KEY IDEA TABLE

Key Idea	Explanation	Chapter (page)
1 Structure-Function	The anatomy (structure) of an element determines its physiology (function).	1 (p. 4)
2 Levels of Organization	All living things are organized from very simple levels (e.g. atoms) to very complex levels (e.g. organisms). Anatomy and physiology can be studied at any of these levels.	1 (p. 4)
3 Homeostasis and Negative Feedback	Despite changing environmental conditions, critical body functions such as blood pressure and temperature are maintained within tight limits. Negative feedback is a control system based on information returning to a source. It reverses any upward or downward shift in a particular body condition.	1 (p. 5)
4 Barriers	Barriers help the body maintain distinct environments. For instance, the skin and mucous membranes separate the inside of the body from the external environment, and the cell's plasma membrane separates the intracellular fluid from the extracellular fluid.	1 (p. 7)
5 Gradients and Resistance	The movement (flow) of a particular substance is promoted by a gradient (a difference in a physical or chemical value between two areas) and opposed by resistance. For instance, resistance can reflect membrane thickness for flow across barriers, and tube width for flow through tubes.	1 (p. 8)
6 Water	Water is of critical importance in many physiological processes.	2 (p. 27)
7 Enzymes	Most chemical reactions require the actions of enzymes. The activity of specific enzymes can be increased or decreased in response to changing body conditions.	2 (p. 32)
8 Energy	Organisms need to generate and use energy. Tracking the flow of energy and matter through systems is a key part of understanding physiology.	2 (p. 34)
9 Genes Code for Proteins	DNA is the cell's master blueprint, determining the body's structures and functions. Mutations, changes in the DNA sequence of a gene, can change the shape of the resulting protein, or even stop it from being synthesized.	3 (p. 54)
10 Adaptation	Organisms adapt to the environment; that is, their structure, function, or behavior changes in response to changes in their environment. Adaptation often results from injury, and helps protect against future damage.	5 (p. 86)
11 Communication	The body uses chemical and electrical signals to convey information. Chemical signals alter the activity of target cells by binding specific receptors.	7 (p. 129)
12 Causation and Correlation	Two events may be correlated (i.e. frequently occur together) without one event causing the other. Establishing causation involves specifying a sequence of interactions linking the cause to the effect.	9 (p. 191)
13 Mass Balance	The amount of a substance stored in a system is determined by inputs and outputs.	13 (p. 271)

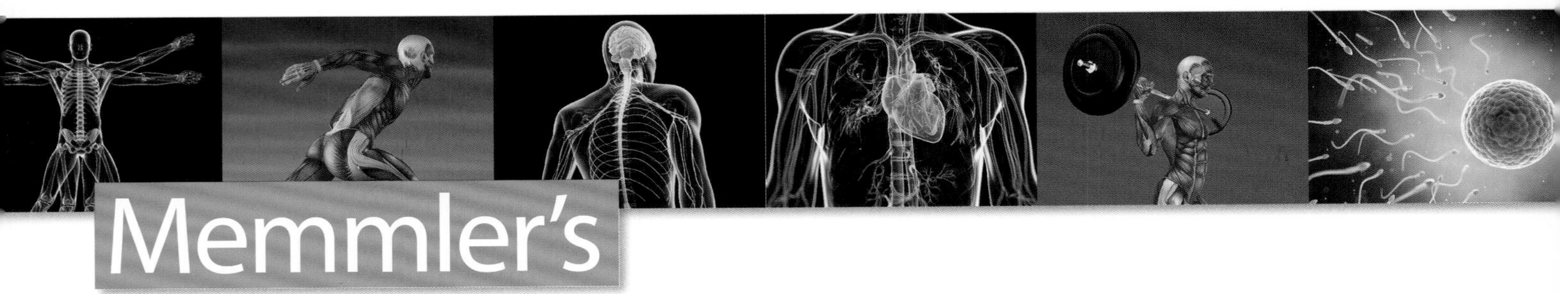

Memmler's Structure and Function of the Human Body

12th Edition

Barbara Janson Cohen, BA, MSEd
Kerry L. Hull, BSC, PhD

Philadelphia • Baltimore • New York • London
Buenos Aires • Hong Kong • Sydney • Tokyo

Vice President and Publisher: Julie K. Stegman
Senior Acquisitions Editor: Jonathan Joyce
Director of Product Development: Jennifer K. Forestieri
Senior Development Editors: Michael Kerns and Julie Vitale
Editorial Coordinator: John Larkin
Marketing Manager: Jason Oberacker
Editorial Assistant: Leo Gray
Design Coordinator: Holly Reid McLaughlin
Art Director, Illustration: Jennifer Clements
Production Project Manager: Alicia Jackson
Manufacturing Coordinator: Karin Duffield
Prepress Vendor: SPi Global

12th Edition

9 8 7 6 5 4 3 2 1

Printed in China

Cataloging-in-Publication Data available on request from the Publisher

ISBN: 978-1-9751-3892-9

---

shop.lww.com

CCS0919

Reviewers

We gratefully acknowledge the generous contributions of the reviewers whose names appear in the list that follows.

Beatrice Avila, BS
Instructor
Palo Alto College
San Antonio, Texas

Christi Blair, MSN, RN
Faculty
Holmes Community College
Goodman, Mississippi

Ernema Boettner, RN, BSN
Instructor
Northwest Technical School
Maryville, Missouri

Beth Gagnon, MSN
Nursing Instructor
Wayne County Schools Career Center
Barberton, Ohio

Michael Hawkes, MEd, RT (R) (ARRT)
Program Director
Pima Medical Institute–Mesa Campus
Mesa, Arizona

Camille Humphreys, BS
Program Director
AmeriTech College
Provo, Utah

Joshua Kramer, MBA
Program Director
Arizona College
Mesa, Arizona

Constance Lieseke, AAS
Faculty and Program Coordinator
Olympic College
Port Orchard, Washington

Maryagnes Luczak, BS, MA
Training Director
Career Training Academy
Pittsburgh, Pennsylvania

Patty Bostwick Taylor, Masters
Instructor
Florence Darlington Technical College
Florence, South Carolina

Robyn Wilhelm, PT, DPT
Assistant Professor
A.T. Still University–Mesa Campus
Mesa, Arizona

Preface

Memmler's Structure and Function of the Human Body is a textbook for introductory-level allied health and nursing students who need a basic understanding of anatomy and physiology and the interrelationships between structure and function.

Like preceding editions, the 12th edition remains true to Ruth Memmler's original vision. The features and content specifically meet the needs of those who may be starting their health career preparation with little or no science background. This book's primary goals are

- To provide the essential knowledge of human anatomy and physiology at an ideal level of detail and in language that is clear and understandable
- To illustrate the concepts discussed with anatomic art of appropriate detail with accuracy, simplicity, and style that is integrated seamlessly with the narrative
- To incorporate the most recent scientific findings into the fundamental material on which Ruth Memmler's classic text is based
- To include pedagogy designed to enhance interest in and understanding of the concepts presented
- To teach the basic anatomic and medical terminology used in health care settings, preparing students to function efficiently in their chosen health careers
- To present an integrated teaching–learning package that includes all of the elements necessary for a successful learning experience

This revision is the direct result of in-depth market feedback solicited to tell us what instructors and students at this level most need. We listened carefully to the feedback, and the results we obtained are integrated into many features of this book and into the ancillary package accompanying it. The text itself has been revised and updated where needed to improve organization of the material and to reflect current scientific thought.

Because visual learning devices are so important to students, this new edition continues to include "The Body Visible," a series of illustrations of the major body systems described in the text with labeled transparent overlays. In addition to being a learning and testing tool, these illustrations provide enrichment and are a valuable general reference.

The 12th edition retains its extensive art program with updated versions of figures from previous editions and many new figures. These features appear in a modified design that makes the content more user-friendly and accessible than ever. Our innovative ancillary package on *thePoint* provides students with a wealth of resources, while the comprehensive package of instructor resources provides instructors with maximum flexibility and efficiency. The online Instructor's Manual describes all of the updates in this new edition and presents teaching and learning strategies for traditional classrooms, flipped classrooms, and online courses.

Organization and Structure

Like previous editions, this 12th edition uses a body systems approach to the study of the normal human body. The book is divided into seven units, grouping related information and body systems together as follows:

- Unit I, The Body as a Whole (Chapters 1–5), focuses on the body's organization; basic chemistry needed to understand body functions; cells and their functions; tissues, glands, and membranes; and the skin.
- Unit II, Movement and Support (Chapters 6 and 7), includes the skeletal and muscular systems.
- Unit III, Coordination and Control (Chapters 8–11), focuses on the nervous system, the sensory system, and the endocrine system.
- Unit IV, Circulation and Body Defense (Chapters 12–15), includes the blood, the heart, blood vessels and circulation, the lymphatic system, and the immune system.
- Unit V, Energy: Supply and Use (Chapters 16–19), includes the respiratory system; the digestive system; metabolism, nutrition, and temperature control; as well as the urinary system and body fluids.
- Unit VI, Perpetuation of Life (Chapters 20 and 21), covers the male and female reproductive systems, as well as development, birth, and heredity.

The main Glossary defines all the chapters' key terms and many additional terms emphasized in the text. An additional Glossary of Word Parts is a reference tool that not only teaches basic medical and anatomic terminology but also helps students learn to recognize unfamiliar terms. Appendices include a variety of supplementary information that students will find useful as they work with the text, including a photographic Dissection Atlas (Appendix 3) and answers to the Chapter Checkpoint questions, Casepoint questions, and Zooming In illustration questions (Appendix 2) that are found in every chapter.

Pedagogic Features

Every chapter contains pedagogy that has been designed with the health professions and nursing student in mind.

- **Key Ideas:** The foundational concepts of Anatomy and Physiology are summarized in the Key Idea Table on page i of the textbook. A large numbered key within a chapter indicates where each key idea is introduced, and small numbered keys indicate when a key idea is used to explain a new concept. These key ideas help students organize and deepen their knowledge.
- **Learning Objectives:** Chapter objectives at the start of every chapter help the student organize and prioritize learning.
- **Ancillaries At-A-Glance:** Learning Tools, Learning Resources, and Learning Activities are highlighted in a one-stop overview of the supplemental materials available for the chapter. These resources are organized for use when the student is preparing to learn, while he or she is learning the material, and when it's time for reviewing the chapter.
- **A & P in Action:** Familiar scenarios transport chapter content into a real-life setting, bringing the information to life for students and showing how the body maintains a state of internal balance.
- **A Look Back:** With the exception of Chapter 1, each chapter starts with a brief review of how its content relates to prior chapters, especially the key ideas.
- **Checkpoint questions:** Brief questions at the end of main sections test and reinforce the student's recall of key information in that section. Answers are in Appendix 2.
- **Casepoint questions:** Critical thinking questions challenge students to apply concepts learned to the A & P in Action case study. Answers are in Appendix 2.
- **Key Points:** Critical information in figure legends spotlights essential aspects of the illustrations.
- **"Zooming In" questions:** Questions in the figure legends test and reinforce student understanding of concepts depicted in the illustration. Answers are in Appendix 2.
- **Phonetic pronunciations:** Easy-to-learn phonetic pronunciations are spelled out in the narrative, appearing in parentheses directly following many terms—no need for students to understand dictionary-style diacritical marks (see the "Guide to Pronunciation").
- **Special interest boxes:** Each chapter contains special interest boxes focusing on topics that augment chapter content. The book includes four kinds of boxes:
    - **One Step at a Time:** Uses the case study to teach scientific literacy and problem solving skills. Each box outlines a step-by-step approach that students can use to answer end of chapter questions or Study Guide problems.
    - **A Closer Look:** Provides additional in-depth scientific detail on topics in or related to the text.
    - **Clinical Perspectives:** Focuses on diseases and disorders relevant to the chapter, exploring what happens to the body when the normal structure–function relationship breaks down.
    - **Hot Topics:** Focuses on current trends and research, reinforcing the link between anatomy and physiology and related news coverage that students may have seen.
- **New! Concept Mastery Alerts:** Text boxes at specific points in the chapters highlight common student misconceptions to help your understanding of potentially confusing topics, as identified by Lippincott's online adaptive learning platform, powered by PrepU. Data from thousands of actual students using this program in courses across the United States have identified common misconceptions that are clarified in this feature.
- **Figures:** The art program includes full-color anatomic line art, many new or revised, with a level of detail that matches that of the narrative. Photomicrographs, radiographs, and other scans give students a preview of what they might see in real-world healthcare settings. Supplementary figures are available on the companion website on *thePoint*.
- **Tables:** The numerous tables in this edition summarize key concepts and information in an easy-to-review form. Additional summary tables are available on the companion website on *thePoint*.
- **Color figure and table callouts:** Figure and table numbers appear in color in the narrative, helping students quickly find their place after stopping to look at an illustration or table.
- **A & P in Action Revisited:** Traces the outcome of the medical story that opens each chapter and shows how the cases relate to material in the chapter and to others in the book.
- **Word Anatomy:** This chart defines and illustrates the various word parts that appear in terms within the chapter. The prefixes, roots, and suffixes presented are grouped according to chapter headings so that students can find the relevant text. This learning tool helps students build vocabulary and promotes understanding even of unfamiliar terms based on a knowledge of common word parts.
- **Chapter Overview:** A graphic outline at the end of each chapter provides a concise overview of chapter content, aiding in study and test preparation.
- **Key Terms:** Selected boldface terms throughout the text are listed at each chapter's end and defined in the book's glossary.
- **Questions for Study and Review:** Study questions are organized hierarchically into three levels. (Note that

answers appear in the student resources on *thePoint*.) This section includes questions that direct students to "The Body Visible" and the various appendices to promote use of these resources. Question levels include the following:

- **Building Understanding:** Includes fill-in-the-blank, matching, and multiple choice questions that test factual recall
- **Understanding Concepts:** Includes short-answer questions (define, describe, compare/contrast) that test and reinforce understanding of concepts
- **Conceptual Thinking:** Includes short-essay questions that promote critical thinking skills. Included are thought questions related to the Disease in Context case stories.

For Students

Look for callouts throughout the chapters for pertinent supplementary material on the companion website on *thePoint*.

This companion website also includes resources that helps students learn faster, remember more, and achieve success. Students may choose from a wealth of materials including a pre-quiz; animations; various types of online learning activities; an audio glossary; and other supplemental materials, such as health professions career information, additional charts and images, and study and test-taking tips. Answers to the chapter Questions for Study and Review are also included.

See the inside front cover of this text for the passcode you will need to gain access to the companion website, and see pages xv–xvii for details about the website and a complete listing of student resources.

Instructor Ancillary Package

All instructor resources are available to approved adopting instructors and can be accessed online at http://thepoint.lww.com/MemmlerSFHB12e.

- Instructor's Manual materials available for each chapter summarize the changes in the new edition, and provide background information and activities relevant to each learning objective.
- Brownstone Test Generator allows you to create customized exams from a bank of questions.
- PowerPoint Presentations use visuals to emphasize the key concepts of each chapter.
- Image Bank includes labels-on and labels-off options.
- Supplemental Image Bank with additional images can be used to enhance class presentations.
- Lesson Plans are organized around the learning objectives and include lecture notes, in-class activities, and assignments, including student activities from the student companion website.
- Answers to "Questions for Study and Review" provide responses to the quiz material found at the end of each chapter in the textbook.
- Strategies for Effective Teaching provide sound, tried-and-true advice for successful instruction in traditional, flipped, and online learning environments.
- WebCT/Blackboard/Angel Cartridge allows easy integration of the ancillary materials into learning management systems.

Instructors also have access to all student ancillary assets, via *thePoint* website.

Guide to Pronunciation

The stressed syllable in each word is shown with capital letters. The vowel pronunciations are as follows:

Any vowel that appears alone or at the end of a syllable is given a long sound, as follows:

a as in say
e as in be
i as in nice
o as in go
u as in true

A vowel followed by a consonant and the letter e (as in rate) also is given a long pronunciation. For example, re-PETE for repeat.

Any vowel followed by a consonant receives a short pronunciation, as follows:

a as in absent
e as in end
i as in bin
o as in not
u as in up

The letter *h* may be added to a syllable to make vowel pronunciation short, as in *vanilla* (vah-NIL-ah).

Summary

The 12th edition of *Memmler's Structure and Function of the Human Body* builds on the successes of the previous 11 editions by offering clear, concise narrative into which accurate, aesthetically pleasing anatomic art has been woven. We have made every effort to respond thoughtfully and thoroughly to reviewers' and instructors' comments, offering the ideal level of detail for students preparing for careers in the health professions and nursing and the pedagogic features that best support them. With the online resources, we have provided students with an integrated system for succeeding in the course. We hope you will agree that the 12th edition of *Memmler's* suits your educational needs.

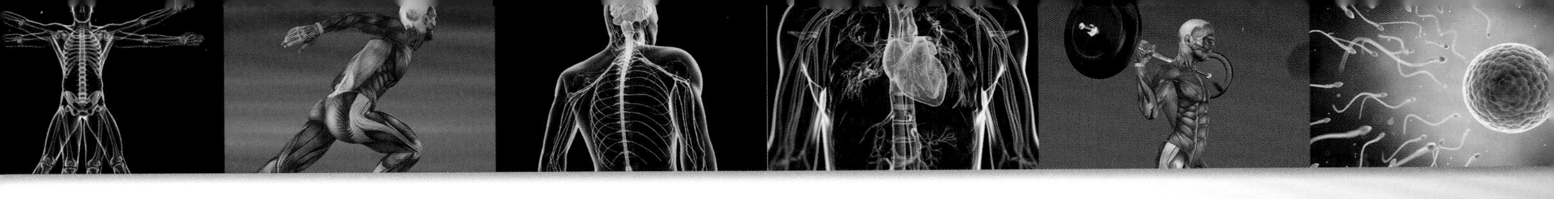

User's Guide

For today's health careers, a thorough understanding of human anatomy and physiology is more important than ever. *Memmler's Structure and Function of the Human Body*, 12th edition, not only provides the conceptual knowledge you'll need but also teaches you how to apply it. This User's Guide introduces you to the features and tools that will help you succeed as you work through the materials.

Your journey begins with your textbook, *Memmler's Structure and function of the Human Body*. Newly updated and fully illustrated, this easy-to-use textbook is filled with resources and activities to help you succeed.

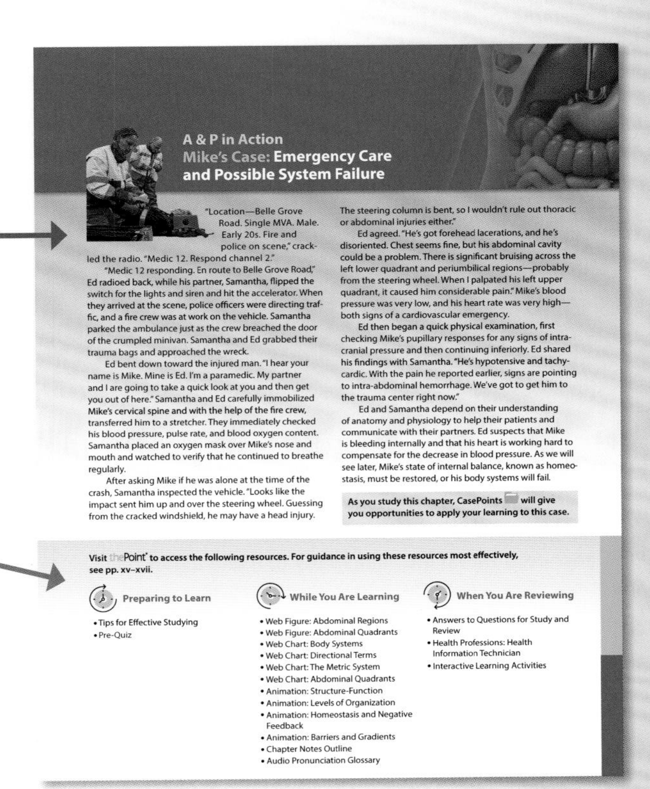

A & P in Action
Mike's Case: **Emergency Care and Possible System Failure**

"Location—Belle Grove Road. Single MVA. Male. Early 20s. Fire and police on scene," crackled the radio. "Medic 12. Respond channel 2."

"Medic 12 responding. En route to Belle Grove Road," Ed radioed back, while his partner, Samantha, flipped the switch for the lights and siren and hit the accelerator. When they arrived at the scene, police officers were directing traffic, and a fire crew was at work on the vehicle. Samantha parked the ambulance just as the crew breached the door of the crumpled minivan. Samantha and Ed grabbed their trauma bags and approached the wreck.

Ed bent down toward the injured man. "I hear your name is Mike. Mine is Ed. I'm a paramedic. My partner and I are going to take a quick look at you and then get you out of here." Samantha and Ed carefully immobilized Mike's cervical spine and with the help of the fire crew, transferred him to a stretcher. They immediately checked his blood pressure, pulse rate, and blood oxygen content. Samantha placed an oxygen mask over Mike's nose and mouth and watched to verify that he continued to breathe regularly.

After asking Mike if he was alone at the time of the crash, Samantha inspected the vehicle. "Looks like the impact sent him up and over the steering wheel. Guessing from the cracked windshield, he may have a head injury. The steering column is bent, so I wouldn't rule out thoracic or abdominal injuries either."

Ed agreed. "He's got forehead lacerations, and he's disoriented. Chest seems fine, but his abdominal cavity could be a problem. There is significant bruising across the left lower quadrant and periumbilical regions—probably from the steering wheel. When I palpated his left upper quadrant, it caused him considerable pain." Mike's blood pressure was very low, and his heart rate was very high—both signs of a cardiovascular emergency.

Ed then began a quick physical examination, first checking Mike's pupillary responses for any signs of intracranial pressure and then continuing inferiorly. Ed shared his findings with Samantha. "He's hypotensive and tachycardic. With the pain he reported earlier, signs are pointing to intra-abdominal hemorrhage. We've got to get him to the trauma center right now."

Ed and Samantha depend on their understanding of anatomy and physiology to help their patients and communicate with their partners. Ed suspects that Mike is bleeding internally and that his heart is working hard to compensate for the decrease in blood pressure. As we will see later, Mike's state of internal balance, known as homeostasis, must be restored, or his body systems will fail.

As you study this chapter, CasePoints will give you opportunities to apply your learning to this case.

Visit thePoint® to access the following resources. For guidance in using these resources most effectively, see pp. xv–xvii.

Preparing to Learn
- Tips for Effective Studying
- Pre-Quiz

While You Are Learning
- Web Figure: Abdominal Regions
- Web Figure: Abdominal Quadrants
- Web Chart: Body Systems
- Web Chart: Directional Terms
- Web Chart: The Metric System
- Web Chart: Abdominal Quadrants
- Animation: Structure-Function
- Animation: Levels of Organization
- Animation: Homeostasis and Negative Feedback
- Animation: Barriers and Gradients
- Chapter Notes Outline
- Audio Pronunciation Glossary

When You Are Reviewing
- Answers to Questions for Study and Review
- Health Professions: Health Information Technician
- Interactive Learning Activities

A & P in Action provides an interesting case story that uses a familiar, real-life scenario to illustrate key concepts in anatomy and physiology. Later in the chapter, the case story is revisited in more detail—improving your understanding and helping you remember the information.

Ancillaries At-A-Glance highlights the Learning Resources and Learning Activities available for the chapter.

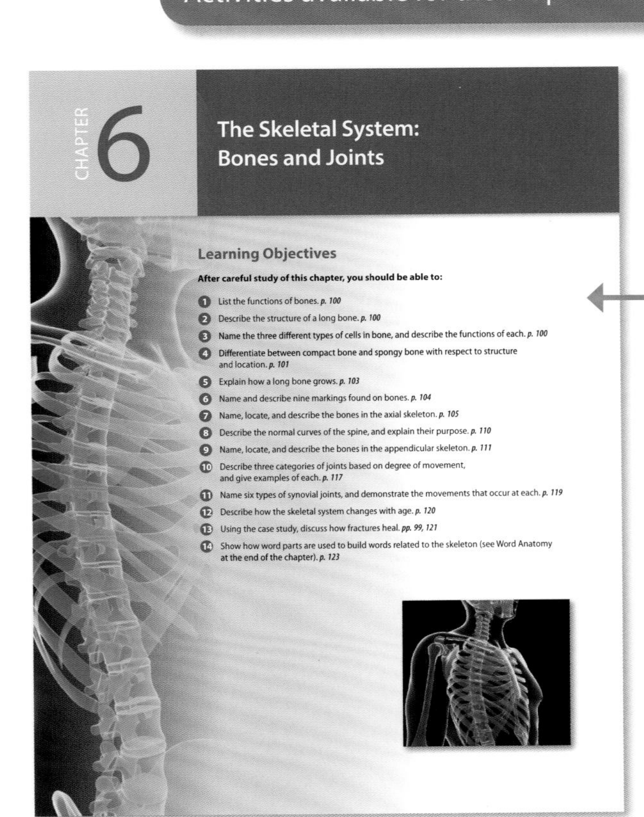

CHAPTER 6

The Skeletal System: Bones and Joints

Learning Objectives

After careful study of this chapter, you should be able to:

1. List the functions of bones. *p. 100*
2. Describe the structure of a long bone. *p. 100*
3. Name the three different types of cells in bone, and describe the functions of each. *p. 100*
4. Differentiate between compact bone and spongy bone with respect to structure and location. *p. 101*
5. Explain how a long bone grows. *p. 103*
6. Name and describe nine markings found on bones. *p. 104*
7. Name, locate, and describe the bones in the axial skeleton. *p. 105*
8. Describe the normal curves of the spine, and explain their purpose. *p. 110*
9. Name, locate, and describe the bones in the appendicular skeleton. *p. 111*
10. Describe three categories of joints based on degree of movement, and give examples of each. *p. 117*
11. Name six types of synovial joints, and demonstrate the movements that occur at each. *p. 119*
12. Describe how the skeletal system changes with age. *p. 120*
13. Using the case study, discuss how fractures heal. *pp. 99, 121*
14. Show how word parts are used to build words related to the skeleton (see Word Anatomy at the end of the chapter). *p. 123*

Learning Objectives help you identify learning goals and familiarize yourself with the materials covered in the chapter. These objectives are referenced to page numbers in the text.

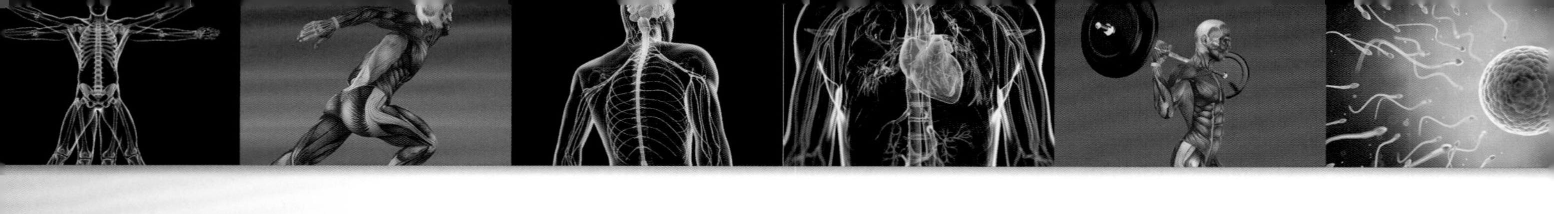

A LOOK BACK

The skin is introduced in Chapter 4 as one of the epithelial membranes, the cutaneous (ku-TA-ne-us) membrane, overlying a connective tissue membrane, the superficial fascia. In this chapter, we describe the skin in much greater detail as it forms the major portion of the integumentary system. 1 *As with all body systems, studying the structure of skin helps us understand its function. The key idea of barriers* 4 *is particularly relevant to this chapter, because the skin provides the major barrier between the external and internal environments.*

A Look Back relates each chapter's content to concepts in the preceding chapters.

CHECKPOINTS

- ☐ **5-1** What is the name of the system that comprises the skin and all its associated structures?
- ☐ **5-2** Moving from the superficial to the deeoer layer, what are the names of the two layers of the skin?
- ☐ **5-3** What is the composition of the subcutaneous layer?

Chapter Checkpoints pose brief questions at the end of main sections that test and reinforce student recall.

CASEPOINTS

- ☐ **5-1** Name the skin layers involved in Hazel's burns in the opening case study.
- ☐ **5-2** Which skin layer will produce new cells to replace Hazel's damaged epidermis?

Case Points Critical thinking questions challenge students to apply concepts learned to the Disease in Context case study. Answers are in Appendix 2.

Concept Mastery Alert

Hair loss involves only the shaft and the root (the small bulblike portion of the hair). The hair follicle is firmly anchored within the skin, ready to make a new hair.

Concept Mastery Alerts highlight common student misconceptions to help understanding of potentially confusing topics, as identified by Lippincott's online adaptive learning platform, powered by PrepU. Data from thousands of actual students using this program in courses across the United States have identified common misconceptions that are clarified in this feature.

12 COMMUNICATION

The negative feedback key idea 3 introduced the concept of signals conveying information between the different components of the feedback loop. However, signals also function outside such loops. **Communication**—the transmission of signals between cells or even within a single cell—is critical to all aspects of body function, ranging from muscle contraction to the development of an entire individual from a single cell. Defects in communication are responsible for many diseases.

Chapter 1 discussed the two types of signals: electri-

New Key Ideas: The foundational concepts of Anatomy and Physiology are summarized in the Key Idea Table on page i of the textbook. A large numbered key within a chapter indicates where each key idea is introduced, and small numbered keys indicate when a key idea is used to explain a new concept. These key ideas help students organize and deepen their knowledge.

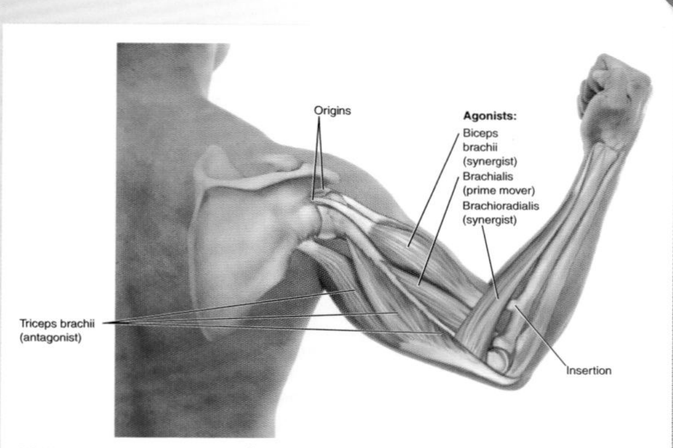

FIGURE 7-6 **Muscle attachments to bones.** KEY POINT Tendons attach muscles to bones. The stable point is the origin; the movable point is the insertion. In this diagram, three attachments are shown—two origins and one insertion. ZOOMING IN Does contraction of the brachialis produce flexion or extension at the elbow?

Key Points in the figure captions spotlight essential aspects of the illustrations.

"Zooming In" questions in the figure captions test and reinforce student understanding of concepts depicted in the illustration.

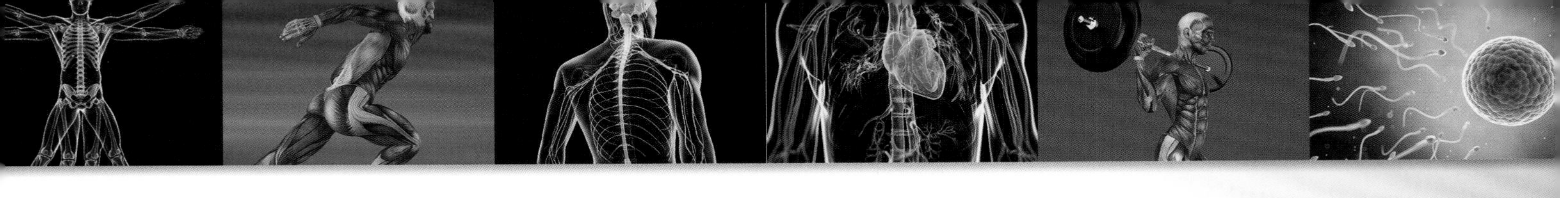

Phonetic pronunciations spelled out in the narrative directly following many terms make learning pronunciation easy—no need to understand dictionary-style diacritical marks.

Two other types of lipids are important in the body. **Phospholipids** (fos-fo-LIP-ids) are complex lipids containing the element phosphorus. Among other functions, phospholipids make up a major part of the membrane around living cells. **Steroids** are lipids that contain rings of carbon atoms. The most important sterol is **cholesterol** (ko-LES-ter-ol), another component of cellular membranes (see FIG. 2-9B). Cholesterol is also used to make steroid hormones, including cortisol, testosterone, and estrogen.

Color figure and table callouts help students quickly find their place after stopping to look at an illustration or table.

number at the bottom of each box. It takes about 1,850 electrons to equal the weight of a single neutron or proton, so electrons are not counted in the determination of atomic weight.

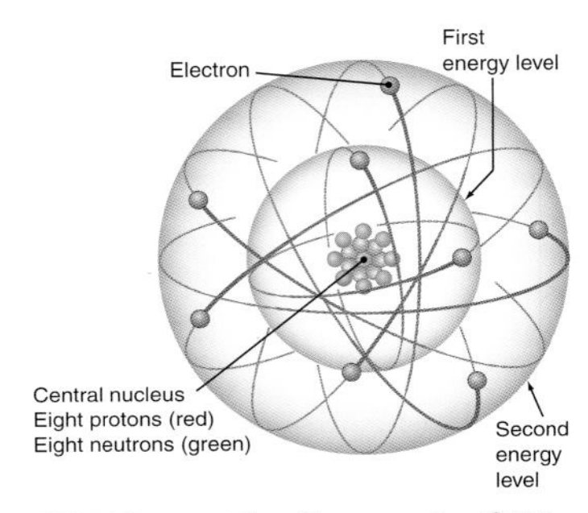

FIGURE 2-2 **Representation of the oxygen atom.** KEY POINT Eight protons and eight neutrons are tightly bound in the central nucleus. The eight electrons are in orbit around the nucleus,

tances from the nucleus in regions called energy levels (see FIG. 2-2). The first energy level, the one closest to the nucleus, can hold only two electrons. The second energy level, the next in distance away from the nucleus, can hold up to eight electrons. FIGURE 2-3A shows the oxygen atom with only the protons in the nucleus and the electrons in fixed positions in their energy levels. It has two electrons in its first energy level and six electrons in its second, outermost, level. The carbon atom, with an atomic number of 6, has four electrons in its outermost energy level (see FIG. 2-3B). More

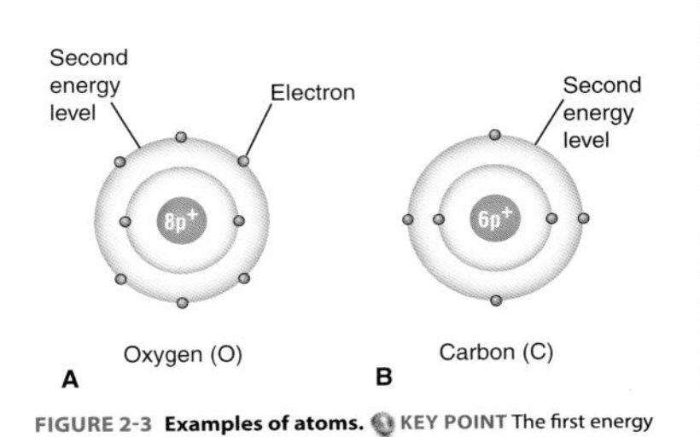

FIGURE 2-3 **Examples of atoms.** KEY POINT The first energy

Blood cell antigens vary among individuals and becom[e] important when blood components are donated from on[e] individual to another in a process called a **transfusion**. Antibodies that recognize red cell antigens are known a[s] *agglutinins*, because they cause red cells to undergo **agglutination** (ah-glu-tih-NA-shun) (clumping). The cells the[n] rupture and release their hemoglobin by a process called **hemolysis** (he-MOL-ih-sis). The resulting condition is dangerous to a patient who has received incompatible blood. There are many types of RBC antigens, but only two groups are particularly likely to cause a transfusion reaction: the so-called A and B antigens and the Rh factor.

THE ABO BLOOD TYPE SYSTEM

There are four blood types involving the A and B antigens: A, B, AB, and O (TABLE 12-3). These letters indicate the type of antigen present on the red cells. If only the A antigen is present, the person has type A blood; if only the B antigen

not react with either A or B antibodies (see TABLE 12-3). People with type O blood are called *universal donors*. Conversely, type AB blood contains no antibodies to agglutinate red cells, and people with this blood type can therefore receive blood from any ABO type donor. Those with AB blood are described as *universal recipients*. Whenever possible, it is safest to give the same blood type as the recipient's blood.

THE RH FACTOR

More than 85% of the US population has another red cell antigen group called the **Rh factor**, named for *Rhesus*

Table 12-3 The ABO Blood TYPE System

Blood Type	Red Blood Cell Antigen	Reacts with Antiserum	Plasma Antibodies	Can Take From	Can Donate To
A	A	Anti-A	Anti-B	A, O	A, AB
B	B	Anti-B	Anti-A	B, O	B, AB
AB	A, B	Anti-A, Anti-B	None	AB, A, B, O	AB
O	None	None	Anti-A, Anti-B	O	O, A, B, AB

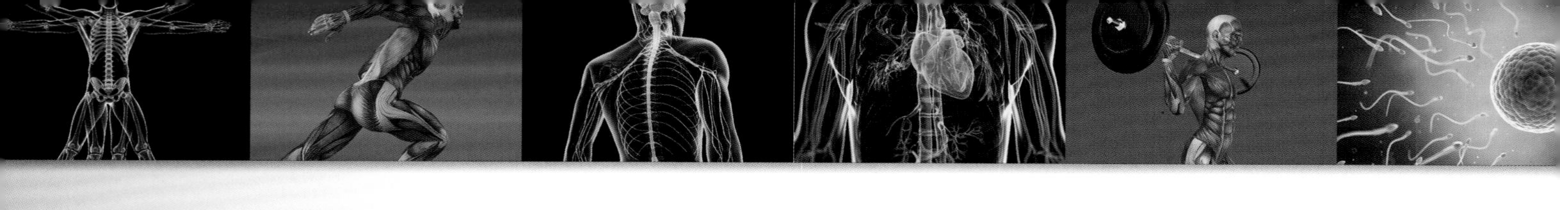

Special interest boxes focus on topics that augment chapter content.

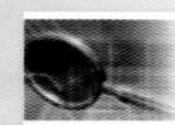

A CLOSER LOOK
Hydrogen Bonds: Strength in Numbers

BOX 2-1

In contrast to ionic and covalent bonds, which hold atoms together, hydrogen bonds hold molecules together. Hydrogen bonds are much weaker than ionic or covalent bonds—in fact, they are more like "attractions" between molecules. While ionic and covalent bonds rely on electron transfer or sharing, hydrogen bonds form bridges between two molecules. A hydrogen bond forms when a slightly positive hydrogen atom in one molecule is attracted to a slightly negative atom in another molecule. Even though a single hydrogen bond is weak, many hydrogen bonds between two molecules can be strong.

Hydrogen bonds hold water molecules together, with the slightly positive hydrogen atom in one molecule attracted to a slightly negative oxygen atom in another. Many of water's unique properties come from its ability to form hydrogen bonds. For example, hydrogen bonds keep water liquid over a wide range of temperatures, which provides a constant environment for body cells.

Hydrogen bonds form not only between molecules but also within large molecules. Hydrogen bonds between regions of the same molecule cause it to fold and coil into a specific shape, as in the process that creates the precise three-dimensional structure of proteins. Because a protein's structure determines its function in the body, hydrogen bonds are essential to protein activity.

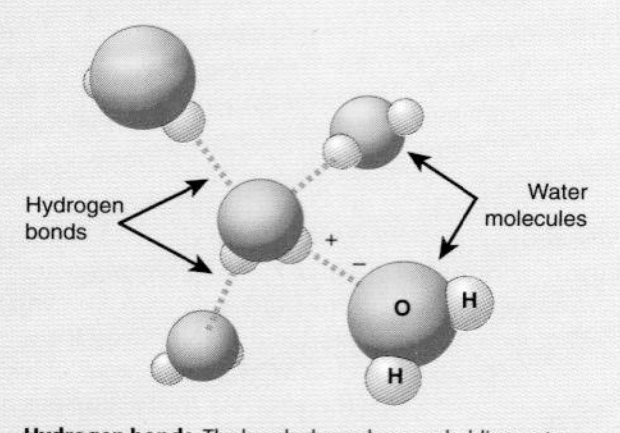

Hydrogen bonds. The bonds shown here are holding water molecules together.

A Closer Look boxes give in-depth scientific detail on topics in or related to the text.

CLINICAL PERSPECTIVES
Anabolic Steroids: Winning at All Costs?

BOX 7-2

Anabolic steroids mimic the effects of the male sex hormone testosterone by promoting metabolism and stimulating growth. These drugs are legally prescribed to promote muscle regeneration and prevent atrophy from disuse after surgery. However, some athletes also purchase them illegally, using them to increase muscle size and strength and improve endurance.

When steroids are used illegally to enhance athletic performance, the doses needed are large enough to cause serious side effects. They increase blood cholesterol levels, which may lead to atherosclerosis, heart disease, kidney failure, and stroke. They damage the liver, making it more susceptible to cancer and other diseases, and suppress the immune system, increasing the risk of infection and cancer. In men, steroids cause impotence, testicular atrophy, low sperm count, infertility, and the development of female sex characteristics, such as breasts (gynecomastia). In women, steroids disrupt ovulation and menstruation and produce male sex characteristics, such as breast atrophy, enlargement of the clitoris, increased body hair, and deepening of the voice. In both sexes, steroids increase the risk for baldness and, especially in men, cause mood swings, depression, and violence.

Clinical Perspectives boxes focus on diseases and disorders relevant to the chapter, exploring what happens to the body when the normal structure–function relationship breaks down.

Hot Topics boxes examine current trends and research.

HOT TOPICS
Radioactive Tracers: Medicine Goes Nuclear

BOX 2-3

Like radiography, computed tomography (CT), and magnetic resonance imaging, **nuclear medicine imaging** (NMI) offers a noninvasive way to look inside the body. An excellent diagnostic tool, NMI not only shows structural details but also provides information about body function. NMI can help diagnose cancer, stroke, and heart disease earlier than can techniques that provide only structural information.

NMI uses **radiotracers**, radioactive substances that specific organs absorb. For example, radioactive iodine is used to image the thyroid gland, which absorbs more iodine than does any other organ. After a patient ingests, inhales, or is injected with a radiotracer, a device called a gamma camera detects the radiotracer in the organ under study and produces a picture, which is used in making a diagnosis. Radiotracers are broken down and eliminated through urine or feces, so they leave the body quickly. A patient's exposure to radiation in NMI is usually considerably lower than with x-ray or CT scan.

Three NMI techniques are positron emission tomography (PET), **bone scanning**, and the **myocardial perfusion imaging** (MPI) stress test. PET is often used to evaluate brain activity by measuring the brain's use of radioactive glucose. PET scans can reveal brain tumors because tumor cells are often more metabolically active than are normal cells and thus absorb more radiotracer. Bone scanning detects radiation from a radiotracer absorbed by bone tissue with an abnormally high metabolic rate, such as a bone tumor. The MPI test is used to diagnose heart disease. A nuclear medicine technologist injects the patient with a radionuclide (e.g., thallium, technetium), and a gamma camera images the heart during exercise and later rest. When compared, the two sets of images help evaluate blood flow to the working, or "stressed," heart.

One Step at a Time boxes expand on basic science concepts introduced in the case studies by walking students step-by-step through higher level critical thinking activities, such as developing scientific literacy and problem-solving skills.

ONE STEP AT A TIME
Muscles and Their Movements

BOX 7-3

Learning the locations of different muscles is the first step to knowing which muscles accomplish which movements. You can use this knowledge for many purposes, from predicting the impact of muscle disorders to optimizing muscle training regimes. This box shows you how to predict muscle actions based on muscle locations and attachments.

Question

Shane has particular weakness in his rectus femoris muscle. Predict which movements will be impacted.

Answer

Step 1. Locate the muscle on your body. The rectus femoris is located on the anterior thigh.

Step 2. Identify the bones where the muscle attaches. You can use a skeleton with labeled muscle origins and insertions, the text, later FIGURE 7-15, or later TABLE 7-5. The rectus femoris attaches to the ilium at one end and the tibia on the other end.

Step 3. Use your knowledge of skeletal anatomy to identify any joints that the muscle crosses. Based on the arrangement of bones identified in step 2, the rectus femoris must cross both the hip joint and the knee joint.

Step 4. Use your body or a model skeleton to shorten the distance between the origin at the hip and insertion on the tibia over the knee. What happens? The rectus femoris most commonly straightens the leg at the knee joint (extension). There might be other possibilities, depending on which bone moves and which bone remains stationary. For instance, this muscle can also act at the hip joint to flex the thigh.

Step 5. If you identified more than one possible outcome in step 4, use your body to figure out when the different movements occur. For the rectus femoris, stabilizing the trunk enables this muscle to flex the thigh and extend the leg. Stabilizing both the trunk and the thigh enables the muscle to extend the leg without flexing the thigh.

Step 6. Which muscles must relax to permit the movement? Remember that antagonist muscles must relax to enable a given movement. Antagonistic muscles in the limbs are usually found on the opposite side of the limb. So, the antagonist of the rectus femoris is the hamstring muscle group. Relaxing the hamstring as much as possible can optimize the function of the rectus femoris.

You can use a similar procedure to answer Question 26 at the end of this chapter, which asks you to design training regimes to strengthen a patient's shoulder and thigh muscles.

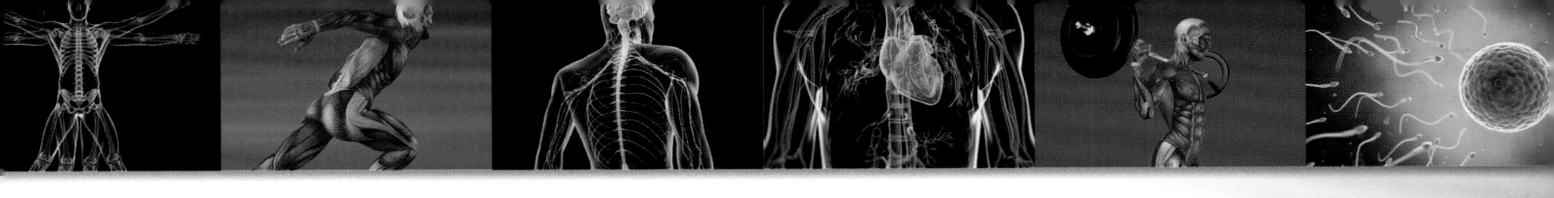

Figures: The art program includes full-color anatomic line art, many new or revised, with a level of detail that matches that of the narrative. Photomicrographs, radiographs, and other scans give students a preview of what they might see in real-world health care settings. Supplementary figures are available on the companion website on thePoint®.

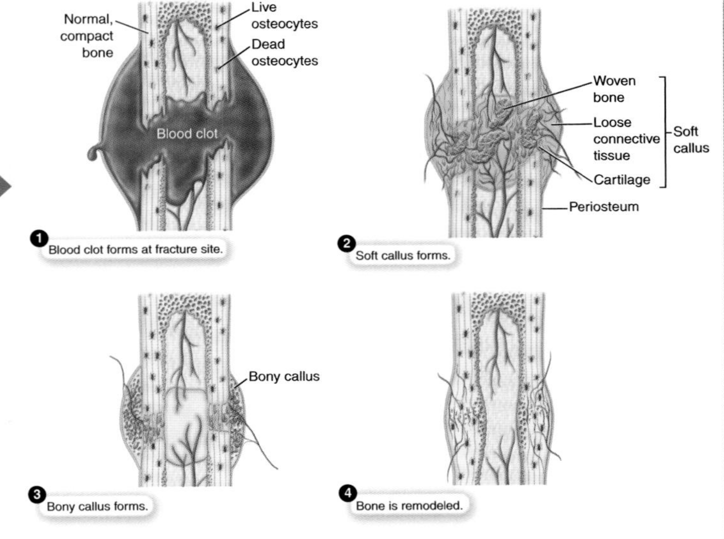

FIGURE 6-4 **Fracture repair.** When bone is fractured, stem cells in the periosteum direct the formation of new bone tissue to repair the injury. ZOOMING IN Which forms first—the soft callus or the bony callus?

Table 7-1 Comparison of the Different Types of the Muscle

	Smooth	Cardiac	Skeletal
Location	Wall of hollow organs, vessels, respiratory, passageways	Wall of the heart	Attached to bones
Cell characteristics	Tapered at each end, branching networks, nonstriated	Branching networks; special membranes (intercalated disks) between cells; single nucleus; lightly striated	Long and cylindrical; multinucleated; heavily striated
Control	Involuntary	Involuntary	Voluntary
Action	Produces peristalsis; contracts and relaxes slowly; may sustain contraction	Pumps blood out of the heart; self-excitatory but influenced by nervous system and hormones	Produces movement at joints; stimulated by nervous system; contracts and relaxes rapidly

Tables: The numerous tables in this edition summarize key concepts and information in an easy-to-review form. Additional summary tables are available on the companion website on thePoint®.

Chapter Overview at the end of each chapter outlines the chapter contents.

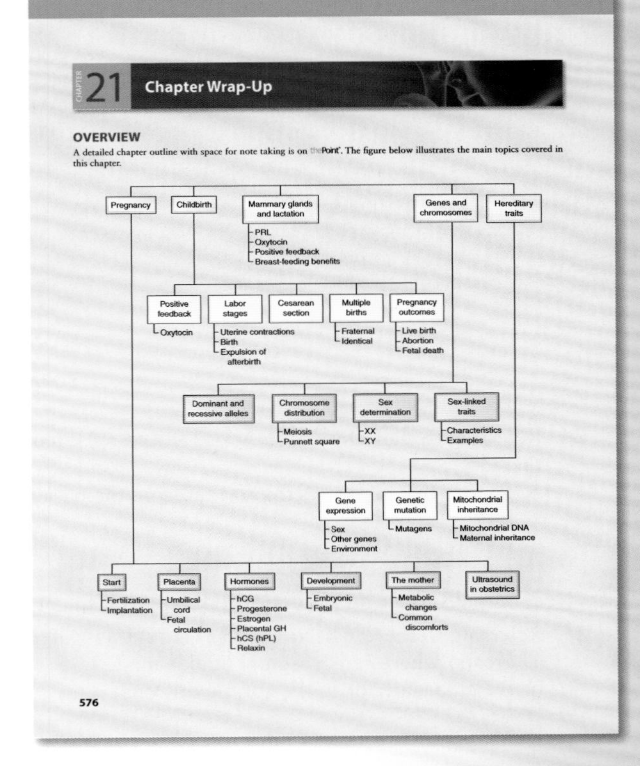
CHAPTER 21 Chapter Wrap-Up

OVERVIEW

A detailed chapter outline with space for note taking is on thePoint®. The figure below illustrates the main topics covered in this chapter.

576

Key Terms sections provide a concise list of selected boldface terms used in the chapter and defined in the book's glossary.

Chapter 21 ▪ Development and Heredity 577

KEY TERMS

The terms listed below are emphasized in this chapter. Knowing them will help you organize and prioritize your learning. These and other boldface terms are defined in the Glossary with phonetic pronunciations.

abortion
allele
amniotic sac
autosome
carrier
chorion
chromosome
colostrum
decidua
dominant
embryo
familial
fertilization
fetus
gene
genetic
genotype
gestation
heredity
heterozygous
homozygous
human chorionic gonadotropin (hCG)
implantation
lactation
meiosis
mutagen
mutation
obstetrics
oxytocin
parturition
phenotype
placenta
progeny
recessive
relaxin
sex-linked
trait
umbilical cord
zygote

Word Anatomy tables define and illustrate the various word parts that constitute the chapter's specialized terminology, helping to build vocabulary and promote understanding of unfamiliar terms.

WORD ANATOMY

Medical terms are built from standardized word parts (prefixes, roots, and suffixes). Learning the meanings of these parts can help you remember words and interpret unfamiliar terms.

WORD PART	MEANING	EXAMPLE
PREGNANCY		
chori/o	membrane, chorion	Human *chorionic* gonadotropin is produced by the chorion (outermost cells) of the embryo.
zyg/o	joined	An ovum and spermatozoon join to form a *zygote*.
CHILDBIRTH		
ox/y	sharp, acute	*Oxytocin* is the hormone that stimulates labor.
toc/o	labor	See preceding example.
THE MAMMARY GLANDS AND LACTATION		
lact/o	milk	The *lactiferous* ducts carry milk from the mammary glands.
mamm/o	breast, mammary gland	A *mammogram* is radiographic study of the breast.
GENES AND CHROMOSOMES		
aut/o-	self	*Autosomes* are all the chromosomes aside from the two that determine sex.
chrom/o	color	*Chromosomes* color darkly with stains.
heter/o	other, different	*Heterozygous* paired genes (alleles) are different from each other.
homo-	same	*Homozygous* paired genes (alleles) are the same.
phen/o	to show	Traits that can be observed or tested for making up a person's *phenotype*.
HEREDITARY TRAITS		
multi-	many	*Multifactorial* traits are determined by multiple pairs of genes.

Questions for Study and Review sections organize study questions hierarchically into three levels.

Building Understanding: Includes fill-in-the-blank, matching, and multiple choice questions that test factual recall.

QUESTIONS FOR STUDY AND REVIEW

BUILDING UNDERSTANDING

Fill in the Blanks

1. Following ovulation, the mature follicle becomes the ______.
2. The first mammary gland secretion is called ______.
3. The basic unit of heredity is a(n) ______.
4. Chromosomes not involved in sex determination are known as ______.
5. The number of chromosomes in each human body cell is ______.

Understanding Concepts: Includes short-answer questions (define, describe, compare/contrast) that test and reinforce understanding of ideas. This section now includes questions pertaining to "The Body Visible" and the diverse information in the appendices.

UNDERSTANDING CONCEPTS

17. What do you study in anatomy? In physiology? Would it be wise to study one without the other?
18. List in sequence the levels of organization in the body from simplest to most complex. Give an example for each level.
19. Compare and contrast the anatomy and physiology of the nervous system with that of the endocrine system.
20. In order of action, name the components of a negative feedback loop.
21. Use The Body Visible overlays at the beginning of this book to name the lateral bone of the lower leg. Name the proximal bone of the arm.
22. Referring to the Dissection Atlas in Appendix 3:
    a. list the figure(s) in which an organ is cut into left and right parts.
    b. list the figure(s) in which an organ is cut into anterior and posterior parts.

Conceptual Thinking: Includes short-essay questions that promote critical thinking skills. This section includes thought questions related to the A & P in Action case stories.

CONCEPTUAL THINKING

23. The human body is organized from very simple levels to more complex levels. With this in mind, describe why a disease at the chemical level can have an effect on organ system function.
24. Use a car operating under cruise control as an example of a negative feedback loop, identifying the set point and the components of the system.
25. In Mike's case, the paramedics discovered bruising of the skin over Mike's left lumbar region and umbilical region. Mike also reported considerable pain in his upper left quadrant. Locate these regions on your own body. Why it is important for health professionals to use medical terminology when describing the human body?
26. If a child swallows a marble, does the marble actually get inside his body? Explain.
27. In Mike's case, blood is flowing out of the blood vessels in his abdomen into his abdominal cavity.
    A. What sort of gradient is driving the flow?
    B. What do you think changed in Mike's case that enabled this flow to occur: the gradient or the resistance?

For more questions, see the Learning Activities on thePoint®.

24. Referring to the "Dissection Atlas" **FIGURE A3-1**, name the:
    a. feature that separates the frontal from the parietal lobe
    b. feature that separates the frontal from the temporal lobe
    c. raised surface area anterior to the central sulcus
25. Referring to the "Dissection Atlas" **FIGURE A3-2**, name:
    a. area(s) where cerebrospinal fluid is made
    b. the area superior to the corpus callosum

Dissection Atlas Questions ask students to respond by examining actual anatomic dissection photographs.

Getting Started with the Student Resources

Your journey begins with your textbook, *Memmler's Structure and Function of the Human Body*, 12th edition. The textbook has callouts that guide you to resources and activities to enhance your learning experience.

Look for these callouts throughout the book for pertinent supplementary material on the companion website.

Here's how to begin:

1. Scratch off the personal access code inside the front cover of your textbook.
2. Log on to http://thepoint.lww.com/MemmlerSFHB12e, the companion website for *Memmler's Structure and Function of the Human Body*, 12th edition, on thePoint®.
3. Click on "Student Resources" and explore the wide variety of learning activities.

thePoint®

Resources and activities available to instructors include the following:

Instructor's Manual
PowerPoints
Image Bank
Answer Key
Customizable Test Generator
WebCT, Angel, and Blackboard-Ready Cartridges

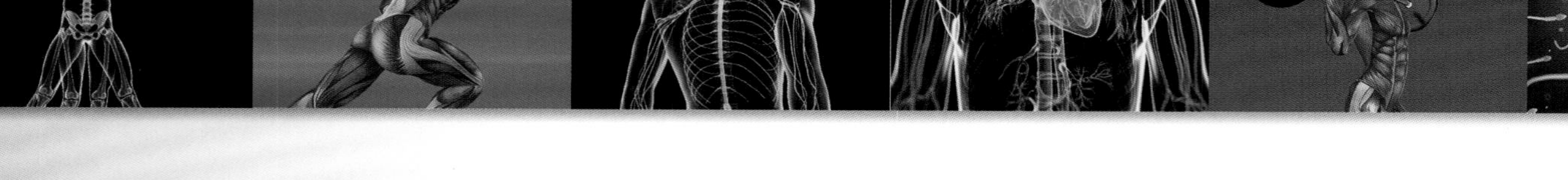

Resources and activities available to students include the following:

- Pre-Quiz
- True or False?
- Key Terms Categories
- Fill-in-the-Blank
- Crossword Puzzle
- Audio Flash Cards
- Word Anatomy
- Supplemental Images
- Audio Pronunciation Glossary
- Health Professions Career Information
- Tips for Effective Studying
- Chapter Notes Outlines and Student Note-Taking Guides

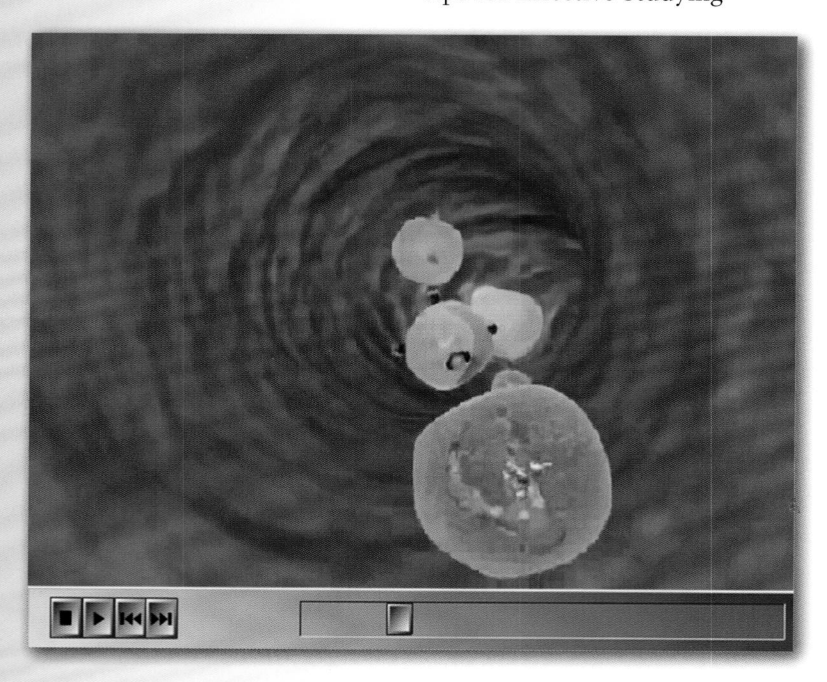

PrepU: An Integrated Adaptive Learning Solution

PrepU

PrepU, Lippincott's adaptive learning system, is an integral component of *Memmler's Structure and Function of the Human Body*.

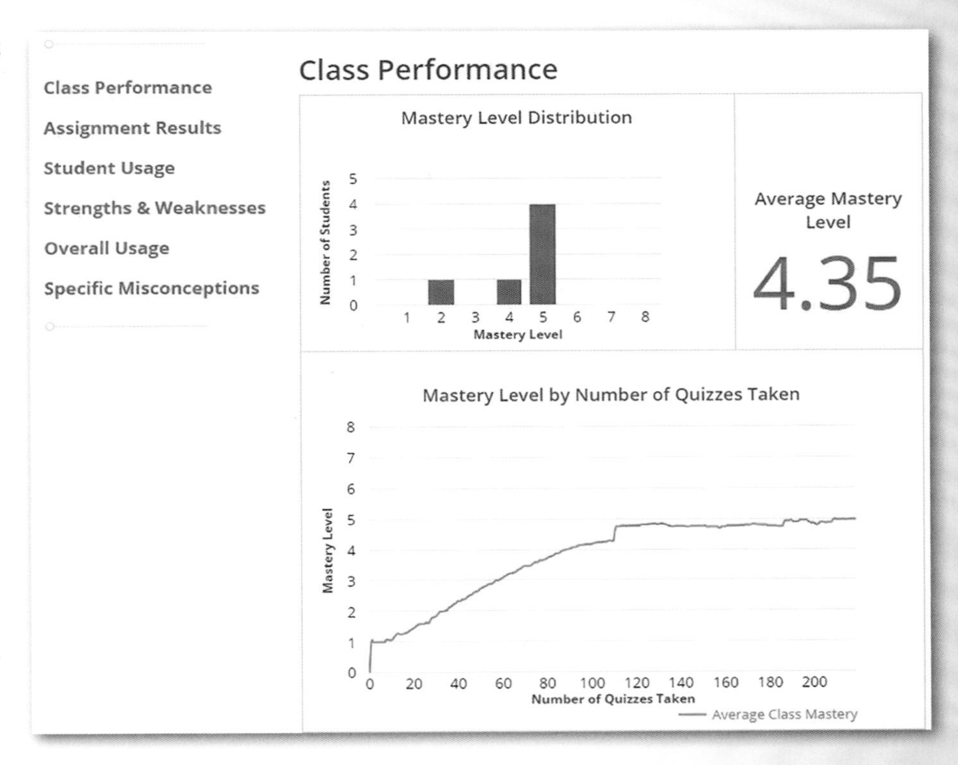

PrepU uses repetitive and adaptive quizzing to build mastery of A & P concepts, helping students to learn more while giving instructors the data they need to monitor each student's progress, strengths, and weaknesses. The hundreds of questions in PrepU offer students the chance to drill themselves on A & P and support their review and retention of the information they have learned. Each question provides not only an explanation for the correct answer but also references of the text page for the student to review the source material. PrepU for *Memmler's Structure and Function of the Human Body* challenges students with questions and activities that coincide with the materials they have learned in the text and gives students a proven tool to learn A & P more effectively. For instructors, PrepU provides tools to identify areas and topics of student misconception; instructors can use this rich course data to assess students' learning and better target their in-class activities and discussions, while collecting data that are useful for accreditation.

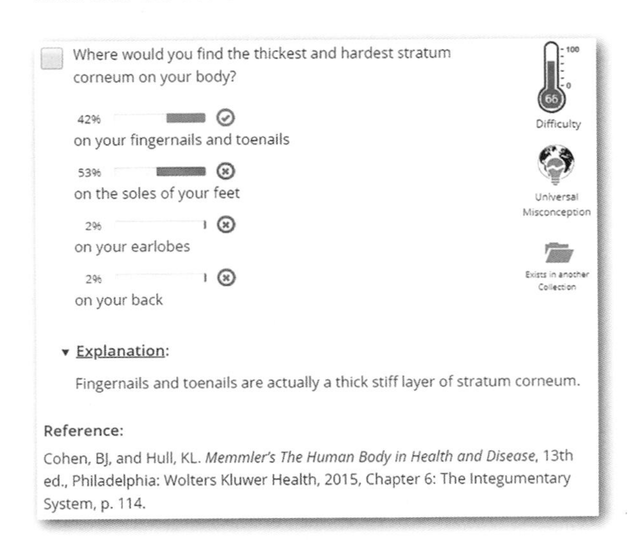

A learning experience individualized to each student. Being an adaptive learning engine, PrepU offers questions customized for each student's level of understanding, challenging students at an appropriate pace and difficulty level, while dispelling common misconceptions. As students review and master PrepU's questions, the system automatically increases the difficulty of questions, effectively driving student understanding of A & P to a mastery level. PrepU not only helps students to improve their knowledge but also helps foster their test-taking confidence.

PrepU works! PrepU works, and not just because we say so. PrepU efficacy is *backed by data*:

1. In an introductory nursing course at Central Carolina Technical College, student course outcomes were positively associated with PrepU usage. The students who answered the most PrepU questions in the class also had the best overall course grades.
2. In a randomized, controlled study at UCLA, students using PrepU (for biology) achieved 62% higher learning gains than those who did not.

To see a video explanation of PrepU, go to http://download.lww.com/wolterskluwer_vitalstream_com/mktg/prepuvid/prepupromo01.html

Study Guide for *Memmler's Structure and Function of the Human Body*, 12th edition

Kerry L. Hull, BSc, PhD
Barbara Janson Cohen, BA, MSEd

Along with the companion website on *thePoint*, this *Study Guide* is the ideal companion to the 12th edition of *Memmler's Structure and Function of the Human Body*. Following the text's organization chapter by chapter, the *Study Guide* provides a full range of self-study aids that actively engage students in learning and enable them to assess and build their knowledge as they advance through the text. Most importantly, the *Study Guide* allows students to get the most out of their study time, with a variety of exercises that meet the needs of all types of learners.

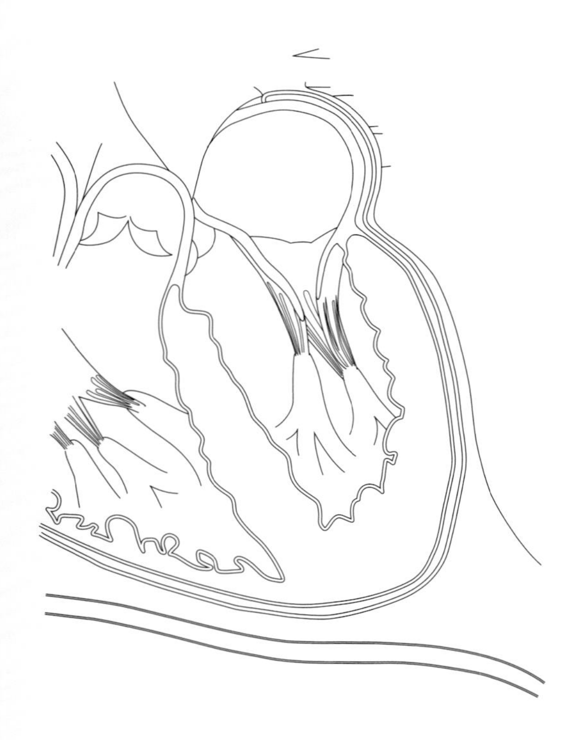

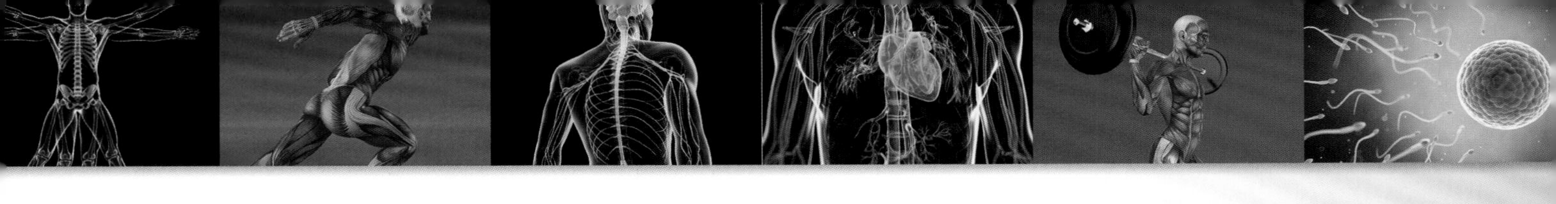

Inside the Study Guide you'll find the following:

- Chapter Overview summarizes the chapter's critical concepts.
- Addressing the Learning Objectives includes labeling, coloring, matching, and short-answer questions, all designed to foster active learning.
- Making the Connections integrates information from each chapter's learning objectives into concept mapping exercises.
- Testing Your Knowledge provides multiple choice, true/false, completion, short-answer, and essay questions to identify areas requiring further study. Practical Applications questions use clinical situations to test your understanding of a subject.
- Expanding Your Horizons helps students learn from the world around them and highlights emerging issues and discoveries in the health professions.

Visit www.lww.com, and reference

ISBN 978-1-9751-3894-3 to order your copy of this important resource.

Acknowledgments

It has been a continuing pleasure to work with Kerry Hull for this 12th edition of *Memmler's Structure and Function of the Human Body.* With her extensive scientific knowledge, familiarity with current pedagogic theory, and teaching experience, Kerry has added immeasurably to this text. She has also coauthored the Study Guide and prepared the Instructor's Manual. I could not have a better coworker or someone more trusted to carry on the traditions of this fine text.

The skilled staff at Wolters Kluwer, as always, has been instrumental in the development of these texts. Consistently striving for improvements and high quality, they have helped achieve the great success of these books over their long history. Thanks to the reviewers, listed separately, who made valuable comments on the text. Their suggestions and insights formed the basis for this revision.

As always, thanks to my husband, Matthew, an instructor in anatomy and physiology, who not only gives consistent support but also contributes advice and suggestions for the text.

—Barbara Janson Cohen

My greatest thanks go to Barbara Cohen. I have enormous respect for her writing skills, deep knowledge base, flexible outlook, and tireless attention to detail. It has been a pleasure and an honor to work with Barbara for the past 15 years, and I hope that our collaboration can continue for many more.

I echo Barbara in her thanks to Marthe Adler, Jay Campbell, Jonathan Joyce, John Larkin, Jennifer Clements, and Staci Wolfson, whose creativity and flexibility enabled us to produce the best book and learning package possible. My thanks also go out to the reviewers for their expertise and careful proofreading.

And, finally, my heartfelt thanks to my husband, Norman, and to my children, Evan and Lauren, for their constant encouragement and support.

—Kerry L. Hull

Brief Contents

Contents

Unit IV Circulation and Body Defense 241

The Body Visible

The Body Visible is a unique study tool designed to enhance your learning of the body's systems in this course and in your future work.

The Body Visible illustrates the systems discussed in the text in the same sequence in which they appear in the text. Each full-color detailed illustration also contains numbers and lines for identifying the structures in the illustration. A transparent overlay with labels for all of the numbered structures in the art accompanies each image.

With the labels in place, *The Body Visible* allows you to study each illustration and helps you learn the body's structures. When you view each system without the overlay in place, *The Body Visible* becomes a self-testing resource. As you test your knowledge and identify each numbered part, you can easily check your answers with the overlay.

Many of the images in *The Body Visible* have somewhat more detail than is covered in the text. We encourage you to keep *The Body Visible* available as a general reference and as a useful study tool as you progress to more advanced levels in your chosen healthcare career.

*The Body Visible** begins on the next page.

*The images in *The Body Visible* are adapted with permission from Anatomical Chart Company, Rapid Review: A Guide for Self-Testing and Memorization, 3rd ed. Philadelphia, PA: Lippincott Williams & Wilkins, 2010.

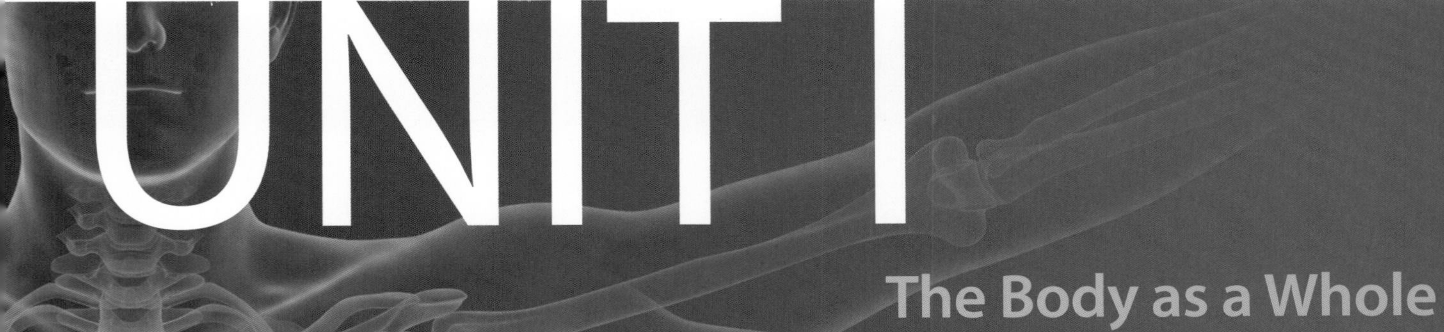

UNIT I

The Body as a Whole

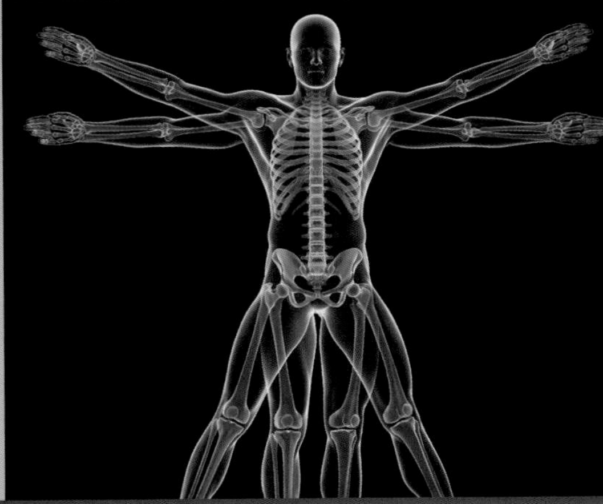

CHAPTER 1

Organization of the Human Body

Learning Objectives

After careful study of this chapter, you should be able to:

1. Define the terms *anatomy* and *physiology*. ***p. 4***
2. Describe the organization of the body from chemicals to the whole organism. ***p. 4***
3. List 11 body systems, and give the general function of each. ***p. 4***
4. Define and give examples of homeostasis. ***p. 5***
5. Using examples, discuss the components of a negative feedback loop. ***p. 6***
6. Explain the importance of barriers in the body, and give several examples of barriers. ***p. 7***
7. Define a gradient, and explain the effect of resistance on flow down a gradient. ***p. 8***
8. List and define the main directional terms for the body. ***p. 9***
9. List and define the three planes of division of the body. ***p. 10***
10. Name the subdivisions of the dorsal and ventral cavities. ***p. 12***
11. Name and locate the subdivisions of the abdomen. ***p. 13***
12. Cite some anterior and posterior body regions along with their common names. ***p. 14***
13. Find examples of anatomic and physiologic terms in the case study. ***pp. 3, 15***
14. Show how word parts are used to build words related to the body's organization (see Word Anatomy at the end of the chapter). ***p. 17***

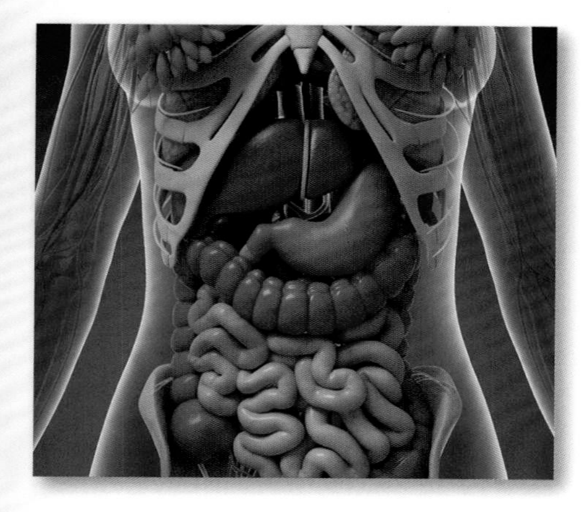

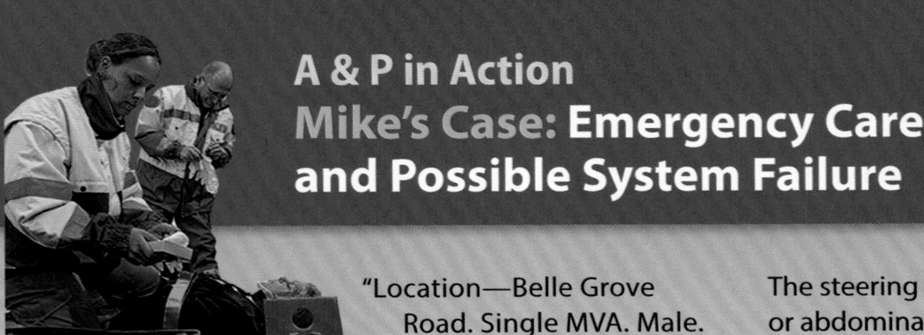

A & P in Action
Mike's Case: Emergency Care and Possible System Failure

"Location—Belle Grove Road. Single MVA. Male. Early 20s. Fire and police on scene," crackled the radio. "Medic 12. Respond channel 2."

"Medic 12 responding. En route to Belle Grove Road," Ed radioed back, while his partner, Samantha, flipped the switch for the lights and siren and hit the accelerator. When they arrived at the scene, police officers were directing traffic, and a fire crew was at work on the vehicle. Samantha parked the ambulance just as the crew breached the door of the crumpled minivan. Samantha and Ed grabbed their trauma bags and approached the wreck.

Ed bent down toward the injured man. "I hear your name is Mike. Mine is Ed. I'm a paramedic. My partner and I are going to take a quick look at you and then get you out of here." Samantha and Ed carefully immobilized Mike's cervical spine and with the help of the fire crew, transferred him to a stretcher. They immediately checked his blood pressure, pulse rate, and blood oxygen content. Samantha placed an oxygen mask over Mike's nose and mouth and watched to verify that he continued to breathe regularly.

After asking Mike if he was alone at the time of the crash, Samantha inspected the vehicle. "Looks like the impact sent him up and over the steering wheel. Guessing from the cracked windshield, he may have a head injury. The steering column is bent, so I wouldn't rule out thoracic or abdominal injuries either."

Ed agreed. "He's got forehead lacerations, and he's disoriented. Chest seems fine, but his abdominal cavity could be a problem. There is significant bruising across the left lower quadrant and periumbilical regions—probably from the steering wheel. When I palpated his left upper quadrant, it caused him considerable pain." Mike's blood pressure was very low, and his heart rate was very high—both signs of a cardiovascular emergency.

Ed then began a quick physical examination, first checking Mike's pupillary responses for any signs of intracranial pressure and then continuing inferiorly. Ed shared his findings with Samantha. "He's hypotensive and tachycardic. With the pain he reported earlier, signs are pointing to intra-abdominal hemorrhage. We've got to get him to the trauma center right now."

Ed and Samantha depend on their understanding of anatomy and physiology to help their patients and communicate with their partners. Ed suspects that Mike is bleeding internally and that his heart is working hard to compensate for the decrease in blood pressure. As we will see later, Mike's state of internal balance, known as homeostasis, must be restored, or his body systems will fail.

As you study this chapter, CasePoints will give you opportunities to apply your learning to this case.

Visit thePoint® to access the following resources. For guidance in using these resources most effectively, see pp. xv–xvii.

Preparing to Learn

- Tips for Effective Studying
- Pre-Quiz

While You Are Learning

- Web Figure: Abdominal Regions
- Web Figure: Abdominal Quadrants
- Web Chart: Body Systems
- Web Chart: Directional Terms
- Web Chart: The Metric System
- Web Chart: Abdominal Quadrants
- Animation: Structure-Function
- Animation: Levels of Organization
- Animation: Homeostasis and Negative Feedback
- Animation: Barriers and Gradients
- Chapter Notes Outline
- Audio Pronunciation Glossary

When You Are Reviewing

- Answers to Questions for Study and Review
- Health Professions: Health Information Technician
- Interactive Learning Activities

Introduction

Studies of the body's normal structure and functions are the basis for all medical sciences. It is only from understanding the normal that we can analyze what is going wrong in cases of disease. These studies give us an appreciation for the design and balance of the human body and for living organisms in general. This chapter discusses several general scientific concepts that provide the foundation for topics to come. These key ideas are indicated with a numbered icon ( ). The table in the front cover of your textbook summarizes all of the key ideas presented in this and later chapters. We will refer back to these key ideas in future chapters. The number on the icon indicates which key idea you should review to understand the topic.

Studies of the Human Body

The scientific term for the study of body structure is **anatomy** (ah-NAT-o-me). The *-tomy* part of this word in Latin means "cutting," because a fundamental way to learn about the human body is to cut it apart, or **dissect** (dis-sekt) it. **Physiology** (fiz-e-OL-o-je) is the term for the study of how the body functions; *physio* is based on a Latin term meaning "nature," and *logy* means "study of."

1 STRUCTURE AND FUNCTION

Anatomy and physiology are closely related—that is, structure and function are intertwined. The stomach, for example, has a pouchlike shape that is well suited for storing food during digestion. The cells in the lining of the stomach are tightly packed. This anatomic feature prevents strong digestive juices from harming underlying tissue.

2 LEVELS OF ORGANIZATION

All living things are organized from very simple levels to more complex levels (FIG. 1-1). Living matter is derived from chemicals, including simple substances, such as water and salts, and more complex materials, such as sugars, fats, and proteins. These chemicals assemble into living **cells**—the basic units of all life. Specialized groups of cells form **tissues**, such as muscle tissue and connective tissue. Tissues function together as **organs**, for example, a muscle contains both muscle tissue and connective tissue. Organs working together for the same general purpose make up the body **systems**, discussed below. All of the systems work together to maintain the body as a whole organism.

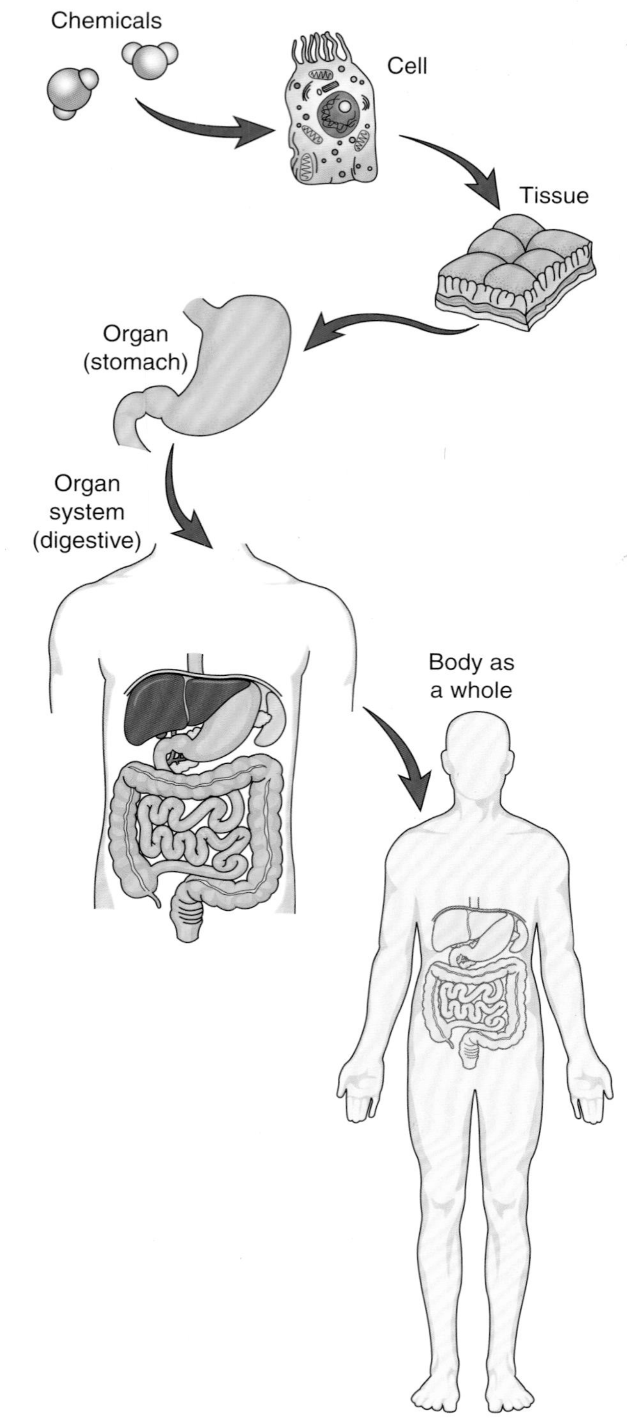

FIGURE 1-1 Levels of organization. **KEY POINT** The body is organized from the level of simple chemicals by increasing levels of complexity to the whole organism. The organ shown here is the stomach, which is part of the digestive system.

See the Student Resources on thePoint® for the animations "Structure-Function" and "Levels of Organization."

BODY SYSTEMS

Although all are interrelated, we can think of the human body as organized into individual systems. Listed according to their descriptions in this text, they are as follows:

- The **integumentary** (in-teg-u-MEN-tar-e) system. The word integument (in-TEG-u-ment) means "skin." The skin with its associated structures is our outermost body system. It protects us from injury and infection, but serves other functions as well. The skin's associated structures include the hair, nails, sweat glands, and oil glands.

The Integumentary System

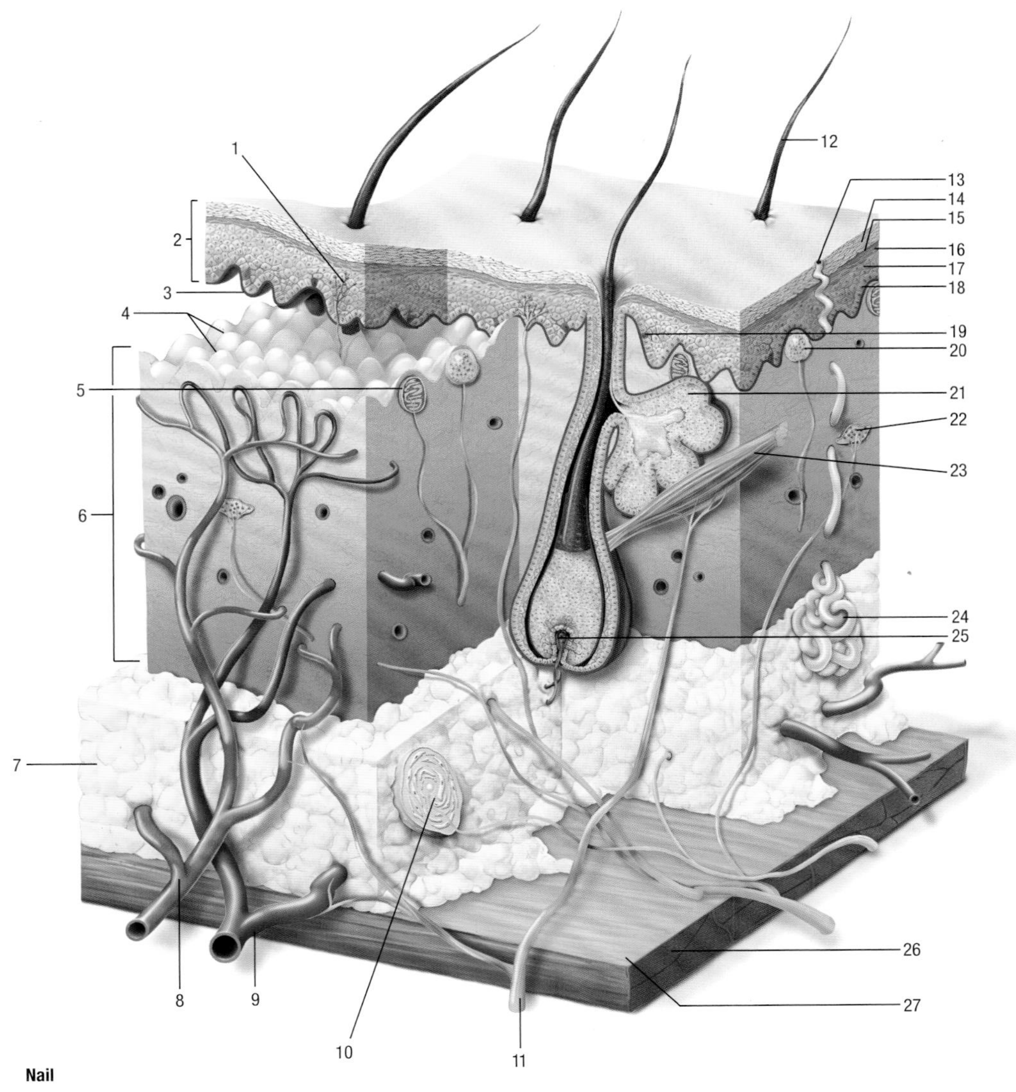

Nail

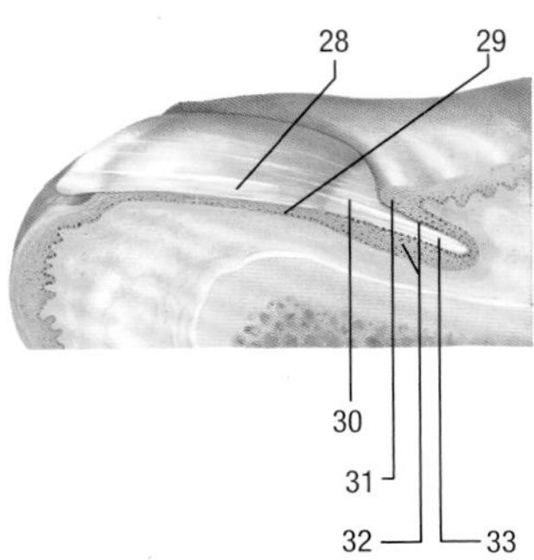

The Skeletal System

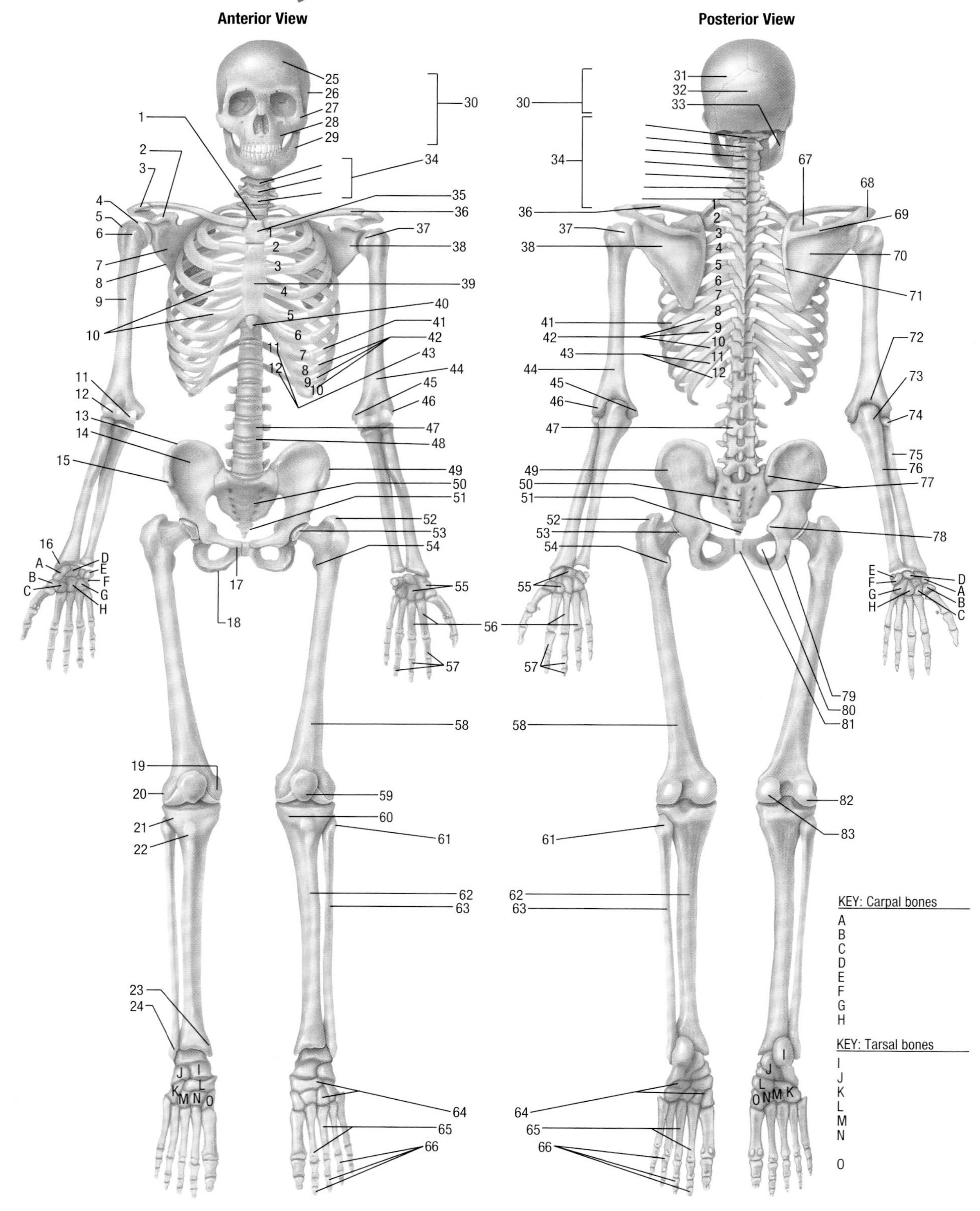

The Muscular System—Anterior View

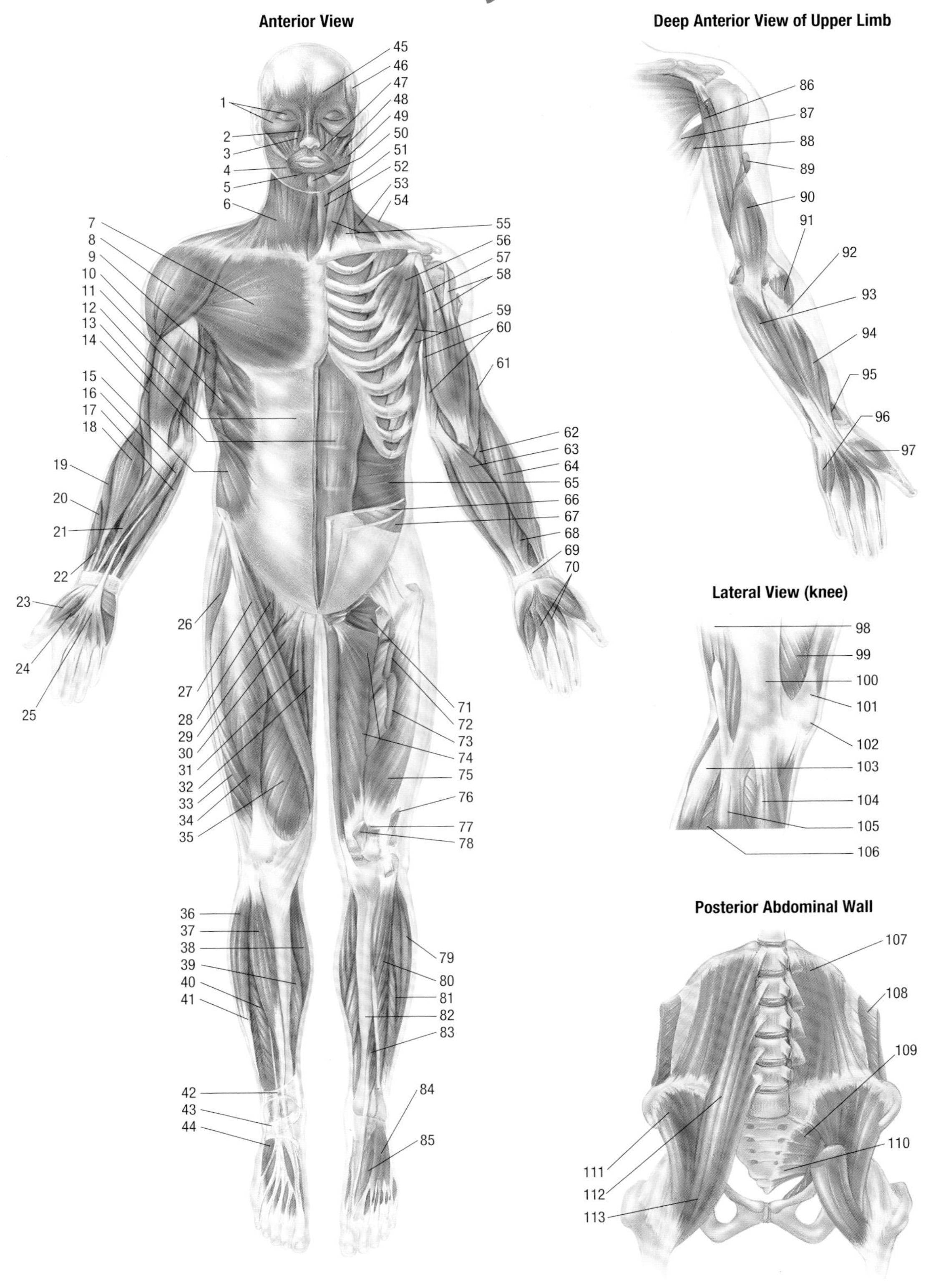

The Muscular System—Posterior View

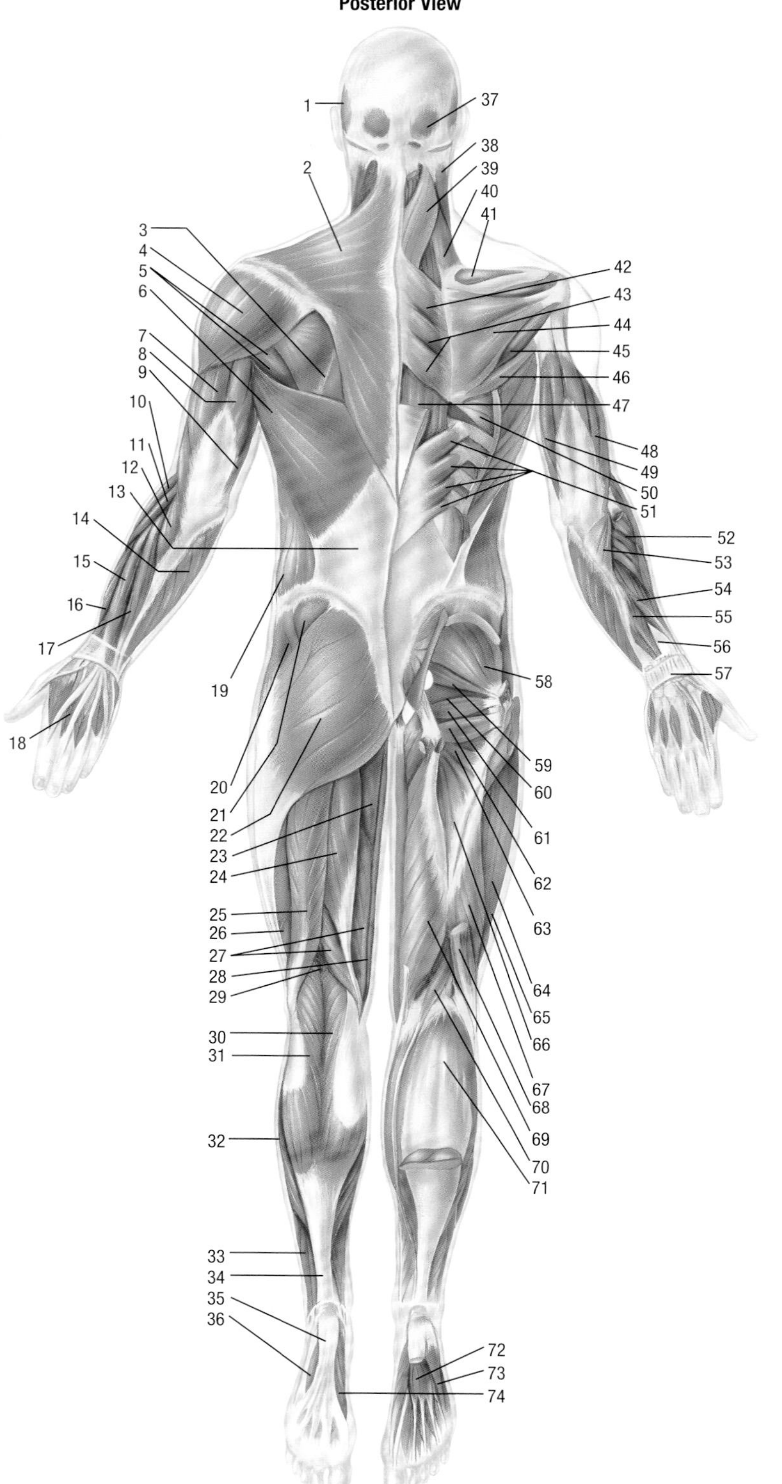

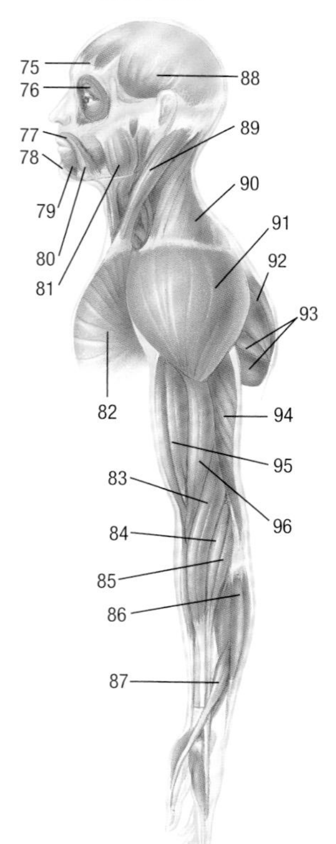

The Peripheral Nervous System

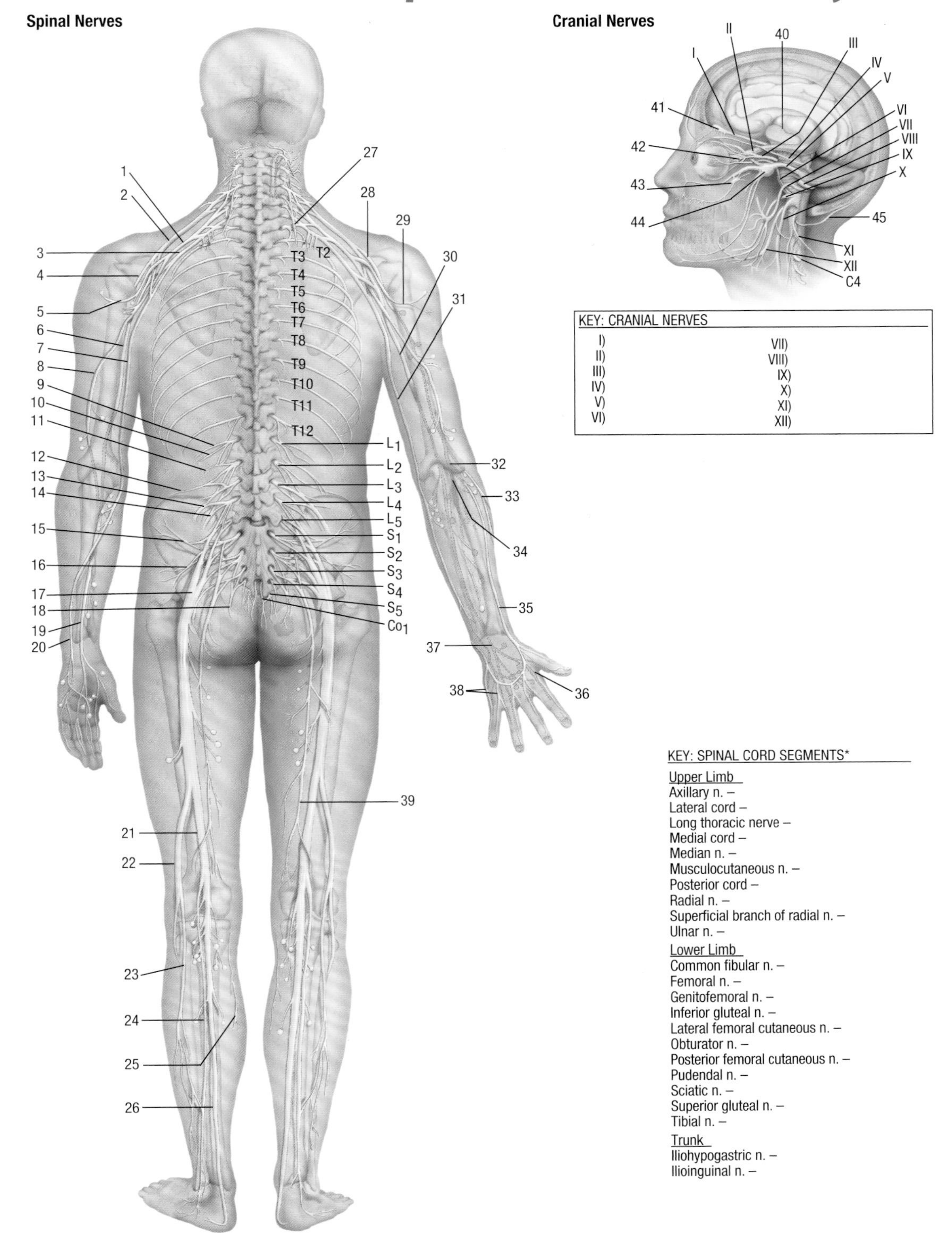

KEY: CRANIAL NERVES	
I)	VII)
II)	VIII)
III)	IX)
IV)	X)
V)	XI)
VI)	XII)

KEY: SPINAL CORD SEGMENTS*

Upper Limb
Axillary n. –
Lateral cord –
Long thoracic nerve –
Medial cord –
Median n. –
Musculocutaneous n. –
Posterior cord –
Radial n. –
Superficial branch of radial n. –
Ulnar n. –

Lower Limb
Common fibular n. –
Femoral n. –
Genitofemoral n. –
Inferior gluteal n. –
Lateral femoral cutaneous n. –
Obturator n. –
Posterior femoral cutaneous n. –
Pudendal n. –
Sciatic n. –
Superior gluteal n. –
Tibial n. –

Trunk
Iliohypogastric n. –
Ilioinguinal n. –

The Brain

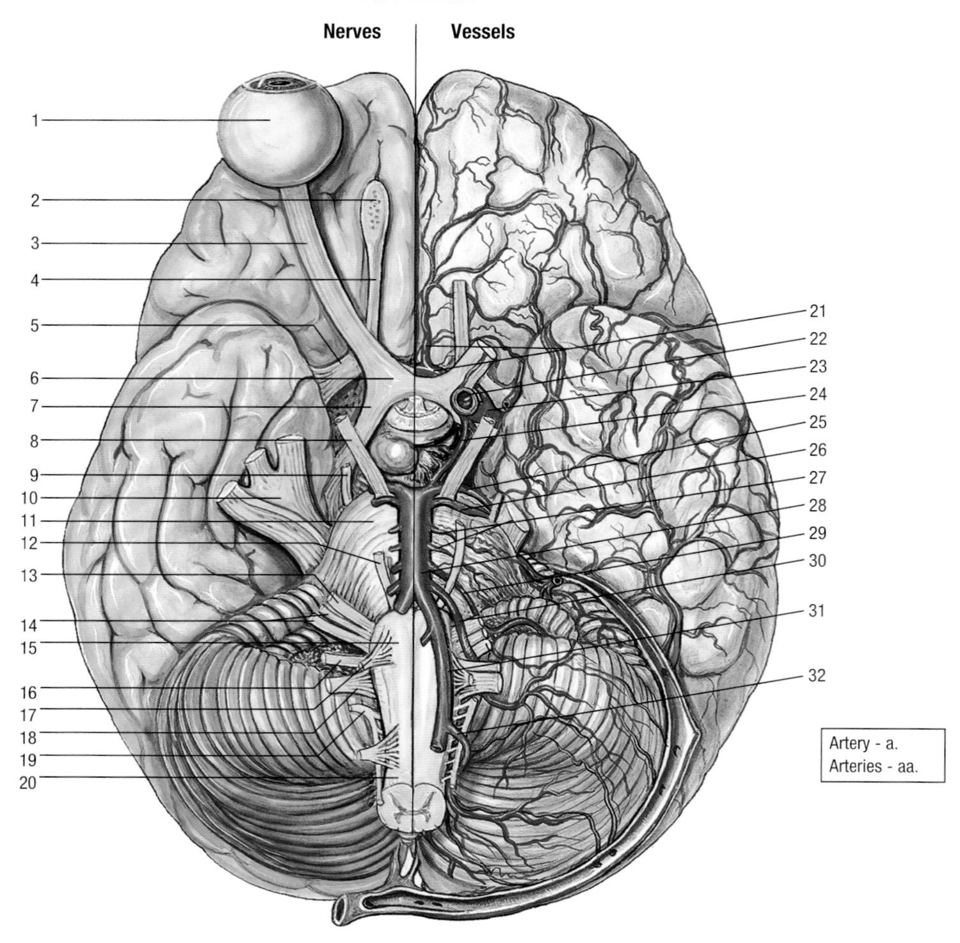

Coronal Section

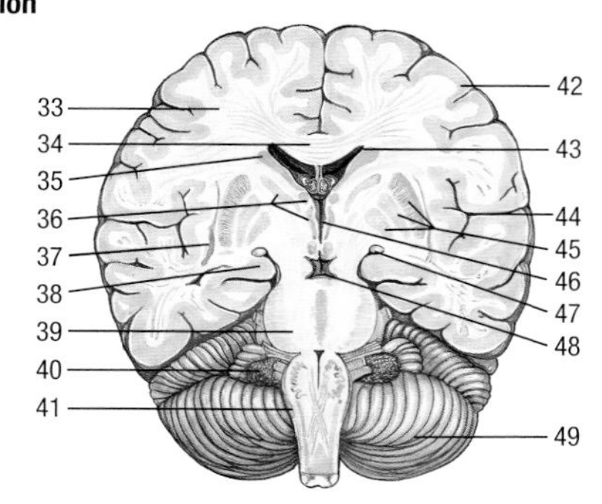

Lobes

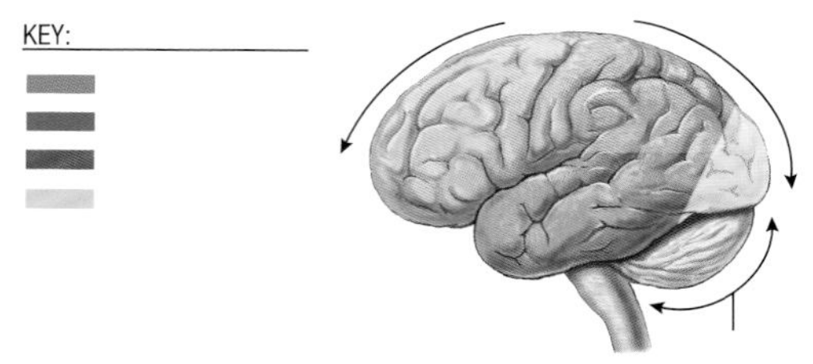

The Eye

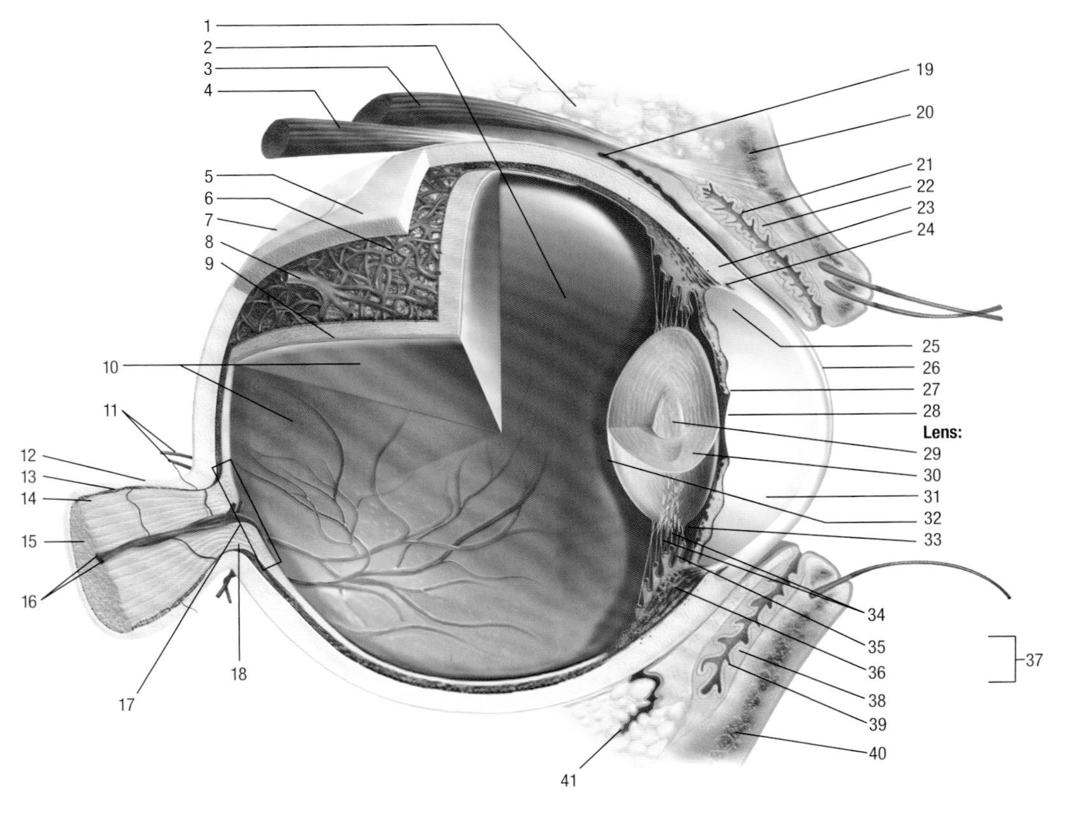

Lacrimal Gland:

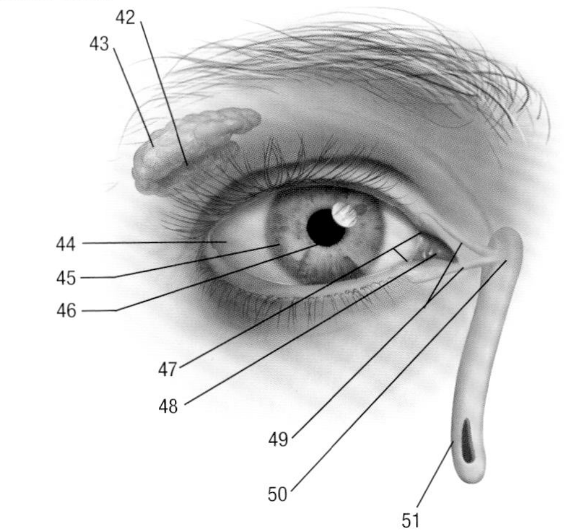

Eye Muscles Superior View

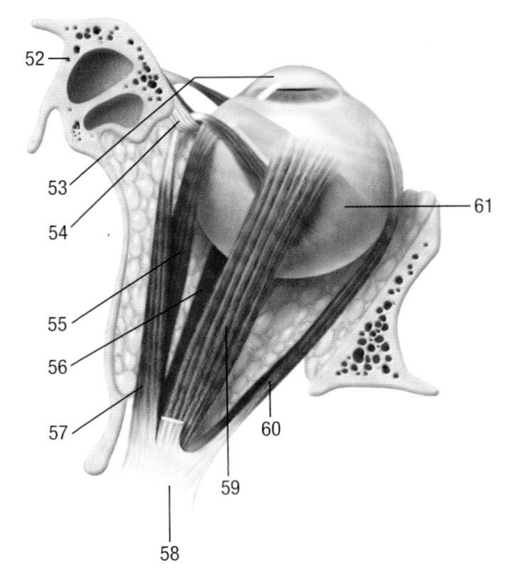

The Ear

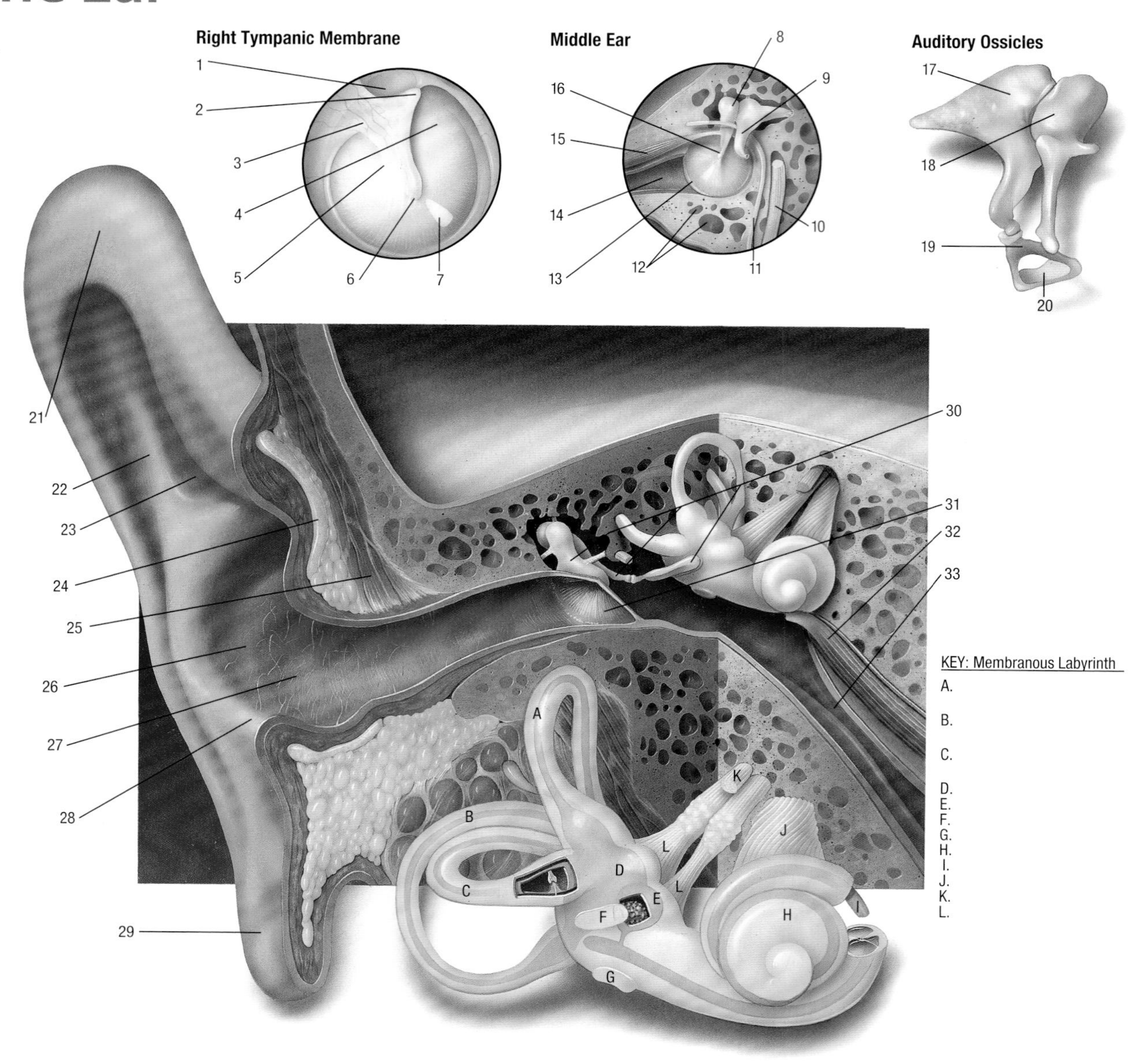

KEY: Membranous Labyrinth

A.

B.

C.

D.
E.
F.
G.
H.
I.
J.
K.
L.

The Heart

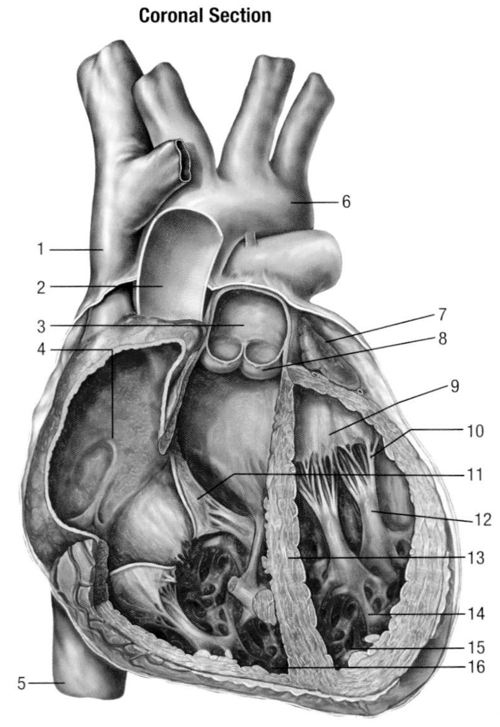

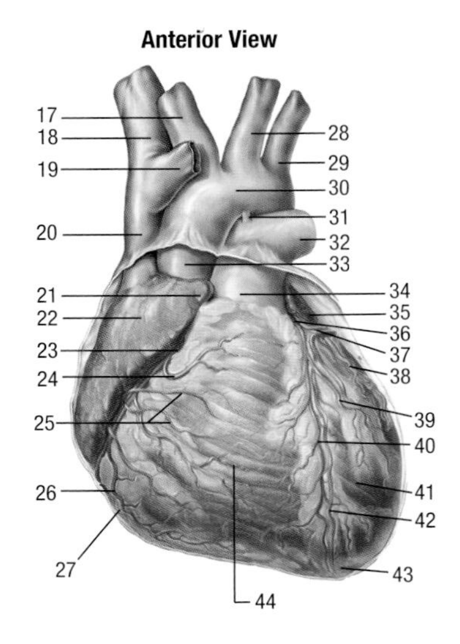

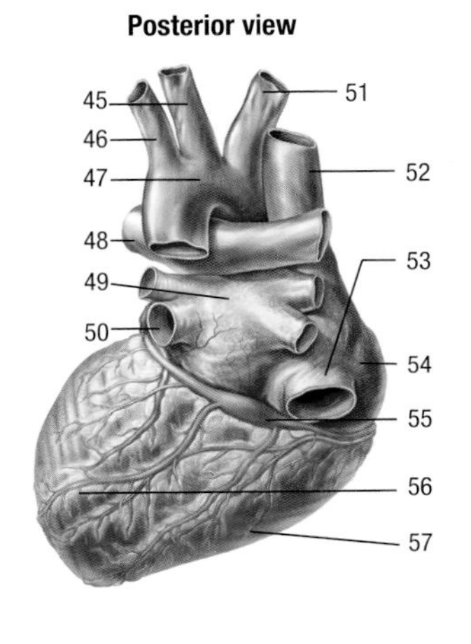

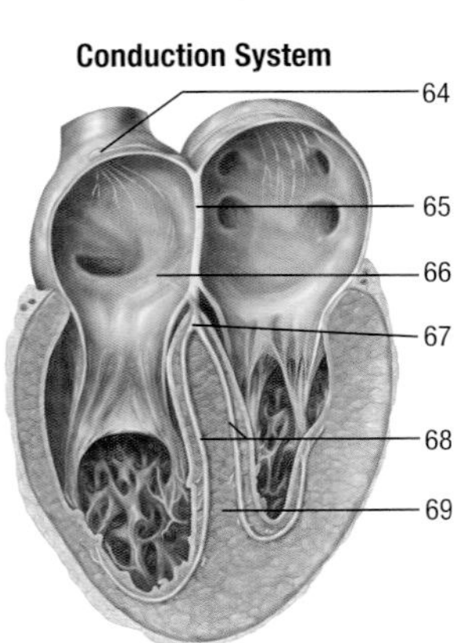

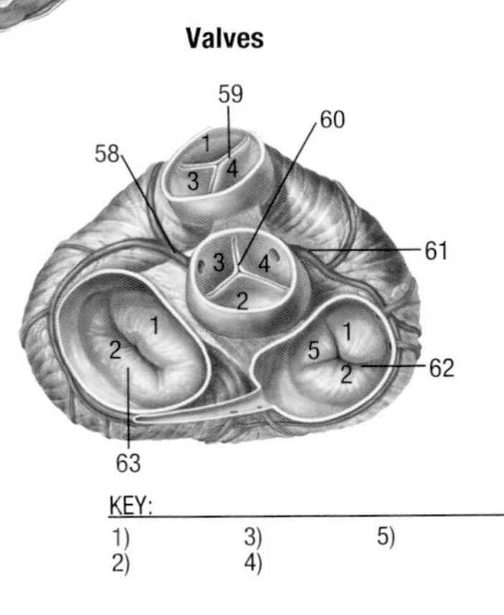

KEY:
1)
2)
3)
4)
5)

The Arteries

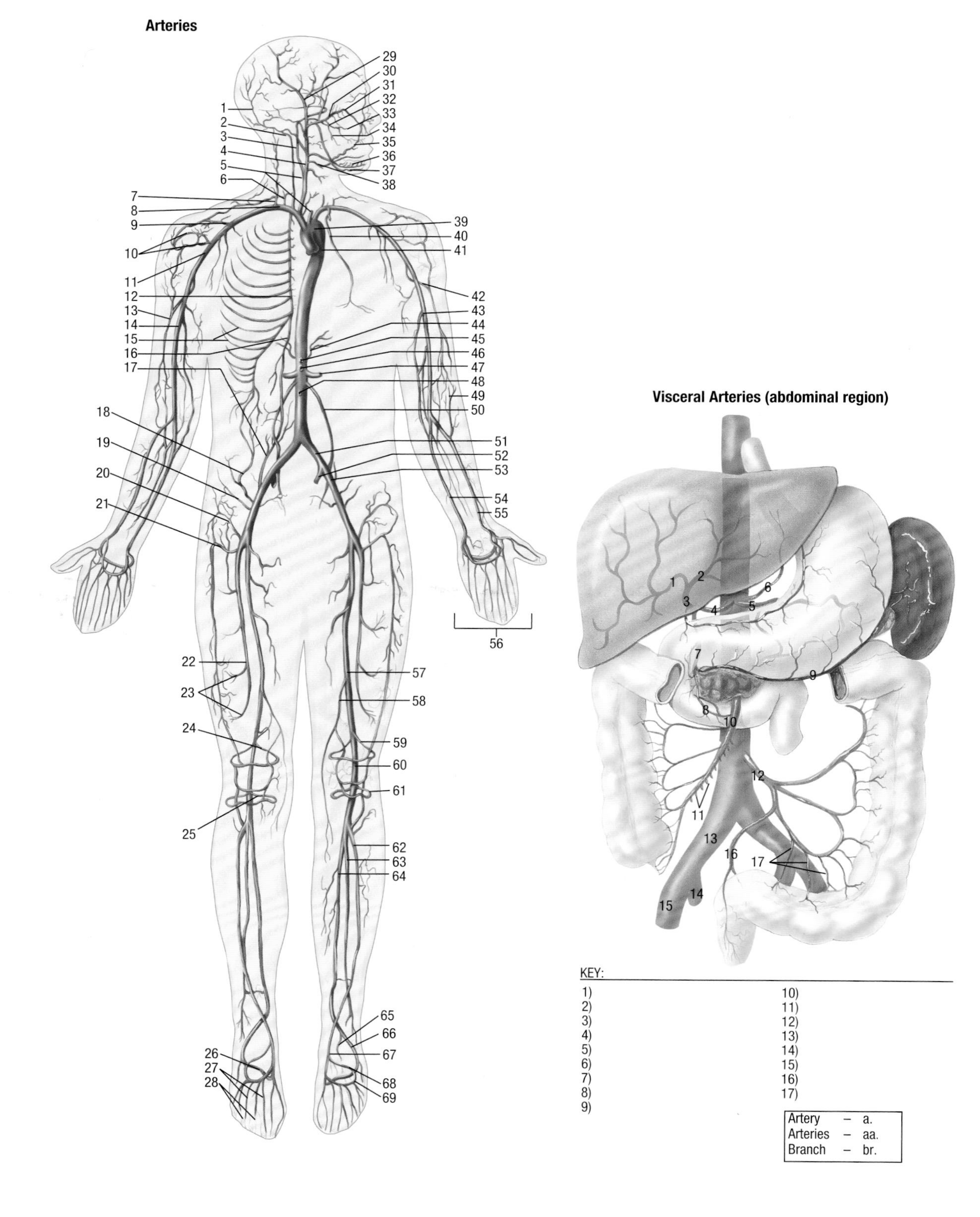

The Veins

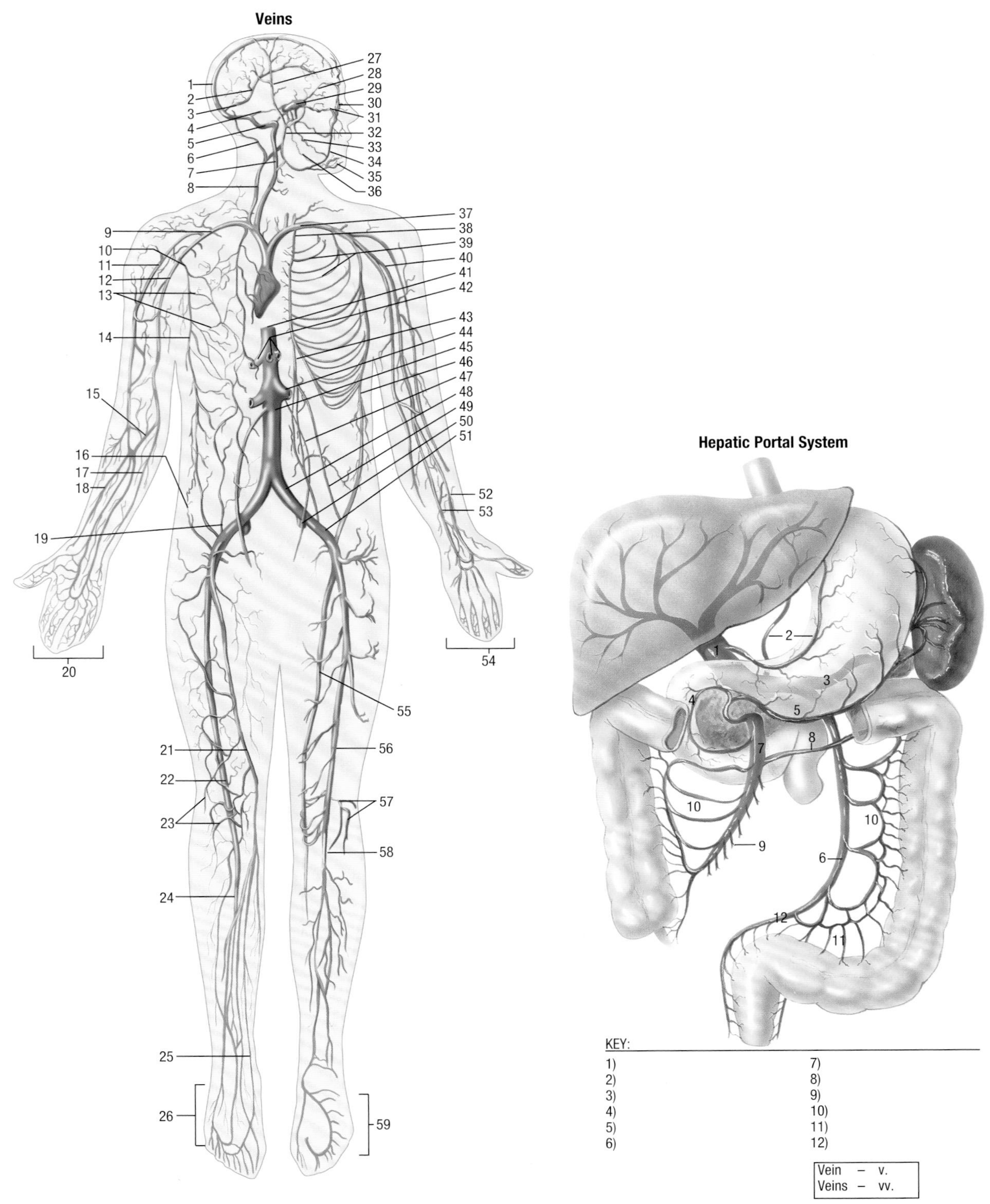

KEY:

1)	7)
2)	8)
3)	9)
4)	10)
5)	11)
6)	12)

Vein	– v.
Veins	– vv.

The Respiratory System

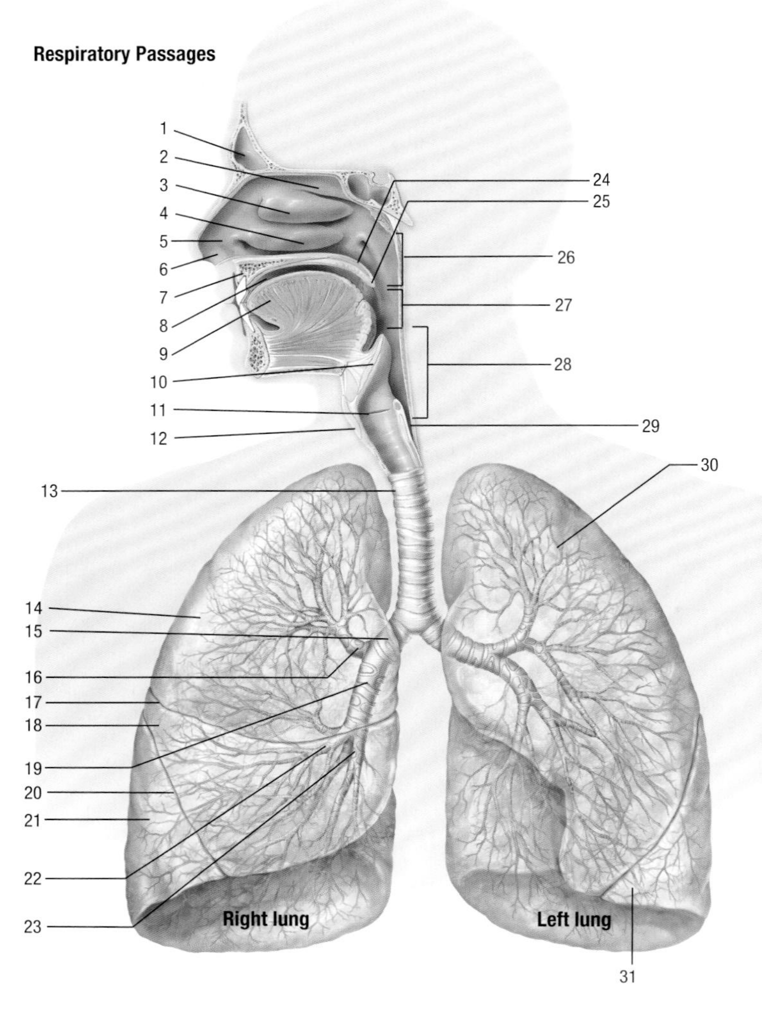

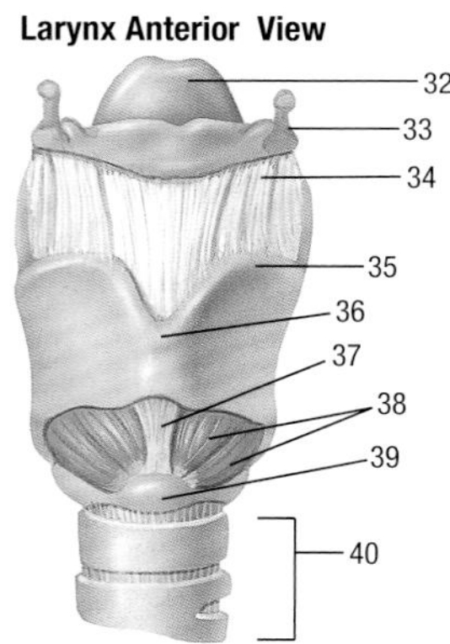

The Digestive System

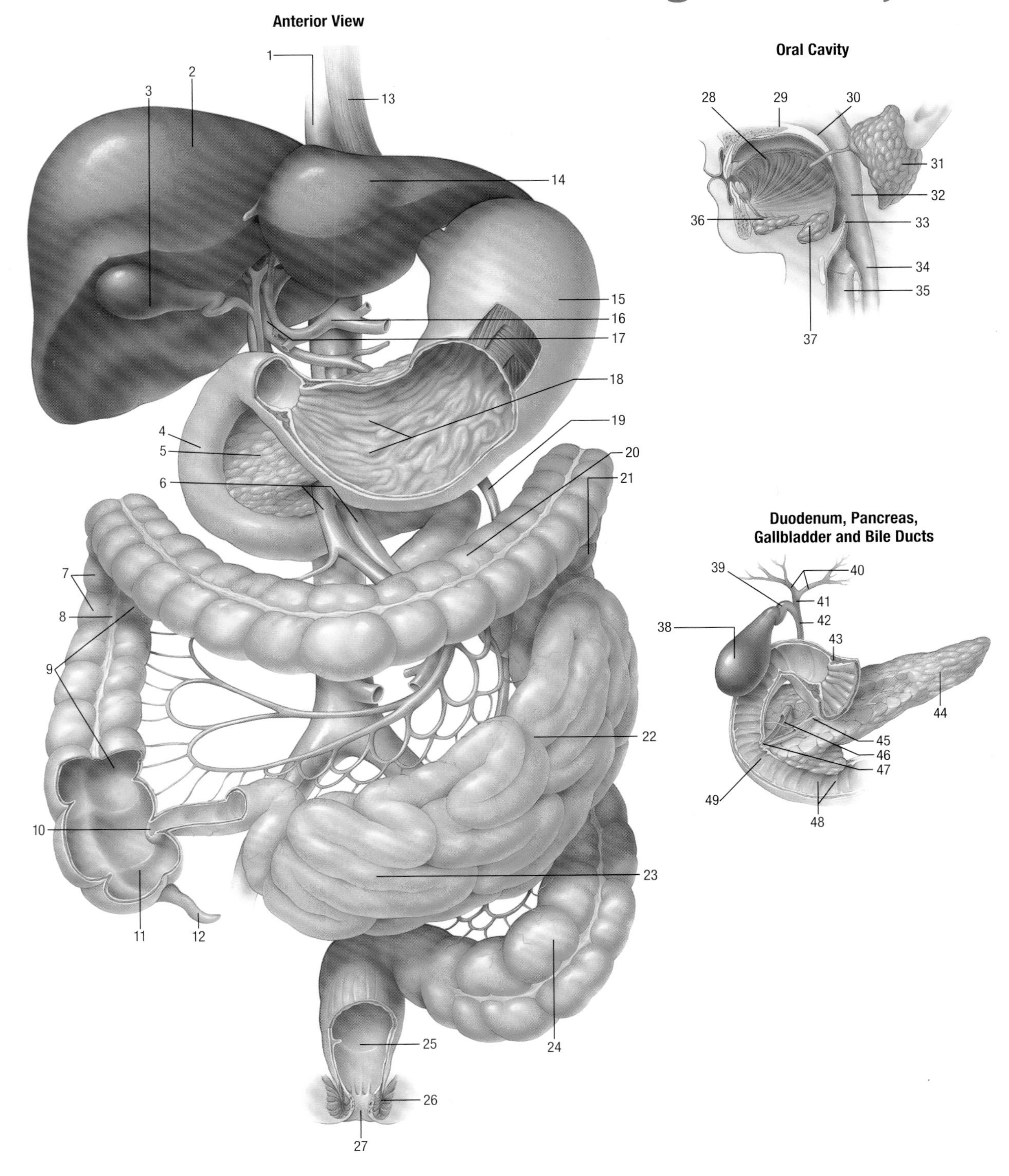

The Urinary System

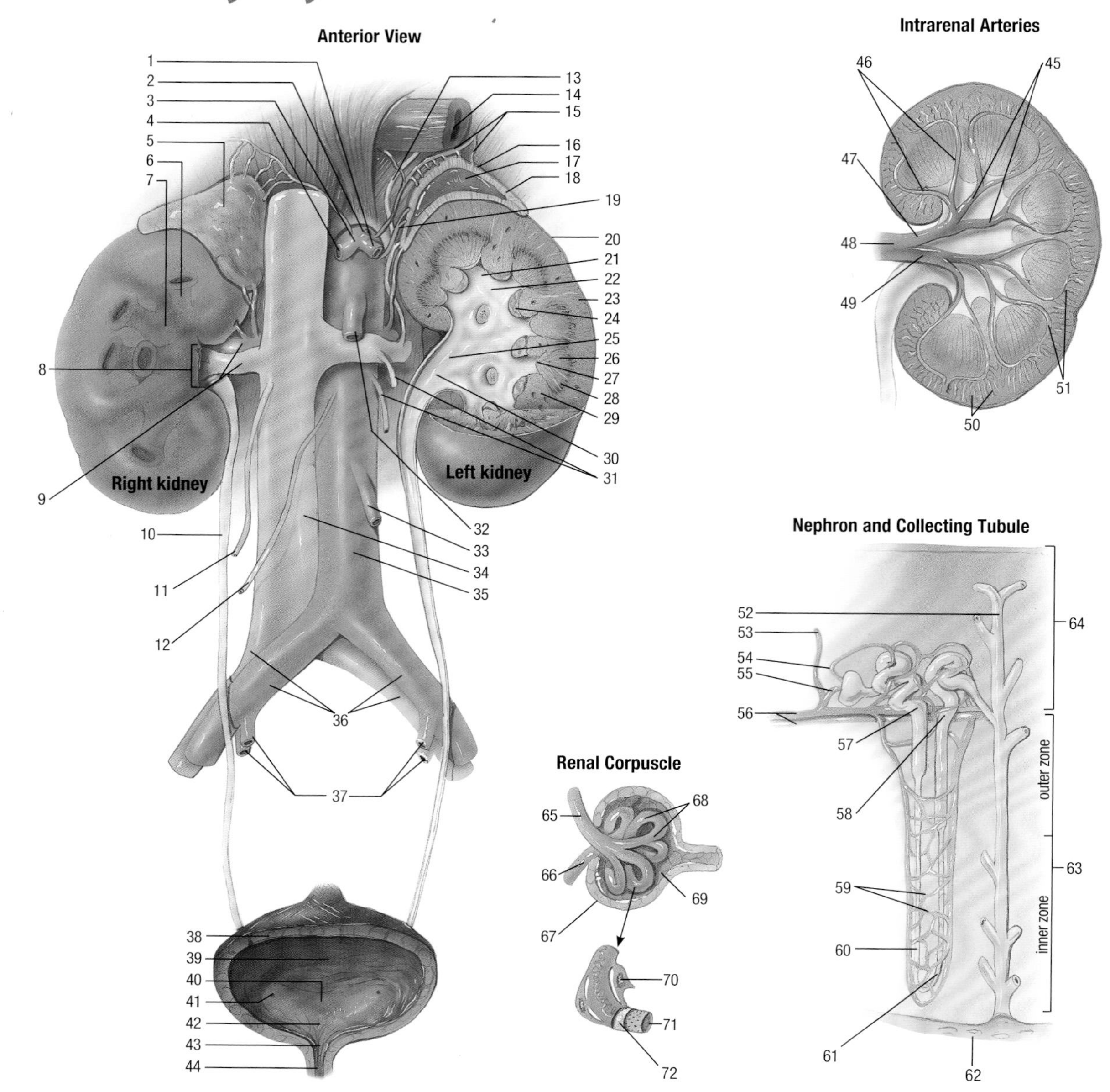

The Male Reproductive System

Pelvic Organs (median section)

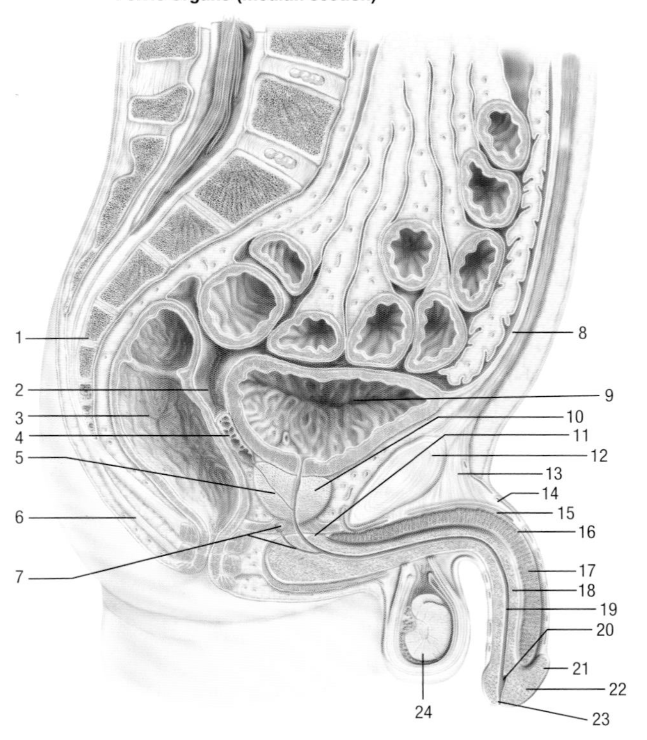

Anterior View (oblique section)

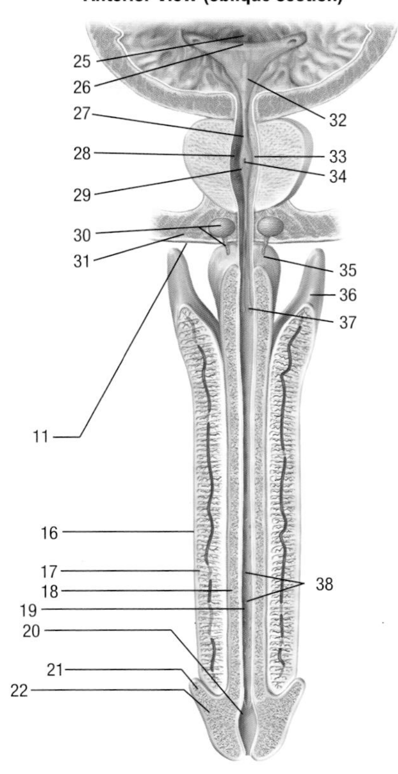

The Female Reproductive System

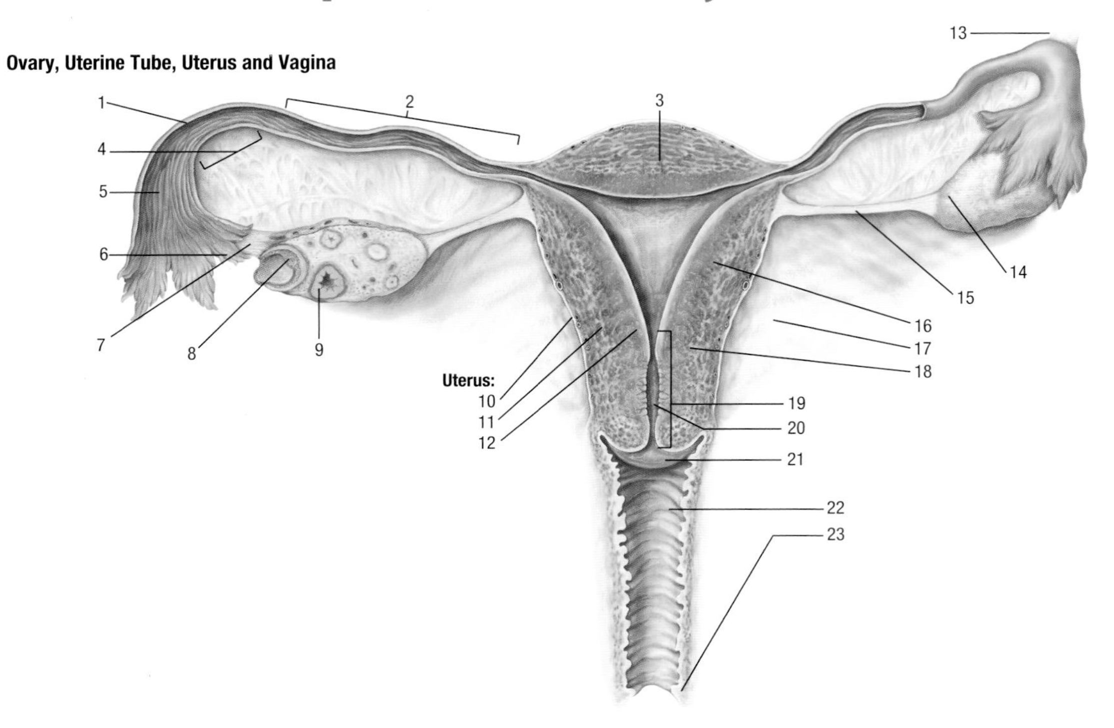

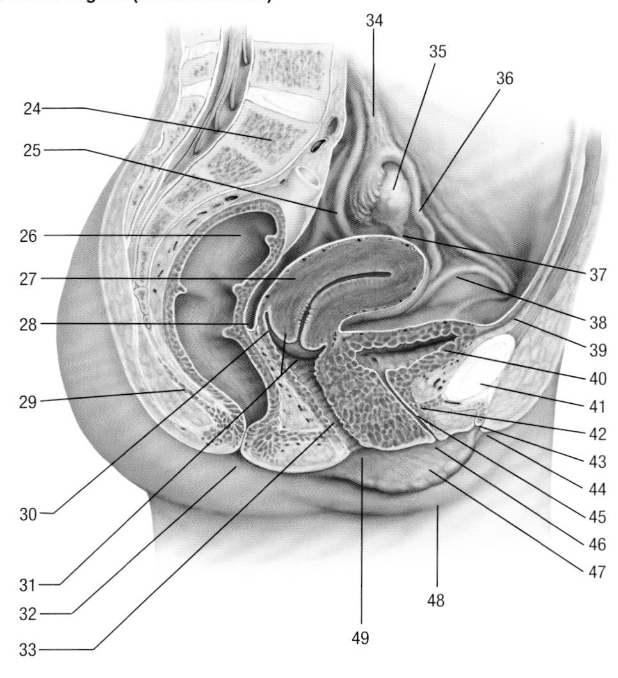

- The **skeletal system**. The body's basic framework is a system of 206 bones and the joints between them, collectively known as the skeleton.
- The **muscular system**. The muscles in this system are attached to the bones and produce movement of the skeleton. These skeletal muscles also give the body structure, protect organs, and maintain posture. Two additional types of muscle contribute to other body systems. Smooth muscle is present in the walls of blood vessels and many organs; cardiac muscle constitutes the bulk of the heart wall.
- The **nervous system**. The brain, spinal cord, and nerves make up this complex system by which the body is controlled and coordinated. The nervous system also includes the special sense organs (the eyes, ears, taste buds, and organs of smell) and the receptors of the general senses, such as pain and touch. When sense organs or receptors detect changes in the external and internal environments, electrical signals are transmitted along nerves to the brain, which directs responses.
- The **endocrine** (EN-do-krin) **system**. The scattered organs known as endocrine glands are grouped together because they share a similar function. All produce special substances called hormones, which regulate such body activities as growth, nutrient utilization, and reproduction. Examples of endocrine glands are the thyroid, pituitary, and adrenal glands. The nervous system and the endocrine system together control bodily functions and responses.
- The **cardiovascular system**. The heart and blood vessels make up the system that pumps blood to all body tissues, bringing with it nutrients, oxygen, and other needed substances. This system then carries waste materials away from the tissues to points where they can be eliminated.
- The **lymphatic system**. Lymphatic vessels assist in circulation by returning fluids from the tissues to the blood. Lymphatic organs, such as the tonsils, thymus, and spleen, play a role in immunity, protecting against disease. The lymphatic system also aids in the absorption of dietary fats. The fluid that circulates in the lymphatic system is called lymph.
- The **respiratory system**. This system includes the lungs and the passages leading to and from the lungs. This system takes in air and conducts it to the areas in the lungs designed for gas exchange. Oxygen passes from the air into the blood and is carried to all tissues by the cardiovascular system. In like manner, carbon dioxide, a gaseous waste product, is taken by the circulation from the tissues back to the lungs to be expelled through the respiratory passages.
- The **digestive system**. This system is composed of all the organs that are involved with taking in nutrients (foods), converting them into a form that body cells can use, and absorbing them into the circulation. Organs of the digestive system include the mouth, esophagus, stomach, small and large intestines, liver, gallbladder, and pancreas. The respiratory system and digestive system together generate energy to fuel all body activities.
- The **urinary system**. The chief purpose of the urinary system is to rid the body of waste products and excess water. This system's components are the kidneys, the ureters, the bladder, and the urethra. (Note that some waste products are also eliminated by the digestive and respiratory systems and by the skin.) The urinary system is the main mechanism for balancing the volume and composition of body fluids.
- The **reproductive system**. This system includes the external sex organs and all related internal structures that are concerned with the production of offspring.

References may vary in the number of body systems cited. For example, some separate the sensory system from the nervous system. Others have a separate entry for the immune system, which protects the body from foreign matter and invading organisms. The immune system is identified by its function rather than its structure and includes elements of both the cardiovascular and lymphatic systems. Bear in mind that even though you will study the systems as separate units, they are interrelated and must cooperate to maintain health.

CASEPOINT

☐ **1-1** The paramedics were concerned about Mike's spinal column and his blood vessels. To which system(s) do these structures belong?

See Appendix 2 for answers to the CasePoint questions

THE EFFECTS OF AGING

With age, changes occur gradually in all body systems. Some of these changes, such as wrinkles and gray hair, are obvious. Others, such as decreased kidney function, loss of bone mass, and formation of deposits within blood vessels, are not visible. However, they may make a person more subject to injury and disease. Changes due to aging are described in chapters on the body systems.

CHECKPOINTS

☐ **1-1** What are the studies of body structure and body function called?

☐ **1-2** What do organs working together combine to form?

See Appendix 2 for answers to the checkpoint questions

See the Student Resources on thePoint® for a chart summarizing the body systems and their functions.

3 Homeostasis

Despite changing environmental conditions, normal body function maintains a state of internal balance or constancy known as **homeostasis** (ho-me-o-STA-sis). Any condition that is subject to change is termed a variable. In the body, certain conditions, described as **regulated variables,** must remain within a somewhat narrow range, or *set point*, if we are to stay healthy. Regulated variables include body

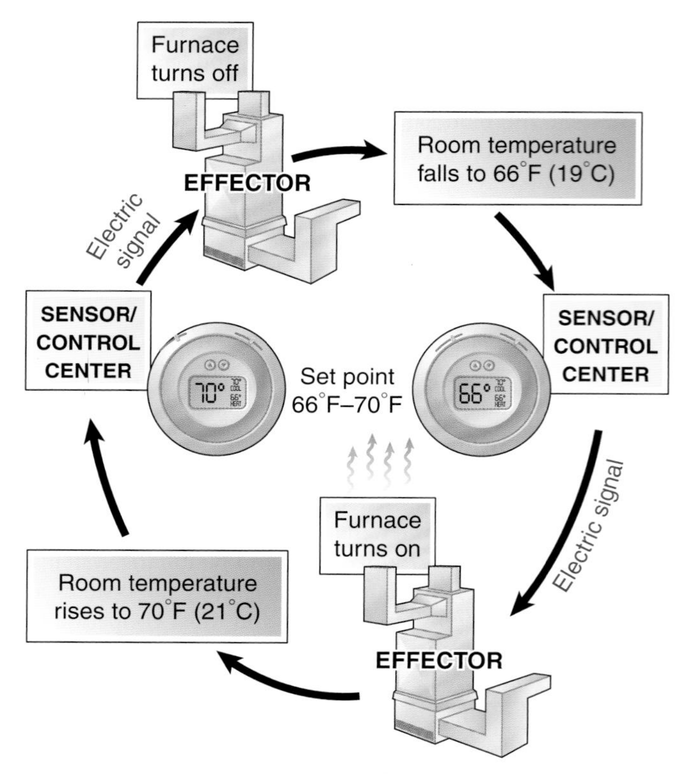

FIGURE 1-2 Negative feedback. **KEY POINT** A home thermostat illustrates how negative feedback keeps temperature within a set range. This thermostat is set to keep the average temperature at 68°F. Note that the different components of the feedback loop are indicated by different colors (regulated variable, *green*; sensors and control centers, *yellow*; signals, *purple*; effectors, *blue*).

temperature, the volume and composition of body fluids, blood gas concentrations, and blood pressure. Disease can be described as a state of homeostatic imbalance.

NEGATIVE FEEDBACK

The main method for maintaining homeostasis is **negative feedback**, a control system based on information returning to a source. We are all accustomed to getting feedback about the results of our actions and using that information to regulate our behavior. Poor marks on tests and assignments, for example, may inspire us to work harder to reverse the downward slide of our grades.

Negative feedback systems keep regulated variables within a set normal range by reversing any upward or downward shift. Note that negative feedback doesn't always mean less response; it just means an opposite response to a stimulus. Any negative feedback loop must contain three components:

- A **sensor** gathers information about a specific variable.
- A **control center** compares the sensor inputs with the set point.
- An **effector** increases or decreases its activity, as necessary, in response to signals from the control center. The actions of the effector restore the normal level of the regulated variable.

A familiar example of negative feedback is the regulation of room temperature (the regulated variable) by means of a thermostat (FIG. 1-2). The user determines the desired room temperature (the set point). Within the thermostat, a thermometer (sensor) measures the actual room temperature, and other components (forming the control center) compare the measured temperature with the set point. If the measured temperature is too low, the control center signals the furnace (effector) to produce heat and increase room temperature. When the room temperature reaches an upper limit (as detected by the sensor), the control center shuts off the furnace. The control center regulates the effector by means of electrical signals traveling through wires.

In the body, sensors in the brain and other organs constantly monitor body temperature and send signals to a specific brain region (the control center). This center activates effectors (such as sweat glands) to cool or warm the body if body temperature deviates above or below the set point of approximately 37°C (98.6°F) (FIG. 1-3). As with our thermostat example, electrical impulses transmitted through the nervous system act as signals between the components of the feedback loop. This example highlights two important

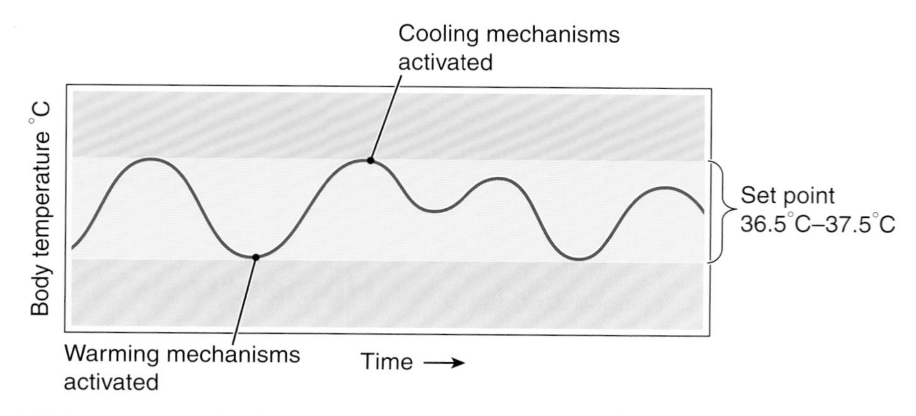

FIGURE 1-3 Negative feedback and body temperature. **KEY POINT** Body temperature is kept within a narrow range by negative feedback acting on a center in the brain. The set point is 37°C.

characteristics of regulated variables: they have specific sensors, and their level cannot vary too much without harming the organism. Many parameters that the body controls are not considered regulated variables. For instance, sweating is controlled. But sweat production is the action of an effector (sweat glands) that helps maintain the regulated variable of body temperature. The body does not directly sense how much sweat is being produced, and sweat production can vary enormously over the course of the day.

As another example, let's say you've just finished eating breakfast—a bowl of cereal and a glass of orange juice. As a result, the level of glucose (a simple sugar) increases in your bloodstream. Blood glucose levels must be tightly regulated to prevent disease, so this increase constitutes a challenge to homeostasis. Endocrine cells in your pancreas act as both sensors and the control center; they sense blood glucose levels and compare them with the set point. When blood glucose levels rise above the set point, pancreatic cells increase the production of a hormone called insulin. Insulin acts as the signal, traveling through the bloodstream from the pancreas to muscle and fat cells (the effectors). Increased glucose uptake and the subsequent drop in the blood glucose level cause the pancreas to reduce insulin secretion, and homeostasis is restored (**FIG. 1-4**). Many more examples of internal regulation by negative feedback will appear throughout this book, so study **FIGURES 1-2 through 1-4** closely. Note that we use a consistent color scheme to indicate the different components:

- Regulated variable: green
- Sensor and control center: yellow
- Signal: purple
- Effector and effector actions: blue

Clearly, negative feedback systems are critical for maintaining our health. **BOX 1-1** uses the case study to explain how to analyze feedback loops and what happens when feedback cannot restore homeostasis.

See the Student Resources on thePoint® for the animation "Homeostasis and Negative Feedback."

4 BARRIERS

The body's ability to maintain homeostasis relies on the barriers separating the relatively constant internal environment from the rapidly changing external environment. During most of our lives, our insides consist of a warm, watery, and sterile environment that is very different from the colder, drier, microbe-riddled external world. Skin forms an effective barrier separating inside from outside—as discussed in Chapter 5, burn patients suffer from fluid loss and frequent infections when they lose the barrier function of skin. The body's tubes—the digestive tract, the airways, the reproductive tract, and the urinary tract—are also considered as "outside." A thinner, moister barrier known as a *mucous membrane* forms the barrier in these locations.

At the microscopic level, the plasma membrane forms a barrier separating the internal environment of cells from their external environment. Chapter 3 discusses the plasma membrane in more detail. The fluid within cells is called **intracellular fluid** (the prefix *intra-* means "within"). The composition of the intracellular fluid differs from that of the fluid outside the cell, known as the **extracellular fluid** (the prefix *extra-* means "outside"). Examples of extracellular fluids are the blood plasma (the fluid portion of blood), lymph, and the

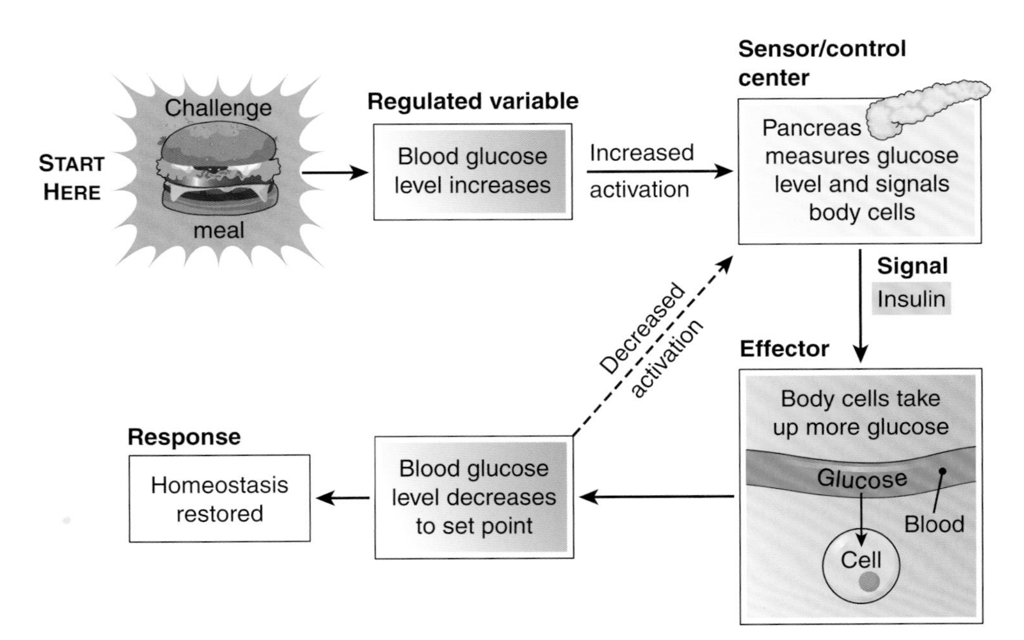

FIGURE 1-4 Negative feedback in the endocrine system. Note that the different components of the feedback loop are indicated by different colors (regulated variable, *green*; sensors and control centers, *yellow*; signals, *purple*; effectors, *blue*). **KEY POINT** The pancreas regulates blood glucose concentration using insulin as the signal.

ONE STEP AT A TIME BOX 1-1

Deciphering Negative Feedback Loops

Homeostasis and negative feedback are key concepts in your study of anatomy and physiology. It can be difficult to identify the components of a negative feedback loop, so here is a worked example to guide your future analysis of such loops. Note that the answer does not identify all of the components. In life, we rarely have all of the information that we need to make a decision. It is important to know what we know and what we do not know!

Question

Read through the opening case study, which discusses a negative feedback loop controlling blood pressure. Focus on these two sentences: "He suspects that Mike is bleeding internally and that his heart is working hard to compensate for the drastic decrease in blood pressure. As we will see later, Mike's state of internal balance, known as homeostasis, must be restored, or his body systems will fail." Using **FIGURE 1-4** as your guide, draw out the negative feedback loop involved. Use question marks for the components that you cannot identify from the case description.

Answer

Step 1: Identify the regulated variable. The question wording tells us the answer to this step—blood pressure. The case study tells us that Mike has a "drastic decrease in blood pressure." Regulated variables are colored *green* in the color-coded textbook diagrams. Remember that regulated variables (unlike other clinical values and body functions) must be kept within tight limits to maintain homeostasis, and they must be directly sensed by the body.

Step 2: Identify the challenge. The challenge is the event that alters the regulated variable. In the case study, bleeding is the challenge that decreases Mike's blood pressure.

Step 3: Identify the sensor/control center. The case study does not give any information about how blood pressure is sensed or controlled, so we will use a question mark for these components. Both the sensor and the control center are colored *yellow* in the textbook diagrams.

Step 4: Identify the effector. The actions of the effector(s) partially or completely reverse the change in the regulated variable. The case study mentions that "his heart is working hard to compensate…" So, the heart is an effector, and we can infer that its actions are increasing blood pressure. The effector and its actions are colored *blue* in the textbook diagrams.

Step 5: Identify the signals. The case study does not explain how signals travel from sensors to the control center to the effectors, so we cannot include this information. Signals are colored *purple* in color-coded textbook diagrams.

Step 6: Is homeostasis restored? The case study tells us that Mike's blood pressure was too low, so the increased heart rate only partially compensated for the challenge. Homeostasis was not completely restored.

Step 7: Create the drawing. Use the components identified in Steps 1 through 6 to illustrate the feedback loop.

fluid between the cells in tissues. Homeostasis requires that the volume and composition of these two different fluids be maintained within narrow limits, as discussed in Chapter 19.

5 GRADIENTS AND RESISTANCE

The previous section discussed the importance of maintaining differences between the inside and outside of the body, and between the inside and the outside of cells. A difference in specific physical or chemical values between two areas is called a **gradient.** For example, there is an altitude gradient between the peak of a roof and the roof edge because the peak is higher than the edge. Water from snow melting on a steep roof will flow downward following this gradient. The steeper the roof, the larger the gradient, and the greater the flow. Indeed, unless they are inhibited from doing so, substances always move down their gradients from high to low.

The term **resistance** describes all factors that inhibit flow down a particular gradient. For example, friction provides resistance to the downward movement of a skateboard on a ramp (**FIG. 1-5**). A smoother surface provides less resistance, and permits a greater amount of flow. Barriers provide significant resistance to flow. Therefore, the barrier formed by skin reduces the amount of heat that moves from the warmer body to the colder surrounding air, down

FIGURE 1-5 Flow, Gradients, and Resistance. KEY POINT The skateboarder moves down the ramp because of an altitude gradient. Friction between the skateboard and the ramp surface provides resistance. ZOOMING IN What will happen to the rate of movement if the angle of the ramp increases? If the surface of the ramp provides more resistance? (See Appendix 2 for answers to the Zooming In questions.)

the temperature gradient. Wearing thicker clothes increases resistance and thus reduces heat flow.

This textbook describes many examples of flow, including the movement of water and solutes across the plasma membrane (Chapter 3), the flow of blood through blood vessels (Chapter 14), and the flow of air through the airways (Chapter 16). The same principle applies in each and every case; a gradient promotes the flow of the substance, and the various factors of resistance oppose the flow.

CASEPOINTS

- ☐ **1-2** In Mike's case, what was the major challenge to his homeostasis?
- ☐ **1-3** Which of these is an effector in the feedback loop controlling blood pressure—the brain or the heart?

See the Student Resources on thePoint® to view the animation "Gradients and Barriers".

CHECKPOINTS

- ☐ **1-3** What is the definition of homeostasis?
- ☐ **1-4** What are the three components of a negative feedback loop?
- ☐ **1-5** What general term refers to any structure that separates differing environments?
- ☐ **1-6** What is formed by a difference in specific physical or chemical values between two regions?
- ☐ **1-7** What term describes a force that opposes flow along a gradient?

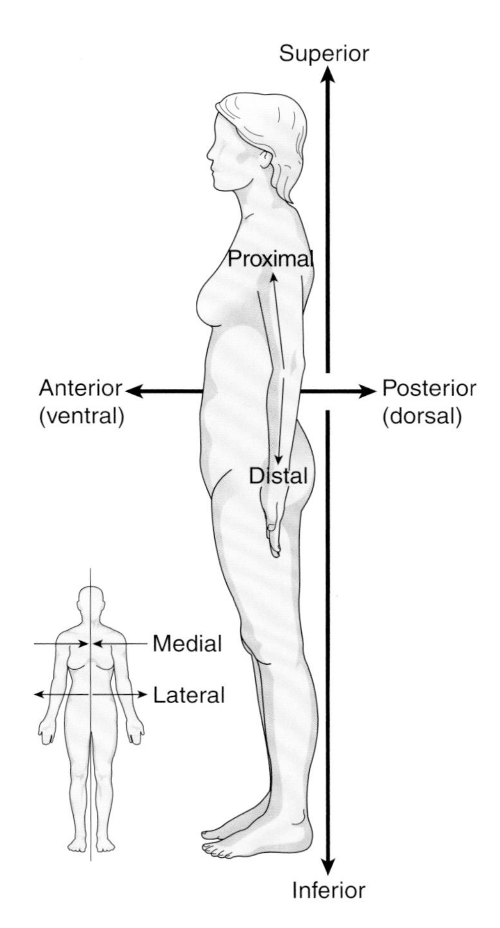

FIGURE 1-6 Directional terms. KEY POINT Healthcare professionals use standardized terms to describe body directions. ZOOMING IN What is the scientific name for the position in which the figures are standing?

Body Directions

Because it would be awkward and inaccurate to speak of bandaging the "southwest part" of the chest, for example, healthcare professionals use standardized terms to designate body positions and directions. For consistency, all descriptions assume that the body is in the **anatomic position.** In this posture, the subject is standing upright with face front, arms at the sides with palms forward, and feet parallel, as shown in **FIGURE 1-6**.

DIRECTIONAL TERMS

The main terms for describing directions in the body are as follows (**see FIG. 1-6**):

- **Superior** is a term meaning above or in a higher position. Its opposite, inferior, means below or lower. The heart, for example, is superior to the intestine.
- **Anterior** and **ventral** have the same meaning in humans: located toward the belly surface or front of the body. Their corresponding opposites, **posterior** and **dorsal**, refer to locations nearer the back.
- **Medial** means nearer to an imaginary plane that passes through the midline of the body, dividing it into left and right portions. **Lateral**, its opposite, means farther away from the midline, toward the side. For example, your nose is medial to your ears.
- **Proximal** means nearer to the origin or attachment point of a structure; **distal** means farther from that point. For example, the part of your thumb where it attaches to your hand is its proximal region; the tip of the thumb is its distal region. Considering the mouth as the beginning (origin) of the digestive tract, the small intestine is distal to the stomach.

Concept Mastery Alert

Students commonly confuse the terms distal and lateral. Both indicate farther away, but the point of reference varies. Lateral means away from the midline and distal means away from an origin or attachment point. For instance, in the anatomic position, the thumb is lateral to the pinkie, but the pinkie is distal to the thumb.*

See the Student Resources on thePoint® for a chart of directional terms with definitions and examples.

*The Concept Mastery Alerts featured in many chapters of this book are derived from common errors students make in responding to questions in PrepU, an online supplemental review program available separately for this text. For information on accessing PrepU, see pp. xviii–xix of the User's Guide at the front of this text.

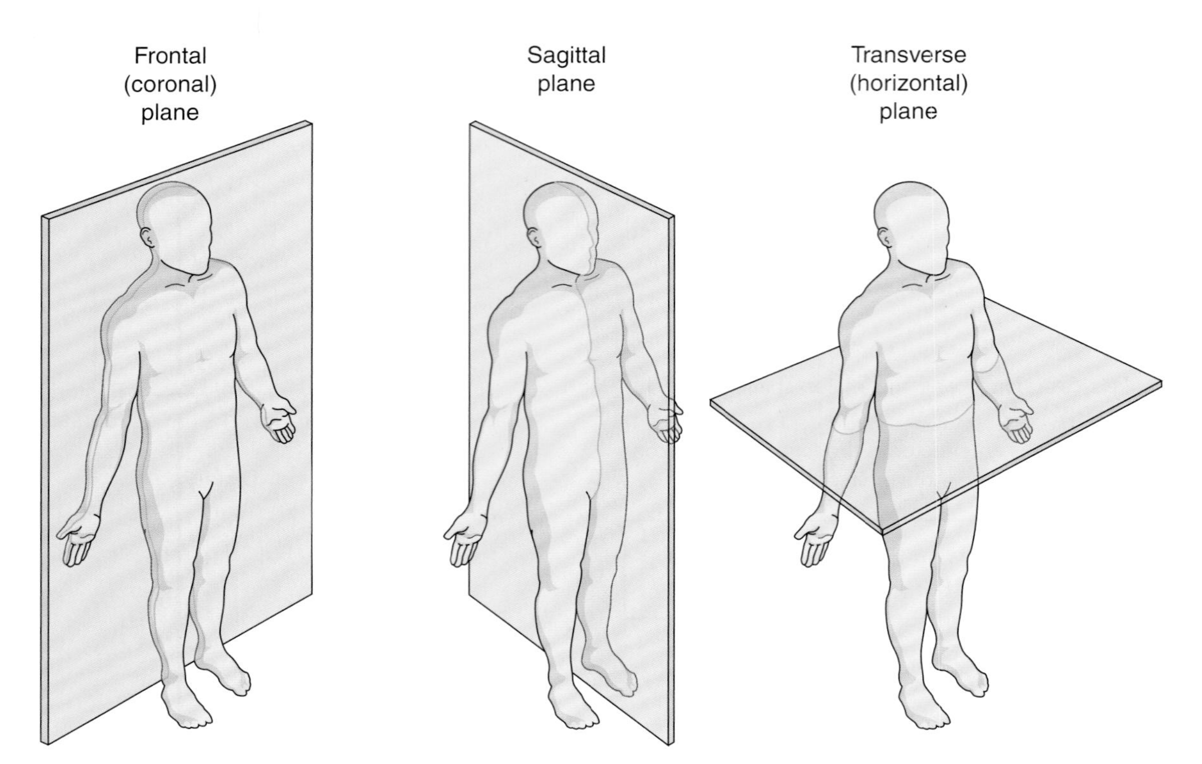

FIGURE 1-7 Planes of division. KEY POINT The body can be divided along three different planes. ZOOMING IN Which plane divides the body into superior and inferior parts? Which plane divides the body into anterior and posterior parts?

PLANES OF DIVISION

To visualize the various internal structures in relation to each other, anatomists can divide the body along three planes, each of which is a cut through the body in a different direction (FIG. 1-7), as follows:

- **Frontal plane.** If the cut were made in line with the ears and then down the middle of the body, you would see an anterior, or ventral (front), section and a posterior, or dorsal (back), section. Another name for this plane is *coronal plane*.
- **Sagittal** (SAJ-ih-tal) **plane.** If you were to cut the body in two from front to back, separating it into right and left portions, the sections you would see would be sagittal sections. A cut exactly down the midline of the body, separating it into equal right and left halves, is a **midsagittal plane.**
- **Transverse plane.** If the cut were made horizontally, across the other two planes, it would divide the body into a superior (upper) part and an inferior (lower) part. A transverse plane is also called a *horizontal plane*.

Some additional terms are used to describe sections (cuts) of tissues, as used to prepare them for study under the microscope. A cross section (FIG. 1-8) is a cut made perpendicular to the long axis of an organ, such as a cut made across a banana to create a small round slice. A longitudinal section is made parallel to the long axis, as in cutting a banana from tip to tip to make a slice for a banana split. An oblique section is made at an angle. The type of section used will determine what is seen under the microscope, as shown with a blood vessel in FIGURE 1-8.

These same terms are used for images taken by techniques such as computed tomography (CT) or magnetic resonance imaging (MRI) (see BOX 1-2). In imaging studies, the term *cross section* is used more generally to mean any two-dimensional view of an internal structure obtained by imaging, as shown in FIGURE 1-9.

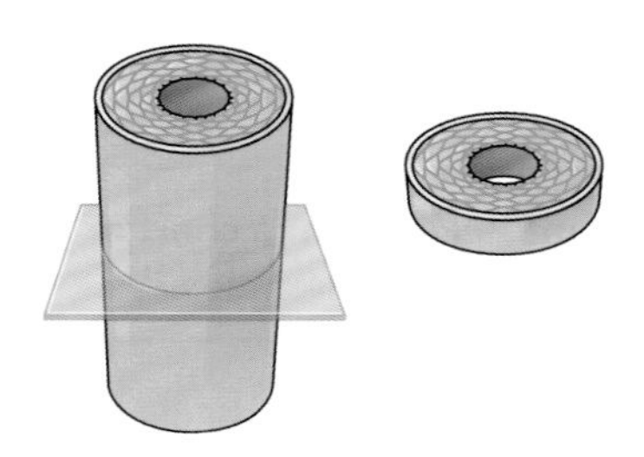

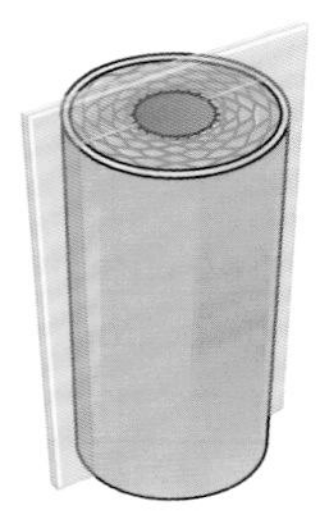

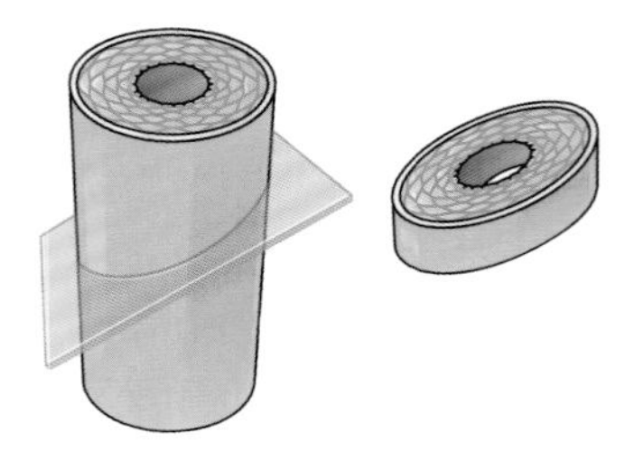

FIGURE 1-8 Tissue sections. KEY POINT The direction in which tissue is cut affects what is seen under the microscope.

HOT TOPICS
Medical Imaging: Seeing without Making a Cut

BOX 1-2

Three imaging techniques that have revolutionized medicine are radiography, computed tomography, and magnetic resonance imaging. With them, physicians today can "see" inside the body without making a single cut. Each technique is so important that its inventor received a Nobel Prize.

The oldest is radiography (ra-de-OG-rah-fe), in which a machine beams x-rays (a form of radiation) through the body onto a piece of film. Like other forms of radiation, x-rays damage body tissues, but modern equipment uses extremely low doses. The resulting picture is called a radiograph. Dark areas indicate where the beam passed through the body and exposed the film, whereas light areas show where the beam did not pass through. Dense tissues (bone, teeth) absorb most of the x-rays, preventing them from exposing the film. For this reason, radiography is commonly used to visualize bone fractures and tooth decay as well as abnormally dense tissues like tumors. Radiography does not provide clear pictures of soft tissues because most of the beam passes through and exposes the film, but contrast media can help make structures like blood vessels and hollow organs more visible. For example, radiologists use ingested barium sulfate (which absorbs x-rays) to coat the digestive tract for imaging.

CT is based on radiography and also uses very low doses of radiation (see FIG. 1-9). During a CT scan, a machine revolves around the patient, beaming x-rays through the body onto a detector. The detector takes numerous pictures of the beam, and a computer assembles them into transverse sections, or "slices." Unlike conventional radiography, CT produces clear images of soft structures such as the brain, liver, and lungs. It is commonly used to visualize brain injuries and tumors and even blood vessels when used with contrast media.

MRI uses a strong magnetic field and radio wave (see FIG. 1-9B). So far, there is no evidence to suggest that MRI causes tissue damage. The MRI patient lies inside a chamber within a very powerful magnet. The molecules in the patient's soft tissues align with the magnetic field inside the chamber. When radio waves beamed at the region to be imaged hit the soft tissue, the aligned molecules emit energy that the MRI machine detects, and a computer converts these signals into a picture. MRI produces even clearer images of soft tissue than does CT and can create detailed pictures of blood vessels without contrast media. MRI can visualize brain injuries and tumors that might be missed using CT.

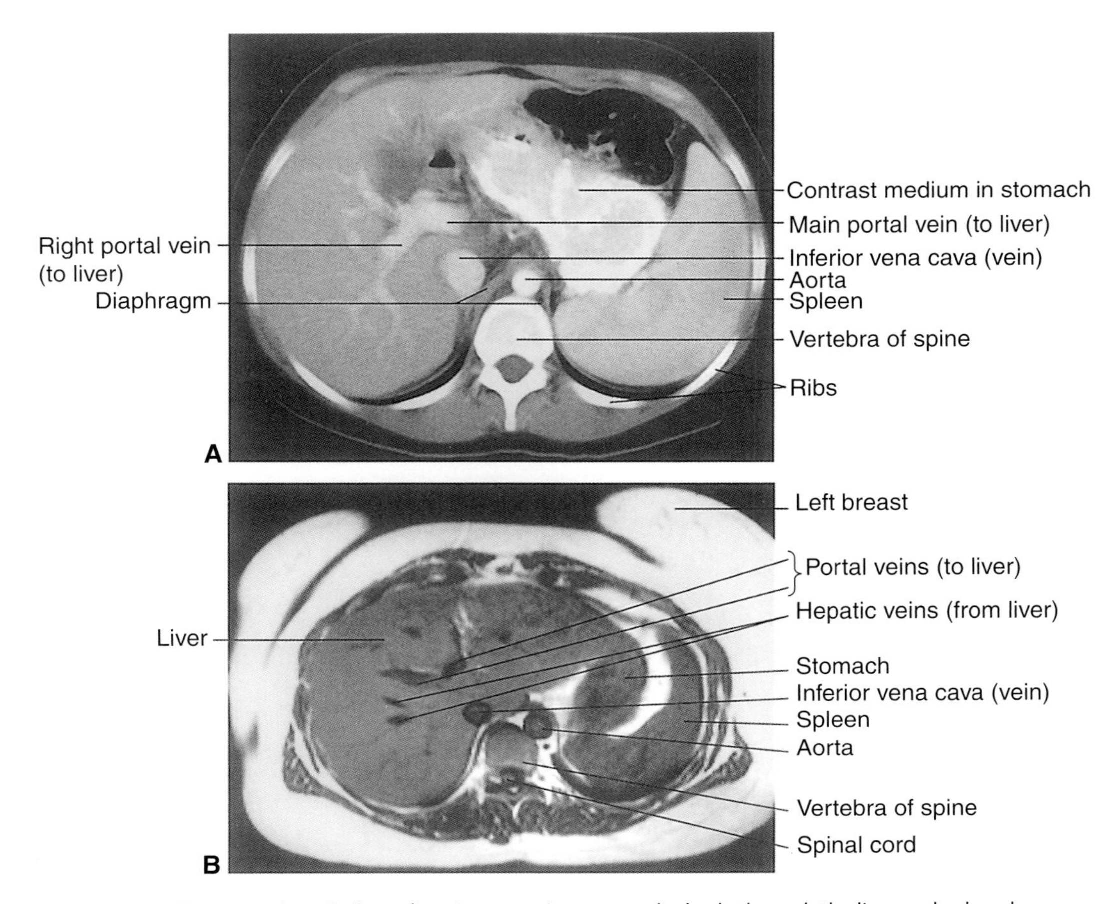

FIGURE 1-9 **Cross sections in imaging.** Images taken across the body through the liver and spleen by **(A)** computed tomography and **(B)** magnetic resonance imaging.

CHECKPOINTS 

- [ ] **1-8** Which term describes a location farther from an origin, such as the wrist in comparison to the elbow?
- [ ] **1-9** What are the three planes in which the body can be cut?

Body Cavities

Internally, the body is divided into a few large spaces, or cavities, which contain the organs. The two main cavities are the **dorsal cavity** and **ventral cavity** (FIG. 1-10).

DORSAL CAVITY

The dorsal body cavity has two subdivisions: the **cranial cavity**, containing the brain, and the **spinal cavity (canal)**, enclosing the spinal cord. These two areas form one continuous space.

VENTRAL CAVITY

The ventral cavity is much larger than the dorsal cavity. It has two main subdivisions, which are separated by the **diaphragm** (DI-ah-fram), a muscle used in breathing. The **thoracic** (tho-RAS-ik) **cavity** is superior to (above) the diaphragm. Its contents include the heart, the lungs, and the large blood vessels that join the heart. The heart is contained in the pericardial cavity, formed by the pericardial sac, the tissue that surrounds the heart; the lungs are in the pleural cavity, formed by the pleurae, the membranes that enclose the lungs (FIG. 1-11). The **mediastinum** (me-de-as-TI-num) is the space between the lungs, including the organs and vessels contained in that space.

 *Concept Mastery Alert*

Remember that the mediastinum is between the lungs; it does not contain them.

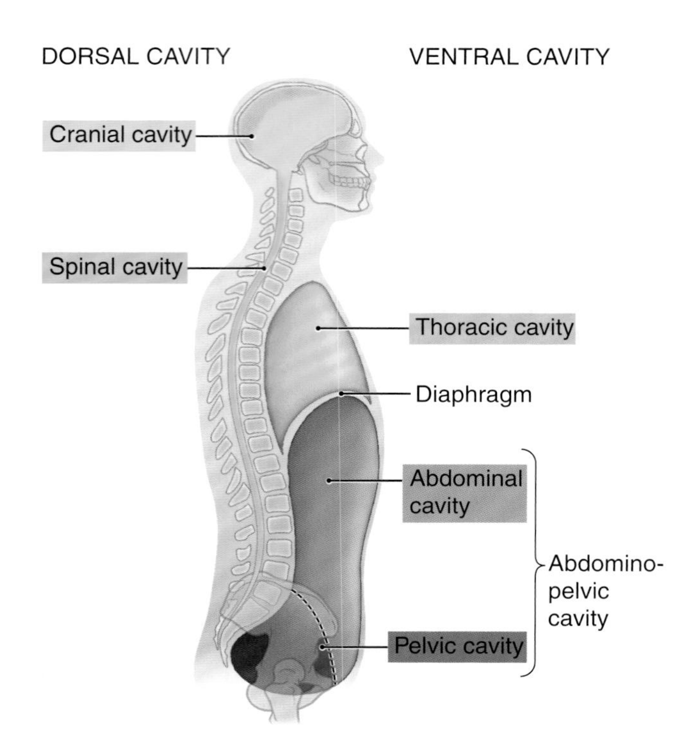

FIGURE 1-10 Body cavities, lateral view. Shown are the dorsal and ventral cavities with their subdivisions. ZOOMING IN Which cavity contains the diaphragm?

See FIGURE A5-8A in Appendix 3, Dissection Atlas, showing the organs in the thoracic cavity.

The **abdominopelvic** (ab-dom-ih-no-PEL-vik) **cavity** (see FIG. 1-11) is inferior to (below) the diaphragm. This space is further subdivided into two regions. The superior portion, the **abdominal cavity**, contains the stomach, most of the intestine, the liver, the gallbladder, the pancreas, and the spleen. The inferior portion, set off by an imaginary line across the top of the hip bones, is the **pelvic cavity**. This cavity contains the urinary bladder, the rectum, and the internal parts of the reproductive system.

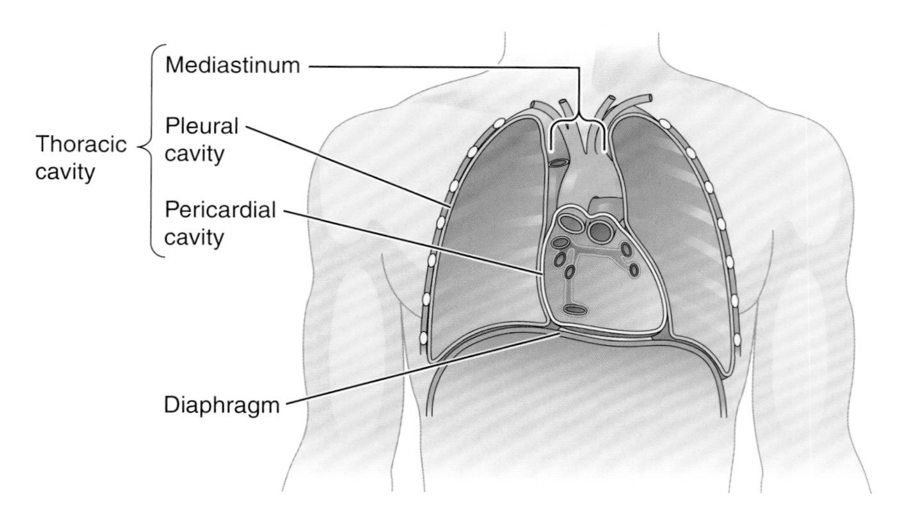

FIGURE 1-11 The thoracic cavity. KEY POINT Among other structures, the thoracic cavity encloses the pericardial cavity, which contains the heart, and the pleural cavity, which contains the lungs.

DIVISIONS OF THE ABDOMEN

It is helpful to divide the abdomen for examination and reference into nine regions (FIG. 1-12).

The three central regions, from superior to inferior, are the following:

- **Epigastric** (ep-ih-GAS-trik) **region**, located just inferior to the breastbone
- **Umbilical** (um-BIL-ih-kal) **region**, around the umbilicus (um-BIL-ih-kus), commonly called the *navel*
- **Hypogastric** (hi-po-GAS-trik) **region**, the most inferior of all the midline regions

The regions on the right and left, from superior to inferior, are the following:

- **Hypochondriac** (hi-po-KON-dre-ak) **regions**, just inferior to the ribs
- **Lumbar regions**, which are on a level with the lumbar regions of the spine
- **Iliac**, or **inguinal** (IN-gwih-nal), **regions**, named for the upper crest of the hip bone and the groin region, respectively

A simpler but less precise division into four quadrants is sometimes used. These regions are the right upper quadrant, left upper quadrant, right lower quadrant, and left lower quadrant (FIG. 1-13).

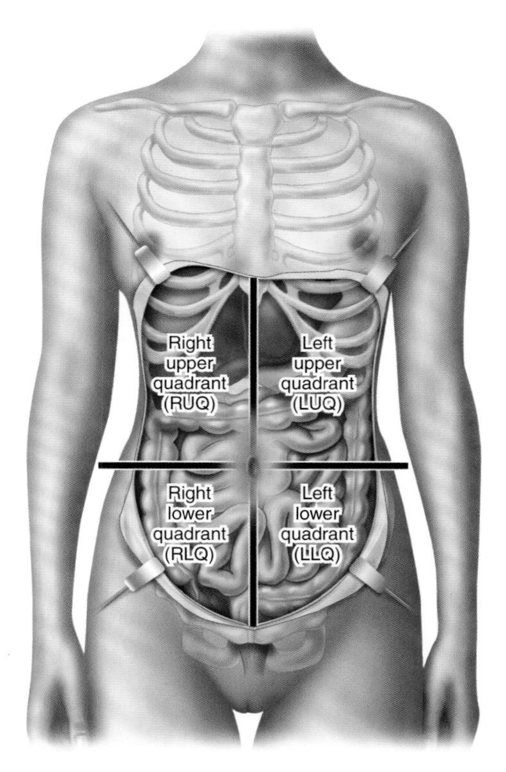

FIGURE 1-13 Quadrants of the abdomen. The organs within each quadrant are shown. ZOOMING IN Which four abdominal regions are represented in the left lower quadrant?

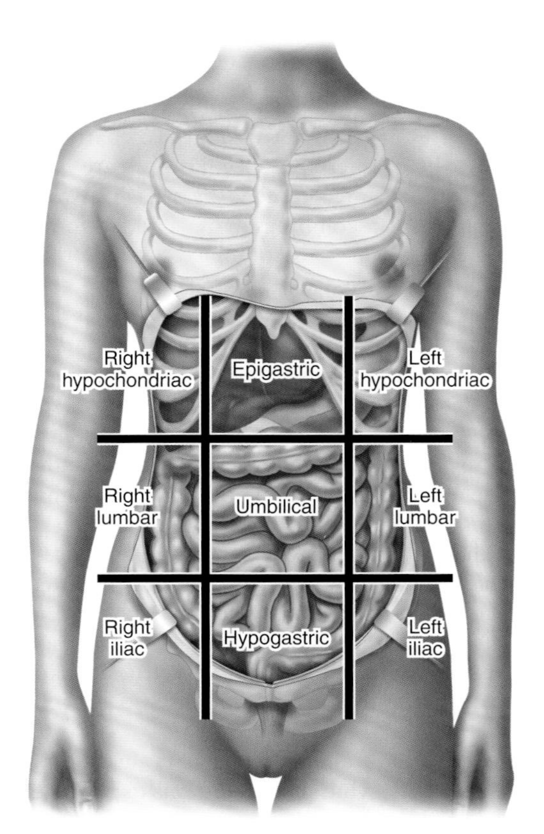

FIGURE 1-12 The nine regions of the abdomen. KEY POINT Internal structures can be localized within nine regions of the abdomen.

For your reference, FIGURES 1-14 and 1-15 give anatomic adjectives for some other body regions along with their common names.

CHECKPOINTS 

- **1-10** Name the two main body cavities.
- **1-11** Name the three central regions and the three left and right lateral regions of the abdomen.

See the Student Resources on thePoint® for photographic versions of FIGURES 1-12 and 1-13 and a list of the organs in each quadrant. You can also find information there on the metric system, which is used for all scientific measurements.

The Language of Healthcare

In Mike's case, we saw that health professionals share a specialized language: medical terminology. This special vocabulary is based on word parts with consistent meanings that are combined to form different words. Each chapter in this book has a section near the end entitled "Word Anatomy." Here, you will find definitions of word parts commonly used in medical terms with examples of their usage.

The main part of a word is the **root**. Some compound words, such as wheelchair, gastrointestinal, and lymphocyte,

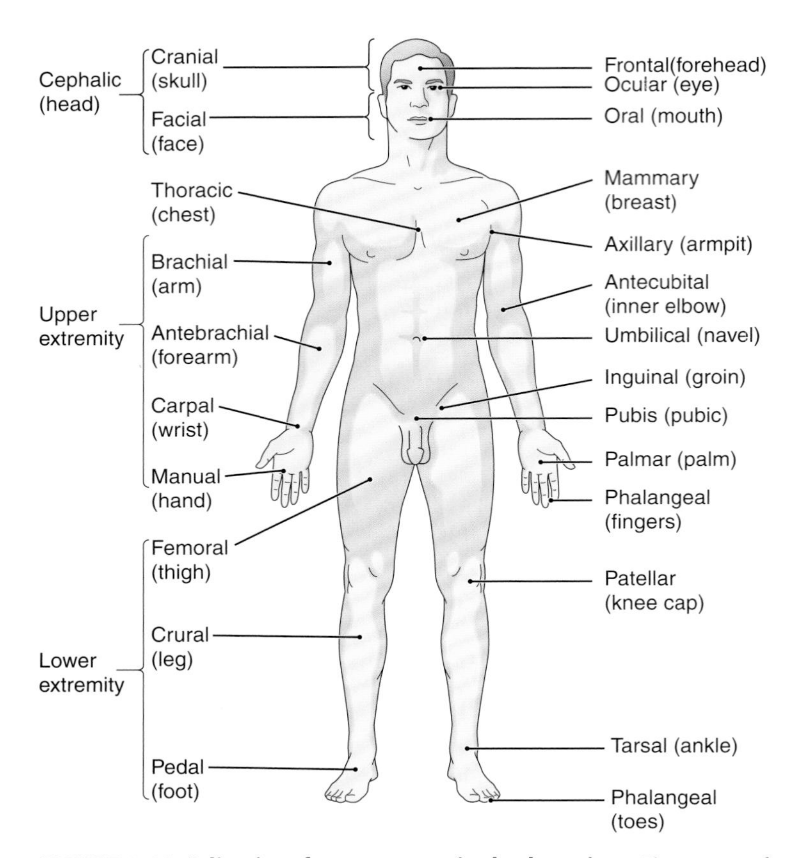

FIGURE 1-14 Adjectives for some anterior body regions. The names of the regions are in parentheses.

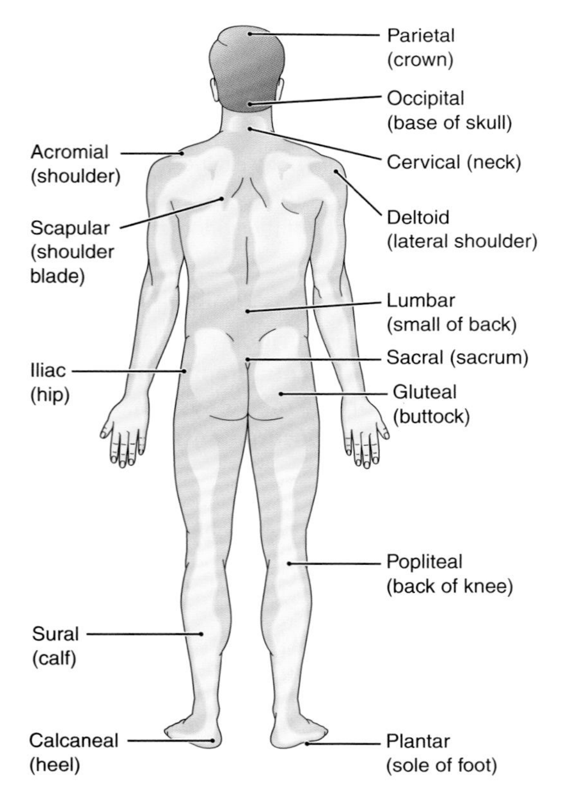

FIGURE 1-15 Adjectives for some posterior body regions. The names of the regions are in parentheses.

use more than one root. A **prefix** is a short part that starts a word and modifies the root. A **suffix** follows the root and also modifies it. In the Word Anatomy charts, the word parts follow the chapter sequence. Prefixes are followed by a dash, and suffixes are preceded by a dash. A root has no dash but often has a combining vowel added to make pronunciation easier when it is combined with another root or a suffix. These vowels are separated from the root with a slash, as in physi/o.

By using the Word Anatomy charts, the Glossary, and the Glossary of Word Parts (the last two found at the back of this text), you too can learn to speak this language.

This text focuses on normal anatomy and physiology. Students in healthcare fields will be adding many more terms relating to disease as they progress in their careers. The study of disease is *pathology*, based on the root path/o. Diseases fall into a number of different categories, including infections, which may be caused by bacteria, viruses, and fungi. Knowing the characteristics of the disease-causing organism can help practitioners know how the disease will spread, potential disease outcomes, and useful treatments.

The diagnosis of disease involves evaluation of accompanying symptoms, which are conditions experienced by the patient, and signs, which are indicators gathered by a health professional. So, fatigue is a symptom, but blood pressure is a sign. Many additional terms are related to treatment.

Although we are concentrating on normal structure and function, the case studies that open each chapter illustrate what happens when conditions go wrong. These cases, and many of the special interest boxes, will introduce some of the additional terms used in today's medical practices.

Mastering the language of health requires more than just learning vocabulary. As healthcare practitioners or consumers, a basic understanding of scientific processes can help us make informed decisions about issues such as vaccination, water quality, environmental safety, healthy habits, and medical treatment. It can be difficult to know if the studies on which we base our decisions were done in a meaningful way and if the conclusions are valid. The Step by Step boxes that follow in the book will discuss some of the elements of good scientific inquiry and how to understand reported results. We will highlight such issues as graph interpretation, the importance of control groups, and the difference between correlation and causation as they relate to the case studies that open each chapter. These boxes will also help you solve other types of problems, as you already saw in **BOX 1-1**.

CASEPOINT 

☐ **1-4** Two important terms in Mike's case are *hypotensive* and *tachycardic*. Name and define the prefix and root for each of these terms, referring to the Word Parts glossary on p. 479.

See the box Health Information Technicians in the Student Resources on thePoint® for a description of a profession that requires knowledge of medical terminology.

A & P in Action Revisited: Mike's Homeostatic Emergency

The dispatch radio crackled to life in the ER. "This is Medic 12. We have Mike, 21 years old. Involved in a head-on collision. Patient is conscious and on oxygen. ETA is 15 minutes."

When they arrived at the ER, Samantha and Ed wheeled their patient into the trauma room. Immediately, the emergency team sprang into action, beginning an IV for administration of fluids. The trauma nurse measured Mike's vital signs while a technician placed an arterial line to monitor blood pressure and draw blood samples for testing in the lab. The emergency physician inserted an endotracheal tube into Mike's pharynx to keep his airway open and then carefully examined his abdominopelvic cavity.

"Blood pressure is 80 over 40. Heart rate is 146. Respirations are shallow and rapid," said the nurse.

"We need to raise his blood pressure—let's start a second IV of plasma. His abdomen is as hard as a board. I think he may have a bleed in there—we need an ultrasound," replied the doctor. The sonographer wheeled the ultrasound machine into position and placed the transducer onto Mike's abdomen. Immediately, she located the cause of Mike's symptoms—blood in the left upper quadrant.

"OK. We have a ruptured spleen here," said the doctor. "Call surgery—they need to operate right now."

CHAPTER 1

Chapter Wrap-Up

OVERVIEW

A detailed chapter outline with space for note-taking is on thePoint®. The figure below illustrates the main topics covered in this chapter.

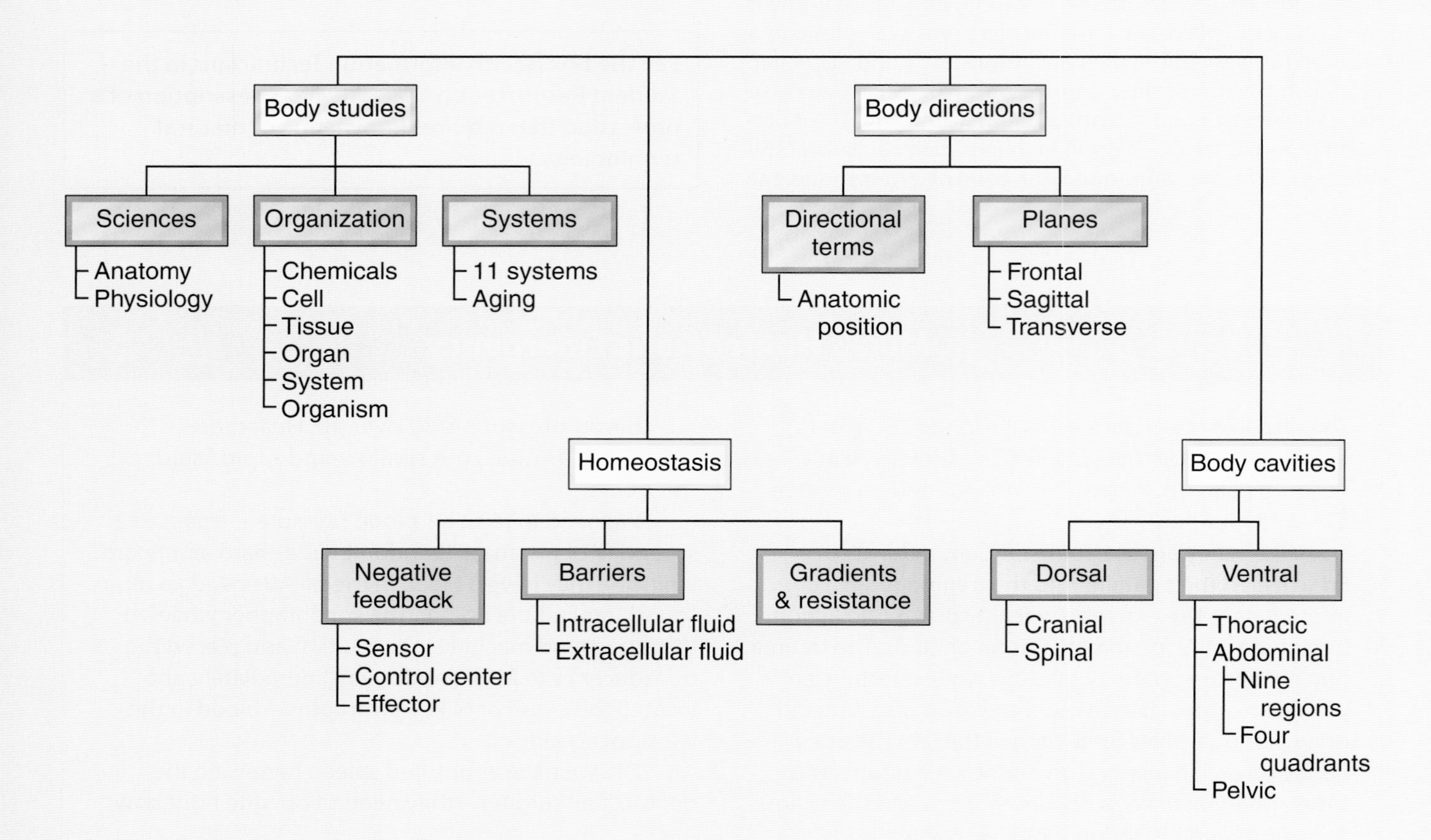

KEY TERMS

The terms listed below are emphasized in this chapter. Knowing them will help you organize and prioritize your learning. These and other boldface terms are defined in the Glossary with phonetic pronunciations.

anatomic position
anatomy
cell
disease
extracellular fluid
gradient
homeostasis
intracellular fluid
negative feedback
organ
pathology
physiology
resistance
system
tissue
regulated variable

WORD ANATOMY

Medical terms are built from standardized word parts (prefixes, roots, and suffixes). Learning the meanings of these parts can help you remember words and interpret unfamiliar terms.

WORD PART	MEANING	EXAMPLE
STUDIES OF THE HUMAN BODY		
dis-	apart, away from	To *dissect* is to cut apart.
-logy	study of	*Radiology* is the study and use of radioactive substances.
path/o	disease	*Pathology* is the study of disease.
physi/o	nature, physical	*Physiology* is the study of how the body functions.
-tomy	cutting, incision of	*Anatomy* can be revealed by cutting the body.
HOMEOSTASIS		
extra-	outside of, beyond	*Extracellular* fluid is outside the cells.
home/o-	same	*Homeostasis* is the steady state (sameness) within an organism.
intra-	within	*Intracellular* fluid is within a cell.
stat, -stasis	stand, stoppage, constancy	In *homeostasis*, "-stasis" refers to constancy.

QUESTIONS FOR STUDY AND REVIEW

BUILDING UNDERSTANDING

Fill in the Blanks

1. Specialized tissues working together for the same general purpose form a(n)_____.
2. In location, the nose is _____ to the eyes.
3. Normal body function maintains a state of internal balance called _____.
4. In the word physiology, *-logy* is an example of a word part called a(n) _____.

Matching Match each numbered item with the most closely related lettered item.

___ 5. The system that uses electrical signals to control and coordinate other systems

___ 6. The system that brings needed substances to the body tissues

___ 7. The system that converts foods into a form that body cells can use

___ 8. The outermost body system

___ 9. The system of glands that produce hormones

a. nervous system
b. integumentary system
c. cardiovascular system
d. endocrine system
e. digestive system

Multiple Choice

___ 10. Which science studies normal body structure?
a. homeostasis
b. anatomy
c. physiology
d. pathology

___ 11. Where is intracellular fluid located?
a. between body cells
b. in blood plasma
c. in lymph
d. inside body cells

___ 12. What is the main way of regulating homeostasis?
a. anabolism
b. biofeedback
c. catabolism
d. negative feedback

___ 13. Which cavity contains the mediastinum?
a. abdominal
b. dorsal
c. thoracic
d. pelvic

___ 14. In location, the ankle is _____ to the knee.
a. distal
b. inferior
c. proximal
d. superior

___ 15. A plane that divides the body into right and left portions is a _____.
a. frontal plane
b. transverse plane
c. sagittal plane
d. horizontal plane

___ 16. The most inferior midline region of the abdomen is the _____.
a. superior region
b. hypogastric region
c. umbilical region
d. epigastric region

UNDERSTANDING CONCEPTS

17. What do you study in anatomy? In physiology? Would it be wise to study one without the other?
18. List in sequence the levels of organization in the body from simplest to most complex. Give an example for each level.
19. Compare and contrast the anatomy and physiology of the nervous system with that of the endocrine system.
20. In order of action, name the components of a negative feedback loop.
21. Use The Body Visible overlays at the beginning of this book to name the lateral bone of the lower leg. Name the proximal bone of the arm.
22. Referring to the Dissection Atlas in Appendix 3:
    a. list the figure(s) in which an organ is cut into left and right parts.
    b. list the figure(s) in which an organ is cut into anterior and posterior parts.

CONCEPTUAL THINKING

23. The human body is organized from very simple levels to more complex levels. With this in mind, describe why a disease at the chemical level can have an effect on organ system function.
24. Use a car operating under cruise control as an example of a negative feedback loop, identifying the set point and the components of the system.
25. In Mike's case, the paramedics discovered bruising of the skin over Mike's left lumbar region and umbilical region. Mike also reported considerable pain in his upper left quadrant. Locate these regions on your own body. Why it is important for health professionals to use medical terminology when describing the human body?
26. If a child swallows a marble, does the marble actually get inside his body? Explain.
27. In Mike's case, blood is flowing out of the blood vessels in his abdomen into his abdominal cavity.
A. What sort of gradient is driving the flow?
B. What do you think changed in Mike's case that enabled this flow to occur: the gradient or the resistance?

For more questions, see the Learning Activities on thePoint®.

CHAPTER 2

Chemistry, Matter, and Life

Learning Objectives

After careful study of this chapter, you should be able to:

1. Define a chemical element. ***p. 22***
2. Describe the structure of an atom. ***p. 22***
3. Differentiate between ionic and covalent bonds. ***p. 24***
4. Define an electrolyte. ***p. 24***
5. Differentiate between molecules and compounds. ***p. 25***
6. Use chemical equations to illustrate different types of chemical reactions. ***p. 26***
7. Define mixture; list the three types of mixtures, and give two examples of each. ***p. 27***
8. Describe roles of water in the body. ***p. 27***
9. Compare acids, bases, and salts. ***p. 28***
10. Explain how the numbers on the pH scale relate to acidity and alkalinity. ***p. 28***
11. Explain why buffers are important in the body. ***p. 29***
12. Define radioactivity, and cite several examples of how radioactive substances are used in medicine. ***p. 29***
13. Name the three main types of organic compounds and the building blocks of each. ***p. 30***
14. Define enzyme; describe how enzymes work. ***p. 32***
15. List the components of nucleotides, and give some examples of nucleotides. ***p. 32***
16. Define metabolism, and name the two types of metabolic reactions. ***p. 34***
17. Distinguish between kinetic and potential energy, and give examples of each. ***p. 34***
18. Use the case study to discuss the importance of regulating body fluid quantity and composition. ***pp. 21, 35***
19. Show how word parts are used to build words related to chemistry, matter, and life (see Word Anatomy at the end of the chapter). ***p. 37***

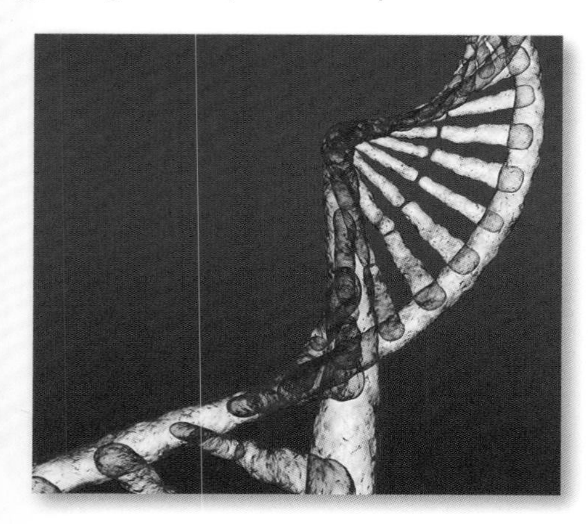

A & P in Action
Margaret's Case: Chemistry's Role in Health Science

"Ugh," sighed Angela as she pulled into her hospital parking spot. The heat wave was into its second week, and she was getting tired of it. It was beginning to take its toll on the city too, especially on its infants and older residents. As Angela walked toward the hospital, she thought back to yesterday's ICU shift. One elderly patient stood out in her mind, probably because she reminded Angela of her own grandmother.

The patient, Margaret Ringland, a 78-year-old widow, lived alone in her apartment on New York's Upper East Side. Yesterday, Margaret's niece found her collapsed on the floor, weak and confused. She called 911, and Margaret was rushed to the emergency room. According to her medical chart, Margaret presented with flushed dry skin, a sticky oral cavity, and a furrowed tongue. She was confused and disoriented. She also had hypotension (low blood pressure) and tachycardia (an elevated heart rate). All were classic signs of dehydration, a severe deficiency of water. Without adequate water, Margaret's body was unable to perform essential metabolic processes, and her tissues and organs were not in homeostatic balance.

Her neurologic symptoms were caused by changes in water volume. Although it was difficult to get a blood sample from Margaret's flattened veins, her blood work confirmed the initial diagnosis. Margaret's electrolyte levels were out of balance; specifically, she had a high blood sodium ion concentration, a condition called hypernatremia. Her hematocrit was also high, indicating low blood volume. This decrease was seriously affecting her cardiovascular system. Margaret's blood pressure had dropped, which forced her heart to beat faster to ensure proper delivery of blood to her tissues.

Because Margaret was still conscious and cooperative, the emergency team started her on oral rehydration therapy with an aqueous solution containing small amounts of glucose and salts. Once stabilized, Margaret was moved to a hospital bed for observation and recovery. Her level of consciousness was monitored, and the staff watched for any signs of cerebral changes due to the dehydration.

Angela depends on her knowledge of chemistry to make sense of the signs and symptoms she observes in her patients. As you read this chapter, keep in mind that a firm understanding of the chemistry presented in this chapter will help you understand the anatomy and physiology of the cells, tissues, and organ systems discussed in subsequent chapters.

As you study this chapter, CasePoints will give you opportunities to apply your learning to this case.

Visit thePoint® to access the following resources. For guidance in using these resources most effectively, see pp. xv–xvii.

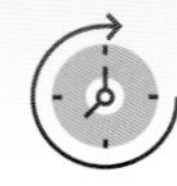

Preparing to Learn

- Tips for Effective Studying
- Pre-Quiz

While You Are Learning

- Animation: Enzymes
- Animation: Energy
- Chapter Notes Outline
- Audio Pronunciation Glossary

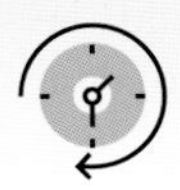

When You Are Reviewing

- Answers to Questions for Study and Review
- Health Professions: Pharmacist and Pharmacy Technician
- Interactive Learning Activities

A LOOK BACK

Chapter 1 introduced the key idea of levels of organization 1. *In this chapter, we explore the most fundamental level—chemicals. Future chapters will build on a solid understanding of how chemicals interact in dynamic ways to produce structures and enable body functions.*

Introduction

Greater understanding of living organisms has come to us through **chemistry**, the science that deals with the composition and properties of matter. Knowledge of chemistry and chemical changes helps us understand the body's normal and abnormal functioning. Food digestion in the intestinal tract, urine production by the kidneys, the regulation of breathing, and all other body activities involve the principles of chemistry. The many drugs used to treat diseases are also chemicals. Chemistry is used for their development and for understanding their actions in the body.

To provide some insights into the importance of chemistry in the life sciences, this chapter briefly describes elements, atoms, molecules, compounds, and mixtures, which are fundamental forms of matter. We also describe the chemicals that characterize organisms—organic chemicals.

Elements

Matter is anything that takes up space, that is, the materials from which the entire universe is made. **Elements** are the unique substances that make up all matter. The food we eat, the atmosphere, and water—everything around us and everything we can see and touch—are made from just 92 naturally occurring elements. (Twenty additional elements have been created in the laboratory.) Examples of elements include various gases, such as hydrogen, oxygen, and nitrogen; liquids, such as mercury used in barometers and other scientific instruments; and many solids, such as iron, aluminum, gold, silver, and zinc. Graphite (the so-called "lead" in a pencil), coal, charcoal, and diamonds are different forms of the element carbon.

Elements can be identified by their names or their chemical symbols, which are abbreviations of their modern or Latin names. Each element is also identified by its own number, which is based on its atomic structure, discussed shortly. The periodic table is a chart used by chemists to organize and describe the elements. Appendix 1 shows the periodic table and gives some information about how it is used.

Of the 92 elements that exist in nature, only 26 have been found in living organisms. Hydrogen, oxygen, carbon, and nitrogen make up about 96% of body weight (**FIG. 2-1**). Nine additional elements—calcium, sodium, potassium, phosphorus, sulfur, chlorine, magnesium, iron, and iodine—make up most of the remaining 4% of the body's elements. The remaining 13, including zinc, selenium, copper, cobalt, chromium, and others, are present in extremely small (trace) amounts totaling about 0.1% of body weight. **TABLE 2-1** lists some of these elements along with their functions.

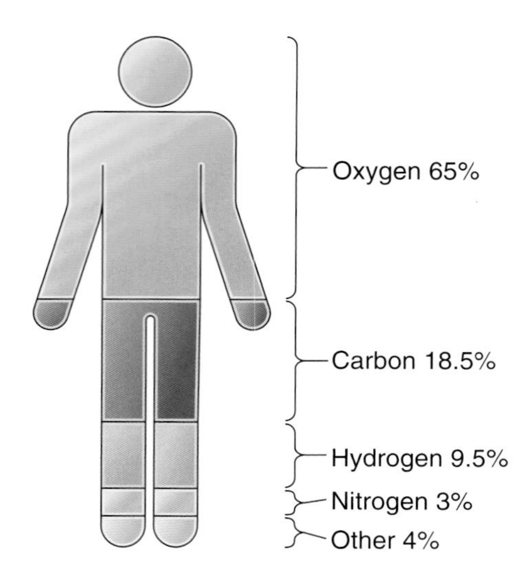

FIGURE 2-1 The body's chemical composition by weight. **KEY POINT** Oxygen, carbon, hydrogen, and nitrogen make up about 96% of body weight.

ATOMIC STRUCTURE

The smallest units of elements are **atoms**. As such, atoms are the smallest complete units of matter. They cannot be broken down or changed into another form by ordinary chemical and physical means. Atoms are so small that millions of them could fit on the sharpened end of a pencil.

Despite the fact that the atom is so tiny, chemists have studied it extensively and have found that it has a definite structure composed of even smaller, or subatomic, particles. These particles differ as to their electric charge. Two types of electric charges exist in nature: positive (+) and negative (−). Particles with the same type of charge repel each other, but particles with different charges attract each other. Therefore, a negatively charged particle would be repelled by another negatively charged particle but attracted by a positively charged particle. At the center of each atom is a nucleus composed of positively charged particles called **protons** (PRO-tonz) and noncharged particles called **neutrons** (NU-tronz) (**FIG. 2-2**). Together, the protons and neutrons contribute nearly all of the atom's weight.

In orbit around the nucleus are **electrons** (e-LEK-tronz). These nearly weightless particles are negatively charged. The protons and electrons of an atom are equal in number so that the atom as a whole is electrically neutral (**see FIG. 2-2**). As discussed later, it is the electrons that determine how (or if) the atom will participate in chemical reactions. Atoms may gain or lose electrons in these reactions and thus become electrically charged.

The **atomic number** of an element is equal to the number of protons that are present in the nucleus of its atoms. Because the number of protons is equal to the number of electrons, the atomic number also represents the number of electrons orbiting the nucleus. As you can see in **FIGURE 2-2**, oxygen has an atomic number of 8. No two elements share the same atomic number. Oxygen is the only element with

Table 2-1 Some Common Elements

Name	Symbol	Function
Oxygen	O	Part of water; needed to metabolize nutrients for energy
Carbon	C	Basis of all organic compounds; component of carbon dioxide, the gaseous byproduct of metabolism
Hydrogen	H	Part of water; participates in energy metabolism; determines the acidity of body fluids
Nitrogen	N	Present in all proteins, ATP (the energy-storing compound), and nucleic acids (DNA and RNA)
Calcium	Ca	Builds bones and teeth; needed for muscle contraction, nerve impulse conduction, and blood clotting
Phosphorus	P	Active ingredient in ATP; builds bones and teeth; component of cell membranes and nucleic acids
Potassium	K	Active in nerve impulse conduction; muscle contraction
Sulfur	S	Part of many proteins
Sodium	Na	Active in water balance, nerve impulse conduction, and muscle contraction
Iron	Fe	Part of hemoglobin, the compound that carries oxygen in red blood cells

The elements are listed in decreasing order by weight in the body.

the atomic number of 8. As another example, a carbon atom has six protons in the nucleus and six electrons orbiting the nucleus, so the atomic number of carbon is 6. In the Periodic Table of the Elements (see Appendix 1), the atomic number is located at the top of the box for each element. The atomic weight (mass), the sum of the protons and neutrons, is the number at the bottom of each box. It takes about 1,850 electrons to equal the weight of a single neutron or proton, so electrons are not counted in the determination of atomic weight.

FIGURE 2-2 Representation of the oxygen atom. **KEY POINT** Eight protons and eight neutrons are tightly bound in the central nucleus. The eight electrons are in orbit around the nucleus, two in the first energy level, and six in the second. **ZOOMING IN** How does the number of protons in this atom compare with the number of electrons? *See Appendix 2 for answers to the Zooming In questions*

The positively charged protons keep the negatively charged electrons in orbit around the nucleus by means of the opposite charges on the particles. Positively charged protons attract negatively charged electrons.

Energy Levels An atom's electrons orbit at specific distances from the nucleus in regions called energy levels (see FIG. 2-2). The first energy level, the one closest to the nucleus, can hold only two electrons. The second energy level, the next in distance away from the nucleus, can hold up to eight electrons. FIGURE 2-3A shows the oxygen atom with only the protons in the nucleus and the electrons in fixed positions in their energy levels. It has two electrons in its first energy level and six electrons in its second, outermost, level. The carbon atom, with an atomic number of 6, has four electrons in its outermost energy level (see FIG. 2-3B). More

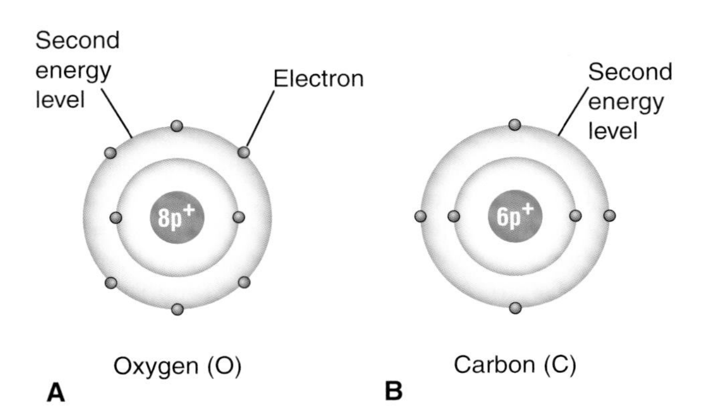

FIGURE 2-3 Examples of atoms. **KEY POINT** The first energy level can hold two electrons, and the second and third can hold eight. The outermost energy level determines chemical reactivity. **ZOOMING IN** How many electrons does oxygen need to complete its outermost energy level?

distant energy levels can hold more than eight electrons, but they are stable (nonreactive) when they have eight.

The electrons in the energy level farthest away from the nucleus determine how the atom will react chemically. Atoms can donate, accept, or share electrons with other atoms to make the outermost level complete. In so doing, they form chemical bonds, as described shortly. Atoms with a stable number of electrons in the outermost energy level do not participate in chemical reactions. Examples are the inert or "noble" gases, including helium, neon, and argon.

CHECKPOINTS

- ☐ **2-1** What are atoms?
- ☐ **2-2** What are three types of particles found in atoms?
- ☐ **2-3** Which of these atoms would be more likely to participate in a chemical reaction—an atom with eight electrons in its outermost energy level or an atom with six electrons in its outermost energy level?

See Appendix 2 for answers to the Checkpoint questions.

Chemical Bonds

When an atom interacts with other atoms to stabilize its outermost energy level, a bond is formed between the atoms. In these chemical reactions, electrons may be transferred from one atom to another or may be shared between atoms.

IONIC BONDS

When electrons are transferred from one atom to another, the type of bond formed is called an **ionic** (i-ON-ik) **bond.** The sodium atom, for example, tends to lose the single electron in its outermost shell leaving the now outermost shell with a stable number of electrons (eight) (**FIG. 2-4**). Removal of a single electron from the sodium atom leaves one more proton than electrons, and the sodium then has a single net positive charge. The sodium in this form is symbolized as Na^+. Calcium loses two electrons when it participates in ionic bonds, so the calcium ion has two positive charges and is abbreviated Ca^{2+}.

Alternately, atoms can gain electrons so that there are more electrons than protons. Chlorine, which has seven electrons in its outermost energy level, tends to gain one electron to fill the level to its capacity. The resultant chlorine is negatively charged (Cl^-) (**see FIG. 2-4**). (Chemists refer to this charged form of chlorine as chloride.) An atom or group of atoms that has acquired a positive or negative charge is called an **ion** (I-on). Any ion that is positively charged is a **cation** (CAT-i-on). Any negatively charged ion is an **anion** (AN-i-on).

Imagine a sodium atom coming in contact with a chlorine atom. The sodium atom gives up its outermost electron to the chlorine and becomes positively charged; the chlorine atom gains the electron and becomes negatively charged. The two newly formed ions (Na^+ and Cl^-), because of their opposite charges, attract each other to produce sodium chloride, ordinary table salt (**see FIG. 2-4**). The attraction between the oppositely charged ions forms an ionic bond. Sodium chloride and other ionically bonded substances tend to form crystals when solid and to dissolve easily in water.

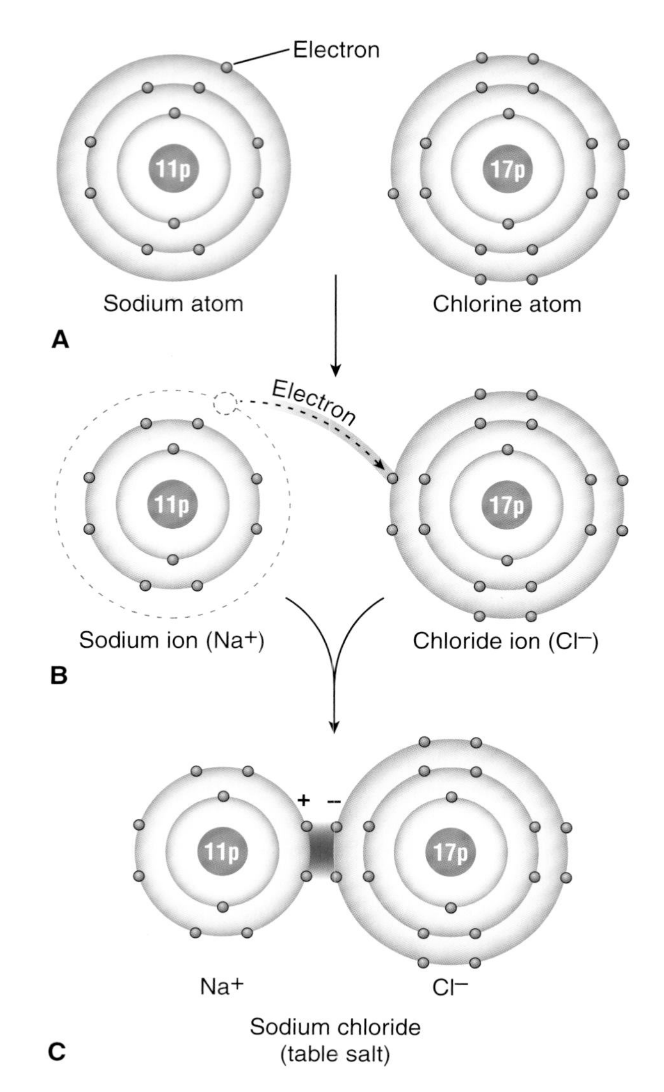

FIGURE 2-4 Ionic bonding. A. A sodium atom has 11 protons and 11 electrons. A chlorine atom has 17 protons and 17 electrons. **B.** A sodium atom gives up one electron to a chlorine atom in forming an ionic bond. The sodium atom now has 11 protons and 10 electrons, resulting in a positive charge of one. The chlorine becomes negatively charged by one, with 17 protons and 18 electrons. **C.** The ionic bond between the sodium ion (Na^+) and the chloride ion (Cl^-) forms the compound sodium chloride (table salt). **ZOOMING IN** How many electrons are in the outermost energy level of a sodium atom? Of a sodium ion?

Electrolytes When ionically bonded substances dissolve in water, the atoms separate as ions. Compounds that release ions when they dissolve in water are called **electrolytes** (e-LEK-tro-lites). Note that in practice, the term electrolytes is also used to refer to the ions themselves in body fluids. Electrolytes include a variety of salts, such as sodium chloride and potassium chloride. They also include acids and bases, which are responsible for the acidity or alkalinity of body fluids, as described shortly.

Concept Mastery Alert

In all cases, a salt consists of a positively charged cation and a negatively charged anion joined by an ionic bond.

Electrolytes must be present in the proper concentrations in the intracellular and extracellular fluids, or damaging effects will result, as seen in Margaret's case study, which opens this chapter.

Ions in the Body Body fluids contain many different ions. Indeed, many of the elements listed in **TABLE 2-1** are only active in their ionic forms. Sodium (Na^+) and potassium (K^+) ions, for instance, play critical roles in the transmission of electric signals by virtue of their positive charges. The concentration of many different ions in body fluids must be kept within narrow limits in order to maintain homeostasis.

Because ions are charged particles, electrolytic solutions can conduct an electric current. Records of electric currents in tissues are valuable indications of the functioning or malfunctioning of tissues and organs. The **electroencephalogram** (e-lek-tro-en-SEF-ah-lo-gram) and the **electrocardiogram** (e-lek-tro-KAR-de-o-gram) are graphic tracings of the electric currents generated by the brain and the heart muscle, respectively (see Chapters 9 and 13).

COVALENT BONDS

Although ionic bonds form some chemicals, many more are formed by another type of chemical bond. This bond involves not the exchange of electrons but a sharing of electrons between the atoms and is called a **covalent bond**. This name comes from the prefix *co-*, meaning "together," and valence, referring to the electrons involved in chemical reactions between atoms. In a covalently bonded substance, the shared electrons orbit around both of the atoms, making both of them stable. Covalent bonds may involve the sharing of one, two, or three pairs of electrons between atoms.

In some covalent bonds, the electrons are equally shared, as in the combination of two identical atoms of hydrogen, oxygen, or nitrogen (**FIG. 2-5**). Electrons may also be shared equally in some bonds involving different atoms—methane (CH_4), for example. If electrons are equally shared in forming a bond, the electric charges are evenly distributed around the atoms and the bond is described as a *nonpolar covalent bond*. That is, no part of the combined particle is more negative or positive than any other part. More commonly, the electrons are held closer to one atom than the other, as in the case of water (H_2O), shown in **FIGURE 2-6**. In water, the shared electrons are actually closer to the oxygen atom than the hydrogen atoms at any one time, making the oxygen region more negative. Such bonds are called *polar covalent bonds*, because one region of the combination is more negative and one part is more positive at any one time.

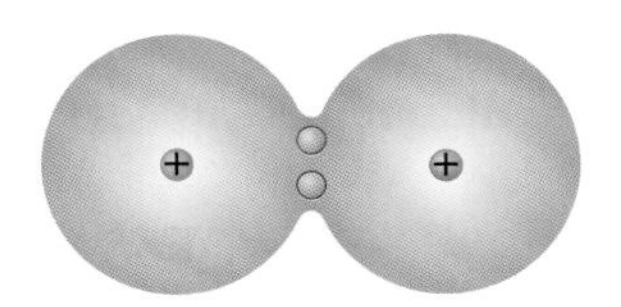

Hydrogen molecule (H_2)

FIGURE 2-5 A nonpolar covalent bond. KEY POINT The electrons involved in the bonding of two hydrogen atoms are equally shared between the two atoms. The electrons orbit evenly around the two. ZOOMING IN How many electrons are needed to complete the energy level of each hydrogen atom in the hydrogen molecule?

Concept Mastery Alert

Remember that electrons are transferred in ionic bonds but shared (equally or unequally) in covalent bonds.

MOLECULES AND COMPOUNDS

When two or more atoms unite covalently, they form a **molecule** (MOL-eh-kule). A molecule is thus the smallest unit of a covalently bonded substance that retains all the properties of that substance. A molecule can be made of like atoms—the oxygen molecule is made of two identical

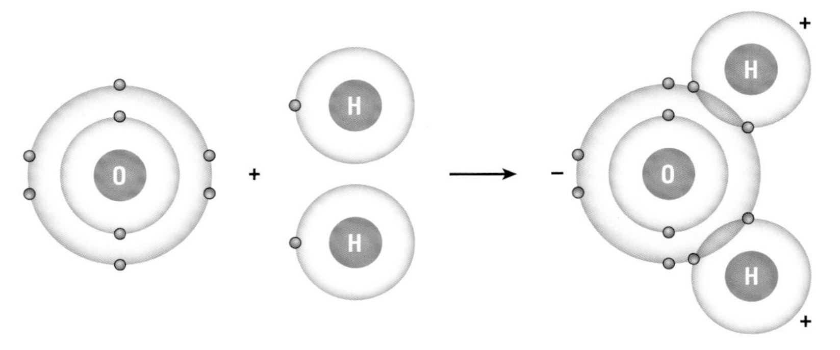

FIGURE 2-6 Formation of water. KEY POINT Polar covalent bonds form water. The unequal sharing of electrons makes the region near the oxygen nucleus more negative and the region near the hydrogen nucleus more positive. ZOOMING IN How many hydrogen atoms bond with an oxygen atom to form water?

atoms, for example—but more often a molecule is made of atoms of two or more different elements. For example, a water molecule (H_2O) contains one atom of oxygen (O) and two atoms of hydrogen (H) (see FIG. 2-6). Any particle resulting from a polar covalent bond is known as a polar molecule. Some polar molecules can gain or lose electrons and thus become ions. Bicarbonate ions (HCO_3^-), for instance, participate in the regulation of body fluid acidity. Polar molecules, whether they are charged or uncharged, can interact using weak bonds called hydrogen bonds, as discussed in **BOX 2-1**.

Note that chemists do not consider ionically bonded substances to be composed of molecules, as their atoms are held together by electrical attraction only. The bonds that hold these atoms together are weak, and the components separate easily in solution into ions, as already described. Thus, unlike ionic bonds and hydrogen bonds, only covalent bonds form molecules.

Concept Mastery Alert

Remember that, ionic bonds, but not covalent bonds, form salts.

Any substance composed of two or more different elements is called a **compound**. This definition includes both ionically and covalently bonded substances. The formula for a compound shows all the elements that make up that compound in their proper ratio, such as NaCl, H_2O, and CO_2. Some compounds are made of a few elements in a simple combination. For example, molecules of the gas carbon monoxide (CO) contain one atom of carbon (C) and one atom of oxygen (O). Other compounds have very large and complex molecules. Such complexity characterizes many of the compounds found in living organisms. Some protein molecules, for example, have thousands of atoms.

It is interesting to observe how different a compound is from any of its constituents. For example, a molecule of liquid water is formed from oxygen and hydrogen, both of which are gases. Another example is the sugar glucose ($C_6H_{12}O_6$). Its constituents include 12 atoms of the gas hydrogen, six atoms of the gas oxygen, and six atoms of the solid element carbon. The component gases and the solid carbon do not in any way resemble the glucose.

CHEMICAL EQUATIONS

In the everyday chemical reactions of life, atoms are not created or destroyed; they are only rearranged by the formation and destruction of chemical bonds. A **chemical equation** is a shorthand way to represent a chemical reaction. Like English sentences, chemical equations are usually read left to right. The molecules to the left of the sideways arrow react together to form the products to the right of the arrow. For instance, in the equation below, A and B react together to form C. The single arrow indicates the reaction is not reversible. That is, C does not break down into A and B.

$$A + B \rightarrow C$$

However, many chemical reactions are reversible. In that case, the chemical equation includes arrows in both directions. The following chemical equation illustrates that D and E combine to form F, but F can also be broken down

A CLOSER LOOK BOX 2-1
Hydrogen Bonds: Strength in Numbers

In contrast to ionic and covalent bonds, which hold atoms together, hydrogen bonds hold molecules together. Hydrogen bonds are much weaker than ionic or covalent bonds—in fact, they are more like "attractions" between molecules. While ionic and covalent bonds rely on electron transfer or sharing, hydrogen bonds form bridges between two molecules. A hydrogen bond forms when a slightly positive hydrogen atom in one molecule is attracted to a slightly negative atom in another molecule. Even though a single hydrogen bond is weak, many hydrogen bonds between two molecules can be strong.

Hydrogen bonds hold water molecules together, with the slightly positive hydrogen atom in one molecule attracted to a slightly negative oxygen atom in another. Many of water's unique properties come from its ability to form hydrogen bonds. For example, hydrogen bonds keep water liquid over a wide range of temperatures, which provides a constant environment for body cells.

Hydrogen bonds form not only between molecules but also within large molecules. Hydrogen bonds between regions of the same molecule cause it to fold and coil into a specific shape, as in the process that creates the precise three-dimensional structure of proteins. Because a protein's structure determines its function in the body, hydrogen bonds are essential to protein activity.

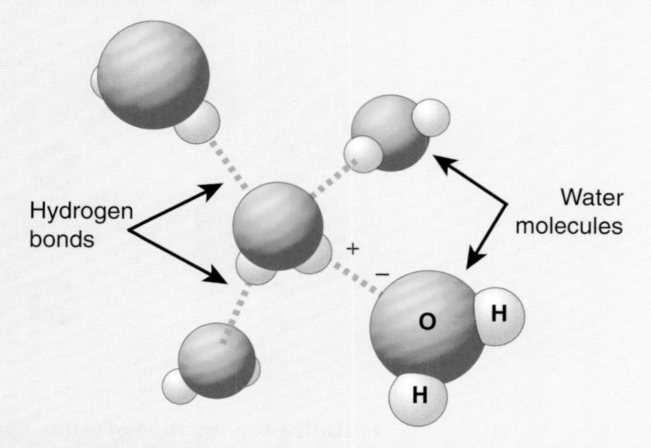

Hydrogen bonds. The bonds shown here are holding water molecules together.

into D and E. If both reactions occur at the same frequency, the chemical reaction is at **equilibrium**. For instance, in this chemical reaction at equilibrium, the rate of F formation is equal to the rate of F breakdown. The concentration of each substance will remain constant.

$$D + E \rightleftarrows F$$

In chemical reactions that are not at equilibrium, one reaction occurs more frequently than the other. For instance, if the rate of F formation exceeds the rate of F breakdown, the concentrations of D and E will decrease and the concentration of F will increase.

These simple illustrations do not cover all the variables that may affect chemical reactions but provide a basis for interpreting chemical equations. More examples will appear later in the text as we explore various physiologic processes.

CHECKPOINTS 

- 2-4 Which type of chemical bond is formed by an exchange of electrons? Which type is formed by a sharing of electrons?
- 2-5 What happens when an electrolyte goes into solution?
- 2-6 What are molecules, and what are compounds?
- 2-7 Write a chemical equation to show that substance X reacts with substance Y to form Z and that the reaction is reversible.

Mixtures

Not all elements or compounds react chemically when brought together. The air we breathe every day is a combination of gases, largely nitrogen, oxygen, and carbon dioxide, along with smaller percentages of other substances. The constituents in the air maintain their identity, although the proportions of each may vary. Blood plasma—the fluid portion of blood—is also a combination in which the various components maintain their identity. The many valuable compounds in the plasma remain separate entities with their own properties. Such combinations are called **mixtures**—blends of two or more substances **(TABLE 2-2)**.

SOLUTIONS AND SUSPENSIONS

A mixture formed when one substance dissolves in another is called a **solution**. One example is salt water. In a solution, the component substances cannot be distinguished from each other and remain evenly distributed throughout; that is, the mixture is homogeneous (ho-mo-JE-ne-us). The dissolving substance, which is water in the body, is the **solvent**. The substance dissolved, table salt in the case of salt water, is the **solute**. An **aqueous** (A-kwe-us) **solution** is one in which water is the solvent. Aqueous solutions of glucose, salts, or both of these together are used for intravenous fluid treatments.

In some mixtures, the substance distributed in the background material is not dissolved and will settle out unless the mixture is constantly shaken. This type of nonuniform, or heterogeneous (het-er-o-JE-ne-us), mixture is called a **suspension**. The particles in a suspension are separate from the material in which they are dispersed, and they settle out because they are large and heavy. Examples of suspensions are milk of magnesia, finger paints, and in the body, red blood cells suspended in blood plasma.

Table 2-2 Mixtures

Type	Definition	Example
Solution	Homogeneous mixture formed when one substance (solute) dissolves in another (solvent)	Table salt (NaCl) dissolved in water; table sugar (sucrose) dissolved in water
Suspension	Heterogeneous mixture in which one substance is dispersed in another but will settle out unless constantly mixed	Red blood cells in blood plasma; milk of magnesia
Colloid	Heterogeneous mixture in which the suspended particles remain evenly distributed based on the small size and opposing charges of the particles	Blood plasma; cytosol

One other type of mixture is important in body function. Some organic compounds form **colloids**, in which the molecules do not dissolve yet remain evenly distributed in the suspending material. The particles have electric charges that repel each other, and the molecules are small enough to stay in suspension. The fluid that fills the cells (cytosol) is a colloid, as is blood plasma.

Many mixtures are complex, with properties of solutions, suspensions, and colloids. For instance, blood plasma has dissolved compounds, making it a solution. The red blood cells and other formed elements give blood the property of a suspension. The proteins in the plasma give it the property of a colloid. Chocolate milk also has all three properties.

CASEPOINTS

- 2-1 What type of solution did the emergency team administer orally to Margaret in the opening case study?
- 2-2 Name the solvent in this solution.

6 THE IMPORTANCE OF WATER

Water is the most abundant compound in the body. No plant or animal can live very long without it. Water is of critical importance in all physiological processes in body tissues. A deficiency of water, or dehydration (de-hi-DRA-shun), can be a serious threat to health, as illustrated by Margaret's case study. Water carries substances to and from the cells and makes possible the essential processes of absorption, exchange, secretion, and excretion. What are some of the properties of water that make it such an ideal medium for living cells?

- Water can dissolve many different substances in large amounts. For this reason, it is called the universal solvent. Many of the body's necessary materials, such as gases and nutrients, dissolve in water to be carried from place to place. Substances, such as salt, that mix with or dissolve in water are described as **hydrophilic** ("water-loving"); substances, such as fats, that do not dissolve in water are described as **hydrophobic** ("water-fearing").

- Water is stable as a liquid at ordinary temperatures. It does not freeze until the temperature drops to 0°C (32°F) and does not boil until the temperature reaches 100°C (212°F). This stability provides a consistent environment for living cells. Water can also be used to distribute heat throughout the body and to cool the body by evaporation of sweat from the body surface.
- Water participates in the body's chemical reactions. It is needed directly in the digestive process and in many of the metabolic reactions that occur in the cells.

CHECKPOINTS 

- 2-8 What is the difference between solutions and suspensions?
- 2-9 What is the most abundant compound in the body?

Acids, Bases, and Salts

An **acid** (AH-sid) is a chemical substance capable of releasing a hydrogen ion (H^+) when dissolved in water. A common example is hydrochloric acid (HCl), the acid found in stomach juices. HCl releases hydrogen ions in solution as follows:

$$\underset{\text{(hydrochloric acid)}}{HCL} \rightarrow \underset{\text{(hydrogen ion)}}{H^+} + \underset{\text{(chloride ion)}}{Cl^-}$$

A **base** is a chemical substance that can accept (react with) a hydrogen ion. A base is also called an **alkali** (AL-kah-li), and bases are described as alkaline. Most bases release a hydroxide ion (OH^-) in solution, and the hydroxide ion subsequently accepts a hydrogen ion to form water. Sodium hydroxide is an example of a base:

$$\underset{\text{(sodium hydroxide)}}{NaOH} \rightarrow \underset{\text{(sodium ion)}}{Na^+} + \underset{\text{(hydroxide ion)}}{OH^-}$$

$$OH^- + H^+ \rightarrow H_2O$$

A reaction between an acid and a base produces a **salt** and also water. In the reaction, the hydrogen of the acid is replaced by the positive ion of the base. A common example of a salt is sodium chloride (NaCl), or table salt, produced by the reaction:

$$HCl + NaOH \rightarrow NaCl + H_2O$$

THE PH SCALE

The greater the concentration of hydrogen ions in a solution, the greater the acidity of that solution. The greater the concentration of hydroxide ion (OH^-), the greater the alkalinity of the solution. The concentrations of H^+ and OH^- in a solution are inversely related; as the concentration of hydrogen ions increases, the concentration of hydroxide ions decreases. Conversely, as the concentration of hydroxide ions increases, the concentration of hydrogen ions decreases.

Acidity and alkalinity are indicated by **pH** units, which represent the relative concentrations of hydrogen and hydroxide ions in a solution. The pH units are listed on a scale from 0 to 14, with 0 being the most acidic and 14 being the most basic (FIG. 2-7). A pH of 7.0 is neutral. At pH 7.0, the solution has an equal number of hydrogen and hydroxide ions. Pure water has a pH of 7.0. Solutions that measure less than 7.0 are acidic; those that measure above 7.0 are alkaline (basic).

Because the **pH scale** is based on multiples of 10, each pH unit on the scale represents a 10-fold change in the number of hydrogen and hydroxide ions present. A solution registering 5.0 on the scale has 10 times the number of hydrogen ions as a solution that registers 6.0. The pH 5.0 solution also has one-tenth the number of hydroxide ions as the solution of pH 6.0. A solution registering 9.0 has one-tenth the number of hydrogen ions and 10 times the number of hydroxide ions as one registering 8.0. Thus, the lower the pH reading, the greater is the acidity, and the higher the pH, the greater is the alkalinity.

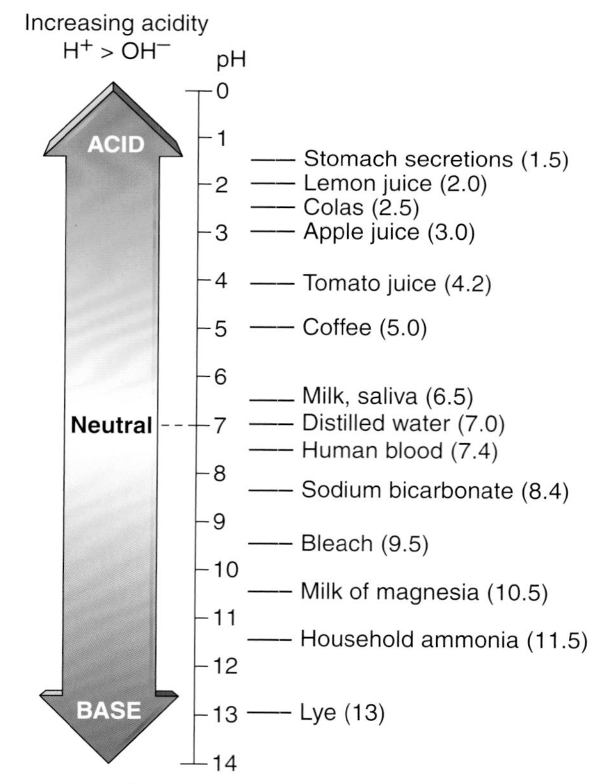

FIGURE 2-7 The pH scale. Degree of acidity or alkalinity is shown in pH units. This scale also shows the pH of some common substances. ZOOMING IN What happens to the amount of hydroxide ion (OH^-) present in a solution when the amount of hydrogen ion (H^+) increases?

ONE STEP AT A TIME
Rituals and the pH Scale

BOX 2-2

If you've ever golfed or skied, you may have experienced this phenomenon—you perform very well on the practice range or an easy slope, but then you revert to bad habits when you hit the golf course or a difficult ski run. The same thing can happen in an exam, when the stress causes us to revert to our ingrained (dominant) habits. Athletes develop rituals to ensure that their practice carries over to competitive situations. You can do the same to maximize your exam performance.

Students frequently make errors in questions about pH, even upper-year and medical students. It may seem logical that higher pH equates with more hydrogen ions, but the opposite is true. To avoid making a mistake on your exam, develop a pH question routine.

Question

The case study focuses on Margaret, an elderly lady with severe dehydration. Margaret's blood pH upon admission was 7.28.

a. Are solutions with this pH considered to be basic (alkaline) or acidic?
b. Does her blood contain more hydrogen ions or more hydroxide (OH^-) ions?
c. The normal pH range for blood is 7.35 to 7.45. Does her blood contain more or fewer hydrogen ions than normal?

Solution

Step 1. Before attempting to solve the problem, it can be handy to write out a mnemonic. You can use "H^+AL" (hydrogen ion, acidity, low pH). If you are more of a visual person, you might want to draw a pH line similar to **FIGURE 2-7** on the question or exam sheet, including changes in hydrogen/hydroxide ion concentrations and pH values. You also need to remember that 7 is neutral.

Step 2. Find the pH value in the question—in this example, 7.28. If you drew the number line, write this value on the line. Note that *7.28* is situated in the basic (alkaline) portion of the line, where hydroxide ions exceed hydrogen ions.

If you used the mnemonic, your starting point is that this number is *higher* than 7—it is a *high* pH. So, based on the mnemonic, high pH is basic (not acidic), and hydroxide (not hydrogen) ions are the most abundant.

Step 3. Answer parts A and B. Based on the work in step 2, her blood is alkaline and contains more hydroxide ions than hydrogen ions.

Step 4. Part C tells us that her blood pH is lower than the normal range. Go back to the mnemonic—*low* pH is associated with acidity and hydrogen ions. So her blood contains more hydrogen ions than normal.

Whatever approach you take, use the same approach in practice problems and exams. Taking the time for your "pre-question ritual" will increase your accuracy and help prevent errors.

Blood and other body fluids are close to neutral but are slightly on the alkaline side, with a pH range of 7.35 to 7.45. Urine averages pH 6.0 but may range from 4.6 to 8.0 depending on body conditions and diet. **FIGURE 2-7** shows the pH of some other common substances.

Because body fluids are on the alkaline side of neutral, the body may be in a relatively acidic state even if the pH does not drop below 7.0. For example, if a patient's pH falls below 7.35 but is still greater than 7.0, the patient is described as being in an acidic state known as *acidosis*. Thus, within this narrow range, physiologic acidity differs from acidity as defined by the pH scale.

An increase in pH to readings greater than 7.45 is termed *alkalosis*. Any shifts in pH to readings above or below the normal range can be dangerous, even fatal. See **BOX 2-2** for more information about interpreting pH values.

CASEPOINT 

☐ **2-3** Margaret's blood pH was 7.28. Is she suffering from acidosis or alkalosis?

BUFFERS

In a healthy person, body fluids are delicately balanced within narrow limits of acidity and alkalinity. This balanced chemical state is maintained in large part by **buffers**. Chemical buffers form a system that prevents sharp changes in hydrogen ion concentration and thus maintains a relatively constant pH. Buffers are important in maintaining stability in the pH of body fluids. More information about body fluids, pH, and buffers can be found in Chapter 19.

CHECKPOINTS 

☐ **2-10** What number is neutral on the pH scale? What kind of compound measures lower than this number? Higher?

☐ **2-11** What is a buffer?

Isotopes and Radioactivity

Elements may exist in several forms, each of which is called an **isotope** (I-so-tope). These forms are alike in their numbers of protons and electrons but differ in their atomic weights because of differing numbers of neutrons in the nucleus. The most common form of oxygen, for example, has eight protons and eight neutrons in the nucleus, giving the atom an atomic weight of 16 atomic mass units (amu). But there are some isotopes of oxygen with only six or seven neutrons in the nucleus and others with 9 to 11 neutrons. The isotopes of oxygen thus range in atomic weight from 14 to 19 amu.

Some isotopes are stable and maintain constant characteristics. Others fall apart and radiate (give off) subatomic particles and/or electromagnetic (energy) waves called *gamma rays*. (Other types of electromagnetic waves are visual light, ultraviolet light, and x-rays.) Isotopes that fall apart easily are said to be **radioactive**. Radioactive elements, also called *radioisotopes*, may occur naturally, as is the case with isotopes of the very heavy elements radium and uranium. Others may be produced artificially by placing the atoms of lighter, nonradioactive elements in accelerators that smash their nuclei together.

The radiation given off by some radioisotopes is used in the treatment of cancer because it can penetrate and destroy tumor cells. A growing tumor contains immature, dividing cancer cells, which are more sensitive to the effects of radiation than are mature body cells. The greater sensitivity of these younger cells allows radiation therapy to selectively destroy them with minimal damage to normal tissues. Modern radiation instruments produce tremendous amounts of energy (in the multimillion electron-volt range) that can destroy deep-seated cancers without causing serious skin reactions.

In addition to its therapeutic values, radiation is extensively used in diagnosis. Radioactive elements that can be administered and then detected internally to identify abnormalities are called *tracers*. Radioactive iodine, for instance, can diagnose problems of the thyroid gland **(see BOX 2-3)**.

When using radiation in diagnosis or therapy, health-care personnel must follow strict precautions to protect themselves and the patient, because the rays can destroy healthy as well as diseased tissues.

CHECKPOINT 

☐ **2-12** What word is used to describe isotopes that give off radiation?

Organic Compounds

The complex molecules that characterize living things are called **organic compounds**. All of these are built on the element **carbon**. Because carbon atoms can form covalent bonds with a variety of different elements and can even covalently bond to other carbon atoms to form long chains, most organic compounds consist of large, complex molecules. The starch found in potatoes, the fat and protein in tissues, hormones, and many drugs are examples of organic compounds. These large molecules are often formed from simpler molecules called *building blocks*, or *monomers* (*mono-* means "one"), which bond together in long chains.

The main types of organic compounds are carbohydrates, lipids, and proteins. All of these organic compounds contain carbon, hydrogen, and oxygen as their main ingredients. Carbohydrates, lipids, and proteins (in addition to minerals, vitamins, and water) must be taken in as part of a normal diet. These nutrients are discussed further in Chapters 17 and 18.

CARBOHYDRATES

The building blocks of **carbohydrates** (kar-bo-HI-drates) are simple sugars, or **monosaccharides** (mon-o-SAK-ah-rides) **(FIG. 2-8)**. (The word root *sacchar/o* means "sugar.") **Glucose** (GLU-kose), a simple sugar that circulates in the blood as a cellular nutrient, is an example of a monosaccharide. Two simple sugars may be linked together to form a **disaccharide** **(see FIG. 2-8B)**, as represented by sucrose, or table sugar. More complex carbohydrates, or **polysaccharides**, consist of many simple sugars linked together **(see FIG. 2-8C)**. (The prefix *di-* means "two," and *poly-* means "many.") Examples of polysaccharides are starch, which is manufactured in plant cells, and **glycogen** (GLI-ko-jen), a storage form of glucose found in liver cells and skeletal muscle cells. Carbohydrates in the form of sugars and starches are important dietary sources of energy.

CASEPOINT 

☐ **2-4** What type of carbohydrate was administered to Margaret in the case study?

HOT TOPICS — BOX 2-3
Radioactive Tracers: Medicine Goes Nuclear

Like radiography, computed tomography (CT), and magnetic resonance imaging, **nuclear medicine imaging** (NMI) offers a noninvasive way to look inside the body. An excellent diagnostic tool, NMI not only shows structural details but also provides information about body function. NMI can help diagnose cancer, stroke, and heart disease earlier than can techniques that provide only structural information.

NMI uses **radiotracers**, radioactive substances that specific organs absorb. For example, radioactive iodine is used to image the thyroid gland, which absorbs more iodine than does any other organ. After a patient ingests, inhales, or is injected with a radiotracer, a device called a gamma camera detects the radiotracer in the organ under study and produces a picture, which is used in making a diagnosis. Radiotracers are broken down and eliminated through urine or feces, so they leave the body quickly. A patient's exposure to radiation in NMI is usually considerably lower than with x-ray or CT scan.

Three NMI techniques are positron emission tomography (PET), **bone scanning**, and the **myocardial perfusion imaging** (MPI) stress test. PET is often used to evaluate brain activity by measuring the brain's use of radioactive glucose. PET scans can reveal brain tumors because tumor cells are often more metabolically active than are normal cells and thus absorb more radiotracer. Bone scanning detects radiation from a radiotracer absorbed by bone tissue with an abnormally high metabolic rate, such as a bone tumor. The MPI test is used to diagnose heart disease. A nuclear medicine technologist injects the patient with a radionuclide (e.g., thallium, technetium), and a gamma camera images the heart during exercise and later rest. When compared, the two sets of images help evaluate blood flow to the working, or "stressed," heart.

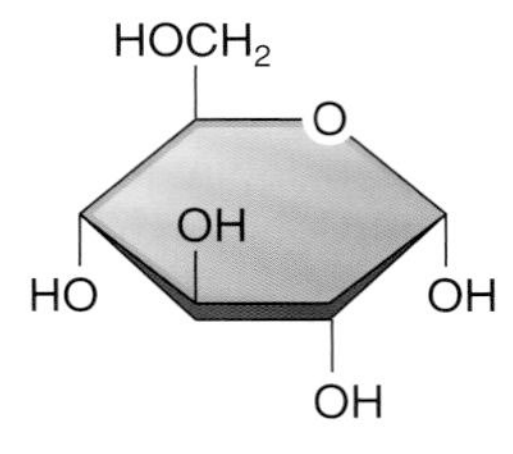

A Monosaccharide

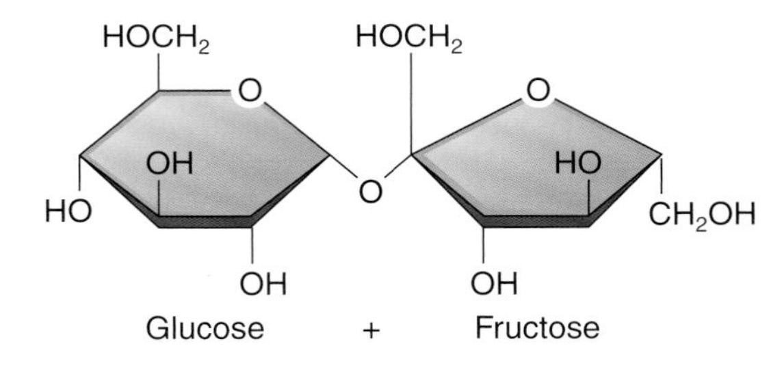

B Disaccharide

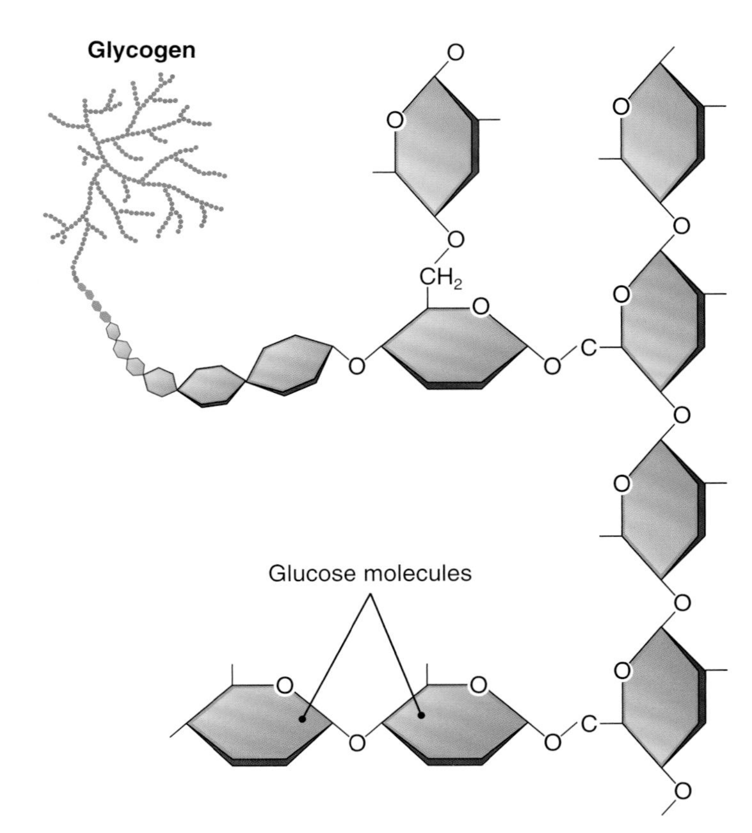

C Polysaccharide

FIGURE 2-8 Examples of carbohydrates. **KEY POINT** A monosaccharide **(A)** is a simple sugar. A disaccharide **(B)** consists of two simple sugars linked together, whereas a polysaccharide **(C)** consists of many simple sugars linked together in chains. **ZOOMING IN** What are the building blocks (monomers) of disaccharides and polysaccharides?

LIPIDS

Lipids are a class of organic compounds that are not soluble in water (hydrophobic). They are mainly found in the body as fat. Simple fats are made from a substance called **glycerol** (GLIS-er-ol), commonly known as glycerin, in combination with three fatty acids (**FIG. 2-9**). One fatty acid is attached to each of the three carbon atoms in glycerol, so simple fats are described as **triglycerides** (tri-GLIS-er-ides) (the prefix *tri-* means "three"). Fats insulate the body and protect internal organs. In addition, fats are the main form in which energy is stored, and most cells use fatty acids for energy.

Two other types of lipids are important in the body. **Phospholipids** (fos-fo-LIP-ids) are complex lipids containing the element phosphorus. Among other functions, phospholipids make up a major part of the membrane around living cells. **Steroids** are lipids that contain rings of carbon atoms. The most important sterol is **cholesterol** (ko-LES-ter-ol), another component of cellular membranes (**see FIG. 2-9B**). Cholesterol is also used to make steroid hormones, including cortisol, testosterone, and estrogen.

PROTEINS

All **proteins** (PRO-tenes) contain, in addition to carbon, hydrogen, and oxygen, the element **nitrogen** (NI-tro-jen). They may also contain sulfur or phosphorus. Proteins are the body's structural materials, found in muscle, bone, and connective tissue. They also make up the pigments that give hair, eyes, and skin their colors. It is proteins that make each individual physically distinct from others. Proteins also serve functional roles. For instance, some act as transporters, moving substances across cell membranes. Other proteins, known as *enzymes,* promote metabolic reactions. Enzymes are discussed further shortly.

Proteins are composed of monomers called **amino** (ah-ME-no) **acids** (**FIG. 2-10**). Although only about 20 different amino acids exist in the body, a vast number of proteins can be made by linking them together in different combinations.

Each amino acid contains an acid group (COOH) and an amino group (NH_2), the part of the molecule that has the nitrogen. These groups are attached to either side of a carbon atom linked to a hydrogen atom. The remainder of the molecule, symbolized by R in **FIGURE 2-10A**, is different in each amino acid, ranging from a single hydrogen atom to a complex chain or ring of carbon and other elements. These variations in the R region of the molecule account for the differences in the amino acids.

In forming proteins, the acid group of one amino acid covalently bonds with the amino group of another amino acid (**see FIG. 2-10B**). This bond is called a *peptide bond.* Many amino acids linked together in this way form a protein, which is essentially a long chain of amino acids. (Shorter

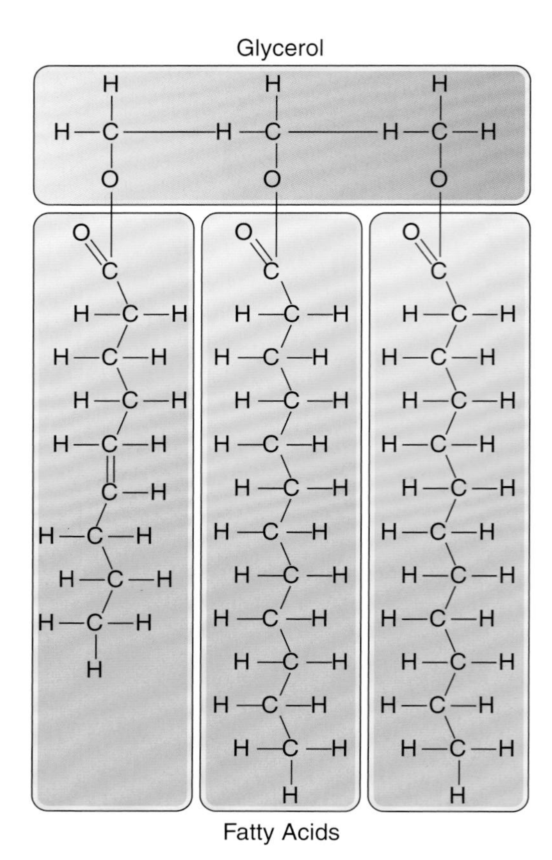

B Cholesterol (a steroid)

FIGURE 2-9 Lipids. A. A triglyceride, a simple fat, contains glycerol combined with three fatty acids. **B.** Cholesterol is a type of steroid, a lipid that contains rings of carbon atoms. ZOOMING IN How many carbon atoms are in glycerol?

chains of amino acids are sometimes called polypeptides.) The linear chain of amino acids can fold into specific shapes because of hydrogen bonding between nonadjacent amino acids. The most common of these simple shapes is a helix (spiral) (FIG. 2-10C). The final, functional form of a protein depends on further interactions between these simple shapes (see FIG. 2-10D). Some proteins consist of multiple protein chains, each folded into a helix, coiled together into rope-like structures (see FIG. 2-10D, left side). These proteins are known as fibrous proteins, and they play important roles in body structure. Collagen, for instance, provides structure to bones and cartilage. Other proteins, known as globular proteins, consist of helices (or other simple shapes) folded back on themselves into complex three-dimensional structures (see FIG. 2-10D, right side). Myoglobin, for example, is a globular protein similar to hemoglobin that stores oxygen in muscle cells. Other globular proteins include hormones, antibodies needed for immunity, and enzymes. The overall three-dimensional shape of a protein is important to its function, as can be seen in the activity of enzymes.

7 Enzymes Enzymes (EN-zimes) are proteins that participate in the hundreds of chemical reactions that take place within cells. They act as **catalysts** (KAT-ah-lists), substances that increase the speed of chemical reactions to a rate sufficient to sustain life. By controlling the activity of enzymes, the body controls which chemical reactions occur under which circumstances. Because each enzyme works only on a specific substance, or **substrate**, and does only one specific chemical job, many different enzymes are needed. Like all catalysts, enzymes take part in reactions only temporarily; they are not used up or changed by the reaction. Therefore, they are needed in small amounts. Many of the vitamins and minerals required in the diet are parts of enzymes.

An enzyme's shape is important in its action. Just as the shape of a key must fit that of its lock, an enzyme's shape must match the shape of the substrate it acts on. This so-called "lock-and-key" mechanism is illustrated in FIGURE 2-11. Because the hydrogen bonds that hold proteins in their shapes are weak, they are easily broken. Harsh conditions, such as extremes of temperature or pH, can alter the shape of any protein, such as an enzyme, and destroy its ability to function. The alteration of a protein's shape so that it can no longer function is termed **denaturation**. Such an event is always harmful to the cells.

You can usually recognize the names of enzymes because, with few exceptions, they end with the suffix *-ase*. Examples are lipase, protease, and oxidase. The first part of the name usually refers to the substance acted on or the type of reaction in which the enzyme is involved.

NUCLEOTIDES

One additional class of organic compounds is composed of building blocks called **nucleotides** (NU-kle-o-tides) (FIG. 2-12). A nucleotide contains:

- A nitrogenous (nitrogen-containing) subunit called a base (not to be confused with an alkali).
- A sugar, usually a sugar called ribose or a related sugar called deoxyribose.
- A phosphate group, which contains phosphorus. There may be more than one phosphate group in the nucleotide.

Out of the four types of organic compounds, nucleotides and carbohydrates contain sugar, but proteins and lipids do not.

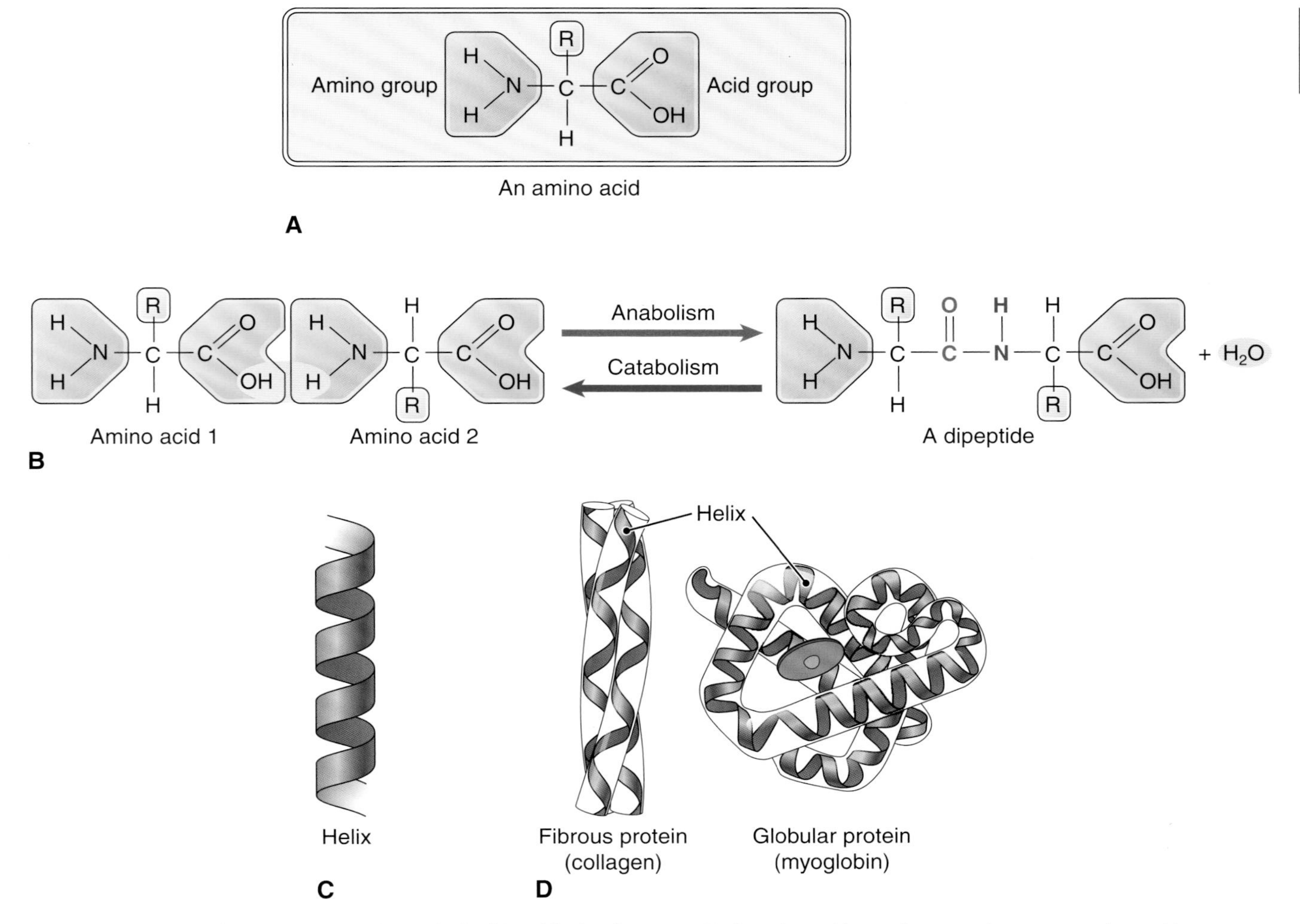

FIGURE 2-10 Proteins. A. Amino acids are the building blocks of proteins. Each amino acid contains an amino group and an acid group attached to a carbon atom. The remainder of the molecule (shown by R) can vary in 20 different ways. **B.** The acid group of one amino acid can react with the amino group of another forming a peptide bond. Further additions of amino acids result in formation of a polypeptide chain. **C.** Chemical attractions between nonadjacent amino acids form simple shapes, such as a helix. **D.** Fibrous proteins consist of multiple protein helices coiled together. Globular proteins consist of helices (or other simple shapes) folded back on themselves into complex three-dimensional structures. The characteristic shape of each protein is critical to its function. ZOOMING IN Which part of an amino acid contains nitrogen?

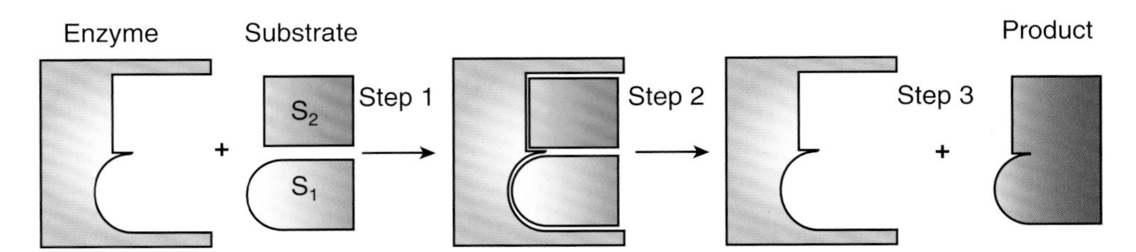

FIGURE 2-11 Diagram of enzyme action. **KEY POINT** An enzyme joins with substrate 1 (S_1) and substrate 2 (S_2) and speeds up the chemical reaction in which the two substrates bond. Once a new product is formed from the substrates, the enzyme is released unchanged. ZOOMING IN How does the shape of the enzyme before the reaction compare with its shape after the reaction?

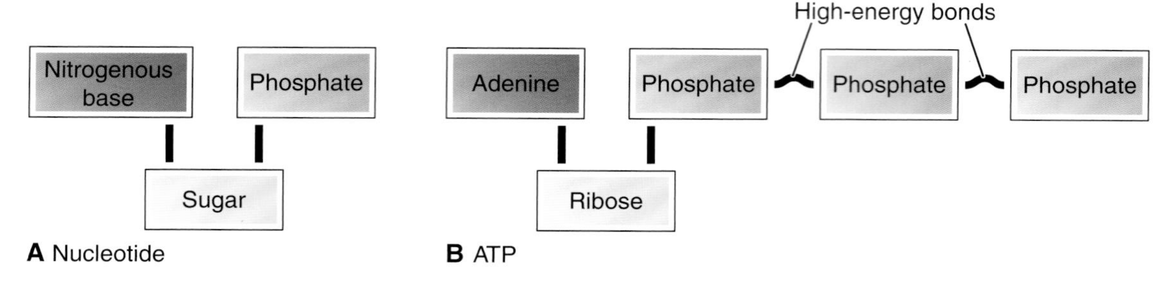

FIGURE 2-12 Nucleotides. A. A nucleotide consists of a nitrogenous base, a sugar, and one or more phosphate groups. **B.** ATP has high-energy bonds between the phosphates. When these bonds are broken, energy is released. ZOOMING IN What does the prefix *tri-* in adenosine triphosphate mean?

The nucleic acids deoxyribonucleic acid (DNA) and ribonucleic acid (RNA), involved in the transmission of genetic traits and their expression in the cell, are composed of nucleotides. These are discussed in further detail in Chapter 3. **Adenosine triphosphate** (ah-DEN-o-sene tri-FOS-fate) (**ATP**), the cell's high-energy compound, is also a nucleotide. The energy in ATP is stored in special bonds between the nucleotide's three phosphates (see FIG. 2-12B).

See the Student Resources on thePoint® to view an animation on enzymes.

Metabolism

All the life-sustaining chemical reactions that occur within the body systems together make up **metabolism** (meh-TAB-o-lizm). Metabolism can be divided into two types of activities:

- In **catabolism** (kah-TAB-o-lizm), complex substances are broken down into simpler components (FIG. 2-13). For instance, proteins are broken down into amino acids. Further catabolic reactions involving these simple substances provide energy to fuel cell processes in the form of ATP. Chapter 18 discusses these reactions in greater detail.
- In **anabolism** (ah-NAB-o-lizm), simple substances are used to manufacture materials needed for growth, function, and tissue repair. Anabolism consists of building (synthesis), reactions, such as constructing proteins from amino acids. These synthesis reactions are fueled by ATP.

8 ENERGY

Every body cell needs energy, and multiple body systems are involved in its generation. Much of physiology involves tracing matter and energy through various bodily processes, so it is helpful to understand its basic forms.

- **Kinetic energy** is the energy of movement. All matter that is in motion possesses kinetic energy, ranging from the vibrations of a single molecule to locomotion in an entire organism. There are also less obvious forms of kinetic energy. One is **radiant energy,** which consists of waves traveling through space. Sunlight, heat, sound, and x-rays are examples. Another type of kinetic energy is **electric energy,** which reflects the movement of electrons, as in electric currents passing through wires or along neurons.
- **Potential energy** is stored energy that can be used to produce activity. An example is **gravitational energy,** the energy of position in a gravitational field. A skier at the top of a hill, for instance, possesses potential energy that can be used to power downhill movement. Another example of potential energy is **chemical energy,** which is stored in covalent bonds. As noted, glucose and ATP contain chemical energy in the bonds that hold the atoms together.

Energy can be converted from one form to another (FIG. 2-14). For example, plants convert radiant energy from the sun into chemical energy within a fructose (fruit sugar) molecule using an anabolic reaction. Body cells use a catabolic reaction to convert the chemical energy of the sugar molecule into the chemical energy of ATP. Breaking the high-energy bonds in ATP can produce kinetic energy in the

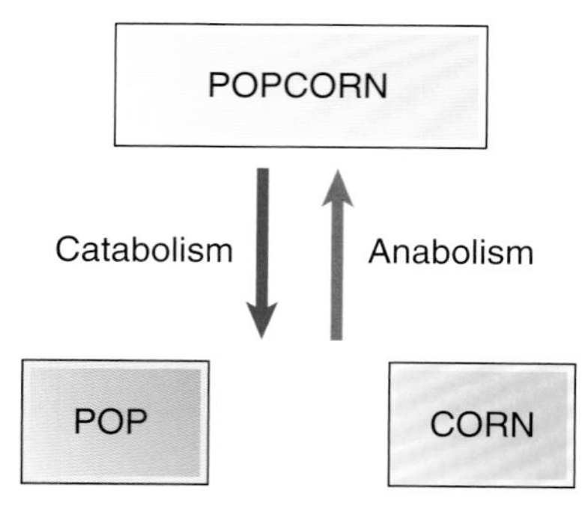

FIGURE 2-13 Metabolism. **KEY POINT** Metabolism includes two types of reactions. In catabolism, substances are broken down into their building blocks. In anabolism, simple components are built into more complex substances. We use the breakdown and building of a simple word here as an example of these reactions.

2

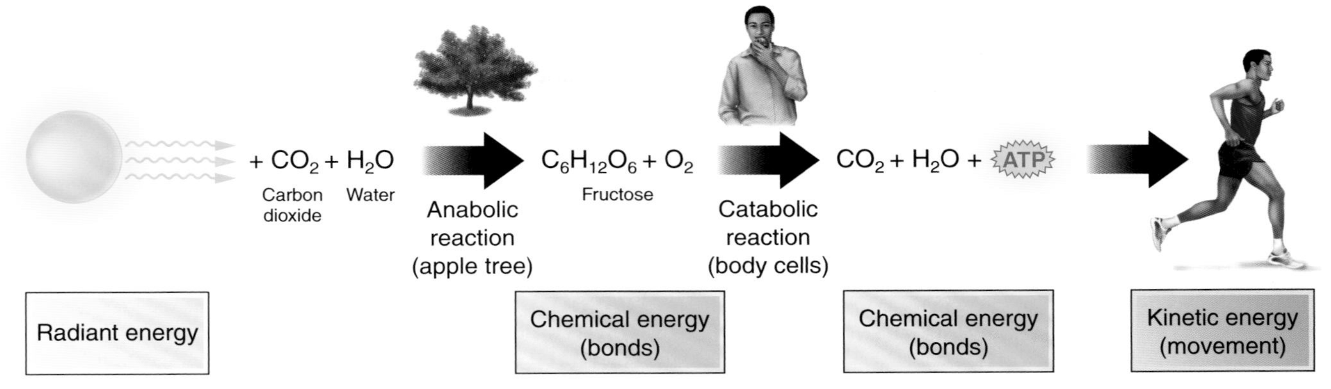

FIGURE 2-14 Energy and chemical reactions. **KEY POINT** Chemical reactions are one way to convert energy from one form to another because chemical bonds are a source of potential energy. **ZOOMING IN** Which chemicals combine to form glucose?

form of body movements. All energy conversions liberate heat, which increases body temperature. Chapter 18 discusses ATP generation in greater detail.

See the Student Resources on thePoint® for an animation on energy. In addition, the Health Professions topic, "Pharmacists and Pharmacy Technicians," describes some professions that require knowledge of chemistry.

CHECKPOINTS 

- ☐ **2-13** What element is the basis of organic chemistry?
- ☐ **2-14** What are the three main categories of organic compounds?
- ☐ **2-15** What is an enzyme?
- ☐ **2-16** What is in a nucleotide, and what compounds are made of nucleotides?
- ☐ **2-17** What are the two main types of metabolic reactions, and what happens during each?
- ☐ **2-18** Use two terms to describe the energy contained in ATP.

A & P in Action Revisited: Margaret: Back in Balance

"Good morning, Mrs. Ringland. How are you feeling today?" asked Angela.

"Much better, thank you," replied Margaret. "I'm so grateful that my niece found me when she did."

"I'm glad too," said Angela. "With the heat wave we're having, dehydration can become a serious problem. Older adults are particularly at risk of dehydration because with age, there is usually a decrease in muscle tissue, which contains a lot of water, and a relative increase in body fat, which does not. So older adults don't have as much water reserve as do younger adults. But," Angela continued as she flipped through Margaret's chart, "it looks like you're well on your way to a full recovery. Your electrolytes are back in balance. Your blood pressure is back to normal, and your heart rate is good too. Your increased urine output tells me that your other organs are recovering as well."

"Does that mean I can go home soon?" asked Margaret.

I'll check with your doctor first," replied Angela. "But when you do return home, you will need to make sure that you drink plenty of fluids."

It was the end of another long shift, and Angela was at her locker, changing into a pair of shorts and a T-shirt. As she closed her locker, she thought of Margaret once again. It always amazed her that chemistry could have such a huge impact on the body as a whole. She grabbed her water bottle, took a long drink, and headed out into the scorching heat.

In this case, we see that health professionals require a background in chemistry to understand how the body works—when healthy and when not. As you learn more about the human body, consider referring back to this chapter when necessary. For more information about the elements that make up every single substance within the body, see Appendix 1: Periodic Table of the Elements at the back of this book.

CHAPTER 2

Chapter Wrap-Up

OVERVIEW

A detailed chapter outline with space for note-taking is on thePoint®. The figure below illustrates the main topics covered in this chapter.

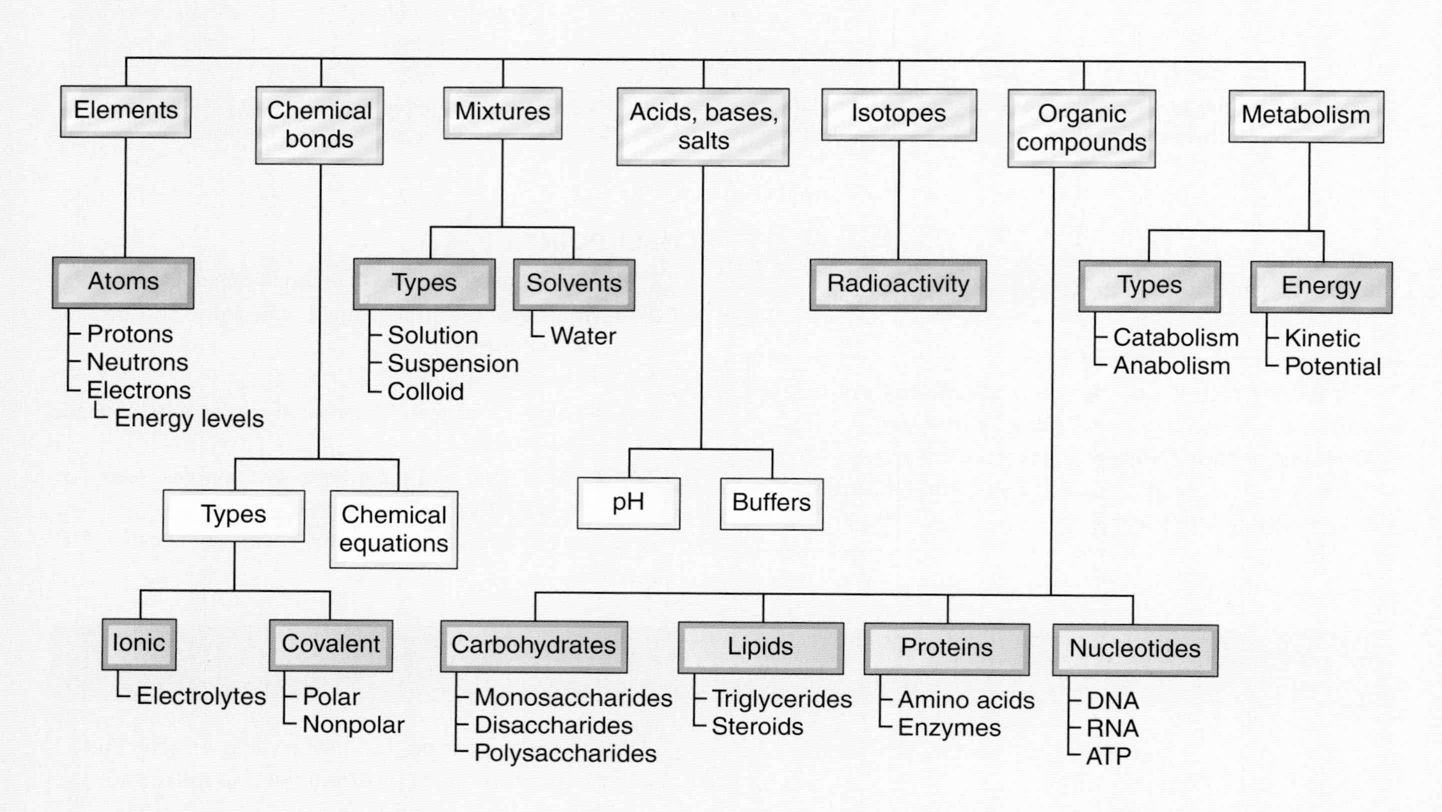

KEY TERMS

The terms listed below are emphasized in this chapter. Knowing them will help you organize and prioritize your learning. These and other boldface terms are defined in the Glossary with phonetic pronunciations.

acid
amino acid
anabolism
anion
aqueous
atom
ATP
base
buffer
carbohydrate
catabolism
catalyst
cation
chemistry
colloid
compound
denaturation
electrolyte
electron
element
enzyme
glucose
glycogen
ion
isotope
kinetic energy
lipid
metabolism
molecule
nucleotide
neutron
pH scale
potential energy
protein
proton
radioactive
salt
solute
solution
solvent
steroid
substrate
suspension

WORD ANATOMY

Medical terms are built from standardized word parts (prefixes, roots, and suffixes). Learning the meanings of these parts can help you remember words and interpret unfamiliar terms.

WORD PART	MEANING	EXAMPLE
CHEMICAL BONDS		
co-	together	*Covalent* bonds form when atoms share electrons.
MIXTURES		
aqu/e	water	In an *aqueous* solution, water is the solvent.
heter/o-	different	*Heterogeneous* solutions are different (not uniform) throughout.
hom/o-	same	*Homogeneous* mixtures are the same throughout.
hydr/o-	water	*Dehydration* is a deficiency of water.
-phil	to like	*Hydrophilic* substances "like" water—they mix with or dissolve in it.
phob/o	fear	*Hydrophobic* substances "fear" water—they repel and do not dissolve in it.
ORGANIC COMPOUNDS		
-ase	suffix used in naming enzymes	A *lipase* is an enzyme that acts on lipids.
de-	remove	*Denaturation* of a protein removes its ability to function (changes its nature).
di-	twice, double	A *disaccharide* consists of two simple sugars.
glyc/o-	sugar, glucose, sweet	*Glycogen* is a storage form of glucose. It breaks down to release glucose.
mon/o-	one	In a *monosaccharide*, "mono-" refers to one.
poly-	many	A *polysaccharide* consists of many simple sugars.
sacchar/o-	sugar	A *monosaccharide* consists of one simple sugar.
tri-	three	*Triglycerides* have one fatty acid attached to each of three carbon atoms.
METABOLISM		
ana-	upward, again, back	*Anabolism* is the building up of simple substances into more complex substances.
cata-	down	*Catabolism* is the breakdown of complex substances into simpler ones.
kine, kinet/o	movement	*Kinetic* energy is the energy of movement.

QUESTIONS FOR STUDY AND REVIEW

BUILDING UNDERSTANDING

Fill in the Blanks

1. The subunits of elements are _____.
2. The atomic number is the number of _____ in an atom's nucleus.
3. A mixture of solute dissolved in a solvent is called a(n) _____.
4. Blood has a pH of 7.35 to 7.45. Gastric juice has a pH of about 2.0. The more alkaline fluid is _____.
5. Proteins that catalyze metabolic reactions are called _____.

Matching Match each numbered item with the most closely related lettered item.

___ 6. A simple carbohydrate such as glucose
___ 7. A complex carbohydrate such as glycogen
___ 8. An important component of cell membranes
___ 9. Examples include DNA, RNA, and ATP
___ 10. The basic building block of protein

a. polysaccharide
b. phospholipid
c. nucleotide
d. amino acid
e. monosaccharide

Multiple Choice

___ 11. Which type of mixture is plasma with red blood cells "floating" in it?
a. compound
b. suspension
c. colloid
d. solution

___ 12. What is the most abundant compound in the body?
a. carbohydrate
b. protein
c. lipid
d. water

___ 13. Which compound releases ions when in solution?
a. solvent
b. electrolyte
c. anion
d. colloid

___ 14. Which substance releases a hydrogen ion when dissolved in water?
a. acid
b. base
c. salt
d. catalyst

___ 15. Which element is found in all organic compounds?
a. oxygen
b. carbon
c. nitrogen
d. phosphorus

UNDERSTANDING CONCEPTS

16. Compare and contrast the following terms:
    a. proton, neutron, and electron
    b. ionic bond and covalent bond
    c. anion and cation
    d. polar and nonpolar covalent bonds
    e. acid and base
17. What are some of the properties of water that make it an ideal medium for living cells?
18. What is pH? Discuss the role of buffers in maintaining a steady pH in the body.
19. Describe some uses of radioactive isotopes in medicine.
20. Compare and contrast carbohydrates, lipids, and proteins, and give examples of each.
21. List the components of nucleotides, and give three examples of nucleotides.
22. Define the term enzyme, and discuss the relationship between enzyme structure and enzyme function.
23. What is the difference between catabolism and anabolism? Give an example of each type of activity.
24. Compare and contrast kinetic energy and potential energy, and give examples of each.

CONCEPTUAL THINKING

25. Explain the statement, "All compounds are composed of molecules, but not all molecules are compounds."
26. Based on your understanding of strong acids and bases, why does the body have to be kept at a close-to-neutral pH?
27. In the opening case study, Margaret's blood tests showed that the percentage of red cells in her blood was high. Explain how the high reading relates to Margaret's condition.
28. Margaret was suffering from dehydration and hypernatremia. Use the glossary of word parts at the back of this book to define the word parts in these two terms.
29. Margaret was rehydrated with an aqueous solution containing 13.5 g/L glucose and several salts, including sodium chloride, potassium chloride, and sodium citrate.
    a. Name the solution's solute(s) and solvent(s).
    b. What is the meaning of the notation g/L? (Consult the metric system chart in the Student Resources for Chapter 1, if needed.)
    c. Give the chemical formulas for sodium chloride and potassium chloride (consult the periodic table of the elements in Appendix 1, if needed).
30. Bacteria break down glucose ($C_6H_{12}O_6$) into ethanol (C_2H_5OH) and carbon dioxide (CO_2).
    a. Write a chemical equation to describe this chemical reaction.
    b. Is this chemical reaction anabolic or catabolic?

For more questions, see the Learning Activities on thePoint®.

CHAPTER 3

Cells and Their Functions

Learning Objectives

After careful study of this chapter, you should be able to:

1. List three types of microscopes used to study cells. ***p. 42***
2. Describe the composition and functions of the plasma membrane. ***p. 43***
3. Describe the cytoplasm of the cell, and cite the names and functions of the main organelles. ***p. 44***
4. Describe methods by which substances enter and leave cells that do not require chemical energy. ***p. 48***
5. Explain what will happen if cells are placed in solutions with concentrations the same as or different from those of the cytoplasm. ***p. 50***
6. Describe methods by which substances enter and leave cells that require chemical energy. ***p. 51***
7. Describe the composition, location, and function of the DNA in a cell. ***p. 53***
8. Compare the functions of three types of RNA in cells. ***p. 55***
9. Explain briefly how cells make proteins. ***p. 55***
10. Name and briefly describe the stages in mitosis. ***p. 57***
11. Use the case study to explain how a small change in DNA sequence can impact the entire organism. ***pp. 41, 59***
12. Show how word parts are used to build words related to cells and their functions (see Word Anatomy at the end of the chapter). ***p. 61***

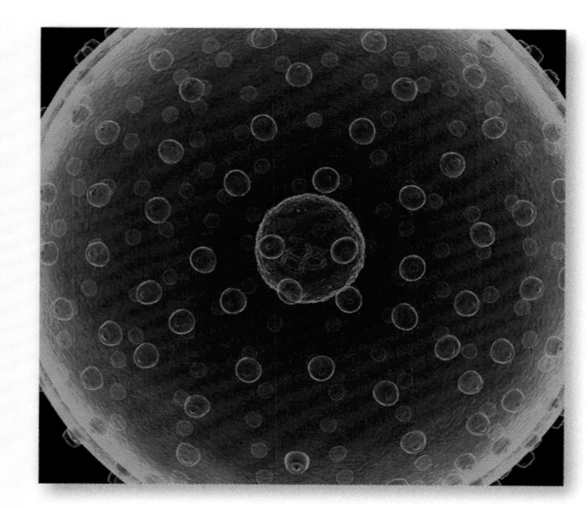

A & P in Action
Ben's Case: How a Cellular Failure Affects the Entire Body

Alison awoke with a start to her baby's coughing. "Not again," she thought as she stumbled out of bed toward Ben's room. For the last few days, Alison's 1-year-old was sick with what appeared to be a nasty chest infection. This wasn't unusual for Ben—he had come down with several lung infections in the past year and often seemed congested, but Alison had chalked this up to normal childhood illnesses. Lately though, Alison had become more worried, especially after taking Ben to their community center, where she noticed that he seemed smaller than the other children of his age and was not as active. "I'll take him in to see the doctor tomorrow," Alison thought as she sat down in the rocking chair beside Ben's crib and began patting his back.

At the medical center, Ben's doctor examined him carefully. Ben was smaller and weighed less than did most boys of his age, despite his mom's observation that he had a good appetite. His recurrent respiratory infections were also cause for worry. In addition, Alison reported that Ben had frequent bowel movements with stools that were often foul smelling and greasy. The doctor's next question caught Alison off guard. "When you kiss your son, does he taste saltier than what you might expect?" The doctor wasn't surprised when Alison answered yes. "I need to run a few more tests before I can make a diagnosis," he said. "In the meantime, let's start Ben on some oral antibiotics for his chest infection."

A few days later, Ben's doctor reviewed his chart and the lab test results. Chest and sinus radiography showed evidence of bacterial infection and thickening of the membrane lining Ben's respiratory passages. The blood test indicated that Ben had elevated levels of the pancreatic enzyme *immunoreactive trypsinogen*. Genetic testing revealed mutations in a specific gene called CFTR. The sweat test revealed that Ben's sweat glands excreted abnormally high concentrations of sodium chloride. With the evidence he had, the doctor was ready to make his diagnosis. Ben had cystic fibrosis (CF).

CF is caused by a mutation in a gene that codes for a channel protein in the plasma membrane of certain types of cells. Its consequences, however, are seen in many different organs and systems—especially the respiratory and digestive systems. We will learn more about the implications of this disease later in the chapter.

As you study this chapter, CasePoints will give you opportunities to apply your learning to this case.

Visit thePoint® to access the following resources. For guidance in using these resources most effectively, **see pp. xv–xvii.**

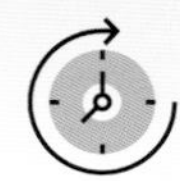

Preparing to Learn

- Tips for Effective Studying
- Pre-Quiz

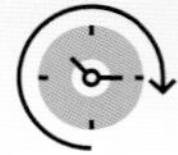

While You Are Learning

- Web Figure: Electron Micrograph of an Animal Cell Magnified over 20,000 Times
- Web Figure: Electron Micrograph of an Animal Cell Magnified over 48,000 Times
- Web Figure: Electron Micrograph of a Replicated Chromosome
- Animation: Osmosis
- Animation: Plasma Membrane and Cell Transport
- Animation: Protein Synthesis Overview
- Animation: The Cell Cycle and Mitosis
- Animation: Phagocytosis
- Chapter Notes Outline
- Audio Pronunciation Glossary

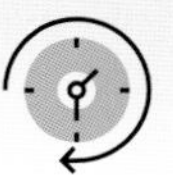

When You Are Reviewing

- Answers to Questions for Study and Review
- Health Professions: Cytotechnologist
- Interactive Learning Activities

A LOOK BACK

The chemicals we learned about in Chapter 2 are the building blocks of cells, the fundamental units of all organisms. In this chapter, we apply five key ideas to understand the organization and activities of cells. Chapter 1 introduced the concepts of barriers 4, flow 5, and the relationship between structure and function 1; Chapter 2 introduced the types of energy 8 and the qualities of water 6.

The cell is the basic unit of all life. It is the simplest structure that shows all the characteristics of life, including organization, metabolism, responsiveness, homeostasis, growth, and reproduction. In fact, it is possible for a single cell to live independently of other cells. Examples of some free-living cells are microscopic organisms such as protozoa and bacteria, some of which produce disease. As we saw in Chapter 1, cells make up all tissues in a multicellular organism; the human body, for example, contains trillions of cells. All of the body's abilities, including thinking, running, and generating energy from food, reflect activities occurring in individual cells. So, understanding how the body accomplishes these complex actions requires first that we understand the structures and abilities of individual cells.

Microscopes

The study of cells is **cytology** (si-TOL-o-je). Scientists first saw the outlines of cells in dried plant tissue almost 350 years ago. They were using a **microscope**, a magnifying instrument that allowed them for the first time to examine structures not visible to the naked eye. Study of a cell's internal structure, however, depended on improvements in the design of the single-lens microscope used in the late 17th century. The following three microscopes, among others, are used today:

- The **compound light microscope** is the microscope most commonly used in laboratories. This instrument, which can magnify an object up to 1,000 times, usually has two lenses and uses visible light for illumination, although some may use other light sources (such as ultraviolet light).
- The **transmission electron microscope (TEM)** uses an electron beam in place of visible light and can magnify an image up to 1 million times.
- The **scanning electron microscope (SEM)** does not magnify as much as does the TEM (100,000 times) and shows only surface features; however, it provides a three-dimensional view of an object.

These microscopes are commonly linked to cameras and computers to record and digitally analyze images. **FIGURE 3-1** shows some cell structures viewed with each of these types of microscopes. The structures are cilia—short, hairlike projections from the cell that move nearby fluids. The metric unit used for microscopic measurements is the **micrometer** (MI-kro-me-ter). This unit is 1/1,000 of a millimeter and is abbreviated as mcm (also µm).

Before scientists can examine cells and tissues under a light microscope, they must usually color them with special dyes called stains to aid in viewing. These stains produce the variety of colors seen in photographs (micrographs) of cells and tissues taken under a microscope.

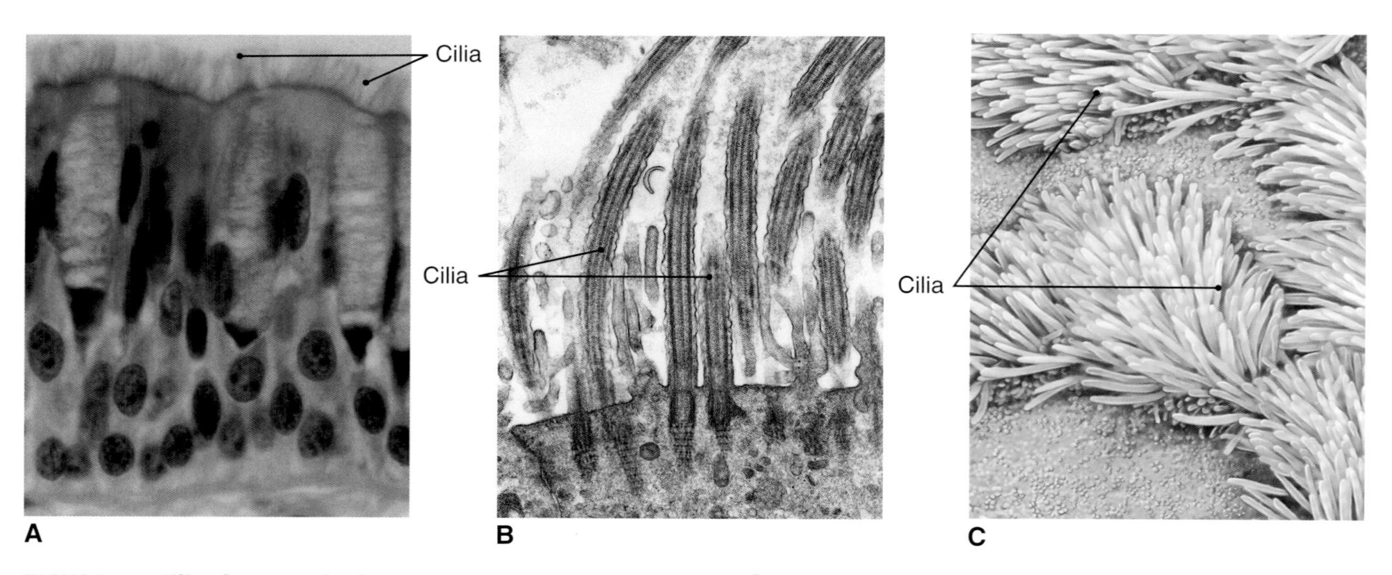

FIGURE 3-1 Cilia photographed under three different microscopes. **KEY POINT** Each type of microscope produces a different type of image that reveals different aspects of structure. **A.** Cilia (hairlike projections) in cells lining the trachea under the highest magnification of a compound light microscope (1,000 times). **B.** Cilia in the bronchial lining viewed with a TEM. Internal components are visible at this much higher magnification. **C.** Cilia on cells lining an oviduct as seen with a SEM (7,000 times). A three-dimensional view is visible. **ZOOMING IN** Which microscope shows the most internal structure of the cilia? Which shows the cilia in three dimensions? *See Appendix 2 for answers to the Zooming In questions.*

CHECKPOINTS

- 3-1 List six characteristics of life shown by cells.
- 3-2 Name three types of microscopes.

See Appendix 2 for answers to the Checkpoint questions.

See the Student Resources on thePoint® for information on careers in cytotechnology, the clinical laboratory study of cells, as well as to view electron micrographs of the cell.

Cell Structure

Just as people may look different but still have certain features in common—two eyes, a nose, and a mouth, for example—all cells share certain characteristics. Refer to FIGURE 3-2 as we describe some of the structures that are common to most animal cells. A summary table follows the descriptions.

PLASMA MEMBRANE

The outer layer of the cell is the **plasma membrane** (FIG. 3-3). (This cell part is still often called the *cell membrane*, although this older term fails to distinguish between the cell's outer membrane and other internal cellular membranes.) The plasma membrane not only encloses the cell contents but also participates in many cellular activities, such as growth, reproduction, and communication between cells, and it is especially important in regulating what can enter and leave the cell.

Some cells specialize in the uptake, or absorption, of materials from the extracellular fluid. The plasma membrane of these cells is often folded into multiple small projections called **microvilli** (mi-kro-VIL-li; see FIG. 3-2). These projections increase the membrane's surface area, allowing for greater absorption, much as a sponge's many holes provide increased surface for absorption. Microvilli are found on cells that line the small intestine, where they promote absorption of digested foods into the circulation. They are also found on kidney cells, where they reabsorb materials that have been filtered out of the blood.

Components of the Plasma Membrane The main substance of the plasma membrane is a double layer—or bilayer—of lipid molecules. Because these lipids contain the element phosphorus, they are called *phospholipids*. We introduced these lipids in Chapter 2, along with cholesterol, another type of lipid found in the plasma membrane. Molecules of cholesterol are located between the phospholipids, and they make the membrane stronger and more flexible.

Carbohydrates are present in small amounts on the outer surface of the membrane, combined either with proteins (glycoproteins) or with lipids (glycolipids). These carbohydrates help cells recognize each other and stick together.

A variety of different proteins float within the lipid bilayer. Some of these proteins extend all the way through the membrane, and some are located near the membrane's inner or outer surface. The importance of these proteins

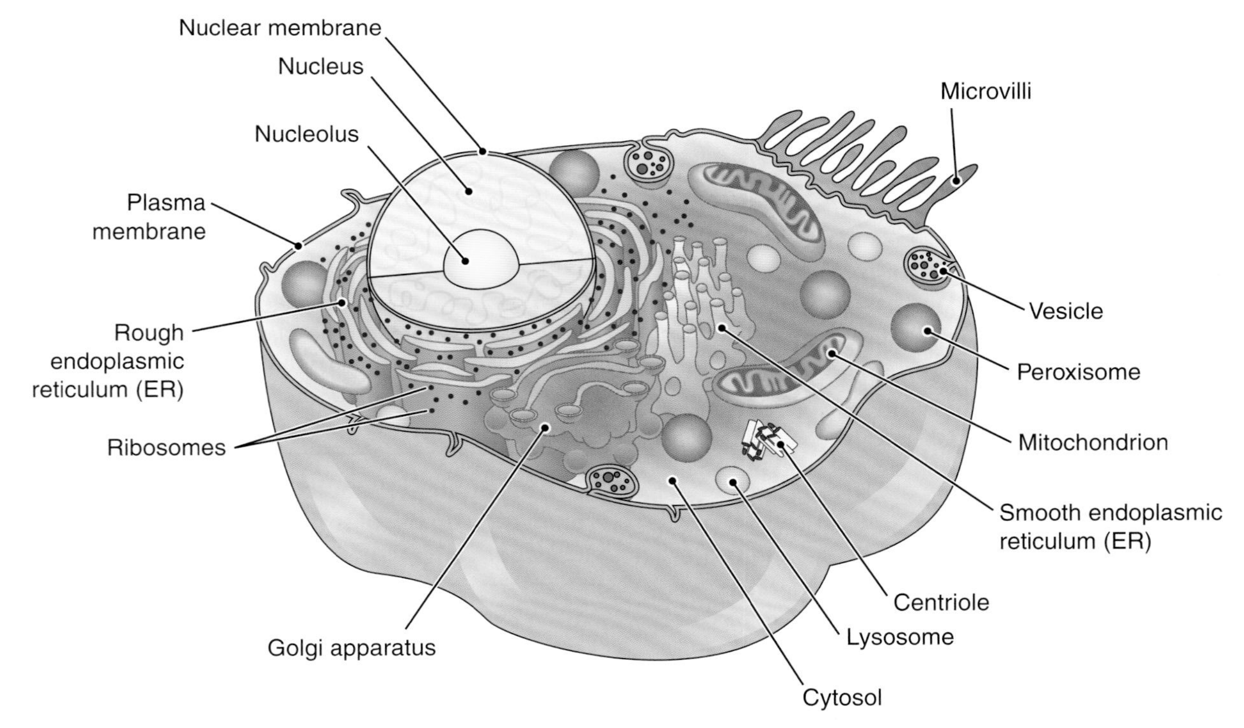

FIGURE 3-2 A generalized animal cell, sectional view. ZOOMING IN What is attached to the ER to make it look rough? What is the liquid part of the cytoplasm called?

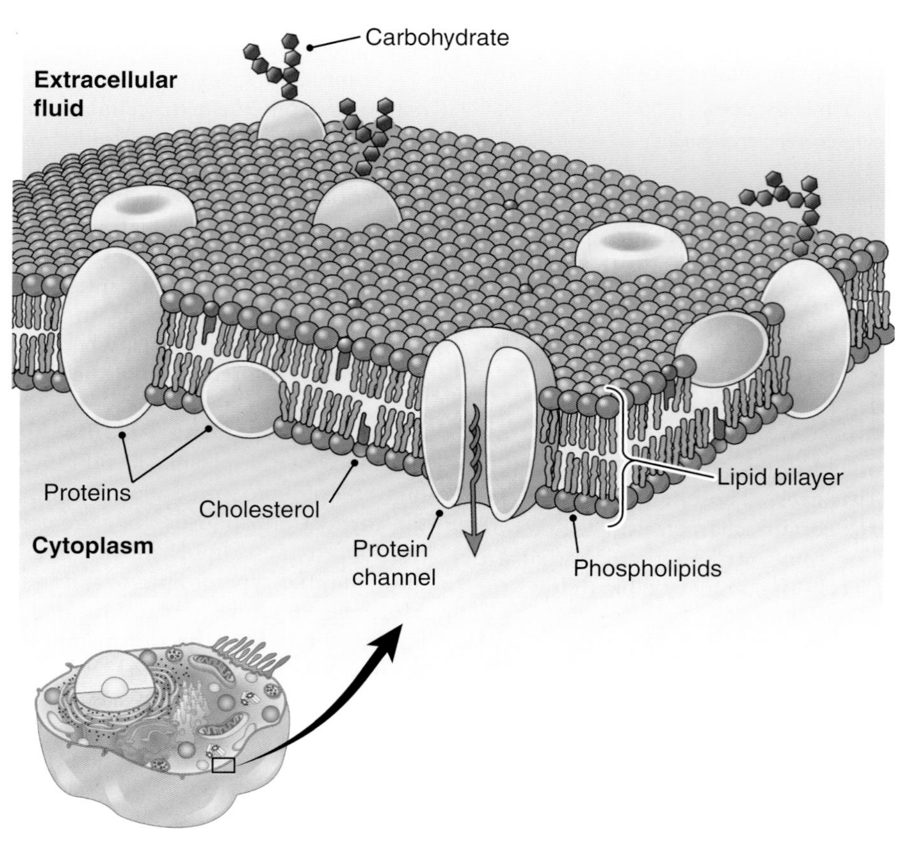

FIGURE 3-3 The plasma membrane. This drawing shows the current concept of its structure. **KEY POINT** The membrane is composed of a double layer of phospholipids with proteins and other materials embedded in it. **ZOOMING IN** Why is the plasma membrane described as a bilayer?

is revealed in later chapters, but they are listed here along with their functions and summarized and illustrated in **TABLE 3-1**.

- Channels—pores in the membrane that allow specific substances to enter or leave. Certain ions and water travel through channels in the membrane.
- Transporters—shuttle substances from one side of the membrane to the other. Unlike channels, transporters change shape during transport. Glucose, for example, is carried into cells using transporters.
- Receptors—specialized proteins that mediate the effects of chemical signals on cells. Chemicals such as hormones or neurotransmitters (chemical signals used by the nervous system) bind to a receptor, which then alters cell function. For example, the hormone insulin binds to a receptor on muscle cells, and the bound receptor stimulates the production of the glucose transporters mentioned above. Some neurotransmitters bind to receptors that open or close membrane channels. Chapter 8 discusses neurotransmitter receptors in greater detail, and Chapter 11 discusses hormone receptors.
- Enzymes—participate in reactions occurring at the plasma membrane.
- Linkers—link to other proteins within the cell to stabilize the membrane and link to membrane proteins of other cells to attach cells together.
- Cell identity markers—proteins unique to an individual's cells. These are important in the immune system.

CASEPOINT

☐ **3-1** Ben's case involves a defect in a protein that allows ions to pass through the plasma membrane. What type of membrane protein is this?

See Appendix 2 for the answers to the CasePoint questions.

See the Student Resources on thePoint® to view an animation on the plasma membrane and cell transport.

THE NUCLEUS

Just as the body has different organs to carry out special functions, the cell contains specialized structures that perform different tasks (**TABLE 3-2**). These structures are called **organelles**, which mean "little organs." The largest of the organelles is

3

Table 3-1 Proteins in the Plasma Membrane and Their Functions

Type of Protein	Function	Illustration
Channels	Pores in the membrane that allow passage of specific substances, such as ions	
Transporters	Proteins that change shape as they shuttle substances, such as glucose, across the membrane	
Receptors	Allow for attachment of substances, such as hormones, to the membrane	
Enzymes	Participate in reactions at the membrane surface	
Linkers	Help stabilize the plasma membrane and attach cells together	
Cell identity markers	Proteins unique to a person's cells; important in the immune system and in transplantation of tissue from one person to another	

the **nucleus** (NU-kle-us), which is surrounded by a membrane, the *nuclear membrane*, that encloses its contents.

The nucleus is often called the *control center* of the cell because it contains the **chromosomes** (KRO-mo-somes), the threadlike structures of heredity that are passed on from parents to their children. It is information contained in the chromosomes that governs all cellular activities, as described later in this chapter. Most of the time, the chromosomes are loosely distributed throughout the nucleus, giving it a uniform, dark appearance when stained and examined under a microscope **(see FIG. 3-2)**. When the cell is dividing, however, the chromosomes tighten into their visible threadlike forms.

Within the nucleus is a darker stained region called the **nucleolus** (nu-KLE-o-lus), which means "little nucleus." The job of the nucleolus is to assemble **ribosomes** (RI-bo-somz), small bodies outside the nucleus that are involved in the manufacture of proteins.

THE CYTOPLASM

The remaining organelles are part of the **cytoplasm** (SI-to-plazm), the material that fills the cell from the nuclear membrane to the plasma membrane. The liquid part of the cytoplasm is the **cytosol**, a suspension of nutrients, electrolytes, enzymes, and other specialized materials in water. The main organelles are described here **(see TABLE 3-2)**.

Recall that ribosomes are small organelles that assemble proteins. Ribosomes begin the process of protein synthesis while floating freely in the cytoplasm. Then, they usually migrate to the surface of a different organelle, the endoplasmic reticulum. The **endoplasmic reticulum** (en-do-PLAS-mik re-TIK-u-lum) is a membranous network located between the nuclear membrane and the plasma membrane. Its name literally means "network" (reticulum) "within the cytoplasm" (endoplasmic), but for ease, it is almost always called simply the **ER**. Sections of the ER studded with

Table 3-2 Cell Parts

Name	Description	Function
Plasma membrane	Outer layer of the cell; composed mainly of lipids and proteins	Encloses the cell contents; regulates what enters and leaves the cell; participates in many activities, such as growth, reproduction, and interactions between cells
Microvilli	Short extensions of the plasma membrane	Absorb materials into the cell
Nucleus	Large, membrane-bound, dark-staining organelle near the center of the cell	Contains the chromosomes, the hereditary structures that direct all cellular activities
Nucleolus	Small body in the nucleus	Makes ribosomes
Cytoplasm	Colloid that fills the cell from the nuclear membrane to the plasma membrane	Site of many cellular activities; consists of cytosol and organelles
Cytosol	The fluid portion of the cytoplasm; contains water, enzymes, nutrients, and other substances	Surrounds the organelles; site of many chemical reactions and nutrient storage
Endoplasmic reticulum (ER)	Network of membranes within the cytoplasm. Rough ER has ribosomes attached to it; smooth ER does not	Rough ER modifies, folds, and sorts proteins; smooth ER participates in lipid synthesis
Ribosomes	Small bodies free in the cytoplasm or attached to the ER; composed of RNA and protein	Manufacture proteins
Golgi apparatus	Layers of membranes	Further modifies proteins; sorts and prepares proteins for transport to other parts of the cell or out of the cell
Mitochondria	Large organelles with internal folded membranes	Convert energy from nutrients into ATP
Lysosomes	Small sacs of digestive enzymes	Digest substances within the cell
Peroxisomes	Membrane-enclosed organelles containing enzymes	Break down harmful substances
Proteasomes	Barrel-shaped organelles	Destroy improperly synthesized proteins
Vesicles	Small membrane-bound sacs in the cytoplasm	Store materials and move materials into or out of the cell
Centrioles	Rod-shaped bodies (usually two) near the nucleus	Help separate the chromosomes during cell division
Surface projections	Structures that extend from the cell	Move the cell or the fluids around the cell
Cilia	Short, hairlike projections from the cell	Move the fluids around the cell
Flagellum	Long, whiplike extension from the cell	Moves the cell

ribosomes have a gritty, uneven surface, causing them to be described as *rough ER*. An attached ribosome feeds the protein into the rough ER, where enzymes add sugar chains and help the protein fold into the correct shape. The part of the ER that is not covered with ribosomes appears to have an even surface and is described as *smooth ER*. This type of ER is involved with the synthesis of lipids.

The rough ER sends proteins to the nearby **Golgi** (GOL-je) **apparatus** (also called the Golgi complex), a large organelle consisting of a stack of membranous sacs. As the proteins pass through this organelle, they are further modified, sorted, and packaged for export from the cell.

The **mitochondria** (mi-to-KON-dre-ah) are large, round, or bean-shaped organelles with folded membranes on the inside. It is within the mitochondria that the chemical energy from nutrients is converted into the chemical energy of adenosine triphosphate (ATP), as described in Chapter 2. These reactions take place on the organelles' internal membranes under the influence of enzymes. Mitochondria are described as the cell's "power plants." Active cells, such as muscle cells or sperm cells, need lots of energy and thus have large numbers of mitochondria.

Several types of organelles appear as small sacs in the cytoplasm. These include **lysosomes** (LI-so-somz), which contain digestive enzymes. (The root *lys/o* means "dissolving" or "separating.") Lysosomes remove waste and foreign materials from the cell. They are also involved in destroying old and damaged cells as needed for repair and remodeling of tissue. **Peroxisomes** (per-OK-sih-somz) have enzymes that destroy harmful substances produced in metabolism. Read **BOX 3-1** to learn about the importance of lysosomes and peroxisomes in health and disease. **Vesicles** (VES-ih-klz) are small, membrane-bound storage sacs. They can be used to move materials into or out of the cell, as described later.

Very small, barrel-shaped protein complexes called **proteasomes** (not shown in FIG. 3-2) also participate in waste removal. They specialize in the destruction of any proteins produced by the ribosomes and ER that do not meet quality control specifications. Sometimes, this quality control system is too sensitive, destroying relatively functional proteins. CF, for instance, results when proteasomes destroy a mutated but relatively functional version of an ion channel. Without this channel, mucus accumulates in the respiratory

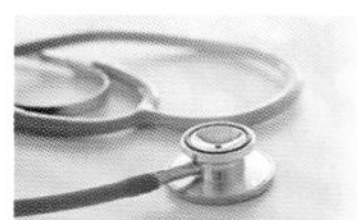

CLINICAL PERSPECTIVES BOX 3-1

Lysosomes and Peroxisomes: Cellular Recycling

Two organelles that play a vital role in cellular disposal and recycling are lysosomes and peroxisomes. **Lysosomes** contain enzymes that break down carbohydrates, lipids, proteins, and nucleic acids. These powerful enzymes must be kept within the lysosome because they would digest the cell if they escaped. In a process called **autophagy** (aw-TOF-ah-je), the cell uses lysosomes to safely recycle cellular structures, fusing with and digesting worn-out organelles. The digested components then return to the cytoplasm for reuse. Lysosomes also break down foreign material, as when cells known as **phagocytes** (FAG-o-sites) engulf bacteria and then use lysosomes to destroy them. The cell may also use lysosomes to digest itself during **autolysis** (aw-TOL-ih-sis), a normal part of development. *Auto-* means "self," and cells that are no longer needed "self-destruct" by releasing lysosomal enzymes into their own cytoplasm.

Peroxisomes are small membranous sacs that resemble lysosomes but contain different kinds of enzymes. They break down toxic substances that may enter the cell, such as drugs and alcohol, but their most important function is to break down free radicals. These substances are byproducts of normal metabolic reactions but can kill the cell if not neutralized by peroxisomes.

Disease may result if either lysosomes or peroxisomes are unable to function. In Tay-Sachs disease, nerve cells' lysosomes lack an enzyme that breaks down certain kinds of lipids. These lipids build up inside the cells, causing malfunction that leads to brain injury, blindness, and death.

and digestive systems, eventually causing death. CF is the subject of Ben's opening case study.

Centrioles (SEN-tre-olz) are rod-shaped bodies near the nucleus that function in cell division. They help to organize the cell and divide the cell contents during this process.

CASEPOINT

☐ **3-2** Cystic fibrosis involves a defect in a membrane protein called CFTR. Name the organelles that synthesize, modify, and transport this protein to the plasma membrane.

SURFACE ORGANELLES

Some cells have structures projecting from their surfaces that are used for motion. **Cilia** (SIL-e-ah) are small, hairlike projections that wave, creating movement of the fluids around the cell (see **FIG. 3-1**). For example, cells that line the passageways of the respiratory tract have cilia that move impurities out of the system. Ciliated cells move the egg cell from the ovary to the uterus in the female reproductive tract.

A long, whiplike extension from a cell is a **flagellum** (flah-JEL-lum). The only type of cell in the human body that has a flagellum is the male sperm cell. Each human sperm cell has a flagellum that propels it toward the egg in the female reproductive tract (**FIG. 3-4E**).

CELLULAR DIVERSITY

1 In Chapter 1, we learned that function is determined by structure. This concept is illustrated by cellular diversity. Although all cells have some fundamental similarities, individual cells may vary widely in size, shape, and composition according to their functions. The average cell size is 10 to

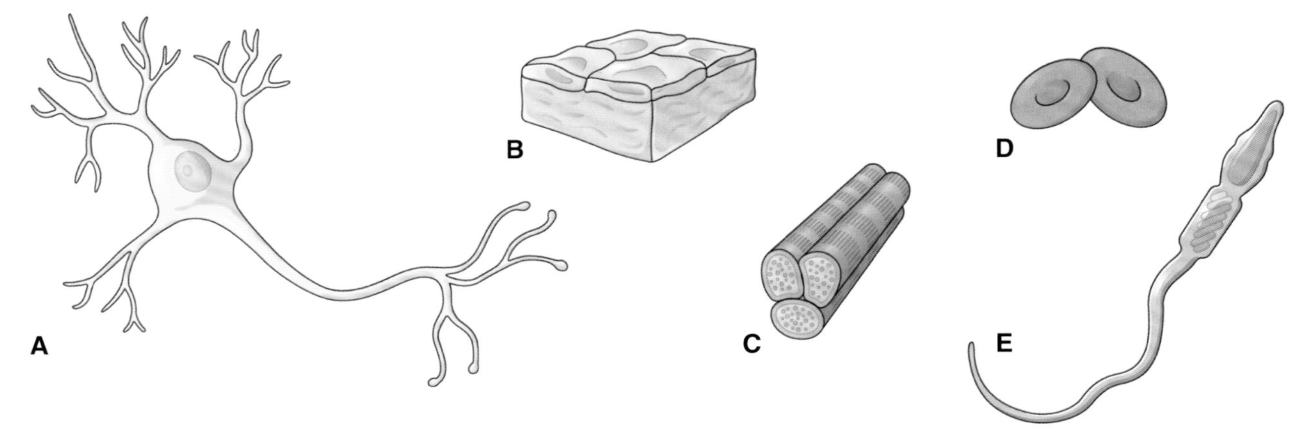

FIGURE 3-4 Cellular diversity. KEY POINT Cells vary in structure according to their functions. **A.** A neuron has long extensions that pick up and transmit electric impulses. **B.** Epithelial cells cover and protect underlying tissue. **C.** Muscle cells have fibers that produce contraction. **D.** Red blood cells lose most organelles and have a round, indented shape to facilitate blood flow through large vessels. **E.** A sperm cell is small and light and swims with a flagellum. ZOOMING IN Which of the cells shown would best cover a large surface area?

15 mcm, but cells may range in size from the 7 mcm of a red blood cell to the 200 mcm or more in the length of a muscle cell.

Cell shape is related to cell function (see FIG. 3-4). A neuron (nerve cell) has long fibers that transmit electric signals over distances up to 1 m (3 ft). Red blood cells are small and flexible, assuming an indented shape in large blood vessels to facilitate blood flow but a cigar-like shape in small vessels to maximize gas exchange. As a red blood cell matures, it loses its nucleus and most other organelles, freeing up space to carry oxygen.

Aside from cilia and flagella, all the organelles described above are present in most human cells. They may vary in number, however. For example, cells producing lipids have lots of smooth ER. Cells that secrete proteins have lots of ribosomes and a prominent Golgi apparatus. All active cells have lots of mitochondria to manufacture the ATP needed for energy.

CHECKPOINTS 

- ☐ **3-3** List four substances found within the plasma membrane.
- ☐ **3-4** What are cell organelles?
- ☐ **3-5** Why is the nucleus called the cell's control center?
- ☐ **3-6** What are the two types of organelles used for movement, and what do they look like?

Movement of Substances across the Plasma Membrane

4 As introduced in Chapter 1, the plasma membrane serves as a barrier between the cell and its environment. Nevertheless, nutrients, oxygen, and many other substances needed by the cell must be taken in, and waste products must be eliminated. Clearly, some substances can be exchanged between the cell and its environment through the plasma membrane. For this reason, the plasma membrane is described at a simple level as **semipermeable** (sem-e-PER-me-ah-bl). It is permeable, or passable, to some molecules but impassable to others. Chapter 2 introduced the importance of water as a solvent. Most of the substances dissolved in intra- and extracellular fluids are soluble in water, that is, they are hydrophilic. However, because the plasma membrane is composed primarily of lipids, only fat-soluble, or hydrophobic, substances can dissolve in and pass freely through it. Steroid hormones and gases (O_2, CO_2, N_2) are examples of lipid-soluble substances. Nutrients, ions, and other hydrophilic substances cannot pass through the lipid bilayer, so they must use transporters or ion channels to cross. For instance, we ingest starch, but intestinal cells only possess transporters for the product of digested starch (glucose). So, the intestinal cell membrane is permeable to glucose but not starch.

Because the permeability of the plasma membrane varies among substances, and over time, the membrane is most accurately described not as simply semipermeable but as **selectively permeable**; it determines what can enter and leave by altering the abundance of specific transporters. Various physical processes are involved in exchanges through the plasma membrane. One way of grouping these processes is according to whether they do or do not require chemical energy.

MOVEMENT THAT DOES NOT REQUIRE CHEMICAL ENERGY

The adjective *passive* describes movement through the plasma membrane that does not require the chemical energy of ATP. Instead, passive mechanisms rely on the kinetic energy of the particles themselves.

5 Chapter 1 introduced the key idea of gradients. Passive mechanisms depend on such gradients, differences in a particular quality between two regions. For instance, a sled moves freely down a gravitational gradient from a higher altitude to a lower altitude. In the body, many substances move because of differences in solute concentrations, but other types of gradients (such as pressure gradients) can also drive transport.

Diffusion **Diffusion** is the net movement of particles from a region of relatively higher concentration to one of lower concentration. Just as couples on a crowded dance floor spread out into all the available space to avoid hitting other dancers, diffusing substances spread throughout their available space until their concentration everywhere is the same—that is, they reach equilibrium (FIG. 3-5). Diffusion uses the particles' kinetic energy and does not directly require ATP. The greater the concentration of a solution, the greater the summed kinetic energy of all the particles together. The particles are said to follow, or move down, their concentration gradient from higher concentration to lower concentration.

Recall from Chapter 1 that the gradient-dependent movement of a substance is opposed by resistance. Particles can only enter or exit the cell by diffusion if they can cross the plasma membrane; in other words, if the membrane does not "resist" their movement. A particle cannot diffuse through, regardless of the gradient strength, if the plasma membrane is impermeable to it. So, hydrophobic substances, such as gases and steroid hormones, diffuse freely in and out of cells whenever a concentration gradient exists. Hydrophilic substances, on the other hand, will only diffuse across the plasma membrane if a suitable ion channel or transporter is available to permit passage through the inhospitable lipid bilayer. FIGURE 3-6 illustrates how glucose uses a transporter to diffuse across the plasma membrane.

CF results from an abnormality in chloride diffusion across the plasma membrane. See BOX 3-2 to learn about a study that treats CF by increasing chloride transport.

FIGURE 3-5 Diffusion of a solid in a liquid. KEY POINT The molecules of the solid tend to spread evenly throughout the liquid as they dissolve.

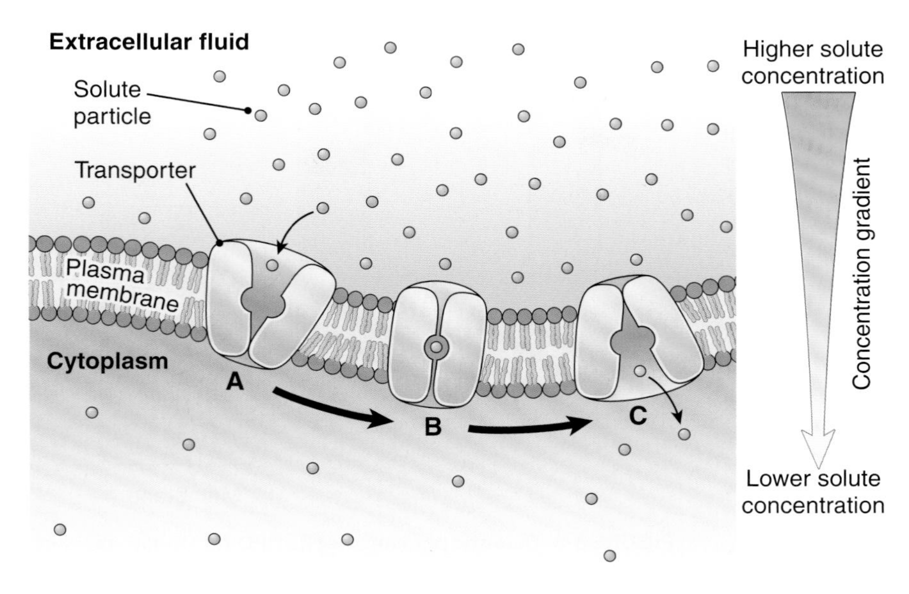

FIGURE 3-6 Diffusion using transporters. **KEY POINT** Protein transporters in the plasma membrane move solute particles through a membrane from an area of higher concentration to an area of lower concentration. **A.** A solute particle enters the transporter. **B.** The transporter changes shape. **C.** The transporter releases the solute particle on the other side of the membrane. **ZOOMING IN** How would a decrease in the number of transporters affect this solute's movement?

ONE STEP AT A TIME
The Relationship Is In the Graph

BOX 3-2

Science is all about relationships. For instance, climate change scientists show the relationship between human activity and ocean temperature, and medical scientists evaluate the relationship between their new drug and improved function. In many cases, scientists use graphs to convince their colleagues that a relationship exists. As a health consumer and perhaps a future health professional, you need to know how to critically evaluate graphs so that you can make informed decisions.

Question

This graph illustrates one effect of a drug being tested for individuals with cystic fibrosis. Gather as much information as you can from the graph to explain the relationship between this drug and changing function in CF patients.

Step 1. Look at the title and the axes. This graph does not have a title, but we can get the needed information from the axes. The X axis (horizontal) uses units such as days and weeks, so it represents time. The Y axis (vertical) is the concentration of chloride in sweat. So, without looking at any data, we know that the graph shows changes in the sweat chloride content overtime.

Step 2. Determine the graph type. This graph shows symbols (data points) joined together by lines, so it is a line graph (see Chapter 19 for an example of a bar graph. This graph actually includes two lines, one representing the placebo group (no active drug) and one representing the treatment group (test drug).

Step 3. Figure out any abbreviations or symbols. Notice that aides (placebo) and squares (drug) are used to represent data points. Each data point represents the average value of all participants. The figure legend tells us that values above the dotted line are diagnostic of cystic fibrosis. The abbreviation "N" indicates the number of values used to calculate the data point. For instance, 65 patients provided data in the placebo group at week 48.

Step 4. Look for trend; in the data, and draw conclusions. The line for the test drug group remains below the diagnostic cutoff for the entire treatment period, but the line for the placebo group remains high above it. Thus, we can conclude that the drug lowers the sweat chloride concentration in CF patients to normal levels within 15 days of treatment. We can also note that the number of participants in the placebo group was consistently lower than that in the treatment group (both groups would have started with equivalent numbers). This trend suggests that the treatment group patients were healthier and better able to provide data for the study.

See the Chapter Review questions to analyze this graph further, and see the Study Guide to get more practice analyzing graphs.

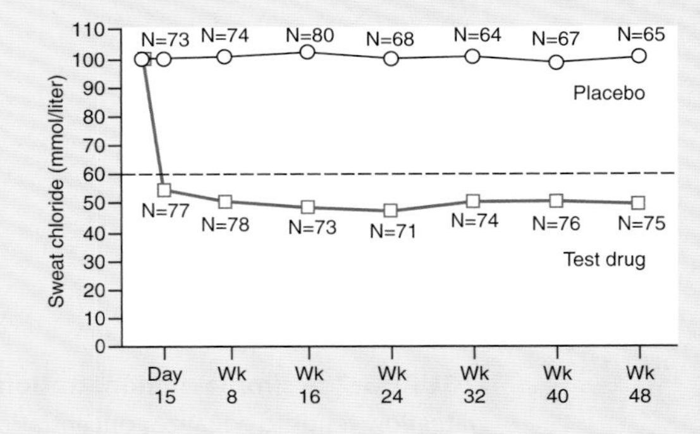

The first data point is the baseline data. The cutoff point for a diagnosis of cystic fibrosis is represented by the *dashed line*. The sample graph and data points are based on the work of Ramsey et al. 2011; NEJM 365:1663–1672. A CFTR potentiator in patients with cystic fibrosis and the G551D mutation.

Osmosis Osmosis (os-MO-sis) is a special type of diffusion. The term applies specifically to the diffusion of water through a semipermeable membrane. Water moves rapidly through the plasma membrane of most cells with the help of channels called *aquaporins* (a-kwa-POR-ins). The water molecules move, as expected, from an area where there are more of them to an area where there are fewer of them. That is, the solvent (the water molecules) moves from an area of lower *solute* concentration to an area of higher *solute* concentration, as demonstrated in FIGURE 3-7.

For a physiologist studying water's flow across membranes, it is helpful to know the direction in which water will flow and at what rate it will move. A measure of the force driving osmosis is called the *osmotic pressure*. This force can be measured, as illustrated in FIGURE 3-8, by applying enough pressure to the surface of a liquid to stop the inward flow of water by osmosis. The pressure needed to counteract osmosis is the osmotic pressure. In practice, the term *osmotic pressure* is used to describe a solution's tendency to draw in water. This force is directly related to concentration; the higher a solution's concentration, the greater is its osmotic pressure.

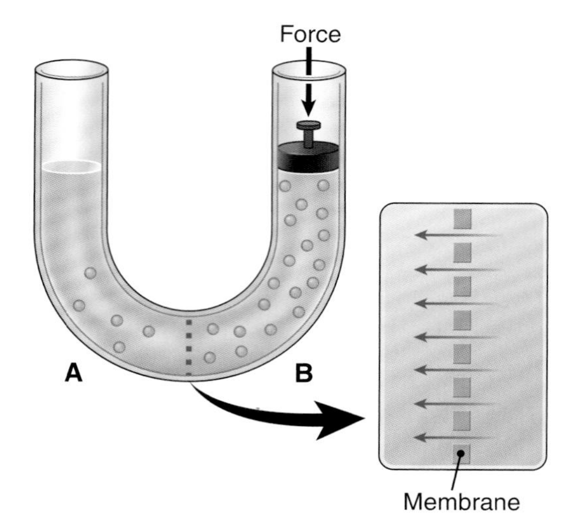

FIGURE 3-8 **Osmotic pressure.** KEY POINT Osmotic pressure is the force needed to stop the flow of water by osmosis. Pressure on the surface of the fluid in side B counteracts the osmotic flow of water from side **A** to side **B.** ZOOMING IN What would happen to osmotic pressure if the concentration of solute were increased on side B of this system?

How Osmosis Affects Cells Because water can move easily through the plasma membrane of most cells, the extracellular fluid must have the same overall concentration of dissolved substances (solutes) as the cytoplasm (intracellular fluid). If this balance is altered, water will move rapidly into or out of the cell by osmosis and change the cell volume (FIG. 3-9). Solutions with concentrations equal to the concentration of the cytoplasm are described as **isotonic** (i-so-TON-ik). Tissue fluids and blood plasma are isotonic for body cells. Manufactured solutions that are isotonic for the cells are available and can be administered intravenously to replace body fluids. One example is 0.9% salt, or normal saline.

A solution that is less concentrated than the cytoplasm is described as **hypotonic.** Based on the principles of osmosis already explained, a cell placed in a hypotonic solution draws water in, swells, and may burst. When a red blood cell draws in water and bursts in this way, the cell is said to undergo

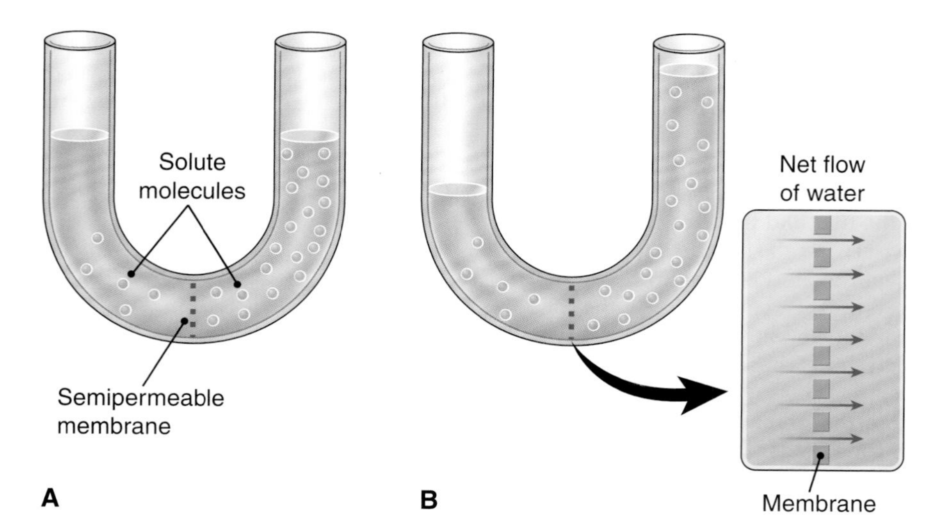

FIGURE 3-7 **A simple demonstration of osmosis.** KEY POINT The direction of water flow tends to equalize concentrations of solutions. Solute molecules are shown in *yellow*. All of the solvent (*blue*) is composed of water molecules. **A.** Two solutions with different concentrations of solute are separated by a semipermeable membrane. Water can flow through the membrane, but the solute cannot. **B.** Water flows into the more concentrated solution, raising the level of the liquid in that side. ZOOMING IN What would happen in this system if the solute could pass through the membrane?

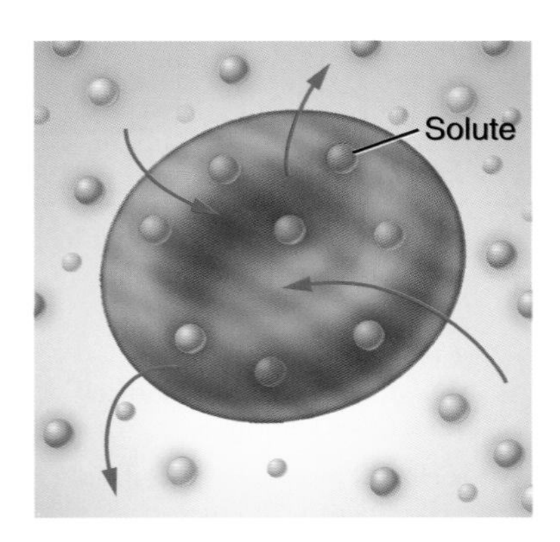

A Isotonic: no volume change

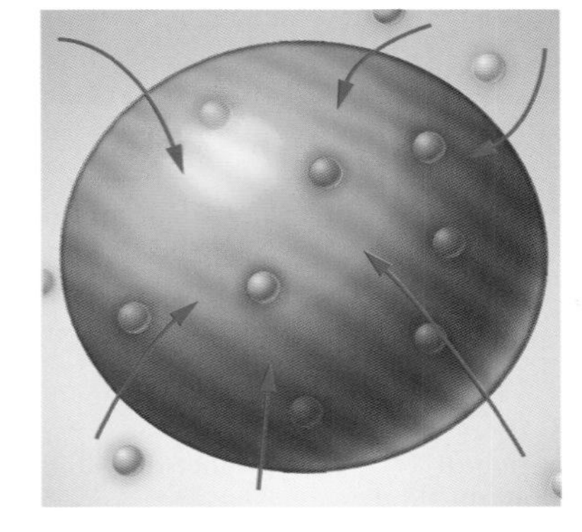

B Hypotonic: cell swells

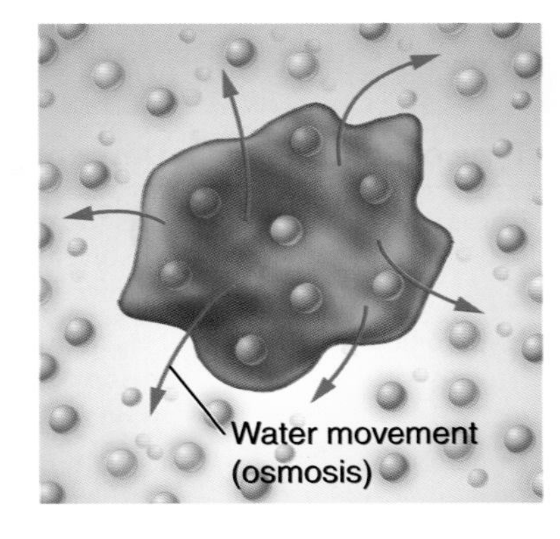

C Hypertonic: cell shrinks

FIGURE 3-9 **The effect of osmosis on cells.** KEY POINT Cells must be kept in fluids that are compatible with the concentration of their cytoplasms. This figure shows how water moves through a red blood cell membrane in solutions with three different concentrations of solute. **A.** The isotonic (normal) solution has the same concentration as the cytoplasm, and water moves into and out of the cell at the same rate. **B.** A cell placed in a hypotonic (more dilute) solution draws water in, causing the cell to swell and perhaps undergo hemolysis (bursting). **C.** The hypertonic (more concentrated) solution draws water out of the cell, causing it to shrink, an effect known as crenation. ZOOMING IN What would happen to red blood cells in the body if blood lost through injury were replaced with pure water?

hemolysis (he-MOL-ih-sis). If a cell is placed in a **hypertonic** solution, which is more concentrated than the cellular fluid, it loses water to the surrounding fluids and shrinks, a process termed **crenation** (kre-NA-shun) (see FIG. 3-9).

Fluid balance is an important facet of homeostasis and must be properly regulated for health. You can figure out in which direction water will move through the plasma membrane if you remember the saying "water follows salt," salt meaning any dissolved material (solute). The total amount and distribution of body fluids is discussed in Chapter 19. TABLE 3-3 summarizes the effects of different solution concentrations on cells.

Concept Mastery Alert

Osmosis is an important topic, so take the time to learn it thoroughly. Remember that it represents the movement of water, not solute. Water moves from the less concentrated solution into the more concentrated solution and changes cell volume as a result.*

CASEPOINT

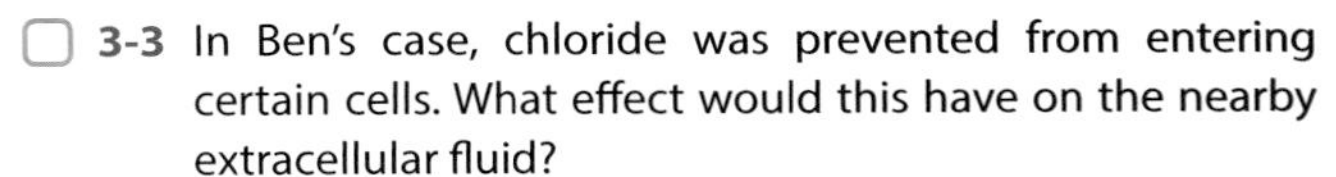

☐ 3-3 In Ben's case, chloride was prevented from entering certain cells. What effect would this have on the nearby extracellular fluid?

Filtration Filtration is the passage of water and dissolved materials through a membrane down a pressure gradient from an area of higher pressure to an area of lower pressure. A mechanical ("pushing") force is usually responsible for the high pressure. The membrane acts as a filter, resisting the movement of larger substances. An everyday example of filtration is an espresso machine, which uses steam to increase the pressure in the machine above atmospheric pressure. This gradient forces water and dissolved chemicals (such as caffeine) into the cup, but the filter retains the grounds. In the body, heart contractions increase the pressure in capillaries (i.e., the blood pressure) above the pressure of the surrounding fluid. The gradient pushes water and electrolytes out of the capillary, but the capillary wall retains the larger proteins and blood cells (see Chapter 14). In the same way, water and dissolved substances are filtered out of blood in the first step of urine formation in the kidney (see Chapter 19).

See the Student Resources on thePoint® to view an animation on osmosis and osmotic pressure.

MOVEMENT THAT REQUIRES CHEMICAL ENERGY

Some materials move across the plasma membrane without depending on a gradient. For instance, intestinal cells import glucose when glucose is more concentrated inside the cell than outside it. Other substances, such as bacteria or complex solutions, are too large or too heterogeneous for channels or transporters to handle. Chemical energy must drive transport in both of these situations. *Active transport* uses transporters and ATP to move ions and nutrients, and *vesicular transport* uses vesicles and ATP to move large amounts of substances at once.

Active Transport While any method that uses chemical energy can be defined as "active," the term **active transport** usually refers to the movement of solutes against their concentration gradients using membrane transporters. These transporter proteins move specific solute particles against

*The Misconception Alerts featured in every chapter of this book are derived from common errors students make in responding to questions in PrepU, an online supplemental review program available separately for this text. For information on accessing PrepU, see pp. xviii–xix of the User's Guide at the front of this text.

Table 3-3 Solutions and Their Effects on Cells

Type of Solution	Description	Examples	Effect on Cells
Isotonic	Has the same concentration of dissolved substances as the fluid in the cell	0.9% salt (normal saline); 5% glucose (dextrose)	None; cell in equilibrium with its environment
Hypotonic	Has a lower concentration of dissolved substances than the fluid in the cell	<0.9% salt or 5%	Cell takes in water, swells, and may burst; red blood cell undergoes hemolysis
Hypertonic	Has a higher concentration of dissolved substances than the fluid in the cell	Higher than 0.9% salt or 5% dextrose	Cell will lose water and shrink; cell undergoes crenation

their gradient, from an area where they are in relatively lower concentration to an area where they are in higher concentration. This movement requires energy, just as getting a sled to the top of a hill requires energy. Instead of the kinetic energy involved in pushing a sled, this process uses the chemical energy of ATP. The nervous system and muscular system, for example, depend on the active transport of sodium, potassium, and calcium ions for proper function. The kidneys also carry out active transport in regulating the composition of urine, and the digestive system uses active transport to absorb virtually all of the nutrients in our ingested food. By means of active transport, the cell can take in what it needs from the surrounding fluids and remove materials from the cell.

Vesicular Transport There are several active (ATP-dependent) methods for moving large quantities of material into or out of the cell. These methods are grouped together as **vesicular transport**, because small sacs, or vesicles, are needed for the processes. These processes are grouped according to whether materials are moved into or out of the cells, as follows:

- **Endocytosis** (en-do-si-TO-sis) is a term that describes the movement of materials into the cell using vesicles. Some examples are:
  - **Phagocytosis** (fag-o-si-TO-sis), in which relatively large particles are engulfed by the plasma membrane and moved into the cell (FIG. 3-10). (The root *phag/o* means "to eat.") Certain white blood cells carry out phagocytosis to rid the body of foreign material and dead cells. Material taken into a cell by phagocytosis is first enclosed in a vesicle made from the plasma membrane and is later destroyed by lysosomes.
  - **Pinocytosis** (pi-no-si-TO-sis), in which the plasma membrane engulfs droplets of fluid. This is a way for large protein molecules in suspension to travel into the cell. The word *pinocytosis* means "cell drinking."
  - **Receptor-mediated endocytosis**, which involves the intake of substances using specific binding sites, or receptors, in the plasma membrane. The bound material, or *ligand* (LIG-and), is then drawn into the cell by endocytosis. Some examples of ligands are lipoproteins (complexes of cholesterol, other lipids, and proteins) and certain vitamins.
- In **exocytosis**, the cell moves materials out in vesicles (FIG. 3-11). One example of exocytosis is the export of neurotransmitters from neurons (neurotransmitters are chemicals that control the activity of the nervous system).

All the transport methods described above are summarized in TABLE 3-4.

See the student resources on thePoint® to view the animation "Phagocytosis."

CHECKPOINTS 

- [ ] **3-7** What types of movement through the plasma membrane do not directly require chemical energy, and what types of movement do require chemical energy?
- [ ] **3-8** What term describes a fluid that is the same concentration as the cytoplasm? What type of fluid is less concentrated? More concentrated?

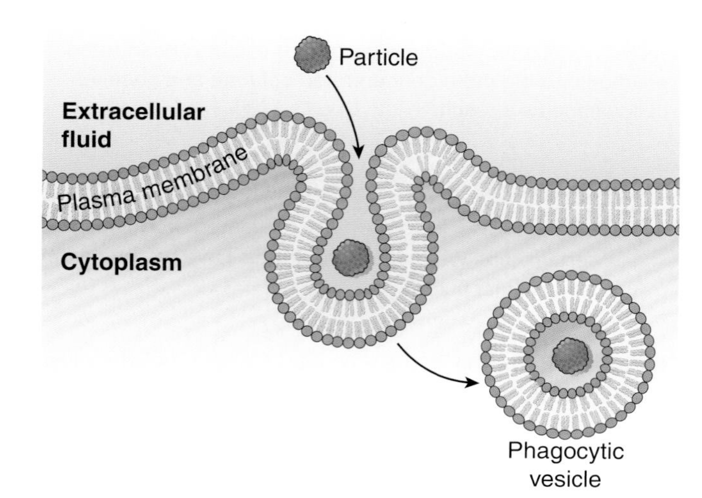

FIGURE 3-10 Phagocytosis. KEY POINT The plasma membrane encloses a particle from the extracellular fluid. The membrane then pinches off, forming a vesicle that carries the particle into the cytoplasm. ZOOMING IN What organelle would likely help to destroy a particle taken in by phagocytosis?

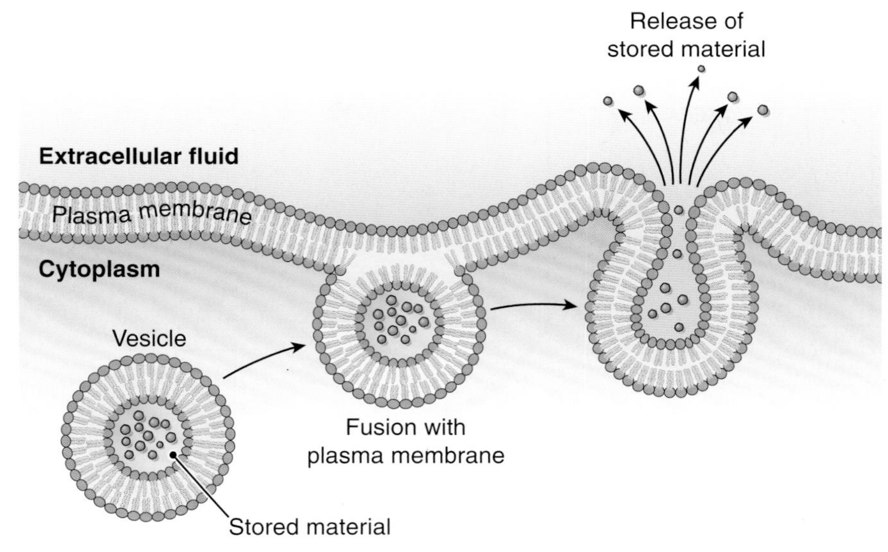

FIGURE 3-11 Exocytosis. KEY POINT A vesicle fuses with the plasma membrane and then ruptures and releases its contents.

Protein Synthesis

Because proteins play an indispensable part in the body's structure and function, we need to identify the cellular substances that direct protein production. As noted earlier, the hereditary structures that govern the cell are the chromosomes in the nucleus. Each chromosome in turn is divided into multiple units, called **genes** (FIG. 3-12). It is the genes that carry the messages for the development of particular inherited characteristics, such as brown eyes, curly hair, or blood type, and they do so by directing protein manufacture in the cell.

STRUCTURE OF DNA AND RNA

Genes are distinct segments of the complex organic chemical that makes up the chromosomes, a substance called **deoxyribonucleic** (de-ok-se-RI-bo-nu-kle-ik) **acid**, or **DNA**.

Table 3-4 Membrane Transport

Process	Definition	Example
Do not require chemical energy (passive)		
Diffusion	Random movement of particles down the concentration gradient (from higher concentration to lower concentration)	Movement of gases through the membrane, ions through an ion channel, or nutrients via transporters
Osmosis	Diffusion of water through a semipermeable membrane	Movement of water across the plasma membrane through aquaporins
Filtration	Movement of materials through a membrane down a pressure gradient	Movement of materials out of the blood under the force of blood pressure
Require chemical energy		
Active transport (pumps)	Movement of materials through the plasma membrane against the concentration gradient using transporters	Transport of ions (e.g., Na^+, K^+, and Ca^{2+}) in neurons
Vesicular transport	Movement of large amounts of material through the plasma membrane using vesicles	
Endocytosis	Transport of materials into the cell using vesicles	Phagocytosis—intake of large particles, as when white blood cells take in waste materials; also pinocytosis (intake of fluid), and receptor-mediated endocytosis, requiring binding sites in the plasma membrane
Exocytosis	Transport of materials out of the cell using vesicles	Release of neurotransmitters from neurons

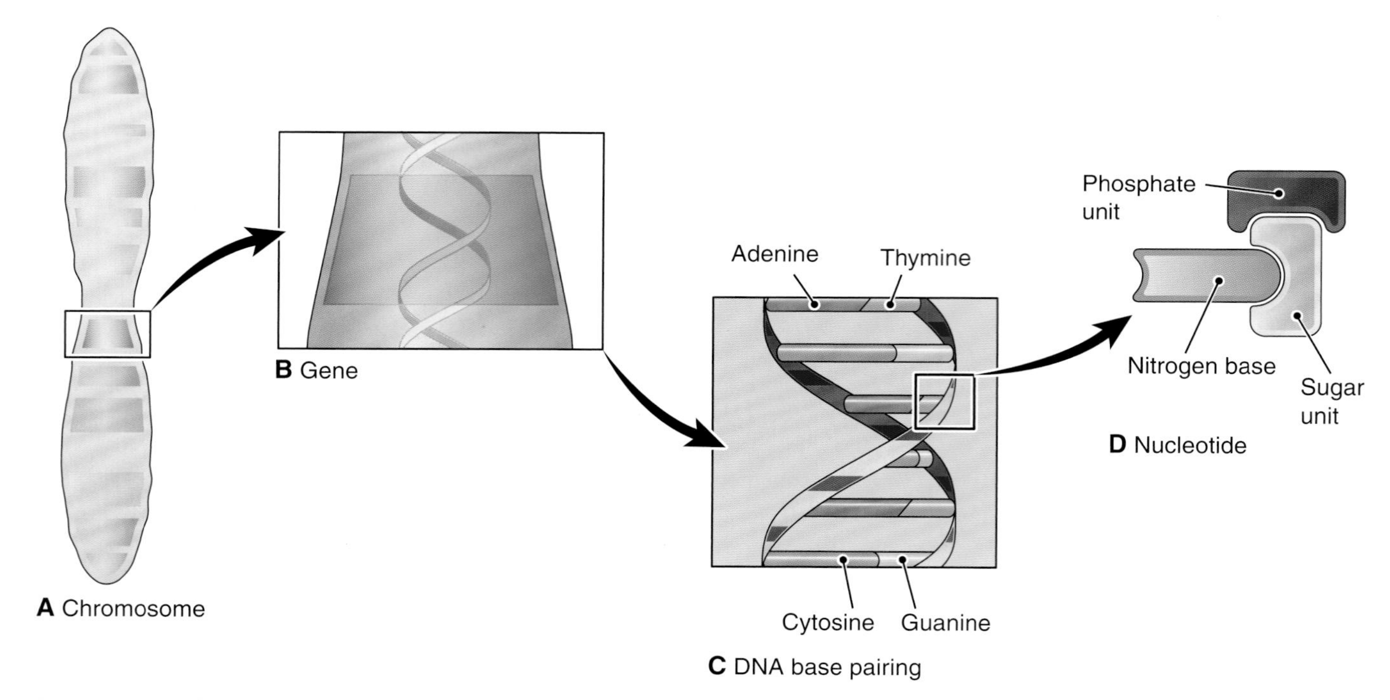

FIGURE 3-12 Chromosomes and DNA. A. A gene is a distinct region of a chromosome. **B.** The DNA making up genes consists of paired nucleic acid strands twisted into a double helix. **C.** The two DNA strands are held together by bonds between the nitrogen bases of complementary nucleotides. **D.** Each structural unit, or nucleotide, consists of a phosphate unit and a sugar unit attached to a nitrogen base. The sugar unit in DNA is deoxyribose. **KEY POINT** There are four different nucleotides in DNA. Their arrangement "spells out" the genetic instructions that control all activities of the cell. ZOOMING IN Two of the DNA nucleotides (A and G) are larger in size than the other two (T and C). How do the nucleotides pair up with regard to size?

DNA is composed of subunits called nucleotides, introduced in Chapter 2 (see FIG. 2-12). A related compound, **ribonucleic** (RI-bo-nu-kle-ik) **acid**, or **RNA**, which participates in protein synthesis but is not part of the chromosomes, is also composed of nucleotides. As noted, a nucleotide contains a sugar, a phosphate, and a nitrogen-containing base. The sugar and phosphate are constant in each nucleotide, although DNA has the sugar deoxyribose and RNA has the sugar ribose. The sugars and phosphates alternate to form a long chain to which the nitrogen bases are attached. The five different nucleotides that appear in DNA and RNA thus differ in the nature of their nitrogen base. Three of the five nucleotides are common to both DNA and RNA. These are the nucleotides containing the nitrogen bases adenine (A), guanine (G), and cytosine (C). However, DNA has one nucleotide containing thymine (T), whereas RNA has one containing uracil (U). **TABLE 3-5** compares the structure and function of DNA and RNA.

9 DNA AND PROTEIN SYNTHESIS

Most of the DNA in the cell is organized into chromosomes within the nucleus (a small amount of DNA is in the mitochondria located in the cytoplasm). **FIGURE 3-12A and B** shows a section of a chromosome and illustrates that the DNA exists as a double strand. Visualizing the complete molecule as a ladder, the sugar and phosphate units of the nucleotides make up the "side rails" of the ladder, and the

Table 3-5 Comparison of DNA and RNA

	DNA	RNA
Location	Almost entirely in the nucleus	Almost entirely in the cytoplasm
Composition	Nucleotides contain adenine (A), guanine (G), cytosine (C), or thymine (T)	Nucleotides contain adenine (A), guanine (G), cytosine (C), or uracil (U)
	Sugar: deoxyribose	Sugar: ribose
Structure	Double-stranded helix formed by nucleotide pairing A–T; G–C	Single strand
Function	Makes up the chromosomes, hereditary units that control all cellular activities; divided into genes that carry the nucleotide codes for the manufacture of proteins	Manufacture proteins according to the codes carried in the DNA; three main types: mRNA, rRNA, and tRNA

nitrogen bases project from the side rails to make up the ladder's "steps" (**FIG. 3-12C and D**). The two DNA strands are paired very specifically according to the identity of the nitrogen bases in the nucleotides. Adenine (A) always pairs with thymine (T); guanine (G) always pairs with cytosine (C). The two strands of DNA are held together by weak bonds (hydrogen bonds; **see BOX 2-1**). The doubled strands then coil into a spiral, giving DNA the descriptive name *double helix*.

Concept Mastery Alert

Remember that chromosomes are composed of DNA, and DNA is composed of nucleotide subunits.

The message of the DNA that makes up the individual genes is actually contained in the varying pattern of the four nucleotides along the strand. Consider the four nucleotides as a small alphabet consisting of four different letters. These "letters" are combined to make different three-letter "words," or triplets, and each word is the code for a specific amino acid (recall that amino acids are the building blocks of proteins). For instance, the sequence CCC is the code for the amino acid glycine (**TABLE 3-6**, first and second columns). Each gene thus consists of a string of three-letter words that codes for a string of amino acids—in other words, an entire protein.

Note that proteins serve diverse and critical roles in the body. They act as chemical signals and receptors; give color to hair, skin, and eyes; and, perhaps most importantly, act as enzymes for cellular reactions. DNA is thus the cell's master blueprint, determining the body's structures and functions. Chapter 21 discusses how an individual's DNA is responsible for the many traits passed down from parents to children. **Mutations**, changes in the DNA sequence of a gene, can change the shape of the resulting protein or even stop it from being synthesized. Some mutations harm cells and may lead to cancer or other disorders, while other mutations are without effect or may actually have a beneficial effect on cell function.

In light of observations on cellular diversity, you may wonder how different cells in the body can vary in appearance and function if they all have the same amount and same kind of DNA. The answer to this question is that only portions of the DNA in a given cell are active at any one time. In some cells, regions of the DNA can be switched on and off, under the influence of hormones, for example. However, as cells differentiate during development and become more specialized, regions of the DNA are permanently shut down, leading to the variations in the different cell types. Scientists now realize that the control of DNA action throughout a cell's life span is a very complex matter involving not only the DNA itself but proteins as well.

Table 3-6 The Genetic Code

Amino Acid	Transcribed DNA Triplet	mRNA	tRNA
Glycine	CCC	GGG	CCC
Proline	GGG	CCC	GGG
Valine	CAC	GUG	CAC
Phenylalanine	AAA	UUU	AAA

The nucleotide triplet code in DNA and RNA is shown for four amino acids.

ROLE OF RNA IN PROTEIN SYNTHESIS

A blueprint is only a guide. The information it contains must be interpreted and acted upon, and RNA is the substance needed for these steps. RNA is much like DNA except that it exists as a single strand of nucleotides and has uracil (U) instead of thymine (T). Thus, when RNA pairs up with another molecule of nucleic acid to manufacture proteins, as explained below, adenine (A) bonds with uracil (U) instead of thymine (T).

A detailed account of protein synthesis is beyond the scope of this book, but a highly simplified description and illustrations of the process are presented. The process begins with the copying of information from DNA to RNA in the nucleus, a process known as *transcription* (**FIG. 3-13**). The RNA copy is called messenger RNA (mRNA) because it carries the DNA message from the nucleus to the cytoplasm. Before transcription begins, the DNA separates into single strands. Then, enzymes assemble a matching strand of RNA along one of the DNA strands by the process of nucleotide pairing. Information on which strand will be used for transcription is contained in the chromosomes themselves. For example, if the DNA strand reads CAC, the corresponding mRNA will read GUG (remember that RNA has U instead of T to bond with A) (**TABLE 3-6**, third column). When complete, this mRNA leaves the nucleus and travels to a ribosome in the cytoplasm (**FIG. 3-14**). Recall that ribosomes are the site of protein synthesis in the cell.

Ribosomes are composed of an RNA type called ribosomal RNA (rRNA) and also protein. At the ribosomes, the genetic message now contained within mRNA is decoded to assemble amino acids into the long chains that form proteins, a process termed *translation*. This final step requires a third RNA type, transfer RNA (tRNA), present in the cytoplasm (**see FIG. 3-14**). Note that both rRNA and tRNA are formed by the transcription process illustrated in **FIGURE 3-13**.

Remember that each amino acid is coded by a nucleotide triplet. Every tRNA contains the complementary nucleotides to one of these sequences and carries the corresponding amino acid (**TABLE 3-6**, fourth column). When the matching triplet is present in the mRNA, the tRNA binds to the mRNA, and the ribosome adds its amino acid to the growing protein chain. After the amino acid chain is formed, it must be coiled and folded into the proper shape for that protein by the endoplasmic reticulum, as discussed above. **TABLE 3-7** summarizes information on the different types of RNA. Also see **BOX 3-3**, "Proteomics: So Many Proteins, So Few Genes."

See the student resources on thePoint® to view the animation "Protein Synthesis Overview."

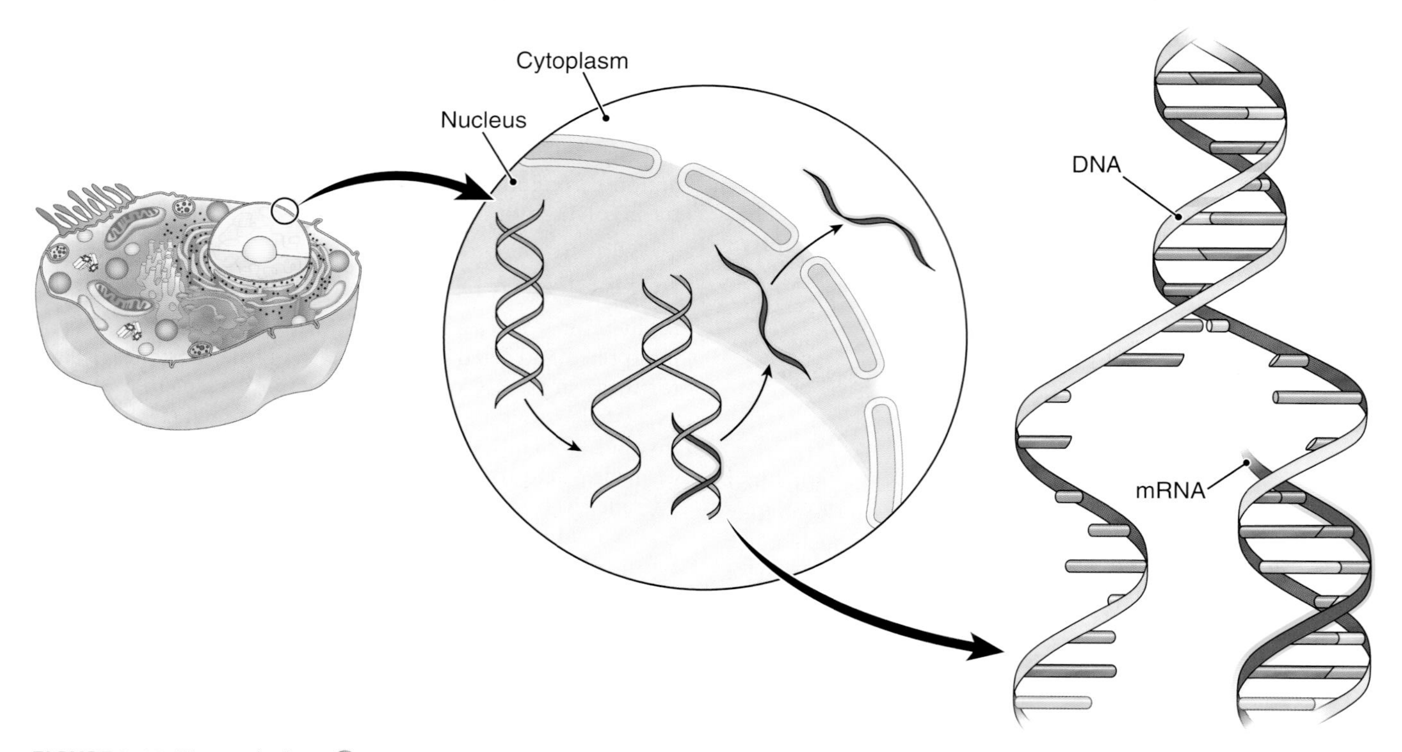

FIGURE 3-13 Transcription. **KEY POINT** In the first step of protein synthesis, the DNA code is transcribed into messenger RNA (mRNA) by nucleotide base pairing. An enlarged view of the nucleic acids during transcription shows how mRNA forms according to the nucleotide pattern of the DNA. Note that adenine (A, *red*) in DNA bonds with uracil (U, *brown*) in RNA.

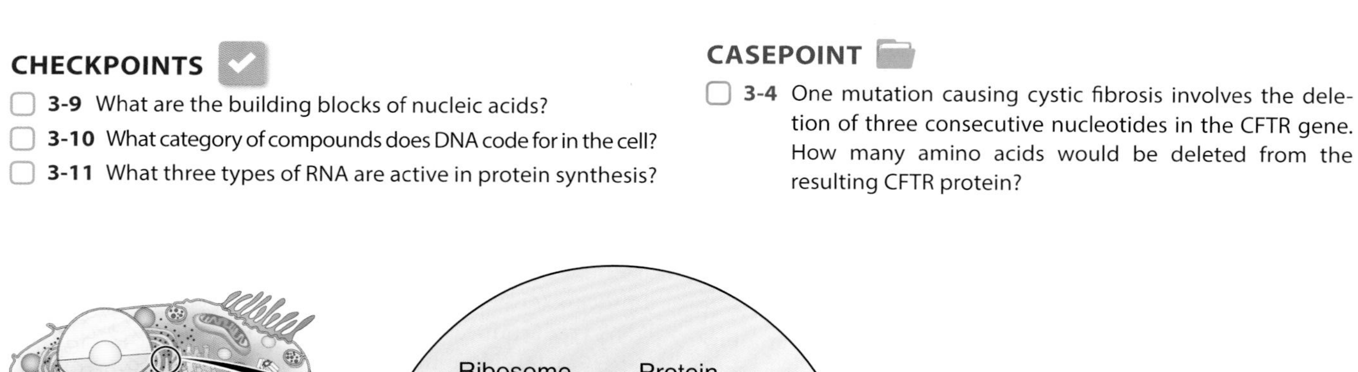

CHECKPOINTS

- ☐ **3-9** What are the building blocks of nucleic acids?
- ☐ **3-10** What category of compounds does DNA code for in the cell?
- ☐ **3-11** What three types of RNA are active in protein synthesis?

CASEPOINT

- ☐ **3-4** One mutation causing cystic fibrosis involves the deletion of three consecutive nucleotides in the CFTR gene. How many amino acids would be deleted from the resulting CFTR protein?

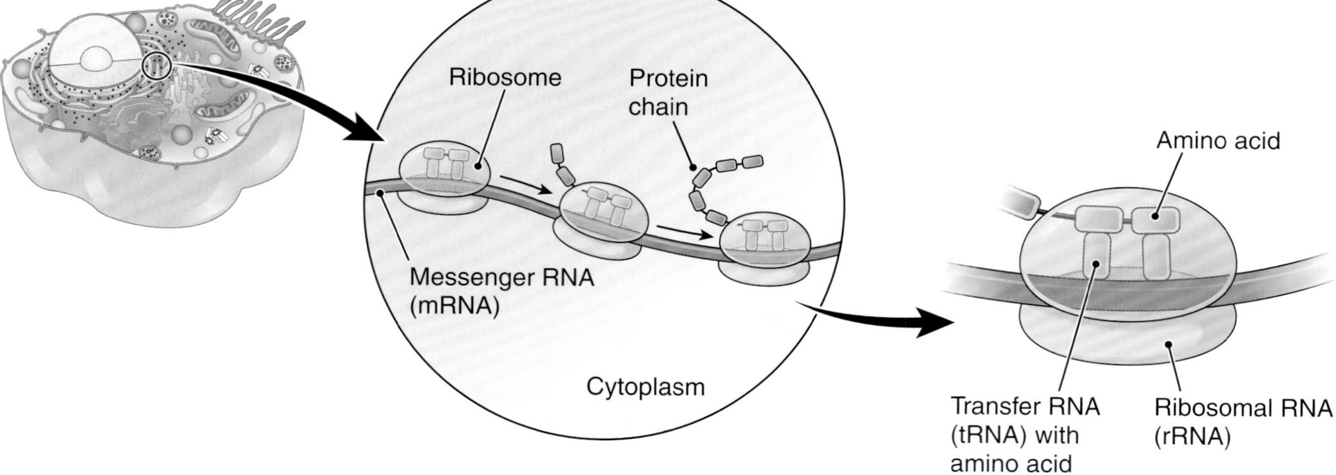

FIGURE 3-14 Translation. **KEY POINT** In protein synthesis, messenger RNA (mRNA) travels to the ribosomes in the cytoplasm. The information in the mRNA codes for the building of proteins from amino acids. Transfer RNA (tRNA) molecules bring amino acids to the ribosomes to build each protein.

Table 3-7 RNA

Types	Function
Messenger RNA (mRNA)	Is built on a strand of DNA in the nucleus and transcribes the nucleotide code; moves to cytoplasm and attaches to a ribosome
Ribosomal RNA (rRNA)	With protein makes up the ribosomes, the sites of protein synthesis in the cytoplasm; involved in the process of translating the genetic message into a protein
Transfer RNA (tRNA)	Works with other forms of RNA to translate the genetic code into protein; each molecule of tRNA carries an amino acid that can be used to build a protein at the ribosome

Cell Division

For growth, repair, and reproduction, cells must multiply to increase their numbers. The cells that form the sex cells (egg and sperm) divide by the process of *meiosis* (mi-O-sis), which cuts the chromosome number in half to prepare for union of the egg and sperm in fertilization. If not for this preliminary reduction, the number of chromosomes in the offspring would constantly double. The process of meiosis is discussed in Chapters 20 and 21. All other body cells, known as *somatic cells*, are formed by a process called **mitosis** (mi-TO-sis). In this process, each original parent cell becomes two identical daughter cells. Somatic cells develop from actively dividing cells called *stem cells*, which we will discuss in more detail in Chapter 4.

PREPARATION FOR MITOSIS

Before mitosis can occur, the genetic information (DNA) in the parent cell must be replicated (doubled), so that each of the two new daughter cells will receive a complete set of chromosomes. For example, a human cell that divides by mitosis must produce two cells with 46 chromosomes each, the same number of chromosomes that are present in the original parent cell. DNA replicates during **interphase**, the stage in the cell's life cycle between one mitosis and the next. During this phase, DNA uncoils from its double-stranded form, and enzymes assemble a matching strand of nucleotides for each old strand according to the pattern of A–T, G–C pairing. There are now two double-stranded DNA molecules, each identical to the original double helix. The two double helices are held together at a region called the *centromere* (SEN-tro-mere) until they separate toward the end of mitosis. A typical stem cell lives in interphase for most of its life cycle and spends only a relatively short period in mitosis. For example, a cell reproducing every 20 hours spends only about one hour in mitosis and the remaining time in interphase. Aside from stem cells, most mature body cells spend their entire lives in interphase and never enter mitosis.

See the Student Resources on thePoint® for a photomicrograph of a replicated chromosome.

STAGES OF MITOSIS

Although mitosis is a continuous process, distinct changes can be seen in the dividing cell at four stages (**FIG. 3-15**).

- In **prophase** (PRO-faze), each replicated chromosome winds up tightly and separates from the other replicated chromosomes. The nucleolus and the nuclear membrane begin to disappear. In the cytoplasm, the two

HOT TOPICS BOX 3-3

Proteomics: So Many Proteins, So Few Genes

To build the many different proteins that make up the body, cells rely on instructions encoded in the genes. Collectively, all the different genes on all the chromosomes make up the **genome**. Genes contain the instructions for making proteins, and proteins perform the body's functions.

Scientists are now studying the human proteome—all the proteins that can be expressed in a cell—to help them understand protein structure and function. Unlike the genome, the proteome changes as the cell's activities and needs change. In 2003, after a decade of intense scientific activity, investigators mapped the entire human genome. We now realize that it probably contains no more than 25,000 genes, far fewer than initially expected. How could this relatively small number of genes code for several million proteins? They concluded that genes were not the whole story.

Gene transcription is only the beginning of protein synthesis. In response to cellular conditions, enzymes can snip newly transcribed mRNA into several pieces, each of which a ribosome can use to build a different protein. After each protein is built, enzymes can further modify the amino acid strands to produce several more different proteins. Other molecules help the newly formed proteins to fold into precise shapes and interact with each other, resulting in even more variations. Thus, while a gene may code for a specific protein, modifications after gene transcription can produce many more unique proteins. There is much left to discover about the proteome, but scientists hope that future research will lead to new techniques for detecting and treating disease.

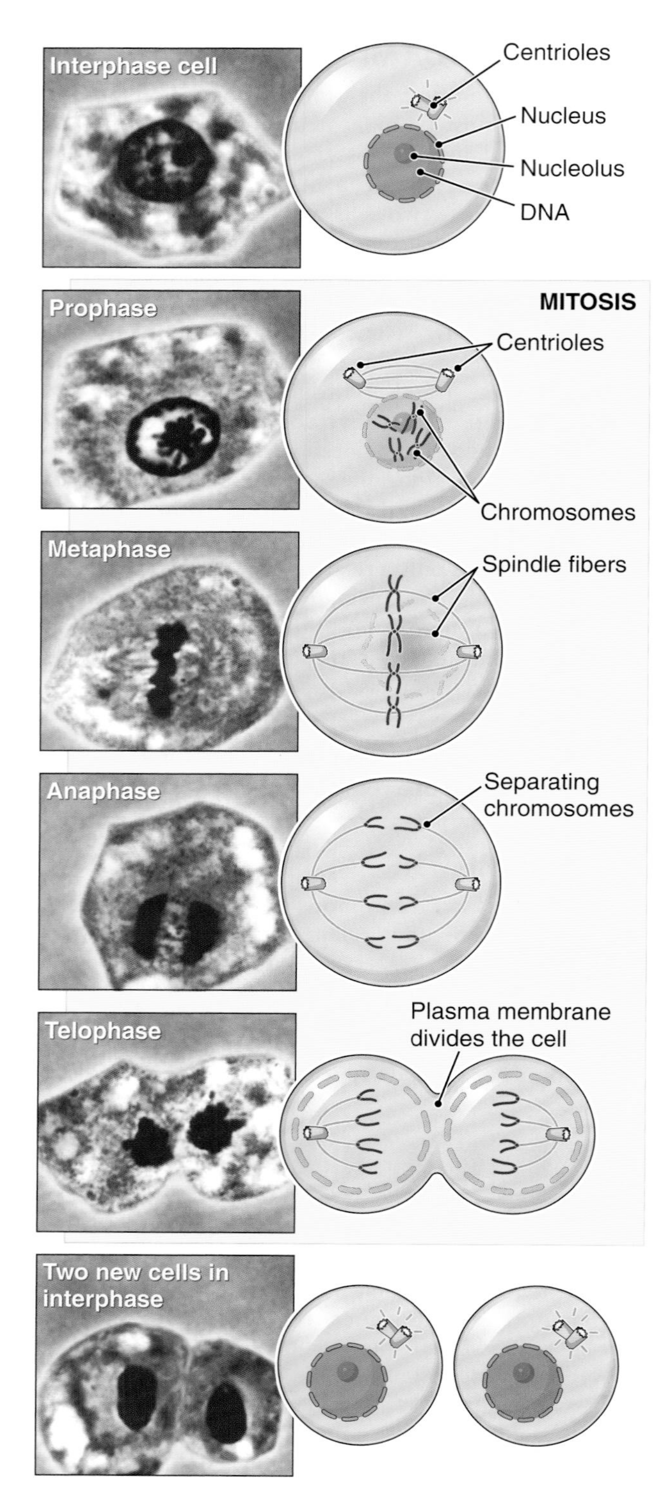

FIGURE 3-15 The stages of mitosis. KEY POINT Although it is a continuous process, mitosis can be seen in four stages. When it is not dividing, the cell is in interphase. The cell shown is for illustration only. It is not a human cell, which has 46 chromosomes. ZOOMING IN If the original cell shown has 46 chromosomes, how many chromosomes will each new daughter cell have?

centrioles move toward opposite ends of the cell, and a spindle-shaped structure made of thin fibers begins to form between them.

- In **metaphase** (MET-ah-faze), the chromosomes line up across the center (equator) of the cell attached to the spindle fibers.
- In **anaphase** (AN-ah-faze), the centromere splits, and the replicated chromosomes separate and begin to move toward opposite ends of the cell.
- As mitosis continues into **telophase** (TEL-o-faze), a membrane appears around each group of separated chromosomes, forming two new nuclei.

Also during telophase, the plasma membrane pinches off to divide the cell. The midsection between the two areas becomes progressively smaller until finally the cell splits into two. There are now two new cells, or daughter cells, each with exactly the same kind and amount of DNA as was present in the parent cell. In just a few types of cells, skeletal muscle cells, for example, the cell itself does not divide following nuclear division. The result, after multiple mitoses, is a giant single cell with multiple nuclei. This pattern is extremely rare in human cells.

CHECKPOINTS 

- ☐ **3-12** What must happen to the DNA in a cell before mitosis can occur? During what stage in the cell life cycle does this occur?
- ☐ **3-13** What are the four stages of mitosis?

Cell Aging

As cells multiply throughout life, changes occur that may lead to their damage and death. Harmful substances known as *free radicals* or *reactive oxygen species (ROS)*, produced in the course of normal metabolism, can injure cells unless they are destroyed. Chapter 18 covers free radicals in more detail. Lysosomes may deteriorate as they age, releasing enzymes that can harm the cell. As noted, mutations sometimes harm cells and may lead to cancer.

As a person ages, stem cells divide less frequently, and mature body cells become less active. These changes slow down repair processes, which rely on the production of new cells and the production of substances from existing cells. A bone fracture, for example, takes considerably longer to heal in an old person than in a young person.

One theory on aging holds that cells are preprogrammed to divide only a certain number of times before they die. Support for this idea comes from the fact that cells taken from a young person divide more times when grown in the laboratory than do similar cells taken from an older individual. This programmed cell death, known as *apoptosis* (ah-pop-TO-sis), is a natural part of growth and remodeling before birth in the developing embryo. For example, apoptosis removes cells from the embryonic limb buds in the development of fingers and toes. Apoptosis also is needed in repair and remodeling of tissue throughout life. Cells subject to wear and tear regularly undergo apoptosis and are replaced. For example, the cells lining the digestive tract are removed and replaced every two to three days. This "cellular suicide" is an orderly, genetically programmed process. The "suicide" genes code for enzymes that destroy the cell quickly without damaging nearby cells. Phagocytes then eliminate the dead cells.

See the Student Resources on thePoint® to view the animation "The Cell Cycle and Mitosis."

A & P in Action Revisited: Ben's CF Diagnosis

Ben's parents were shocked when the doctor diagnosed their 1-year-old with CF. Apparently, the condition had not been detected when Ben was tested at birth, as required by state law. The doctor asserted that they were not to blame for Ben's condition. CF is an inherited disease—Ben's parents each carried a defective gene in their DNA and both had, by chance, passed copies to Ben. As a result, Ben was unable to synthesize a channel protein found in the plasma membranes of certain cells. Normally, this channel regulates the movement of chloride into the cell. Because the channels did not work in Ben's case, chloride was trapped outside the cells. The negatively charged chloride ions attract positively charged sodium ions normally found in extracellular fluid. These two ions form the salt, sodium chloride, which is lost in high amounts in the sweat of individuals with CF.

Abnormal chloride channel function causes cells in many organs to produce thick, sticky mucus. In the lungs, this mucus causes difficulty breathing, inflammation, and frequent bacterial infections. The thick mucus also decreases the ability of the large and small intestines to absorb nutrients, resulting in low weight gain, poor growth, and vitamin deficiencies. This problem is compounded by damage to the pancreas, preventing production of essential digestive enzymes.

Alison and her husband's immediate concern was, of course, for their son. The doctor reassured them that with proper treatment, Ben could lead a relatively normal life. As he explained, "New oral medications can increase membrane channels for people with certain CF mutations, which fortunately include Ben's F508del mutation. Inhaled therapies can improve pulmonary function, and replacement enzymes can help overcome the pancreatic enzyme deficiency. Future therapies," he concluded, "might even offer a cure for CF."

In this case, we saw that defective plasma membrane channels in some of Ben's cells had widespread effects on his whole body. In later chapters, as you learn about the body's organs, remember that their structure and function are closely related to the condition of their cells and tissues.

CHAPTER 3

Chapter Wrap-Up

OVERVIEW

A detailed chapter outline with space for note-taking is on thePoint®. The figure below illustrates the main topics covered in this chapter.

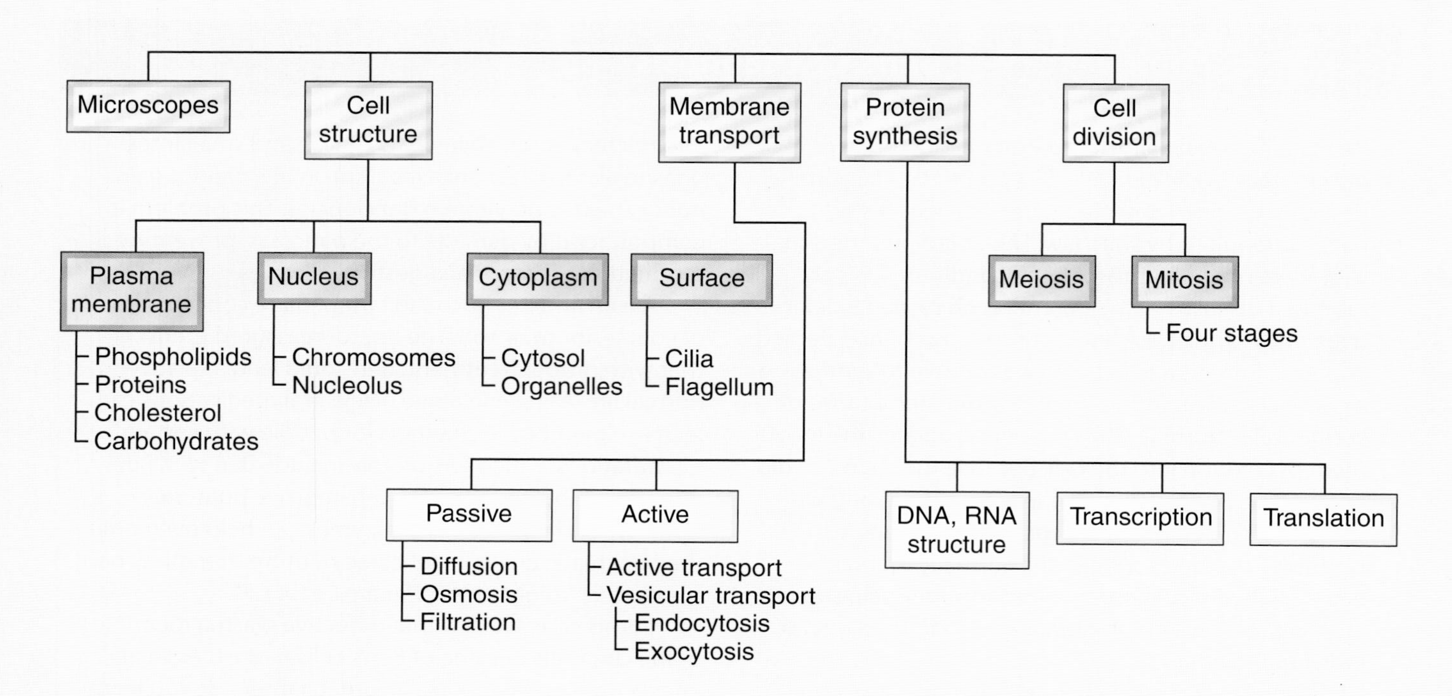

KEY TERMS

The terms listed below are emphasized in this chapter. Knowing them will help you organize and prioritize your learning. These and other boldface terms are defined in the Glossary with phonetic pronunciations.

active transport
chromosome
cytology
cytoplasm
diffusion
DNA
endocytosis
exocytosis
filtration
gene
hemolysis
hypertonic
hypotonic
interphase
isotonic
micrometer
microscope
mitochondria
mitosis
mutation
nucleus
organelle
osmosis
phagocytosis
plasma membrane
ribosome
RNA

WORD ANATOMY

Medical terms are built from standardized word parts (prefixes, roots, and suffixes). Learning the meanings of these parts can help you remember words and interpret unfamiliar terms.

WORD PART	MEANING	EXAMPLE
MICROSCOPES		
cyt/o	cell	*Cytology* is the study of cells.
micr/o	small	*Microscopes* are used to view structures too small to see with the naked eye.
CELL STRUCTURE		
bi-	two	The lipid *bilayer* is a double layer of lipid molecules.
chrom/o-	color	*Chromosomes* are small, threadlike bodies that stain darkly with basic dyes.
end/o-	in, within	The *endoplasmic* reticulum is a membranous network within the cytoplasm.
lys/o	loosening, dissolving, separating	*Lysosomes* are small bodies (organelles) with enzymes that dissolve materials (*see also hemolysis*).
-some	body	*Ribosomes* are small bodies in the cytoplasm that help make proteins.
MOVEMENT ACROSS THE PLASMA MEMBRANE		
ex/o-	outside, out of, away	In *exocytosis*, the cell moves material out from vesicles.
hem/o	blood	*Hemolysis* is the destruction of red blood cells.
hyper-	above, over, excessive	A *hypertonic* solution's concentration is higher than that of the cytoplasm.
hypo-	deficient, below, beneath	A *hypotonic* solution's concentration is lower than that of the cytoplasm.
iso-	same, equal	An *isotonic* solution has the same concentration as that of the cytoplasm.
phag/o	to eat, ingest	In *phagocytosis*, the plasma membrane engulfs large particles and moves them into the cell.
pin/o	to drink	In *pinocytosis*, the plasma membrane "drinks" (engulfs) droplets of fluid.
semi-	partial, half	A *semipermeable* membrane lets some molecules pass through but not others.
CELL DIVISION		
ana-	upward, back, again	In the *anaphase* stage of mitosis, chromosomes move to opposite sides of the cell.
inter-	between	*Interphase* is the stage between one cell division (mitosis) and the next.
meta-	change	*Metaphase* is the second stage of mitosis when the chromosomes change position and line up across the equator.
pro-	before, in front of	*Prophase* is the first stage of mitosis.
tel/o-	end	*Telophase* is the last stage of mitosis.

QUESTIONS FOR STUDY AND REVIEW

BUILDING UNDERSTANDING

Fill in the Blanks

1. The part of the cell that regulates what can enter or leave is the _____.
2. The cytosol and organelles make up the _____.
3. If Solution A has more solute and less water than does Solution B, then Solution A is _____ to Solution B.
4. Mechanisms that require cellular energy to move substances across the plasma membrane are called _____ transport mechanisms.
5. Distinct segments of DNA that code for specific proteins are called _____.

Matching Match each numbered item with the most closely related lettered item.

___ 6. DNA replication occurs.

___ 7. DNA is tightly wound into chromosomes.

___ 8. Chromosomes line up along the cell's equator.

___ 9. Chromosomes separate and move toward opposite ends of the cell.

___ 10. Cell membrane pinches off, dividing the cell into two new daughter cells.

a. metaphase
b. anaphase
c. telophase
d. interphase
e. prophase

Multiple Choice

___ 11. The nucleus is called the cell's control center because it contains the
a. nucleolus
b. chromosomes
c. cytosol
d. cilia

___ 12. Where in the cell does ATP synthesis occur?
a. endoplasmic reticulum
b. Golgi apparatus
c. mitochondria
d. nucleus

___ 13. The movement of solute from a region of high concentration to one of lower concentration is called _____.
a. diffusion
b. endocytosis
c. exocytosis
d. osmosis

___ 14. A DNA sequence reads: TGAAC. What is its mRNA sequence?
a. ACTTG
b. ACUUG
c. CAGGT
d. CAGGU

___ 15. Rupture of red blood cells placed in a hypotonic solution is called _____.
a. crenation
b. hemolysis
c. permeability
d. mitosis

UNDERSTANDING CONCEPTS

16. List the components of the plasma membrane, and state a function for each.
17. Compare and contrast the following cellular components and processes:
    a. microvilli and cilia
    b. rough ER and smooth ER
    c. lysosome and peroxisome
    d. endocytosis and exocytosis
    e. DNA and RNA
    f. chromosome and gene
18. List and define five methods by which materials cross the plasma membrane. Which of these requires chemical energy?
19. Why is the plasma membrane described as selectively permeable?
20. What will happen to a body cell placed in a 5.0% salt solution? In distilled water?
21. Describe the role of each of the following in protein synthesis: DNA, nucleotide, RNA, ribosomes, rough ER, and Golgi apparatus.

CONCEPTUAL THINKING

22. Kidney failure causes a buildup of waste and water in the blood. A procedure called hemodialysis removes these substances from the blood. During this procedure, the patient's blood passes over a semipermeable membrane within the dialysis machine. Waste and water from the blood diffuse across the membrane into dialysis fluid on the other side. Based on this information, compare the osmotic concentration of the blood with that of the dialysis fluid.
23. CF can result from a change in the CFTR gene on chromosome number 7. This gene codes for a membrane protein that regulates chloride. Describe the process that produces such an abnormal protein beginning with a change in the DNA of the CFTR gene.
24. Changes at the cellular level can ultimately affect the entire organism. Using Ben's case, explain why this is so.
25. Look at the graph in the One Step at a Time box on p. 49, and answer these questions.
    a. How many children in the placebo group provided data in week 24?
    b. When were drug effects first noted, according to this graph?
    c. Give the graph a title (what does it show?).

For more questions, see the Learning Activities on thePoint®.

CHAPTER 4

Tissues, Glands, and Membranes

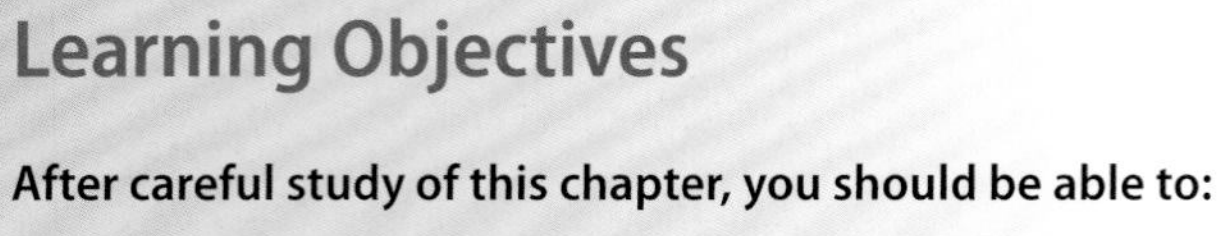

Learning Objectives

After careful study of this chapter, you should be able to:

1. Define stem cells, and describe their role in development and repair of tissue. ***p. 66***
2. Name the four main groups of tissues, and give the location and general characteristics of each. ***p. 66***
3. Classify the different types of epithelial tissue based on the shape and arrangement of the cells. ***p. 67***
4. Describe the difference between exocrine and endocrine glands, and give examples of each. ***p. 68***
5. Classify the different types of connective tissue based on the cells and the material separating the cells. ***p. 69***
6. Describe the structure, location, and function of three types of epithelial membranes. ***p. 73***
7. List six types of connective tissue membranes. ***p. 74***
8. Using examples from the case study, define histology and explain the role of histology in medical diagnosis and treatment. ***p. 76***
9. Show how word parts are used to build words related to tissues, glands, and membranes (see Word Anatomy at the end of the chapter). ***p. 79***

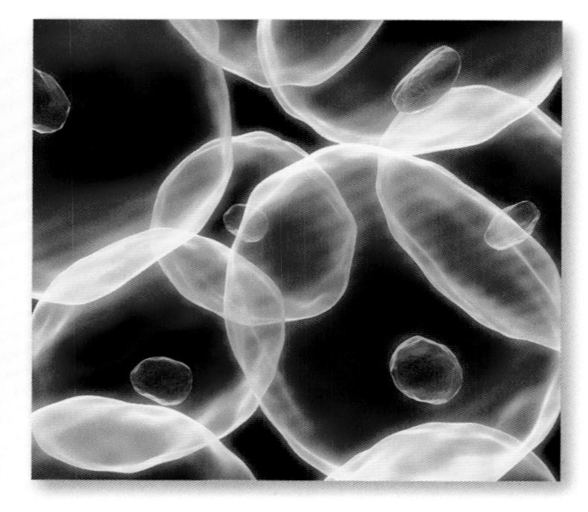

A & P in Action
Paul's Case: Sun-Damaged Skin

"Wait a minute," Paul said to his reflection in the mirror as he examined his face after shaving. He had noticed a small nodule to the side of his left nostril. The lump was mostly pink with a pearly white border and painless to the touch. *I haven't seen that before. Probably just a pimple or maybe a small cyst*, Paul thought, although he couldn't help thinking back to the many hours he had spent as a kid sailing competitively at the seashore. *I know sun exposure isn't great for your skin, even dangerous, and I wasn't real careful about wearing sunscreen. Even if I did, it would have washed off anyway while I was sailing*, he thought. Paul finished his trimming and decided the lump was probably nothing.

Despite his attempts to forget about the lump, Paul was concerned. Over the next several days, he showed the small, rounded mass to several people to get their opinions. No one had an answer when he asked, "What do you think this is?" When several weeks produced no change, except maybe a little depression in the center of the mass, worry led him to make an appointment with a dermatologist.

"Well Paul, I'm not sure. It could be nothing, but we'd better look a little closer," said Dr. Nielsen. "It could be benign, meaning it's a simple tissue overgrowth that never spreads to other tissues. But we have to be sure that it's not a small skin cancer. This is a very common site for such a lesion. Basal cell carcinomas arise from the epithelial portion of skin, especially in sun-exposed areas. UV rays from the sun damage DNA, causing the cells to divide more rapidly than normal. We usually completely cure this type of cancer, because it is commonly diagnosed before it spreads to other tissues. We'll remove this and send it to the pathology lab to see what's going on." Paul left Dr. Nielsen's office with a small bandage over the site of excision, some ointment to apply, and instructions to call the office in three days.

Paul's dermatologist suspects that he may have skin cancer—a disease affecting the cutaneous membrane. Later in the chapter, we revisit Paul and learn the final diagnosis of that lump on his nose.

As you study this chapter, CasePoints will give you opportunities to apply your learning to this case.

Visit thePoint® to access the following resources. For guidance in using these resources most effectively, see pp. xv–xvii.

Preparing to Learn

- Tips for Effective Studying
- Pre-Quiz

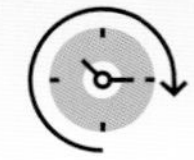

While You Are Learning

- Web Chart: Epithelial Tissue
- Web Chart: Connective Tissue
- Chapter Notes Outline
- Audio Pronunciation Glossary

When You Are Reviewing

- Answers to Questions for Study and Review
- Health Professions: Histotechnologist
- Interactive Learning Activities

In Chapter 3, we learned about cells and their structures and functions, including the principles of cell division and how cancer can result from abnormal cell division. In this chapter, we apply these concepts to the next level of organization, cells working together to form tissues. Important key ideas include barriers 4 and the relationship between structure and function 1 from Chapter 1 and the relationship between genes and proteins 9 from Chapter 3.

Introduction

Tissues are groups of cells similar in structure, arranged in a characteristic pattern, and specialized for the performance of specific tasks. The tissues in our bodies might be compared with the different materials used to construct a building. Think for a moment of the great variety of building materials used according to need—wood, stone, steel, plaster, insulation, and others. All the functions of the building depend upon the properties and organization of the individual building materials. Similarly, an organ's ability to accomplish its functions depends on the organization, structure, and abilities of its tissues. The study of tissues is known as **histology** (his-TOL-o-je).

Tissue Origins

During development, all tissues derive from young, actively dividing cells known as **stem cells** (BOX 4-1). Most stem cells gradually differentiate into the mature, functioning cells that make up different body tissues. These mature body cells no longer undergo mitosis but remain in interphase, as described in Chapter 3. Dividing cells that are active during development in the uterus are described as *embryonic stem (ES) cells*. These cells have the greatest potential for multiplication. Stem cells that persist after birth are commonly called *adult stem cells*, but as they are also present in babies and children, they are more accurately called *postnatal stem cells*.

A variable number of stem cells persist in each tissue, producing new cells that can differentiate into mature cells. Tissues subject to wear and tear, such as the skin and the lining of the digestive and respiratory tracts, maintain a large population of stem cells that continually divide in order to replace lost or damaged cells. These tissues can repair themselves relatively easily. Other tissues, especially nervous tissue and muscle tissue, maintain few stem cells that divide infrequently, so these tissues repair themselves slowly, if at all. Brain tissue injured by a stroke or heart muscle tissue injured by a heart attack has limited regenerative ability. Between these two extremes are organs, such as the liver, that maintain enough stem cells to replace the entire organ within months or years. So, for example, a portion of the liver can be transplanted from one person to another, and the donor's organ will be restored.

Stem cells give rise to four main tissue groups, presented as follows:

- **Epithelial** (ep-ih-THE-le-al) tissue covers surfaces, lines cavities, and forms glands.
- **Connective tissue** supports and forms the framework of all parts of the body.
- **Muscle tissue** contracts and produces movement.
- **Nervous tissue** conducts nerve impulses.

This chapter concentrates mainly on epithelial and connective tissues, the less specialized of the four types. As discussed later in the chapter, epithelial and connective tissues often form thin sheets of tissues called **membranes** that line and cover organs. Muscle and nervous tissues receive more attention in later chapters.

HOT TOPICS — BOX 4-1
Stem Cells: So Much Potential

At least 200 different cell types are found in the human body, each with its own unique structure and function. All originate from unspecialized precursors called stem cells, which exhibit two important characteristics: they can divide repeatedly and have the potential to become specialized cells.

Stem cells come in two types. **Embryonic stem (ES) cells** found in early embryos are the source of all body cells and can potentially differentiate into any cell type. **Postnatal stem cells**, also called adult stem cells, are found in babies and children as well as adults. These are stem cells that remain in the body after birth and can differentiate into different cell types. They assist with tissue growth and repair. For example, in the red bone marrow, these cells differentiate into blood cells, whereas in the skin, they differentiate into new skin cells to replace cells in surface layers that are shed continually or cells that are damaged by a cut, scrape, or other injuries.

The potential healthcare applications of stem cell research are numerous. In the near future, stem cell transplants may be used to repair damaged tissues in treating illnesses such as diabetes, cancer, heart disease, Parkinson disease, and spinal cord injury. This research may also help explain how cells develop and why some cells develop abnormally, causing birth defects and cancer. Scientists may also use stem cells to test drugs before trying them on animals and humans.

But stem cell research is controversial. Some argue that it is unethical to use embryonic stem cells because they are obtained from aborted fetuses or fertilized eggs left over from in vitro fertilization. Others argue that these cells would be discarded anyway and have the potential to improve lives. A possible solution is the use of postnatal stem cells. However, these cells are less abundant than are embryonic stem cells and lack their potential to differentiate.

Epithelial Tissue

4 Epithelial tissue, or **epithelium** (ep-ih-THE-le-um), forms the barrier separating the body's relatively constant internal environment from the rapidly changing external environment. It is the main tissue of the skin's outer layer. Epithelium lining the body's tubes and hollow organs serves a similar role, separating the tube's contents (which are considered "outside") from body cells. Within the body, epithelium also helps protect individual organs and makes up specialized structures called *glands*, discussed further below.

Epithelium repairs itself quickly after it is injured. In areas of the body subject to normal wear and tear, such as the skin, the inside of the mouth, and the lining of the intestinal tract, epithelial stem cells reproduce frequently, replacing dead or damaged cells. Certain areas of the epithelium that form the outer layer of the skin are capable of modifying themselves for greater strength whenever they are subjected to unusual wear and tear; the growth of calluses is a good example of this response.

STRUCTURE OF EPITHELIAL TISSUE

Epithelial cells are tightly packed to better protect underlying tissue or form barriers between systems. 1 The cells vary in shape and arrangement according to their functions. In shape, the cells may be described as follows:

- **Squamous** (SKWA-mus)—flat and irregular
- **Cuboidal**—square
- **Columnar**—long and narrow

The cells may be arranged in a single layer, in which case the epithelium is described as **simple** (FIG. 4-1). Simple epithelium functions as a thin barrier through which materials can pass fairly easily. For example, simple epithelium allows for absorption of materials from the lining of the digestive tract to the blood and allows for passage of oxygen from the blood to body tissues. Areas subject to wear and tear that require protection are covered with epithelial cells in multiple layers, an arrangement described as **stratified** (FIG. 4-2). If the cells are staggered so that they appear to be in multiple layers but really are not, they are termed *pseudostratified*.

Terms for both shape and arrangement are used to describe epithelial tissue. Thus, a single layer of flat, irregular cells would be described as *simple squamous epithelium*, whereas tissue with many layers of these same cells would be described as *stratified squamous epithelium*. Regardless of cell shape, the overall thickness of the epithelial layer determines its protective ability.

Some organs, such as the urinary bladder, must vary a great deal in size as they work. These organs are lined with **transitional epithelium**, which is capable of great expansion but returns to its original form once tension is relaxed—as when, in this case, the urinary bladder is emptied.

CASEPOINT

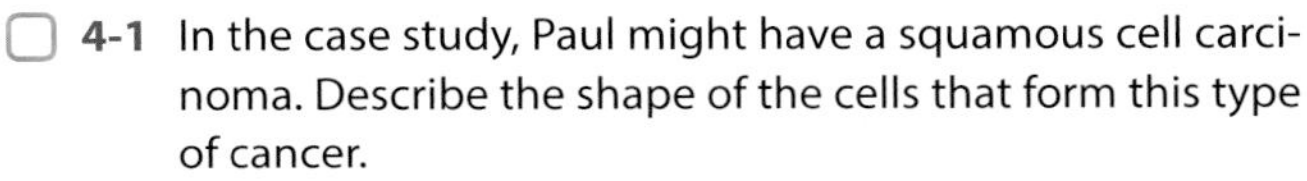

☐ **4-1** In the case study, Paul might have a squamous cell carcinoma. Describe the shape of the cells that form this type of cancer.

See Appendix 2 for the answers to the Casepoint questions.

See the Student Resources on thePoint® for a summary chart on epithelial tissue.

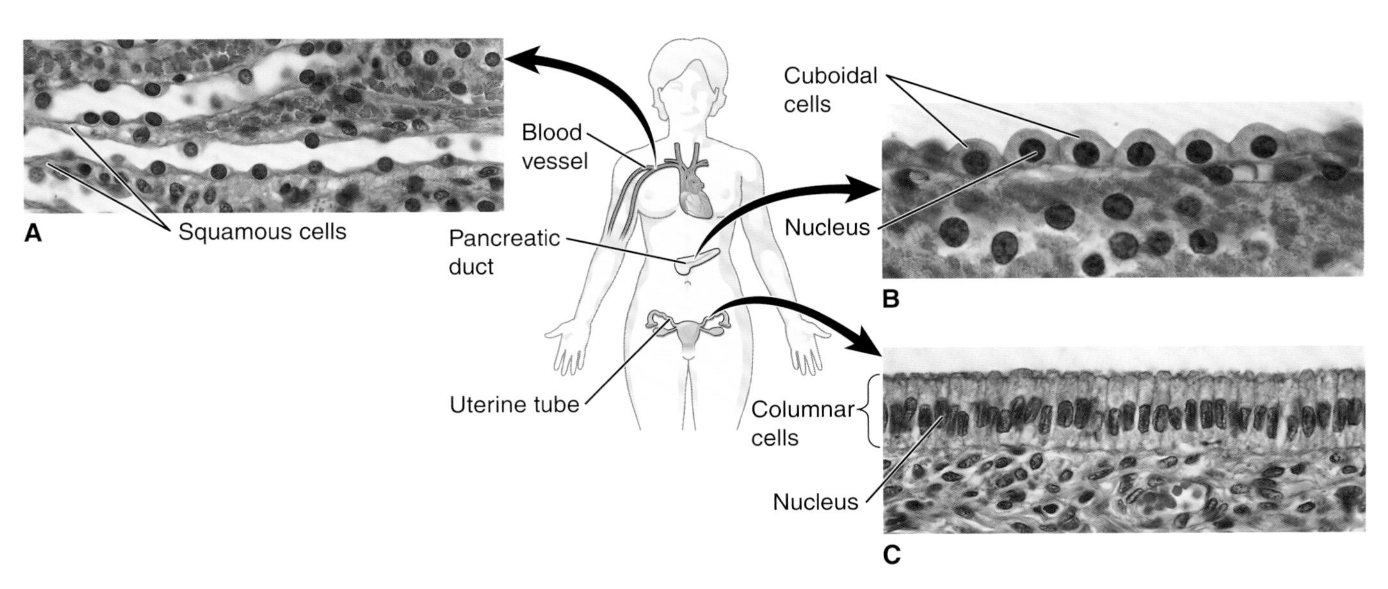

FIGURE 4-1 **Simple epithelial tissues.** KEY POINT Epithelial tissue can be described by the shape of its cells. **A.** Simple squamous epithelium has flat, irregular cells with flat nuclei. **B.** Cuboidal epithelial cells are square in shape with round nuclei. **C.** Columnar epithelial cells are long and narrow with ovoid nuclei. ZOOMING IN How many layers are these epithelial cells? *See Appendix 2 for answers to the Zooming In questions.*

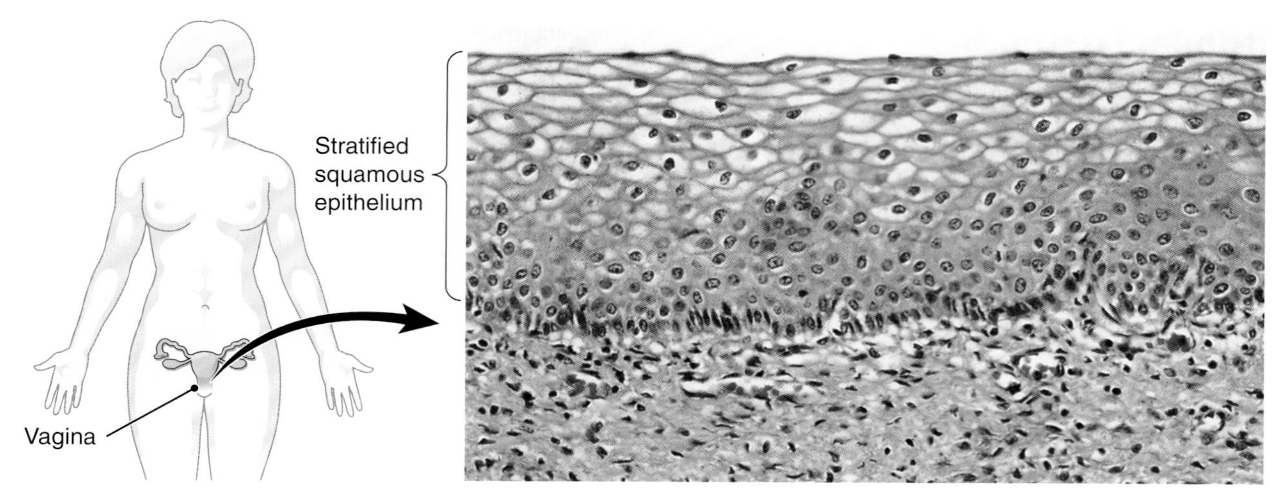

FIGURE 4-2 **Stratified squamous epithelium.** KEY POINT Stratified epithelium has multiple layers of cells. ZOOMING IN What is the function of stratified epithelium?

GLANDS

Epithelial tissue produces the many secretions necessary for health, including **mucus** (MU-kus) (a clear, sticky fluid), digestive juices, sweat, and hormones. A **gland** is an organ or cell specialized to produce a substance that is sent out to other parts of the body. The gland manufactures these secretions from materials removed from the blood. Glands are divided into two categories based on how they release their secretions: exocrine glands and endocrine glands.

Exocrine Glands **Exocrine** (EK-so-krin) **glands** produce secretions that are carried out of the body (recall that *ex/o* means "outside" or "away from"). The exocrine glands usually have ducts or tubes to carry their secretions away from the glands. Their secretions are delivered into an organ, to a cavity, or to the body surface and act in a limited area near their sources. Examples of exocrine glands include the glands in the stomach and intestine that secrete digestive juices, the salivary glands, the sweat and sebaceous (oil) glands of the skin, and the lacrimal glands that produce tears.

Most exocrine glands are composed of multiple cells in various arrangements, including tubular, coiled, or saclike formations. **Goblet cells**, in contrast, are single-celled exocrine glands that secrete mucus. Goblet cells are scattered among the epithelial cells lining the respiratory and digestive passageways (FIG. 4-3). As discussed further, the mucus lubricates the passageways and protects the underlying tissue.

Endocrine Glands **Endocrine** (EN-do-krin) **glands** secrete not through ducts but directly into surrounding tissue fluid. Most often, the secretions are then absorbed into the bloodstream, which distributes them internally, as indicated by the prefix *end/o*, meaning "within." These secretions, called **hormones**, have effects on specific tissues known as the *target tissues*.

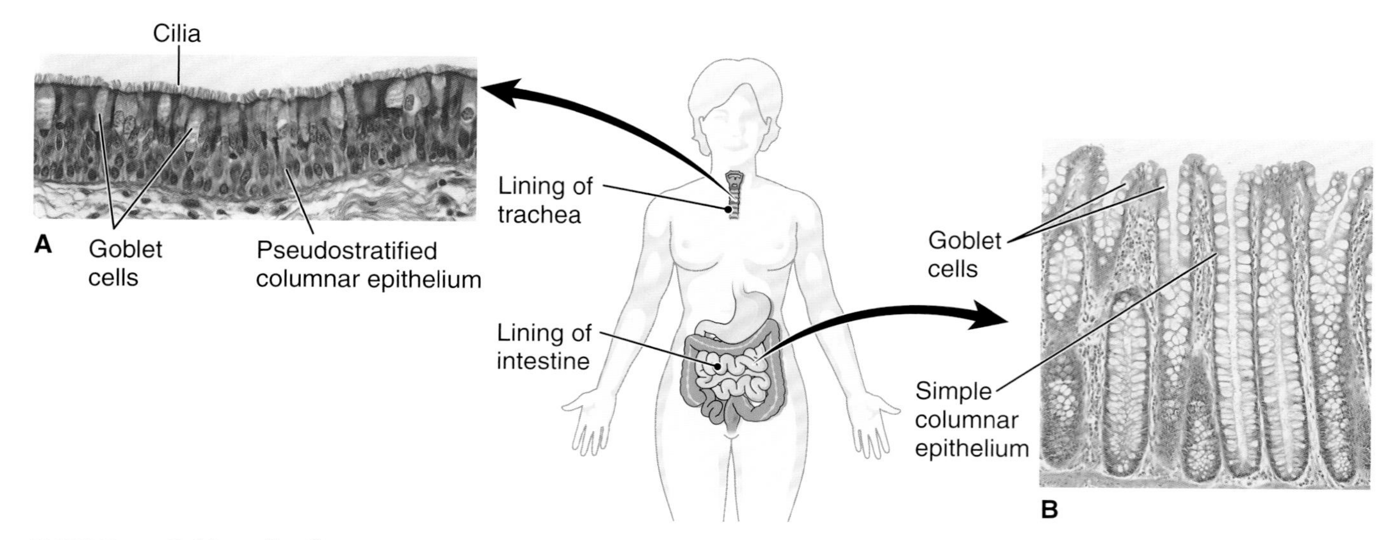

FIGURE 4-3 **Goblet cells.** KEY POINT Goblet cells in epithelium secrete mucus. **A.** The lining of the trachea showing cilia and goblet cells that secrete mucus. **B.** The lining of the intestine showing goblet cells.

Concept Mastery Alert

Note that hormones pass through the interstitial fluid before they reach the bloodstream.

Endocrine glands have an extensive network of blood vessels. These so-called ductless glands include the pituitary, thyroid, adrenal, and other glands described in greater detail in Chapter 11.

CHECKPOINTS 

- [ ] **4-1** What are the three basic shapes of epithelial cells?
- [ ] **4-2** What are the two categories of glands based on their methods of secretion?

See Appendix 2 for answers to the Checkpoint questions.

Connective Tissue

The supporting fabric everywhere in the body is the connective tissue. This is so extensive and widely distributed that if we were able to dissolve all the tissues except the connective tissue, we would still be able to recognize the entire body. The connective tissue has large amounts of nonliving material between the cells. This intercellular background material or **matrix** (MA-trix) contains varying amounts of water, protein fibers, and hard minerals.

Histologists, specialists in the study of tissues, have numerous ways of classifying connective tissues based on their structure or function. Here, we describe the different types in order of increasing hardness.

- **Circulating connective tissue** has a fluid consistency; its cells are suspended in a liquid matrix. The two types are blood, which circulates in blood vessels (**FIG. 4-4**), and lymph, a fluid derived from blood that circulates in lymphatic vessels
- **Loose connective tissue** has a soft consistency, similar to jelly.
- **Dense connective tissue** contains many fibers and is quite strong, similar to a rope or canvas fabric.
- **Cartilage** has a very firm consistency. The gristle at the end of a chicken bone is an example of this tissue type.
- **Bone tissue**, the hardest type of connective tissue, is solidified by minerals in the matrix.

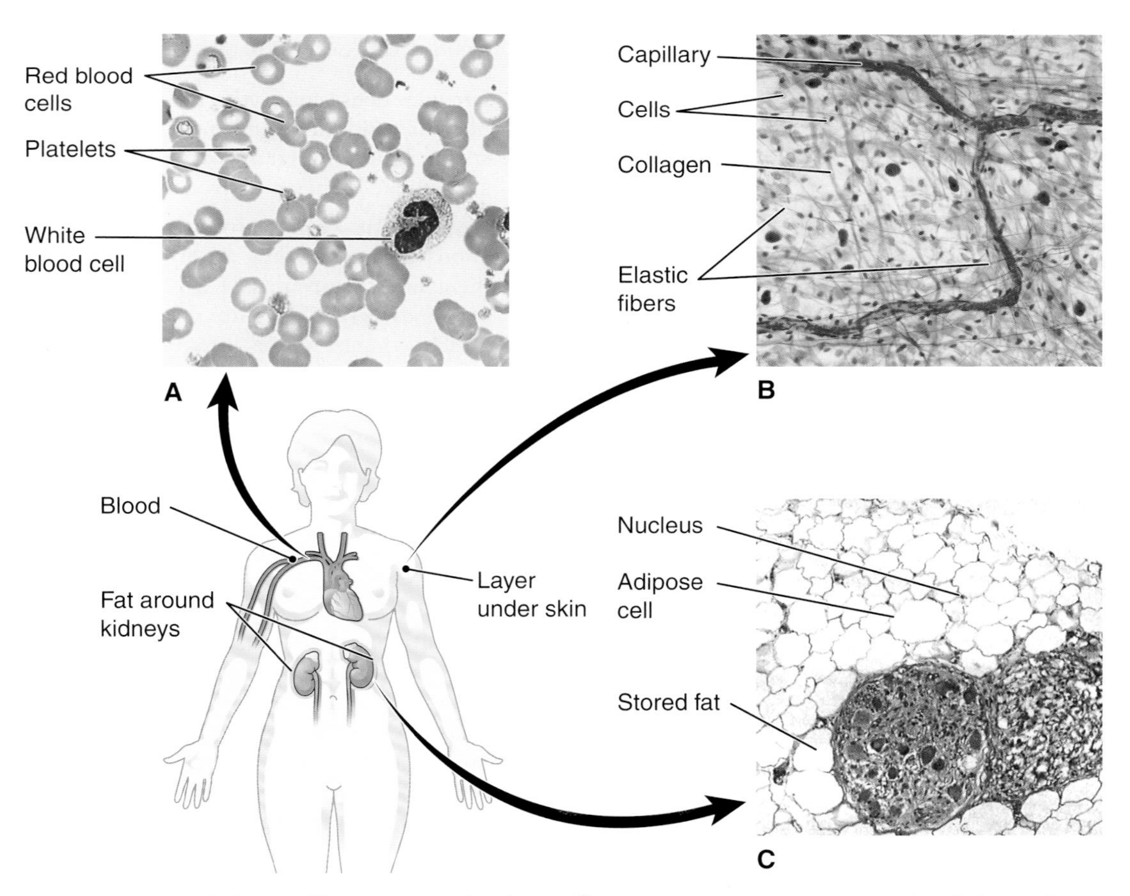

FIGURE 4-4 Circulating and loose connective tissue. **KEY POINT** Connective tissue is classified according to its distribution and the consistency of its matrix. **A.** Blood smear showing various blood cells in a liquid matrix. **B.** Areolar connective tissue, a mixture of cells and fibers in a jelly-like matrix. **C.** Adipose tissue shown here surrounding dark-stained glandular tissue. The micrograph shows areas where fat is stored and nuclei are at the edge of the cells. **ZOOMING IN** Which of these tissues has the most fibers? Which of these tissues is modified for storage?

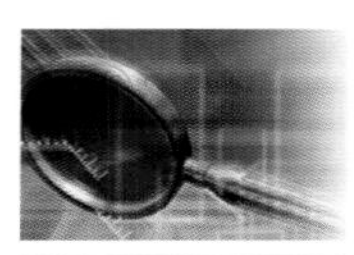

A CLOSER LOOK
Collagen: The Body's Scaffolding

BOX 4-2

The most abundant protein in the body, making up about 25% of total protein, is collagen. Its name, derived from a Greek word meaning "glue," reveals its role as the main structural protein in connective tissue.

Fibroblasts secrete collagen molecules into the surrounding matrix, where the molecules are then assembled into fibers. These fibers give the matrix its strength and its flexibility. Collagen fibers' high tensile strength makes them stronger than steel fibers of the same size, and their flexibility confers resilience on the tissues that contain them. For example, collagen in the skin, bone, tendons, and ligaments resists pulling forces, whereas collagen found in joint cartilage and between vertebrae resists compression. Based on amino acid structure, there are at least 19 types of collagen, each of which imparts a different property to the connective tissue containing it.

The arrangement of collagen fibers in the matrix reveals much about the tissue's function. In the skin and membranes covering muscles and organs, collagen fibers are arranged irregularly, with fibers running in all directions. The result is a tissue that can resist stretching forces in many different directions. In tendons and ligaments, collagen fibers have a parallel arrangement, forming strong ropelike cords that can resist longitudinal pulling forces. In bone tissue, collagen fibers' meshlike arrangement promotes deposition of calcium salts into the tissue, which gives bone strength while also providing flexibility.

Collagen's varied properties are also evident in the preparation of a gelatin dessert. Gelatin is a collagen extract made by boiling animal bones and other connective tissues. It is a viscous liquid in hot water but forms a semisolid gel on cooling.

Chapters 12 and 15 have more information on circulating connective tissue, which is highly specialized in both composition and function. The other types of connective tissue are discussed in greater detail below.

See the Student Resources on thePoint® for a summary chart of the connective tissue types.

LOOSE CONNECTIVE TISSUE

As the name implies, loose connective tissue has a soft or semiliquid consistency. There are two types:

- **Areolar** (ah-RE-o-lar) **tissue** is named from a word that means "space" because of its open composition (see FIG. 4-4B). It contains cells and fibers in a soft, jelly-like matrix. The main cell type is the **fibroblast**, which produces the protein fibers and other components of the matrix (the word ending in *blast* refers to a young and active cell). Fibroblasts produce **collagen** (KOL-ah-jen), a flexible white protein (see BOX 4-2), as well as elastic fibers. Areolar tissue forms an important component of many tissue membranes (discussed later) and is the most common type of connective tissue.

- **Adipose** (AD-ih-pose) **tissue** is primarily composed of fat cells (adipocytes) with minimal intercellular matrix. Adipocytes are able to store large amounts of fat that serves as a reserve energy supply for the body (see FIG. 4-4C). Adipose tissue underlying the skin acts as a heat insulator, and adipose tissue surrounding organs and joints provides protective padding.

DENSE CONNECTIVE TISSUE

Dense connective tissue, like areolar tissue, contains fibroblasts that synthesize a collagen-rich matrix. However, dense connective tissue contains significantly more protein fibers, so it is stronger, firmer, and more flexible than areolar tissue. The different types of dense connective tissue vary in the arrangement of the collagen fibers:

- **Irregular dense connective tissue** has mostly collagenous fibers in random arrangement. This tissue makes up the strong membranes that cover joints and various organs, such as the kidney and liver, and strengthen the skin.
- **Regular dense connective tissue** also has mostly collagenous fibers, but they are in a regular, parallel alignment like the strands of a cable. This tissue can pull in one direction. Examples are the cordlike **tendons**, which connect muscles to bones, and the **ligaments**, which connect bones to other bones (FIG. 4-5A). The regular dense connective tissue in the walls of blood vessels, the respiratory passageways, and the vocal cords contains large amounts of elastic fibers. As a result, these structures can stretch and return to their original dimensions.

CARTILAGE

Because of its strength and flexibility, cartilage is a structural material and provides reinforcement. It is also a shock absorber and a bearing surface that reduces friction between moving parts, as at joints. The cells that produce cartilage are **chondrocytes** (KON-dro-sites), a name derived from the root word *chondro*, meaning "cartilage," and the root word *cyto*, meaning "cell." There are three forms of cartilage:

Concept Mastery Alert

Students frequently confuse chondrocytes, which are cells, and collagen, which is a protein between the cells.

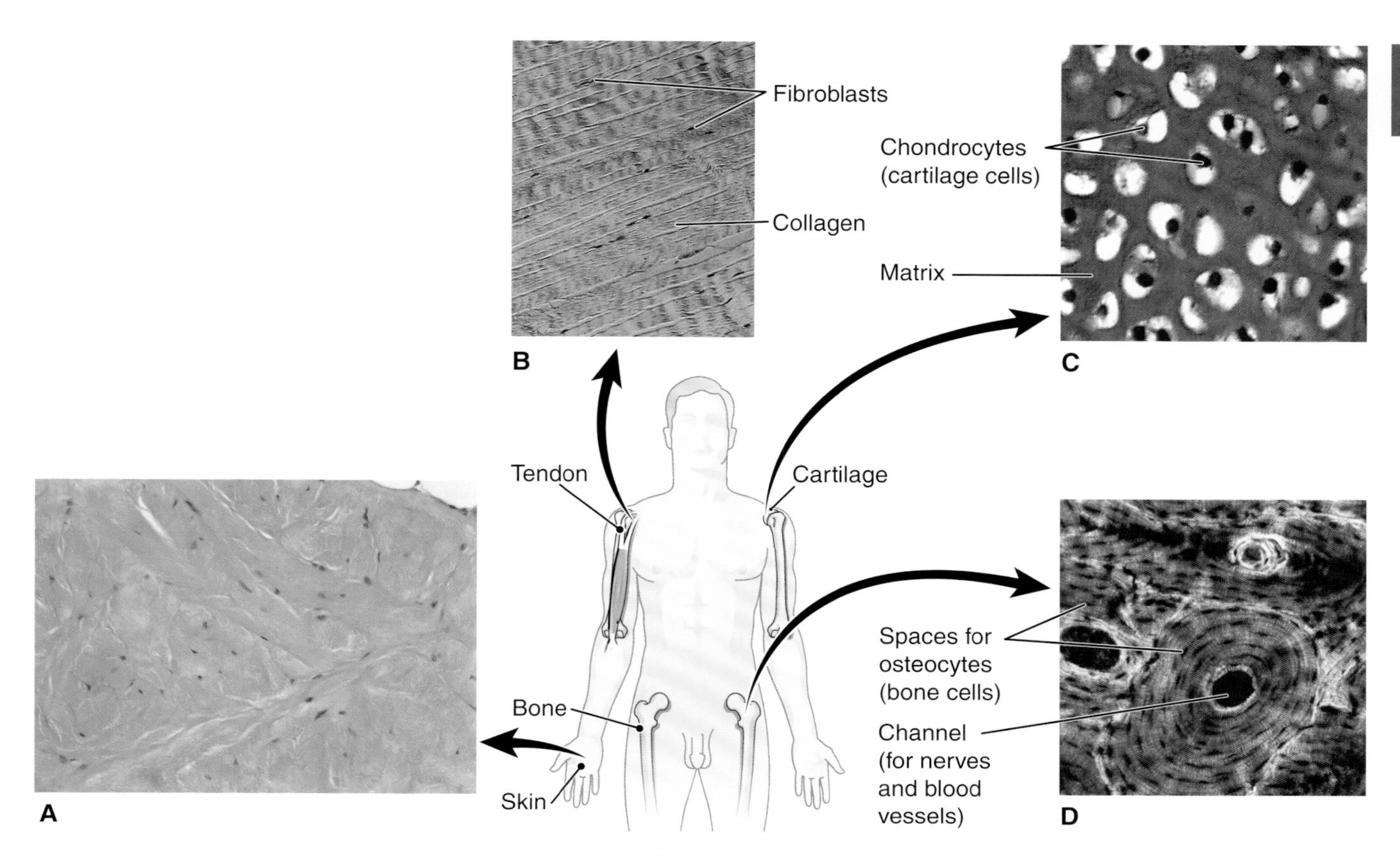

FIGURE 4-5 Dense connective tissue, cartilage, and bone. KEY POINT Fibers are a key component of connective tissue. **A.** Dense irregular connective tissue. **B.** Dense regular connective tissue. In tendons and ligaments, collagenous fibers are arranged in the same direction. **C.** In cartilage, the cells (chondrocytes) are enclosed in a firm matrix. **D.** The bone is the hardest connective tissue. The cells (osteocytes) are within the hard matrix.

- **Hyaline** (HI-ah-lin) **cartilage** is the tough translucent material, popularly called gristle, that covers the ends of the long bones (see FIG. 4-5B). You can feel hyaline cartilage at the tip of your nose and along the front of your throat, where rings of this tissue reinforce the trachea ("windpipe"). Hyaline cartilage also reinforces the larynx ("voice box") at the top of the trachea and can be felt anteriorly at the top of the throat as the "Adam's apple."
- **Fibrocartilage** (fi-bro-KAR-tih-laj) is firm and rigid and is found between the vertebrae (segments) of the spine, at the anterior joint between the pubic bones of the hip, and in the knee joint.
- **Elastic cartilage** can spring back into shape after it is bent. An easy place to feel the properties of elastic cartilage is in the outer portion of the ear. It is also located in the larynx.

BONE

The tissue that composes bones, called **osseous** (OS-e-us) **tissue**, is much like cartilage in its cellular structure (see FIG. 4-5C). In fact, the fetal skeleton in the early stages of development is made almost entirely of cartilage. This tissue gradually becomes impregnated with salts of calcium and phosphorus that make bone characteristically solid and hard. The cells that form bone are called **osteoblasts** (OS-te-o-blasts), a name that combines the root for bone (*osteo*) with the ending *blast*. As these cells mature, they are referred to as **osteocytes** (OS-te-o-sites). Within the osseous tissue are nerves and blood vessels. A specialized type of tissue, the bone marrow, is enclosed within bones. The red bone marrow contained in certain regions produces blood cells. Chapter 6 has more information on bones.

CHECKPOINTS 

- ☐ **4-3** What is the general name for the intercellular material in connective tissue?
- ☐ **4-4** What protein makes up the most abundant fibers in connective tissue?
- ☐ **4-5** What type of cell characterizes dense connective tissue? Cartilage? Bone tissue?

Muscle Tissue

Muscle tissue is capable of producing movement by contraction of its cells, which are called *muscle fibers* because most of them are long and threadlike. If you were to pull apart a piece of well-cooked meat, you would see small groups of these muscle fibers. Muscle tissue is usually classified as follows:

- **Skeletal muscle**, which works with tendons and bones to move the body (**FIG. 4-6A**). This type of tissue is described as **voluntary muscle** because we make it contract by conscious thought. The cells in the skeletal muscle are very large and are remarkable in having multiple nuclei and a pattern of dark and light banding described as **striations**. For this reason, skeletal muscle is also called *striated muscle*. Chapter 7 has more details on skeletal muscles.
- **Cardiac muscle**, which forms the bulk of the heart wall and is known also as **myocardium** (mi-o-KAR-de-um) (**see FIG. 4-6B**). This muscle produces the regular contractions known as *heartbeats*. Cardiac muscle is described as **involuntary muscle** because it typically contracts independently of thought. Most of the time, we are not aware of its actions at all. Cardiac muscle has branching cells that form networks. The heart and cardiac muscles are discussed in Chapter 13.
- **Smooth muscle** is also an involuntary muscle (**see FIG. 4-6C**). It forms the walls of the hollow organs in the ventral body cavities, including the stomach, intestines, gallbladder, and urinary bladder. Together, these organs are known as viscera (VIS-eh-rah), so smooth muscle is sometimes referred to as *visceral muscle*. Smooth muscle is also found in the walls of many tubular structures, such as the blood vessels and the tubes that carry urine from the kidneys to the bladder. A smooth muscle is also attached to the base of each body hair. Contraction of these muscles causes the condition of the skin that we call *gooseflesh*. Smooth muscle cells are of a typical size and taper at each end. They are not striated and have only one nucleus per cell. Structures containing smooth muscle are discussed in the chapters on the various body systems.

Muscle tissue, like nervous tissue, repairs itself only with difficulty or not at all once a major injury has been sustained. When severely injured, muscle tissue is frequently replaced with connective tissue.

CHECKPOINT 

☐ **4-6** What are the three types of muscle tissue?

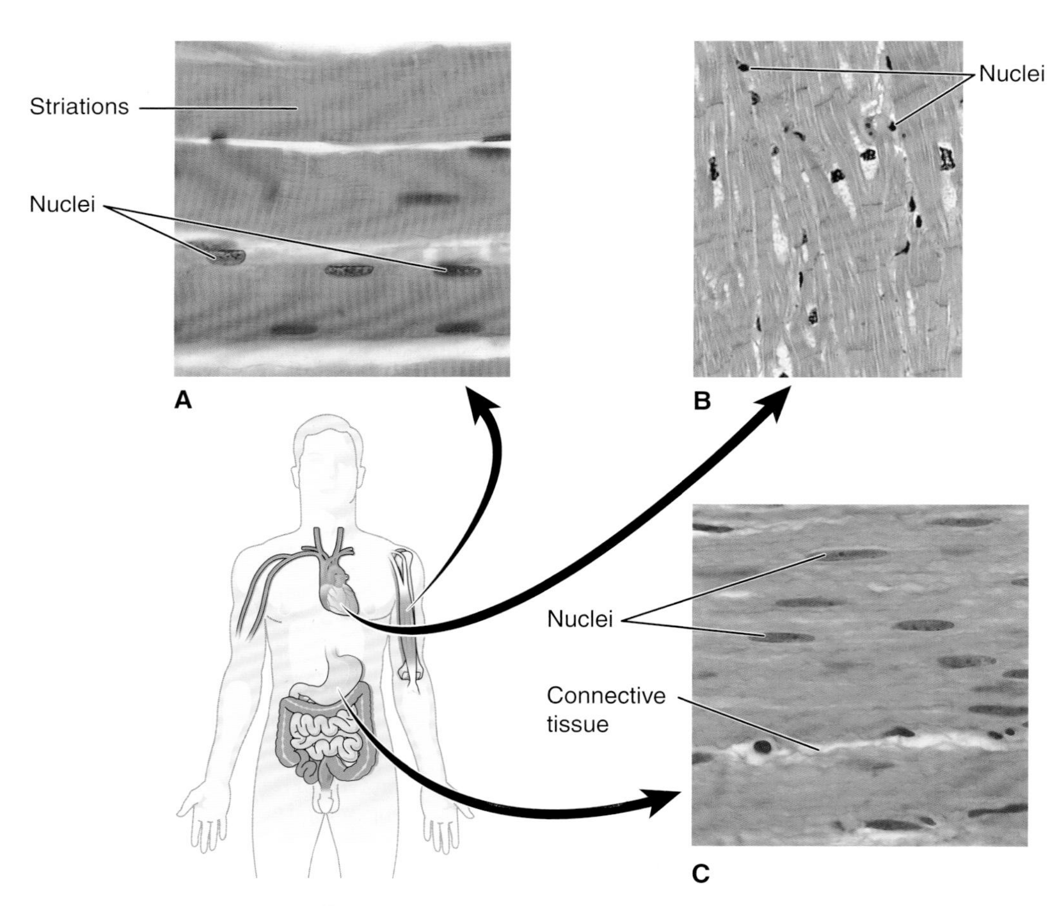

FIGURE 4-6 Muscle tissue. **KEY POINT** There are three types of muscle tissue. **A.** Skeletal muscle cells have bands (striations) and multiple nuclei. **B.** Cardiac muscle makes up the wall of the heart. **C.** Smooth muscle is found in soft body organs and in vessels.

4

Nervous Tissue

The human body is made up of countless structures, each of which contributes to the action of the whole organism. This aggregation of structures might be compared to a large corporation. For all the workers in the corporation to coordinate their efforts, there must be some central control, such as the president or CEO. In the body, this central agent is the **brain** (FIG. 4-7A). Each body structure is in direct communication with the brain by means of its own set of "wires," called **nerves** (see FIG. 4-7B). Nerves from even the most remote parts of the body come together and feed into a great trunk cable called the **spinal cord**, which in turn leads into the central switchboard of the brain. Here, messages come in and orders go out 24 hours a day. Some nerves, the cranial nerves, connect directly with the brain and do not communicate with the spinal cord. This entire control system, including the brain, is made of nervous tissue.

THE NEURON

The basic unit of nervous tissue is the **neuron** (NU-ron), or nerve cell (see FIG. 4-7C). A neuron consists of a nerve cell body plus small branches from the cell called *fibers*. These fibers carry nerve impulses to and from the cell body. Neurons may be quite long; their fibers can extend for several feet. A nerve is a bundle of such nerve cell fibers held together with connective tissue (see FIG. 4-7B).

NEUROGLIA

Nervous tissue is supported and protected by specialized cells known as **neuroglia** (nu-ROG-le-ah) or *glial* (GLI-al) *cells*, which are named from the Greek word *glia* meaning "glue." Some of these cells protect the brain from harmful substances; others get rid of foreign organisms and cellular debris; still, others form the myelin sheath around axons. They do not, however, transmit nerve impulses.

A more detailed discussion of nervous tissue and the nervous system appears in Chapters 8 and 9.

CHECKPOINTS 

- ☐ **4-7** What is the basic cell of the nervous system, and what is its function?
- ☐ **4-8** What are the nonconducting support cells of the nervous system called?

Membranes

Recall that membranes are thin sheets of tissue. Their properties vary: some are fragile, and others are tough; some are transparent, and others are opaque (i.e., you cannot see through them). Membranes may cover a surface, may be a dividing partition, may line a hollow organ or body cavity, or may anchor an organ. They may contain cells that secrete lubricants to ease the movement of organs, such as the heart and lung, and the movement of joints. Membranes are classified based on the tissues they contain. Epithelial membranes contain epithelium and supporting tissues, but connective tissue membranes consist exclusively of connective tissue.

EPITHELIAL MEMBRANES

An **epithelial membrane** is so named because its outer surface is made of epithelium. Underneath, however, there is a

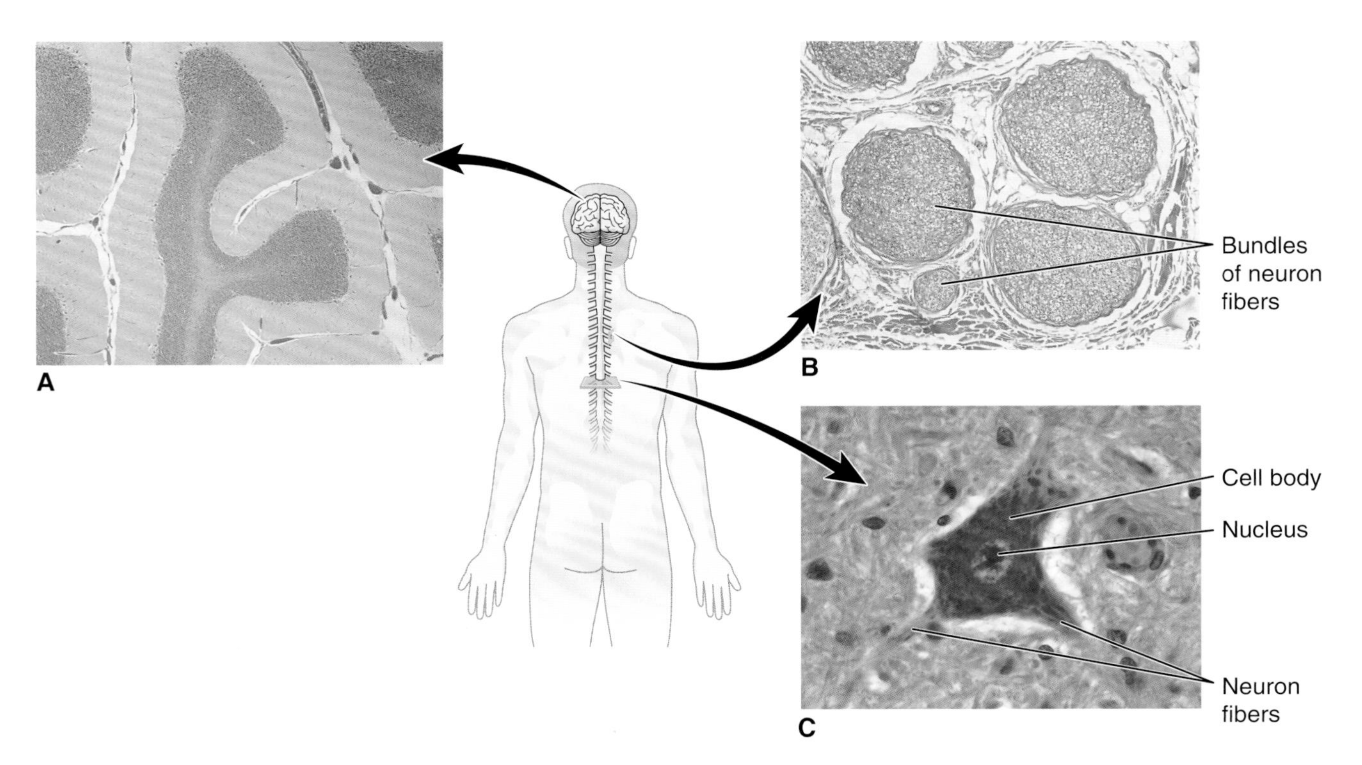

FIGURE 4-7 Nervous tissue. A. Brain tissue. **B.** Cross section of a nerve. **C.** A neuron or nerve cell.

layer of areolar and/or dense irregular connective tissue that strengthens the membrane, and in some cases, there is a thin layer of smooth muscle under that. Epithelial membranes are made of closely packed active cells that manufacture lubricants and protect the deeper tissues from invasion by microorganisms. Epithelial membranes are of several types:

- **Serous** (SE-rus) **membranes** line the walls of body cavities and are folded back onto the surface of internal organs, forming their outermost layer.
- **Mucous** (MU-kus) **membranes** line tubes and other spaces that open to the outside of the body.
- The **cutaneous** (ku-TA-ne-us) **membrane**, commonly known as the skin, has an outer layer of stratified squamous epithelium and an inner layer of dense irregular connective tissue. This membrane is complex and is discussed in detail in Chapter 5.

Serous Membranes Serous membranes line the closed ventral body cavities and do not connect with the outside of the body. They secrete a thin, watery lubricant, known as serous fluid, that allows organs to move with a minimum of friction. The thin epithelium of serous membranes is a smooth, glistening kind of tissue called *mesothelium* (meso-THE-le-um) overlying areolar tissue. The membrane itself may be referred to as the **serosa** (se-RO-sah).

There are three serous membranes:

- The **pleurae** (PLU-re), or *pleuras* (PLU-rahs), line the thoracic cavity and cover each lung.
- The **serous pericardium** (per-ih-KAR-de-um) forms part of a sac that encloses the heart, which is located in the chest between the lungs.
- The **peritoneum** (per-ih-to-NE-um) is the largest serous membrane. It lines the walls of the abdominal cavity, covers the abdominal organs, and forms supporting and protective structures within the abdomen (see FIG. 17-2 in Chapter 17).

Serous membranes are arranged so that one portion forms the lining of a closed cavity, while another part folds back to cover the surface of the organ contained in that cavity. The relationship between an organ and the serous membrane around it can be visualized by imagining your fist punching into a large, soft balloon (FIG. 4-8). Your fist is the organ and the serous membrane around it is in two layers, one against your fist and one folded back to form an outer layer. Although in two layers, each serous membrane is continuous.

The portion of the serous membrane attached to the wall of a cavity or sac is known as the **parietal** (pah-RI-eh-tal) **layer**; the word *parietal* refers to a wall. In the example above, the parietal layer is represented by the outermost layer of the balloon. Parietal pleura lines the thoracic (chest) cavity, and parietal pericardium lines the fibrous sac (the fibrous pericardium) that encloses the heart (see FIG. 4-8B).

Because internal organs are called *viscera*, the portion of the serous membrane attached to an organ is the **visceral layer**. Visceral pericardium is on the surface of the heart, and each lung surface is covered by visceral pleura. Portions of the peritoneum that cover organs in the abdomen are named according to the particular organ involved. The visceral layer in our balloon example is in direct contact with your fist.

A serous membrane's visceral and parietal layers normally are in direct contact with a minimal amount of lubricant between them. The area between the two layers forms a potential space. That is, it is *possible* for a space to exist there, although normally one does not. Only if substances accumulate between the layers, as when inflammation causes the production of excessive amounts of fluid, is there an actual space.

Mucous Membranes Mucous membranes are so named because they contain goblet cells that produce mucus. (Note that the adjective *mucous* contains an "o," whereas the noun *mucus* does not.) These membranes form extensive continuous linings in the digestive, respiratory, urinary, and reproductive systems, all of which are connected with the outside of the body. The membranes vary somewhat in both structure and function, but they all have an underlying layer of areolar tissue known as the *lamina propria*. The epithelial cells that line the nasal cavities and the respiratory passageways have cilia. The microscopic cilia move in waves that force secretions outward. In this way, foreign particles, such as bacteria, dust, and other impurities trapped in the sticky mucus, are prevented from entering the lungs and causing harm. Ciliated epithelium is also found in certain tubes of both the male and the female reproductive systems.

The mucous membranes that line the digestive tract have special functions. For example, the stomach's mucous membrane protects its deeper tissues from the action of powerful digestive juices. If for some reason a portion of this membrane is injured, these juices begin to digest a part of the stomach itself—as happens in cases of peptic ulcers. Mucous membranes located farther along in the digestive system are designed to absorb nutrients, which the bloodstream then transports to all cells.

The noun **mucosa** (mu-KO-sah) refers to the mucous membrane of an organ.

CASEPOINT 

☐ **4-2** Which type of epithelial membrane was involved in Paul's case study?

CONNECTIVE TISSUE MEMBRANES

The following list is an overview of membranes that consist of connective tissue with no epithelium. These membranes are described in greater detail in later chapters.

- **Synovial** (sin-O-ve-al) **membranes** are thin layers of areolar tissue that line the joint cavities. They secrete a lubricating fluid that reduces friction between the ends of bones, thus permitting free movement of the joints. Synovial membranes also line small cushioning sacs near the joints called **bursae** (BUR-se).

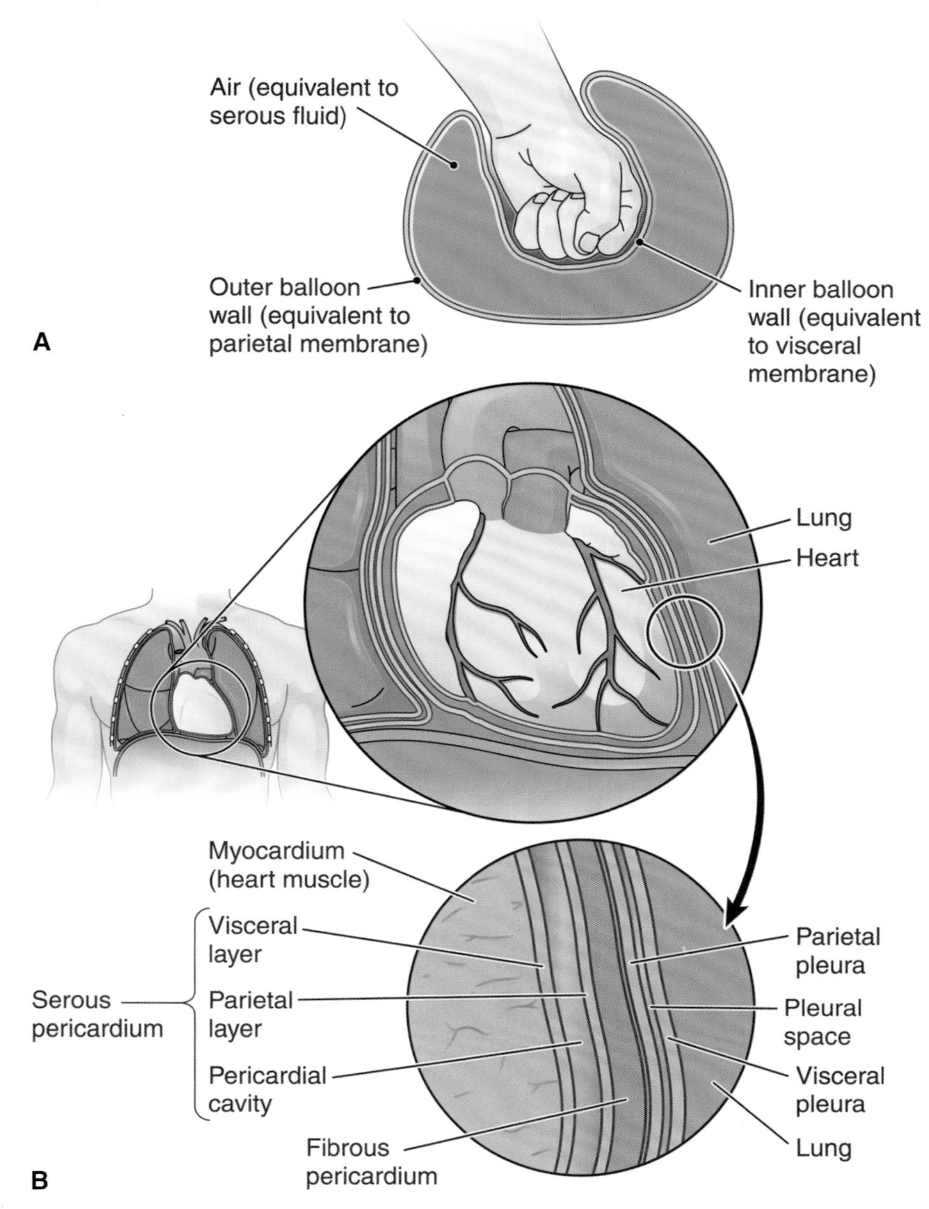

FIGURE 4-8 Organization of serous membranes. KEY POINT A serous membrane that encloses an organ has a visceral and a parietal layer. **A.** An organ fits into a serous membrane like a fist punching into a soft balloon. **B.** The outer layer of a serous membrane is the parietal layer. The inner layer is the visceral layer. The fibrous pericardium reinforces the serous pericardium around the heart. The pleura is the serous membrane around the lungs.

- The **meninges** (men-IN-jeze) are several membranous layers covering the brain and the spinal cord.

Fascia (FASH-e-ah) refers to fibrous bands or sheets that support organs and hold them in place. Fascia is found in two regions:

- **Superficial fascia** is the continuous sheet of tissue that underlies the skin. Composed of areolar and adipose tissue, this membrane insulates the body and cushions the skin. This tissue is also called *subcutaneous fascia* because it is located beneath the skin.
- **Deep fascia** covers, separates, and protects skeletal muscles, nerves, and blood vessels. It consists of dense connective tissue.

Finally, there are membranes with names that all start with the prefix *peri* because they are around organs. These tough, protective coverings are made of dense irregular connective tissue.

- The **fibrous pericardium** (per-e-KAR-de-um) forms the cavity that encloses the heart, the pericardial cavity. This fibrous sac and the serous pericardial membranes described above are often described together as the pericardium (see FIG. 4-8B).

- The **periosteum** (per-e-OS-te-um) is the membrane covering a bone.
- The **perichondrium** (per-e-KON-dre-um) is the membrane covering cartilage.

Now that our discussion of tissues is complete, try analyzing histology elsewhere in the textbook and on the internet. BOX 4-3, "The Histology Challenge," can help you get started.

ONE STEP AT A TIME

The Histology Challenge

BOX 4-3

In the case study opening this chapter, Paul's skin cancer was diagnosed by histology—the microscopic examination of tissue. Histology relies on close observation and a systematic approach, skills that will facilitate many aspects of learning anatomy and physiology. The histology of skin will be discussed in detail in Chapter 5, so we have chosen a mystery structure (below) to analyze. We've used a table to summarize our observations.

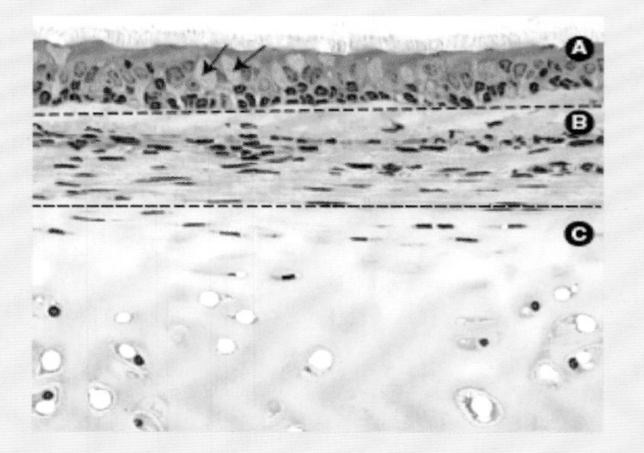

Step 1. Divide the image into different tissues. Remember that each tissue consists of similar cells. So divide your image into different regions based on appearance. This image contains at least three different tissues—we used dotted lines to separate them. Focus on architecture rather than color, as tissues may be stained with different dyes for examination.

Step 2. Write your overall impression of each tissue. Don't be afraid to say "It looks like…" Loose connective tissue looks like a mish mash of cells and fibers; dense bone tissue is organized in rings; the endocrine cells of the pancreas look like islands in a sea of different-appearing cells; kidney tissue has areas that look like filled balls.

Step 3. Look at the nuclei for shape and spacing. For example, nuclei are segmented in certain red blood cells, close together in epithelium, multiple in skeletal muscle cells, and far apart in connective tissue.

Step 4. Look at the shape and arrangement of the cells: flat, cuboidal, or columnar in epithelial cells; spindle-shaped in muscle cells; in clusters, as in the islet cells of the pancreas; or in tubules, as in much of the kidney. See if the cells are densely packed, as in epithelial tissue or far apart as are the white cells in blood or the cells in bone tissue. The plasma membranes separating cells may not be visible, but the arrangement of the nuclei can provide some clues.

Step 5. Look for surface features, such as microvilli in the cells lining the intestine, cilia in the cells lining the respiratory passageways, or flagella in sperm cells.

Step 6. Examine the appearance of the material between the cells (the matrix). Is it loosely or tightly packed? Are there visible thin or thick fibers?

Step 7. Identify the tissue. Based on your observations, classify the tissue as specifically as possible.

Step 8. Hypothesize about function. Cilia produce currents in fluid; epithelial cells line body surfaces and cavities; connective tissues serve various structural roles.

Step 9. If possible, identify the location. Where in the body do you have a tube, lined with cilia and fluid, supported by cartilage? The trachea!

	Tissue A	Tissue B	Tissue C
Overall impression	Fringed carpet	Mish mash	Cell islands in a sea of pink
Nuclei	Round, close together, at different levels	Flattened and separated	Small, round, and widely separated
Cells	Tall and thin; tightly packed; contain goblet cells (arrows)	Thin, somewhat regular spacing	Small cells in spaces
Surface features	Cilia	None	None
Matrix	None	Fibrous	Very homogeneous appearance
Identification	Pseudostratified ciliated epithelium	Loose connective tissue	Cartilage
Function	Lines and protects a body structure, produces currents in a fluid	Reinforcement or support	Support

CHECKPOINTS

- ☐ **4-9** What are the three types of epithelial membranes?
- ☐ **4-10** What is the difference between a parietal and a visceral serous membrane?
- ☐ **4-11** What is fascia, and where is it located?

See the Student Resources on thePoint® for information on careers in histotechnology—the laboratory study of tissues.

Tissues and Aging

With aging, connective tissues lose elasticity, and collagen becomes less flexible. These changes affect the skin most noticeably, but internal changes occur as well. The blood vessels, for example, have a reduced capacity to expand. Less blood supply, lower metabolism, and decline in hormone levels slow the healing process. Tendons and ligaments stretch, causing a stooped posture and joint instability. Bones may lose calcium salts, becoming brittle and prone to fracture. With age, muscles and other tissues waste from loss of cells, a process termed *atrophy* (AT-ro-fe) (**FIG. 4-9**). Changes that apply to specific organs and systems are described in later chapters.

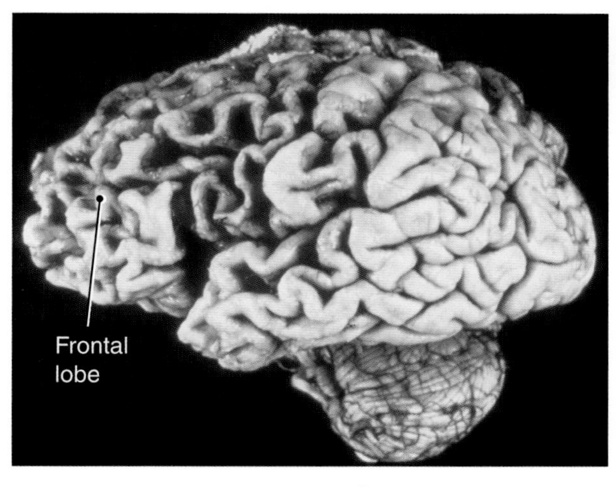

FIGURE 4-9 **Atrophy of the brain.** **KEY POINT** Brain tissue has thinned, and large spaces appear between sections of tissue, especially in the frontal lobe.

A & P in Action Revisited: Sun-damaged Skin

Paul was edgy during the three days before he telephoned the dermatologist's office. *What if I have skin cancer? Even if it's treatable, I may have a scar—and right smack in the center of my face!* Finally, he made the call and learned from Dr. Nielsen that he did indeed have a small basal cell carcinoma.

"I recommend that you consult Dr. Morris, a local surgeon who specializes in a procedure that guarantees the removal of all abnormal cells," the dermatologist advised. "Mohs surgery is done in stages, with the surgeon first removing just the visible lesion and then checking microscopically to be sure that the margins of the excised tissue are free of cancerous cells. If not, additional tissue is removed by degrees until the margins are clean."

Fortunately, Dr. Morris had to repeat the procedure only once after the first pathology examination to be sure of success. Paul left reassured after several hours. Dr. Morris was confident that Paul was safe from the cancer and that scarring would be minimal. "Let's make an appointment for a follow-up visit on when I'll remove the stitches and you can see for yourself," he said.

That evening, Paul described his day to his wife and told her about his additional instructions. "I need to see my regular dermatologist every six months now, as I may be prone to these types of carcinomas." Paul can't undo earlier damage, but he can prevent further insult to his skin by wearing sunscreen outdoors and reapplying it often. The doctor also advised him to cover up in the sun and avoid times of high sun intensity. "Come to think of it," Paul said to his wife, "that's good advice for you too!"

In this case, we saw the cancer-causing effect of sun damage on skin tissue. Paul was fortunate in that he saw his physician promptly and avoided the spread of the abnormal cells into other regions of the skin or into other tissues. Cancer cells disrupt homeostasis by interfering with the structure and function of the host tissue. The science of histology was important in the diagnosis and treatment of his disorder.

CHAPTER 4

Chapter Wrap-Up

OVERVIEW

A detailed chapter outline with space for note-taking is available on thePoint®. The figure below illustrates the main topics covered in this chapter.

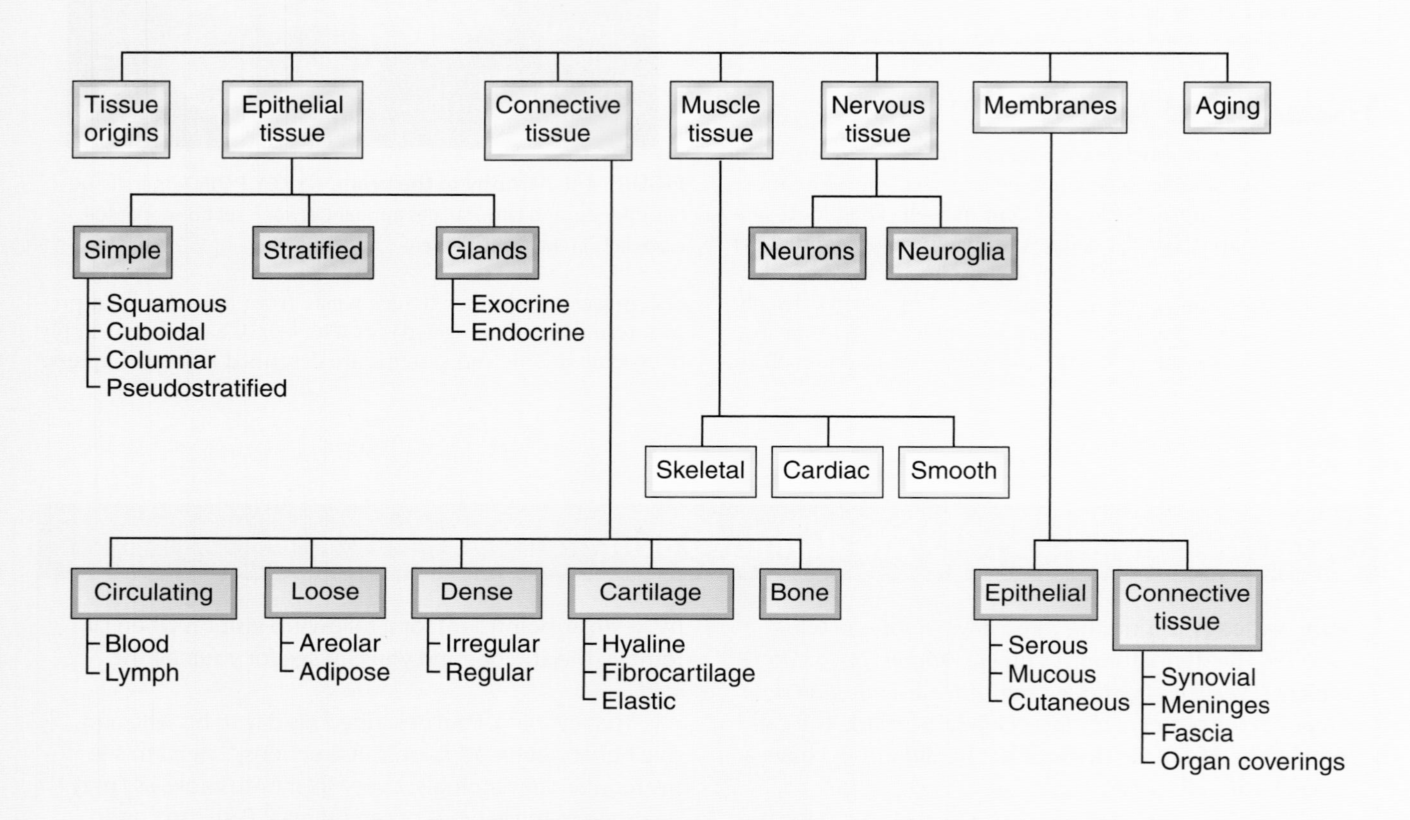

KEY TERMS

The terms listed below are emphasized in this chapter. Knowing them will help you organize and prioritize your learning. These and other boldface terms are defined in the Glossary with phonetic pronunciations.

adipose
areolar
cartilage
chondrocyte
collagen
endocrine
epithelium
exocrine
fascia
fibroblast
histology
matrix
membrane
mucosa
mucus
neuroglia
neuron
osteocyte
parietal
serosa
stem cell
visceral

WORD ANATOMY

Medical terms are built from standardized word parts (prefixes, roots, and suffixes). Learning the meanings of these parts can help you remember words and interpret unfamiliar terms.

WORD PART	MEANING	EXAMPLE
hist/o	tissue	*Histology* is the study of tissues.
EPITHELIAL TISSUE		
epi-	on, upon	*Epithelial* tissue covers body surfaces.
pseud/o-	false	*Pseudostratified* epithelium appears to be in multiple layers but is not.
CONNECTIVE TISSUE		
blast/o	immature cell, early stage of cell	A *fibroblast* is a cell that produces fibers.
chondr/o	cartilage	A *chondrocyte* is a cartilage cell.
oss, osse/o	bone, bone tissue	*Osseous* tissue is bone tissue.
oste/o	bone, bone tissue	An *osteocyte* is a mature bone cell.
MUSCLE TISSUE		
cardi/o	heart	The *myocardium* is the heart muscle.
my/o	muscle	See preceding example.
NERVOUS TISSUE		
neur/o	nerve, nerve system	A *neuron* is a nerve cell.
MEMBRANES		
peri-	around	The *peritoneum* wraps around the abdominal organs.
pleur/o	side, rib	The *pleurae* are membranes that line the chest cavity.

QUESTIONS FOR STUDY AND REVIEW

BUILDING UNDERSTANDING

Fill in the Blanks

1. A group of similar cells arranged in a characteristic pattern is called a(n) _____.
2. Glands that secrete their products directly into the blood are called _____ glands.
3. Tissue that supports and forms the framework of the body is called _____ tissue.
4. Skeletal muscle is also described as _____ muscle.
5. Nervous tissue is supported by specialized cells known as _____.

Matching

Match each numbered item with the most closely related lettered item.

___ 6. Membrane around the heart
___ 7. Membrane around each lung
___ 8. Membrane around the bone
___ 9. Membrane around cartilage
___ 10. Membrane around abdominal organs

a. perichondrium
b. pericardium
c. peritoneum
d. periosteum
e. pleura

Multiple Choice

___ 11. You look under the microscope and see tissue composed of a single layer of long and narrow cells. What is it?
a. simple cuboidal epithelium
b. simple columnar epithelium
c. stratified cuboidal epithelium
d. stratified columnar epithelium

___ 12. What tissue forms tendons and ligaments?
a. areolar connective tissue
b. loose connective tissue
c. regular, dense connective tissue
d. cartilage

___ 13. Which tissue is composed of long striated cells with multiple nuclei?
a. smooth muscle
b. cardiac muscle
c. skeletal muscle
d. nervous tissue

___ 14. A bundle of nerve cell fibers held together with connective tissue is a(n) _____.
a. dendrite
b. axon
c. nerve
d. neuroglia

___ 15. Which membrane is formed from connective tissue?
a. cutaneous
b. mucous
c. serous
d. fascia

UNDERSTANDING CONCEPTS

16. Define stem cells, and explain how stem cells function in healing.
17. Explain how epithelium is classified, and discuss at least three functions of this tissue type.
18. Compare the structure and function of exocrine and endocrine glands, and give two examples of each type.
19. Put the following terms in a logical order: cartilage, dense connective tissue, circulating connective tissue, bone, loose connective tissue.
20. Compare and contrast the three different types of muscle tissue.
21. Compare the three types of epithelial membranes in terms of their structure and their ability to form a barrier.
22. Referring to the Dissection Atlas in Appendix 3:
    a. List the figure(s) in which nervous tissue is shown.
    b. List the figure(s) in which cardiac muscle is shown.

CONCEPTUAL THINKING

23. Prolonged exposure to cigarette smoke causes damage to ciliated epithelium that lines portions of the respiratory tract. Discuss the implications of this damage.
24. The hereditary disease osteogenesis imperfecta is characterized by abnormal collagen fiber synthesis. Which tissue type would be most affected by this disorder? Based on your knowledge of this tissue's functions, list some possible symptoms of this disease.
25. In Paul's case, sun damage caused skin cancer. Name the type of membrane that was damaged. Why is this tissue more likely to become cancerous than muscle tissue, for example?

For more questions, see the Learning Activities on thePoint®.

CHAPTER 5

The Integumentary System

Learning Objectives

After careful study of this chapter, you should be able to:

1. Name and describe the layers of the skin. *p. 84*
2. Describe the subcutaneous layer. *p. 86*
3. Give the locations and functions of the accessory structures of the integumentary system. *p. 86*
4. List the main functions of the integumentary system. *p. 88*
5. Discuss the factors that contribute to skin color. *p. 90*
6. Cite the steps in repair of skin wounds and the factors that affect healing. *p. 91*
7. Describe how the skin changes with age. *p. 91*
8. Using information in the case study and the text, explain how a burn disrupts skin function. *pp. 83, 92*
9. Show how word parts are used to build words related to the integumentary system (see Word Anatomy at the end of the chapter). *p. 94*

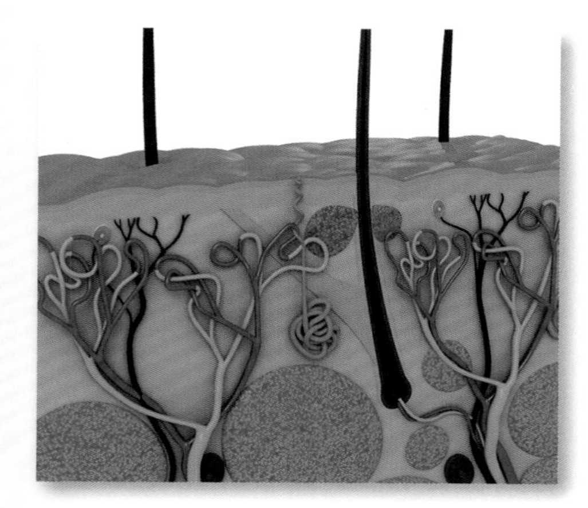

A & P in Action
Hazel's Deep Burn

"Grandma, I'm here!" Tiffany announced as she walked into her grandmother's farm kitchen. It was time for their fall ritual of canning tomatoes and making dill pickles. Tiffany had just started her nursing degree and was glad to get away from studying and classes to spend time with 75-year-old Hazel. Hazel's tomato plants had produced a bountiful yield, so it took all morning to harvest the fruit.

While Tiffany prepared the tomatoes, Hazel sterilized the jars by immersing them into the hot water bath in the canner. As she used the special tongs to lift out the jars, one slipped and splashed the steaming water directly onto her forearm and hand. "Blast it!" Hazel exclaimed and hurried to the sink to cool her arm under running water. Tiffany raced over and saw that Hazel's skin was reddened over her hand and forearm. Blisters filled with clear fluid were beginning to form in a small area on her forearm. Tiffany pressed lightly on a reddened area and noticed that it blanched. When she removed her finger from her grandmother's skin, it promptly turned an angry red again.

"Grandma, those blisters mean that your burn is pretty deep. We better get you to the urgent care clinic," Tiffany said.

"Just let me get some painkillers," Hazel replied. "It hurts already, so I think you better drive."

At the clinic, Dr. Stanford took a history of the event and examined the damaged tissue. "You were smart to come in now," Dr. Stanford told Hazel and Tiffany as he applied an antimicrobial cream to the burned areas. "The most serious is this blistered area on your forearm. It involves the top layer of skin, called the epidermis, and portions of the next layer, the dermis. The fact that the area hurts is actually a good sign, because it means that the nerve supply to the skin wasn't destroyed. The other burns aren't as serious, but there is still damage to the epidermis. Infections are the main complication in this situation. You will have to be meticulous in keeping the area clean, so the burns can heal without delay. I also want you to apply this antimicrobial cream daily and keep the area covered with a special nonstick sterile dressing. We can give you a supply to take home for your dressing changes. The red areas should heal in about six days, but the blistered area may take up to three weeks to heal. You will be in pain for a while, so I'll give you a prescription for painkillers."

The body's first line of defense, intact skin, was compromised when Hazel suffered these burns. In this chapter, we will learn the functions of the skin and the importance of this protective barrier.

As you study this chapter, CasePoints will give you opportunities to apply your learning to this case.

Visit thePoint® to access the following resources. For guidance in using these resources most effectively, see pp. xv–xvii.

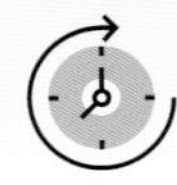

Preparing to Learn

- Tips for Effective Studying
- Pre-Quiz

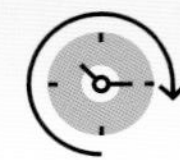

While You Are Learning

- Web Figure: Repair of the Integument
- Web Chart: Skin Structure
- Web Chart: Accessory Skin Structures
- Animation: Wound Healing
- Chapter Notes Outline
- Audio Pronunciation Glossary

When You Are Reviewing

- Answers to Questions for Study and Review
- Health Professions: Registered Nurse
- Interactive Learning Activities

A LOOK BACK

The skin is introduced in Chapter 4 as one of the epithelial membranes, the cutaneous (ku-TA-ne-us) membrane, overlying a connective tissue membrane, the superficial fascia. In this chapter, we describe the skin in much greater detail as it forms the major portion of the integumentary system. 1 *As with all body systems, studying the structure of skin helps us understand its function. The key idea of barriers* 4 *is particularly relevant to this chapter, because the skin provides the major barrier between the external and internal environments.*

Introduction

Although the skin may be viewed simply as a membrane enveloping the body, it is far more complex than are the other epithelial membranes previously described. The skin is associated with accessory structures, also known as appendages, which include glands, hair, and nails. Together with blood vessels, nerves, and sensory organs, the skin and its associated structures form the **integumentary** (in-teg-u-MEN-tar-e) **system**. This name is from the word *integument* (in-TEG-u-ment), which means "covering," but the skin has many functions and reflects an individual's health and emotional state.

See the Student Resources on thePoint® for a summary chart of skin structure.

Structure of the Skin

Recall from Chapter 4 that the skin, or cutaneous membrane, is an epithelial membrane. Like all epithelial membranes, it contains an outer layer of epithelium and an inner layer of connective tissue (**FIG. 5-1**):

- The **epidermis** (ep-ih-DER-mis), the outermost portion, which itself is subdivided into thin layers called **strata** (STRA-tah) (sing. stratum). The epidermis is composed entirely of epithelial cells and contains no blood vessels.

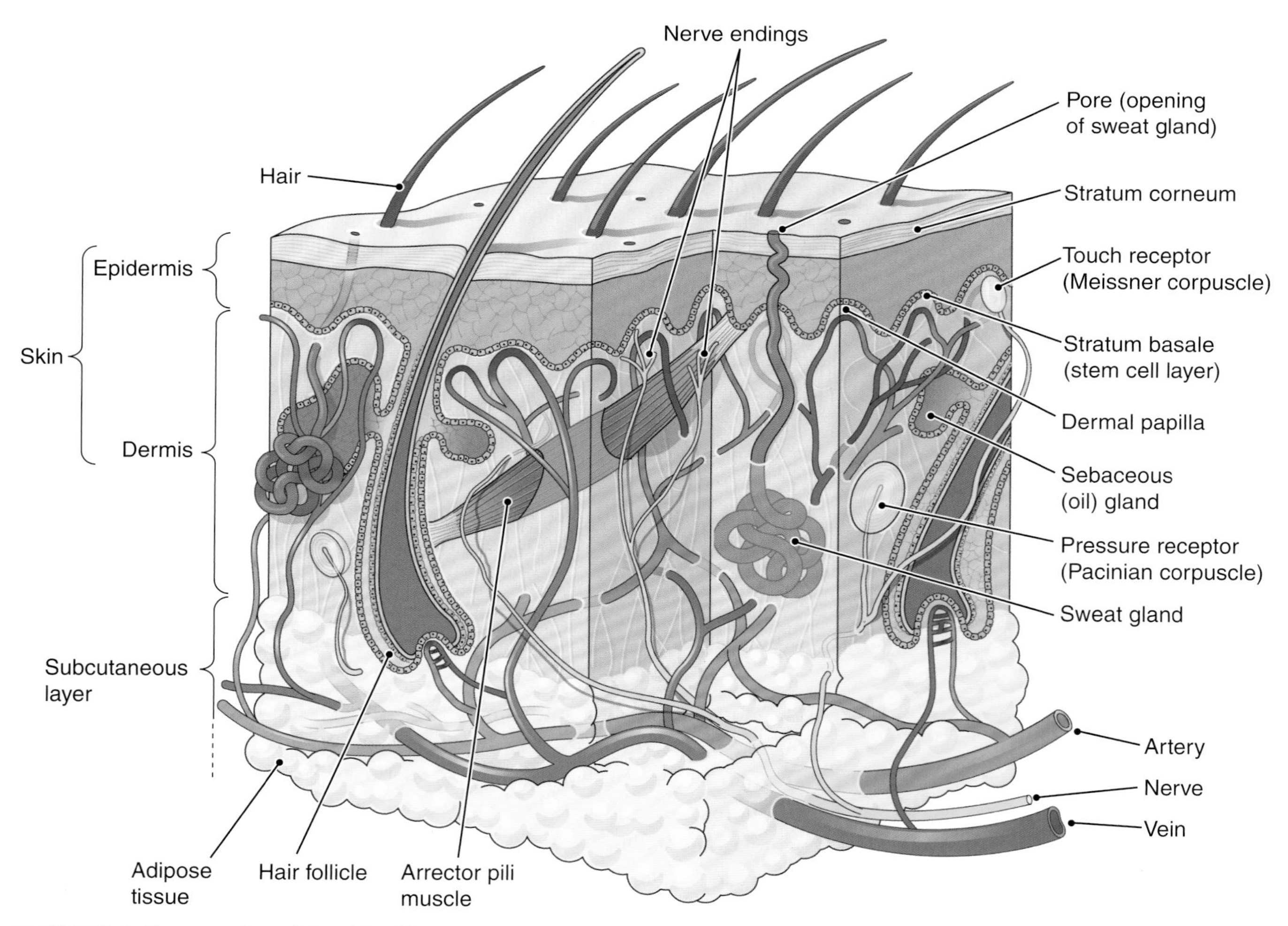

FIGURE 5-1 Cross section of the skin. **KEY POINT** The dermis and epidermis make up the skin. The skin and its associated structures make up the integumentary system. **ZOOMING IN** How is the epidermis supplied with oxygen and nutrients? What tissue is located beneath the skin?

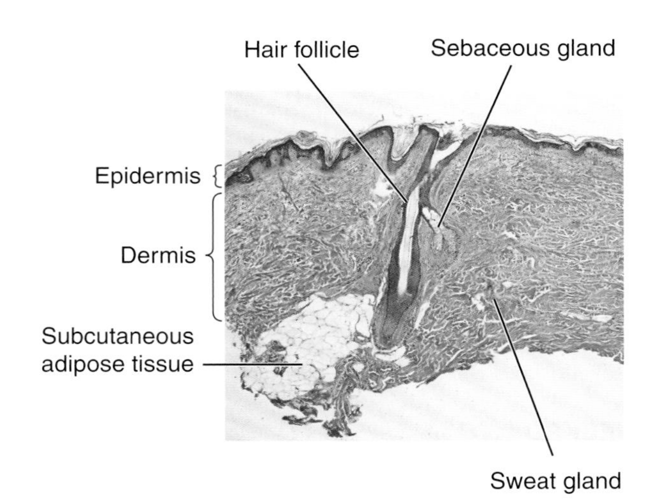

FIGURE 5-2 Microscopic view of thin skin. Tissue layers and some accessory structures are labeled.

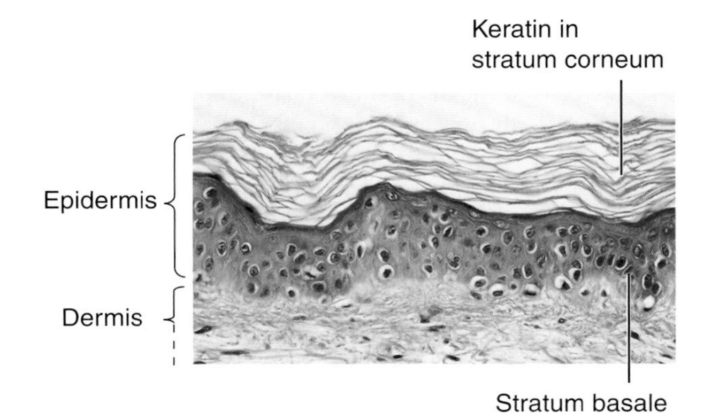

FIGURE 5-3 Upper portion of the skin. KEY POINT Layers of keratin in the stratum corneum are visible at the surface. Below are layers of stratified squamous epithelium making up the remainder of the epidermis.

- The **dermis**, which is the connective tissue portion of the cutaneous membrane, contains many blood vessels, nerve endings, and glands.

FIGURE 5-2 is a photograph of the skin as seen through a microscope showing the layers and some accessory structures.

EPIDERMIS

The epidermis is the skin's surface portion, the outermost cells of which are constantly lost through wear and tear. Because there are no blood vessels in the epidermis, the cells must be nourished by capillaries in the underlying dermis. New epidermal cells are produced from stem cells in the deepest layer, which is closest to the dermis. The cells in this layer, the **stratum basale** (bas-A-le), or *stratum germinativum* (jer-min-a-TI-vum), are constantly dividing and producing new cells, which are then pushed upward toward the skin surface. As the epidermal cells die from the gradual loss of nourishment, they undergo changes. Mainly, their cytoplasm is replaced by large amounts of a protein called **keratin** (KER-ah-tin), which thickens and protects the skin (FIG. 5-3).

By the time epidermal cells approach the surface, they have become flat and keratinized, or cornified, forming the uppermost layer of the epidermis, the **stratum corneum** (KOR-ne-um). The stratum corneum is a protective layer and is more prominent in thick skin than in thin skin. Cells at the surface are constantly being lost and replaced from below, a process that gets rid of accumulated environmental toxins and microbes. Although this process of **exfoliation** (eks-fo-le-A-shun) occurs naturally at all times, many cosmetics companies sell products to promote exfoliation, promising to "enliven" and "refresh" the skin.

Between the stratum basale and the stratum corneum are additional layers of stratified epithelium that vary in number and quantity depending on the skin's thickness. Some places, such as the soles of the feet and the palms of the hands, are covered with very thick, hairless skin, such as illustrated in FIGURE 5-3. Other regions, such as the eyelids, are covered with very thin and delicate layers. BOX 5-1 describes the structural and functional differences between

A CLOSER LOOK
Thick and Thin Skin: Getting a Grip on Their Differences

BOX 5-1

The skin is the largest organ in the body, weighing about 4 kg. Though it appears uniform in structure and function, its thickness in fact varies. Many of the functional differences between skin regions reflect the thickness of the epidermis and not the skin's overall thickness. Based on epidermal thickness, skin can be categorized as **thick** (about 1 mm deep) or **thin** (about 0.1 mm deep).

Areas of the body exposed to significant friction (the palms, fingertips, and bottoms of the feet and toes) are covered with thick skin. It is composed of a thick stratum corneum and an extra layer not found in thin skin, the stratum lucidum, both of which make thick skin resistant to abrasion. Thick skin is also characterized by epidermal ridges (e.g., fingerprints) and numerous sweat glands but lacks hair and sebaceous (oil) glands. These adaptations make the thick skin covering the hands and feet effective for grasping or gripping. The dermis of thick skin also contains many sensory receptors, giving the hands and feet a superior sense of touch.

Thin skin covers body areas not exposed to much friction. It has a very thin stratum corneum and lacks a distinct stratum lucidum. Thin skin lacks epidermal ridges and has fewer sensory receptors than does thick skin. It also contains numerous sebaceous glands, making it supple and free of cracks that might let infectious organisms enter.

thick and thin skin. Like the epidermis, the dermis varies in thickness in different areas.

Some cells in the deepest layer of the epidermis produce **melanin** (MEL-ah-nin), a dark pigment that colors the skin and protects it from sunlight's harmful rays. The cells that produce this pigment are **melanocytes** (MEL-ah-no-sites), which are scattered throughout the skin surface. Irregular patches of melanin form freckles.

DERMIS

The **dermis** has a framework of dense irregular connective tissue and is well supplied with blood vessels and nerves. It is called the "true skin" because it carries out the skin's vital functions. Because of dermal elasticity, the skin can stretch, even dramatically as in pregnancy, with little damage. Most of the skin's accessory structures, including the sweat glands, the oil glands, and the hair, are located in the dermis and may extend into the **subcutaneous layer** under the skin.

Portions of the dermis extend upward into the epidermis, allowing blood vessels to get closer to the superficial cells (see FIG. 5-1). These extensions, or dermal papillae, can be seen on the surface of thick skin, such as at the tips of the fingers and toes. Here, they form a distinct pattern of ridges that help to prevent slipping, as when grasping an object. The unchanging patterns of the ridges are determined by heredity. Because they are unique to each person, fingerprints and footprints can be used for identification.

CASEPOINTS 

- ☐ **5-1** Name the skin layers involved in Hazel's burns in the opening case study.
- ☐ **5-2** Which skin layer will produce new cells to replace Hazel's damaged epidermis?

10 ADAPTATION

The skin's responses to various stimuli illustrate the principle of adaptation. Organisms *adapt* to the environment; that is, their structure, function, or behavior changes in response to changes in their environment. Adaptation often results from injury and helps protect against future damage. For instance, excessive sunlight damages skin but also stimulates melanocyte activity. The increased melanin helps protect skin from further sunlight-induced damage. Similarly, skin thickens into calluses in areas subject to physical damage. Adaptations persist only as long as the stimulus is present. For example, the suntan resulting from a winter tropical vacation only lasts for a few weeks, and the calluses resulting from a summer landscaping job disappear after a few months back at school.

SUBCUTANEOUS LAYER

The skin rests on a connective tissue membrane, the **subcutaneous** (sub-ku-TA-ne-us) **layer**, sometimes referred to as the hypodermis or the superficial fascia (see FIG. 5-1). This layer connects the skin to the deep fascia covering the underlying muscles. It consists of areolar connective tissue and variable amounts of adipose (fat) tissue. The fat serves as insulation and as a reserve energy supply. Continuous bundles of elastic fibers connect the subcutaneous tissue with the dermis, so there is no clear boundary between the two.

The blood vessels that supply the skin with nutrients and oxygen and help to regulate body temperature run through the subcutaneous layer and send branches into the dermis. This tissue is also rich in nerves and nerve endings, including those that supply nerve impulses to and from the dermis and epidermis. The thickness of the subcutaneous layer varies in different parts of the body; it is thinnest on the eyelids and thickest on the abdomen.

CHECKPOINTS 

- ☐ **5-1** What is the name of the system that comprises the skin and all its associated structures?
- ☐ **5-2** Moving from the superficial to the deeper layer, what are the names of the two layers of the skin?
- ☐ **5-3** What is the composition of the subcutaneous layer?

Accessory Structures of the Skin

The integumentary system includes some structures associated with the skin—glands, hair, and nails—that protect the skin and serve other functions.

See the Student Resources on thePoint® for a chart summarizing the skin's accessory structures.

SEBACEOUS (OIL) GLANDS

The **sebaceous** (se-BA-shus) **glands** are saclike in structure, and their oily secretion, **sebum** (SE-bum), lubricates the skin and hair and prevents drying. The ducts of the sebaceous glands open into the hair follicles (FIG. 5-4A).

Babies are born with a covering produced by these glands that resembles cream cheese; this secretion is called the **vernix caseosa** (VER-niks ka-se-O-sah), which literally means "cheesy varnish." Modified sebaceous glands, **meibomian** (mi-BO-me-an) **glands**, are associated with the eyelashes and produce a secretion that lubricates the eyes.

SWEAT GLANDS

The **sweat glands**, also known as *sudoriferous* (su-do-RIF-er-us) *glands*, are coiled, tubelike structures located in the dermis and the subcutaneous tissue (see FIG. 5-4B). They release sweat, or perspiration, onto the body surface. They are thus all classified as exocrine glands.

The smaller-sized **eccrine** (EK-rin) sweat glands are widely distributed throughout the skin and function to cool the body. As discussed in Chapter 18, the evaporation of sweat from the body surface carries heat away with it. Sweat also contains antimicrobial proteins that help to prevent skin infections and moisturizing factors that help keep skin hydrated. Because sweat contains small amounts of dissolved salts and other wastes in addition to water, these glands also

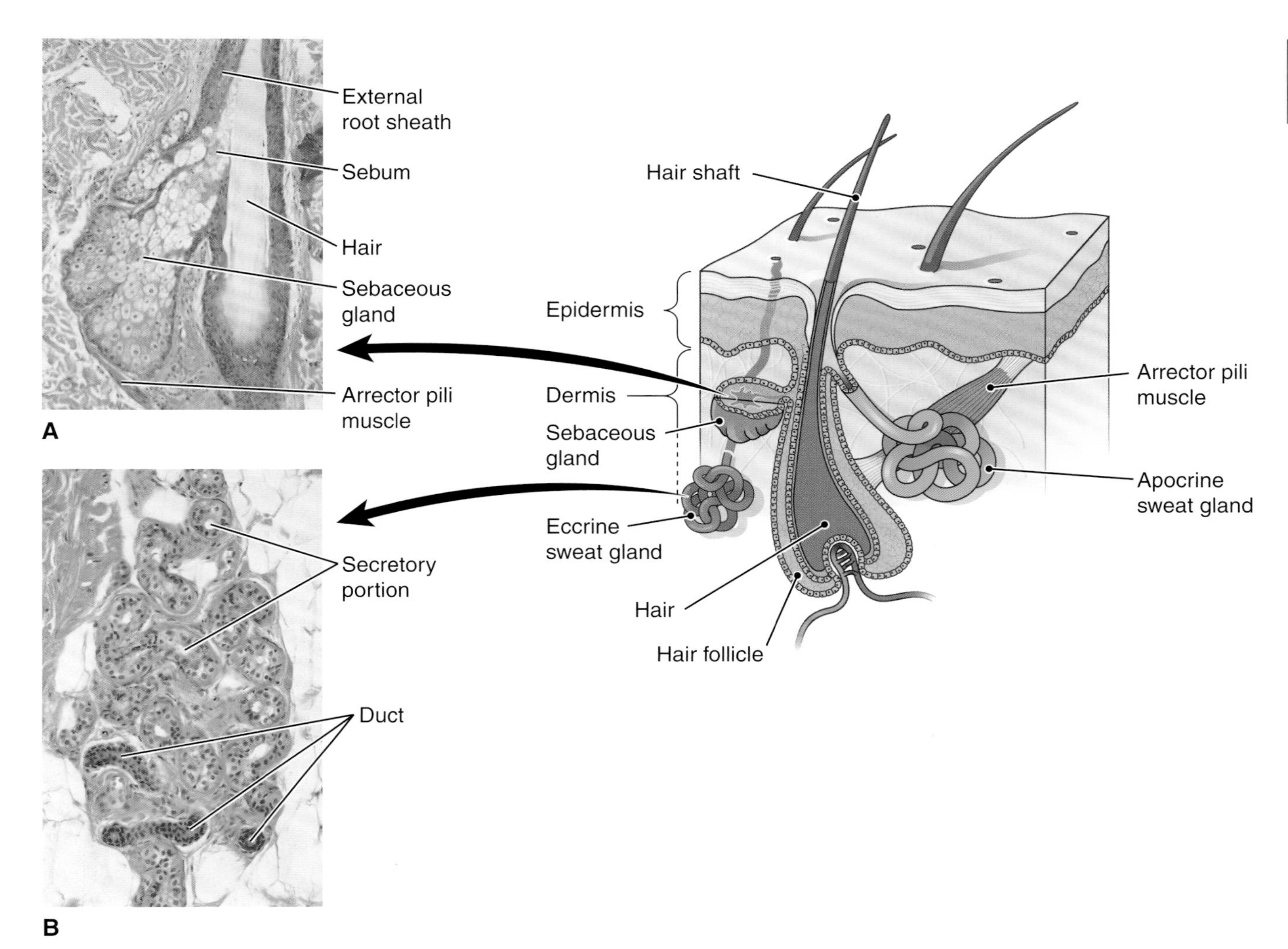

FIGURE 5-4 Portion of the skin showing associated glands and hair. A. A sebaceous (oil) gland and its associated hair follicle. **B.** An eccrine (temperature-regulating) sweat gland. ZOOMING IN How do the sebaceous glands and apocrine sweat glands secrete to the outside? What kind of epithelium makes up the sweat glands?

serve a minor excretory function. Each eccrine gland has a secretory portion and an excretory tube that extends directly to the surface and opens at a pore (see FIG. 5-1).

The less numerous **apocrine** (AP-o-krin) sweat glands are located mainly in the armpits (axillae) and groin area.

Concept Mastery Alert

Unlike eccrine glands, apocrine glands are always associated with hair follicles.

These glands become active at puberty and release their secretions through the hair follicles in response to emotional stress and sexual stimulation. The apocrine glands release some cellular material in their secretions. Body odor develops from the action of bacteria in breaking down these organic cellular materials.

Several other types of glands associated with the skin are also classified as apocrine glands. These are the **ceruminous** (seh-RU-min-us) **glands** in the ear canal that produce ear wax, or **cerumen**; the **ciliary** (SIL-e-er-e) **glands** at the edges of the eyelids; and the **mammary glands** in the breasts.

HAIR

Almost all of the body is covered with hair, which in most areas is soft and fine. Hairless regions are the palms of the hands, soles of the feet, lips, nipples, and parts of the external genitalia. Hair is composed mainly of keratin and is not living. Each hair develops, however, from stem cells located in a bulb at the base of the **hair follicle**, a sheath of epithelial and connective tissue that encloses the hair (see FIG. 5-4). Melanocytes in this growth region add pigment to the developing hair. Different shades of melanin produce the various hair colors we see in the population. The part of the hair that projects above the skin is the **shaft**; the portion below the skin is the hair's **root**.

Concept Mastery Alert

Hair loss involves only the shaft and the root (the small bulblike portion of the hair). The hair follicle is firmly anchored within the skin, ready to make a new hair.

Attached to most hair follicles is a thin band of involuntary muscle (see FIG. 5-4). When a person is frightened or cold, this muscle contracts, raising the hair and forming "goose bumps" or "chicken skin." The name of this muscle is **arrector pili** (ah-REK-tor PI-li), which literally means "hair raiser." This response is not important in humans but is a warning sign in animals and helps animals with furry coats to conserve heat. As the arrector pili contracts, it presses on the sebaceous gland associated with the hair follicle, causing the release of sebum to lubricate the skin.

NAILS

Nails protect the fingers and toes and also help in grasping small objects with the hands. They are made of hardened keratin formed by the epidermis (FIG. 5-5).

Concept Mastery Alert

Remember that collagen and keratin are very different proteins. Keratin is an intracellular protein in epithelial cells, but collagen is an extracellular protein in connective tissue.

New cells develop continuously in a growth region (nail matrix) located under the nail's proximal end, a portion called the **nail root**. The remainder of the **nail plate** rests on a **nail bed** of epithelial tissue. The color of the dermis below the nail bed can be seen through the clear nail. The pale **lunula** (LU-nu-lah), literally "little moon," at the nail's proximal end appears lighter because it lies over the nail's thicker growing region. The **cuticle**, an extension of the stratum corneum, seals the space between the nail plate and the skin above the root.

Nails of both the toes and the fingers are affected by general health. Changes in nails, including abnormal color, thickness, shape, or texture (e.g., grooves or splitting), occur in chronic diseases such as heart disease, peripheral vascular disease, malnutrition, and anemia.

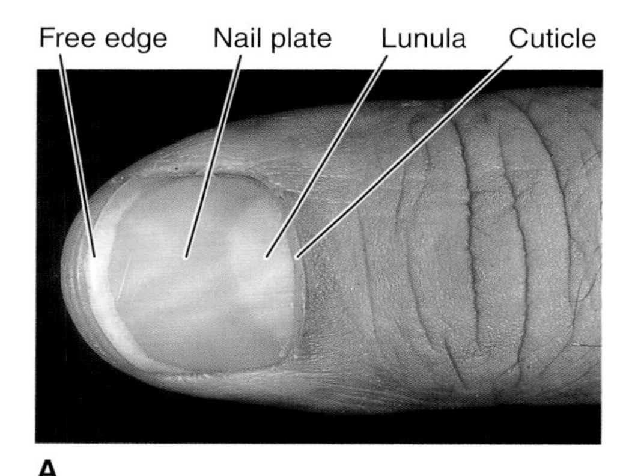

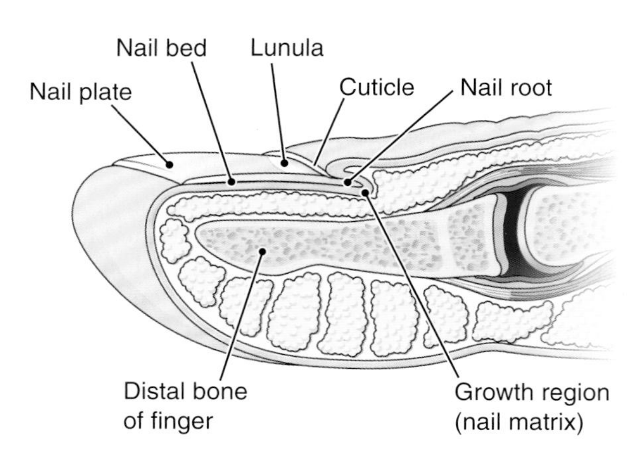

FIGURE 5-5 Nail structure. **KEY POINT** Nails protect the fingertips and toes. They form from epidermal cells at the nail root. **A.** Photograph of a nail, superior view. **B.** Midsagittal section of a fingertip.

See the Student Resources on thePoint® to view some of these conditions affecting nails.

CHECKPOINTS

- ☐ **5-4** What is the name of the skin glands that produce an oily secretion?
- ☐ **5-5** What is the scientific name for the sweat glands?
- ☐ **5-6** What is the name of the sheath in which a hair develops?
- ☐ **5-7** Where are the active cells that produce a nail located?

Functions of the Integumentary System

4 Recall from Chapter 1 that skin provides the major barrier separating our internal cells and fluids from the external environment. As such, skin plays a critical role in maintaining homeostasis, preventing the loss of heat and fluids, and blocking the entry of harmful pathogens. Among the main functions of the integumentary system are the following:

- Protection against infection
- Protection against dehydration (drying)
- Regulation of body temperature
- Collection of sensory information

PROTECTION AGAINST INFECTION

Intact skin forms a primary barrier against invasion of pathogens. The cells of the stratum corneum form a tight interlocking pattern that resists penetration. The surface cells are constantly shed, thus mechanically removing pathogens. The antimicrobial properties of sebum and sweat also help protect against infection. Rupture of this barrier, as in cases of wounds or burns, invites infection of deeper tissues. The skin also protects against bacterial toxins (poisons) and some harmful chemicals in the environment.

PROTECTION AGAINST DEHYDRATION

Both keratin in the epidermis and the oily sebum released to the skin's surface from the sebaceous glands help to waterproof the skin and keep it moist and supple, even in dry environments. These substances also prevent excessive water loss by evaporation. The skin itself forms a boundary that encloses body fluids, limiting water loss. When the skin is burned, fluid losses are significant, and burn patients commonly complain of intense thirst.

REGULATION OF BODY TEMPERATURE

Both the loss of excess heat and protection from cold are important functions of the integumentary system. Indeed, most of the blood flow to the skin is concerned with temperature regulation. In cold conditions, vessels in the skin constrict (become narrower) to reduce blood flow to the surface and diminish heat loss. The skin may become visibly pale under these conditions.

To cool the body, the skin forms a large surface area for radiating body heat to the surrounding air. When the blood vessels dilate (widen), more blood is brought to the surface so that heat can dissipate. The other mechanism for cooling the body involves the sweat glands, as noted previously. The evaporation of perspiration draws heat from the skin. Sebum also participates in this function, because it causes sweat to spread thinly over the skin instead of forming droplets that will fall off without cooling the body. A person feels uncomfortable on a hot and humid day because water does not evaporate as readily from the skin into the surrounding air. A dehumidifier makes one more comfortable even when the temperature remains high.

As is the case with so many body functions, temperature regulation is complex and involves other areas, including certain centers in the brain. See Chapter 18 for more information.

COLLECTION OF SENSORY INFORMATION

Because of its many nerve endings and other special receptors, the integumentary system may be regarded as one of the body's chief sensory organs. Free nerve endings detect pain and moderate changes in temperature. Other types of sensory receptors in the skin respond to light touch and deep pressure. **FIGURE 5-1** shows some free nerve endings, a touch receptor (Meissner corpuscle), and a deep pressure receptor (Pacinian corpuscle) in a section of skin.

Many of the reflexes that make it possible for humans to adjust themselves to the environment begin as sensory impulses from the skin. For example, touching a hot stove activates skin receptors, which initiate a reflex that ends with withdrawing your finger. As elsewhere in the body, the skin receptors work with the brain and the spinal cord to accomplish these important functions.

CASEPOINTS 

- ☐ **5-3** Why did Hazel and her doctor use antibacterial cream in treating her burns?
- ☐ **5-4** Why was pain a positive sign for in Hazel's case?

OTHER ACTIVITIES OF THE INTEGUMENTARY SYSTEM

Substances can be absorbed through the skin in limited amounts. Some drugs—for example, estrogens, other steroids, anesthetics, and medications to control motion sickness—can be absorbed from patches placed on the skin **(see BOX 5-2)**. Most medicated ointments used on the skin, however, are for the treatment of local conditions only. Even medication injected into the subcutaneous tissue is absorbed very slowly.

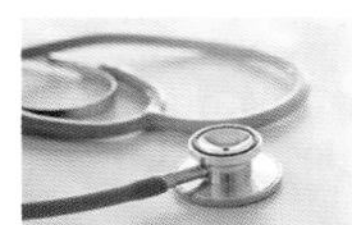

CLINICAL PERSPECTIVES — BOX 5-2
Medication Patches: No Bitter Pill to Swallow

For most people, pills are a convenient way to take medication, but for others, they have drawbacks. Pills must be taken at regular intervals to ensure consistent dosing, and they must be digested and absorbed into the bloodstream before they can begin to work. For those who have difficulty swallowing or digesting pills, **transdermal (TD) patches** offer an effective alternative to some oral medications.

TD patches deliver a consistent dose of medication that diffuses at a constant rate through the skin into the bloodstream. There is no daily schedule to follow, nothing to swallow, and no stomach upset. TD patches can also deliver medication to unconscious patients, who would otherwise require intravenous drug delivery. TD patches are used in hormone replacement therapy, to treat heart disease, to manage pain, and to suppress motion sickness. Nicotine patches are also used as part of programs to quit smoking.

TD patches must be used carefully. Drug diffusion through the skin takes time, so it is important to know how long the patch must be in place before it is effective. It is also important to know how long the medication's effects will persist after the patch is removed. Because the body continues to absorb what has already diffused into the skin, removing the patch does not entirely remove the medicine. Also, increased heat may elevate drug absorption to dangerous levels.

A recent advance in TD drug delivery is **iontophoresis**. Based on the principle that like charges repel each other, this method uses a mild electric current to move ionic drugs through the skin. A small electrical device attached to the patch uses positive current to "push" positively charged drug molecules through the skin and a negative current to push negatively charged ones. Even though very low levels of electricity are used, people with pacemakers should not use iontophoretic patches. Another disadvantage is that they can move only ionic drugs through the skin.

There is also a minimal amount of excretion through the skin. Water and electrolytes are excreted in sweat (perspiration). Some nitrogen-containing wastes are eliminated through the skin, but even in disease, the amount of waste products excreted by the skin is small. Therefore, claims that saunas and hot yoga "detoxify" the body are not substantiated by science.

Vitamin D needed for the development and maintenance of bone tissue is manufactured in the skin under the effects of ultraviolet (UV) radiation in sunlight.

Note that the human skin does not "breathe." The pores of the epidermis serve only as outlets for perspiration from the sweat glands and sebum (oil) from the sebaceous glands. They are not used for exchange of gases.

CHECKPOINTS 

- 5-8 Which two substances produced in the skin help to prevent dehydration?
- 5-9 Which two mechanisms involving the skin are used to regulate temperature?

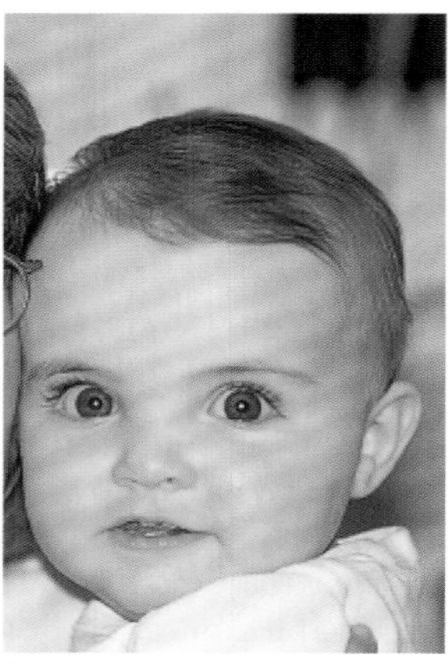

A B

FIGURE 5-6 Hemoglobin and skin color. KEY POINT Hemoglobin in blood gives skin a reddish coloration. **A.** Increased blood flow causes redness. **B.** Decreased blood flow results in pallor.

SKIN COLOR

Skin color is determined by pigments present in the skin itself and in blood circulating through the skin. The three main pigments that impart color to the skin are:

Melanin Melanin is the skin's main pigment. As discussed earlier, it is produced by melanocytes in the stratum germinativum of the epithelium. In addition to its presence in the skin, it is found in the hair, the middle coat of the eyeball, the iris of the eye. It is common to all races, but darker people have a much larger quantity in their tissues because their melanocytes are more active. The melanin in the skin helps protect against sunlight's damaging UV radiation. Thus, skin that is exposed to the sun shows a normal increase in this pigment, a response we call tanning.

Hemoglobin Hemoglobin (he-mo-GLO-bin) is the pigment that carries oxygen in red blood cells (further described in Chapters 12 and 16). It gives blood its color and is visible in the skin through vessels in the dermis. Circulation increases, and so does skin redness, when body temperature rises or in response to excitement or embarrassment (FIG. 5-6A). Conversely, shock or cold decreases circulation and causes skin to become pale (FIG. 5-6B).

Carotene Carotene (KAR-o-tene) is a skin pigment obtained from carrots and other orange and yellow vegetables. Excessive intake of these vegetables can result in carotene accumulation in blood, a condition known as **carotenemia** (kar-o-te-NE-me-ah) (the suffix *-emia* refers to blood). The excess carotene is deposited in the stratum corneum, resulting in a yellowish red skin discoloration known as carotenoderma.

CHECKPOINT 

- 5-10 Name three pigments that give color to the skin.

CASEPOINT 

- 5-5 What pigment gave Hazel's skin its color when she suffered a burn?

Repair of the Integument

The integumentary system, our outer covering, is more prone to damage than any other system. Superficial wounds involving only the epidermis do not bleed, but they still hurt. Wounds that penetrate the dermis do bleed because the blood vessels in that layer are damaged. Large skin wounds can result in life-threatening infection and dehydration, because the skin's function as a barrier has been lost. More extensive injuries penetrate the subcutaneous layer and may extend down to the underlying adipose and muscle tissues. Interestingly, deeper skin wounds sometimes hurt less than do more superficial injuries, because they destroy skin's pain-detecting nerve supply. Hazel's opening case study describes a thermal injury, that is, a burn.

Regardless of the injury's cause, repair begins with inflammation. Blood brings growth factors that promote the activity of restorative cells and agents that break down tissue debris and fight infection. New vessels branch from damaged capillaries and grow into the injured tissue. Fibroblasts (connective tissue cells) manufacture collagen and other substances to close the gap made by the wound. Stem cells in neighboring undamaged skin (specifically the stratum basale of the epidermis) produce epidermal cells that migrate to cover the new connective tissue. In severe wounds, stem cells in dermal hair follicles can be activated to produce new epidermal cells. However, if both the dermis and epidermis are destroyed, new epidermal tissue cannot be made and skin grafts may be required.

The new connective tissue (dermis) at the center of the healed wound is usually different from normal tissue, forming a **scar**, also called a **cicatrix** (SIK-ah-trix), which may continue to show at the surface as a white line. Scar tissue is strong but is not as flexible as normal tissue and does not function like the tissue it replaces. Suturing (sewing) the edges of a clean wound together, as is done for operative wounds, decreases the amount of connective tissue needed for repair and thus minimizes scarring.

FACTORS THAT AFFECT HEALING

Wound healing is a complex process involving multiple body systems. It is affected by the following:

- Nutrition—A complete and balanced diet will provide the nutrients needed for cell regeneration. All required vitamins and minerals are important, especially vitamins A and C, which are needed for collagen production.
- Blood supply—The blood brings oxygen and nutrients to the tissues and also carries away waste materials and toxins (poisons) that might form during the healing process. White blood cells attack invading bacteria at the site of the injury. Poor circulation, as occurs in cases of diabetes, for example, will delay wound healing.
- Infection—Contamination prolongs inflammation and interferes with the formation of materials needed for wound repair.
- Age—Healing is generally slower among the elderly reflecting their slower rate of cell replacement. The elderly also may have lowered immune responses to infection.

CHECKPOINT 

☐ **5-11** Name four factors that affect skin healing.

See the Student Resources on thePoint® for a figure on integument repair and to view an animation on wound healing.

Effects of Aging on the Integumentary System

As people age, wrinkles, or crow's feet, develop around the eyes and mouth owing to the loss of fat, elastic fibers, and collagen in the underlying tissues. The dermis becomes thinner, and the skin may become transparent and lose its elasticity, an effect sometimes called "parchment skin." Exposure to the UV radiation in sunlight degrades collagen and elastic fibers, thereby accelerating these changes. Pigment formation decreases with age. However, there may be localized areas of extra pigmentation in the skin with the formation of brown spots ("liver spots"), especially on areas exposed to the sun (e.g., the backs of the hands). Circulation to the dermis decreases, so white skin looks paler. Wounds heal more slowly and are more susceptible to infection.

The hair does not replace itself as rapidly as before and thus becomes thinner on the scalp and elsewhere on the body. Decreased melanin production leads to gray or white hair. Hair texture changes as the hair shaft becomes less dense, and hair, like the skin, becomes drier as sebum production decreases.

The eccrine sweat glands decrease in number, so there is less output of perspiration and lowered ability to withstand heat. The elderly are also more sensitive to cold because of having less fat in the skin and poor circulation. The fingernails may flake, become brittle, or develop ridges, and toenails may become discolored or abnormally thickened.

Observation and care of the skin are important in nursing as well as other healthcare professions. The Student Resources on thePoint® have information on nursing careers.

A & P in Action Revisited: Hazel's Healing Process

The following weekend, Tiffany returned to check up on her grandmother. "How have you been getting along?" she asked.

"Well, it hasn't been fun," Hazel groused, "but Mavis from next door has been coming over every few days to change my dressing, and I gave her all of the tomatoes. I only have the bandage on my forearm now, but it's taking forever to heal!"

"I've been practicing dressing changes this week," Tiffany replied, "so let me take care of it today." She donned sterile gloves and carefully peeled back the dressing. Part of the burned forearm was still moist and red, but she saw very little pus or other signs of infection, and the blisters were gone. Small islands of new skin had appeared over the injured dermis, indicating skin regrowth from epidermal stem cells in the hair shafts. Tiffany gently cleaned the wound and then applied more antibiotic cream and a new dressing. The hand was dry, pink, and sensitive to touch, so she left it alone. The unaffected skin was quite papery and had some brown spots, but she knew that these were normal indications of aging. "As long as the wound doesn't get infected, you might not even have a scar," Tiffany said, "and applying sunscreen to the healed skin for the next year whenever you go outside can help prevent permanent color changes." Hazel's story highlights the ability of skin to heal after injury, even in the elderly, as long as the damage is not overly deep.

Chapter Wrap-Up

OVERVIEW

A detailed chapter outline with space for note-taking is on thePoint®. The figure below illustrates the main topics covered in this chapter.

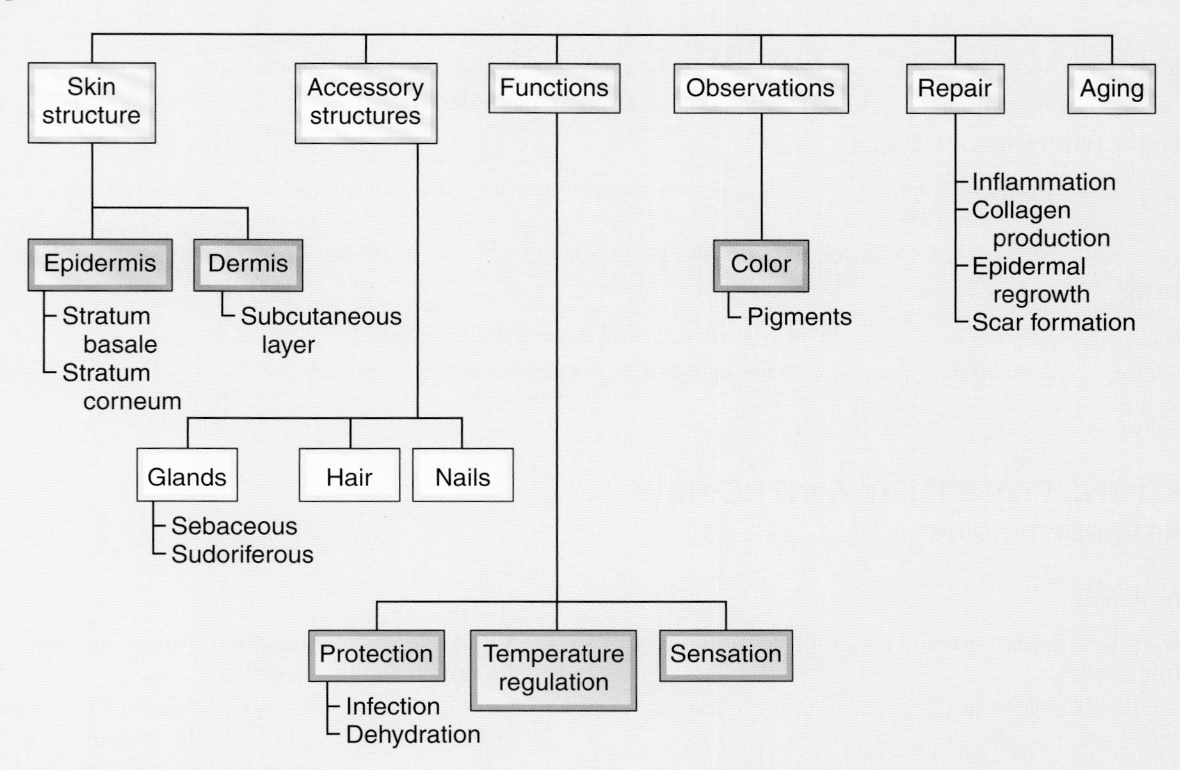

KEY TERMS

The terms listed below are emphasized in this chapter. Knowing them will help you organize and prioritize your learning. These and other boldface terms are defined in the Glossary with phonetic pronunciations.

apocrine
arrector pili
cerumen
dermis
eccrine
epidermis
hair follicle
integumentary system
keratin
melanin
melanocyte
sebaceous gland
sebum
stratum basale
subcutaneous layer
sweat gland

WORD ANATOMY

Medical terms are built from standardized word parts (prefixes, roots, and suffixes). Learning the meanings of these parts can help you remember words and interpret unfamiliar terms.

WORD PART	MEANING	EXAMPLE
STRUCTURE OF THE SKIN		
corne/o	cornified, keratinized	The stratum *corneum* is the outermost thickened, keratinized layer of the skin.
derm/o	skin	The *epidermis* is the outermost layer of the skin.
melan/o	dark, black	A *melanocyte* is a cell that produces the dark pigment melanin.
sub-	under, below	The *subcutaneous* layer is under the skin.
ACCESSORY STRUCTURES OF THE SKIN		
ap/o-	separation from, derivation from	The *apocrine* sweat glands release some cellular material in their secretions.
pil/o	hair	The *arrector pili* muscle raises the hair to produce "goose bumps."
SKIN COLOR		
hemo	blood	*Hemoglobin* is a substance found in blood.
melan/o	dark, black	A *melanocyte* is a cell that contains a dark pigment.

QUESTIONS FOR STUDY AND REVIEW

BUILDING UNDERSTANDING

Fill in the Blanks

1. Cells of the stratum corneum contain large amounts of a protein called _____.
2. Sweat glands located in the axillae and groin are classified as _____ glands.
3. The name of the muscle that raises the hair is the _____.
4. A dark-colored pigment that protects the skin from the rays in sunlight is called _____.
5. Refer to the integumentary system in "The Body Visible" overlays to find that the Meissner and Ruffini corpuscles are receptors for the sense of _____.

Matching Match each numbered item with the most closely related lettered item.

___ 6. Accessory skin structure that lubricates the eye
___ 7. Deepest, dividing epithelial layer of the skin
___ 8. A deep pressure receptor in the skin
___ 9. Modified sweat gland that produces ear wax
___ 10. Superficial layer of the epidermis

a. stratum basale
b. stratum corneum
c. meibomian gland
d. Pacinian corpuscle
e. ceruminous gland

Multiple Choice

___ 11. The dermis is _____ to the epidermis.
a. superficial
b. deep
c. lateral
d. medial

___ 12. The layer of the skin that contains blood vessels is the
a. epidermis
b. hypodermis
c. dermis
d. subcutaneous layer

___ 13. Fingerprints and footprints are formed by
a. Pacinian corpuscles
b. melanocytes
c. Meissner corpuscles
d. dermal papillae

___ 14. Which glands are involved in temperature regulation?
a. ceruminous
b. sudoriferous
c. ciliary
d. papillary

___ 15. Nails grow from which area?
a. lunula
b. cuticle
c. nail root
d. nail bed

UNDERSTANDING CONCEPTS

16. Compare and contrast the epidermis, dermis, and hypodermis. How are the outermost cells of the epidermis replaced?
17. Referring to the integumentary system in "The Body Visible" overlays and information in Chapter 4, name the type of tissue that constitutes number 27.
18. Describe the location and function of the two types of skin glands.
19. What are the four most important functions of the skin?
20. Describe the events associated with skin wound healing.
21. What changes may occur in the skin with age?

CONCEPTUAL THINKING

22. Why is the skin described as a membrane? An organ? A system?
23. Hazel's deep burn was painful, moist, warm, and prone to infection. Which of the four skin functions have been compromised, and which are still functioning?

For more questions, see the Learning Activities on thePoint®.

UNIT II

Movement and Support

CHAPTER 6

The Skeletal System: Bones and Joints

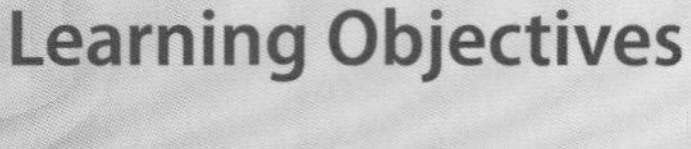

Learning Objectives

After careful study of this chapter, you should be able to:

1. List the functions of bones. ***p. 100***
2. Describe the structure of a long bone. ***p. 100***
3. Name the three different types of cells in bone, and describe the functions of each. ***p. 100***
4. Differentiate between compact bone and spongy bone with respect to structure and location. ***p. 101***
5. Explain how a long bone grows. ***p. 103***
6. Name and describe nine markings found on bones. ***p. 104***
7. Name, locate, and describe the bones in the axial skeleton. ***p. 105***
8. Describe the normal curves of the spine, and explain their purpose. ***p. 110***
9. Name, locate, and describe the bones in the appendicular skeleton. ***p. 111***
10. Describe three categories of joints based on degree of movement, and give examples of each. ***p. 117***
11. Name six types of synovial joints, and demonstrate the movements that occur at each. ***p. 119***
12. Describe how the skeletal system changes with age. ***p. 120***
13. Using the case study, discuss how fractures heal. ***pp. 99, 121***
14. Show how word parts are used to build words related to the skeleton (see Word Anatomy at the end of the chapter). ***p. 123***

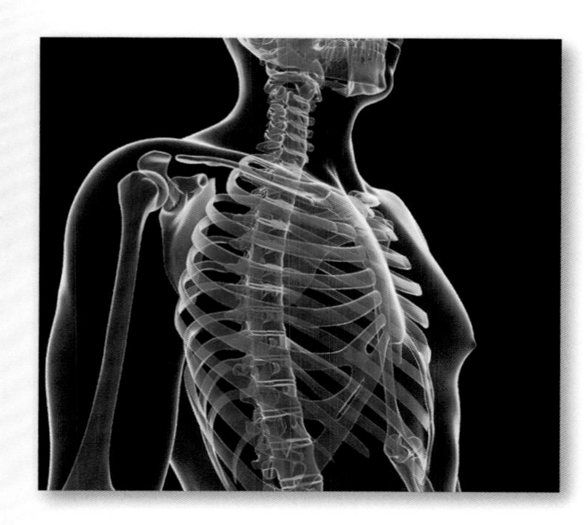

A & P in Action

Reggie's Case: A Footballer's Fractured Femur

"Donnelly throws deep for a touchdown. Wilson makes a beautiful catch! Ooh, a nasty hit from number 26." The crowd roared their approval for the wide receiver. On the ground, Reggie Wilson knew that something was wrong with his hip. In fact, he thought he had actually heard the bone break. It didn't take long for the coaches and medical staff to realize that Reggie needed help. And it didn't take long for the ambulance to get him to the trauma center closest to the stadium.

At the hospital, the emergency team examined Reggie. His injured leg appeared shorter than the other and was adducted and laterally rotated—all signs of a hip fracture. An x-ray confirmed the team's suspicions; Reggie had sustained an intertrochanteric fracture of his right femur. His tibia and fibula were intact. He would need surgery, but luckily for Reggie, the fracture line extended from the greater trochanter to the lesser trochanter and didn't involve the femoral neck. This meant that the blood supply to the femoral head was not in danger, so the surgery would be more straightforward.

In the operating room, the surgical team applied traction to Reggie's right leg, pulling on it to reposition the broken ends of his proximal femur back into anatomic position (verified with another x-ray). Then, the orthopedic surgeon made an incision beginning at the tip of the greater trochanter and continuing distally along the lateral thigh through the skin, subcutaneous fat, and vastus lateralis muscle. After exposing the proximal femur, the surgeon drilled a hole and installed a titanium screw through the greater trochanter and neck and into the femoral head. He then positioned a titanium plate over the screw and fastened it to the femoral shaft with four more screws. Confident that the broken ends of the femur were firmly held together, the surgeon closed the wound with sutures and skin staples. Reggie was then wheeled into the recovery room.

The surgical team successfully realigned the fractured ends of Reggie's femur. Now Reggie's body will begin the healing process. In this chapter, we learn more about bones and joints. Later in the chapter, we see how Reggie's skeletal system is repairing itself.

As you study this chapter, CasePoints will give you opportunities to apply your learning to this case.

Visit thePoint® to access the following resources. For guidance in using these resources most effectively, **see pp. xv–xvii.**

Preparing to Learn

- Tips for Effective Studying
- Pre-Quiz

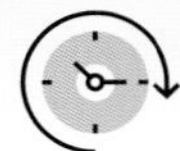

While You Are Learning

- Web Figure: Bone Markings and Formations
- Web Figure: Skeletal Features of the Head and Neck
- Web Figure: Skeletal Features of the Shoulder and Torso
- Web Figure: Skeletal Features of the Upper Extremity
- Web Figure: Skeletal Features of the Lower Extremity
- Web Chart: Bones of the Skull
- Animation: Bone Growth
- Chapter Notes Outline
- Audio Pronunciation Glossary

When You Are Reviewing

- Answers to Questions for Study and Review
- Health Professions: Radiologic Technologist
- Interactive Learning Activities

A LOOK BACK

Illustrating a principle introduced in Chapter 1, we will see that the skeletal system can be studied at multiple levels of organization 2, including bone cells, bone tissue (as introduced in Chapter 4), and bone organs (individual bones). Together with the joints, these bones constitute the skeletal system. The key ideas of homeostasis 3 (Chapter 1) and adaptation 10 (Chapter 5) are also relevant to this chapter.

Introduction

The skeleton is the strong framework on which the body is constructed. Much like the frame of a building, the skeleton must be strong enough to support and protect all the body structures. Bones work with muscles to produce movement at the joints.

Bones

Bones have a number of functions, several of which are not evident in looking at the skeleton. They

- form a sturdy framework for the entire body
- protect delicate structures, such as the brain and the spinal cord
- work as levers with attached muscles to produce movement
- store calcium salts, which may be resorbed into the blood if calcium is needed
- produce blood cells (in the red marrow)

BONE STRUCTURE

The complete bony framework of the body, known as the **skeleton** (FIG. 6-1), consists of approximately 206 bones (the precise number can vary somewhat among individuals). The axial skeleton includes the bones of the head and torso, and the appendicular skeleton includes the bones of the extremities. The individual bones in these two divisions are described in detail later in this chapter. The bones of the skeleton can be of several different shapes. They may be flat (ribs, cranium), short (carpals of wrist, tarsals of ankle), or irregular (vertebrae, facial bones). The most familiar shape, however, is the **long bone**, the type of bone that makes up most of the appendicular skeleton. The long narrow shaft of this type of bone is called the **diaphysis** (di-AF-ih-sis). At the center of the diaphysis is a **medullary** (MED-u-lar-e) **cavity**, which contains bone marrow. The long bone also has two irregular ends, a proximal and a distal **epiphysis** (eh-PIF-ih-sis) (FIG. 6-2).

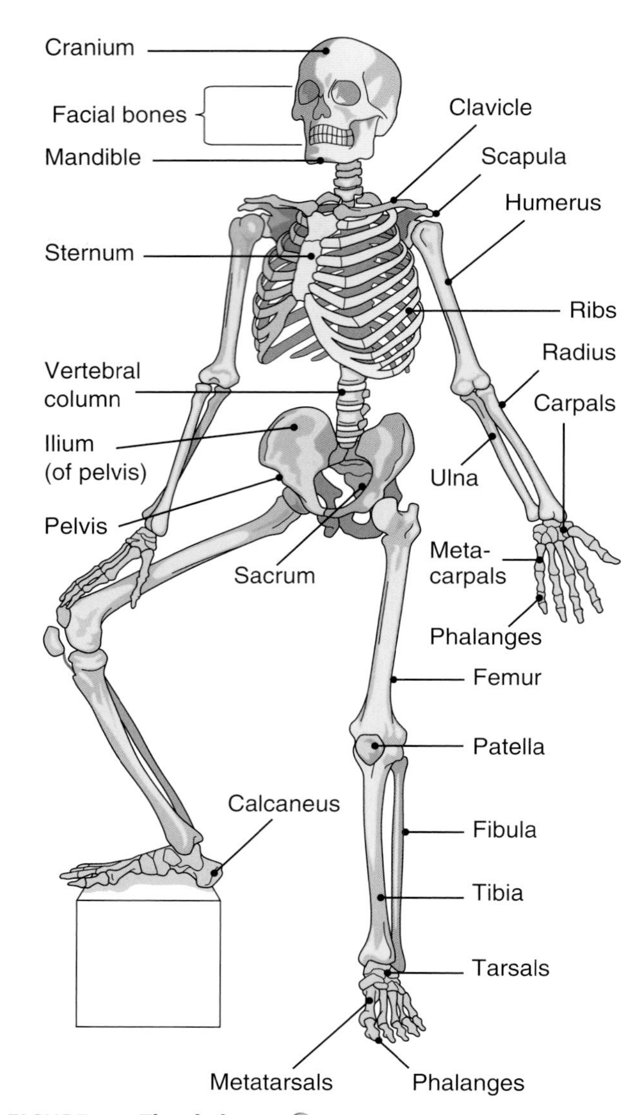

FIGURE 6-1 The skeleton. KEY POINT The skeleton is divided into two portions. The axial skeleton is shown here in *yellow*; the appendicular in *blue*.

Bone Tissue 2 Bones are living organs with their own systems of blood vessels and nerves. The bulk of each bone is composed of bone tissue, also known as **osseous** (OS-e-us) **tissue,** the hardest form of connective tissue. Bone tissue contains three types of cells:

- **Osteoblasts** (OS-te-o-blasts) build bone tissue. (You can use the mnemonic "**B**lasts **B**uild" to remember the role of these cells.)
- **Osteocytes** (OS-te-o-sites) are mature osteoblasts that become trapped in the bone matrix. They maintain bone tissue.
- **Osteoclasts** (OS-te-o-klasts) are large, multinucleated cells responsible for the process of **resorption,** which is the breakdown of bone tissue. Osteoclasts develop from a type of white blood cell (monocyte). (You can use the mnemonic "**C**lasts **C**leave.")

Bone's hardness and strength reflect the components of the **matrix**, the material between the living bone cells. This material is rich in the protein **collagen** (KOL-ah-jen) and in calcium salts. Both substances are necessary for healthy bones; without minerals, bones would bend easily, but

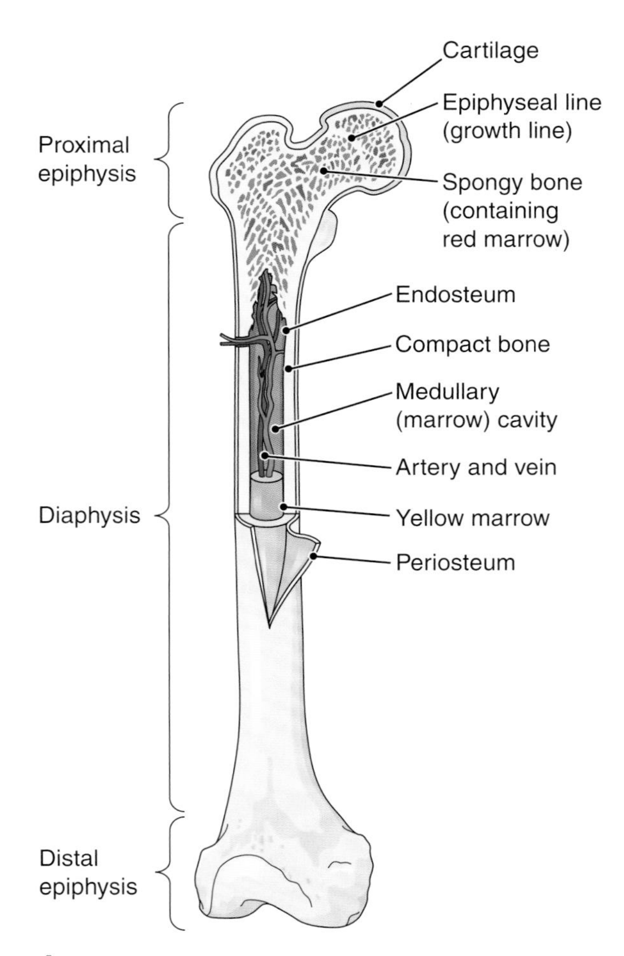

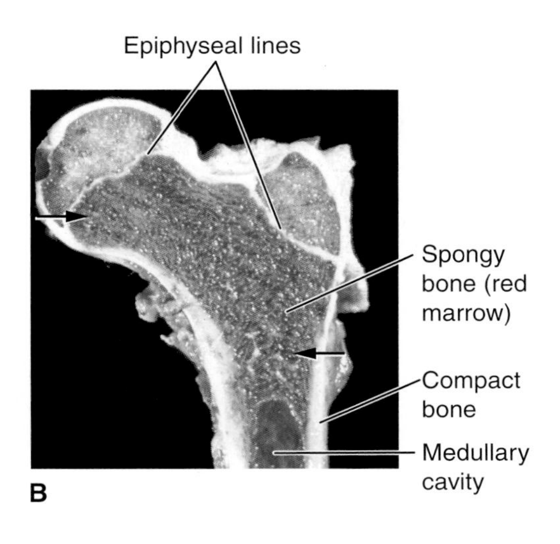

FIGURE 6-2 **The structure of a long bone.** KEY POINT A long bone has a long, narrow shaft, the diaphysis, and two irregular ends, the epiphyses. The medullary cavity has yellow marrow. Red marrow is located in spongy bone. **A.** Diagram of a long bone. **B.** Photograph of bone tissue. This longitudinal section of a long bone shows an outer layer of compact bone. Spongy bone tissue is indicated by the arrows. Transverse growth lines are also visible. ZOOMING IN What are the membranes on the outside and the inside of a long bone called?

without collagen, the bones would shatter easily, like sticks of chalk.

Types of Osseous Tissue There are two types of osseous tissue: compact and spongy. **Compact bone** is hard and dense (FIG. 6-3). This tissue makes up the main shaft of a long bone and the outer layer of other bones. The osteocytes (mature bone cells) in this type of bone are located in rings of bone tissue around a **central canal**, also called a *haversian* (ha-VER-shan) *canal*, containing nerves and blood vessels. Osteocytes live in spaces (lacunae) between the rings and extend out into many small radiating channels so that they can be in contact with nearby cells. Each ringlike unit with its central canal makes up an **osteon** (OS-te-on) or haversian system (see FIG. 6-3B). **Perforating** (Volkmann) **canals** form channels across the bone from one side of the shaft to the other, permitting the passage of blood vessels and nerves.

The second type of bone tissue, called **spongy bone**, or cancellous bone, has more spaces than does compact bone. It is made of a meshwork of small, bony plates filled with red marrow. Spongy bone is found at the epiphyses (ends) of the long bones and at the center of other bones. It also lines the medullary cavity of long bones. FIGURE 6-2B shows a photograph of both compact and spongy bone tissue.

Bone Marrow Bones contain two kinds of marrow. In adults, **red marrow** is found in the spongy bone at the ends of the long bones and at the center of other bones (see FIG. 6-2). Red bone marrow manufactures blood cells. **Yellow marrow** is found chiefly in the central cavities of the long bones. Yellow marrow is composed largely of fat. The long bones of babies and children contain mostly red marrow, reflecting their greater need for new blood cells.

Bone Membranes Bones are covered (except at the joint region) by a membrane called the **periosteum** (per-e-OS-te-um) (see FIG. 6-2), composed of dense irregular connective tissue. This membrane's inner layer contains osteoblasts that build bone tissue and osteoclasts that break down bone tissue. The coordinated actions of these

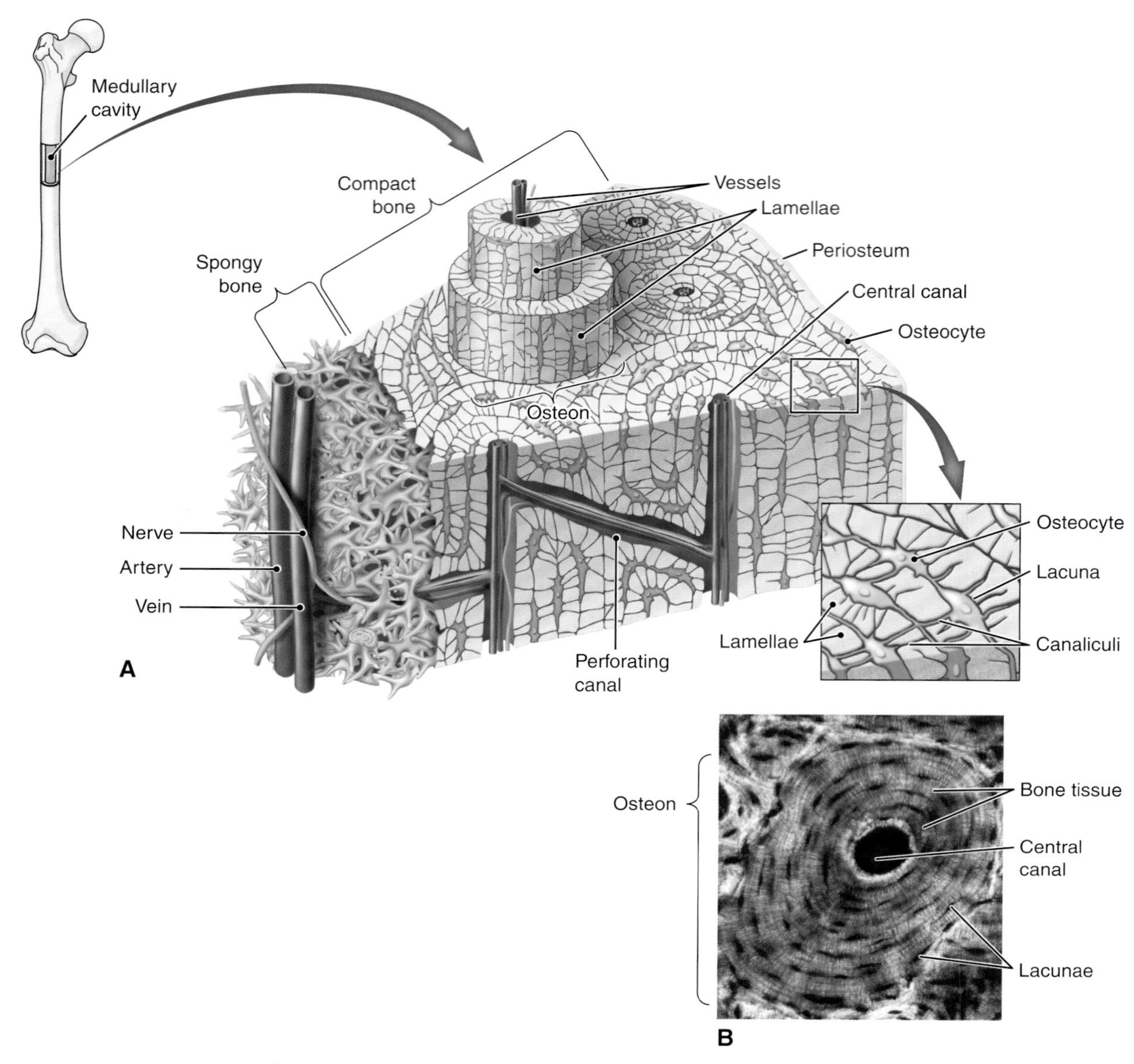

FIGURE 6-3 Bone tissue. KEY POINT There are two types of bone tissue—compact and spongy. **A.** This section shows osteocytes (bone cells) within osteons (haversian systems) in compact bone. It also shows the canals that penetrate the tissue. **B.** Microscopic view of compact bone in cross-section (×300) showing a complete osteon. In living tissue, osteocytes (bone cells) reside in spaces (lacunae) and extend out into channels that radiate from these spaces. ZOOMING IN Which cells are located in the spaces of compact bone?

cells build, repair, and maintain bone throughout life. Blood vessels in the periosteum play an important role in the nourishment of bone tissue. Nerve fibers in the periosteum make their presence known when a person suffers a fracture or receives a blow, such as on the shinbone. A thinner membrane, the **endosteum** (en-DOS-te-um), lines the bone's marrow cavity; it too contains osteoblasts and osteoclasts.

CHECKPOINTS 

- ☐ **6-1** What are the scientific names for the shaft and the ends of a long bone?
- ☐ **6-2** What compounds are deposited in the intercellular matrix of the embryonic skeleton to harden it?
- ☐ **6-3** What are the three types of cells found in bone, and what is the role of each?
- ☐ **6-4** What are the two types of osseous (bone) tissue, and where is each type found?

CASEPOINTS 

- ☐ **6-1** Reggie fractured the proximal end of a long bone. What is the scientific name for the bone region involved?
- ☐ **6-2** Would Reggie's fracture line contact red marrow or yellow marrow?

BONE GROWTH, MAINTENANCE, AND REPAIR

The process of bone formation begins in the earliest weeks of embryonic life and continues until young adulthood.

Fetal Ossification During early development, the long bones of the embryonic skeleton are composed primarily of hyaline cartilage. The conversion of cartilage to bone, a process known as **ossification**, begins during the second and third months of embryonic life. At this time, osteoclast-like cells remove the cartilage, and osteoblasts deposit bone tissue in place of the cartilage.

Once this intercellular material has hardened, the cells remain enclosed within the lacunae (small spaces) in the matrix. These cells, now known as osteocytes, are still living and continue to maintain the existing bone matrix, but they do not produce new bone tissue. Osteoblasts and osteoclasts in the bone membranes are responsible for bone growth, repair, and remodeling later in life. You will see the importance of these cells in Reggie's case study.

The flat bones of the skull and other regions develop from dense, irregular connective tissue membranes instead of from cartilage. Osteoblasts deposit bone tissue within these fibrous membranes.

Formation of a Long Bone In a long bone, the transformation of cartilage into bone begins at the center of the shaft during fetal development. As bone synthesis continues, osteoclasts (the bone removers) degrade the bone tissue at the center of the bone, producing the medullary cavity. Around the time of birth, secondary bone-forming centers, or **epiphyseal** (ep-ih-FIZ-e-al) **plates**, develop across the ends of the bones. The long bones continue to grow in length at these centers by the production of new cartilage within the plate and calcification of older cartilage. The large amount of cartilage in a child's bones renders them more pliable and tougher to break.

Finally, by the late teens or early 20s, the bones stop growing in length. Bone tissue replaces all of the cartilage in the epiphyseal plate, which can be seen in x-ray films as a thin line, known as the epiphyseal line (see FIG. 6-2). Physicians can use the presence of the epiphyseal plate or line on x-rays to evaluate a patient's age.

As a bone grows in length, it also grows in width. To prevent bones from becoming too heavy, osteoclasts remove bone tissue from the shaft to enlarge the central marrow cavity as osteoblasts add bone tissue to the outside.

Bone Tissue Regulation Even after skeletal growth is complete, osteoblasts and osteoclasts actively maintain and repair bone tissue and remodel it according to the stresses placed upon it. 10 For instance, a right-handed person uses the right arm more than the left arm and the right arm bones adapt by becoming larger and stronger. Astronauts (in the absence of gravity) lose bone mass because their bones are no longer subjected to stress. Resorption is also necessary for repair of bone injury.

3 Bone tissue also plays a key role in the regulation of the blood calcium concentration, which is a regulated variable. That is, a negative feedback loop keeps the levels of blood calcium within tight limits by altering bone deposition or absorption. The important signal in this loop is called parathyroid hormone, which is produced by the parathyroid glands in the neck (posterior to the thyroid gland). The parathyroid glands act as sensors and control centers in this loop. As blood calcium levels decline, parathyroid hormone synthesis increases. Parathyroid hormone stimulates osteoclast activity, resulting in bone resorption and release of calcium into the blood. Thus, bone is an effector in this feedback loop. The body will weaken bones as needed to maintain adequate blood calcium concentrations.

The sex hormones, estrogen and testosterone, also contribute to bone growth and maintenance. Vitamin D, consumed in the diet and produced by the skin, promotes calcium absorption from the intestine. Chapter 11 provides more information about these and other hormones.

The balance between osteoblast and osteoclast activity dictates changes in bone mass. Bones increase in density until the early 20s in females and the late 20s in males, at which point bones are at peak density and strength. Most people maintain peak bone density until about age 40. As people age, there is a slowing of bone tissue renewal. As a result, the bones become weaker, and damage heals more slowly.

Fracture Repair When a bone is fractured, nearby blood vessels immediately constrict to prevent further bleeding, and the blood then forms a clot (FIG. 6-4). Some cells around the injury die off, but fibroblasts in the area survive, multiply, and contribute to the formation of new connective tissue. In a few days, stem cells in the periosteum near the fracture site develop into chondroblasts and produce hyaline cartilage. Meanwhile, periosteal cells farther from the fracture point develop into osteoblasts capable of building a type of loosely organized primitive bone tissue, called *woven bone*. Eventually, these new tissues join to form a *soft* callus that closes the fracture gap. The cartilage and woven bone then ossify into spongy bone, forming a *bony callus* and restoring some bone strength. Over the next several years, osteoclasts resorb the spongy bone and osteoblasts replace it with compact bone. To summarize, the steps in fracture repair are first *reaction*, involving hemostasis and an inflammatory response; second, the *repair* of the damage; and finally, *remodeling* of new tissue into compact bone.

Treatment of a serious fracture, as in Reggie's case, requires repositioning of the bone and stabilization. Factors that affect healing include the nature and extent of the break as well as a person's nutritional status, age, and general health. Healing may be promoted by injectable, synthetic bone cements or by application of an external magnetic field.

CASEPOINT 

☐ **6-3** During the healing process of Reggie's fracture, which cell type would produce new bone tissue?

See the Student Resources on thePoint® to view the animation "Bone Growth," showing the growth process in a long bone.

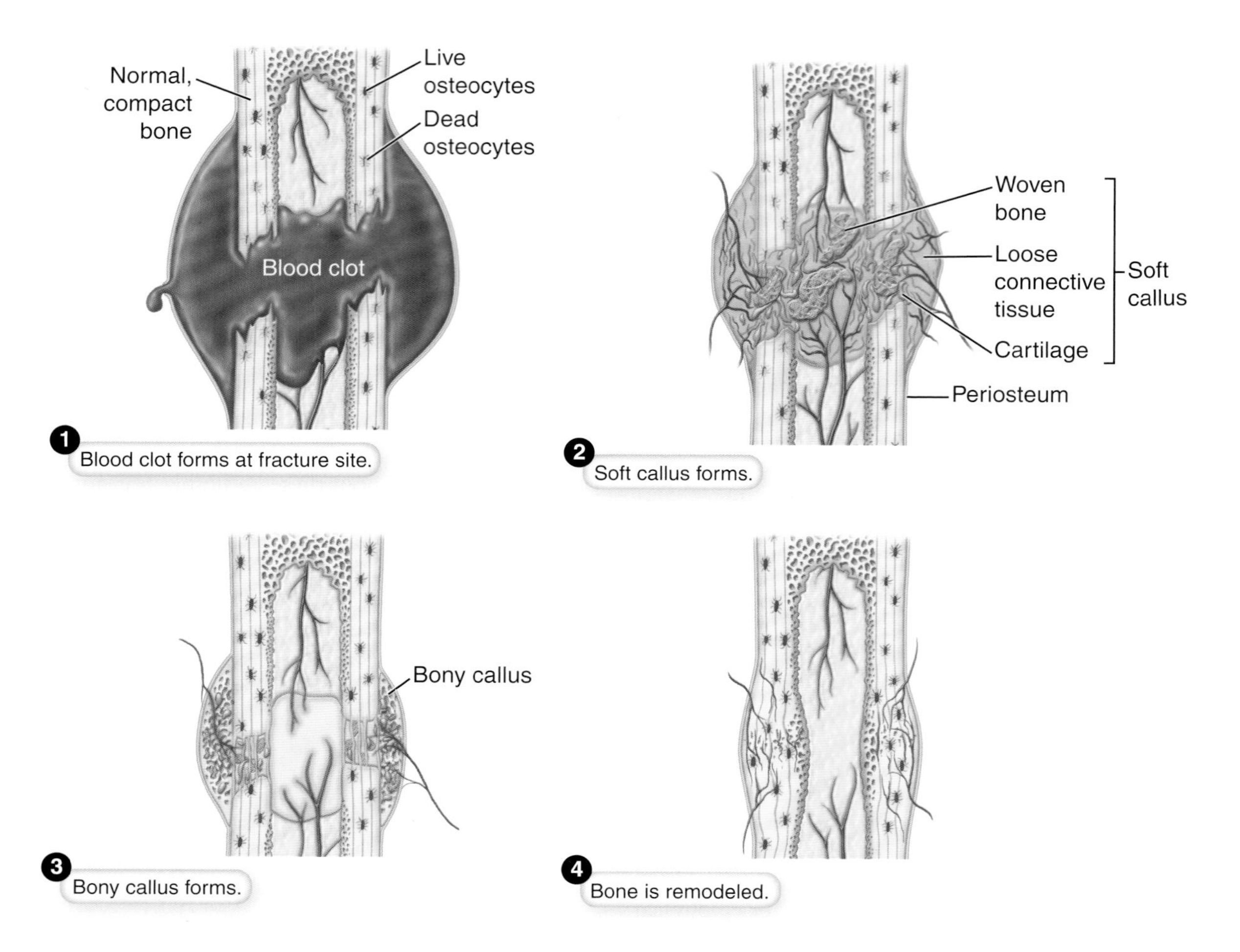

FIGURE 6-4 Fracture repair. When bone is fractured, stem cells in the periosteum direct the formation of new bone tissue to repair the injury. ZOOMING IN Which forms first—the soft callus or the bony callus?

BONE MARKINGS

In addition to their general shape, bones have other distinguishing features, or **bone markings**. These markings include raised areas and depressions, which help form joints or serve as points for muscle attachments, and various holes, which allow the passage of nerves and blood vessels. Some of these identifying features are described next.

Projections

- **Head**—a rounded, knoblike end separated from the rest of the bone by a slender region, the neck
- **Process**—a large projection of a bone
- **Condyle** (KON-dile)—a rounded projection; a small projection above a condyle is an epicondyle
- **Crest**—a distinct border or ridge, often rough, such as over the top of the hip bone
- **Spine**—a sharp projection from the surface of a bone, such as the spine of the scapula (shoulder blade)

Depressions or Holes

- **Foramen** (fo-RA-men)—a hole that allows a vessel or a nerve to pass through or between bones. The plural is foramina (fo-RAM-ih-nah).
- **Sinus** (SI-nus)—A cavity or hollow space. Most commonly, an air-filled chamber found in some skull bones (FIG. 6-5).
- **Fossa** (FOS-sah)—a depression on a bone surface. The plural is fossae (FOS-se).
- **Meatus** (me-A-tus)—a short channel or passageway, usually the external opening of a canal. An example is the channel in the skull that leads to the inner ear.

CHECKPOINTS

- ☐ **6-5** What are the centers for secondary growth of a long bone called?
- ☐ **6-6** Which hormone stimulates osteoclast activity?
- ☐ **6-7** What are some functions of bone markings?

See the Student Resources on thePoint to view bone markings on an illustration of a whole skeleton.

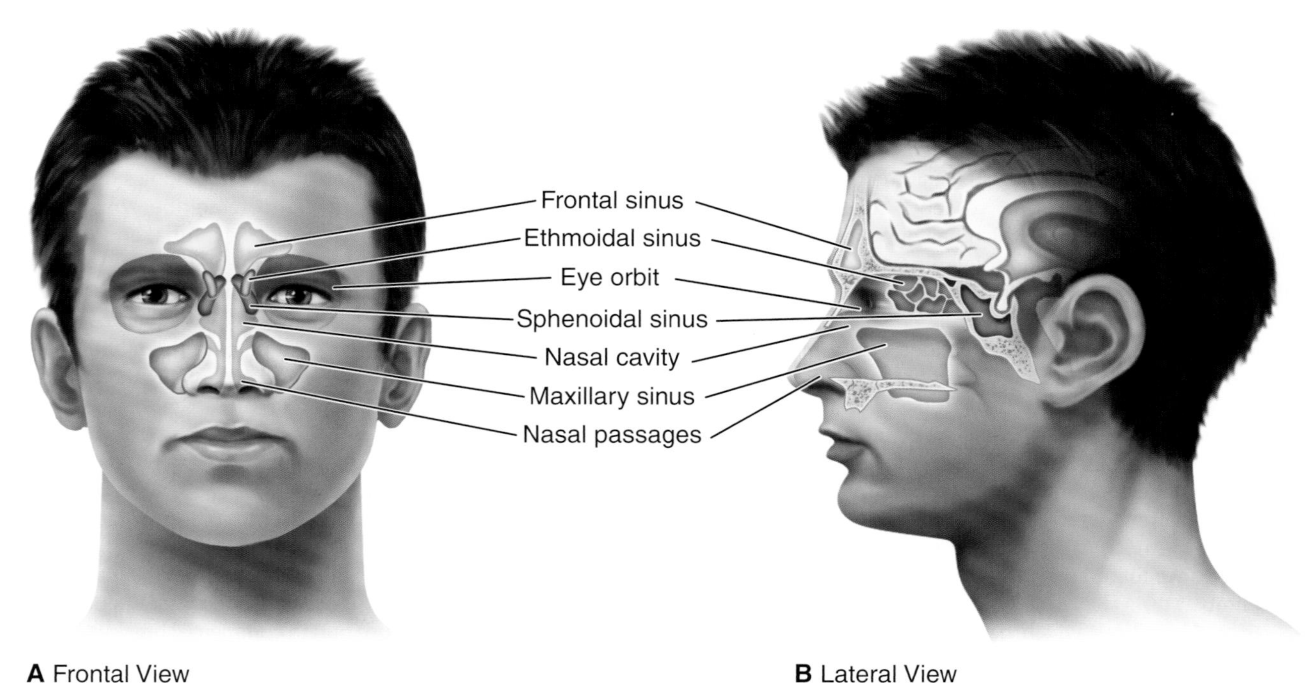

FIGURE 6-5 **Sinuses.** KEY POINT A sinus is a cavity or hollow space, such as the air-filled chambers in certain skull bones. View of the skull showing the sinuses from frontal **(A)** and lateral **(B)** aspects.

Bones of the Axial Skeleton

As noted earlier, the skeleton may be divided into two main groups of bones (see FIG. 6-1):

- The **axial** (AK-se-al) **skeleton** consists of 80 bones and includes the bony framework of the head and the trunk. Think of the axial skeleton as the body's "axis."
- The **appendicular** (ap-en-DIK-u-lar) **skeleton** consists of 126 bones and forms the framework for the **extremities** (limbs) and for the shoulders and hips. Think of the appendicular skeleton as the body's "appendages."

We describe the axial skeleton first and then proceed to the appendicular skeleton. A table at the end of this section summarizes all of the bones described. Also refer back to FIGURE 6-1 as you study this chapter. See BOX 6-1, "So Many Bones, So Little Time" for tips about learning bone names and features.

FRAMEWORK OF THE SKULL

The bony framework of the head, called the **skull**, is subdivided into two parts: the cranium and the facial portion. Refer to FIGURE 6-6, which shows different views of the skull, as you study the following descriptions. The individual bones are color coded to help you identify them as you study the skull in different views.

Note the many features of the skull bones as you examine these illustrations. For example, openings in the base of the skull provide spaces for the entrance and exit of many blood vessels, nerves, and other structures. Bone projections and fossae (depressions) provide for muscle attachment. Some portions protect delicate structures, for example, the eye orbit (socket) and the part of the temporal bone at the lateral skull that encloses the inner ear. The sinuses provide lightness and serve as resonating chambers for the voice (which is why your voice sounds better to you as you are speaking than it sounds when you hear it played back as a recording).

Cranium This rounded chamber that encloses the brain is composed of eight distinct cranial bones.

- The **frontal bone** forms the forehead, the anterior of the skull's roof, and the roof of the eye orbit. The **frontal sinuses** communicate with the nasal cavities (see FIG. 6-5). These sinuses and others near the nose are described as paranasal sinuses.
- The two **parietal** (pah-RI-eh-tal) bones form most of the top and the side walls of the cranium.
- The two **temporal bones** contribute to the sides and the base of the skull.

Concept Mastery Alert

Remember that the temporal bones are inferior to the two parietal bones, which form most of the top and sides of the skull.

- Each contains one ear canal, eardrum, and the ear's entire middle and inner portions. The **mastoid process** of the temporal bone projects downward immediately behind the outer ear (see FIG. 6-6B). It is a place for muscle

ONE STEP AT A TIME
So Many Bones, So Little Time

BOX 6-1

Learning skeletal anatomy is the first major memorization feat required of most anatomy students. Your body contains at least 208 bones, each of which has numerous distinguishing marks. The responsibility for learning all of these bones is lightened somewhat by the fact that we are bilaterally symmetrical; the skeleton can figuratively be split down the center, with equal structures on both sides of the midline. This cuts the learning in half. And chances are, your instructor will assign you a specific subset of bones and features of these 208 bones. Nevertheless, learning bone anatomy remains a daunting task for any student. Reading the textbook and looking at the images are only the first steps. You don't need to employ all of the strategies we propose, but using a variety of learning methods can maximize your retention. We begin with the bones featured in the case study—the bones of the lower limb.

Step 1. Find out which details you need to know. Your instructor might provide a list of required bones and bone markings.

Step 2. Using FIGURE 6-17 as a model, draw a rough sketch of the femur, tibia, and fibula, and label the features you need to know. If you don't like to draw, consider using a coloring exercise (see the available exercises in the Study Guide). As you label or color each bone and feature, say the term aloud. Using your visual, auditory, and tactile senses together will help you remember.

Step 3. Make up or look online for mnemonics for hard-to-remember details. For instance, to remember which leg bone is larger, you can use "Tibia is big like a Tuba," and "Fibula is small like a Flute." Also, "the fibuLA is LAteral."

Step 4. Find the bones and features on your own body or on a model skeleton. Make sure that you say the feature aloud as you find it. For instance, you can palpate the lateral malleolus (mah-LE-o-lus) of the fibula and potentially the greater trochanter (TRO-kan-ter) of the femur. You can get inexpensive skeletons online (try "Tiny Tim").

Step 5. Use flashcards to test your learning. Studies show that flashcards are most effective when you make them yourself! Draw the structure on an index card, using numbers for each bone and bone feature. Write the answers on the other side of the card. You can use images from the Study Guide or from the Internet if you don't like to draw. Review your flashcards often—sitting on the bus, waiting for a friend, or during commercial breaks!

Step 6. Use the online resources provided with your textbook to determine which areas need further study.

Learning skeletal gross anatomy is a critical step in your A&P education, because you often will learn the components of other systems (muscles, arteries, nerves) in relation to the bones. For instance, the radial artery, radial vein, and radial nerve are all located along the radius of the forearm. Take the time to discover which active learning techniques work best for you, and they will serve you well in future studies.

attachments and contains air cells (spaces) that make up the **mastoid sinus** (not illustrated).

- The **ethmoid** (ETH-moyd) **bone** is a light, fragile bone located between the eyes (see FIG. 6-6A and C). It forms a part of the medial wall of the eye orbit, a small portion of the cranial floor, and most of the nasal cavity roof. It also forms the superior and middle nasal conchae (KON-ke), bony plates that extend into the nasal cavity (the name *concha* means "shell"). The mucous membranes covering the conchae help filter, warm, and moisten air as it passes through the nose. The ethmoid houses several air cells, constituting some of the paranasal sinuses. A thin, platelike, downward extension of this bone (the perpendicular plate) forms much of the **nasal septum**, the midline partition in the nose (see FIG. 6-6A).
- The **sphenoid** (SFE-noyd) **bone**, when seen from a superior view, resembles a bat with its wings extended (see FIG. 6-6D). It lies at the base of the skull anterior to the temporal bones and forms part of the eye orbit. It contains the sphenoid sinuses. It also contains a depression called the **sella turcica** (SEL-ah TUR-sih-ka), literally "Turkish saddle," that holds and protects the pituitary gland like a saddle.
- The **occipital** (ok-SIP-ih-tal) **bone** forms the skull's posterior portion and a part of its base. The **foramen magnum,** located at the base of the occipital bone, is a large opening through which the spinal cord attaches to the brain (see FIG. 6-6C and D).

Uniting the skull bones is a type of flat, immovable joint known as a **suture** (SU-chur) (see FIG. 6-6B). Some of the most prominent cranial sutures are as follows:

- The coronal (ko-RO-nal) suture joins the frontal bone with the two parietal bones along the coronal plane.
- The squamous (SKWA-mus) suture joins the temporal bone to the parietal bone on the cranium's lateral surface (named because it is in a flat portion of the skull).
- The lambdoid (LAM-doyd) suture joins the occipital bone with the parietal bones in the posterior cranium (named because it resembles the Greek letter lambda).
- The sagittal (SAJ-ih-tal) suture joins the two parietal bones along the superior midline of the cranium, along the sagittal plane. Although this suture is not visible in FIGURE 6-6B, you can feel it if you press your fingertips along the top center of your skull.

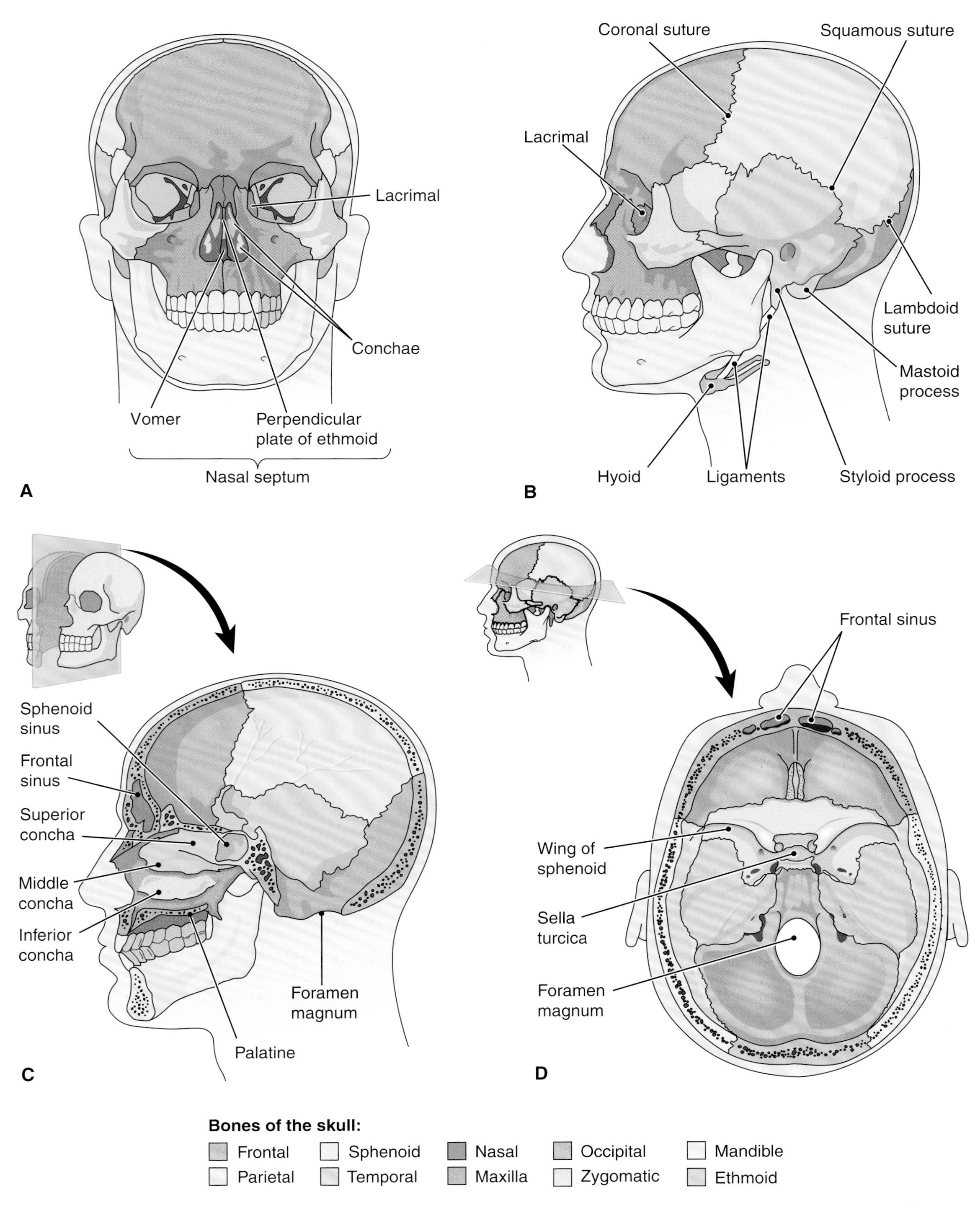

FIGURE 6-6 The skull. A. Anterior view. **B.** Left lateral view. **C.** Inferior view. The mandible (lower jaw) has been removed. **D.** Sagittal section. ZOOMING IN Which two bones make up each side of the hard palate? What is a foramen? Which bone makes up the superior and middle conchae?

Facial Bones The facial portion of the skull is composed of 14 bones (see FIG. 6-6):

- The **mandible** (MAN-dih-bl), or lower jaw bone, is the skull's only movable bone.
- The two **maxillae** (mak-SIL-e) fuse in the midline to form the upper jaw bone, including the anterior part of the hard palate (roof of the mouth). Each maxilla contains a large air space, called the **maxillary sinus**, that communicates with the nasal cavity.
- The two **zygomatic** (zi-go-MAT-ik) **bones**, one on each side, form the prominences of the cheeks. The zygomatic forms an arch over the cheek with a process of the temporal bone (see FIG. 6-6B).
- Two slender **nasal bones** lie side by side, forming the bridge of the nose.
- The two **lacrimal** (LAK-rih-mal) **bones**, each about the size of a fingernail, form the anterior medial wall of each orbital cavity.
- The **vomer** (VO-mer), shaped like the blade of a plow, forms the inferior part of the nasal septum (see FIG. 6-6A).
- The paired **palatine** (PAL-ah-tine) **bones** form the posterior part of the hard palate (see FIG. 6-6C).
- The two **inferior nasal conchae** (KON-ke) extend horizontally along the lateral wall (side) of the nasal cavities. (As noted, the paired superior and middle conchae are part of the ethmoid bone, as shown in FIG. 6-5C.)

In addition to the cranial and facial bones, there are three tiny bones, or **ossicles** (OS-sik-ls), in each middle ear (see Chapter 10), and just below the mandible (lower jaw), a single horseshoe, or U-shaped, bone called the **hyoid** (HI-oyd) **bone,** to which the tongue and other muscles are attached (see FIG. 6-6B).

Infant Skull The infant's skull has areas in which the bone formation is incomplete, leaving membranous "soft spots," properly called **fontanels** (fon-tah-NELS) (also spelled *fontanelles*) (FIG. 6-7). These flexible regions allow the skull to compress and change shape during the birth process. They also allow for rapid brain growth during infancy. Although there are a number of fontanels, named for their location or the bones they border, the largest and most recognizable is near the front of the skull at the junction of the two parietal bones and the frontal bone. This anterior fontanel usually does not close until the child is about 18 months old.

See the Student Resources on thePoint® for a summary table of the cranial and facial bones and figures on the skeletal features of the head and neck.

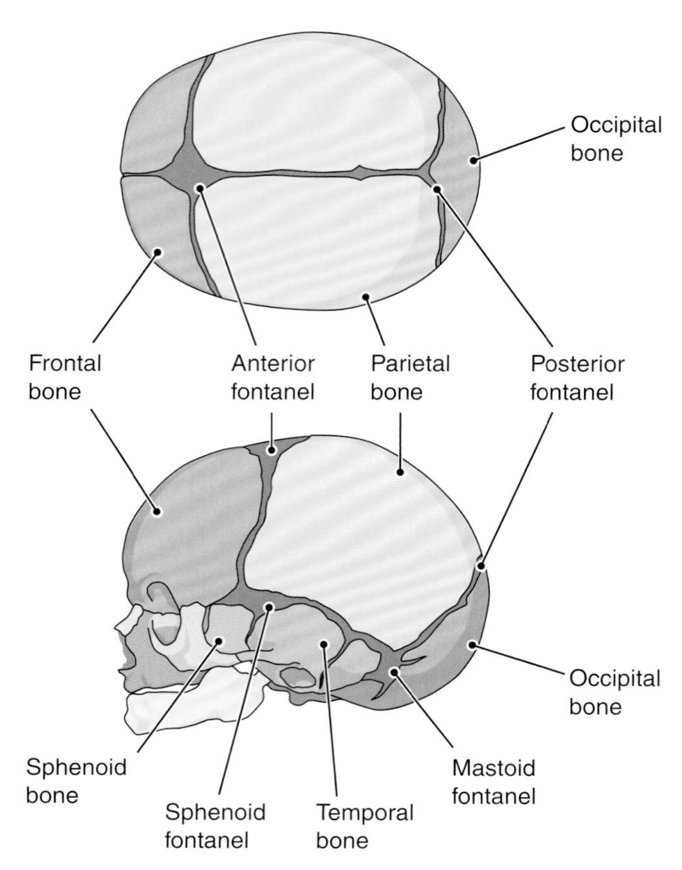

FIGURE 6-7 **Infant skull, showing fontanels.** KEY POINT Fibrous membranes between the skull bones allow the skull to compress during childbirth. Sutures later form in these areas. ZOOMING IN Which is the largest fontanel?

FRAMEWORK OF THE TRUNK

The bones of the trunk include the spine, or **vertebral** (VER-teh-bral) **column**, and the bones of the chest, or **thorax** (THO-raks).

Vertebral Column This bony sheath for the spinal cord is made of a series of irregularly shaped bones. These number 33 or 34 in the child, but because of fusions that occur later in the lower part of the spine, there usually are just 26 separate bones in the adult spinal column. FIGURE 6-8 shows a lateral view of the vertebral column.

Each **vertebra** (VER-teh-brah) (aside from the first two) has a drum-shaped **body** located anteriorly (toward the front) that serves as the weight-bearing part; disks of cartilage between the vertebral bodies absorb shock and provide flexibility (see FIG. 6-8A and B). In the center of each vertebra is a large hole, the vertebral foramen. When all the vertebrae are linked in series by strong connective tissue bands (ligaments), these spaces form the spinal canal, a bony cylinder that protects the spinal cord. Projecting posteriorly (toward the back) from the bony arch that encircles the spinal cord is the **spinous process**, which usually can be felt just under the skin of the back. Projecting

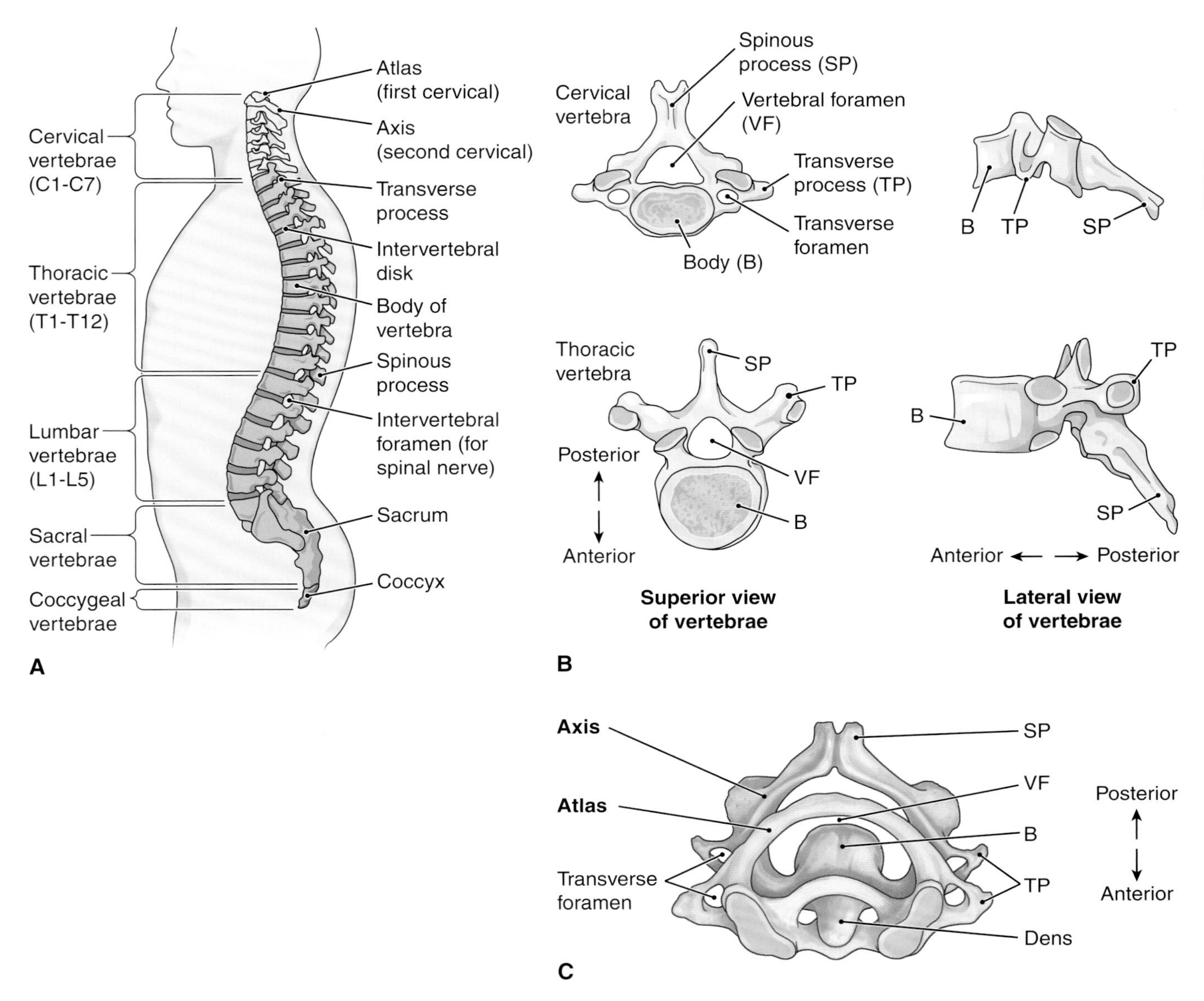

FIGURE 6-8 The vertebral column and vertebrae. A. Vertebral column, left lateral view. **B.** Features of the vertebrae. The blue areas on the vertebrae show points of contact with other bones. **C.** Atlas and axis, superior view. The first two cervical vertebrae are adapted to support the skull and allow for movements of the head in different directions. **KEY POINT** The adult spine has five regions and four curves. **ZOOMING IN** From an anterior view, which group(s) of vertebrae form a convex curve? Which vertebrae are the largest and heaviest? Why?

laterally is a **transverse process** on each side. These processes are attachment points for muscles. Other processes form joints with adjacent vertebrae. A lateral view of the vertebral column shows a series of **intervertebral foramina**, formed between the vertebrae as they join together. Spinal nerves emerge from the spinal cord through these openings (see FIG. 6-8A).

The bones of the vertebral column are named and numbered from superior to inferior and according to location. There are five groups:

- The **cervical** (SER-vih-kal) **vertebrae**, seven in number (C1 to C7), are located in the neck. The first vertebra, called the **atlas**, supports the head (see FIG. 6-8C). (This vertebra is named for the mythologic character who was able to support the world in his hands.) When you nod your head, the skull rocks on the atlas at the occipital bone. The second cervical vertebra, the **axis** (see FIG. 6-8C), serves as a pivot when you turn your head from side to side. It has an upright toothlike part, the dens, that projects into the atlas as a pivot point. The absence of a body in these vertebrae allows for the extra movement. Only the cervical vertebrae have a hole in the transverse process on each side (see FIG. 6-8B and C). These **transverse foramina** accommodate blood vessels and nerves that supply the neck and head.

- The **thoracic vertebrae,** 12 in number (T1 to T12), are located in the chest. They are larger and stronger than the cervical vertebrae and have a longer spinous process that points downward (see FIG. 6-8B). The posterior ends of the 12 pairs of ribs are attached to the transverse processes of these vertebrae.
- The **lumbar vertebrae,** five in number (L1 to L5), are located in the small of the back. They are larger and heavier than the vertebrae superior to them and can support more weight (see FIG. 6-8A). All of their processes are shorter and thicker.
- The **sacral** (SA-kral) **vertebrae** are five separate bones in the child. They eventually fuse to form a single bone, called the **sacrum** (SA-krum), in the adult. Wedged between the two hip bones, the sacrum completes the posterior part of the bony pelvis.
- The **coccygeal** (kok-SIJ-e-al) **vertebrae** consist of four or five tiny bones in the child. These later fuse to form a single bone, the **coccyx** (KOK-siks), or tail bone, in the adult.

Spinal Curves When viewed from the side, the adult vertebral column shows four curves, corresponding to the four vertebral groups (see FIG. 6-8A). In the fetus, the entire column is concave forward (like a letter "C" and your spine when you assume a "fetal position"). This is the primary curve.

When an infant begins to assume an erect posture, secondary curves develop. The cervical curve is convex and appears as the baby holds its head up at about 3 months of age. The lumbar curve is also convex and appears when the child begins to walk. The thoracic and sacral curves remain the two primary concave curves. These curves of the vertebral column provide some of the resilience and spring so essential in balance and movement.

Thorax The bones of the thorax form a cone-shaped cage (FIG. 6-9). Twelve pairs of **ribs** form the bars of this cage, completed anteriorly by the **sternum** (STER-num), or breastbone. These bones enclose and protect the heart, lungs, and other organs contained in the thorax.

The superior portion of the sternum is a roughly triangular **manubrium** (mah-NU-bre-um) that joins laterally on the right and left with a clavicle (collarbone). (The name manubrium comes from a Latin word meaning "handle.") The point on the manubrium where the clavicle joins can be seen on FIGURE 6-9 is the clavicular notch. Laterally and inferiorly, the manubrium joins with the anterior ends of the first pair of ribs. The sternum's body is long and bladelike. It joins along each side with ribs 2 through 7. Where the manubrium joins the body of the sternum, there is a slight elevation, the **sternal angle,** which easily can be felt as a surface landmark.

The inferior end of the sternum consists of a small tip that is made of cartilage in youth but becomes bone in the adult. This is the **xiphoid** (ZIF-oyd) **process.** It is used as a landmark for cardiopulmonary resuscitation (CPR) to locate the region for chest compression.

All 12 ribs on each side are attached to the vertebral column posteriorly. However, variations in the anterior attachment of these slender, curved bones have led to the following classification:

- **True ribs,** the first seven pairs, are those that attach directly to the sternum by means of individual extensions called *costal* (KOS-tal) *cartilages.*

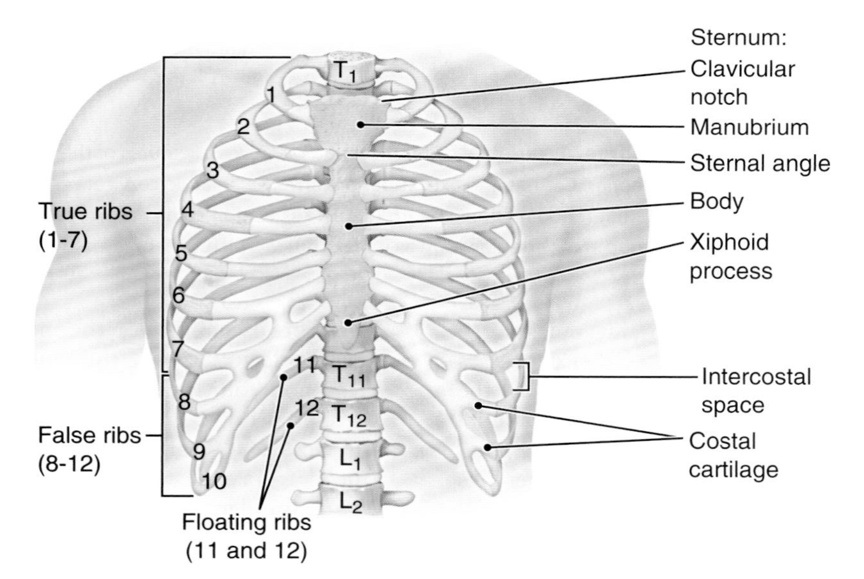

FIGURE 6-9 Bones of the thorax, anterior view. KEY POINT The first seven pairs of ribs are the true ribs; pairs 8 through 12 are the false ribs, of which the last two pairs are also called floating ribs. ZOOMING IN To what bones do the costal cartilages attach?

- **False ribs** are the remaining five pairs. Of these, the eighth, ninth, and 10th pairs attach to the cartilage of the rib above. The last two pairs have no anterior attachment at all and are known as **floating ribs**.

The spaces between the ribs, called *intercostal spaces*, contain muscles, blood vessels, and nerves.

CHECKPOINTS

- [ ] **6-8** Which bones make up the skeleton of the trunk?
- [ ] **6-9** What are the five regions of the vertebral column?

Bones of the Appendicular Skeleton

The appendicular skeleton includes an upper division and a lower division. The upper division on each side includes the shoulder, the arm (between the shoulder and the elbow), the forearm (between the elbow and the wrist), the wrist, the hand, and the fingers. The lower division includes the hip (part of the pelvic girdle), the thigh (between the hip and the knee), the leg (between the knee and the ankle), the ankle, the foot, and the toes.

THE UPPER DIVISION OF THE APPENDICULAR SKELETON

The bones of the upper division may be divided into two groups, the shoulder girdle and the upper extremity.

The Shoulder Girdle The shoulder girdle consists of two bones (FIG. 6-10).

- The **clavicle** (KLAV-ih-kl), or collarbone, is a slender bone with two shallow curves. It joins the sternum anteriorly and the scapula laterally and helps to support the shoulder. Because it often receives the full force of falls on outstretched arms or of blows to the shoulder, it is the most frequently broken bone.
- The **scapula** (SKAP-u-lah), or shoulder blade, is shown from anterior and posterior views in FIGURE 6-10. The spine of the scapula is the posterior raised ridge that can be felt behind the shoulder in the upper portion of the back. Muscles that move the arm attach to fossae (depressions), known as the **supraspinous fossa** and the **infraspinous fossa**, superior and inferior to the scapular spine. The **acromion** (ah-KRO-me-on) is the process that joins the clavicle. You can feel this as the highest point of your shoulder. Below the acromion, there is a shallow socket, the **glenoid cavity**, that forms a ball-and-socket joint with the arm bone (humerus). Medial to the glenoid cavity is the **coracoid** (KOR-ah-koyd) **process**, to which arm and chest muscles and ligaments attach.

The Upper Extremity The upper extremity is also called the upper limb, or simply the arm, although technically, the arm is only the region between the shoulder and the elbow. The region between the elbow and wrist is the forearm. The upper extremity consists of the following bones:

- The proximal bone is the **humerus** (HU-mer-us), or arm bone (FIG. 6-11). The head of the humerus articulates (forms a joint) with the glenoid cavity of the scapula. The distal end has a projection on each side, the medial and lateral **epicondyles** (ep-ih-KON-diles), to which tendons attach, and a midportion, the **trochlea** (TROK-le-ah), that forms a joint with the ulna of the forearm. (The name comes from a word that means "pulley wheel" because of its shape.)
- The forearm bones are the **ulna** (UL-nah) and the **radius** (RA-de-us). In the anatomic position, the ulna lies on the medial side of the forearm in line with the little finger, and the radius lies laterally, above the thumb (see FIG. 6-11). When the forearm is supine, with the palm up or forward, the two bones are parallel; when the

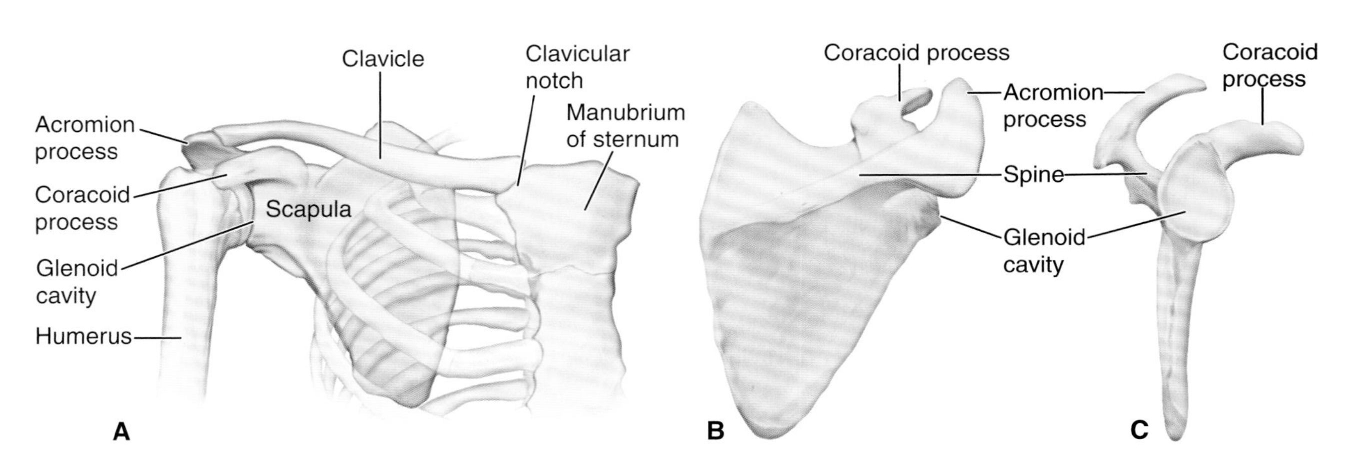

FIGURE 6-10 **The shoulder girdle.** KEY POINT The shoulder girdle consists of the clavicle and scapula. **A.** Bones of the left shoulder girdle, anterior view. **B.** Bones of the left shoulder girdle, posterior view. **C.** Scapula, lateral view. ZOOMING IN What does the prefix *supra-* mean?

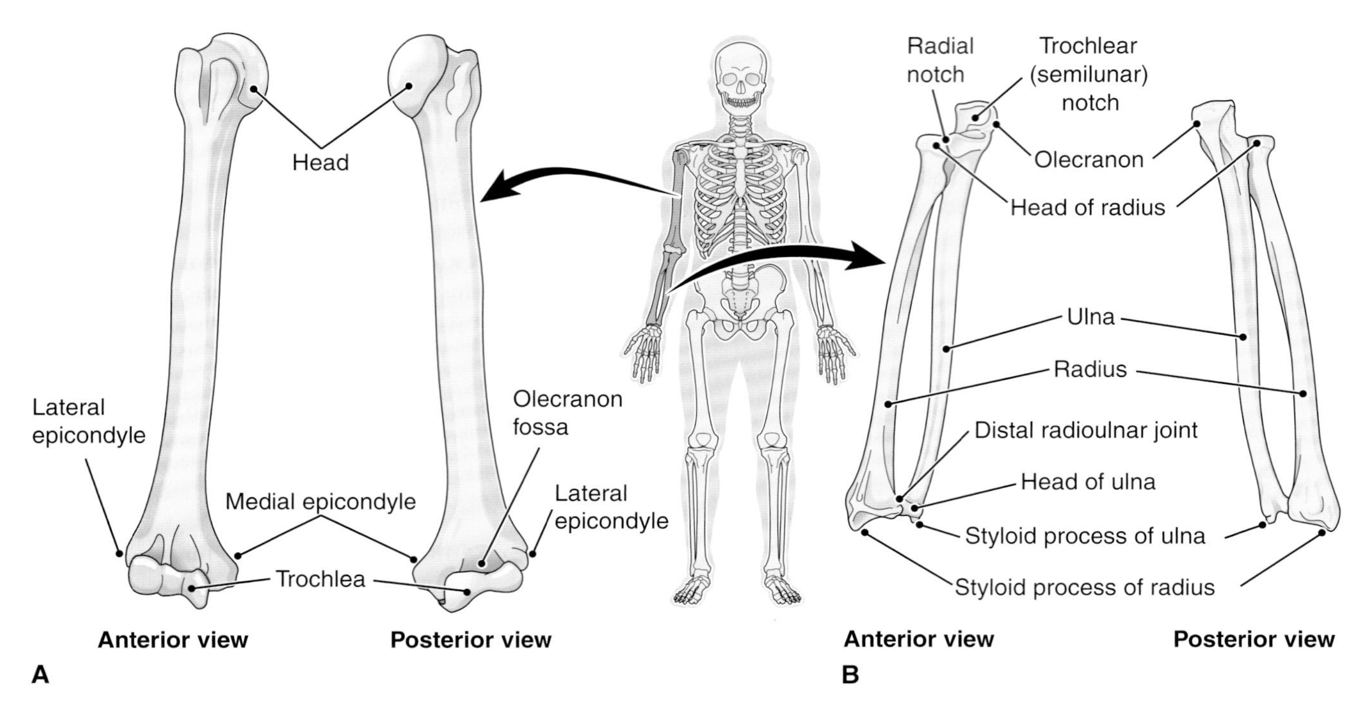

FIGURE 6-11 Bones of the upper extremity. KEY POINT The upper extremity consists of the arm and forearm **A.** The humerus of the right arm in anterior and posterior view. **B.** The radius and ulna of the right forearm in anterior and posterior view. ZOOMING IN What is the medial bone of the forearm?

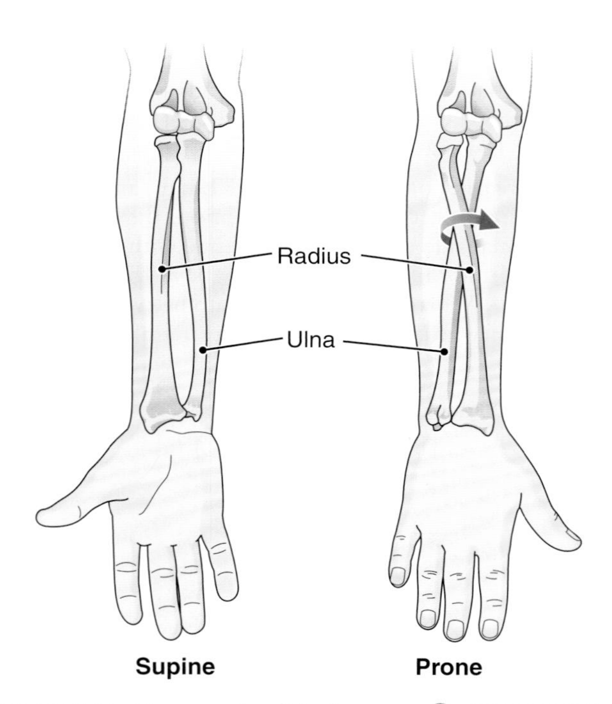

FIGURE 6-12 Movements of the forearm. KEY POINT When the palm is supine (facing up or forward), the radius and ulna are parallel. When the palm is prone (facing down or to the rear), the radius crosses over the ulna.

forearm is prone, with the palm down or back, the distal end of the radius rotates around the ulna so that the shafts of the two bones are crossed (FIG. 6-12). In this position, a distal projection (styloid process) of the ulna shows at the outside of the wrist.

The proximal end of the ulna has the large **olecranon** (o-LEK-rah-non), a process that forms the point of the elbow (FIG. 6-13). At the posterior elbow joint, the olecranon fits into a depression of the distal humerus, the **olecranon fossa**. The trochlea of the distal humerus fits into the ulna's deep **trochlear notch**, allowing a hinge action at the elbow joint. This ulnar depression, because of its deep

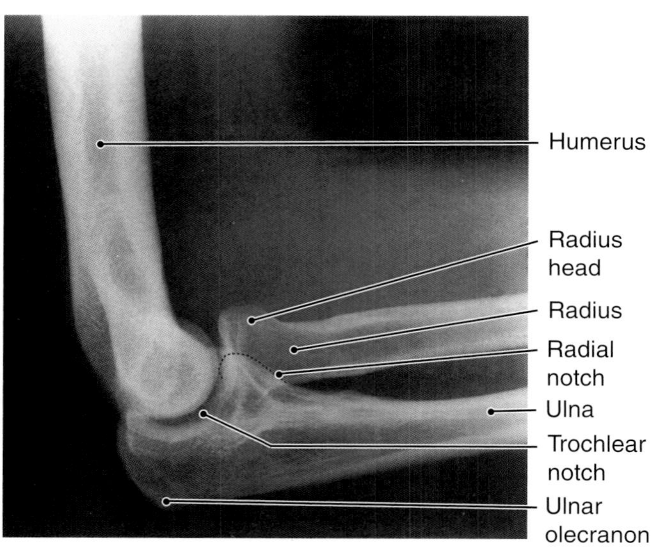

FIGURE 6-13 Left elbow, lateral view. ZOOMING IN What part of what bone forms the bony prominence of the elbow?

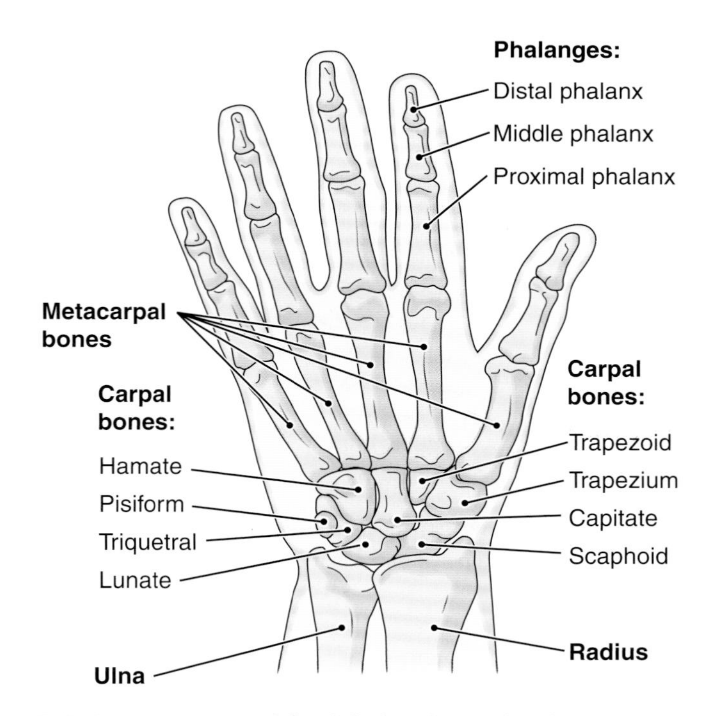

FIGURE 6-14 Bones of the right hand, anterior view.
ZOOMING IN How many phalanges are there on each hand?

half-moon shape, is also known as the semilunar notch (see FIG. 6-13).

- The wrist contains eight small **carpal** (KAR-pal) **bones** arranged in two rows of four each. The names of these eight different bones are given in FIGURE 6-14. Note that the anatomic wrist, composed of the carpal bones, is actually the heel of the hand. We wear a "wristwatch" over the distal ends of the radius and ulna.
- Five **metacarpal bones** are the framework for the palm of each hand. Their rounded distal ends form the knuckles.
- There are 14 **phalanges** (fah-LAN-jeze), or finger bones, in each hand, two for the thumb and three for each finger. Each of these bones is called a **phalanx** (FA-lanx). They are identified as the proximal, which is attached to a metacarpal; the middle; and the distal. Note that the thumb has only two phalanges, a proximal and a distal (see FIG. 6-14).

See the Student Resources on thePoint® for figures on the skeletal features of the shoulder, torso, and upper extremity.

THE LOWER DIVISION OF THE APPENDICULAR SKELETON

The bones of the lower division also fall into two groups, the pelvis and the lower extremity.

The Pelvic Bones The hip bone, or **os coxae**, begins its development as three separate bones that later fuse (FIG. 6-15). These individual bones are the following:

- The **ilium** (IL-e-um) forms the upper, flared portion. The **iliac** (IL-e-ak) **crest** is the curved rim along the ilium's superior border. It can be felt just below the waist. At either end of the crest are two bony projections. The most prominent of these is the **anterior superior iliac spine**, which is often used as a surface landmark in diagnosis and treatment.

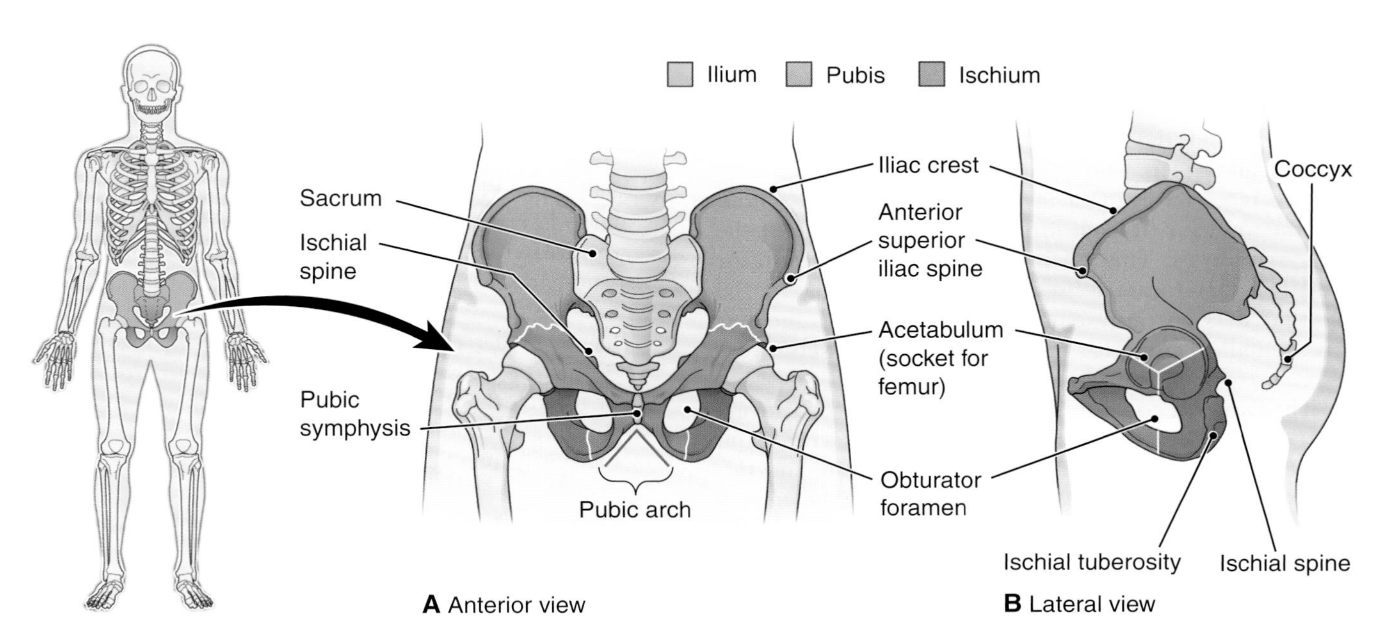

FIGURE 6-15 The pelvic bones. KEY POINT The hip bone, or os coxae, is formed of three fused bones. **A.** Anterior view. **B.** Lateral view showing the joining of the three pelvic bones to form the acetabulum. ZOOMING IN What bone is nicknamed the "sit bone?"

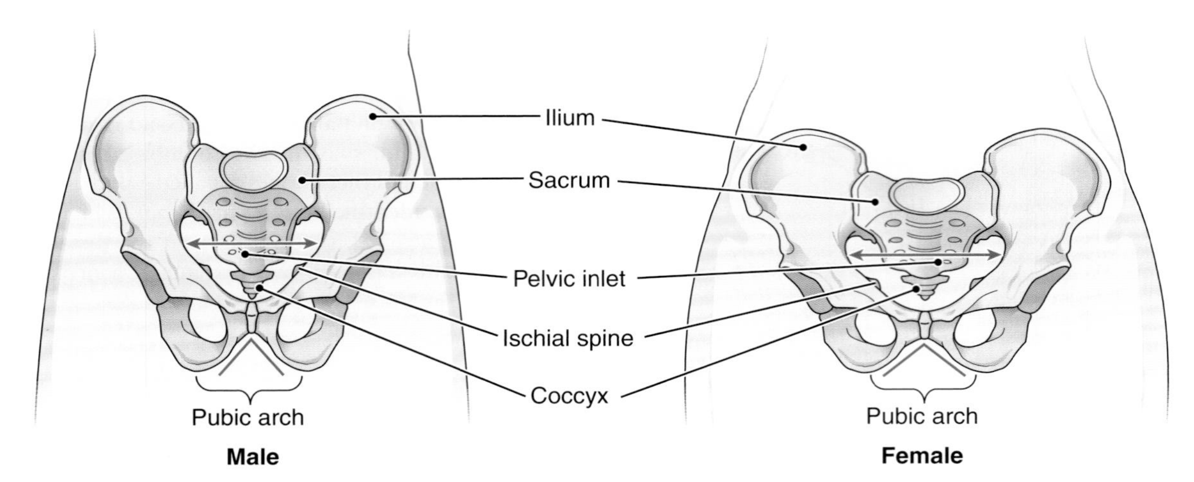

FIGURE 6-16 Comparison of male and female pelvis, anterior view. **KEY POINT** The female pelvis is adapted for pregnancy and childbirth. Note the broader angle of the pubic arch and the wider pelvic outlet in the female. Also, the ilia are wider and more flared; the sacrum and coccyx are shorter and less curved.

- The **ischium** (IS-ke-um) is the lowest and strongest part. The **ischial** (IS-ke-al) **spine** at the posterior of the pelvic outlet is used as a reference point during childbirth to indicate the progress of the presenting part (usually the baby's head) down the birth canal. Just inferior to this spine is the large **ischial tuberosity**, which helps support the trunk's weight when a person sits down. You may sometimes be aware of this ischial projection when sitting on a hard surface for a while.
- The **pubis** (PU-bis) forms the anterior part of the os coxae. The joint formed by the union of the two hip bones anteriorly is called the **pubic symphysis** (SIM-fih-sis). This joint becomes more flexible late in pregnancy to allow for passage of the baby's head during childbirth.

Portions of all three pelvic bones contribute to the formation of the **acetabulum** (as-eh-TAB-u-lum), the deep socket that holds the head of the femur (thigh bone) to form the hip joint (see FIG. 6-15).

The largest foramina in the entire body are found near the anterior of each hip bone on either side of the pubic symphysis. This opening is named the **obturator** (OB-tu-ra-tor) **foramen** (see FIG. 6-15), referring to the fact that it is partially closed by a membrane and has only a small opening for passage of blood vessels and a nerve.

The two ossa coxae join in forming the pelvis, a strong bony girdle completed posteriorly by the sacrum and coccyx of the spine. The pelvis supports the trunk and surrounds the organs in the pelvic cavity, including the urinary bladder, the internal reproductive organs, and parts of the intestine.

The female pelvis is adapted for pregnancy and childbirth (FIG. 6-16). Some ways in which the female pelvis differs from that of the male are as follows:

- It is lighter in weight.
- The ilia are wider and more flared.
- The pubic arch, the anterior angle between the pubic bones, is wider.
- The pelvic inlet, the upper opening, bordered by the pubic joint and sacrum, is wider and more rounded.
- The pelvic outlet, the lower opening, bordered by the pubic joint and coccyx, is larger.
- The sacrum and coccyx are shorter and less curved.

The Lower Extremity The lower extremity is also called the lower limb, or simply the leg, although technically the leg is only the region between the knee and the ankle. The portion of the extremity between the hip and the knee is the thigh. The lower extremity consists of the following bones:

- The **femur** (FE-mer), the thigh bone, is the longest and strongest bone in the body. Proximally, it has a large ball-shaped head that joins the os coxae (FIG. 6-17). The large lateral projection near the head of the femur is the **greater trochanter** (tro-KAN-ter), used as a surface landmark. Movements of the greater trochanter can indicate the degree of hip mobility. The **lesser trochanter**, a smaller elevation, is located on the medial side. A fracture between the greater and lesser trochanters, or intertrochanteric fracture, is the injury described in Reggie's opening case study. On the posterior surface, there is a long central ridge, the **linea aspera** (literally "rough line"), which is a point for attachment of hip muscles. The distal anterior **patellar surface** articulates with the kneecap.
- The **patella** (pah-TEL-lah), or kneecap (see FIG. 6-1), is embedded in the tendon of the large anterior thigh muscle, the quadriceps femoris, where it crosses the knee

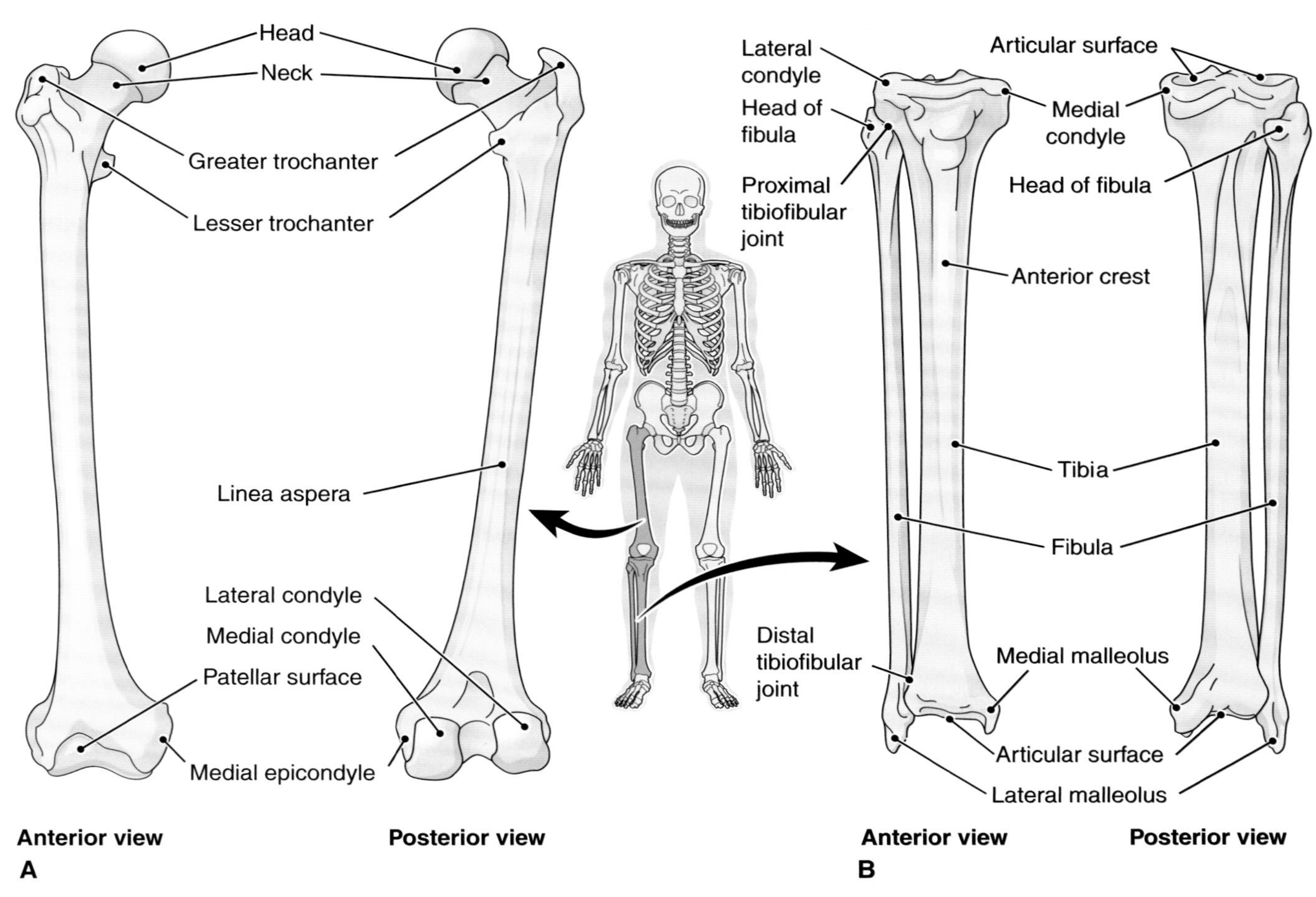

FIGURE 6-17 Bones of the lower extremity. KEY POINT The lower extremity consists of the thigh and leg. **A.** The femur of the right thigh. **B.** The tibia and fibula of the right leg. ZOOMING IN What is the lateral bone of the leg? Which bone of the leg is weight bearing?

joint. It is an example of a **sesamoid** (SES-ah-moyd) **bone**, a type of bone that develops within a tendon or a joint capsule.

- There are two bones in the leg (see FIG. 6-17). Medially (on the great toe side), the **tibia**, or shin bone, is the longer, weight-bearing bone. Its proximal surface articulates with the distal femur. The tibia has a sharp anterior crest that can be felt as the shin bone. Laterally, the slender **fibula** (FIB-u-lah) does not reach the knee joint; thus, it is not a weight-bearing bone. The **medial malleolus** (mal-LE-o-lus) is a downward projection at the tibia's distal end; it forms the prominence on the inner aspect of the ankle. The **lateral malleolus**, at the fibula's distal end, forms the prominence on the outer aspect of the ankle. Most people think of these projections as their "ankle bones," whereas, in truth, they are features of the tibia and fibula.
- The structure of the foot is similar to that of the hand. However, the foot supports the body's weight, so it is stronger and less mobile than the hand. There are seven **tarsal bones** associated with the ankle and foot. These are named and illustrated in FIGURE 6-18. The

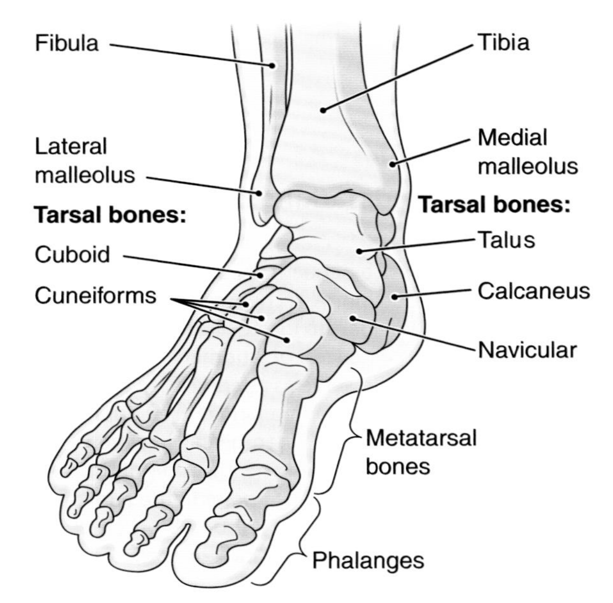

FIGURE 6-18 Bones of the right foot. ZOOMING IN Which tarsal bone is the heel bone? Which tarsal bone forms a joint with the tibia?

Table 6-1 Bones of the Skeleton

Region	Bones	Description
Axial Skeleton		
Skull		
Cranium	Cranial bones (8)	Chamber enclosing the brain; houses the ear and forms part of the eye socket
Facial portion	Facial bones (14)	Form the face and chambers for sensory organs
Hyoid		U-shaped bone under lower jaw; used for muscle attachments
Ossicles	Ear bones (3)	Transmit sound waves through middle ear
Trunk		
Vertebral column	Vertebrae (26)	Enclose the spinal cord
Thorax	Sternum	Anterior bone of the thorax
	Ribs (12 pairs)	Enclose the organs of the thorax
Appendicular Skeleton		
Upper division		
Shoulder girdle	Clavicle	Anterior; between sternum and scapula
	Scapula	Posterior; anchors muscles that move arm
Upper extremity	Humerus	Arm bone
	Ulna	Medial bone of forearm
	Radius	Lateral bone of forearm
	Carpals (8)	Wrist bones
	Metacarpals (5)	Bones of palm
	Phalanges (14)	Bones of fingers
Lower division		
Pelvis	Os coxae (2)	Join sacrum and coccyx of vertebral column to form the bony pelvis
Lower extremity	Femur	Thigh bone
	Patella	Kneecap
	Tibia	Medial bone of leg
	Fibula	Lateral bone of leg
	Tarsal bones (7)	Ankle bones
	Metatarsals (5)	Bones of instep
	Phalanges (14)	Bones of toes

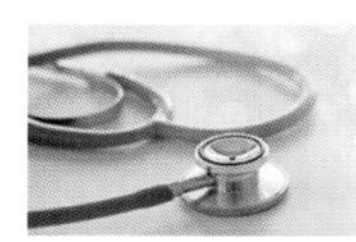

CLINICAL PERSPECTIVES

BOX 6-2

Landmarking: Seeing with Your Fingers

Most body structures lie beneath the skin, hidden from direct view except in dissection. A technique called **landmarking** allows healthcare providers to locate hidden structures simply and easily. Bony prominences, or landmarks, can be palpated (felt) beneath the skin to serve as reference points for locating other internal structures. Landmarking is used during physical examinations and surgeries, when giving injections, and for many other clinical procedures. The lower tip of the sternum, the xiphoid process, is a reference point in the administration of cardiopulmonary resuscitation (CPR).

Practice landmarking by feeling for some of the other bony prominences. You can feel the joint between the mandible and the temporal bone of the skull (the temporomandibular joint, or TMJ) anterior to the ear canal as you move your lower jaw up and down. Feel for the notch in the sternum (breast bone) between the clavicles (collar bones). Approximately 4 cm below this notch, you will feel a bump called the sternal angle. This prominence is an important landmark because its location marks where the trachea splits to deliver air to both lungs. Move your fingers lateral to the sternal angle to palpate the second ribs, important landmarks for locating the heart and lungs. Feel for the most lateral bony prominence of the shoulder, the acromion process of the scapula (shoulder blade). Two to three fingerbreadths down from this point is the correct injection site into the deltoid muscle of the shoulder. Place your hands on your hips and palpate the iliac crest of the hip bone. Move your hands forward until you reach the anterior end of the crest, the anterior superior iliac spine (ASIS). Feel for the part of the bony pelvis that you sit on. This is the ischial tuberosity. This and the ASIS are important landmarks for locating safe injection sites in the gluteal region.

largest of these is the **calcaneus** (kal-KA-ne-us), or heel bone. The **talus** above it forms the ankle joint with the tibia.

- Five **metatarsal bones** form the framework of the instep, and the heads of these bones form the ball of the foot (see FIG. 6-18).
- The phalanges of the toes are counterparts of those in the fingers. There are three of these in each toe except for the great toe, which has only two.

See **TABLE 6-1** for a summary outline of all the bones of the skeleton. It can be helpful to locate some of these bones and bone markings on your own body. To find out how these markings can be used in healthcare, see **BOX 6-2**, "Landmarking: Seeing with your Fingers."

CHECKPOINTS 

- ☐ **6-10** What are the four regions of the appendicular skeleton?
- ☐ **6-11** Where would you find phalanges?

CASEPOINTS 

- ☐ **6-4** Reggie's fracture extended from the greater trochanter to the lesser trochanter. Does it cross the proximal or distal end of the bone?
- ☐ **6-5** See **FIGURE 6-2** to revisit the structure of a long bone. Would this fracture pass through the medullary cavity?

See the Student Resources on thePoint® for figures on the skeletal features of the lower extremity.

Table 6-2 Joints

Type	Material between the Bones	Examples
Immovable (synarthrosis)	Fibrous: No joint cavity; fibrous connective tissue between bones	Sutures between skull bones
Slightly movable (amphiarthrosis)	No joint cavity; cartilage (or sometimes fibrous tissue) between bones	Pubic symphysis; joints between vertebral bodies
Freely movable (diarthrosis)	Joint cavity containing synovial fluid	Gliding, hinge, pivot, condyloid, saddle, ball-and-socket joints

The Joints

An **articulation**, or **joint**, is an area of junction or union between two or more bones. Joints are classified into three main types according to the degree of movement permitted. The joints also differ in the type of material between the adjoining bones (TABLE 6-2):

- **Synarthrosis** (sin-ar-THRO-sis). The bones in this type of joint are held together so tightly that they cannot move in relation to one another. Most synarthroses use fibrous tissue to join the bones, so they are often described as **fibrous joints**. An example is a suture between bones of the skull.
- **Amphiarthrosis** (am-fe-ar-THRO-sis). This type of joint is slightly moveable. For example, the radius and ulna are joined by a large band of fibrous tissue that permits slight movement, so this joint is a fibrous amphiarthrosis. Most amphiarthroses, however, use cartilage to join the bones and are thus described as **cartilaginous joints**. The joint between the pubic bones of the pelvis—the pubic symphysis—and the joints between the bodies of the vertebrae are examples.
- **Diarthrosis** (di-ar-THRO-sis). Diarthroses are freely moveable joints. The bones in this type of joint have a potential space between them called the **joint cavity**, which contains a small amount of thick, colorless fluid. This lubricant, **synovial fluid**, resembles uncooked egg white (*ov* is the root, meaning "egg") and is secreted by the membrane that lines the joint cavity. For this reason, diarthroses are also called **synovial** (sin-O-ve-al) **joints.** Most of the body's joints are synovial joints; they are described in more detail next.

MORE ABOUT SYNOVIAL JOINTS

The bones in freely movable joints are held together by **ligaments**, bands of dense regular connective tissue. Additional ligaments reinforce and help stabilize the joints at various points (FIG. 6-19). Also, for strength and protection, there is a **joint capsule** of fibrous connective tissue that encloses each joint and is continuous with the periosteum of the bones (see FIG. 6-19B). A smooth layer of hyaline cartilage called the **articular** (ar-TIK-u-lar) **cartilage** protects the bone surfaces in synovial joints. Some complex joints may have additional cushioning cartilage between the bones, such as the crescent-shaped medial menisci (meh-NIS-si) and lateral menisci in the knee joint (FIG. 6-20). Fat may also appear as padding around a joint.

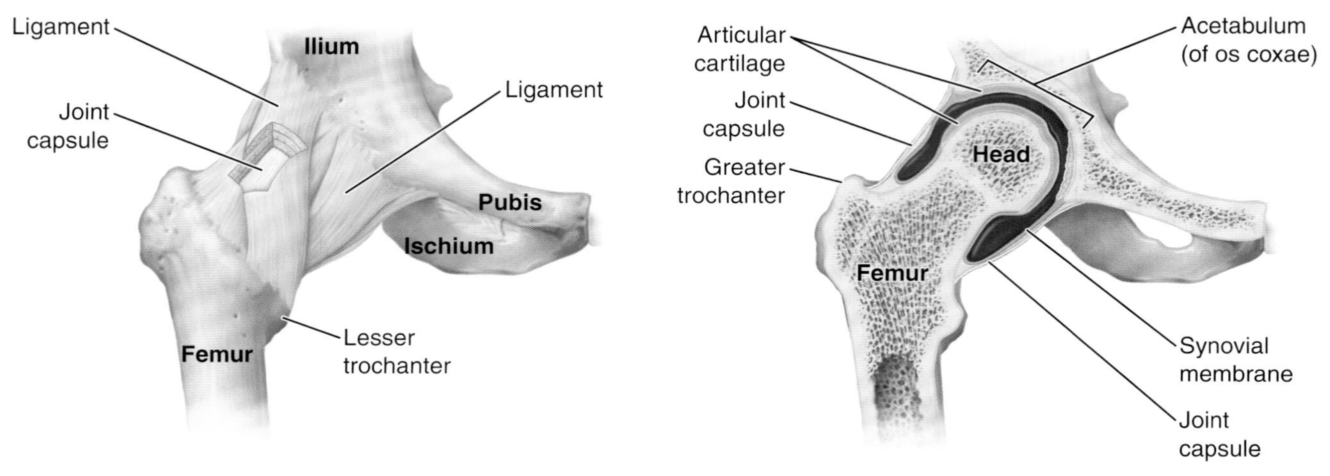

FIGURE 6-19 Structure of a synovial joint. KEY POINT Connective tissue structures stabilize and protect synovial joints. **A.** Anterior view of the hip joint showing ligaments that reinforce and stabilize the joint. **B.** Frontal section through right hip joint showing protective structures. ZOOMING IN What is the purpose of the greater trochanter of the femur? What type of tissue covers and protects the ends of the bones?

Near some joints are small sacs called **bursae** (BER-se), which are filled with synovial fluid (see FIG. 6-20). These lie in areas subject to stress and help ease movement over and around the joints.

Movement at Synovial Joints Freely movable joints allow the articulating bones to move in relation to each other. For instance, bending the knee joint moves the leg in relation to the thigh. The specific terms describing movements assume the anatomic position introduced in Chapter 1 (see FIG. 1-6). Experiment with moving your body in these different directions as you read these descriptions and examine the illustrations. There are four kinds of angular movement, or movement that changes the angle between bones (FIG. 6-21):

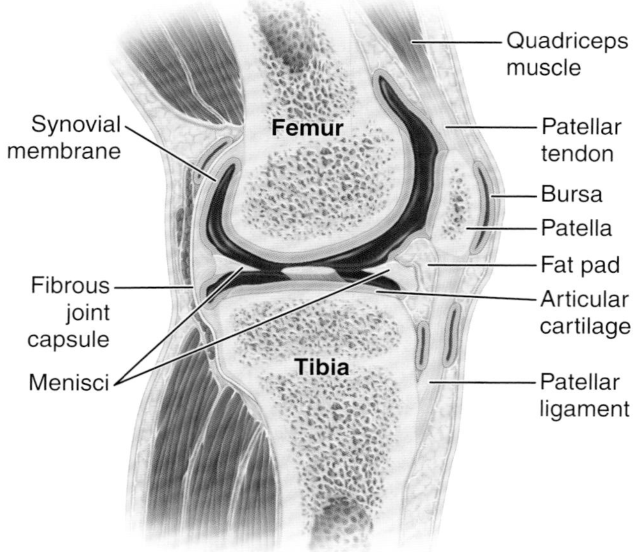

- **Flexion** (FLEK-shun) is a bending motion that decreases the angle between bones away from the anatomic position, as in bending the fingers to close the hand. Arm flexion at the shoulder involves raising the upper limb in front of the body, as in raising your hand to ask a question. Specialized terms describe flexion at the ankle:
  - **Dorsiflexion** (dor-sih-FLEK-shun) bends the foot upward at the ankle, narrowing the angle between the leg and the top of the foot.
  - **Plantar flexion** bends the foot so that the toes point downward, as in toe dancing.
- **Extension** is a straightening motion that increases the angle between bones and returns the joint toward the anatomic position, as in straightening the fingers to open the hand. Arm extension lowers the arm from the flexed position. In **hyperextension**, a part is extended beyond its anatomic position, as in opening the hand to its maximum by hyperextending the fingers or hyperextending the thigh at the hip in preparation for kicking a ball from a standing position.
- **Abduction** (ab-DUK-shun) is movement away from the midline of the body, as in moving the arm straight out to the side.
- **Adduction** is movement toward the midline of the body, as in bringing the arm back to its original position beside the body.

Specialized terms describe movements of the foot in the lateral plane:

- **Inversion** (in-VER-zhun) is the act of turning the sole inward, so that it faces the opposite foot.

FIGURE 6-20 The knee joint, sagittal section. Protective structures are also shown.

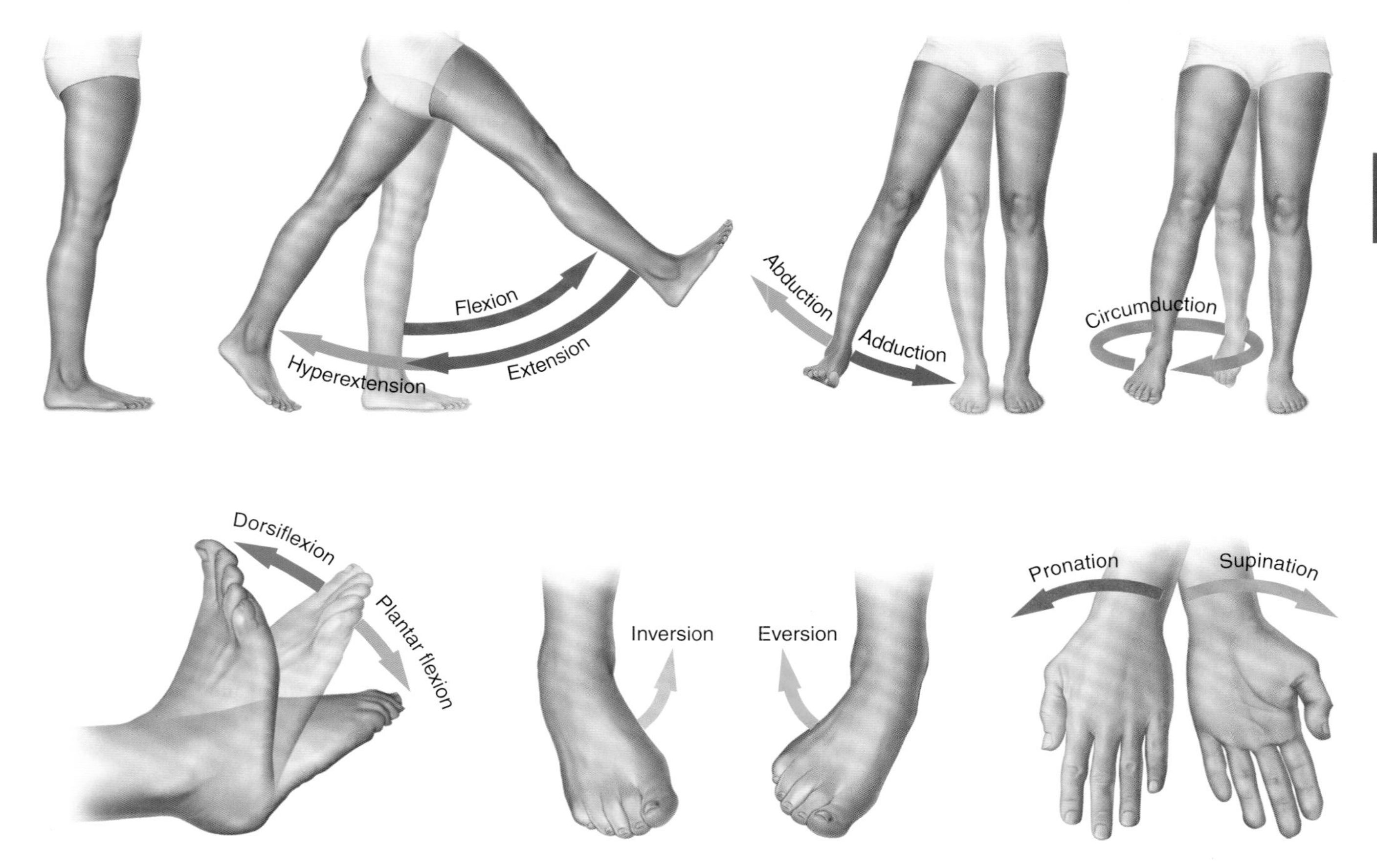

FIGURE 6-21 Movements at synovial joints. Specific movements are labeled on the diagrams and explained more fully in the text. All movements are in reference to the anatomic position (shown first). **KEY POINT** Synovial joints allow the greatest range of motion.

- **Eversion** (e-VER-zhun) turns the sole outward, away from the body.

A combination of angular movements enables one to execute a movement referred to as **circumduction** (ser-kum-DUK-shun). To perform this movement, stand with your arm outstretched and draw a large imaginary circle in the air. Note the smooth combination of flexion, abduction, extension, and adduction that makes circumduction possible.

Rotation refers to a twisting or turning of a bone on its own axis, as in turning the head from side to side to say no. Specialized terms describe rotation of the forearm:

- **Supination** (su-pin-A-shun) is the act of turning the palm up or forward.
- **Pronation** (pro-NA-shun) turns the palm down or backward.

Types of Synovial Joints Synovial joints are classified according to the types of movement they allow, as described and illustrated in **TABLE 6-3**. Locate these types of joints on your body, and demonstrate the different movements they allow. Listed in order of increasing range of motion, they are

- Gliding joint—two relatively flat bone surfaces slide over each other with little change in the joint angle. Examples are the joints between the tarsal and carpal bones.
- Hinge joint—a convex (curving outward) surface of one bone fits into the concave (curving inward) surface of another bone, allowing movement in one direction. Hinge joints allow flexion and extension only. Examples are the elbow joint and the joints between the phalanges.
- Pivot joint—a rounded or pointed portion of one bone fits into a ring in another bone. This joint allows rotation only, as in the joint between the atlas and axis of the cervical spine or the proximal joint between the radius and ulna that allows supination and pronation of the forearm.
- Condyloid joint—an oval-shaped projection of one bone fits into an oval-shaped depression on another bone. This joint allows movement in two directions: flexion and extension and abduction and adduction. Examples are the joints between the metacarpal bones and the proximal phalanges of the fingers.
- Saddle joint—similar to the condyloid joint, but deeper and allowing greater range of motion. One bone fits into

Table 6-3 Synovial Joints

Type of Joint	Type of Movement	Examples
Gliding joint	Flat bone surfaces slide over one another with little change in the joint angle	Joints in the wrist and ankles **(see FIGS. 6-14 and 6-18)**
Hinge joint	Allows movement in one direction, changing the angle of the bones at the joint, as in flexion and extension	Elbow joint; joints between phalanges of fingers and toes **(see FIGS. 6-13, 6-14, and 6-18)**
Pivot joint	Allows rotation around the length of the bone	Joint between the first and second cervical vertebrae; joint at proximal ends of the radius and ulna **(see FIGS. 6-8 and 6-12)**
Condyloid joint	Allows movement in two directions: flexion and extension, abduction and adduction	Joint between the occipital bone of the skull and the first cervical vertebra (atlas) **(see FIG. 6-8)**; joint between the metacarpal and the first phalanx of the finger (knuckle) **(see FIG. 6-14)**
Saddle joint	Like a condyloid joint, but with deeper articulating surfaces and movement in three directions, rotation in addition to flexion and extension, abduction and adduction	Joint between the wrist and the metacarpal bone of the thumb **(see FIG. 6-14)**
Ball-and-socket joint	Allows the greatest range of motion. Permits movement in three directions around a central point, as in circumduction	Shoulder joint and hip joint **(see FIGS. 6-10, 6-15, and 6-23)**

a saddle-like depression on another bone. It allows movement in three directions: flexion and extension, abduction and adduction, and rotation. An example is the joint between the wrist and the metacarpal of the thumb.

- Ball-and-socket joint—a ball-like surface of one bone fits into a deep cuplike depression in another bone. It allows the greatest range of motion in three directions, as in circumduction. Examples are the shoulder and hip joints.

CHECKPOINTS

- ☐ **6-12** What are three types of joints based on the degree of movement they allow?
- ☐ **6-13** Which is the most freely moveable type of joint?

CASEPOINTS

- ☐ **6-6** Give three terms that describe the joint proximal to Reggie's fracture.
- ☐ **6-7** Reggie's lower limb was moved away from the midline of his body. What term describes this movement?

Effects of Aging on the Skeletal System

The aging process includes significant changes in all connective tissues, including bone. There is a loss of calcium salts and a decreased ability to form the protein framework on which calcium salts are deposited. Cellular metabolism slows, so bones are weaker, less dense, and more fragile; fractures and other bone injuries heal more slowly. Muscle tissue is also lost throughout adult life. Loss of balance and diminished reflexes may lead to falls. Thus, there is a tendency to decrease the exercise that is so important to the maintenance of bone tissue.

Changes in the vertebral column with age lead to a loss in height. Approximately 1.2 cm (about 0.5 in) are lost every 20 years beginning at 40 years of age, owing primarily to a thinning of the intervertebral disks (between the bodies of the vertebrae). Even the vertebral bodies themselves may lose height in later years. The costal (rib) cartilages become calcified and less flexible, and the chest may decrease in diameter by 2 to 3 cm (about 1 in), mostly in the lower part.

At the joints, reduction of collagen in bone, tendons, and ligaments contributes to the diminished flexibility so often experienced by older people. Thinning of articular cartilage and loss of synovial fluid may contribute to joint damage. By the process of calcification, minerals may be deposited in and around the joints, especially at the shoulder, causing pain and limiting mobility.

See the Student Resources on thePoint® for information on careers in radiology, a method used to detect bone injuries and disorders.

A & P in Action Revisited: Reggie's Fracture Begins to Heal

"So, Doc, what's the chance my leg's going to heal up enough to play football again?" asked Reggie.

"Well," replied the doctor, "you've had a complicated injury, and it's going to take some time to heal. When that happens, it may be better than new, but we'll have to wait and see what you're your prospects are for catching footballs again."

The surgeon knew that even before the surgery to realign its broken ends, Reggie's femur had already begun to heal itself. Immediately after the injury occurred on the football field, a blood clot formed around the fracture. A day or two later, chemical messengers within the clot would stimulate blood vessels from the periosteum and endosteum to invade the clot, bringing connective tissue cells with them. Over the next several weeks, fibroblasts and chondroblasts in the clot would secrete collagen and cartilage, converting it into a soft callus. Meanwhile, macrophages would remove the remains of the blood clot and osteoclasts would digest dead bone tissue. Soon after, osteoblasts in the callus would convert it into spongy bone called a hard callus. Months after the injury, osteoclasts and osteoblasts would work together to remodel the outer layers of the hard callus into compact bone, resulting in a repair even stronger than the original bone tissue in Reggie's femur.

During this case, we saw how fractured bones are repaired using screws and plates. We also saw that the body has its own "orthopedic surgeons"—cells like osteoblasts and osteoclasts, which can engineer a bone repair that is even stronger than the original.

CHAPTER 6

Chapter Wrap-Up

OVERVIEW

A detailed chapter outline with space for note-taking is on thePoint®. The figure below illustrates the main topics covered in this chapter.

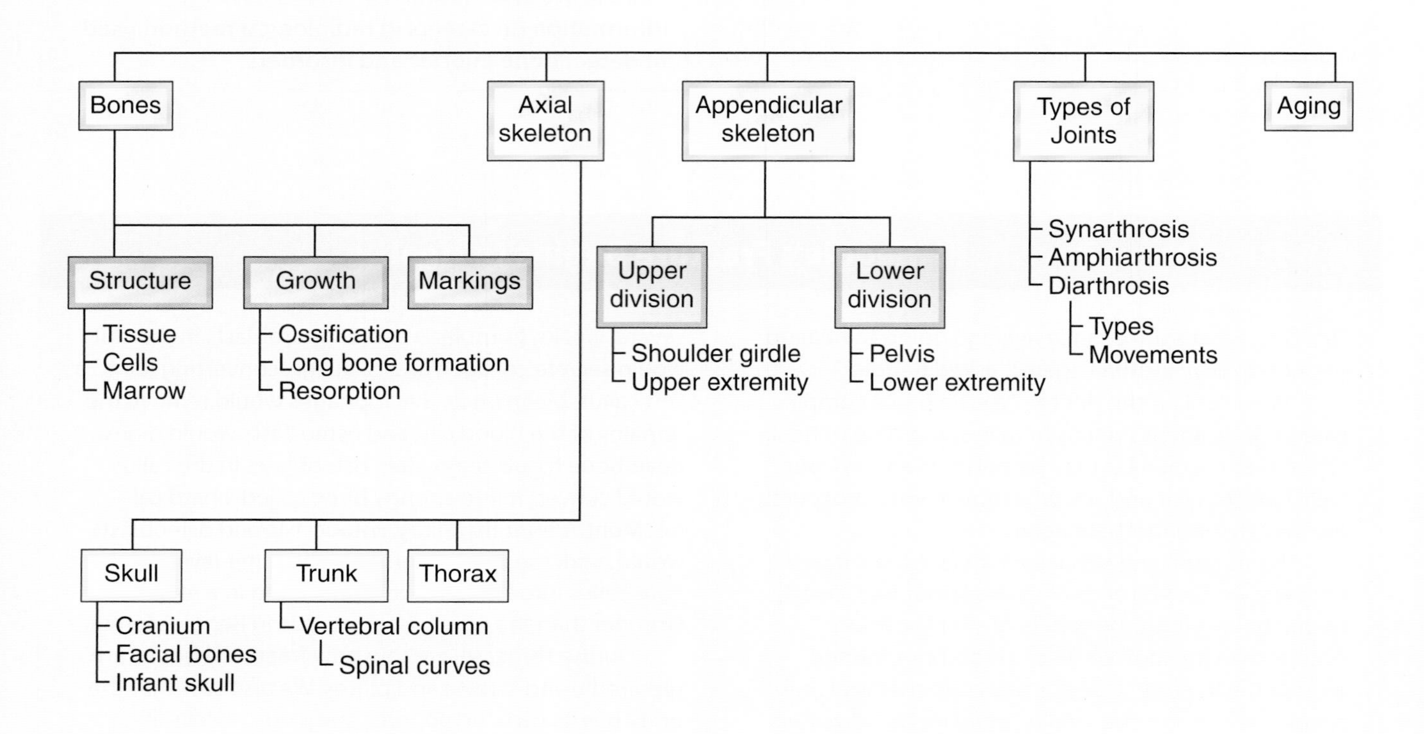

KEY TERMS

The terms listed below are emphasized in this chapter. Knowing them will help you organize and prioritize your learning. These and other boldface terms are defined in the Glossary with phonetic pronunciations.

amphiarthrosis
articulation
bone marrow
bursa
circumduction
diaphysis
diarthrosis
endosteum
epiphysis
extremity
fontanel
joint
osteoblast
osteoclast
osteocyte
osteon
periosteum
resorption
skeleton
synarthrosis
synovial

WORD ANATOMY

Medical terms are built from standardized word parts (prefixes, roots, and suffixes). Learning the meanings of these parts can help you remember words and interpret unfamiliar terms.

WORD PART	MEANING	EXAMPLE
BONES		
-clast	break	An *osteoclast* breaks down bone in the process of resorption.
dia-	through, between	The *diaphysis*, or shaft, of a long bone is between the two ends, or epiphyses.
oss, osse/o	bone, bone tissue	*Osseous* tissue is another name for bone tissue.
oste/o	bone, bone tissue	The *periosteum* is the fibrous membrane around a bone.
BONES OF THE AXIAL SKELETON		
cost/o	rib	*Intercostal* spaces are located between the ribs.
para-	near	The *paranasal* sinuses are near the nose.
pariet/o	wall	The *parietal* bones are the side walls of the skull.
BONES OF THE APPENDICULAR SKELETON		
infra-	below, inferior	The *infraspinous* fossa is a depression inferior to the spine of the scapula.
meta-	near, beyond	The *metacarpal* bones of the palm are near and distal to the carpal bones of the wrist.
supra-	above, superior	The *supraspinous* fossa is a depression superior to the spine of the scapula.
THE JOINTS		
ab-	away from	*Abduction* is movement away from the midline of the body.
ad-	toward, added to	*Adduction* is movement toward the midline of the body.
amphi-	on both sides, around, double	An *amphiarthrosis* is a slightly movable joint.
arthr/o	joint, articulation	A *synarthrosis* is an immovable joint, such as a suture.
circum-	around	*Circumduction* is movement around a joint in a circle.

QUESTIONS FOR STUDY AND REVIEW

BUILDING UNDERSTANDING

Fill in the Blanks

1. The shaft of a long bone is called the _____.
2. The structural unit of compact bone is the _____.
3. Red bone marrow manufactures _____.
4. Bones are covered by a connective tissue membrane called _____.
5. Bone cells active in resorption are _____.

Matching Match each numbered item with the most closely related lettered item.

___ 6. A rounded bony projection
___ 7. A sharp bony prominence
___ 8. An air-filled space in bone
___ 9. A bony depression
___ 10. A short channel or passageway in bone

a. A condyle
b. A sinus
c. A fossa
d. A meatus
e. A spine

Multiple Choice

___ 11. Where is the growth plate of a long bone located?
a. epiphysis
b. articular cartilage
c. marrow cavity
d. endosteum

___ 12. Which bone contains the foramen magnum?
a. temporal
b. hyoid
c. occipital
d. fibula

___ 13. Your anatomy teacher exclaims "Ouch! I bruised my olecranon!" Which part did she injure?
a. big toe
b. elbow
c. ankle
d. tailbone

___ 14. Which type of joint is freely moveable?
a. arthrotic
b. synarthrotic
c. diarthrotic
d. amphiarthrotic

___ 15. What kind of synovial joint is between the atlas and axis?
a. gliding
b. hinge
c. saddle
d. pivot

UNDERSTANDING CONCEPTS

16. List five functions of bone, and describe how a long bone's structure enables it to carry out each of these functions.
17. Explain the differences between the terms in each of the following pairs:
    a. osteoblast and osteocyte
    b. red marrow and yellow marrow
    c. compact bone and spongy bone
    d. synarthrosis and amphiarthrosis
    e. periosteum and endosteum
18. Discuss the process of long bone formation during fetal development and childhood. What role does resorption play in bone formation?
19. Name the five groups of vertebrae. Explain how the different structures of the different vertebrae correspond to their functions.
20. Referring to the "The Body Visible" at the beginning of the book, give the numbers of the following:
    a. large proximal projection of the thigh bone
    b. lower jaw
    c. projection that forms the elbow
    d. crest of the os coxae
    e. inferior process of the sternum
    f. medial and distal projection of the tibia
21. Describe the structure of a synovial joint. Explain how the structure of synovial joints relates to their function.
22. What is circumduction? Which type of joint allows for circumduction, and where are such joints located?
23. Name three effects of aging on the skeletal system.
24. Describe the structural differences between the male pelvis and the female pelvis.
25. Differentiate between the terms in each of the following pairs:
    a. flexion and extension
    b. abduction and adduction
    c. supination and pronation
    d. dorsiflexion and plantar flexion

CONCEPTUAL THINKING

26. Nine-year-old Alek is admitted to the emergency room with a closed fracture of the right femur. Radiography reveals that the fracture crosses the distal epiphyseal plate. What concerns should Alek's healthcare team have about the location of his injury?
27. The vertebral bodies are much larger in the lower back than the neck. What is the functional significance of this structural difference?

For more questions, see the Learning Activities on thePoint®.

CHAPTER 7

The Muscular System

Learning Objectives

After careful study of this chapter, you should be able to:

1. Compare the three types of muscle tissue. *p. 128*
2. Describe three functions of skeletal muscle. *p. 129*
3. Describe the structure of a skeletal muscle to the level of individual cells. *p. 129*
4. Outline the steps in skeletal muscle contraction. *p. 129*
5. List compounds stored in muscle cells that are used to generate energy. *p. 134*
6. Explain what happens in muscle cells contracting anaerobically. *p. 135*
7. Cite the effects of exercise on muscles. *p. 135*
8. Compare isotonic and isometric contractions. *p. 136*
9. Explain how muscles work together to produce movement. *p. 137*
10. Compare the workings of muscles and bones to lever systems. *p. 138*
11. Explain how muscles are named. *p. 139*
12. Name some of the major muscles in each muscle group, and describe the locations and functions of each. *p. 139*
13. Describe how muscles change with age. *p. 150*
14. Using information in the text, list the major muscles involved in walking and breathing, muscles that are typically affected in cases of muscular dystrophy. *pp. 127, 150*
15. Show how word parts are used to build words related to the muscular system (see Word Anatomy at the end of the chapter). *p. 152*

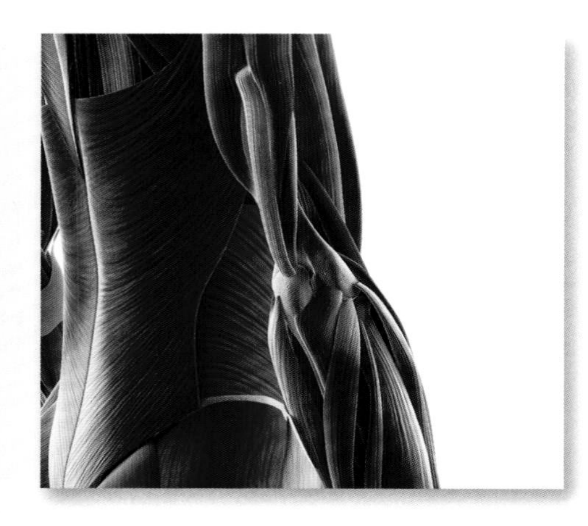

A & P in Action
Shane's Daycare Incident

"Mrs. Anderson. It's Annie Beaumont at the daycare center. We think everything is OK, but Shane fell off a small plastic slide this morning, and we think you should take a look at him and maybe have him checked by his doctor."

The 2-year-old Shane had started daycare a few days earlier. His mother Kathy, a single working parent, had been using babysitters to watch him since he was 2 months old. This year, she felt it was time to enroll him in a daycare program and was able to register him at a reputable center near the city, not too far from where she worked. She was excited that he would be able to play with other children, as there weren't many social opportunities for him where they lived and with her work schedule. She was hoping the teachers and a more structured daytime environment would benefit Shane until she got home from work and could care for him herself.

"Everything is probably fine," Ms. Beaumont told Kathy when she arrived at the center. "We're just concerned that Shane had a difficult time standing up when he fell, and just to be safe, we think he should be evaluated by his physician."

Later that afternoon, Shane's pediatrician, Dr. Schroeder, listened to Kathy as she described the daycare center incident as it had been related to her. Then, he examined Shane for any injuries that may have occurred from the fall. During the evaluation, he observed Shane's voluntary movements. He noticed that the boy's calf muscles were enlarged (pseudohypertrophic), and his thighs were thin. He asked Shane to sit on the floor and then stand up. Shane had to use his hands and arms to "walk" up his own body (Gowers sign), reflecting weak thigh muscles. Dr. Schroeder took Kathy into his office and talked to her privately.

"I haven't seen Shane in some time for his well checkups," the doctor said.

"He has seemed healthy so I didn't think it was necessary," Kathy replied.

"The fall didn't result in any injury, but after evaluating Shane I am concerned that his muscles are underdeveloped," Dr. Schroeder said. "This may have led to the fall. I want to do some more tests to figure out why Shane's muscles are weak. Let's start with a blood test and go from there."

Dr. Schroeder suspected that Shane had a condition called Duchenne muscular dystrophy or DMD, a hereditary disease that causes damage to muscle cells. In this chapter, we will learn about muscle tissue and how it interacts with the nervous system to produce movement. Later in the chapter, we will find out more about the progression of Shane's disease.

As you study this chapter, CasePoints will give you opportunities to apply your learning to this case.

Visit thePoint® to access the following resources. For guidance in using these resources most effectively, see pp. xv–xvii.

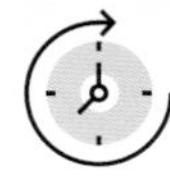

Preparing to Learn

- Tips for Effective Studying
- Pre-Quiz

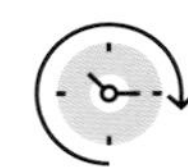

While You Are Learning

- Web Figure: Muscles of the Head and Neck
- Web Figure: Muscles of the Shoulder and Upper Torso
- Web Figure: Muscles of the Upper Extremity
- Web Figure: Muscles of the Lower Extremity
- Animation: The Neuromuscular Junction
- Animation: Skeletal Muscle Contraction
- Chapter Notes Outline
- Audio Pronunciation Glossary

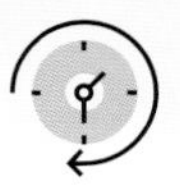

When You Are Reviewing

- Answers to Questions for Study and Review
- Health Professions: Physical Therapist
- Interactive Learning Activities

A LOOK BACK

The voluntary muscles discussed in this chapter attach to the skeleton to create the movements described in Chapter 6. As with the skeletal system, understanding the muscular system requires study at many levels of organization 2, including the muscle tissue introduced in Chapter 4. We also reference additional key ideas from Chapter 1 (structure–function 1, negative feedback 3), Chapter 2 (enzymes 7, energy 8), and Chapter 5 (adaptation 10).

Muscle Tissue

There are three kinds of muscle tissue, smooth, cardiac, and skeletal muscle, as introduced in Chapter 4. After a brief description of all three types **(TABLE 7-1)**, this chapter concentrates on skeletal muscle.

SMOOTH MUSCLE

Smooth muscle makes up most of the walls of the hollow body organs as well as those of the blood vessels and respiratory passageways. It contracts involuntarily and produces the wavelike motions of peristalsis that move substances through a system. Smooth muscle can also regulate the diameter of an opening, such as the central opening of blood vessels, or produce contractions of hollow organs, such as the uterus.

Smooth muscle fibers (cells) are tapered at each end and have a single, central nucleus. The cells appear smooth under the microscope because they do not contain the visible bands, or **striations**, that are seen in the other types of muscle cells. Smooth muscle may contract in response to a nerve impulse, hormonal stimulation, stretching, and other stimuli. The muscle contracts and relaxes slowly and can remain contracted for a long time.

CARDIAC MUSCLE

Cardiac muscle, also involuntary, makes up the heart's wall and creates the pulsing action of that organ. The cells of cardiac muscle are striated, like those of skeletal muscle. They differ in having one nucleus per cell and branching interconnections. The membranes between the cells are specialized to allow electric impulses to travel rapidly through them, so that contractions can be better coordinated. These specialized membrane regions appear as dark lines between the cells **(see TABLE 7-1)** and are called intercalated (in-TER-kah-la-ted) disks, because they are "inserted between" the cells. The electric impulses that produce cardiac muscle contractions are generated within the muscle itself but can be modified by nervous stimuli and hormones.

SKELETAL MUSCLE

When viewed under the microscope, skeletal muscle cells appear heavily striated. The arrangement of protein threads within the cell that produces these striations is described later. The cells are very long and cylindrical, and because of their great length compared to other cells, they are often described as muscle *fibers*. They have multiple

Table 7-1 Comparison of the Different Types of the Muscle

	Smooth	Cardiac	Skeletal
Location	Wall of hollow organs, vessels, respiratory, passageways	Wall of the heart	Attached to bones
Cell characteristics	Tapered at each end, branching networks, nonstriated	Branching networks; special membranes (intercalated disks) between cells; single nucleus; lightly striated	Long and cylindrical; multinucleated; heavily striated
Control	Involuntary	Involuntary	Voluntary
Action	Produces peristalsis; contracts and relaxes slowly; may sustain contraction	Pumps blood out of the heart; self-excitatory but influenced by nervous system and hormones	Produces movement at joints; stimulated by nervous system; contracts and relaxes rapidly

nuclei per cell because during development, groups of precursor cells called myoblasts fuse to form large multinucleated cells.

Skeletal muscle is under the control of the nervous system division known as the voluntary, or somatic, nervous system. Because it is under conscious control, skeletal muscle is described as voluntary. This muscle tissue usually contracts and relaxes rapidly.

Skeletal muscle is so named because most of these muscles are attached to bones and produce movement at the joints. There are a few exceptions. The muscles of the abdominal wall, for example, are partly attached to other muscles, and the muscles of facial expression are attached to the skin. Skeletal muscles constitute the largest amount of the body's muscle tissue, making up about 40% of the total body weight. This muscular system is composed of more than 600 individual skeletal muscles. Although each one is a distinct structure, muscles usually act in groups to execute body movements.

CHECKPOINT 

- ☐ **7-1** What are the three types of muscle?

CASEPOINTS 

- ☐ **7-1** Shane's disorder involved voluntary movements. What are the effectors in voluntary movements?
- ☐ **7-2** Which division of the nervous system controls voluntary movement?

The Muscular System

The three primary functions of skeletal muscles are as follows:

- Movement of the skeleton. Muscles are attached to bones and contract to change the position of the bones at a joint.
- Maintenance of posture. A steady partial contraction of the muscle, known as **muscle tone**, keeps the body in position. Some of the muscles involved in maintaining posture are the large muscles of the thighs, back, neck, and shoulders as well as the abdominal muscles.
- Generation of heat. Muscles generate most of the heat needed to keep the body at 37°C (98.6°F). Heat is a natural byproduct of muscle cell metabolism. When we are cold, muscles can boost their heat output by the rapid small contractions we know of as shivering.

MUSCLE STRUCTURE

In forming whole muscles, individual muscle fibers (cells) are arranged in bundles, or **fascicles** (FAS-ih-kls), held together by dense connective tissue **(FIG. 7-1A and C)**. These layers are as follows:

- The **endomysium** (en-do-MIS-e-um) is the deepest layer of this connective tissue and surrounds the individual fibers within fascicles.
- The **perimysium** (per-ih-MIS-e-um) is a connective tissue layer around each fascicle.
- The **epimysium** (ep-ih-MIS-e-um) is a connective tissue sheath that encases the entire muscle. The epimysium forms the innermost layer of the **deep fascia**, the tough, fibrous connective tissue membrane that encloses and defines a muscle.

Note that all these layers are named with prefixes that describe their position with regard to the fascicle: *endo-* meaning "within," *peri-* meaning "around," and *epi-* meaning "above." (These prefixes are added to the root *my/o*, meaning "muscle.") All of these supporting tissues merge to form the **tendon**, a band of dense regular connective tissue that attaches a muscle to a bone **(see FIG. 7-1)**.

Muscle cells are among the most specialized cell types in the body, so many of their organelles are given special names. Several of these names include the word root *sarco*, which literally means "flesh" but here means muscle. The muscle fiber's cytoplasm is called the **sarcoplasm**, and its plasma membrane is the **sarcolemma** **(see FIG. 7-1B)**. Extensions of the sarcolemma tunnel deep in the interior of the muscle fiber as a network of **T-tubules**, which are important in muscle cell stimulation. Muscle fibers contain large amounts of smooth endoplasmic reticulum, known as the **sarcoplasmic reticulum (SR)**. This organelle stores calcium, an important element in muscle contraction that will be discussed later. The vast majority of the muscle fiber's volume is taken up by **myofibrils**, which are bundles of protein filaments. It is these myofibrils that accomplish the work of muscle contraction. **FIGURE 7-1B** also shows **satellite cells**, stem cells that can produce new myoblasts. These myoblasts can then fuse with an existing muscle fiber, making it (and thus the muscle) larger and stronger.

11 COMMUNICATION

The negative feedback key idea 3 introduced the concept of signals conveying information between the different components of the feedback loop. However, signals also function outside such loops. **Communication**—the transmission of signals between cells or even within a single cell—is critical to all aspects of body function, ranging from muscle contraction to the development of an entire individual from a single cell. Defects in communication are responsible for many diseases.

Chapter 1 discussed the two types of signals: electrical and chemical. Electrical signals travel down cells much like electricity travels down a wire. Chemical signals alter cell activity by binding to specific proteins called **receptors**. The signal that binds the receptor is known as the **ligand**. A ligand fits into its receptor like a key in a lock. Once the ligand binds, the receptor initiates events that change the target cell's activity. A cell's response may vary depending

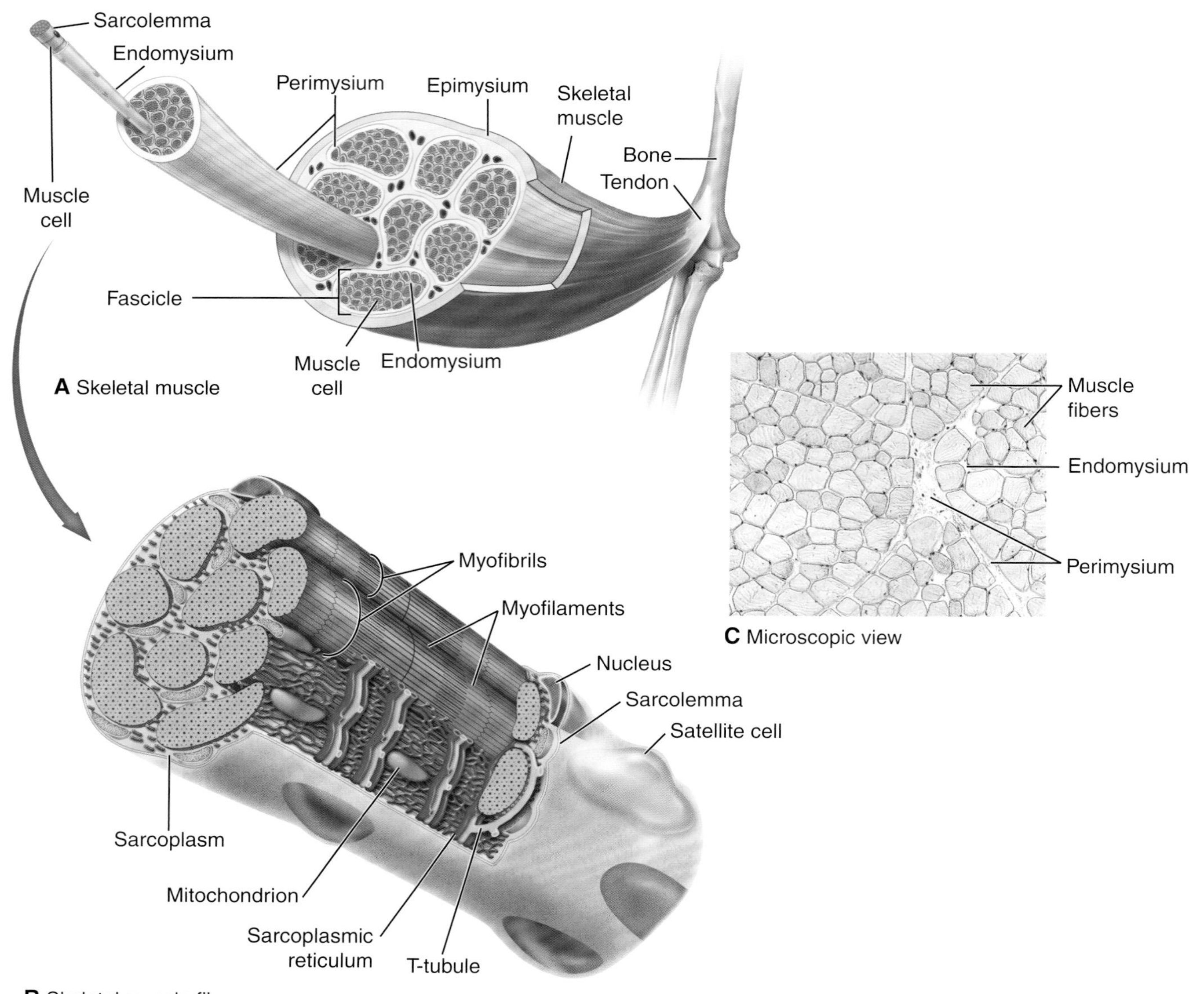

FIGURE 7-1 Structure of a skeletal muscle. KEY POINT Muscles consist of specialized cells called muscle fibers held together by multiple layers of connective tissue. **A.** Numerous muscle fibers are packed together to form a fascicle, and many fascicles are packed together to form a muscle. **B.** Muscle cells (fibers) contain specialized organelles. **C.** Muscle tissue seen under a microscope. Portions of several fascicles are shown with connective tissue coverings. ZOOMING IN What is the innermost layer of connective tissue in a muscle? What layer of connective tissue surrounds a fascicle?

on the receptors it contains. As we will see in this chapter, both electrical and chemical signals are required for muscle contraction.

MUSCLE CELLS IN ACTION

Signals coming from the brain and the spinal cord stimulate skeletal muscle fiber contractions (see Chapter 8). Because these signals often stimulate movement, the neurons (nerve cells) that carry these signals are described as motor neurons. In contrast, sensory signals travel in sensory neurons from the periphery toward the central nervous system. Each neuron has a long extension called an axon. The axons of motor neurons branch to supply from a few to hundreds of individual muscle cells or in some cases more than 1,000 (FIG. 7-2A).

A single neuron and all the muscle fibers it stimulates constitute a **motor unit**. Stimulation of the neuron activates all of the associated muscle fibers, so stronger or weaker contractions use more or fewer motor units (respectively). Muscles containing small motor units (with few muscle fibers per neuron) provide more control because they can change contraction strength in small increments. Muscles controlling the hand and the eye, for example, contain small motor units and perform very precise movements. Muscles containing larger motor units are used for maintaining posture or for broad movements, such as walking or swinging a tennis racquet.

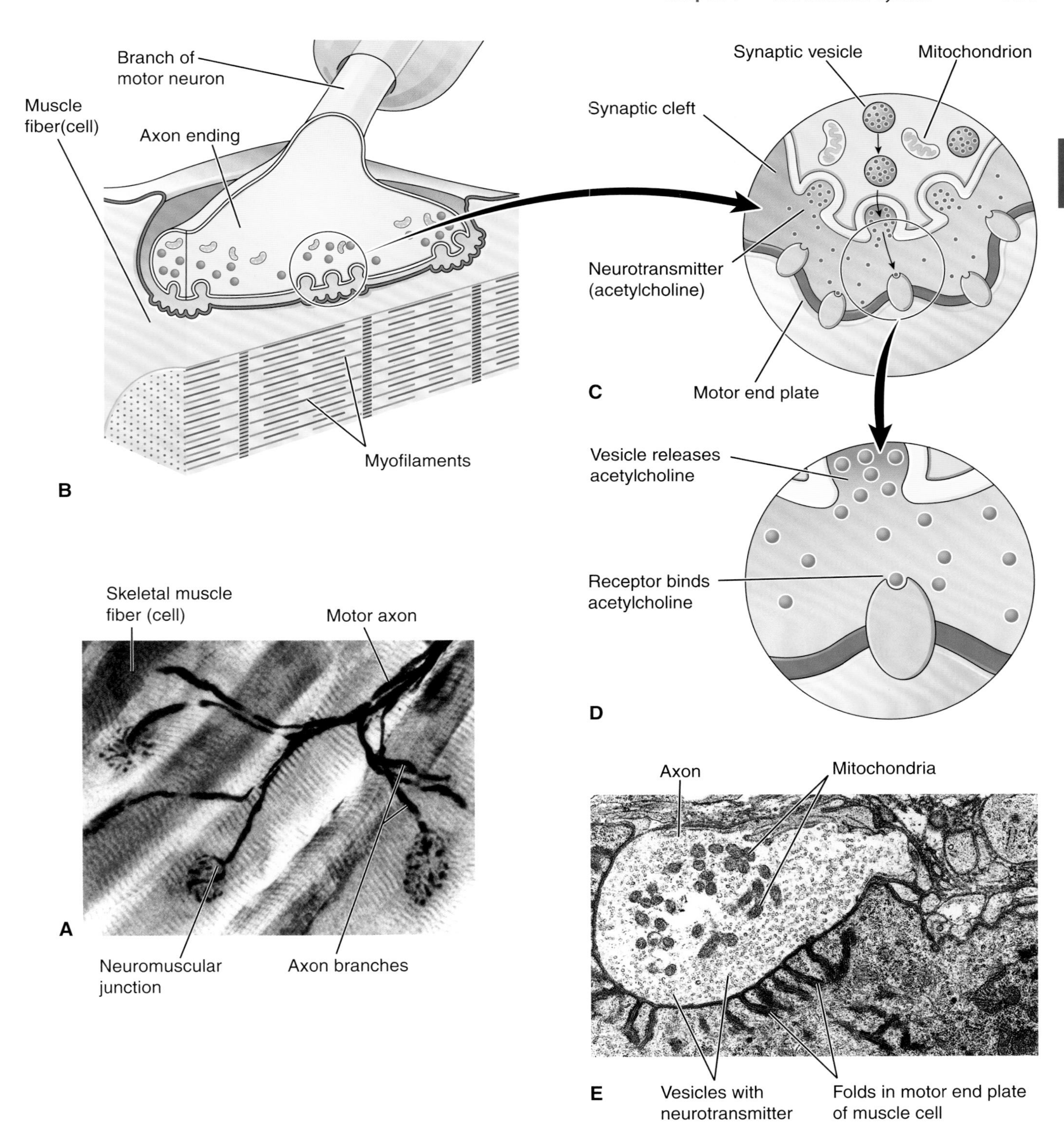

FIGURE 7-2 The neuromuscular junction (NMJ). **KEY POINT** Motor neurons stimulate skeletal muscle cells at the NMJ. **A.** A motor axon branches to stimulate multiple muscle fibers (cells). **B.** An axon branch makes contact with the membrane of a muscle fiber at the NMJ. **C.** Enlarged view of the NMJ showing release of neurotransmitter (acetylcholine) into the synaptic cleft. **D.** Acetylcholine attaches to receptors in the motor end plate, whose folds increase surface area. **E.** Electron microscope photograph of the NMJ.

The Membrane Potential Understanding how electrical signals travel through neurons and muscle cells requires us to revisit the chemical properties of ions. Both the extracellular and intracellular fluids contain large numbers of positive and negative ions. For the most part, each negative ion can pair with a positive ion. Just as when we add "+1" to "−1" to get zero, equal numbers of positive and negative ions cancel each other out. But in a resting cell, the intracellular fluid contains a small excess of negative ions, and the extracellular fluid contains a small excess of positive ions. As a result of these unpaired charges, the plasma membrane of a living cell carries a difference in electric charge (voltage) on either side that is known as a **membrane** (or transmembrane) **potential** (po-TEN-shal). Membrane potential is measured inside the cell, so in resting cells, it is negative (about −70 millivolts or mV).

Muscle cells and neurons show the property of **excitability**, because their membrane potential can change. For instance, the membrane potential becomes more negative if negative ions enter the cell. It becomes less negative (more positive) if positive ions enter the cell to neutralize the unpaired negative ions. These changes create electric signals, because they spread along the membrane, much like an electric current spreads along a wire. This spreading wave of electric current is called the **action potential** because it calls the cell into action.

The Neuromuscular Junction The point at which a nerve fiber contacts a muscle cell is called the **neuromuscular junction** (NMJ) (see FIG. 7-2). It is here that a chemical signal classified as a **neurotransmitter** is released from the neuron to stimulate the muscle fiber. The specific neurotransmitter released here is **acetylcholine** (as-e-til-KO-lene), abbreviated ACh, which is found elsewhere in the body as well. A great deal is known about the events that occur at this junction, and this information is important in understanding muscle action.

The NMJ is an example of a **synapse** (SIN-aps), a point of communication between a neuron and another cell (the term comes from a Greek word meaning "to clasp"). At every synapse, there is a tiny space, the **synaptic cleft**, across which the neurotransmitter must travel. Until its release, the neurotransmitter is stored in tiny membranous sacs, called vesicles, in the axon ending. Once released, the neurotransmitter crosses the synaptic cleft and attaches to a receptor, which is a protein embedded in the muscle cell membrane. The muscle cell membrane forms multiple folds at this point, and these serve to increase surface area and hold a maximum number of receptors. The muscle cell's receiving membrane is known as the **motor end plate**.

Once acetylcholine binds the receptor in the motor end plate, the bound receptor initiates an action potential in the muscle cell that spreads rapidly along the sarcolemma and the T-tubules throughout the muscle cell. We will see later how this action potential results in muscle contraction. Chapter 8 provides more information on synapses and the action potential.

See the Student Resources on thePoint® to view the animation "The Neuromuscular Junction."

Contraction Another important property of muscle tissue is **contractility**. This is a muscle fiber's capacity to undergo shortening, becoming thicker. Studies of muscle chemistry and observation of cells under the powerful electron microscope have increased our understanding of how muscle cells work. These studies reveal the structure of the myofibrils shown in FIGURE 7-1. Each myofibril contains many threads, or myofilaments, made primarily of two kinds of proteins, called **actin** (AK-tin) and **myosin** (MI-o-sin). Filaments made of actin are thin and light; those made of myosin are thick and dark. The filaments are present in alternating bundles within the myofibril (FIG. 7-3). It is the alternating bands of light actin and heavy myosin filaments that give skeletal muscle its striated appearance. They also give a view of what occurs when muscles contract.

Note that the actin and myosin filaments overlap where they meet, just as your fingers overlap when you fold your hands together. A contracting subunit of skeletal muscle is

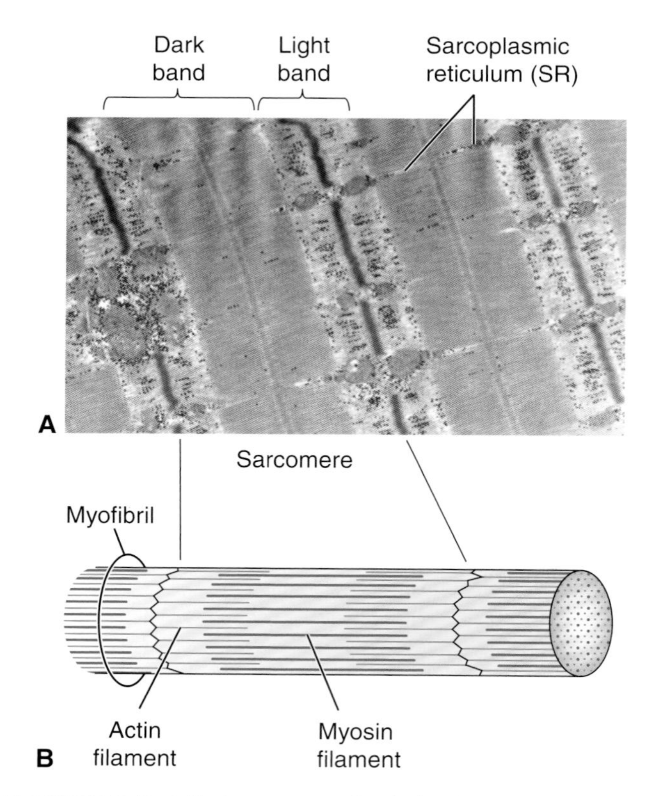

FIGURE 7-3 **Detailed structure of a skeletal muscle cell.** **KEY POINT** Myofibrils are composed of overlapping contractile bands of protein filaments. **A.** Photomicrograph of skeletal muscle cell (×6,500). Actin makes up the light band, and myosin makes up the dark band. The dark line in the actin band marks points where actin filaments are held together. A sarcomere is a contracting subunit of skeletal muscle. The sarcoplasmic reticulum is the ER of muscle cells. **B.** Diagram of the photographic image.

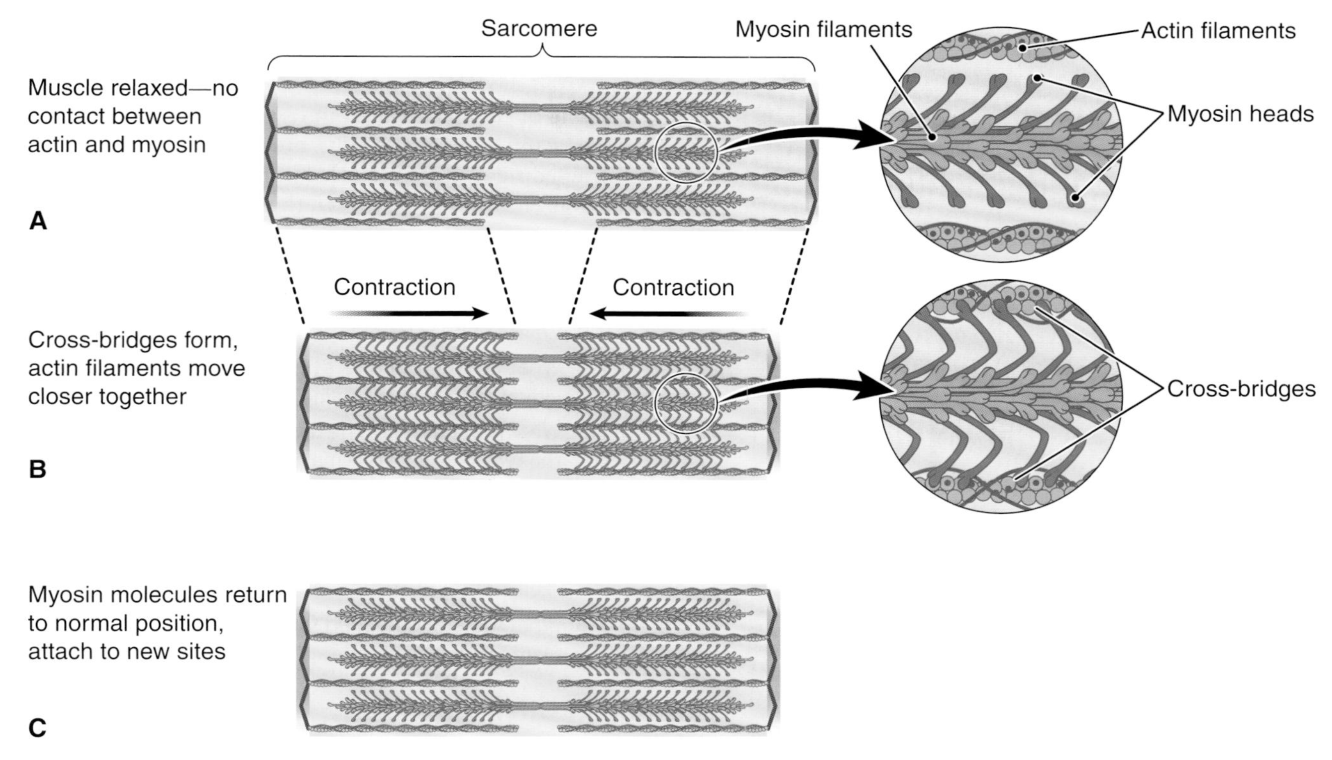

FIGURE 7-4 Sliding filament mechanism of skeletal muscle contraction. KEY POINT Muscle contraction depends on the interaction of actin and myosin filaments within the cell. **A.** Muscle is relaxed, and there is no contact between the actin and myosin filaments. **B.** Cross-bridges form, and the actin filaments are moved closer together as the muscle fiber contracts. **C.** The cross-bridges break, and the myosin heads attach to new sites to prepare for another pull on the actin filaments and further contraction. ZOOMING IN Do the actin or myosin filaments change in length as contraction proceeds?

called a **sarcomere** (SAR-ko-mere). It consists of a band of myosin filaments and the actin filaments on each side (see FIG. 7-3). The myosin molecules are shaped like two golf clubs twisted together with their paddlelike heads projecting away from the sarcomere's center. The actin molecules are twisted together like two strands of beads, each bead having a myosin-binding site (FIG. 7-4).

8 By studying events at the molecular level, we can understand how muscles convert the potential energy of adenosine triphosphate (ATP) into the kinetic energy of muscle movement (see FIG. 7-4). In movement, the myosin heads "latch on" firmly to the actin filaments in their overlapping region, forming attachments between the filaments that are described as *cross-bridges*. 7 The myosin head acts as an enzyme to break a phosphate bond in the ATP molecule and liberate the chemical energy it contains (see FIG. 2-12). This release converts ATP to adenosine diphosphate (ADP), a compound with just two phosphates:

$$\underset{\text{adenosine triphosphate}}{\text{ATP}} \rightarrow \underset{\text{adenosine diphosphate}}{\text{ADP}} + \underset{\text{phosphate}}{\text{P}} + \text{energy}$$

Then, the myosin heads, like the oars of a boat moving water, use the released energy to pull all the actin strands closer together within each sarcomere, producing movement. New ATP molecules trigger the release of the myosin heads and move them back to position for another "power stroke." With repeated movements, the overlapping filaments slide together, and the muscle fiber contracts, becoming shorter and thicker. This action is aptly named the *sliding filament mechanism* of muscle contraction. Note that the filaments overlap increasingly as the cell contracts. (In reality, not all the myosin heads are moving at the same time. About one-half are forward at any time, and the rest are preparing for another swing.) During contraction, each sarcomere becomes shorter, but the individual filaments do not change in length. As in shuffling a deck of cards, as you push the cards together, the deck becomes smaller, but the cards do not change in length.

The Role of Calcium In addition to actin, myosin, and ATP, calcium is needed for muscle contraction. It enables cross-bridges to form between actin and myosin so the sliding filament action can begin. When muscles are at rest, two additional proteins called **troponin** (tro-PO-nin) and **tropomyosin** (tro-po-MI-o-sin) block the sites on actin filaments where cross-bridges can form (FIG. 7-5). When calcium attaches to the troponin, these proteins move aside, uncovering the binding sites. In resting muscles, the calcium is not available because it is stored within the cell's SR. Calcium is released into the sarcoplasm in response to the action potential along the sarcolemma and T-tubules, as discussed earlier. Muscles relax when nervous stimulation stops, and the calcium is pumped back into the SR, ready for the next contraction.

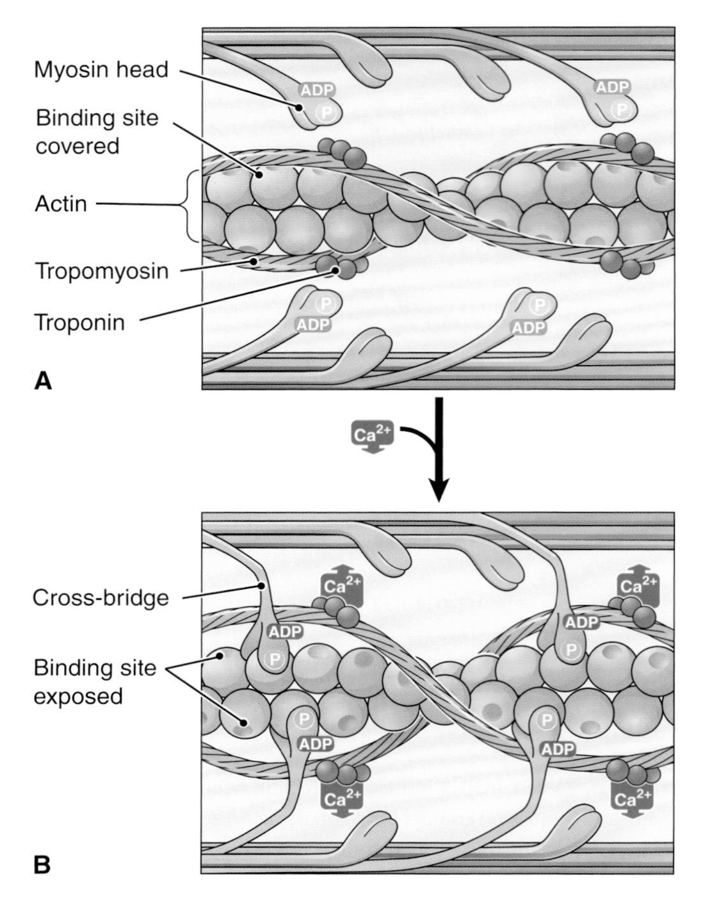

FIGURE 7-5 **Role of calcium in muscle contraction.**
KEY POINT Calcium unblocks sites where cross-bridges can form between actin and myosin filaments to begin muscle contraction. **A.** Troponin and tropomyosin cover the binding sites where cross-bridges can form between actin and myosin. **B.** Calcium shifts troponin and tropomyosin away from binding sites so cross-bridges can form.

A summary of the events in a muscle contraction is as follows:

1. ACh is released from a neuron ending into the synaptic cleft at the NMJ.
2. ACh binds to receptors in the muscle's motor end plate and stimulates an action potential.
3. The action potential travels down the sarcolemma and T-tubules and activates calcium channels in the SR.
4. The SR releases calcium into the cytoplasm.
5. Calcium shifts troponin and tropomyosin so that binding sites on actin are exposed.
6. Myosin heads bind to actin, forming cross-bridges.
7. Using the chemical energy contained in ATP, myosin heads pull actin filaments together within the sarcomeres, and the cell shortens.
8. New ATP is used to detach myosin heads and move them back to position for another "power stroke."
9. The muscle relaxes when stimulation ends, and the calcium is pumped back into the SR.

The phenomenon of rigor mortis, a state of rigidity that occurs after death, illustrates ATP's crucial role in muscle contraction. Shortly after death, muscle cells begin to degrade. Calcium escapes into the cytoplasm, stimulating cross-bridge formation and muscle contraction. Metabolism has ceased, however, and there is no ATP to disengage the filaments or pump calcium out of the cytoplasm. The cross-bridges cannot detach, so the muscle remains locked in a contracted rigor mortis that lasts about 24 hours, gradually fading as enzymes break down the muscle filaments.

CHECKPOINTS 

- ☐ **7-2** What are the three main functions of skeletal muscle?
- ☐ **7-3** What are bundles of muscle fibers called?
- ☐ **7-4** What are the bundles of protein molecules within individual muscle fibers called?
- ☐ **7-5** What is the term for the difference in electrical charge on the two sides of a plasma membrane?
- ☐ **7-6** What is the name of the special synapse where a nerve cell makes contact with a muscle cell?
- ☐ **7-7** Which neurotransmitter is involved in the stimulation of skeletal muscle cells?
- ☐ **7-8** What mineral is needed for interaction of the contractile filaments?

CASEPOINT

- ☐ **7-3** Shane's disorder affects a protein that links groups of myofibrils to the muscle cell membrane. This change damages the cells when they contract. What are the main myofilaments in the myofibrils of muscle cells

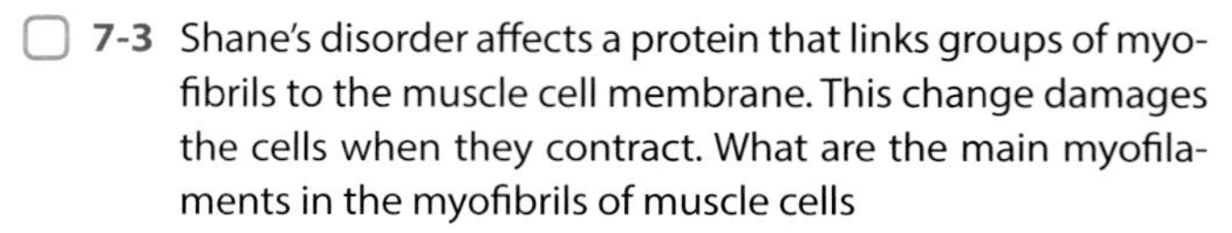
See the Student Resources on thePoint® for the animation "Skeletal Muscle Contraction."

ENERGY SOURCES

8 As noted earlier, all muscle contraction requires chemical energy in the form of ATP. Most of this energy is produced by the oxidation (commonly called "burning") of nutrients within the cells' mitochondria, especially the oxidation of glucose and fatty acids. Chapter 18 discusses this process in greater detail. Metabolism that requires oxygen is described as aerobic (the root *aer/o* means "air" or "gas" but in this case refers to oxygen).

Storage Compounds The circulating blood constantly brings nutrients and oxygen to the cells, but muscle cells also store a small supply of each for rapid ATP generation, such as during vigorous exercise. For example:

- **Myoglobin** (mi-o-GLO-bin) stores oxygen. This compound is similar to the hemoglobin in blood, but it is located specifically in muscle cells as indicated by the root *my/o* in its name.
- **Glycogen** (GLI-ko-jen) is the storage form of glucose. It is a polysaccharide made of multiple glucose molecules, and it can be broken down into glucose when needed by the muscle cells.
- Fatty acids are stored as triglycerides formed into fat droplets. These droplets can be broken down into fatty acids when needed by the muscle cells.

Anaerobic Metabolism Oxidation is very efficient and yields a large amount of ATP per nutrient molecule, but it has some limitations. First, it takes a while to start generating ATP, so oxidation cannot supply enough energy for the first few seconds of muscle contraction. Second, during strenuous exercise, mitochondrial activity cannot keep up with the demands of hardworking muscles. These two situations rely on alternate, rapid mechanisms of ATP production that do not require oxygen. They are described as *anaerobic* (*an-* means "not" or "without"). Note that these anaerobic processes are always occurring, but they are particularly important at the beginning of exercise or during very strenuous exercise.

1. Breakdown of **creatine** (KRE-ah-tin) **phosphate**. Creatine phosphate is a compound similar to ATP in that it has a high-energy bond that breaks down to release energy. This energy is used to make ATP for muscle contraction. It generates ATP very rapidly, but its supply is limited **(BOX 7-1)**.
2. **Anaerobic glycolysis**. This process breaks glucose down incompletely without using oxygen (*glyc/o* means "glucose" and *-lysis* means "separation"). A few ATPs are generated in these reactions, as is a byproduct called lactate, which is later oxidized for energy when oxygen is available.

When a person stops exercising, the body must generate enough ATP to reestablish a resting state by replenishing stored materials. ATP is also needed to restore normal body temperature. The person must take in extra oxygen by continued rapid breathing, known as *excess postexercise oxygen consumption*. Chapter 18 has more details on metabolism.

Muscle Fatigue It is commonly thought that muscles tire because they are out of ATP or because lactate accumulates. In fact, fatigue in nonathletes frequently originates in the nervous system, not in the muscles. People unaccustomed to strenuous exercise find the sensations it generates unpleasant and consciously or unconsciously reduce the nervous impulses to skeletal muscles. It is difficult to overcome the brain's inhibition and truly fatigue a muscle, that is, take it to the point that it no longer responds to stimuli. True muscle fatigue has many causes and may depend on individual factors, fitness and genetic makeup, for example, and the type of exercise involved. These causes include depletion of glycogen reserves, inadequate oxygen supply, or the accumulation of phosphates from ATP breakdown.

EFFECTS OF EXERCISE

10 Regular exercise results in a number of changes in muscle tissue as the muscle cells adapt to the increased workload. The changes depend on the type of exercise. Resistance training, such as weight lifting, causes muscle cells to increase in size, a condition known as hypertrophy (hi-PER-tro-fe). Larger muscle cells contain more myofibrils and can form more cross-bridges, so they can generate more force. Resistance training also increases muscle stores of creatine phosphate and glycogen, so that muscle cells can use anaerobic metabolism to generate a large amount of ATP in a short time. Muscle hypertrophy is stimulated by hormones, especially the male sex steroids. **BOX 7-2** has information on how some athletes abuse these steroids to increase muscle size and strength at the expense of their health.

A CLOSER LOOK — BOX 7-1
Creatine Kinase: Muscle's Backup Energy Enzyme

At rest, muscle cells store some of the ATP they produce in the sarcoplasm. But the amount stored is sufficient for only a few seconds of contraction, and it takes several more seconds before anaerobic glycolysis and oxidation can replenish it. So, how do muscle fibers power their contractions in the meantime? They have a backup energy source called creatine phosphate to tide them over.

When muscle cells are resting, they manufacture creatine phosphate by transferring energy from ATP to creatine, a substance produced by the liver, pancreas, and kidneys. When muscle cells are exercising actively, they transfer that energy to ADP to create ATP. There is enough creatine phosphate in the sarcoplasm to produce four or five times the original amount of stored ATP—enough to power the cell until the other ATP-producing reactions take over.

Creatine kinase (CK) catalyzes the transfer of energy from creatine phosphate to ADP. CK is found in all muscle cells and other metabolically active cells, such as neurons. It is composed of two subunits, which can be either *B* (brain type) or *M* (muscle type). Therefore, there are three forms of CK, CK-BB, CK-MM, and CK-MB, which are present at different levels in various tissues. CK-BB is found mainly in nervous and smooth muscle tissue. CK-MM is found predominantly in skeletal muscle. Cardiac muscle contains both CK-MM and CK-MB.

Normally, the blood level of CK is low, but damage to CK-containing tissues increases it. Thus, clinicians can use blood CK levels in diagnosis. Elevated blood CK-BB may indicate a nervous system disorder, such as stroke or amyotrophic lateral sclerosis (ALS, Lou Gehrig disease). Elevated blood levels of CK-MM may indicate a muscular disorder, such as myositis or muscular dystrophy, as seen in Shane's opening case. Elevated blood levels of both CK-MM and CK-MB may indicate cardiac muscle damage following a myocardial infarction (heart attack).

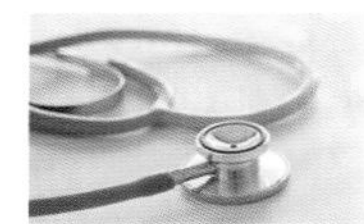

CLINICAL PERSPECTIVES

Anabolic Steroids: Winning at All Costs?

BOX 7-2

Anabolic steroids mimic the effects of the male sex hormone testosterone by promoting metabolism and stimulating growth. These drugs are legally prescribed to promote muscle regeneration and prevent atrophy from disuse after surgery. However, some athletes also purchase them illegally, using them to increase muscle size and strength and improve endurance.

When steroids are used illegally to enhance athletic performance, the doses needed are large enough to cause serious side effects. They increase blood cholesterol levels, which may lead to atherosclerosis, heart disease, kidney failure, and stroke. They damage the liver, making it more susceptible to cancer and other diseases, and suppress the immune system, increasing the risk of infection and cancer. In men, steroids cause impotence, testicular atrophy, low sperm count, infertility, and the development of female sex characteristics, such as breasts (gynecomastia). In women, steroids disrupt ovulation and menstruation and produce male sex characteristics, such as breast atrophy, enlargement of the clitoris, increased body hair, and deepening of the voice. In both sexes, steroids increase the risk for baldness and, especially in men, cause mood swings, depression, and violence.

Aerobic exercise, that is, exercise that increases oxygen consumption, such as running, biking, or swimming, leads to improved muscular endurance. Endurance training increases the muscle cells' blood supply and number of mitochondria, improving their ability to generate ATP aerobically and to get rid of waste products. Endurance-trained muscles can contract more frequently and for longer periods without fatiguing.

Cardiovascular changes are perhaps the most important physical benefits of endurance exercise. You can think of endurance exercise as "strength training for the heart." That organ has to pump up to five times as much blood during endurance exercise as it does at rest. The heart muscle (especially the left ventricle) adapts to its increased workload by growing larger and stronger. Its increased pumping efficiency means that the heart doesn't have to work very hard at rest, so the resting heart rate declines. Endurance exercise also benefits the body by decreasing the amount of less healthy (LDL) cholesterol in the blood, reducing blood pressure, and improving blood glucose control.

These beneficial changes may derive in part from the reduced body fat and improved psychological well-being that result from regular physical activity. Whatever the cause, it's well established that aside from not smoking, exercise is the most important thing you can do to improve your health. Recent studies show that a few short bouts of high-intensity exercise are equally or more effective than longer, low-intensity workouts. Even people such as the elderly those with type II diabetics show significant improvements when they exercise for 10 intervals of 60 seconds, each at maximum intensity.

The benefits of one form of exercise do not significantly carry over to the other—endurance exercise does not significantly increase muscle strength, and resistance training does not significantly increase muscle endurance. An exercise program thus should include both methods with periods of warm-up and cooldown before and after working out. Stretching generally improves the range of motion at the joints and improves balance. However, studies show that static stretching (holding an extended position for 60 seconds to two minutes) just before a strenuous workout actually decreases muscle strength and increases the risk of injury.

TYPES OF MUSCLE CONTRACTIONS

Muscle tone refers to a muscle's partially contracted state that is normal even when the muscle is not in use. The maintenance of this tone, or **tonus** (TO-nus), is due to the action of the nervous system in keeping the muscles in a constant state of readiness for action. Muscles that are little used soon become flabby, weak, and lacking in tone.

In addition to the partial contractions that are responsible for muscle tone, there are two other types of contractions on which the body depends:

- In **isotonic** (i-so-TON-ik) **contractions**, the tone or tension within the muscle remains the same, but muscle length changes, and the muscle bulges as it accomplishes work (*iso-* means "same" or "equal" and *ton* means "tension"). Within this category, there are two forms of contractions:
  - **Concentric contractions.** These contractions are more familiar, as they produce more obvious changes in position. In concentric contractions, a muscle as a whole shortens to produce movement. Try flexing your arm at the elbow to pick up a dumbbell or heavy can. The anterior arm flexors, the biceps brachii and brachialis, move the forearm at the elbow, lifting the weight, and you can see that the muscles change shape and bulge outward.
  - **Eccentric contractions.** In these contractions, the muscle lengthens as it exerts force. Think of gradually lowering that dumbbell or heavy can. The arm flexors tense as they lengthen. These contractions strengthen muscles considerably but are more likely to cause soreness later, perhaps because of microscopic tears in the muscle fibers.
- In **isometric** (i-so-MET-rik) **contractions**, there is no change in muscle length, but there is a great increase in muscle tension (*metr/o* means "measure"). Pushing against an immovable force produces an isometric contraction, as in trying to lift a weight that is too heavy to move. Try pushing the palms of your hands hard against each other. There is no movement or change in muscle length, but you can feel the increased tension in your arm muscles.

Most body movements involve a combination of both isotonic and isometric contractions. When walking, for example, some muscles contract isotonically to propel the body forward, but at the same time, other muscles are contracting isometrically to keep your body in position.

CHECKPOINTS

- 7-9 Which compound is formed in oxidation of nutrients that supplies the energy for muscle contraction?
- 7-10 Which compound stores reserves of oxygen in muscle cells?
- 7-11 What are the two main types of muscle contraction?

The Mechanics of Muscle Movement

Most muscles have two or more points of attachment to the skeleton. The muscle is attached to a bone at each end by means of a cordlike extension called a tendon (FIG. 7-6). All of the connective tissue within and around the muscle merges to form the tendon, which then attaches directly to the bone's periosteum (see FIG. 7-1). In some instances, a broad sheet called an **aponeurosis** (ap-o-nu-RO-sis) may attach muscles to bones or to other muscles.

In moving the bones, one end of a muscle is attached to a more freely movable part of the skeleton, and the other end is attached to a relatively stable part. The less movable (more fixed) attachment is called the **origin**; the attachment to the body part that moves is called the **insertion** (see FIG. 7-6). When a muscle contracts, it pulls on both attachment points, bringing the more movable insertion closer to the origin and thereby causing movement of the body part. FIGURE 7-6 shows the action of the brachialis (in the arm) in flexing the arm at the elbow. The muscle's insertion on the ulna of the forearm is brought toward its origin on the humerus of the arm.

MUSCLES WORK TOGETHER

Many of the skeletal muscles function in pairs. The main muscle that performs a given movement is the **prime mover**. For instance, the brachialis is the prime mover for flexion of the arm at the elbow (see FIG. 7-6). Because any muscle that performs a given action is technically called an **agonist** (AG-on-ist), the muscle that produces an opposite action is termed the **antagonist** (an-TAG-on-ist) (the prefix *anti-* means "against"). Clearly, for any given movement, the antagonist must relax when the agonist contracts. For example, when the brachialis at the anterior arm contracts to flex the arm, the triceps brachii at the back must relax;

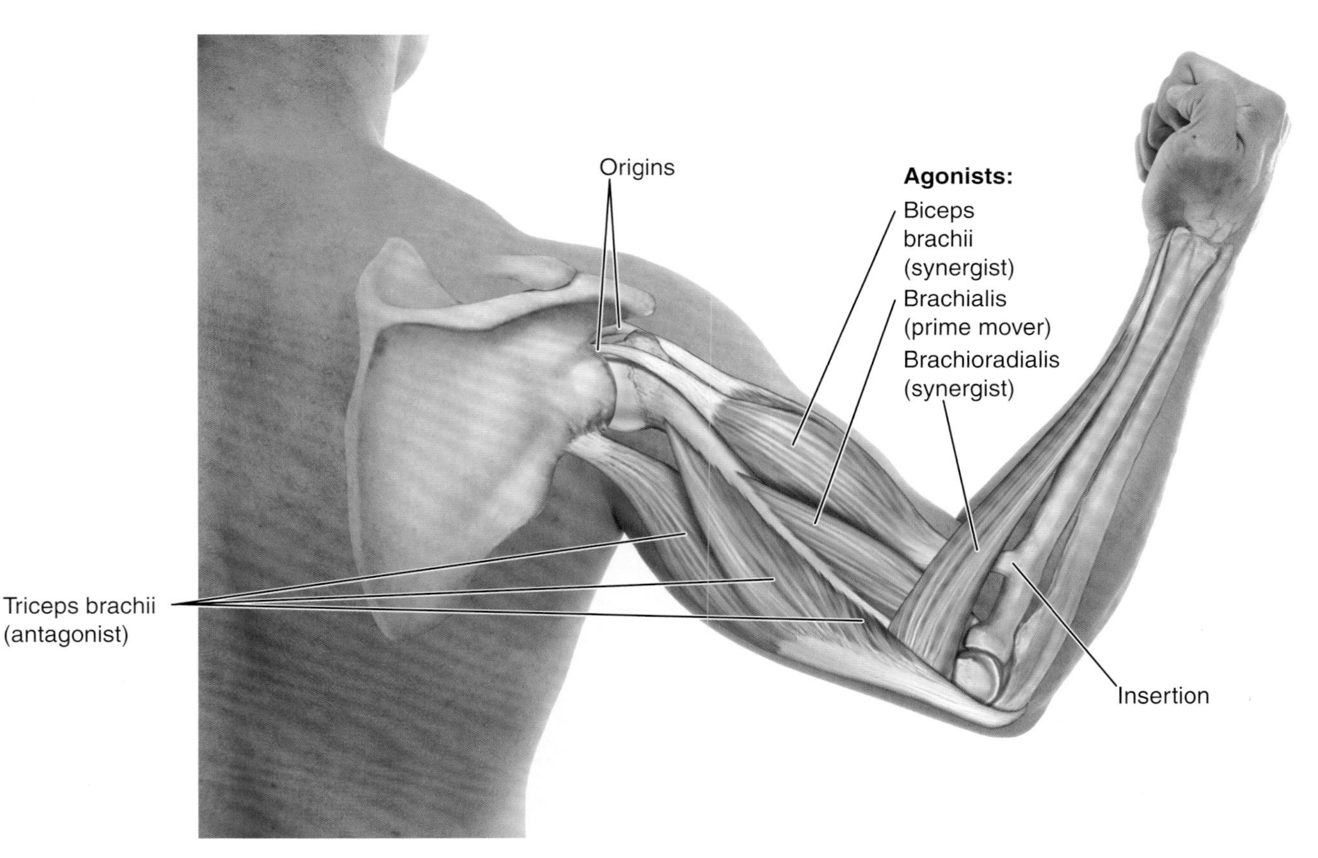

FIGURE 7-6 **Muscle attachments to bones.** KEY POINT Tendons attach muscles to bones. The stable point is the origin; the movable point is the insertion. In this diagram, three attachments are shown—two origins and one insertion. ZOOMING IN Does contraction of the brachialis produce flexion or extension at the elbow?

when the triceps brachii contracts to extend the arm, the brachialis must relax. In addition to prime movers and antagonists, there are also muscles that steady body parts or assist in an action. These "helping" muscles are called **synergists** (SIN-er-jists), because they work with the prime mover to accomplish a movement (*syn-* means "together" and *erg/o* means "work"). For example, the biceps brachii and the brachioradialis are synergists to the brachialis in flexing the arm (see FIG. 7-6).

As the muscles work together, actions are coordinated to accomplish many complex movements. Note that during development, the nervous system must gradually begin to coordinate our movements. A child learning to walk or to write, for example, may use muscles unnecessarily at first or fail to use appropriate muscles when needed.

LEVERS AND BODY MECHANICS

Proper body mechanics help conserve energy and ensure freedom from strain and fatigue; conversely, such ailments as lower back pain—a common complaint—can be traced to poor body mechanics. Body mechanics have special significance to healthcare workers, who frequently must move patients and handle cumbersome equipment. Maintaining the body segments in correct alignment also affects the vital organs that are supported by the skeleton.

If you have had a course in physics, recall your study of levers. A lever is simply a rigid bar that moves about a fixed pivot point, the fulcrum. There are three classes of levers, which differ only in the location of the fulcrum (F); the effort (E), or force; and the resistance (R), the weight or load:

- In a first-class lever, the fulcrum is located between the resistance and the effort; a seesaw or a pair of scissors is an example of this class (FIG. 7-7A).
- The second-class lever has the resistance located between the fulcrum and the effort; a wheelbarrow or a mattress lifted at one end is an illustration of this class. However, there are no significant examples of second-class levers in the body.
- In the third-class lever, the effort is between the resistance and the fulcrum. A forceps or a tweezers is an example of this type of lever. The effort is applied in the tool's center, between the fulcrum, where the pieces join, and the resistance at the tip (see FIG. 7-7B).

The musculoskeletal system can be considered a system of levers, in which the bone is the lever, the joint is the fulcrum, and the force is applied by a muscle. An example of a first-class lever in the body is using the muscles at the back of the neck to lift the head at the joint between the skull's occipital bone and the first cervical vertebra (atlas) (see FIG. 7-7A). However, there are very few examples of first-class levers in the body. Most lever systems in the body are of the third-class type. A muscle usually inserts past a joint and exerts force between the fulcrum and the resistance. That is, the fulcrum is behind both the point of effort and the weight. As shown in FIGURE 7-7B, when the biceps brachii helps flex the forearm at the elbow, the muscle exerts its force at its insertion on the radius. The weight of the hand and forearm creates the resistance, and the fulcrum is the elbow joint, which is behind the point of effort.

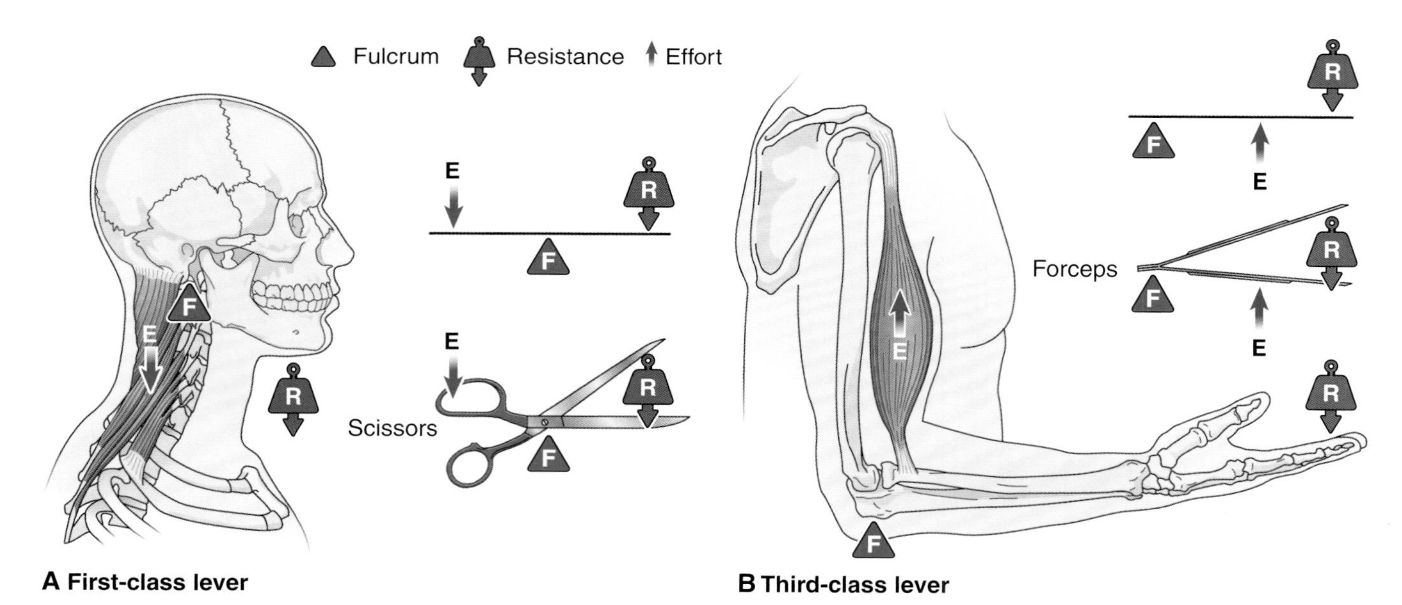

FIGURE 7-7 Levers. **KEY POINT** Muscles work with bones as lever systems to produce movement. Two classes of levers are shown along with tools and anatomic examples that illustrate each type. R, resistance (weight); E, effort (force); F, fulcrum (pivot point). A first-class lever **(A)** and third-class lever **(B)** are shown. ZOOMING IN In a third-class lever system, where is the fulcrum with regard to the effort and the resistance?

By understanding and applying knowledge of levers to body mechanics, healthcare workers can reduce the risk of musculoskeletal injury while carrying out their numerous clinical tasks.

CHECKPOINTS 

- ☐ **7-12** What are the names of the two attachment points of a muscle, and how do they function?
- ☐ **7-13** What is the name of the muscle that produces a movement as compared with the muscle that produces an opposite movement?
- ☐ **7-14** Of the three classes of levers, which one represents the action of most muscles?

Skeletal Muscle Groups

The study of muscles is made simpler by grouping them according to body regions. Knowing how muscles are named can also help in remembering them. A number of different characteristics are used in naming muscles, including the following:

- Location, using the name of a nearby bone, for example, or a position, such as lateral, medial, internal, or external
- Size, using terms such as maximus, major, minor, longus, or brevis
- Shape, such as circular (orbicularis), triangular (deltoid), or trapezoidal (trapezius)
- Direction of fibers, including straight (rectus) or angled (oblique)
- Number of heads (attachment points), as indicated by the suffix *-ceps*, as in biceps, triceps, and quadriceps
- Action, as in flexor, extensor, adductor, abductor, or levator

Often, more than one feature is used in naming. Refer to **FIGURES 7-8 and 7-9** as you study the locations and functions of the superficial skeletal muscles, and try to figure out the basis for each name. Although they are described in the singular, most of the muscles are present on both sides of the body. The origins and insertions given refer to the manner in which a muscle is ordinarily used. If the muscle action is reversed, the origin and insertion are reversed as well. 1 Knowing the attachment points of a muscle can help you predict its action. See **BOX 7-3** "Muscles and Their Movements," for hints about linking muscles to different movements.

MUSCLES OF THE HEAD

The principal muscles of the head are those of facial expression and of mastication (chewing) (**FIG. 7-10**; **TABLE 7-2**).

The muscles of facial expression include ring-shaped ones around the eyes and the lips, called the **orbicularis** (or-bik-u-LAH-ris) **muscles** because of their shape (think of "orbit"). The muscle surrounding the eye is called the **orbicularis oculi** (OK-u-li), whereas the lip muscle is the **orbicularis oris**. These muscles all have antagonists. For example, the **levator palpebrae** (PAL-pe-bre) **superioris**, or lifter of the upper eyelid, is the antagonist for the orbicularis oculi.

One of the largest muscles of expression forms the fleshy part of the cheek and is called the **buccinator** (BUK-se-na-tor). Used in whistling or blowing, it is sometimes referred to as the trumpeter's muscle. You can readily think of other muscles of facial expression: for instance, the zygomaticus muscles produce a smile, while the depressor anguli oris muscles turn a smile into a grimace. There are a number of scalp muscles that lift the eyebrows or draw them together into a frown.

There are four pairs of mastication (chewing) muscles, all of which insert on and move the mandible. The largest are the **temporalis** (TEM-po-ral-is), which is superior to the ear, and the **masseter** (mas-SE-ter) at the angle of the jaw.

The tongue has two muscle groups. The first group, called the *intrinsic muscles*, is located entirely within the tongue. The second group, the *extrinsic muscles*, originates outside the tongue. Note that *intrinsic* is a generic term to describe muscles located within the moving structure, whereas *extrinsic* describes muscles that connect to the moving structure by tendons. It is because of these many muscles that the tongue has such remarkable flexibility and can perform so many different functions. Consider the intricate tongue motions involved in speaking, chewing, and swallowing. **FIGURE 7-10** shows some additional muscles of the face.

MUSCLES OF THE NECK

The neck muscles tend to be ribbonlike and extend vertically or obliquely in several layers and in a complex manner (**see FIG. 7-10; TABLE 7-2**). The one you will hear of most frequently is the **sternocleidomastoid** (ster-no-kli-do-MAS-toyd), sometimes referred to simply as the sternomastoid. This strong muscle extends superiorly from the sternum across the lateral neck to the mastoid process of the temporal bone. When the left and right muscles work together, they bring the head forward on the chest (flexion). Working alone, each muscle tilts and rotates the head so as to orient the face toward the side opposite that muscle. If the head is abnormally fixed in this position, the person is said to have **torticollis** (tor-tih-KOL-is), or wryneck; this condition may be caused by muscle injury or spasm.

A portion of the trapezius muscle (described later) is located at the posterior neck, where it helps hold the head up (extension). Other back muscles, discussed later, extend the entire vertebral column, including the neck.

See the Student Resources on thePoint® to view additional pictures of head and neck musculature.

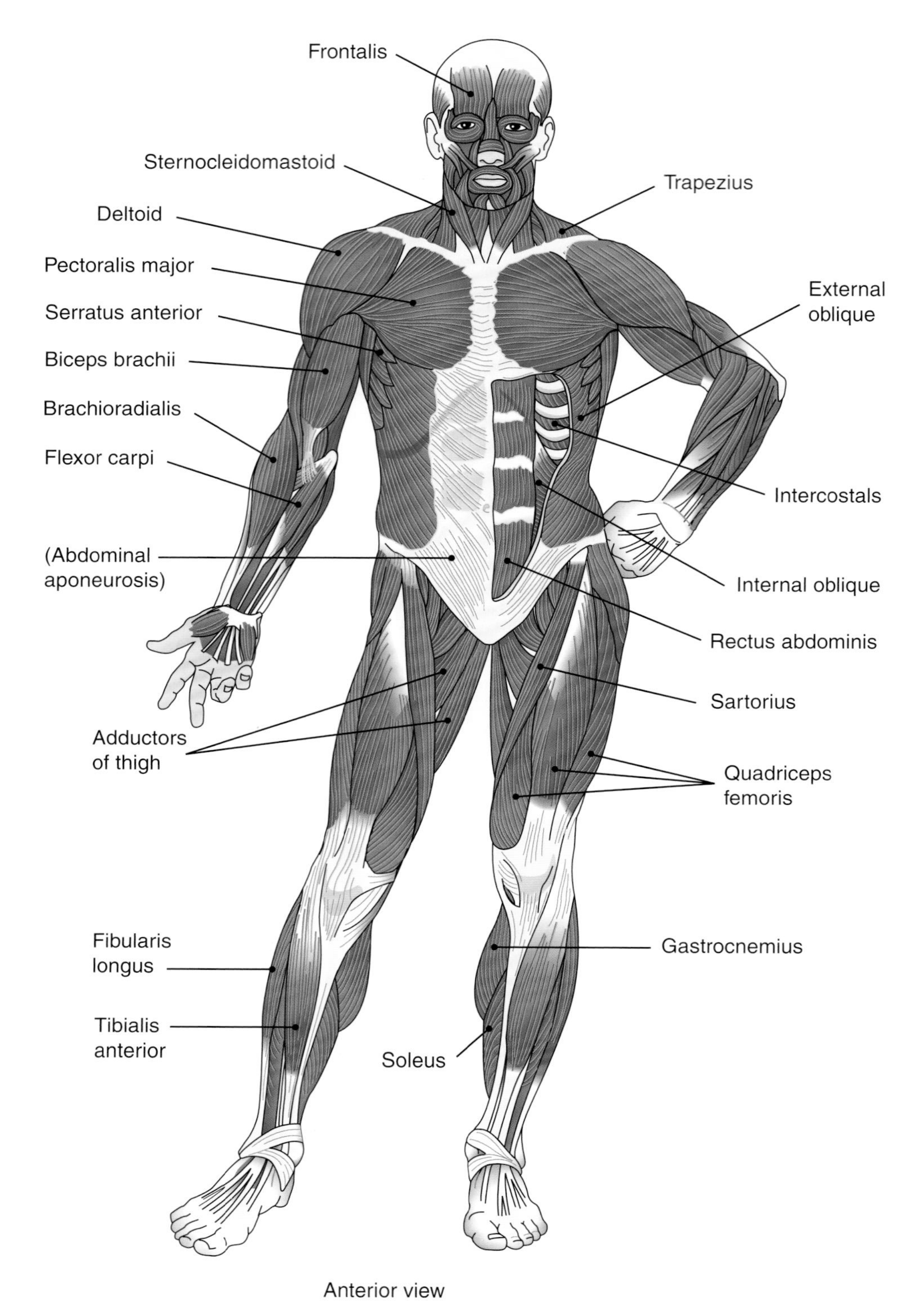

FIGURE 7-8 Superficial muscles, anterior view. An associated structure is labeled in parentheses. An aponeurosis is a broad, sheetlike tendon.

MUSCLES OF THE UPPER EXTREMITIES

Muscles of the upper extremities include the muscles that determine the position of the shoulder, the anterior and posterior muscles that move the arm, and the muscles that move the forearm and hand.

Muscles that Move the Shoulder and Arm The position of the shoulder depends to a large extent on the degree of contraction of the **trapezius** (trah-PE-ze-us), a triangular muscle that covers the posterior neck and extends across the posterior shoulder to insert on the clavicle and scapula (see FIGS. 7-8 and 7-9; TABLE 7-3). The trapezius muscles enable

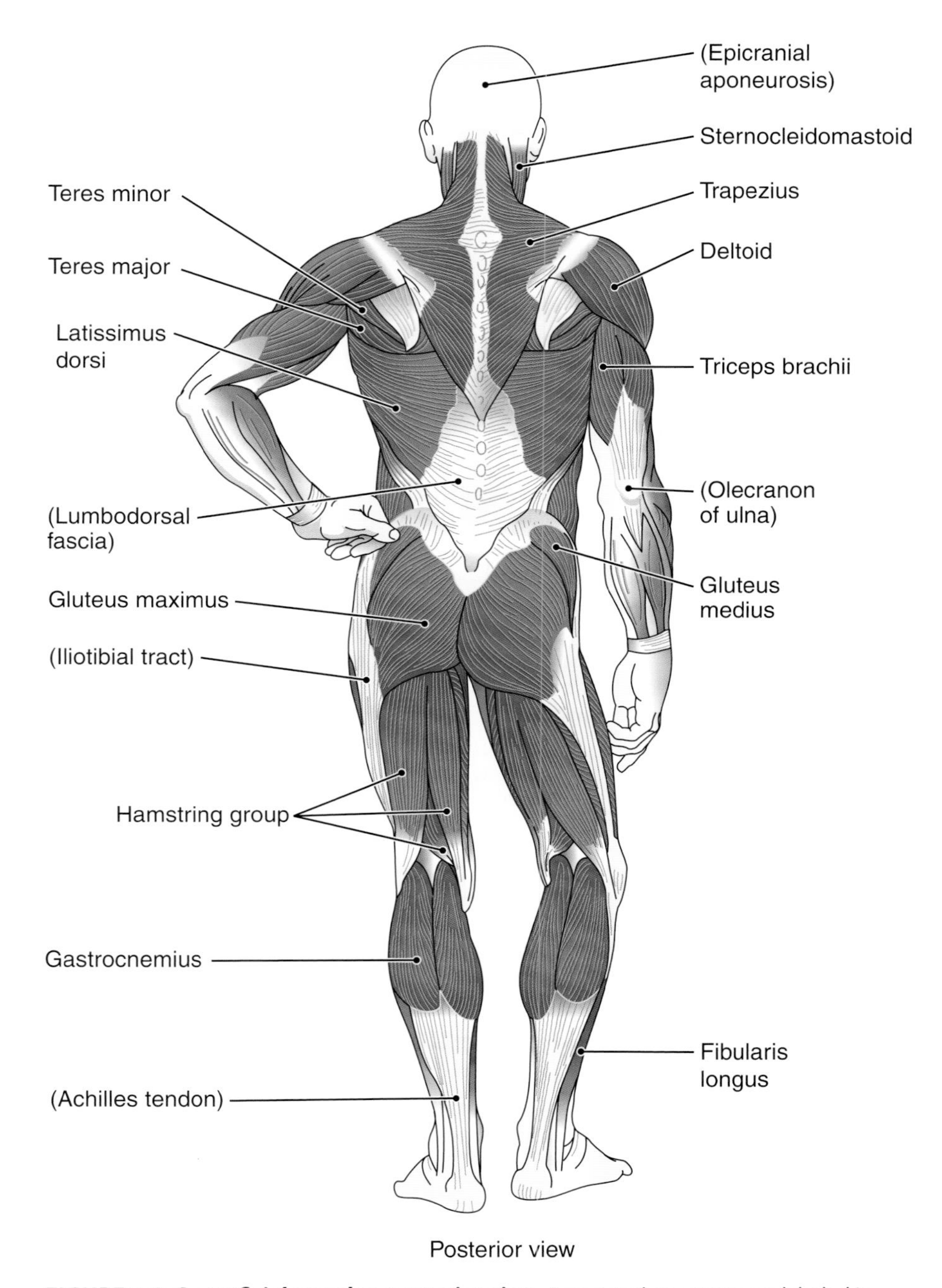

FIGURE 7-9 Superficial muscles, posterior view. Associated structures are labeled in parentheses.

one to raise the shoulders and pull them back. The superior portion of each trapezius can also extend the head and turn it from side to side.

The **latissimus** (lah-TIS-ih-mus) **dorsi** is the wide muscle of the back and lateral trunk (see FIG. 7-9). It originates from the vertebral spine in the middle and lower back and covers the inferior half of the thoracic region, forming the posterior portion of the axilla (armpit). The fibers of each muscle converge to a tendon that inserts on the humerus. The latissimus dorsi powerfully extends the arm, bringing it down forcibly as, for example, in swimming. It is also the prime mover in arm adduction.

A large **pectoralis** (pek-to-RAL-is) **major** is located on either side of the superior chest (see Fig. 7-8). This muscle arises from the sternum, the upper ribs, and the clavicle and forms the anterior "wall" of the axilla; it inserts on the superior humerus. The superior part of the pectoralis major flexes the arm (raising it overhead). Other portions of the muscle medially rotate the arm, pulling it across the chest, and synergize with the latissimus dorsi to adduct the arm.

ONE STEP AT A TIME

Muscles and Their Movements

BOX 7-3

Learning the locations of different muscles is the first step to knowing which muscles accomplish which movements. You can use this knowledge for many purposes, from predicting the impact of muscle disorders to optimizing muscle training regimes. This box shows you how to predict muscle actions based on muscle locations and attachments.

Question

Shane has particular weakness in his rectus femoris muscle. Predict which movements will be impacted.

Answer

Step 1. Locate the muscle on your body. The rectus femoris is located on the anterior thigh.

Step 2. Identify the bones where the muscle attaches. You can use a skeleton with labeled muscle origins and insertions, the text, FIGURE 7-15, or TABLE 7-5. The rectus femoris attaches to the ilium at one end and the tibia on the other end.

Step 3. Use your knowledge of skeletal anatomy to identify any joints that the muscle crosses. Based on the arrangement of bones identified in step 2, the rectus femoris must cross both the hip joint and the knee joint.

Step 4. Use your body or a model skeleton to shorten the distance between the origin at the hip and insertion on the tibia over the knee. What happens? The rectus femoris most commonly straightens the leg at the knee joint (extension). There might be other possibilities, depending on which bone moves and which bone remains stationary. For instance, this muscle can also act at the hip joint to flex the thigh.

Step 5. If you identified more than one possible outcome in step 4, use your body to figure out when the different movements occur. For the rectus femoris, stabilizing the trunk enables this muscle to flex the thigh and extend the leg. Stabilizing both the trunk and the thigh enables the muscle to extend the leg without flexing the thigh.

Step 6. Which muscles must relax to permit the movement? Remember that antagonist muscles must relax to enable a given movement. Antagonistic muscles in the limbs are usually found on the opposite side of the limb. So, the antagonist of the rectus femoris is the hamstring muscle group. Relaxing the hamstring as much as possible can optimize the function of the rectus femoris.

You can use a similar procedure to answer Question 26 at the end of this chapter, which asks you to design training regimes to strengthen a patient's shoulder and thigh muscles.

The **serratus** (ser-RA-tus) **anterior** is below the axilla, on the lateral chest (see FIG. 7-8). It originates on the upper eighth or ninth ribs on the lateral and anterior thorax and inserts in the scapula on the side toward the vertebrae. The serratus anterior moves the scapula forward and stabilizes it when, for example, one is pushing or punching something. It also aids in flexing and abducting the arm above the horizontal level.

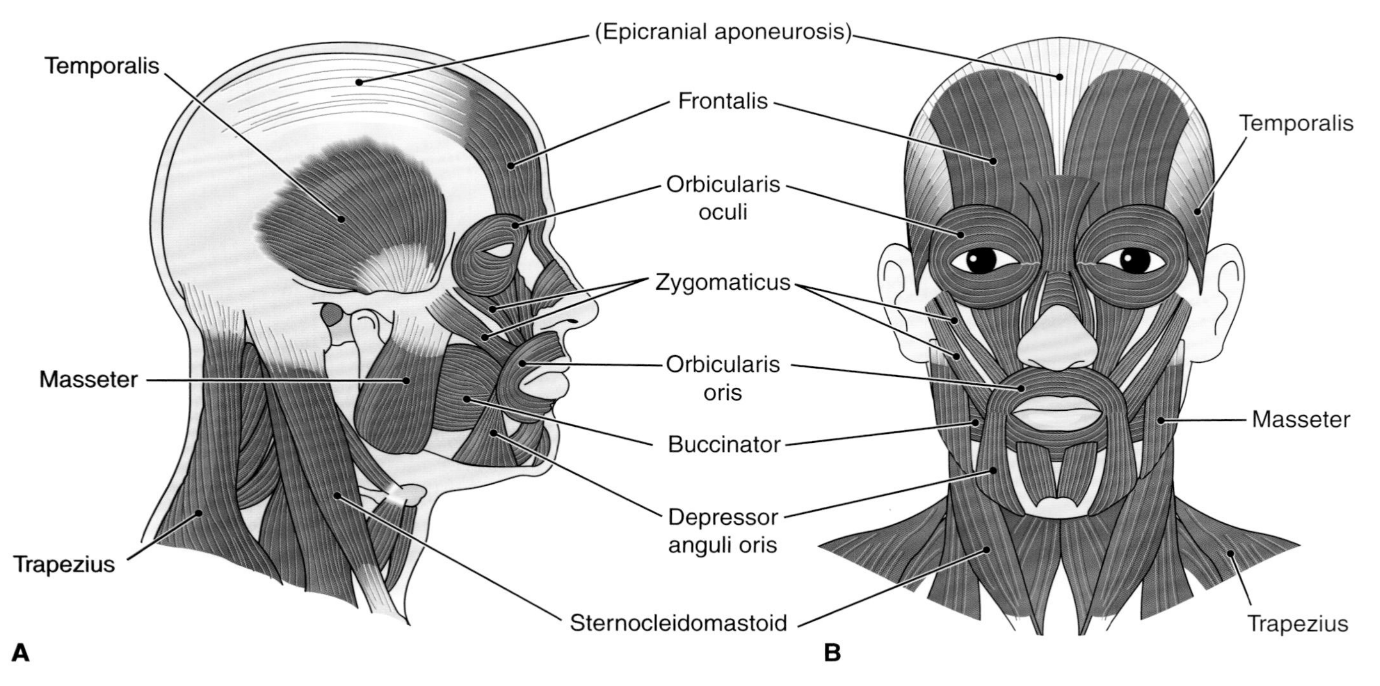

FIGURE 7-10 Muscles of the head. An associated structure is labeled in parentheses **A.** Lateral view **B.** Anterior view. ZOOMING IN Which of the muscles in this illustration are named for a nearby bone?

Table 7-2 Muscles of the Head and Neck*

Name	Origin	Insertion	Function
Orbicularis oculi	Frontal bone, maxilla	Skin encircling eye	Closes the eye
Levator palpebrae superioris (deep muscle; not shown)	Posterior orbit	Upper eyelid	Opens the eye
Orbicularis oris	Mandible, maxilla	Skin and muscle around mouth	Closes lips
Buccinator	Maxilla, mandible	Orbicularis oris	Flattens the cheek; helps in eating, whistling, and blowing wind instruments
Zygomaticus major and minor	Zygomatic bone	Orbicularis oris	Raises mouth corners upward and laterally (smile)
Depressor anguli oris	Mandible	Skin and muscle at mouth corners	Lowers mouth corners (grimace)
Temporalis	Temporal bone	Mandible	Closes the jaw
Masseter	Temporal bone	Mandible	Closes the jaw
Sternocleidomastoid	Sternum, clavicle	Temporal bone (mastoid process)	Flexes the head; rotates the head toward opposite side from the muscle

*These and other muscles of the face are shown in **FIGURE 7-10**.

Table 7-3 Muscles of the Upper Extremities*

Name	Origin	Insertion	Function
Trapezius	Occipital bone, cervical and thoracic vertebrae	Clavicle, scapula (acromion, spine)	Raises the shoulder and pulls it back; superior portion extends and turns the head
Latissimus dorsi	Inferior vertebrae, iliac crest	Humerus	Extends and adducts the arm (prime mover)
Pectoralis major	Sternum, clavicle, superior ribs	Humerus	Flexes and adducts the arm; medially rotates the arm across the chest; pulls the shoulder forward and downward
Serratus anterior	Superior ribs	Scapula	Moves the shoulder forward; synergist in arm flexion and abduction
Deltoid	Clavicle, sternum (acromion, spine)	Humerus	Abducts the arm; synergist in arm flexion, rotation, and extension
Biceps brachii	Scapula	Radius	Supinates the forearm and hand; synergist in forearm flexion
Brachialis	Humerus	Ulna	Primary flexor of the forearm
Brachioradialis	Humerus	Radius	Synergist in forearm flexion
Triceps brachii	Scapula, humerus	Olecranon of ulna	Extends the forearm to straighten upper extremity
Flexor carpi group	Humerus, ulna	Metacarpals (anterior)	Flexes the hand
Extensor carpi group	Humerus, ulna	Metacarpals (posterior)	Extends the hand
Flexor digitorum group	Humerus, ulna, radius	Middle phalanges (anterior)	Flexes fingers
Extensor digitorum group	Humerus	Distal and middle phalanges (posterior)	Extends fingers

*These and other muscles of the upper extremities are shown in **FIGURES 7-8, 7-9, and 7-11**.

The **deltoid** covers the shoulder joint and is responsible for the roundness of the upper arm just inferior to the shoulder **(see FIGS. 7-8 and 7-9)**. This muscle is named for its triangular shape, which resembles the Greek letter delta. The deltoid is often used as an injection site. Arising from the shoulder girdle (clavicle and scapula), the deltoid fibers converge to insert on the lateral surface of the humerus. Contraction of the middle portion of this muscle abducts the arm, raising it laterally to the horizontal position. The anterior portion synergizes with the pectoralis major to flex and rotate the arm anteriorly, whereas the posterior portion works with latissimus dorsi to extend and rotate the arm posteriorly.

The shoulder joint allows for a wide range of movements. This freedom of movement is possible because the humerus fits into a shallow scapular socket, the glenoid cavity. This joint requires the support of four deep muscles and their tendons, which compose the **rotator cuff**. The four muscles are the **s**upraspinatus, **i**nfraspinatus, **t**eres minor, and **s**ubscapularis, known together as SITS, based on the first letters of their names. In certain activities, such as swinging a golf club, playing tennis, or pitching a baseball, the rotator cuff muscles may be injured, even torn, and may require surgery for repair.

Muscles that Move the Forearm and Hand The **biceps brachii** (BRA-ke-i), located at the anterior arm along the humerus, is the muscle you usually display when you want to "flex your muscles" to show your strength **(FIG. 7-11)**. The root *brachi* means "arm" and is found in the names of several arm muscles. The biceps brachii inserts on the radius. It supinates the hand, as in turning a screwdriver, and acts as a synergist in forearm flexion. The **brachialis** (bra-ke-AL-is) lies deep to the biceps brachii and inserts distally over the anterior elbow joint. It flexes the forearm forcefully in all positions, sustains flexion, and steadies the forearm's slow extension.

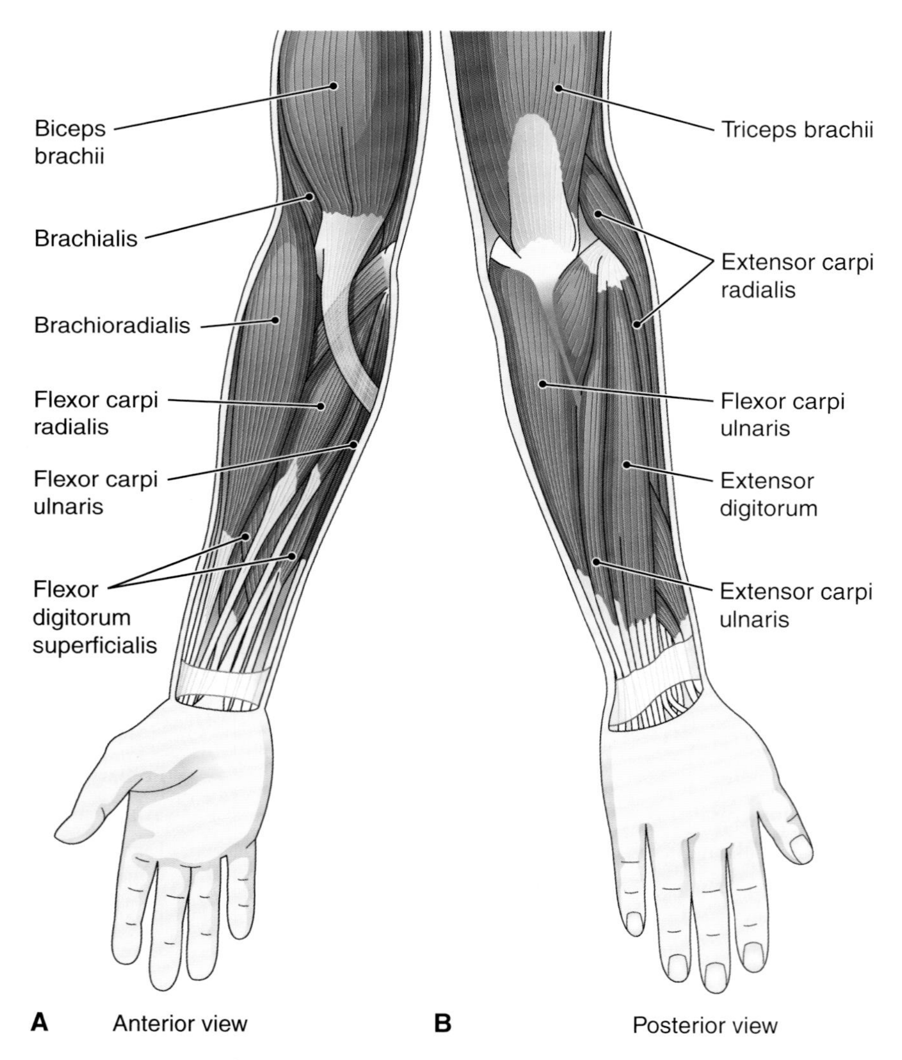

FIGURE 7-11 Muscles that move the forearm and hand. The muscles are shown in anterior **(A)** and posterior **(B)** views. ZOOMING IN What does carpi refer to in the names of muscles? Digitorum?

Another forearm flexor at the elbow is the **brachioradialis** (bra-ke-o-ra-de-A-lis), a prominent forearm muscle that originates at the distal humerus and inserts on the distal radius (see FIG. 7-11).

The **triceps brachii**, located on the posterior arm, inserts on the olecranon of the ulna (see FIG. 7-11B). It is used to extend the forearm and thus straighten the arm, as in lowering a weight from an arm curl. It is also important in pushing because it converts the arm and forearm into a sturdy rod.

Most of the muscles that move the hand and fingers originate from the distal humerus. Their muscle bellies are usually in the forearm, and they control the hand and fingers by long tendons that are held in place at the wrist by a connective tissue band, the transverse carpal ligament. This arrangement enables precise hand and finger movements unencumbered by bulky muscles.

The action and sometimes location of each forearm muscle can be inferred by its name. The **flexors** in the anterior forearm and the **extensors** in the posterior forearm act on the hand (carpi muscles) and fingers (digitorum muscles). The flexor carpi ulnaris is located posteriorly to the other flexors but still inserts on the anterior surface of the carpal bones and accomplishes hand flexion (see FIG. 7-11). The flexor carpi and extensor carpi that insert on the radius (thumb side) work together to abduct the hand, and the flexor carpi and extensor carpi that insert on the ulna adduct the hand.

Tiny intrinsic muscle groups in the fleshy parts of the hand fine-tune the intricate movements that can be performed with the thumb and the fingers. The thumb's freedom of movement has been one of humankind's most useful capacities.

> **See the Student Resources on thePoint® for additional pictures of the rotator cuff and the muscles of the upper extremity.**

MUSCLES OF THE TRUNK

The trunk muscles include the muscles involved in breathing, the thin muscle layers of the abdomen, and the muscles of the pelvic floor. The following discussion also includes the deep muscles of the back that support and move the vertebral column.

Muscles of Respiration The most important muscle involved in the act of breathing is the **diaphragm**. This dome-shaped muscle forms the partition between the thoracic cavity above and the abdominal cavity below (FIG. 7-12). When the diaphragm contracts, the central dome-shaped portion is pulled downward, thus enlarging the thoracic cavity from top to bottom. This action results in inhalation (breathing in).

The **intercostal muscles**, which also act to change thoracic volume, are attached to the ribs and fill the spaces between them. Additional muscles of the neck, chest, and abdomen are employed in forceful breathing. The mechanics of breathing are described in Chapter 16.

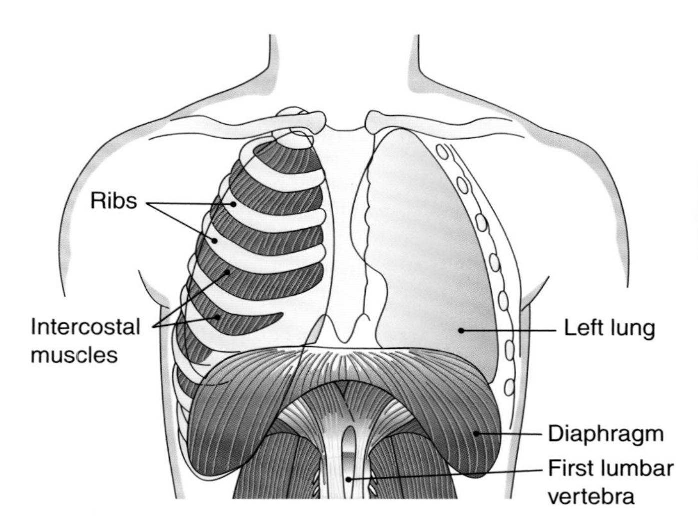

FIGURE 7-12 Muscles of respiration. **KEY POINT** The diaphragm is the main muscle of respiration. The left lung and ribs are also shown.

Muscles of the Abdomen and Pelvis The abdominal wall has three muscle layers that extend from the back (dorsally) and around the sides (laterally) to the front (ventrally) (FIG. 7-13; TABLE 7-4). They are the **external oblique** on the exterior, the **internal oblique** in the middle, and the

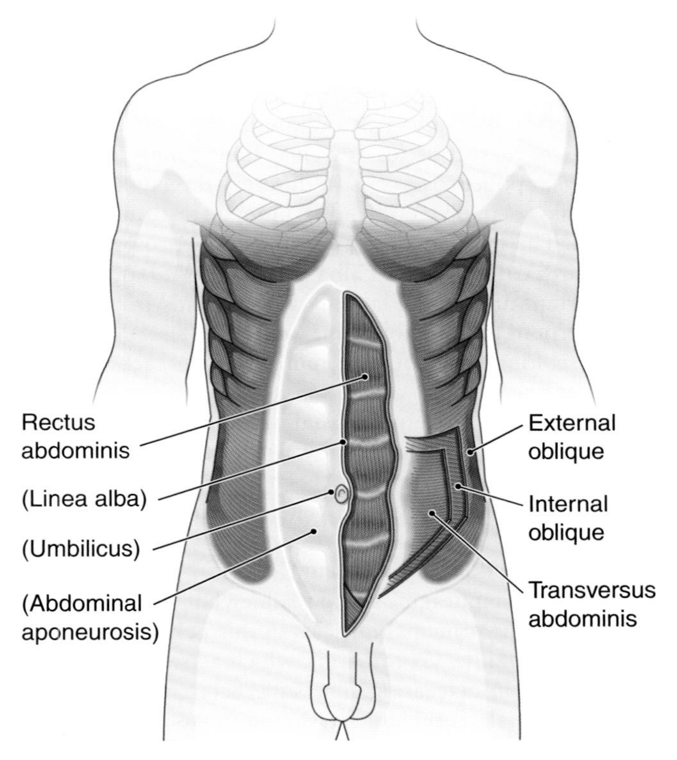

FIGURE 7-13 Muscles of the abdominal wall. **KEY POINT** Thin layers of muscle tissue with fibers running in different directions give strength to the abdominal wall. Surface tissue is removed here on the right side to show deeper muscles. Associated structures are labeled in parentheses. **ZOOMING IN** What does rectus mean? Oblique?

Table 7-4 Muscles of the Trunk*

Name	Origin	Insertion	Function
Diaphragm	Xiphoid process, cartilage of inferior ribs, lumbar vertebrae	Central tendon	Contracts to enlarge thorax causing inhalation; relaxes to shrink the thorax causing exhalation
Internal intercostals	Superior border of ribs	Inferior border of the rib above	Depress ribs during active exhalation
External intercostals	Inferior border of ribs	Superior border of the rib below	Elevate ribs during inhalation
Rectus abdominis	Pubis	Xiphoid process of the sternum, ribs	Flexes spinal column, compresses abdomen
External oblique	Fifth–12th ribs	Ilium, pubis, linea alba	Both: Flex spinal column, compress abdomen One: Rotates, flexes the vertebral column laterally
Internal oblique	Iliac crest	10th–12th ribs, linea alba	Same as above
Transverse abdominis	Iliac crest, cartilage of Seventh–12th ribs	Xiphoid process, linea alba, pubis	Compresses the abdomen
Levator ani	Pubis, ischial spine	Coccyx, urethra, rectum, perineum	Aids in defecation; stabilizes the perineum
Erector spinae (deep; not shown)	Ribs and vertebrae	Occipital bone, temporal bone, ribs, vertebrae	Extends the neck; moves the vertebral column

*These and other muscles of the trunk are shown in **FIGURES 7-12 through 7-14**.

transversus abdominis, the innermost. The connective tissue from these muscles extends anteriorly and encloses the vertical **rectus abdominis** of the anterior abdominal wall. The fibers of these muscles as well as their connective tissue extensions (aponeuroses) run in different directions, resembling the layers in plywood and resulting in a strong abdominal wall. The midline meeting of the aponeuroses forms a whitish area called the **linea alba** (LIN-e-ah AL-ba), which is an important abdominal landmark. It extends from the tip of the sternum to the pubic joint (**see FIG. 7-13**).

These four pairs of abdominal muscles act together to protect the internal organs and compress the abdominal cavity, as in forcefully exhaling, coughing, emptying the bladder (urination) and bowel (defecation), sneezing, vomiting, and childbirth (labor). The two oblique muscles and the rectus abdominis help bend the trunk forward and sideways.

The pelvic floor, or **perineum** (per-ih-NE-um), has its own form of diaphragm, shaped somewhat like a shallow dish. One of the principal muscles of this pelvic diaphragm is the **levator ani** (le-VA-tor A-ni), which acts on the rectum and thus aids in defecation. The superficial and deep muscles of the female perineum are shown in **FIGURE 7-14** along with some associated structures.

Deep Muscles of the Back The deep muscles of the back, which act on the vertebral column itself, are thick vertical masses that lie under the trapezius and latissimus dorsi and thus are not illustrated. The **erector spinae** muscles make up a large group located between the sacrum and the skull. These muscles extend the spine and maintain the vertebral column in an erect posture. They can be strained in lifting heavy objects if the spine is flexed while lifting. One should bend at the hip and knee instead and use the thigh and buttock muscles to help in lifting.

Even deeper muscles lie beneath the lumbodorsal fascia. These small muscles extend the vertebral column in the lumbar region and are also easily strained.

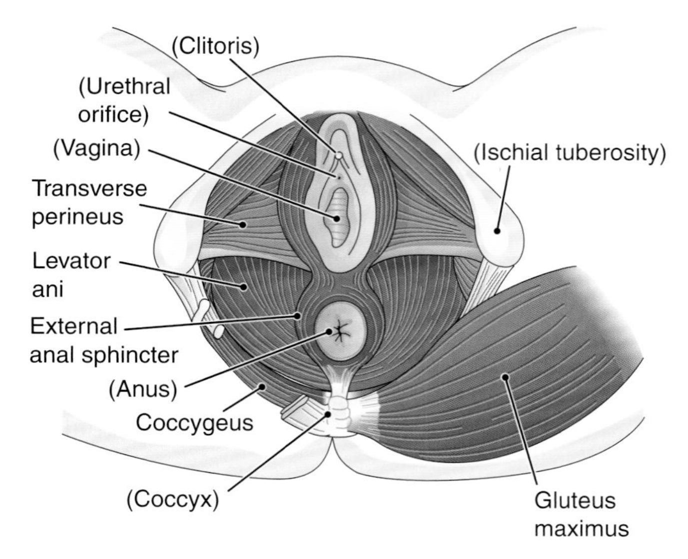

FIGURE 7-14 Muscles of the female perineum (pelvic floor). Associated structures are labeled in parentheses.

MUSCLES OF THE LOWER EXTREMITIES

The muscles in the lower extremities, among the longest and strongest muscles in the body, are specialized for locomotion and balance. They include the muscles that move the thigh and leg and those that control movement of the foot.

Concept Mastery Alert

Before moving on, review the terms describing movements of the lower extremity. Remember that flexion bends the limb at the hip or knee, and extension straightens it.

Muscles that Move the Thigh and Leg The **gluteus maximus** (GLU-te-us MAK-sim-us), which forms much of the buttock's fleshy part, is relatively large in humans because of its support function when a person is standing erect (see **FIGS. 7-9 and 7-14; TABLE 7-5**). This muscle extends the thigh and is important in walking and running by providing a pushing force at the end of a stride. The **gluteus medius**, which is partially covered by the gluteus maximus, abducts the thigh. This movement moves the foot slightly outward and upward with every step. It is one of the sites used for intramuscular injections.

The **iliopsoas** (il-e-o-SO-as) arises from the ilium and the bodies of the lumbar vertebrae; it crosses the anterior

Table 7-5 Muscles of the Lower Extremities*

Name	Origin	Insertion	Function
Gluteus maximus	Iliac crest, sacrum, coccyx	Iliotibial tract, linea aspera of femur	Extends and laterally rotates the hip
Gluteus medius	Ilium	Greater trochanter of femur	Abducts and laterally rotates the hip
Iliopsoas	Ilium and lumbar vertebrae	Femur	Flexes and laterally rotates the hip; flexes the vertebral column
Adductor group (e.g., adductor longus, adductor magnus)	Pubis, ischium	Femur	Adducts, medially rotates the hip
Sartorius	Iliac spine	Tibia	Flexes, abducts, and laterally rotates the hip and flexes the knee in order to sit cross-legged
Gracilis	Pubis	Tibia	Adducts and medially rotates the hip; flexes the knee
Quadriceps femoris:		All four muscles: patella and proximal tibia	
Rectus femoris	Iliac spine		Extends the knee; flexes the hip
Vastus medialis	Femur (greater trochanter, linea aspera)		Extends the knee
Vastus lateralis	Femur (greater trochanter, linea aspera)		Extends the knee
Vastus intermedius (deep; not shown)	Femur		Extends the knee
Hamstring group:			Flexes the knee and extends the hip
Biceps femoris	Ischial tuberosity, femur (linea aspera)	Fibula (head) and tibia (lateral condyle)	
Semimembranosus	Ischial tuberosity	Tibia (medial condyle)	
Semitendinosus	Ischial tuberosity	Proximal tibia	
Gastrocnemius	Femur (lateral and medial condyles)	Calcaneus by Achilles tendon	Plantar flexes the ankle; raises the heel in walking; flexes the knee
Soleus	Fibula (head) and proximal tibia	Calcaneus by Achilles tendon	Plantar flexes the ankle; maintains balance when walking
Tibialis anterior	Tibia	Ankle and first metatarsal	Dorsiflexes foot inverts the ankle
Fibularis (peroneus) longus	Fibula, tibia	Ankle and first metatarsal	Everts the ankle; steadies the leg
Flexor digitorum group	Posterior tibia	Distal phalanges	Flexes the toes; plantar flexes the ankle
Extensor digitorum group	Tibia	Distal phalanges	Extends the toes; dorsiflexes the ankle

*These and other muscles of the lower extremities are shown in **FIGURES 7-15 and 7-16**.

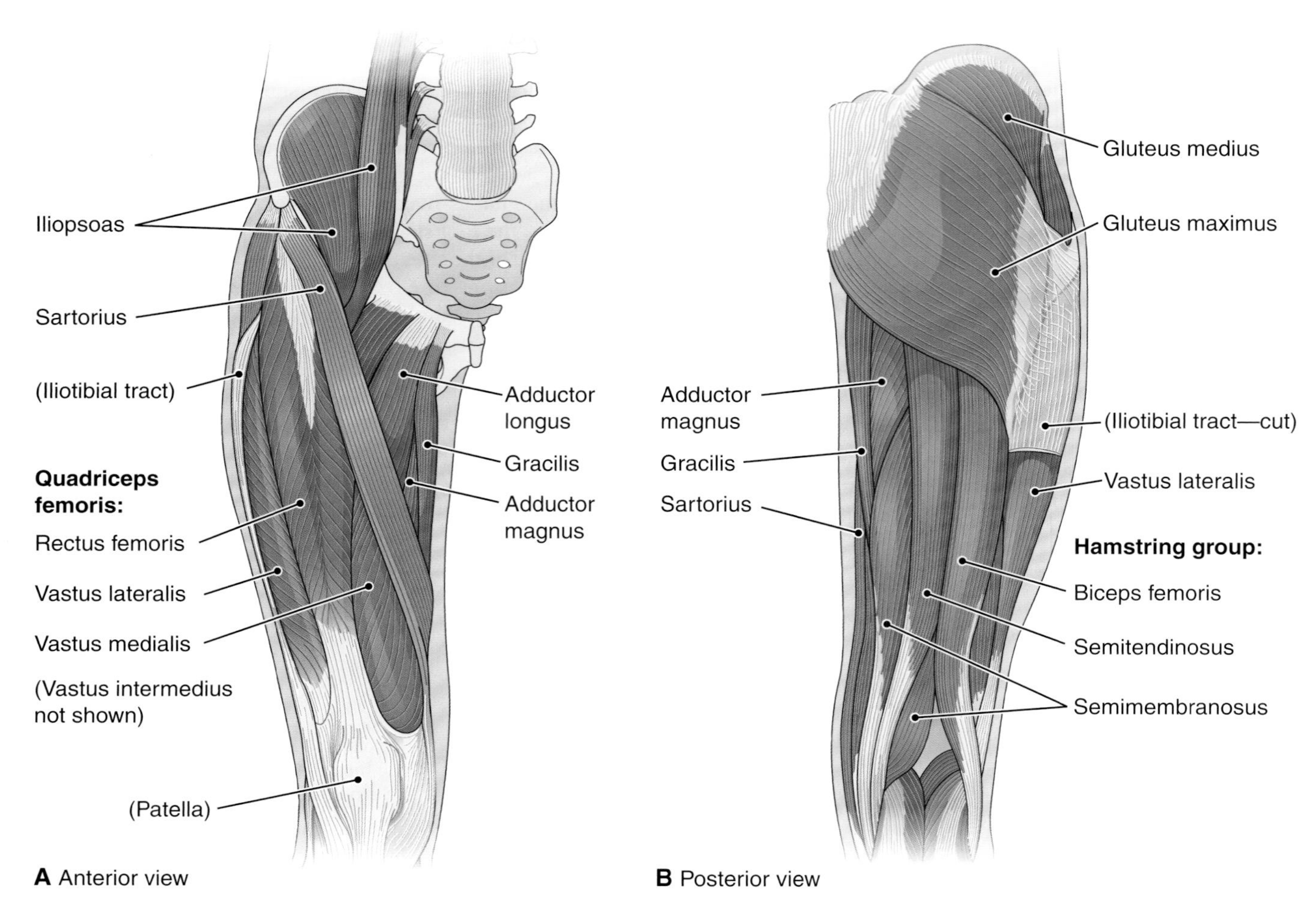

FIGURE 7-15 Muscles of the thigh. Associated structures are labeled in parentheses. Anterior **(A)** and posterior **(B)** views are shown. ZOOMING IN How many muscles make up the quadriceps femoris?

hip joint to insert on the femur (FIG. 7-15A). It is a powerful thigh flexor and helps keep the trunk from falling backward when one is standing erect. When the thighs are immobilized, the iliopsoas flexes the trunk, as in doing a sit-up.

The **adductor muscles** are located on the medial thigh (see FIG. 7-15). They arise from the pelvis and insert on the femur. These strong muscles press the thighs together, as in grasping a saddle between the knees when riding a horse. They include the **adductor longus** and **adductor magnus**. The **gracilis** (grah-SIL-is) is also located in the medial thigh, but it crosses both the hip joint and the knee joint. It too adducts the thigh and also flexes the leg at the knee.

The **sartorius** (sar-TO-re-us) is a long, narrow muscle that begins at the iliac spine, winds downward and medially across the anterior thigh, and ends on the tibia's superior medial surface (see FIG. 7-15). It is called the tailor's muscle because it is used in crossing the legs in the manner of tailors, who, in days gone by, sat cross-legged on the floor.

The anterior and lateral femur are covered by the **quadriceps femoris** (KWOD-re-seps FEM-or-is), a large muscle that has four heads of origin (see FIG. 7-15A). The individual parts are as follows: in the center, covering the anterior thigh, the **rectus femoris**; on either side, the **vastus medialis** and **vastus lateralis**; and deeper in the center, the **vastus intermedius**. One of these muscles (rectus femoris) originates from the ilium, and the other three are from the femur, but all four have a common tendon of insertion on the tibia. You may remember that this is the tendon that encloses the patella (kneecap). This muscle extends the leg and straightens the lower limb, as in kicking a ball. The vastus lateralis is also a site for intramuscular injections.

The iliotibial tract is a thickened band of fascia that covers the lateral thigh muscles. It extends from the ilium of the hip to the superior tibia and reinforces the fascia of the thigh (the fascia lata) (see FIG. 7-15).

The **hamstring muscles** are located in the posterior thigh (see FIG. 7-15B). They originate on the ischium and femur, and you can feel their tendons behind the knee as they descend to insert on the tibia and fibula. The hamstrings flex the leg at the knee as in kneeling. They also extend and rotate the thigh at the hip. Individually, moving from lateral to medial position, they are the **biceps femoris**, the **semimembranosus**, and the **semitendinosus**. The name of this muscle group refers to the tendons at the posterior of the knee by which these muscles insert on the leg.

Muscles that Move the Foot The **gastrocnemius** (gas-trok-NE-me-us) is the chief muscle of the calf of the leg (its name means "belly of the leg") (**FIG. 7-16**). It has been called the toe dancer's muscle because it is used in standing on tiptoe. It ends near the heel in a prominent cord called the **Achilles tendon** (**see FIG. 7-16B**), which attaches to the calcaneus (heel bone). The Achilles tendon is the largest tendon in the body. According to the Greek mythology, the region above the heel was the only place on his body where the hero Achilles was vulnerable, and if the Achilles tendon is cut, it is impossible to walk. The **soleus** (SO-le-us) is a flat muscle deep to the gastrocnemius. It also inserts by means of the Achilles tendon and, like the gastrocnemius, flexes the foot at the ankle.

Another leg muscle that acts on the foot is the **tibialis** (tib-e-A-lis) **anterior**, located on the anterior region of the leg (**see FIG. 7-16A**). This muscle performs the opposite function of the gastrocnemius. Walking on the heels uses the tibialis anterior to raise the rest of the foot off the ground (dorsiflexion). This muscle is also responsible for inversion of the foot. The muscle for the foot's eversion is the **fibularis** (fib-u-LA-ris) **longus**, also called the peroneus (per-o-NE-us) longus, located on the lateral leg. This muscle's long tendon crosses under the foot, forming a sling that supports the transverse (metatarsal) arch.

The toes, like the fingers, are provided with flexor and extensor muscles. The tendons of the extensor muscles are located in the superior part of the foot and insert on the superior surface of the phalanges (toe bones). The flexor digitorum tendons cross the sole of the foot and insert on the undersurface of the phalanges (**see FIG. 7-16**).

Concept Mastery Alert

Confusingly, both the fingers and toes are moved by digitorum muscle groups.

CHECKPOINTS 

- [ ] **7-15** Which muscle is most important in breathing?
- [ ] **7-16** Which structural feature gives strength to the muscles of the abdominal wall?

CASEPOINTS 

- [ ] **7-4** Shane's physician found that his calf muscles were enlarged. Name the two largest muscles of the calf.
- [ ] **7-5** Shane's thigh muscles were weak. Name the large muscle group of the anterior thigh.
- [ ] **7-6** Name the large muscle group of the posterior thigh.

See the Student Resources on thePoint® for additional pictures of the muscles that move the lower extremity.

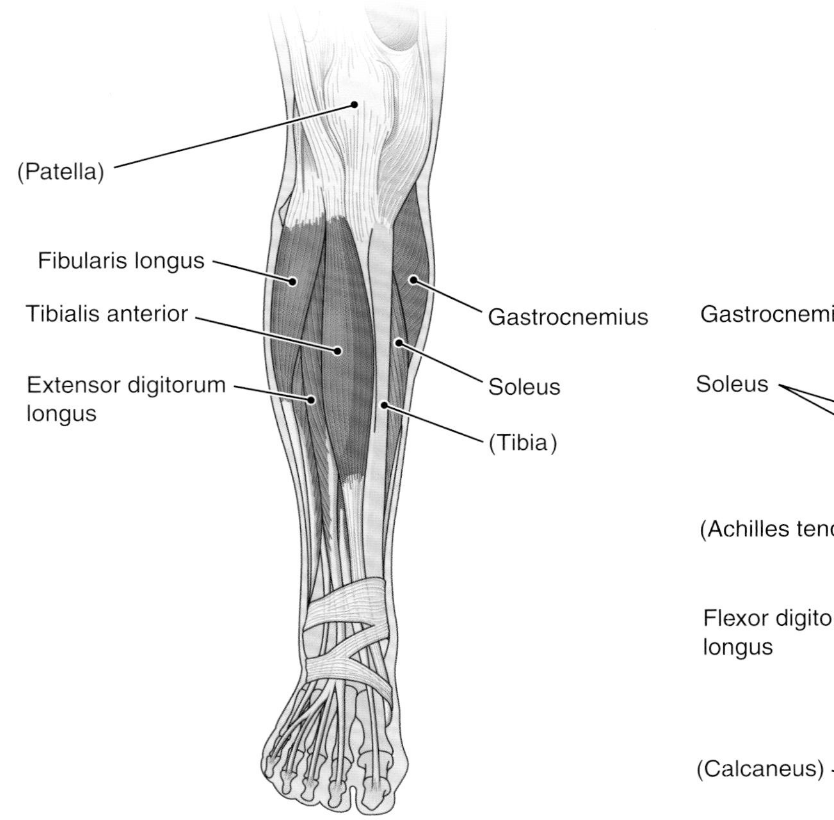

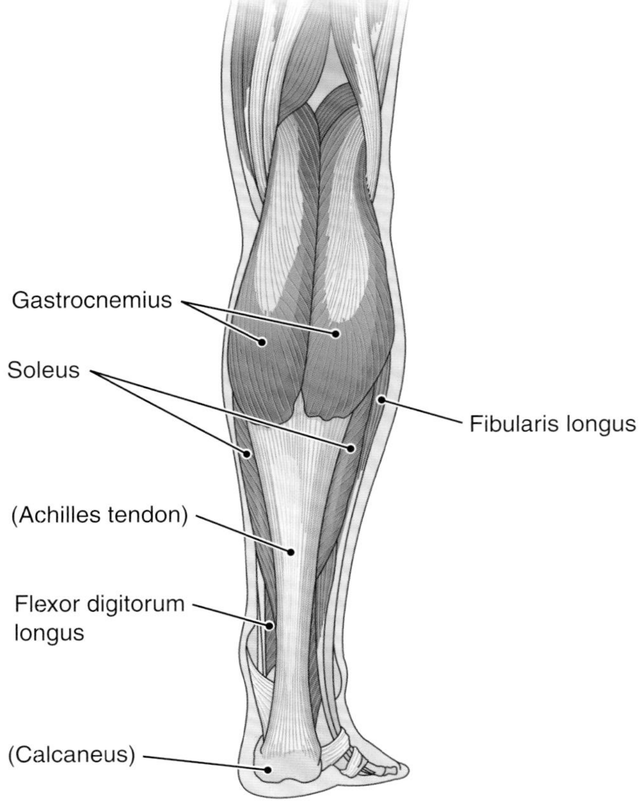

FIGURE 7-16 Muscles that move the foot. Associated structures are labeled in parentheses. Anterior **(A)** and posterior **(B)** views are shown. ZOOMING IN On what bone does the Achilles tendon insert?

Effects of Aging on Muscles

Beginning at about 40 years of age, there is a gradual loss of muscle cells with a resulting decrease in the size of each individual muscle. The muscle fibers that enable quick, explosive movements die first. The loss of these fibers makes it more difficult for the elderly to recover their balance, leading to more frequent falls. Loss of power in the extensor muscles, such as the large sacrospinalis near the vertebral column, causes the "bent over" appearance of a hunchback. Sometimes, there is a tendency to bend (flex) the hips and knees. In addition to the previously noted changes in the vertebral column (see Chapter 6), these effects on the extensor muscles result in a further decrease in the elderly person's height. Activity and exercise throughout life delay and decrease these undesirable effects of aging. Even among the elderly, resistance exercise, such as weight lifting, increases muscle strength and function.

See the Student Resources on thePoint® for information on careers in physical therapy and how physical therapists participate in treatment of muscular disorders.

A & P in Action Revisited

Looking at Shane's blood test results, Dr. Schroeder saw that the level of creatine phosphokinase was elevated. Usually, this protein is confined to muscle cells, so this finding indicated that substances were leaking from the weakened walls of his muscle cells. At the next visit, Dr. Schroeder used a small needle to extract a sample of Shane's gastrocnemius (a muscle biopsy). The sample revealed evidence of muscle degeneration and abnormally low levels of dystrophin (the protein that links myofibril bundles to the sarcolemma). A genetic test of Shane's initial blood sample showed that Shane had a gene mutation that causes Duchenne muscular dystrophy.

Dr. Schroeder discussed these results with Kathy. "Kathy, the tests conclusively show that Shane has Duchenne muscular dystrophy. As Shane grows older, his muscles will become weaker, and this disease will considerably shorten his life. With therapy, we can preserve his ability to walk for as long as possible and decrease the development of deformities. I'm going to refer you to a rehabilitation medicine specialist, called a physiatrist, who can further evaluate Shane's condition and formulate a management program. Your family is going to need some help dealing with Shane's condition, so here is the contact information for a DMD support group and some government agencies that can help. While we don't have any treatments for Shane right now, research efforts are ongoing. Don't give up hope."

During this case, we learned that structural components of the muscle cells determine the ability of muscles to contract. Weakness in limb muscles (such as the gastrocnemius) will affect Shane's ability to walk and perform everyday tasks. Eventually, weakness in his respiratory muscles will likely cause his death.

CHAPTER 7

Chapter Wrap-Up

OVERVIEW

A detailed chapter outline with space for note-taking is on thePoint®. The figure below illustrates the main topics covered in this chapter.

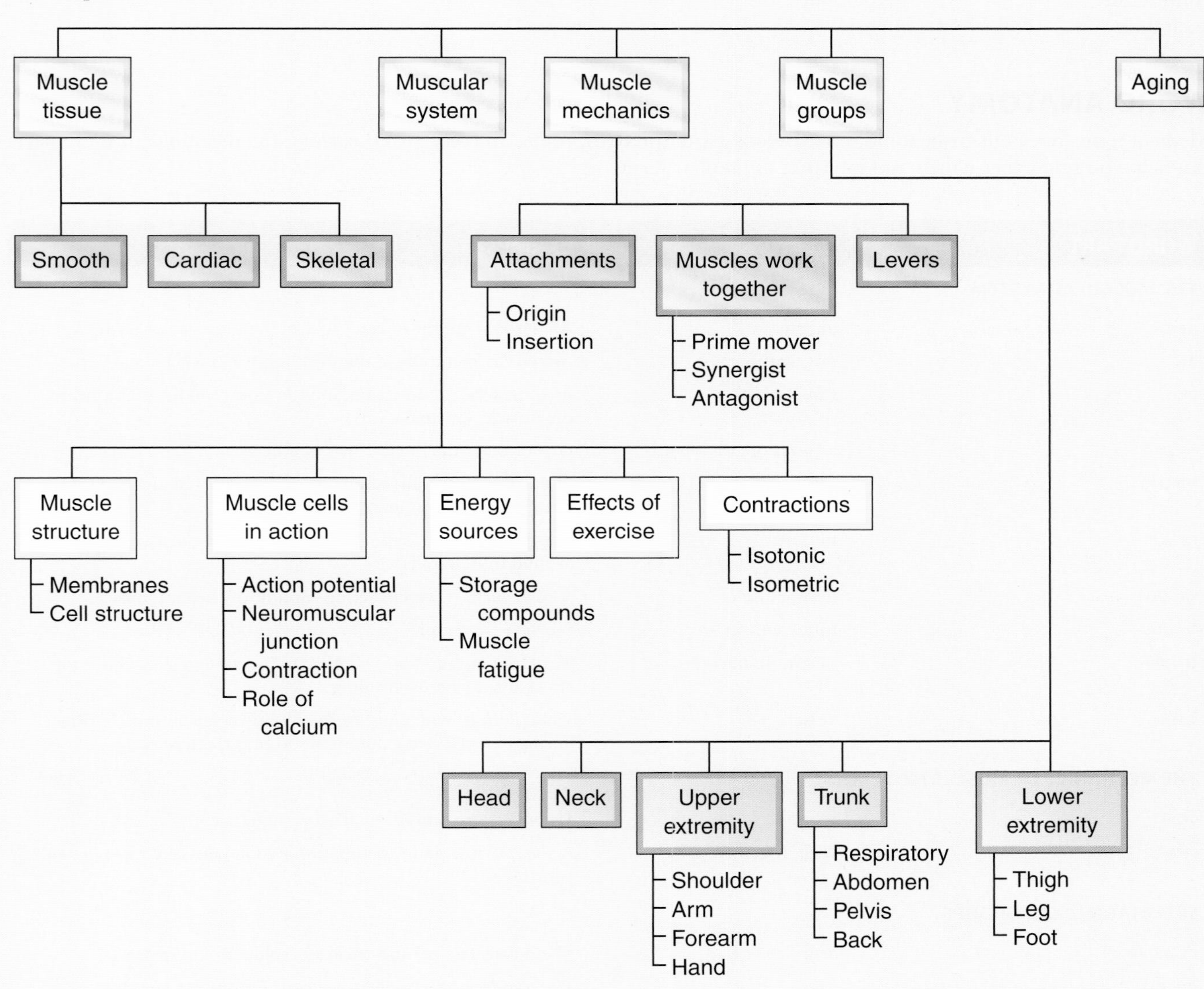

KEY TERMS

The terms listed below are emphasized in this chapter. Knowing them will help you organize and prioritize your learning. These and other boldface terms are defined in the Glossary with phonetic pronunciations.

acetylcholine
actin
action potential
agonist
antagonist
contractility
excitability
fascicle
glycogen
insertion
membrane potential
motor unit
myoglobin
myosin
neuromuscular junction
neurotransmitter
origin
prime mover
receptor
sarcomere
synapse
synergist
tendon
tonus
tropomyosin
troponin

WORD ANATOMY

Medical terms are built from standardized word parts (prefixes, roots, and suffixes). Learning the meanings of these parts can help you remember words and interpret unfamiliar terms.

WORD PART	MEANING	EXAMPLE
THE MUSCULAR SYSTEM		
aer/o	air, gas	An *aerobic* organism can grow in the presence of air (oxygen).
an-	not, without	*Anaerobic* metabolism does not require oxygen.
iso-	same, equal	In an *isotonic* contraction, muscle tone remains the same, but the muscle shortens.
-lysis	separation, dissolving	*Glycolysis* is the breakdown of glucose.
metr/o	measure	In an *isometric* contraction, muscle length remains the same, but muscle tension increases.
my/o	muscle	The *endomysium* is the deepest layer of connective tissue around muscle cells.
sarc/o	flesh, muscle	A *sarcomere* is a contracting subunit of skeletal muscle.
ton/o	tone, tension	See “iso-” example.
troph/o	nutrition, nurture	Muscles undergo *hypertrophy*, an increase in size, under the effects of resistance training.
vas/o	vessel	*Vasodilation* (widening) of the blood vessels in muscle tissue during exercise brings more blood into the tissue.
THE MECHANICS OF MUSCLE MOVEMENT		
erg/o	work	*Synergists* are muscles that work together.
syn-	with, together	A *synapse* is a point of communication between a neuron and another cell.
SKELETAL MUSCLE GROUPS		
brachi/o	arm	The biceps *brachii* and triceps brachii are in the arm.
quadr/i	four	The *quadriceps* muscle group consists of four muscles.

QUESTIONS FOR STUDY AND REVIEW

BUILDING UNDERSTANDING

Fill in the Blanks

1. The point at which a nerve fiber contacts a muscle cell is called the _____.
2. A single neuron and all the muscle fibers it stimulates make up a(n) _____.
3. A wave of electric current that spreads along a plasma membrane is called a(n) _____.
4. A contraction in which there is no change in muscle length but there is a great increase in muscle tension is _____.
5. A skeletal muscle's partially contracted state that is normal even when the muscle is not in use is called _______.

7

Matching Match each numbered item with the most closely related lettered item.

___ 6. Raises and pulls back the shoulder
___ 7. Active in breathing
___ 8. Closes jaw
___ 9. Muscle of the perineum
___ 10. Closes lips

a. levator ani
b. masseter
c. orbicularis oris
d. trapezius
e. diaphragm

Multiple Choice

___ 11. From superficial to deep, the correct order of muscle structure is
a. deep fascia, epimysium, perimysium, and endomysium
b. epimysium, perimysium, endomysium, and deep fascia
c. deep fascia, endomysium, perimysium, and epimysium
d. endomysium, perimysium, epimysium, and deep fascia

___ 12. In order of decreasing size, which of the following is correct?
a. myofibril, myofilament, myofiber
b. myofilament, myofiber, myofibril
c. myofiber, myofibril, myofilament
d. myofibril, myofiber, myofilament

___ 13. What is the function of calcium ions in skeletal muscle contraction?
a. bind to receptors on the motor end plate to stimulate muscle contraction
b. cause a pH change in the cytoplasm to trigger muscle contraction
c. block the myosin-binding sites on actin
d. bind to regulatory proteins to expose myosin-binding sites on actin

___ 14. Which structure is a broad flat extension that attaches muscle to a bone or other muscles?
a. tendon
b. fascicle
c. aponeurosis
d. motor end plate

___ 15. Which type of lever is responsible for forearm flexion?
a. first class
b. second class
c. third class
d. fourth class

16. Which muscle is most involved in the act of quiet breathing?
a. sternocleidomastoid
b. diaphragm
c. pectoralis major
d. internal intercostal

UNDERSTANDING CONCEPTS

17. Compare smooth, cardiac, and skeletal muscle with respect to location, structure, and function. Briefly explain how each type of muscle is specialized for its function.
18. Explain what is meant by the statement "The membrane potential in resting muscle cells is about −70 mV."
19. Explain the concept of excitability in muscle cells and how excitability is necessary for muscle function.
20. Describe four substances stored in skeletal muscle cells that are used to manufacture a constant supply of ATP.
21. The first sign of Shane's muscular dystrophy was weakness of his thigh muscles. Name the muscles that are involved in movement of the thigh.

22. Name the following muscle(s):
    a. antagonist of the orbicularis oculi
    b. prime mover in arm adduction
    c. prime mover in hand extension
    d. antagonist of zygomaticus major
    e. prime mover in dorsiflexion
    f. antagonist of the brachialis

23. During a Cesarean section, a transverse incision is made through the abdominal wall. Name the muscles incised, and state their functions.

24. What effect does aging have on muscles? What can be done to resist these effects?

CONCEPTUAL THINKING

25. Margo recently began working out and jogs three times a week. After her jog, she is breathless, and her muscles ache. From your understanding of muscle physiology, describe what has happened inside of Margo's skeletal muscle cells. How do Margo's muscles recover from this? If Margo continues to exercise, what changes would you expect to occur in her muscles?

26. Alfred suffered a mild stroke, leaving him partially paralyzed on his left side. Physical therapy was ordered to prevent left-sided weakness. Prescribe some exercises for Alfred's shoulder and thigh.

For more questions, see the Learning Activities on thePoint®.

UNIT III

Coordination and Control

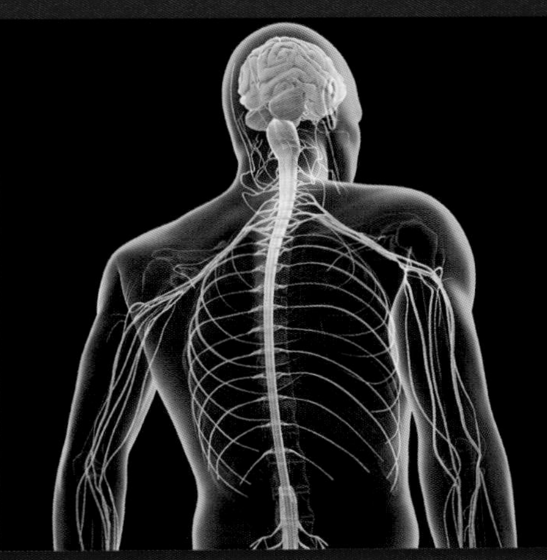

CHAPTER 8

The Nervous System: The Spinal Cord and Spinal Nerves

Learning Objectives

After careful study of this chapter, you should be able to:

1. Outline the organization of the nervous system according to structure and function. *p. 158*
2. Describe the structure of a neuron. *p. 159*
3. Explain the construction and function of the myelin sheath. *p. 160*
4. Describe how neuron fibers are built into a nerve. *p. 161*
5. List four types of neuroglia in the central nervous system, and cite the functions of each. *p. 162*
6. Diagram and describe the steps in an action potential. *p. 162*
7. Explain the role of neurotransmitters in impulse transmission at a synapse. *p. 165*
8. Describe the distribution of gray and white matter in the spinal cord. *p. 167*
9. Describe and name the spinal nerves and three of their main plexuses. *p. 168*
10. List the components of a reflex arc. *p. 170*
11. Define a simple reflex, and give several examples of reflexes. *p. 171*
12. Compare the locations and functions of the sympathetic and parasympathetic nervous systems. *p. 171*
13. Explain the role of cellular receptors in the action of neurotransmitters in the autonomic nervous system. *p. 174*
14. Using the case study, describe the effects of demyelination on motor and sensory function. *pp. 157, 175*
15. Show how word parts are used to build words related to the nervous system (see Word Anatomy at the end of the chapter). *p. 177*

A & P in Action
Sue's Case: The Importance of Myelin

Dr. Jensen glanced at her patient's chart as she stepped into the consulting room. Sue Pritchard, a seemingly healthy 26-year-old Caucasian, had presented to her family doctor with right-hand weakness and difficulty in walking, a feeling that she was off-balance. She also had experienced an episode of double vision (diplopia) within the past year. Some preliminary tests of her muscular strength and reflexes had led her to suspect multiple sclerosis (MS), a disease that affects the nerves. In addition to the referral, she had ordered a magnetic resonance imaging (MRI) of her brain and spinal cord.

"Hi Sue. My name is Dr. Jensen. I'm a neurologist, which means I specialize in the diagnosis and treatment of nervous system disorders. Let's start with a few tests to determine how well your brain and spinal cord communicate with the rest of your body. Then, we'll take a look at your MRI results."

Using a reflex hammer, Dr. Jensen tapped on the tendons of several muscles in Sue's arms and legs to elicit stretch reflexes. Her responses indicated damage to areas of the spinal cord that control reflexes. The doctor also detected muscle weakness in Sue's limbs—an indication of damage to the descending tracts in the spinal cord, which carry motor nerve impulses from the brain to skeletal muscle. In addition, she discovered that Sue's sense of touch was impaired—an indication of damage to the spinal cord's ascending tracts, which carry sensory impulses from receptors in the skin to the brain. Sue was exhibiting several of the most common clinical signs of MS.

Dr. Jensen then showed Sue the results of her MRI scan. "Here's the MRI of your spinal cord. The nervous tissue making up the spinal cord is organized into two regions—this inner region of gray matter and this outer one of white matter. If you look closely at the white matter, you can see several damaged areas, which we call lesions. They are causing many of your symptoms because they prevent your spinal cord from transmitting impulses between your brain and the rest of your body. These lesions, or scleroses, are what give multiple sclerosis its name. There is also a lesion in your brain that may be responsible for your earlier incident of double vision."

The evidence shows that Sue has MS, a disease of neurons in the central nervous system (CNS). In this chapter, we learn more about neurons and the spinal cord, one part of the CNS.

As you study this chapter, CasePoints will give you opportunities to apply your learning to this case.

Visit thePoint® to access the following resources. For guidance in using these resources most effectively, see pp. xv–xvii.

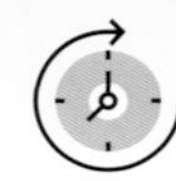

Preparing to Learn

- Tips for Effective Studying
- Pre-Quiz

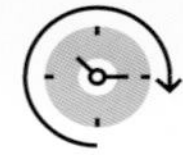

While You Are Learning

- Web Figure: The Cauda Equina
- Web Chart: Neuroglia
- Animation: The Neuromuscular Junction
- Animation: The Action Potential
- Animation: Propagation of the Action Potential
- Animation: The Reflex Arc
- Chapter Notes Outline
- Audio Pronunciation Glossary

When You Are Reviewing

- Answers to Questions for Study and Review
- Health Professions: Occupational Therapist
- Interactive Learning Activities

A LOOK BACK

In Chapter 7, we discussed how action potentials and neurotransmitters are involved in muscle contraction. Now, we broaden our outlook to see how the nervous system uses these same signals to transmit information and coordinate responses to changes in the environment. Key ideas relevant to this chapter include Gradients and Flow 5, Homeostasis 3, and Structure–Function 1 from Chapter 1 and Communication 11 from Chapter 7.

Overview of the Nervous System

No body system is capable of functioning alone. All are interdependent and work together as one unit to promote health and well-being. The nervous system serves as the chief coordinating agency for most body functions. Conditions both within and outside the body are constantly changing. The nervous system must detect and respond to these changes (known as *stimuli*) so that the body can adapt itself to new conditions.

The nervous system can be compared with a large corporation in which market researchers (sensory receptors) feed information into middle management (the spinal cord), who then transmit information to the chief executive officer or CEO (the brain). The CEO organizes and interprets the information and then sends instructions out to workers (effectors) who carry out appropriate actions for the good of the company. Information (signals) between the researchers, management, and workers are communicated by e-mail through a network, which, like the body's nerves, carries information throughout the system.

3 In Chapter 1, we discussed the role of the nervous system in negative feedback loops. These loops help maintain homeostasis—the maintenance of a relatively constant internal environment—by keeping the levels of specific regulated variables relatively constant. This chapter discusses a different type of involuntary response called a **reflex**. Reflexes protect us by initiating rapid responses to damaging stimuli, such as pulling your hand away from a hot stove. The actions of the nervous system, though, are not only concerned with our physical well-being. Chapter 9 introduces "higher" brain functions, such as the planning of complex movements and speech, and Chapter 10 explains how we gather and process sensory information to form an integrated picture of the world around us.

Although all parts of the nervous system work in coordination, portions may be grouped together on the basis of either structure or function.

DIVISIONS OF THE NERVOUS SYSTEM

The entire nervous system is classified into two divisions (**FIG. 8-1**):

- The **central nervous system** (CNS) includes the brain and spinal cord.
- The **peripheral** (per-IF-er-al) **nervous system** (PNS) is made up of all the nerves outside the CNS. It includes all the **cranial nerves** that carry impulses to and from the brain and all the **spinal nerves** that carry messages to and from the spinal cord.

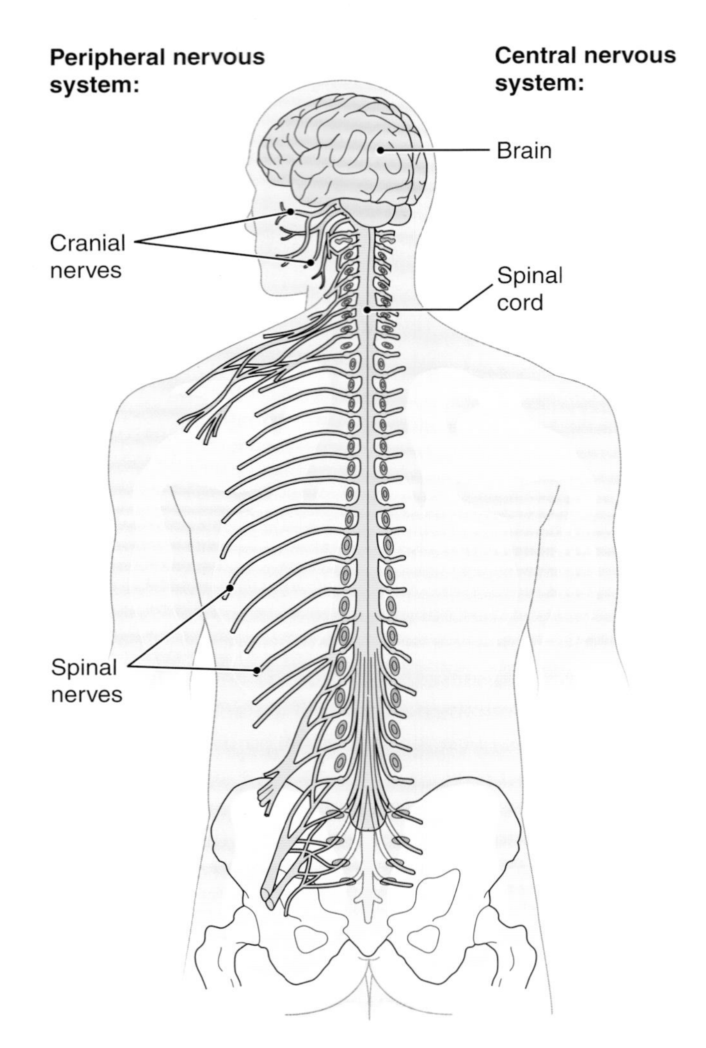

FIGURE 8-1 Anatomic divisions of the nervous system, posterior view. **KEY POINT** The nervous system is divided structurally into a central nervous system and a peripheral nervous system. **ZOOMING IN** What structures make up the central nervous system? The peripheral nervous system?

 *Concept Mastery Alert*

Remember, even though the cranial nerves attach to the brain, they are still part of the PNS.

The CNS and PNS together include all of the nervous tissue in the body.

Functional Divisions of the PNS The PNS is split by function into two subdivisions. The somatic nervous system controls voluntary functions, whereas the autonomic nervous system (ANS) controls functions we cannot consciously control (**TABLE 8-1**). Any tissue or organ that carries out a nervous system command is called an **effector**, all of which are muscles or glands.

Table 8-1 Functional Divisions of the Peripheral Nervous System

Division	Control	Effectors
Somatic nervous system	Voluntary	Skeletal muscle
Autonomic nervous system	Involuntary	Smooth muscle, cardiac muscle, and glands

The **somatic nervous system** is voluntarily controlled (by conscious will), and all its effectors are skeletal muscles (described in Chapter 7). The nervous system's involuntary division is called the **autonomic nervous system**, making reference to its automatic activity. It is also called the **visceral nervous system** because its effectors are smooth muscle, cardiac muscle, and glands, which are found in the soft body organs, the viscera. The ANS is described in more detail later in this chapter.

Although these divisions are helpful for study purposes, dividing the PNS and its effectors by function can be misleading. Although skeletal muscles *can* be controlled voluntarily, they may function commonly without conscious control. The diaphragm, for example, a skeletal muscle, typically functions in breathing without conscious thought. In addition, we have certain rapid reflex responses involving skeletal muscles—drawing the hand away from a hot stove, for example—that do not involve the brain. In contrast, people can be trained to consciously control involuntary functions, such as blood pressure and heart rate, by training techniques known as *biofeedback*.

CHECKPOINTS

- 8-1 What are the two divisions of the nervous system based on structure?
- 8-2 What division of the PNS is voluntary and controls skeletal muscles? What division is involuntary and controls smooth muscle, cardiac muscle, and glands?

Neurons and Their Functions

The functional cells of the nervous system are highly specialized cells called **neurons** (FIG. 8-2). 1 These cells have a unique structure related to their function.

STRUCTURE OF A NEURON

The main portion of each neuron, the cell body, contains the nucleus and other organelles typically found in cells. A distinguishing feature of the neurons, however, is the long, threadlike fibers that extend out from the cell body and carry impulses across the cell (FIG. 8-3).

Dendrites and Axons Two kinds of fibers extend from the neuron cell body: dendrites and axons.

- **Dendrites** are neuron fibers that conduct impulses *to* the cell body. Most dendrites have a highly branched, treelike appearance (see FIG. 8-2). In fact, the name comes from a Greek word meaning "tree." Some dendrites

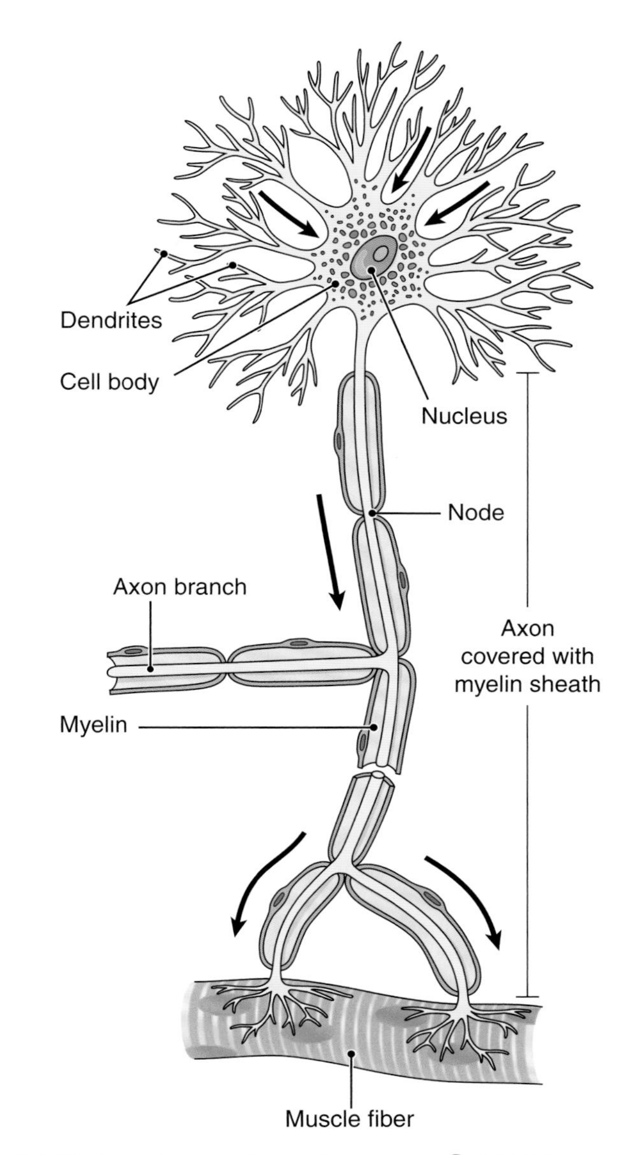

FIGURE 8-2 Diagram of a motor neuron. KEY POINT A neuron has fibers extending from the cell body. Dendrites carry impulses toward the cell body; axons carry impulses away from the cell body. The break in the axon denotes length. The *arrows* show the direction of the nerve impulse. ZOOMING IN How do you know the neuron shown here is a motor neuron? Is it part of the somatic or visceral nervous system? Explain.

function as **receptors** in the nervous system. That is, they receive a stimulus that begins a neural pathway. In Chapter 10, we describe how the dendrites of the sensory system may be modified to respond to a specific type of stimulus, such as pressure or taste.

- **Axons** (AK-sons) are neuron fibers that conduct impulses *away from* the cell body (see FIG. 8-2). These impulses may be delivered to another neuron, to a muscle, or to a gland. A neuron has only one axon, which can extend up to 1 m and give off many branches.

The Myelin Sheath Some axons are covered with a whitish, fatty substance called **myelin** (MI-eh-lin) that insulates

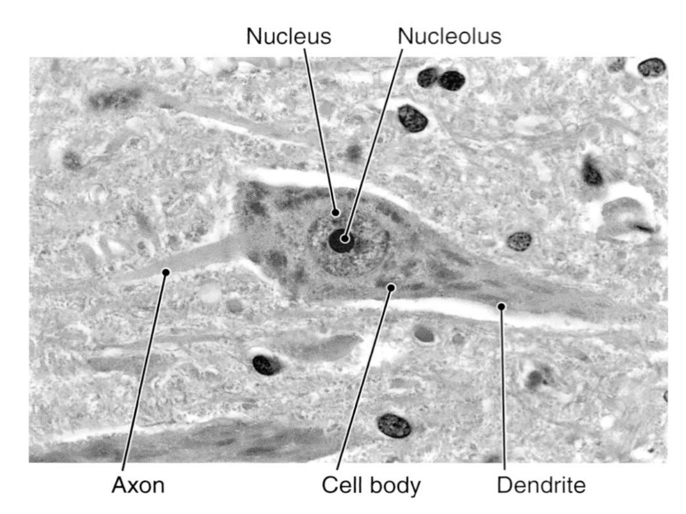

FIGURE 8-3 Microscopic view of a neuron. Based on staining properties and structure, the fiber on the *left* is identified as an axon; the fiber on the *right* is a dendrite. The clear space around the axon is caused by the staining procedure.

and protects the axon (see FIG. 8-2). In the PNS, myelin is actually entire cells, known as **Schwann** (shvahn) **cells**, that are wrapped many times around the axon like a jelly roll (FIG. 8-4). The cell bodies and most of the cytoplasm get squeezed into the outermost layer of the Schwann cell wrapping, known as the **neurilemma** (nu-rih-LEM-mah).

When the sheath is complete, small spaces remain between the individual Schwann cells. These tiny gaps, called **nodes** (originally, nodes of Ranvier), are important in speeding nerve impulse conduction. In the CNS, the myelin sheath is formed by another type of cell, the oligodendrocyte (ol-ih-go-DEN-dro-site) (literally meaning "cell with few dendrites"). One oligodendrocyte sends cellular extensions to myelinate several neighboring axons (see FIG. 8-6 and the later discussion of oligodendrocytes). The cytoplasm and nucleus remain with the oligodendrocyte, so CNS neurons do not have a neurilemma.

Schwann cells help injured neurons regenerate. Undamaged Schwann cells near the injury divide to produce replacement Schwann cells. These new cells remove the damaged tissue and provide a mold (tube) to guide the extension of the new axon. Unfortunately, oligodendrocytes in the CNS do not have the same capabilities. While they can partially restore the myelin coating if it gets damaged, they cannot guide axon regeneration to the same extent as Schwann cells. If CNS neurons are injured, the damage is almost always permanent. Even in the peripheral nerves, however, repair is a slow and uncertain process.

Myelinated axons, because of myelin's color, are called **white fibers** and are found in the **white matter** of the brain and spinal cord as well as in most nerves throughout the body. The fibers and cell bodies of the **gray matter** are not covered with myelin. Sue, the subject of the case study, was diagnosed with multiple sclerosis. This disorder results from the loss of myelin in the CNS, and at present, there is no cure. See BOX 8-1, "Multiple sclerosis and experimental design" to learn how scientists test potential new treatments.

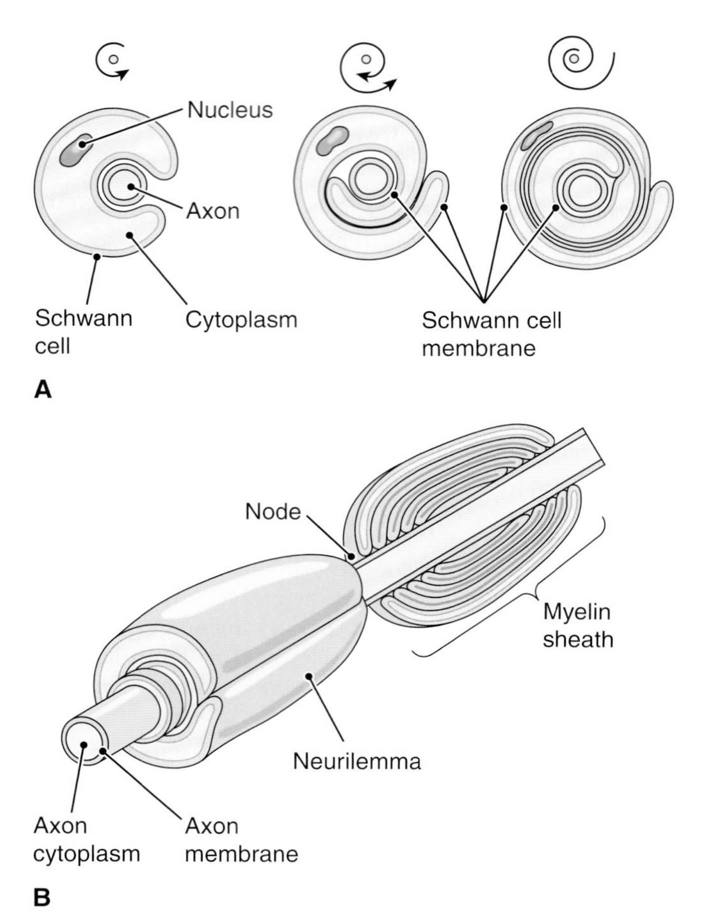

FIGURE 8-4 Formation of a myelin sheath. KEY POINT The myelin sheath is formed by Schwann cells in the peripheral nervous system. **A.** Schwann cells wrap around the axon, creating a myelin coating. **B.** The outermost layer of the Schwann cell forms the neurilemma. Spaces between the cells are the nodes (of Ranvier).

TYPES OF NEURONS

The job of neurons is to relay information to or from the CNS or to different places within the CNS itself. There are three functional categories of neurons:

- **Sensory neurons**, also called *afferent neurons*, conduct impulses *to* the spinal cord and brain. For example, if you touch a sharp object with your finger, sensory neurons will carry impulses generated by the stimulus to the CNS for interpretation.
- **Motor neurons**, also called *efferent neurons*, carry impulses *from* the CNS to muscles and glands (effectors). For example, the CNS responds to the pain of touching a sharp object by directing skeletal muscles in your arm to flex and withdraw your hand.
- **Interneurons**, also called *central* or *association neurons*, relay information from place to place within the CNS. Following our original example, in addition to immediate withdrawal from pain, impulses may travel to other parts of the CNS to help retain balance as you withdraw your hand or to help you learn how to avoid sharp objects!

ONE STEP AT A TIME

BOX 8-1

Multiple Sclerosis and Experimental Design

Sue, the case study subject, suffers from multiple sclerosis (MS). Scientists have not yet identified an effective drug for this disorder, so research continues. Imagine that you developed a new drug to treat MS. How would you design a trial to investigate your drug? Since it is untested, a good starting point would be testing your drug in mice that have been treated with a virus that demyelinates neurons.

Step 1. The point of your study is to investigate the effects of your drug. However, you want to minimize the possibility that the injections themselves have an effect. So, your study requires two groups. The experimental group consists of mice treated with the drug. The control group consists of mice injected with a **placebo**, a solution that does not contain any medication. Placebos are particularly important in human trials, because believing that something might help can actually produce positive effects. Any factor that can change is called a **variable**. So, the type of injection (placebo or drug) is the **experimental variable** (also known as the **Independent variable**).

Step 2. In order to isolate the effect of your drug, the experimental variable should be the only difference between your two groups. You need to minimize all other differences, called **controlled variables**. So, your mice should be of the same age, gender, and genetic strain. They should also eat the same diet, be housed in identical cages, and have the same light: dark cycle. Even cage position has been shown to modify a mouse's physiology. Controlled variables in human studies are far more complicated, but can include age, gender, socioeconomic status, education, and life habits such as smoking, among others.

Step 3. Group size is important. Large groups will provide more trustworthy statistical analysis of results (see the One Step at a Time box in Chapter 21). The size of your groups may be limited by manpower and finances, but aim for at least 20 mice per group. Large group sizes (in the hundreds or thousands) are particularly important in human studies, because they can minimize the impact of individual differences among participants. Studies with few participants do not produce reliable data.

Step 4. Decide how you will measure your drug's effectiveness. For instance, you could examine the mice's ability to run on a rodent wheel or take a brain sample to directly measure the extent of demyelination or remyelination.

Scientists must take great care in designing and carrying out their experiments, because a poorly designed or interpreted experiment can have a huge negative impact. For instance, a study using only 13 participants and few controlled variables was responsible for the fallacy that vaccines cause autism. Armed with a basic understanding of variables and experimental design, you can start to critically evaluate the claims you hear about through the media and online. See the Study Guide (available separately) for practice identifying good and bad experimental designs.

NERVES AND TRACTS

Everywhere in the nervous system, neuron fibers (dendrites and axons) are collected into bundles of varying size (FIG. 8-5). A fiber bundle is called a **nerve** in the PNS but a **tract** in the CNS. Tracts are located both in the brain, where they conduct impulses between regions, and in the spinal cord, where they conduct impulses to and from the brain.

A nerve or tract contains many neuron fibers, just like an electric cable contains many wires. As in muscles, the individual fibers are organized into groups called fascicles and are bound together by connective tissue. The names of the connective tissue layers are similar to their names in muscles, but the root *neur/o*, meaning "nerve," is substituted for the muscle root *my/o*, as follows:

- Endoneurium surrounds each individual fiber.
- Perineurium surrounds each fascicle.
- Epineurium surrounds the whole nerve.

A nerve may contain all sensory fibers, all motor fibers, or a combination of both types of fibers. A few of the cranial nerves contain only sensory fibers, so they only conduct impulses toward the brain. These are described as **sensory** (**afferent**) **nerves.** A few of the cranial nerves, called **motor** (**efferent**) **nerves,** contain only motor fibers, so they only conduct impulses away from the brain. However, most of the cranial nerves and *all* of the spinal nerves contain both sensory *and* motor fibers and are referred to as **mixed nerves.** Note that in a mixed nerve, impulses may be traveling in two directions (toward or away from the CNS), but each individual fiber in the nerve is carrying impulses in one direction only. Think of the nerve as a large highway. Traffic may be going north and south, for example, but each lane carries cars traveling in only one direction.

CHECKPOINTS 

- ☐ **8-3** What is the name of the neuron fiber that carries impulses toward the cell body? What is the name of the fiber that carries impulses away from the cell body?
- ☐ **8-4** What color describes myelinated fibers? What color describes the nervous system's unmyelinated tissue?
- ☐ **8-5** What name is given to nerves that convey impulses toward the CNS? What name is given to nerves that transport away from the CNS?
- ☐ **8-6** What is a nerve? What is a tract?

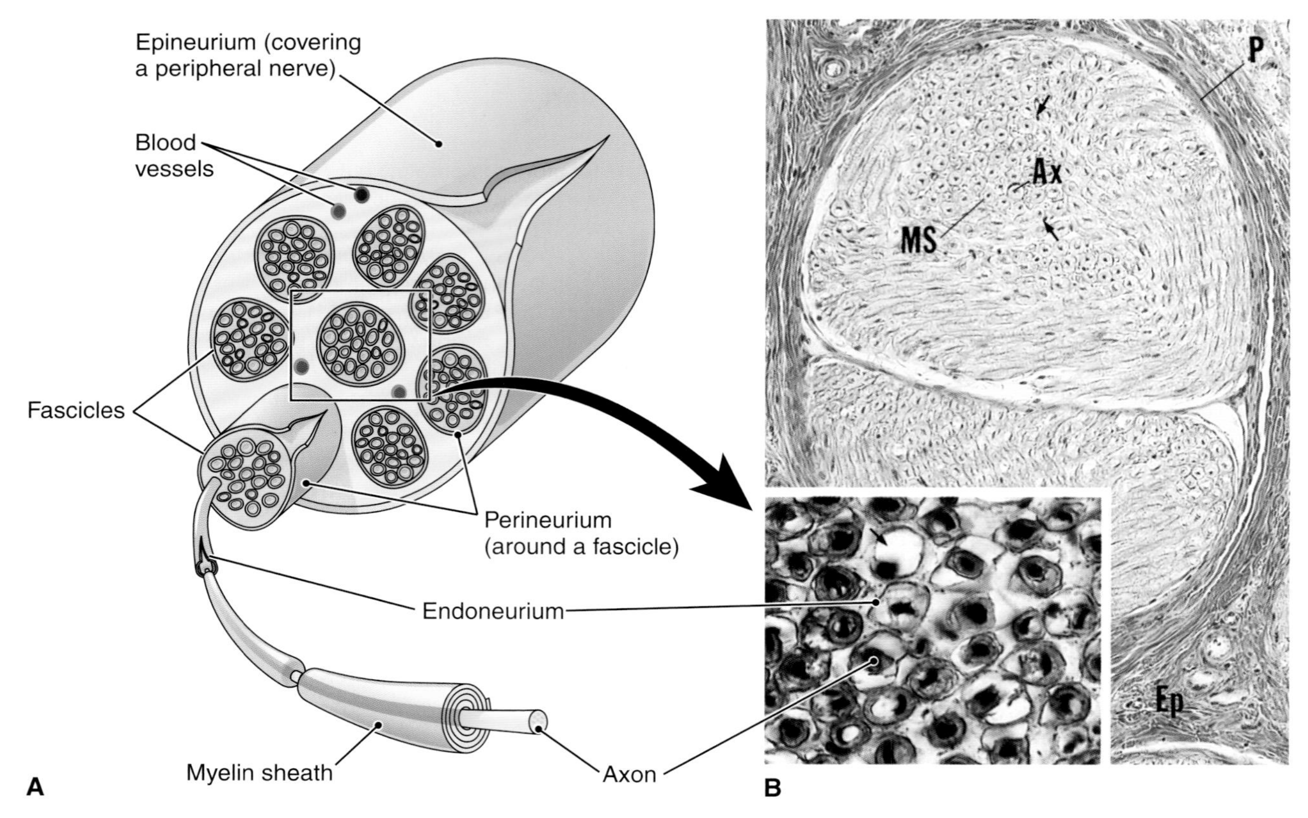

FIGURE 8-5 Structure of a nerve. **KEY POINT** Neuron fibers are collected in bundles called fascicles. Groups of fascicles make up a nerve. Connective tissue holds all components of the nerve together. **A.** Structure of a nerve showing neuron fibers and fascicles. **B.** Micrograph of a nerve (X132). Two fascicles are shown. Perineurium (P) surrounds each fascicle. Epineurium (Ep) is around the entire nerve. Individual axons (Ax) are covered with a myelin sheath (MS). Inset shows myelinated axons surrounded by endoneurium. **ZOOMING IN** What is the deepest layer of connective tissue in a nerve? What is the outermost layer?

Neuroglia

Neurons make up only 10% of nervous tissue. The remaining 90% of the CNS and the PNS consists of support cells known as **neuroglia** (nu-ROG-le-ah) or **glial** (GLI-al) **cells,** from a Greek word meaning "glue." The Schwann cell that forms the myelin sheath in the PNS is one type of neuroglia. Other types are located exclusively in the CNS (**FIG. 8-6**). Some of these and their functions are as follows:

- **Astrocytes** (AS-tro-sites) are star-shaped cells serving many functions. They nourish and physically support and anchor neurons; they regulate the composition of the extracellular fluid by absorbing and degrading neurotransmitters and excess ions; they form a barrier between blood and brain tissue; and they act as stem cells to make new neurons. These new neurons are important in the generation of memories. Astrocytes can also contribute new neurons to repair damaged areas, but their ability to completely repair brain damage is limited.
- **Oligodendrocytes** form the myelin sheath of CNS neurons.

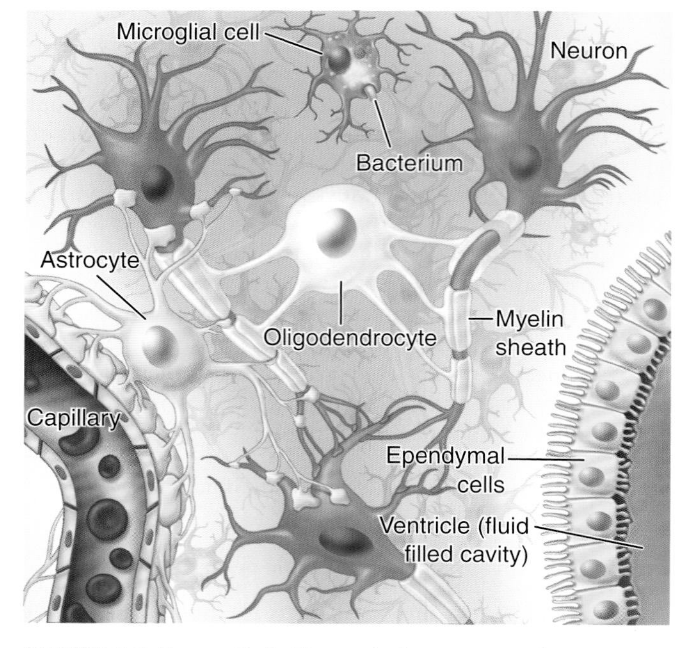

FIGURE 8-6 Neuroglia in the central nervous system. **KEY POINT** Neuroglia serve multiple roles in the nervous system.

- **Microglia** (mi-KROG-le-ah) act as phagocytes to remove pathogens, impurities, and dead neurons.
- **Ependymal** (ep-EN-dih-mal) **cells** line ventricles in the brain, fluid-filled cavities discussed in Chapter 9. These cells form a barrier between the nervous tissue of the CNS and the fluid filling the ventricles, cerebrospinal fluid (CSF). In a modified form, these cells also synthesize CSF and promote its movement in the ventricles.

Stem cells produce new neuroglia throughout life. Because neuroglia multiply more frequently than do neurons, most tumors of nervous tissue are glial tumors.

CHECKPOINT 

☐ **8-7** What is the name of the nervous system's nonconducting cells, which protect, nourish, and support the neurons?

See the Student Resources on thePoint® for a summary of the different neuroglial types.

The Nervous System at Work

We presented a snapshot of how the nervous system works in Chapter 7, when we explained how neurons stimulate muscle contraction. Here, we revisit and expand upon the mechanisms by which signals pass down neurons and are transmitted from a neuron to a neighboring cell.

THE NERVE IMPULSE

Recall that neuron fibers can extend over large distances—up to 1 m (3 ft). Specialized electrical signals called **action potentials**, or *nerve impulses*, pass from one end of the neuron to the other, much like an electric current spreads along a wire.

The Resting State Before we discuss the action potential, though, we need to review the basics of membrane potential discussed in Chapter 7. In living cells, positive and negative ions are unequally distributed in the cytoplasm and extracellular fluid. This unequal distribution creates an electric charge across the membrane, known as a membrane (or transmembrane) potential. In this state, the membrane is said to be *polarized*, because oppositely charged particles are separated. In resting cells, the membrane potential is negative (about –70 millivolts, or mV) because of the relative excess of negative ions in the cytoplasm.

Also important for the generation of a nerve impulse are large concentration gradients for sodium and potassium ions across the plasma membrane. Sodium ions are more concentrated along the extracellular side of the plasma membrane than they are along the intracellular side of the membrane. Conversely, potassium ions are in higher concentration on the inside than on the outside of the membrane. **5** Remember that substances flow by diffusion from an area where they are in higher concentration to an area where they are in lower concentration. Because of these concentration gradients, ions are constantly diffusing across the membrane in small amounts through channels known as *leak channels* and, as discussed shortly, during nerve impulse transmission.

Left unchecked, Na^+ and K^+ diffusion would eventually eliminate the gradients needed for nerve impulse conduction. In living cells, however, the Na^+ and K^+ gradients stay relatively constant due to specialized membrane proteins called sodium–potassium pumps (Na^+–K^+ pumps or Na–K–ATPases). These carriers use the chemical energy of ATP (see Chapter 2) to export sodium ions and import potassium ions against their concentration gradients.

Changes in Membrane Potential The movement of ions across the plasma membrane changes the membrane potential (**FIG. 8-7A**). It becomes less negative (more positive) if positive ions enter the cell to neutralize the unpaired negative ions. This change is known as **depolarization**, because it reduces membrane polarity closer to zero. If positive ions leave the cell (or if negative ions enter the cell), the membrane becomes more negative (or less positive). If the membrane potential falls below its resting value, the change is known as **hyperpolarization**. Any change that returns the membrane potential closer to its resting value is known as **repolarization**.

The Action Potential Action potentials result from ion movement across the plasma membrane. A simple description of the events in an action potential is as follows (**FIG. 8-7B**):

- **Rising phase.** A stimulus, such as an electrical, chemical, or mechanical signal of adequate force, causes specific channels in the membrane to open and allow Na^+ ions to flow into the cell. These newly arrived positive ions pair with and eventually outnumber the negative ions, so the membrane potential rises from –70 mV to about +55 mV.
- **Falling phase.** In the next step of the action potential, K^+ channels open to allow K^+ to leave the cell. The departure of positively charged potassium ions causes the membrane potential to fall from about +55 back to –70 mV, so it is described as the falling phase. Since the membrane potential returns to its resting value, it is also known as repolarization. Note that the ions moving in this step are not the same ones that caused the rising phase. During the falling phase, the membrane does not respond to further stimulation.

The action potential occurs rapidly, in less than one-thousandth of a second, and is followed by a rapid return to the resting state. However, this local electrical change in the membrane opens the sodium channels in the adjacent membrane region, causing a new action potential (**FIG. 8-8**). And so, the action potential spreads along the membrane as a wave of electric current.

8

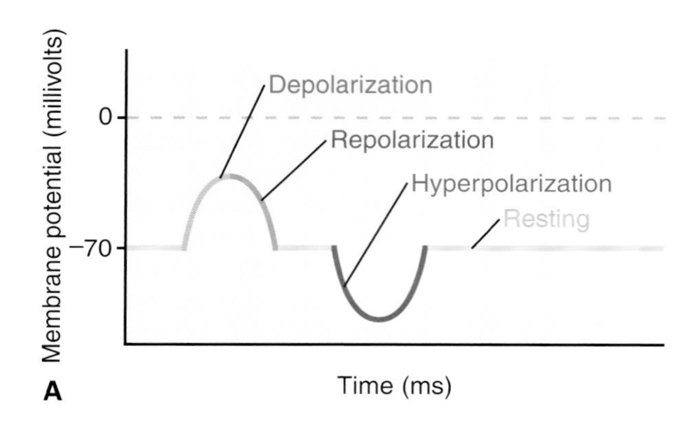

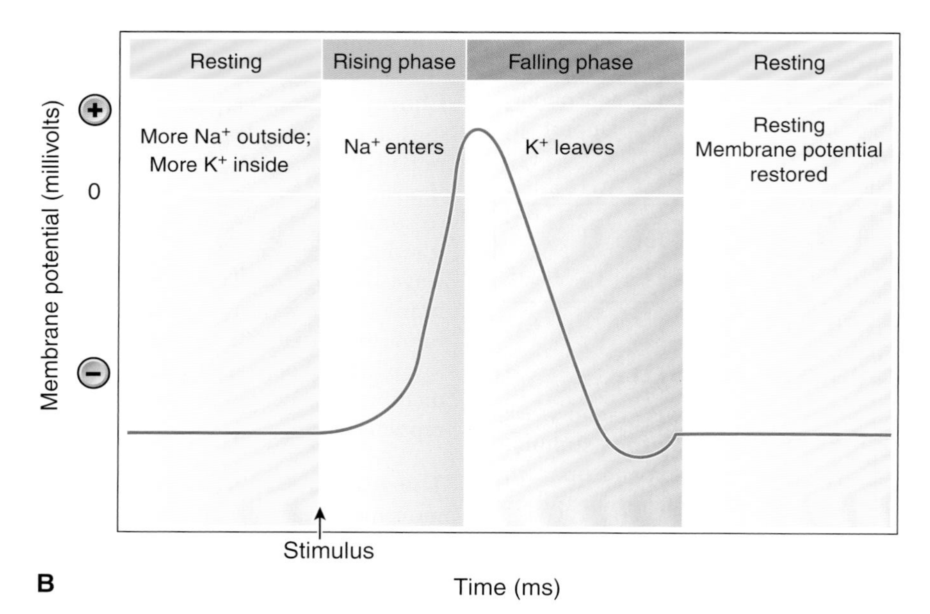

FIGURE 8-7 The membrane potential. **KEY POINT** The membrane potential changes when ions cross the plasma membrane. **A.** Possible changes in a membrane's potential. **B.** Recording of a single action potential. In depolarization, Na^+ membrane channels open, and Na^+ enters the cell. In repolarization, K^+ membrane channels open, and K^+ leaves the cell.

Concept Mastery Alert

Students often confuse the action potential with the membrane potential. The membrane potential is the electric charge that exists across the plasma membrane of all cells. The action potential describes a particular change in the membrane potential, consisting of a rising phase and a falling phase.

The Role of Myelin in Conduction As previously noted, some axons are coated with the fatty material myelin (see FIG. 8-4). If a fiber is not myelinated, the action potential spreads continuously along the cell's membrane (see FIG. 8-8). When myelin is present on an axon, however, it insulates the fiber against the spread of current. This would appear to slow or stop conduction along these fibers, but in fact, the myelin sheath speeds conduction. The reason is that the myelin causes the action potential to "jump" like a spark from node to node along the sheath (FIG. 8-9). This type of conduction, called **saltatory** (SAL-tah-to-re) **conduction** (from the Latin verb meaning "to leap"), is actually faster than continuous conduction, because fewer action potentials are needed for an impulse to travel a given distance. It is this type of conduction that is impaired in Sue's case of MS.

See the Student Resources on thePoint® to view the animations "The Action Potential," "Propagation of the Action Potential," and "The Neuromuscular Junction."

THE SYNAPSE

Neurons do not work alone; impulses must be transferred between neurons to convey information within the nervous system. The point of junction for transmitting the nerve impulse is the **synapse** (FIG. 8-10). (There are also synapses between neurons and effector organs. We studied synapses

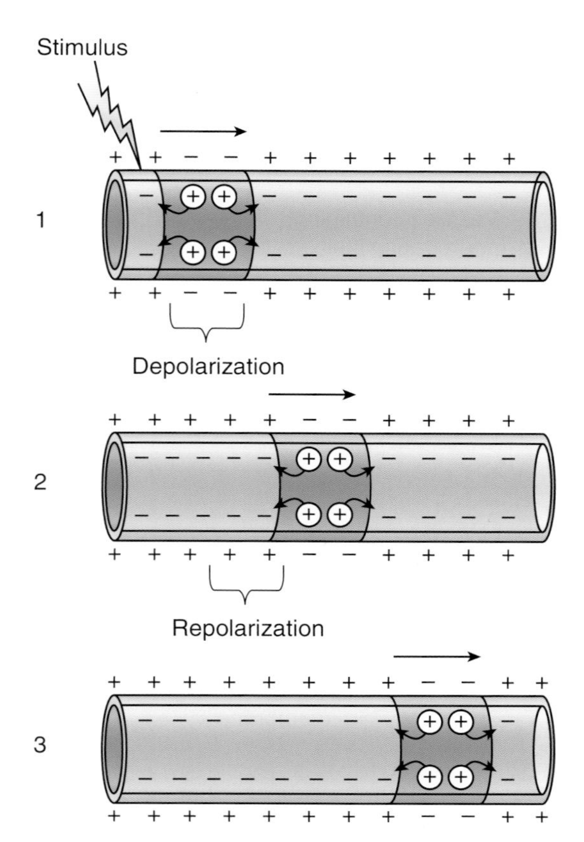

FIGURE 8-8 A nerve impulse. KEY POINT From a point of stimulation, a wave of depolarization followed by repolarization travels along the membrane of a neuron. This spreading action potential is a nerve impulse. ZOOMING IN What happens to the charge on the membrane at the point of an action potential?

between neurons and muscle cells in Chapter 7.) At a nerve-to-nerve synapse, transmission of an impulse often occurs from the axon of one cell, the **presynaptic cell**, to the dendrite of another cell, the **postsynaptic cell**.

As described in Chapter 7, information must be passed from one cell to another at the synapse across a tiny gap between the cells, the synaptic cleft. Information usually

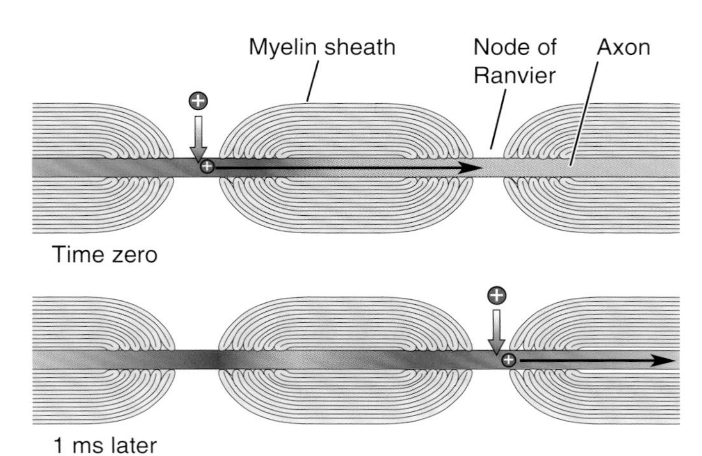

FIGURE 8-9 Saltatory conduction. KEY POINT The action potential along a myelinated axon jumps from node to node, speeding conduction.

crosses this gap by means of a chemical signal called a **neurotransmitter**. While the cells at a synapse are at rest, the neurotransmitter is stored in many small vesicles (sacs) within the enlarged axon endings, usually called *end bulbs* or *terminal knobs* but known by several other names as well.

When an action potential traveling along a neuron membrane reaches the end bulb, some of these vesicles fuse with the membrane and release their neurotransmitter into the synaptic cleft (an example of exocytosis, as described in Chapter 3). The neurotransmitter then acts as a chemical signal to the postsynaptic cell. 11 Recall that chemical signals alter cell function by binding to specialized membrane receptors. Neurotransmitter receptors on the postsynaptic cell's membrane alter the membrane potential when the neurotransmitter binds to them.

Neurotransmitters Although there are many known neurotransmitters, some common ones are **norepinephrine** (nor-ep-ih-NEF-rin), **serotonin** (ser-o-TO-nin), **dopamine** (DO-pah-mene), and **acetylcholine** (as-e-til-KO-lene). Acetylcholine (ACh) is the neurotransmitter released at the neuromuscular junction.

It is common to think of neurotransmitters as stimulating the cells they reach; in fact, they have been described as such in this discussion. Stimulatory neurotransmitters depolarize neurons, increasing the chance that an action potential will occur. Note, however, that some of these chemicals inhibit the postsynaptic cell and keep it from reacting, as will be demonstrated later in discussions of the ANS. These inhibitory neurotransmitters hyperpolarize the cell, making action potentials less likely.

The connections between neurons can be quite complex. One cell can branch to stimulate many receiving cells, or a single cell may be stimulated by a number of different axons. The cell's response is based on the total effects of all the neurotransmitters it receives over a short period of time.

After its release into the synaptic cleft, the neurotransmitter may be removed by several methods:

- It may slowly diffuse away from the synapse.
- It may be rapidly destroyed by enzymes in the synaptic cleft.
- It may be taken back into the presynaptic cell to be used again, a process known as *reuptake*.
- It may be taken up by neuroglial cells, specifically astrocytes.

The method of removal helps determine how long a neurotransmitter will act.

Many drugs that act on the mind, substances known as *psychoactive drugs*, function by affecting neurotransmitter activity in the brain. Prozac, for example, increases the level of the neurotransmitter serotonin by blocking its reuptake into presynaptic cells at synapses. This and other selective serotonin reuptake inhibitors prolong the neurotransmitter's activity and produce a mood-elevating effect. They are used to treat depression, anxiety, and obsessive–compulsive disorder. Similar psychoactive drugs prevent the reuptake of the neurotransmitters norepinephrine and dopamine. Another

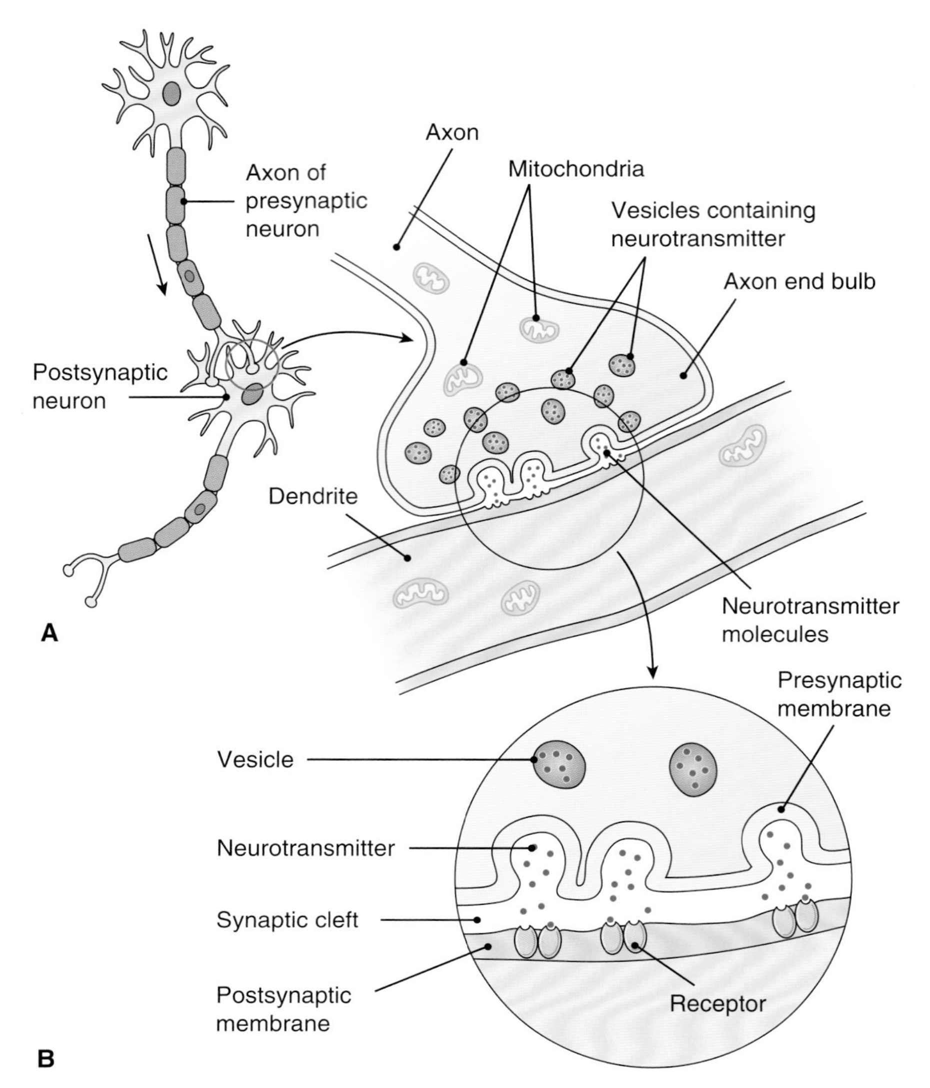

FIGURE 8-10 A synapse. **KEY POINT** Neurotransmitters carry impulses across a synaptic cleft. **A.** The end bulb of the presynaptic (transmitting) axon has vesicles containing neurotransmitter, which is released into the synaptic cleft to the membrane of the postsynaptic (receiving) cell. **B.** Close-up of a synapse showing receptors for neurotransmitter in the postsynaptic cell membrane.

class of antidepressants prevents serotonin's enzymatic breakdown in the synaptic cleft, thus extending its action.

Electrical Synapses Not all synapses are chemically controlled. In smooth muscle, in cardiac muscle, and also in the CNS, there is a type of synapse in which electrical signals travel directly from one cell to another. The membranes of the presynaptic and postsynaptic cells are close together, and an electric charge can spread directly between them through an intercellular bridge.

Concept Mastery Alert

Note that electrical synapses can use action potentials but not neurotransmitters to convey a signal.

These electrical synapses allow more rapid and coordinated communication. In the heart, for example, it is important that large groups of cells contract together for effective pumping action.

CHECKPOINTS

- ☐ **8-8** What are the two stages of an action potential, and what happens during each?
- ☐ **8-9** What ions are involved in generating an action potential?
- ☐ **8-10** How does the myelin sheath affect conduction along an axon?
- ☐ **8-11** What is the junction between two neurons called?
- ☐ **8-12** As a group, what are all the chemicals that carry information across the synaptic cleft called?

The Spinal Cord

The spinal cord is the link between the spinal nerves and the brain. It also helps coordinate some simple actions that do not involve the brain. The spinal cord is contained in and protected by the vertebrae, which fit together to form a continuous tube extending from the occipital bone to the coccyx (FIG. 8-11). In the embryo, the spinal cord occupies the entire spinal canal, extending down into the tail portion of the vertebral column. The bony column grows much more rapidly than does the nervous tissue of the cord, however, and eventually, the end of the spinal cord no longer reaches the lower part of the spinal canal. This disparity in growth continues to increase so that, in adults, the spinal cord ends in the region just below the area to which the last rib attaches (between the first and second lumbar vertebrae). Individual nerves fan out from this point within the vertebral canal, hence its name as the **cauda equina** (KAW-dah eh-KWI-nah) or horse's tail.

STRUCTURE OF THE SPINAL CORD

The spinal cord has a small, irregularly shaped core of gray matter (unmyelinated axons and cell bodies) surrounded by white matter (myelinated axons) (see FIG. 8-11B and C). The internal gray matter is arranged so that a column of gray matter extends up and down posteriorly (dorsally), one on each side; another column is found in the anterior (ventral) region

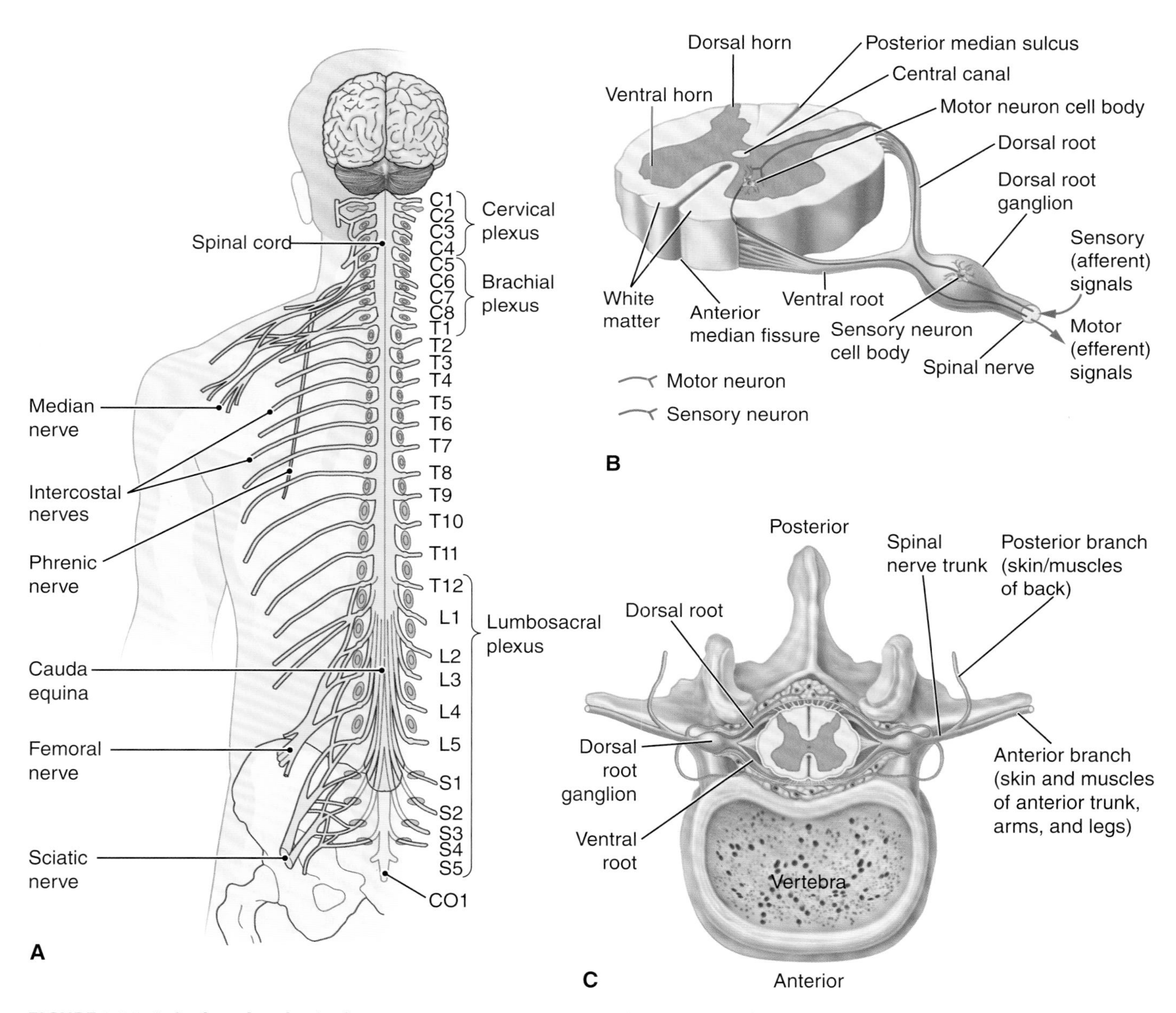

FIGURE 8-11 Spinal cord and spinal nerves. A. Posterior view. Nerve plexuses (networks) are shown. Nerves extending from the distal cord form the cauda equina. **B.** Cross section of the spinal cord showing the organization of the gray and white matter. The roots of the spinal nerves are also shown. **C.** Branches of the spinal nerves. ZOOMING IN Is the spinal cord the same length as the spinal column? How does the number of cervical vertebrae compare to the number of cervical spinal nerves?

on each side. These two pairs of columns, called the **dorsal horns** and **ventral horns**, give the gray matter an H-shaped appearance in cross section. The bridge of gray matter that connects the right and left horns is the **gray commissure** (KOM-ih-shure). In the center of the gray commissure is a small channel, the **central canal**, containing CSF, the liquid that circulates around the brain and spinal cord. A narrow groove, the **posterior median sulcus** (SUL-kus), divides the right and left portions of the posterior white matter. A deeper groove, the **anterior median fissure** (FISH-ure), separates the right and left portions of the anterior white matter.

The spinal cord neurons do not regenerate easily when damaged. So, spinal cord injuries are frequently permanent. See **BOX 8-2**, "Spinal cord injuries: crossing the divide," for more information.

Spinal Tracts The spinal cord is the pathway for sensory and motor impulses traveling to and from the brain. Most impulses are carried in the thousands of myelinated axons in the spinal cord's white matter, which are subdivided into tracts (fiber bundles). Sensory impulses entering the spinal cord are transmitted toward the brain in **ascending tracts** of the white matter. Motor impulses traveling from the brain are carried in **descending tracts** toward the PNS.

CHECKPOINTS 

- ☐ **8-13** How are the gray and white matter arranged in the spinal cord?
- ☐ **8-14** What is the purpose of the tracts in the spinal cord's white matter?

CASEPOINTS 

- ☐ **8-1** Sue's lesions were in the white matter of her spinal cord. Are the neuron fibers in the white matter myelinated or unmyelinated?
- ☐ **8-2** Sue's doctor studied transfer of information from the brain to her muscles. Which tracts in the spinal cord carry these motor impulses?

See the Student Resources on thePoint® to view an illustration of the cauda equina.

The Spinal Nerves

There are 31 pairs of spinal nerves, usually named after the vertebra superior to their point of emergence (see FIG. 8-11A). The exception to this rule is the cervical nerves; nerves C1 through C7 emerge above the corresponding vertebra, and C8 arises below vertebra C7. Each nerve passes through an intervertebral foramen between adjacent vertebrae.

Each nerve is attached to the spinal cord by two roots residing within the spinal canal: the **dorsal root** and the **ventral root** (see FIG. 8-11B). The **dorsal root ganglion**, a swelling on each dorsal root, contains the cell bodies of the sensory neurons. A **ganglion** (GANG-le-on) is any collection of nerve cell bodies located outside the CNS.

A spinal nerve's ventral root contains motor fibers that supply muscles and glands (effectors). The cell bodies of these neurons are located in the cord's ventral gray matter (ventral horns). Because the dorsal (sensory) and ventral (motor) roots combine to form the spinal nerve, all spinal nerves are mixed nerves.

BRANCHES OF THE SPINAL NERVES

Each spinal nerve continues only a short distance away from the spinal cord and then branches into small posterior divisions and larger anterior divisions (see FIG. 8-11C). The posterior divisions distribute branches to the back. The anterior branches of the thoracic nerves 2 through 11 become the intercostal nerves supplying the regions between the ribs. The remaining anterior branches interlace to form networks called **plexuses** (PLEK-sus-eze), which then distribute branches to the body parts

HOT TOPICS
Spinal Cord Injury: Crossing the Divide

BOX 8-2

Approximately 13,000 new cases of traumatic spinal cord injury occur each year in the United States, the majority involving males aged 16 to 30 years. More than 80% of these injuries are due to motor vehicle accidents, acts of violence, and falls. Because neurons show little, if any capacity to repair themselves, spinal cord injuries almost always result in a loss of sensory or motor function (or both), and therapy has focused on injury management rather than cure. However, scientists are investigating four improved treatment approaches:

- *Minimizing spinal cord trauma after injury.* Intravenous injection of the steroid methylprednisolone shortly after injury reduces swelling at the site of injury and improves recovery.
- *Using neurotrophins to induce repair in damaged nerve tissue.* Certain types of neuroglia produce chemicals called neurotrophins (e.g., nerve growth factor) that have promoted nerve regeneration in experiments.
- *Regulation of inhibitory factors that keep neurons from dividing.* "Turning off" these factors (produced by neuroglia) in the damaged nervous system may promote tissue repair. The factor called Nogo is an example.
- *Nervous tissue transplantation.* Successfully transplanted donor tissue may take over the damaged nervous system's functions.

(see FIG. 8-11A). The three main plexuses are described as follows:

- The **cervical plexus** supplies motor impulses to the neck muscles and receives sensory impulses from the neck and the back of the head. The phrenic nerve, which activates the diaphragm, arises from this plexus.
- The **brachial** (BRA-ke-al) **plexus** sends numerous branches to the shoulder, arm, forearm, wrist, and hand. For example, the median nerve emerges from the brachial plexus.
- The **lumbosacral** (lum-bo-SA-kral) **plexus** supplies nerves to the pelvis and legs. The femoral nerve to the thigh is part of this plexus. The largest branch in this plexus is the sciatic (si-AT-ik) nerve, which leaves the dorsal part of the pelvis, passes beneath the gluteus maximus muscle, and extends down the posterior thigh. At its beginning, it is nearly 1 in thick, but it soon branches to the thigh muscles. Near the knee, it forms two subdivisions that supply the leg and the foot.

DERMATOMES

Sensory neurons from all over the skin, except for the skin of the face and scalp, feed information into the spinal cord through the spinal nerves. The skin surface can be mapped into distinct regions that are supplied by a single spinal nerve. Each of these regions is called a **dermatome** (DER-mah-tome) (FIG. 8-12).

Sensation from a given dermatome is carried over its corresponding spinal nerve. This information can be used to identify the spinal nerve or spinal segment that is involved in an injury, as sensation from its corresponding skin surface will be altered. In some areas, the dermatomes are not absolutely distinct. Some dermatomes may share a nerve supply with neighboring regions. For this reason, it is necessary to

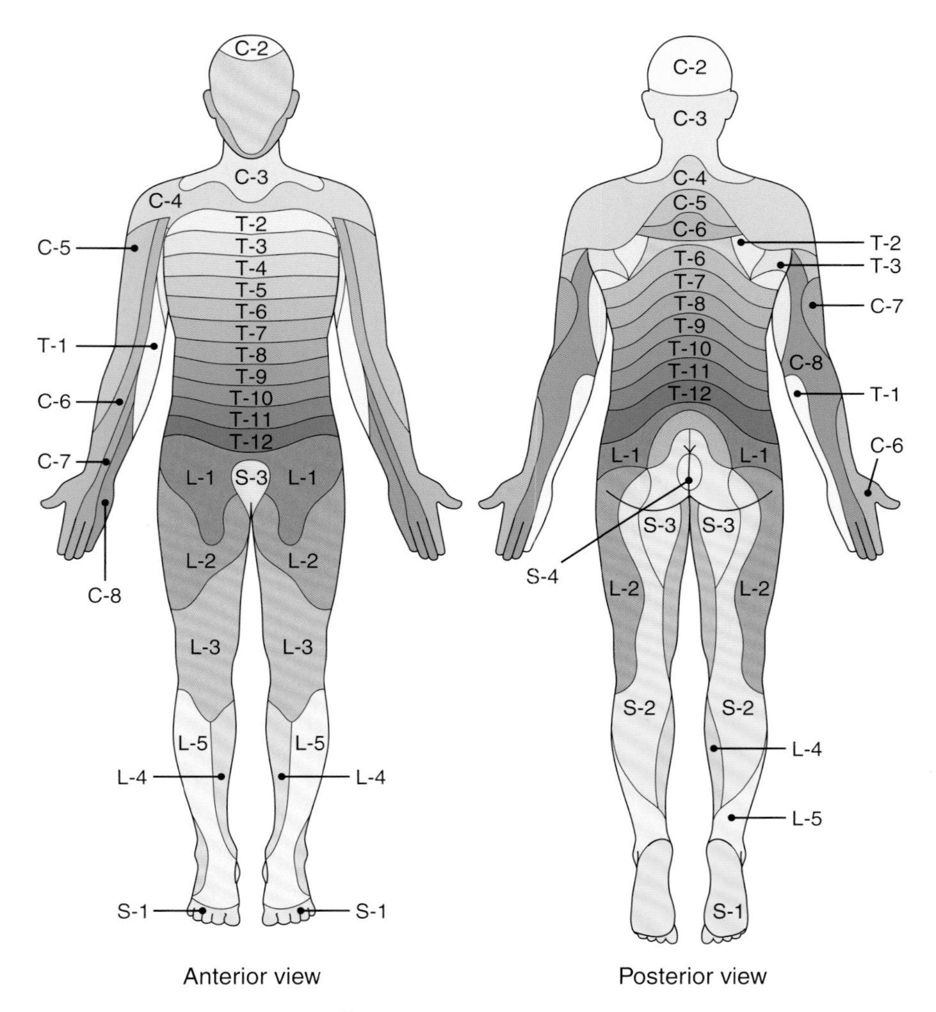

FIGURE 8-12 Dermatomes. **KEY POINT** A dermatome is a region of the skin supplied by a single spinal nerve. **ZOOMING IN** Which spinal nerves carry impulses from the skin of the toes? From the anterior hand and fingers?

8

numb several adjacent dermatomes to achieve successful anesthesia.

CHECKPOINTS 

- [ ] **8-15** How many pairs of spinal nerves are there?
- [ ] **8-16** What types of fibers are in a spinal nerve's dorsal root? What types are in its ventral root?
- [ ] **8-17** What is the term for a network of spinal nerves?

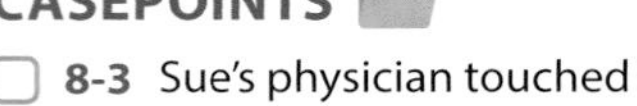

CASEPOINTS

- [ ] **8-3** Sue's physician touched the top of her foot to investigate her sense of touch. Which spinal nerve carried the sensory signal?
- [ ] **8-4** What CNS tracts would conduct these touch stimuli to the brain?

Reflexes

As the nervous system develops, neural pathways are formed to coordinate responses. Most of these pathways are very complex, involving multiple neurons and interactions between different regions of the nervous system. Easier to study are simple pathways involving a minimal number of neurons. Responses controlled by some of these simpler pathways are useful in neurologic studies.

THE REFLEX ARC

A complete pathway through the nervous system from stimulus to response is termed a **reflex arc**. This is the nervous system's basic functional pathway. **FIGURE 8-13** illustrates the components of a reflex arc, using the example of the

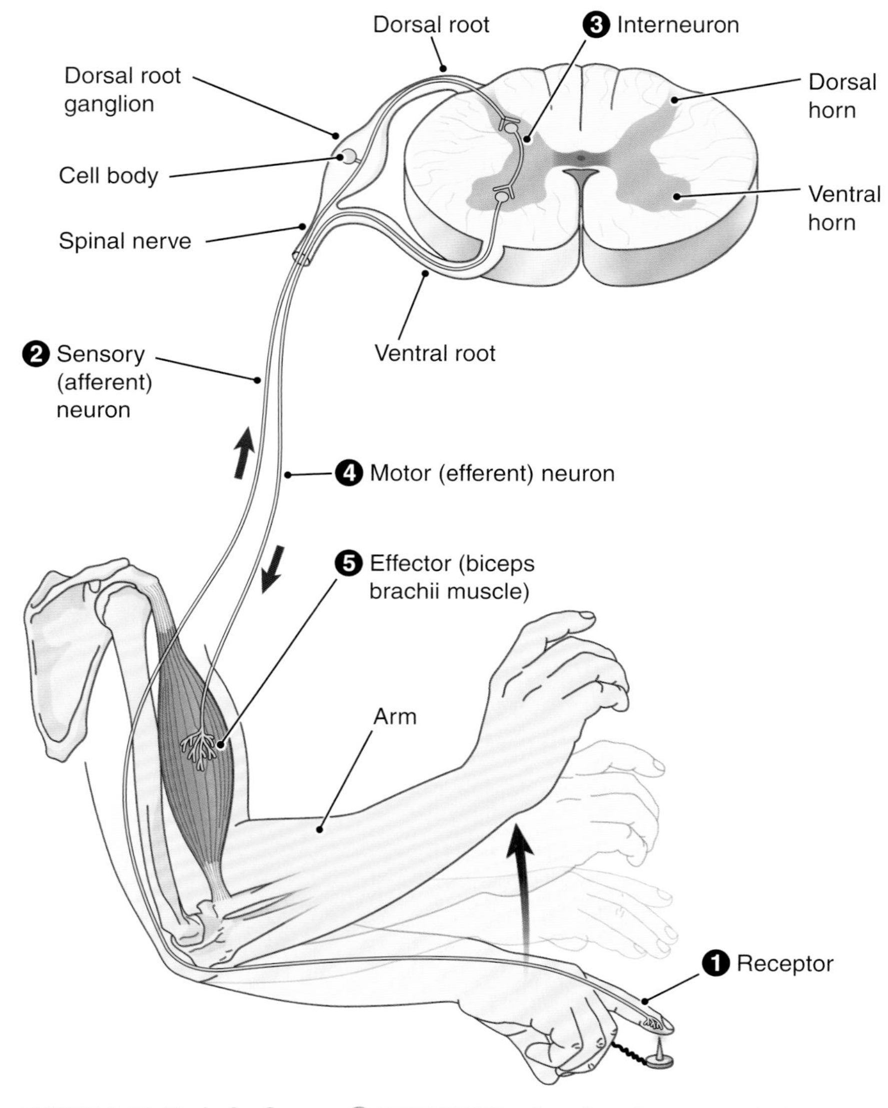

FIGURE 8-13 Typical reflex arc. **KEY POINT** Numbers show the sequence in the pathway of impulses through the spinal cord (*solid arrows*). Contraction of the biceps brachii results in flexion of the arm at the elbow. **ZOOMING IN** Is this a somatic or an autonomic reflex arc? What type of neuron is located between the sensory and motor neurons in the CNS?

pinprick reflex. Try it yourself—if you touch a tack or pin with your finger, your finger will automatically pull away. The fundamental parts of a reflex arc are the following:

1. **Receptor**—the end of a dendrite or some specialized receptor cell, as in a special sense organ, that detects a stimulus. In Step 1 of **FIGURE 8-13**, a pinprick activates a sensory receptor in the finger.
2. **Sensory neuron**—a cell that transmits impulses toward the CNS. In Step 2, a dendrite carries the signal from the sensory receptor to the cell body in the dorsal root ganglion. An axon carries the signal from the cell body through the dorsal root and into the CNS (in this case, the spinal cord).
3. **Central nervous system**—where impulses are coordinated and a response is organized. In Step 3, the sensory neuron synapses with an interneuron within the CNS. Each interneuron can receive signals from multiple neurons.
4. **Motor neuron**—a cell that carries impulses away from the CNS. The motor neuron in Step 4 conveys a signal from the interneuron to the skeletal muscle. Motor impulses leave the cord through the ventral horn of the spinal cord's gray matter.
5. **Effector**—a muscle or a gland outside the CNS that carries out a response. Step 5 of the figure shows how the biceps brachii responds to the motor signal by contracting, thereby removing the finger.

At its simplest, a reflex arc can involve just two neurons, one sensory and one motor, with a synapse in the CNS. Few reflex arcs require only this minimal number of neurons. (The knee-jerk reflex described below is one of the few examples in humans.) Most reflex arcs involve many more, even hundreds, of connecting neurons within the CNS. The many intricate patterns that make the nervous system so responsive and adaptable also make it difficult to study, and investigation of the nervous system is one of the most active areas of research today.

REFLEX ACTIVITIES

Although reflex pathways may be quite complex, a **simple reflex** is a rapid, uncomplicated, and automatic response involving very few neurons. Reflexes are specific; a given stimulus always produces the same response. When you fling out an arm or leg to catch your balance, withdraw from a painful stimulus, or blink to avoid an object approaching your eyes, you are experiencing reflex behavior. A simple reflex arc that passes through the spinal cord alone and does not involve the brain is termed a **spinal reflex**. Returning to our opening corporation analogy, it's as if middle management makes a decision independently without involving the CEO.

The **stretch reflex**, in which a muscle is stretched and responds by contracting, is one example of a spinal reflex. If you tap the tendon below the kneecap (the patellar tendon), the muscle of the anterior thigh (quadriceps femoris) contracts, eliciting the knee-jerk reflex. Such stretch reflexes may be evoked by appropriate tapping of most large muscles (such as the triceps brachii in the arm and the gastrocnemius in the calf of the leg). Because reflexes are simple and predictable, they are used in physical examinations to test the condition of the nervous system. In the case study, Dr. Jensen tested Sue's stretch reflexes to help in diagnosis.

CHECKPOINT 

☐ **8-18** What is the name for a pathway through the nervous system from a stimulus to an effector?

CASEPOINT 

☐ **8-5** The physician in Sue's case tested the stretch reflexes in her limbs. What type of reflexes was she studying?

See the Student Resources on thePoint® to view the animation "The Reflex Arc."

The Autonomic Nervous System

The autonomic (visceral) nervous system regulates the action of the glands, the smooth muscles of hollow organs and vessels, and the heart muscle. These actions are carried out automatically; whenever a change occurs that calls for a regulatory adjustment, it is made without conscious awareness. The ANS consists of the **sympathetic** and **parasympathetic** divisions. These two divisions have distinct functional and structural characteristics (**FIG. 8-14**), as described below and summarized in **TABLES 8-2 and 8-3**.

FUNCTIONS OF THE AUTONOMIC NERVOUS SYSTEM

Most visceral organs are supplied by both sympathetic and parasympathetic fibers, and the two systems generally have opposite effects (**see TABLE 8-2**). The sympathetic system tends to act as an accelerator for those organs needed to meet a stressful situation. It promotes what is called the **fight-or-flight response** because in the most primitive terms, the person must decide to stay and "fight it out" with the enemy or to run away from danger. The times when the sympathetic nervous system comes into play can be summarized by the four "Es," that is, times of emergency, excitement, embarrassment, and exercise. If you think of what happens to a person who is in any of these situations, you can easily remember the effects of the sympathetic nervous system:

- Increase in the rate and force of heart contractions.
- Increase in blood pressure due partly to the more effective heartbeat and partly to constriction of most small arteries everywhere except the brain.
- Dilation of the bronchial tubes to allow more oxygen to enter and more carbon dioxide to leave.
- Stimulation of the central portion of the adrenal gland. This gland produces hormones that prepare the body to meet emergency situations in many ways (see Chapter 11). The sympathetic nerves and hormones from the adrenal gland reinforce each other.

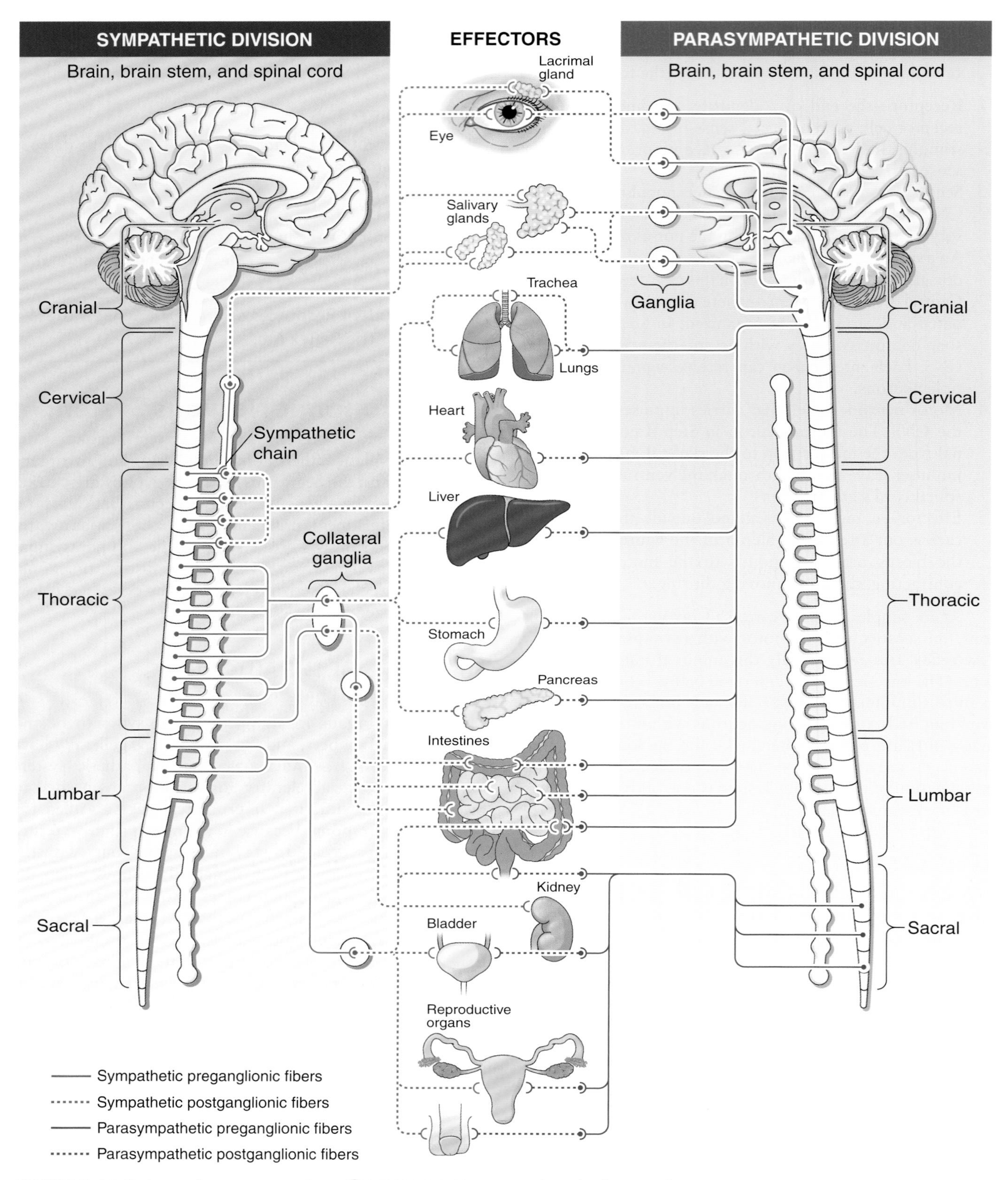

FIGURE 8-14 Autonomic nervous system. KEY POINT Most organs have both sympathetic and parasympathetic fibers. The diagram shows only one side of the body for each division. ZOOMING IN Which division of the autonomic nervous system has ganglia closer to the effector organ?

Table 8-2 Effects of the Sympathetic and Parasympathetic Systems on Selected Organs

Effector	Sympathetic System	Parasympathetic System
Pupils of eye	Dilation	Constriction
Lacrimal glands	None	Secretion of tears
Sweat glands	Stimulation	None
Digestive glands	Inhibition	Stimulation
Heart	Increased rate and strength of beat	Decreased rate of beat
Bronchi of lungs	Dilation	Constriction
Muscles of digestive system	Decreased contraction	Increased contraction
Kidneys	Decreased activity	None
Urinary bladder	Relaxation	Contraction and emptying
Liver	Increased release of glucose	None
Penis	Ejaculation	Erection
Adrenal medulla	Stimulation	None
Blood vessels	Constriction	Dilation, penis and clitoris only

- Increase in basal metabolic rate.
- Dilation of the eye's pupil and increase in distance focusing ability.

The sympathetic system also acts as a brake on those systems not directly involved in the stress response, such as the urinary and digestive systems. If you try to eat while you are angry, you may note that your saliva is thick and so small in amount that you can swallow only with difficulty. Under these circumstances, when food does reach the stomach, it seems to stay there longer than usual.

The parasympathetic system normally acts as a balance for the sympathetic system once a crisis has passed. It is the "rest and digest" system. It causes constriction of the pupils, slowing of the heart rate, and constriction of the bronchial tubes. However, the parasympathetic nervous system also stimulates certain activities needed for maintenance of homeostasis. Among other actions, it promotes the formation and release of urine and activity of the digestive tract. Saliva, for example, flows more easily and profusely under its effects. These stimulatory actions are summarized by the acronym SLUDD: salivation, lacrimation (tear formation), urination, digestion, and defecation.

STRUCTURE OF THE AUTONOMIC NERVOUS SYSTEM

All autonomic pathways contain two motor neurons connecting the spinal cord with the effector organ **(see TABLE 8-3)**. The two neurons synapse in ganglia that serve as relay stations. The first neuron, the preganglionic neuron, extends from the spinal cord to the ganglion. The second neuron, the postganglionic neuron, travels from the ganglion to the effector. This arrangement differs from the voluntary (somatic) nervous system, in which each motor nerve fiber extends all the way from the spinal cord to the skeletal muscle with no

Table 8-3 Divisions of the Autonomic Nervous System

	Divisions	
Characteristics	**Sympathetic Nervous System**	**Parasympathetic Nervous System**
Origin of fibers	Thoracic and lumbar regions of the spinal cord; thoracolumbar	Brain stem and sacral regions of the spinal cord; craniosacral
Location of ganglia	Sympathetic chains and three single collateral ganglia (celiac, superior mesenteric, and inferior mesenteric)	Terminal ganglia in or near the effector organ
Neurotransmitter at effector	Mainly norepinephrine; adrenergic	Acetylcholine; cholinergic
Effects	Response to stress; fight-or-flight response	Reverses fight-or-flight (stress) response **(see TABLE 8-2)**: stimulates some activities

intervening synapse. Some of the autonomic fibers are within the spinal nerves; some are within the cranial nerves (see Chapter 9).

Sympathetic Nervous System Pathways Sympathetic motor neurons originate in the thoracic and lumbar regions of the spinal cord, the **thoracolumbar** (tho-rah-ko-LUM-bar) area. The cell bodies of the preganglionic fibers are located within the cord. The axons travel via spinal nerves T1-T5 and L1-L2 to ganglia near the cord, where they synapse with postganglionic neurons, which then extend to the effectors.

Many of the sympathetic ganglia form the **sympathetic chains**, two cordlike strands of ganglia that extend along either side of the spinal column from the lower neck to the upper abdominal region. (Note that FIGURE 8-14 shows only one side for each division of the ANS.)

In addition, the nerves that supply the abdominal and pelvic organs synapse in three single **collateral ganglia** farther from the spinal cord. These are:

- the celiac ganglion, which sends fibers mainly to the digestive organs
- the superior mesenteric ganglion, which sends fibers to the large and small intestines
- the inferior mesenteric ganglion, which sends fibers to the distal large intestine and organs of the urinary and reproductive systems

The postganglionic neurons of the sympathetic system, with few exceptions, act on their effectors by releasing the neurotransmitter norepinephrine (noradrenaline), a compound similar in chemical composition and action to the hormone epinephrine (adrenaline). This system is therefore described as **adrenergic**, which means "activated by adrenaline."

Parasympathetic Nervous System Pathways The parasympathetic motor pathways begin in the **craniosacral** (kra-ne-o-SA-kral) areas, with fibers arising from cell bodies in the brain stem (midbrain and medulla) and the lower (sacral) part of the spinal cord. From these centers, the first fibers extend to autonomic ganglia that are usually located near or within the walls of the effector organs and are called **terminal ganglia.** The pathways then continue along postganglionic neurons that stimulate involuntary muscles and glands.

The neurons of the parasympathetic system release the neurotransmitter ACh, leading to the description of this system as **cholinergic** (activated by ACh).

Autonomic Nervous System Receptors 11 Recall from Chapter 7 that chemical signals, including neurotransmitters, exert their effects by binding to specific receptors. Once the neurotransmitter binds, the receptor initiates events that change the postsynaptic cell's activity. Different receptors' responses to the same neurotransmitter may vary, and a cell's response depends on the receptors it contains.

Concept Mastery Alert

Note that cellular receptors are individual proteins, whereas sensory receptors are usually an entire cell or grouping of cells.

Among the many different classes of identified receptors, two are especially important and well studied. The first is the cholinergic receptors, which bind ACh. Cholinergic receptors are further subdivided into two types, each named for drugs that researchers have discovered bind to them and mimic ACh's effects. Nicotinic receptors, which bind nicotine, are found on skeletal muscles and stimulate muscle contraction when ACh is present. These receptors are involved only in the somatic nervous system and were discussed in Chapter 7. The cholinergic receptors of the ANS are classified as muscarinic receptors, because they bind muscarine, a poison. These receptors are found on effector cells of the parasympathetic nervous system. Depending on the type of muscarinic receptor in a given effector organ, ACh can either stimulate or inhibit a response. For example, ACh stimulates digestive organs but inhibits the heart.

The second class of receptors is the adrenergic receptors, which bind norepinephrine. They are found on effector cells of the sympathetic nervous system. They are further subdivided into alpha (α) and beta (β) categories. Depending on the type of adrenergic receptor in a given effector organ, norepinephrine can either stimulate or inhibit a response. For example, norepinephrine stimulates the heart and inhibits the digestive organs.

Some drugs block specific receptors. For example, "beta-blockers" regulate the heart in cardiac disease by preventing β receptors from binding norepinephrine, the neurotransmitter that increases the rate and strength of heart contractions.

CHECKPOINTS 

- ☐ **8-19** How many neurons are there in each motor pathway of the ANS?
- ☐ **8-20** Which division of the ANS stimulates a stress response? Which division reverses the stress response?

Occupational therapists often help care for people with nervous system disorders. See the Student Resources on thePoint® for more information about this career.

A & P in Action Revisited: Sue Learns about Her MS

"Sue, I can't really answer the question of why you developed multiple sclerosis," Dr. Jensen explained to her patient. "There is evidence that the disease has a genetic component but the environment, and perhaps even a virus, might be involved. We do know that MS affects women more frequently than it does men and is more prevalent in areas like the Northern United States and Canada. We also know that MS is an autoimmune disease. Normally, immune cells travel through the brain and spinal cord looking for pathogens. In MS, the immune cells make a mistake and cause inflammation in healthy nervous tissue. This inflammatory response damages neuroglial cells called oligodendrocytes. These cells form the myelin sheath that covers and insulates the axons of neurons much like the plastic covering on an electrical wire does. When the oligodendrocytes are damaged, they are unable to make this myelin sheath, and the axons can't properly transmit nerve impulses. Right now, it appears that the largest areas of demyelination are in the white matter tracts of your spinal cord."

"Is there a medication I can take to stop the disease?" asked Sue.

"Unfortunately," replied the doctor, "there isn't a cure for MS yet. But we can slow down the disease's progress using antiinflammatory drugs to decrease the inflammation and drugs called interferons that depress the immune response. There are also newer drugs that help prevent antibodies from accessing the central nervous system and causing additional lesions.

During this case, we saw that neurons carrying information to and from the CNS require myelin sheaths. Inflammation and subsequent damage of the myelin sheath in diseases like MS have profound effects on sensory and motor function. For more information about the inflammatory response and interferons, see Chapter 15.

8

CHAPTER 8 Chapter Wrap-Up

OVERVIEW

A detailed chapter outline with space for note-taking is on thePoint®. The figure below illustrates the main topics covered in this chapter.

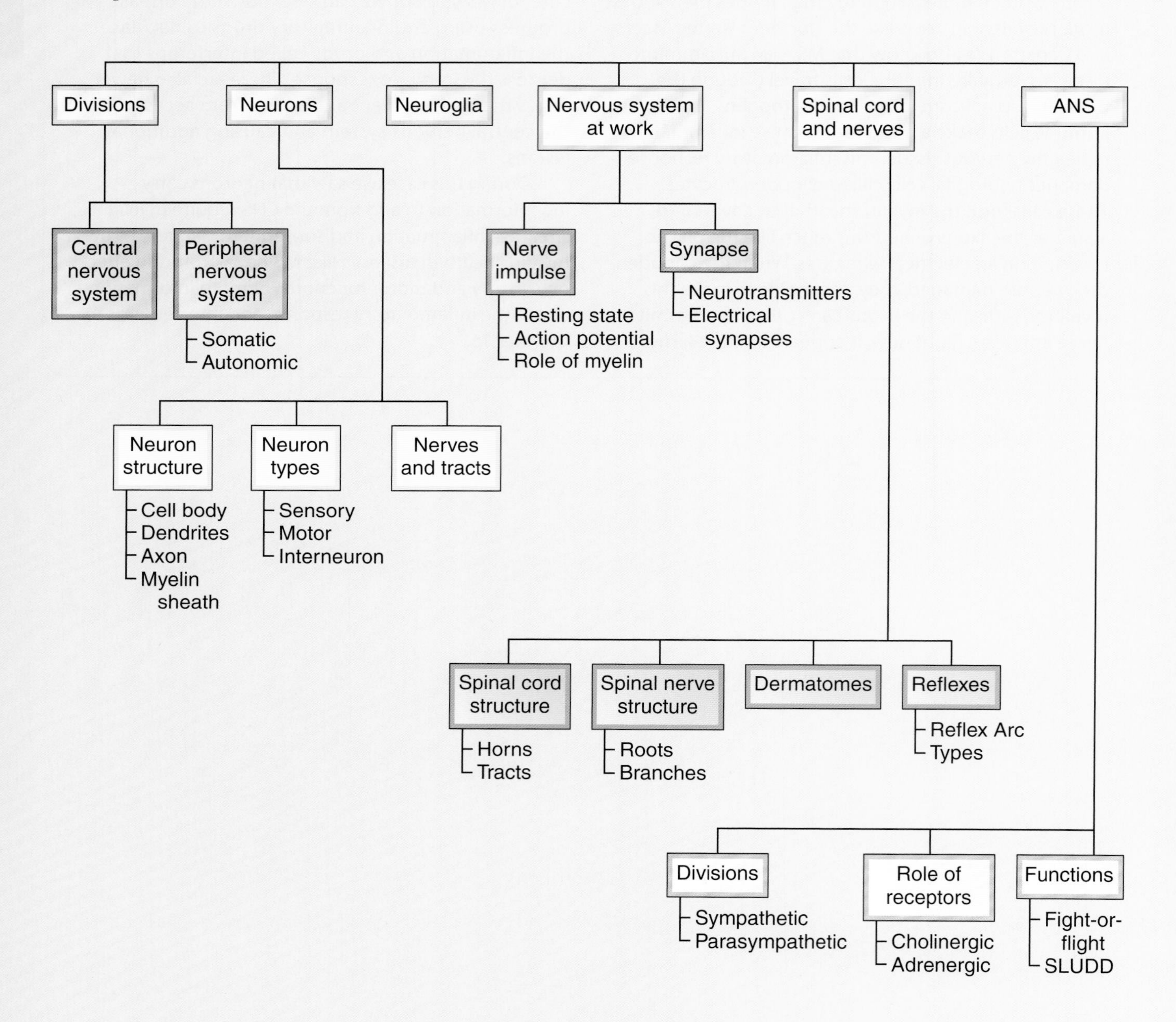

KEY TERMS

The terms listed below are emphasized in this chapter. Knowing them will help you organize and prioritize your learning. These and other boldface terms are defined in the Glossary with phonetic pronunciations.

acetylcholine
action potential
afferent
autonomic nervous system
axon
dendrite
depolarization
effector
efferent
ganglion
interneuron
motor
nerve
nerve impulse
neuroglia
neuron
neurotransmitter
norepinephrine
parasympathetic nervous system
plexus
postsynaptic
presynaptic
receptor
reflex
repolarization
saltatory conduction
sensory
somatic nervous system
sympathetic nervous system
synapse
tract

WORD ANATOMY

Medical terms are built from standardized word parts (prefixes, roots, and suffixes). Learning the meanings of these parts can help you remember words and interpret unfamiliar terms.

WORD PART	MEANING	EXAMPLE
OVERVIEW OF THE NERVOUS SYSTEM		
aut/o	self	The *autonomic* nervous system is automatically controlled and is involuntary.
NEURONS AND THEIR FUNCTIONS		
dendr/o	tree	The *dendrite* of a neuron has treelike branched fibers.
hyper-	over, excess, increased	A *hyperpolarized* neuron is more polarized than in the resting state.
inter-	between	An *interneuron* carries nerve impulses between regions of the central nervous system.
-lemma	sheath	See example below.
neur/i	nerve, nervous tissue	The *neurilemma* is the outer membrane of the myelin sheath around an axon.
olig/o	few, deficiency	An *oligodendrocyte* has few dendrites.
soma-	body	The *somatic* nervous system controls skeletal muscles that move the body.
THE NERVOUS SYSTEM AT WORK		
de-	remove	*Depolarization* removes the charge on the plasma membrane of a cell.
post-	after	The *postsynaptic* cell is located after the synapse and receives neurotransmitter from the presynaptic cell.
pre-	before, in front of	A *presynaptic* neuron is located before the synapse and delivers neurotransmitter into the synaptic cleft.
re-	again, back	*Repolarization* restores the charge on the plasma membrane of a cell.

QUESTIONS FOR STUDY AND REVIEW

BUILDING UNDERSTANDING

Fill in the Blanks

1. The brain and spinal cord make up the ______ nervous system.
2. The ion that enters a cell to cause depolarization is ______.
3. The term that describes conduction along a myelinated axon is ______.
4. In the spinal cord, sensory information travels in ______ tracts.
5. With few exceptions, the sympathetic nervous system uses the neurotransmitter ______ to act on effector organs.

Matching Match each numbered item with the most closely related lettered item.

___ 6. Cells that carry impulses from the CNS

___ 7. Cells that carry impulses to the CNS

___ 8. Cells that carry impulses within the CNS

___ 9. Cells that detect a stimulus

___ 10. Cells that carry out a response to a stimulus

a. receptors
b. effectors
c. sensory neurons
d. motor neurons
e. interneurons

Multiple Choice

___ 11. Which system directly innervates skeletal muscles?
  a. central nervous system
  b. somatic nervous system
  c. parasympathetic nervous system
  d. sympathetic nervous system

___ 12. What cells are involved in most nervous system tumors?
  a. motor neurons
  b. sensory neurons
  c. interneurons
  d. neuroglia

___ 13. What is the correct order of synaptic transmission?
  a. postsynaptic neuron, synapse, and presynaptic neuron
  b. presynaptic neuron, synapse, and postsynaptic neuron
  c. presynaptic neuron, postsynaptic neuron, and synapse
  d. postsynaptic neuron, presynaptic neuron, and synapse

___ 14. Where do afferent nerve fibers enter the spinal cord?
  a. dorsal horn
  b. ventral horn
  c. gray commissure
  d. central canal

___ 15. What system promotes the "fight-or-flight" response?
  a. sympathetic nervous system
  b. parasympathetic nervous system
  c. somatic nervous system
  d. reflex arc

UNDERSTANDING CONCEPTS

16. Differentiate between the terms in each of the following pairs:
  a. axon and dendrite
  b. gray matter and white matter
  c. nerve and tract
  d. dorsal and ventral spinal nerve root
17. Describe an action potential. How does conduction along a myelinated fiber differ from conduction along an unmyelinated fiber?
18. What are neuroglia, and what are their functions?
19. Explain the reflex arc using stepping on a tack as an example.
20. Describe the anatomy of a spinal nerve. How many pairs of spinal nerves are there?
21. Define a *plexus*. Name the three main spinal nerve plexuses.
22. What is a dermatome? How did Sue's physician use dermatomes to identify her damaged nerves?
23. Differentiate between the functions of the sympathetic and parasympathetic divisions of the ANS.
24. Explain how a single neurotransmitter can stimulate some cells and inhibit others.

CONCEPTUAL THINKING

25. Clinical depression is associated with abnormal serotonin levels. Medications that block the removal of this neurotransmitter from the synapse can control the disorder. Based on this information, is clinical depression associated with increased or decreased levels of serotonin? Explain your answer.
26. Mr. Hayward visits his dentist for a root canal and is given novocaine, a local anesthetic, at the beginning of the procedure. Novocaine reduces membrane permeability to Na^+. What effect does this have on action potential?
27. In Sue's case, her symptoms were caused by demyelination in her CNS. Would her symptoms be the same or different if her spinal nerves were involved? Explain why or why not.

For more questions, see the Learning Activities on thePoint®.

CHAPTER 9

The Nervous System: The Brain and Cranial Nerves

Learning Objectives

After careful study of this chapter, you should be able to:

1. Give the locations of the four main divisions of the brain. *p. 182*
2. Name and describe the three meninges. *p. 184*
3. Cite the function of cerebrospinal fluid, and describe where and how this fluid is formed. *p. 184*
4. Name and locate the lobes of the cerebral hemispheres. *p. 186*
5. Cite one function of the cerebral cortex in each lobe of the cerebrum. *p. 187*
6. Name two divisions of the diencephalon, and cite the functions of each. *p. 188*
7. Locate the three subdivisions of the brain stem, and give the functions of each. *p. 189*
8. Describe the cerebellum, and identify its functions. *p. 190*
9. Name three neuronal networks that involve multiple regions of the brain, and describe the function of each. *p. 190*
10. Describe four techniques used to study the brain. *p. 191*
11. List the names and functions of the 12 cranial nerves. *p. 192*
12. Using information in the case study, list the possible effects of mild traumatic brain injury. *pp. 181, 195*
13. Show how word parts are used to build words related to the nervous system (see "Word Anatomy" at the end of the chapter). *p. 197*

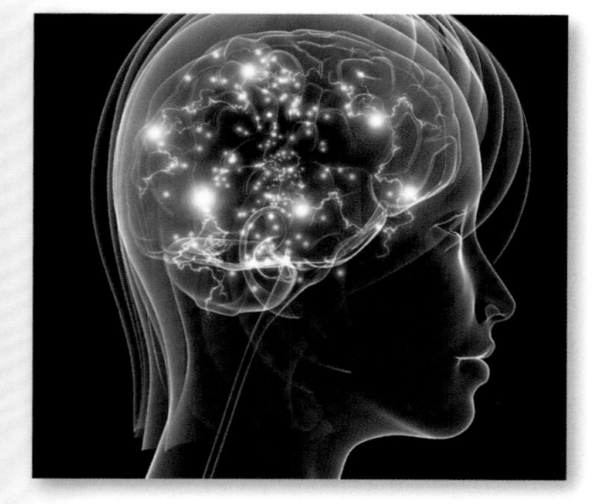

A & P in Action
Natalie's Cerebral Concussion

Lacey was agitated and a little panicky as she helped her sister into her SUV. Despite Natalie's protests, she insisted on driving her to Mount Desert Island Hospital to be checked. Lacey was an active outdoor person who enjoyed hiking in nearby Acadia National Park. She had been eager to share her favorite trail with her sister, who lived in New York City and rarely strayed from her Wall Street Office.

They reached the trail head at dawn and began their hike up Cadillac Mountain. As they climbed, the path became more precarious and lined with boulders. Lacey had experience with climbing and started to scramble up a small boulder. Natalie took a few steps up the rock and fell, striking her head on the ground. Lacey feared her sister had suffered head trauma. Fortunately, they had not climbed very far, and Natalie was conscious, so the two sisters carefully picked their way back down the mountain and drove to the hospital.

Dr. Erickson, the resident on call, performed Natalie's neurologic examination, starting with an evaluation of her mental status and then moving on to a cranial nerve exam. He noted sluggish eye movements during the ocular exam. He proceeded with a motor exam of the extremities and then assessed her balance and coordination. When questioned, Natalie admitted she felt dizzy and complained of a headache, nausea, and blurred vision. Twice, she vomited small amounts. She kept repeating that she was tired and wanted to go to sleep. Dr. Erickson ordered a computed tomography (CT) scan to determine if injury was present and if so its extent.

Dr. Erickson explained his findings to the sisters. "As a result of hitting your head, Natalie, you have incurred a cerebral concussion, or mild traumatic brain injury (MTBI). The dizziness, headaches, nausea, and vomiting you are experiencing are caused by injury and swelling in your brain. Thankfully, the CT scan did not show more serious damage, such as bleeding within your brain, but I'm concerned about your vision problems. Just to be safe, we're going to admit you and get an MRI. The neurologist will be in to see you shortly."

Natalie's visual symptoms could have resulted from brain swelling and injury to cranial nerves II, III, IV, or VI. In this chapter, we will learn about the structure and function of the brain and cranial nerves. We will revisit Natalie later in the chapter to see how she is progressing.

As you study this chapter, CasePoints will give you opportunities to apply your learning to this case.

Visit thePoint® to access the following resources. For guidance in using these resources most effectively, see pp. xv–xvii.

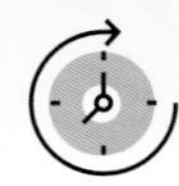

Preparing to Learn

- Tips for Effective Studying
- Pre-Quiz

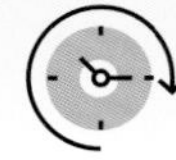

While You Are Learning

- Chapter Notes Outline
- Audio Pronunciation Glossary

When You Are Reviewing

- Answers to Questions for Study and Review
- Health Professions: Speech Therapist
- Interactive Learning Activities

A LOOK BACK

Having discussed the basics of nerve impulse conduction and the reflex arc, we now apply these fundamentals to look at how the various brain regions coordinate information and orchestrate responses. As you might guess, this is a very complex topic, spanning activities from the cellular level to the highest abstract brain function. The homeostasis key idea 3 introduced in Chapter 1 is relevant to this chapter.

Overview of the Brain

The brain is the control center of the nervous system, where sensory information is processed; responses are coordinated; and the higher functions of reasoning, learning, and memory occur. The brain occupies the cranial cavity and is surrounded by membranes, fluid, and skull bones.

DIVISIONS OF THE BRAIN

For study purposes, the brain can be divided into four regions with specific activities. These divisions are in constant communication as they work together to regulate body functions (FIG. 9-1):

- The **cerebrum** (SER-e-brum) is the most superior and largest part of the brain. It consists of left and right hemispheres, each shaped like a small boxer's glove.
- The **diencephalon** (di-en-SEF-ah-lon) sits in the center of the brain between the two hemispheres and superior to the brain stem. It includes the thalamus and the hypothalamus.
- The **brain stem** spans the region between the diencephalon and the spinal cord. The superior portion of the brain stem is the **midbrain**. Inferior to the midbrain is the **pons** (ponz), followed by the **medulla oblongata** (meh-DUL-lahob-long-GAH-tah). The medulla connects with the spinal cord through a large opening in the base of the skull (foramen magnum).
- The **cerebellum** (ser-eh-BEL-um) is posterior to the brain stem and is connected with the cerebrum, brain stem, and spinal cord by means of the pons. The word *cerebellum* means "little brain."

Each of these divisions is described in greater detail later in this chapter and summarized in **TABLE 9-1**. Also, see Dissection Atlas **FIGURES A3-1 and A3-2**.

PROTECTIVE STRUCTURES OF THE BRAIN AND SPINAL CORD

The protective structures of the brain include the meninges and the cerebrospinal fluid (CSF). Both the meninges and the CSF also protect the spinal cord.

Meninges The **meninges** (men-IN-jez) are three layers of connective tissue that surround both the brain and spinal

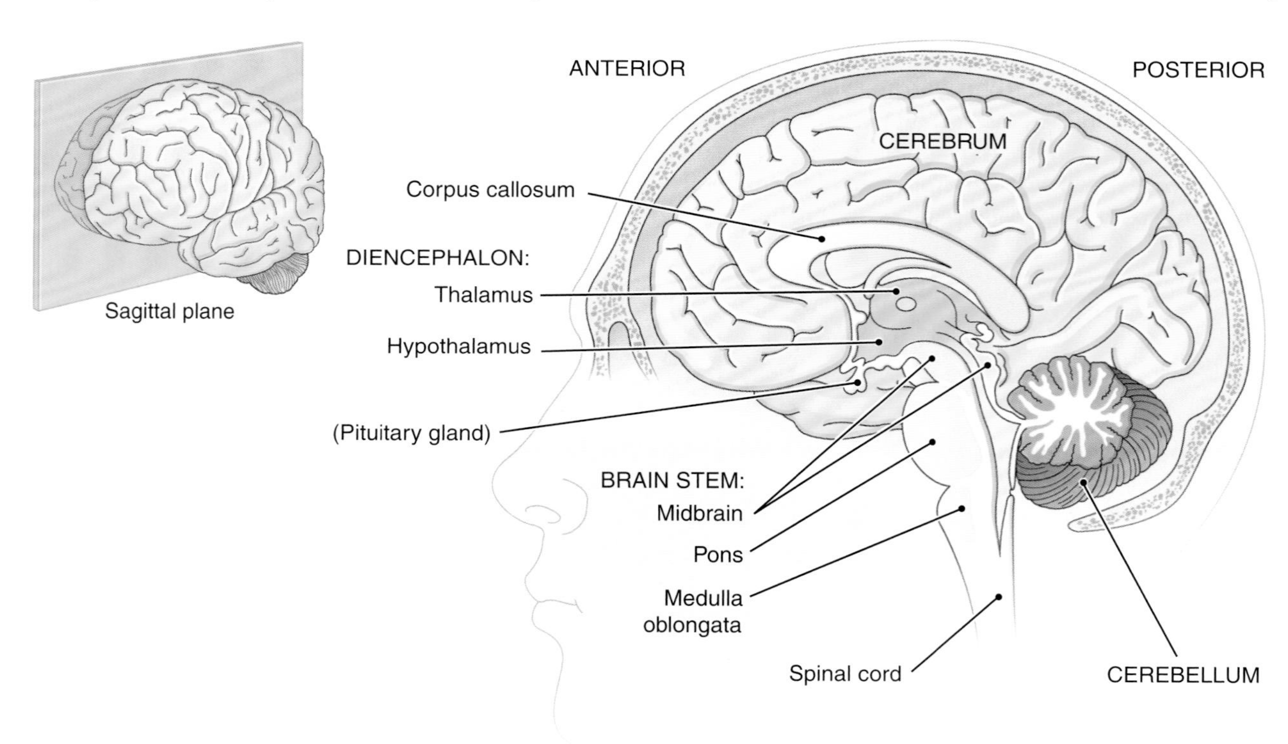

FIGURE 9-1 Brain, sagittal section. **KEY POINT** The four main divisions of the brain are the cerebrum, diencephalon, brain stem, and cerebellum. The pituitary gland is closely associated with the brain. **ZOOMING IN** What is the largest part of the brain? What part connects with the spinal cord?

Table 9-1 Organization of the Brain

Division	Description	Functions
CEREBRUM	Largest and most superior portion of the brain Divided into two hemispheres, each subdivided into lobes	Cortex (outer layer) is the site for conscious thought, memory, reasoning, perception, and abstract mental functions, all localized within specific lobes.
DIENCEPHALON	Between the cerebrum and the brain stem Contains the thalamus and hypothalamus	Thalamus sorts and redirects sensory input. Hypothalamus maintains homeostasis and controls autonomic nervous system and pituitary gland.
BRAIN STEM	Anterior region below the cerebrum	Connects the cerebrum and diencephalon with the spinal cord
Midbrain	Below the center of the cerebrum	Has reflex centers concerned with vision and hearing. Connects the cerebrum with lower portions of the brain
Pons	Anterior to the cerebellum	Connects the cerebellum with other portions of the brain Helps regulate respiration
Medulla oblongata	Between the pons and the spinal cord	Links the brain with the spinal cord. Has centers for control of vital functions, such as respiration and the heartbeat
CEREBELLUM	Below the posterior portion of the cerebrum Divided into two hemispheres	Coordinates voluntary muscles Maintains balance and muscle tone

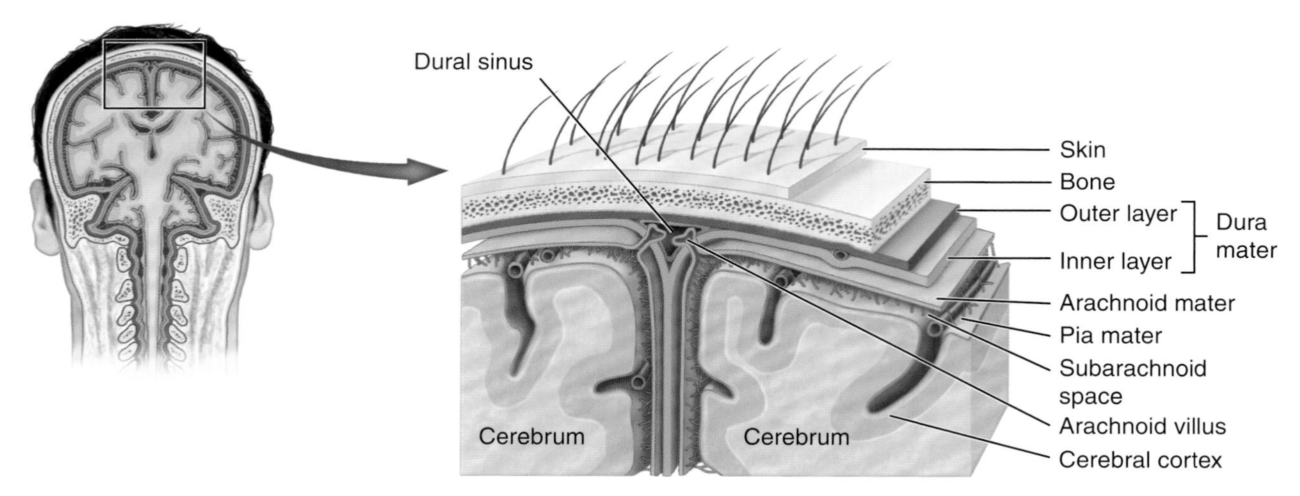

FIGURE 9-2 The meninges. KEY POINT The skull and meninges protect fragile brain tissue. The brain has many layers of protective substances. ZOOMING IN What are the channels formed where the dura mater divides into two layers? How many layers of meninges are there?

cord to form a complete enclosure (FIG. 9-2 and Dissection Atlas FIG. A3-3). The innermost layer around the brain, the **pia mater** (PI-ah MA-ter), is attached to the nervous tissue of the brain and spinal cord and follows all the contours of these structures. The pia is made of a delicate connective tissue (*pia* meaning "tender" or "soft"). It holds blood vessels that supply nutrients and oxygen to the brain and spinal cord.

The middle layer of the meninges is the **arachnoid** (ah-RAK-noyd). This membrane is loosely attached to the pia mater by weblike fibers, forming a space (the subarachnoid space) where CSF circulates. Blood vessels also pass through this space. The arachnoid is named from the Latin word for spider because of its weblike appearance.

The outermost **dura mater** (DU-rah MA-ter) is the thickest and toughest of the meninges. (*Mater* is from the Latin meaning "mother," referring to the protective function of the meninges; *dura* means "hard.") Around the brain, the dura mater is in two layers, and the outer layer is fused to the cranial bones. In certain places, these two layers separate to provide venous channels, called **dural sinuses**, for the drainage of blood coming from brain capillaries. Extensions of the arachnoid membrane called *arachnoid villi* project into this space. The dura is in a single layer around the spinal cord.

Cerebrospinal Fluid **Cerebrospinal** (ser-e-bro-SPI-nal) **fluid** (CSF) is a clear liquid that circulates in and around the brain and spinal cord (FIG. 9-3). The function of the CSF is to support nervous tissue and to cushion shocks that would otherwise injure these delicate structures. This fluid also carries nutrients to the cells and transports waste products from the cells.

CSF forms in four spaces within the brain called **ventricles** (VEN-trih-klz) (see FIG. 9-3A). A network of ependymal (specialized neuroglial) cells and blood vessels, known as the **choroid** (KOR-oyd) **plexus,** makes CSF within all four ventricles. You can see a choroid plexus in the third ventricle in FIGURE 9-3B. The journey of CSF can be summarized as follows:

1. CSF formed by the choroid plexuses in the lateral ventricles (ventricles 1 and 2) flows into the third ventricle (see FIG. 9-3A and B). This cavity forms a midline space within the diencephalon (see FIG. 9-3B).
2. Next, CSF passes through a small canal in the midbrain, the **cerebral aqueduct,** into the fourth ventricle. The fourth ventricle is located between the brain stem and the cerebellum.
3. A small volume of CSF passes from the fourth ventricle into the central canal of the spinal cord and travels down the cord. The rest passes through small openings in the roof of the fourth ventricle into the subarachnoid space of the meninges. CSF travels in the subarachnoid space around the spinal cord and brain. Nutrients and oxygen pass from the CSF, across the pia mater, and into the brain and spinal cord tissue, and waste products follow the opposite route.
4. Some of the CSF volume filters out of the arachnoid villi into blood in the dural sinus. CSF loss by this route occurs at the same rate as CSF synthesis by the choroid plexuses of the four ventricles, so the overall CSF volume remains constant.

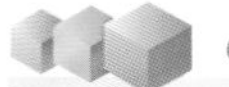

Concept Mastery Alert

Remember that the subarachnoid space contains CSF, but the dural sinuses and other cerebral vessels contain blood.

Many blood substances cannot access the CSF and thus fragile brain tissue. See BOX 9-1 for more information about the benefits and costs of this arrangement.

CHECKPOINTS

- ☐ **9-1** What are the main divisions of the brain?
- ☐ **9-2** What are the names of the three layers of the meninges from the outermost to the innermost?
- ☐ **9-3** Where is CSF produced?

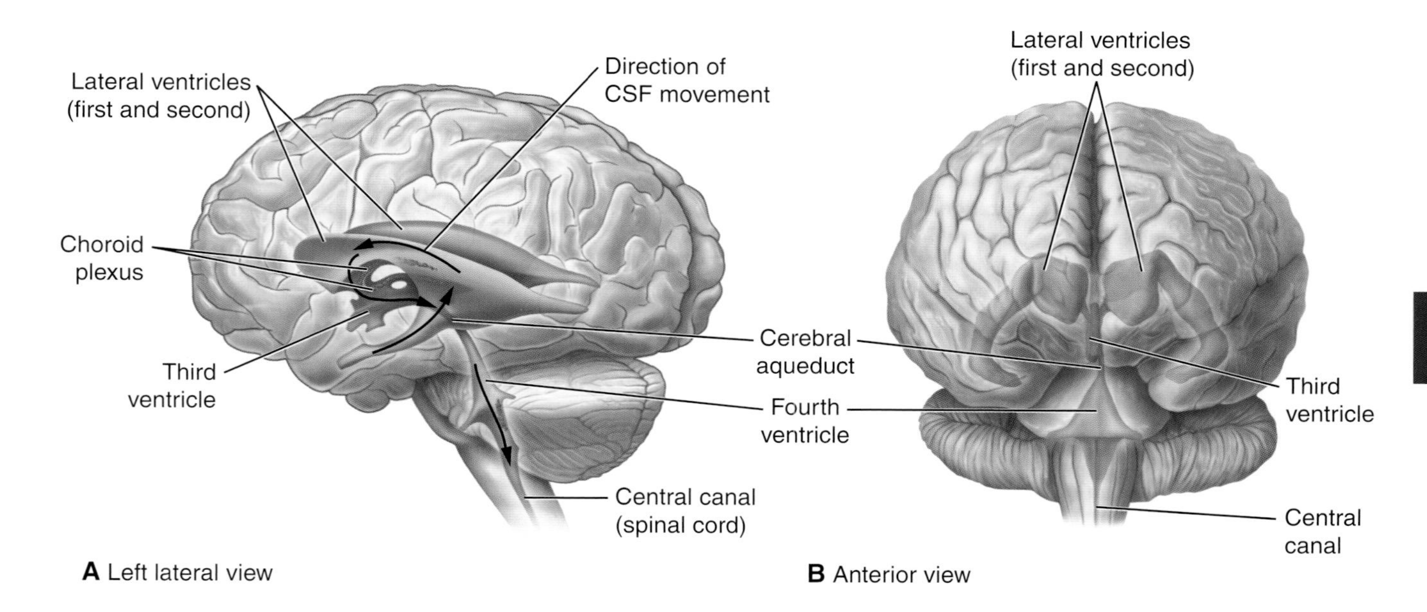

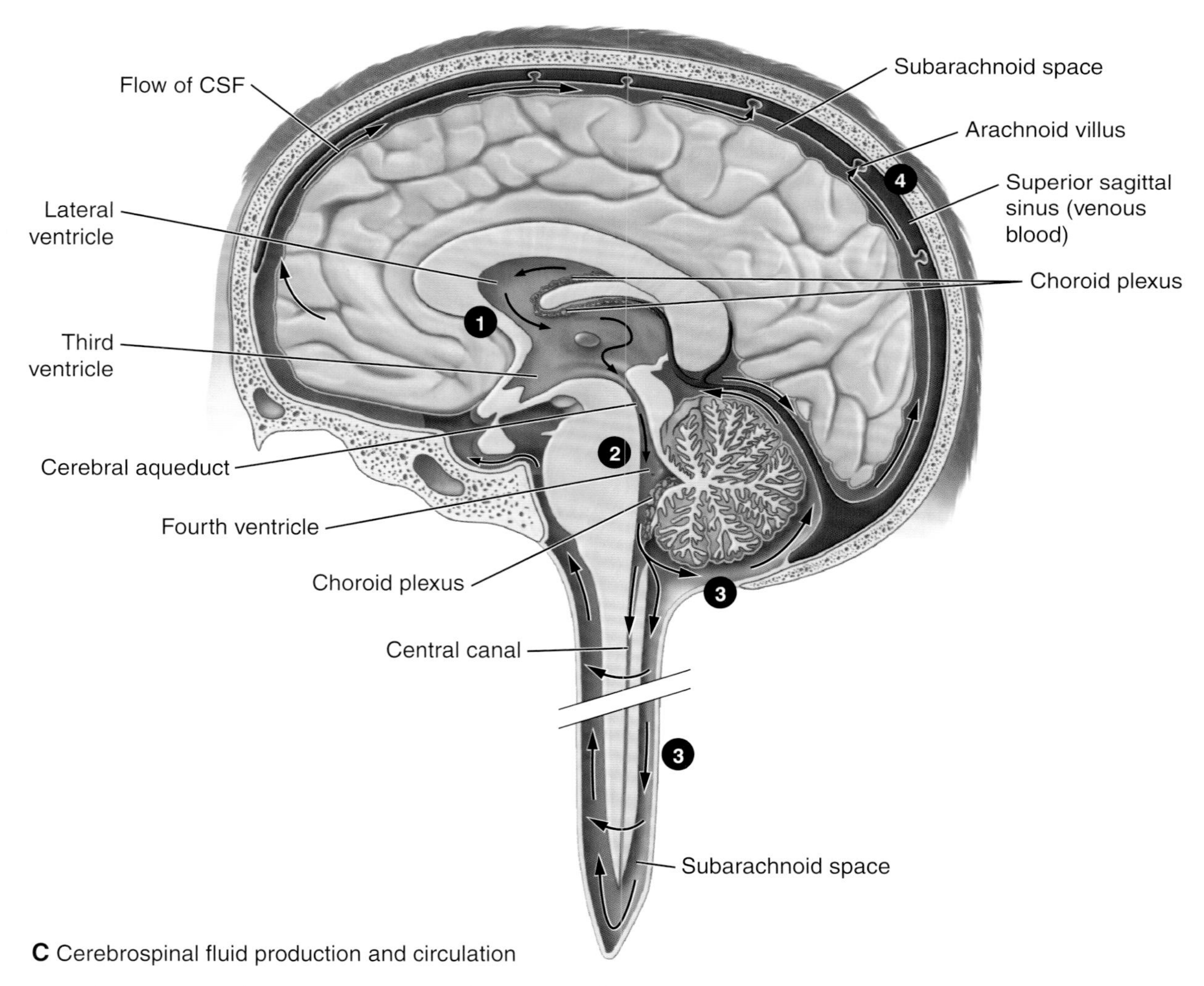

FIGURE 9-3 Cerebrospinal fluid (CSF) and the cerebral ventricles. **KEY POINT** Each ventricle contains a choroid plexus, which synthesizes CSF. **A.** Left lateral view. **B.** Anterior view. **C.** CSF circulates through the ventricular system into the arachnoid villi, where it is filtered into the blood of the superior sagittal sinus. The numbers correspond to the numbers in the narrative. (The actual passageways through which the CSF flows are narrower than those shown here, which have been enlarged for visibility.) **ZOOMING IN** Which ventricle is continuous with the central canal of the spinal cord?

A CLOSER LOOK
The Blood–Brain Barrier: Access Denied

BOX 9-1

Neurons in the central nervous system (CNS) function properly only if the composition of the extracellular fluid bathing them is carefully regulated. The semipermeable blood–brain barrier helps maintain this stable environment by allowing some substances to cross it while blocking others. Whereas it allows glucose, amino acids, and some electrolytes to cross, it prevents passage of hormones, drugs, neurotransmitters, and other substances that might adversely affect the brain.

Structural features of CNS capillaries create this barrier. In most parts of the body, capillaries are lined with simple squamous epithelial cells that are loosely attached to each other. The small spaces between cells let materials move between the bloodstream and the tissues. In CNS capillaries, the simple squamous epithelial cells are joined by tight junctions that limit passage of materials between them.

The blood–brain barrier excludes pathogens, although some viruses, including poliovirus and herpesvirus, can bypass it by traveling along peripheral nerves into the CNS. Some streptococci also can breach the tight junctions. Disease processes, such as hypertension, ischemia (lack of blood supply), and inflammation, can increase the blood–brain barrier's permeability.

The blood–brain barrier is an obstacle to delivering drugs to the brain. Some antibiotics can cross it, whereas others cannot. Neurotransmitters also pose problems. In Parkinson disease, the neurotransmitter dopamine is deficient in the brain. Dopamine itself will not cross the barrier, but a related compound, L-dopa, will. L-Dopa crosses the blood–brain barrier and is then converted to dopamine. Mixing a drug with a concentrated sugar solution and injecting it into the bloodstream are other effective delivery methods. The solution's high osmotic pressure causes water to osmose out of capillary cells, shrinking them and opening tight junctions through which the drug can pass.

The Cerebrum

The cerebrum, the brain's largest portion, is divided into right and left cerebral (SER-e-bral) hemispheres by a deep groove called the longitudinal fissure (FIG. 9-4). The two hemispheres have overlapping functions and are similarly subdivided.

DIVISIONS OF THE CEREBRAL HEMISPHERES

Each cerebral hemisphere is divided into four visible lobes named for the overlying cranial bones. These are the frontal, parietal, temporal, and occipital lobes (see FIG. 9-4B). In addition, there is a small fifth lobe deep within each hemisphere that cannot be seen from the surface. Not much is known about this lobe, which is called the **insula** (IN-su-lah).

Each cerebral hemisphere is covered by a thin (2 to 4 mm) layer of gray matter known as the **cerebral cortex** (see FIG. 9-2). The neuronal cell bodies and synapses in this region are responsible for conscious thought, reasoning, and abstract mental functions. Specific functions are localized in the cortex of the different lobes, as described in greater detail later.

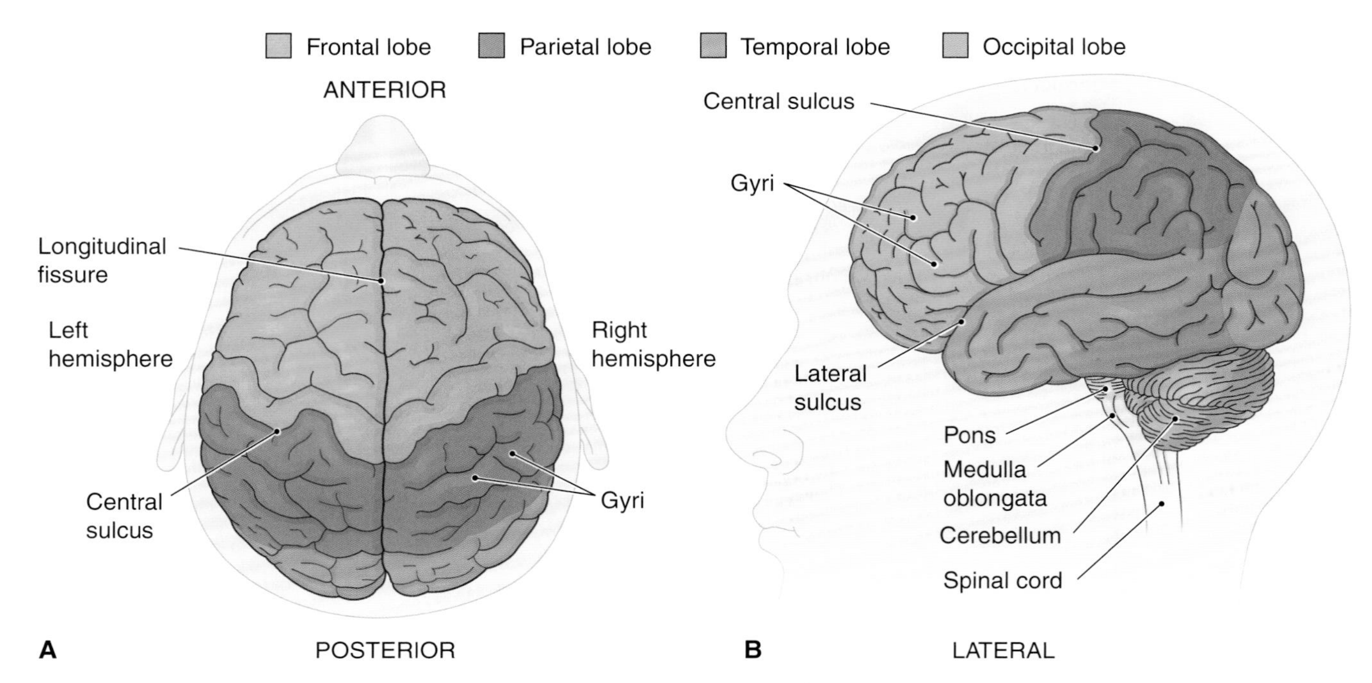

FIGURE 9-4 External surface of the brain. A. Superior view. **B.** Lateral view. **KEY POINT** The brain is divided into two hemispheres by the longitudinal fissure. Each hemisphere is subdivided into lobes. ZOOMING IN What structure separates the frontal from the parietal lobe? The temporal lobe from the frontal and parietal lobes?

The cortex is arranged in folds forming elevated portions known as **gyri** (JI-ri), singular *gyrus*. These raised areas are separated by shallow grooves called **sulci** (SUL-si), singular **sulcus**. Although there are many sulci, the following two are especially important landmarks:

- The **central sulcus** lies between the frontal and parietal lobes of each hemisphere at right angles to the longitudinal fissure (see FIG. 9-4).
- The **lateral sulcus** curves along the side of each hemisphere and separates the temporal lobe from the frontal and parietal lobes (see FIG. 9-4B).

The interior of the cerebral hemispheres consists primarily of white matter—myelinated fibers that connect the cortical areas with each other and with other parts of the nervous system. The **corpus callosum** (kah-LO-sum) is an important band of white matter located at the bottom of the longitudinal fissure (see FIG. 9-1). This band is a bridge between the right and left hemispheres, permitting impulses to cross from one side of the brain to the other.

FUNCTIONS OF THE CEREBRAL CORTEX

The cerebral cortex houses our consciousness—our awareness of the world around us and our ability to voluntarily interact with it. The cerebral cortex "stores" information, much of which can be recalled on demand by means of the phenomenon called *memory*. It is in the cerebral cortex that thought processes such as association, judgment, and discrimination take place.

Although the various brain areas act in coordination to produce behavior, particular functions are localized in the cortex of each lobe (FIG. 9-5). Some of these are described below.

Frontal Lobe The **frontal lobe**, which is relatively larger in humans than in any other organism, lies anterior to the central sulcus. The gyrus just anterior to the central sulcus in this lobe contains a **primary motor area**, which provides conscious control of skeletal muscles. Specific segments of the primary motor area control the muscles in different body regions. Relatively, larger portions of the cortex are devoted to muscles requiring precise control, such as those of the hand.

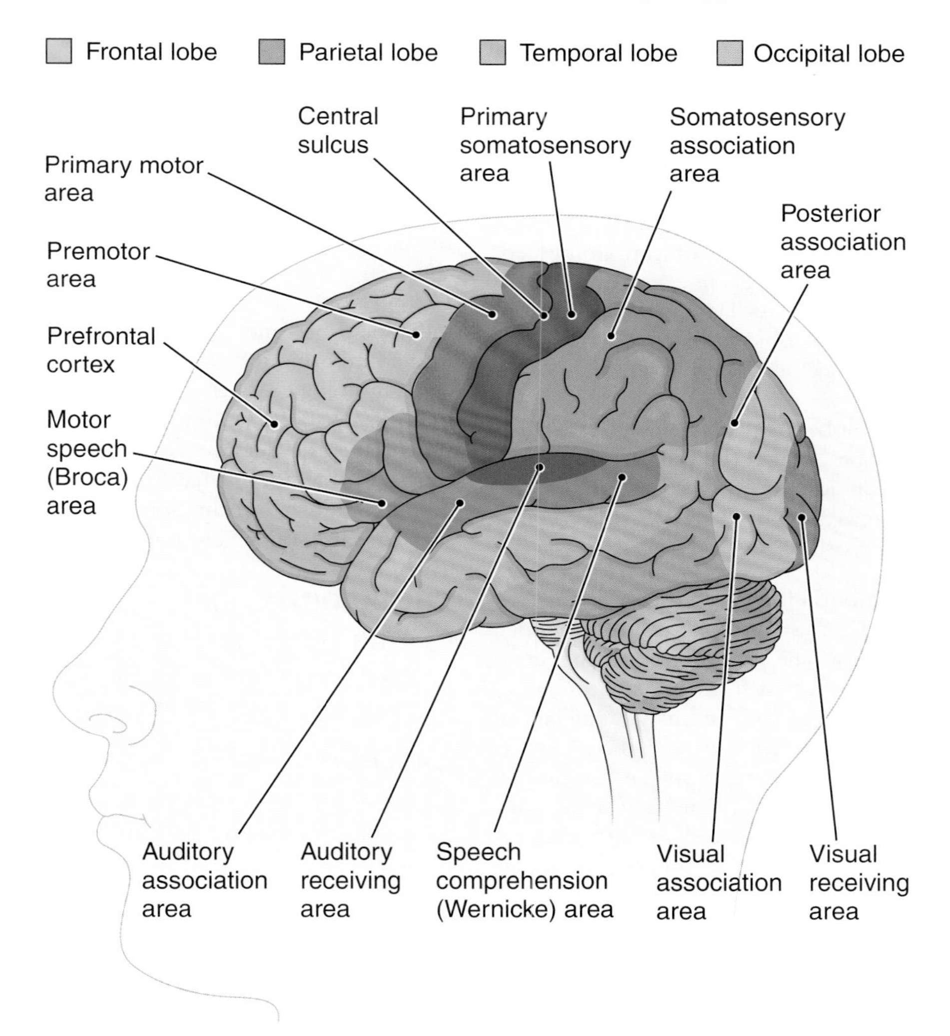

FIGURE 9-5 Functional areas of the cerebral cortex. KEY POINT Regions of the cerebral cortex are specialized for specific functions. ZOOMING IN What cortical area is posterior to the central sulcus? What area is anterior to the central sulcus?

9

Just anterior to the primary motor area is the premotor cortex, which helps plan complex movements. It receives sensory information from other parts of the brain to assist it in this task, and it sends most of its commands to the primary motor area. Anterior to the premotor cortex is the **prefrontal cortex**, involved in memory, problem-solving, and conscious thought. Within the prefrontal cortex in one cerebral hemisphere lies the *motor speech area*, or **Broca** (bro-KAH) **area**. This region, which is usually (but not always) found in the left hemisphere, plans the sequences of muscle contractions in the tongue, larynx, and soft palate required to form meaningful sentences. People with damage in this area can understand sentences but have trouble expressing their ideas in words.

Parietal Lobe The **parietal lobe** occupies the superior part of each hemisphere and lies posterior to the central sulcus. The gyrus just posterior to the central sulcus in this lobe contains the **primary somatosensory area**, where impulses from the skin, such as touch, pain, and temperature, are received. As with the motor cortex, the greater the intensity of sensation from a particular area, the tongue or fingers, for example, the more area of the cortex is involved.

Just posterior to this region is the somatosensory association area, which integrates somatosensory input with memories to identify physical sensations. For instance, imagine that you are reaching for a water glass at night, but your hand encounters your pet cat instead. The primary somatosensory area sends information regarding the softness, warmth, and shape of the sensed object. Combined with your memories of what a cat feels like, you are able to identify the object as a cat, not a glass. Damage to this region makes it impossible to identify objects by touch alone, so you would know that the object is fuzzy and warm, but not that it was a cat.

Much of the parietal lobe, as well as portions of the temporal and occipital lobes, forms the **posterior association area.** This brain region accepts information from all of the sensory association areas and our memories to construct an integrated view of the world.

Temporal Lobe The **temporal lobe** lies inferior to the lateral sulcus and folds under the hemisphere on each side. This lobe processes sounds. The **auditory receiving area** detects sound impulses transmitted from the environment, whereas the surrounding area, the **auditory association area**, interprets the sounds. Another region of the auditory cortex, located on the left side in most people, is the *speech comprehension area*, or **Wernicke** (VER-nih-ke) **area**. This area functions in speech recognition and the meaning of words. Someone who suffers damage in this region of the brain, as by a stroke, will have difficulty in understanding the meaning of speech. The **olfactory area**, concerned with the sense of smell, is located in the medial part of the temporal lobe and is not visible from the surface; it is stimulated by impulses arising from receptors in the nose.

Occipital Lobe The **occipital lobe** lies posterior to the parietal lobe and extends over the cerebellum. This lobe contains the **visual receiving area**, which collects sensory information from the retina, such as brightness and color, and the **visual association area**, which interprets the impulses into a mental "picture." Additional information processing by the posterior association area is necessary for us to label the mental picture as a flower, an understandable word, or a gesturing friend.

There is a functional relationship among areas of the brain. Many neurons must work together to enable a person to receive, interpret, and respond to verbal and written messages as well as to touch (tactile stimulus) and other sensory stimuli.

Speech therapists treat patients with language or communication problems from any cause. See the Student Resources on thePoint® for more information on this career.

MEMORY

Memory is the mental faculty for recalling ideas. In the initial stage of the memory process, sensory signals (e.g., visual, auditory) are retained for a very short time, perhaps only fractions of a second. Nevertheless, they can be used for further processing. **Short-term memory** refers to the retention of bits of information for a few seconds or perhaps a few minutes, after which the information is lost unless reinforced. **Long-term memory** refers to the storage of information that can be recalled at a later time. There is a tendency for a memory to become more fixed the more often a person repeats the remembered experience; thus, short-term memory signals can lead to long-term memories. Furthermore, the more often a memory is recalled, the more indelible it becomes; such a memory can be so deeply fixed in the brain that it can be recalled immediately.

Physiologic studies show that rehearsal (repetition) of the same information again and again accelerates and potentiates the degree of short-term memory transfer into long-term memory. It has also been noted that the brain is able to organize information so that new ideas are stored in the same areas in which similar ones had been stored before.

CHECKPOINTS 

- ☐ **9-4** Name the four surface lobes of each cerebral hemisphere.
- ☐ **9-5** Name the thin outer layer of gray matter where higher brain functions occur.

CASEPOINTS 

- ☐ **9-1** Natalie complains of blurred vision. Which brain lobe processes visual stimuli?
- ☐ **9-2** Natalie's sister later complains that Natalie loses the thread of a conversation. What form of her memory has been affected?

The Diencephalon

The diencephalon, or interbrain, is located between the cerebral hemispheres and the brain stem. One can see it by cutting into the central and inferior section of the brain. The diencephalon includes the **thalamus** (THAL-ah-mus) and the **hypothalamus** (FIG. 9-6).

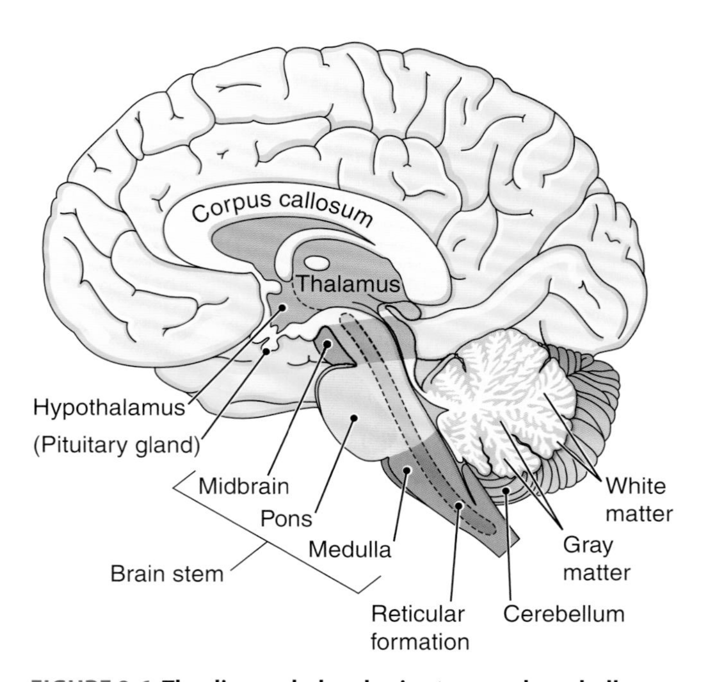

FIGURE 9-6 The diencephalon, brain stem, and cerebellum. **KEY POINT** The diencephalon consists of the thalamus, hypothalamus, and pituitary gland (hypophysis). The brain stem has three divisions: the midbrain, pons, and medulla oblongata. The white matter of the cerebellum is in a treelike pattern. **ZOOMING IN** To what part of the brain is the pituitary gland attached?

The two parts of the thalamus form the lateral walls of the third ventricle (see **FIGS. 9-1 and 9-3**). Nearly, all sensory impulses travel through the masses of gray matter that form the thalamus. The role of the thalamus is to sort out the impulses and direct them to particular areas of the cerebral cortex.

The hypothalamus is located in the midline area inferior to the thalamus and forms the floor of the third ventricle. 3 It acts as the control center for negative feedback loops controlling body temperature, food intake, and water balance, so it plays a critical role in homeostasis. It also controls other body functions necessary for well-being, such as sleep, and some emotions, such as fear and pleasure. Both the sympathetic and parasympathetic divisions of the autonomic nervous system are under hypothalamic control (see Chapter 8), as is the pituitary gland (see Dissection Atlas **FIG. A3-3A**). Chapter 11 discusses the important endocrine roles of the hypothalamus and pituitary gland.

CHECKPOINT

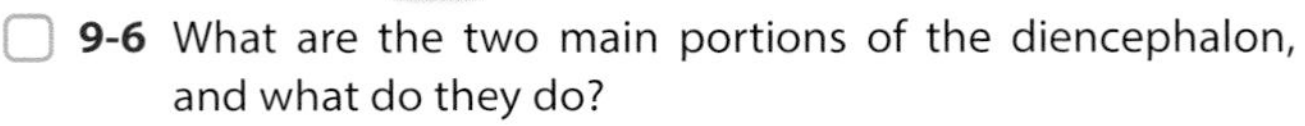

- [ ] **9-6** What are the two main portions of the diencephalon, and what do they do?

The Brain Stem

The brain stem is composed of the midbrain, the pons, and the medulla oblongata (see **FIG. 9-6**). These structures connect the cerebrum and diencephalon with the spinal cord.

THE MIDBRAIN

The **midbrain**, inferior to the center of the cerebrum, forms the superior part of the brain stem (see **FIG. 9-6**). Four rounded masses of gray matter that are hidden by the cerebral hemispheres form the superior part of the midbrain. These four bodies act as centers for certain reflexes involving the eye and the ear, for example, moving the eyes in order to track an image or to read. The white matter at the anterior of the midbrain conducts impulses between the higher centers of the cerebrum and the lower centers of the pons, medulla, cerebellum, and spinal cord.

THE PONS

The pons lies between the midbrain and the medulla, anterior to the cerebellum (see **FIG. 9-6**). It is composed largely of myelinated nerve fibers, which connect the two halves of the cerebellum with the brain stem as well as with the cerebrum above and the spinal cord below. Its name means "bridge," and it is an important connecting link between the cerebellum and the rest of the nervous system. It also contains nerve fibers that carry impulses to and from the centers located above and below it. Certain reflex (involuntary) actions, such as some of those regulating respiration, are integrated in the pons.

THE MEDULLA OBLONGATA

The medulla oblongata of the brain stem is located between the pons and the spinal cord (see **FIG. 9-6**). It appears white externally because like the pons, it contains many myelinated nerve fibers. Internally, it contains collections of cell bodies (gray matter) called **nuclei**, or *centers*. 3 Some of these nuclei act as control centers for the homeostatic control of blood gas levels and blood pressure, which are both regulated variables.

- The **respiratory center** controls the muscles of respiration in order to maintain consistent blood gas levels.
- The **cardiac center** helps regulate the rate and force of the heartbeat, which helps determine blood pressure.
- The **vasomotor** (vas-o-MO-tor) **center** regulates the contraction of smooth muscle in the blood vessel walls and thus controls blood flow and blood pressure.

The ascending sensory fibers that carry messages through the spinal cord up to the brain travel through the medulla, as do descending motor fibers. These groups of fibers form tracts (bundles) and are grouped together according to function.

The motor fibers from the motor cortex of the cerebral hemispheres extend down through the medulla, and most of them cross from one side to the other (decussate) while going through this part of the brain. The crossing of motor fibers in the medulla results in contralateral (opposite side) control—the right cerebral hemisphere controls muscles in the left side of the body and the left cerebral hemisphere controls muscles in the right side of the body.

The medulla is an important reflex center; here, certain neurons end, and impulses are relayed to other neurons.

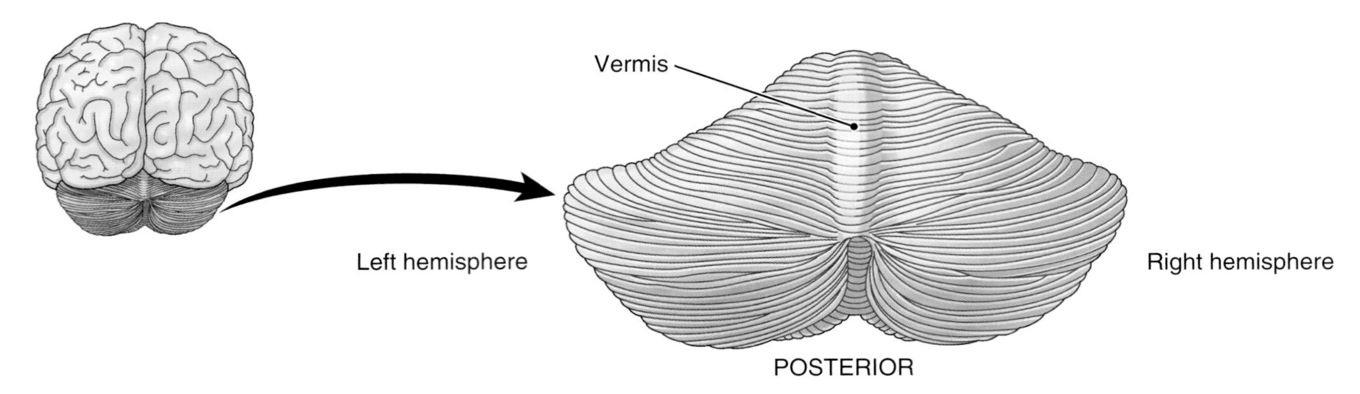

FIGURE 9-7 **The cerebellum.** Posterior view. KEY POINT The cerebellum is divided into two hemispheres.

CHECKPOINT 

- [ ] **9-7** What are the three subdivisions of the brain stem?

The Cerebellum

The cerebellum is made up of three parts: the middle portion (vermis) and two lateral hemispheres, the left and right (FIG. 9-7). Like the cerebral hemispheres, the cerebellum has an outer area of gray matter and an inner portion that is largely white matter (see FIG. 9-6). However, the white matter is distributed in a treelike pattern. The functions of the cerebellum are as follows:

- Helps coordinate voluntary muscles to ensure smooth, orderly function. Disease of the cerebellum causes muscular jerkiness and tremors.
- Helps maintain balance in standing, walking, and sitting as well as during more strenuous activities. Messages from the internal ear and from sensory receptors in tendons and muscles aid the cerebellum.
- Helps maintain muscle tone so that all muscle fibers are slightly tensed and ready to produce changes in position as quickly as necessary.

CHECKPOINT 

- [ ] **9-8** What are some functions of the cerebellum?

Widespread Neuronal Networks

Some coordinating networks involve select regions of the diencephalon and brain stem or extend throughout the entire brain. The limbic system, basal nuclei, and reticular formation are three such networks; the limbic system helps control behavior, the basal nuclei participate in motor control, and the reticular formation helps govern awareness.

THE LIMBIC SYSTEM

The cerebrum and the diencephalon contribute structures to the **limbic system**, a diffuse collection of neurons involved in emotional states and behavior (FIG. 9-8). The limbic system has extensive connections with all brain regions and can be considered the interface between the "thinking" brain of the prefrontal cortex and the "autonomic" brain gathering sensory information and controlling motor output. While there is some disagreement regarding which structures are part of the limbic system, most agree that it includes:

- The **cingulate** (SIN-gu-late) **gyrus**, the portion of the cerebral cortex looping over the corpus callosum. This region of the cortex associates emotions with memories.
- The **hippocampus** (hip-o-KAM-pus), shaped like a sea horse and located under the lateral ventricles. The hippocampus enables us to store new memories—that is, to learn new things. Lesions in the hippocampus impair one's ability to form new memories but leave old memories intact.
- The **amygdale** (ah-MIG-dah-lah), two clusters of nuclei deep in the temporal lobes. This brain region coordinates our emotional responses to stimuli. It receives extensive

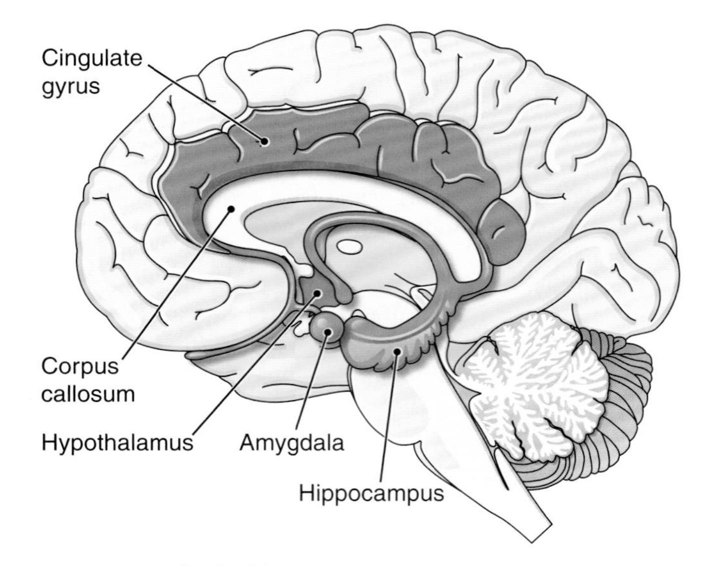

FIGURE 9-8 **The limbic system.** The system is shown in *red* in this figure. KEY POINT The limbic system consists of regions in the cerebrum and diencephalon that are involved in memory and emotion. ZOOMING IN Which part of the cerebral cortex contributes to the limbic system?

input from the olfactory lobe, which is why we often have strong emotional responses to smells.

- Parts of the hypothalamus and nearby nuclei. These regions control our motor responses to emotional stimuli, for instance, activation of the sympathetic nervous system.

BASAL NUCLEI

The **basal nuclei**, also called *basal ganglia*, modulate motor inputs and facilitate practiced, routine motor tasks. The basal nuclei consist of masses of gray matter spread throughout the brain that are extensively interconnected. They include three regions deep within the cerebrum, part of the midbrain called the *substantia nigra*, and a small collection of cell bodies in the diencephalon. As we will see later, death of different basal nuclei cells can cause Huntington or Parkinson disease.

RETICULAR FORMATION

The **reticular** (reh-TIK-u-lar) **formation** is a sausage-shaped network of neuronal cell bodies spanning the length of the brain stem (see FIG. 9-6). The reticular activating system (RAS) within this formation sends impulses to the cerebral cortex that keep us awake and attentive. In conjunction with the cortex, the RAS also screens out unnecessary sensory input (such as regular traffic noise) to increase the impact of novel stimuli (such as a car horn). Sleep centers in other brain regions inhibit the RAS, and thus arousal, when we sleep.

CHECKPOINTS 

- ☐ **9-9** What are four structures in the limbic system?
- ☐ **9-10** What is the function of the basal nuclei?
- ☐ **9-11** What is the function of the reticular activating system?

CASEPOINT 

- ☐ **9-3** Natalie had trouble staying awake and was irritable. Which neuronal networks are implicated by these two symptoms?

Brain Studies

Some of the imaging techniques used to study the brain are described in **BOX 1-2** in Chapter 1. These techniques include the following:

- CT scan, which provides photographs of the bone, soft tissue, and cavities of the brain (FIG. 9-9A). Anatomic lesions, such as tumors or scar tissue accumulations, are readily seen.
- MRI (magnetic resonance imaging), which gives a more detailed view of the brain tissue than does CT and may reveal tumors, scar tissue, and hemorrhaging not shown by CT (see FIG. 9-9B).
- PET (positron emission tomography), which visualizes the brain in action (see FIG. 9-9C).

The interactions of the brain's billions of nerve cells give rise to measurable electric currents. These may be recorded using an instrument called the **electroencephalograph** (e-lek-tro-en-SEF-ah-lo-graf). Electrodes placed on the head pick up the electrical signals produced as the brain functions. These signals are then amplified and recorded to produce the tracings, or brain waves, of an electroencephalogram (EEG).

The electroencephalograph is used to study sleep patterns; to diagnose disease, such as epilepsy; to locate tumors; to study the effects of drugs; and to determine brain death. FIGURE 9-10 shows some typical tracings.

CASEPOINT 

- ☐ **9-4** Why did Dr. Erickson choose a CT scan to evaluate Natalie's injuries?

12 Causation and Correlation

A CT scan helped Dr. Erickson diagnose Natalie's mild traumatic brain injury (concussion). Frequent concussions are associated with long-lasting changes in behavior, such as increased aggression. But, this association may be an example of correlation; two events that frequently occur

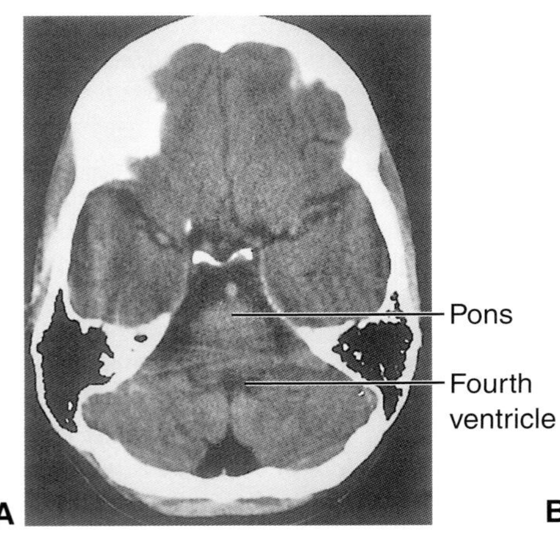

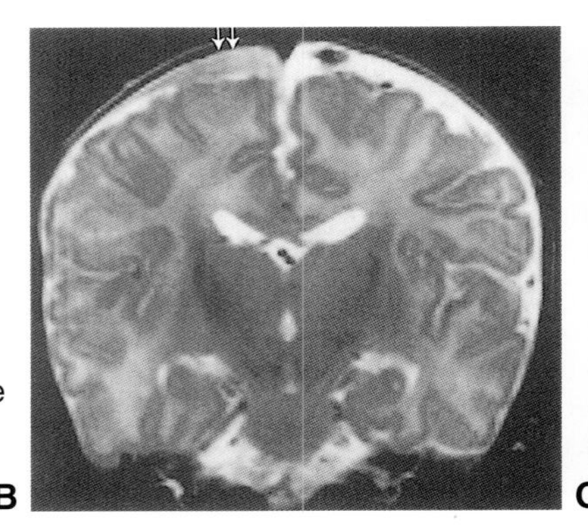

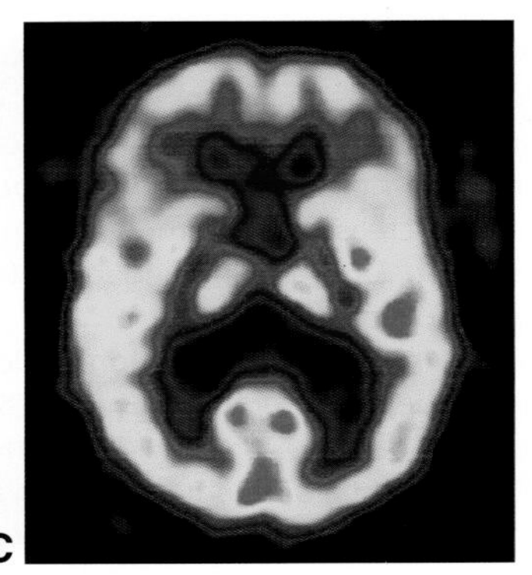

FIGURE 9-9 Imaging the brain. A. CT scan of a normal adult brain at the level of the fourth ventricle. **B.** MRI of the brain showing a point of injury (*arrows*). **C.** PET scan.

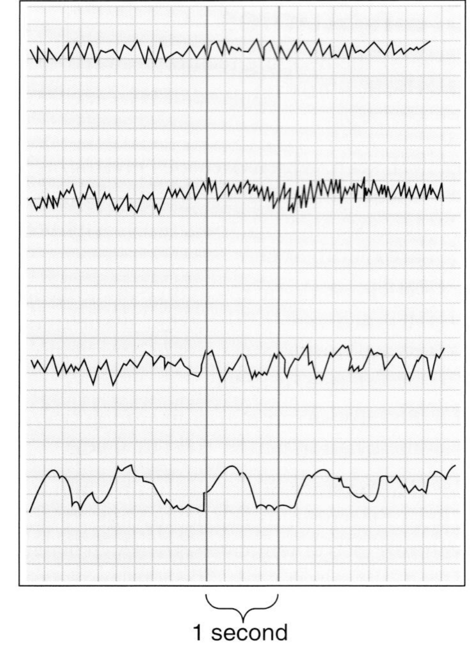

FIGURE 9-10 **Electroencephalography.** Normal brain waves.

together without one event causing the other. **BOX 9-2** in this chapter (Head Trauma and Violence: Correlation or Causation) outlines some strategies used by scientists to differentiate between correlation and causation. Identifying the underlying cause of a physiologic or pathologic process allows us to make predictions and design cures. Establishing causation involves specifying a sequence of interactions linking the cause to the effect. The One Step at a Time Box in Chapter 16 explains how to do this for the events of breathing.

Cranial Nerves

There are 12 pairs of cranial nerves (in this discussion, when a cranial nerve is identified, a pair is meant). They are numbered, usually in Roman numerals, according to their connection with the brain, beginning anteriorly and proceeding posteriorly (**FIG. 9-11**). Except for the first two pairs, which connect with the cerebrum and diencephalon, respectively, the cranial nerves connect with the brain stem. Also, note that anatomists now have found that cranial nerve XI connects with the cervical spinal cord. The first nine pairs and the 12th pair supply structures in the head.

ONE STEP AT A TIME

BOX 9-2

Head Trauma and Violence: Correlation or Causation?

Natalie, the subject of this chapter's case study, suffered a MTBI or concussion. The occurrence of concussions in professional sports, especially football, raises many issues/questions. For instance, NFL players suffer from frequent concussions, and players also commit more violent crimes compared with other groups of a similar income level. A logical conclusion might be that brain injuries are responsible for the violent behavior. But, an alternate explanation also exists that concussions and violent behavior are correlated, but the concussions do not cause the behavior. To understand the difference between correlation and causation, let's turn to a simpler observation.

A study in New York City showed that ice cream consumption is positively correlated with violent crime. A positive correlation means that both factors (ice cream consumption and violent crimes) are more prevalent in the same period or in a particular population. (Decreased ice cream consumption and increased violent behavior would be a negative correlation, as would increase ice cream consumption and decrease violent behavior). Does this finding indicate that ice cream makes us more violent?

Step 1. Experimentation. Following the rules of experimental design outlined in Chapter 8 (**BOX 8-1**), we could compare the behavior of individuals after they eat ice cream and after they eat a different food (such as salad).

Step 2. Postulate underlying causes. Is there an obvious, logical, common cause for both observations? The underlying cause of our ice cream example is relatively clear—ambient temperature. Hot and muggy weather makes us want to eat ice cream but also might increase aggression and violence.

The issue of concussions and violence can also be examined by the same process.

Step 1. Experiments are ideal for demonstrating causation. But, it would be unethical to give people concussions. One researcher proposed following the behavior of children diagnosed with concussions over decades to examine changes in their brain development and behavior compared with children without concussions.

Step 2. A tendency to engage in high-risk behavior could be responsible for both concussions and violent crime. Professional football is a high-risk undertaking, and risk-takers are more prone to alcohol, steroid, and drug abuse, all of which contribute to violent behavior.

As you can see, it is much easier to establish correlation than causation. This distinction reflects reality, because many separate events can occur sequentially or at the same time without one event actually causing another. The events could be caused by another factor common to all or by completely unrelated factors. One infamous example is the proposed link between frequent vaccinations and autism. The two events are correlated, because autism is usually diagnosed during, or just after, children receive most of their vaccinations. However, studies of over 14 million children have not been able to find any causative link between these two factors (see the "Study Guide" (available separately) for more practice at distinguishing between correlation and causation.

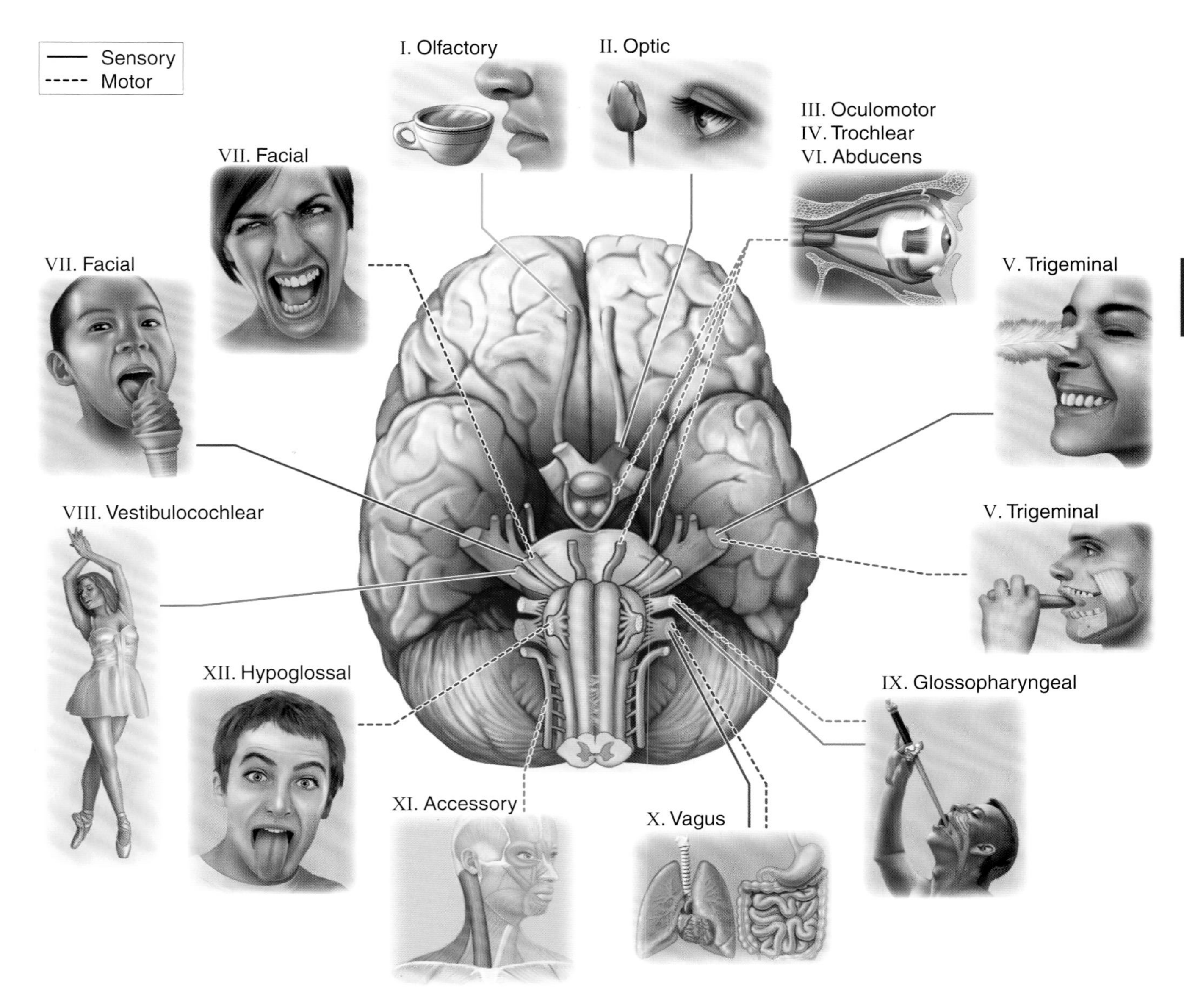

FIGURE 9-11 Cranial nerves. **KEY POINT** There are 12 pairs of cranial nerves, each designated by name and Roman numeral. They are shown here from the base of the brain.

From a functional viewpoint, the cranial nerves carry four types of signals. Note that each neuron carries only one signal type, but one cranial nerve can contain neurons carrying different signal types:

- **Special sensory impulses**, such as those for smell, taste, vision, and hearing, originating in special sense organs in the head.
- **General sensory impulses**, such as those for pain, touch, temperature, deep muscle sense, pressure, and vibrations. These impulses come from receptors that are widely distributed throughout the body.
- **Somatic motor impulses** that control the skeletal muscles.
- **Visceral motor impulses** that control glands, smooth muscle, and cardiac muscle. These motor pathways are part of the autonomic nervous system, parasympathetic division.

NAMES AND FUNCTIONS OF THE CRANIAL NERVES

A few of the cranial nerves (I, II, and VIII) contain only sensory fibers; some (III, IV, VI, XI, and XII) contain all or mostly motor fibers. The remainders (V, VII, IX, and X) contain both sensory and motor fibers; they are known as *mixed nerves*. All 12 nerves are listed below and summarized in **TABLE 9-2**:

I. The **olfactory nerve** carries smell impulses from receptors in the nasal mucosa to the brain.
II. The **optic nerve** carries visual impulses from the eye to the brain.
III. The **oculomotor nerve** controls the contraction of most of the eye muscles. These muscles are skeletal muscles and are under voluntary control. This nerve also carries the parasympathetic fibers that constrict the pupil.

Table 9-2 The Cranial Nerves and Their Functions

Nerve (Roman Numeral Designation)	Name	Function
I	Olfactory	Carries impulses for the sense of smell toward the brain
II	Optic	Carries visual impulses from the eye to the brain
III	Oculomotor	Controls contraction of eye muscles
IV	Trochlear	Supplies one eyeball muscle
V	Trigeminal	Carries sensory impulses from eye, upper jaw, and lower jaw toward the brain
VI	Abducens	Controls an eyeball muscle
VII	Facial	Controls muscles of facial expression, carries sensation of taste, stimulates small salivary glands and lacrimal (tear) gland
VIII	Vestibulocochlear	Carries sensory impulses for hearing and equilibrium from the inner ear toward the brain
IX	Glossopharyngeal	Carries sensory impulses from the tongue and pharynx (throat), controls swallowing muscles, and stimulates the parotid salivary gland
X	Vagus	Supplies most of the organs in the thoracic and abdominal cavities; carries motor impulses to the larynx (voice box) and pharynx
XI	Accessory	Controls muscles in the neck and larynx
XII	Hypoglossal	Controls muscles of the tongue

IV. The **trochlear** (TROK-le-ar) **nerve** supplies one eyeball muscle.

V. The **trigeminal** (tri-JEM-in-al) **nerve** is the great sensory nerve of the face and head. It has three branches that carry general sensory impulses (e.g., pain, touch, temperature) from the eye, the upper jaw, and the lower jaw. Motor fibers to the muscles of mastication (chewing) join the third branch. A dentist anesthetizes branches of the trigeminal nerve to work on the teeth without causing pain.

VI. The **abducens** (ab-DU-senz) **nerve** is another nerve sending motor impulses to an eyeball muscle.

VII. The **facial nerve** is largely motor nerve, controlling the muscles of facial expression. This nerve also includes special sensory fibers for taste (anterior two-thirds of the tongue), and it contains secretory fibers to the smaller salivary glands (the submandibular and sublingual) and to the lacrimal (tear) gland.

VIII. The **vestibulocochlear** (ves-tib-u-lo-KOK-le-ar) **nerve** carries sensory impulses for hearing and equilibrium from the inner ear. This nerve was formerly called the auditory or acoustic nerve.

IX. The **glossopharyngeal** (glos-o-fah-RIN-je-al) **nerve** contains general sensory fibers from the posterior tongue and the pharynx (throat). This nerve also contains sensory fibers for taste from the posterior third of the tongue, visceral motor fibers that supply the largest salivary gland (parotid), and motor nerve fibers to control the swallowing muscles in the pharynx.

X. The **vagus** (VA-gus) **nerve** is the longest cranial nerve. (Its name means "wanderer.") It carries autonomic motor impulses to most of the organs in the thoracic and abdominal cavities, including the muscles and glands of the digestive system. This nerve also contains somatic motor fibers supplying the larynx (voice box).

XI. The **accessory nerve** (also called the *spinal accessory nerve*) is a motor nerve with two branches. One branch controls two muscles of the neck, the trapezius and sternocleidomastoid; the other supplies muscles of the larynx.

XII. The **hypoglossal nerve,** the last of the 12 cranial nerves, carries impulses controlling the tongue muscles.

Concept Mastery Alert

Note that three nerves supply the tongue and five nerves supply the eyeball. Make sure that you can identify the information carried by each nerve.

It has been traditional for students in medical fields to use mnemonics (ne-MON-iks), or memory devices, to remember lists of terms. These devices are usually words (real or made-up) or sayings formed from the first letter of each item. We've used the example SLUDD for the actions of the parasympathetic system; for the cranial nerves, students use "On Occasion Our Trusty Truck Acts Funny. Very Good Vehicle Any How." Can you and your classmates make up any other mnemonic phrases for the cranial nerves? You can also check the Internet for sites where medical mnemonics are shared, but be forewarned, students often enjoy making them raunchy!

CHECKPOINTS

- ☐ **9-12** How many pairs of cranial nerves are there?
- ☐ **9-13** What are the three types of cranial nerves? What is a mixed nerve?

CASEPOINT

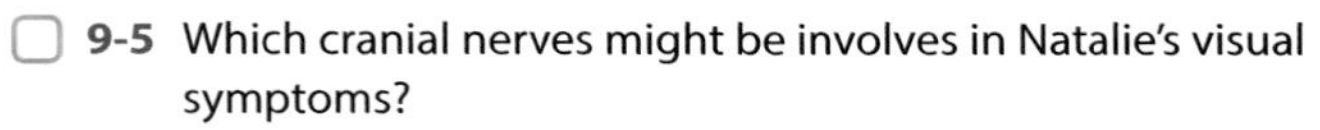

- ☐ **9-5** Which cranial nerves might be involves in Natalie's visual symptoms?

Effects of Aging on the Nervous System

The nervous system is one of the first systems to develop in the embryo. By the beginning of the third week of development, the rudiments of the CNS have appeared. Beginning with maturity, the nervous system begins to undergo degenerative changes. Neurons and glial cells die and are not replaced, decreasing the size and weight of the brain. Neuron loss in the cerebral cortex, accompanied by decreased neurotransmitter production and fewer synapses, results in slower information processing and movements. Memory diminishes, especially for recent events. Narrowing of the cerebral arteries reduces the brain's blood flow. Vascular degeneration increases the likelihood of stroke.

Much individual variation is possible, however, with regard to location and severity of changes. Although age might make it harder to acquire new skills, tests have shown that practice enhances skill retention. As with other body systems, the nervous system has vast reserves, and most elderly people are able to cope with life's demands.

A & P in Action Revisited: Natalie's Progress

The neurologist discharged Natalie the next morning, after an MRI did not show any brain lesions. Lacey was instructed to watch her sister carefully over the next few days. Natalie was to return to the hospital if her headache worsens, her speech becomes slurred, or she becomes difficult to arouse.

"And," said the neurologist, "take away Natalie's cell phone, laptop computer, and any other electronics, as well as all reading material. She needs complete quiet and rest."

At the follow-up appointment later that week, it was obvious that Lacey's patience was wearing thin.

"Natalie is irritable, keeps forgetting where she left her tea, and can't hold a coherent conversation for more than 2 minutes," Lacey complained.

"Natalie, you are experiencing postconcussion syndrome," said the neurologist. "Although the brain is encased within the skull and cushioned by CSF, it can still be damaged by the force of the brain colliding with the skull. It can take a few weeks for the cerebral swelling to decrease. Until that time, you can expect cognitive, physical, emotional, and behavioral changes."

"I expect you to recover fully," he continued. "But for the time being, you still need rest. Avoid overstimulation and alcohol. Your symptoms should resolve in the next few weeks."

CHAPTER 9

Chapter Wrap-Up

OVERVIEW

A detailed chapter outline with space for note-taking is on thePoint®. The figure below illustrates the main topics covered in this chapter.

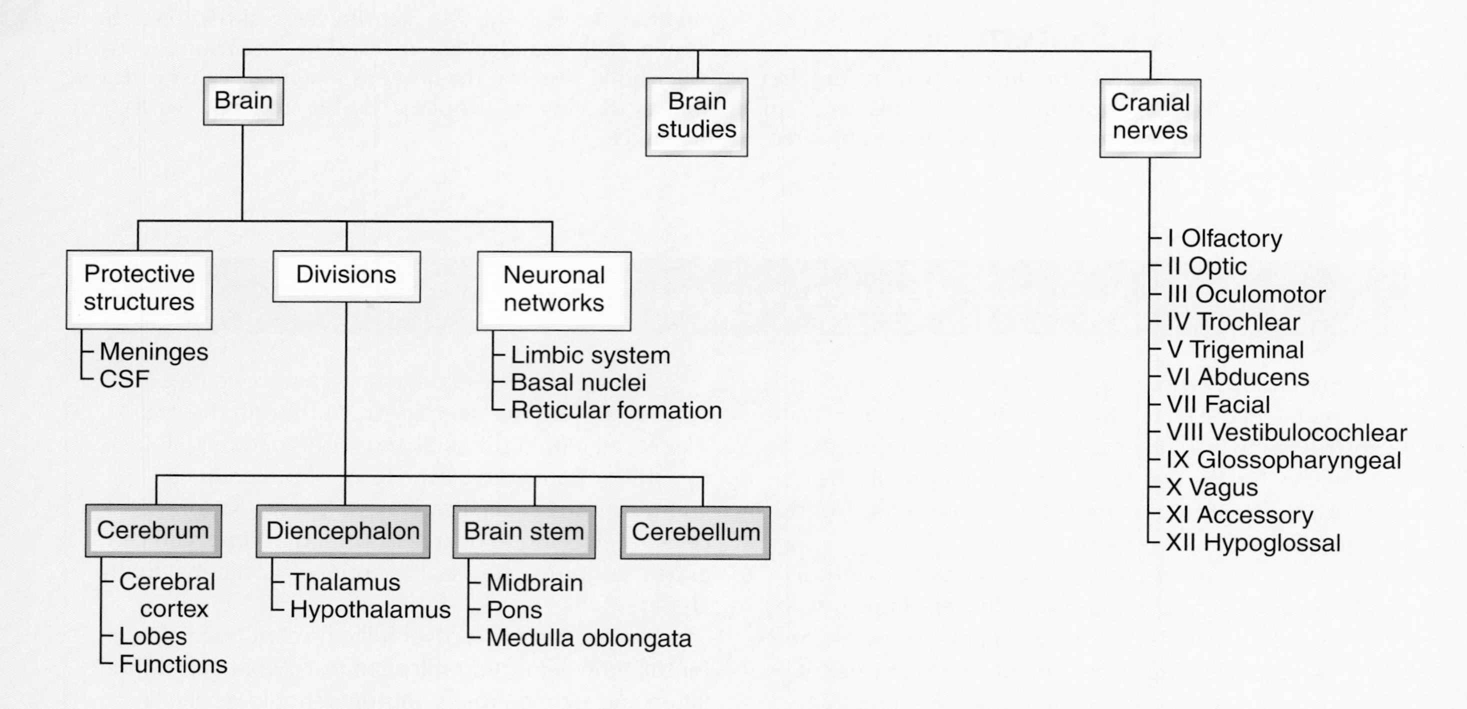

KEY TERMS

The terms listed below are emphasized in this chapter. Knowing them will help you organize and prioritize your learning. These and other boldface terms are defined in the "Glossary" with phonetic pronunciations.

basal nuclei
brain stem
cerebellum
cerebral cortex
cerebrospinal fluid (CSF)
cerebrum
concussion
corpus callosum
diencephalon
electroencephalograph (EEG)
gyrus (pl. gyri)
hypothalamus
limbic system
medulla oblongata
meninges
midbrain
pons
reticular formation
sulcus (pl. sulci)
thalamus
ventricle

WORD ANATOMY

Medical terms are built from standardized word parts (prefixes, roots, and suffixes). Learning the meanings of these parts can help you remember words and interpret unfamiliar terms.

WORD PART	MEANING	EXAMPLE
PROTECTIVE STRUCTURES OF THE BRAIN AND SPINAL CORD		
cerebr/o	brain	*Cerebrospinal* fluid circulates around the brain and spinal cord.
chori/o	membrane	The *choroid* plexus is the vascular membrane in the ventricle that produces CSF.
contra-	opposed, against	The cerebral cortex has *contralateral* control of motor function.
encephal/o	brain	The *diencephalon* is the part of the brain located between the cerebral hemispheres and the brain stem.
gyr/o	circle	A *gyrus* is a circular raised area on the surface of the brain.
later/o	lateral, side	See "contra-" example.
BRAIN STUDIES		
tom/o	cut	*Tomography* is a method for viewing sections as if cut through the body.
CRANIAL NERVES		
gloss/o	tongue	The *hypoglossal* nerve controls muscles of the tongue.
tri-	three	The *trigeminal* nerve has three branches.

QUESTIONS FOR STUDY AND REVIEW

BUILDING UNDERSTANDING

Fill in the Blanks

1. The delicate innermost layer of the meninges is the _____.
2. The large band of white matter that connects the right and left hemispheres is the _____.
3. Sound is processed in the _____ lobe of the brain.
4. The thalamus and hypothalamus are parts of the brain division named the _____.
5. The third and fourth ventricles are connected by a small canal called the _____.

Matching Match each numbered item with the letter of the most closely related cranial nerve.

___ 6. The nerve involved with the sense of smell

___ 7. The large sensory nerve of the face and head

___ 8. The nerve that controls muscles of the tongue

___ 9. The sensory nerve for hearing and equilibrium

___ 10. The long nerve that carries autonomic impulses to the thorax and abdomen

a. olfactory nerve
b. vestibulocochlear nerve
c. trigeminal nerve
d. vagus nerve
e. hypoglossal nerve

Multiple Choice

___ 11. What divides the cerebrum into left and right hemispheres?
a. central sulcus
b. insula
c. lateral sulcus
d. longitudinal fissure

___ 12. Which lobe interprets impulses arising from the retina of the eye?
a. frontal
b. occipital
c. parietal
d. temporal

___ 13. What is the name of the neuronal network in the brain stem that maintains wakefulness and screens out unimportant sensory stimuli?
a. reticular formation
b. basal nuclei
c. limbic system
d. corpus callosum

___ 14. Dr. H diagnosed her patient with a brain injury but cannot identify its location. Which diagnostic technique did she use?
a. computed tomography
b. electroencephalography
c. positron emission tomography
d. radiography

___ 15. What type of impulses is involved in the sense of touch?
a. special sensory
b. general sensory
c. somatic motor
d. visceral motor

UNDERSTANDING CONCEPTS

16. Based on your knowledge of their functions, predict the effects of injury to the following brain areas:
a. Broca area
b. hypothalamus
c. medulla oblongata
d. cerebellum
17. A neurosurgeon has drilled a hole through her patient's skull and is preparing to remove a cerebral tumor. In order, list the membranes she must cut through to reach the cerebral cortex.
18. What is the function of the limbic system? Describe the effect of damage to the hippocampus.
19. Explain the working of an electroencephalograph. What kind of information does the electroencephalograph provide?
20. Compare and contrast the functions of the following structures:
a. frontal lobe and parietal lobe
b. temporal lobe and occipital lobe
c. thalamus and hypothalamus
21. Referring to the 12 cranial nerves and their functions, make a list of the ones that are sensory, motor, and mixed.
22. Explain the function of the four cranial nerves that might have been involved in causing Natalie's visual symptoms discussed in the opening case study.
23. Referring to the brain overlays in The Body Visible at the beginning of the book, give the numbers of the following:
a. a network involved in the manufacture of CSF
b. myelinated fibers
c. a portion of the limbic system
d. a shallow groove in the surface of the cerebral cortex
e. a chamber where CSF is made
24. Referring to the Dissection Atlas **FIGURE A3-1**, name the:
a. feature that separates the frontal from the parietal lobe
b. feature that separates the frontal from the temporal lobe
c. raised surface area anterior to the central sulcus
25. Referring to the "Dissection Atlas" **FIGURE A3-2**, name:
a. area(s) where cerebrospinal fluid is made
b. the area superior to the corpus callosum
26. Referring to the "Dissection Atlas" **FIGURE A3-3B**:
a. Name the area that is external to the dura mater.
b. Name the area that is deep to the middle layer of the meninges.

CONCEPTUAL THINKING

27. The parents of Molly R., a 2-month-old girl, are informed that their daughter requires a shunt to drain excess CSF from her brain. What would happen to Molly if the shunt were not put into place?
28. Natalie's fall resulted in swelling in her brain. Explain the areas of the brain that might be affected as shown by her symptoms, which included drowsiness, irritability, loss of balance, lack of coordination, and slurred speech.

For more questions, see the Learning Activities on thePoint®.

CHAPTER 10

The Sensory System

Learning Objectives

After careful study of this chapter, you should be able to:

1. Describe the functions of the sensory system. *p. 202*
2. Differentiate between the different types of sensory receptors, and give examples of each. *p. 202*
3. Describe sensory adaptation, and explain its value. *p. 202*
4. List and describe the structures that protect the eye. *p. 203*
5. Cite the location and the purpose of the extrinsic eye muscles. *p. 203*
6. Identify the three tunics of the eye. *p. 204*
7. Describe the processes involved in vision. *p. 205*
8. Differentiate between the rods and the cones of the eye. *p. 207*
9. Describe the three divisions of the ear. *p. 209*
10. Describe the receptor for hearing, and explain how it functions. *p. 212*
11. Compare the locations and functions of the equilibrium receptors. *p. 213*
12. Discuss the locations and functions of the sense organs for taste and smell. *p. 214*
13. Describe five general senses. *p. 215*
14. Referring to the case study, discuss the purpose and mechanism of a cochlear implant. *pp. 201, 218*
15. Show how word parts are used to build words related to the sensory system (see "Word Anatomy" at the end of the chapter). *p. 220*

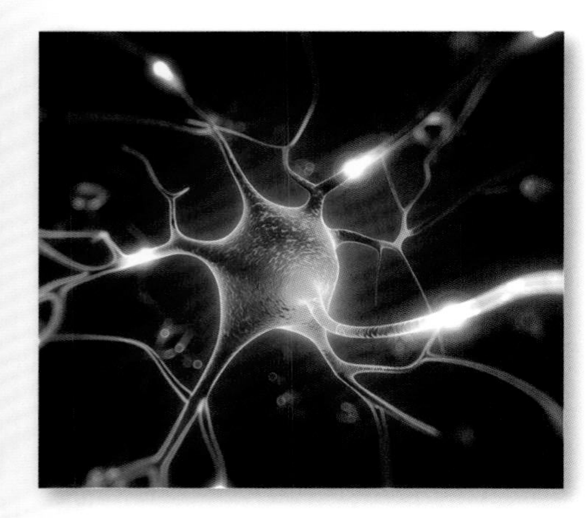

A & P in Action
Evan Needs a Cochlear Implant

"Bacterial meningitis," the ER resident told Evan's worried parents. Evan, 20 months old, had been sick for a day with vomiting, a high fever, and a rash on his chest. When he had a seizure, his parents realized that this was no ordinary childhood illness and brought him immediately to the hospital.

The resident continued, "He needs to be admitted right away for treatment and observation. Strong antibiotics should cure the infection, and steroids should limit the damage caused by Evan's immune system fighting the bacteria. Meningitis is an infection of the central nervous system and can be very dangerous, but we caught it early. Be prepared to stay in the hospital at least a week."

It was 10 days until Evan was home and feeling better. The entire family was so exhausted by the ordeal that it took them another week to notice that Evan wasn't responding to sounds. A follow-up appointment with his pediatrician and subsequent hearing test revealed that Evan had 95% hearing loss in the right ear and 60% in the left.

"I'm referring you to Dr. Sanchez, a specialist in cochlear implants for young children," the pediatrician said. "A cochlear implant can potentially restore some of Evan's hearing."

"A multidisciplinary approach is used for each patient," Dr. Sanchez explained. "A complete audiological, speech, and medical evaluation will be conducted along with a CT scan to determine if Evan is a candidate. If approved, you and Evan will need to be enrolled into the preimplant program, where you will learn about how a prosthetic implant works. The device will stimulate the cochlear nerve directly, bypassing the receptor cells, and it may restore hearing for medium to loud sounds."

Evan's parents were hopeful and eager to proceed with the evaluation. Dr. Sanchez fully expected Evan to fit the criteria, as his hearing loss was recent. We will check later to see how Evan's case progressed.

As you study this chapter, CasePoints will give you opportunities to apply your learning to this case.

Visit thePoint® to access the following resources. For guidance in using these resources most effectively, **see pp. xv–xvii.**

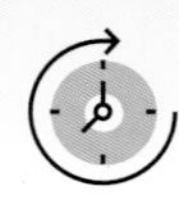

Preparing to Learn

- Tips for Effective Studying
- Pre-Quiz

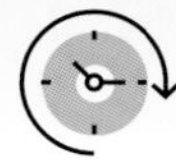

While You Are Learning

- Animation: Eye Structure and Function
- Animation: Hearing
- Chapter Notes Outline
- Audio Pronunciation Glossary

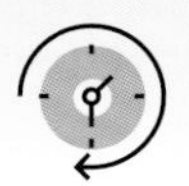

When You Are Reviewing

- Answers to Questions for Study and Review
- Health Professions: Audiologist
- Interactive Learning Activities

A LOOK BACK

In describing the basic organization of the nervous system, we included both sensory and motor functions. Now, we concentrate on just the sensory portion of the nervous system and the special receptors that detect environmental changes. These specialized structures initiate the reflex pathways described in the previous chapters. The key ideas of homeostasis 3*, adaptation* 10*, communication* 11*, and causation* 12 *are particularly relevant to this chapter.*

The Senses

The sensory system provides us with an awareness of our external and internal environments. An environmental change becomes a stimulus when it initiates a nerve impulse, which then travels to the central nervous system (CNS) by way of a sensory neuron. A stimulus becomes a sensation—something we experience—only when a specialized area of the cerebral cortex interprets the nerve impulse received. Many stimuli arrive from the external environment and are detected at or near the body surface. 3 Others originate internally and help maintain homeostasis; they act as the sensors in some negative feedback loops.

SENSORY RECEPTORS

The part of the nervous system that detects a stimulus is the **sensory receptor.** 11 Note that sensory receptors are different from the protein receptors involved in binding chemical signals. In structure, a sensory receptor may be one of the following:

- The free dendrite of a sensory neuron, such as the receptors for pain and temperature
- A modified ending on the dendrite of a sensory neuron, such as those for touch
- A specialized cell associated with a sensory neuron, such as the rods and cones of the eye's retina

Receptors can be classified according to the type of stimulus to which they respond:

- Chemoreceptors, such as receptors for taste and smell, detect chemicals in solution.
- Photoreceptors, located in the retina of the eye, respond to light.
- Thermoreceptors detect changes in temperature. Many of these receptors are located in the skin.
- Mechanoreceptors respond to movement, such as stretch, pressure, or vibration. These include pressure receptors in the skin, receptors that monitor body position, and the receptors of hearing and equilibrium in the ear, which are activated by the movement of cilia on specialized receptor cells.

Any receptor must receive a stimulus of adequate intensity, that is, at least a **threshold stimulus**, in order to respond and generate a nerve impulse.

SPECIAL AND GENERAL SENSES

Another way of classifying the senses is according to the distribution of their receptors. A **special sense** is localized in a special sense organ; a **general sense** is widely distributed throughout the body.

- Special senses:
  - **Vision** from receptors in the eye
  - **Hearing** from receptors in the inner ear
  - **Equilibrium** (balance) from receptors in the inner ear
  - **Taste** from receptors in the tongue
  - **Smell** from receptors in the upper nasal cavities
- General senses:
  - **Pressure, temperature, pain,** and **touch** from receptors in the skin and internal organs
  - Sense of **position** from receptors in the muscles, tendons, and joints

SENSORY ADAPTATION

10 When sensory receptors are exposed to a continuous and unimportant stimulus, they often adjust so that the sensation becomes less acute. The term for this phenomenon is **sensory adaptation.** For example, when you first put on a watch, you may be aware of its pressure on your wrist. Soon, you do not notice it at all. If you are rinsing dishes in very warm water, you may be aware of the temperature at first, but you soon adapt and stop noticing the water's temperature. Similarly, both delicious and horrible odors weaken the longer you smell them. As these examples show, both special and general senses are capable of adaptation. However, different receptors adapt at different rates. Those for warmth, cold, and light pressure adapt rapidly. In contrast, receptors for pain do not adapt. In fact, the sensations from receptors for slow, chronic pain tend to increase over time. This variation in receptors allows us to save energy by not responding to unimportant stimuli while always heeding the warnings of pain.

CHECKPOINTS

- ☐ **10-1** What is a sensory receptor?
- ☐ **10-2** What are some categories of sensory receptors based on the type of stimulus?
- ☐ **10-3** How do the special and general senses differ in location?
- ☐ **10-4** What happens when a sensory receptor adapts to a stimulus?

CASEPOINT

- ☐ **10-1** In the case study, what type of sensory receptor was damaged?

The Eye and Vision

Vision is arguably the most important of the special senses, contributing more than half of the information we use to perceive the world. Before we discuss the eye itself, we begin with the structures that protect, move, and control the eye.

PROTECTIVE STRUCTURES OF THE EYE

The eye is a delicate organ and is protected by a number of structures:

- The skull bones form the walls of the eye orbit (cavity) and protect the posterior part of the eyeball (see FIG. 6-6).
- The upper and lower eyelids aid in protecting the eye's anterior portion (FIGS. 10-1 and 10-2). The eyelids can be closed to keep harmful materials out of the eye, and blinking helps to lubricate the eye. An eyelid is technically called a palpebra (PAL-peh-brah). A muscle, the levator palpebra, is attached to the upper eyelid (see FIG. 10-2). When this muscle contracts, it keeps the eye open. If the muscle becomes weaker with age, the eyelids may droop and interfere with vision, a condition called *ptosis*.
- The eyelashes and eyebrow help keep foreign matter out of the eye.
- Tears, produced by the **lacrimal** (LAK-rih-mal) **glands** (see FIG. 10-1), lubricate the eye and contain an enzyme that protects against infection. As tears flow across the eye from the lacrimal gland located in the orbit's upper lateral part, they carry away small particles that may have entered the eye. The tears then flow into canals near the eye's nasal corner where they drain into the nose by way of the **nasolacrimal** (na-zo-LAK-rih-mal) **duct**. The lacrimal glands, ducts, and canals together make up the **lacrimal apparatus**. An excess of tears causes a "runny nose"; the overproduction causes tears to spill onto the cheeks. With age, the lacrimal glands secrete less, but tears still may overflow if the nasolacrimal ducts become plugged.

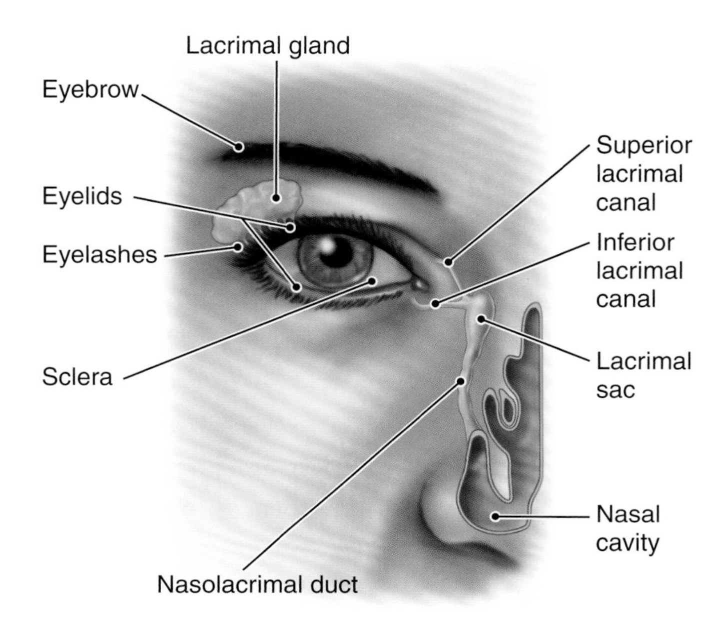

FIGURE 10-1 The eye's protective structures. KEY POINT Tears are produced in the lacrimal gland, located laterally, and flow across the eye to the lacrimal canals, located medially. The eyelid, eyelashes, and eyebrow also protect the eye.

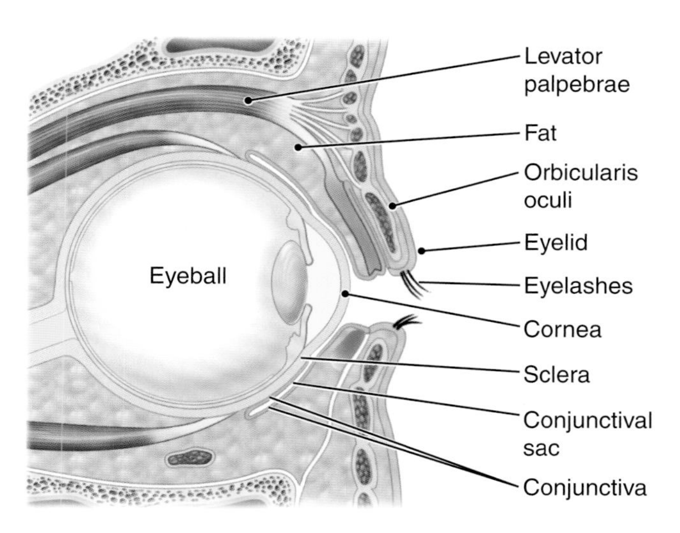

FIGURE 10-2 A sagittal view of the eye orbit. KEY POINT The eye is a delicate organ well guarded by a bony socket and other protective structures.

- A thin membrane, the **conjunctiva** (kon-junk-TI-vah), lines the inner surface of the eyelids and folds back to cover the visible portion of the white of the eye (sclera) (see FIG. 10-1).

Concept Mastery Alert

The conjunctiva does not cover the cornea or the eye's posterior surface.

Cells within the conjunctiva produce mucus that aids in lubricating the eye. The pocket formed by the folded conjunctiva, known as the conjunctival sac, can be used to instill medication drops (see FIG. 10-2). With age, the conjunctiva often thins and dries, resulting in inflammation and dilated blood vessels.

THE EXTRINSIC EYE MUSCLES

The **extrinsic muscles** move the eyeball within the eye socket and are attached to the eyeball's outer surface. The six ribbonlike extrinsic muscles connected with each eye originate on the orbital bones and insert on the surface of the sclera (FIG. 10-3). They are named for their location and the direction of the muscle fibers. These muscles pull on the eyeball in a coordinated fashion so that both eyes center on one visual field. This process of **convergence** is necessary to the production of a clear retinal image. Having the image come from a slightly different angle from each retina is believed to be important for three-dimensional (stereoscopic) vision, a characteristic of primates.

NERVE SUPPLY TO THE EYE

Two sensory cranial nerves supply the eye (see FIG. 9-11):

- The optic nerve (cranial nerve II) carries visual impulses from the eye's photoreceptors to the brain.
- The ophthalmic (of-THAL-mik) branch of the trigeminal nerve (cranial nerve V) carries impulses of pain, touch, and temperature from the eye and surrounding parts to the brain.

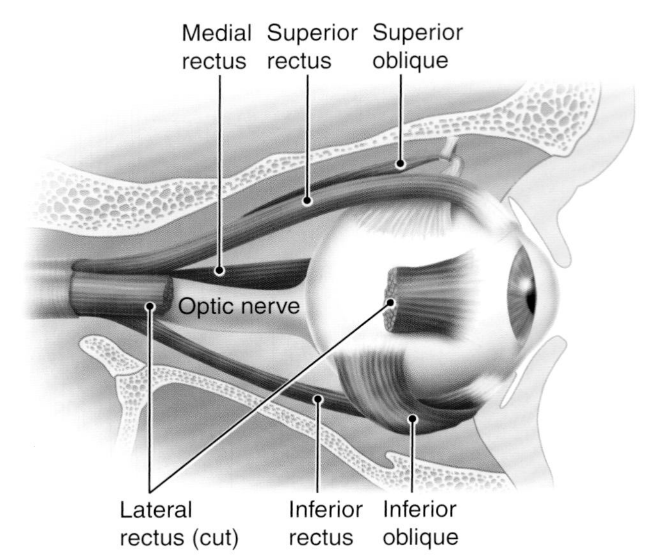

FIGURE 10-3 Extrinsic muscles of the eye. KEY POINT The extrinsic muscles coordinate eye movements for proper vision. ZOOMING IN Describe the fiber directions of a rectus muscle and of an oblique muscle.

Three cranial nerves carry motor impulses to the eyeball muscles (see FIG. 9-11):

- The oculomotor nerve (cranial nerve III) is the largest nerve; it supplies voluntary and involuntary motor impulses to all but two eye muscles.
- The trochlear nerve (cranial nerve IV) supplies the superior oblique extrinsic eye muscle.
- The abducens nerve (cranial nerve VI) supplies the lateral rectus extrinsic eye muscle.

CHECKPOINTS 

- ☐ **10-5** What are five structures that protect the eye?
- ☐ **10-6** What is the function of the extrinsic eye muscles?
- ☐ **10-7** Which cranial nerve carries impulses from the retina to the brain?

STRUCTURE OF THE EYEBALL

In the embryo, the eye develops as an outpocketing of the brain, a process that begins at about 22 days of development. The eyeball has three separate coats or tunics. The components of these tunics are shown in **FIGURE 10-4**.

1. The outermost tunic is the fibrous tunic. It consists mainly of the **sclera** (SKLE-rah), which is made of tough connective tissue and is commonly referred to as the *white of the eye*. It appears white because of the collagen it contains and because it has no blood vessels to add color. (Reddened or "bloodshot" eyes result from inflammation and swelling of blood vessels in the conjunctiva.) The anterior portion of the fibrous tunic is the forward-curving, transparent, and colorless **cornea** (KOR-ne-ah).
2. The middle tunic is the vascular tunic, consisting mainly of the **choroid** (KO-royd). This layer is composed of a delicate network of connective tissue interlaced with many blood vessels. Its dark coloration reflects the presence of **melanin**, a dark brown pigment. Melanin absorbs light rays and prevents them from reflecting

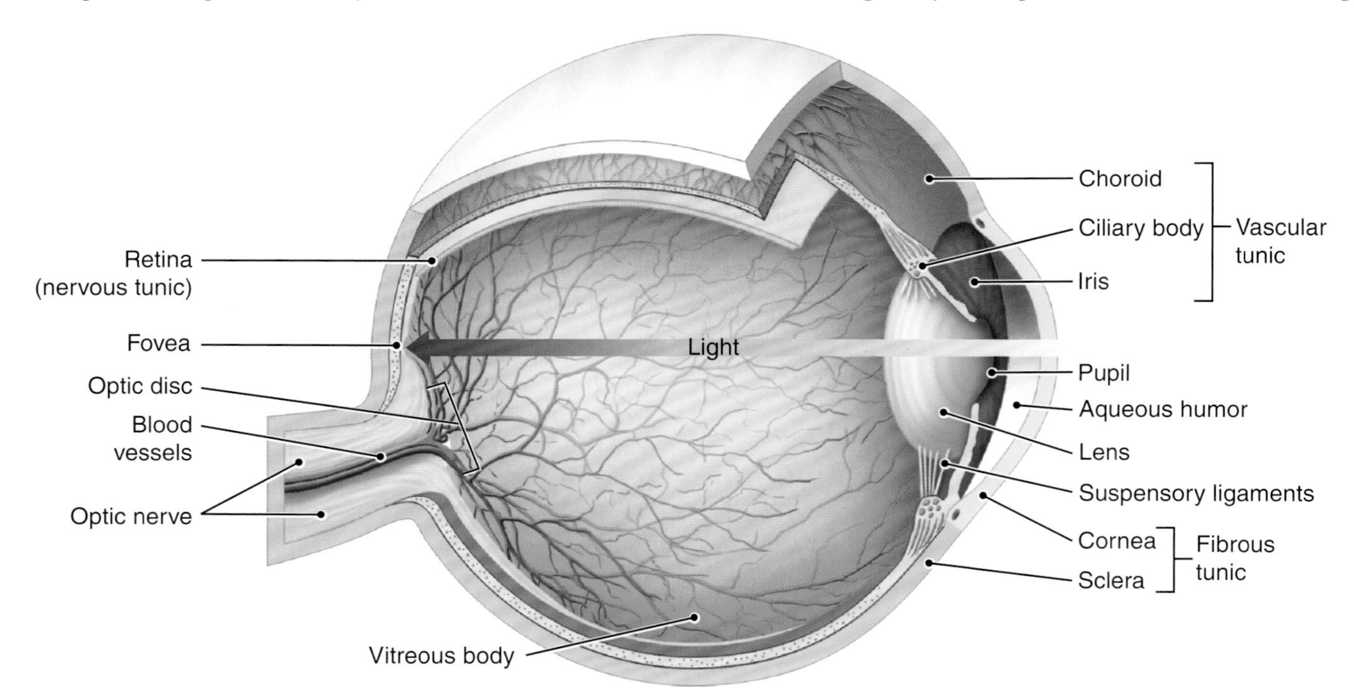

FIGURE 10-4 The eye. KEY POINT The eye has three tunics or coats—the sclera, choroid, and retina. Its refractive parts are the cornea, aqueous humor, lens, and vitreous body. These and other structures involved in vision are shown. ZOOMING IN What anterior structure is continuous with the sclera?

within the eye, much like the black grease athletes used under their eyes to reduce sun glare. At the eye's anterior, the vascular tunic continues as the **ciliary** (SIL-e-ar-e) **muscle** and **suspensory ligaments** (which control the shape of the **lens**, described shortly) and the **iris** (I-ris), the colored, ringlike portion of the eye.

Concept Mastery Alert

Remember that the iris is anterior to the lens but is not part of it.

3. The innermost coat is the nervous tunic, consisting of the **retina** (RET-ih-nah), the eye's actual receptor layer. The retina contains light-sensitive cells known as **rods** and **cones**, which generate the nerve impulses associated with vision. The neural tunic covers only the posterior surface of the eye.

PATHWAY OF LIGHT RAYS AND REFRACTION

As light rays pass through the eye toward the retina, they travel through a series of transparent, colorless parts described below and seen in **FIGURE 10-4**. On the way, they undergo a process known as **refraction**, which is the bending of light rays as they pass from one substance to another substance of different densities. (For a simple demonstration of refraction, place a spoon into a glass of water, and observe how the handle appears to bend at the surface of the water.) Because of refraction, light from a very large area can be focused on a very small area of the retina. As light travels from the environment to the retina, it passes through the following refractory structures:

1. The transparent cornea curves forward slightly and is the eye's main refracting structure. The cornea has no blood vessels; it is nourished by the fluids that constantly bathe it.
2. The **aqueous** (A-kwe-us) **humor**, a watery fluid that fills much of the eyeball anterior to the lens, helps maintain the cornea's convex curve. The aqueous humor is constantly produced and drained from the eye.
3. The lens, technically called the *crystalline lens*, is a clear, circular structure made of a firm, elastic material. The lens has two outward-curving surfaces and is thus described as biconvex. The lens is important in light refraction because its thickness can be adjusted to focus light for near or far vision.
4. The **vitreous** (VIT-re-us) **body** is a soft jelly-like substance that fills the entire space posterior to the lens (the adjective *vitreous* means "glasslike"). Like the aqueous humor, it helps maintain the shape of the eyeball.

Accommodation **Accommodation** is the process of adjusting lens thickness to allow for vision at near and far distances. It involves three structures of the vascular tunic: the ciliary muscle, the suspensory ligaments, and the lens. The doughnut-shaped ciliary muscle sits posterior to the iris, surrounding the lens. Its central hole is about the same size as the entire iris. Suspensory ligaments extend from the lens periphery to the inner surface of the ciliary muscle, similar to the springs joining a trampoline to its wire frame (**FIG. 10-5**).

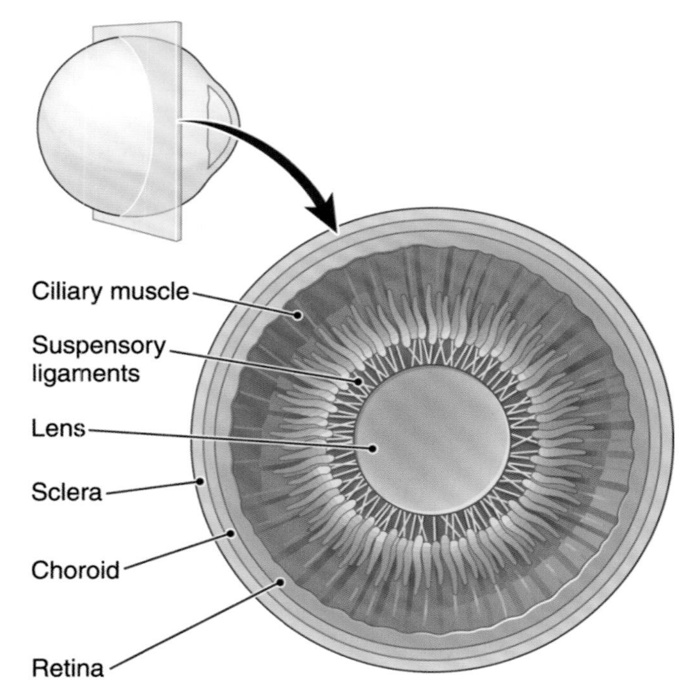

FIGURE 10-5 The ciliary muscle and lens (posterior view). **KEY POINT** Contraction of the ciliary muscle relaxes tension on the suspensory ligaments, allowing the lens to become rounder for near vision. **ZOOMING IN** What structures hold the lens in place?

Accommodation occurs as follows (**FIG. 10-6**). The ciliary muscle is a smooth muscle, so the cells are thin and long when relaxed but short and wide when contracted. For distant vision, the ciliary muscle relaxes and thins, enlarging the central opening. This enlarged opening pulls on the suspensory ligaments, keeping the lens in a more flattened shape. Light rays from a distant object do not require much refraction, so the flattened lens perfectly focuses the light rays on the retina (**see FIG. 10-6A**). For close vision, the ciliary muscle contracts and fattens, relaxing tension on the suspensory ligaments. The elastic lens then recoils and becomes thicker, in much the same way that a rubber band thickens when we release the pull on it. The thickened lens refracts the light to a greater extent, focusing the image from a near object on the retina (**see FIG. 10-6B**). Eyeglasses and contact lenses are used to correct problems with refraction. See **BOX 10-1** for more information.

Function of the Iris The iris is the pigmented ring that gives an eye its distinctive color. It is composed of two sets of muscle fibers that govern the size of the iris's central opening, the **pupil** (PU-pil) (**FIG. 10-7**). One set of fibers is arranged in a circular fashion, and the other set extends radially like the spokes of a wheel. The iris regulates the amount of light entering the eye. In bright light, the iris's circular muscle fibers contract (and the radial fibers relax), reducing the size of the pupil. This narrowing is termed *constriction*. In contrast, in dim light, the radial fibers contract (and the circular fibers relax), pulling the opening outward and enlarging it. This enlargement of the pupil is known as *dilation*.

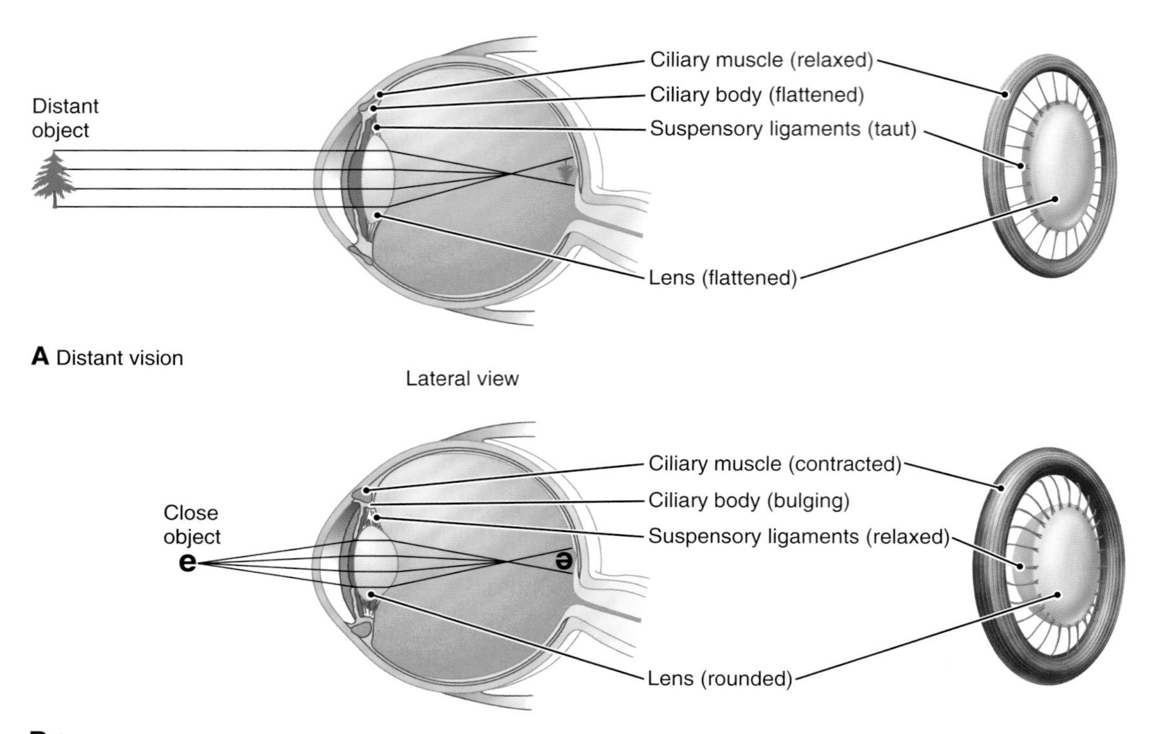

FIGURE 10-6 Accommodation. KEY POINT When viewing a close object, the lens must become more rounded to focus light rays on the retina. **A.** Distant vision. The lens is flattened to lessen refraction when viewing distant objects. **B.** Close vision. The lens is rounded to increase refraction when viewing close objects. ZOOMING IN When the lens is rounded, what is the position of the ciliary muscle?

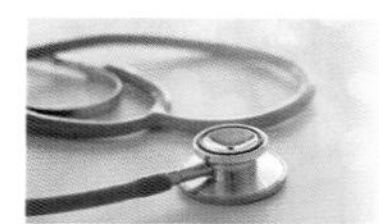

CLINICAL PERSPECTIVES
Corrective Lenses

BOX 10-1

An important part of routine health care is the vision test, in which the patient reads off letters from a poster about 20 feet away. This test investigates one aspect of vision—the ability of the lens to refract (bend) light rays just enough so that they converge on the retina **(see FIG. A)**. The point of convergence is known as the *focal point*. The focusing of light rays in front of or behind the retina results in a blurry image.

Farsightedness, or *hyperopia*, usually results from an abnormally short eyeball. In this situation, light rays focus behind the retina because the lens is unable to bend them sharply enough to focus on the retina **(see FIG. B)**. To see a near object clearly, a person must move it away from the eye so that the image falls on the retina. A common age-related change, known as *presbyopia* (literally "old eye disorder"), also impacts close vision. In this case, the lens loses elasticity and cannot curve enough to refract light properly.

In the case of nearsightedness, or *myopia*, the eyeball is too long or the cornea bends light rays too sharply, and the focal point falls in front of the retina **(see FIG. D)**. Distant objects appear blurry, and only near objects can be seen clearly.

Farsightedness or presbyopia can be corrected with a convex eyeglass or contact lens that increases the angle of light refraction to focus on the retina. Nearsightedness can be corrected with a concave eyeglass or contact lens that causes light rays to diverge and moves the focal point farther back to focus on the retina **(see FIGS. C and E)**.

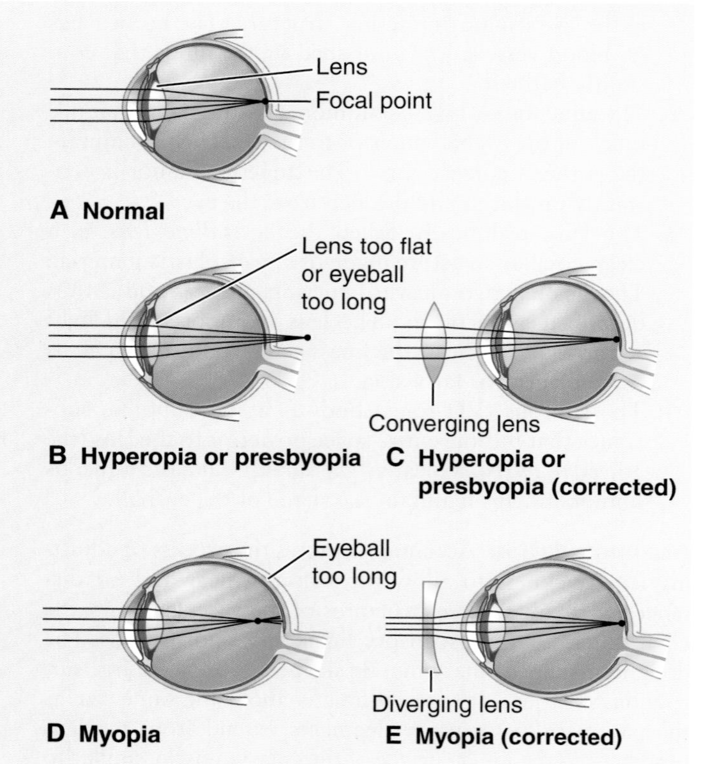

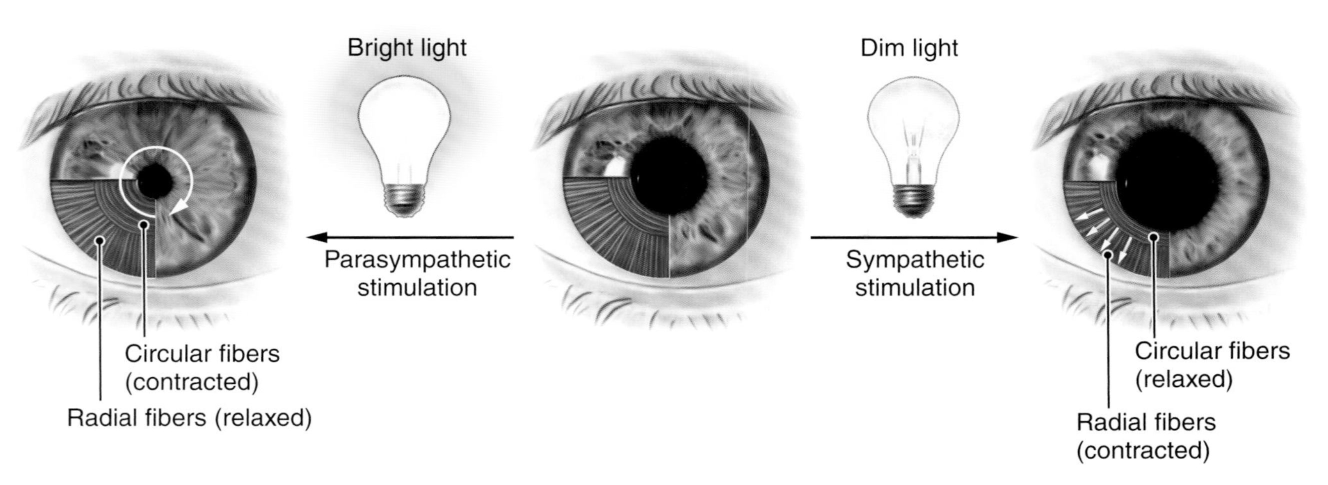

FIGURE 10-7 Function of the iris. KEY POINT In bright light, circular muscles contract and constrict the pupil, limiting the light that enters the eye. In dim light, the radial muscles contract and dilate the pupil, allowing more light to enter the eye. ZOOMING IN What muscle fibers of the iris contract to make the pupil smaller and to make the pupil larger?

Concept Mastery Alert

Students often confuse the functions of the iris and the lens. The lens regulates accommodation, that is, light refraction. The iris controls how much light reaches the lens, but does not function in accommodation.

FUNCTION OF THE RETINA

The retina has a complex structure with multiple layers of cells (FIG. 10-8). The deepest layer is a pigmented layer just superficial to the choroid. Next are the rods and cones, the eye's receptor cells, named for their respective shapes.

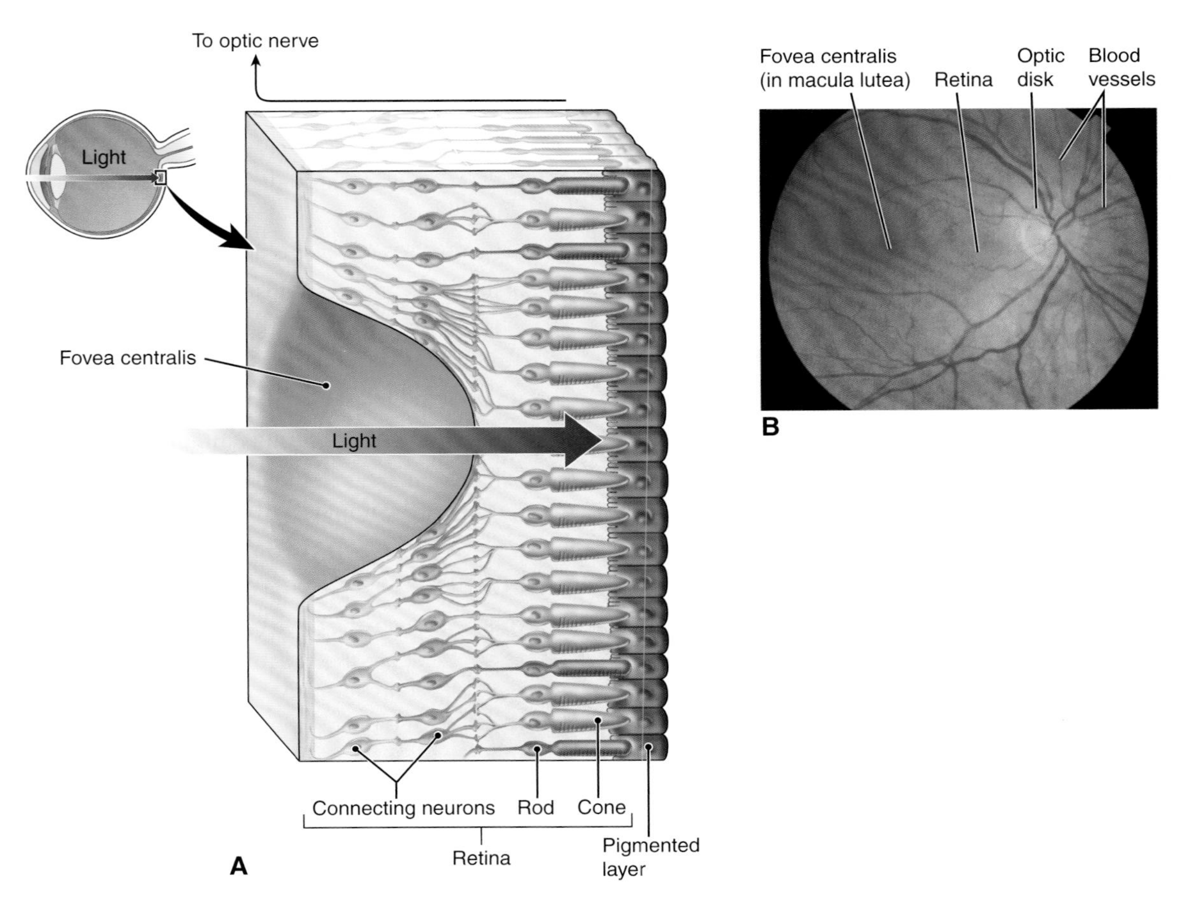

FIGURE 10-8 The retina. A. Structure of the retina. Rods and cones form a deep layer of the retina near the choroid. Connecting neurons carry visual impulses toward the optic nerve. **B.** The fundus (back) of the eye as seen through an ophthalmoscope. An abnormal appearance of the fundus can indicate disease.

Table 10-1 Comparison of the Rods and Cones of the Retina

Characteristic	Rods	Cones
Shape	Cylindrical	Flask shaped
Number	About 120 million in each retina	About 6 million in each retina
Distribution	Toward the periphery (anterior) of the retina	Concentrated at the center of the retina
Stimulus	Dim light	Bright light
Visual acuity (sharpness)	Low	High
Pigments	Rhodopsin	Pigments sensitive to red, green, or blue
Color perception	None; shades of gray	Respond to color

TABLE 10-1 lists the differences between these two cell types. Superficial to the rods and cones are connecting neurons that carry impulses toward the optic nerve. Light rays must pass through these connecting neurons before they can activate the photoreceptors of the retina.

The optic nerve arises from the retina a little toward the medial or nasal side of the eye. There are no photoreceptors in the area of the optic nerve. Consequently, no image can form on the retina at this point, which is known as the blind spot or **optic disk** (see **FIG. 10-4**).

The optic nerve transmits impulses from the photoreceptors to the thalamus (part of the diencephalon), from which they are directed to the occipital cortex. Note that the light rays passing through the eye are actually over-refracted (overly bent) so that an image falls on the retina upside down and backward (see **FIG. 10-6**). It is the job of the brain's visual centers to invert the images.

The rods are highly sensitive to light and thus function best in dim light, but they do not provide a sharp image or differentiate colors. They are more numerous than the cones and are distributed more toward the periphery (anterior portion) of the retina. (If you visualize the retina as the inside of a bowl, the rods would be located toward the bowl's lip.) When you enter into dim light, such as a darkened movie theater, you cannot see for a short period. It is during this time that the rods are beginning to function well, a change that is described as **dark adaptation**. When you are able to see again, images are indistinct and appear only in shades of gray.

The cones function best in bright light, are sensitive to color, and give sharp images. The cones are localized at the retinal center, especially in a tiny depressed area near the optic nerve that is called the **fovea centralis** (FO-ve-ah sen-TRA-lis) (see **FIG. 10-8**; also **FIG. 10-4**). (Note that *fovea* is a general term for a pit or depression.) Because this area contains the highest concentration of cones, it is the point of the sharpest vision. In addition, all of the neurons that connect the rods and cones to the optic nerve are displaced away from this region so that the maximum amount of light reaches the cones. The fovea is contained within a yellowish spot, the **macula lutea** (MAK-u-lah LU-te-ah).

There are three types of cones, each sensitive to red, green, or blue light. Color blindness results from a deficiency of retinal cones. People who completely lack cones are totally color blind; those who lack one type of cone are partially color blind. For instance, individuals without green cones cannot distinguish green from red (red–green color blindness). This disorder, because of its pattern of inheritance, occurs much more commonly in males.

The rods and cones function by means of pigments that are sensitive to light. The light-sensitive pigment in rods is **rhodopsin** (ro-DOP-sin), a complex molecule synthesized from vitamin A. If a person lacks vitamin A, and thus rhodopsin, he or she may have difficulty seeing in dim light, because the rods cannot be activated; this condition is termed **night blindness**. Nerve impulses from the rods and cones flow into sensory neurons that gather to form the optic nerve (cranial nerve II) at the eye's posterior (see **FIG. 10-4**). The impulses travel to the visual center in the brain's occipital cortex.

When an **ophthalmologist** (of-thal-MOL-o-jist), a physician who specializes in treatment of the eye, examines the retina with an **ophthalmoscope** (of-THAL-mo-skope), he or she can see abnormalities in the retina and in the retinal blood vessels (see **FIG. 10-8B**). Some of these changes may signal more widespread diseases that affect the eye, such as diabetes and high blood pressure (hypertension).

THE VISUAL PROCESS

To summarize, the events required for proper vision (some of which may be occurring simultaneously) are as follows:

- The extrinsic eye muscles produce convergence.
- Light refracts through the cornea and the aqueous humor.
- The muscles of the iris adjust the pupil.
- The ciliary muscle adjusts the lens (accommodation).
- The light continues to refract through the vitreous body and passes through the superficial layers of the retina.
- Light stimulates retinal receptor cells (rods and cones).
- The optic nerve transmits impulses to the brain.
- The visual areas in the occipital lobe cortex receive and interpret the impulses.

CHECKPOINTS 

- ☐ **10-8** What are the three tunics of the eyeball?
- ☐ **10-9** What are the structures that refract light as it passes through the eye?
- ☐ **10-10** What is the function of the ciliary muscle?
- ☐ **10-11** What is the function of the iris?
- ☐ **10-12** What are the receptor cells of the retina?

See the "Student Resources" on thePoint® to view the animation "Eye Structure and Function."

The Ear

The ear is the sense organ for both hearing and equilibrium (FIG. 10-9). It is divided into three main sections:

- The **outer ear** includes an outer projection and a canal ending at a membrane.
- The **middle ear** is an air space containing three small bones.
- The **inner ear** is the most complex and contains the sensory receptors for hearing and equilibrium.

THE OUTER EAR

The external portion of the ear consists of a visible projecting portion, the **pinna** (PIN-nah), also called the *auricle* (AW-rih-kl), and the **external auditory canal**, or *meatus* (me-A-tus), that leads into the ear's deeper parts. The pinna directs sound waves into the ear, but it is probably of little importance in humans. The external auditory canal extends medially from the pinna for about 2.5 cm or more, depending on which wall of the canal is measured. The skin lining this tube is thin and, in the first part of the canal, contains many wax-producing **ceruminous** (seh-RU-mih-nus) **glands**. The wax, or **cerumen** (seh-RU-men), may become dried and impacted in the canal and must then be removed. The same kinds of disorders that involve the skin elsewhere—atopic dermatitis, boils, and other infections—may also affect the skin of the external auditory canal.

The **tympanic** (tim-PAN-ik) **membrane**, or *eardrum*, is at the end of the external auditory canal and separates this canal from the middle ear cavity. The tympanic membrane vibrates freely when struck with sound waves that enter the ear.

THE MIDDLE EAR AND OSSICLES

The middle ear cavity is a small, flattened space that contains three small bones, or **ossicles** (OS-ih-klz) (see FIG. 10-9). The three ossicles are joined in such a way that they amplify the sound waves received by the tympanic membrane as they transmit the sounds to the inner ear. The first bone is shaped like a hammer (or mallet) and is called the **malleus** (MAL-e-us). The handle-like part of the malleus is attached to the tympanic membrane, whereas the headlike part is connected to the second bone, the **incus** (ING-kus). The incus is shaped like an anvil, an iron block used by blacksmiths to shape metal. The innermost ossicle is shaped somewhat like the stirrup of a saddle and is called the **stapes** (STA-peze), which is Latin for *stirrup*. The base of the stapes is in contact with the inner ear.

The **auditory tube**, also called the *eustachian* (u-STA-shun) *tube*, connects the middle ear cavity with the throat or pharynx (FAR-inks) (see FIG. 10-9). This tube opens to allow pressure to equalize on the two sides of the tympanic membrane. A valve that closes the tube can be forced open by swallowing hard, yawning, or blowing with the

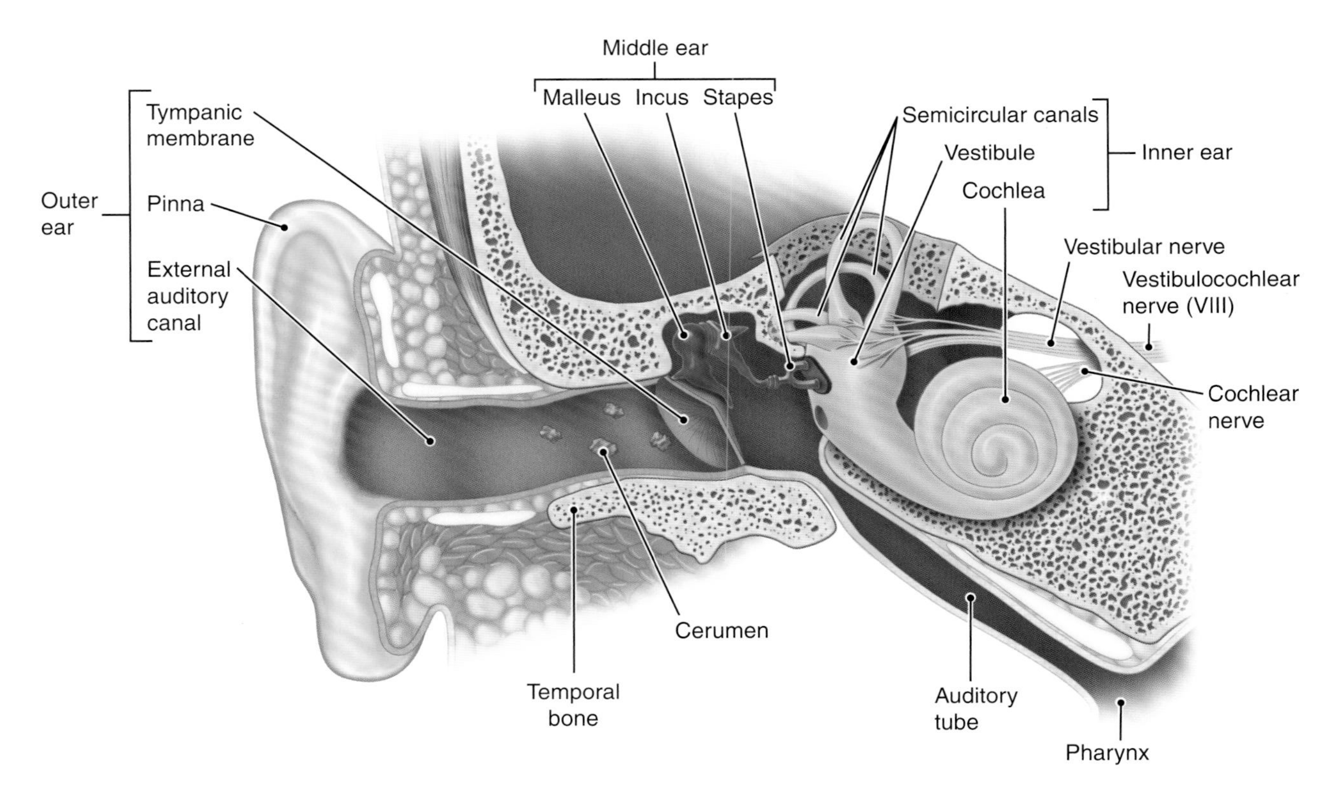

FIGURE 10-9 The ear. KEY POINT Structures of the outer, middle, and inner ear amplify and detect sound waves. ZOOMING IN What structure separates the outer ear from the middle ear?

nose and mouth sealed, as a person often does when experiencing pain from pressure changes in an airplane.

The mucous membrane of the pharynx is continuous through the auditory tube into the middle ear cavity. The posterior wall of the middle ear cavity contains an opening into the mastoid air cells, which are spaces inside the temporal bone's mastoid process **(see FIG. 6-6B)**.

THE INNER EAR

The ear's most complicated and important part is the internal portion, which is described as a *labyrinth* (LAB-ih-rinth) because it has a complex, mazelike construction **(FIG. 10-10)**. The outer shell of the inner ear is composed of hollow bone comprising the **bony labyrinth**. This outer portion is filled with a fluid called **perilymph** (PER-e-limf).

Within the bony labyrinth is an exact replica of this bony shell made of membrane, much like an inner tube within a tire. The tubes and chambers of this **membranous labyrinth** are filled with a fluid called **endolymph** (EN-do-limf) **(see FIG. 10-10A)**. Thus, the endolymph is within the membranous labyrinth, and the perilymph surrounds the membranous labyrinth. These fluids are important to the sensory functions of the inner ear. The inner ear has three divisions:

- The **vestibule** consists of two chambers (the utricle and the saccule) that contain some of the receptors for equilibrium.

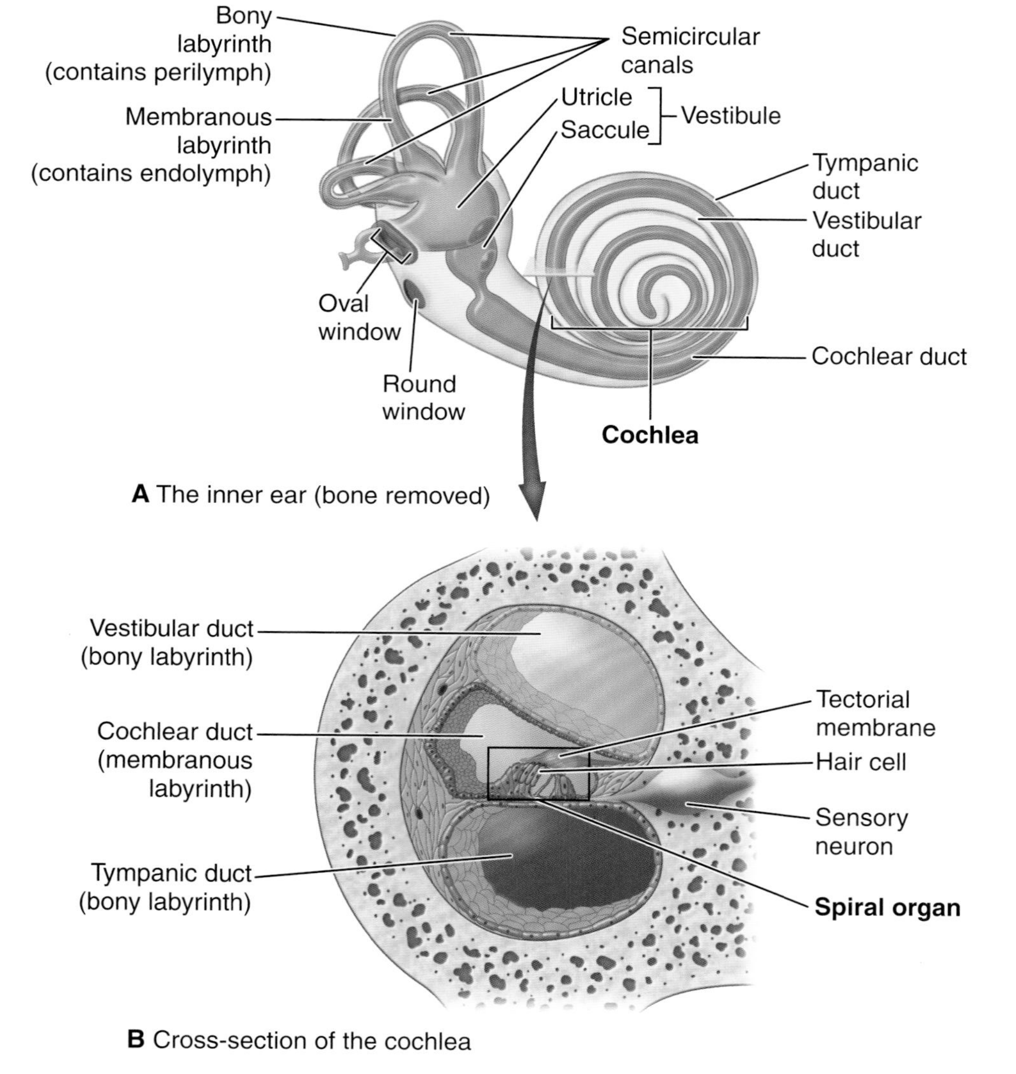

FIGURE 10-10 The inner ear. **KEY POINT** The vestibule, semicircular canals, and cochlea are made of a bony shell, described as a bony labyrinth, with an interior membranous labyrinth. Endolymph fills the membranous labyrinth, and perilymph surrounds it in the bony labyrinth. The cochlea is the organ of hearing. The semicircular canals and vestibule are concerned with equilibrium. **ZOOMING IN** What type of fluid is in contact with the membrane lining the vestibular duct?

- The **semicircular canals** are three projecting tubes located toward the posterior. The area at the base of each semicircular canal contains receptors for equilibrium.
- The **cochlea** (KOK-le-ah) is coiled like a snail shell (*cochlea* is Latin for "snail"). It contains the receptors for hearing.

HEARING

Within the cochlea, the membranous labyrinth is known as the **cochlear duct** (see FIG. 10-10A and B). It bisects the bony labyrinth into a superior portion, the **vestibular duct**, and an inferior portion, the **tympanic duct. Hair cells,** the receptors for hearing, sit on the lower membrane of the cochlear duct. The long strip of hair cells is also known as the **spiral organ** (organ of Corti [KOR-te]). The tips of the hair cells are embedded in a gelatinous membrane called the **tectorial membrane.** (The membrane is named from a Latin word that means "roof.")

The numbers below match the numbers in FIGURE 10-11. The cochlea is pictured as unrolled to more easily show how

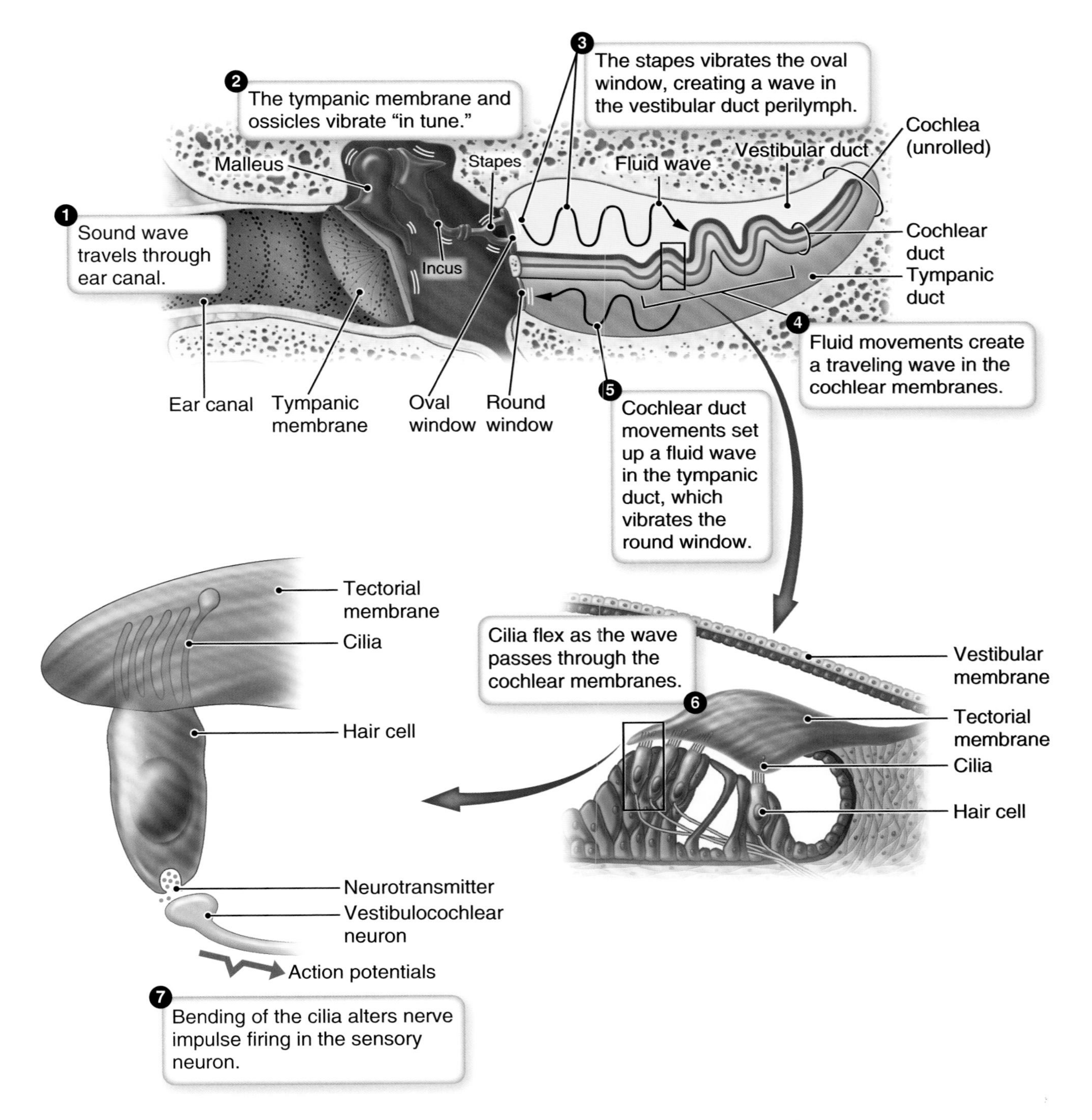

FIGURE 10-11 The mechanics of hearing. The bullets refer to numbered steps in the narrative. ZOOMING IN What membrane contains the cilia of the hair cells?

sound waves progress through the inner ear chambers. The steps in hearing are as follows:

1. Sound waves first enter the external auditory canal. The pitch (high or low) and amplitude (loud or soft) depend on the characteristic of the wave.
2. The sound waves set up vibrations in the tympanic membrane. The ossicles amplify these vibrations, which can be large (loud) or small (soft) and fast (high pitch) or slow (low pitch).
3. The stapes transmits the vibrations to the **oval window** of the inner ear. This membrane then transmits the sound waves to the perilymph within the vestibular duct.
4. This fluid wave in the vestibular duct sets up vibrations in the membranes of the cochlear duct. High-pitched sounds cause the proximal portions of the cochlear membranes to vibrate, whereas low-pitched sounds initiate vibrations in the distal portion.
5. The cochlear duct vibrations initiate a second fluid wave in the tympanic duct. This wave dissipates when it hits the round window.
6. Recall that the hair cells perch upon the lower cochlear duct membrane with their tips (cilia) embedded in the tectorial membrane. Vibrations in these membranes move the cilia back and forth.
7. Ciliar movement produces an electrical signal in the hair cell.
8. This motion sets up nerve impulses that travel to the brain via neurons of the **vestibulocochlear nerve**, a branch of the eighth cranial nerve shown in **FIGURE 10-9** (formerly called the *auditory* or *acoustic nerve*).

See **BOX 10-2** for hints about mastering the steps of the hearing process.

Hearing receptors respond to both the pitch (tone) of sound and its intensity (loudness). Loud sounds stimulate more cells and produce more vibrations, sending more nerve impulses to the brain. Exposure to loud noises, such as very loud music, jet plane noise, or industrial noises, can damage the receptors for particular pitches of sound and lead to hearing loss for those tones.

See the Student Resources on thePoint® for an animation on hearing and for information on how audiologists treat hearing disorders.

CASEPOINTS 

- ☐ **10-2** What cells were damaged in Evan's case?
- ☐ **10-3** Would a cochlear implant have worked for Evan if the eighth cranial nerve was damaged? Explain.

ONE STEP AT A TIME — BOX 10-2
Mastering Multistep Pathways

This chapter discusses two of the most complex processes in the body—hearing and vision. Understanding these processes is important, because it enables us to design effective treatment for sensory disorders. This box provides some strategies that you can use to analyze any multistep pathway.

Question

Hearing aids amplify sound waves. Explain why a hearing aid would be of limited use to correct Evan's deafness.

Answer

Step 1. Learn the steps of the pathway. Start by reading the textbook and looking at the diagrams, but you'll need to employ active learning techniques. For instance:

a. Write out the steps on separate pieces of paper, and practice rearranging them in the correct order.
b. Using an unlabeled drawing of the ear, explain the steps of the hearing pathway to a classmate (or a dog or a wall). Speaking the steps aloud will help you remember. Use language such as "Step X causes step Y" so that you can be sure of the cause–effect relationships.
c. Gather a group of students to act out the steps of the pathway. A student representing a sound wave can shake a student representing the tympanic membrane, who in turn shakes the student representing the stapes, and so on. Perhaps a blanket can represent the sound wave in the perilymph. Be creative in developing your role-playing simulation, and chances are you will remember the pathway for a very long time.

Step 2. Identify the damaged component. In the case study, meningitis damaged Evan's hair cells. Find the hair cells in your list of steps or in your role-playing simulation. They are responsible for the final step of transmitting the signal to the vestibulocochlear nerve.

Step 3. Determine the impact of the damaged component. Remove the slip of paper or actor representing hair cells, and see if your pathway still works. Remember that each step in a pathway depends on the previous step. If one step fails, any subsequent (downstream) step will not occur. In Evan's case, the lack of hair cells prevents movements of the tectorial membrane from activating the cochlear nerve.

Step 4. Identify the action of the medical intervention. A hearing aid amplifies the sound wave, so all of the mechanical changes are bigger. Use your role-playing simulation to visualize the impact of bigger sound waves. All of the events up to tectorial membrane movements are magnified, but the signal cannot reach the nerve. In reality, a hearing aid would help if a few hair cells escaped destruction. However, it would have no effect if all of the hair cells were gone. This is why cochlear implants bypass all of the mechanical events to directly stimulate the nerve.

See the "Study Guide" (available separately) for more practice in understanding and analyzing multistep pathways.

EQUILIBRIUM

The other sensory receptors in the inner ear are those related to equilibrium (balance). They are located in the vestibule and the semicircular canals. Receptors for the sense of equilibrium respond to acceleration and, like the hearing receptors, are ciliated hair cells. As the head moves, a shift in the position of the cilia within a thick material around them generates a nerve impulse.

Receptors located in the vestibule's two small chambers sense the position of the head relative to the force of gravity and also respond to acceleration in a straight line, as in a forward-moving vehicle or an elevator. Each receptor is called a **macula**. (There is also a macula in the eye, but *macula* is a general term that means "spot.") The macular hair cells are embedded in a gelatinous material, the **otolithic** (o-to-LITH-ik) **membrane**, which is weighted with small crystals of calcium carbonate, called **otoliths** (O-to-liths). The force of gravity pulls the membrane downward, which bends the cilia of the hair cells, generating a nerve impulse (**FIG. 10-12**). In linear acceleration, the otolithic membrane lags behind the forward motion, bending the cilia in a direction opposite to the direction of acceleration. Imagine sweeping mud off a garden path with a broom. As you move forward, the thick mud is dragging the broom straws in the opposite direction.

The receptors for detecting rotation, such as when you shake your head or twirl in a circle, are located at the bases

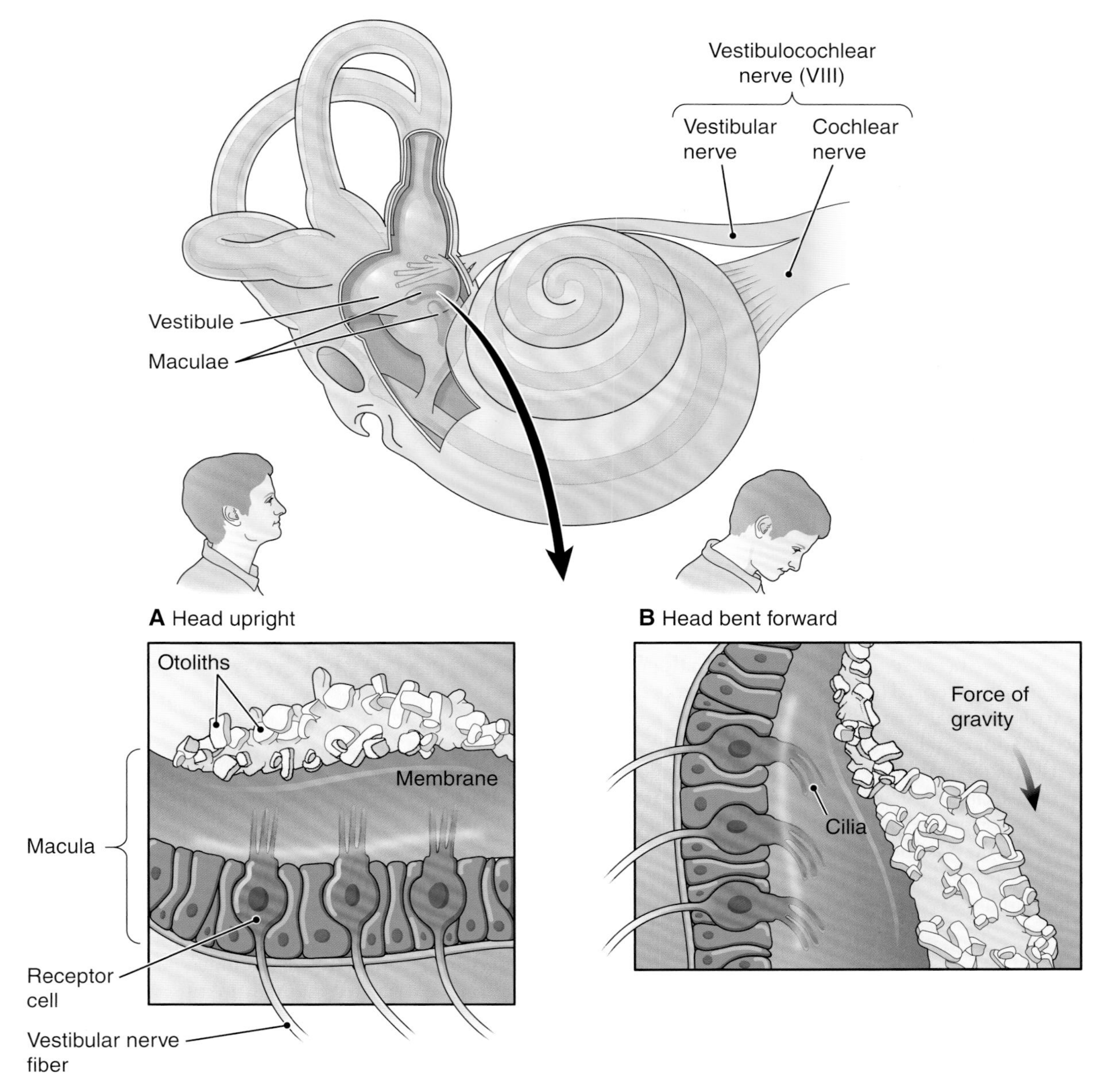

FIGURE 10-12 Action of the vestibular equilibrium receptors (maculae). **KEY POINT** As the head moves, the otolithic membrane, weighted with otoliths, pulls on the receptor cells' cilia, generating a nerve impulse. These receptors also function in linear acceleration. **A.** Position of the hair cells and otoliths when the head is upright. **B.** Their position when the head bends forward. **ZOOMING IN** What happens to the cilia of the macular cells when the otolithic membrane moves?

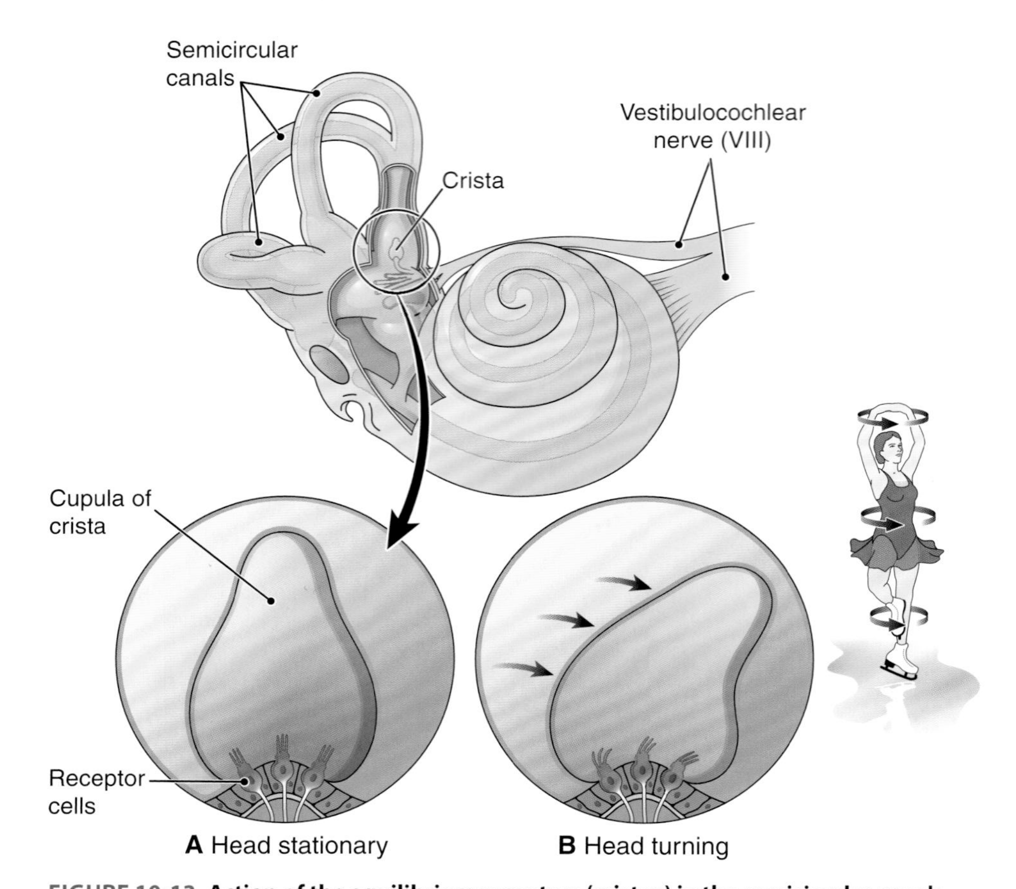

FIGURE 10-13 Action of the equilibrium receptors (cristae) in the semicircular canals. KEY POINT As the body spins or moves in different directions, the receptor cells' cilia bend, generating nerve impulses. **A.** Position of a crista when the head is stationary. **B.** Position when the head is turning.

of the semicircular canals (FIG. 10-13). These receptors, called **cristae** (KRIS-te), are hair cells embedded in a gelatinous material called the *cupula* (KU-pu-lah). As with the maculae, when the head moves, the cupula lags behind a bit, bending the cilia in the opposite direction. It's easy to remember what these receptors do, because the semicircular canals go off in three different directions. The crista in the horizontal canal responds to horizontal rotation, as in a dancer's spin; the one in the superior canal responds to forward and backward rotation, as in somersaulting; and the one in the posterior canal responds to left–right rotation, as in doing a cartwheel.

Nerve fibers from the vestibule and from the semicircular canals form the **vestibular** (ves-TIB-u-lar) **nerve**, which joins the cochlear nerve to form the vestibulocochlear nerve, the eighth cranial nerve (see FIG. 10-13).

CHECKPOINTS 

- ☐ **10-13** What are the three divisions of the ear?
- ☐ **10-14** What are the names of the ear ossicles, and what do they do?
- ☐ **10-15** What are the two fluids found in the inner ear, and where are they located?
- ☐ **10-16** What is the name of the hearing organ, and where is it located?
- ☐ **10-17** Where are the receptors for equilibrium located?

CASEPOINT 

- ☐ **10-4** Would Evan's equilibrium be affected by his disease? Explain.

Other Special Sense Organs

The sense organs of taste and smell respond to chemical stimuli.

SENSE OF TASTE

The sense of taste, or **gustation** (gus-TA-shun), involves receptors in the tongue and two different nerves that carry taste impulses to the brain (FIG. 10-14). The gustatory sensory organs, known as **taste buds**, are located mainly on the superior surface of the tongue. They are enclosed in raised projections called **papillae** (pah-PIL-e), which give the tongue's surface a rough texture and help manipulate food when chewing. Taste buds are stimulated only if the substance to be tasted is in solution or dissolves in the fluids of the mouth. Within each taste bud, modified epithelial cells (the gustatory cells) respond to one of the five basic tastes:

- **Sweet** receptors respond to simple sugars.
- **Salty** receptors respond to sodium.

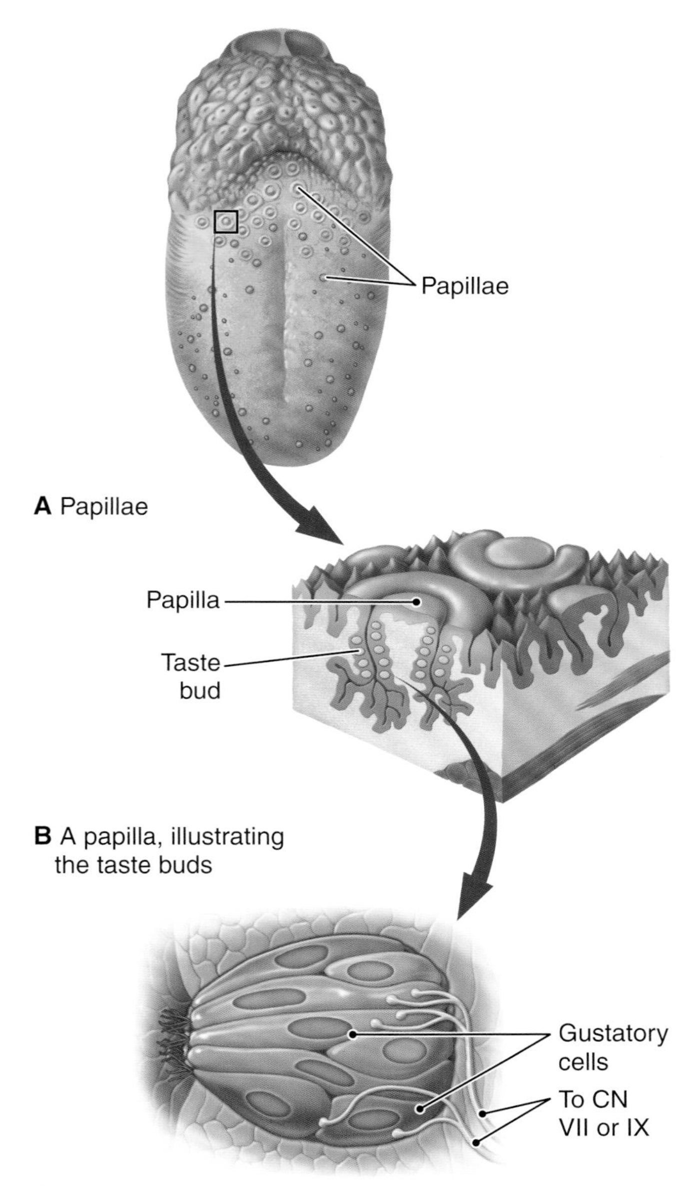

FIGURE 10-14 Taste. KEY POINT **A.** Papillae are small bumps on the tongue. **B.** A papilla containing taste buds. **C.** Anatomy of a taste bud. Gustatory cells in taste buds act as the taste receptors.

- **Sour** receptors detect hydrogen ions.
- **Bitter** receptors respond to various organic compounds.
- **Umami** (u-MOM-e) is a pungent or savory taste based on a response to the amino acids glutamate and aspartate, which add to the meaty taste of protein. Glutamate is found in MSG (monosodium glutamate), a flavor enhancer used in some processed foods and some restaurants.

Some investigators consider spiciness to be a sixth taste, but the chemicals involved (such as capsaicin) activate pain/touch receptors, namely, the trigeminal nerve, rather than specialized gustatory cells. Other tastes are a combination of these five with additional smell sensations. The nerves of taste include the facial and the glossopharyngeal cranial nerves (VII and IX) (see FIG. 10-14). The interpretation of taste impulses is probably accomplished by the brain's lower frontal cortex, although there may be no sharply separate gustatory center.

SENSE OF SMELL

The importance of the sense of smell, or **olfaction** (ol-FAK-shun), is often underestimated. This sense helps detect gases and other harmful substances in the environment and helps warn of spoiled food. Smells can trigger memories and other psychological responses. Smell is also important in sexual behavior.

The olfactory receptor cells are neurons embedded in the epithelium of the nasal cavity's superior region (FIG. 10-15). These neurons extend dendrites into the nasal cavity that interact with smell chemicals (odorants). Again, the chemicals detected must be dissolved in the mucus that lines the nose. Because these receptors are high in the nasal cavity, you must "sniff" to bring odors upward in your nose.

The axons of the olfactory receptor cells pass through the ethmoid bone to synapse with other neurons in the olfactory bulb, the enlarged ending of the olfactory nerve (cranial nerve I). The olfactory nerve carries smell impulses directly to the olfactory center in the brain's temporal cortex as well as to the limbic system.

The interpretation of smell is closely related to the sense of taste, but a greater variety of dissolved chemicals can be detected by smell than by taste. We have hundreds of different types of odor receptors; FIGURE 10-15 illustrates two types. Different odors can also activate specific combinations of receptors so that we can detect over 10,000 different smells. The smell of foods is just as important in stimulating appetite and the flow of digestive juices as is the sense of taste. When you have a cold, food often seems tasteless and unappetizing because nasal congestion reduces your ability to smell the food.

The olfactory receptors deteriorate with age, and food may become less appealing. It is important when presenting food to elderly people that the food look inviting so as to stimulate their appetites.

CHECKPOINT 

☐ **10-18** What are the special senses that respond to chemical stimuli?

The General Senses

Unlike the special sensory receptors, which are localized within specific sense organs and are limited to a relatively small area, the general sensory receptors are scattered throughout the body. These include receptors for touch, pressure, temperature, position, and pain (FIG. 10-16).

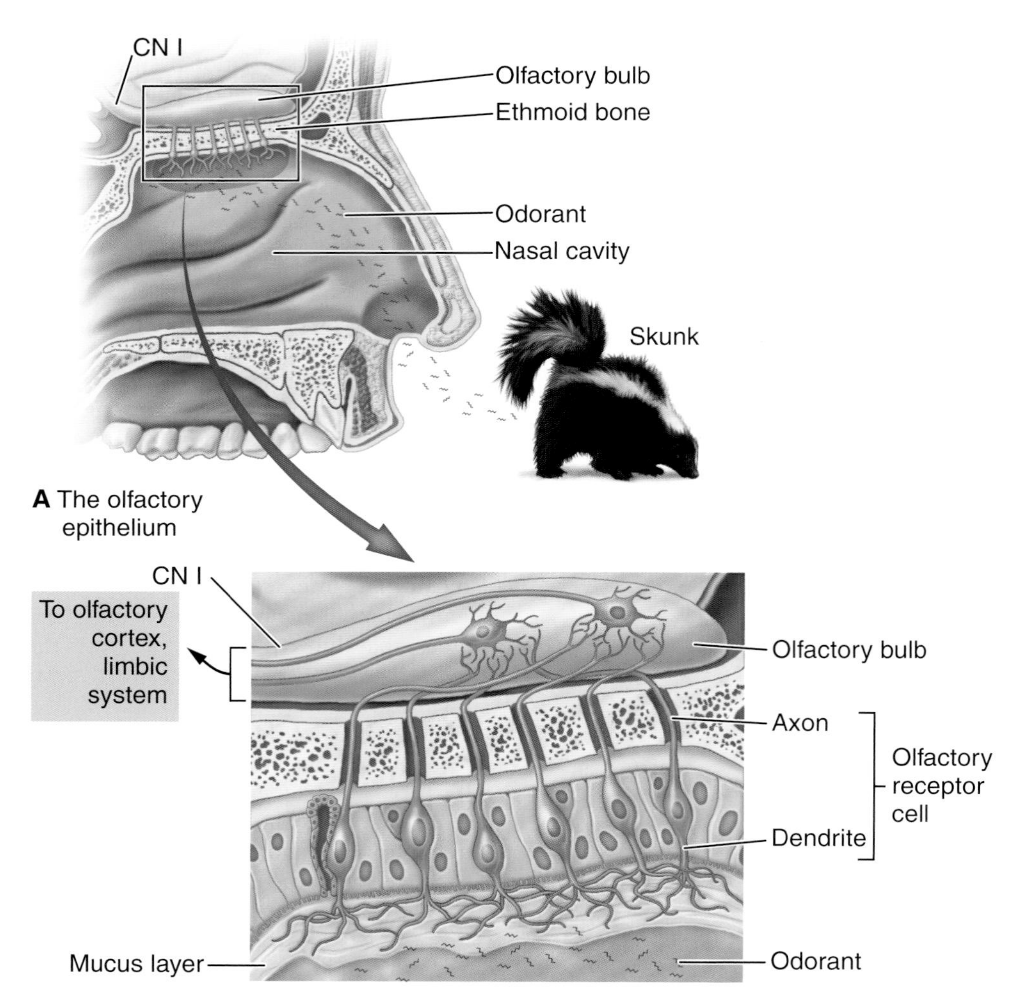

FIGURE 10-15 Smell. A. Olfactory cells in the superior portion of the nasal cavity detect smells. **B.** Olfactory receptor cells detect an odorant and convey a signal to a neuron in the olfactory bulb. ZOOMING IN What part of an olfactory receptor cell interacts with an odorant?

SENSE OF TOUCH

The touch receptors, **tactile** (TAK-til) **corpuscles**, are found mostly in the dermis of the skin and around hair follicles. Touch sensitivity varies with the number of touch receptors in different areas. They are especially numerous and close together in the tips of the fingers and the toes. The lips and the tip of the tongue also contain many of these receptors and are very sensitive to touch. Other areas, such as the back of the hand and the back of the neck, have fewer receptors and are less sensitive to touch.

The sensation of tickle is related to the sense of touch but is still something of a mystery. Tickle receptors are free nerve endings associated with the tactile mechanoreceptors. No one knows the value of tickling, but it may be a form of social interaction. Oddly, we experience tickling only when touched by someone else. Apparently, the brain inhibits these sensations when you are trying to tickle yourself and know the tickling site, eliminating the element of surprise.

SENSE OF PRESSURE

Even when the skin is anesthetized, it can still respond to pressure stimuli. These sensory receptors for deep pressure are located in the subcutaneous tissues beneath the skin and also near joints, muscles, and other deep tissues. They are sometimes referred to as *receptors for deep touch*.

SENSE OF TEMPERATURE

The temperature receptors are **free nerve endings**, receptors that are not enclosed in capsules but are simply branchings of nerve fibers. Temperature receptors are widely distributed in the skin, and there are separate receptors for heat and cold. A warm object stimulates only the heat receptors, and a cool object affects only the cold receptors. Internally, there are temperature receptors in the brain's hypothalamus, which help adjust body temperature according to the temperature of the circulating blood.

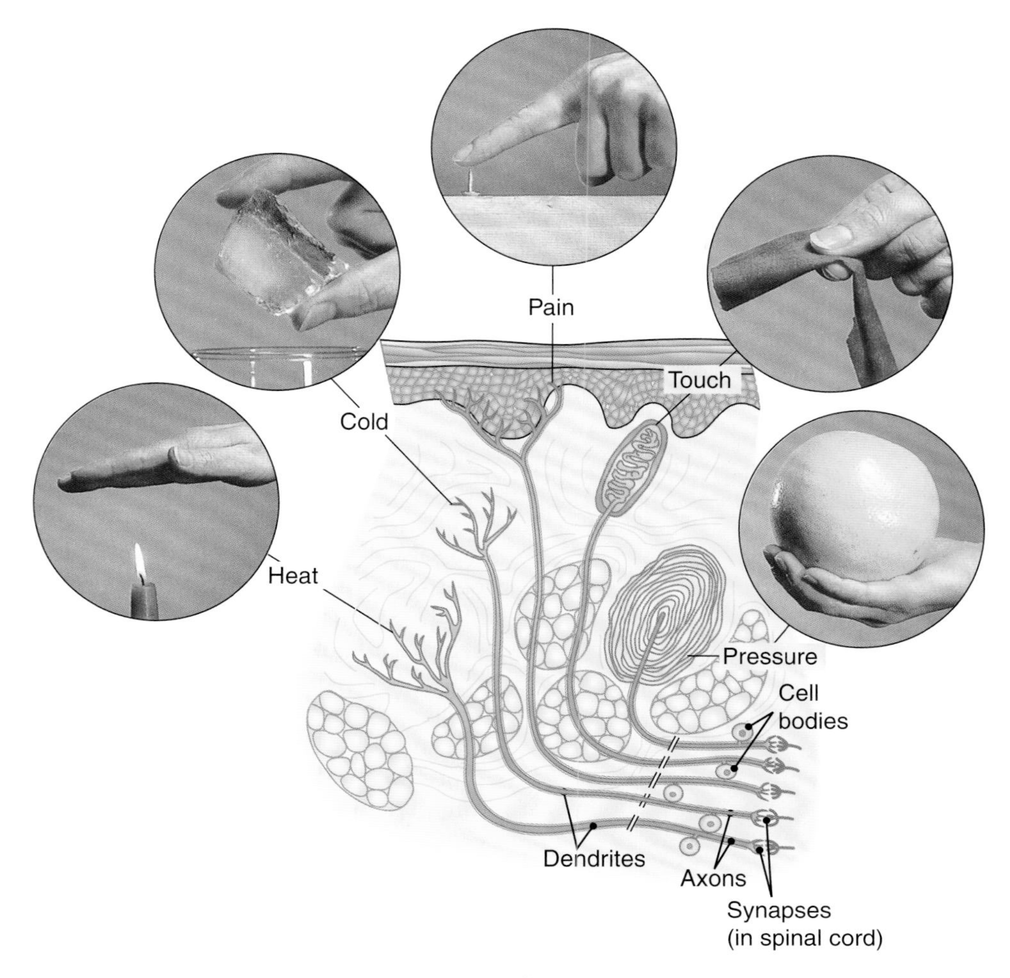

FIGURE 10-16 Sensory receptors in the skin. KEY POINT The skin has a variety of sensory receptors. Synapses with interneurons are in the spinal cord.

SENSE OF POSITION

Receptors located in muscles, tendons, and joints relay impulses that aid in judging body position and relative changes in the locations of body parts. They also inform the brain of the amount of muscle contraction and tendon tension. These rather widespread receptors, known as **proprioceptors** (pro-pre-o-SEP-tors), are aided in this function by the internal ear's equilibrium receptors. The term **kinesthesia** (kin-es-THE-ze-ah) is sometimes used to describe dynamic, or movement-associated, aspects of proprioception.

Information received by proprioceptors is needed for muscle coordination and is important in such activities as walking, running, and many more complicated skills, such as playing a musical instrument. Proprioceptors play an important part in maintaining muscle tone and good posture. They also help assess the weight of an object to be lifted so that the right amount of muscle force is used.

The nerve fibers that carry impulses from these receptors enter the spinal cord and ascend to the brain in the posterior part of the cord. The cerebellum is a main coordinating center for these impulses.

SENSE OF PAIN

Pain is the most important protective sense. The receptors for pain are widely distributed free nerve endings. They are found in the skin, muscles, and joints and to a lesser extent in most internal organs (including the blood vessels and viscera). Two pathways transmit pain to the CNS. One is for acute, sharp pain, and the other is for slow, chronic pain. Thus, a single strong stimulus can produce an immediate sharp pain, followed in a second or so by a slow, diffuse pain that increases in severity with time.

Referred Pain Sometimes, pain that originates in an internal organ is experienced as coming from a more superficial part of the body, particularly the skin. This phenomenon is known as *referred pain*. Liver and gallbladder diseases often cause referred pain in the skin over the right shoulder. Spasm of the coronary arteries that supply the heart may cause pain in the left shoulder and arm. Infection of the appendix can be felt as pain of the skin covering the lower right abdominal quadrant.

Apparently, some interneurons in the spinal cord have the twofold duty of conducting impulses from visceral pain receptors in the chest and abdomen and from somatic pain receptors in neighboring areas of the skin, resulting in referred pain. The brain cannot differentiate between these two possible sources, but because most pain sensations originate in the skin, the brain automatically assigns the pain to this more likely place of origin. Knowing where visceral pain is referred to in the body is of great value in diagnosing chest and abdominal disorders.

Itch Itch receptors are free nerve endings that may be specific for that sensation or may share pathways with other receptors, such as those for pain. There are multiple causes for itching, including skin disorders, allergies, kidney disease, infection, and a host of chemicals. Usually, itching is a mild, short-lived annoyance, but for some, it can be chronic and debilitating. No one knows why scratching helps alleviate itch. It may replace the itch sensation with pain or send signals to the brain to relieve the sensation.

CHECKPOINTS 

- ☐ **10-19** What are the five examples of general senses?
- ☐ **10-20** What are proprioceptors, and where are they located?

A & P in Action Revisited: Evan's Cochlear Implant

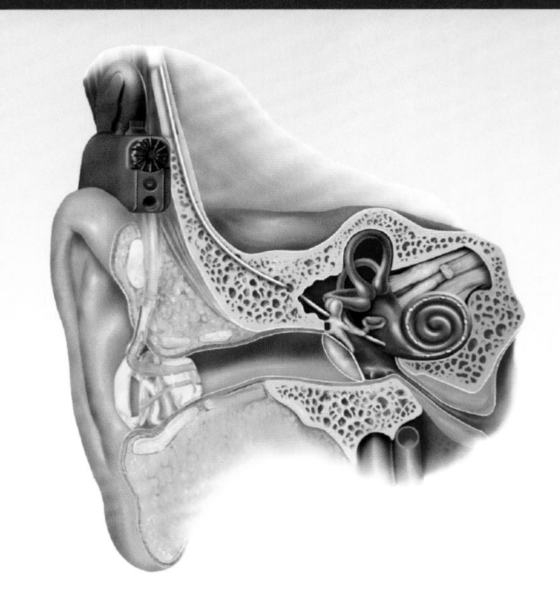

"Hi. How are we doing today?" asked Dr. Sanchez. Evan had undergone a right cochlear implant 4 weeks ago and was in for another follow-up visit. He appeared to be a normal, happy 20-month-old sitting on his mother's lap. A couple of months earlier, meningitis had destroyed many of the essential hair cells in Evan's cochlea. The drugs used to treat the meningitis may also have been ototoxic, contributing to further destruction of these cells.

Dr. Sanchez explained, "Evan is ready to have his external fitting. If you recall, the implant has three main components: the internal receiver that we implanted behind Evan's ear with electrodes going to the inner ear, and the external parts, a transmitter and a sound processor. These components allow the brain to interpret the frequency of sound as it would if the hair cells were functioning properly. Today, we are going to activate the implant by connecting the processor to the internal device. Please understand that an implant is not as sensitive as one's normal hearing. Evan will need some help from a speech therapist and audiologist, but we're sure he will make progress. If all goes as expected, the device will complement the limited hearing in his left ear, and he will not need a second implant."

"This entire process has been very difficult for us all," said Evan's father. "We understand the limitations you have explained, but to think that Evan will be able to hear again is truly thrilling for us!"

In this case, we saw how the sense of hearing can be compromised by damaged hair cells in the cochlea, but new and advancing technology is helping to treat sensorineural hearing loss.

CHAPTER 10

Chapter Wrap-Up

OVERVIEW

A detailed chapter outline with space for note-taking is on thePoint®. The figure below illustrates the main topics covered in this chapter.

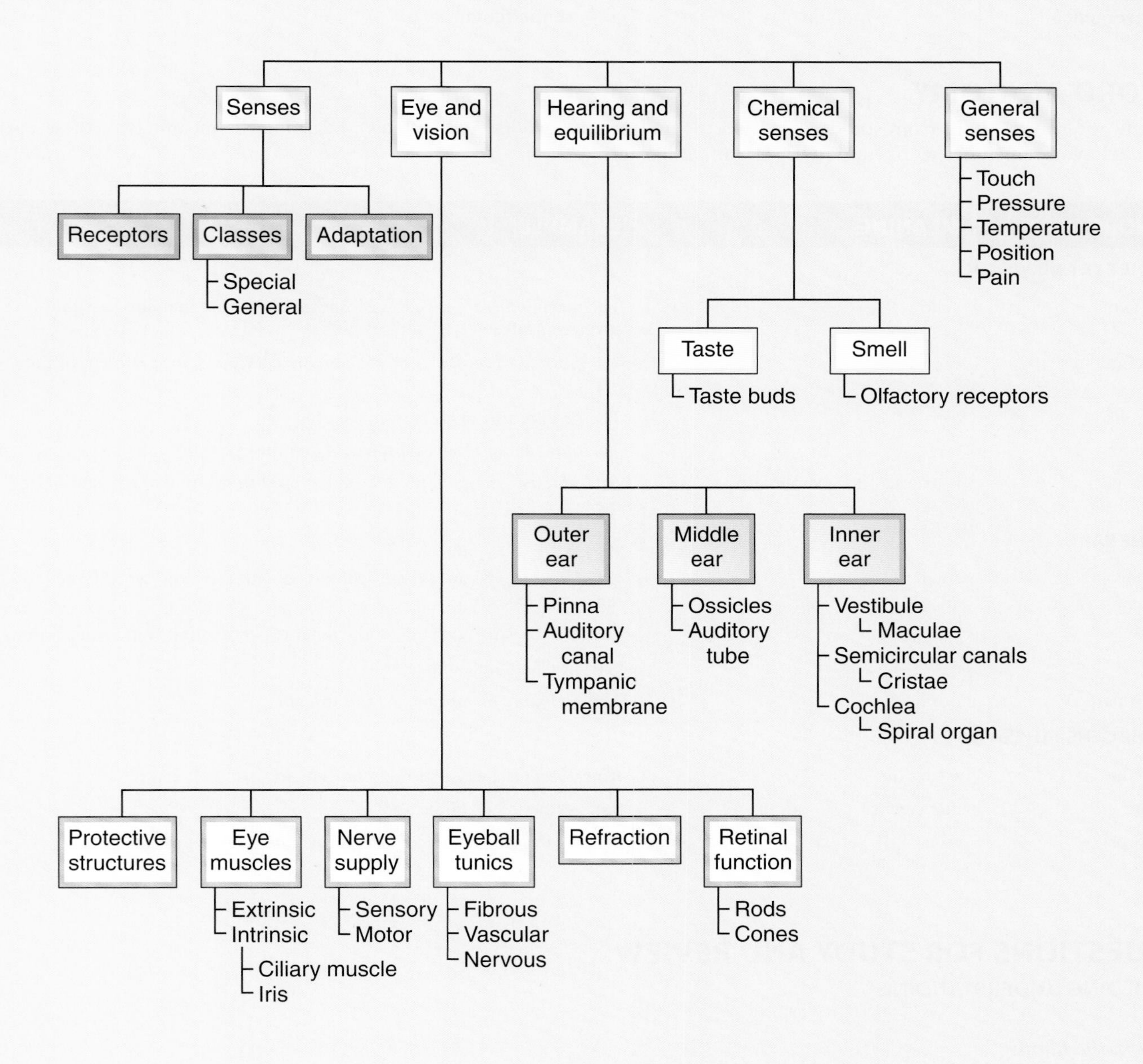

KEY TERMS

The terms listed below are emphasized in this chapter. Knowing them will help you organize and prioritize your learning. These and other boldface terms are defined in the "Glossary" with phonetic pronunciations:

accommodation
auditory tube
aqueous humor
choroid
cochlea
conjunctiva
convergence
cornea
gustation
iris
kinesthesia
lacrimal apparatus
lens (crystalline lens)
macula
olfaction
ossicle
proprioceptor
refraction
retina
sclera
semicircular canal
sensory adaptation
sensory receptor
spiral organ
tympanic membrane
vestibule
vitreous body

WORD ANATOMY

Medical terms are built from standardized word parts (prefixes, roots, and suffixes). Learning the meanings of these parts can help you remember words and interpret unfamiliar terms.

WORD PART	MEANING	EXAMPLE
THE EYE AND VISION		
lute/o	yellow	The macula *lutea* is a yellowish spot in the retina that contains the fovea centralis.
ophthalm/o	eye	An *ophthalmologist* is a physician who specializes in treatment of the eye.
ocul/o	eye	The *oculomotor* nerve moves the eye.
opt/o	eye	The *optic* nerve receives impulses from the eye.
-scope	instrument for examination	An *ophthalmoscope* is an instrument used to examine the posterior of the eye.
THE EAR		
equi-	equal	*Equilibrium* is balance (*equi-* combined with the Latin word *libra* meaning "balance").
lith	stone	*Otoliths* are small crystals in the inner ear that aid in static equilibrium.
ot/o	ear	See "lith" example above
tympan/o	drum	The *tympanic* membrane is the eardrum.
THE GENERAL SENSES		
-esthesia	sensation	*Kinesthesia* is a sense of body movement.
kine	movement	See "-esthesia" example above.
propri/o-	own	*Proprioception* is a perception of one's own body position.

QUESTIONS FOR STUDY AND REVIEW

BUILDING UNDERSTANDING

Fill in the Blanks

1. The part of the nervous system that detects a stimulus is a(n) _____.
2. The bending of light rays as they pass from one substance to another is called _____.
3. Nerve impulses are carried from the ear to the brain by the _____ nerve.
4. A receptor that senses body position is a(n) _____.
5. A receptor's loss of sensitivity to a continuous stimulus is called _____.

Matching Match each numbered item with the most closely related lettered item.

___ 6. Contains ciliated receptors sensitive to vibration
___ 7. Contains receptors sensitive to light
___ 8. Location of equilibrium receptors
___ 9. A touch receptor
___ 10. A pain receptor

a. retina
b. free nerve ending
c. vestibule
d. spiral organ
e. tactile corpuscle

Multiple Choice

___ 11. Which of the following is a general sense?
a. taste
b. smell
c. equilibrium
d. touch

___ 12. From superficial to deep, what is the order of the eyeball's tunics?
a. nervous, vascular, fibrous
b. fibrous, nervous, vascular
c. vascular, nervous, fibrous
d. fibrous, vascular, nervous

___ 13. Which eye structure has the greatest effect on light refraction?
a. cornea
b. lens
c. vitreous body
d. retina

___ 14. Which nerve carries sensory signals from the retina to the brain?
a. ophthalmic
b. optic
c. oculomotor
d. abducens

___ 15. What do receptors in the vestibule sense?
a. muscle tension
b. sound
c. light
d. acceleration

UNDERSTANDING CONCEPTS

16. Differentiate between the terms in each of the following pairs:
    a. special sense and general sense
    b. extrinsic and intrinsic eye muscles
    c. rods and cones
    d. endolymph and perilymph
    e. maculae and cristae
17. Trace the path of a light ray from the outside of the eye to the retina.
18. Define convergence and accommodation, and describe the eye structures involved in each.
19. List in order the structures that sound waves pass through in traveling through the ear to the receptors for hearing.
20. Name the five basic tastes. Where are the taste receptors? Name the nerves of taste.
21. Trace the pathway of a nerve impulse from the olfactory receptors to the olfactory center in the brain.

CONCEPTUAL THINKING

22. The heroine in the wartime story starts to cry when she hears her boyfriend is believed to be missing in action. At the same time, she starts to sniffle and reaches for a tissue. Why?
23. You and a friend have just finished riding the roller coaster at the amusement park. As you walk away from the ride, your friend stumbles and comments that the ride has affected her balance. How do you explain this?
24. Referring to the case study, imagine you were Dr. Sanchez speaking to Evan's parents. How would you explain the role of hair cells in hearing and how the cochlear implant can overcome their loss?

For more questions, see the Learning Activities on thePoint®.

CHAPTER 11

The Endocrine System: Glands and Hormones

Learning Objectives

After careful study of this chapter, you should be able to:

1. Compare the effects of the nervous system and the endocrine system in controlling the body. *p. 224*
2. Describe the functions of hormones. *p. 224*
3. Identify the glands of the endocrine system on a diagram. *p. 224*
4. Discuss the chemical composition of hormones. *p. 225*
5. Explain how hormones are regulated. *p. 225*
6. List the hormones produced by each endocrine gland, and describe the effects of each on the body. *p. 225*
7. Describe how the hypothalamus controls the posterior and anterior pituitary. *p. 225*
8. List seven tissues other than the endocrine glands that produce hormones. *p. 234*
9. Explain the origin and function of prostaglandins. *p. 234*
10. List eight medical uses of hormones. *p. 234*
11. Explain how the endocrine system responds to stress. *p. 235*
12. Referring to the case study, discuss the signs, symptoms, and treatment of type 1 diabetes mellitus. *pp. 223, 236*
13. Show how word parts are used to build words related to the endocrine system (see "Word Anatomy" at the end of the chapter). *p. 238*

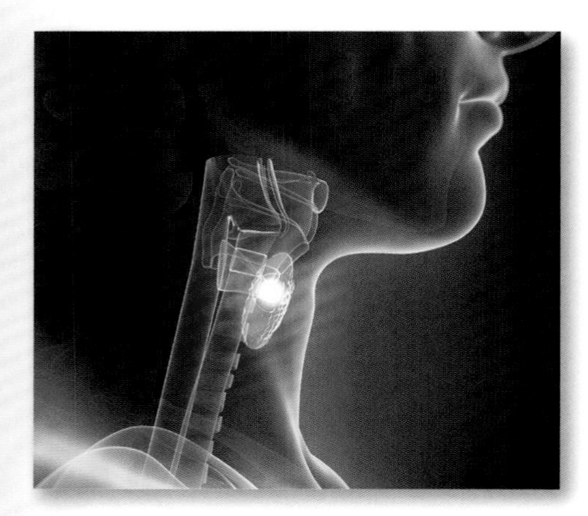

A & P in Action
Becky's Case: When an Endocrine Organ Fails

Becky stumbled down the stairs, hoping that Max hadn't finished all the pancakes that she could smell cooking.

"How was your sleep last night?" asked Becky's mother.

"Awful," sighed Becky, drowning the pancakes she was served in a lake of syrup. "I woke up a bunch of times to go to the bathroom."

"Were you actually able to make it this time?" chimed Becky's little brother. Becky wished Max hadn't brought *that* up. She hoped he wasn't blabbing to his friends that she was wetting the bed again.

"You know, if you didn't drink so much, you wouldn't have to pee so much," explained Max, as his sister gulped down her orange juice. Becky pretended that she didn't care about Max's comment. But he was right. She was so thirsty—and hungry!

It had been a long day when the bell rang and Becky boarded the bus for home. Math class had been a disaster, because she couldn't concentrate. During gym, she was tired and had a stomach ache. And, she had to keep asking for permission to go to the bathroom! Now, she was exhausted and her head hurt. Fighting tears, she remembered that during breakfast, her mom had mentioned that she'd made an appointment for Becky to see her doctor. She hadn't been too keen on the idea, but now she was relieved.

Later that week, Becky's pediatrician weighed and measured her and asked her a bunch of questions.

"So, Becky," said Dr. Carter. "For the past couple of weeks, you say you've felt lethargic and sick to your stomach. You've been really thirsty and have needed to go to the bathroom a lot. You've also been really hungry. You've had headaches and some difficulty concentrating at school and have felt tired when playing sports." Becky wasn't too sure what lethargic meant, but other than that he seemed to have gotten the facts right. So, Becky nodded her head yes.

Turning to Becky's mother, Dr. Carter said, "Checking her chart, it appears that she's lost several pounds since her last appointment despite her appetite. I'm going to order urine and blood tests. I'd like to see what her glucose levels are." Becky didn't enjoy the tests one bit. Having to pee in a cup was gross, and as for the blood test, that was the worst.

The next day, Dr. Carter called Becky's mother. "The urinalysis was positive for glucose and ketones, suggesting that Becky is not metabolizing glucose correctly. Her blood test revealed that she's hyperglycemic; her blood sugar is too high. My diagnosis so far is that Becky has type 1 diabetes mellitus (T1DM) and needs insulin."

Dr. Carter suspects that Becky's pancreas does not produce enough insulin, a hormone needed to utilize glucose. As we will see later, diabetes has a dramatic effect on Becky's health.

As you study this chapter, CasePoints will give you opportunities to apply your learning to this case.

Visit thePoint® to access the following resources. For guidance in using these resources most effectively, see pp. xv–xvii.

Preparing to Learn

- Tips for Effective Studying
- Pre-Quiz

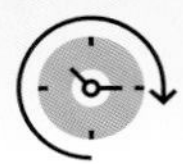

While You Are Learning

- Animation: Negative Feedback in the Hypothalamo-Pituitary Axis
- Animation: Hormonal Control of Blood Glucose
- Chapter Notes Outline
- Audio Pronunciation Glossary

When You Are Reviewing

- Answers to Questions for Study and Review
- Health Professions: Exercise and Fitness Specialist
- Interactive Learning Activities

The past several chapters have described the nervous system and its role in regulating body responses. The endocrine system is also viewed as a controlling system, exerting its effects through hormones. These hormones act as signals in the homeostatic feedback loops described in Chapter 1 and also regulate other body processes such as growth and reproduction. So, the key ideas of homeostasis 3 *and communication* 11 *are particularly relevant to this chapter.*

Introduction

The **endocrine system** consists of a group of glands that produces regulatory chemicals called **hormones**. These glands specialize in hormone secretion and are illustrated in **FIGURE 11-1**. The endocrine system and the nervous system work together to control and coordinate all other body systems. The nervous system controls such rapid actions as muscle movement and intestinal activity by means of electrical and chemical signals. The effects of the endocrine system occur more slowly and over a longer period. They involve chemical signals only, and these chemical messengers have widespread effects on the body.

Although the nervous and endocrine systems differ in some respects, the two systems are closely related. For example, the activity of the pituitary gland, which in turn regulates other glands, is controlled by the brain's hypothalamus. You can see both structures in **FIGURE 11-1**. The connections between the nervous system and the endocrine system enable endocrine function to adjust to the demands of a changing environment.

Hormones

Hormones are chemical messengers that have specific regulatory effects on certain cells or organs. Hormones from the endocrine glands are released, not through ducts, but directly into surrounding tissue fluids. Most then diffuse into the bloodstream, which carries them throughout the body. The specific tissue acted on by each hormone is the **target tissue.** 11 Recall from Chapter 7 that chemical signals act on target tissues by binding to **receptors** in the plasma membrane or within the cytoplasm. Once a hormone binds to a receptor on or in a target cell, the bound receptor affects cell activities such as regulating the

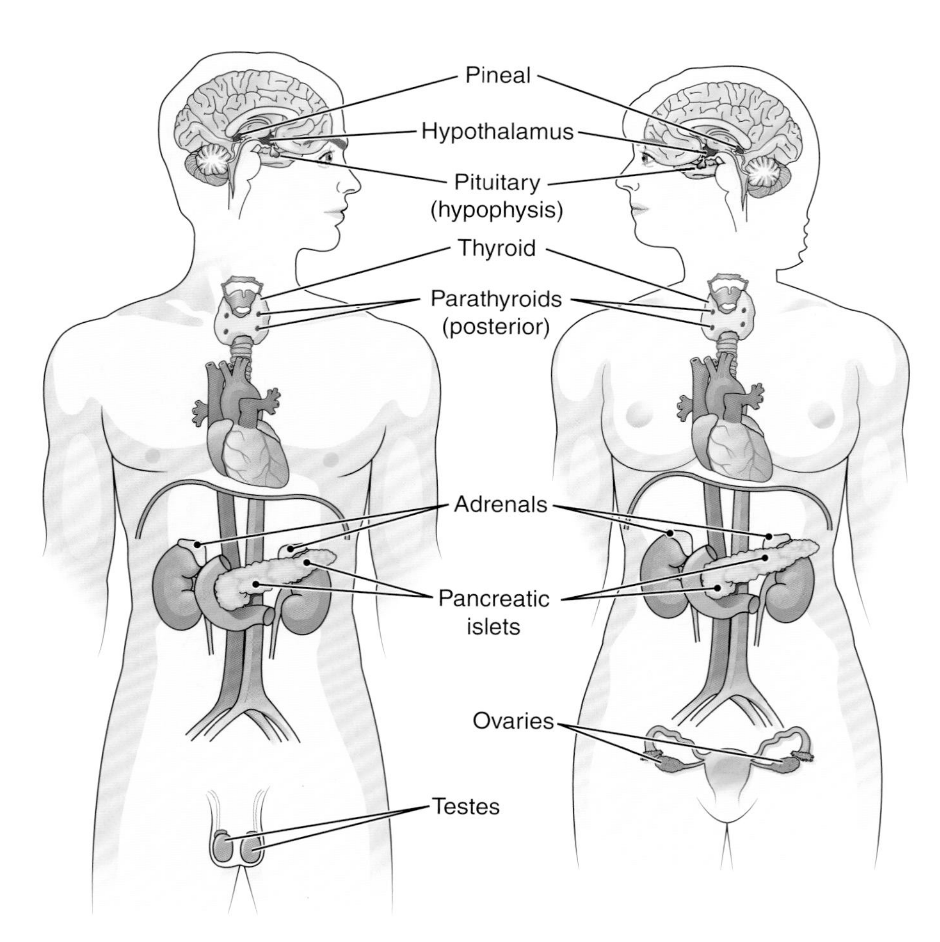

FIGURE 11-1 The endocrine glands. **KEY POINT** The endocrine system comprises glands with a primary function of hormone secretion.

manufacture of specific proteins, changing the membrane's permeability to specific substances, or affecting metabolic reactions. Since blood carries hormones throughout the body, any cell possessing receptors for a specific hormone will respond to the hormone, be it a close neighbor of the secreting cell or not.

HORMONE CHEMISTRY

Chemically, hormones fall into two main categories:

- **Amino acid compounds.** These hormones are proteins or related compounds also made of amino acids. All hormones except those of the adrenal cortex and the sex glands fall into this category.
- **Steroids.** These hormones are derived from the steroid cholesterol, a type of lipid (see FIG. 2-9). Steroid hormones are produced by the adrenal cortex and the sex glands. Many can be recognized by the ending *sterone*, as in progesterone and testosterone.

HORMONE REGULATION

3 As discussed in Chapter 1, the process of negative feedback keeps the level of a regulated variable within a specific range. FIGURE 1-4 illustrates how the hormone insulin is a signal in the negative feedback loop regulating blood glucose concentrations. When blood glucose increases, the pancreas (the sensor/control center) secretes more insulin. Insulin prompts effectors, such as the liver and muscle cells, to utilize more glucose, thus reducing blood glucose and restoring homeostasis.

Hormone release may fall into a rhythmic pattern. Hormones of the adrenal cortex follow a 24-hour cycle related to a person's sleeping pattern, with the secretion level greatest just before arising and least at bedtime. Hormones of the female menstrual cycle follow a monthly pattern.

CHECKPOINTS 

- ☐ **11-1** What are hormones, and what are some effects of hormones?
- ☐ **11-2** What name is given to the specific tissue that responds to a hormone?
- ☐ **11-3** Hormones belong to what two chemical categories?
- ☐ **11-4** What is the most common mechanism used to regulate hormone secretion?

The Endocrine Glands and Their Hormones

The remainder of this chapter discusses hormones and the tissues that produce them. Although most of the discussion centers on the endocrine glands, which specialize in hormone production, it is important to note that many tissues—other than the endocrine glands—also secrete hormones. These tissues include the brain, digestive organs, and kidneys. Some of these other tissues are discussed later in the chapter. TABLE 11-1 summarizes the information on the endocrine glands and their hormones.

CASEPOINT 

- ☐ **11-1** Referring to FIGURE 1-4, what step in the negative feedback cycle is interrupted in Becky's case?

THE PITUITARY

The **pituitary** (pih-TU-ih-tar-e), or *hypophysis* (hi-POF-ih-sis), is a gland about the size of a cherry. It is located in a saddle-like depression of the sphenoid bone just posterior to the point where the optic nerves cross. It is surrounded by a bone except where it connects with the brain's **hypothalamus** by a stalk called the **infundibulum** (in-fun-DIB-u-lum). The gland is divided into two parts: the **anterior lobe** and the **posterior lobe** (FIG. 11-2 and Dissection Atlas FIGS. A3-2 and A3-3A). The anterior lobe is a true endocrine gland, composed of epithelial tissue. The posterior lobe, however, is not a true gland. It consists of the axons and axon terminals of neurons that originate in the hypothalamus. The two lobes are considered separately below. A small band of tissue between the two lobes secretes a protein called melanocyte-stimulating hormone (MSH). See BOX 11-1 for more information.

Posterior Lobe The two hormones of the posterior pituitary (antidiuretic hormone, or ADH, and oxytocin) are actually produced in the hypothalamus and only stored in the posterior pituitary (see FIG. 11-2). Their release is controlled by nerve impulses that travel over pathways (tracts) between the hypothalamus and the posterior pituitary. Their actions are as follows:

- **Antidiuretic** (an-ti-di-u-RET-ik) **hormone** (**ADH,** also known as *vasopressin*) promotes the reabsorption of water from the kidney tubules and thus decreases water excretion (see Chapter 19). It also stimulates contraction of smooth muscle in blood vessel walls. These two actions raise blood pressure.
- **Oxytocin** (ok-se-TO-sin) causes uterine contractions and triggers milk ejection from the breasts (see Chapter 21). Under certain circumstances, commercial preparations of this hormone are administered during childbirth to promote uterine contraction.

Anterior Lobe The hormone-producing cells of the anterior pituitary are controlled by secretions called **releasing hormones** produced in the hypothalamus (see FIG. 11-2). These releasing hormones travel to the anterior pituitary by way of a special type of circulatory pathway called a **portal system**. By this circulatory "detour," some of the blood that leaves the hypothalamus travels to capillaries in the anterior pituitary before returning to the heart. Each pituitary cell produces a particular hormone and is stimulated by specific

Table 11-1 The Endocrine Glands and Their Hormones

Gland	Hormone	Principal Functions
Hypothalamus	Releasing hormones	Control the release of anterior pituitary hormones
Hypothalamus and posterior pituitary	ADH (antidiuretic hormone)	Promotes water reabsorption in kidney tubules; at high concentration, stimulates constriction of blood vessels
	Oxytocin	Causes uterine muscle contraction; causes milk ejection from mammary glands
Anterior pituitary	GH (growth hormone)	Promotes growth of all body tissues
	TSH (thyroid-stimulating hormone)	Stimulates thyroid gland to produce thyroid hormones
	ACTH (adrenocorticotropic hormone)	Stimulates growth and hormonal activity of the adrenal cortex
	PRL (prolactin)	Stimulates milk production by mammary glands
	FSH (follicle-stimulating hormone)	Stimulates growth and hormonal activity of ovarian follicles; stimulates growth of the testes; promotes sperm cell development
	LH (luteinizing hormone)	Initiates ovulation, corpus luteum formation, and progesterone production in the female; stimulates testosterone secretion in males
Thyroid	Thyroxine (T_4) and triiodothyronine (T_3)	Increase metabolic rate, influencing both physical and mental activities; required for normal growth
Parathyroids	PTH (parathyroid hormone)	Regulates exchange of calcium between blood and bones; increases the calcium level in blood
Adrenal medulla	Epinephrine	Increases blood pressure and heart rate; activates cells influenced by sympathetic nervous system plus many not supplied by sympathetic nerves
Adrenal cortex	Cortisol (95% of glucocorticoids)	Increases blood glucose concentration in response to stress
	Aldosterone (95% of mineralocorticoids)	Promotes salt (and thus water) retention and potassium excretion
	Weak androgens	Contribute to some secondary sex characteristics in women
Pancreatic islets	Insulin	Reduces blood glucose concentrations by promoting glucose uptake into cells and glucose storage; promotes fat and protein synthesis
	Glucagon	Stimulates the liver to release glucose, thereby increasing blood glucose levels
Testes	Testosterone	Stimulates growth and development of sexual organs (testes and penis) plus development of secondary sexual characteristics, such as hair growth on the body and face and deepening of voice; stimulates sperm cell maturation
Ovaries	Estrogens (e.g., estradiol)	Stimulates growth of primary sexual organs (uterus and tubes) and development of secondary sexual organs, such as breasts; stimulates development of ovarian follicles
	Progesterone	Stimulates development of mammary glands' secretory tissue; prepares uterine lining for implantation of fertilized ovum; aids in maintaining pregnancy
Pineal	Melatonin	Regulates mood, sexual development, and daily cycles in response to the amount of light in the environment

hypothalamic releasing hormones. Hypothalamic releasing hormones are indicated with the abbreviation *RH* added to an abbreviation for the name of the hormone stimulated. For example, the releasing hormone (RH) that controls growth hormone (GH) is GHRH. Inhibitory hormones from the hypothalamus also regulate the anterior pituitary hormones.

Concept Mastery Alert

Remember that releasing hormones are produced in the hypothalamus and regulate the production of specific anterior pituitary hormones.

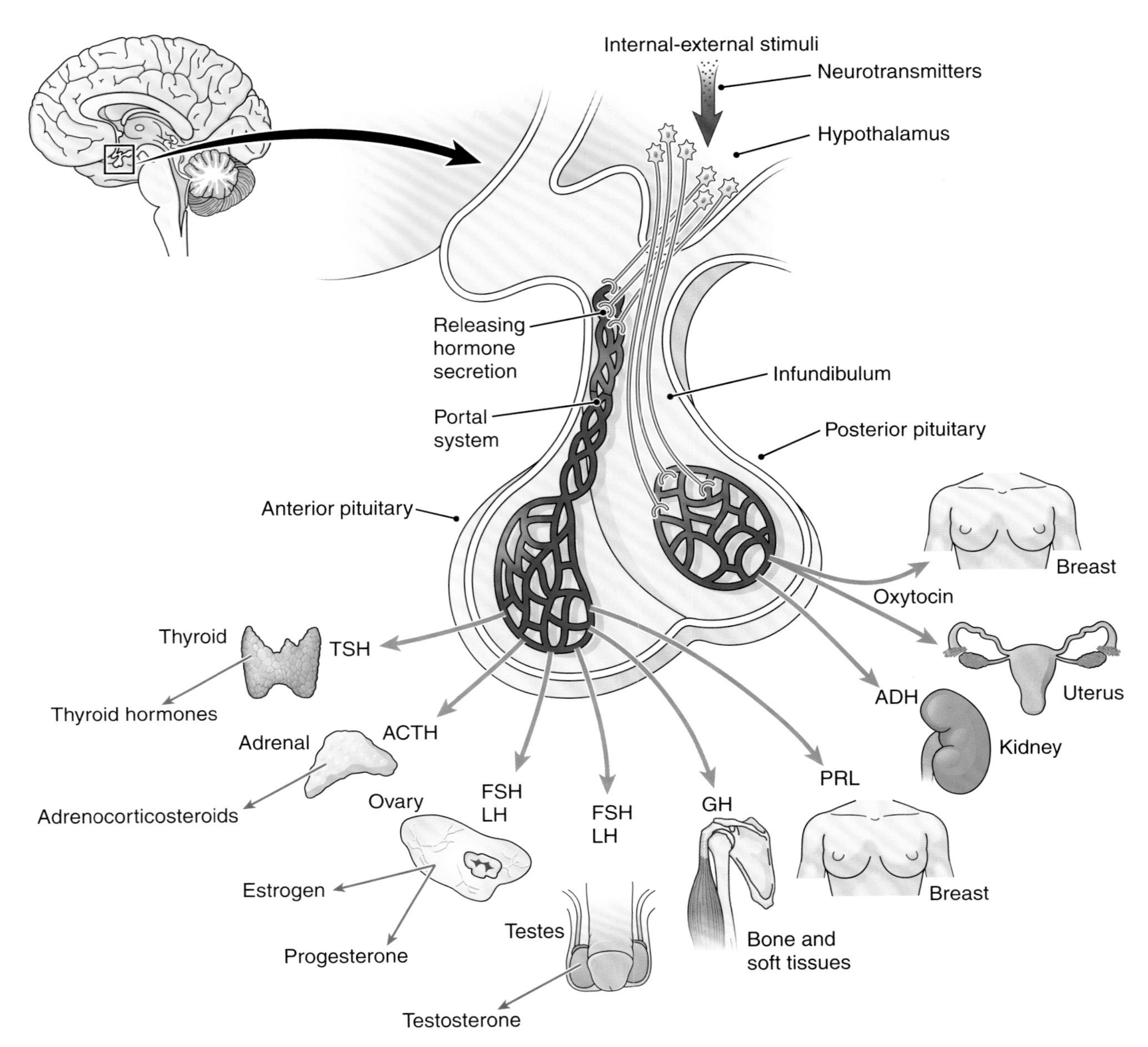

FIGURE 11-2 The hypothalamus, pituitary gland, and target tissues. KEY POINT The hypothalamus synthesizes hormones secreted by the posterior pituitary and synthesizes releasing hormones that regulate the anterior pituitary. ZOOMING IN What two structures does the infundibulum connect?

A CLOSER LOOK

BOX 11-1

Melanocyte-Stimulating Hormone: More Than a Tan?

In amphibians, reptiles, and certain other animals, melanocyte-stimulating hormone (MSH) darkens skin and hair by stimulating melanocytes to manufacture the pigment melanin. In humans, though, MSH levels are usually so low that its role as a primary regulator of skin pigmentation and hair color is questionable. What, then, is its function in the human body?

Recent research suggests that MSH is probably more important as a neurotransmitter in the brain than as a hormone in the rest of the body. A narrow region between the anterior and posterior pituitary, the intermediate lobe, produces MSH. When the pituitary gland secretes adrenocorticotropic hormone (ACTH), it secretes MSH as well. This is so because pituitary cells do not produce ACTH directly but produce a large precursor molecule, proopiomelanocortin (POMC), which enzymes cut into ACTH and MSH.

MSH's other roles include helping the brain regulate food intake, fertility, and even the immune response. Interestingly, despite MSH's relatively small role in regulating pigmentation, women do produce more MSH during pregnancy and often have darker skin.

Anterior Lobe Hormones The anterior pituitary is often called the *master gland* because it releases hormones that affect the working of other glands, such as the thyroid, gonads (ovaries and testes), and adrenal glands. (Hormones that stimulate other glands may be recognized by the ending *tropin* as in *thyrotropin*, which means "acting on the thyroid gland.") The major hormones are as follows (see FIG. 11-2):

- **Growth hormone (GH)**, or *somatotropin* (so-mah-to-TRO-pin), acts directly on most body tissues, promoting protein manufacture that is essential for growth. GH causes increases in size and height to occur in youth, before the closure of long bone epiphyses. A young person with a GH deficiency will remain small, though relatively well-proportioned unless treated with adequate hormone. GH is produced throughout life. It stimulates protein synthesis and is needed for cellular maintenance and repair. It also stimulates the liver to release fatty acids and glucose for energy in time of stress.
- **Thyroid-stimulating hormone (TSH)**, or *thyrotropin* (thi-ro-TRO-pin), stimulates the thyroid gland (a large gland found in the neck) to produce thyroid hormones.
- **Adrenocorticotropic hormone** (ad-re-no-kor-tih-ko-TRO-pik) (ACTH) stimulates hormone production in the cortex of the adrenal glands.
- **Prolactin** (pro-LAK-tin) (**PRL**) stimulates milk production in the breasts.
- **Follicle-stimulating hormone (FSH)** stimulates the development of ovarian follicles in which egg cells mature and the development of sperm cells in the testes.
- **Luteinizing** (LU-te-in-i-zing) **hormone (LH)** causes ovulation in females and promotes progesterone secretion in females and testosterone secretion in males.

FSH and LH are classified as **gonadotropins** (gon-ah-do-TRO-pinz), hormones that act on the gonads to regulate growth, development, and reproductive function in both males and females.

3 Negative Feedback and the Anterior Lobe Most negative feedback loops control the level of a regulated variable that can be monitored internally, such as body temperature or blood pressure. However, the anterior pituitary hormones control variables that are not easily monitored, such as metabolic rate or growth. For this reason, it can be convenient to consider the levels of certain hormones as regulated variables. So, in this instance, a hormone itself inhibits further hormone secretion. An example is the secretion of thyroid hormones (FIG. 11-3). Thyrotropin-releasing hormone (TRH) from the hypothalamus stimulates the production of TSH from the anterior pituitary gland. TSH promotes the release of thyroid hormones from the thyroid gland. The hypothalamus and anterior pituitary gland sense any increase in thyroid hormone levels and reduce their production of TRH and TSH, respectively. As a result, thyroid hormone levels decline back to normal. Conversely, TRH and TSH production increases if thyroid hormone levels decline below normal levels. As mentioned above, TSH signals the thyroid gland (a gland in the neck) to secrete more hormones.

GH, ACTH, and, to a certain extent, the gonadal steroids are all subject to similar negative feedback loops. These self-regulating systems keep hormone levels within a set normal range.

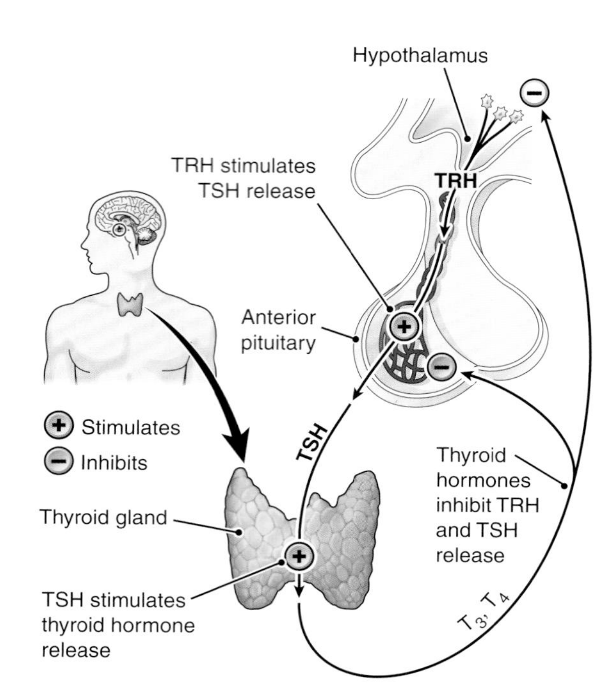

FIGURE 11-3 Negative feedback control of thyroid hormones. KEY POINT Thyroid hormone levels are kept constant by negative feedback. ZOOMING IN What gland controls the thyroid gland?

Concept Mastery Alert

Students sometimes confuse the roles of negative feedback and receptors. Negative feedback maintains hormone concentrations within relatively narrow limits. Receptors ensure that these hormones exert their effects exclusively at the intended target tissue.

See the Student Resources on thePoint® for the animation "Negative Feedback in the Hypothalamo-Pituitary Axis."

CHECKPOINTS 

- ☐ **11-5** What part of the brain controls the pituitary?
- ☐ **11-6** What hormones are released from the posterior pituitary?
- ☐ **11-7** What hormones does the anterior pituitary secrete?

THE THYROID GLAND

The thyroid, located in the neck, is the largest of the endocrine glands (FIG. 11-4). The thyroid has two roughly oval lateral lobes on either side of the larynx (voice box) connected by a narrow band called an *isthmus* (IS-mus). A connective tissue capsule encloses the entire gland.

Thyroid Hormones The thyroid produces two hormones that regulate metabolism. It produces large amounts of the hormone **thyroxine** (thi-ROK-sin), which is symbolized as T_4, based on the four iodine atoms contained in each molecule. It produces smaller amounts of **triiodothyronine** (tri-i-o-do-THI-ro-nene), or T_3, which contains three iodines. T_4 is much less potent than T_3, but target tissues convert T_4 to T_3 by removing one iodine. These thyroid hormones increase the metabolic rate in body cells. That is, they increase the rate at which cells use nutrients to generate ATP and heat. Both thyroid hormones and GH are needed for normal growth. As we saw in FIGURE 11-3, thyroid hormone production is under the control of TSH from the anterior pituitary, and thyroid hormones feed back to inhibit TSH production.

THE PARATHYROID GLANDS

The four tiny **parathyroid glands** are embedded in the thyroid's posterior capsule or in the surrounding connective tissue (see FIG. 11-4). 3 As discussed in Chapter 6, the secretion of these glands, **parathyroid hormone (PTH)**, acts as a signal in the negative feedback loop controlling the blood calcium concentration. When calcium levels are low, the parathyroid glands increase PTH production. PTH promotes calcium release from bone tissue and calcium retention by the kidney, and blood calcium levels increase.

Calcium Metabolism Calcium balance is required not only for the health of bones and teeth but also for the proper function of the nervous system and muscles. Another hormone, in addition to PTH, controls calcium balance. This hormone is **calcitriol** (kal-sih-TRI-ol), technically called dihydroxycholecalciferol (di-hi-drok-se-ko-le-kal-SIF-eh-rol), the active form of vitamin D. Calcitriol is produced by modification of vitamin D in the liver and then the kidney, a process stimulated by PTH. Calcitriol increases intestinal absorption of calcium to raise blood calcium levels.

PTH and calcitriol work together to regulate the amount of calcium in the blood and provide calcium for bone maintenance and other functions.

CHECKPOINTS 

- ☐ **11-8** What is the effect of thyroid hormones on cells?
- ☐ **11-9** What mineral is needed to produce thyroid hormones?
- ☐ **11-10** Which two hormones control calcium levels in blood?

THE ADRENAL GLANDS

The **adrenal glands** are two small glands located atop the kidneys. Each adrenal gland has two parts that act as separate glands. The inner area is called the **medulla**, and the outer portion is called the **cortex** (FIG. 11-5).

Hormones from the Adrenal Medulla The hormones of the adrenal medulla are released in response to stimulation

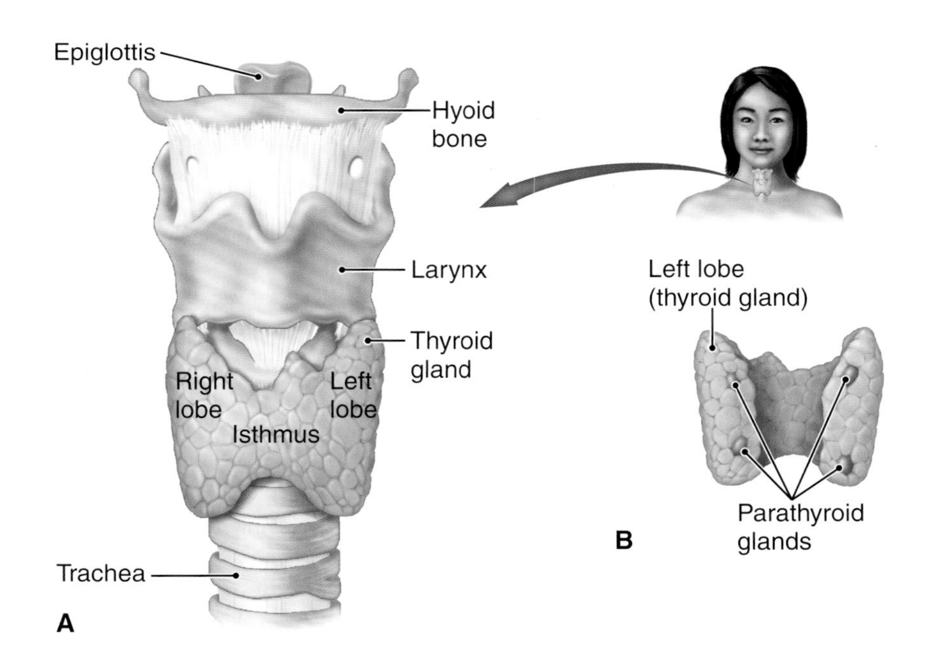

FIGURE 11-4 **Thyroid and parathyroid glands. A.** The thyroid has two lobes connected by an isthmus. These are shown here in relation to other structures in the throat. The epiglottis is a cartilage of the larynx. **B.** The parathyroid glands are embedded in the posterior surface of the thyroid gland. ZOOMING IN What structure is superior to the thyroid? Inferior to the thyroid?

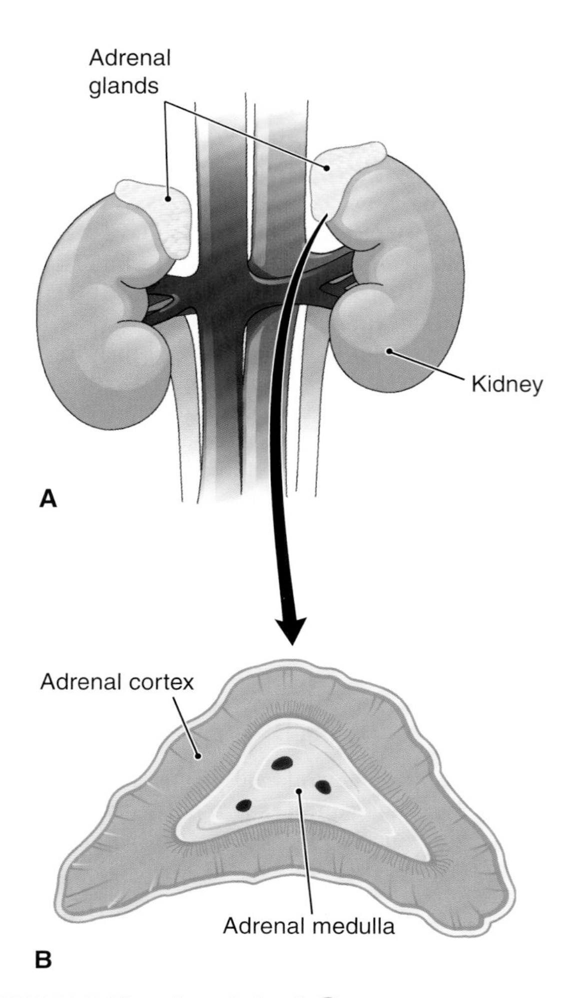

FIGURE 11-5 The adrenal gland. KEY POINT The gland has two distinct regions. The medulla secretes epinephrine. The cortex secretes steroid hormones. **A.** Location of the adrenal glands on the kidneys. **B.** Structure of the adrenal gland. ZOOMING IN What is the outer region of the adrenal gland called? The inner region?

by the sympathetic nervous system. The principal hormone produced by the medulla is **epinephrine**, also called *adrenaline*. Epinephrine is chemically and functionally similar to norepinephrine, the neurotransmitter active in the sympathetic nervous system, as described in Chapter 8. However, epinephrine is generally considered to be a hormone because it is released into the bloodstream instead of being released locally at synapses. Both epinephrine and norepinephrine are responsible for *fight-or-flight* responses during emergency situations. Some of their effects are as follows:

- Stimulation of smooth muscle contraction in the walls of some arterioles, causing them to constrict, and blood pressure to rise accordingly.
- Increase in the heart rate.
- Increase in the metabolic rate of body cells.
- Conversion of glycogen stored in the liver into glucose; the glucose enters the blood and travels throughout the body, allowing the voluntary muscles and other tissues to do an increased amount of work.
- Dilation of the bronchioles through relaxation of the smooth muscle in their walls.

Hormones from the Adrenal Cortex There are three main groups of hormones secreted by the adrenal cortex:

- **Glucocorticoids** (glu-ko-KOR-tih-koyds) help the body respond to unfavorable conditions such as starvation. They maintain blood glucose levels in times of stress by stimulating the liver to convert amino acids into glucose instead of protein (as indicated by *gluco* in the name). In addition, they raise the levels of other nutrients in the blood, including amino acids from tissue proteins and fatty acids from fats stored in adipose tissue. Glucocorticoids also have the ability to suppress the inflammatory response and are often administered as medication for this purpose. The major hormone of this group is **cortisol**, which is also called *hydrocortisone*. Cortisol levels are controlled by negative feedback. See **BOX 11-2** to learn about the impact of cortisol injections on this negative feedback loop.
- **Mineralocorticoids** (min-er-al-o-KOR-tih-koyds) are important in the regulation of electrolyte balance. They control sodium reabsorption and potassium secretion by the kidney tubules. The major hormone of this group is **aldosterone** (al-DOS-ter-one), which is discussed further in Chapter 19.
- **Androgens** ("male" sex hormones) are secreted in small amounts. Whereas these hormone quantities are insignificant in males (the testes produce large amounts of androgens), they contribute significant amounts of androgens in both premenopausal and postmenopausal women. In normal amounts, they promote some bone and muscle growth and stimulate libido (sexual desire).

CHECKPOINTS 

- **11-11** What is the main hormone produced by the adrenal medulla?
- **11-12** What three categories of hormones are released by the adrenal cortex?
- **11-13** What effect does cortisol have on blood glucose levels?

THE ENDOCRINE PANCREAS

The **pancreas** is located in the left upper quadrant of the abdomen, inferior to the liver and gallbladder and lateral to the first portion of the small intestine, the duodenum. It is posterior to the stomach, in the space behind the peritoneum, with its head to the right against the duodenum and its tail pointed to the left (**FIG. 11-6** and Dissection Atlas **FIGS. A3-9C and A3-10**). It has two main types of cells that perform different functions. The most abundant type forms small clusters called *acini* (AS-ih-ni) (singular *acinus*) that resemble blackberries (*acinus* comes from a Latin word meaning "berry"). Acini secrete digestive enzymes through

ONE STEP AT A TIME

BOX 11-2

Negative Feedback, One More Time

In Chapter 1, **BOX 1-1** (One Step at a Time: Deciphering Negative Feedback Loops) discussed the components of relatively simple feedback loops. As you saw in **FIGURE 11-3**, negative feedback involving the hypothalamus and anterior pituitary is somewhat more complicated! Answering questions such as the one below will help cement your knowledge of these important endocrine organs and enrich your understanding of homeostasis and feedback.

Question

You have been getting weekly injections of cortisol to treat a persistent athletic injury, and your physician requests a blood test. Normally, corticotrophin-releasing hormone (CRH) from the hypothalamus stimulates ACTH release from the anterior pituitary gland. In turn, ACTH stimulates cortisol release from the adrenal cortex. Cortisol inhibits CRH release as the messenger in a negative feedback loop. On the basis of this information, what would your blood test show in terms of CRH, ACTH, and cortisol levels?

Step 1. Identity the hormones involved and draw out the feedback loop. The question gives you all of the necessary information for this step. You may want to sketch out a simple feedback loop such as this one:

CRH → ACTH → cortisol
⊖ ↑ (cortisol feeds back to CRH)

Step 2. Identify the intervention. You are receiving cortisol injections, so your cortisol levels will be high.

Step 3. Map changes in the other hormones involved. The injected cortisol will act at the hypothalamus to reduce CRH production. If CRH production decreases, ACTH production will decrease as well.

Step 4. Remember that negative feedback only regulates "normal" hormone production. It cannot reduce (or increase) the hormone concentration if:

1. The hormone is administered therapeutically
2. A tumor is producing unregulated levels of a hormone
3. The gland that normally produces the hormone is deficient

Thus, even though negative feedback will act to reduce cortisol production from the adrenal gland, the constant cortisol injections ensure that blood cortisol levels remain high. So, the answer to the question is elevated cortisol but reduced CRH and ACTH. (Incidentally, CRH inhibits appetite. So, patients receiving large amounts of cortisol generally get very hungry!)

See Question 26 at the end of this chapter for more practice with this question type.

ducts directly into the small intestine (see Chapter 17), thus making up the exocrine portion of the pancreas. In addition, scattered throughout the pancreas are specialized cells that form "little islands" called islets (I-lets), originally the *islets of Langerhans* (LAHNG-er-hanz) **(see FIG. 11-6)**. These cells produce hormones that diffuse into the bloodstream, thus making up the endocrine portion of the pancreas, which we discuss further here.

Pancreatic Hormones The most important hormone secreted by the islets is **insulin** (IN-su-lin), which is produced by beta (β) cells. 3 As illustrated in **FIGURE 1-4**, insulin is an important signal in the regulation of glucose homeostasis by negative feedback. Increased blood glucose levels stimulate insulin production by the pancreas, and insulin stimulates glucose uptake and use by body cells. All cells use more glucose for energy, and liver and muscle cells convert excess glucose into glycogen, the storage form of glucose. As a result of these actions, blood glucose levels decline, and insulin secretion declines.

In addition to its role in glucose homeostasis, insulin also promotes overall tissue building. Under its effects, tissues store fatty acids as triglycerides (fats) and use amino acids to build proteins.

The second islet hormone, produced by alpha (α) cells, is **glucagon** (GLU-kah-gon), which works with insulin to regulate blood glucose levels. When blood glucose levels decrease, such as during an overnight fast, glucagon secretion increases. The liver responds to glucagon by increasing glucose production from amino acids and from glycogen. The increased hepatic glucose production increases blood glucose levels.

To summarize, insulin is known as the *hormone of feasting*, because food intake stimulates its release and insulin acts to stimulate glucose use and storage. Glucagon, on the other hand, is the *hormone of fasting*, because starvation stimulates its release and glucagon promotes glucose production by the liver. The activities of insulin and glucagon are summarized in **FIGURE 11-7**.

CASEPOINTS

- ☐ **11-2** Is the insulin involved in Becky's case composed of amino acids or is it a steroid?
- ☐ **11-3** Why is Becky losing weight?
- ☐ **11-4** Would Becky's glucagon secretion be increased or decreased compared to normal?

See the "Student Resources" on thePoint® to view the animation "Hormonal Control of Blood Glucose."

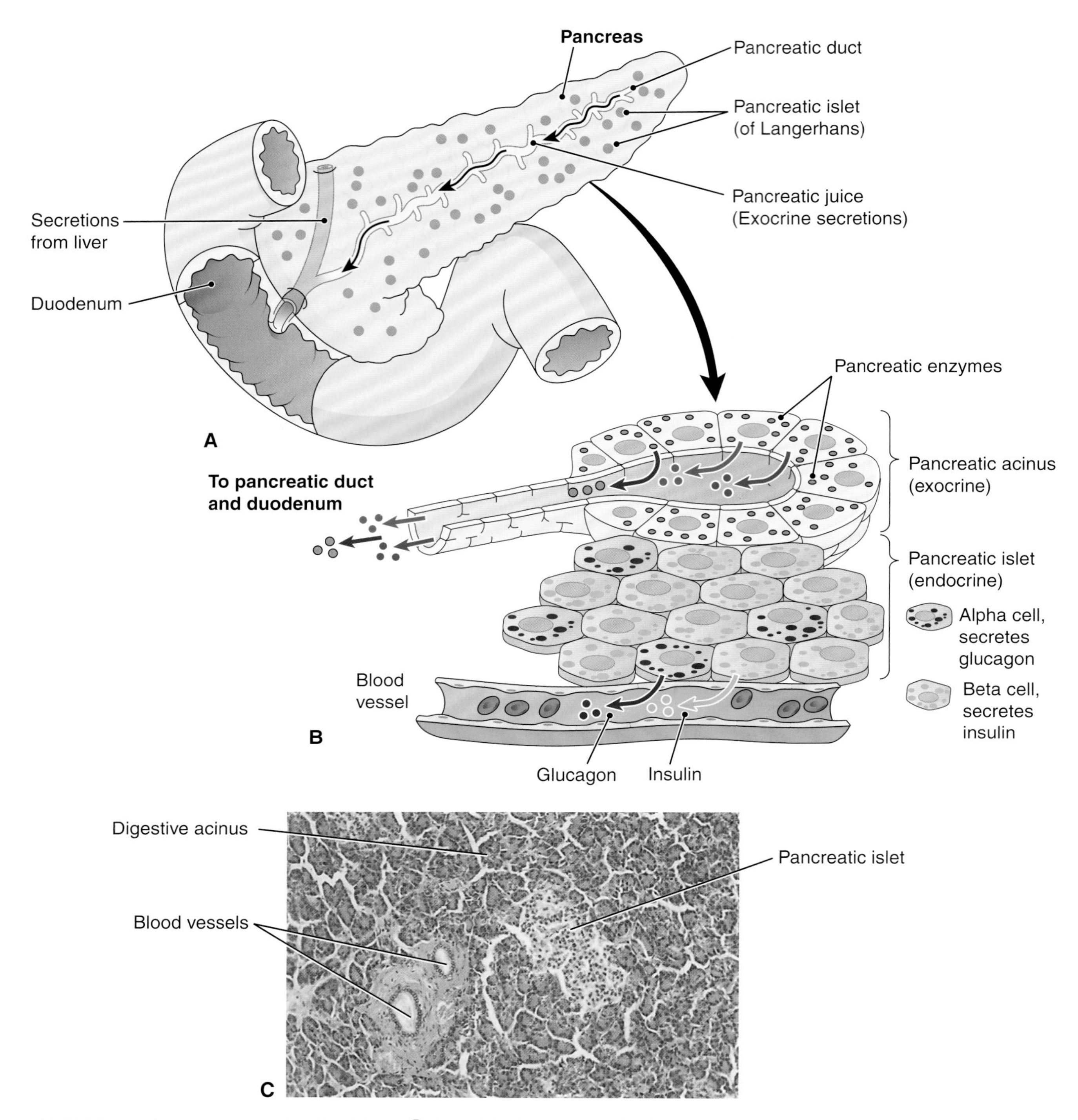

FIGURE 11-6 The pancreas and its functions. KEY POINT The pancreas has both exocrine and endocrine functions. **A.** The pancreas in relation to the duodenum (small intestine). **B.** Diagram of an acinus, which secretes digestive juices into ducts, and an islet, which secretes hormones into the bloodstream. **C.** Photomicrograph of pancreatic cells. Light-staining islets are visible among the many acini that produce digestive juices.

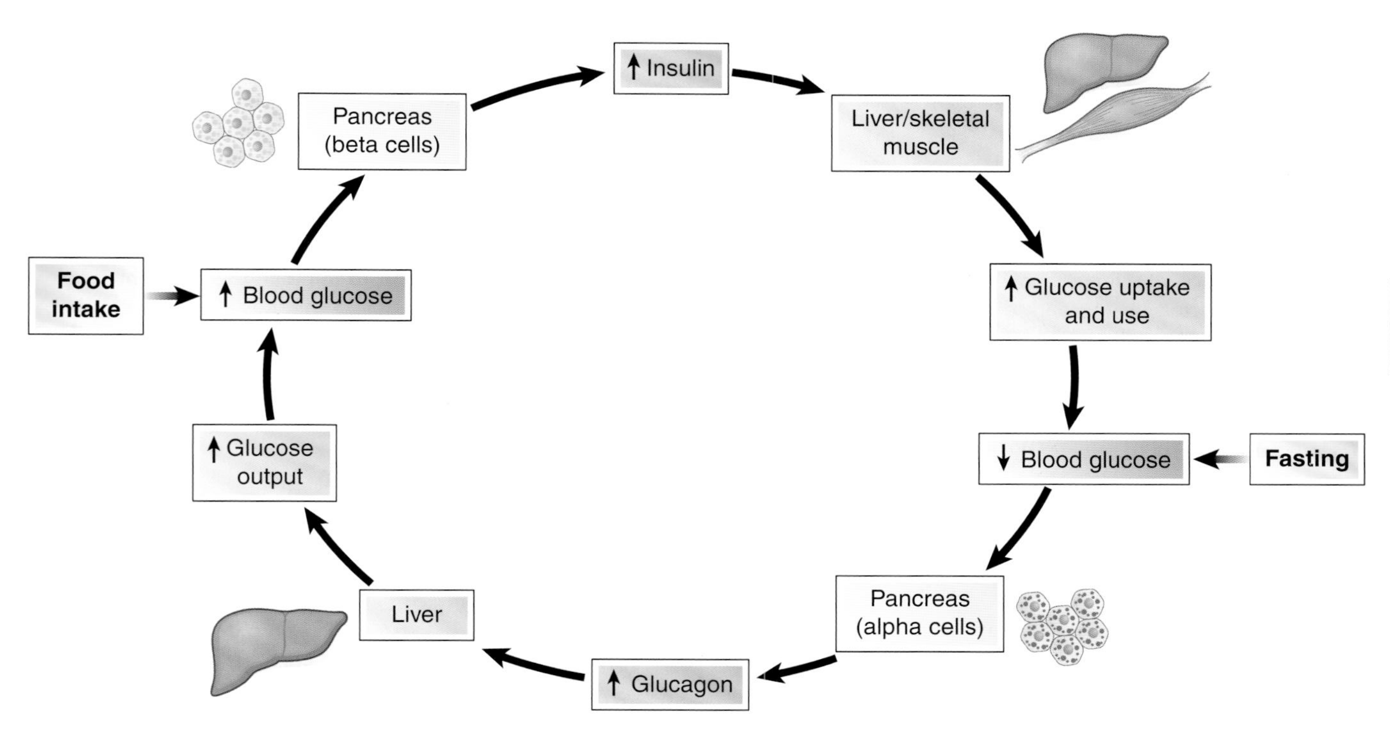

FIGURE 11-7 The effects of insulin and glucagon. KEY POINT These two hormones work together to regulate blood glucose levels. Note that the different components of the feedback loop are indicated by different colors (regulated variable, green; sensors and control centers, yellow; signals, purple; effectors, blue). ZOOMING IN Which effector organ responds to insulin but not glucagon?

THE SEX GLANDS

The sex glands, the female ovaries, and the male testes not only produce the sex cells (sperm and ova) but are also important endocrine organs. The hormones produced by these organs are needed in the development of the sexual characteristics, which usually appear in the early teens, and for the maintenance of the reproductive organs once full development has been attained. They also control nonreproductive characteristics, such as bone and muscle growth and repair. Those features that typify a male or female other than the structures directly concerned with reproduction are termed *secondary sex characteristics*. They include a deep voice and facial and body hair in males and wider hips, breast development, and a greater ratio of fat to muscle in females.

All male sex hormones are classified as **androgens** (AN-dro-jens). The main androgen produced by the testes is **testosterone** (tes-TOS-ter-one). In the females, the hormones that most nearly parallel to testosterone in their actions are the **estrogens** (ES-tro-jens), produced by the ovaries. Estrogens contribute to the development of the female secondary sex characteristics and stimulate mammary gland development, the onset of menstruation, and the development and functioning of the reproductive organs.

The other hormone produced by the ovaries, called **progesterone** (pro-JES-ter-one), assists in the normal development of pregnancy (gestation). All the sex hormones are discussed in more detail in Chapter 20.

THE PINEAL GLAND

The **pineal** (PIN-e-al) **gland** is a small, flattened, cone-shaped structure located posterior to the midbrain and connected to the roof of the third ventricle (see **FIG. 11-2** and Dissection Atlas **FIG. A3-2**).

The pineal gland produces the hormone **melatonin** (mel-ah-TO-nin) during dark periods; little hormone is produced during daylight hours. This pattern of hormone secretion influences the regulation of sleep–wake cycles (**BOX 11-3**). Melatonin also appears to delay the onset of puberty.

CHECKPOINTS 

- ☐ **11-14** What two hormones produced by the islets of the pancreas regulate blood glucose levels?
- ☐ **11-15** Sex hormones confer certain features associated with male and female gender. What are these features called as a group?
- ☐ **11-16** What hormone does the pineal gland secrete?

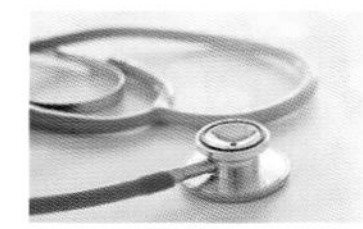

CLINICAL PERSPECTIVES BOX 11-3

Seasonal Affective Disorder: Seeing the Light

Most of us find that long, dark days make us blue and sap our motivation. Are these learned responses or is there a physical basis for them? Studies have shown that the amount of light in the environment does have a physical effect on behavior. Evidence that light alters mood comes from people who are intensely affected by the dark days of winter—people who suffer from **seasonal affective disorder**, aptly abbreviated SAD. When days shorten, these people feel sleepy, depressed, and anxious. They tend to overeat, especially carbohydrates. Research suggests that SAD has a genetic basis and may be associated with decreased levels of the neurotransmitter serotonin.

As light strikes the retina of the eye, it sends impulses that decrease the amount of melatonin produced by the pineal gland in the brain. Because melatonin depresses mood, the final effect of light is to elevate mood. Daily exposure to bright lights has been found to improve the mood of most people with SAD. Exposure for 15 minutes after rising in the morning may be enough, but some people require longer sessions both morning and evening. Other aids include aerobic exercise, stress management techniques, and antidepressant medications.

Other Hormone Sources

Originally, the word *hormone* is applied to the secretions of the endocrine glands only. The term now includes various body substances that have regulatory actions, either locally or at a distance from where they are produced. Many body organs and tissues produce such regulatory substances. Some of these are:

- Adipose tissue (fat) produces **leptin**, a hormone that controls appetite, as well as androgens and estrogens.
- The small intestine secretes hormones that control appetite and help regulate the digestive process.
- The kidneys produce a hormone called **erythropoietin** (e-rith-ro-POY-eh-tin) (EPO), which stimulates red blood cell production in the bone marrow. Production of this hormone increases when there is a decreased oxygen supply in the blood.
- **Osteocalcin**, produced in bone, stimulates such diverse processes as bone formation and insulin secretion.
- The atria (upper chambers) of the heart produce a substance called **atrial natriuretic** (na-tre-u-RET-ik) **peptide** (**ANP**) in response to their increased filling with blood. ANP increases sodium excretion by the kidneys and lowers blood pressure.
- The **placenta** (plah-SEN-tah) produces several hormones during pregnancy. These cause changes in the uterine lining, and later in pregnancy, they help prepare the breasts for lactation. Pregnancy tests are based on the presence of placental hormones (see Chapter 21).

PROSTAGLANDINS

Prostaglandins (pros-tah-GLAN-dins) are a group of hormone-like substances derived from fatty acids. The name prostaglandin comes from the fact that they were first discovered in semen and thought to be derived from the male prostate gland. We now know that prostaglandins are synthesized by almost all body cells. One reason that they are not strictly classified as hormones is that they are produced, act, and are rapidly inactivated in or close to where they are produced. In addition, they are produced not at a defined location, but throughout the body.

A bewildering array of functions has been ascribed to prostaglandins. Some cause constriction of blood vessels, bronchial tubes, and the intestine, whereas others cause dilation of these same structures. Prostaglandins are active in promoting inflammation; certain antiinflammatory drugs, such as aspirin, act by blocking prostaglandin production. Some prostaglandins have been used to induce labor or abortion and have been recommended as possible contraceptive agents.

Overproduction of prostaglandins by the uterine lining (endometrium) can cause painful cramps of the uterine muscle during menstruation. Treatment with prostaglandin inhibitors has been successful in some cases. Much has been written about these substances, and extensive research on them continues.

Hormones and Treatment

Hormones used for medical treatment are obtained from several different sources. Some are extracted from animal tissues. Some hormones and hormone-like substances are available in synthetic form, meaning that they are manufactured in commercial laboratories. A few hormones are produced by the genetic engineering technique of recombinant DNA. In this method, a gene for the cellular manufacture of a given product is introduced in the laboratory into a harmless strain of the common bacterium *Escherichia coli*. The organisms are then grown in quantity, and the desired substance is harvested and purified.

A few examples of natural and synthetic hormones used in treatment are as follows:

- GH is used for the treatment of children and adults with a deficiency of this hormone. Adequate supplies are produced by recombinant DNA techniques.

- Insulin is used in the treatment of diabetes mellitus. Pharmaceutical companies now produce "human" insulin by recombinant DNA methods.
- Adrenal steroids, primarily the glucocorticoids, are used for the relief of inflammation in such diseases as rheumatoid arthritis, lupus erythematosus, asthma, and cerebral edema, for immunosuppression after organ transplantation, and for relief of symptoms associated with circulatory shock.
- Epinephrine (adrenaline) has many uses, including stimulation of the heart muscle when rapid response is required, treatment of asthmatic attacks by relaxation of the small bronchial tubes, and treatment of the acute allergic reaction called **anaphylaxis** (an-ah-fi-LAK-sis).
- Thyroid hormones are used in treatment for an underactive thyroid (hypothyroidism) and as replacement therapy after surgical removal of the thyroid gland.
- Oxytocin is used to cause uterine contractions and induce labor.
- Androgens, including testosterone and androsterone, are used in severe chronic illness to aid tissue building and promote healing.
- Estrogen and progesterone are used as oral contraceptives (e.g., birth control pills, "the pill"). They are highly effective in preventing pregnancy. Occasionally, they give rise to unpleasant side effects, such as nausea. More rarely, they cause serious complications, such as thrombosis (blood clots) or hypertension (high blood pressure). These adverse side effects are more common among women who smoke. Any woman taking birth control pills should have a yearly medical examination.

In women experiencing menopause, levels of estrogen and progesterone begin to decline. Thus, preparations of synthetic estrogen and progesterone, called *hormone replacement therapy (HRT)* or *menopausal hormone therapy (MHT)*, have been developed to treat the unpleasant symptoms associated with this decline. Studies on the most popular forms of MHT have raised questions about their benefits and revealed some risks associated with their use (see Chapter 20 for more information).

CASEPOINT

☐ **11-5** Some forms of diabetes mellitus can be treated with drugs that increase the effectiveness of insulin. Why does Becky have to inject insulin daily?

Hormones and Stress

Stress in the form of physical injury, disease, emotional anxiety, and even pleasure calls forth a specific physiologic response that involves both the nervous system and the endocrine system. The nervous system response, the "fight-or-flight" response, is mediated by parts of the brain, especially the hypothalamus, and by the sympathetic nervous system, which releases norepinephrine. The hypothalamus also coordinates the endocrine response to stress, stimulating the production of some hormones and inhibiting the production of others. Some of these hormones include:

- Epinephrine released from the adrenal medulla facilitates the fight-or-flight response.
- Cortisol released from the adrenal gland helps deal with the stress of starvation, increasing nutrient availability in blood, and inhibiting inflammation.
- ADH released from the posterior pituitary promotes water conservation.
- GH released from the anterior pituitary also increases nutrient availability and helps repair damaged tissues. GH is only released in response to physical stress, such as exercise.

Stress inhibits the production of insulin in order to maximize nutrient availability. It also suppresses the production of thyroid hormones and sex hormones, because these hormones do not facilitate short-term survival.

These stimulatory and inhibitory changes in hormone levels help the body meet stressful situations. Unchecked, however, they are harmful and may lead to such stress-related disorders as hyperglycemia, high blood pressure, heart disease, ulcers, insomnia, back pain, and headaches. Cortisol decreases the immune response, leaving the body more susceptible to infection.

Although no one would enjoy a life totally free of stress in the form of stimulation and challenge, unmanaged stress, or "distress," has negative physical effects. For this reason, techniques such as biofeedback and meditation to control stress are useful. The simple measures of setting priorities, getting adequate periods of relaxation, and getting regular physical exercise are important in maintaining total health.

CHECKPOINTS 

☐ **11-17** What organs produce each of the following: erythropoietin, osteocalcin, ANP?

☐ **11-18** What are four hormones released in times of stress?

See the "Student Resources" on thePoint® for information on careers in exercise and fitness.

Effects of Aging on the Endocrine System

The incidence of endocrine diseases, particularly hypothyroidism and type 2 diabetes mellitus, increases with age. In addition, some of the changes associated with healthy aging, such as loss of muscle and bone tissue, can be linked to changes in the endocrine system. For instance, GH declines, accounting for some losses in strength, immunity, skin thickness, and healing. Sex hormones decline during the middle-age years in both males and females. These changes come from decreased activity of the gonads and also decreased anterior pituitary activity, resulting in decline of gonadotropic hormone secretion. Production of adrenal androgens also declines. Decrease in bone mass leading to osteoporosis is one result of declining sex steroid production. HRT has shown some beneficial effects on mucous membranes and bone mass, but as noted above, its use is controversial. In contrast, the production of thyroid hormone, cortisol, and pancreatic hormones remains relatively constant in healthy old age.

A & P in Action Revisited: Becky's New "Normal"

Becky made her way to the kitchen, hoping she was still in time for pancakes. "Good morning, sleepyhead," greeted her mother as she handed Becky the glucose monitor and lancet. "How was your sleep last night?"

"Great," yawned Becky as she lanced the side of her finger and squeezed a tiny drop of blood onto the monitor's test strip. After a few seconds, the monitor beeped and displayed her blood glucose concentration. "I'm normal," said Becky, half-expecting a wisecrack from her little brother, but he kept on eating.

Since Dr. Carter's diagnosis, Becky had been getting used to her new "normal." It wasn't easy being diabetic. She had to be really careful about what she ate and when. She had to measure her glucose and inject herself with insulin before meals. Monitoring her glucose wasn't too bad, but Becky didn't think she would ever get used to the needles. She was also a little worried about what the kids at school were saying about her and her disease. One unexpected benefit was that Max seemed to have a newfound respect for her and her ability to inject herself. "What a weirdo!" she thought as she carefully poured a little bit of syrup on her pancakes.

During this case, we saw that the lack of the hormone insulin had negative effects on Becky's whole body. In this and later chapters, we learn more about the endocrine system's role in regulating body functions.

CHAPTER 11

Chapter Wrap-Up

OVERVIEW

A detailed chapter outline with space for note-taking is on thePoint®. The figure below illustrates the main topics covered in this chapter.

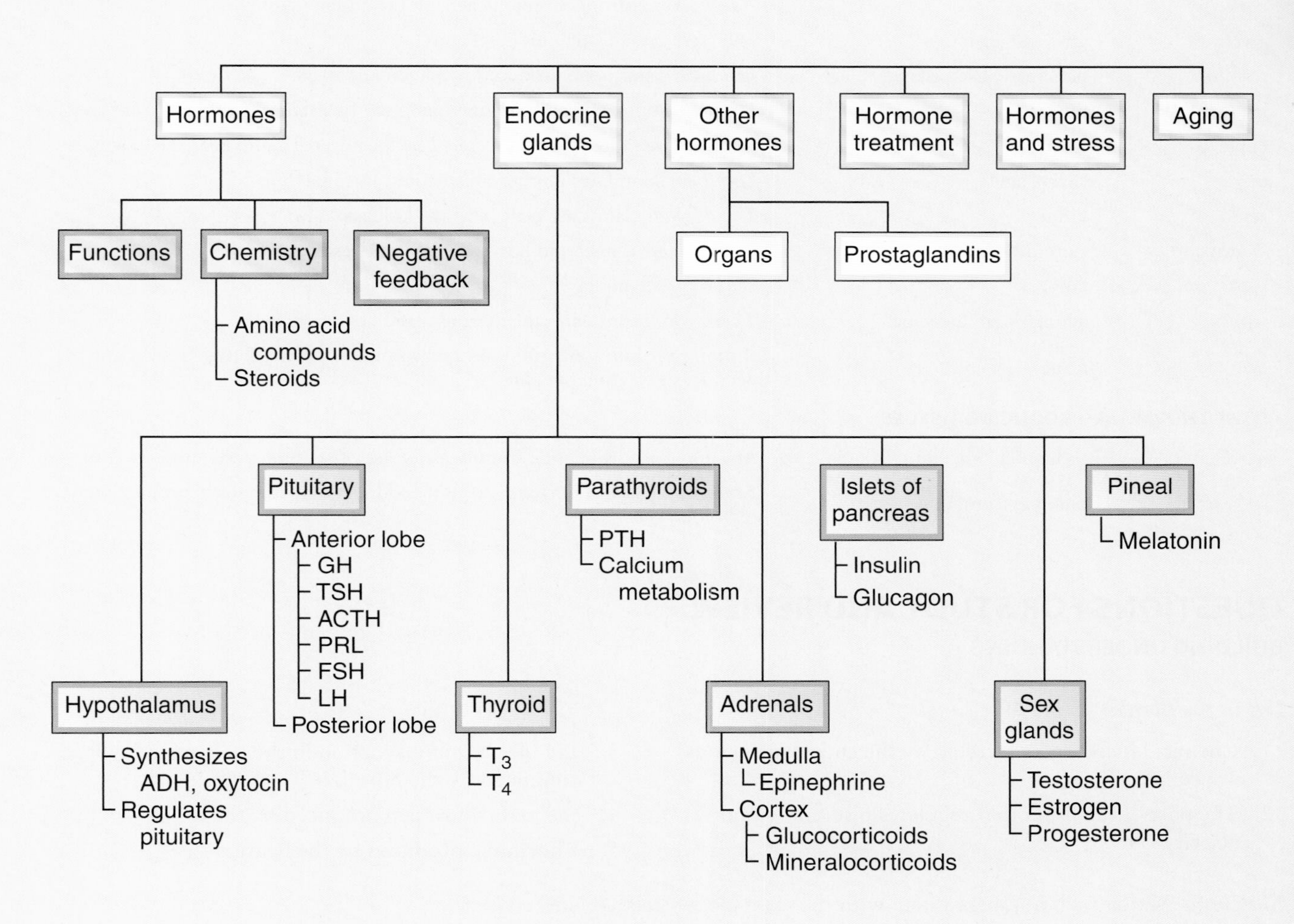

KEY TERMS

The terms listed below are emphasized in this chapter. Knowing them will help you organize and prioritize your learning. These and other boldface terms are defined in the "Glossary" with phonetic pronunciations.

endocrine system
hormone
hypothalamus
pituitary (hypophysis)
prostaglandin
receptor
steroid
target tissue

WORD ANATOMY

Medical terms are built from standardized word parts (prefixes, roots, and suffixes). Learning the meanings of these parts can help you remember words and interpret unfamiliar terms.

WORD PART	MEANING	EXAMPLE
THE ENDOCRINE GLANDS AND THEIR HORMONES		
andr/o	male	An *androgen* is any male sex hormone.
cortic/o	cortex	*Adrenocorticotropic* hormone acts on the adrenal cortex.
glyc/o	glucose, sugar	*Hyperglycemia* is high blood glucose.
insul/o	pancreatic islet, island	*Insulin* is a hormone produced by pancreatic islets.
lact/o	milk	*Prolactin* stimulates milk production in the breasts.
nephr/o	kidney	*Epinephrine* (adrenaline) is secreted by the adrenal gland near the kidney.
oxy	sharp, acute	*Oxytocin* stimulates uterine contractions during labor.
ren/o	kidney	The *adrenal* glands are near (ad-) the kidneys.
-sterone	steroid hormone	*Testosterone* is a steroid hormone from the testes.
toc/o	labor	See "oxy" example.
trop/o	acting on, influencing	*Thyrotropin* stimulates the thyroid gland.
ur/o	urine	*Antidiuretic* hormone promotes reabsorption of water in the kidneys and decreases excretion of urine.
OTHER HORMONE-PRODUCING TISSUES		
natri	sodium (*L. natrium*)	Atrial *natriuretic* peptide stimulates release of sodium in the urine.
-poiesis	making, forming	*Erythropoietin* is a hormone from the kidneys that stimulates production of red blood cells.

QUESTIONS FOR STUDY AND REVIEW

BUILDING UNDERSTANDING

Fill in the Blanks

1. Chemical messengers secreted by the endocrine glands are called ______.
2. The part of the brain that regulates pituitary gland activity is the ______.
3. Red blood cell production in the bone marrow is stimulated by the hormone ______.
4. The main androgen produced by the testes is ______.
5. A hormone produced by the heart is ______.

Matching Match each numbered item with the most closely related lettered item.

___ 6. An anterior pituitary lobe hormone

___ 7. An adrenal cortex hormone

___ 8. A pancreatic hormone

___ 9. A posterior pituitary lobe hormone

___ 10. An ovarian hormone

a. aldosterone
b. estrogen
c. glucagon
d. growth hormone
e. oxytocin

Multiple Choice

___ 11. To what do hormones bind?
a. lipid bilayer
b. transporters
c. ion channels
d. receptors

___ 12. Which hormone promotes uterine contractions and milk ejection?
a. prolactin
b. oxytocin
c. estrogen
d. luteinizing hormone

___ 13. Choose the principal hormonal regulator of the basal metabolic rate.
a. cortisol
b. triiodothyronine
c. aldosterone
d. progesterone

___ 14. Which structure secretes epinephrine?
a. adrenal cortex
b. adrenal medulla
c. kidneys
d. pancreas

___ 15. Which organ regulates sleep–wake cycles?
a. pituitary
b. thyroid
c. thymus
d. pineal

UNDERSTANDING CONCEPTS

16. With regard to regulation, what are the main differences between the nervous system and the endocrine system?
17. Explain how the hypothalamus and pituitary gland regulate certain endocrine glands. Use the thyroid as an example.
18. Name the two divisions of the pituitary gland. List the hormones released from each division, and describe the effects of each.
19. Compare and contrast the following hormones:
a. thyroxine and triiodothyronine
b. cortisol and aldosterone
c. insulin and glucagon
d. testosterone and estrogen
20. Describe the anatomy of the following endocrine glands:
a. thyroid
b. pancreas
c. adrenals
21. Name the hormone released by the kidneys and by the pineal gland. What are the effects of each?
22. List several hormones released during stress. What is the relationship between prolonged stress and disease?
23. Referring to the Dissection Atlas, Appendix 3, name all the endocrine glands shown and give the number of the figure(s) in which they appear.

CONCEPTUAL THINKING

24. In the case study, Dr. Carter noted that Becky presented with the three cardinal signs of type 1 diabetes mellitus. What are they? What causes them?
25. How is type 1 diabetes mellitus similar to starvation?
26. One consequence of decreased blood levels of thyroid hormone is increased blood levels of thyroid-stimulating hormone (TSH). Why?

For more questions, see the "Learning Activities" on thePoint®.

UNIT IV

Circulation and Body Defense

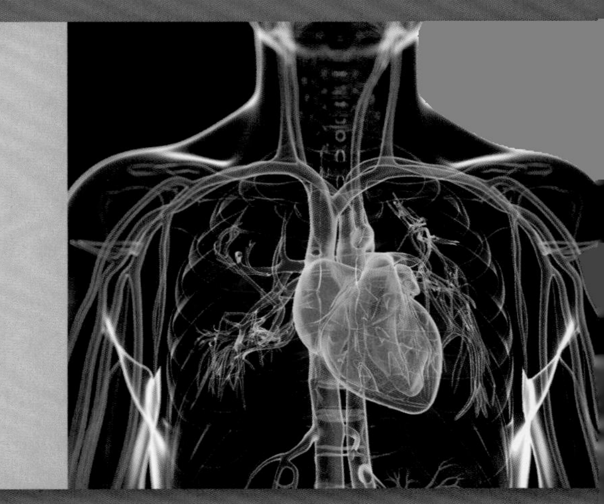

CHAPTER 12

The Blood

Learning Objectives

After careful study of this chapter, you should be able to:

1. List the functions of the blood. ***p. 244***
2. Identify the main components of plasma. ***p. 245***
3. Describe the formation of blood cells. ***p. 245***
4. Name and describe the three types of formed elements in the blood, and give the functions of each. ***p. 246***
5. Characterize the five types of leukocytes. ***p. 247***
6. Define hemostasis, and cite three steps in hemostasis. ***p. 250***
7. Briefly describe the steps in blood clotting. ***p. 250***
8. Define *blood type*, and explain the relation between blood type and transfusions. ***p. 251***
9. Explain the basis of Rh incompatibility and its possible consequences. ***p. 251***
10. List four possible reasons for transfusions of whole blood and blood components. ***p. 253***
11. Identify six types of tests used to study blood. ***p. 254***
12. Referring to the case study, discuss the adverse effects of abnormal blood cell production. ***pp. 243, 257***
13. Show how word parts are used to build words related to the blood (see Word Anatomy at the end of the chapter). ***p. 259***

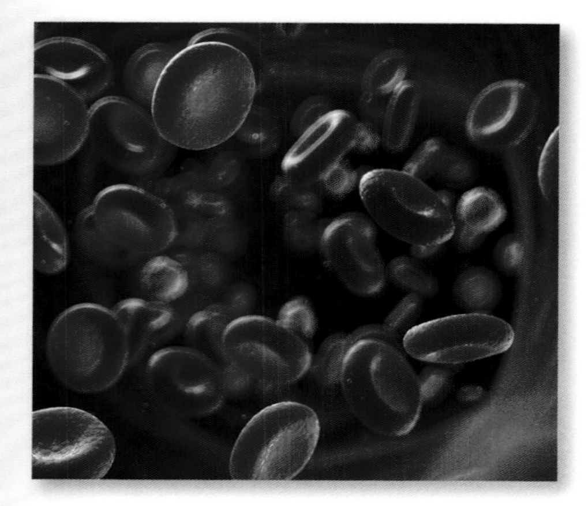

A & P in Action
Allen's Blood Disorder

Allen, a 62-year-old retired chemist, presented to the emergency department of his local hospital. "I don't know what to do, doctor," he stated to the emergency physician on call, "I blew my nose this morning and it started to bleed, and now it just won't stop!" Allen had tried pinching his nose for 20 minutes and placing a bag of frozen peas over his nose, but when the blood just kept coming, his wife had driven him to the ED. While a nurse and the physician staunched the bleeding and packed Allen's left nostril, a resident took Allen's history and performed a physical exam.

Allen, a former smoker, had no significant medical history. "I've been meaning to schedule a check-up, though," Allen said, "because I've been feeling so tired lately - I get winded doing the littlest things." Allen also revealed that he seemed to run fevers and woke up shivering and sweating sometimes at night. The resident noticed that the roof of Allen's mouth was pale and that his lower legs were covered in pinpoint red spots that did not blanch when pressed. His blood work showed a low platelet count; a low hemoglobin and hematocrit, indicating anemia; and an elevated white blood cell count. His blood type was O positive. Allen was admitted to the hospital for further workup and consultation with a hematologist.

When Dr. Howard, the hematologist, examined Allen's blood smear under the microscope the next morning, he noticed that the white cells were not only too numerous but also abnormally large with enlarged nuclei. He ordered a bone marrow biopsy.

Several days later, Dr. Howard sat down with Allen and his wife to discuss his condition. "All of our tests unfortunately showed that you have developed acute myelogenous leukemia, or AML," the doctor explained. "This is the most common acute leukemia in adults, and it means that one type of stem cell in your bone marrow is multiplying out of control and producing non-helpful cells that don't mature properly. This is causing your symptoms of fatigue, fevers, and bleeding. We will need to start chemotherapy right away and follow your blood cell counts closely. We can give you transfusions of whole blood and platelets to help with your symptoms, and hormones to stimulate normal blood cell production. The nurse practitioner will be discussing good general hygiene to help you prevent infections going forward."

This case study shows the importance of the bone marrow in producing all of the cells found in blood. We'll learn more about the functions of the different blood cells in this chapter and follow up on Allen's case.

As you study this chapter, CasePoints will give you opportunities to apply your learning to this case.

Visit thePoint® to access the following resources. For guidance in using these resources most effectively, see pp. xv–xvii.

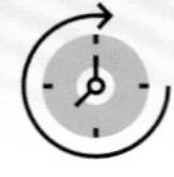

Preparing to Learn

- Tips for Effective Studying
- Pre-Quiz

While You Are Learning

- Web Figure: Hematopoiesis
- Web figure: Erythrocyte Life Cycle
- Web Figure: Hemostasis
- Animation: Hemostasis
- Chapter Notes Outline
- Audio Pronunciation Glossary

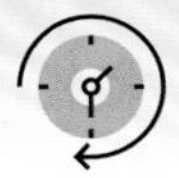

When You Are Reviewing

- Answers to Questions for Study and Review
- Health Professions: Hematology Specialist
- Interactive Learning Activities

A LOOK BACK

Blood is one of the two types of circulating connective tissue introduced in Chapter 4. The other is lymph, discussed more fully in Chapter 15. In this chapter, we fully discuss the structure and function of this all-important body fluid. Important key ideas include homeostasis and negative feedback 3, the function of water 6 and of enzymes 7, and the relationship between genes and proteins 9.

Introduction

Blood is a life-giving fluid that brings nutrients and oxygen to the cells and carries away waste. The heart pumps blood continuously through a closed system of vessels. The heart and blood vessels are described in Chapters 13 and 14, respectively.

Blood is classified as a connective tissue because it consists of cells suspended in an extracellular background material, or matrix. However, blood differs from other connective tissues in that its cells are not fixed in position; instead, they move freely in the plasma, the blood's liquid matrix.

Whole blood is a viscous (thick) fluid that varies in color from bright scarlet to dark red, depending on how much oxygen it is carrying. (It is customary in drawings to color blood high in oxygen as red and blood low in oxygen as blue.) The blood volume accounts for approximately 8% of total body weight. The actual quantity of circulating blood differs with a person's size and gender; adult men have about 75 mL/kg body weight, while adult women have about 65 mL/kg body weight. So, a 150-lb (68-kg) male contains about 5.1 L (5.4 qt) of blood, and a 120-lb (54-kg) woman contains about 3.5 L (3.7 qt).

Blood Functions

The blood participates in three important body functions: transportation, protection, and regulation.

Blood serves as a transport medium for many important substances, including gases, nutrients, waste products, and hormones. It transports heat from warmer regions to cooler regions, for example, from the heat-producing muscles to the cooler extremities. As discussed in Chapter 18, modifying the amount of blood traveling to different regions is one method used to equalize body temperature.

The cells and proteins found in the blood also serve protective functions. This chapter explains how blood components protect against blood loss from the site of an injury. This chapter also discusses how blood proteins and cells protect against pathogens, a topic that is further explained in Chapter 15.

The body regulates both the volume and composition of the blood in order to maintain homeostasis. For instance, the kidneys and the hypothalamic thirst center control the overall blood volume in order to maintain normal blood pressure (see Chapter 14). They do so by stimulating fluid intake and regulating fluid loss in the urine (see Chapter 19). The kidneys and other organs control the concentration of many blood components, such as sodium, potassium, and calcium. The blood also contains substances (proteins and electrolytes) that act as *buffers,* substances that help keep the pH of body fluids steady at about 7.4. (The actual range of blood pH is 7.35 to 7.45.) Recall that pH is a measure of a solution's acidity or alkalinity. At an average pH of 7.4, blood is slightly alkaline (basic).

Water and most dissolved substances (except proteins) can move freely between the blood and the fluid surrounding cells (the interstitial fluid). So, by regulating the composition of the blood, the body also regulates the interstitial fluid.

CHECKPOINTS 

- 12-1 What are four types of substances transported in the blood?
- 12-2 What is the pH range of the blood?

See the Student Resources on thePoint® for information on careers in hematology, the study of blood.

Blood Constituents

The blood is divided into two main components (FIG. 12-1). The liquid portion is the **plasma**. The **formed elements**,

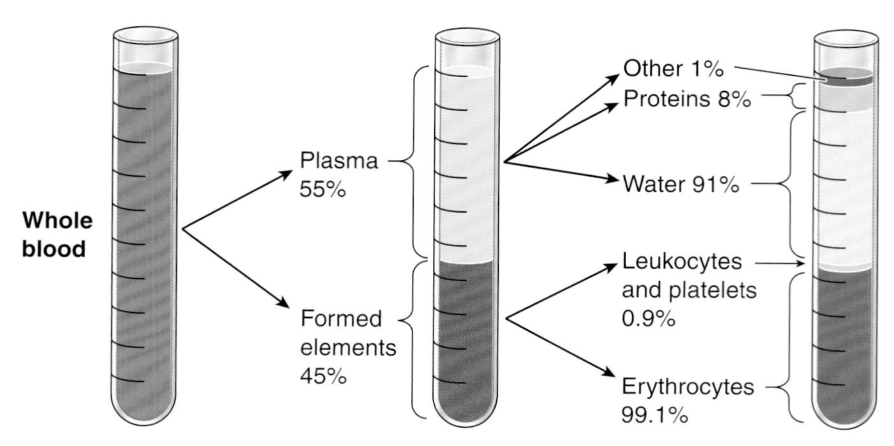

FIGURE 12-1 Composition of whole blood. KEY POINT Percentages show the relative proportions of the different components of plasma and formed elements.

Table 12-1 Formed Elements of Blood

Elements	Number per mcL of Blood	Description	Function
Erythrocytes (red blood cells)	5 million	Tiny (7 mcm diameter), biconcave disks without nucleus (anuclear)	Carry oxygen bound to hemoglobin; also carry some carbon dioxide and buffer blood
Leukocytes (white blood cells)	5,000–10,000	Larger than red cells with prominent nucleus that may be segmented (granulocyte) or unsegmented (agranulocyte); vary in staining properties	Active in immunity; located in blood, tissues, and lymphatic system
Platelets	150,000–450,000	Fragments of large cells (megakaryocyte)	Hemostasis; form a platelet plug and start blood clotting (coagulation)

which include cells and cell fragments, can be further divided into three categories:

- **Erythrocytes** (eh-RITH-ro-sites), from *erythro*, meaning "red," are the red blood cells (RBCs), which transport oxygen and, to a lesser extent, carbon dioxide.
- **Leukocytes** (LU-ko-sites), from *leuko*, meaning "white," are the several types of white blood cells (WBCs), which help defend against infections and cancer.
- **Platelets**, also called **thrombocytes** (THROM-bo-sites), are cell fragments that participate in blood clotting.

TABLE 12-1 summarizes information on the different types of formed elements. **FIGURE 12-2** shows all the categories of formed elements in a blood smear, that is, a blood sample spread thinly over the surface of a glass slide, as viewed under a microscope.

BLOOD PLASMA

About 55% of the total blood volume is plasma. The plasma itself is 91% water. 6 So, many of the properties of blood reflect the properties of water discussed in Chapter 2, such as its ability to transport heat and to act as a solvent for water-soluble (hydrophilic) substances. Many different substances, dissolved or suspended in the water, make up the other 9% by weight (see FIG. 12-1). The plasma content may vary somewhat because substances are removed and added as the blood circulates to and from the tissues. However, the body tends to maintain a fairly constant level of most substances. For example, the level of glucose, a simple sugar, is maintained at a remarkably constant level of about one-tenth of 1% (0.1%) in solution.

After water, the next largest percentage (about 8%) of material in the plasma is protein. The liver synthesizes most plasma proteins. They include the following:

- **Albumin** (al-BU-min), the most abundant protein in plasma, is important for maintaining the blood's osmotic pressure. Recall from Chapter 3 that osmotic pressure reflects the ability of a solution to attract water. Albumin is thus necessary to maintain normal blood volume, as discussed further in Chapter 14.
- **Clotting factors**, necessary for preventing blood loss from damaged vessels, are discussed later.
- **Antibodies**, substances that combat infection and are made by certain WBCs involved in immunity (see Chapter 15).
- **Complement** consists of a group of enzymes that participate in immunity (see Chapter 15).

The remaining 1% of the plasma consists of nutrients, electrolytes, and other materials that must be transported. With regard to the nutrients, glucose is the principal carbohydrate found in the plasma. This simple sugar is absorbed from digested foods in the intestine. It can also be released from the liver. Amino acids, the products of protein digestion, also circulate in the plasma. Lipids constitute a small percentage of blood plasma. Lipid components include cholesterol and fats. As lipids are not soluble in plasma, they are packaged together with proteins to form lipoproteins. The electrolytes in the plasma include sodium, potassium, calcium, magnesium, chloride, carbonate, and phosphate. These electrolytes have a variety of functions, including the formation of bone (calcium and phosphorus), the production of certain hormones (such as iodine for the production of thyroid hormones), and maintenance of the acid–base balance (such as sodium and potassium carbonates and phosphates present in buffers).

Other materials transported in plasma include hormones, waste products, drugs, and small amounts of dissolved gases such as oxygen and carbon dioxide.

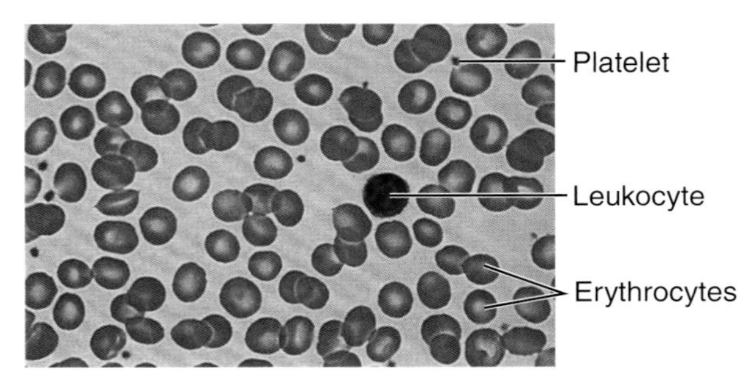

FIGURE 12-2 Blood cells as viewed under the microscope.
KEY POINT All three types of formed elements are visible.
ZOOMING IN Which cells are the most numerous in the blood?

THE FORMED ELEMENTS

All of the blood's formed elements are produced in the red bone marrow, which is located in the ends of long bones and in the inner portion of all other bones. The ancestors of all the blood cells are called **hematopoietic** (blood-forming) **stem cells**. These cells reproduce frequently, and their offsprings differentiate into the different blood cell types discussed below.

In comparison with other cells, most blood cells are short-lived. The need for constant blood cell replacement means that normal activity of the red bone marrow is absolutely essential to life. In the opening case study, Allen has lost the protective functions of the blood cells due to cancerous stem cells interfering with the function of the red bone marrow.

See the Student Resources on thePoint® for a figure on hematopoiesis detailing the development of all the formed elements and a figure on the erythrocyte life cycle.

Erythrocytes Erythrocytes, the red blood cells (RBCs or red cells), measure about 7 mcm (micrometers) in diameter (a micrometer is one-millionth of a meter). They are disk-shaped bodies with a depression on both sides. This biconcave shape creates a central area that is thinner than the edges (**FIG. 12-3**). Erythrocytes are different from other cells in that the mature form found in the circulating blood lacks a nucleus (is anuclear) and also lacks most of the other organelles commonly found in cells. As red cells mature, these components are lost, providing more space for the cells to carry oxygen. This vital gas is bound in the red cells to **hemoglobin** (he-mo-GLO-bin), a protein that contains iron. (**BOX 12-1** on the structure and function of hemoglobin.) Hemoglobin, combined with oxygen, gives the blood its characteristic red color. The more oxygen carried by the hemoglobin, the brighter is the blood's red color. Therefore, the blood that goes from the lungs to the tissues is a bright red because it carries a great supply of oxygen; in contrast, the blood that returns to the lungs is a much darker red because it has given up some of its oxygen to the tissues.

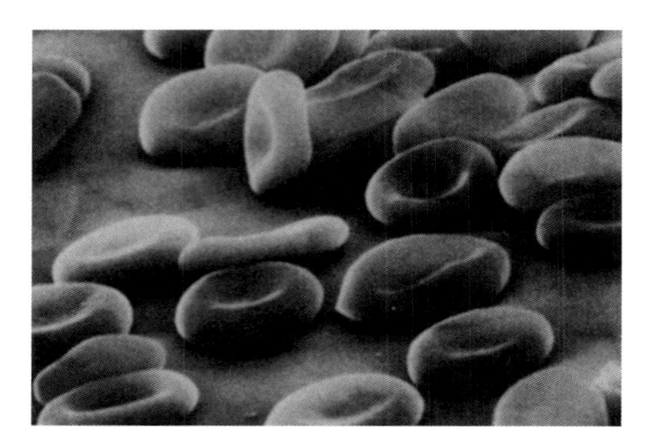

FIGURE 12-3 Red blood cells as seen under a scanning electron microscope. KEY POINT This type of microscope provides a three-dimensional view of the cells, revealing their shape. ZOOMING IN Why are these cells described as biconcave?

Hemoglobin has two lesser functions in addition to the transport of oxygen. Hemoglobin can carry hydrogen ions, so it acts as a buffer and plays an important role in acid–base balance (see Chapter 19). Hemoglobin also carries some carbon dioxide.

A CLOSER LOOK
Hemoglobin: Door-to-Door Oxygen Delivery
BOX 12-1

The hemoglobin molecule is a protein made of four amino acid chains (the globin part of the molecule), each of which holds an iron-containing heme group. Each of the four hemes can bind one molecule of oxygen.

Hemoglobin allows the blood to carry much more oxygen than it could were the oxygen simply dissolved in the plasma. A red blood cell contains about 250 million hemoglobins, each capable of binding four molecules of oxygen. So a single red blood cell can carry about 1 billion oxygen molecules! Hemoglobin reversibly binds oxygen, picking it up in the lungs and releasing it in the body tissues. Active cells need more oxygen and also generate heat and acidity. These changing conditions promote the release of oxygen from hemoglobin into metabolically active tissues.

Immature red blood cells (erythroblasts) produce hemoglobin as they mature into erythrocytes in the red bone marrow. When the liver and spleen destroy old erythrocytes, they break down the released hemoglobin. Some of its components are recycled, and the remainder leaves the body as a brown fecal pigment called stercobilin. Despite some conservation, dietary protein and iron are still essential to maintain hemoglobin supplies.

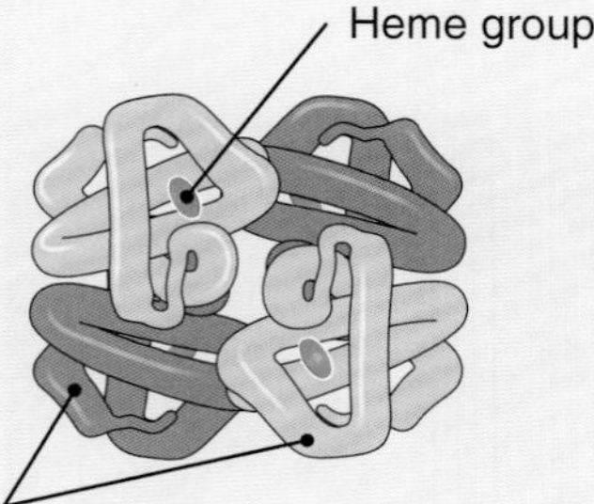

Hemoglobin. This protein in red blood cells consists of four amino acid chains (globins), each with an oxygen-binding heme group.

Hemoglobin's ability to carry oxygen can be blocked by carbon monoxide. This odorless and colorless but harmful gas combines with hemoglobin to form a stable compound that can severely restrict the erythrocytes' ability to carry oxygen. Carbon monoxide is a byproduct of the incomplete burning of fuels, such as gasoline and other petroleum products and coal, wood, and other carbon-containing materials. It also occurs in cigarette smoke and automobile exhaust.

Erythrocytes are by far the most numerous of the blood cells, averaging from 4.5 to 5 million per microliter (mcL) of blood. (A microliter is one-millionth of a liter.) Because mature red cells have no nucleus and cannot divide or repair themselves, they must be replaced constantly. After leaving the bone marrow, they circulate in the bloodstream for about 120 days before their membranes deteriorate, and they are destroyed by the liver and spleen.

3 Red cells act in a negative feedback loop controlling the level of blood oxygen, which is a regulated variable. The signal in this loop is the hormone **erythropoietin** (eh-rith-ro-POY-eh-tin) (**EPO**), which is released from the kidney (the sensor/control center) in response to decreased oxygen. EPO acts on the bone marrow (the effector) to increase red cell production. Once the body's oxygen transport capacity returns to normal, EPO production declines.

Constant red cell production requires an adequate supply of nutrients, particularly protein; the vitamins B_{12} and folic acid, required for the production of DNA; and the minerals iron and copper for the production of hemoglobin. Vitamin C is also important for the proper absorption of iron from the small intestine.

CHECKPOINTS 

- ☐ **12-3** What are the two main components of blood?
- ☐ **12-4** Next to water, what is the most abundant type of substance in plasma?
- ☐ **12-5** Where do blood cells form, and from which stem cell type are they produced?
- ☐ **12-6** What is the main function of hemoglobin?
- ☐ **12-7** What is the signal in the feedback loop controlling red cell production? What is the regulated variable?

Leukocytes The leukocytes, or white blood cells (WBCs or white cells), differ from the erythrocytes in appearance, quantity, and function. The cells themselves are round, but they contain prominent nuclei of varying shapes and sizes. Occurring at a concentration of 5,000 to 10,000/mcL of blood, leukocytes are outnumbered by red cells by about 700 to 1. Although the red cells have a definite color, the leukocytes are colorless and must be stained if we are to study them under the microscope.

Types of Leukocytes The different types of white cells are identified by their size, the shape of the nucleus, and the appearance of granules in the cytoplasm when the cells are stained **(TABLE 12-2)**. The stain commonly used for blood is the Wright stain, which is a mixture of dyes that differentiates the various blood cells. The "granules" in the white cells are actually lysosomes and secretory vesicles. They are present in all WBCs, but they are more easily stained and more visible in some cells than in others. Leukocytes are active in immunity. As we discuss later in this chapter, the relative percentage of the different types of leukocytes is a valuable clue in arriving at a medical diagnosis.

The granular leukocytes, or **granulocytes** (GRAN-u-lo-sites), are so named because they show visible granules in the cytoplasm when stained. Each has a very distinctive, highly segmented nucleus **(see TABLE 12-2)**. The different types of granulocytes are named for the type of dyes they take up when stained. They include the following:

- **Neutrophils** (NU-tro-fils) stain with either acidic or basic dyes
- **Eosinophils** (e-o-SIN-o-fils) stain with acidic dyes (eosin is one)
- **Basophils** (BA-so-fils) stain with basic dyes

Neutrophils are the most numerous of the white cells, constituting approximately 60% of all circulating leukocytes. Because the nuclei of the neutrophils have various shapes, these cells are also called **polymorphs** (meaning "many forms") or simply *polys*. Other nicknames are *segs*, referring to the segmented nucleus, and *PMNs*, an abbreviation of **p**oly**m**orphonuclear **n**eutrophils. Before reaching full maturity and becoming segmented, a neutrophil's nucleus looks like a thick, curved band **(FIG. 12-4)**. An increase in the number of these **band cells** (also called *stab* or *staff cells*) is a sign of infection and active neutrophil production. Eosinophils and basophils make up a small percentage of the white cells but increase in number during allergic reactions.

The agranular leukocytes, or **agranulocytes**, are so named because they lack easily visible granules **(see TABLE 12-2)**. Their nuclei are round or curved and are not segmented. There are two types of agranular leukocytes:

- **Lymphocytes** (LIM-fo-sites) are the second most numerous of the white cells. Although lymphocytes originate in the red bone marrow, they develop to maturity in lymphoid tissue and can multiply in this tissue as well. They are more abundant in the lymphatic system than in blood (see Chapter 15).
- **Monocytes** (MON-o-sites) are the largest of all white cells. They average about 5% of the leukocytes.

Functions of Leukocytes Leukocytes clear the body of foreign material and cellular debris and help defend against cancer. Most importantly, they destroy pathogens that may invade the body. Neutrophils and monocytes engage in **phagocytosis** (fag-o-si-TO-sis), the engulfing of foreign matter **(FIG. 12-5)**. Whenever pathogens enter the tissues, as through a wound, phagocytes are attracted to the area. They squeeze between the cells of the capillary walls and proceed by ameboid (ah-ME-boyd), or ameba-like, motion to the area of infection where they engulf the invaders. Lysosomes in the cytoplasm then digest the foreign organisms, and the cells eliminate the waste products.

When foreign organisms invade, the bone marrow and lymphoid tissue go into emergency production of white

Table 12-2 Leukocytes (White Blood Cells)

Cell Type	Relative Percentage (Adult)	Description	Function
Neutrophils	54%–62%	Stain with either acidic or basic dyes; show lavender granules when stained	Phagocytosis
Eosinophils	1%–3%	Stain with acidic dyes; show beadlike, bright pink granules when stained	Allergic reactions; defense against parasites
Basophils	Less than 1%	Stain with basic dyes; have large, dark blue granules that can obscure the nucleus	Allergic reactions; inflammatory reactions
Lymphocytes	25%–38%	Mature and can multiply in lymphoid tissue	Immunity (T cells and B cells)
Monocytes	3%–7%	Largest of leukocytes	Phagocytosis

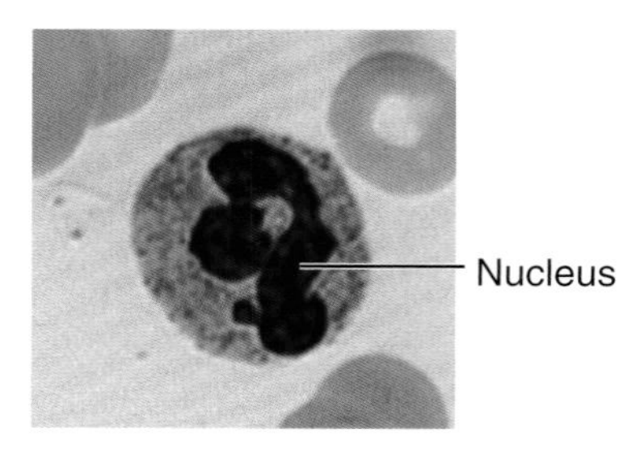

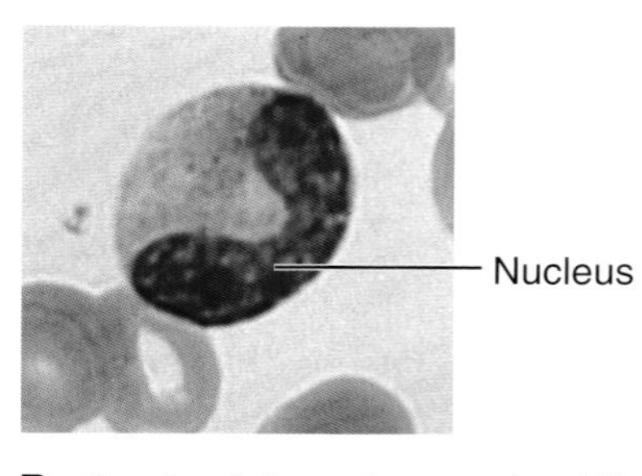

FIGURE 12-4 Stages in neutrophil development. A. A mature neutrophil has a segmented nucleus. **B.** An immature neutrophil is called a band cell because the nucleus is shaped like a thick, curved band.

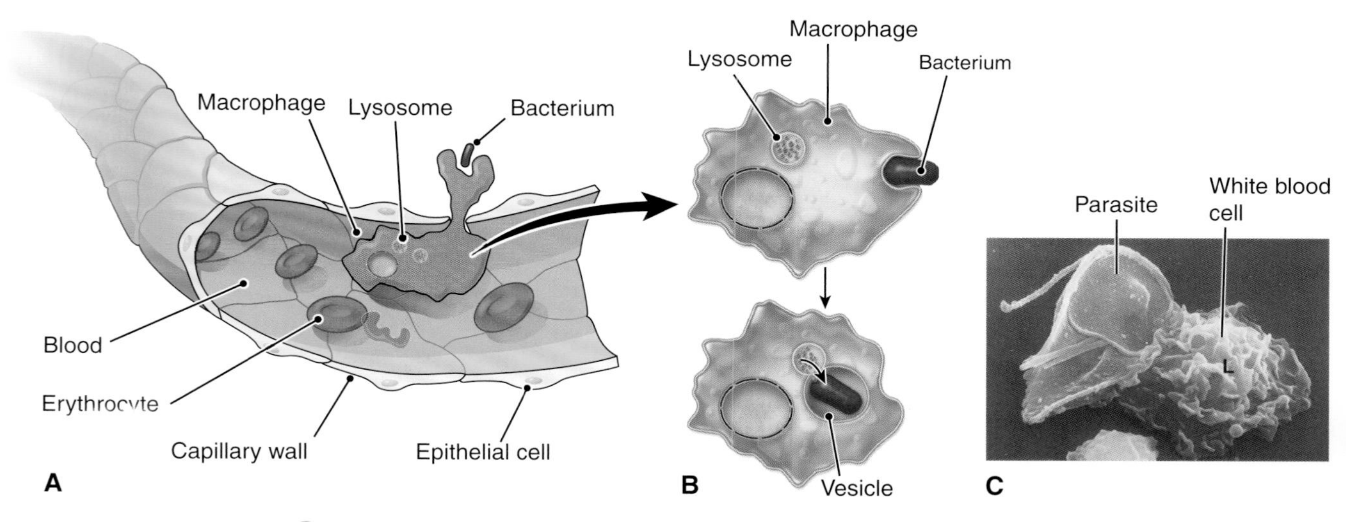

FIGURE 12-5 Phagocytosis. KEY POINT Phagocytosis is the engulfing of foreign matter by white cells. **A.** A phagocytic leukocyte (white blood cell) squeezes through a capillary wall in the region of an infection and engulfs a bacterium. **B.** The bacterium is enclosed in a vesicle and digested by a lysosome. **C.** A scanning electron microscope image of a phagocyte ingesting a parasite. ZOOMING IN What type of epithelium makes up the capillary wall?

12

cells, and their number increases enormously as a result. Detection of an abnormally large number of white cells in the blood is an indication of infection. In battling pathogens, leukocytes themselves are often destroyed. A mixture of dead and living bacteria, together with dead and living leukocytes, forms **pus**. A collection of pus localized in one area is known as an **abscess**.

When monocytes enter the tissues, they enlarge and mature into **macrophages** (MAK-ro-faj-ez). These phagocytic superstars are highly active in disposing of invaders and foreign material. Although most circulating lymphocytes live only six to eight hours, those that enter the tissues may survive for longer periods—days, months, or even years.

Some lymphocytes become **plasma cells**, active in the production of circulating antibodies needed for immunity. The activities of the various white cells are further discussed in Chapter 15.

Platelets **Blood platelets** (thrombocytes) are the smallest of all the formed elements **(FIG. 12-6A)**. These tiny structures are not cells in themselves but rather fragments constantly released from giant bone marrow cells called **megakaryocytes** (meg-ah-KAR-e-o-sites) **(see FIG. 12-6B)**. Platelets do not have nuclei or DNA, but they do contain active enzymes and mitochondria. The number of platelets in the circulating blood normally varies from 150,000 to 450,000/mcL. They have a life span of about ten days.

Platelets are essential for the prevention of blood loss (hemostasis), discussed next. When blood comes in contact with any tissue other than the smooth lining of the blood vessels, as in the case of injury, the platelets stick together and form a plug that seals the wound. The platelets then release chemicals that participate in the formation of a clot to stop blood loss. More details on these reactions follow:

CHECKPOINTS

- ☐ **12-8** What are the three types of granular leukocytes? What are the two types of agranular leukocytes?
- ☐ **12-9** What is the most important function of leukocytes?
- ☐ **12-10** What is the function of blood platelets?

CASEPOINTS

- ☐ **12-1** What explains Allen's pale-colored palate and extreme tiredness?
- ☐ **12-2** Why was he subject to repeated infections, as evidenced by his frequent fevers and chills?

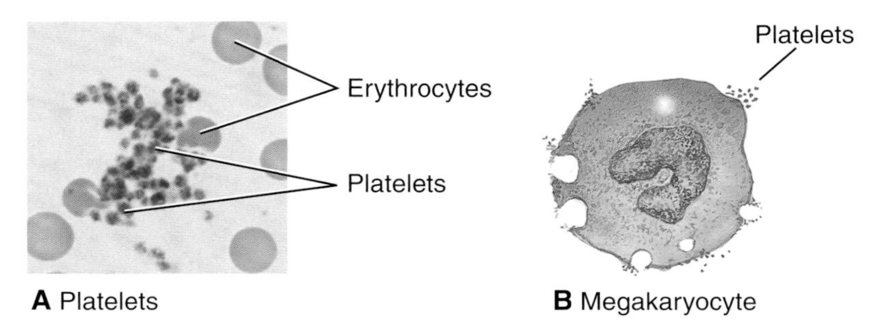

FIGURE 12-6 Platelets (thrombocytes). KEY POINT Platelets are fragments of larger cells. **A.** Platelets in a blood smear. **B.** A megakaryocyte releases platelets.

Hemostasis

Hemostasis (he-mo-STA-sis) is the process that prevents blood loss from the circulation when a blood vessel is ruptured by an injury. Events in hemostasis include the following:

1. **Contraction** of the smooth muscles in the blood vessel wall. The resulting reduction in the vessel's diameter, known as *vasoconstriction*, reduces blood flow and loss from the defect in the vessel.
2. Formation of a **platelet plug.** Activated platelets become sticky and adhere to the defect to form a temporary plug.
3. Formation of a **blood clot**, by the process of **coagulation** (ko-ag-u-LA-shun).

Once initiated, the clotting process proceeds rapidly. This quick response is possible because all of the substances needed for clotting are already present in the blood but usually in their inactive forms. Think of a racecar driver at the beginning of the race. They maintain a state of readiness by pressing on both the brake and the accelerator and can achieve maximum speed within a short time interval. Similarly, a balance is maintained between compounds that promote clotting, known as **procoagulants**, and those that prevent clotting, known as **anticoagulants**. Under normal conditions, the substances that prevent clotting (the anticoagulants) prevail. When an injury occurs, however, the procoagulants are activated, and a clot is formed (FIG. 12-7).

The clotting process is a well-controlled series of separate events involving 12 different clotting factors, each designated by a Roman numeral. Calcium ion (Ca^{2+}) is one such factor. Others are released from damaged tissue and activated platelets 7. Still others are enzymes, made in the liver and released into the bloodstream in their inactive form, which can be activated in the clotting process. To manufacture these enzymes, the liver requires vitamin K. We obtain some of this vitamin in food from green vegetables and grains, but a large proportion is made by bacteria living symbiotically in the large intestine. The final step in the clotting reaction is the conversion of a plasma protein called **fibrinogen** (fi-BRIN-o-jen) into solid threads of **fibrin**, in which blood cells are trapped to form the clot. The final steps involved in blood clot formation are described below and illustrated in FIGURE 12-7.

1. Substances released from damaged tissue and sticky platelets initiate a reaction sequence that leads to the formation of an active enzyme called **prothrombinase** (pro-THROM-bih-nase).
2. Prothrombinase converts prothrombin in the blood to **thrombin**. Calcium is needed for this step.
3. Thrombin, in turn, converts soluble fibrinogen into insoluble fibrin. Threads of fibrin form a meshwork that entraps plasma and blood cells to form a clot.

Blood clotting occurs in response to injury. Blood also clots when it comes into contact with some surface other than the lining of a blood vessel, for example, a glass or plastic tube used for a blood specimen. In this case, the preliminary steps of clotting are somewhat different and require more time, but the final steps are the same as those illustrated in FIGURE 12-7. The fluid that remains after clotting has occurred is called **serum** (plural, *sera*). Serum contains all the components of blood plasma *except* the clotting factors, as expressed in the formula:

Plasma = serum + clotting factors

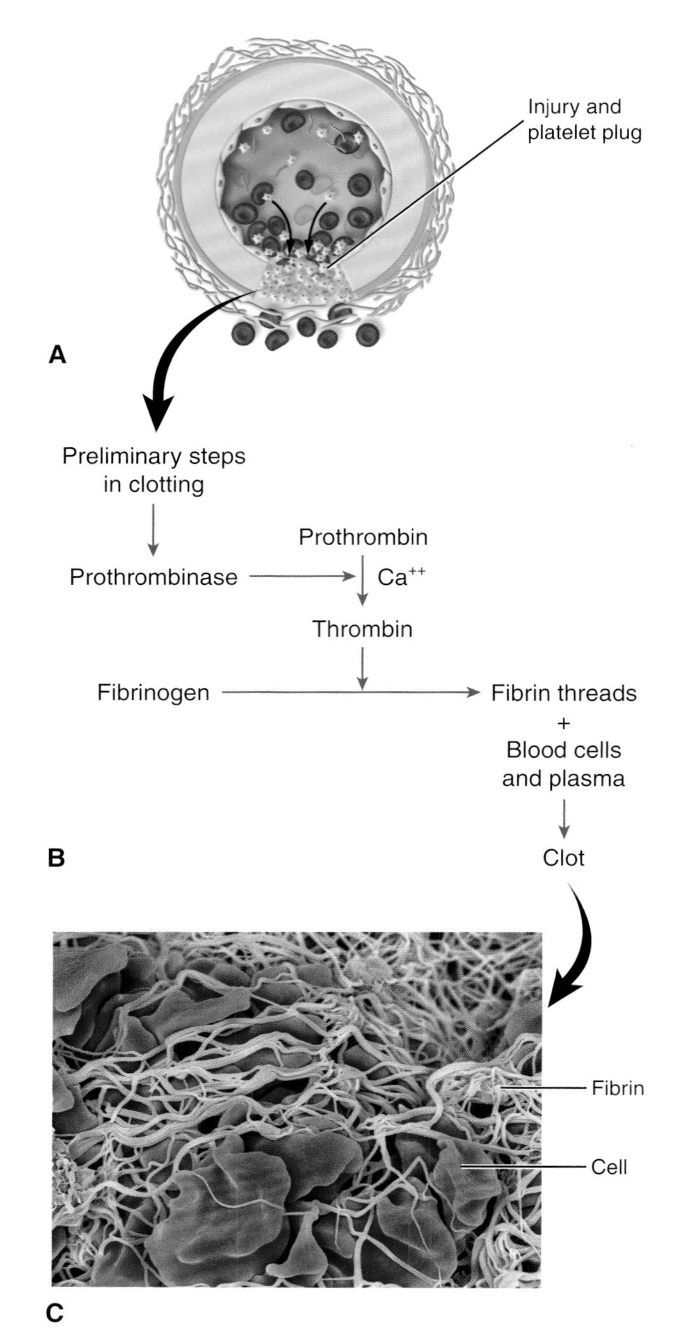

FIGURE 12-7 Blood clotting (coagulation). KEY POINT Blood coagulation requires a complex series of reactions that lead to the formation of fibrin, an insoluble protein. Fibrin threads trap blood cells to form a clot. **A.** Substances released from damaged tissue and sticky platelets initiate the preliminary steps in clotting. **B.** The final coagulation reactions lead to formation of a fibrin clot. **C.** Scanning electron micrograph of blood cells trapped in fibrin. ZOOMING IN What part of the word *prothrombinase* indicates that it is an enzyme? What part of the word *prothrombin* indicates that it is a precursor?

Several methods used to measure the body's ability to coagulate blood are described later in this chapter.

See the Student Resources on thePoint® for a summary diagram and an animation on hemostasis.

CHECKPOINTS 

- ☐ **12-11** What is the general term for the process that stops blood loss?
- ☐ **12-12** What substance in the blood forms a clot?
- ☐ **12-13** How does serum differ from blood plasma?

CASEPOINT 

- ☐ **12-3** What symptoms indicated that Allen was lacking in platelets?

Blood Types

Like all body cells, blood cells contain substances (usually proteins) capable of activating an immune response. These substances, known as **antigens** (AN-ti-jens), trigger the immune system to make specialized proteins called *antibodies* that help destroy any cell with the offending antigen, as discussed more fully in Chapter 15.

Blood cell antigens vary among individuals and become important when blood components are donated from one individual to another in a process called a **transfusion**. Antibodies that recognize red cell antigens are known as *agglutinins*, because they cause red cells to undergo **agglutination** (ah-glu-tih-NA-shun) (clumping). The cells then rupture and release their hemoglobin by a process called **hemolysis** (he-MOL-ih-sis). The resulting condition is dangerous to a patient who has received incompatible blood. There are many types of RBC antigens, but only two groups are particularly likely to cause a transfusion reaction: the so-called A and B antigens and the Rh factor.

THE ABO BLOOD TYPE SYSTEM

There are four blood types involving the A and B antigens: A, B, AB, and O (**TABLE 12-3**). These letters indicate the type of antigen present on the red cells. If only the A antigen is present, the person has type A blood; if only the B antigen is present, he or she has type B blood. Type AB red cells have both antigens, and type O blood has neither. Of course, no one has antibodies to his or her own blood type antigens, or their plasma would destroy their own cells. Each person does, however, produce antibodies that react with the AB antigens he or she is lacking. (These antibodies are produced early in life from exposure to A and B antigens in the environment.) It is these antibodies in the patient's plasma that can react with antigens on the donor's red cells to cause a transfusion reaction.

Testing for Blood Type Blood type can be tested using blood sera containing antibodies to the A or B antigens. These **antisera** are prepared in animals using either the A or the B antigens to induce a response. Blood serum containing antibodies that recognize the A antigen is called **anti-A serum**; blood serum containing antibodies that recognize the B antigen is called **anti-B serum**. When combined with a blood sample in the laboratory, each antiserum causes the corresponding red cells to agglutinate. The blood's agglutination pattern when mixed with these two sera *one at a time* reveals its blood type (**FIG. 12-8**). Type A reacts with anti-A serum only; type B reacts with anti-B serum only. Type AB agglutinates with both, and type O agglutinates with neither A nor B.

Blood Compatibility An individual's genes determine which antigens he or she will produce, so heredity determines a person's blood type. The percentage of people with each of the different blood types varies in different populations. For example, about 45% of the white population of the United States have type O blood, 40% have A, 11% have B, and only 4% have AB. The percentages vary within other population groups.

In an emergency, type O blood can be given to any ABO type because the cells lack both A and B antigens and will not react with either A or B antibodies (**see TABLE 12-3**). People with type O blood are called *universal donors*. Conversely, type AB blood contains no antibodies to agglutinate red cells, and people with this blood type can therefore receive blood from any ABO type donor. Those with AB blood are described as *universal recipients*. Whenever possible, it is safest to give the same blood type as the recipient's blood.

THE RH FACTOR

More than 85% of the US population has another red cell antigen group called the **Rh factor**, named for *Rhesus*

Table 12-3 The ABO Blood TYPE System

Blood Type	Red Blood Cell Antigen	Reacts with Antiserum	Plasma Antibodies	Can Take From	Can Donate To
A	A	Anti-A	Anti-B	A, O	A, AB
B	B	Anti-B	Anti-A	B, O	B, AB
AB	A, B	Anti-A, Anti-B	None	AB, A, B, O	AB
O	None	None	Anti-A, Anti-B	O	O, A, B, AB

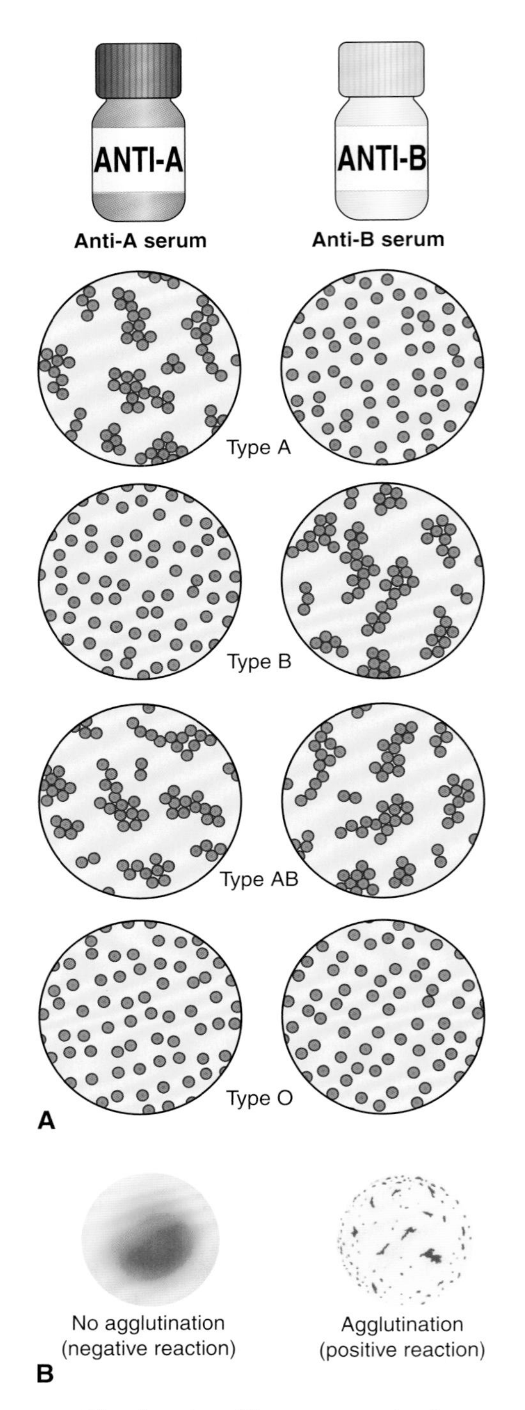

FIGURE 12-8 Blood typing. **KEY POINT** Blood type can be determined by mixing small volumes with antisera prepared against the different red cell antigens (proteins). Agglutination (clumping) with an antiserum indicates the presence of the corresponding antigen. **A.** Labels at the top of each column denote the kind of antiserum added to the blood samples. Anti-A serum agglutinates red cells in type A blood, but anti-B serum does not. Anti-B serum agglutinates red cells in type B blood, but anti-A serum does not. Both sera agglutinate type AB blood cells, and neither serum agglutinates type O blood. **B.** Photographs of blood typing reactions. **ZOOMING IN** Can you tell from these reactions whether these cells are Rh positive or Rh negative?

monkeys, in which it was first found. Rh is also known as the *D antigen*. People with this antigen are said to be **Rh positive**; those who lack this protein are said to be **Rh negative**. If Rh-positive blood is given to an Rh-negative person, he or she may produce antibodies to the "foreign" Rh antigens. The blood of this "Rh-sensitized" person will then destroy any Rh-positive cells received in a later transfusion.

Concept Mastery Alert

Remember that the Rh status of the donated blood must be taken into consideration. Thus, O positive blood could cause an incompatibility reaction.

Rh incompatibility is a potential problem in certain pregnancies (FIG. 12-9). A mother who is Rh negative may develop antibodies to the Rh protein of an Rh-positive fetus (the fetus having inherited this factor from the father). Red cells from the fetus that enter the mother's circulation during pregnancy and childbirth evoke the response. In a subsequent pregnancy with an Rh-positive fetus, some of the anti-Rh antibodies may pass from the mother's blood into the blood of her fetus and destroy the fetus' red cells. This reaction is now prevented by administration of immune globulin Rh_o(D), trade name RhoGAM, to the mother during pregnancy and shortly after delivery. These preformed antibodies clear the mother's circulation of Rh antigens and prevent stimulation of an immune response.

CHECKPOINTS 

- ☐ **12-14** What is the term for any substance that activates an immune response?
- ☐ **12-15** What are the four ABO blood types?
- ☐ **12-16** What blood factor is associated with incompatibility during pregnancy?

CASEPOINTS 

- ☐ **12-4** Allen had type O positive blood. What ABO blood type antigens were present on his red cells?
- ☐ **12-5** What blood type antibodies were in his blood plasma?
- ☐ **12-6** Did he have Rh antibodies in his plasma? Explain.

Uses of Blood and Blood Components

Blood can be packaged and kept in blood banks for emergencies. To keep the blood from clotting, a solution such as citrate–phosphate–dextrose–adenine (CPDA-1) is added. The blood may then be stored for up to 35 days. The blood supplies in the bank are dated with an expiration date to prevent the use of blood in which red cells may have disintegrated. Blood banks usually have all types of blood and blood products available. It is important that there be an extra supply of type O, Rh-negative blood because in an emergency, this type can be used for any patient. It is a normal procedure, however, to test the recipient and give blood of the same type. In this chapter's case study, Allen's

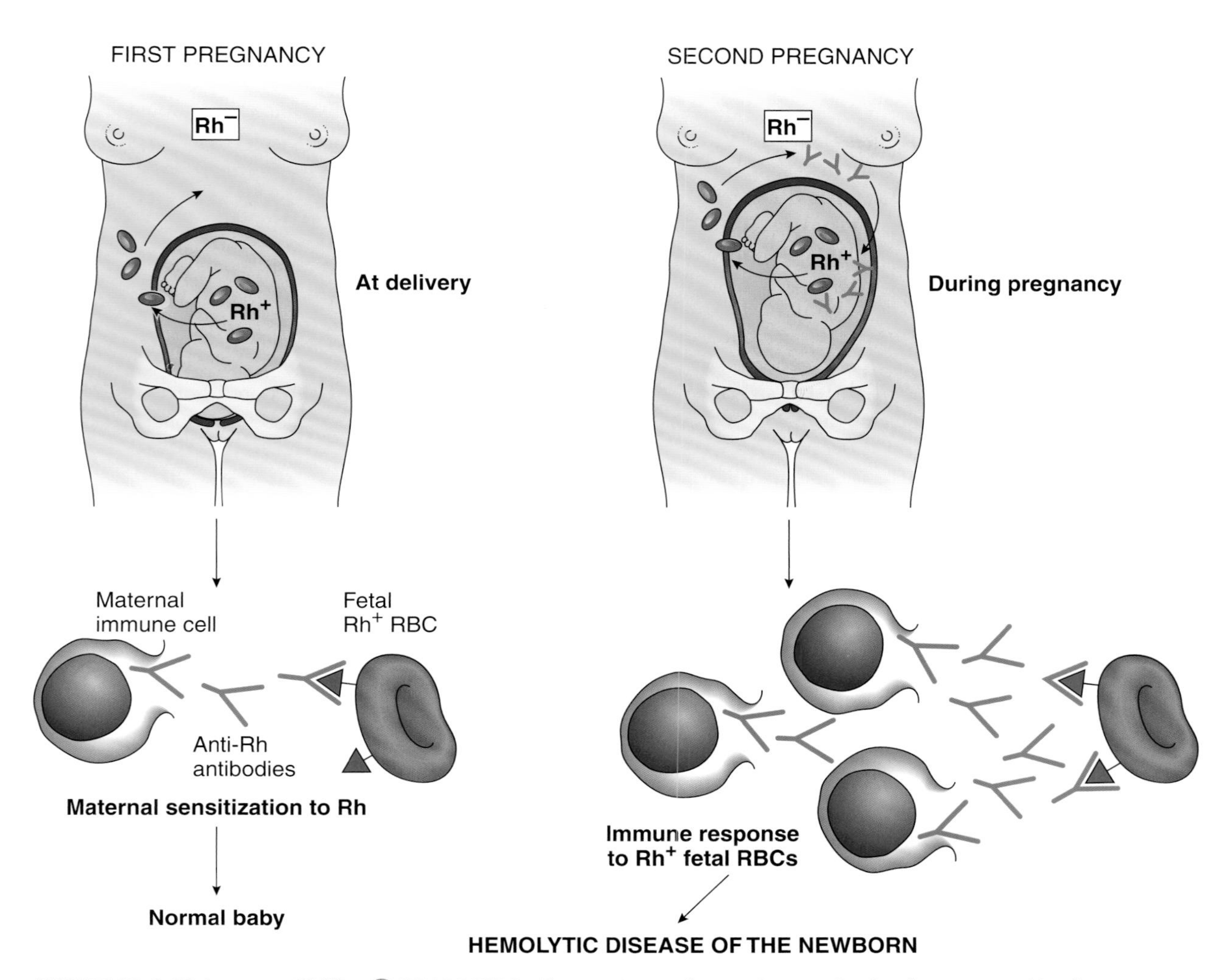

FIGURE 12-9 **Rh incompatibility.** KEY POINT An Rh-negative mother can form antibodies (become sensitized) to an Rh-positive fetus' red cells when exposed to the antigen during delivery. Unless she is treated with RhoGAM to prevent a response, her Rh antibodies can cross the placenta in a subsequent pregnancy and destroy fetal red cells if they are Rh positive. The result is hemolytic disease of the newborn (HDN).

blood was typed, and he was transfused with whole blood and platelets to overcome the effects of his bone marrow failure.

A person can donate his or her own blood before undergoing elective (planned) surgery to be used during surgery if needed. This practice eliminates the possibility of incompatibility and of disease transfer as well. Such **autologous** (aw-TOL-o-gus) (self-originating) blood is stored in a blood bank only until the surgery is completed.

WHOLE BLOOD TRANSFUSIONS

The transfer of whole human blood from a healthy person to a patient is often a life-saving process. Whole blood transfusions may be used for any condition in which there is loss of a large volume of blood, known as a **hemorrhage** (HEM-eh-rij). We may be familiar with the hemorrhages resulting from external wounds, but less evident are internal hemorrhages resulting from ruptures in deeper vessels. Regardless of the cause, severe hemorrhage starves body cells of oxygen and nutrients. Serious mechanical injuries, surgical operations, and internal injuries such as bleeding ulcers commonly require whole blood transfusions.

Caution and careful evaluation of the need for a blood transfusion is the rule, however, because of the risk for transfusion reactions and the possible transmission of viral diseases, particularly hepatitis and AIDS. (In developed countries, careful screening has virtually eliminated transmission of these viruses in donated blood.) Ideally, the compatibility of donor and recipient blood is tested prior to transfusion, a process called *cross matching*. Both the red cells and the serum are tested separately for any possible cross-reactions with donor blood. This procedure is particularly important in individuals who have previously received transfused blood.

USE OF BLOOD COMPONENTS

Blood can be separated into its various parts, which may be used for different purposes.

Preparation of Blood Components A common method for separating the blood plasma from the formed elements is by the use of a **centrifuge** (SEN-trih-fuje), a machine that spins in a circle at high speed to separate a mixture's components according to density. When a container of blood is spun rapidly, all the blood's heavier formed elements are pulled to the bottom of the container. They are thus separated from the plasma, which is less dense. The formed elements may be further separated, for example, packed red cells alone for the treatment of hemorrhage or platelets alone for the treatment of clotting disorders.

Blood losses to the donor can be minimized if the blood is removed, the desired components are separated, and the remainder is returned to the donor. The general term for this procedure is **hemapheresis** (hem-ah-fer-E-sis) (from the Greek word *apheresis* meaning "removal"). If the plasma is removed and the formed elements returned to the donor, the procedure is called **plasmapheresis** (plas-mah-fer-E-sis). Autoimmune diseases can be treated by plasmapheresis. A patient's blood is removed and separated, and the plasma portion containing the autoantibodies is discarded and replaced with normal saline. The blood components are then returned to the patient.

Uses of Plasma Blood plasma alone may be given in an emergency to replace blood volume and prevent circulatory failure (shock). Plasma is especially useful when blood typing and the use of whole blood are not possible, such as in natural disasters or in emergency rescues. Because the red cells have been removed from the plasma, there should be no incompatibility problems. (In the rare instances when they occur, they are related to high levels of blood type antibodies in the plasma, which could react with the recipient's blood cells). Plasma separated from the cellular elements is usually further separated by chemical means into various components, such as plasma protein fraction, serum albumin, immune serum, and clotting factors.

The packaged plasma that is currently available is actually plasma protein fraction. Further separation yields serum albumin that is available in solutions of 5% or 25% concentration. In addition to its use in treatment of circulatory shock, these solutions are given when plasma proteins are deficient. They increase the blood's osmotic pressure and thus draw fluids back into circulation. The use of plasma proteins and serum albumin has increased because these blood components can be treated with heat to prevent transmission of viral diseases.

In emergency situations, healthcare workers may administer fluids known as *plasma expanders*. These are cell-free isotonic solutions used to maintain blood fluid volume to prevent circulatory shock.

Fresh plasma may be frozen and saved. Plasma frozen when it is less than six hours old contains all the factors needed for clotting. When frozen plasma is thawed, a white precipitate called **cryoprecipitate** (kri-o-pre-SIP-ih-tate) forms in the bottom of the container. Cryoprecipitate is especially rich in fibrinogen and clotting factors. It may be given when there is a special need for these substances.

A portion of the plasma called the *gamma globulin fraction* contains antibodies produced by lymphocytes when they come in contact with foreign agents, such as bacteria and viruses. Antibodies play an important role in immunity (see Chapter 15). Commercially prepared immune sera are available for administration to patients in immediate need of antibodies, such as infants born to mothers with active hepatitis.

CHECKPOINT 

☐ **12-17** How is blood commonly separated into its component parts?

CASEPOINT 

☐ **12-7** Allen was treated with transfusions of red cells for his anemia. What other blood component did he need to treat his bleeding problem?

Blood Studies

Many kinds of studies can be done on blood, and some of these have become standard parts of a routine physical examination. Machines that are able to perform several tests at the same time have largely replaced manual procedures, particularly in large institutions.

THE HEMATOCRIT

The **hematocrit** (he-MAT-o-krit) (Hct) measures how much of the blood volume is taken up by red cells. It reflects both the size and the number of cells and can provide an estimate of the oxygen-carrying capacity of blood.

The hematocrit is determined by spinning a blood sample in a high-speed centrifuge for three to five minutes to separate the cellular elements from the plasma (**FIG. 12-10**). Results are expressed as the volume of packed red cells per unit volume of whole blood. For example, "hematocrit, 38%" in a laboratory report means that the patient has 38 mL red cells/dL (deciliter; 100 mL) of blood; red cells comprise 38% of the total blood volume. For adult men, the normal range is 42% to 54%, whereas for adult women, the range is slightly lower, 36% to 46%. These normal ranges, like all normal ranges for humans, may vary depending on the method used and the interpretation of the results by an individual laboratory. Some values change over the course of the day, month (for menstrual cycling women), and entire lifespan. Thus, two healthy individuals may have different test results. See **BOX 12-2** for more information about interpreting normal and abnormal test values.

Usually, a decreased hematocrit is characteristic of a hemoglobin deficiency known as *anemia*. This disorder can result from a decreased RBC synthesis or increased RBC destruction. A chronically elevated hematocrit may indicate excess production of red cells, technically known as *polycythemia rubra*. This condition may indicate an abnormality, but may also occur normally, as in people living at high altitudes. However, an elevated hematocrit can also reflect dehydration, because plasma volume decreases, but the RBC volume remains constant.

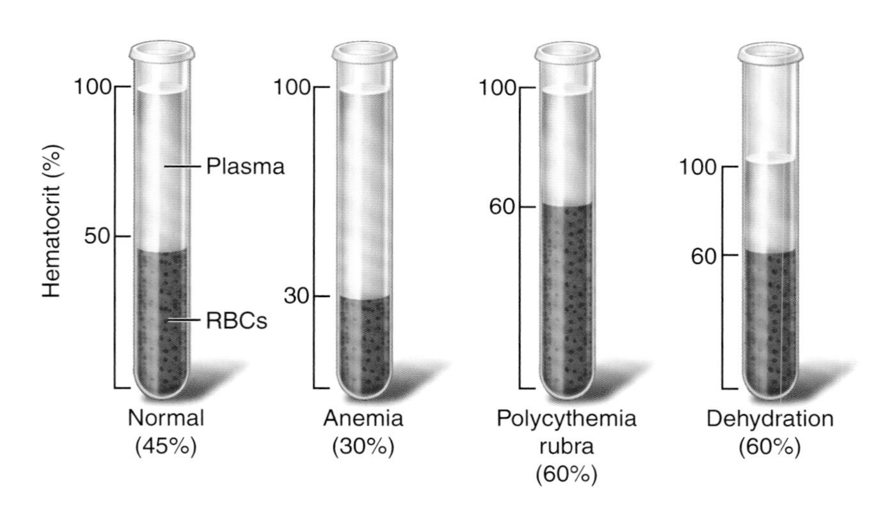

FIGURE 12-10 Hematocrit. KEY POINT The hematocrit tests the volume percentage of red cells in whole blood. The **tube on the left** shows a normal hematocrit. Abnormal hematocrit results can indicate disease (**two middle tubes**) or simply dehydration (**far right tube**).

12

HEMOGLOBIN TESTS

Hemoglobin carries the oxygen in blood cells, and the best estimate of the blood's oxygen-carrying capacity is the hemoglobin concentration. New techniques can measure the concentration directly in a blood sample. Older techniques require lysing the cells to release the hemoglobin into solution, then quantifying the color intensity. Hemoglobin (Hb) is expressed in grams per deciliter of whole blood. Normal hemoglobin concentrations for adult males range from 14 to 17 g/dL blood. Values for adult women are in a somewhat lower range, at 12 to 15 g/dL blood. The hemoglobin reading can also be expressed as a percentage

ONE STEP AT A TIME

Reading Laboratory Test Results

BOX 12-2

Results of medical lab tests are reported in a variety of units and usually include a normal range for each particular test. This range is needed because in any living population, there are variations in all characteristics among individuals. Even for a normal healthy person, variations occur at different times and under different circumstances. So homeostasis is always achieved within a given set of parameters. In addition, results may vary somewhat depending on the type of test used and instrumentation.

Question

Using the printout of selected blood test results in Allen's case, list which measurements are normal and which are abnormal.

LAKESIDE MEDICAL CENTER

47 Medical Drive Chicago, IL 60604 PHONE 312-333-3333 FAX 312-333-3323

NAME : ALLEN.MCKINLEY DOB: 22/11/1956 AGE: 62
MR#: DBTE-333 GENDER: M
ACCT#: 155534243
COLL: 22/12/2018xx 14:23

TEST	RESULT	REFERENCE RANGE	UNITS
RBC	3.7	[4.3-5.7]	M/mcL
WBC	15	[5-10]	K/mcL
HGB	9	[13.5-17.5]	g/dL
HCT	29	[38.8-50]	%
PLT	27	[150-450]	K/mcL
ALB	3.9	[3.8-5.5]	g/dL
Na	140	[136-145]	mEq/L
GLU	105	[70-110]	mg/dL

Answer

Step 1. Determine the person's age and gender. Reference ranges for some tests differ between men and women. In some cases, age is also a factor. For example, the normal range for cholesterol levels increases with age. Values may be different for children than adults. Information at the top of the page indicates that Allen is a 62-year-old man.

Step 2. Identify abbreviations. Printouts often use abbreviations for the test name (in the first column) and the units (in the fourth column). Look up unknown abbreviations in a medical dictionary or the Internet. For instance, ALB stands for albumin, K stands for thousands, and M stands for millions.

Step 3. Determine whether or not each result conforms to its normal range. The numbers in the third column indicate the lowest and highest values considered "normal," and the second column shows Allen's results. For instance, the normal WBC count should be between 5 and 10 thousand/microliter of blood. Allen's value is higher than 10, so it is abnormal. Note that the normal spread is much larger for some measurements, such as platelet count, than for others, such as sodium. Deviations would be more significant for parameters with a narrow range.

Step 4. Summarize your findings. Looking at these results, it appears that all of Allen's blood cell readings are abnormal, whereas all of his measurements of substances in blood plasma are within the normal range.

See the Study Guide (available separately) for a similar question that you can answer yourself.

of a given standard, usually the average male normal of 15.6 g Hb/dL. Thus, a reading of 90% would mean 90% of 15.6 or 14 g Hb/dL. A decrease in hemoglobin to below normal levels signifies anemia.

Normal and abnormal types of hemoglobin can be separated and measured by the process of **electrophoresis** (e-lek-tro-fo-RE-sis). In this procedure, an electric current is passed through the liquid that contains the hemoglobin to separate different components based on their size. This test is useful in the diagnosis of sickle cell anemia and other disorders caused by abnormal types of hemoglobin. The case study in Chapter 21 features a young boy with sickle cell anemia.

CASEPOINT 

☐ **12-8** Which of these is a likely hematocrit value for Allen 30%, 42%, or 54%?

BLOOD CELL COUNTS

Laboratories use automated methods for obtaining the data for blood counts. The values are then compared to normal ranges for an individual of the same gender and age, as shown in **BOX 12-2**. Values outside the normal range indicate a blood disorder. Visual counts are sometimes done using a **hemocytometer** (he-mo-si-TOM-eh-ter), a ruled slide used to count the cells in a given volume of blood under the microscope. A blood cell count includes the following information:

- **Red cell count:** the normal red cell count varies from 4.5 to 5.5 million cells/mcL of blood. An increase in the red cell count is termed polycythemia rubra (pol-e-si-THE-me-ah RU-brah), the "rubra" referring to red cells. People who live at high altitudes develop higher red cell counts. As discussed earlier, low red cell counts are usually indicative of anemia.
- **White cell count:** the leukocyte count varies from 5,000 to 10,000 cells/mcL of blood. A **differential white count** is an estimation of the percentage of each white cell type in the smear. Because each type has a specific function, changes in their proportions can be a valuable diagnostic aid **(see TABLE 12-2)**. WBC counts above normal usually indicate that the body is fighting a bacterial infection. In this chapter's case study, Allen's elevated white cell count resulted from myelogenous leukemia, as cancer arising in the bone marrow.
- **Platelet count:** it is difficult to count platelets visually because they are so small. Laboratories can obtain more accurate counts with automated methods. The normal platelet count ranges from 150,000 to 450,000/mcL of blood, but counts may fall to 100,000 or less without causing serious bleeding problems. If a count is very low, a platelet transfusion may be given.

THE BLOOD SMEAR (SLIDE)

In addition to the above tests, blood studies include the examination of a stained blood slide **(see FIG. 12-2)**. In this procedure, a drop of blood is spread thinly and evenly over a glass slide, and a special stain (Wright) is applied to differentiate the otherwise colorless white cells. The slide is then studied under the microscope. The red cells are examined for abnormalities in size, color, or shape and for variations in the percentage of immature red cell forms, known as reticulocytes. (See **BOX 12-3** to learn about

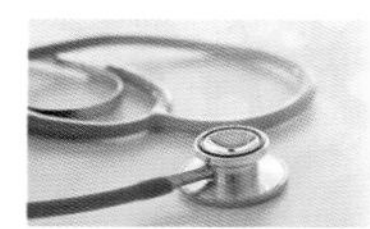

CLINICAL PERSPECTIVES — BOX 12-3
Counting Reticulocytes to Diagnose Disease

As erythrocytes mature in the red bone marrow, they go through a series of stages in which they lose their nucleus and most other organelles, maximizing the space available to hold hemoglobin. In one of the last stages of development, small numbers of ribosomes and some rough endoplasmic reticulum remain in the cell. These appear as a network (or reticulum) when stained. Cells at this stage are therefore called **reticulocytes**. Reticulocytes leave the red bone marrow and enter the bloodstream where they become fully mature erythrocytes in about 24 to 48 hours. The average number of red cells maturing through the reticulocyte stage at any given time is about 1% to 2%. Changes in these numbers can be used in diagnosing certain blood disorders.

When erythrocytes are lost or destroyed, as from chronic bleeding or some form of hemolytic anemia, red blood cell production is "stepped up" to compensate for the loss. Greater numbers of reticulocytes are then released into the blood before reaching full maturity, and counts increase above normal. On the other hand, a decrease in the number of circulating reticulocytes suggests a problem with red blood cell production, as in cases of deficiency anemias or suppression of bone marrow activity.

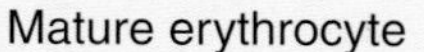
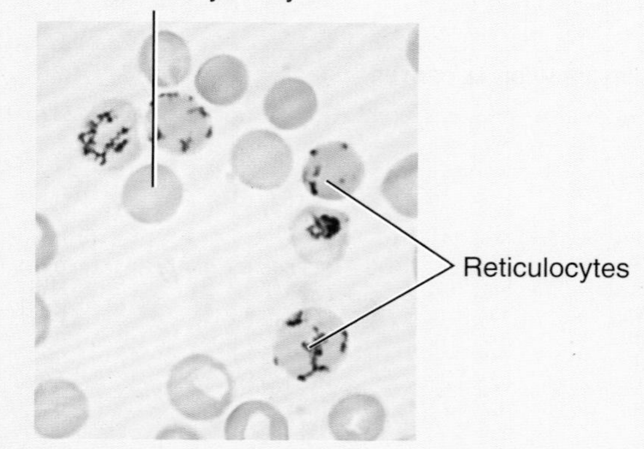

Reticulocytes. Some ribosomes and rough ER appear as a network in a late stage of erythrocyte development.

reticulocytes and how their counts are used to diagnose disease.) Parasites, such as the malarial organism and others, may be found.

BLOOD CHEMISTRY TESTS

Batteries of tests on blood serum are often done by machine. The "Chem-7" test quantifies levels of four electrolytes (sodium, potassium, chloride, and bicarbonate), blood glucose, blood urea nitrogen, and **creatinine** (kre-AT-in-in).

Other tests check for enzymes. Increased levels of **creatine kinase** (CK), **lactic dehydrogenase** (LDH), and other enzymes indicate tissue damage, such as that resulting from heart disease.

Blood can be tested for amounts of lipids, such as cholesterol, triglycerides (fats), and lipoproteins or for amounts of plasma proteins. For example, the presence of more than the normal amount of glucose in the blood indicates uncontrolled diabetes mellitus. The list of blood chemistry tests is extensive and constantly increasing. We may now obtain values for various hormones, vitamins, antibodies, and toxic or therapeutic drug levels.

COAGULATION STUDIES

Before surgery and during treatment of certain diseases, hemophilia, for example, it is important to know that coagulation will take place within normal time limits. Because clotting is a complex process involving many reactants, a delay may result from a number of different causes, including lack of certain hormones, calcium, or vitamin K. The amounts of the various clotting factors are measured to aid in the diagnosis and treatment of bleeding disorders.

Additional tests for coagulation include tests for bleeding time, clotting time, capillary strength, and platelet function.

CHECKPOINTS 

- ☐ **12-18** What test measures the relative volume of red cells in the blood?
- ☐ **12-19** What are two ways of expressing hemoglobin level?

A & P in Action Revisited: Allen's Follow-Up

After his diagnosis, Allen received a complete workup in the hospital. This included a detailed evaluation of his cancer cells, including a cytogenetic analysis, looking at the mutations in his cancer cell chromosomes, and immunophenotyping, looking at the pattern of proteins, or antigenic markers, present on the surfaces of his cancer cells. In addition, imaging studies and a lumbar puncture were done to look for the spread of leukemic cells outside the bone marrow. Fortunately, Allen's leukemia had not spread to his spleen, liver, or lymph nodes. There was no evidence of leukemic cells in his central nervous system or skin. Allen was started on chemotherapy and targeted therapy, injections of anticancer antibodies tailored to his specific cancer cells. Because of his age over 50, the oncologist then decided on a mini-bone marrow transplant designed to mop up any remaining cancer cells. This procedure involves a transplant from a compatible donor without destruction of the patient's own bone marrow. The hope is that the donor tissue will destroy any remaining cancer cells and gradually replace the blood-forming cells in the patient's bone marrow. Medical researchers are currently developing new clinical approaches and targeted therapies for blood cancers, but there is still a long way to go in treating these insidious diseases.

CHAPTER 12

Chapter Wrap-Up

OVERVIEW

A detailed chapter outline with space for note-taking is on thePoint®. The figure below illustrates the main topics covered in this chapter.

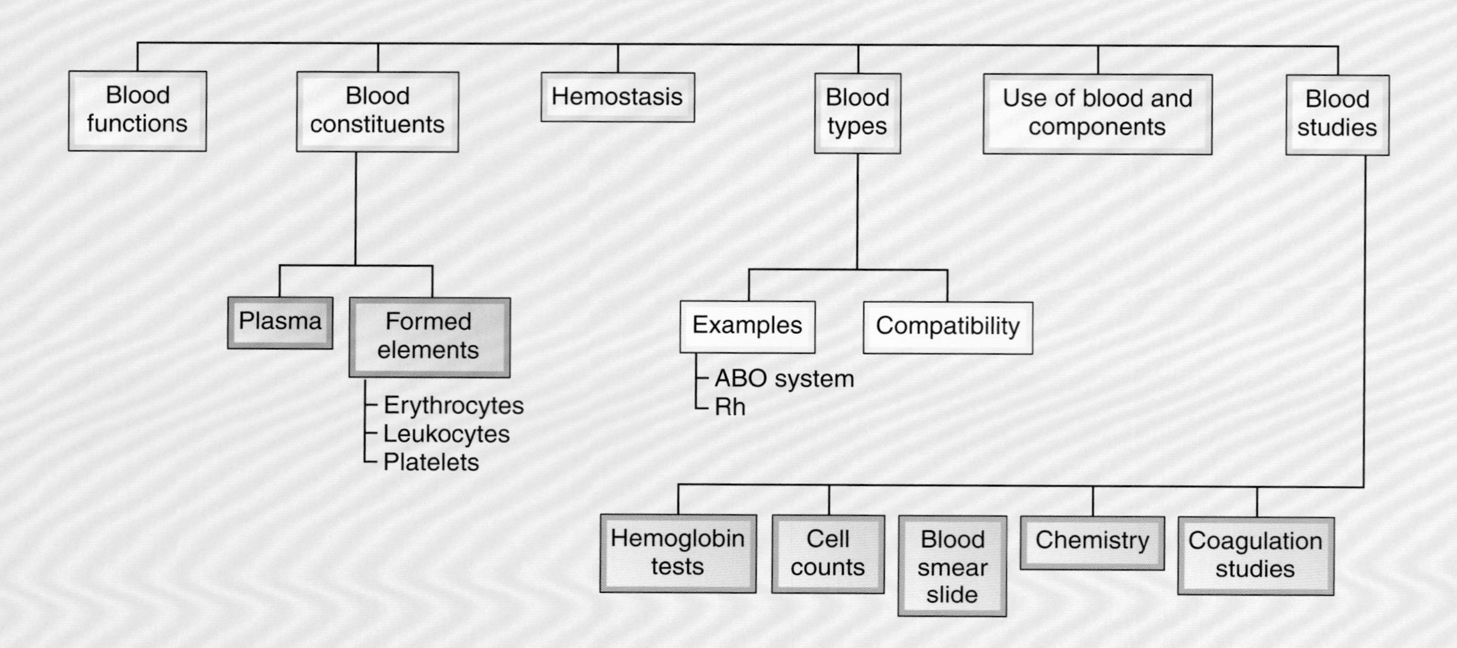

KEY TERMS

The terms listed below are emphasized in this chapter. Knowing them helps you organize and prioritize your learning. These and other boldface terms are defined in the Glossary with phonetic pronunciations.

agglutination
albumin
antigen
antiserum
basophil
centrifuge
coagulation
cryoprecipitate
eosinophil
erythrocyte
fibrin
hematocrit
hematopoietic
hemoglobin
hemolysis
hemostasis
leukocyte
lymphocyte
megakaryocyte
monocyte
neutrophil
plasma
platelet (thrombocyte)
serum
thrombin
transfusion

WORD ANATOMY

Medical terms are built from standardized word parts (prefixes, roots, and suffixes). Learning the meanings of these parts can help you remember words and interpret unfamiliar terms.

WORD PART	MEANING	EXAMPLE
BLOOD CONSTITUENTS		
erythr/o-	red, red blood cell	An *erythrocyte* is a red blood cell.
hemat/o	blood	*Hematopoietic* stem cells form (–poiesis) all of the blood cells.
hem/o	blood	*Hemoglobin* is a protein that carries oxygen in the blood.
kary/o	nucleus	A *megakaryocyte* has a very large nucleus.
leuk/o-	white, colorless	A *leukocyte* is a white blood cell.
lymph/o	lymph, lymphatic system	*Lymphocytes* are white blood cells that circulate in the lymphatic system.
macr/o	large	A *macrophage* takes in large amounts of foreign matter by phagocytosis.
mon/o	single, one	A *monocyte* has a single, unsegmented nucleus.
morph/o	shape	The nuclei of *polymorphs* have many shapes.
phag/o	eat, ingest	Certain leukocytes take in foreign matter by the process of *phagocytosis*.
thromb/o	blood clot	A *thrombocyte* is a cell fragment that is active in blood clotting.
HEMOSTASIS		
-gen	producing, originating	*Fibrinogen* converts to fibrin in the formation of a blood clot.
pro-	before, in front of	Prothrombinase is an enzyme (–ase) that converts *prothrombin* to thrombin.
BLOOD TYPES		
-lysis	loosening, dissolving, separating	A recipient's antibodies to donated red cells can cause hemolysis of the cells.
USES OF BLOOD AND BLOOD COMPONENTS		
cry/o	cold	*Cryoprecipitate* forms when blood plasma is frozen and then thawed.
BLOOD STUDIES		
-hemia, -emia (hem+ -ia)	condition of blood	*Anemia* is a lack (an-) of red cells or hemoglobin.

QUESTIONS FOR STUDY AND REVIEW

BUILDING UNDERSTANDING

Fill in the Blanks

1. The liquid portion of blood is called ________.
2. The ancestors of all blood cells are called ________cells.
3. Platelets are produced by certain giant cells called ________.
4. Some monocytes enter the tissues and mature into phagocytes called ________.
5. Erythrocytes have a life span of approximately ________ days.

Matching Match each numbered item with the most closely related lettered item:

___ 6. Type A blood
___ 7. Type B blood
___ 8. Type O blood
___ 9. Type AB blood
___ 10. Rh-positive blood

a. its plasma contains anti-A antibody only
b. it belongs to a universal donor
c. only A antigen is present on its erythrocytes
d. it belongs to a universal recipient
e. Rh antigen are present on its erythrocytes

Multiple Choice

___ 11. What iron-containing protein transports oxygen?
a. erythropoietin
b. complement
c. hemoglobin
d. thrombin

___ 12. What is the correct sequence for hemostasis?
a. vessel contraction, plug formation, blood clot
b. blood clot, plug formation, vessel contraction
c. plug formation, blood clot, vessel contraction
d. vessel contraction, blood clot, plug formation

___ 13. The hematology specialist needs to measure the number of eosinophils in a blood sample. Which test should she conduct?
a. hematocrit
b. electrophoresis
c. bone marrow biopsy
d. differential white blood cell count

___ 14. What vitamin is needed for blood clotting?
a. vitamin A
b. vitamin K
c. biotin
d. vitamin E

UNDERSTANDING CONCEPTS

15. List the three main functions of blood. What is the average volume of circulating blood in the body?
16. Compare and contrast the following:
a. formed elements and plasma
b. erythrocyte and leukocyte
c. hemorrhage and transfusion
d. hemapheresis and plasmapheresis
17. List four main types of proteins in blood plasma, and state their functions. What are some other substances carried in blood plasma?
18. Describe the structure and function of erythrocytes. State the normal blood cell count for erythrocytes.
19. Construct a chart that compares the structure and function of the five types of leukocytes. State the normal blood cell count for leukocytes.
20. Diagram the three final steps in blood clot formation.
21. Name the four blood types in the ABO system. What antigens and antibodies (if any) are found in people with each type?
22. Is an Rh-negative fetus of an Rh-negative mother in any danger of hemolytic disease (HDN)? Explain.
23. Using the glossary of word parts at the back of the book, define each of the components of the word polycythemia listed below, and write a definition of the term.
a. poly
b. cyt/o
c. hem/o
d. -ia

CONCEPTUAL THINKING

24. In the feedback loop controlling red blood cell production, name the role (regulated variable, sensor, signal, or effector) of each of these components:
a. kidney
b. blood oxygen level
c. bone marrow
d. erythropoietin
25. J. Regan, a 40-year-old firefighter, has just had his annual physical. He is in excellent health, but his red blood cell count is above the normal range. How might Mr. Regan's job explain his polycythemia?
26. List the signs and symptoms Allen experienced in the opening case study, and relate them to the problem in his bone marrow.

For more questions, see the Learning Activities on thePoint®.

CHAPTER 13

The Heart

Learning Objectives

After careful study of this chapter, you should be able to:

1. Describe the three tissue layers of the heart wall. ***p. 264***
2. Describe the location and structure of the pericardium, and cite its functions. ***p. 265***
3. Compare the functions of the right and left chambers of the heart. ***p. 265***
4. Name the valves at the entrance and exit of each ventricle, and identify the function of each. ***p. 266***
5. Briefly describe blood circulation through the myocardium. ***p. 267***
6. Briefly describe the cardiac cycle. ***p. 269***
7. Name and locate the components of the heart's conduction system. ***p. 270***
8. Define and explain common variations in heart rate. ***p. 271***
9. Explain the effects of the autonomic nervous system (ANS), hormones, and ions on cardiac output. ***p. 271***
10. Explain what produces each of the two normal heart sounds, and identify the usual cause of a murmur. ***p. 273***
11. Briefly describe five methods used to study the heart. ***p. 273***
12. List four risk factors for coronary artery disease that cannot be modified. ***p. 274***
13. List seven risk factors for coronary artery disease that can be modified. ***p. 274***
14. Describe four changes that may occur in the heart with age. ***p. 274***
15. Referring to the case study, list the emergency and surgical procedures commonly performed following a myocardial infarction, and explain why they are done. ***pp. 263, 275***
16. Show how word parts are used to build words related to the heart (see Word Anatomy at the end of the chapter). ***p. 277***

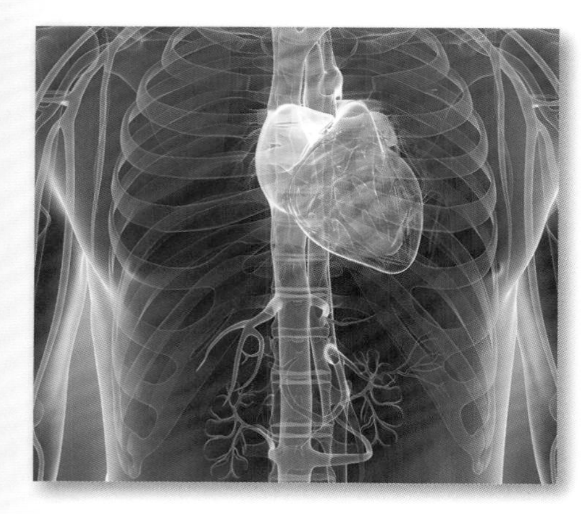

A & P in Action
Jim's Coronary Emergency

The emergency room's dispatch radio echoed from the triage desk.

"This is Medic 5 en route with Jim, a 58-year-old Caucasian male. Suspected acute myocardial infarction with subsequent cardiac arrest while playing basketball. Cardiopulmonary resuscitation was initiated on scene with return of spontaneous circulation. Portable electrocardiography (ECG) indicates ST segment elevation consistent with MI. Patient is receiving high-flow oxygen through a bag valve mask. We have established IV access and administered 10 milligrams' morphine" the EMS medic continued. "Before the patient arrested, we were able to administer 325 mg of aspirin and sublingual nitroglycerin therapy. Estimated time of arrival (ETA) 10 minutes."

When Jim arrived at the ER, the emergency team rushed to stabilize him. An anesthetist intubated him and inserted a central IV line. A trauma nurse measured his vital signs—he was hypotensive—and attached an oxygen mask to his endotracheal tube. Meanwhile, a phlebotomist drew blood from Jim's other arm for lab tests. The cardiac catheterization team was standing by to rush Jim to the cath lab. A cardiology technician attached ECG leads to his chest and began to record his cardiac muscle's electrical activity. Looking at the cardiac monitor, the doctor confirmed that Jim's heart showed signs of a heart attack. The doctor knew that one or more of the coronary arteries feeding Jim's heart muscle were blocked with a thrombus (blood clot). He was aware that Jim had already received aspirin, an anticoagulant, to inhibit the formation of any more thrombi and nitroglycerin to reduce strain on the heart. Morphine had been administered to manage Jim's pain and reduce his heart rate.

In the cardiac catheterization lab, a coronary angiography revealed a blockage of Jim's left main coronary artery. A specialized catheter was then inserted through an artery and quickly extended to the left main coronary artery to dilate the blockage with a tiny balloon (balloon angioplasty). A metal stent then was left in place to hold the artery open. Jim was transferred to the cardiac intensive care unit for monitoring and scheduled for an echocardiogram the next day. Jim's nurses knew to check his cardiac enzyme levels every six hours for the next 24 hours to evaluate heart muscle damage.

Thanks to the quick action of the paramedics, emergency team, and cath lab personnel, Jim was resting comfortably in the intensive care unit a few hours after his heart attack. He was lucky to be alive! Later in the chapter, we will visit Jim again and learn more about how doctors can restore blood flow to coronary arteries in cases of coronary vascular disease.

As you study this chapter, CasePoints will give you opportunities to apply your learning to this case.

Visit thePoint® to access the following resources. For guidance in using these resources most effectively, see pp. xv–xvii.

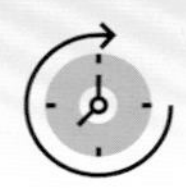

Preparing to Learn

- Tips for Effective Studying
- Pre-Quiz

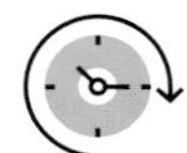

While You Are Learning

- Web Figure: Interior View of the Left Atrium and Ventricle
- Web Chart: Layers of the Heart Wall
- Web Chart: Layers of the Pericardium
- Web Chart: Chambers of the Heart
- Web Chart: Valves of the Heart
- Animation: Blood Circulation
- Animation: Heart Anatomy
- Animation: Cardiac Cycle
- Animation: Myocardial Blood Flow
- Animation: Heart Sounds
- Chapter Notes Outline
- Audio Pronunciation Glossary

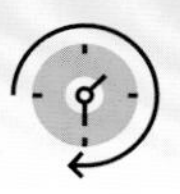

When You Are Reviewing

- Answers to Questions for Study and Review
- Health Professions: Surgical Technologist
- Interactive Learning Activities

A LOOK BACK

In Chapter 4, we learned that cardiac muscle is one of the three types of muscles in the body. Now, it is time to study this tissue and the organ where it is found—the heart. The heart propels blood through the body, so the key idea of gradients and flow 5 *provides a foundation for understanding heart function. Even though the heart can work on its own, the nervous and endocrine systems influence its actions in order to maintain adequate blood pressure, so reviewing the key ideas of homeostasis and negative feedback* 3 *and communication* 11 *is also recommended. The relationship between structure and function* 1 *and the importance of adaptation* 10 *are also relevant to this chapter.*

Introduction

The next two chapters investigate how the blood delivers oxygen and nutrients to the cells and carries away the waste products of cellular metabolism. The continuous one-way circuit of blood through the blood vessels is known as **circulation**. The prime mover that propels blood throughout the body is the **heart**. This chapter examines the heart's structure and function as a foundation for the detailed discussion of blood vessels that follows.

The heart's importance has been recognized for centuries. Strokes (contractions) of this pump average about 72 per minute and continue unceasingly for a lifetime. The beating of the heart is affected by the emotions, which may explain the frequent references to it in song and poetry. However, the heart's vital functions and its disorders are of more practical concern.

Heart Structure

The heart is slightly bigger than a person's fist. It is located between the lungs in the center and a bit to the left of the body's midline (**FIG. 13-1A**). It occupies most of the **mediastinum** (me-de-as-TI-num), the central region of the thorax. The heart's **apex**, the pointed, inferior portion, is directed

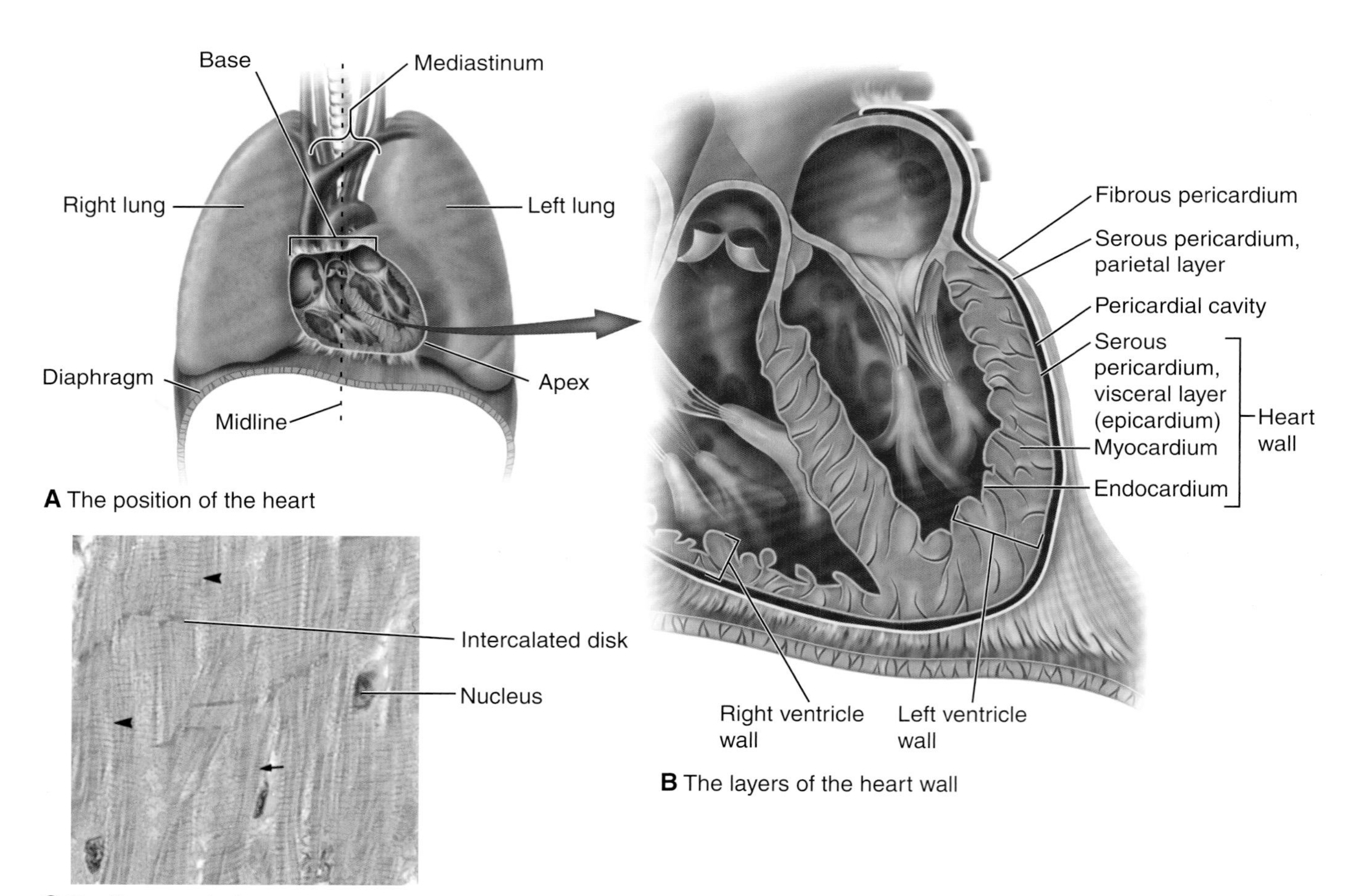

FIGURE 13-1 The heart. A. The position of the heart in the thorax (anterior view). The heart is located between the lungs and just superior to the diaphragm in a region known as the mediastinum. **B.** Layers of the heart wall and pericardium. The serous pericardium covers the heart and lines the fibrous pericardium. **C.** Cardiac muscle tissue viewed under the microscope (×540). The sample shows light striations (see the *arrowheads*), intercalated disks, and branching fibers (*arrow*). ZOOMING IN Which layer of the heart wall is the thickest?

toward the left. The broad, superior **base**, directed toward the right, is the area of attachment for the large vessels carrying blood into and out of the heart. See Dissection Atlas **FIGURE A3-8A** for a photograph of the heart position in the thorax.

TISSUE LAYERS OF THE HEART WALL

The heart is a hollow organ, with walls formed of three different layers. Just as a warm coat might have a smooth lining, a thick interlining, and an outer covering of a third fabric, so the heart wall has three tissue layers **(see FIG. 13-1B)**. Starting with the innermost layer, these are as follows:

1. The **endocardium** (en-do-KAR-de-um) is a thin, smooth layer of epithelial cells that lines the heart's interior. The endocardium provides a smooth surface for easy flow as blood travels through the heart. Extensions of this membrane cover the flaps (cusps) of the heart valves.
2. The **myocardium** (mi-o-KAR-de-um), the heart muscle, is the thickest layer and pumps blood through the vessels. The cardiac muscle's unique structure will be described in more detail shortly.
3. The **epicardium** (ep-ih-KAR-de-um) is a serous membrane that forms the thin, outermost layer of the heart wall. It is also considered the visceral layer of the pericardium, discussed next.

THE PERICARDIUM

The **pericardium** (per-ih-KAR-de-um) is the sac that encloses the heart **(see FIG. 13-1B)**. The formation of the pericardial sac was described and illustrated in Chapter 4 under the discussion of membranes **(see FIG. 4-8)**. This sac's outermost and heaviest layer is the fibrous pericardium, a connective tissue membrane. Additional connective tissue anchors this pericardial layer to the diaphragm, located inferiorly; to the sternum, located anteriorly; and to other structures surrounding the heart, thus holding the heart in place. A serous membrane forms the inner layer of the pericardium. This membrane, known as the serous pericardium, consists of an outer, parietal layer that lines the fibrous pericardium and an inner, visceral layer (the epicardium) that covers the myocardium. A thin film of fluid between these two layers reduces friction as the heart moves within the pericardium. Normally, the visceral and parietal layers are very close together, but fluid may accumulate in the region between them, the pericardial cavity, under certain disease conditions.

SPECIAL FEATURES OF THE MYOCARDIUM

Cardiac muscle cells are lightly striated (striped) based on alternating actin and myosin filaments, as seen in skeletal muscle cells **(see TABLE 7-1)**. Unlike skeletal muscle cells, however, cardiac muscle cells have a single nucleus instead of multiple nuclei. Also, cardiac muscle tissue is involuntarily controlled; it typically contracts independently of conscious thought. There are specialized partitions between cardiac muscle cells that show faintly under a microscope **(see FIG. 13-1C)**. These **intercalated** (in-TER-cah-la-ted) **disks** are actually plasma membranes of adjacent cells that are tightly joined together by specialized membrane proteins. Other membrane proteins within the disks permit electric impulses to travel between adjacent cells. Such electrical synapses, mentioned in Chapter 8, provide rapid and coordinated communication between cells.

Another feature of cardiac muscle tissue is the branching of the muscle fibers (cells). These branched fibers are interwoven so that the stimulation that causes the contraction of one fiber results in the contraction of a whole group. The intercalated disks between the fibers and the branching cellular networks allow cardiac muscle cells to contract in a coordinated manner for effective pumping.

HEART DIVISIONS

Healthcare professionals often refer to the *right heart* and the *left heart*, because the human heart is really a double pump **(FIG. 13-2)**. The right side receives blood low in oxygen content that has already passed through the body and pumps it to the lungs through the pulmonary circuit. The left side receives highly oxygenated blood from the lungs and pumps it throughout the body via the systemic circuit. Each side of the heart is divided into two chambers. See Dissection Atlas **FIGURES A3-4 and A3-5** for photographs of the human heart showing the chambers and the vessels that connect to the heart.

Four Chambers The upper chambers on the right and left sides, the **atria** (A-tre-ah), are mainly blood-receiving chambers **(see FIG. 13-2)**. The lower chambers on the right and left sides, the **ventricles** (VEN-trih-klz), are forceful pumps. 1 The thickness of the wall in each chamber indicates the force it can generate. The atria have the thinnest walls and the weakest contractions. The stronger ventricles have thicker walls, with the left ventricular wall being the thickest of all. The chambers, listed in the order in which blood originating in the body tissues flows through them, are as follows:

1. The **right atrium** (A-tre-um) is a thin-walled chamber that receives the blood returning from the body tissues. This blood, which is comparatively low in oxygen, is carried in veins, the blood vessels leading back to the heart. The superior vena cava brings blood from the head, chest, and arms; the inferior vena cava delivers blood from the trunk and legs. A third vessel that opens into the right atrium brings blood from the heart muscle itself, as described later in this chapter.
2. The **right ventricle** receives blood from the right atrium and pumps it to the lungs. Blood passes from the right ventricle to a large pulmonary trunk, which then divides into right and left pulmonary arteries. Branches of these arteries carry blood to the lungs. An artery is a vessel that takes blood from the heart to the tissues. Note that the pulmonary arteries in **FIGURE 13-2** are colored blue because they are carrying blood low in oxygen, unlike other arteries, which carry blood high in oxygen.
3. The **left atrium** receives oxygen-rich blood as it returns from the lungs in pulmonary veins. Note that the

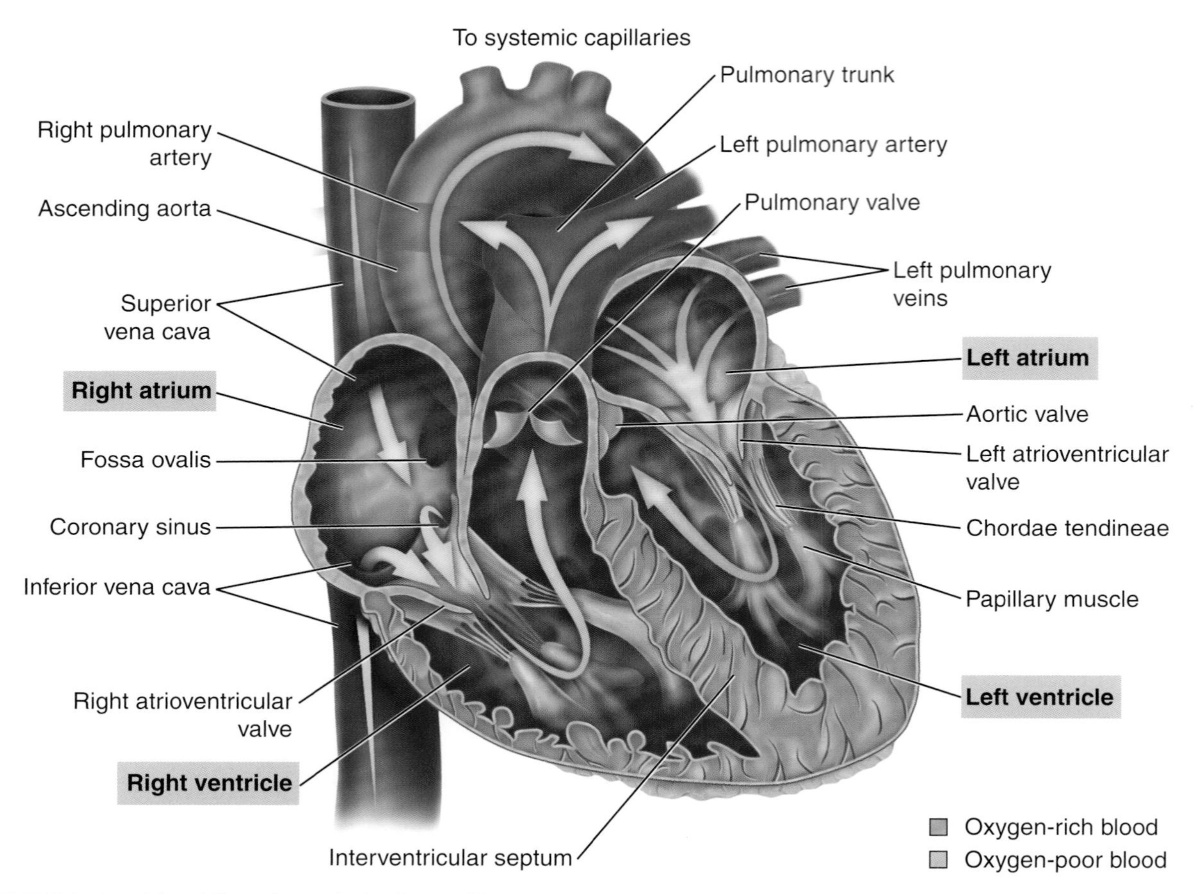

FIGURE 13-2 Blood flow through the heart. **KEY POINT** The right heart has blood low in oxygen; the left heart has blood high in oxygen. The *arrows* show the direction of blood flow through the heart. **ZOOMING IN** Which heart chamber has the thickest wall?

pulmonary veins in **FIGURE 13-2** are colored red because they are carrying blood high in oxygen content, unlike other veins, which carry blood low in oxygen.

4. The **left ventricle**, the chamber with the thickest wall, pumps highly oxygenated blood to all parts of the body, including the lung tissues. This blood goes first into the aorta (a-OR-tah), the largest artery, and then into the branching systemic arteries that take blood to the tissues. The heart's apex, the lower pointed region, is formed by the wall of the left ventricle.

The heart's right and left chambers are completely separated from each other by partitions, each of which is called a **septum.** The **interatrial** (in-ter-A-tre-al) **septum** separates the two atria, and the **interventricular** (in-ter-ven-TRIK-u-lar) **septum** separates the two ventricles. The septa, like the heart wall, consist largely of myocardium.

Four Valves One-way valves that direct blood flow through the heart are located at the entrance and exit of each ventricle (see **FIG. 13-2**). The entrance valves are the **atrioventricular** (a-tre-o-ven-TRIK-u-lar) (AV) **valves,** so named because they are between the atria and ventricles. The exit valves are the **semilunar** (sem-e-LU-nar) **valves,** so named because each flap of these valves resembles a half-moon. 5 Recall from Chapter 1 that flow is promoted by a gradient and opposed by resistance. Blood flow through the heart is governed by this principle, with contractions of the myocardium or accumulating blood volume creating a pressure gradient. The heart valves provide resistance when the pressure gradient would promote flow in the wrong direction. So, the atrioventricular valves open when the pressure is greater in the atria than in the ventricles, and flow occurs. However, they close when ventricular pressure exceeds atrial pressure, preventing flow. The semilunar valves similarly govern flow between the ventricles and great vessels.

Each valve has a specific name as follows:

- The **right atrioventricular** (AV) **valve** is also known as the **tricuspid** (tri-KUS-pid) **valve** because it has three cusps, or flaps, that open and close (**FIG. 13-3**). When this valve is open, blood flows freely from the right atrium into the right ventricle. When the right ventricle begins to contract, however, the valve is closed by blood pressing against the cusps. With the valve closed, blood cannot return to the right atrium but must flow forward into the pulmonary trunk.
- The **left atrioventricular** (AV) **valve** is the bicuspid valve, but it is commonly referred to as the **mitral** (MI-tral)

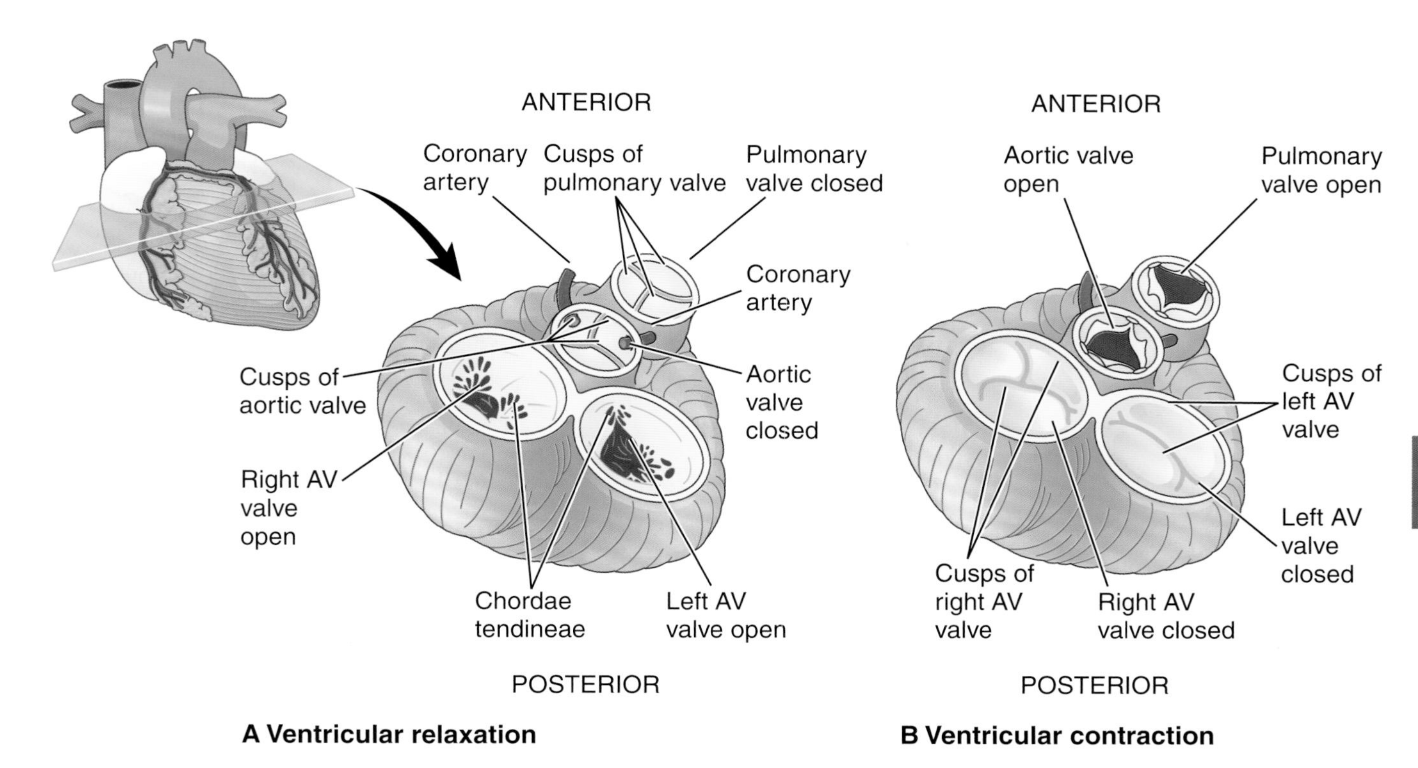

FIGURE 13-3 Heart valves (superior view from posterior, atria removed). **KEY POINT** Valves keep blood flowing in a forward direction through the heart. **A.** When the ventricles are relaxed, the AV valves are open and blood flows freely from the atria to the ventricles. The pulmonary and aortic valves are closed. **B.** When the ventricles contract, the AV valves close, and blood pumped out of the ventricles opens the pulmonary and aortic valves. The abbreviation AV means atrioventricular. **ZOOMING IN** How many cusps does the right AV valve have? The left?

valve (named for a miter, the pointed, two-sided hat worn by bishops). It has two heavy cusps that permit blood to flow freely from the left atrium to the left ventricle. The cusps close when the left ventricle begins to contract; this closure prevents blood from returning to the left atrium and ensures the forward flow of blood into the aorta. Both the right and left AV valves are attached by means of thin fibrous threads to **papillary** (PAP-ih-lar-e) **muscles** arising from the walls of the ventricles. The function of these threads, called the **chordae tendineae** (KOR-de ten-DIN-e-e) (**see FIG. 13-2**), is to stabilize the valve flaps when the ventricles contract so that the blood's force will not push the valves up into the atria. In this manner, they help prevent a backflow of blood when the heart beats.

- The **pulmonary** (PUL-mon-ar-e) **valve** is a semilunar valve located between the right ventricle and the pulmonary trunk that leads to the lungs. When the right ventricle relaxes, pressure in that chamber drops. The higher pressure in the pulmonary artery, described as *back pressure*, closes the valve and prevents blood from returning to the ventricle.
- The **aortic** (a-OR-tik) **valve** is a semilunar valve located between the left ventricle and the aorta. When the left ventricle relaxes, back pressure closes the aortic valve and prevents the backflow of blood from the aorta to the ventricle.

Note that blood passes through the heart twice in making a trip from the heart's right side through the pulmonary circuit to the lungs and back to the heart's left side to start on its way through the systemic circuit. However, it is important to bear in mind that the heart's two sides function in unison to pump identical volumes of blood through both circuits at the same time.

See the Student Resources on thePoint® for charts summarizing the structure of the heart and pericardium; for a detailed picture of the heart's interior; and the animation "Heart Anatomy." See also the animation "Blood Circulation."

BLOOD SUPPLY TO THE MYOCARDIUM

Only the endocardium comes into contact with the blood that flows through the heart chambers. Therefore, the myocardium must have its own blood vessels to provide oxygen and nourishment and to remove waste products. Together, these blood vessels form the **coronary** (KOR-o-na-re) **circulation**. Like vessels elsewhere in the body, the coronary arteries with time undergo degenerative changes (known as coronary artery disease). These

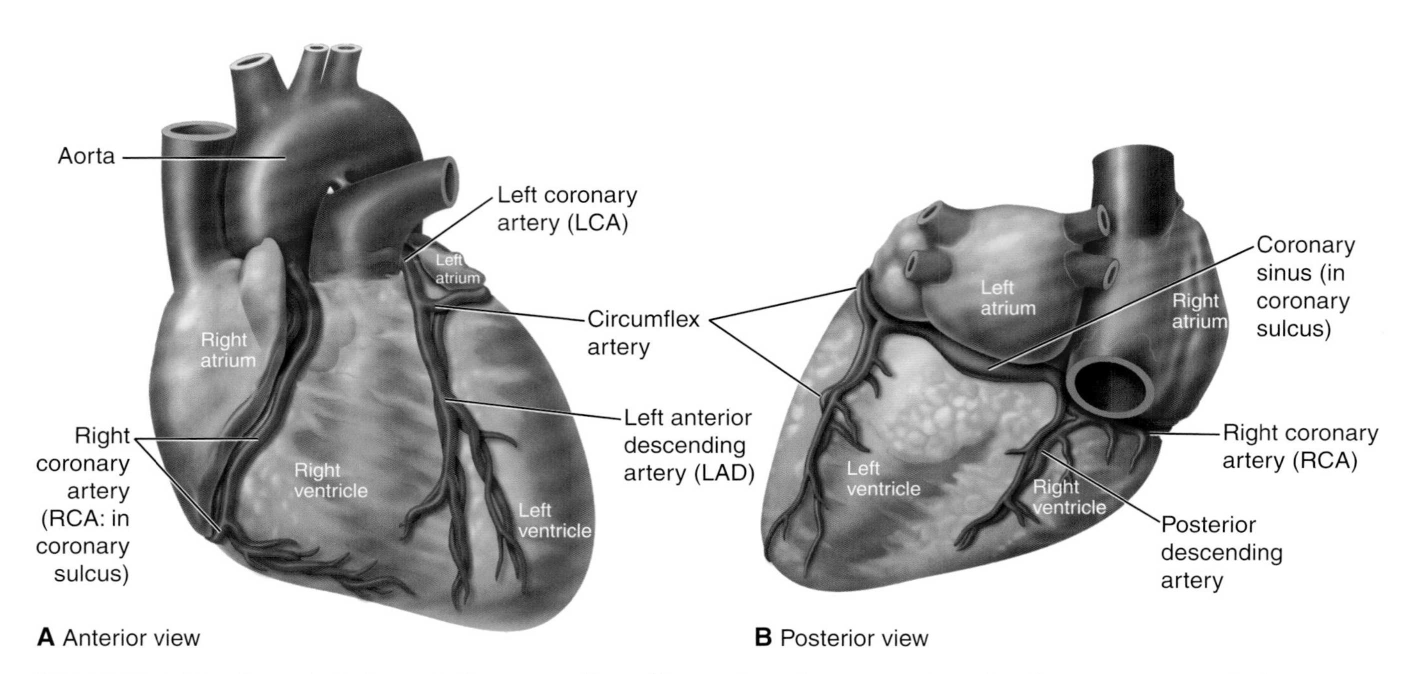

FIGURE 13-4 Blood vessels that supply the myocardium. KEY POINT Coronary arteries and cardiac veins constitute the heart's circulatory pathways. **A.** Anterior view. **B.** Posterior view. ZOOMING IN What is the largest cardiac vein, and where does it lead?

changes narrow the lumen (space) inside the vessel, reducing blood flow to the cardiac muscle. Inadequate blood supply reduces the heart's ability to convey electrical signals and generate force. To make matters worse, the narrowed vessels facilitate the formation of thrombi (blood clots) that completely block the vessel, resulting in a heart attack (myocardial infarction). Unless the vessel is reopened quickly, tissue death results in the heart region supplied by the blocked vessel. In the case study, Jim suffered a heart attack. In this instance, blood flow through his coronary circulation was restored before the heart attack took his life.

The main arteries that supply blood to the heart muscle are the right and left coronary arteries (**FIG. 13-4**), named because they encircle the heart like a crown. These arteries, which are the first to branch off the aorta, arise just above the cusps of the aortic valve and branch to all regions of the heart muscle. They receive blood only when the ventricles relax because the aortic valve must be closed to expose the entrance to these vessels (**FIG. 13-5**). The left coronary artery

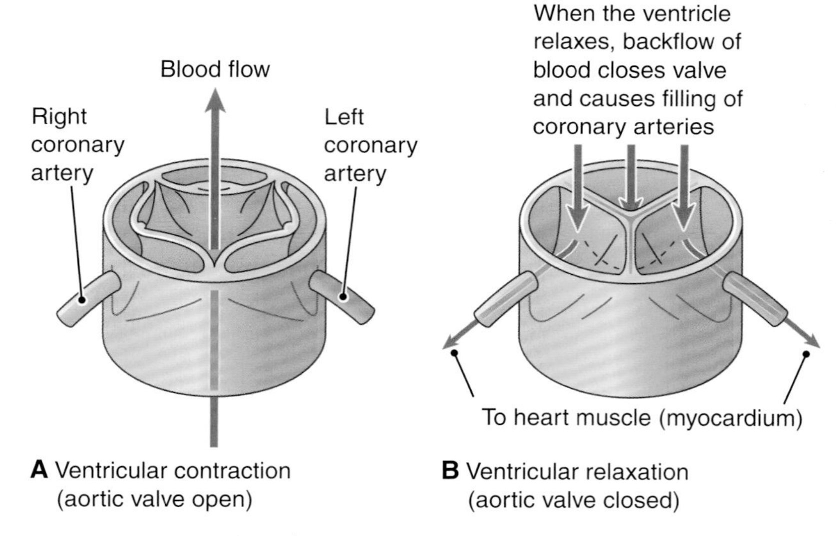

FIGURE 13-5 Opening of coronary arteries in the aortic valve (anterior view). A. When the left ventricle contracts, the aortic valve opens. The valve cusps prevent filling of the coronary arteries. **B.** When the left ventricle relaxes, backflow of blood closes the aortic valve, and the coronary arteries fill.

(LCA) branches into the circumflex artery and the left anterior descending (LAD) artery (also known as the anterior interventricular branch of the LCA). The right coronary artery (RCA) snakes around the heart just inferior to the right atrium, giving off a major branch called the posterior descending artery (also known as the posterior interventricular artery). After passing through the capillaries in the myocardium, blood drains into a system of cardiac veins that brings blood back toward the right atrium. Blood finally collects in the **coronary sinus**, a dilated vein that opens into the right atrium near the inferior vena cava (SEE FIG. 13-4).

CHECKPOINTS 

- ☐ **13-1** What are the names of the innermost, middle, and outermost layers of the heart wall?
- ☐ **13-2** What is the name of the sac that encloses the heart?
- ☐ **13-3** What is the heart's upper receiving chamber on each side called? What is the lower pumping chamber called?
- ☐ **13-4** What is the purpose of the valves in the heart?
- ☐ **13-5** What is the name of the system that supplies blood to the myocardium?

CASEPOINT 

- ☐ **13-1** In the case study, Jim had a blockage of his left coronary artery. Which two arteries would no longer receive blood because of this blockage?

Heart Function

Although the heart's right and left sides are separated from each other, they work together. A heart muscle contraction begins in the thin-walled upper chambers, the atria, and is followed by a contraction of the thick muscle of the lower chambers, the ventricles. In each case, the active phase, called **systole** (SIS-to-le), is followed by a resting phase known as **diastole** (di-AS-to-le). One complete sequence of heart contraction and relaxation is called the **cardiac cycle** (FIG. 13-6). Each cardiac cycle represents a single heartbeat. At rest, one cycle takes an average of 0.8 seconds.

The cardiac cycle begins with contraction of both atria, which forces blood through the AV valves into the ventricles. The atrial walls are thin, and their contractions are not very powerful. However, they do improve the heart's efficiency by forcing blood into the ventricles before these lower chambers contract. Atrial contraction ends before ventricular contraction begins. Thus, atrial diastole begins at the same time ventricular systole begins. While the ventricles are contracting, forcing blood through the semilunar valves, the atria are relaxed and again are filling with blood (see FIG. 13-6).

After the ventricles have contracted, all the chambers are relaxed for a short period. During this period of complete relaxation, blood enters the atria from the great veins and passively drains into the ventricles.

Concept Mastery Alert

Note that the semilunar valves are closed during atrial systole to prevent blood from flowing backward from the great vessels to the ventricles.

Then, another cycle begins with an atrial contraction followed by a ventricular contraction. Although both upper and lower chambers have a systolic and diastolic phase in each cardiac cycle, discussions of heart function usually refer to these phases as they occur in the ventricles, because

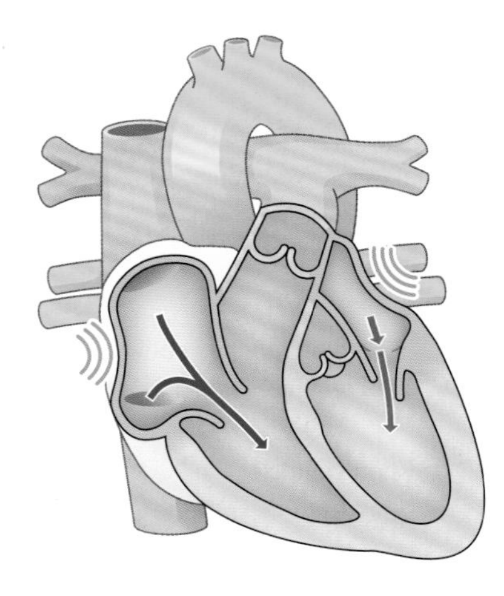

Atrial systole
Contraction of atria pumps additional blood into the ventricles.

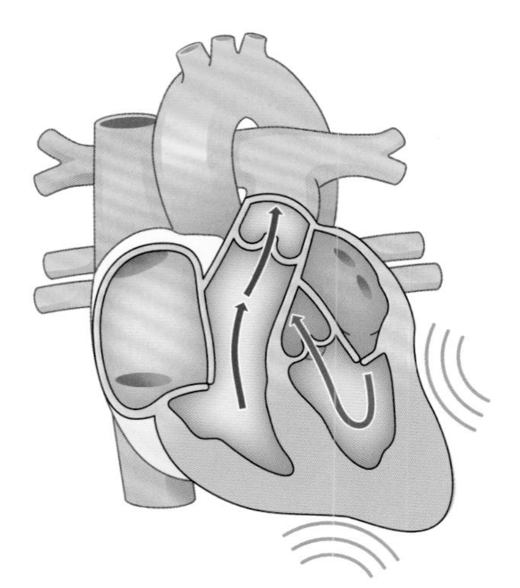

Ventricular systole
Contraction of ventricles pumps blood into aorta and pulmonary arteries.

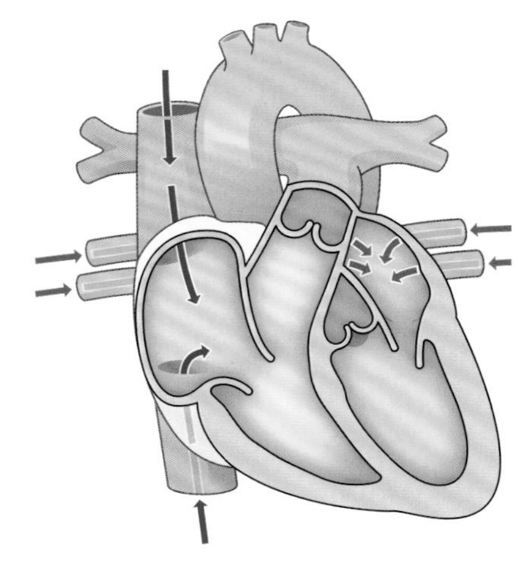

Complete diastole
Atria fill with blood, which flows directly into the relaxed ventricles.

FIGURE 13-6 **The cardiac cycle.** KEY POINT In one cardiac cycle, contraction of both atria is followed by contraction of both ventricles. The entire heart relaxes briefly before the next cardiac cycle begins. The green shading indicates that the chamber is contracting. ZOOMING IN When the ventricles contract, what valves close? What valves open?

13

these chambers contract more forcefully and drive blood into the arteries.

See the Student Resources on thePoint® for the animations "Myocardial Blood Flow" and "The Cardiac Cycle."

THE HEART'S CONDUCTION SYSTEM

Like other muscles, the heart muscle is stimulated to contract by a wave of electric energy that passes along the cells. This action potential is generated by specialized tissue within the heart and spreads over structures that form the heart's conduction system (FIG. 13-7). Two of these structures are tissue masses called **nodes**, and the remainder consists of specialized fibers that branch through the myocardium.

The **sinoatrial** (SA) **node** is located in the upper wall of the right atrium in a small depression described as a sinus. This node initiates the heartbeats by generating an action potential at regular intervals. Because the SA node sets the rate of heart contractions, it is commonly called the **pacemaker.** The second node, located in the interatrial septum at the bottom of the right atrium, is called the **atrioventricular** (AV) **node.**

The **atrioventricular** (AV) **bundle,** also known as the *bundle of His,* is located at the top of the interventricular septum. Fibers travel first down both sides of the interventricular septum in groups called the right and left bundle branches. Smaller **Purkinje** (pur-KIN-je) **fibers** then travel in a branching network throughout the myocardium of the ventricles. Intercalated disks allow the rapid flow of impulses throughout the heart muscle.

The order in which impulses travel through the heart is as follows:

1. The SA generates the electric impulse that begins the heartbeat (see FIG. 13-7).
2. The excitation wave travels throughout the myocardium of each atrium, causing the atria to contract. At the same time, impulses also travel directly to the AV node by means of fibers in the wall of the atrium that make up the **internodal pathways.**
3. The atrioventricular node is stimulated. A relatively slower rate of conduction through the AV node allows time for the atria to contract and complete the filling of the ventricles before the ventricles contract.
4. The excitation wave rapidly travels through the AV bundle and then throughout the ventricular walls by means of the bundle branches and Purkinje fibers. The entire ventricular musculature contracts in a wave, beginning at the apex and squeezing the blood upward toward the aorta and pulmonary artery.

A normal heart rhythm originating at the SA node is termed a **sinus rhythm.** As a safety measure, a region of the conduction system other than the SA node can generate a heartbeat if the SA node fails, but it does so at a slower rate.

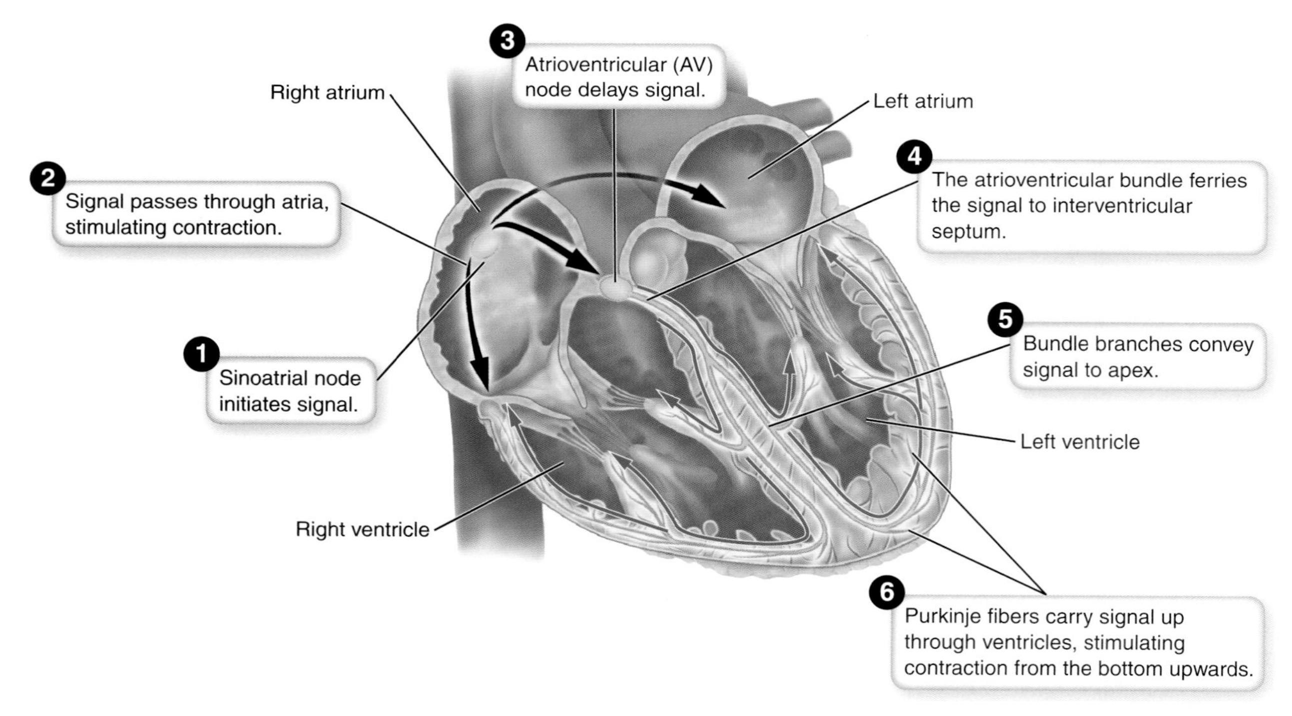

FIGURE 13-7 Conduction system of the heart. KEY POINT The sinoatrial (SA) node, the atrioventricular (AV) node, and specialized fibers conduct the electric signal that stimulates the heart muscle to contract. ZOOMING IN Atrial muscle fibers carry signals between which two parts of the conducting system?

CARDIAC OUTPUT

13 **Mass Balance: Output Balances Input** The key idea of mass balance is not confined to physiology—it is a basic principle of life that also applies to your bank account. Your stored money (savings) will stay constant if the input (income) matches the output (expenses). If income exceeds expenses, your savings increase and vice versa. The concept of mass balance explains that the amount of material in a system depends on both inputs and outputs. With few exceptions, the adult body keeps storage pools of most substances relatively constant by matching input and output. For instance, increased fluid consumption results in increased urine production, thus keeping the amount of "stored" body water constant.

Determinants of Cardiac Output A unique property of heart muscle is its ability to adjust the ventricles' output to their input, thereby keeping ventricular blood volume at the end of systole relatively constant. This property reflects changes in the strength of contraction. When the heart chamber is filled and the wall stretched (within limits), the contraction is strong. As less blood enters the heart, contractions become less forceful. Thus, as more blood enters the heart, the muscle contracts with greater strength to push the larger volume of blood out into the blood vessels. The heart's ability to pump out all of the blood it receives prevents blood from pooling in the chambers.

The volume of blood pumped by each ventricle in one minute is termed the **cardiac output** (CO). It is the product of the **stroke volume** (SV)—the volume of blood ejected from the ventricle with each beat—and the **heart rate** (HR), the number of times the heart beats per minute. To summarize:

$$CO = HR \times SV$$

Based on a heart rate of 75 bpm and a stroke volume of 70 mL/beat, the average cardiac output for an adult at rest is about 5 L/min. This means that at rest, the heart pumps the equivalent of the total blood volume each minute. But like many other organs, the heart has great reserves of strength. The cardiac reserve is a measure of how many times more than resting output the heart can produce when needed.

During mild exercise, cardiac output might double. During strenuous exercise, it might double again. In other words, for most people, the cardiac reserve is four to five times the resting output but can be much higher in trained athletes. This increase is achieved by an increase in stroke volume and heart rate. In athletes exercising vigorously, the ratio may reach six to seven times the resting volume. In contrast, those with heart disease may have little or no cardiac reserve. They may be fine at rest but quickly become short of breath or fatigued when exercising or even when carrying out the simple tasks of daily living.

Variations in Heart Rate The resting heart rate varies significantly among individuals. Generally speaking, the heart rate is generally faster in small people than large people and slightly faster in women than in men. Resting heart rate also declines with age, from 120 to 140 bpm in a newborn to 60 to 80 bpm in adults. 10 The heart responds to the demands of regular exercise by getting stronger and ejecting more blood with each beat. Consequently, the body's circulatory needs at rest can be met with a lower heart rate. Trained athletes usually have a low resting heart rate.

Specific terms describe some variations in heart rate. Note that these variations do not necessarily indicate pathology:

- **Bradycardia** (brad-e-KAR-de-ah) is a relatively slow heart rate of less than 60 bpm. During rest and sleep, the heart may beat less than 60 bpm, but the rate usually does not fall below 50 bpm.
- **Tachycardia** (tak-e-KAR-de-ah) refers to a heart rate of more than 100 bpm. Tachycardia is normal during exercise or stress or with excessive caffeine intake but may also occur with certain disorders.
- **Sinus arrhythmia** (ah-RITH-me-ah) is a regular variation in heart rate caused by changes in the rate and depth of breathing. It is a normal phenomenon.
- **Premature ventricular contraction** (PVC), also called *ventricular extrasystole*, is a ventricular contraction initiated by the Purkinje fibers rather than the SA node. It can be experienced as a palpitation between normal heartbeats or as a skipped beat. PVCs may be initiated by caffeine, nicotine, or psychological stresses. They are also common in people with heart disease.

13

CASEPOINT

☐ **13-2** Jim received morphine, which lowered his heart rate to 80 bpm. His stroke volume remained constant at 50 mL/beat. What is his cardiac output?

Control of Cardiac Output Although the heart's fundamental beat originates within the heart itself, the heart rate and stroke volume (and thus cardiac output) can be influenced by the nervous system, hormones, drugs, and ions.

3 Recall from Chapter 1 that blood pressure is a regulated variable that must be kept within narrow limits. The body controls cardiac output in order to maintain adequate blood pressure as conditions and activities alter. Chapter 14 discusses this topic in greater detail, but for now, we need to know that increasing the cardiac output increases blood pressure. When blood pressure decreases, heart rate and stroke volume increase in order to restore normal blood pressure. Similarly, heart rate and stroke volume increase when the set point for blood pressure is raised due to stressors such as excitement, increased body temperature, and exercise. It is important to remember that cardiac output, heart rate, and stroke volume are not regulated variables, because they vary significantly over time without causing disease and the body cannot directly sense them. Instead, changing heart activity reflects the actions of an effector (the heart) in the negative feedback loop controlling blood pressure.

The signals in this negative feedback loop are carried by the autonomic nervous system (**FIG. 13-8**). Sympathetic

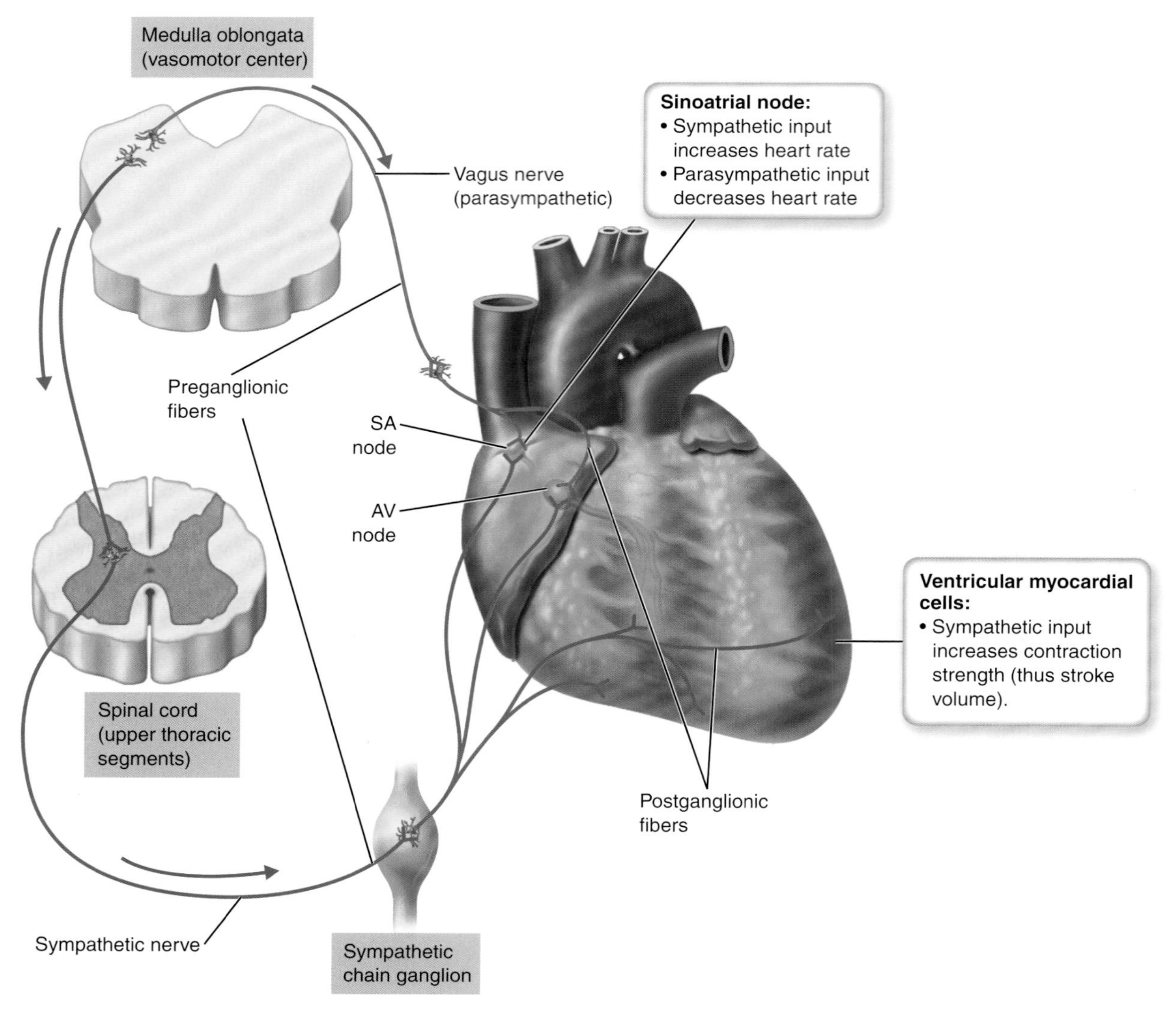

FIGURE 13-8 Autonomic nervous system (ANS) regulation of the heart. KEY POINT The ANS affects the rate and force of heart contractions by acting on the SA and AV nodes and the myocardium itself. ZOOMING IN Which cranial nerve carries parasympathetic impulses to the heart?

fibers increase the heart rate by stimulating the SA and AV nodes. They also increase the contraction force and thus the stroke volume by acting directly on the fibers of the myocardium. Parasympathetic stimulation decreases cardiac output by lowering the heart rate. The parasympathetic nerve that supplies the heart is the vagus nerve (cranial nerve X). It slows the heart rate by acting on the SA and AV nodes but does not influence the stroke volume (see FIG. 13-8).

Concept Mastery Alert

Remember that the sympathetic nervous system increases both the heart rate and the stroke volume, but the parasympathetic nervous system reduces only the heart rate.

The heart rate is also affected by substances circulating in the blood, including hormones, drugs, and ions. The most important of these is epinephrine, a hormone released from the adrenal gland. 11 Recall from Chapter 8 that chemical signals affect their target sites by binding to receptors and that sympathetic nerve fibers use norepinephrine as their neurotransmitter. Epinephrine also binds to norepinephrine receptors, so it too increases cardiac output. Thyroid hormones increase the heart rate by increasing the number of epinephrine receptors and magnifying epinephrine's effects. Some drugs, conversely, lower heart rate (and thus blood pressure) by blocking epinephrine receptors and preventing epinephrine's actions. These "beta-blockers" (named after a type of epinephrine receptor) are important for the treatment of high blood pressure, as discussed in Chapter 14. Pathologic changes in blood concentrations of Na^+, K^+, and Ca^{2+} can also change heart rate, because they alter the membrane potential of cardiac cells. For example, excessive potassium hyperpolarizes cardiac cells, so they fire fewer (if any) action potentials and contract less often (if at all). Surgeons use this technique to stop the heartbeat during surgeries.

NORMAL AND ABNORMAL HEART SOUNDS

The normal heart sounds are usually described by the syllables "lub" and "dup." The first heart sound (S_1), the "lub," is a longer, lower-pitched sound that occurs at the start of ventricular systole. It is caused by a combination of events, mainly closure of the AV valves. This action causes vibrations in the blood passing through the valves and in the tissue surrounding the valves. The second heart sound (S_2), the "dup," is shorter and sharper. It occurs at the beginning of ventricular relaxation and is caused largely by sudden closure of the semilunar valves.

Any abnormal heart sound is called a **murmur.** Anything that disrupts the smooth flow of blood through the heart will cause a murmur. For example, if a valve fails to close tightly and blood leaks back, a murmur is heard. An abnormal sound caused by any structural change in the heart or the vessels connected with the heart is called an **organic murmur**. Certain normal sounds heard while the heart is working may also be described as murmurs, such as the sound heard during rapid filling of the ventricles. To differentiate these from abnormal sounds, they are more properly called **functional murmurs**.

CHECKPOINTS 

- ☐ **13-6** What name is given to the contraction phase of the cardiac cycle? To the relaxation phase?
- ☐ **13-7** What is cardiac output? What two factors determine cardiac output?
- ☐ **13-8** What is the scientific name of the heart's pacemaker?
- ☐ **13-9** What system exerts the main influence on the rate and strength of heart contractions?
- ☐ **13-10** What is a heart murmur?

CASEPOINTS 

- ☐ **13-3** A trauma nurse measured Jim's vital signs when he arrived at the hospital. Vital signs include temperature, heart rate, blood pressure, and respiration rate. What part of the brain stem has vital centers that control heart rate and respiration rate?
- ☐ **13-4** Jim had tachycardia upon admission. Which heart rate would be considered as tachycardia, 160 bpm or 90 bpm?

See the Student Resources on thePoint® for the animation "Heart Sounds."

Heart Studies

Experienced listeners can gain important information about the heart using a **stethoscope** (STETH-o-skope). This relatively simple instrument is used to convey sounds from within the patient's body to an examiner's ear.

The **electrocardiograph** (ECG or EKG) is used to record the electrical activity of the heart as it functions. (The abbreviation EKG comes from the German spelling of the word.) This activity corresponds to the depolarization and repolarization that occur during an action potential, as described in Chapters 7 and 8. The ECG may reveal certain myocardial injuries. Electrodes (leads) placed on the skin surface pick up electric activity, and the ECG tracing, or electrocardiogram, represents this activity as waves (FIG. 13-9). These waves are identified by consecutive letters of the alphabet. The P wave corresponds to depolarization of the atria; the QRS wave corresponds to depolarization of the ventricles. The T wave shows ventricular repolarization, but atrial repolarization is hidden by the QRS wave. Cardiologists use changes in the waves and the intervals between them to diagnose heart damage and arrhythmias.

Many people with heart disease undergo **catheterization** (kath-eh-ter-i-ZA-shun). In right heart catheterization, an extremely thin tube (catheter) is passed through the veins of the right arm or right groin and then into the right side of the heart. This procedure gives diagnostic information and monitors heart function. A **fluoroscope** (flu-OR-o-skope), an instrument for examining deep structures with x-rays, is used to show the route taken by the catheter. In left heart

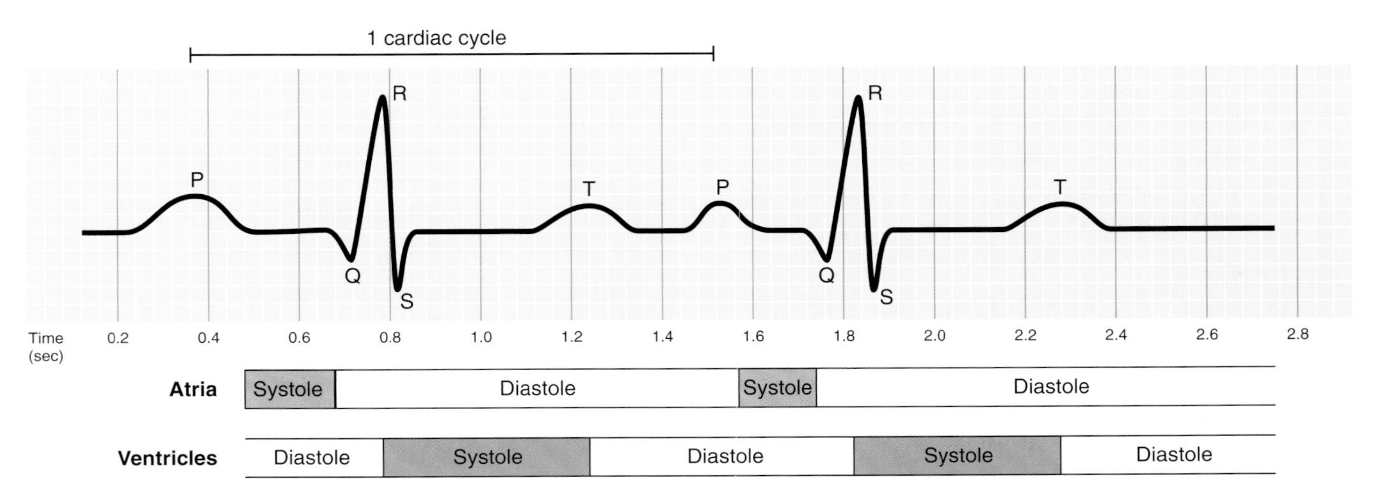

FIGURE 13-9 Normal electrocardiography (ECG) tracing. KEY POINT Electric activity in the myocardium produces ECG waves. The P wave represents atrial depolarization, the QRS ventricular depolarization and atrial repolarization, and the T wave ventricular repolarization. Changes in the wave patterns indicate a disorder. ZOOMING IN What is the length of one cardiac cycle shown in this diagram?

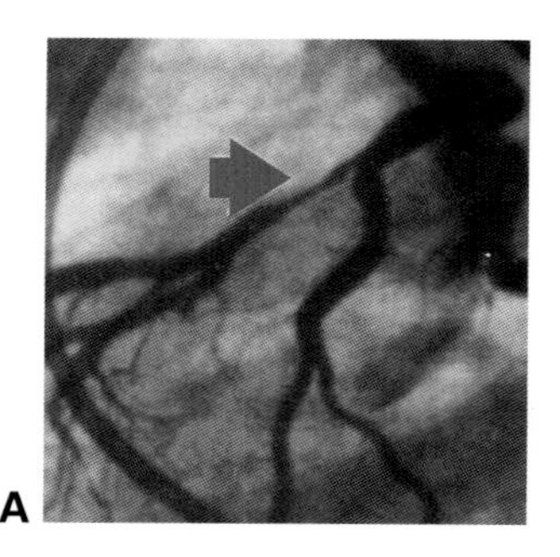

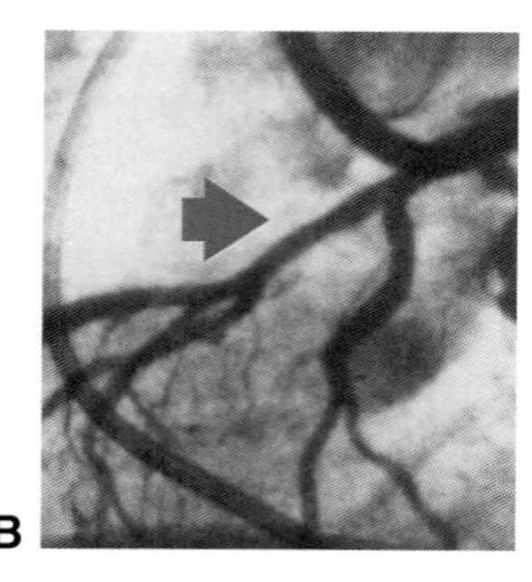

FIGURE 13-10 Coronary angiography. **KEY POINT** The coronary vessels are imaged following administration of a dye. **A.** Coronary angiography shows narrowing in the mid–left anterior descending artery (LAD) (*arrow*). **B.** The same vessel (*arrow*) after a procedure to remove plaque. Note the improved blood flow through the artery.

catheterization, a catheter is passed through an artery in the left groin or arm to the heart. The tube may be passed through the aortic valve into the left ventricle for studies of pressure and volume in that chamber. During catheterization, dye can be injected into the coronary arteries to map vascular damage, a procedure known as **coronary angiography** (an-je-OG-rah-fe) (**FIG. 13-10**). The root *angi/o* means "vessel."

Ultrasound consists of sound waves generated at a frequency above the human ear's range of sensitivity. In **echocardiography** (ek-o-kar-de-OG-rah-fe), also known as *ultrasound cardiography*, high-frequency sound waves are sent to the heart from a small instrument on the chest surface. The ultrasound waves bounce off the heart and are recorded as they return, showing the heart in action. Movement of the echoes is traced on an electronic instrument called an *oscilloscope* and recorded on film. (The same principle is employed by submarines to detect ships.) The method is safe and painless, and it does not use x-rays. It provides information on the size and shape of heart structures, on cardiac function, and on possible heart defects.

CHECKPOINTS 

- 13-11 What do ECG and EKG stand for?
- 13-12 What is the general term for using a thin tube threaded through a vessel for diagnosis or repair?
- 13-13 What techniques use a dye and x-rays to visualize the coronary arteries?

CASEPOINT 

- 13-5 Jim's ECG in the ambulance showed an ST segment elevation. What event in the cardiac cycle is occurring during the T wave?

Maintaining Heart Health

Earlier in this chapter we highlighted the dangerous consequences of degenerative changes in the coronary arteries. Some factors contributing to arterial disease are out of one's control. These include:

- Age. The risk of coronary heart disease increases with age.
- Gender. Until middle age, men have greater risk than do women. Women older than 50 years or past menopause have risk equal to that of men.
- Heredity. Those with immediate family members with heart disease are at greater risk.
- Body type. In particular, the hereditary tendency to deposit fat in the abdomen or on the chest surface increases risk.

Other risk factors for heart disease can be reduced by modifying one's lifestyle. These include:

- Smoking and other forms of tobacco use, which lead to spasm and hardening of the arteries. These arterial changes result in decreased blood flow and poor supply of oxygen and nutrients to the myocardium.
- Physical inactivity. Lack of exercise weakens the heart muscle and decreases the heart's efficiency. It also decreases the efficiency of the skeletal muscles, which further taxes the heart.
- Weight over the ideal increases risk.
- Saturated fat in the diet. Elevated fat levels in the blood lead to blockage of the coronary arteries by plaque (fatty deposits) (**BOX 13-1**).
- Hypertension damages heart muscle. Smoking cessation; regular physical activity; a healthful, low-sodium diet; and appropriate medication, if needed, are all important in reducing this risk factor.
- Diabetes causes damage to small blood vessels. Type 2 diabetes can be managed with diet, exercise, and proper medication, if needed.
- Individuals suffering from sleep apnea, that is, people who frequently stop breathing for short periods when they sleep, have a higher risk of coronary artery disease. Sleep apnea can be treated with devices to aid breathing during the night and sometimes with surgery to remove obstructions of air passageways.

Efforts to prevent heart disease should include having regular physical examinations and minimizing the controllable risk factors. Jim's physicians will undoubtedly discuss these lifestyle changes with him during the course of his treatment.

See the Student Resources on thePoint® for career information on surgical technologists, who assist in all types of surgical operations.

Effects of Aging on the Heart

There is a great deal of individual variation in the way the heart ages, depending on heredity, environmental factors, diseases, and personal habits such as diet, exercise patterns,

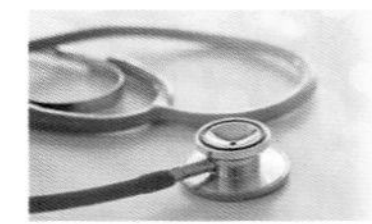

CLINICAL PERSPECTIVES
Lipoproteins: What's the Big DL?

BOX 13-1

Although cholesterol has received a lot of bad press in recent years, it is a necessary substance in the body. It is found in bile salts needed for digestion of fats, in hormones, and in the cell's plasma membrane. However, high levels of cholesterol in the blood have been associated with atherosclerosis (hardening of the arteries) and heart disease.

It now appears that the total amount of blood cholesterol is not as important as the form in which it occurs. Cholesterol is transported in the blood in combination with other lipids and with protein, forming compounds called lipoproteins. These compounds are distinguished by their relative density. High-density lipoprotein (HDL) is composed of a high proportion of protein and relatively little cholesterol. HDLs remove cholesterol from the tissues, including the arterial walls, and carry it back to the liver for reuse or disposal. In contrast, low-density lipoprotein (LDL) contains less protein and a higher proportion of cholesterol. LDLs carry cholesterol from the liver to the tissues, making it available for membrane or hormone synthesis. However, excess LDLs can deposit cholesterol along the lining of arterial walls. Thus, high levels of HDLs (60 mg/dL and above) indicate efficient removal of arterial plaques, whereas high levels of LDLs (130 mg/dL and above) suggest that arteries will become clogged.

Diet is an important factor in regulating lipoprotein levels. Saturated fatty acids (found primarily in animal fats) raise LDL levels, while unsaturated fatty acids (found in most vegetable oils) lower LDL levels and stimulate cholesterol excretion. Thus, a diet lower in saturated fat and higher in unsaturated fat may reduce the risk of atherosclerosis and heart disease. Other factors that affect lipoprotein levels include cigarette smoking, caffeine, and stress, which raise LDL levels, and exercise, which lowers LDL levels.

and tobacco use. However, many changes still commonly occur with age. The heart chambers become smaller, and myocardial tissue atrophies and gets replaced with connective tissue. These changes significantly reduce cardiac output. The valves become less flexible, and incomplete closure may produce an audible murmur. By 70 years of age, the cardiac output may decrease by as much as 35%. Damage within the conduction system can produce abnormal rhythms, including extra beats, rapid atrial beats, and slowing of ventricular contraction rate. Temporary failure of the conduction system (heart block) can cause periodic loss of consciousness. Because of the decrease in the heart's reserve strength, elderly people may be less able to respond efficiently to physical or emotional stress.

A & P in Action Revisited: Jim's Heart Treatment

After Jim's heart attack, his cardiac enzymes peaked and then began to decline. An echocardiogram showed that his left ventricle's anterior wall was still moving effectively, a sign that the quick action of the medical team had preserved his heart muscle! A few days later, the cardiologist sat down with Jim and his family to discuss what to expect in the future.

"You will need to follow a diet with lots of fresh fruits and vegetables and low fat and low salt and also get regular exercise for good heart health," the cardiologist told Jim. "To help keep your stent and coronary arteries open, we will start you on aspirin and another drug to keep your platelets from sticking together. We will also start you on a "statin," a drug that lowers the levels of bad cholesterol in your blood and reduces the risk of developing additional coronary vascular disease. If, in the future, another clot forms and closes the blood flow through your stent, we would need to consider a coronary artery bypass graft. This involves open heart surgery and taking a piece of a vein from your leg to sew between your aorta and your left coronary artery below the blockage. However, with good care, you should be fine, and there's no need to be concerned about that at present."

Although this chapter concentrates on medical terms related to the heart, Jim's case also contains terminology about blood vessels. In Chapter 14, Blood Vessels and Blood Circulation, you will examine these terms in more detail.

CHAPTER 13

Chapter Wrap-Up

OVERVIEW

A detailed chapter outline with space for note-taking is on thePoint®. The figure below illustrates the main topics covered in this chapter.

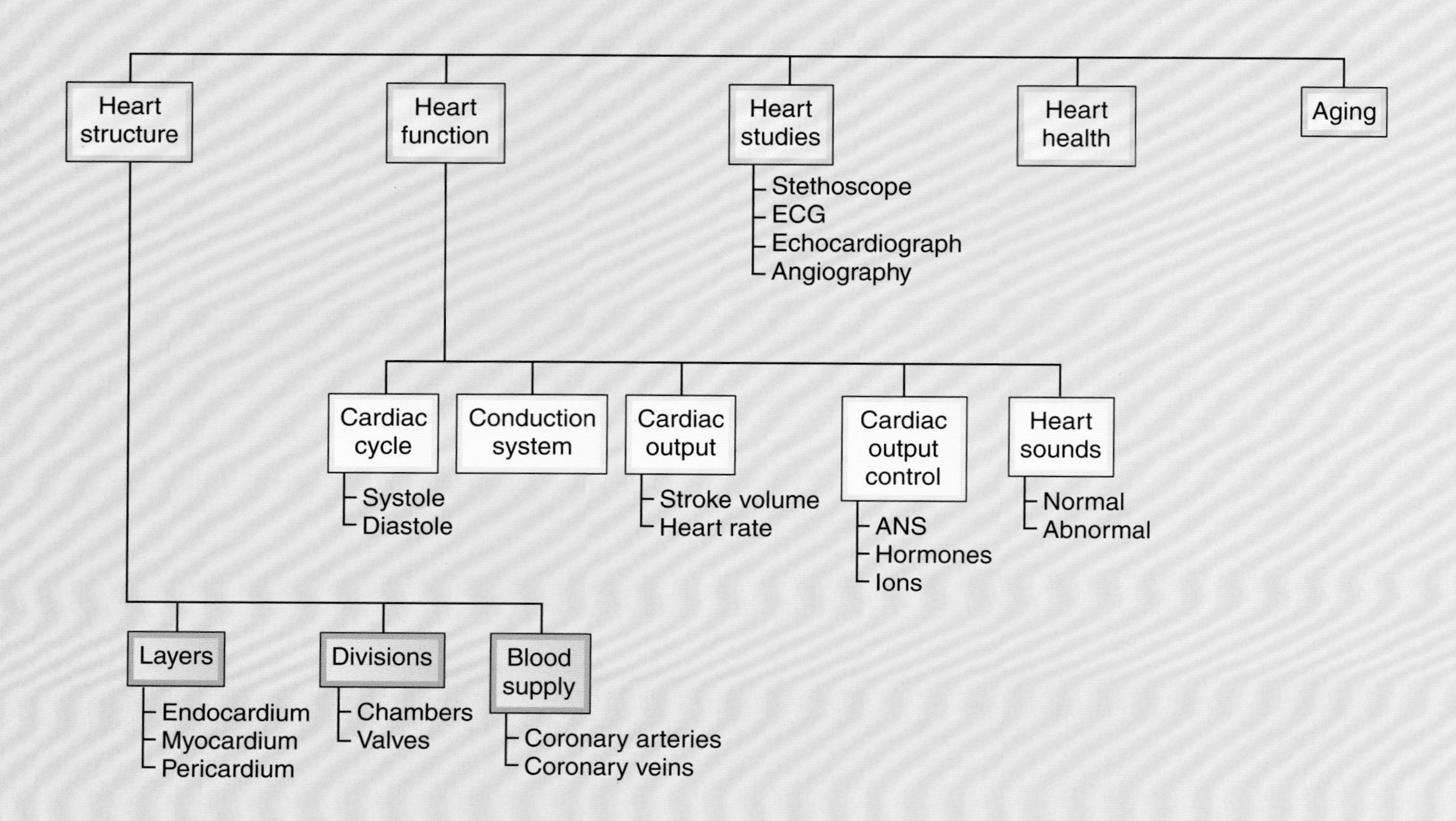

KEY TERMS

The terms listed below are emphasized in this chapter. Knowing them will help you organize and prioritize your learning. These and other boldface terms are defined in the Glossary with phonetic pronunciations.

angiography
arrhythmia
atrium
bradycardia
cardiac output
coronary
diastole
echocardiography
electrocardiograph
endocardium
epicardium
mediastinum
murmur
myocardium
pacemaker
pericardium
septum
systole
tachycardia
valve
ventricle

WORD ANATOMY

Medical terms are built from standardized word parts (prefixes, roots, and suffixes). Learning the meanings of these parts can help you remember words and interpret unfamiliar terms.

WORD PART	MEANING	EXAMPLE
HEART STRUCTURE		
cardi/o	heart	The *myocardium* is the heart muscle.
pulmon/o	lung	The *pulmonary* circuit carries blood to the lungs.
HEART FUNCTION		
brady-	slow	*Bradycardia* is a slow heart rate.
sin/o	sinus	The *sinoatrial* node is in a space (sinus) in the wall of the right atrium.
tachy-	rapid	*Tachycardia* is a rapid heart rate.
HEART STUDIES		
angi/o	vessel	*Angiography* is a radiographic study of vessels.
steth/o	chest	A *stethoscope* is used to listen to body sounds.

QUESTIONS FOR STUDY AND REVIEW

BUILDING UNDERSTANDING

Fill in the Blanks

1. The heart's pointed inferior portion is the ______.
2. The layer of the heart wall that pumps blood is the ______.
3. The heartbeat is initiated by electrical impulses from the ______.
4. The sac that encloses the heart is the ______.
5. Any partition that separates chambers of the heart is called a(n) ______.

Matching Match each numbered item with the most closely related lettered item.

___ 6. receives blood low in oxygen from the body
___ 7. receives blood high in oxygen from the lungs
___ 8. sends blood low in oxygen to the lungs
___ 9. sends blood high in oxygen to the body

a. right ventricle
b. left ventricle
c. right atrium
d. left atrium

Multiple Choice

___ 10. Which structural characteristic promotes rapid transfer of electrical signals between cardiac muscle cells?
a. the striated nature of the cells
b. branching of the cells
c. the abundance of mitochondria within the cells
d. intercalated disks between the cells

___ 11. What separates the upper chambers of the heart from each other?
a. intercalated disk
b. interatrial septum
c. interventricular septum
d. ductus arteriosus

___ 12. Which term describes one complete sequence of heart contraction and relaxation?
a. systole
b. diastole
c. cardiac cycle
d. cardiac output

___ 13. Cardiac output is the product of which factors?
a. stroke volume and heart rate
b. cardiac reserve and atrial systole
c. heart rate and ventricular diastole
d. stroke volume and dysrhythmia

___ 14. Which of the following sends impulses to the Bundle of His?
a. Purkinje fibers
b. atrioventricular bundle
c. atrioventricular node
d. sinoatrial node

___ 15. Which variation in heart rate can be due to changes in the rate and depth of breathing?
a. murmur
b. tachycardia
c. sinus arrhythmia
d. premature ventricular contraction

UNDERSTANDING CONCEPTS

16. Referring to *The Body Visible* at the beginning of this book, give the name and number of the following:
    a. two structures that keep the AV valves from opening into the atria
    b. the heart's pacemaker
    c. the two vessels that carry blood into the coronary circulation
    d. the vein that drains blood from the coronary circulation and empties into the right atrium
17. Differentiate between the terms in each of the following pairs:
    a. serous pericardium and fibrous pericardium
    b. atrium and ventricle
    c. coronary artery and coronary sinus
    d. systole and diastole
    e. bradycardia and tachycardia
18. Explain the purpose of the four heart valves, and describe their structures and locations.
19. Trace a drop of blood from the superior vena cava to the lungs and then from the lungs to the aorta.
20. Describe the order in which electrical impulses travel through the heart.
21. Compare the effects of the sympathetic and parasympathetic nervous systems on heart function.
22. List some age-related changes to the heart.

CONCEPTUAL THINKING

23. Certain risk factors for coronary artery disease may have contributed to Jim's heart attack. What can Jim do to lower his risk of having it happen again? What risk factors is he unable to change? Apply your knowledge of these factors to your own life or the life of someone you know.
24. Jorge's new watch records his heart rate. To his alarm, he notices that his rate varies from 65 bpm to 180 bpm over the course of the day. Is this a failure of homeostasis? Explain.

For more questions, see the Learning Activities on thePoint®.

CHAPTER 14

Blood Vessels and Blood Circulation

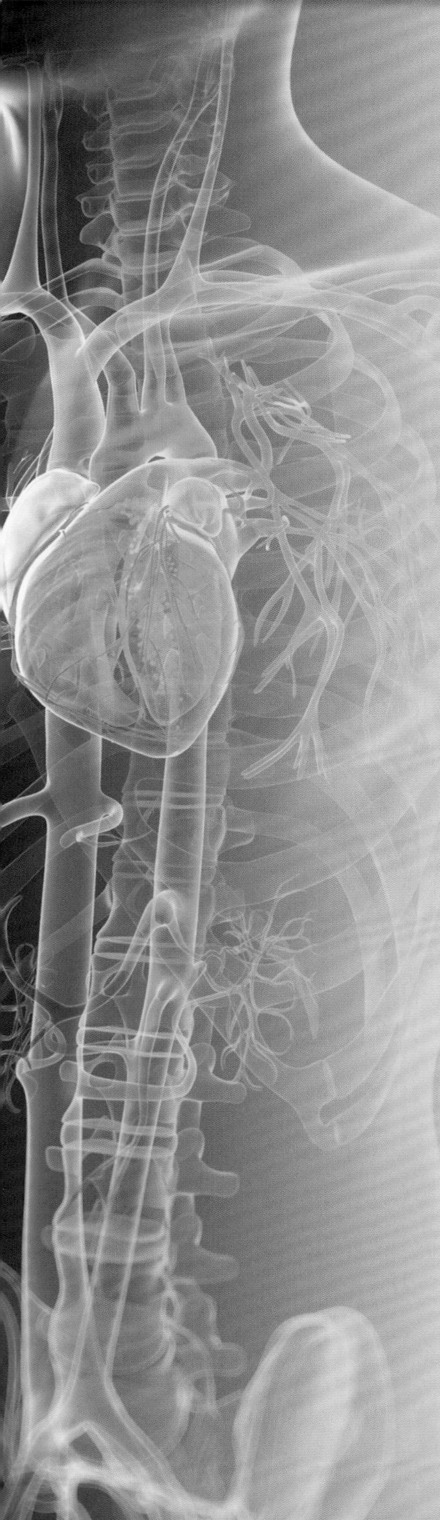

Learning Objectives

After careful study of this chapter, you should be able to:

1. Differentiate among the five types of blood vessels with regard to structure and function. *p. 282*
2. Compare the pulmonary and systemic circuits relative to location and function. *p. 282*
3. Name the four sections of the aorta, and list the main branches of each section. *p. 285*
4. Trace the pathway of blood through the main arteries of the upper and lower limbs. *p. 288*
5. Define anastomosis, cite its function, and give four examples of anastomoses. *p. 289*
6. Compare superficial and deep veins, and give examples of each type. *p. 289*
7. Name the main vessels that drain into the superior and inferior venae cavae. *p. 289*
8. Define venous sinus, and give four examples of venous sinuses. *p. 291*
9. Describe the structure and function of the hepatic portal system. *p. 291*
10. Explain the forces that affect bulk flow across the capillary wall. *p. 293*
11. Discuss mechanisms to control blood flow to different organs and then propel blood back to the heart. *p. 295*
12. Describe the negative feedback loop controlling blood pressure. *p. 297*
13. Explain how changes in overall blood volume, blood viscosity, and vessel compliance impact blood pressure. *p. 298*
14. Explain how blood pressure is commonly measured. *p. 298*
15. Based on the opening case study, discuss the dangers of thrombosis, and describe one approach to its treatment. *pp. 281, 300*
16. Show how word parts are used to build words related to the blood vessels and circulation (see Word Anatomy at the end of the chapter). *p. 302*

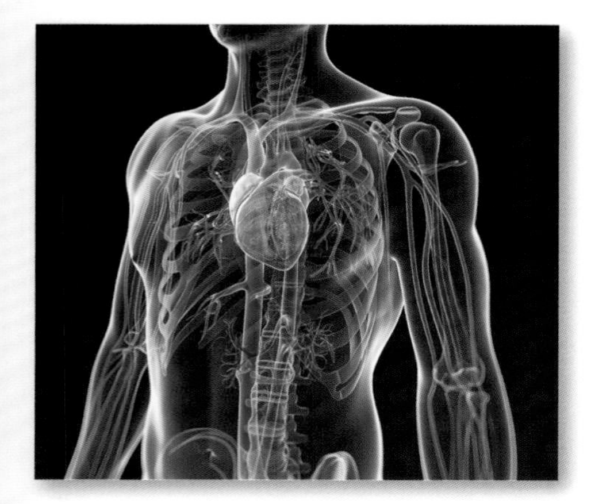

A & P in Action
Jocelyn's: Circulation Crisis

"I feel like I'm getting old," Jocelyn lamented to her husband, John, as they relaxed after dinner. "This left knee that I strained at work has been bothering me for a few weeks, and it doesn't seem to be getting much better. The elastic brace and the antiinflammatory drug the orthopedist prescribed are not helping much. My leg looks swollen, and the pain actually seems to be moving into my calf."

Jocelyn, age 52, works at a day care center for the elderly and thinks she may have injured herself in working with her clients. Taking John's advice to see the doctor again, she visited Dr. Rennard's office, and considering her continuing pain, he ordered a venous ultrasound of her leg.

"Get this," Jocelyn said, in tears, as she hung up the phone. "The report came back that I have a blood clot in my leg. The condition might be really dangerous, and the doctor wants me to come in this afternoon."

Dr. Rennard explained to Jocelyn that she had a deep vein thrombosis, or DVT, in the popliteal vein behind her knee. She would need injections and medication daily until the clot resolved. He told her to watch for any signs of phlebitis, such as pain, redness, or swelling in the affected limb or pulmonary embolism, that is, a blot clot breaking loose and traveling to her lungs. Signs of this dangerous development include shortness of breath, chest pain, cough, or fainting.

"Your body should develop alternate blood routes to compensate for the clot, but we still have to get rid of it," he explained.

"I really appreciate your giving me these shots," Jocelyn told her husband, "but you know how I hate needles, and I might be a really bad patient."

John was to give her twice-daily injections of Lovenox into her lateral abdomen for 11 days. In addition, she took Coumadin orally once a day, adjusting the dose based on the drug's activity as measured with twice-weekly blood tests. Lovenox is a form of the anticoagulant heparin, which inhibits certain clotting factors, and Coumadin interferes with the action of vitamin K needed for clotting. To everyone's relief, she was able to stop the injections after 11 days but had to continue the Coumadin for a total of six months. She also required regular tests to be sure that her blood would clot properly if necessary.

"Your ultrasound shows no blood clot," Dr. Rennard was pleased to tell Jocelyn at the end of her treatment. "However, I highly recommend that you see a hematologist. Maybe we can find out why this clot formed and how you might be able to prevent recurrences."

Dr. Rennard recognizes the dangers of a blockage anywhere in the circulatory system. In this chapter, we'll learn about the normal blood routes and the physiology of capillary exchange and blood pressure.

As you study this chapter, CasePoints will give you opportunities to apply your learning to this case.

Visit thePoint® to access the following resources. For guidance in using these resources most effectively, see pp. xv–xvii.

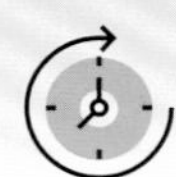

Preparing to Learn

- Tips for Effective Studying
- Pre-Quiz

While You Are Learning

- Web Figure: Capillary Micrograph
- Animation: Blood Circulation
- Animation: Measuring Blood Pressure
- Chapter Notes Outline
- Audio Pronunciation Glossary

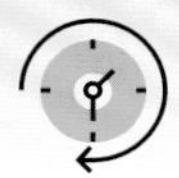

When You Are reviewing

- Answers to Questions for Study and Review
- Health Professions: Vascular Technologist
- Interactive Learning Activities

A LOOK BACK

The story of circulation continues with a discussion of the structure and function 1 *of the vessels that carry blood away from and then back to the heart. Along with the heart, the blood vessels are under the control of the nervous and endocrine systems* 11*, because they act as effectors in the feedback loop controlling blood pressure homeostasis* 3*. Both blood flow through the circulatory system and the flow of materials between the tissues and the bloodstream rely on gradients* 5 *and follow the principles of mass balance* 13*.*

Introduction

The blood vessels, together with the four chambers of the heart, form a closed system in which blood is carried to and from the tissues. Although whole blood does not leave the vessels, components of the plasma and tissue fluids can be exchanged through the walls of the tiniest vessels—the capillaries (FIG. 14-1).

The vascular system is easier to understand if you refer to the appropriate illustrations in this chapter as the vessels are described. When this information is added to what you already know about the blood and the heart, a picture of the cardiovascular system as a whole will emerge.

Blood Vessels

Blood vessels may be divided into five groups, named according to the sequence of blood flow from the heart:

1. **Arteries** carry blood away from the heart and toward the tissues. The heart's ventricles pump blood into the arteries (see FIG. 14-1).
2. **Arterioles** (ar-TE-re-olz) are small subdivisions of the arteries (FIG. 14-2). They divide into the capillaries.
3. **Capillaries** are tiny, thin-walled vessels that allow for exchanges between systems. These exchanges occur between the blood and the body cells and between the blood and the air in the lung tissues. The capillaries connect the arterioles and venules.
4. **Venules** (VEN-ulz) are small vessels that receive blood from the capillaries and begin its transport back toward the heart (see FIG. 14-2).
5. **Veins** are vessels formed by the merger of venules. They continue blood's transport until it is returned to the heart.

BLOOD CIRCUITS

The vessels together may be subdivided into two groups, or circuits: pulmonary and systemic. FIGURE 14-1 diagrams blood flow through these two circuits.

The Pulmonary Circuit The **pulmonary circuit** delivers blood to the lungs, where some carbon dioxide is eliminated and oxygen is replenished. The pulmonary vessels that carry blood to and from the lungs include the following:

1. The pulmonary trunk and its arterial branches, which carry blood low in oxygen from the right ventricle to the lungs

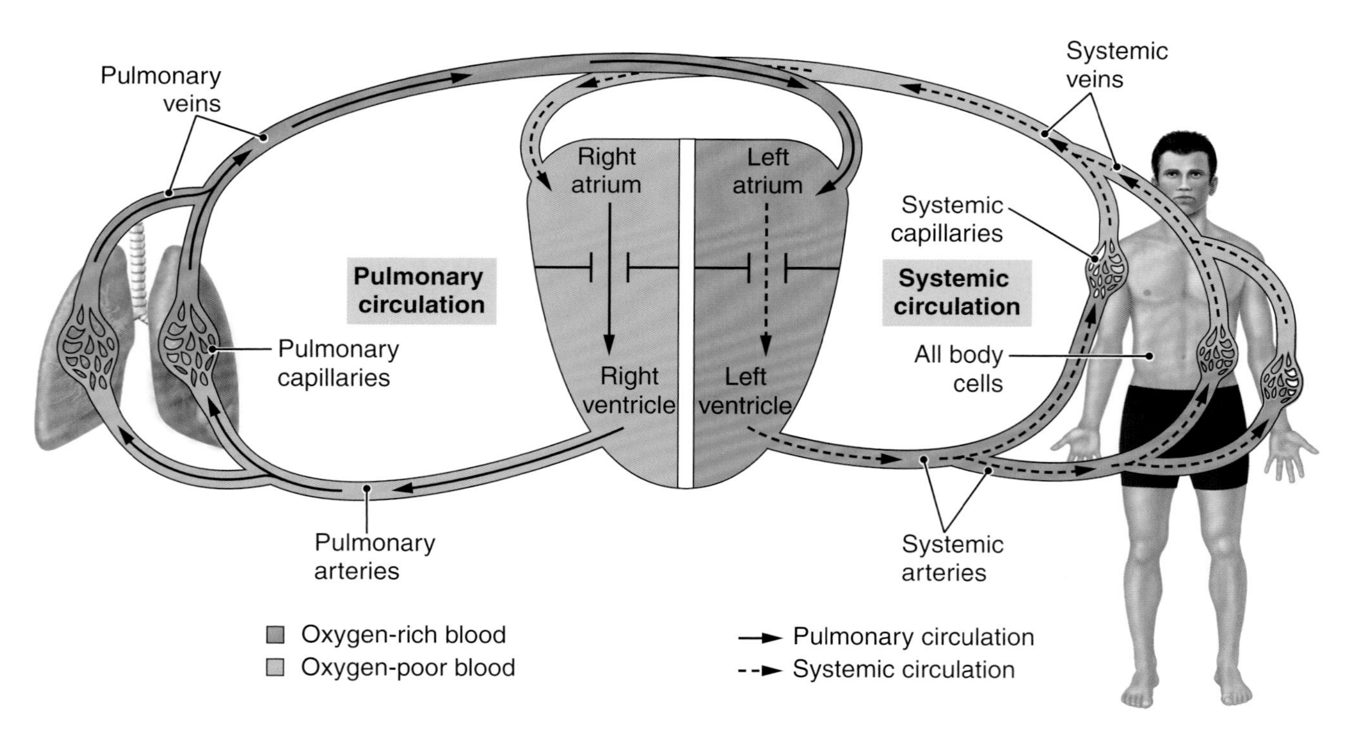

FIGURE 14-1 The cardiovascular system. KEY POINT Blood flows in a closed system with exchanges of material between the blood and tissues through the capillary walls. There are two circuits, pulmonary and systemic. ZOOMING IN Which arteries contain oxygen-poor blood? Which veins contain oxygen-rich blood?

2. The pulmonary capillaries in the lungs, through which gases, nutrients, and wastes are exchanged
3. The pulmonary veins, which carry freshly oxygenated blood back to the left atrium

Note that the pulmonary vessels differ from those in the systemic circuit in that the pulmonary arteries carry blood that is *low* in oxygen content, and the pulmonary veins carry blood that is *high* in oxygen content. In contrast, the systemic arteries carry highly oxygenated blood, and the systemic veins carry blood that is low in oxygen.

The Systemic Circuit The **systemic** (sis-TEM-ik) **circuit** supplies nutrients and oxygen to all the tissues and carries waste materials away from the tissues for disposal. The systemic vessels include the following:

1. The **aorta** (a-OR-tah) receives freshly oxygenated blood from the left ventricle and then branches into the systemic arteries carrying blood to the tissues.
2. The systemic capillaries are the blood vessels through which materials are exchanged.
3. The systemic veins carry blood low in oxygen back toward the heart. The venous blood flows into the right atrium of the heart through the superior vena cava and inferior vena cava.

VESSEL STRUCTURE

1 The structure of the different types of blood vessels provides clues to their function. For instance, the arteries have thick walls because they must be strong enough to receive blood pumped under pressure from the heart's ventricles (FIG. 14-2).

The three tunics (coats) of the arteries resemble the heart's three tissue layers. Named from internal to external, they are as follows:

1. The inner tunic consists of a smooth lining of simple, squamous epithelial cells, the **endothelium** (en-do-THE-le-um), over which the blood flows easily. These fragile endothelial cells are supported by a mat of fibrous connective tissue called the *basement membrane*.
2. The middle tunic, the thickest layer, is made up of smooth (involuntary) muscle, which is under the control of the autonomic nervous system.
3. The outer tunic is made of supporting connective tissue.

Elastic tissue between the layers of the arterial wall allows these vessels to stretch when receiving blood and then return to their original size. The ease with which vessels expand to receive blood is termed their **compliance** (kom-PLI-ans). The term **elasticity** (e-las-TIH-sih-te) describes the ability of blood vessels to return to their original size after being stretched.

14

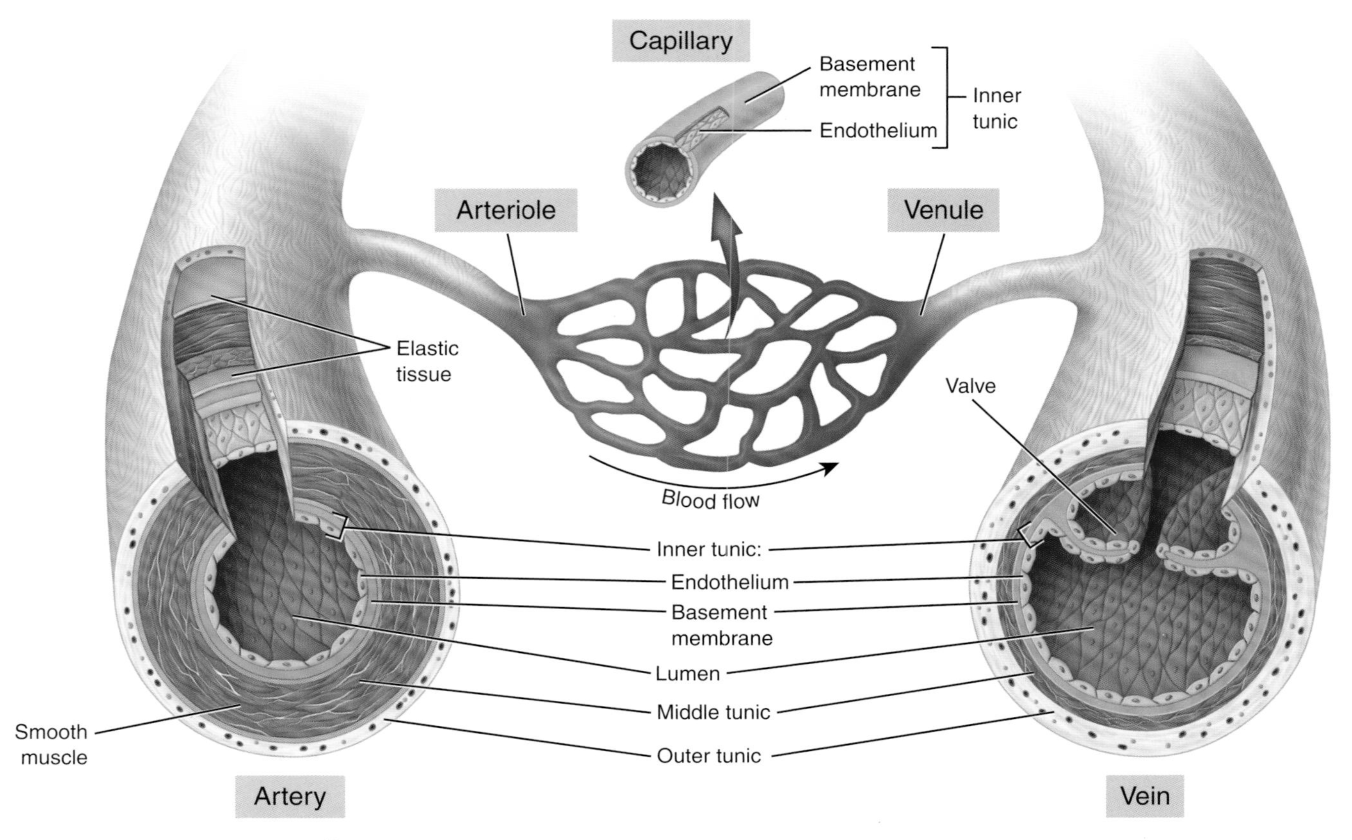

FIGURE 14-2 Blood vessels. KEY POINT Drawings show the thick wall of an artery, the thin wall of a vein, and the single-layered wall of a capillary. A valve is also shown. ZOOMING IN Which vessels have valves that control the direction of blood flow?

This elastic force propels blood forward between heartbeats, ensuring continuous blood flow. The amount of elastic tissue diminishes as the arteries branch and become smaller.

The small subdivisions of the arteries, the arterioles, have thinner walls in which there is little elastic connective tissue but relatively more smooth muscle. The autonomic nervous system controls this involuntary muscle. The vessel lumens (openings) become narrower (constrict) when the muscle contracts, and they widen (dilate) when the muscle relaxes. In this manner, the arterioles regulate the amount of blood that enters the various tissues at a given time. As discussed later, change in the diameter of many arterioles at once alters blood pressure.

The microscopic capillaries that connect arterioles and venules reach a maximum diameter of 10 mcm, just wide enough for a blood cell to pass through. They have the thinnest walls of any vessels: one cell layer (see **FIG. 14-2**). The transparent capillary walls are a continuation of the smooth endothelium that lines the arteries. The thinness of these walls allows for exchanges between the blood and the body cells and between the lung tissue and the outside air. The capillary boundaries are the most important center of activity for the entire circulatory system. Their function is explained later in this chapter. Capillary structure varies according to function, as described in **BOX 14-1**.

See the Student Resources on to review the animation "Blood Circulation" and for a micrograph of a capillary in cross-section.

The smallest veins, the venules, are formed by the union of capillaries, and their walls are only slightly thicker than are those of the capillaries. As the venules merge to form veins, the smooth muscle in the vessel walls becomes thicker, and the venules begin to acquire the additional layers found in the larger vessels.

A CLOSER LOOK
Capillaries: The Body's Free Trade Zones

BOX 14-1

The exchange of substances between body cells and the blood occurs along about 50,000 miles (80,000 km) of capillaries. Exchange rates vary because based on their structure, different types of capillaries vary in permeability.

Continuous capillaries (top) are the most common type and are found in muscle, connective tissue, the lungs, and the central nervous system (CNS). These capillaries are composed of a continuous layer of endothelial cells. Adjacent cells are loosely attached to each other with small openings, called intercellular clefts, between them. Although continuous capillaries are the least permeable, water and small molecules can diffuse easily through their walls. Large molecules, such as plasma proteins and blood cells, cannot. In certain body regions like the CNS, adjacent endothelial cells are joined tightly together, making the capillaries impermeable to many substances (see **BOX 9-1**, "The Blood–Brain Barrier," in Chapter 9).

Fenestrated (FEN-es-tra-ted) *capillaries* (middle) are much more permeable than are continuous capillaries, because they have many holes, or fenestrations, in the endothelium (the word is derived from Latin meaning "window"). These sieve-like capillaries are permeable to water and solutes as large as peptides. In the digestive tract, fenestrated capillaries permit rapid absorption of water and nutrients into the bloodstream. In the kidneys, they permit rapid filtration of blood plasma, the first step in urine formation.

Sinusoidal capillaries (bottom) are the most permeable. In addition to fenestrations, they have large spaces between endothelial cells that allow the exchange of water; large solutes, such as plasma proteins; and even blood cells. Sinusoidal capillaries, also called sinusoids, are found in the liver and red bone marrow, for example. Albumin, clotting factors, and other proteins formed in the liver enter the bloodstream through sinusoidal capillaries. In red bone marrow, newly formed blood cells travel through sinusoidal capillary walls to join the bloodstream.

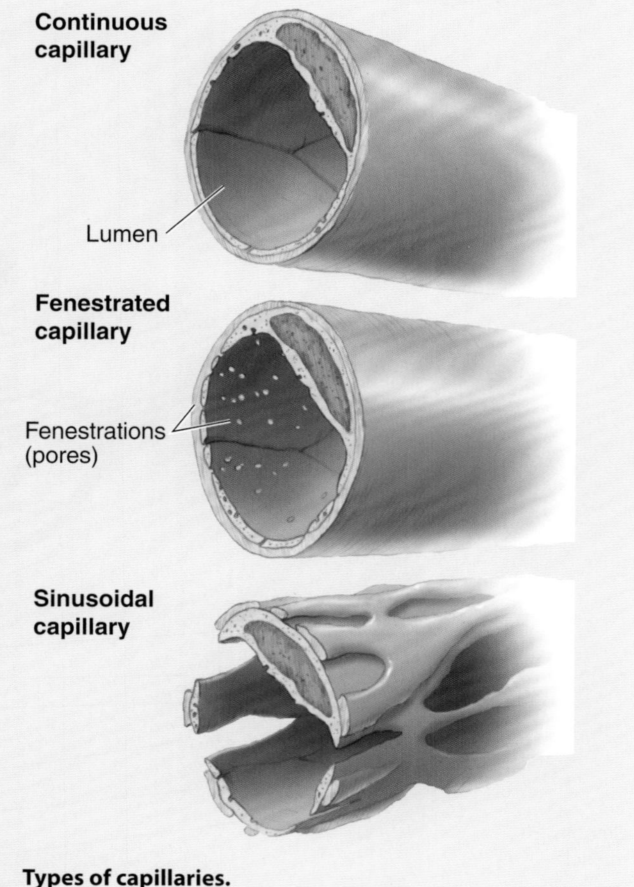

Types of capillaries.

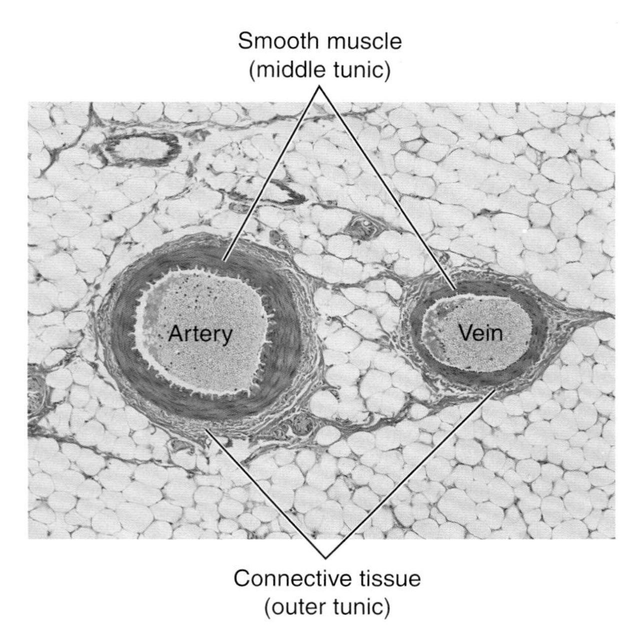

FIGURE 14-3 Cross-section of an artery and vein. The smooth muscle and connective tissue of the vessels are visible in this photomicrograph. ZOOMING IN Which type of vessel shown has a thicker wall?

The walls of the veins have the same three layers as those of the arteries. However, the middle smooth muscle tunic is relatively thin in the veins. A vein wall is much thinner than is the wall of a comparably sized artery (see FIG. 14-2). These vessels also have less elastic tissue between the layers, so they expand easily and carry blood under much lower pressure. Because of their thinner walls, the veins collapse easily. Even a slight pressure on a vein by a tumor or other mass may interfere with blood flow.

Most veins are equipped with one-way valves that permit blood to flow in only one direction: toward the heart (see FIG. 14-2). Such valves are most numerous in the veins of the extremities. FIGURE 14-3 is a cross-section of an artery and a vein as seen through a microscope.

CHECKPOINTS 

- [ ] **14-1** What are the five types of blood vessels?
- [ ] **14-2** What are the two blood circuits, and what areas does each serve?
- [ ] **14-3** What type of tissue makes up the middle tunic of arteries and veins, and how is this tissue controlled?
- [ ] **14-4** How many cell layers make up the wall of a capillary?

CASEPOINTS 

- [ ] **14-1** In the case study, Jocelyn developed a blood clot in a leg vein. Would the clot block blood flow toward the tissues or toward the heart?
- [ ] **14-2** Clots form more readily if the inner blood vessel layer is rough or damaged. Name this layer.

Systemic Arteries

The systemic arteries begin with the aorta, the largest artery, which measures about 2.5 cm (1 in) in diameter. This vessel receives blood from the left ventricle, ascends from the heart, and then arches back to travel downward through the body, branching to all organs.

THE AORTA AND ITS PARTS

The aorta is a thick-walled vessel about the diameter of your thumb (FIG. 14-4). It is one continuous artery, but its regions are named as follows:

1. The **ascending aorta** extends upward and slightly to the right from the left ventricle. It lies within the pericardial sac.
2. The **aortic arch** curves from the right to the left and also extends posteriorly.
3. The **thoracic aorta** descends just anterior to the vertebral column posterior to the heart in the mediastinum.
4. The **abdominal aorta** is the longest section of the aorta, beginning at the diaphragm and spanning the abdominal cavity.

The thoracic aorta and abdominal aorta together make up the descending aorta.

Branches of the Ascending Aorta and Aortic Arch

The aorta's ascending part has two branches near the heart, called the *left* and *right coronary arteries*, which supply the heart muscle (see FIGS. 13-4 and 13-5). As noted in Chapter 13, these arteries form a crown around the heart's base and give off branches to all parts of the myocardium. Some of these vessels can be seen in FIGURE A3-4 of the Dissection Atlas.

The aortic arch, located immediately past the ascending aorta, gives rise to three large branches, shown in part in FIGURE A3-8 of the Dissection Atlas.

1. The first, the **brachiocephalic** (brak-e-o-seh-FAL-ik) **artery**, is a short vessel that supplies the arm and the head on the right side (see FIG. 14-4). After extending upward about 5 cm (2 in), it divides into the **right subclavian** (sub-KLA-ve-an) **artery**, which extends under the right clavicle (collar bone) and supplies the right upper extremity (arm) and part of the brain, and the **right common carotid** (kah-ROT-id) **artery**, which supplies the right side of the neck, head, and brain. Note that the brachiocephalic artery is unpaired.
2. The second, the **left common carotid artery**, extends upward from the highest part of the aortic arch. It supplies the left side of the neck and the head.
3. The third, the **left subclavian artery**, extends under the left clavicle and supplies the left upper extremity and part of the brain. This is the aortic arch's last branch.

Branches of the Descending Aorta The thoracic aorta supplies branches to the chest wall and esophagus (e-SOF-ah-gus), the bronchi (subdivisions of the trachea), the lungs,

14

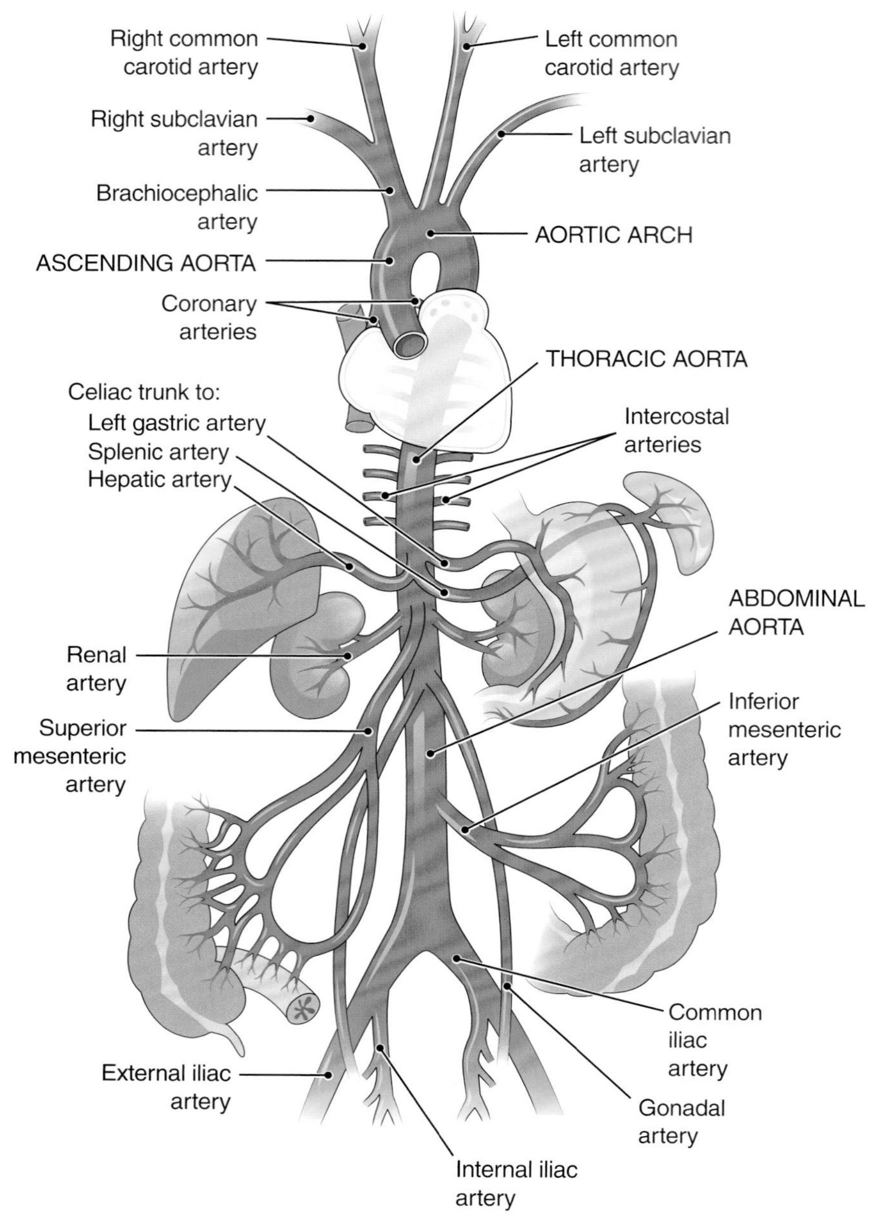

FIGURE 14-4 The aorta and its branches. KEY POINT As the aorta travels from the heart through the body, it branches to all tissues. ZOOMING IN How many brachiocephalic arteries are there?

and the muscles of the chest wall (FIG. 14-5). There are usually nine to ten pairs of **intercostal** (in-ter-KOS-tal) **arteries** that extend between the ribs, sending branches to the muscles and other structures of the chest wall.

The abdominal aorta has unpaired branches extending anteriorly and paired branches extending laterally. The unpaired vessels are large arteries that supply the abdominal viscera. The most important of these visceral branches are as follows:

1. The **celiac** (SE-le-ak) **trunk** is a short artery, about 1.25 cm (1/2 in) long, that subdivides into three branches: the **left gastric artery** goes to the stomach, the **splenic** (SPLEN-ik) **artery** goes to the spleen, and the **hepatic** (heh-PAT-ik) **artery** goes to the liver (see FIG. 14-4 and Dissection Atlas FIG. A3-7).
2. The large **superior mesenteric** (mes-en-TER-ik) **artery** carries blood to most of the small intestine and to the first half of the large intestine.
3. The much smaller **inferior mesenteric artery**, located below the superior mesenteric artery and near the end of the abdominal aorta, supplies the second half of the large intestine.

The abdominal aorta's paired lateral branches include the following right and left vessels:

1. The **superior** and **inferior phrenic** (FREN-ik) **arteries** supply the diaphragm. (These are not shown in the figures.)
2. The **renal** (RE-nal) **arteries**, the largest in this group, carry blood to the kidneys.

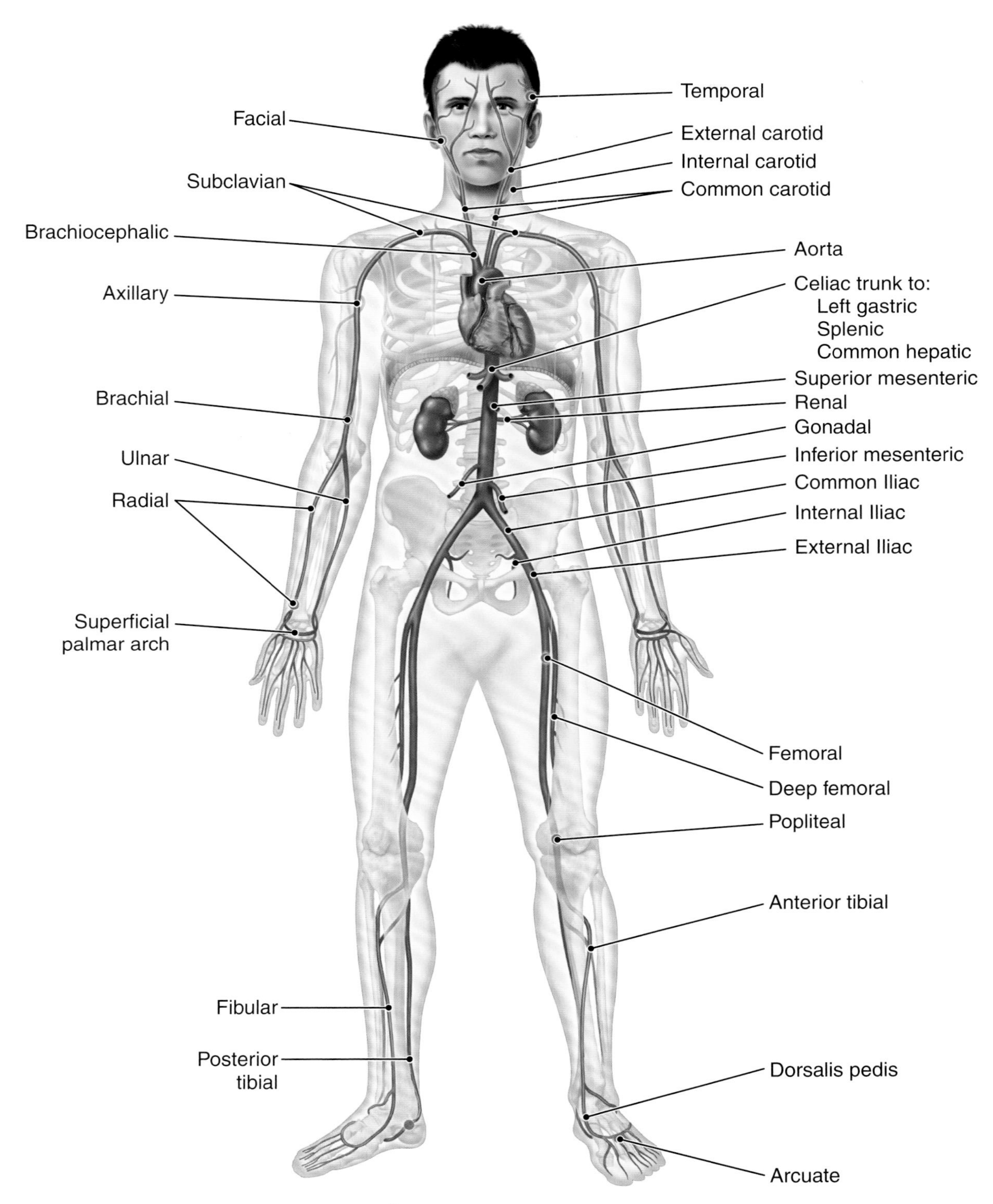

FIGURE 14-5 Principal systemic arteries. ZOOMING IN What large vessels branch from the terminal aorta?

3. The **gonadal** (go-NAD-al) **arteries**—the **ovarian** (o-VAR-e-an) **arteries** in females and **testicular** (tes-TIK-u-lar) **arteries** in males—supply the sex glands.
4. Four pairs of **lumbar** (LUM-bar) **arteries** extend into the musculature of the abdominal wall. (These are not shown in the figures.)

THE ILIAC ARTERIES AND THEIR SUBDIVISIONS

The abdominal aorta finally divides into two **common iliac** (IL-e-ak) **arteries** (see FIG. 14-5). These vessels, which are about 5 cm (2 in) long, extend into the pelvis, where each subdivides into an **internal** and an **external iliac artery**. The internal iliac vessels then send branches to the pelvic organs, including the urinary bladder, the rectum, and reproductive organs other than the gonads.

Each external iliac artery continues into the thigh as the **femoral** (FEM-or-al) **artery**. This vessel gives rise to the **deep femoral artery** in the thigh and then becomes the **popliteal** (pop-LIT-e-al) **artery**, which subdivides below the knee into the anterior and posterior **tibial arteries**, supplying the leg and foot. The anterior tibial artery terminates as the **dorsalis pedis** (dor-SA-lis PE-dis) at the foot. The posterior tibial artery gives rise to the **fibular** (FIB-u-lar) **artery** (peroneal artery) in the leg.

ARTERIES THAT BRANCH TO THE ARM AND HEAD

The **subclavian** (sub-KLA-ve-an) **artery** supplies blood to the arm and hand. Its first branch, however, is the **vertebral** (VER-te-bral) **artery**, which passes through the transverse processes of the first six cervical vertebrae and supplies blood to the posterior brain **(FIG. 14-6)**. The subclavian artery

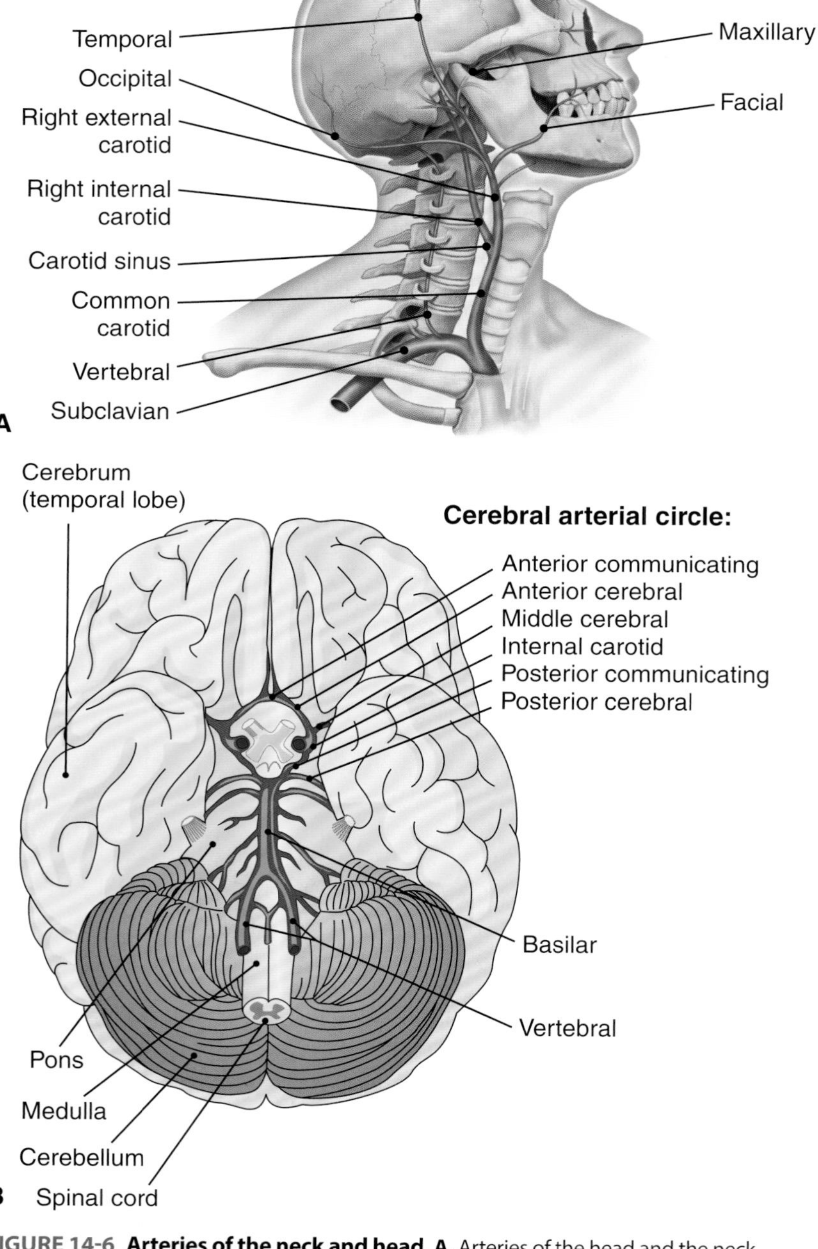

FIGURE 14-6 Arteries of the neck and head. A. Arteries of the head and the neck, lateral view. **B.** Arteries supplying the brain. The arteries of the cerebral arterial circle (CAC) are shown. **KEY POINT** Arterial anastomoses help preserve blood supply to the brain.

changes names as it travels through the arm and branches to the forearm and hand (see FIG. 14-5). It first becomes the **axillary** (AK-sil-ar-e) **artery** in the axilla (armpit). The longest part of this vessel, the **brachial** (BRA-ke-al) **artery**, is in the arm proper. The brachial artery subdivides into two branches near the elbow: the **radial artery**, which continues down the thumb side of the forearm and wrist, and the **ulnar artery**, which extends along the medial or little finger side into the hand. These two arteries unite in the **palmar arches**, which give off smaller **digital arteries** that supply the hand and fingers.

The right and left common carotid arteries travel along either side of the trachea enclosed in a sheath with the internal jugular vein and the vagus nerve. Just anterior to the angle of the mandible (lower jaw), each branches into the **external** and **internal carotid arteries** (see FIG. 14-6). You can feel the pulse of the carotid artery just anterior to the large sternocleidomastoid muscle in the neck and below the jaw. The internal carotid artery travels into the head and branches to supply the eye, the anterior portion of the brain, and other structures in the cranium. The external carotid artery branches to the thyroid gland and to other structures in the head and upper part of the neck.

Just as the larger branches of a tree divide into limbs of varying sizes, so the arterial tree has a multitude of subdivisions. Hundreds of names might be included. We have mentioned only some of them.

ANASTOMOSES

A communication between two vessels is called an **anastomosis** (ah-nas-to-MO-sis). By means of arterial anastomoses, blood reaches vital organs by more than one route. Some examples of such end-artery unions are as follows:

- The **cerebral arterial circle** (circle of Willis) (see FIG. 14-6 and **Dissection Atlas** FIG. A3-6) receives blood from the two internal carotid arteries and from the **basilar** (BAS-il-ar) **artery**, which is formed by the union of the two vertebral arteries. This arterial circle lies just under the brain's center and sends branches to the cerebrum and other parts of the brain.
- The **superficial palmar arch** is formed by the union of the radial and ulnar arteries in the hand. It sends branches to the hand and the fingers (see FIG. 14-5).
- The **mesenteric** arches are made up of communications between branches of the vessels that supply blood to the intestinal tract.
- One of several arterial arches in the foot is the **arcuate artery**, formed by the union of the dorsalis pedis artery and a second branch of the anterior tibial artery (the lateral tarsal artery).

CHECKPOINTS 

- [ ] **14-5** What are the subdivisions of the aorta, the largest artery?
- [ ] **14-6** What are the three branches of the aortic arch?
- [ ] **14-7** What areas are supplied by the brachiocephalic artery?
- [ ] **14-8** What is an anastomosis?

Systemic Veins

Whereas most arteries are located in protected and rather deep areas of the body, many of the principal systemic veins are found near the surface (FIG. 14-7). The most important of the **superficial veins** are in the extremities and include the following:

- The veins on the back of the hand and at the front of the elbow. Those at the elbow are often used for drawing blood for test purposes, as well as for intravenous injections. The largest of this venous group are the **cephalic** (seh-FAL-ik), the **basilic** (bah-SIL-ik), and the **median cubital** (KU-bih-tal) **veins**.
- The **saphenous** (sah-FE-nus) **veins** of the lower extremities, which are the body's longest veins. The great saphenous vein begins in the foot and extends up the medial side of the leg, the knee, and the thigh. It finally empties into the femoral vein near the groin.

The **deep veins** tend to parallel arteries and usually have the same names as the corresponding arteries (see FIG. 14-7). Examples of these include the **femoral** and the **external** and **internal iliac vessels** of the lower body, and the **radial**, **ulnar**, **brachial**, **axillary**, and **subclavian vessels** of the upper extremities. (A deep vein was involved in Jocelyn's thrombosis in the opening case study.) The veins of the head and the neck, however, have different names than the arteries. The two jugular (JUG-u-lar) veins on each side of the neck drain the areas supplied by the carotid arteries (*jugular* is from a Latin word meaning "neck"). The larger of the two veins, the **internal jugular**, receives blood from the large veins (cranial venous sinuses) that drain the head and also from regions of the face and neck (FIG. 14-8). The smaller **external jugular** drains the areas supplied by the external carotid artery. Both veins empty directly into a **subclavian vein**. On each side, the subclavian, external jugular, and internal jugular veins join to form a **brachiocephalic vein**. (Remember, there is only *one* brachiocephalic artery.)

Concept Mastery Alert

Students often confuse the jugular with the carotid. Remember that the former is a vein and the latter is an artery.

THE VENAE CAVAE AND THEIR TRIBUTARIES

Two large veins receive blood from the systemic vessels and empty directly into the heart's right atrium. The veins of the head, neck, upper extremities, and chest all drain into the **superior vena cava** (VE-nah KA-vah). This vessel is formed by the union of the right and left brachiocephalic veins. The unpaired **azygos** (AZ-ih-gos) **vein** drains the veins of the chest

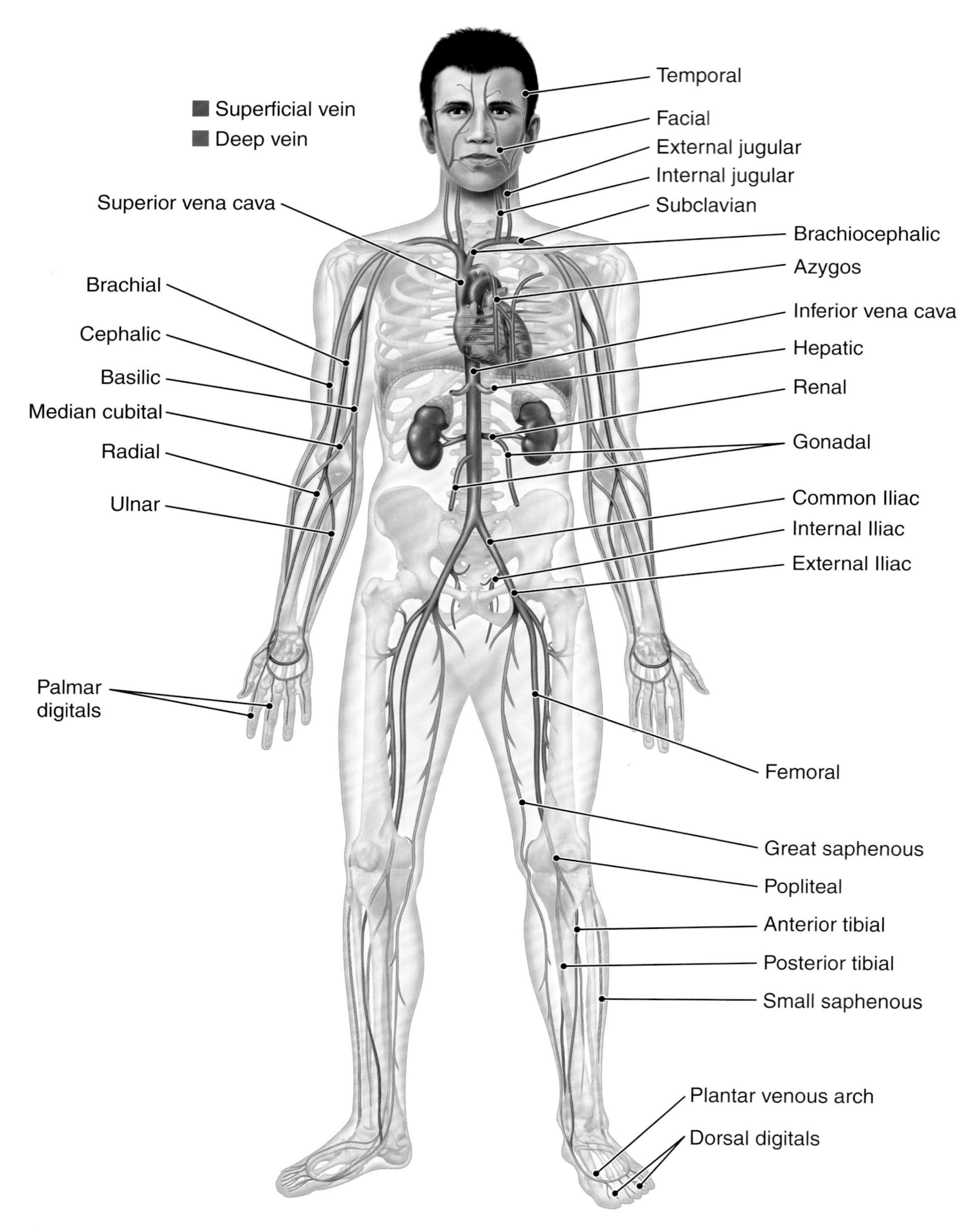

FIGURE 14-7 Principal systemic veins. Anterior view. **KEY POINT** Deep veins (*in blue*) usually parallel arteries and carry the same names. Superficial veins are shown *in purple*. **ZOOMING IN** How many brachiocephalic veins are there?

wall and empties into the superior vena cava just before the latter empties into the right atrium of the heart (see FIG. 14-7) (*azygos* is from a Greek word meaning "unpaired").

The **inferior vena cava**, which is much longer than is the superior vena cava, returns blood from areas below the diaphragm (see Dissection Atlas FIG. A3-4). It begins in the lower abdomen with the union of the two common iliac veins. It then ascends along the abdomen's posterior wall, through a groove in the posterior part of the liver, through the diaphragm, and finally through the lower thorax to empty into the heart's right atrium.

The large veins below the diaphragm may be divided into two groups:

1. The right and left veins that drain paired parts and organs. They include the **external** and **internal iliac veins** from near the groin that join to form the **common iliac veins**; four pairs of **lumbar veins** from the dorsal trunk and from the spinal cord; the **gonadal veins**—the **testicular veins** from the male testes and the **ovarian veins** from the female ovaries; the **renal veins** from the kidneys; and, finally, the large **hepatic veins** from the liver. For the most part, these vessels empty directly into the inferior vena cava. The left testicular in the male and the left ovarian in the female empty into the left renal vein, which carries this blood to the inferior vena cava; these veins thus constitute exceptions to the rule that the paired veins empty directly into the vena cava.
2. Unpaired veins that drain the spleen and parts of the digestive tract (stomach and intestine) empty into the **hepatic portal vein**, discussed shortly and shown in FIGURE 14-9. Unlike other lower veins, which empty directly into the inferior vena cava, the hepatic portal vein is part of a special system that enables blood to circulate through the liver before returning to the heart. This system, the hepatic portal system, will be described in more detail later.

VENOUS SINUSES

The word *sinus* means "space" or "hollow." A **venous sinus** is a large channel that drains blood low in oxygen but does not have a vein's usual tubular structure. One example of a venous sinus is the **coronary sinus**, which receives most of the blood from the heart wall (see FIG. 13-2 in Chapter 13). It lies between the left atrium and the left ventricle on the heart's posterior surface and empties directly into the right atrium, along with the two venae cavae.

Other important venous sinuses are the **cranial venous sinuses**, which are spaces between the two layers of the dura mater. Veins from throughout the brain drain into these channels (see FIG. 14-8). They also collect cerebrospinal fluid from the CNS and return it to the bloodstream. The largest of the cranial venous sinuses are the following:

- The **superior sagittal** (SAJ-ih-tal) **sinus** is a single long space located in the midline above the brain and in the fissure between the two cerebral hemispheres. It ends in an enlargement called the **confluence** (KON-flu-ens) **of sinuses**.
- The **inferior sagittal sinus** parallels the superior sagittal sinus and merges with the straight sinus.
- The **straight sinus** receives blood from the inferior sagittal sinus and a large cerebral vein and flows into the confluence of sinuses.
- The two **transverse sinuses**, also called the **lateral sinuses**, begin posteriorly from the confluence of sinuses and then extend laterally. As each sinus extends around the skull's interior, it picks up additional blood. Nearly all of the blood leaving the brain eventually empties into one of the transverse sinuses. Each of these extends anteriorly to empty into an internal jugular vein, which then passes through a channel in the skull to continue downward in the neck.

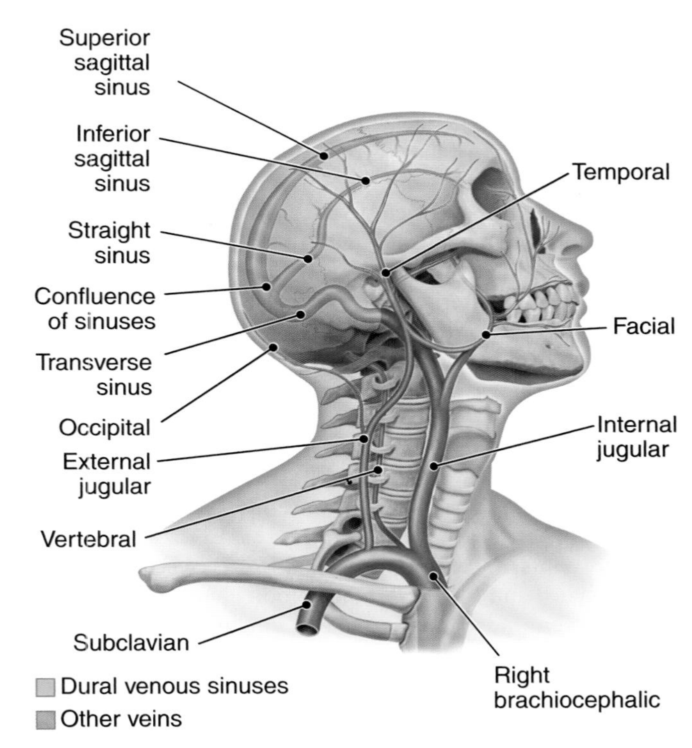

FIGURE 14-8 Veins of the head and the neck and the cranial sinuses, lateral view. KEY POINT The cranial sinuses are spaces between the two layers of the dura mater. ZOOMING IN Which vein receives blood from the transverse sinus?

THE HEPATIC PORTAL SYSTEM

Almost always, when blood leaves a capillary bed, it flows through venules and veins directly back to the heart. In a portal system, however, blood circulates through a second capillary bed in a second organ before it returns to the heart. Portal circulations enable all of the products of one organ to pass directly to another organ. Chapter 11 described the small local portal system that carries secretions from the hypothalamus to the pituitary gland. A much larger portal system is the **hepatic portal system**, which carries blood from the abdominal organs to the liver to be processed before it returns to the heart (**FIG. 14-9**).

The hepatic portal system includes the veins that drain blood from capillaries in the spleen, stomach, pancreas, and intestine. Instead of emptying their blood directly into the inferior vena cava, they deliver it through the hepatic portal vein to the liver. The portal vein's largest tributary is the **superior mesenteric vein**, which drains blood from the proximal portion of the intestine. It is joined by the **splenic vein** just under the liver. Other tributaries of the portal circulation are the **gastric, pancreatic**, and **inferior mesenteric veins**. As it enters the liver, the portal vein divides and subdivides into ever smaller branches.

Eventually, the portal blood flows into a vast network of sinus-like capillaries called **sinusoids** (SI-nus-oyds) (**see BOX 14-1**). These leaky vessels allow free exchange of proteins, nutrients, and dissolved substances between liver cells (hepatocytes) and blood. (Similar blood channels are found in the spleen and endocrine glands, including the thyroid and adrenals.) After leaving the sinusoids, blood is finally collected by the hepatic veins, which empty into the inferior vena cava.

The hepatic portal system ensures that most substances absorbed from the intestines can be processed by the liver before they encounter body cells. For example, the liver can inactivate some of the ingested toxins (such as alcohol and drugs) before they reach the general circulation. The liver processes nutrients and can store or release them according to body needs. The liver also receives arterial blood and breaks down alcohol, certain drugs, and various other toxins that have already reached the general circulation.

Remember that all of the blood vessels we have discussed are linked together with the heart into the circulatory system. See **BOX 14-2** to practice mapping routes through the many elements of the vasculature.

CHECKPOINTS 

- ☐ **14-9** What is the difference between superficial and deep veins?
- ☐ **14-10** What two large veins drain the systemic blood vessels and empty into the right atrium?
- ☐ **14-11** What is a venous sinus?
- ☐ **14-12** The hepatic portal system takes blood from the abdominal organs to which organ?

CASEPOINTS 

- ☐ **14-3** If the clot in Jocelyn's popliteal vein broke loose, what is the next vessel it would enter?
- ☐ **14-4** What vessel would carry the clot directly into the heart?
- ☐ **14-5** Which chamber of the heart would it enter?
- ☐ **14-6** Would the clot enter the hepatic portal system? Why or why not?

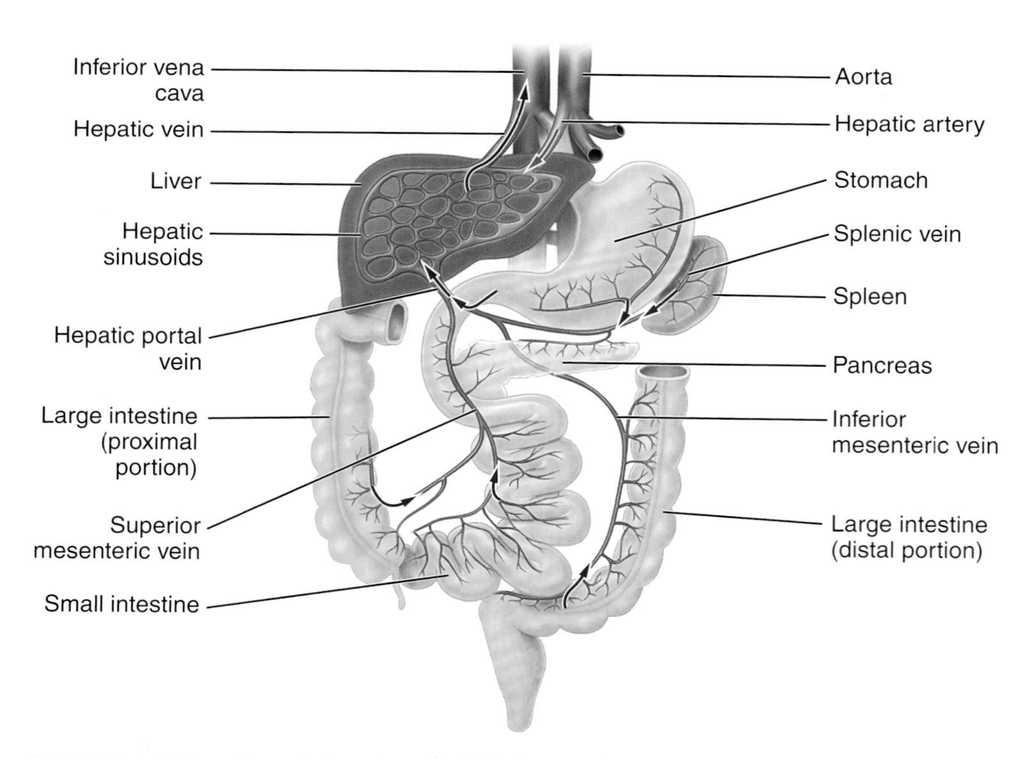

FIGURE 14-9 Hepatic portal system. **KEY POINT** Veins from the abdominal organs carry blood to the hepatic portal vein leading to the liver. *Arrows* show the direction of blood flow. **ZOOMING IN** What vessel do the hepatic veins drain into?

ONE STEP AT A TIME

The Vascular Highway: Mapping Blood Vessels

BOX 14-2

Now that you've learned the locations of bones and muscles, mastering blood vessel anatomy is the final challenge. Your skeletal system knowledge will help, since some vessels are named after nearby bones. A fun and proven way to learn vascular anatomy is to map the route taken by a blood cell through the body.

Question

A blood cell travels from the popliteal artery to the toe capillaries and then back to the femoral vein. Map out a route it could take based on the diagrams provided in your textbook.

Answer

Step 1. Go to the starting point. Make sure you are on the correct diagram. You will need FIGURE 14-5, since it shows the arteries. FIGURE 14-7 shows the veins.

Step 2. Determine your starting direction. The starting point is a systemic artery, so blood will flow away from the heart. Systemic veins flow toward the heart.

Step 3. Travel through the first set of vessels. Use a pencil to follow the artery to a branch point. Unless specified, you can take either branch. A potential routing through the arteries to the toes could be popliteal a., anterior tibial a., dorsalis pedis a., arcuate a., toes.

Step 4. If necessary, transition to the second set of vessels. Although not visible on the diagrams, the arcuate artery would give off arterioles that would branch into capillaries. The capillaries would merge into venules, which would drain into nearby veins. So, we need to find a vein draining the toes. FIGURE 14-7 shows the veins. Mapping a route between a vein and an artery would require you to travel through the heart and the pulmonary circulation.

Step 5. Travel through the second set of vessels. A potential route from the toes to the femoral vein could be dorsal digital v., plantar venous arch, anterior tibial v., popliteal v., femoral v. See the Chapter Review questions and the Study Guide (available separately) for more practice in vessel mapping.

Circulation Physiology

Circulating blood might be compared to a bus that travels around the city, picking up and delivering passengers at each stop on its route. Take gases, for example. As blood flows through capillaries surrounding the air sacs in the lungs, it picks up oxygen and unloads carbon dioxide. Later, when this oxygen-rich blood is pumped to systemic capillaries, it unloads some of the oxygen and picks up carbon dioxide and other substances generated by the cells. The microscopic capillaries are of fundamental importance in these activities. It is only through the cells of these thin-walled vessels that the necessary exchanges can occur.

CAPILLARY EXCHANGE

All living cells are immersed in a slightly salty liquid, the **interstitial** (in-ter-STISH-al) **fluid**, or *tissue fluid*. Looking at FIGURE 14-10A, one can see how this fluid serves as a "middleman" between the capillary membrane and the neighboring cells. As water, oxygen, electrolytes, and other necessary cellular materials pass through the capillary walls, they enter the tissue fluid and subsequently the cells. At the same time, carbon dioxide and other metabolic end products leave the cells and move in the opposite direction. These substances enter the capillaries and are carried away in the bloodstream for processing in other organs or elimination from the body.

4 The capillary wall forms a barrier between the blood and the interstitial fluid, but the effectiveness of the barrier varies between tissues. For instance, capillaries in the bone marrow are very leaky so that newly synthesized blood cells can pass through the spaces between endothelial cells. The endothelial cells of brain capillaries, conversely, are joined together very tightly so as to restrict the passage of substances. BOX 14-1 contains more information about the structure and function of different capillary types.

Materials can cross the capillary wall using the various means of transport described in Chapter 3 (see FIG. 14-10A). 5 Gases and lipids, for example, can diffuse through or between the endothelial cells based on their concentration gradients. Sugars, ions, and amino acids can diffuse between the cells or can cross the cells using specific transporters. Vesicular transport also allows large amounts of certain substances to cross endothelial cells enclosed in vesicles.

Bulk Flow In addition to diffusion and vesicular transport, there is a third mechanism involved in capillary exchange. **Bulk flow** describes the movement of fluid across the capillary wall down a pressure gradient. This gradient reflects the balance between the hydrostatic pressure of blood pushing fluid *out* and the osmotic pressure of blood pulling fluid *in* (see FIG. 14-10B). The hydrostatic pressure is created by blood pressure. The osmotic pressure reflects the presence of proteins in the blood, mainly albumin. (Note that the hydrostatic and osmotic pressures of interstitial fluid are very low and can usually be ignored). Bulk flow plays an important role in water balance between different fluid compartments.

When the hydrostatic pressure of the blood pushing fluid out of the capillary exceeds the osmotic pressure pulling fluid into the capillary, water and dissolved materials move out of the capillary into the interstitial fluid. We described this activity as filtration in Chapter 3. Filtration happens as blood enters the capillary bed. However, hydrostatic pressure declines as blood moves through the capillary bed, because of water loss (see FIG. 14-10C). In contrast,

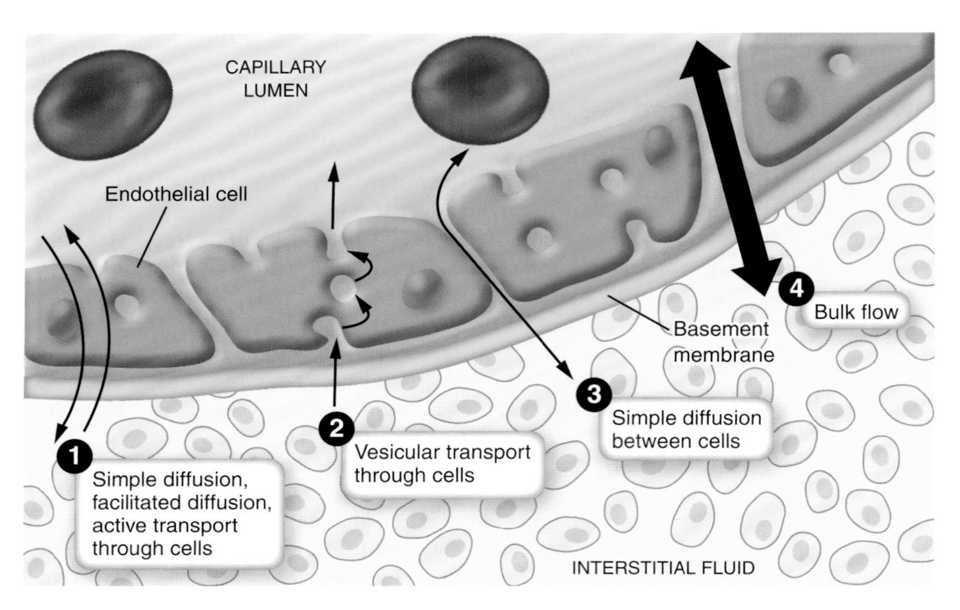

A Transport through the capillary wall

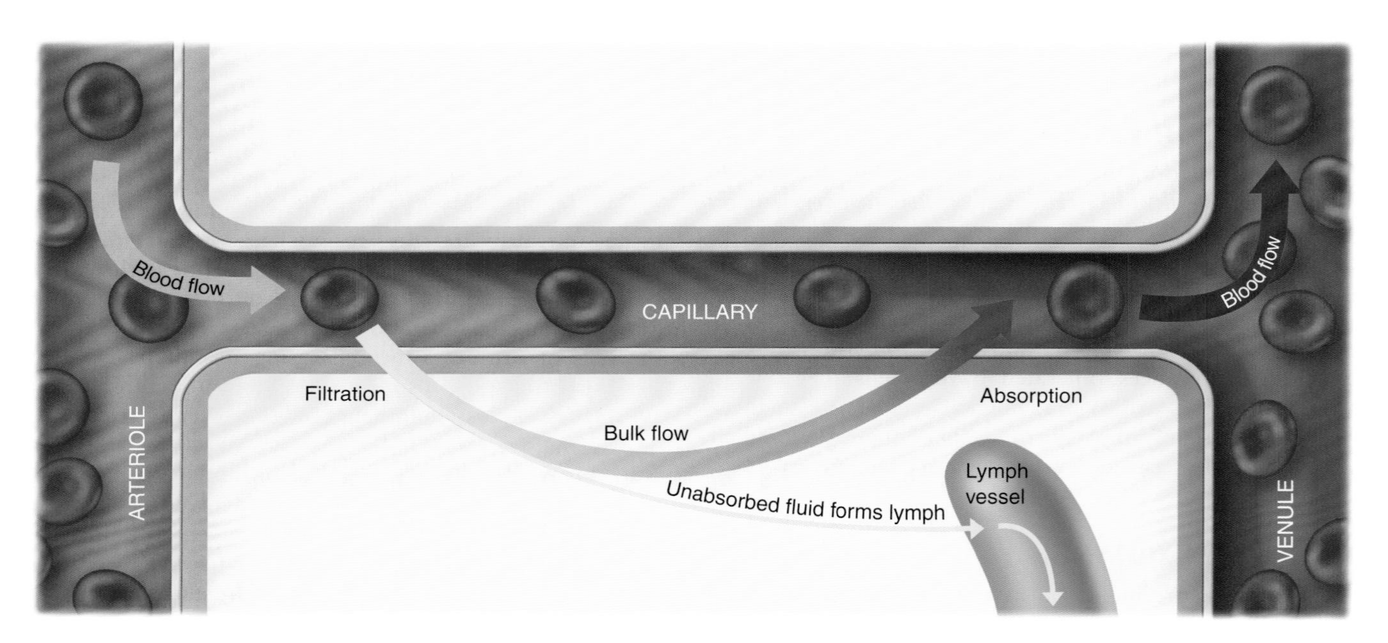

B Bulk flow

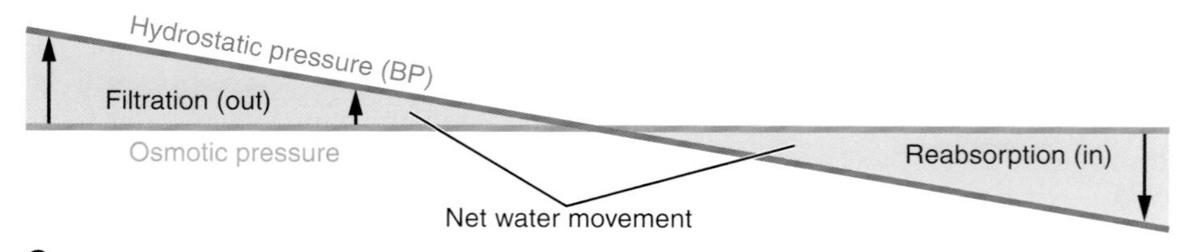

C Changing pressures over the length of the capillary

FIGURE 14-10 The role of capillaries. **KEY POINT** Capillaries are the point of exchanges between the bloodstream and the tissues. **A.** Substances can move between the blood and the tissues using four different mechanisms. **B.** Net fluid movement as blood flows through a capillary bed. **C.** The balance between hydrostatic and osmotic pressure of blood determines the direction of bulk flow.

osmotic pressure remains relatively constant, because proteins are retained within the capillary. Thus, near the end of most capillary beds, osmotic pressure exceeds blood pressure and fluid is absorbed into the capillary.

In most capillary beds, filtration exceeds absorption, and some fluid is left behind in the tissues. Also, some proteins escape from the capillaries into the tissues. The lymphatic system, discussed in Chapter 15, collects this extra fluid and proteins and returns them to the circulation.

Concept Mastery Alert

Students often struggle with the concept of bulk flow, so study FIGURE 14-10B closely. Don't forget that blood pressure pushes fluid out of the capillary. Osmotic pressure pulls fluid back into the capillary.

The movement of blood through the capillaries is relatively slow, owing to the much overall larger cross-sectional area of the capillaries compared with that of the vessels from which they branch. This slow progress through the capillaries allows time for exchanges to occur.

CASEPOINT

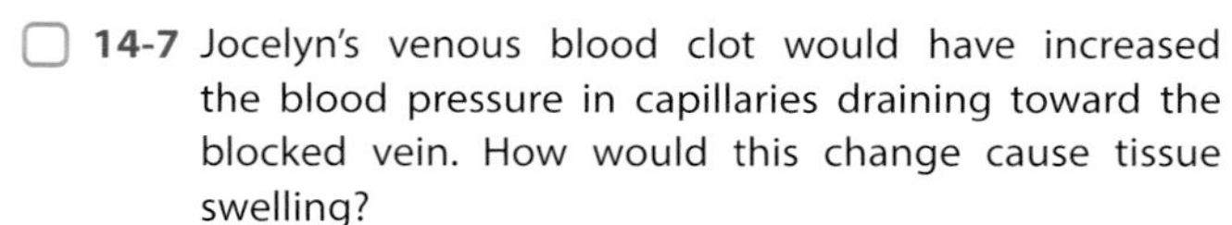

☐ **14-7** Jocelyn's venous blood clot would have increased the blood pressure in capillaries draining toward the blocked vein. How would this change cause tissue swelling?

BLOOD FLOW DYNAMICS

Blood flow is carefully regulated to supply tissue needs without unnecessary burden on the heart. Some organs, such as the brain, liver, and kidneys, require large quantities of blood even at rest. The requirements of some tissues, such as the skeletal muscles and digestive organs, increase greatly during periods of activity. For example, the blood flow in muscle can increase up to 25 times during exercise. 13 This increase reflects increased cardiac output. It also requires decreased blood flow to other organs, because the total blood volume is limited. The volume of blood flowing to a particular organ can be regulated by changing the diameter of the blood vessels supplying that organ.

Vasomotor Changes An increase in a blood vessel's internal diameter is called **vasodilation.** This change allows for the delivery of more blood to an area. **Vasoconstriction** is a decrease in a blood vessel's internal diameter, causing a decrease in blood flow. These *vasomotor activities* result from the contraction or relaxation of smooth muscle in the walls of the blood vessels, mainly the arterioles. A **vasomotor** center in the medulla of the brain stem regulates changes in vessel diameter, sending its messages through the autonomic nervous system.

A **precapillary sphincter** of smooth muscle encircles the entrance to each capillary, controlling its blood supply. This sphincter widens to allow more blood to enter when the nearby cells need more oxygen.

Blood's Return to the Heart Blood leaving the capillary networks returns in the venous system to the heart and even picks up some speed along the way, despite factors that work against its return. Blood flows in a closed system and must continually move forward, whether the heart is contracting or relaxing. However, by the time blood arrives in the veins, little force remains from the heart's pumping action. Also, because the veins expand easily under pressure, blood tends to pool in the veins. Considerable amounts of blood are normally stored in these vessels. Finally, the force of gravity works against upward flow from regions below the heart. Several mechanisms help overcome these forces and promote blood's return to the heart in the venous system. These are as follows:

- **Contraction of skeletal muscles.** As skeletal muscles contract, they compress the veins and squeeze blood forward (FIG. 14-11).
- **Valves in the veins.** They prevent backflow and keep blood flowing toward the heart.
- **Breathing.** Pressure changes in the abdominal and thoracic cavities during breathing also promote blood return in the venous system. During inhalation, the

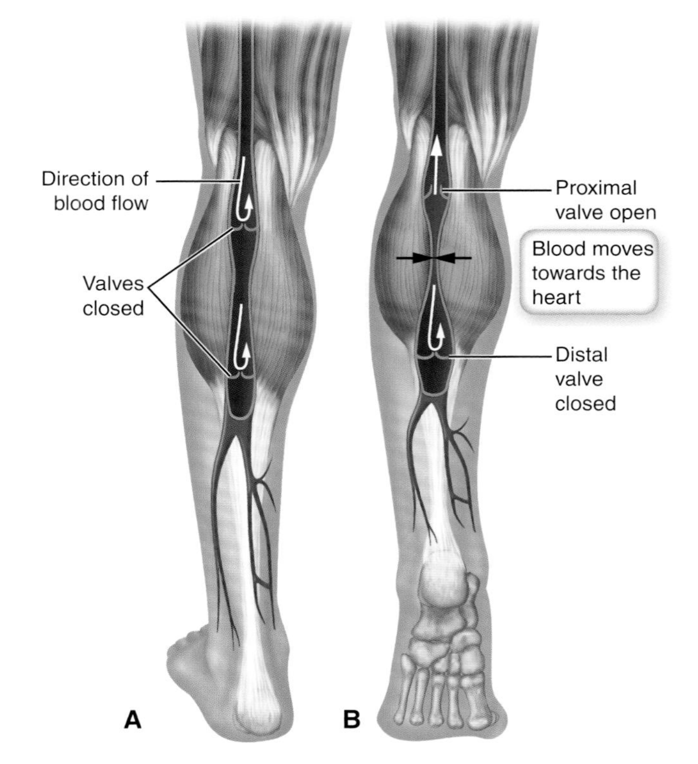

FIGURE 14-11 The skeletal muscle pump. KEY POINT Muscle contraction and valves keep blood flowing back toward the heart. **A.** Both proximal and distal valves are closed in the relaxed muscle, as gravity propels blood distally. **B.** Contracting skeletal muscle compresses the vein and drives blood forward, opening the proximal valve, while the distal valve closes to prevent backflow of blood. ZOOMING IN Which of the two valves shown is closer to the heart?

diaphragm flattens and puts pressure on the large abdominal veins. At the same time, chest expansion causes pressure to drop in the thorax. Together, these actions serve to both push and pull blood through the abdominal and thoracic cavities and return it to the heart.

As evidence of these effects, if a person stands completely motionless, especially on a hot day when the superficial vessels dilate, enough blood can accumulate in the lower extremities to cause fainting from insufficient oxygen to the brain.

CHECKPOINTS 

- 14-13 What force helps push materials out of a capillary? What force helps draw materials into a capillary?
- 14-14 Name the two types of vasomotor changes.
- 14-15 Where are vasomotor activities regulated?

BLOOD PRESSURE

Blood pressure is the force exerted by the blood against the walls of the vessels and is the force propelling blood to the tissues and back to the heart.

The force of ventricular contraction starts a wave of increased pressure, the **pulse,** that begins at the heart and travels along the arteries. Each heart contraction increases left ventricle pressure to about 120 mm Hg, but pressure falls to about 0 between contractions because the ventricle wall is very stretchy (compliant) but not very elastic (**FIG. 14-12**). The arteries, conversely, are very elastic—they stretch slightly to accommodate the new blood ejected from the ventricles but then rapidly return to their original size. This elastic recoil squeezes blood forward and keeps arterial pressure at about 80 mm Hg even though the ventricle has completely relaxed. The pressure in the blood vessels declines with increasing distance from the left ventricle, falling to just above zero in the vena cava and right atrium. **FIGURE 14-12** also illustrates four important values in blood pressure measurement:

- **Systolic pressure**, the maximum pressure that develops in the arteries after heart muscle contraction.
- **Diastolic pressure**, the lowest pressure measured in the arteries after relaxation of the heart muscle.
- **Mean arterial pressure** (**MAP**), the average pressure in the arteries. This value is closer to value for diastolic pressure than systolic pressure, because diastole lasts much longer than systole.
- **Pulse pressure,** the difference between the systolic and the diastolic pressures. It represents the pressure increase in the vessels created by ventricular contraction. Pulse pressure declines to about zero in the declines to virtually zero by the time it reaches the capillaries.

The pulse beat can be felt in any artery that is relatively close to the surface, particularly if the vessel can be pressed down against a bone. At the wrist, the radial artery passes over the bone on the forearm's thumb side, and the pulse

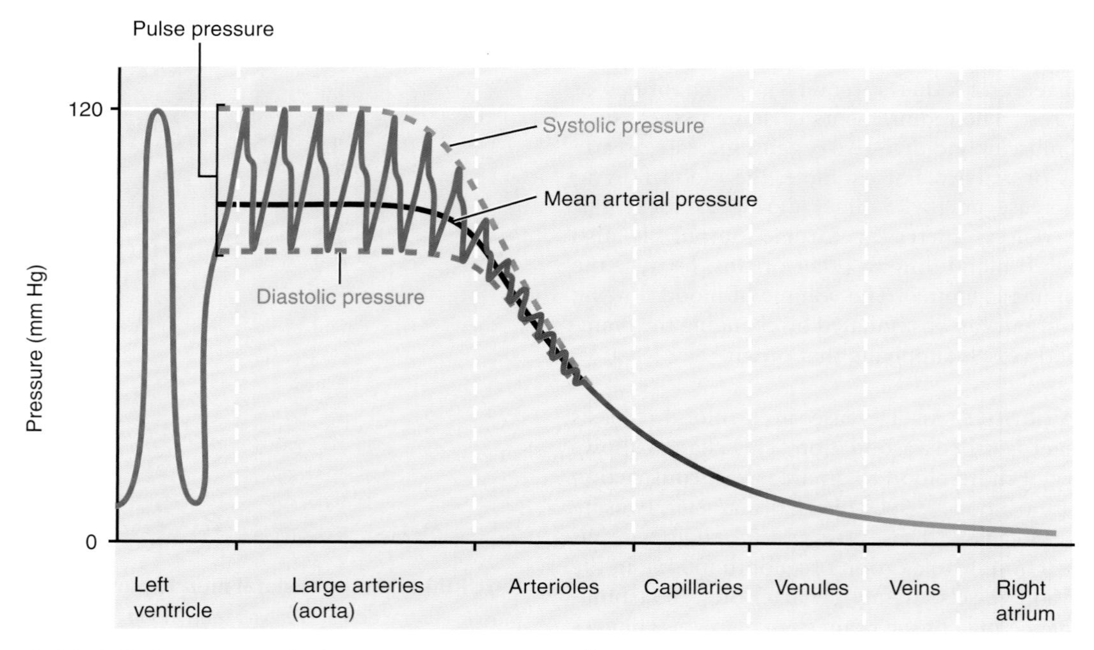

FIGURE 14-12 Pressures in the systemic circulation. **KEY POINT** Blood pressure declines as blood flows farther from the heart. Systolic pressure is the maximum pressure that develops in the arteries after heart muscle contraction; diastolic pressure is the lowest pressure in the arteries after relaxation of the heart. The difference between the two pressures (pulse pressure) declines as blood flows through the arteries. **ZOOMING IN** In which vessels does the pulse pressure drop to zero?

is most commonly obtained here. Other vessels sometimes used for taking the pulse are the carotid artery in the neck and the dorsalis pedis on the top of the foot.

Since each pulse beat represents one cardiac cycle, the pulse rate is usually equal to the heart rate. However, the heartbeat may not be detected as a pulse if the heartbeat is abnormally weak or if the artery is obstructed. In checking another person's pulse, it is important to use your second or third finger. If you use your thumb, you may be feeling your own pulse. When taking a pulse, it is important to gauge its strength as well as its regularity and rate.

Short-Term Blood Pressure Regulation 3 Blood pressure is a regulated variable—it must be kept within relatively narrow limits in order to guarantee adequate blood flow to the tissues without damaging blood vessels or overworking the heart. Specifically, arterial blood pressure is the regulated variable. So, unless stated otherwise, blood pressure refers to arterial pressure.

Just as pressure will increase within a water balloon as it fills, converting a floppy sac into a ball, increasing the volume of blood in the arteries increases blood pressure. As shown in FIGURE 14-13, the body can increase arterial blood volume in two ways.

- First, it can increase the blood *entering* the arteries from the heart by changing the cardiac output. As discussed in Chapter 13, the cardiac output is the amount of blood leaving each ventricle each minute. Cardiac output is the product of the heart rate (number of heart beats per minute) and the stroke volume (the volume pumped by each ventricle in each beat). Increased cardiac output increases blood pressure.
- Second, it can decrease the amount of blood *leaving* the arteries and entering the arterioles by reducing the diameter of many arterioles at once. This factor is known as *peripheral resistance*, because narrow vessels offer more resistance to blood flow than do wider vessels. Thus, generalized vasoconstriction increases peripheral resistance, which increases arterial volume, which increases blood pressure. Vasodilation does the opposite, reducing peripheral resistance and making it easier for blood to leave the arteries. Arterial volume, and thus pressure, decreases.

A negative feedback loop enables the body to maintain relatively constant blood pressure despite changes in body position and overall blood volume (FIG. 14-14). The sensors for blood pressure consist of **baroreceptors** (bar-o-re-SEP-torz) in the walls of the carotid arteries and the aorta. Baroreceptors are stretch receptors; increased pressure stretches the vessel wall, increasing the signal rate from the baroreceptors. Decreased pressure decreases the signal rate. For instance, hemorrhage reduces blood pressure and thus the rate of signals traveling to the cardiovascular control center in the medulla oblongata. The control center responds by increasing sympathetic activation. Recall from Chapter 13 that the sympathetic nervous system increases cardiac output by increasing both heart rate and stroke volume. Sympathetic neurons also travel to arterioles, where they stimulate vasoconstriction. Thus, both cardiac output and peripheral resistance increase, and blood pressure rises. On the other hand, high blood pressure increases baroreceptor signaling. The control center responds by sending signals via the parasympathetic nervous system that result in a slower heart rate and dilation of peripheral vessels, and blood pressure decreases.

Long-Term Blood Pressure Regulation The long-term regulation of blood pressure involves controlling the body's fluid balance, because changes in this parameter directly alter total blood volume. Arterial blood is about 13% of total blood volume in a resting person; therefore, if total blood volume increases, arterial blood volume (and thus blood pressure) increases with it. For instance, a high-salt diet increases water retention and thus total blood volume. In the absence of compensatory changes, arterial volume and thus blood pressure will also increase. As discussed in

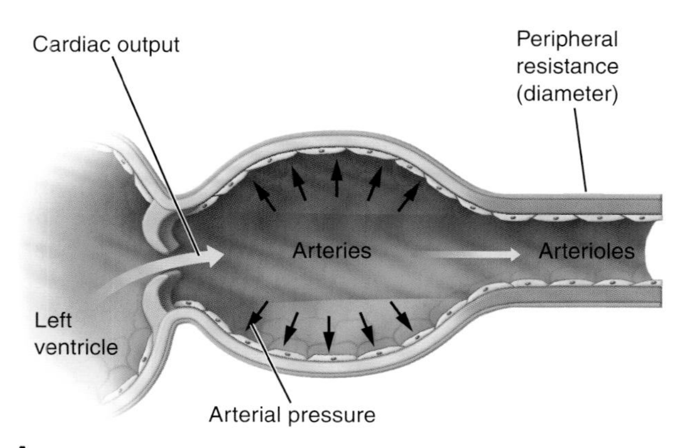

A Lower blood pressure

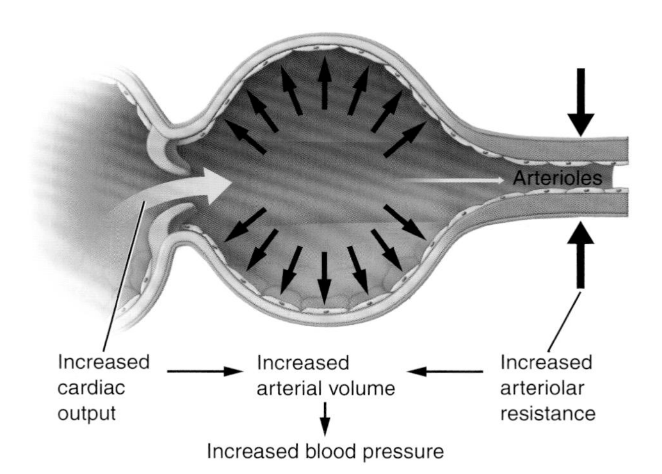

B Higher blood pressure

FIGURE 14-13 **Determinants of blood pressure. A.** Arterial blood volume is a major determinant of arterial blood pressure. **B.** An increase in cardiac output and/or arteriolar resistance increases arterial blood volume and thus blood pressure. ZOOMING IN If arteriolar diameter increases, will blood pressure rise or fall?

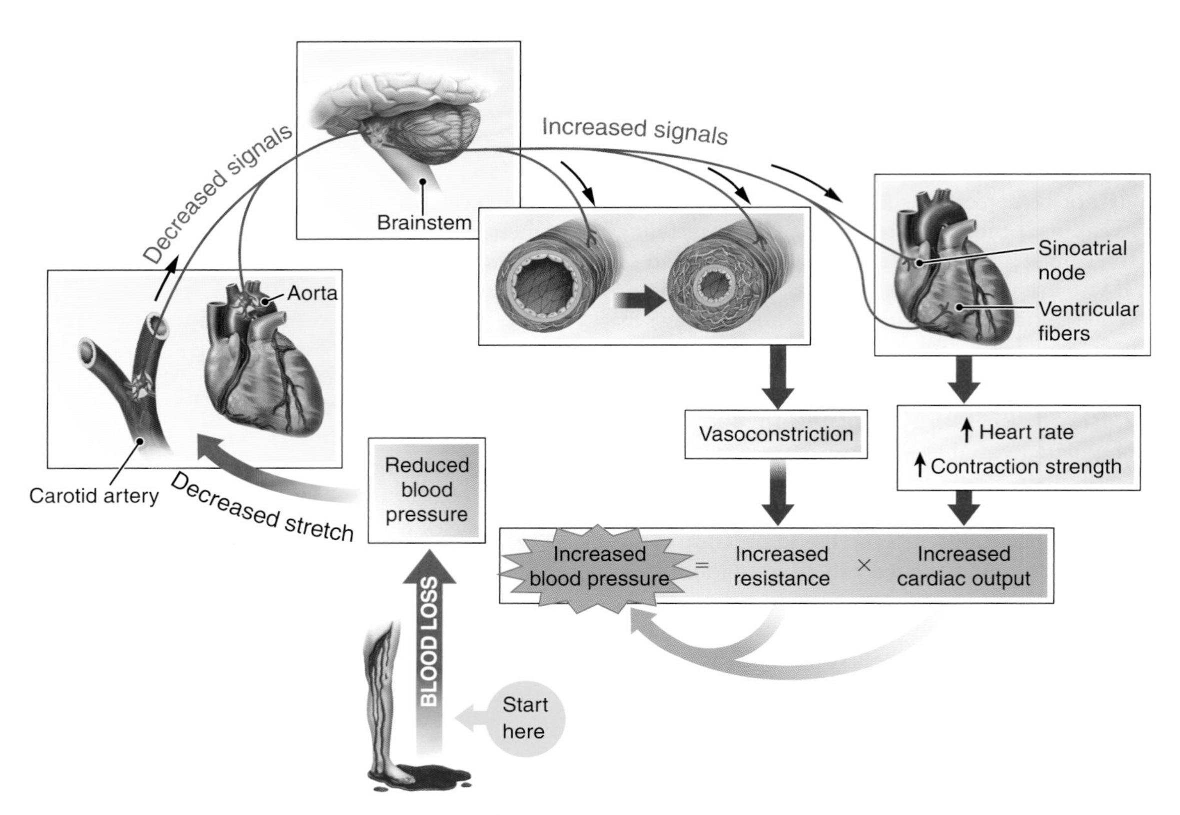

FIGURE 14-14 The baroreceptor response. KEY POINT The baroreceptors act as sensors in a negative feedback loop that keeps blood pressure relatively constant. Note that the different components of the feedback loop are indicated by different colors (regulated variable, *green*; sensors and control centers, *yellow*; signals, *purple*; effectors and their actions, *blue*). ZOOMING IN What happens to the heart rate when blood pressure falls?

Chapter 19, the kidneys modify the total blood volume by altering how much fluid is released in the urine. So, the kidneys play a critical role in the long-term regulation of blood pressure.

Other Factors Affecting Blood Pressure Blood pressure also depends on some additional factors, most importantly blood viscosity (thickness) and arterial compliance. While the body does not regulate these factors directly, they can cause abnormal changes in blood pressure.

Blood Viscosity Increased blood viscosity increases blood pressure. Under normal circumstances, blood viscosity remains within a constant range. However, loss of red cells, or loss of plasma proteins will decrease viscosity. Conversely, increased numbers of red cells or a loss of plasma volume, as by dehydration, will increase viscosity. The hematocrit test described in Chapter 12 is one measure of blood viscosity; it measures the relative percentage of packed cells in whole blood.

Blood Vessel Compliance Recall that compliance describes the ease with which vessels expand to receive blood. If vessels lose their capacity for expansion, the same volume of blood entering the arteries creates greater pressure because it is confined to a smaller space. Blood vessels lose compliance with aging, thus increasing blood pressure.

As you can see by this discussion, blood pressure regulation is quite complex, involving changes in heart rate, stroke volume, arteriolar diameter, overall blood volume, blood viscosity, and blood vessel compliance. We will return to the subject of blood pressure regulation in Chapter 19.

Blood Pressure Measurement The careful measurement and interpretation of blood pressure readings may prove a valuable guide in the care and evaluation of a person's health. Healthcare providers ordinarily measure arterial pressure only, since this is the homeostatically regulated variable. The most common measurement site is the brachial artery of the arm. In taking blood pressure, both systolic and diastolic pressures are measured.

The instrument used to measure blood pressure is a **sphygmomanometer** (sfig-mo-mah-NOM-eh-ter) (FIG. 14-15), more simply called a blood pressure cuff or blood pressure apparatus. The sphygmomanometer is an inflatable cuff attached to a pressure gauge. Pressure is expressed in millimeters mercury (mm Hg), that is, the height to which the pressure can push a column of mercury in a tube. The examiner wraps the cuff around the patient's upper arm and inflates it with air until the brachial artery is compressed

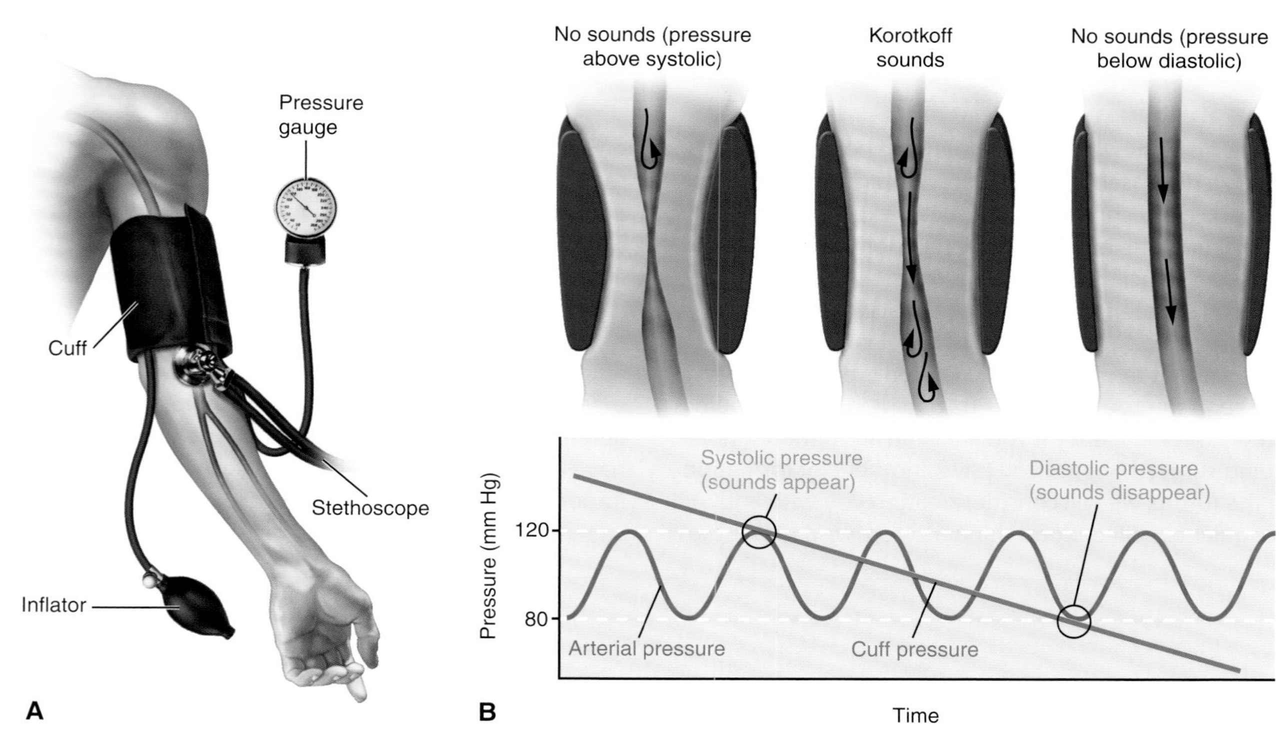

FIGURE 14-15 Measurement of blood pressure. A. A sphygmomanometer, or blood pressure cuff. **B.** As the clinician lowers the cuff pressure, Korotkoff sounds begin at the systolic pressure and disappear at the diastolic pressure. ZOOMING IN What is the systolic pressure in part B?

and the blood flow is cut off. Then, listening with a stethoscope, he or she slowly lets air out of the cuff until the first pulsations (Korotkoff sounds) are heard. At this point, the pressure in the cuff is equal to the systolic pressure, and this pressure is read. Then, more air is let out gradually until a characteristic muffled sound indicates that the vessel is open and the diastolic pressure is read off of a gauge or digital display. Newer devices measure blood pressure electronically: the examiner simply applies the cuff, which self-inflates and provides a digital reading. A typical normal systolic pressure is less than 120 mm Hg; a typical normal diastolic pressure is less than 80 mm Hg. Blood pressure is reported as systolic pressure first, then diastolic pressure, separated by a slash, such as 120/80. This reading would be reported verbally as "120 over 80."

Considerable experience is required to ensure an accurate blood pressure reading. Often, it is necessary to repeat measurements. Note also that blood pressure varies throughout the day and under different conditions, so a single reading does not give a complete picture. Some people typically have a higher reading in a doctor's office because of stress. People who experience such "white coat hypertension" may need to take their blood pressure at home while relaxed to get a more accurate reading. For more specific diagnoses, hospital clinicians can use complex methods involving catheterization or echocardiography to measure internal blood pressures directly.

See the Student Resources thePoint® for an animation on measuring blood pressure.

CHECKPOINTS 

- ☐ **14-16** What is the definition of a pulse?
- ☐ **14-17** What is the definition of blood pressure?
- ☐ **14-18** What are the effectors involved in the short-term regulation of blood pressure?
- ☐ **14-19** What are the sensors for blood pressure called?
- ☐ **14-20** What two components of blood pressure are measured?

See the Student Resources on thePoint® for career information on vascular technology. Vascular technologists collect information on the blood vessels and circulation to aid in diagnosis.

A & P in Action Revisited: Jocelyn Sees a Hematologist

In her first visit with Dr. Schuman, the hematologist, Jocelyn discussed the history of the blood clot in her left leg. In a second visit, the doctor explained the meaning of the blood test she had done for hereditary protein S deficiency.

"This protein normally helps block blood clotting," she said. "You have a mild deficiency, so your blood might tend to clot easily. I don't recommend any further treatment for you at this point, but you should report this deficiency if you are in an accident or undergoing surgery. Should you have another clot formation in the future, we will consider treating you with anticoagulants, called blood thinners. If so, we'd prescribe one of the newer oral anticoagulants, that don't require regular blood tests. Realize that you will probably always have poor circulation in that leg, so try not to sit or stand in one position for long. If, for example, you are traveling for a long time, wear a compression stocking on that leg and move around at least once every hour. I also recommend that you have your two young adult daughters tested for this deficiency. Definitely no menopausal hormone replacement therapy for you, and no hormonal contraceptives for the girls if they have this genetic trait. And of course, not smoking is important."

CHAPTER 14

Chapter Wrap-Up

OVERVIEW

A detailed chapter outline with space for note taking is on thePoint®. The figure below illustrates the main topics covered in this chapter.

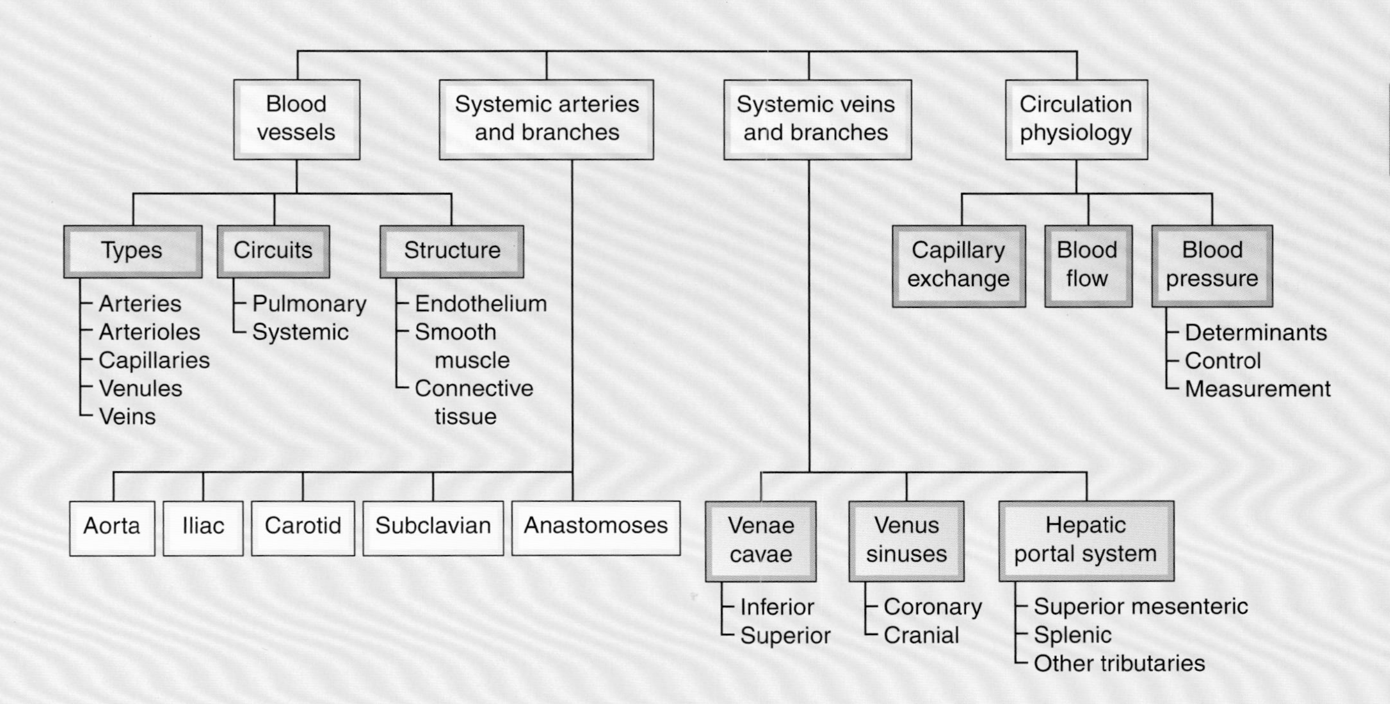

KEY TERMS

The terms listed below are emphasized in this chapter. Knowing them will help you organize and prioritize your learning. These and other boldface terms are defined in the Glossary with phonetic pronunciations.

aorta
arteriole
artery
baroreceptor
bulk flow
compliance
capillary
elasticity
endothelium
pulse
pulse pressure
sinusoid
sphygmomanometer
vasoconstriction
vasodilation
vasomotor
vein
vena cava
venous sinus
venule

WORD ANATOMY

Medical terms are built from standardized word parts (prefixes, roots, and suffixes). Learning the meanings of these parts can help you remember words and interpret unfamiliar terms.

WORD PART	MEANING	EXAMPLE
SYSTEMIC ARTERIES		
brachi/o	arm	The *brachiocephalic* artery supplies blood to the arm and head on the right side.
celi/o	abdomen	The *celiac* trunk branches to supply blood to the abdominal organs.
cephal/o	head	See "brachi/o" example.
clav/o	clavicle	The *subclavian* artery extends under the clavicle on each side.
cost/o	rib	The *intercostal* arteries are between the ribs.
enter/o	intestine	The *mesenteric* arteries supply blood to the intestines.
gastr/o	stomach	The *gastric* artery goes to the stomach.
hepat/o	liver	The *hepatic* artery supplies blood to the liver.
ped/o	foot	The dorsalis *pedis* artery supplies blood to the foot.
phren/o	diaphragm	The *phrenic* artery supplies blood to the diaphragm.
splen/o	spleen	The *splenic* artery goes to the spleen.
stoma	mouth	An *anastomosis* is a communication between two vessels.
CIRCULATION PHYSIOLOGY		
bar/o	pressure	A *baroreceptor* responds to changes in pressure.
man/o	pressure	See next example.
sphygm/o	pulse	A *sphygmomanometer* is used to measure blood pressure.

QUESTIONS FOR STUDY AND REVIEW

BUILDING UNDERSTANDING

Fill in the Blanks

1. Capillaries receive blood from vessels called ______.
2. The specific part of the medulla oblongata that regulates blood flow and blood pressure is the ______.
3. The flow of blood into an individual capillary is regulated by a(n) ______.
4. Blood is delivered to the lungs by the ______ circuit.
5. The technical name for a blood pressure cuff is ______.

Matching Match each numbered item with the most closely related lettered item.

___ 6. Supplies blood from the heart to the right arm and head

___ 7. Supplies blood from the heart to the kidney

___ 8. Returns blood from the brain to the heart

___ 9. Returns blood from the small intestine to the heart

___ 10. Supplies oxygen-rich blood to the liver

a. jugular vein
b. superior mesenteric vein
c. brachiocoephalic artery
d. renal artery
e. hepatic artery

Multiple Choice

___ 11. Which tissue makes up a blood vessel's inner tunic?
a. smooth muscle
b. epithelium
c. connective tissue
d. nervous tissue

___ 12. What is the name of either large vein that drains into the right atrium?
a. vena cava
b. jugular
c. carotid
d. iliac

___ 13. Which term describes the movement of fluid down a pressure gradient?
a. endocytosis
b. exocytosis
c. bulk flow
d. diffusion

___ 14. Which vessel supplies oxygenated blood to the stomach, spleen, and liver?
a. hepatic portal system
b. superior mesenteric artery
c. inferior mesenteric artery
d. celiac trunk

___ 15. Which structure regulates vasomotor activities?
a. medulla oblongata
b. cerebellum
c. cerebrum
d. spinal cord

UNDERSTANDING CONCEPTS

___ 16. Differentiate between the terms in each of the following pairs:
a. artery and vein
b. arteriole and venule
c. anastomosis and venous sinus
d. vasoconstriction and vasodilation
e. systolic and diastolic pressure

17. How does the structure of the blood vessels correlate with their function?

18. Trace a drop of blood from the left ventricle to the:
a. right side of the head and the neck
b. lateral surface of the left hand
c. right foot
d. liver
e. small intestine

19. Trace a drop of blood from capillaries in the wall of the small intestine to the right atrium. What is the purpose of going through the liver on this trip?

20. Describe three mechanisms that promote the return of blood to the heart in the venous system.

21. What physiological factors influence blood pressure?

22. Name the blood vessels that contribute to the hepatic portal system.

23. Referring to the Dissection Atlas in Appendix 3:
a. Give the number of any figure that shows an anastomosis.
b. Name the artery leading to the gallbladder.
c. Name two branches of the common hepatic artery.
d. Name the artery that supplies the stomach and the first portion of the small intestine.
e. Name an artery that supplies the reproductive organs.

CONCEPTUAL THINKING

24. Kidney disease usually results in the loss of protein from the blood into the urine. One common sign of kidney disease is edema. From your understanding of capillary exchange, explain why edema is often associated with kidney disease.

25. Cliff C., a 49-year-old self-described "couch potato," has high blood pressure. His doctor suspects that Cliff's lifestyle has led to fatty deposits in the lining of his arteries, resulting in narrowing and hardening of these vessels. How has this disorder contributed to Cliff's high blood pressure?

26. Jocelyn's blood clot has blocked the popliteal vessel behind her knee. If this clot broke loose, what structures would it pass through before it reached a lung?

For more questions, see the Learning Activities on thePoint®.

CHAPTER 15

The Lymphatic System and Immunity

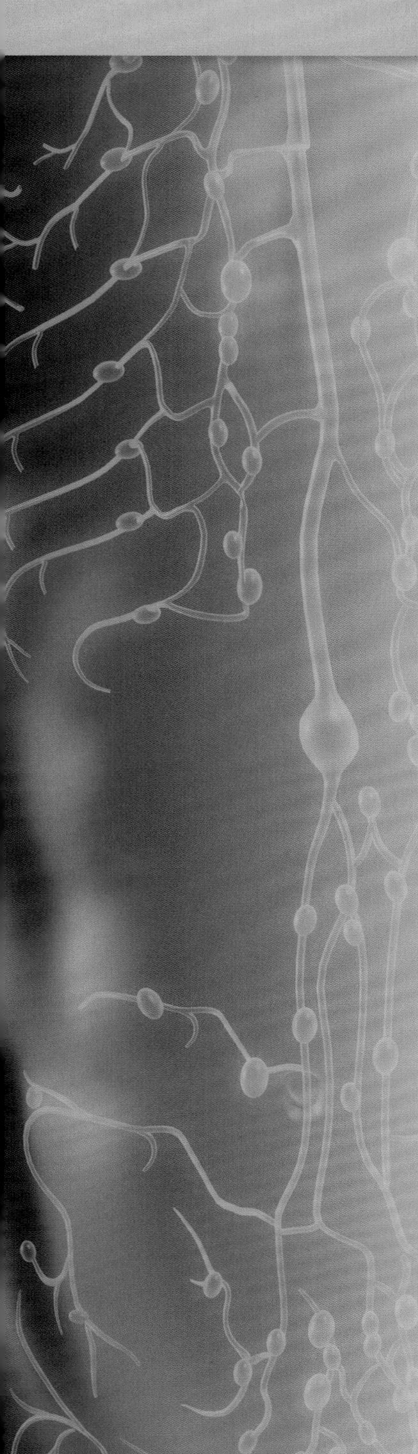

Learning Objectives

After careful study of this chapter, you should be able to:

1. List the functions of the lymphatic system. *p. 306*
2. Compare and contrast the structure of lymphatic capillaries and blood capillaries. *p. 306*
3. Explain how fluid moves into lymphatic capillaries and through lymphatic vessels. *p. 306*
4. Name the two main lymphatic ducts, and describe the area drained by each. *p. 307*
5. Name and give the locations of five types of lymphoid tissue, and list the functions of each. *p. 309*
6. Differentiate between innate and adaptive immunity and give examples of each. *p. 312*
7. Name three types of cells and three types of chemicals active in innate immunity. *p. 312*
8. Briefly describe the inflammatory reaction. *p. 315*
9. Define antigen and antibody. *p. 316*
10. Compare and contrast T cells and B cells with respect to development and type of activity. *p. 316*
11. Describe the activities of four types of T cells. *p. 316*
12. Explain the role of antigen-presenting cells in adaptive immunity. *p. 316*
13. Differentiate between natural and artificial adaptive immunity. *p. 319*
14. Differentiate between active and passive immunity. *p. 319*
15. Define the term vaccine and give three examples of vaccine types. *p. 320*
16. Define the term antiserum. *p. 320*
17. Using information from the case study and narrative, explain how the body rids itself of virus-infected cells. *pp. 305, 323*
18. Show how word parts are used to build words related to the lymphatic system and immunity (see Word Anatomy at the end of the chapter). *p. 325*

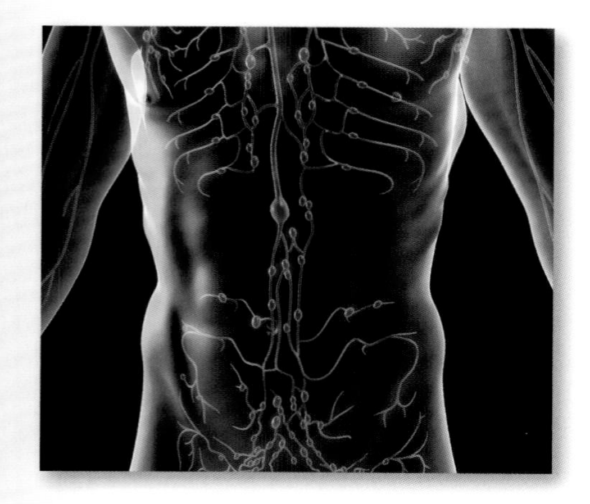

A & P in Action
Lucas's Mononucleosis

Lucas was looking forward to the upcoming high school marching band season. His band had been putting in many hours practicing new tunes and marching sequences. Lucas recently had been feeling pretty tired, which he attributed to the late hours and twice-a-day practices. His 30-lb sousaphone seemed to weigh more these days.

"Come on Lucas, you have to eat dinner before going to band practice tonight," his mom urged one evening.

"I don't feel like eating, and besides, my throat is kind of sore," Lucas replied.

Thinking about her son's answer, his mom recalled that Lucas had been sleeping more than usual the past few days. She is worried about his hectic schedule, the frequent crowded bus trips, and the intense classroom sessions with the 125-plus band members. She was aware that two of the band members had come down with mono the past week, and that illness was already a topic of discussion at the parent booster club meeting. She decided to call Lucas's physician the next day and schedule an appointment.

"Hi Lucas, how's the band season coming along?" Dr. Fischer asked when he saw Lucas later that week.

"Pretty good; we've got some new songs and I think we are going to score well in the band competition. We have a new tuba player, and he's got one of those new lightweight sousaphones, pretty cool" Lucas said. "Everyone was trying it out, it's really neat."

Dr. Fischer took a history and then asked Lucas to lie down on the examination table. "Let's take a look and see why you are feeling so tired lately and what might be causing the sore throat."

Dr. Fischer considered the symptoms: general malaise for 7 to 10 days, fever, loss of appetite, and a sore throat. He began the physical examination by observing Lucas's throat and noted it was red and swollen. He palpated the lymph nodes in the cervical, axillary, and inguinal regions. They were all enlarged. He also palpated the left upper quadrant (LUQ) of the abdomen and noted that the spleen was enlarged.

"It looks like you might have come down with a viral infection called mononucleosis," Dr. Fischer told Lucas. "I'm going to take a throat culture and blood sample to confirm my suspicions."

Later, we will check on the results of Lucas's laboratory tests and his diagnosis.

In this chapter, you will learn about the lymphatic system and how, along with other components of the immune system, it helps protect us from infections.

As you study this chapter, CasePoints will give you opportunities to apply your learning to this case.

Visit thePoint® to access the following resources. For guidance in using these resources most effectively, see pp. xv–xvii.

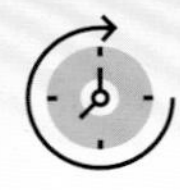

Before Begin

- Tips for Effective Studying
- Pre-Quiz

While You Are Learning

- Web Figure: Chain of Events in Inflammation
- Web Chart: Lymphoid Tissue
- Animation: Phagocytosis
- Animation: Acute Inflammation
- Animation: Immune Response
- Chapter Notes Outline
- Audio Pronunciation Glossary

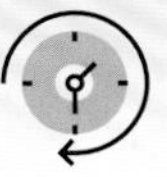

When You Are Reviewing

- Answers to Questions for Study and Review
- Health Professions: Nurse Practitioner
- Interactive Learning Activities

A LOOK BACK

In Chapter 14, we learned that the blood leaves some fluid behind in the tissues as it travels through the capillary networks. The lymphatic system collects this fluid and returns it to the circulation. Revisiting the key ideas of barriers [4]*, flow and gradients* [5]*, and mass balance* [13] *will help you understand how lymphatic vessels accomplish this task. The lymphatic system has other functions besides aiding in circulation. These include its role in immunity, as discussed in this chapter, and its participation in digestion, as presented in Chapter 17. These processes involve negative feedback* [3]*, barriers* [4]*, and communication* [11]*.*

Introduction

The lymphatic system is a widespread system of tissues and vessels. Its organs are not grouped together but are scattered throughout the body, and it services almost all regions.

Functions of the Lymphatic System

The lymphatic system's functions are just as varied as its locations. These functions fall into three categories:

- **Fluid balance.** As blood circulates through the capillaries in the tissues, water and dissolved substances are constantly exchanged between the bloodstream and the interstitial (in-ter-STISH-al) fluids that bathe the cells. [13] In order to prevent fluid accumulation, an equal volume of fluid must enter and leave a particular tissue. But, as shown in **FIGURE 14-10**, the volume of fluid that leaves the blood at the proximal end of the capillary is not quite matched by the amount that returns to the blood at the distal end. This extra volume is returned to the bloodstream via the lymphatic vessels. Proteins that leaked out of the capillaries are also returned to the circulation via the lymphatic system (**FIG. 15-1**).

  In addition to the blood-carrying capillaries, the tissues also contain microscopic lymphatic capillaries (**FIG. 15-2**). These small vessels pick up excess fluid and protein from the tissues. They then drain into larger vessels, which eventually return these materials to the venous system near the heart.

  The fluid that circulates in the lymphatic system is called **lymph** (limf), a clear fluid similar in composition to interstitial fluid. Although lymph is formed from the components of blood plasma, it differs from the plasma in that it has much less protein.
- **Protection.** The lymphatic system is an important component of the immune system, which fights infection and helps prevent cancer. One group of white blood cells, the lymphocytes, can live and multiply in the lymphatic system, where they attack and destroy foreign organisms. Lymphoid tissue scattered throughout the body filters out pathogens, other foreign matters, tumor cells, and cellular debris found in body fluids. More will be said about the lymphocytes and immunity later in this chapter.
- **Absorption of fats.** Following the chemical and mechanical breakdown of food in the digestive tract, most nutrients are absorbed into the blood through intestinal capillaries. Many digested fats, however, are too large to enter the blood capillaries and are instead absorbed into specialized lymphatic capillaries in the lining of the small intestine. Fats taken into these **lacteals** (LAK-te-als) are transported in lymphatic vessels until the lymph is added to the blood. More information on the lymphatic system's role in digestion is found in Chapter 17.

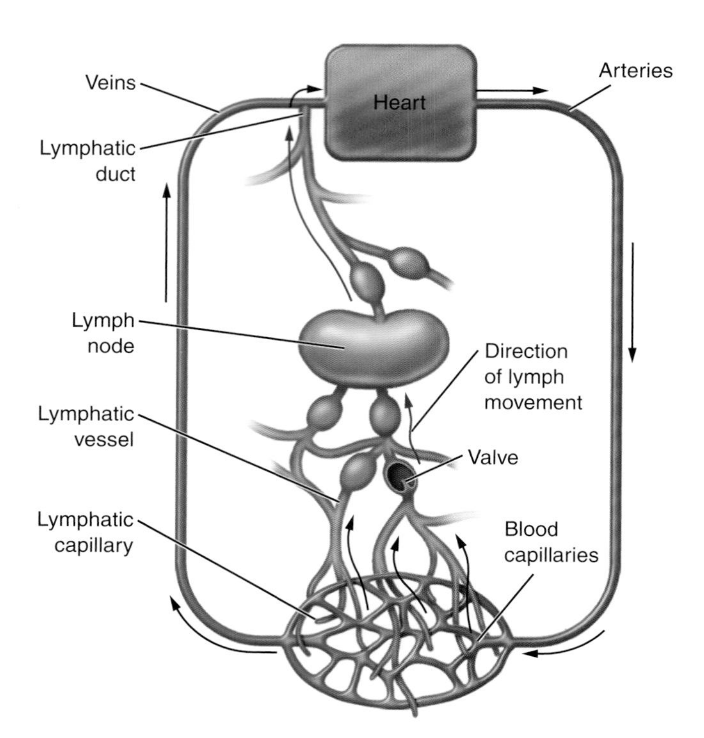

FIGURE 15-1 The lymphatic system in relation to the cardiovascular system. KEY POINT Lymphatic vessels pick up fluid in the tissues and return it to the blood in vessels near the heart. ZOOMING IN What type of blood vessel receives lymph collected from the body?

CHECKPOINT 

☐ **15-1** What are the three functions of the lymphatic system?

Lymphatic Circulation

Lymph travels through a network of small and large channels that are in some ways similar to the blood vessels. However, the system is not a complete circuit. It is a one-way system that begins in the tissues and ends when the lymph joins the blood (see **FIG. 15-1**).

LYMPHATIC CAPILLARIES

The walls of the lymphatic capillaries resemble those of the blood capillaries in that they are made of one layer

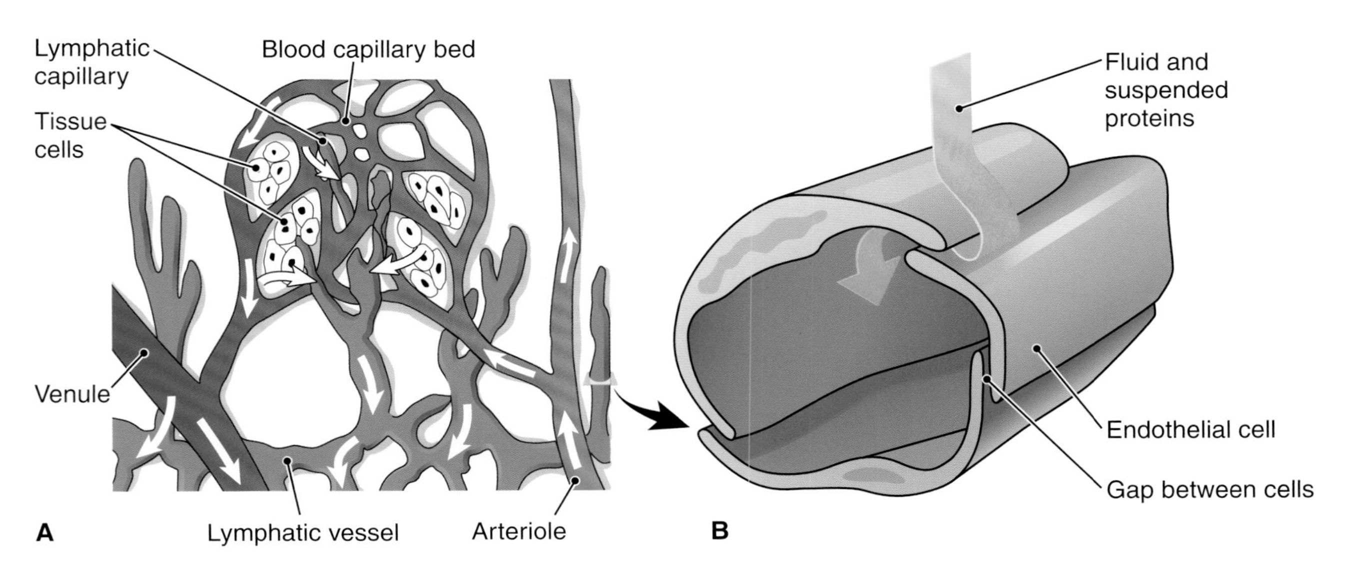

FIGURE 15-2 Lymphatic drainage in the tissues. KEY POINT Lymphatic capillaries pick up fluid and proteins from the tissues for return to the heart in lymphatic vessels. **A.** Blind-ended lymphatic capillaries in relation to blood capillaries. *Arrows* show the direction of flow. **B.** Structure of a lymphatic capillary. Fluid and proteins can enter the capillary with ease through gaps between the endothelial cells. Overlapping cells act as valves to prevent the material from leaving.

15

of flattened (squamous) epithelial cells (the endothelium). However, the gaps between the endothelial cells in the lymphatic capillaries are larger than those of the blood capillaries. 4 The barrier formed by the lymphatic capillary wall is thus more permeable than that of blood capillaries. This "leakiness" facilitates the entry of water and larger molecules, including proteins, into the lymphatic system.

Interstitial fluid enters the lymphatic capillaries through these gaps using bulk flow. 5 As introduced in Chapter 14, bulk flow is the movement of fluid down a pressure gradient (see FIG. 14-10). The pressure gradient across the lymphatic capillary wall is created by an increase in the interstitial fluid volume. When this happens, the excess fluid is absorbed into the lymphatic capillaries (see FIG. 15-2B).

Unlike the blood capillaries, the lymphatic capillaries arise blindly; that is, they are closed at one end and do not form a bridge between two larger vessels. Instead, one end simply lies within a lake of tissue fluid, and the other joins with a larger lymphatic vessel that transports the lymph toward the heart (see FIG. 15-2A).

LYMPHATIC VESSELS

The lymphatic vessels are thin walled and delicate and have a beaded appearance because of indentations where valves are located (see FIG. 15-1). These valves prevent backflow in the same way as do those found in veins (see FIG. 14-11).

Lymphatic vessels include superficial and deep sets (FIG. 15-3). The surface lymphatics are immediately below the skin, often lying near the superficial veins. The deep vessels are usually larger and accompany the deep veins.

Lymphatic vessels are named according to location. For example, those in the breast are called mammary lymphatic vessels (see FIG. 15-3A), those in the thigh are called femoral lymphatic vessels, and those in the leg are called tibial lymphatic vessels. At certain points, the vessels drain through **lymph nodes**, small masses of lymphoid tissue that filter the lymph. The nodes are in groups that serve a particular region, as will be described shortly.

Lymphatic vessels carrying lymph away from the regional nodes eventually drain into one of two terminal vessels, the right lymphatic duct or the thoracic duct, both of which empty into the bloodstream near the heart.

The Right Lymphatic Duct The **right lymphatic duct** is a short vessel, approximately 1.25-cm (1/2 in) long, that receives only the lymph that comes from the body's superior right quadrant: the right side of the head, neck, and thorax, as well as the right upper extremity. The right lymphatic duct empties into the right subclavian vein near the heart (see FIG. 15-3B). Its opening into this vein is guarded by two pocket-like semilunar valves to prevent blood from entering the duct. The rest of the body is drained by the thoracic duct.

The Thoracic Duct The **thoracic duct**, or left lymphatic duct, is the larger of the two terminal vessels, measuring approximately 40 cm (16 in) in length. As shown in FIGURE 15-3C, the thoracic duct receives lymph from all parts of the body except those superior to the diaphragm on the right side. It then drains into the left subclavian vein. This duct begins in the posterior part of the abdominal cavity, inferior to the attachment of the diaphragm. The duct's first part is enlarged to form a cistern, or temporary storage pouch, called the **cisterna chyli** (sis-TER-nah KI-li). **Chyle** (kile) is the milky fluid that drains from the intestinal lacteals; it is formed by the combination of fat globules and lymph. Chyle passes through the intestinal lymphatic

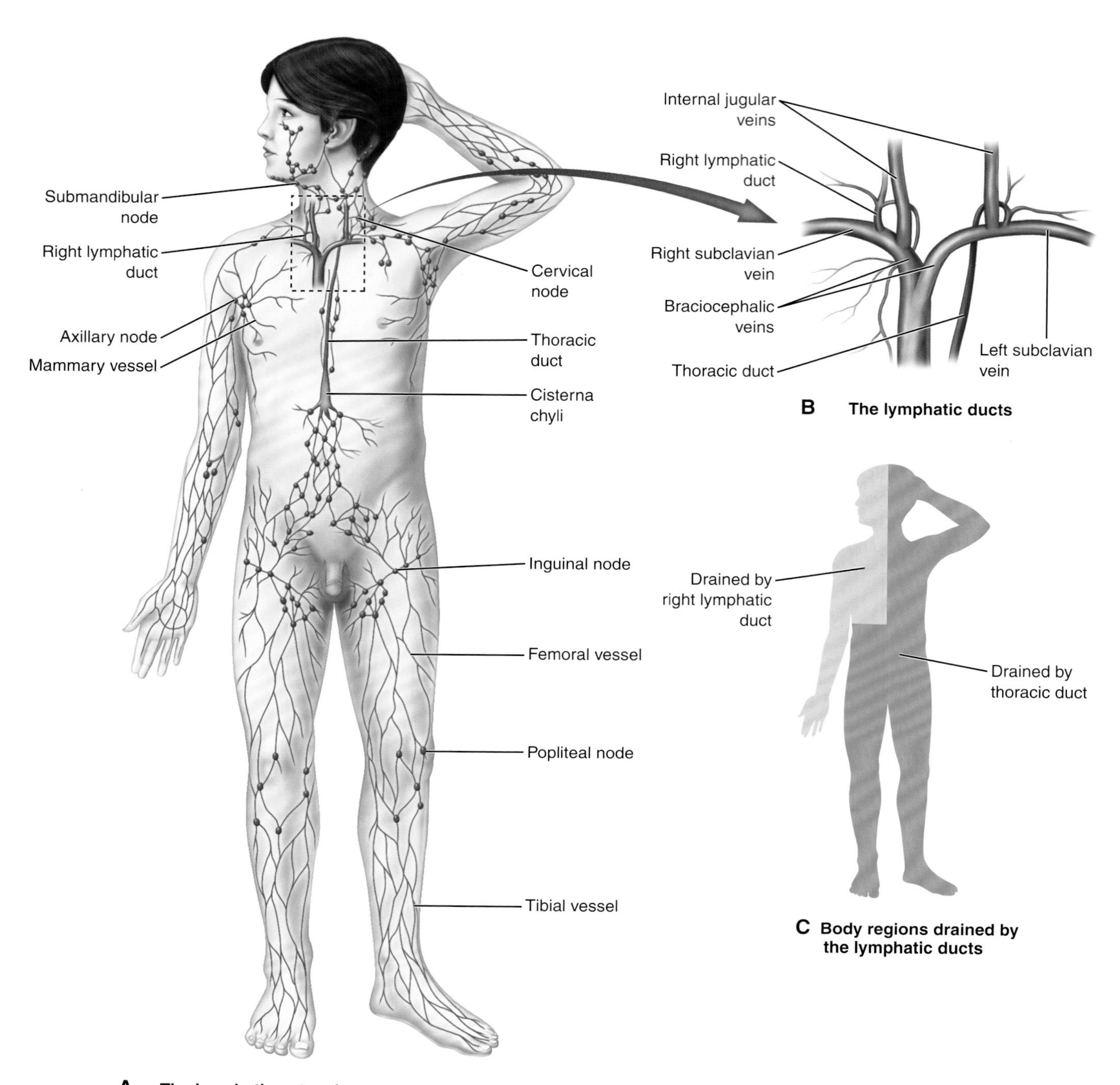

FIGURE 15-3 Vessels and nodes of the lymphatic system. KEY POINT Lymphatic vessels serve almost every area in the body. Lymph nodes are distributed along the path of the vessels. **A.** Lymph nodes and vessels. **B.** The thoracic and right lymphatic ducts drain into the subclavian veins. **C.** Body regions drained by the two lymphatic ducts. ZOOMING IN What are some nodes that receive lymph drainage from the mammary vessels? From the leg vessels?

vessels and the lymph nodes of the mesentery (the membrane around the intestines), finally entering the cisterna chyli. In addition to chyle, all the lymph from below the diaphragm empties into the cisterna chyli and subsequently the thoracic duct.

The thoracic duct extends upward from the cisterna chyli through the diaphragm and along the posterior thoracic wall into the base of the neck on the left side. Here, it receives the left jugular lymphatic vessels from the head and neck and the left subclavian vessels from the left upper extremity. In addition to the valves along the duct, there are two valves at its opening into the left subclavian vein to prevent the passage of blood into the duct.

LYMPH MOVEMENT

The segments of lymphatic vessels located between the valves contract rhythmically, propelling the lymph forward. The contraction rate is related to the fluid volume in the vessel—the more fluid, the more rapid the contractions.

The same mechanisms that promote venous return of blood to the heart also move lymph. As skeletal muscles contract during movement, they compress the lymphatic vessels and drive lymph forward. Changes in pressures within the abdominal and thoracic cavities caused by breathing aid lymphatic movement from the abdomen to the thorax. When lymph does not flow properly, a condition called lymphedema results. This is discussed later in this chapter.

CHECKPOINTS 

- ☐ **15-2** What are the two differences between blood capillaries and lymphatic capillaries?
- ☐ **15-3** What are the two main lymphatic vessels?

CASEPOINTS 

- ☐ **15-1** In the case study, Lucas's physician noted swelling of his cervical, axillary, and inguinal lymph nodes. Using FIGURE 15-3 as reference, where are these located?
- ☐ **15-2** What is the path of lymph from the inflamed nodes on his left side to his bloodstream?

Lymphoid Tissue

Lymphoid (LIM-foyd) **tissue** is distributed throughout the body and makes up the lymphatic system's specialized organs. As previously mentioned, the lymph nodes are part of the network of lymphatic vessels. In contrast, the spleen, thymus, tonsils, and other lymphoid organs do not encounter lymph.

See the Student Resources on thePoint® for a quick study chart on lymphoid tissue.

LYMPH NODES

The lymph nodes, as noted, filter the lymph as it travels through the lymphatic vessels (FIG. 15-4). They are also sites where lymphocytes of the immune system multiply and work to combat foreign organisms. The lymph nodes are small, rounded masses varying from pinhead size to as long as 2.5 cm (1 in.). Each has a fibrous connective tissue capsule from which partitions called *trabeculae* extend into the node's substance. At various points in the node's surface, afferent lymphatic vessels pierce the capsule to carry lymph toward open channels, or sinuses, in the node. An indented area, the hilum (HI-lum), is the exit point for efferent lymphatic vessels carrying lymph out of the node. At this location, other structures, including blood vessels and nerves, link with the node. (*Hilum* is a general term for an indented region of an organ where vessels and nerves connect.)

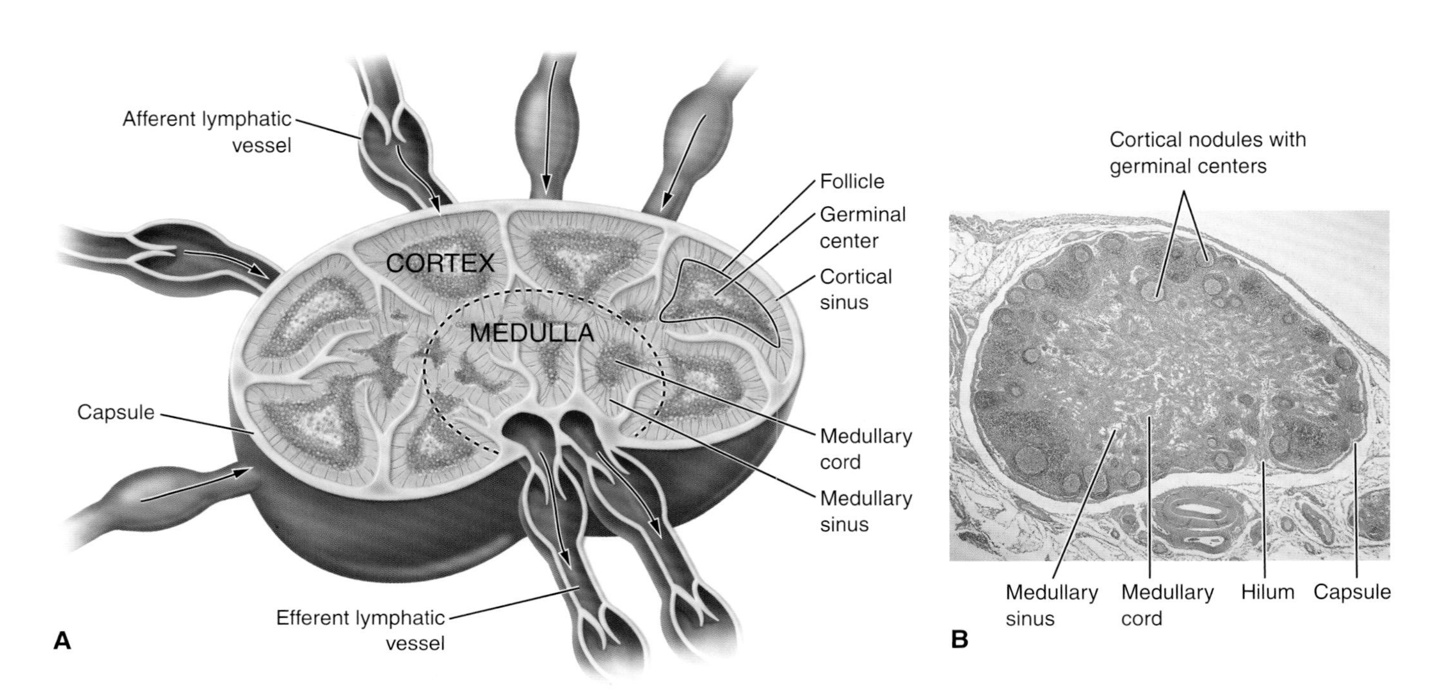

FIGURE 15-4 Lymph nodes. **KEY POINT** Lymph is filtered as it travels through a lymph node. Cells of the immune system also multiply in the nodes. **A.** Structure of a lymph node. *Arrows* indicate the flow of lymph. **B.** Section of a lymph node as seen under the microscope (low power). **ZOOMING IN** What type of lymphatic vessel carries lymph into a node? What type of lymphatic vessel carries lymph out of a node?

A lymph node is divided into two regions: an outer cortex and an inner medulla. Each of these regions includes lymph-filled sinuses and cords of lymphatic tissue. The cortex contains pulplike *cortical nodules*, each of which has a *germinal center* where certain lymphocytes multiply. The medulla has populations of immune cells, including lymphocytes and macrophages (phagocytes) along the *medullary sinuses*, which drain into the efferent lymphatic vessels.

Lymph nodes are seldom isolated. As a rule, they are grouped together in numbers varying from 2 or 3 to well over 100. Some of these groups are placed deeply, whereas others are superficial. The main groups include the following:

- **Cervical nodes**, located in the neck in deep and superficial groups, drain various parts of the head and neck. They often become enlarged during upper respiratory infections.
- **Axillary nodes**, located in the axillae (armpits), may become enlarged after infections of the upper extremities and the breasts. Cancer cells from the breasts often metastasize (spread) to the axillary nodes.
- **Tracheobronchial** (tra-ke-o-BRONG-ke-al) **nodes** are found near the trachea and around the larger bronchial tubes. In people who smoke, are exposed to smoke, or live in highly polluted areas, these nodes become filled with airborne contaminants.
- **Mesenteric** (mes-en-TER-ik) **nodes** are found between the two layers of peritoneum that form the mesentery. There are some 100 to 150 of these nodes.
- **Inguinal nodes**, located in the groin region, receive lymph drainage from the lower extremities and from the external reproductive organs. When they become enlarged, they are often referred to as **buboes** (BU-bose), from which bubonic plague got its name.

THE SPLEEN

The **spleen** filters blood, much like the lymph nodes filter lymph. It is located in the superior left hypochondriac region of the abdomen, high up under the dome of the diaphragm, and normally is protected by the lower part of the rib cage (**FIG. 15-5**). The spleen is a soft, purplish, and somewhat flattened organ, measuring approximately 12.5- to 15-cm (5 to 6 in) long and 5- to 7.5-cm (2 to 3 in.) wide. The spleen's capsule, as well as its framework, is more elastic than that of the lymph nodes. It contains an involuntary muscle, which enables the splenic capsule to contract and also to withstand some swelling.

Not surprisingly, considering its role in blood filtration, the spleen has an unusually rich blood supply. The organ is filled with a soft pulp rich in phagocytes and lymphocytes. As blood slowly percolates through this tissue, connective tissue fibers trap cellular debris and other impurities for destruction by phagocytes. The spleen is classified as part of the lymphatic system because it contains prominent masses

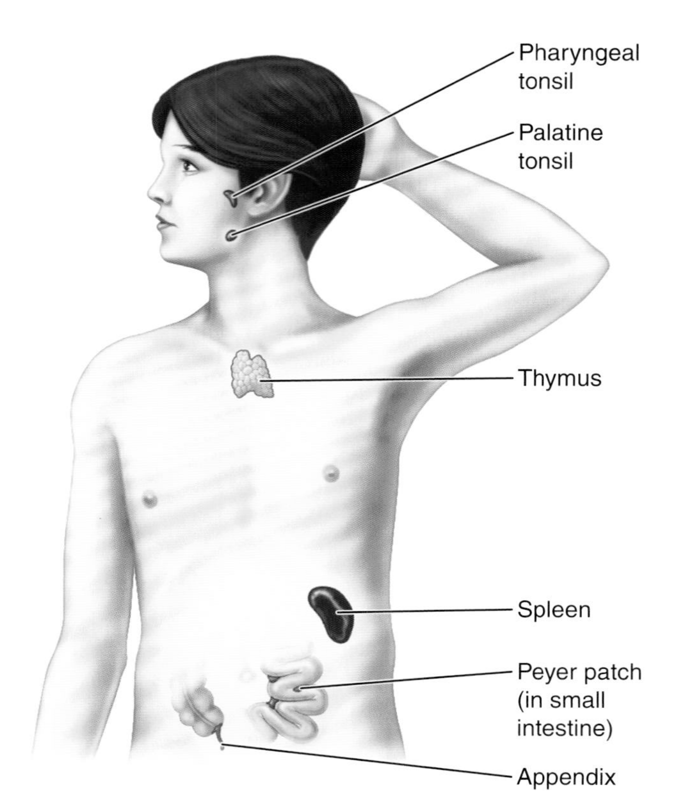

FIGURE 15-5 Location of lymphoid organs. KEY POINT In addition to lymphatic vessels and nodes, the lymphatic system includes the thymus, spleen, and mucosa-associated lymphoid tissue (MALT). MALT includes the tonsils, Peyer patches, and appendix.

of lymphoid tissue. However, it has wider functions than other lymphatic structures, including the following:

- Destroying old, worn-out red blood cells. The iron and other breakdown products of hemoglobin are then carried to the liver by the hepatic portal system to be reused or eliminated from the body.
- Producing red blood cells before birth.
- Serving as a reservoir for blood, which can be returned to the bloodstream in case of hemorrhage or other emergencies.

Splenectomy (sple-NEK-to-me), or surgical removal of the spleen, is usually a well-tolerated procedure. Although the spleen is the body's largest lymphoid organ, other lymphoid tissues can take over its functions. There is evidence that splenectomy carries a risk of developing certain infections, especially in younger patients. Physicians might prescribe prophylactic antibiotics, educate patients about the risk, or perform a partial splenectomy if possible.

THE THYMUS

The **thymus** (THI-mus), located in the superior thorax deep to the sternum, plays a role in immune system development during fetal life and infancy (**see FIG. 15-5**). Certain lymphocytes, T cells, must mature in the thymus gland before they can perform their functions in the immune system. Removal of the thymus causes a generalized decrease in the production of T cells as well as a decrease in the size of the spleen and of lymph nodes throughout the body.

The thymus is most active during early life. After puberty, the tissue undergoes changes; it shrinks in size and is gradually replaced by connective tissue and fat.

MUCOSA-ASSOCIATED LYMPHOID TISSUE

In the mucous membranes lining portions of the digestive, respiratory, and urogenital tracts, there are areas of lymphoid tissue that help destroy foreign contaminants. Leukocytes within this **m**ucosa-**a**ssociated **l**ymphoid **t**issue, or **MALT**, attack microorganisms and environmental contaminants using phagocytosis and antibodies specialized proteins that counteract infectious agents. Some leukocytes travel to other lymphoid organs, where they stimulate the production of more leukocytes targeted against the invading microorganism. So, along with the mucous membranes themselves, MALT helps prevent microorganisms from invading deeper tissues.

Some of the largest aggregations of MALT are found in the digestive tract and are known as **GALT**, or **g**ut-**a**ssociated **l**ymphoid **t**issue. **Peyer** (PI-er) **patches** are clusters of lymphoid nodules located in the mucous membranes lining the distal small intestine. The **appendix** (ah-PEN-diks) is a finger-like tube of lymphoid tissue, measuring approximately 8-cm (3 in) long. It is attached, or "appended," to the first portion of the large intestine **(see FIG. 15-5)**. Like the tonsils, the appendix seems to be noticed only when it becomes infected, causing appendicitis. However, it may, like the tonsils, figure in the development of immunity.

TONSILS

The **tonsils** are unencapsulated masses of GALT located in the vicinity of the pharynx (throat). They help protect against ingested or inhaled contaminants **(FIG. 15-6)**. The tonsils have deep grooves lined with lymphatic nodules. Lymphocytes attack pathogens trapped in these grooves. *Like the spleen and thymus, the tonsils are not connected with the lymphatic vessel network*. The tonsils are located in the following three areas:

- The **palatine** (PAL-ah-tine) **tonsils** are oval bodies located at each side of the soft palate. These are generally what is meant when one refers to "the tonsils."
- The single **pharyngeal** (fah-RIN-je-al) **tonsil** is commonly referred to as the **adenoid** (AD-eh-noyd) (a general term that means "glandlike"). It is located behind the nose on the posterior wall of the upper pharynx.
- The **lingual** (LING-gwal) **tonsils** are little mounds of lymphoid tissue posterior of the tongue.

Any of these tonsils may become so loaded with bacteria that they become reservoirs for repeated infections, and their removal is advisable. In children, a slight enlargement of any of them is not an indication for surgery, however, because all lymphoid tissue masses tend to be larger in childhood. A physician must determine whether these masses are abnormally enlarged taking the patient's age into account, because the tonsils are important in immune function during early childhood. Surgery may be advisable in cases of recurrent infection or if the enlarged tonsils make swallowing or breathing difficult. Their removal may also help children suffering from otitis media, because bacteria infecting the tonsils may travel to the middle ear. The surgery to remove the palatine tonsils is a tonsillectomy; an adenoidectomy is removal of the adenoid. Often,

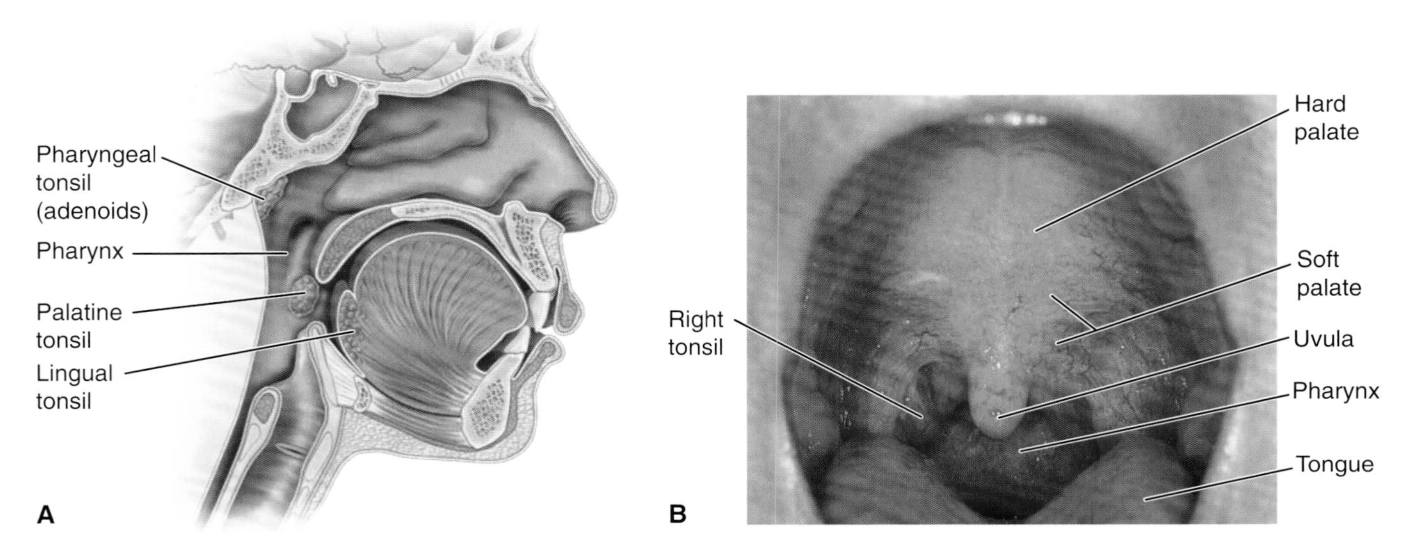

FIGURE 15-6 The tonsils. **KEY POINT** All of the tonsils are in the vicinity of the pharynx (throat) where they filter impurities. **A.** The tonsils in a sagittal section. **B.** Healthy tonsils in a healthy adult.

15

these two procedures are done together and abbreviated as T&A. Most tonsillectomies are performed by electrocautery, which uses an electric current to burn the tissue away. A newer technique, which allows faster recovery and fewer complications, uses radio waves to break down the tonsillar tissue.

CHECKPOINTS

- [ ] **15-4** What is the function of the lymph nodes?
- [ ] **15-5** What does the spleen filter?
- [ ] **15-6** What kind of immune system cells develop in the thymus?
- [ ] **15-7** What is the meaning of the acronyms MALT and GALT?
- [ ] **15-8** What is the general location of the tonsils?

CASEPOINTS

- [ ] **15-3** The germinal centers of Lucas's lymph nodes would have been enlarged because of his infection. What does this change indicate?
- [ ] **15-4** Which tonsils would the doctor have seen in Lucas's throat?
- [ ] **15-5** Of all the lymphoid organs mentioned in Lucas's case, which filter lymph and which filter blood?

Immunity

The body is constantly exposed to potentially harmful organisms, such a bacteria and viruses. Fortunately, most of us survive contact with these invaders and even become more resistant to disease in the process. All of the defenses that protect us against disease constitute **immunity**. These defenses protect us against any harmful agent that enters the body, such as an infectious organism. They also protect us against abnormal cells that arise within the body, as described later in the chapter. Some immune defenses combat any foreign agent, while others are effective against a particular pathogen type or a specific pathogen. Scientists use the term *immune system* to describe all the cells and tissues that protect us against foreign organisms or any cells different from our own normal cells.

Innate Immunity

The features that protect the body against disease are usually considered successive "lines of defense" beginning with the relatively simple outer barriers and proceeding through progressively more complicated, internal responses. Our defenses can be categorized as *innate* or *adaptive*. Innate defenses are inborn, that is, they are inherited along with all of a person's other characteristics. They include barriers and certain internal cellular and metabolic responses that protect against any foreign invaders or harmful substances. Innate responses are very rapid, are relatively nonspecific, and can help prevent or slow infections. Adaptive defenses, discussed later, develop *after* exposure to a particular pathogen and are specific to that pathogen. Adaptive defenses are slower to emerge, but can completely eliminate a specific pathogen from the body as well as prevent future infections.

THE FIRST LINE OF DEFENSE: INNATE BARRIERS

4 The first defense against invading organisms includes chemical and mechanical barriers, such as the following (FIG. 15-7):

- The skin serves as a mechanical barrier as long as it remains intact. A serious danger to burn victims, for example, is the risk of infection resulting from skin destruction.
- The mucous membranes that line the passageways leading into the body also act as barriers, trapping foreign material in their sticky secretions. The cilia in membranes in the upper respiratory tract help sweep impurities out of the body.
- Body secretions, such as tears, perspiration, and saliva, wash away microorganisms and may contain acids, enzymes, or other chemicals that destroy invaders. Digestive juices destroy many ingested bacteria and their toxins.
- Certain reflexes aid in the removal of pathogens. Sneezing and coughing, for instance, tend to remove foreign matter including microorganisms from the upper respiratory tract. Vomiting and diarrhea are ways in which ingested toxins and bacteria may be expelled from the digestive tract.

THE SECOND LINE OF DEFENSE: INNATE CELLS AND CHEMICALS

If an organism has overcome initial defenses, we have a number of internal activities that constitute a second line of defense (see FIG. 15-7). Although we present them in a separate category, you will see later in this chapter that many participate in and promote adaptive immune responses. Innate immunity was previously described as *nonspecific immunity*, but we now know that bacteria, viruses, and cancer cells induce distinct innate responses.

For many years, scientists did not know how the cells of the innate immune system were able to differentiate between body cells and foreign cells. Recently, it was discovered that innate immune cells possess a specific class of receptors called **tolllike receptors** (TLRs). (*Toll* means "great" in German.) 11 Recall that chemical signals induce their effects by binding to specific receptors. TLRs use a similar strategy to recognize pathogens. For instance, one TLR binds to specific substances in bacterial cell walls, whereas other TLRs bind to viral DNA. While TLRs cannot identify specific pathogens, they can identify specific patterns that strongly suggest that a cell is "nonself," and they can also distinguish between different pathogen types. These harmful cells can then be eliminated by various pathways, including phagocytosis.

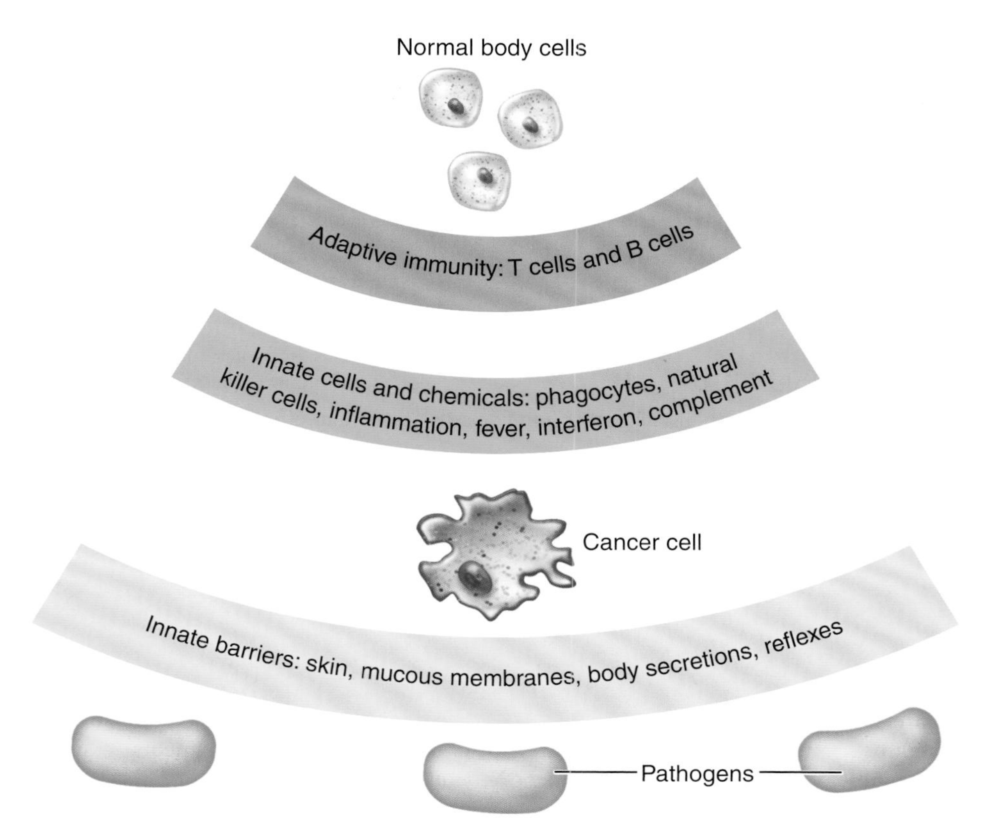

FIGURE 15-7 Lines of defense. KEY POINT Physical barriers are the first line of defense against pathogens. The innate and adaptive arms of the immune system are the second and third lines.

Phagocytosis In the process of **phagocytosis,** white blood cells take in and destroy waste (including worn-out cells), cancer cells, and pathogens. (see FIG. 12-5 in Chapter 12.) Neutrophils, a category of granular leukocytes, are important phagocytic white blood cells. (see TABLE 12-2 in Chapter 12.) Another active phagocyte is the **macrophage** (MAK-ro-faj). (The name *macrophage* means "big eater.") Macrophages are large white blood cells derived from monocytes, a type of agranular leukocyte. Monocytes develop into macrophages upon entering the tissues. Some macrophages remain fixed in the tissues, for example, in the skin, liver, lungs, lymphoid tissue, bone marrow, and soft connective tissue throughout the body. In some organs, macrophages are given special names. For example, Kupffer (KOOP-fer) cells are macrophages located in the lining of the liver sinusoids (blood channels). In the lungs, where they ingest solid particles, macrophages are called *dust cells.*

Natural Killer Cells The **natural killer (NK) cell** is a type of lymphocyte different from those active in adaptive immunity, which are described later. NK cells possess TLRs and other receptors that detect antigen patterns found in pathogens and cancerous body cells but not in healthy cells. As the name indicates, NK cells kill foreign and abnormal cells on contact. NK cells are found in the lymph nodes, spleen, bone marrow, and blood. They destroy abnormal cells by secreting a protein that breaks down the plasma membrane.

Cytokines and Other Chemicals 11 Scientists have identified a huge number of chemical mediators involved in innate immunity, a few of which are summarized in TABLE 15-1. A **cytokine (SI-to-kine)** is a peptide produced by immune cells or other body cells that is used for cellular signaling (the name means "cell activator"). Many cytokines modulate the immune response; these include **interleukins** (in-ter-LU-kinz) (IL) and **interferons** (in-ter-FERE-onz). Immune cells also produce many noncytokine molecules that activate innate immune responses, including histamine (HIS-tah-mene) and prostaglandins. Finally, **complement** (KOM-ple-ment) is a group of proteins that circulate in blood in inactive forms. They are activated by tissue damage or the presence of pathogens. These chemicals work together with immune cells to promote inflammation and fever, two generalized innate immune responses that fight pathogens. Before we explain these important processes, we must first discuss interferons and complement in greater detail.

Interferons Certain cells infected with a virus release a substance that prevents nearby cells from producing more viruses. This substance was first found in cells infected with influenza virus, and it was called "interferon" because it "interferes" with multiplication and spread of the virus.

Table 15-1 Selected Chemicals of Innate Immunity*

Chemical	Producing Cells	Actions
Cytokines		
Interleukins	Activated macrophages and other immune cells	Various actions; some promote inflammation and fever
Interferons (alpha and beta)	Released by virus-infected cells	Block viruses from infecting other cells; activate natural killer cells
Noncytokines		
Histamine	Mast cells	Dilates arterioles and increases capillary leakiness, resulting in redness and swelling
Prostaglandins	Mast cells, neutrophils, and other immune cells	Similar effects as histamine; activate pain receptors
Complement	Synthesized in the liver; found in blood in their inactive forms	Activates mast cells; lyses pathogens; stimulates phagocytosis

*Note that many of these chemicals are also active in adaptive immunity.

Interferon is now known to be a group of substances. Each is abbreviated as IFN with a Greek letter, alpha (α) or beta (β), to indicate different categories.

Pure interferons are now being produced by genetic engineering in microorganisms, making adequate quantities for medical therapy available. They are used to treat certain viral infections, such as hepatitis. Interferons are also of interest because they act nonspecifically on cells of the immune system. They have been used with varying success to boost the immune response in the treatment of disease.

Complement The destruction of foreign cells sometimes requires the activity of a group of nonspecific proteins in the blood, together called complement (**FIG. 15-8**). Complement proteins are always present in the blood, but they must be activated by contact with foreign cell surfaces or by specific immune complexes (described shortly). Complement is so named because it complements (assists with) immune reactions. Some of complement's actions are to:

- Bind to foreign cells to help phagocytes recognize and engulf them (**FIG. 15-8A**).
- Destroy cells by forming channels called **membrane attack complexes** (MACs) in pathogens' membranes (water enters the pathogen, resulting in cell rupture) (**FIG. 15-8B**).

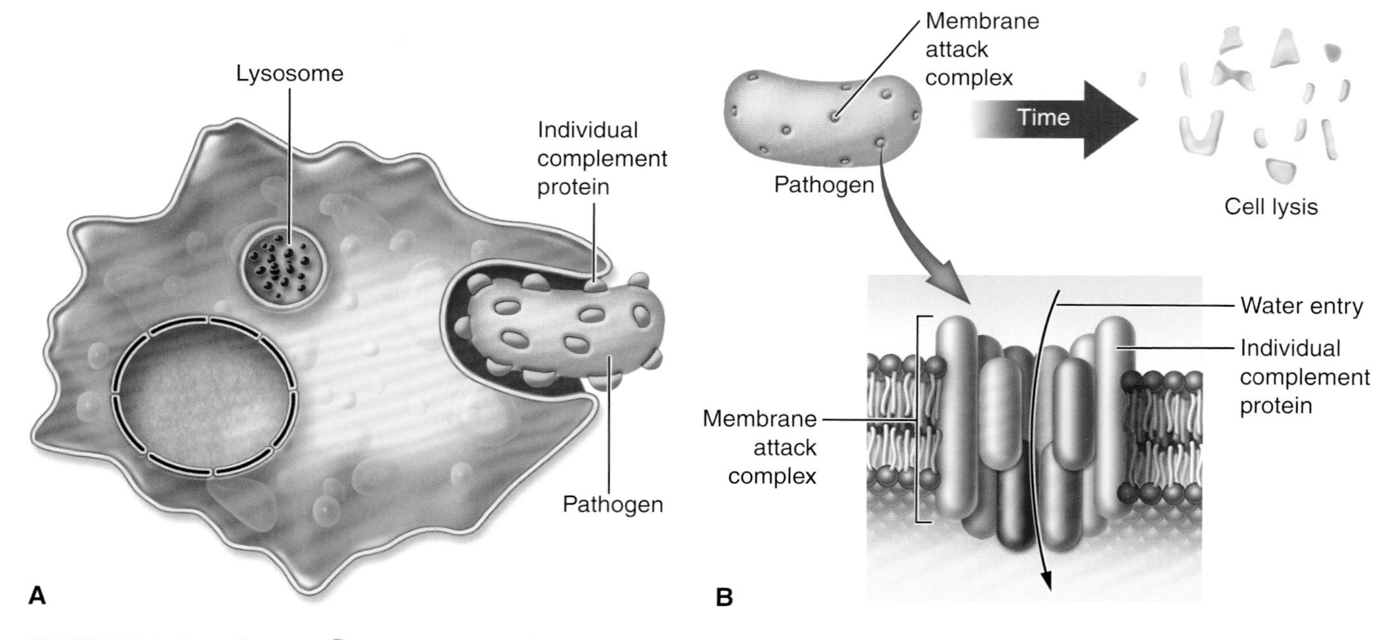

FIGURE 15-8 Complement. **KEY POINT** Complement participates in the destruction of pathogens. **A.** Complement proteins attract phagocytes to an area of inflammation and help them attach to and engulf a foreign organism. The organism is enclosed in a phagocytic vesicle, which then merges with a lysosome. Lysosomal enzymes digest the organism. **B.** A complex of specific complement proteins forms the membrane attack complex (MAC).

- Promote inflammation by increasing capillary permeability.
- Attract phagocytes to an area of inflammation.

Inflammation **Inflammation** is a nonspecific defensive response to a tissue-damaging irritant. Examples of irritants include friction, x-rays, extreme temperatures, caustic chemicals, and allergens. Often, however, inflammation results from irritation caused by infection. With the entrance and multiplication of pathogens, a whole series of defensive processes begins (**FIG. 15-9**). This inflammatory reaction is accompanied by four classic symptoms: heat, redness, swelling, and pain, as described below.

Tissue injury or pathogens activate **mast cells**, basophil-like immune cells in tissues. Activated mast cells produce histamine, which causes local blood vessels to dilate (widen) and become leaky. Blood plasma leaks out of the vessels into the tissues and begins to clot, thus limiting the spread of infection to other areas. The increased blood flow causes heat and redness.

Pathogens also activate resident immune cells, which release cytokines and other inflammatory chemicals. Some of these chemicals attract leukocytes to the area, especially neutrophils and monocytes. These cells enter the tissue through gaps in the capillary wall, and monocytes convert into macrophages. The resident immune cells and the new arrivals work together to phagocytose pathogens and damaged cells.

The mixture of leukocytes and fluid, the **inflammatory exudate**, contributes to swelling and puts pressure on nerve endings, causing pain. Substances secreted from the activated immune cells, including prostaglandins, cytokines, and histamine, also contribute to the pain of inflammation. Phagocytes are destroyed in large numbers as they work, and dead cells gradually accumulate in the area. The mixture of exudate, living and dead white blood cells, pathogens, and destroyed tissue cells is called "pus."

Meanwhile, the lymphatic vessels drain fluid from the inflamed area and carry it toward the lymph nodes. The regional lymph nodes become enlarged and tender (lymphadenitis), a sign that they are working overtime to produce immune cells to fight the infection (**see FIG. 15-3**).

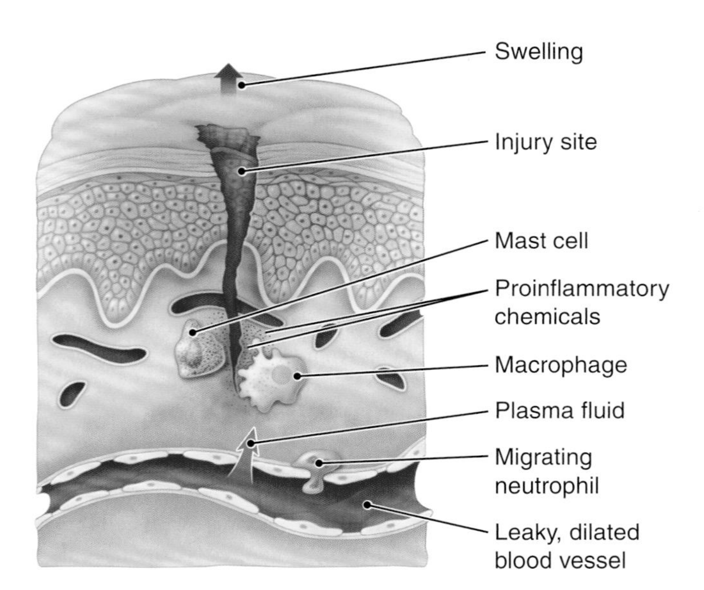

FIGURE 15-9 Inflammation. KEY POINT Acute inflammation involves components in blood and tissues. ZOOMING IN What causes the heat, redness, swelling, and pain characteristic of inflammation?

See the Student Resources on thePoint® for a diagram summarizing the events in inflammation and for the animations "Phagocytosis" and "Acute Inflammation."

Fever 3 As introduced in Chapter 1, negative feedback maintains body temperature within a narrow range—the set point. When phagocytes are exposed to infecting organisms, they release substances that raise this set point and body temperature increases. This increase, known as a **fever**, boosts the immune system. For instance, it stimulates phagocytes, increases metabolism, and decreases certain disease organisms' ability to multiply.

A common misperception is that fever is a dangerous symptom that should always be eliminated. Control of fever in itself does little to alter the course of an illness. Healthcare workers, however, should always be alert to fever development as a possible sign of a serious disorder and should recognize that an increased metabolic rate may have adverse effects on a weak patient's heart.

CHECKPOINTS 

- ☐ **15-9** What constitutes the first line of defense against the invasion of pathogens?
- ☐ **15-10** What are two types of components in the second line of defense against infection?
- ☐ **15-11** What are the four signs of inflammation?
- ☐ **15-12** What are three ways that fever boosts the immune system?

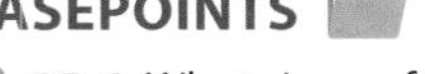

CASEPOINTS

- ☐ **15-6** What signs of inflammation are seen in Lucas's opening case study?
- ☐ **15-7** What other signs were there in Lucas's case that his body was fighting an infection?

Adaptive Immunity: The Final Line of Defense

Adaptive immunity to disease can be defined as an individual's power to resist or overcome the effects of a *particular* disease agent or its harmful products. These defense mechanisms also recognize and attack potentially useful, but foreign, materials, such as transplanted organs. Adaptive immunity is a selective process (i.e., immunity to one disease does not necessarily cause immunity to another). This selective characteristic is called **specificity** (spes-ih-FIS-ih-te).

Adaptive immunity is also termed **acquired immunity**, because it develops during a person's lifetime as he or she encounters various specific harmful agents. If the following description of adaptive immunity seems complex, bear in mind that from infancy onward, your immune system is able to protect you from millions of foreign substances, even synthetic substances not found in nature. All the while, the system is kept in check so that it does not usually overreact to produce allergies or mistakenly attack and damage your own body tissues.

ANTIGENS

Specific immunity is based on the body's ability to recognize a particular foreign substance. Any substance that induces an immune response is called an **antigen** (AN-te-jen) **(Ag)**. (The word is formed from *anti*body + *gen*, because an antigen generates antibody production.) Most antigens are large protein molecules, but carbohydrates and some lipids also may act as antigens. Normally, only nonself antigens stimulate an immune response. Such antigens can be found on the surfaces of pathogenic organisms, transfused blood cells, transplanted tissues, and cancerous cells and also on pollens, in toxins, and in foods. The critical feature of any substance described as an antigen is that it stimulates the activity of certain lymphocytes classified as T or B cells.

T CELLS

Both T and B cells come from hematopoietic (blood-forming) stem cells in bone marrow, as do all blood cells. The T and B cells differ, however, in their development and their methods of action. Some of the immature stem cells migrate to the thymus and become T cells, which constitute about 80% of the lymphocytes in the circulating blood. While in the thymus, these T lymphocytes multiply and become capable of combining with specific foreign antigens, at which time they are described as *sensitized*. These thymus-derived cells produce an immunity that is said to be **cell-mediated immunity**.

There are several types of T cells, each with different functions. The different types and some of their functions are as follows:

- **Cytotoxic T cells** (T_c) destroy certain abnormal cells directly. They recognize cells infected with viruses or other intracellular pathogens, cancer cells, and foreign antigens present in transplanted tissue. They produce substances that cause the cells to "self-destruct" by apoptosis (programmed death).
- **Helper T cells** (T_h) control immune responses by releasing interleukins. These signaling molecules stimulate the production and activity of cytotoxic T cells as well as B cells and macrophages. (Interleukins are so named because they act between white blood cells.) There are several subtypes of helper T cells, one of which is infected and destroyed by the AIDS virus (HIV). The HIV-targeted T cells have a special surface receptor (CD4) to which the virus attaches.
- **Regulatory T cells** (T_{reg}) suppress the immune response in order to prevent overactivity. These T cells may inhibit or destroy active lymphocytes.
- **Memory T cells** remember an antigen and start a rapid response if that antigen is contacted again.

The T-cell portion of the immune system is generally responsible for defense against cancer cells, certain viruses, and other pathogens that grow within cells, as well as for the rejection of tissue transplanted from another person.

Antigen-Presenting Cells T cells cannot respond to foreign antigens directly. Instead, the antigen must be "presented" to them by an antigen-presenting cell (APC). The most important APCs are **dendritic** (den-DRIT-ik) **cells**, large phagocytic cells derived from monocytes or lymphocytes and named for their many fibrous processes. Macrophages and some lymphocytes can also act as APCs.

An APC mobilizes the immune system against a particular foreign antigen. First, it ingests the foreign material, such as a disease organism, by enclosing it in a vesicle. As is typical in phagocytosis, this vesicle then merges with a lysosome filled with digestive enzymes that break down the organism into smaller fragments (FIG. 15-10). Specialized proteins then display the antigens on the plasma membrane—in a sense, advertising the presence of the invader to other immune cells. These proteins are known as major histocompatibility complex (MHC) molecules or human leukocyte antigens (HLAs).

11 Each T_h cell expresses a receptor that recognizes a specific foreign antigen bound to the MHC protein (see FIG. 15-10). When a T_h cell encounters an APC displaying its particular foreign antigen, the MHC/foreign antigen complex binds to the receptor and activates the T_h cell. Activated T_h cells produce cytokines that stimulate the production and activity of other lymphocytes.

CHECKPOINTS

- ☐ **15-13** What is adaptive immunity?
- ☐ **15-14** What is an antigen?
- ☐ **15-15** List four types of T cells.
- ☐ **15-16** What is the role of APCs in immunity?

CASEPOINT

- ☐ **15-8** Which type of T cell would be directly responsible for eliminating Lucas's virus-infected cells?

B CELLS AND ANTIBODIES

B cells (B lymphocytes) are the second main class of lymphocytes active in immunity. Whereas T cells mature in the thymus, B cells mature in the red bone marrow. B cells function in immunity by producing Y-shaped proteins called **antibodies (Ab)**, also known as *immunoglobulins* (Ig), in response to a foreign antigen. (Globulins, in general, are a class of folded proteins described in Chapter 2.)

Like T cells, B cells have surface receptors that bind with a specific type of antigen (FIG. 15-11). Exposure to the antigen stimulates the cells to rapidly multiply and produce

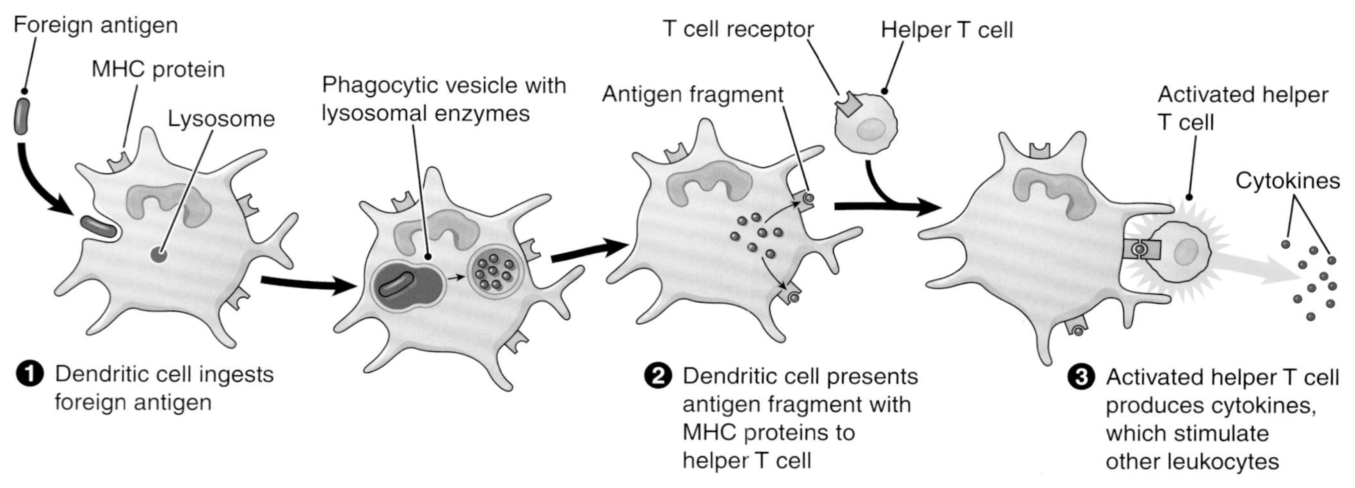

FIGURE 15-10 Activation of a helper T cell by a dendritic cell (antigen-presenting cell). KEY POINT An antigen-presenting cell (APC) displays digested foreign antigen on its surface along with self major histocompatibility complex (MHC) antigen. A helper T cell is activated by contact with this complex and produces stimulatory interleukins. ZOOMING IN What is contained in the lysosome that joins the phagocytic vesicle?

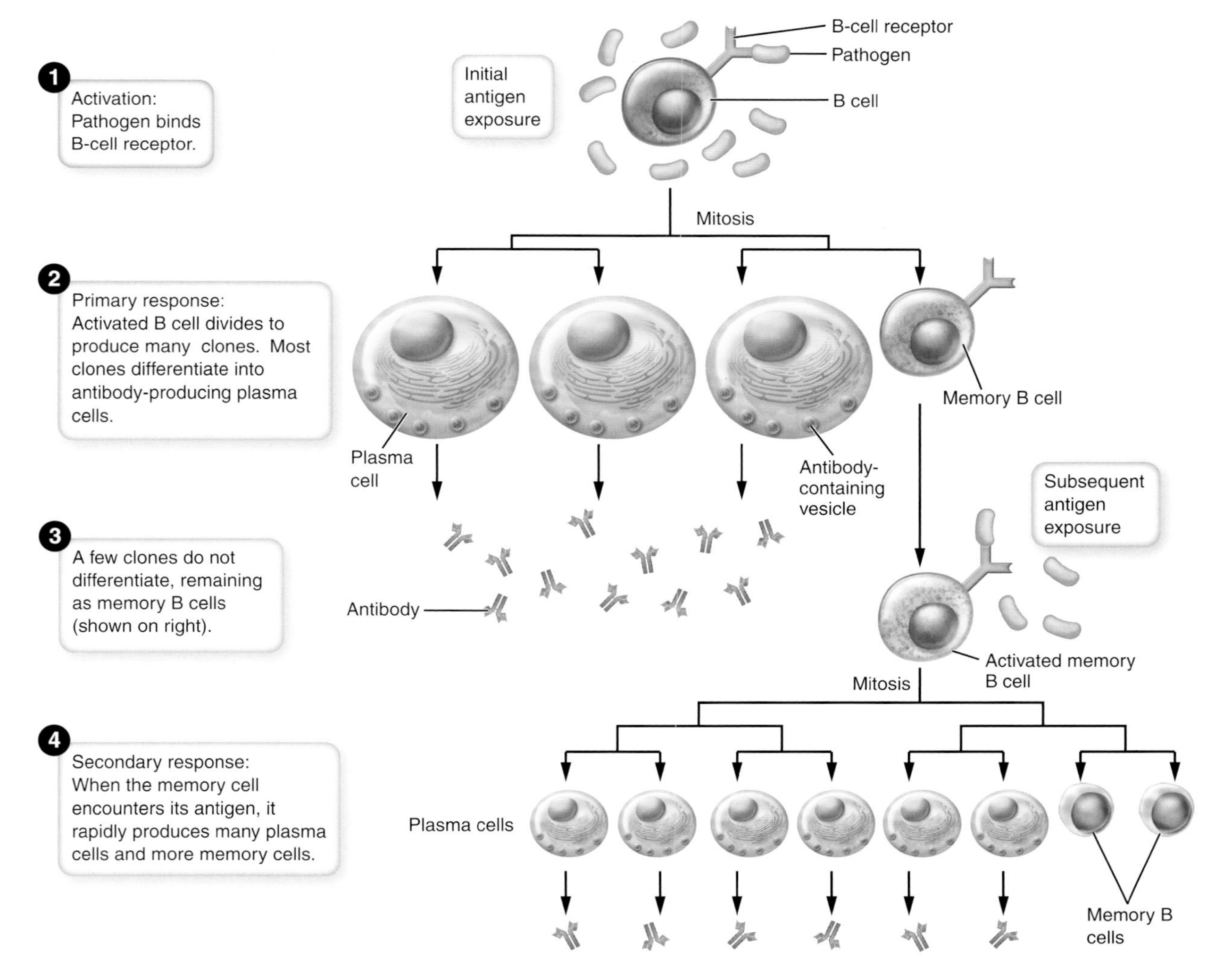

FIGURE 15-11 Activation of B cells. KEY POINT The B cell combines with a specific antigen. The cell divides to form plasma cells, which produce antibodies. Some of the cells develop into memory cells, which protect against reinfection. ZOOMING IN What two types of cells develop from activated B cells?

Table 15-2 Antigen–Antibody Interactions and Their Effects

Interaction	Effects
Prevention of attachment	A pathogen coated with antibody is prevented from attaching to a cell.
Clumping of antigen	Antibodies can link antigens together, forming a cluster that phagocytes can ingest.
Neutralization of toxins	Antibodies bind to toxin molecules to prevent them from damaging cells.
Help with phagocytosis	Phagocytes can attach more easily to antigens that are coated with antibody.
Activation of complement	When complement attaches to antibody on a cell surface, a series of reactions begins that activates complement to destroy cells.
Activation of NK cells	NK cells respond to antibody adhering to a cell surface and attack the cell.

large numbers (clones) of **plasma cells**. These mature cells produce antibodies against the original antigen and release them into the blood providing a form of immunity described as **humoral immunity** (the term humoral refers to body fluids). Humoral immunity is long term and generally protects against circulating antigens and bacteria that grow outside the cells. All antibodies are contained in a portion of the blood plasma called the **gamma globulin** fraction, obtained when plasma is separated in a laboratory into its different protein components.

The antibody that is produced in response to a specific antigen, such as a bacterial protein or a toxin, has a shape that matches some part of that antigen, much in the same way that a key's shape matches the shape of its lock. The antibody can bind only to the antigen that caused its production. Antibodies do not destroy cells directly; rather, they assist in the immune response. For example, they prevent a pathogen's attachment to a host cell; help with phagocytosis; activate NK cells; and neutralize toxins. The antigen–antibody complex may activate the complement system, which helps in immunity, as previously described. These antigen–antibody interactions are illustrated, and their protective effects are described in **TABLE 15-2**.

Notice in **FIGURE 15-11** that some of the activated B cells do not become plasma cells but, like certain T cells, become memory cells. These do not immediately produce antibodies but live on in lymphoid tissue and the bloodstream. Upon repeated contact with an antigen, they immediately begin dividing to produce many active plasma cells. Because of this "immunologic memory," one is usually immune to a childhood disease, such as chickenpox, after having it. Note, however, that these B cells do not live forever, which explains the need for booster vaccines to maintain immunity.

FIGURES 15-11 and 15-12 illustrate both the primary response, following the first encounter with an antigen, and the secondary response, which follows each successive encounter. There are five different classes of antibodies distinguished by their locations and functions. The antibodies in this figure are designated as IgM (immunoglobulin M) and IgG (immunoglobulin G). IgM is the first type of antibody produced in an immune response, followed shortly by IgG. A second encounter with the antigen stimulates production of both types of antibodies but has a much greater effect on IgG production. These and the other classes of Igs are described in **BOX 15-1**.

CHECKPOINTS 

- 15-17 What is an antibody?
- 15-18 What type of cells produce antibodies?

CASEPOINT 

- 15-9 Which cells of the immune system should make Lucas immune to a second mono infection?

See the Student Resources on thePoint® for the animation "Immune Response."

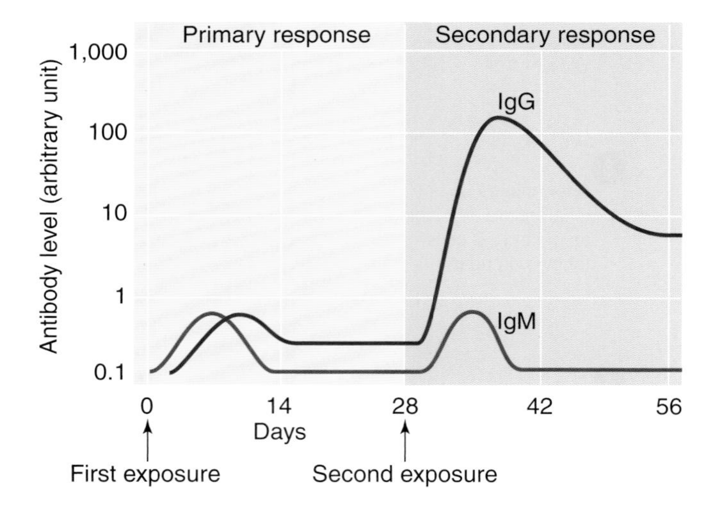

FIGURE 15-12 Production of antibodies (Ab). KEY POINT Antibodies or immunoglobulins (Ig) are produced in response to a first encounter with a foreign antigen. A second exposure produces a greater response. Immunoglobulin M (IgM) and immunoglobulin G (IgG) are two of the five classes of antibodies.

A CLOSER LOOK

Antibodies: A Protein Army that Fights Disease

BOX 15-1

Antibodies are proteins secreted by plasma cells (activated B cells) in response to specific antigens. They are all contained in a fraction of the blood plasma known as gamma globulin. Because the plasma contains other globulins as well, antibodies have become known as Igs. Immunologic studies have shown that there are several classes of Igs that vary in molecular size and in function (see below). Studies of these antibody fractions can be helpful in making a diagnosis. For example, high levels of IgM antibodies, because they are the first to be produced in an immune response, indicate a recent infection.

Class	Abundance	Characteristics and Function
IgG	75%	Found in the blood, lymph, and intestines Enhances phagocytosis, neutralizes toxins, and activates complement Crosses the placenta and confers passive immunity from mother to fetus
IgA	15%	Found in glandular secretions such as sweat, tears, saliva, mucus, and digestive juices Provides local protection in mucous membranes against bacteria and viruses Also found in breast milk, providing passive immunity to newborn
IgM	5%–10%	Found in the blood and lymph The first antibody to be secreted after infection Stimulates agglutination and activates complement
IgD	less than 1%	Located on the surface of B cells
IgE	less than 0.1%	Located on basophils and mast cells Active in allergic reactions and parasitic infections

TYPES OF ADAPTIVE IMMUNITY

As we have just seen, adaptive immunity may develop naturally through contact with a specific disease organism. In this case, the infected person's T cells and antibodies act against the infecting agent or its toxins. The infection that triggers the immunity may be so mild as to cause no symptoms (subclinical). Nevertheless, it stimulates the host's immune response. Moreover, each time a person is invaded by the disease organism, his or her cells will respond to the infection. Such immunity may last for years and in some cases for life. Because the host is actively involved in generating protection, this type of immunity is described as *active*. Because the immunity is formed against harmful agents encountered in the normal course of life, it is called **natural active immunity** (FIG. 15-13).

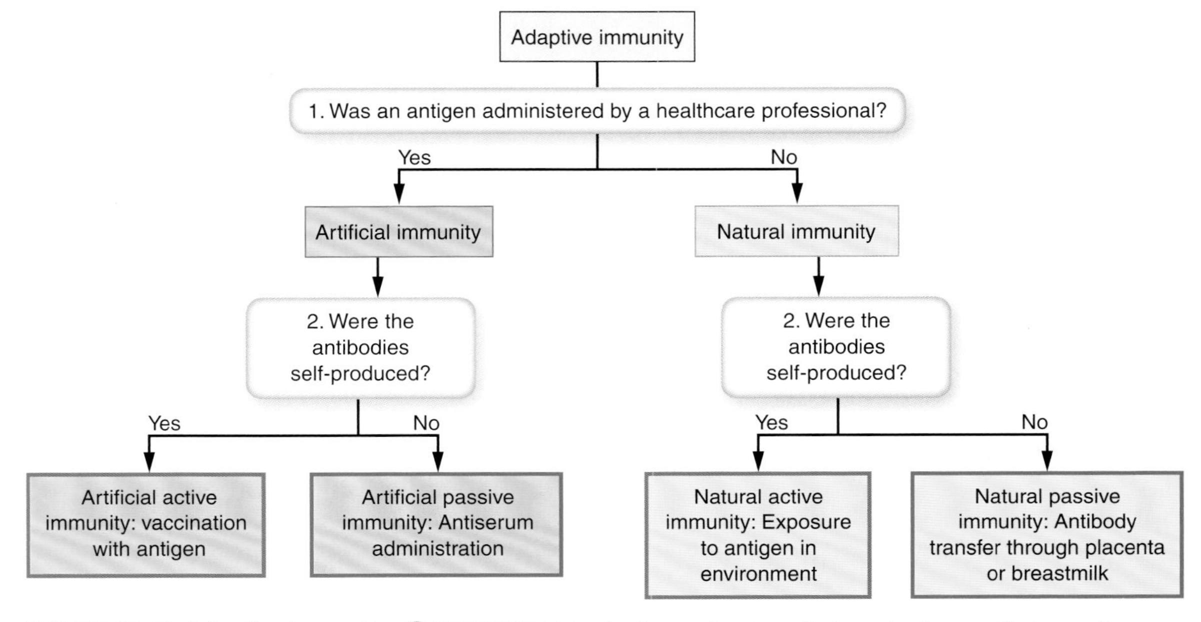

FIGURE 15-13 Adaptive immunity. **KEY POINT** Adaptive immunity, also called acquired or specific immunity, is a response to a particular antigen. Adaptive immunity may develop naturally (as by contact with the pathogen) or by artificial means (as by vaccination). Both natural and artificial adaptive immunity may be active (generated by the individual) or passive (provided from an outside source).

Immunity also may be acquired naturally by the passage of antibodies from a mother to her fetus through the placenta. Because these antibodies come from an outside source, this type of immunity is called **natural passive immunity**. The antibodies obtained in this way do not last as long as actively produced antibodies, but they do help protect the infant for about six months, by which time the child's own immune system begins to function. Nursing an infant can lengthen this protective period because the mother's specific antibodies are present in her breast milk and colostrum (the first breast secretion). These are the only known examples of naturally acquired passive immunity.

A person who has not been exposed to a particular pathogen has no antibodies or T cells against that organism and may be defenseless against infection. Therefore, medical personnel may use artificial measures to establish immunity. As with naturally acquired immunity, artificially acquired immunity may be active or passive (see FIG. 15-13). Artificial active immunity results from the use of a **vaccine** (vak-SENE), a prepared substance that initiates an immune response against a particular pathogen. Administration of a vaccine is described as **immunization** or *vaccination* (vak-sih-NA-shun). This type of immunity is described as active because the recipient's own immune system is at work. Artificial passive immunity, on the other hand, involves the administration of antibodies obtained from an outside source, a preparation known as an antiserum. Vaccines and antisera are discussed in greater detail next.

CASEPOINT

- [ ] **15-10** What type of immunity will Lucas develop against the virus that causes mono?

VACCINES

The administration of a vaccine is a preventive measure designed to provide protection in anticipation of contact with a specific disease organism. Originally, the word *vaccination* meant inoculation against smallpox. (The term even comes from the Latin word for cow, referring to cowpox, which was used to vaccinate against smallpox.) According to the World Health Organization, however, smallpox has now been eliminated as a result of widespread immunization programs. Mandatory vaccination against smallpox has been discontinued because the chance of adverse side effects from the vaccine is thought to be greater than the probability of contracting the disease.

All vaccines carry a small risk of adverse side effects and may be contraindicated in some cases. In general, however, vaccines are extensively tested for safety, and for most people, their potential benefits far outweigh their risks. Moreover, the vaccination of most people in a given area protects individuals who cannot be vaccinated for some reason because the pathogen will lack a susceptible population in which to spread. Declining levels of pertussis (whooping cough) vaccinations, for instance, have recently led to increased infection rates.

Types of Vaccines The administration of virulent pathogens to stimulate immunity obviously would be dangerous. So, vaccines are designed to activate the immune system without causing a serious illness.

Vaccines can be made with live organisms or with organisms killed by heat or chemicals. If live organisms are used, they must be nonvirulent for humans, such as the cowpox virus used for smallpox immunization, or they must be treated in the laboratory to weaken them as human pathogens. An organism weakened for use in vaccines is described as **attenuated**. People who are immunosuppressed should not be given vaccines that contain a live virus. Also, pregnant women should not receive live virus vaccines because the virus could cross the placenta and harm the fetus.

Another type of vaccine is made from the toxin produced by a disease organism. The toxin is altered with heat or chemicals to reduce its harmfulness, but it can still function as an antigen to induce immunity. Such an altered toxin is called a **toxoid**.

The newest types of vaccines are produced from antigenic components of pathogens or by genetic engineering. By techniques of recombinant DNA, the genes for specific disease antigens are inserted into the genetic material of harmless organisms. The antigens produced by these organisms are extracted and purified and used for immunization. The hepatitis B vaccine is produced in this manner.

Boosters In many cases, an active immunity acquired by artificial (or even natural) means does not last a lifetime. Circulating antibodies can decline with time. To help maintain a high titer (level) of antibodies in the blood, repeated inoculations, called *booster shots*, are administered at intervals. The number of booster injections recommended varies with the disease and with an individual's environment or degree of exposure. On occasion, epidemics in high schools or colleges may prompt recommendations for specific boosters. TABLE 15-3 lists the vaccines currently recommended in the United States. The number and timing of doses vary with the different vaccines.

Nurse practitioners often prescribe and administer vaccines. See the Student Resources on thePoint® to read about this career and specifically about pediatric nurse practitioners.

ANTISERUM

Active immunity, either natural or artificial, requires several weeks or longer to fully develop. Therefore, a person who receives a large dose of virulent organisms and has no established immunity to them is in great danger. To prevent illness, the person must quickly receive counteracting antibodies from an outside source. This is accomplished through the administration of an **antiserum** or *immune serum*. The "ready-made" serum gives short-lived but effective protection against the organisms in the form of an artificially acquired passive immunity.

Table 15-3 Immunizations*

Vaccine	Disease(s)	Schedule
DTaP	Diphtheria, tetanus, pertussis (whooping cough)	2, 4, 6, and 15–18 months; 4–6 years; pregnant women Diphtheria and tetanus toxoid (Tdap) at 11–12 years; booster every 10 years
Hib	*Haemophilus influenza* type b (spinal meningitis)	2 and 4 months or 2, 4, and 6 months depending on type used
HepA	Hepatitis A virus	12–23 months; second dose 6 months later
HepB	Hepatitis B	Birth, 1–2 months, 6–18 months
Influenza	Influenza ("flu")	Yearly from 6 months to 6 years
MMR	Measles, mumps, rubella	15 months and 4–6 years
PCV	Pneumococcus (pneumonia, meningitis)	PCV13 at 2, 4, 6, and 12–15 months; also for immunocompromised adults and as first vaccine for adults over 65; PPSV23 for adults over 65 after PCV13
Polio vaccine (IPV)	Poliomyelitis	2 and 4 months, 6–18 months, and 4–6 years
Rotavirus (RV)	Rotavirus gastroenteritis	2 and 4 months or 2, 4, and 6 months, depending on the version used
Varicella	Chickenpox	12–15 months and 4–6 years
HPV	Human papillomavirus	Three doses administered between 11 and 26 years (sooner is better)
Meningococcal	Meningococcus	High risk: 9 months to 10 years Routine: 11–12 years, 16 years
Zoster	Shingles	After 50 years of age

*Recommended by the Advisory Committee on Immunization Practices (www.cdc.gov/vaccines/recs/acip), the American Academy of Pediatrics (www.aap.org), and the American Academy of Family Physicians (www.aafp.org). Information is also available through the National Immunization Program website (www.cdc.gov/vaccines).

Preparation of Antisera Traditionally, immune sera have been prepared in animals, mainly horses. These animals produce large quantities of antibodies in response to the injection of organisms or their toxins. Injecting humans with serum derived from animals has its problems, however, as the foreign proteins in animal sera may cause an often serious sensitivity reaction. Immunologists are instead promoting the use of antibodies prepared in the laboratory using genetic recombination techniques. Genes that direct antibody production are inserted into cell cultures that yield large quantities of these desired immunizing agents. The problem of animal sensitivity can also be avoided by using human antibodies contained in the gamma globulin fraction of the blood.

Concept Mastery Alert

Remember that antibodies are also called immunoglobulins. So, immunoglobulin administration results in passive immunity.

CHECKPOINTS 

- ☐ **15-19** What is the difference between active and passive adaptive immunity?
- ☐ **15-20** What is the difference between natural and artificial adaptive immunity?
- ☐ **15-21** What is a vaccine?
- ☐ **15-22** What is a booster?
- ☐ **15-23** What is an antiserum, and when are antisera used?

The Immune System and Cancer

Cancer cells differ slightly from normal body cells, and therefore, the immune system should recognize them as "nonself." The fact that people with AIDS and other immune deficiencies develop cancer at a higher rate than normal suggests that this is true. Cancer cells probably form

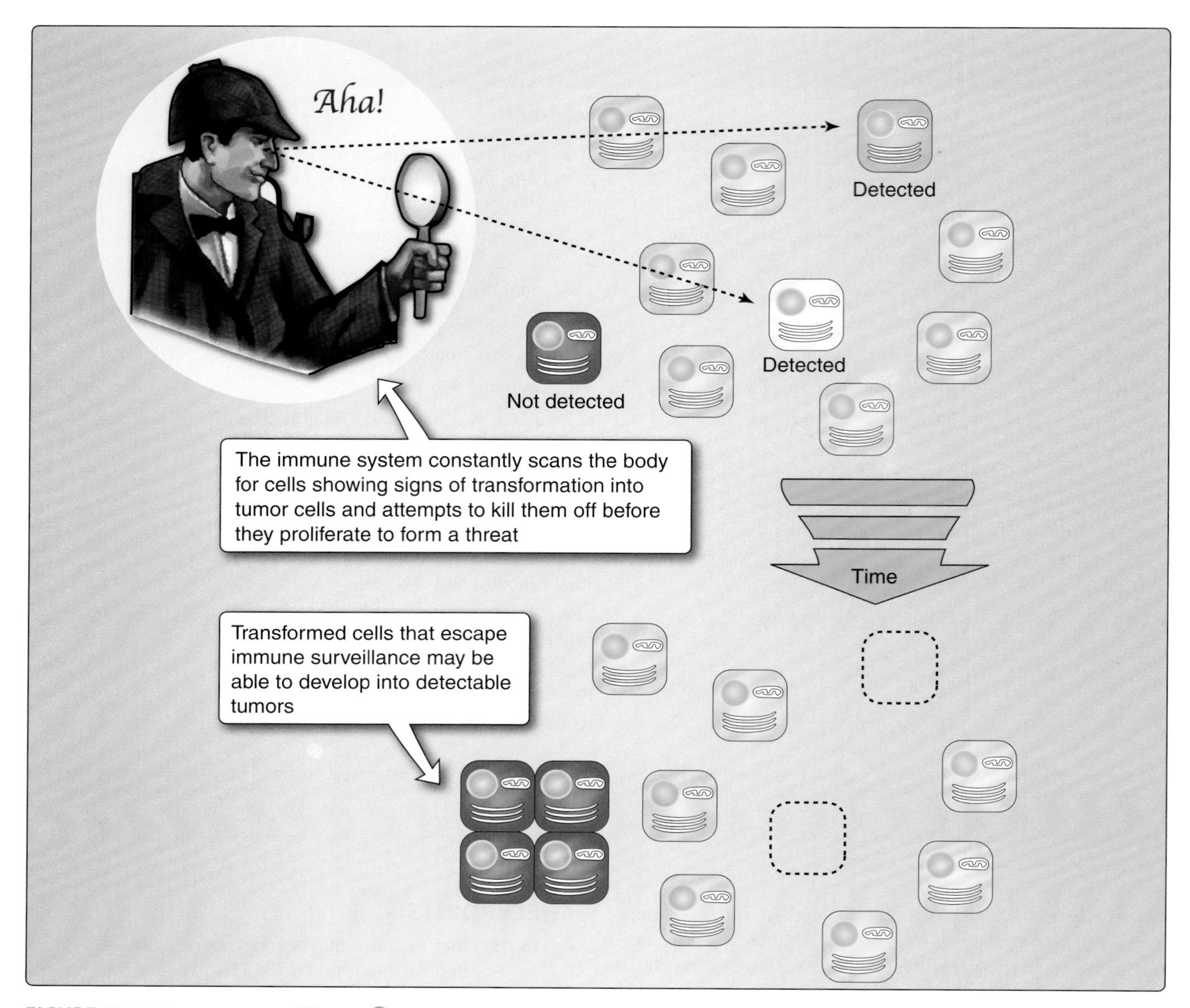

FIGURE 15-14 Immune surveillance. KEY POINT Tumors can occur when abnormal cells evade immune surveillance.

continuously in the body but normally are destroyed by NK cells, cytotoxic T cells, and macrophages in a process called **immune surveillance** (sur-VAY-lans) (FIG. 15-14). As a person ages, cell-mediated immunity declines, and cancer is more likely to develop.

Medical scientists are attempting to treat cancer by stimulating the patient's immune responses, a practice called **immunotherapy**. In one approach, T cells are removed from the patient, activated with cytokines, and then reinjected. This method has given some positive results, especially in treatment of melanoma, a highly malignant form of skin cancer. In the future, a vaccine against cancer may become a reality. Vaccines that target specific proteins produced by cancer cells have already been tested in a few forms of cancer. Read more about using the immune system to combat cancer in BOX 15-2.

HOT TOPICS
Harnessing the Immune System to Fight Cancer

BOX 15-2

Cancer therapy is beginning to emerge from its long history of treating the disease by harming cells, both normal and diseased, with chemotherapy and radiation. Investigators are trying to make treatment more individualized, for example, by selecting a chemotherapy drug based on the genetic makeup of a person's tumor cells. Other avenues of research involve using the immune system in a variety of approaches:

- Nonspecific boosters of the immune system involving substances active in immunity, such as cytokines, interleukins, and interferons, to promote a general increase in the immune response.
- Monoclonal antibodies (mAbs) made in the lab specifically target the antigens on cancer cells or mark them for immune attack. They can also deliver chemotherapeutic drugs or radiation directly to cancer cells.
- Immune checkpoint inhibitors block points on T cells that cancer cells use to inhibit an immune response.
- Vaccines can prevent or treat cancer. Vaccines for human papilloma virus (HPV) and hepatitis, for example, can prevent cervical and liver cancers, among others. Vaccines made from cancer cell components can increase the body's immune response to these abnormal cells. The process may be amplified by exposing patients' T cells to a cancer antigen in the lab and reintroducing them into the body.
- Genetically engineered T cells. These patient-derived blood cells are altered in the lab to produce special receptors on their surfaces called chimeric antigen receptors (CARs). The added proteins allow T cells to recognize a specific cancer cell antigen. The modified CAR-T cells are multiplied in the lab and then reintroduced to attack malignant cells. Further, the cells can continue to multiply throughout the body to combat any recurrences of the cancer.

Immunotherapy has been most successful so far in treating blood cancers, such as leukemia, cancer of white blood cells, and lymphoma, cancer of the lymphatic system and also the skin cancer, melanoma. Solid tumors present more difficulties. A danger in these approaches is overstimulating the immune system so that it attacks normal body cells. The trick is finding an antigen that is unique to the cancer cells, which after all, arise from normal body cells. Future medical studies hold the promise of killing unwanted cells while leaving normal cells intact.

15

A & P in Action Revisited: Lucas's Mononucleosis

Lucas's throat culture came back positive for streptococcal pharyngitis. His blood tests showed a high white cell count with an increase in the number of lymphocytes. Lucas also tested positive in the monospot test for antibodies to the Epstein-Barr virus (EBV), the usual cause of infectious mononucleosis. That result in combination with the sore throat (pharyngitis), enlarged lymph nodes and spleen, fever, fatigue, and loss of appetite all pointed to mono. Dr. Fischer told Lucas that this illness is frequently spread through contact with oral secretions and that sharing the new band member's tuba might not have been a good idea.

"We can treat your symptoms and offer support, but there is no cure for mono. We just have to let it run its course" Dr. Fischer said. "You will be tired for a couple of weeks. That means you need to get lots of rest and drink lots of fluids." He explained that Lucas would have to avoid strenuous exercise and marching band practice for three weeks and contact sports for an additional week due to the enlarged spleen. Once an ultrasound showed that this condition was resolved, he could return to his normal activities. Lucas was disappointed when he learned that he could not participate in practice for a while but was reassured that he would recover completely and most likely be able to participate in the band competitions the following month.

CHAPTER 15

Chapter Wrap-Up

OVERVIEW

A detailed chapter outline with space for note-taking is on thePoint®. The figure below illustrates the main topics covered in this chapter.

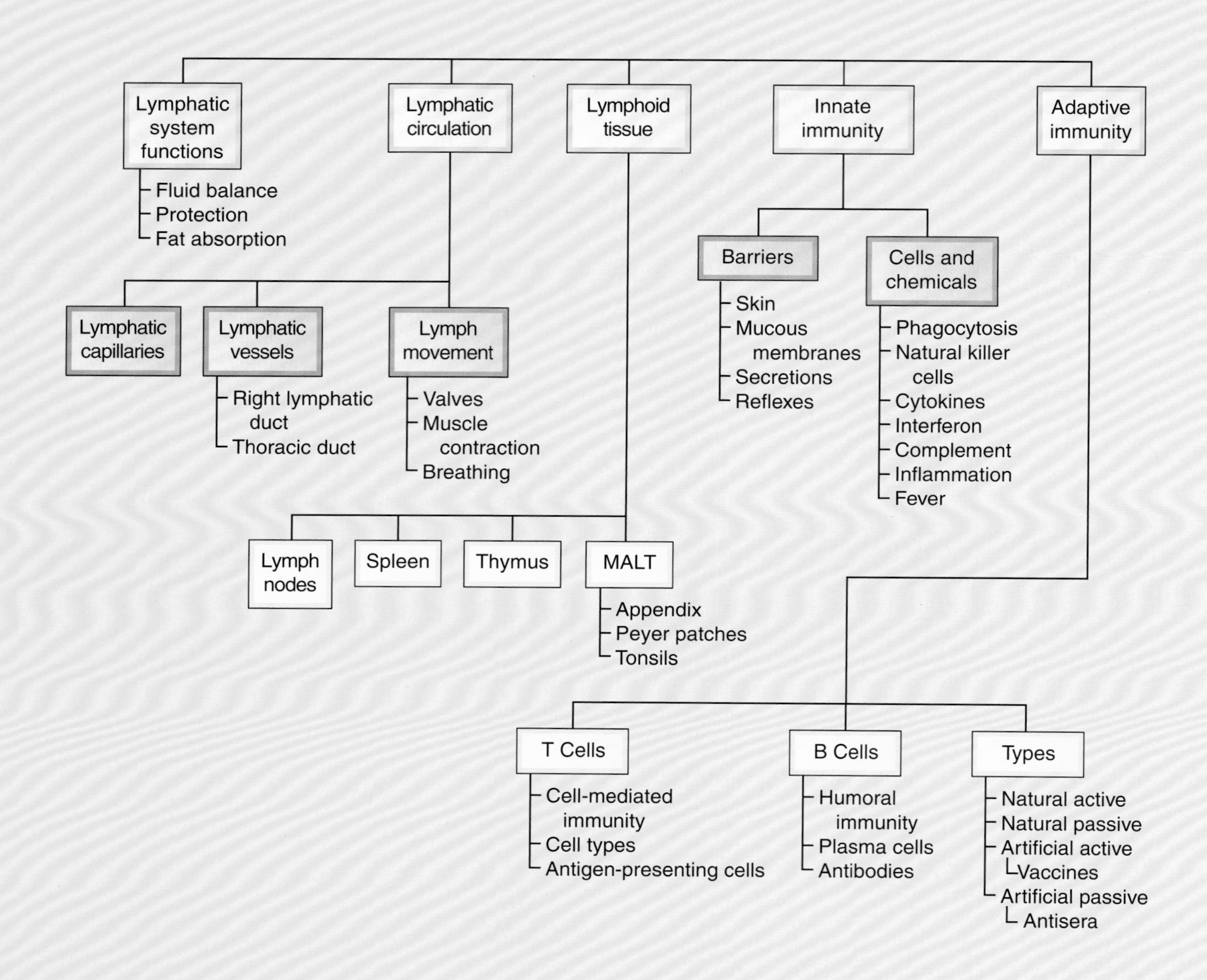

KEY TERMS

The terms listed below are emphasized in this chapter. Knowing them will help you organize and prioritize your learning. These and other boldface terms are defined in the Glossary with phonetic pronunciations.

adenoid
antibody
antigen
antiserum
antitoxin
attenuated
B cell
chyle
complement
cytokine
dendritic cells
fever
GALT
gamma globulin
immunity
immunization
immunoglobulin
inflammation
interferon
interleukin
lymph
lymphadenopathy
lymphatic duct
lymph node
macrophage
MALT
mast cell
natural killer (NK) cell
phagocytosis
plasma cell
spleen
T cell
thymus
tonsil
toxin
toxoid
vaccine

WORD ANATOMY

Medical terms are built from standardized word parts (prefixes, roots, and suffixes). Learning the meanings of these parts can help you remember words and interpret unfamiliar terms.

WORD PART	MEANING	EXAMPLE
LYMPHOID TISSUE		
aden/o	gland	The *adenoids* are glandlike tonsils.
lingu/o	tongue	The *lingual* tonsils are at the back of the tongue.
-oid	like, resembling	*Lymphoid* tissue makes up the specialized organs of the lymphatic system.

QUESTIONS FOR STUDY AND REVIEW

BUILDING UNDERSTANDING

Fill in the Blanks

1. The fluid that circulates in the lymphatic system is called _____.
2. Digested fats enter the lymphatic circulation through vessels called _____.
3. Digested fats and lymph combine to form a milky fluid called _____.
4. All antibodies are contained in a portion of the blood plasma termed the _____.
5. Any substance that induces an immune response is called a(n) _____.

Matching Match each numbered item with the most closely related lettered item.

___ 6. Destroy foreign cells directly
___ 7. Release interleukins, which stimulate other cells to join the immune response
___ 8. Suppress the immune response in order to prevent overactivity
___ 9. Recognize an antigen and start a rapid response if the antigen is contacted again
___ 10. Produce antibodies when activated by antigens

a. regulatory T cells
b. memory cells
c. cytotoxic T cells
d. B cells
e. helper T cells

Multiple Choice

___ 11. Compared to plasma, lymph contains much less:
a. fat
b. carbohydrate
c. protein
d. water

___ 12. Which vessel returns lymph from the lower extremities to the cardiovascular system?
a. appendix
b. lacteal
c. right lymphatic duct
d. thoracic duct

___ 13. Which tonsil is located behind the nose on the posterior wall of the upper pharynx?
a. appendix
b. lingual
c. palatine
d. pharyngeal

___ 14. Which of the following is NOT part of barriers, the first line of defense?
a. tears
b. saliva
c. complement
d. skin

___ 15. Which of the following is an active phagocyte?
a. NK cell
b. plasma cell
c. macrophage
d. toll-like receptor

___ 16. Which cell matures in the thymus?
a. T cell
b. B cell
c. dendritic cell
d. mast cell

UNDERSTANDING CONCEPTS

17. How does the structure of lymphatic capillaries correlate with their function? List some differences between lymphatic and blood capillaries.
18. Describe three mechanisms that propel lymph through the lymphatic vessels.
19. Trace a globule of fat from a lacteal in the small intestine to the right atrium.
20. Describe the structure of a typical lymph node.
21. Name two examples of GALT.
22. Define cytokine. List some roles of cytokines, and give two examples.
23. What are the symptoms of inflammation, and what causes them?
24. Differentiate between the terms in each of the following pairs:
    a. first and second lines of defense
    b. innate immunity and adaptive immunity
    c. cell-mediated immunity and humoral immunity
    d. active immunity and passive immunity
25. Describe the events that must occur for a T cell to react with a foreign antigen. Once activated, what do the T cells do?
26. What role do antibodies play in immunity? How are they produced? How do they work?

CONCEPTUAL THINKING

27. Explain the absence of arteries in the lymphatic circulatory system.
28. Name several locations of MALT, and explain the significance of those locations.
29. Maria's young son wakes during the night with a low-grade fever. Give some reasons why she should take steps to limit the fever. Give some reasons why she might not want to eliminate his fever.

For more questions, see the Learning Activities on thePoint®.

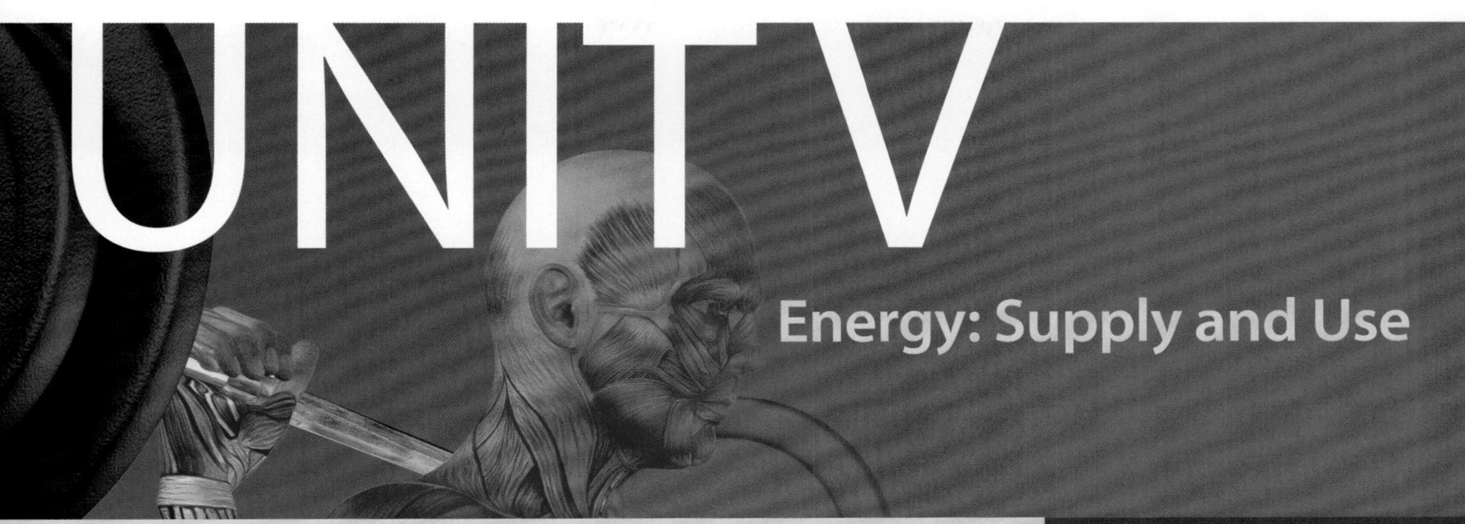

UNIT V

Energy: Supply and Use

CHAPTER 16

The Respiratory System

Learning Objectives

After careful study of this chapter, you should be able to:

1. Define respiration, and describe its four phases. ***p. 330***
2. Name and describe all the structures of the respiratory system. ***p. 331***
3. Explain the mechanism for pulmonary ventilation. ***p. 335***
4. Discuss the processes of internal and external gas exchange. ***p. 338***
5. List the ways in which oxygen and carbon dioxide are transported in the blood. ***p. 340***
6. Describe factors that control ventilation. ***p. 341***
7. Describe the negative feedback loop regulating arterial CO_2 levels. ***p. 342***
8. Describe normal and altered breathing patterns. ***p. 343***
9. Compare hyperventilation and hypoventilation. ***p. 344***
10. Referring to the case study, discuss how asthma can be diagnosed and treated. ***pp. 329, 345***
11. Show how word parts are used to build words related to respiration (see "Word Anatomy" at the end of the chapter). ***p. 347***

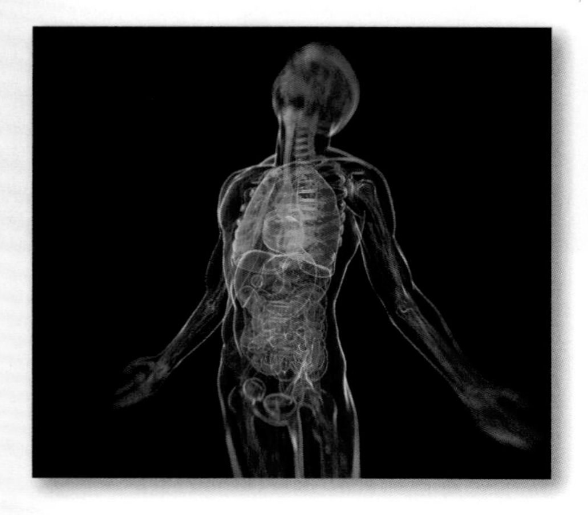

A & P in Action
Emily's Case: Advances in Asthma Therapy

"Remind me to mention to Dr. Martinez that Emily still has that nagging cough," Nicole told her husband.

"I've been worried about that," he replied. "You know, I had asthma as a kid—I hope she doesn't. I could hardly do any sports without taking a couple puffs of my inhaler."

Later that week, Dr. Martinez listened carefully to 3-year-old Emily's lungs. He knew that the common symptoms of asthma—coughing, wheezing, and shortness of breath—were due to swelling of the airway tissues and spasm of the smooth muscle around them.

"I don't hear any wheezing, but given the family history, we can't rule out asthma," he said. "In addition to the genetic component, asthma can have several environmental triggers such as respiratory infections, allergies, cold air, and exercise."

"Well," replied Nicole, "Emily did have a cold right before the coughing began. I haven't noticed any allergies, but now that I think about it, she did have a persistent cough last winter too. And she is getting lots of exercise at preschool and dance class. If Emily does have asthma, will that limit her activities?"

"The asthma drugs we have now are much improved since your husband's youth," the doctor replied. "But first we need to figure out what is causing Emily's cough. I'm going to order a chest x-ray to rule out infection. She's a bit young, but we'll try to measure her lung function using a spirometry test. I know a clinic that specializes in preschool children. I'm also going to ask you to monitor her for the next few weeks and see if anything exacerbates her symptoms. If she has asthma, we'll start daily treatment with a short-acting inhaler to relax Emily's airway muscles and ease breathing. If needed, we can add inhaled corticosteroids to control airway inflammation. This is jumping ahead, but we could also go with a long-term oral medication that prevents the lungs from producing leukotrienes, substances that cause constriction of smooth muscle in the airways."

Asthma is the most common chronic respiratory disease of childhood. In this chapter, we'll examine the respiratory system and its involvement in this disease. Later in the chapter, we'll check in on Emily and learn about other medications used to treat asthma.

As you study this chapter, CasePoints will give you opportunities to apply your learning to this case.

Visit thePoint® to access the following resources. For guidance in using these resources most effectively, see pp. xv–xvii.

Preparing to Learn

- Tips for Effective Studying
- Pre-Quiz

While You Are Learning

- Web Figure: Principal Muscles of Breathing and Lateral Chest
- Animation: The Lungs
- Animation: Pulmonary Ventilation
- Animation: Oxygen Transport
- Animation: Carbon Dioxide Exchange
- Chapter Notes Outline
- Audio Pronunciation Glossary

When You Are Reviewing

- Answers to Questions for Study and Review
- Health Professions: Respiratory Therapist
- Interactive Learning Activities

A LOOK BACK

The respiratory system works together with the cardiovascular system to move gases between the atmosphere and body cells. The key idea of flow down gradients 5, in relation to both diffusion (Chapter 3) and flow through tubes (Chapter 14), is thus fundamental to this chapter. The overall purpose of gas flow is to maintain homeostasis of body fluids 3. Revisiting the key ideas of barriers 4, water 6, enzymes 7, causation 12, and structure–function 1 can help you master the intricacies of this important body system.

Phases of Respiration

Most people think of respiration simply as the process by which air moves into and out of the lungs, that is, *breathing*. By scientific definition, **respiration** is the process by which oxygen is obtained from the environment and delivered to the cells. Carbon dioxide is transported to the outside in a reverse pathway (FIG. 16-1).

The coordinated actions of the respiratory and cardiovascular systems accomplish the four phases of respiration:

- **Pulmonary ventilation**, which is the exchange of air between the atmosphere and the air sacs (alveoli) of the lungs. This is normally accomplished by the inhalation and exhalation of breathing.
- **External gas exchange**, which occurs in the lungs as oxygen (O_2) diffuses from the air sacs into the blood and carbon dioxide (CO_2) diffuses out of the blood to be eliminated.
- **Gas transport in the blood.** The circulating blood carries gases between the lungs and the tissues, supplying oxygen to the cells and bringing back carbon dioxide.
- **Internal gas exchange**, which occurs in the tissues as oxygen diffuses from the blood to the cells, whereas carbon dioxide travels from the cells into the blood.

The term *respiration* is also used to describe a related process that occurs at the cellular level. In *cellular*

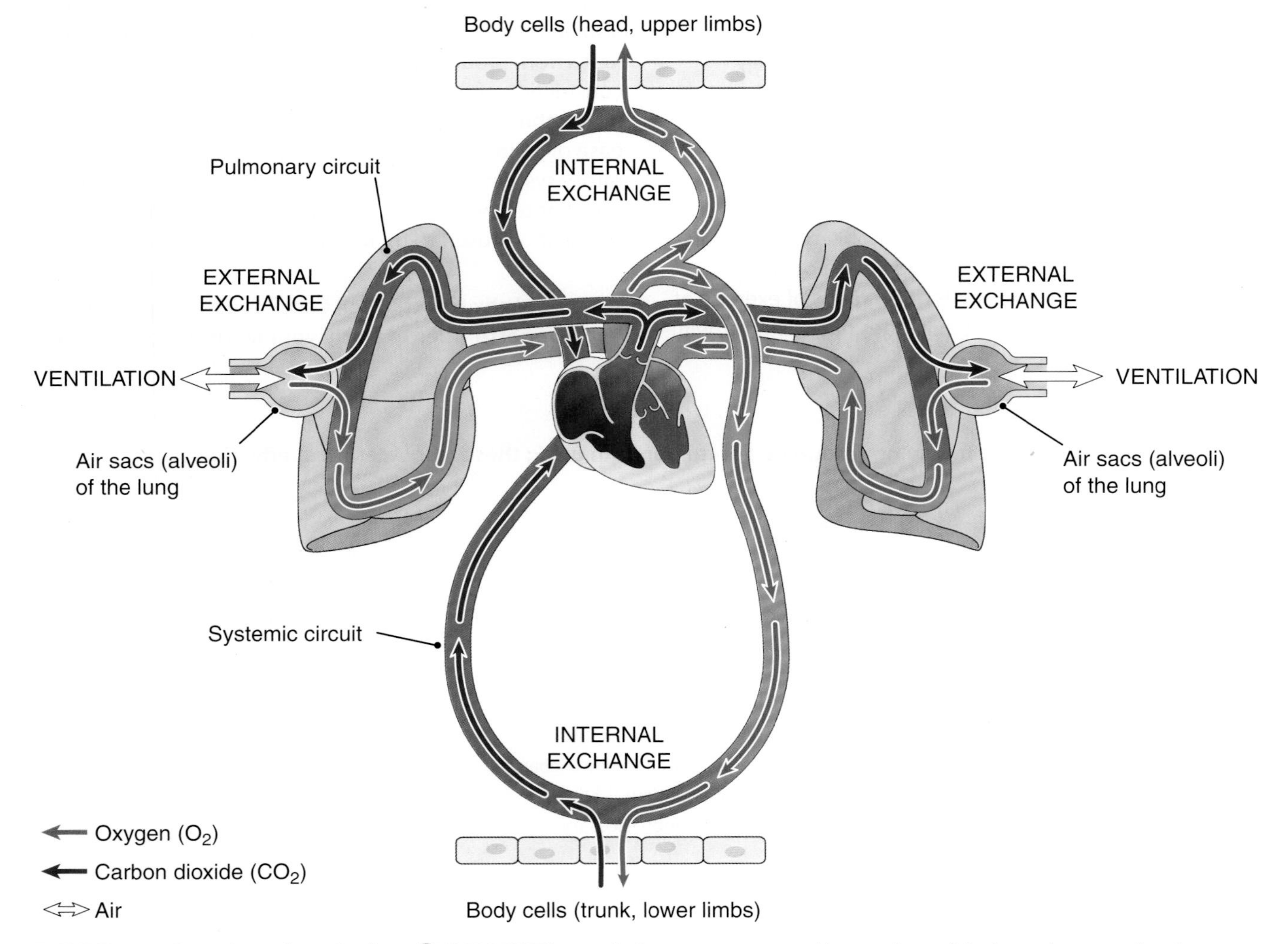

FIGURE 16-1 Overview of respiration. KEY POINT In ventilation, gases are moved into and out of the lungs. In external exchange, gases move between the air sacs (alveoli) of the lungs and the blood. In internal exchange, gases move between the blood and body cells. The circulation transports gases in the blood. ZOOMING IN From which side of the heart does blood leave to travel to the lungs? To which side does it return?

respiration, the cells take in oxygen and use it in the breakdown of nutrients. In this process, the cells release energy and carbon dioxide, a waste product of cellular respiration, as described in Chapter 18.

CHECKPOINT 

☐ **16-1** What are the four phases of respiration?

CASEPOINT 

☐ **16-1** What phase of respiration is affected by Emily's asthma?

Structure of the Respiratory System

The respiratory system is a complex series of spaces and passageways that conduct air into and through the lungs (**FIG. 16-2**). These spaces include the nasal cavities; the pharynx, which is common to the digestive and respiratory systems; the voice box or larynx; the windpipe or trachea; and the lungs themselves, with their conducting tubes and air sacs.

THE NASAL CAVITIES

Air enters the respiratory system through the openings in the nose called the **nostrils** or *nares* (NA-reze) (sing. naris). Immediately, inside the nostrils, located between the roof of the mouth and the cranium, are the two spaces known as the **nasal cavities**. These two spaces are separated from each other by a partition, the **nasal septum**. The septum's superior portion is formed by a thin plate of the ethmoid bone, and the inferior portion is formed by the vomer (**see FIG. 6-6A** in Chapter 6). An anterior extension of the septum is made of hyaline cartilage.

The nasal cavity is lined by a mucous membrane containing stratified squamous epithelium. The many layers of this epithelium guard against abrasion, microbes, and other environmental insults.

On the lateral walls of each nasal cavity are three projections called the **conchae** (KONG-ke) (**see FIG. 6-6A and C** and later **FIG. 16-12**). The shelllike conchae greatly increase the surface area of the mucous membrane over which air travels on its way through the nasal cavities. This membrane contains many blood vessels that deliver heat and moisture. The membrane's cells secrete a large amount of fluid—up to 1 qt each day. Breathing through the nose (instead of the mouth) enables the nasal mucosa to warm and moisten inhaled air. Moreover, foreign bodies, such as dust and pathogens, can be trapped in the nasal hairs or the surface mucus.

As noted in Chapter 6, the paranasal sinuses are small cavities in the skull bones near the nose (**see FIG. 6-5**). They are resonating chambers for the voice and lessen

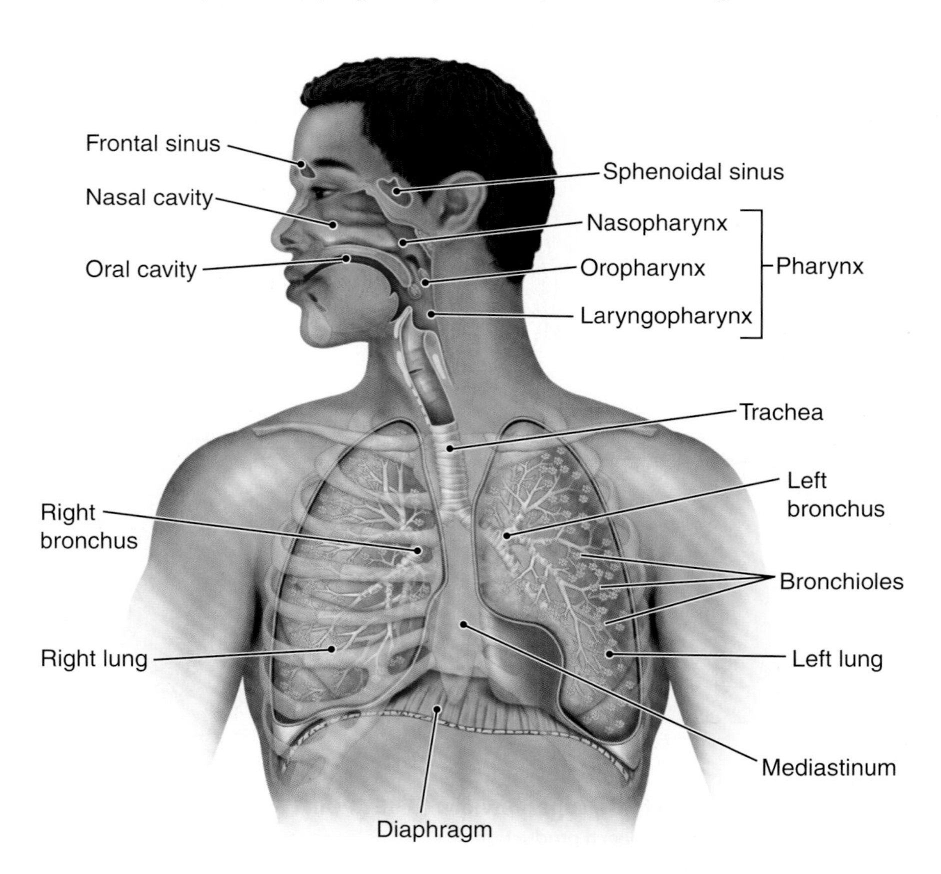

FIGURE 16-2 The respiratory system. KEY POINT The respiratory system consists of a series of airways that finally branch through the lungs. ZOOMING IN What organ is located in the medial depression of the left lung?

16

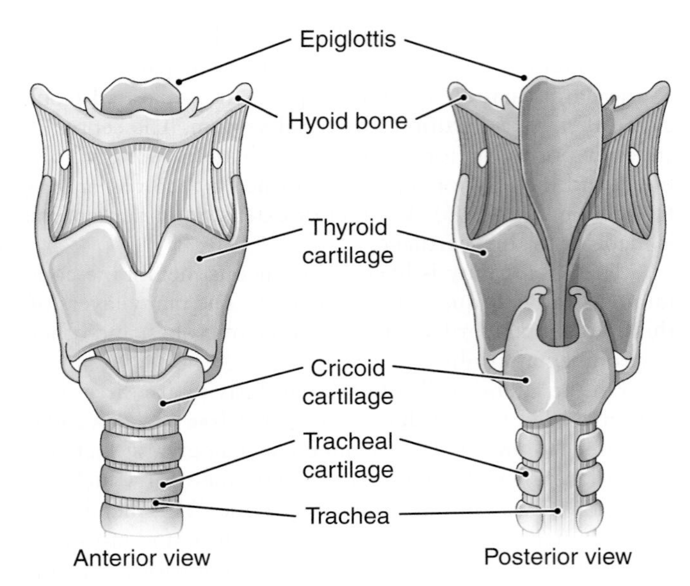

FIGURE 16-3 The larynx. KEY POINT The larynx is reinforced by hyaline cartilage, as is the trachea. The laryngeal cartilages include the epiglottis, thyroid cartilage, and cricoid cartilage.

the skull's weight. These sinuses are lined with mucous membrane and communicate with the nasal cavities. They are highly susceptible to infection traveling from the nose and throat.

THE PHARYNX

The muscular **pharynx** (FAR-inks), or throat, carries air into the respiratory tract and carries foods and liquids into the digestive system (see FIG. 16-2). The superior portion, located immediately behind the nasal cavity, is called the **nasopharynx** (na-zo-FAR-inks); the middle section, located posterior to the mouth, is called the **oropharynx** (o-ro-FAR-inks); and the most inferior portion is called the (lah-rin-go- FAR-inks). This last section opens into the larynx toward the anterior and into the esophagus toward the posterior.

THE LARYNX

The **larynx** (LAR-inks), commonly called the *voice box* (FIG. 16-3), connects the pharynx with the trachea. Its rigid framework consists of nine portions of hyaline cartilage. The anterior part, the thyroid cartilage, protrudes at the anterior of the neck. This projection is commonly called the *Adam's apple* because it is considerably larger in men than in women. The inferior cricoid (KRI-koyd) cartilage forms a ring below the thyroid cartilage. It is used as a landmark for medical procedures involving the trachea.

Folds of mucous membrane used in producing speech are located centrally in the superior larynx. These are the **vocal folds**, or *vocal cords* (FIG. 16-4), which vibrate as air flows over them. You can feel this vibration by placing your

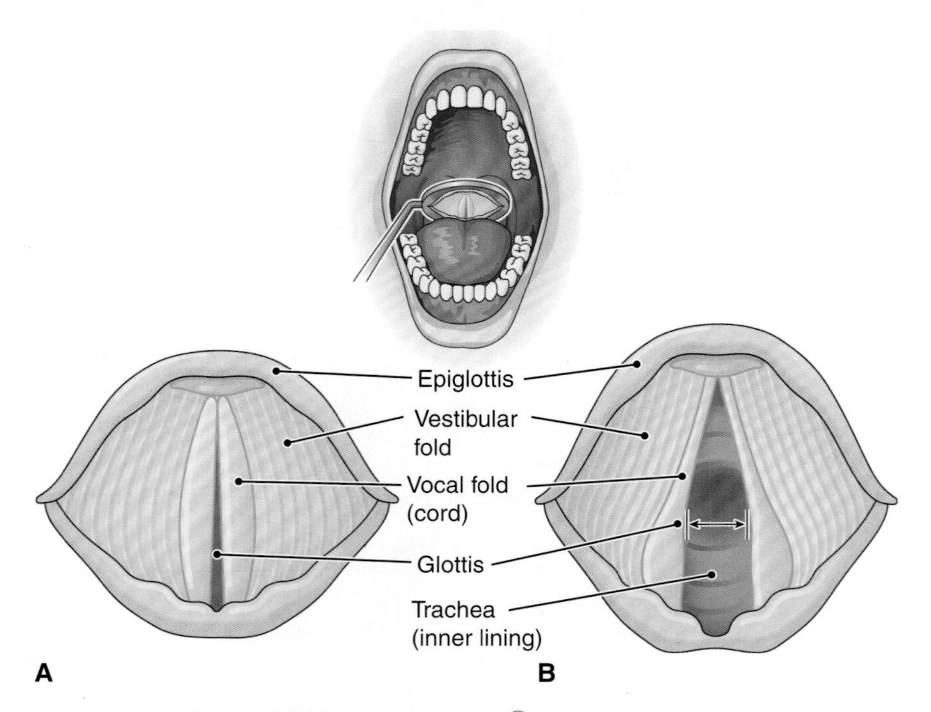

FIGURE 16-4 The vocal folds, superior view. KEY POINT The larynx contains the vocal folds (cords), used in speech production. The glottis is the space between the vocal folds. The vestibular folds help close off the glottis when necessary. **A.** The glottis in closed position. **B.** The glottis in open position. ZOOMING IN What cartilage is named for its position above the glottis?

fingertips over the larynx at the center of your anterior throat and saying "ah." Variations in the length and tension of the vocal folds and the distance between them regulate the pitch of sound. The amount of air forced over them regulates volume. 1 A difference in the size of the larynx and the vocal folds is what accounts for the difference between adult male and female voices. In general, a man's larynx is larger than a woman's. His vocal folds are thicker and longer, so they vibrate more slowly, resulting in a lower range of pitch. Muscles of the pharynx, tongue, lips, and face also are used to articulate words. The mouth, nasal cavities, paranasal sinuses, and the pharynx all serve as resonating chambers for speech, just as does the cabinet for an audio speaker. The space between the vocal folds is called the **glottis** (GLOT-is). This is partially open during normal breathing but widely open during forced breathing (see FIG. 16-4).

Superior to the vocal folds are additional folds in the laryngeal mucous membrane. These are known as the *vestibular folds* (see FIG. 16-4), sometimes called the "false vocal folds" because they do not contribute to speech production. Muscles in the larynx can bring these folds together to close off the glottis and help keep materials out of the respiratory tract during swallowing. They are also closed to help in holding one's breath against pressure in the thoracic cavity, as when straining to lift a heavy weight or to defecate.

The little leaf-shaped cartilage that covers the larynx during swallowing is called the **epiglottis** (ep-ih-GLOT-is). (The name means "above the glottis.") The glottis and epiglottis help keep food and liquids out of the remainder of the respiratory tract. As the larynx moves upward and forward during swallowing, the epiglottis moves downward, covering the opening into the larynx. You can feel the larynx move upward toward the epiglottis during this process by placing the pads of your fingers on your larynx as you swallow.

Despite laryngeal protections, it is possible to inhale or **aspirate** (AS-pih-rate) material into the respiratory tract. This might occur when someone is laughing or talking vigorously while eating, or an incapacitated individual inhales vomited gastric contents. Children can aspirate small objects or pieces of slippery food, such as hot dogs. Objects that enter the respiratory tract may cut off air supply completely, resulting in suffocation, or become lodged in the respiratory passageways. If the object is not removed, infection and inflammation due to irritation are likely to result.

THE TRACHEA

The **trachea** (TRA-ke-ah), commonly called the *windpipe*, is a tube that extends from the inferior edge of the larynx to the mediastinum, just superior to the heart (see FIG. 16-2). The trachea conducts air between the larynx and the lungs.

A framework of separate cartilages reinforces the trachea and keeps it open. These cartilages, each shaped somewhat like a tiny horseshoe or the letter C, are found along the trachea's entire length (see FIG. 16-3). The open sections in the cartilages are lined up at their posterior so that the esophagus can expand into this region during swallowing.

Concept Mastery Alert

Remember that cartilage rings brace only the trachea's anterior surface.

CHECKPOINTS 

- ☐ **16-2** What happens to air as it passes over the nasal mucosa?
- ☐ **16-3** What are the three regions of the pharynx?
- ☐ **16-4** What are the scientific names for the throat, voice box, and windpipe?

THE BRONCHI

At its inferior end, the trachea divides into two mainstem, or primary, **bronchi** (BRONG-ki), which enter the lungs (FIG. 16-5 and Dissection Atlas FIG. A3-8B). Cartilage rings stabilize the bronchi and keep them open to permit air passage. The right bronchus is considerably larger in diameter than is the left and extends downward in a more vertical direction (see FIG. 16-19B later in the chapter). Therefore, if a foreign body is inhaled, it is likely to enter the right lung. Each bronchus enters the lung at a depression called the *hilum* (HI-lum). Blood vessels and nerves also connect with the lung here and, together with the bronchus, make up a region known as the *root of the lung*.

THE LINING OF THE AIR PASSAGEWAYS

The trachea, bronchi, and other conducting passageways of the respiratory tract are lined with a mucous membrane (see FIG. 4-3A). Basically, it is simple, ciliated columnar epithelium with an underlying lamina propria, but the cells are arranged in such a way that they appear stratified. The tissue is thus described as *pseudostratified*, meaning "falsely stratified." Goblet cells within this epithelial membrane secrete mucus to trap impurities. The cilia on nearby cells sweep the mucus upward toward the throat, where it can be swallowed or eliminated by coughing, sneezing, or blowing the nose.

CASEPOINT

- ☐ **16-2** Which of the four characteristics of inflammation affects Emily's breathing?

THE LUNGS

The **lungs** contain both air passageways and minute, thin-walled sacs called *alveoli* (described later). They are the sites of external gas exchange, that is, the exchange of gases between air and blood. The two lungs are set side by side in the thoracic (chest) cavity (see FIG. 16-5A). Between them are the heart, the great blood vessels, and other organs of the mediastinum (the space and organs between the lungs) (Dissection Atlas FIG. A3-8A shows the lungs in position in the thoracic cavity). The left lung has an indentation on its medial side that accommodates the heart.

Divisions of the Lungs The right lung is subdivided by a horizontal and an oblique fissure into three lobes (superior,

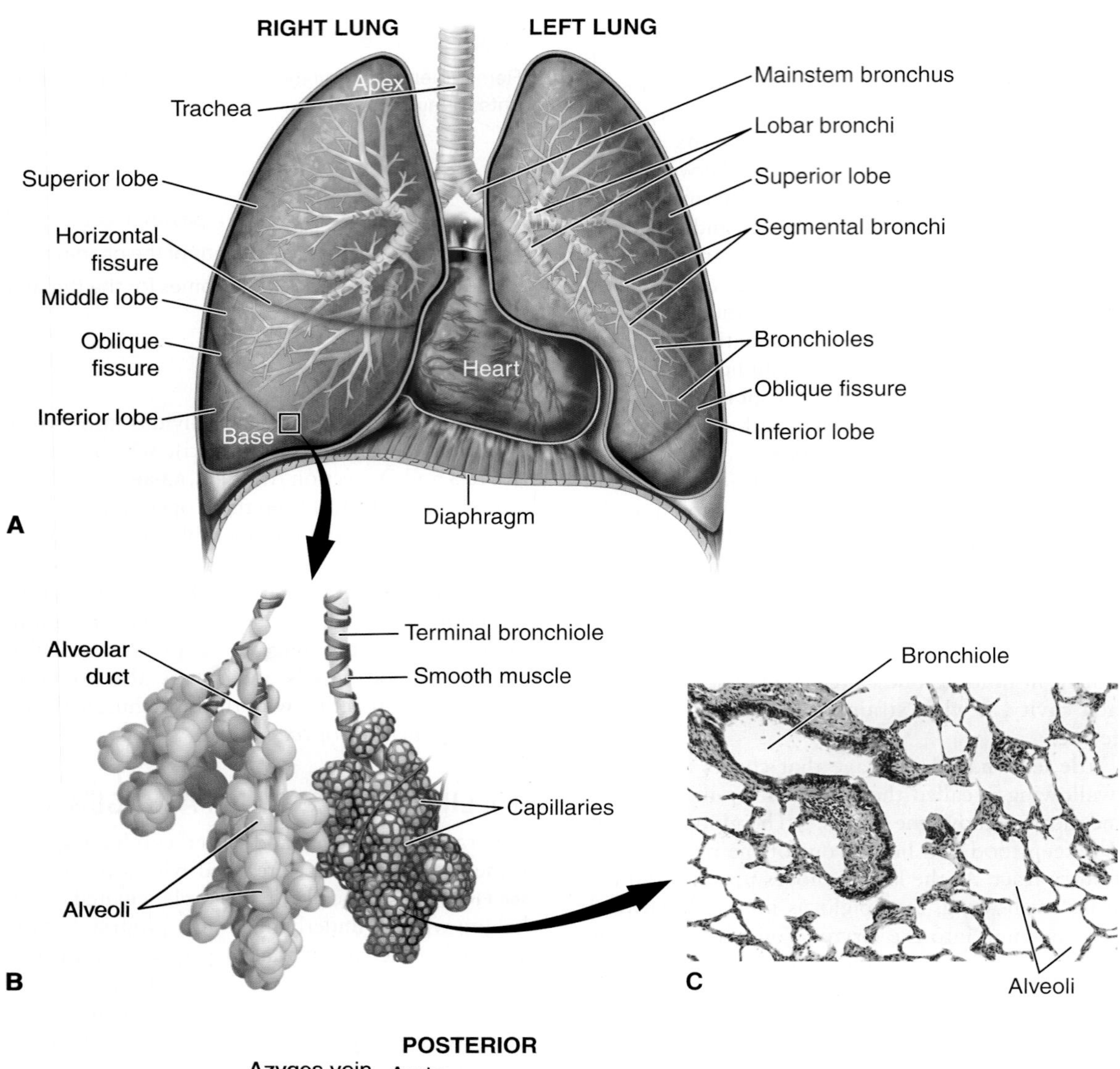

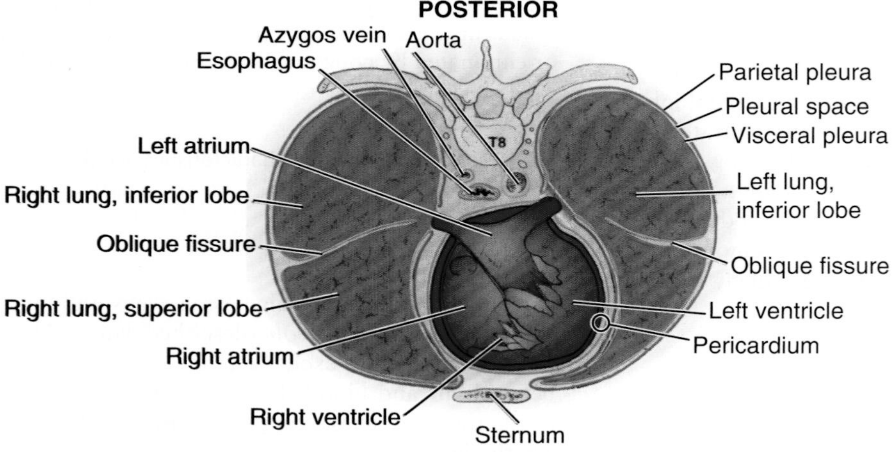

FIGURE 16-5 The lungs. KEY POINT The lungs are divided into lobes and segments that correspond to divisions of the bronchial tree. **A.** Position and structure of the lungs. **B.** Alveoli (air sacs). The left section of the diagram shows alveoli with capillaries removed. **C.** Histology of lung tissue. **D.** Cross-section of the thorax through the lungs (superior view), showing the visceral and parietal pleurae and the pleural space.

middle, and inferior); the left lung is divided by a single oblique fissure into two lobes (superior and inferior). Each lobe is then further subdivided into segments and then lobules. These subdivisions correspond to subdivisions of the bronchi as they branch throughout the lungs.

Each mainstem bronchus enters the lung at the hilum and immediately subdivides. The right bronchus divides into three lobar or secondary bronchi, each of which enters one of the right lung's three lobes. The left bronchus gives rise to two lobar bronchi, which enter the left lung's two lobes. The bronchi subdivide again and again, becoming progressively smaller as they branch through the lung tissue. The lobar bronchi become segmental bronchi as they branch into smaller segments of the lung. Because the bronchial subdivisions resemble the branches of a tree, they have been given the common name *bronchial tree*.

The Bronchioles The smallest of these conducting tubes are called **bronchioles** (BRONG-ke-oles). The histology of the tubes gradually changes as they become smaller. The amount of cartilage decreases until it is totally absent in the bronchioles; what remains is mostly smooth muscle, which is under the control of the autonomic (involuntary) nervous system.

The Alveoli At the end of the **terminal bronchioles**, the smallest subdivisions of the bronchial tree, there are clusters of tiny air sacs in which most external gas exchange takes place. These sacs, known as the **alveoli** (al-VE-o-li) (sing. alveolus), are covered by capillaries (see FIG. 16-5B). 4 Recall from Chapter 1 that a mucous membrane forms a barrier between outside air in the respiratory system and the blood and cells considered inside the body. This barrier thins to a single layer of flattened, squamous epithelium in the alveoli, in order to permit the exchange of gases between air and blood.

There are about 300 million alveoli in the human lungs. The resulting surface area in contact with gases approximates 60 m^2 (some sources say even more). This area is equivalent, as an example, to the floor surface of a classroom that measures about 24 by 24 ft. As with many other body systems, there is great functional reserve; we have about three times as much lung tissue as is minimally necessary to sustain life. Because of the many air spaces, the lung is light in weight; normally, a piece of lung tissue dropped into a glass of water will float. FIGURE 16-5C shows a microscopic view of the lung tissue.

The pulmonary circuit brings blood to and from the lungs for the purpose of external gas exchange—the transfer of oxygen to pulmonary blood and the removal of carbon dioxide. The systemic circuit also travels to the lungs for the same purposes as in other tissues—providing oxygen-rich blood and carrying away waste products for the lung cells themselves.

The Lung Cavities and Pleura The lungs occupy a considerable portion of the thoracic cavity, which is separated from the abdominal cavity by the muscular partition known as the **diaphragm** (DI-ah-fram). A continuous doubled sac, the **pleura** (PLU-ra), covers each lung (see FIG. 16-5D). As discussed in Chapter 4, the pleura is a serous membrane composed of an epithelial layer overlying areolar tissue. The two layers of the pleura are named according to location. The portion that is attached to the chest wall is the parietal pleura, and the portion that is attached to the lung surface is the visceral pleura. Each closed sac completely surrounds the lung, except at the hilum, where the bronchus and blood vessels enter the lung.

Between the two layers of the pleura is the **pleural space**, containing a thin film of fluid that lubricates the membranes. The effect of this fluid is the same as between two flat pieces of glass joined by a film of water; that is, the surfaces slide easily on each other but strongly resist separation. Thus, the lungs are able to expand and contract effortlessly during breathing, but the pleural fluid keeps them from separating from the chest wall.

CHECKPOINTS 

- ☐ **16-5** What structures does the inferior branching of the trachea form?
- ☐ **16-6** What feature of the cells lining the respiratory passageways enables them to move impurities away from the lungs?
- ☐ **16-7** In what structures does gas exchange occur in the lung?
- ☐ **16-8** What is the name of the membrane that encloses the lung?

CASEPOINT 

- ☐ **16-3** Emily's inhaler activates adrenergic receptors on the airways' smooth muscle cells. Would these receptors cause muscle contraction or relaxation?

16

The Process of Respiration

Respiration involves ventilation of the lungs, exchange of gases at the lungs and the body tissues, and their transport in the blood. 3 By controlling ventilation, the body maintains relatively constant levels of O_2 and CO_2 in arterial blood.

PULMONARY VENTILATION

Ventilation is the movement of air into and out of the lungs, normally accomplished by breathing. There are two phases of ventilation: **inhalation**, which is the drawing of air into the lungs, and **exhalation**, or expiration, which is the expulsion of air from the lungs. 5 Air flows down a pressure gradient from an area of higher pressure to an area of lower pressure. Thus, inhalation occurs when lung pressure drops below atmospheric pressure, and exhalation occurs when lung pressure rises above atmospheric pressure. As discussed next, we change lung pressure by altering lung volume.

Inhalation In inhalation, the active phase of quiet breathing, respiratory muscles of the thorax and diaphragm contract to enlarge the thoracic cavity (FIG. 16-6). During quiet breathing, the diaphragm's movement accounts for most of the increase in thoracic volume. The diaphragm is a strong, dome-shaped muscle attached to the body wall around the base of the rib cage. The diaphragm's contraction

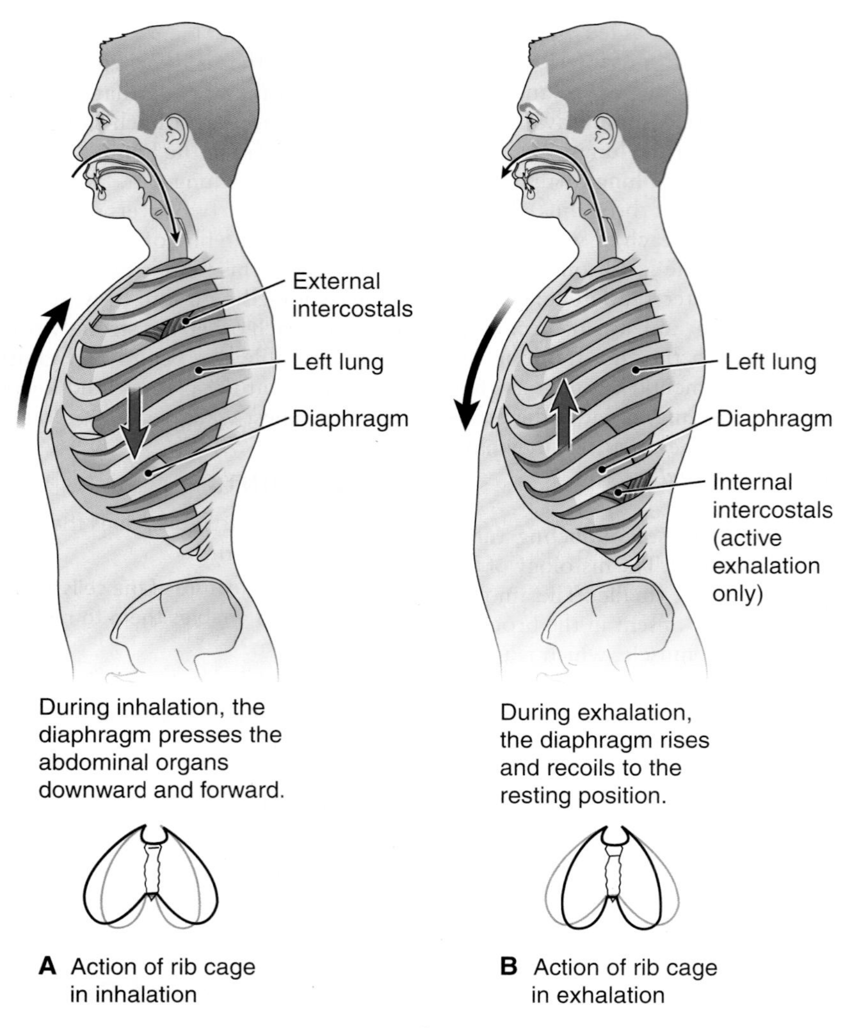

FIGURE 16-6 Pulmonary ventilation. KEY POINT The diaphragm and external intercostal muscles are involved in inhalation, the active phase of quiet breathing. These muscles relax in exhalation, the passive phase of quiet breathing. The *black arrow* shows the movement of the ribs, and the *green arrow* shows the movement of the diaphragm. **A.** Inhalation. **B.** Exhalation. ZOOMING IN What muscles are located between the ribs?

and flattening cause a piston-like downward motion that increases the chest's vertical dimension. Other muscles that participate in breathing are the intercostal muscles located between the ribs. The external intercostal muscles contract during inhalation, lifting the rib cage upward and outward. Put the palms of your hands on either side of the rib cage to feel this action as you inhale. During forceful inhalation, the rib cage is moved further up and out by contraction of muscles in the neck and chest wall. Paralysis or weakness in the respiratory muscles thus interferes with one's ability to breathe.

As the thoracic cavity increases in size, gas pressure within the cavity decreases. This phenomenon follows a law in physics stating that when the volume of a given amount of gas increases, the pressure of the gas decreases. Conversely, when the volume decreases, the pressure increases. If you push down on the plunger of a capped syringe, the gas particles will hit the wall of the syringe more frequently, creating greater pressure (FIG. 16-7A). 5 In the middle syringe, pressure is greater inside the syringe than outside. But, the resistance provided by the syringe cap prevents flow. If the resistance is removed by uncapping the syringe, air flows down the pressure gradient (bottom syringe).

FIGURE 16-7B illustrates the pressure gradients that enable ventilation. During inhalation, pressure in the chest cavity drops as the thorax expands. When the pressure drops to slightly below the air pressure outside the lungs, air moves down the pressure gradient into the lungs by suction. Air flows out of the lungs (exhalation) when pressure in the lungs exceeds that of the atmosphere. 12 When learning about ventilation, it can be difficult to understand if the change in lung volume is the cause or the result of air flow (see BOX 16-1 for an explanation).

Air enters the respiratory passages and flows through the ever-dividing tubes of the bronchial tree. As the air travels this route, it moves more and more slowly through

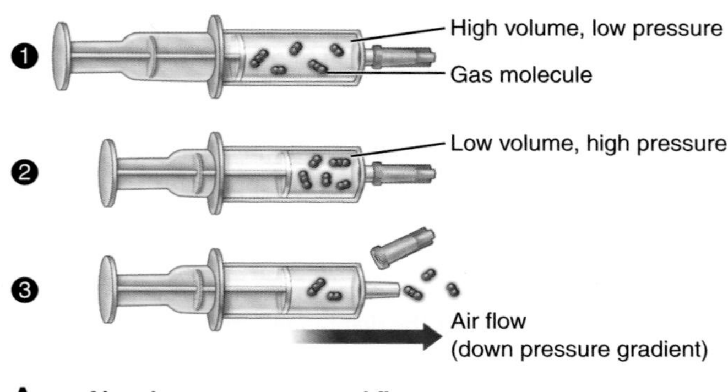

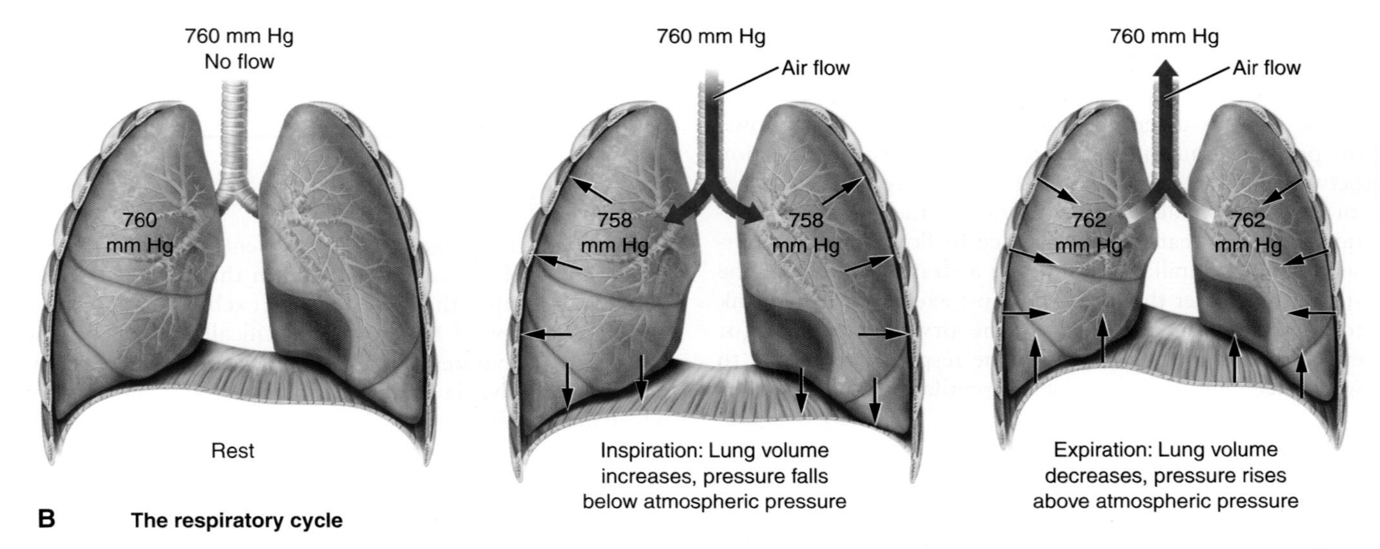

FIGURE 16-7 The relationship of gas pressure to volume. **KEY POINT** Volume changes create pressure gradients, which cause airflow. **A.** Pressure increases if the same number of gas molecules is confined to a smaller space. **B.** Air does not flow at rest, because there is no pressure gradient between the lungs and the atmosphere. During inspiration, alveolar pressure decreases as the thorax and lungs expand, so air flows into the lungs. Lung size decreases during expiration, increasing alveolar pressure and forcing air out of the lungs. **ZOOMING IN** When is alveolar pressure greater than atmospheric pressure—during inhalation or exhalation?

the many bronchial tubes until there is virtually no forward flow as it reaches the alveoli. The incoming air mixes with the residual air remaining in the respiratory passageways so that the gases soon are evenly distributed. Each breath causes relatively little change in the gas composition of the alveoli, but normal continuous breathing ensures the presence of adequate oxygen and the removal of carbon dioxide.

As with expansion of blood vessels discussed in Chapter 15, the ease with which one can expand the lungs and thorax in inhalation is called **compliance**. The health of lung tissue helps determine compliance; scar tissue makes lung tissue less compliant and can thus interfere with inhalation. 6 To understand some other conditions that can affect lung compliance, we must look again at the properties of water, first introduced in Chapter 2. A thin film of water lines the alveoli. This moisture is necessary, because gases in the lungs must go into solution before they can diffuse across the membrane into capillary blood. Water has the property of high *surface tension* based on water molecules' attraction for each other as a result of hydrogen bonding **(see BOX 2-1)**. You may have noticed that water will climb a short way up the inside of a narrow glass tube—a result of surface tension. A kind of "skin" forms at the surface of water that will support a light object or small insect, for example. Water's surface tension exerts an inward "pull" on the alveoli, causing them to resist expansion. To counteract this force, certain alveolar cells produce **surfactant** (sur-FAK-tant), a substance that reduces the surface tension of the fluids that line the alveoli. Surfactant is a lipoprotein mixture that behaves much like a dish detergent, which reduces surface tension to aid in breaking down fats. Thus, normal compliance of the lung tissue, aided by surfactant, allows the lungs to expand and fill adequately with air during inhalation. Conditions that can decrease compliance include diseases that damage or scar the lung tissue or deficiency of surfactant.

See the "Student Resources" on thePoint® for illustrations of the breathing muscles and for the animations "The Lungs" and "Pulmonary Ventilation."

Exhalation In exhalation, the passive phase of quiet breathing, the respiratory muscles relax, allowing the ribs and diaphragm to return to their original positions. The lung tissues are elastic and recoil to their original size during exhalation. Surface tension within the alveoli also aids in this return to resting size.

During forced exhalation, the internal intercostal muscles contract, pulling the bottom of the rib cage in and down. The muscles of the abdominal wall contract, pushing the abdominal viscera upward against the relaxed diaphragm. Even with maximum exhalation, however, you cannot expel all the air from your lungs. There is always a certain residual volume left to fill the airways and keep the lungs partially inflated.

Airway Radius 5 The pressure gradients created by the respiratory muscles drive air movement. But, the radius of the airways determines how easily the air flows down the pressure gradient—it determines the *resistance*. Friction between the air molecules and the airway walls opposes air flow. The smaller the airway radius, the greater the friction and the greater the resistance to flow. Imagine yourself drinking a milkshake through a straw. The thinner the straw, the greater the force you must exert to get the drink to flow. Similarly, narrowing of the airways in asthma (or even a bad chest cold) requires the respiratory muscles to work much harder to accomplish ventilation.

Spirometry The effectiveness of ventilation is measured by a **spirometer** (spi-ROM-eh-ter), an instrument for recording volumes of air inhaled and exhaled (FIG. 16-8). The tracing is a **spirogram** (SPI-ro-gram). TABLE 16-1 gives the definitions and average values for some of the breathing volumes and capacities that are important in evaluating respiratory function. A lung *capacity* is a sum of volumes. These same values are shown on a graph as they might appear on a spirogram.

CASEPOINTS 

- ☐ **16-4** Emily's narrowed airways make it particularly difficult to exhale. Does the diaphragm contract or relax during exhalation?
- ☐ **16-5** Which muscles could Emily use to aid in exhalation?

Respiratory therapists evaluate and treat breathing disorders. See the "Student Resources" on thePoint® for a description of this career.

GAS EXCHANGE

External gas exchange is the movement of gases between the alveoli and the capillary blood in the lungs (FIG. 16-9). 5 The principle that governs gas exchange is the now familiar concept of flow, more specifically diffusion. The barrier that separates alveolar air from the blood is composed of the alveolar wall and the capillary wall, both of which are extremely thin in order to minimize the resistance to diffusion. This respiratory membrane is not only very thin, but it is also moist. As previously noted, the moisture is

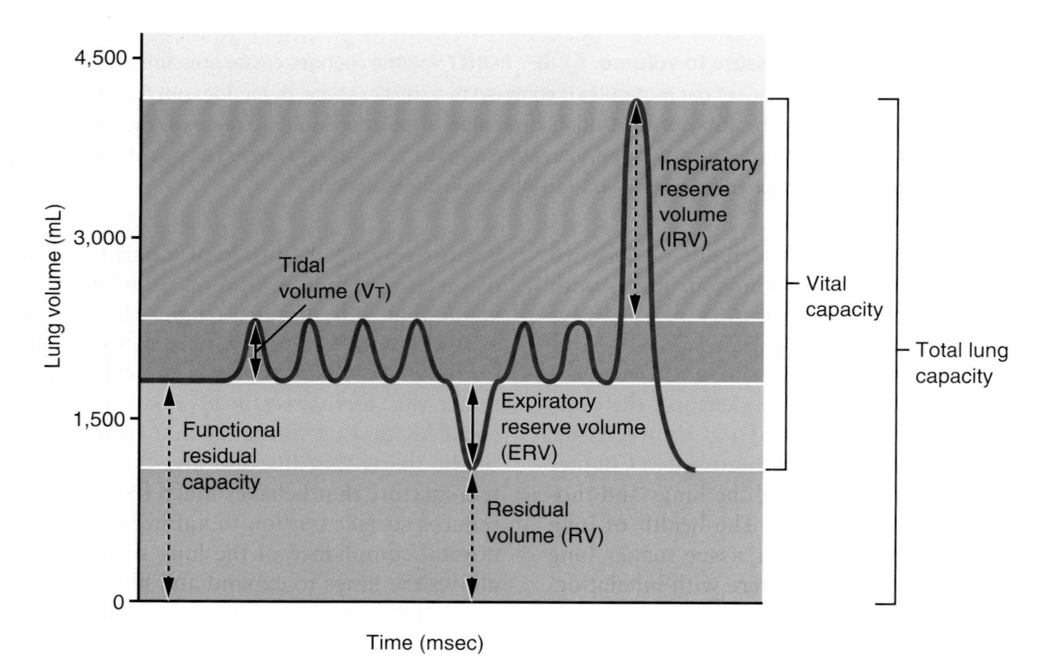

FIGURE 16-8 A spirogram. Lung volumes and capacities of an average adult female, as measured by spirometry. Upward deflections are inhalations; downward deflections are exhalations. The *solid arrows* indicate values that can be directly measured by spirometry. The *dotted arrows* indicate values that must be calculated or measured using specialized equipment. ZOOMING IN What lung volume cannot be measured with a spirometer? Which lung capacities cannot be measured with a spirometer?

Table 16-1 Lung Volumes and Capacities

Volume	Definition	Average Value (mL)
Tidal volume	The amount of air moved into or out of the lungs in quiet, relaxed breathing	500
Residual volume	The volume of air that remains in the lungs after maximum exhalation	1,100
Inspiratory reserve volume	The additional amount that can be breathed in by force after a normal inhalation	1,900
Expiratory reserve volume	The additional amount that can be breathed out by force after a normal exhalation	700
Vital capacity	The volume of air that can be expelled from the lungs by maximum exhalation after maximum inhalation	3,100
Functional residual capacity	The amount of air remaining in the lungs after normal exhalation	1,800
Total lung capacity	The total volume of air that can be contained in the lungs after maximum inhalation	4,200

important because the oxygen and carbon dioxide must go into a solution before they can diffuse across the membrane.

Recall that solutes diffuse down concentration gradients. Gases diffuse instead down pressure gradients, and the pressure of a gas is expressed in millimeters of mercury (mm Hg), as is blood pressure. The pressure of a gas within a gas mixture (such as air) is called its **partial pressure**, symbolized with a P and a subscript of its formula. Thus, the partial pressures of oxygen and carbon dioxide are symbolized as PO_2 and PCO_2, respectively. Note that even though the gases are in a mixture, each diffuses independently of any other gas in that mixture.

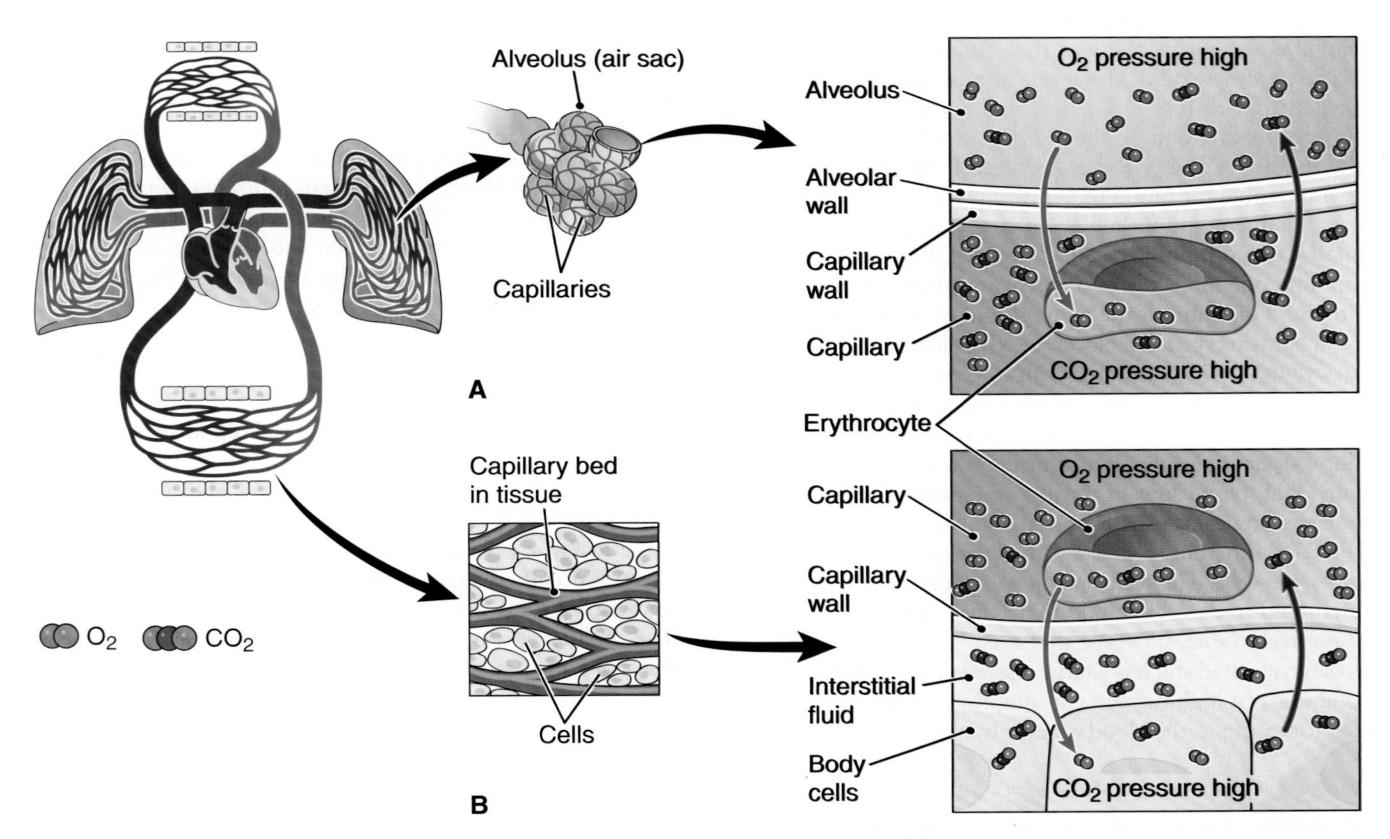

FIGURE 16-9 Gas exchange. KEY POINT Gas exchanges are based on relative partial pressures of oxygen and carbon dioxide on either side of a membrane. **A.** External exchange between the alveoli and the blood. Oxygen diffuses into the blood, and carbon dioxide diffuses out based on pressures of the two gases in the alveoli and in the blood. **B.** Internal exchange between the blood and the cells. Oxygen diffuses out of the blood and into tissues, while carbon dioxide diffuses from the cells into the blood.

Gases diffuse from the area of higher partial pressure to the area of lower partial pressure (see FIG. 16-9). Recall that blood in the pulmonary capillaries has already encountered body cells, so its Po_2 is lower than that in inspired air. In the lungs, therefore, oxygen will diffuse across the alveolar wall and into the capillaries. Because the cells generate carbon dioxide in cellular respiration, the Pco_2 in the pulmonary capillaries is higher than the Pco_2 in inspired air, and carbon dioxide diffuses out of the blood and into the alveoli.

In contrast, internal gas exchange takes place between the blood and the tissues, again based on partial pressure gradients for each gas. The blood arriving in the tissues has received additional oxygen in the lungs, and oxygen will pass into the oxygen-poor tissues. The lower partial oxygen pressure in highly active tissues will create a larger gradient, so these tissues will receive more oxygen than less active tissues. Carbon dioxide will diffuse out of the tissues and into the blood. Blood returning from the tissues and entering the lung capillaries through the pulmonary circuit is relatively low in oxygen and high in carbon dioxide. Again, the blood will pick up oxygen and give up carbon dioxide. After a return to the left side of the heart, it starts once more on its route through the systemic circuit.

OXYGEN TRANSPORT

A very small amount (1.5%) of the oxygen in the blood is carried in solution in the plasma. (Oxygen does dissolve in water, as shown by the fact that aquatic animals get their oxygen from water.) However, almost all (98.5%) of the oxygen that diffuses into the capillary blood in the lungs binds to hemoglobin in the red blood cells. If not for hemoglobin and its ability to hold oxygen in the blood, it would be impossible for blood to supply enough oxygen to the tissues. The hemoglobin molecule is a large protein with four small iron-containing "heme" regions. Each heme portion can bind one molecule of oxygen. See more about hemoglobin and a diagram of the hemoglobin molecule in BOX 12-1.

Highly oxygenated blood (in systemic arteries and pulmonary veins) is 97% saturated with oxygen. That is, only 3% of all the oxygen-binding sites on all of the hemoglobin molecules are unoccupied. Many hemoglobin molecules are carrying four oxygen molecules, but some carry only three. In a resting individual, blood low in oxygen (in systemic veins and pulmonary arteries) is usually about 70% saturated with oxygen. This 27% difference represents the oxygen that has been taken up by the cells. The terms *oxygenated* and *deoxygenated* are often used to describe blood that is high and low in oxygen, respectively. Note, however, that even blood that is described as deoxygenated still has a large reserve of oxygen. Even under conditions of high oxygen consumption, as in vigorous exercise, for example, the blood is never totally depleted of oxygen.

Concept Mastery Alert

Remember that oxygen saturation is always higher in pulmonary veins, which carry blood from the lungs to the heart, than in the pulmonary arteries.

To enter the cells, oxygen must separate from hemoglobin. Normally, the bond between oxygen and hemoglobin is easily broken, and oxygen is released as blood travels into tissues where the oxygen pressure is relatively low. Cells are constantly using oxygen in cellular respiration and obtaining fresh supplies by diffusion from the blood. In addition, some conditions increase the rate of oxygen's release from hemoglobin. Increasing body temperature and increasing blood acidity, both generated by cellular activity, promote its release to the tissues.

The poisonous gas carbon monoxide (CO), at low partial pressure, binds with hemoglobin at the same molecular sites as does oxygen. However, it binds more tightly and displaces oxygen. Even a small amount of CO causes a serious reduction in the blood's ability to carry oxygen.

CARBON DIOXIDE TRANSPORT

Carbon dioxide is produced continuously in the tissues as a by-product of cellular respiration. It diffuses from the tissue cells into the blood and is transported to the lungs in three ways:

- About 10% is dissolved in the plasma and in the fluid within red blood cells. (In carbonated beverages, the bubbles represent carbon dioxide that was previously dissolved in the solution.)
- About 15% is combined with the protein portion of hemoglobin and with plasma proteins.
- About 75% is transported as an ion, known as a **bicarbonate** (bi-KAR-bon-ate) **ion**, which is formed when carbon dioxide undergoes a chemical change after it enters red blood cells. It first combines with water to form **carbonic** (kar-BON-ik) **acid**, which then separates (ionizes) into hydrogen and bicarbonate ions. 7 This reaction is catalyzed by an enzyme called **carbonic anhydrase** (an-HI-drase) (CA).

$$\underset{\text{Carbon dioxide}}{CO_2} + \underset{\text{Water}}{H_2O} \underset{}{\overset{CA}{\rightleftharpoons}} \underset{\text{Carbonic acid}}{H_2CO_3} \rightleftharpoons \underset{\text{Hydrogen ion}}{H^+} + \underset{\text{Bicarbonate ion}}{HCO_3^-}$$

The bicarbonate formed in the red blood cells moves to the plasma and then is carried to the lungs. The process is reversed as bicarbonate re-enters the red blood cells, joins with a hydrogen ion to form carbonic acid, and again under the effects of CA, releases carbon dioxide and water. The carbon dioxide diffuses into the alveoli and is exhaled. The amount of CO_2 determines the direction of the reaction. The high level of CO_2 in the tissues promotes the forward reactions indicated by the upper arrows. The low level of CO_2 in the lungs promotes the backward reactions indicated by the lower arrows.

Carbon dioxide is important in regulating the blood's pH (acid–base balance). As a bicarbonate ion is formed from carbon dioxide in the plasma, a hydrogen ion (H^+) is also produced. Therefore, the blood becomes more acidic as the amount of carbon dioxide in the blood increases. Eliminating more carbon dioxide than the body produces

ONE STEP AT A TIME — BOX 16-1
Distinguishing Cause and Effect

In the case study, Emily's lung function was measured by asking her to breathe in and out of a spirometer. The goal of this test was to see how easily Emily moved air between the atmosphere and her lungs. To truly understand the mechanics of ventilation, as well as any other physiological process, we need to differentiate between cause and effect. The question below asks you to distinguish between changes in chest size and air flow: which is the cause, and which is the effect?

Question

Which of these statements is correct?

A. My chest gets bigger when I inhale because air enters my lungs.
B. Air enters my lungs when I inhale because my chest gets bigger.

Answer

Step 1. Look for the factor that the body can directly change. The body creates change by various means, such as muscle contraction or altering enzyme activity. If we look at our two possible answers to the question, we see that the two factors are "the chest gets bigger" and "air enters my lungs." The body can directly change chest size by contracting the ventilator muscles. So, step 1 suggests that the change in chest size is the cause, and airflow is the result.

Step 2. List all of the steps linking the two factors. In other words, try to expand the sentence and see if it still makes sense. Consider option B. We could expand this sentence to read "*Air enters my lungs when I inhale* because air moves down a pressure gradient. The pressure is lower in the lungs in the atmosphere, because I expand the size of my lungs. The size of my lungs *increases because my chest gets bigger.*" We cannot add any logical links to option B.

So, although it is somewhat counterintuitive, air flows because our chest (and thus lungs) changes in size, not the other way around. Anything interfering with that change in volume, such as decreased lung compliance, will reduce ventilation. And what about Emily's case? She must create an exceptionally strong pressure gradient to overcome the problem of her constricted airways. So, during asthma attacks, she may need to recruit additional muscles to change her chest volume enough to create the stronger gradient.

See the Study Guide (available separately) for more practice at distinguishing cause and effect.

16

shifts the blood's pH more toward the alkaline (basic) range. Chapter 19 has more information on acid–base balance.

CHECKPOINTS 

- ☐ **16-9** What are the two phases of quiet breathing? Which is active and which is passive?
- ☐ **16-10** What substance produced by lung cells aids in compliance?
- ☐ **16-11** What type of gradient drives diffusion across the respiratory membrane?
- ☐ **16-12** What substance in red blood cells holds almost all of the oxygen carried in the blood?
- ☐ **16-13** What is the main form in which carbon dioxide is carried in the blood?

See the "Student Resources" on thePoint® to view the animation "Oxygen Transport" and "Carbon Dioxide Exchange."

REGULATION OF VENTILATION

As mentioned above, arterial levels of O_2 and CO_2 are regulated variables. They are kept remarkably constant, despite large variations in cellular oxygen requirements and carbon dioxide production, by controlling the activity of the ventilatory muscles. In addition to maintaining arterial gas homeostasis, breathing must also be coordinated with complex behaviors such as laughing, swallowing, coughing, and singing. Centers in the central nervous system control the fundamental respiratory pattern and also respond to varying respiratory demands.

The Respiratory Control Center The respiratory control center is a complex network of neurons located in the medulla and pons of the brain stem. The control center's main part, located in the medulla, sets the basic pattern of respiration (**FIG. 16-10**). At rest, these centers fire about 12 times per minute, so we breathe about every five seconds. This pattern changes in response to input from other brain regions, as discussed later.

From the respiratory center in the medulla, motor nerve fibers extend into the spinal cord. From the cervical (neck) part of the cord, these fibers continue through the **phrenic** (FREN-ik) **nerve** (a branch of the vagus nerve) to the diaphragm and also to the intercostal muscles. The diaphragm and the other respiratory muscles are voluntary in the sense that they can be regulated consciously by messages from the higher brain centers, notably the cerebral cortex (**see FIG. 16-10**). It is possible for you to deliberately breathe more rapidly or more slowly or to hold your breath and not breathe at all for a while. In a short time, however, the respiratory center in the brain stem will override the voluntary desire to hold your breath, and breathing will resume. Most of the time, we breathe without thinking about it, and the respiratory center is in control.

Factors Regulating Breathing As shown by the yellow neurons in FIGURE 16-10, the respiratory center receives inputs from other brain regions as well as peripheral receptors:

- Central **chemoreceptors** (ke-mo-re-SEP-tors) in the medulla oblongata and peripheral chemoreceptors in the carotid artery and aorta sense arterial blood gas levels. Like the receptors for taste and smell, they are sensitive to chemicals dissolved in body fluids. Their contribution will be discussed shortly.
- The hypothalamus increases ventilation in response to pain, increased heat, or intense emotions such as excitement.
- Stretch receptors in the lung airways stop inhalation to prevent overexpansion of the lungs.
- Proprioceptors in muscles and joints (not shown in FIG. 16-10) respond to movement. Along with inputs from the motor cortex itself, these inputs enable the respiratory center to match ventilation with exercise intensity.

Feedback Control of Arterial CO_2 and O_2 Levels 3 A homeostatic feedback loop maintains relatively constant blood gas levels in arterial, but not venous, blood (FIG. 16-11). The chemoreceptors mentioned earlier act as sensors, the respiratory center in the medulla acts as the control center, and the ventilatory muscles are the effectors.

The central chemoreceptors are on either side of the brain stem near the medullary respiratory center. These receptors respond to the CO_2 level in circulating blood, but the gas acts indirectly. CO_2 is capable of diffusing through the capillary blood–brain barrier. It dissolves in medullary interstitial fluid and separates into hydrogen ion and bicarbonate ion, as explained previously. It is the presence of hydrogen ion that actually stimulates the central chemoreceptors.

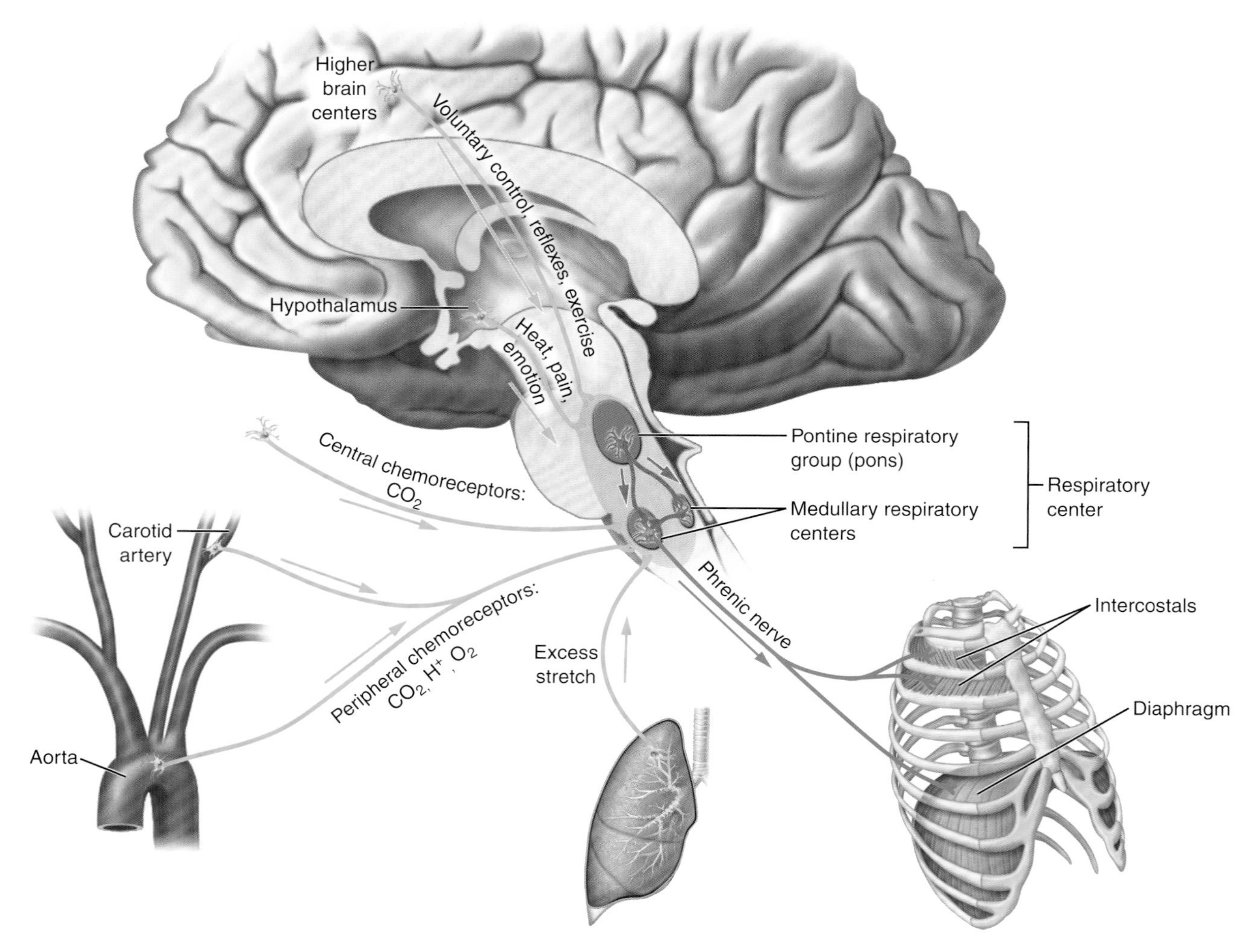

FIGURE 16-10 **Regulation of ventilation.** KEY POINT The medullary respiratory center sets the basic respiratory rhythm. It sends signals to the diaphragm and intercostal muscles through the vagus nerve (X). Input from other brain regions, chemoreceptors, and mechanoreceptors are shown in *yellow*. These inputs modify the rate and depth of breathing.

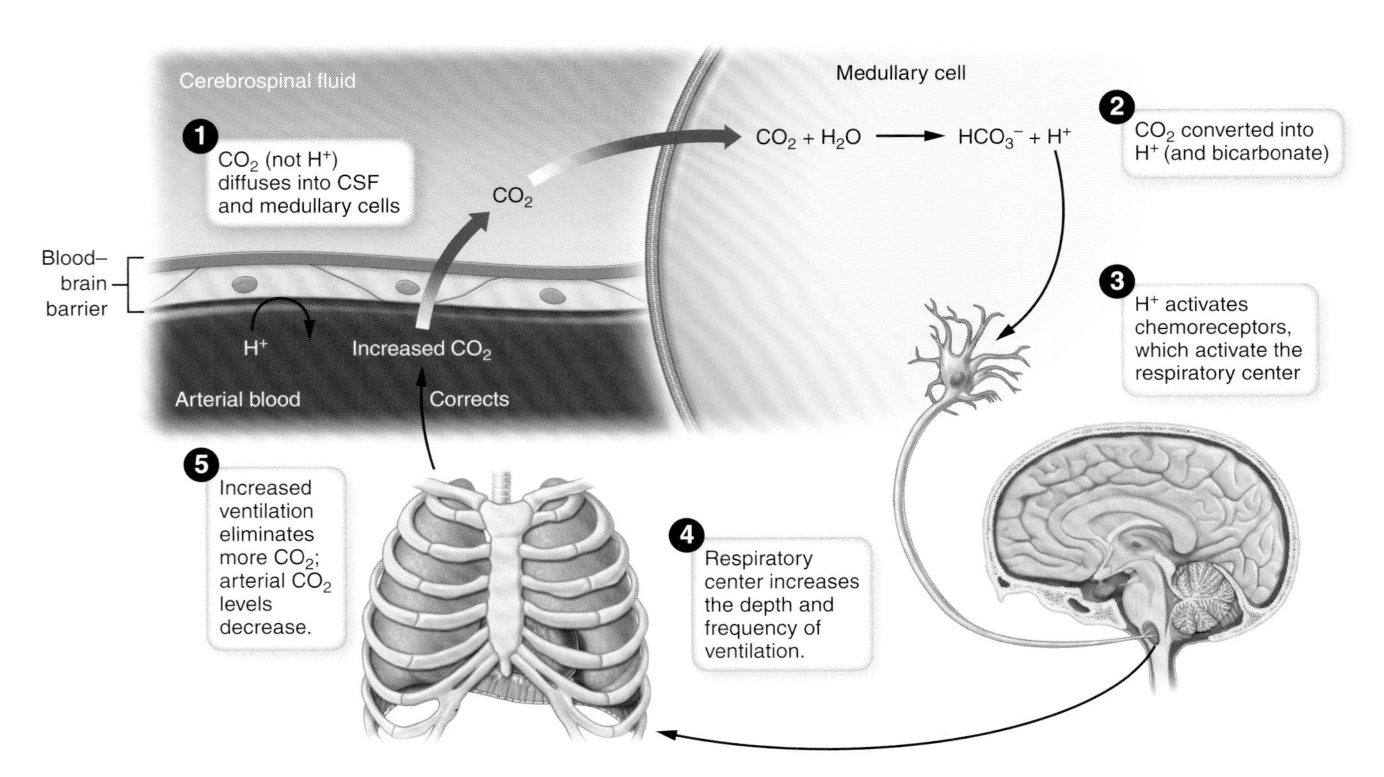

FIGURE 16-11 The central chemoreceptors. **KEY POINT** A negative feedback loop keeps arterial levels of CO_2 constant. The central chemoreceptors sense CO_2 indirectly, by measuring the amount of H^+ ions in interstitial fluid.

Increased arterial blood CO_2 is known as **hypercapnia** (hi-per-KAP-ne-ah). The respiratory center responds to hypercapnia by sending more signals to the ventilatory muscles. The resulting increase in the rate and depth of breathing eliminates more CO_2, and arterial levels return to normal.

Concept Mastery Alert

Remember that the central chemoreceptors measure the acidity of the medullary interstitial fluid, not the CO_2 concentration.

The peripheral chemoreceptors that detect arterial gas levels are found in structures called the *carotid* and *aortic bodies* (see FIG. 16-10). The carotid bodies are located near the bifurcation (forking) of the common carotid arteries in the neck, whereas the aortic bodies are located in the aortic arch. These bodies contain sensory neurons that respond mainly to a decrease in oxygen supply. They are not usually involved in regulating breathing because they do not act until oxygen drops to a very low level. Because there is usually an ample reserve of oxygen in the blood, carbon dioxide has the most immediate effect in regulating respiration at the level of the central chemoreceptors as discussed above. Oxygen only becomes a controlling factor when its level falls considerably below normal, as in cases of lung disease or in high-altitude environments where oxygen partial pressure is low.

Concept Mastery Alert

It may surprise you that arterial P_{CO_2} is under much tighter controls than P_{O_2}.

CASEPOINT 

☐ **16-6** If Emily has trouble exhaling carbon dioxide as a result of her asthma, what will happen to her breathing rate?

BREATHING PATTERNS

Normal breathing rates vary from 12 to 20 breaths per minute for adults. In children, rates may vary from 20 to 40 breaths per minute, depending on age and size. In infants, the respiratory rate may be more than 40 breaths per minute. Changes in respiratory rates are important in various disorders, and a healthcare provider should record them carefully. To determine the respiratory rate, the clinician counts the client's breathing for at least 30 seconds, usually by watching the chest rise and fall with each inhalation and exhalation. The count is then multiplied by 2 to obtain the rate in breaths per minute. It is best if the person does not realize that he or she is being observed because awareness of the measurement may cause a change in the breathing rate.

Some Terms for Altered Breathing The following is a list of terms designating various respiratory abnormalities. These are symptoms, not diseases. Note that the word ending in *-pnea* refers to breathing:

- **Hyperpnea** (hi-PERP-ne-ah) refers to an increase in the depth and rate of breathing to meet the body's metabolic needs, as in exercise.
- **Hypopnea** (hi-POP-ne-ah) is a decrease in the rate and depth of breathing.

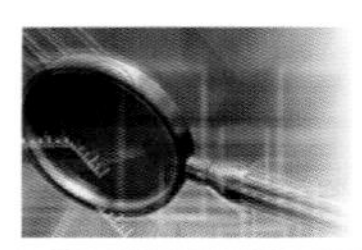

A CLOSER LOOK
Adaptations to High Altitude: Living with Hypoxia

BOX 16-2

Our bodies work best at low altitudes where oxygen is plentiful. However, people are able to live at high altitudes where oxygen is scarce and can even survive climbing Mount Everest, the tallest peak on our planet, showing that the human body can adapt to both short-term and long-term reductions in available oxygen. This adaptation process compensates for decreased atmospheric oxygen by increasing the efficiency of the respiratory and cardiovascular systems.

The body's immediate response to high altitude is to increase the rate of ventilation (hyperventilation) and raise the heart rate to increase cardiac output. Hyperventilation makes more oxygen available to the cells and increases blood pH (alkalosis), which boosts hemoglobin's capacity to bind oxygen. Over time, the body adapts in additional ways. Hypoxia (low tissue oxygen) stimulates the kidneys to secrete erythropoietin, prompting red bone marrow to manufacture more erythrocytes and hemoglobin. Also, capillaries proliferate, increasing blood flow to the tissues. Some people are unable to adapt to high altitudes, and for them, hypoxia and alkalosis lead to potentially fatal **altitude sickness**.

Successful adaptation to high altitude illustrates the principle of homeostasis and also helps to explain how the body adjusts to hypoxia associated with disorders such as chronic obstructive pulmonary disease (COPD).

- **Tachypnea** (tak-IP-ne-ah) is an excessive rate of breathing that may be normal, as in exercise.
- **Apnea** (AP-ne-ah) is a temporary cessation of breathing. Short periods of apnea occur normally during deep sleep. More severe sleep apnea can result from obstruction of the respiratory passageways or, less commonly, by failure in the central respiratory center.

ABNORMAL VENTILATION

In **hyperventilation** (hi-per-ven-tih-LA-shun), the rate and depth of breathing increase above optimal levels. This condition can occur during anxiety attacks or when a person is experiencing pain or other forms of stress. Hyperventilation increases the exhalation of carbon dioxide and decreases the level of that gas in the blood, a condition called **hypocapnia** (hi-po-KAP-ne-ah). Recall the equation that links levels of carbon dioxide with blood pH. Hyperventilation shifts the equation to the left, removing acidic products from the blood and increasing its pH. This condition is referred to as *alkalosis*, excess alkalinity of body fluids, and can result in dizziness and tingling sensations.

In the absence of continued overriding inputs from other brain centers prompting hyperventilation, the respiratory center responds to decreased carbon dioxide levels by decreasing the rate and depth of respiration. Gradually, the carbon dioxide level returns to normal, and a regular breathing pattern is resumed. In extreme cases, a person may faint, and then breathing will involuntarily return to normal. In assisting a person who is hyperventilating, one should speak calmly, reassure him or her that the situation is not dangerous, and encourage even breathing using the diaphragm.

In **hypoventilation**, an amount of air insufficient to meet the body's metabolic needs enters the alveoli. The many possible causes of this condition include respiratory obstruction; lung disease; injury to the respiratory center; depression of the respiratory center, as by drugs; and chest deformity. Hypoventilation increases the blood's carbon dioxide concentration, shifting the equation cited previously to the right and decreasing the blood's pH. This condition is called *acidosis*, excess acidity of body fluids. The respiratory center responds to this condition by attempting to increase the rate and depth of respiration.

BOX 16-2 offers information on adjusting to high altitudes where the availability of oxygen in the air is lower than usual.

CHECKPOINTS

- ☐ **16-14** Where in the brain stem are the centers that set the basic pattern of respiration?
- ☐ **16-15** What is the name of the motor nerve that controls the diaphragm?
- ☐ **16-16** What gas is the main chemical controller of respiration?
- ☐ **16-17** What is the meaning of the word ending *-pnea*?

Effects of Aging on the Respiratory Tract

With age, the tissues of the respiratory tract lose elasticity and become more rigid. Similar rigidity in the chest wall, combined with arthritis and loss of strength in the breathing muscles, results in an overall decrease in compliance and in lung capacity. However, there is a great deal of variation among individuals, and regular aerobic activity throughout adulthood (walking, running, swimming, etc.) can contribute significantly to maintaining respiratory function.

Reduction in phagocytosis and other protective mechanisms in the lungs, leads to increased susceptibility to infection. The incidence of lung disease increases with age but is hastened by cigarette smoking and by exposure to other environmental irritants.

A & P in Action Revisited: Emily's Asthma

It had been about a month since Emily's appointment with Dr. Martinez. During that time, her parents had monitored her breathing carefully and had noticed patterns in her coughing bouts.

"It does seem that Emily gets out of breath sooner than the other kids when she exercises," Emily's mother reported to Dr. Martinez during her follow-up appointment. "And now that the weather has cooled off, I do notice that she coughs more, but her cough improves when she uses the rescue inhaler you prescribed."

"Let's take a listen to Emily's lungs," replied the doctor as he placed his stethoscope on the little girl's chest. "Yes, I hear wheezing today, which suggests that Emily's airways are narrowed. The wheezing and spirometry results, along with your observations regarding the inhaler and the family history, all point to an asthma diagnosis."

Although Nicole had expected the doctor's diagnosis, it was still a shock to hear him say it. Seeing her look of alarm, Dr. Martinez continued.

"Most kids with asthma lead very normal, active lives. The medications available today target asthma right at its source—inflammation. In fact, with proper treatment, many kids maintain near-normal pulmonary function. First of all, you should take steps to eliminate allergens in Emily's environment. Let's continue with the current inhaler and add an anti-inflammatory in the form of a low-dose corticosteroid inhaler if Emily's symptoms do not improve in a few weeks. Let's go slowly here and see if we can help Emily with conservative measures. Down the road, if needed, we can turn to an oral leukotriene modifier that she will take daily to control airway inflammation".

In this case, we learned that Emily's asthma was caused by airway inflammation. Medications that limit the inflammatory response in the respiratory passages can prevent the symptoms of asthma. For a review of the role of inflammation in normal body defense mechanisms, see Chapter 15.

CHAPTER 16

Chapter Wrap-Up

OVERVIEW

A detailed chapter outline with space for note-taking is on thePoint®. The figure below illustrates the main topics covered in this chapter.

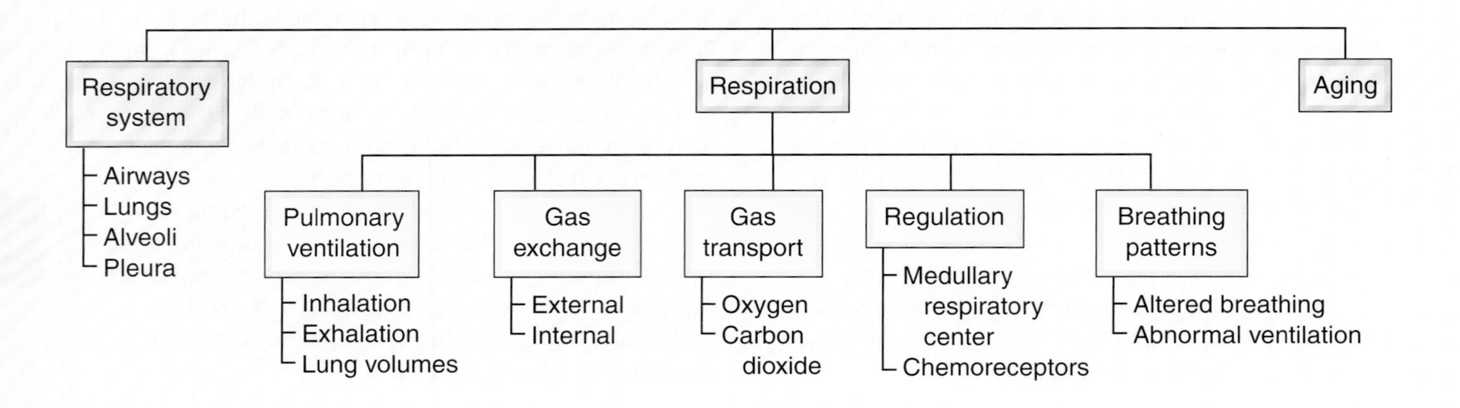

KEY TERMS

The terms listed below are emphasized in this chapter. Knowing them will help you organize and prioritize your learning. These and other boldface terms are defined in the "Glossary" with phonetic pronunciations.

alveoli (sing. alveolus)
bicarbonate ion
bronchiole
bronchus (pl. bronchi)
carbonic acid
carbonic anhydrase
chemoreceptor
compliance
diaphragm
epiglottis
hypercapnia
larynx
lung
pharynx
phrenic nerve
pleura
pulmonary ventilation
respiration
spirometer
surfactant
trachea

WORD ANATOMY

Medical terms are built from standardized word parts (prefixes, roots, and suffixes). Learning the meanings of these parts can help you remember words and interpret unfamiliar terms.

WORD PART	MEANING	EXAMPLE
STRUCTURE OF THE RESPIRATORY SYSTEM		
laryng/o	larynx	The *laryngeal* pharynx opens into the larynx.
nas/o	nose	The *nasopharynx* is behind the nasal cavity.
or/o	mouth	The *oropharynx* is behind the mouth.
pleur/o	side, rib	The *pleura* covers the lung and lines the chest wall (rib cage).
THE PROCESS OF RESPIRATION		
capn/o	carbon dioxide	*Hypercapnia* is a rise in the blood level of carbon dioxide.
-pnea	breathing	*Hypopnea* is a decrease in the rate and depth of breathing.
spir/o	breathing	A *spirometer* is an instrument used to record breathing volumes.

QUESTIONS FOR STUDY AND REVIEW

BUILDING UNDERSTANDING

Fill in the Blanks

1. The exchange of air between the atmosphere and the lungs is called ________.
2. The space between the vocal cords is the ________.
3. The ease with which the lungs and thorax can be expanded is termed ________.
4. A lower-than-normal level of oxygen in the tissues is called ________.
5. The serous membrane around the lung is the ________.

Matching Match each numbered item with the most closely related lettered item.

___ 6. The amount of air remaining in the lungs after a normal exhalation

___ 7. The amount of additional air that can be breathed out by force after a normal exhalation

___ 8. The amount of air moved into or out of the lungs in quiet, relaxed breathing

___ 9. The amount of air remaining in the lungs after the maximum exhalation

___ 10. The amount of air that can be expelled from the lungs by maximum exhalation after maximum inhalation

a. vital capacity
b. functional residual capacity
c. tidal volume
d. residual volume
e. expiratory reserve volume

Multiple Choice

___ 11. Which of the following are bony projections in the nasal cavity?
a. nares
b. septa
c. conchae
d. sinuses

___ 12. Which structure contains the vocal folds?
a. pharynx
b. larynx
c. trachea
d. lungs

___ 13. What covers the larynx during swallowing?
a. epiglottis
b. glottis
c. conchae
d. sinus

___ 14. Which structure has the centers that regulate respiration?
a. cerebral cortex
b. diencephalon
c. brain stem
d. cerebellum

___ 15. All of the following are true during forced expiration except
a. the diaphragm relaxes
b. the rib cage is elevated
c. the abdominal muscles contract
d. the volume of the thoracic cavity decreases

UNDERSTANDING CONCEPTS

16. Differentiate between the terms in each pair:
    a. internal and external gas exchange
    b. carbonic acid and carbonic anhydrase
    c. visceral and parietal pleura
    d. hyperpnea and tachypnea
17. Describe the characteristics of the tissue that lines the respiratory passageways and explain how this tissue protects us.
18. Trace the path of air from the nostrils to the lung capillaries.
19. What is the relationship between gas pressure and volume? What happens to gas pressure in the thoracic cavity when the diaphragm contracts?
20. Describe the direction in which oxygen and carbon dioxide diffuse during external and internal gas exchange.
21. Compare and contrast oxygen and carbon dioxide transport in the blood.
22. Define hyperventilation and hypoventilation. What is the effect of each on blood CO_2 levels and blood pH?
23. What are chemoreceptors, and how do they function to regulate breathing?

CONCEPTUAL THINKING

24. Jake, a sometimes exasperating 4-year-old, threatens his mother that he will hold his breath until "he dies." Should his mother be concerned that he might succeed?
25. Why is it important that airplane interiors are pressurized? If the cabin lost pressure, what physiological adaptations to respiration might occur in the passengers?

For more questions, see the Learning Activities on thePoint®.

CHAPTER 17

The Digestive System

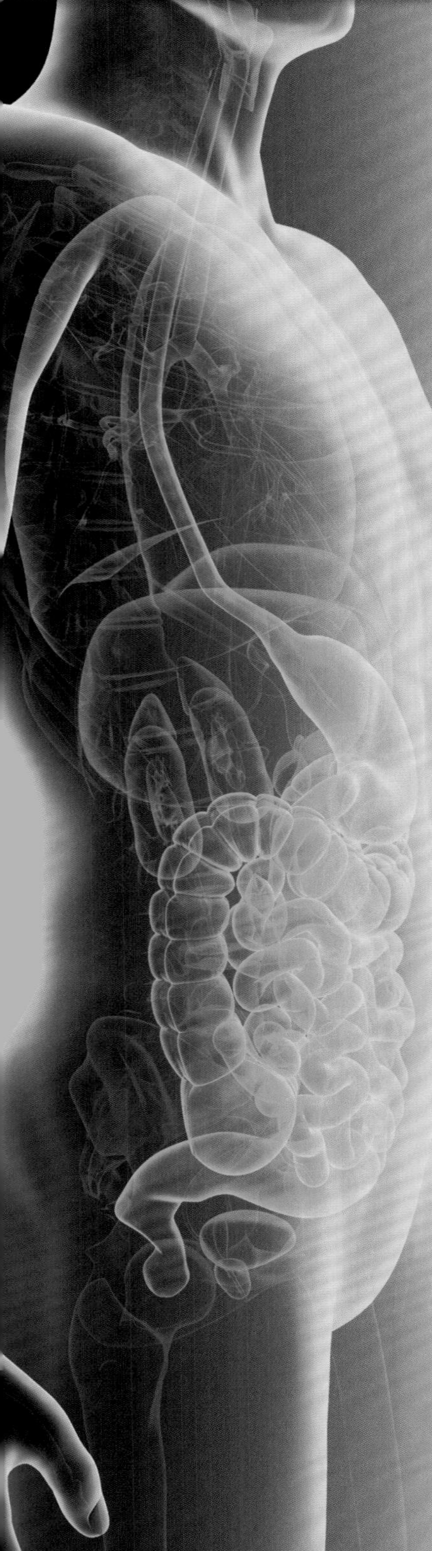

Learning Objectives

After careful study of this chapter, you should be able to:

1. Describe the three main functions of the digestive system. ***p. 352***
2. Trace the path of food from the mouth to the anus. ***p. 352***
3. Name and locate the two main layers and the subdivisions of the peritoneum. ***p. 352***
4. Describe the four layers of the digestive tract wall. ***p. 353***
5. Describe the two types of muscular contractions important in the digestive process. ***p. 354***
6. Describe the structure and function of the oral cavity, including the different types of teeth. ***p. 355***
7. Name and describe the parts and structural specializations of the esophagus and stomach. ***p. 356***
8. Describe the components and function of gastric juice. ***p. 357***
9. Identify the parts and structural specializations of the small and large intestines. ***p. 358***
10. Name and describe the functions of the accessory organs of digestion. ***p. 360***
11. List the steps involved in the digestion and absorption of carbohydrates, proteins, and fats. ***p. 362***
12. Define the term hydrolysis, and explain its role in digestion. ***p. 363***
13. Explain nervous control of digestion and the role of the enteric nervous system. ***p. 364***
14. Explain feedback loops involving the hormones gastrin, secretin, cholecystokinin (CCK), and gastric inhibitory peptide (GIP). ***p. 364***
15. Explain how hormones regulate hunger and appetite. ***p. 365***
16. Using information in the case study and the text, explain possible adverse effects associated with acidic gastric juice. ***pp. 351, 367***
17. Show how word parts are used to build words related to digestion (see Word Anatomy at the end of the chapter). ***p. 369***

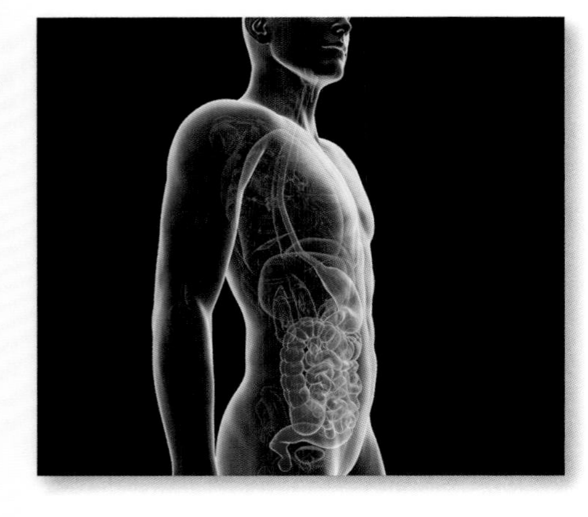

A & P in Action

Benjamin's: Gastroesophageal Reflux Disease (GERD)

Benjamin, a 51-year-old traveling salesman, was preparing for yet another night away from home when he experienced his familiar epigastric pain. He had experienced intermittent heartburn for the last 10 years, but it seemed to be getting worse. "More job stress, and age, I imagine" he told himself, but made a mental note to check in with his physician when he returned home.

Benjamin told Dr. Eliason, his primary care physician, about his daily episodes of discomfort. "This heartburn occurs after meals and at bedtime. It often interrupts my sleep and I feel exhausted sometimes, especially when I travel for business," he related. "My eating habits and sleep patterns are definitely not the best at those times. On occasion, I feel that food is backing up into my throat. I've taken antacids and other over-the-counter drugs, which help a little, but I follow the package recommendation to never take them for more than 2 weeks."

Dr. Eliason performed a brief physical exam and reviewed Benjamin's file. "First, the good news," he began. Your colonoscopy last year was completely normal, and your blood pressure is within the normal range. My notes state that you do not have a history of smoking or alcohol abuse, and, aside from your gastric pain, you haven't had any health problems. Are these observations still valid?" Benjamin nodded his agreement, so the physician continued.

"You're experiencing classic esophageal symptoms that suggest gastroesophageal reflux disease (GERD). This condition might be associated with erosive esophagitis, which is best diagnosed by endoscopy in an esophagogastroduodenoscopy (EGD) procedure. We really should take care of this, because it can lead to some serious complications with time. I'll refer you for the procedure. In the meantime, I'll prescribe a medication called a proton pump inhibitor to reduce stomach acid. It may take a while to arrange for the EGD test, but see me again in 4 weeks."

As you study this chapter, CasePoints will give you opportunities to apply your learning to this case.

Visit thePoint® to access the following resources. For guidance in using these resources most effectively, see pp. xv–xvii.

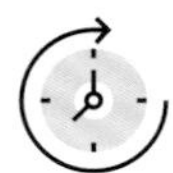

Preparing to Learn

- Tips for Effective Studying
- Pre-Quiz

While You Are Learning

- Animation: General Digestion
- Animation: Digestion of Carbohydrates
- Animation: Enzymes
- Animation: The Liver in Health and Disease
- Chapter Notes Outline
- Audio Pronunciation Glossary

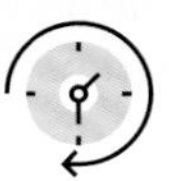

When You Are Reviewing

- Answers to Questions for Study and Review
- Health Professions: Dental Hygienist
- Interactive Learning Activities

A LOOK BACK

Recall from Chapter 1 that the contents of the digestive tract are considered "outside" the body. They are separated from the body's cells by the mucous membrane described in Chapter 4 4. Converting food into components that can cross this barrier requires both enzymes 7 and water 6. The key ideas of structure-function 1, energy 8, and homeostasis 3 and the structure of nutrients from Chapter 2 are also highly relevant to the topics in this chapter.

Overview of the Digestive System

Every body cell needs a constant supply of nutrients. 8 Cells use the potential energy contained in nutrients to do their work. In addition, they rearrange the nutrients' chemical building blocks to manufacture materials the body needs for metabolism, growth, and repair.

Most ingested nutrients are too large to enter cells. They must first be broken down into smaller components, a process known as **digestion**. The transfer of nutrients from the digestive tract to the circulation is called **absorption**. The circulatory system carries nutrients to body cells for use and storage. Finally, undigested waste material must be eliminated. Digestion, absorption, and elimination are the three chief functions of the digestive system.

For study purposes, the digestive system may be divided into two groups of organs (FIG. 17-1):

- The **digestive tract**, which is a muscular tube extending through the body. It begins at the **mouth**, where food is taken in, and terminates at the **anus**, where the solid waste products of digestion are expelled. The remainder of the tract consists of the **pharynx**, **esophagus**, **stomach**, **small intestine**, and **large intestine**. The digestive tract is also described as the **gastrointestinal** (GI) **tract** because of the major importance of the stomach and intestine in the digestive process.
- The **accessory organs**, which are necessary for the digestive process but are not a direct part of the digestive tract. The teeth and the tongue physically manipulate food prior to digestion. The other accessory organs release substances into the digestive tract through ducts. They include the **salivary glands, liver, gallbladder**, and **pancreas**.

Within this chapter, we will discuss each of these structures in detail.

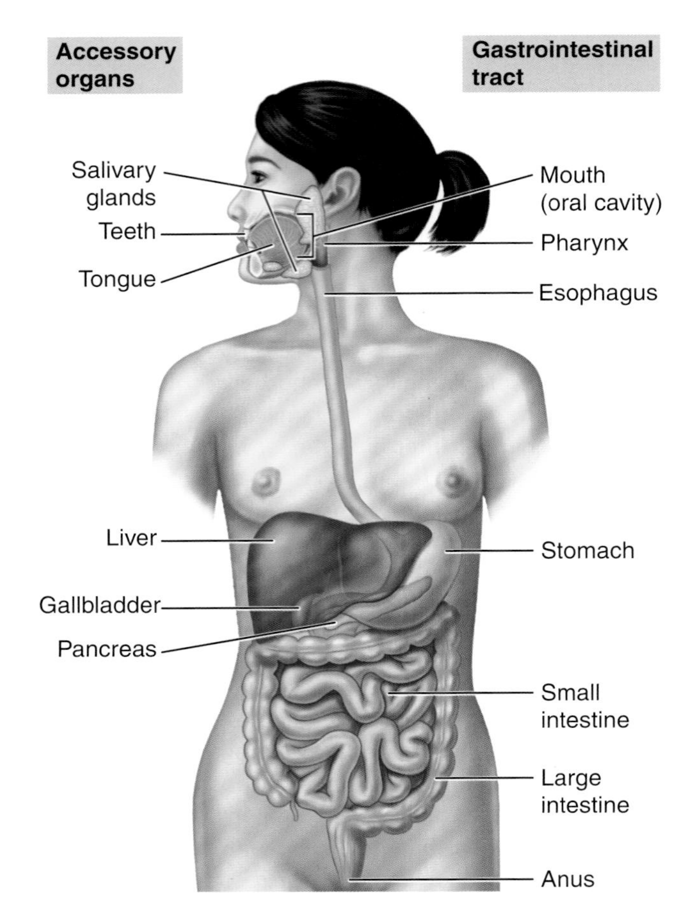

FIGURE 17-1 The digestive system. KEY POINT The digestive system extends from the mouth to the anus. Accessory organs secrete into the digestive tract. ZOOMING IN Which accessory organs of digestion secrete into the mouth?

CASEPOINT

☐ **17-1** In the EGD test described in the case study, a tube passes from the lips to the small intestine. List the structures that the tube will encounter.

THE PERITONEUM

The **peritoneum** (per-ih-to-NE-um) is a thin, shiny serous membrane that lines the abdominopelvic cavity and also folds back to cover most of the organs contained within the cavity (FIG. 17-2). As noted in Chapter 4, the outer portion of this membrane, the layer in contact with the body wall, is called the *parietal* (pah-RI-eh-tal) *peritoneum*; the layer that covers the organs is called the *visceral* (VIS-eh-ral) *peritoneum*. This slippery membrane allows the organs to slide over each other as they function. The peritoneum also carries blood vessels, lymphatic vessels, and nerves between the two layers. In some places, it supports the organs and binds them to each other. The peritoneal cavity is the potential space between the membrane's two layers and contains serous fluid (peritoneal fluid).

Subdivisions of the peritoneum around the various organs have special names. The **mesentery** (MES-en-ter-e) is a double-layered portion of the peritoneum shaped somewhat like a fan (see FIG. 17-2). The handle portion is attached to the posterior abdominal wall, and the expanded long edge is attached to the small intestine. Between the two membranous layers of the mesentery are the vessels and nerves that supply the intestine. The section of the peritoneum that extends from the colon to the posterior abdominal wall is the mesocolon (mes-o-KO-lon).

A large double layer of the peritoneum containing much fat hangs like an apron over the front of the intestine. This **greater omentum** (o-MEN-tum) extends from the lower border of the stomach into the pelvic cavity and then loops back up to the transverse colon.

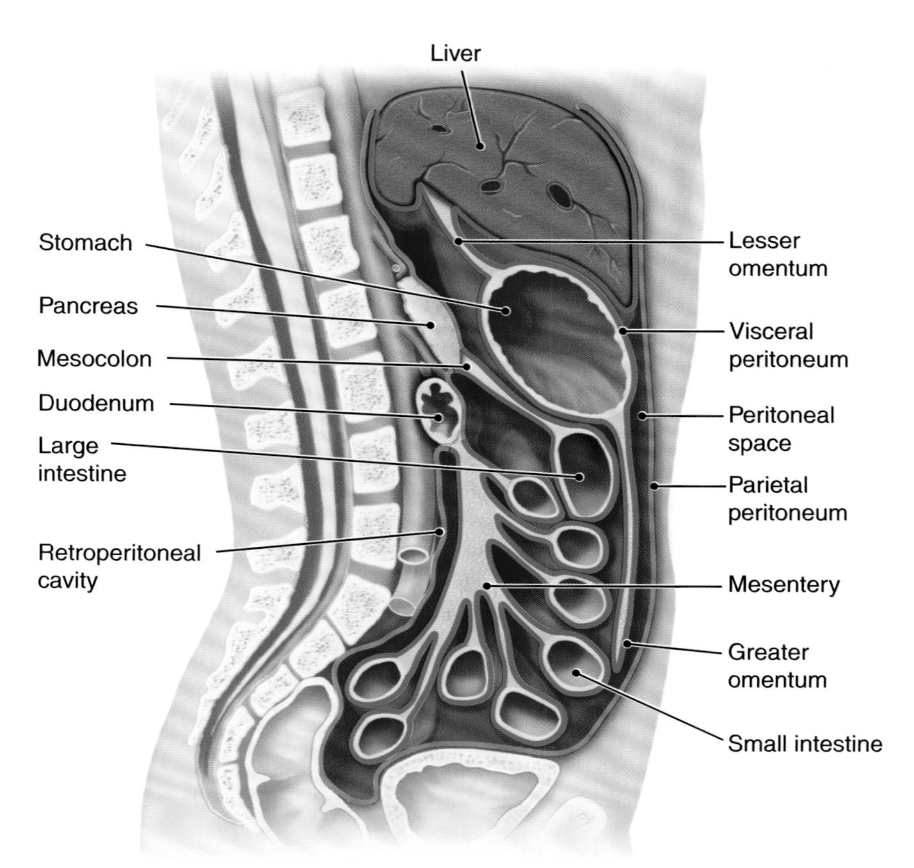

A The peritoneum

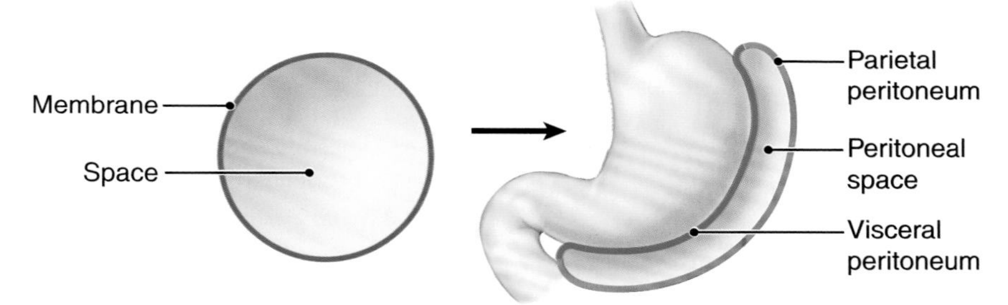

B Formation of the peritoneum

FIGURE 17-2 The abdominopelvic cavity and peritoneum. KEY POINT **A.** The parietal peritoneum lines the abdominopelvic cavity. **B.** Subdivisions of the visceral peritoneum fold over, supporting and separating individual organs. ZOOMING IN Which part of the peritoneum is around the small intestine?

Concept Mastery Alert

Note that the greater omentum does not actually enclose the transverse colon.

A smaller membrane, called the **lesser omentum,** extends between the stomach and the liver.

THE WALL OF THE DIGESTIVE TRACT

Although modified for specific tasks in different organs, the wall of the digestive tract, from the esophagus to the anus, is similar in structure throughout (**FIG. 17-3**). The general pattern consists of four layers, which are, from innermost to outermost:

1. Mucous membrane, or mucosa
2. Submucosa
3. Smooth muscle, the muscularis externa
4. Serous membrane, or serosa

Lining the tube's lumen is the mucous membrane, or **mucosa,** a type of epithelial membrane described in Chapter 4. 1 From the mouth through the esophagus, and also in the anus, the mucosal epithelium consists of stratified squamous cells, which help protect deeper tis-

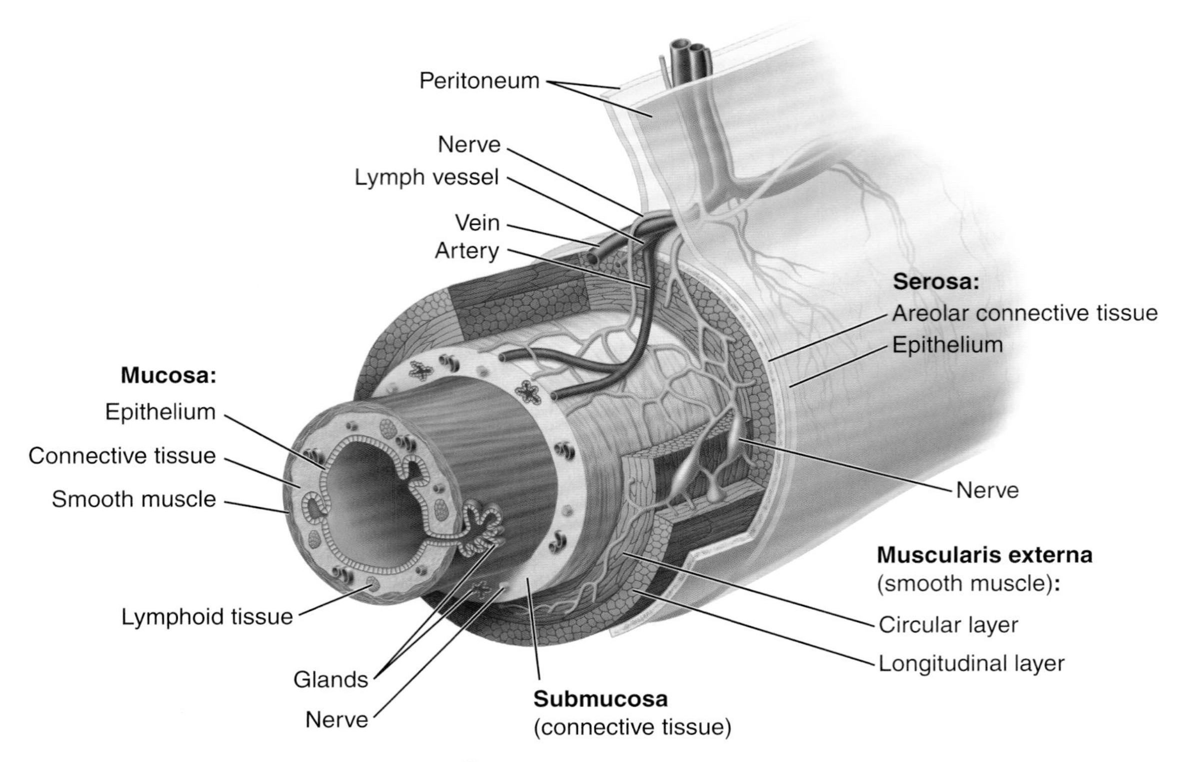

FIGURE 17-3 Wall of the digestive tract. KEY POINT The four layers are the mucosa, submucosa, muscularis externa (smooth muscle), and serosa. There is some variation in different organs. ZOOMING IN What type of tissue is between the submucosa and the serous membrane in the digestive tract wall?

sues. Throughout the remainder of the digestive tract, the mucosa contains simple columnar epithelium, which maximizes the absorption of nutrients. Goblet cells within this epithelium secrete mucus to protect the system's lining. Adjacent to the epithelium is a thin layer of connective tissue containing gut-associated lymphoid tissue (GALT; see Chapter 15), followed by a very thin layer of smooth muscle.

The thick layer of connective tissue beneath the mucosa is the **submucosa**, which contains blood vessels and some of the nerves that help regulate digestive activity. The next layer, the **muscularis externa**, is composed of smooth muscle. Most of the digestive organs have two layers of smooth muscle: an inner layer of circular fibers and an outer layer of longitudinal fibers. When a section of the circular muscle contracts, the organ's lumen narrows; when the longitudinal muscle contracts, a section of the wall shortens. A wave of circular muscle contractions propels food through parts of the digestive tract, a movement known as **peristalsis** (per-ih-STAL-sis) (FIG. 17-4). Alternatively, rhythmic contractions of the circular muscle mix food with digestive juices, a process known as **segmentation**. Different regions of the digestive tract use these different forms of motility, as discussed later. Nerves that control motility of the digestive organs are located in this smooth muscle layer.

The digestive organs in the abdominopelvic cavity have an outermost layer of serous membrane, or **serosa**, a thin, moist tissue composed of simple squamous epithelium and areolar (loose) connective tissue (see FIG. 17-3). This membrane forms the inner layer of the peritoneum.

CHECKPOINTS

- 17-1 Why does food have to be digested before cells can use it?
- 17-2 What is the name of the large serous membrane that lines the abdominopelvic cavity and covers the organs it contains?
- 17-3 What are the four layers of the digestive tract wall?

CASEPOINT

- 17-2 What layer of the digestive tract wall is contacted by the endoscope in Benjamin's EGD exam?

Digestive Tract Organs

Although the digestive tract is a continuous tube, each section is modified to accomplish particular functions. Next we discuss the structure and function of each digestive organ followed by descriptions of the secretory accessory organs. These discussions are followed by an overview of how all the organs work together in digestion.

THE MOUTH

The mouth, also called the *oral cavity*, is where a substance begins its travels through the digestive tract (FIG. 17-5A). The mouth has the following digestive functions:

- It receives food, a process called **ingestion**.
- It breaks food into small portions. This is done mainly by the teeth in the process of chewing or **mastication**

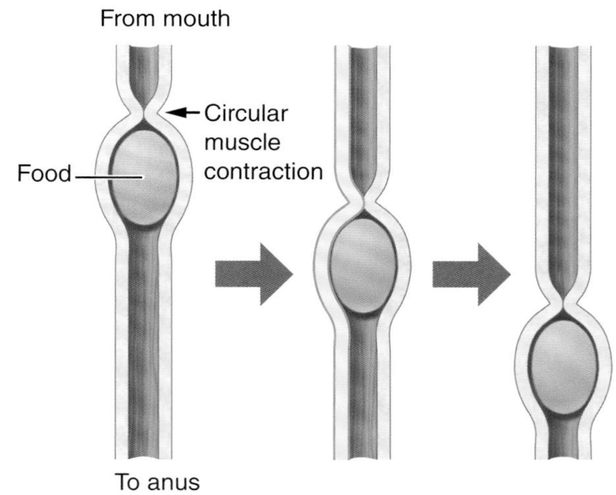

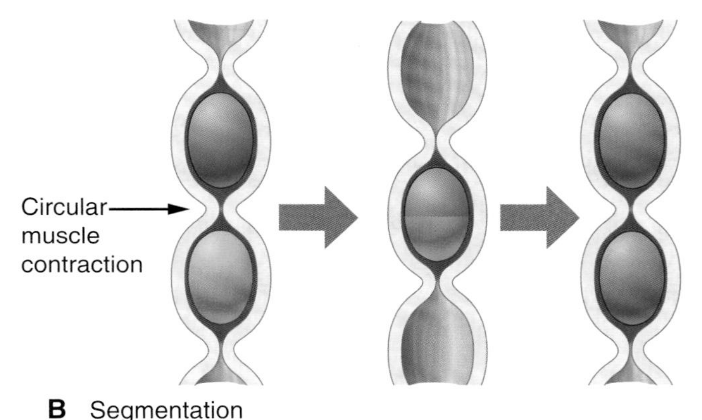

FIGURE 17-4 Gastrointestinal motility. KEY POINT **A.** Two types of contractions aid in digestion. Peristalsis moves the tube's contents ahead of the wave of contraction. **B.** Segmentation squishes the contents back and forth to mix them. ZOOMING IN Which type of motility would be most useful in the esophagus where the contents should move quickly?

(mas-tih-KA-shun), but the tongue, cheeks, and lips are also used.

- It mixes the food with **saliva** (sah-LI-vah), which is produced by the salivary glands and secreted into the mouth. The salivary glands will be described with the other accessory organs.
- It moves controlled amounts of food toward the throat to be swallowed, a process called **deglutition** (deg-lu-TISH-un).

The **tongue** (tung), a muscular organ that projects into the mouth, aids in chewing and swallowing and is one of the principal organs of speech. The tongue has a number of special surface receptors, called *taste buds*, which can differentiate taste sensations (e.g., bitter, sweet, sour, or salty) (see Chapter 10).

THE TEETH

The oral cavity also contains the teeth (see FIG. 17-5A). Young children have 20 teeth, known as the baby teeth or **deciduous** (de-SID-u-us) teeth. (The word deciduous means "falling off at a certain time," such as the leaves that fall off the trees in autumn.) A complete set of adult permanent teeth numbers 32. The cutting teeth, or **incisors** (in-SI-sors), occupy the anterior part of the oral cavity. The **cuspids** (KUS-pids), commonly called the *canines* (KA-nines) or *eyeteeth*, are lateral to the incisors. They are pointed teeth with deep roots that are used for more forceful gripping and tearing of food. The posterior **molars** (MO-lars) are the larger grinding teeth. There are two premolars and three molars. In an adult, each quadrant (quarter) of the mouth, moving from anterior to posterior, has two incisors, one cuspid, and five molars.

The first eight deciduous (baby) teeth to appear through the gums are usually the incisors. Later, the cuspids and molars appear. Usually, the 20 baby teeth all have appeared by the time a child has reached the age of 2 to 3 years. During the first 2 years, the permanent teeth develop within the upper jaw (maxilla) and lower jaw (mandible) from buds that are present at birth. The first permanent teeth to appear are the four 6-year molars, which usually come in before any baby teeth are lost. Because decay and infection of deciduous molars may spread to new, permanent teeth, deciduous teeth need proper care.

As a child grows, the jawbones grow, making space for additional teeth. After the 6-year molars have appeared, the baby incisors loosen and are replaced by permanent incisors. Next, the baby canines (cuspids) are replaced by permanent canines, and finally, the baby molars are replaced by the permanent bicuspids (premolars).

At this point, the larger jawbones are ready for the appearance of the 12-year, or second, permanent molar teeth. During or after the late teens, the third molars, or so-called *wisdom teeth*, may appear. In some cases, the jaw is not large enough for these teeth, or there are other abnormalities, so that the third molars may not erupt or may have to be removed. FIGURE 17-5B shows the parts of a molar.

The main substance of the tooth is **dentin**, a calcified substance harder than bone. Within the tooth is a soft pulp containing blood vessels and nerves. The tooth's *crown* projects above the gum, the **gingiva** (JIN-jih-vah), and is covered with **enamel**, the hardest substance in the body. The *root* or roots of the tooth, located below the gum line in a bony socket, are covered with a rigid connective tissue (cementum) that helps hold the tooth in place. A connective tissue sheet called the *periodontal ligament* joins the cementum to the tooth socket. Each root has a canal containing extensions of the pulp.

Dental hygienists are concerned with oral health. See the Student Resources on thePoint® for information on this career.

THE PHARYNX

The pharynx (FAR-inks) is commonly referred to as the throat. It is a combined pathway for the respiratory and digestive systems and was described in Chapter 16 (FIG. 16-2). The oral part of the pharynx, the oropharynx, is visible when you look into an open mouth and depress the

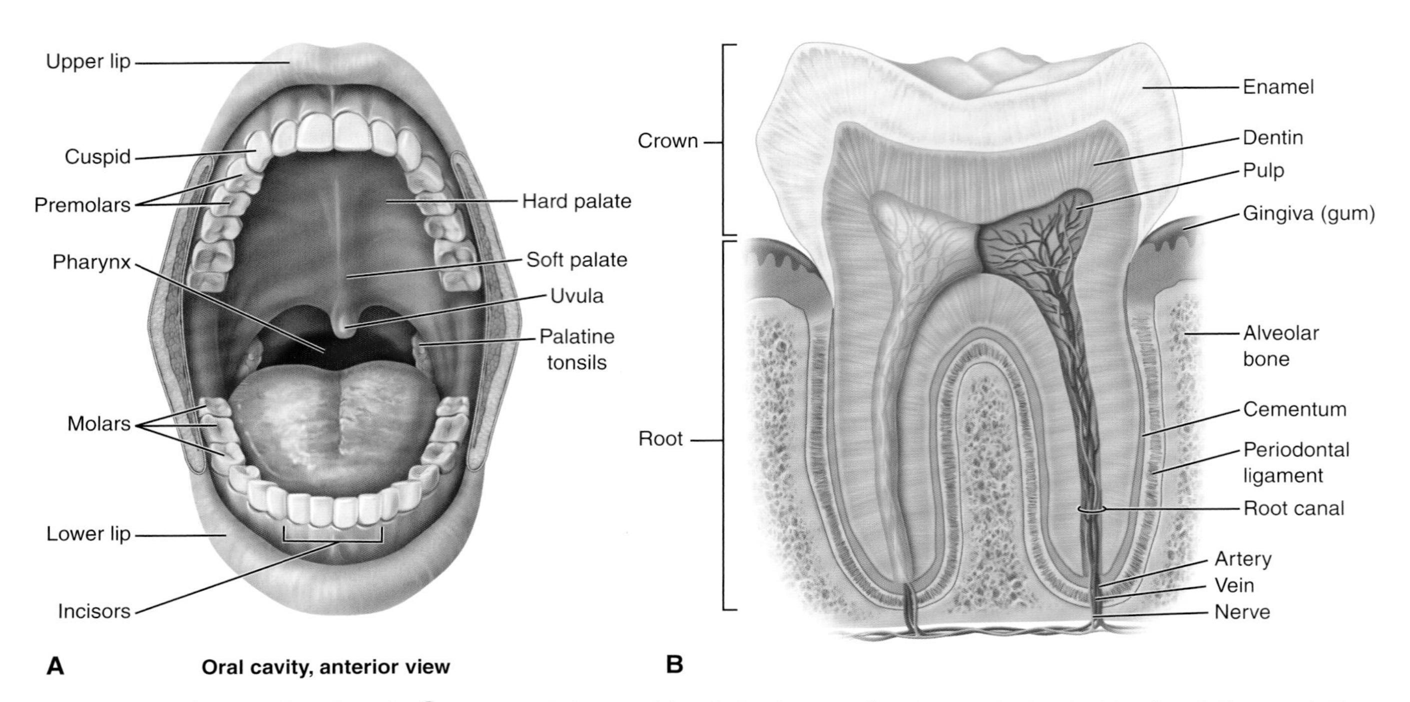

FIGURE 17-5 The mouth and teeth. KEY POINT The mouth breaks food into smaller pieces and mixes it with saliva. **A.** The mouth. The teeth and tonsils are visible in this view. **B.** A molar tooth. ZOOMING IN What is the common name for the gingiva?

tongue. The palatine tonsils may be seen at either side of the oropharynx. The pharynx also extends upward to the nasal cavity, where it is referred to as the nasopharynx, and downward to the larynx, where it is called the laryngopharynx. The **soft palate** forms the posterior roof of the oral cavity. From it hangs a soft, fleshy, V-shaped mass called the **uvula** (U-vu-lah).

In swallowing, the tongue pushes a **bolus** (BO-lus) of food, a small portion of chewed food mixed with saliva, into the pharynx. Once the food reaches the pharynx, swallowing occurs rapidly by an involuntary reflex action. At the same time, the soft palate and uvula are raised to prevent food and liquid from entering the nasal cavity, and the tongue is raised to seal the back of the oral cavity. As described in Chapter 16, the entrance of the trachea is guarded during swallowing by the leaf-shaped cartilage, the epiglottis, which covers the opening of the larynx (see FIG. 16-3). The swallowed food is then moved into the esophagus.

THE ESOPHAGUS

The esophagus (eh-SOF-ah-gus) is a muscular tube about 25 cm (10 in) long. Its musculature differs slightly from that of the other digestive organs because it has voluntary striated muscle in its upper portion, which gradually changes to smooth muscle along its length. In the esophagus, food is lubricated with mucus, and peristalsis moves it into the stomach. No additional digestion occurs in the esophagus.

Before joining the stomach, the esophagus must pass through the diaphragm. It travels through an opening in the diaphragm called the **esophageal hiatus** (eh-sof-ah-JE-al hi-A-tus) (FIG. 17-6). If there is a weakness in the diaphragm at this point, a portion of the stomach or other abdominal organ may protrude through the space, a condition called *hiatal hernia*, discussed later.

THE STOMACH

The stomach is an expanded J-shaped organ in the superior left region of the abdominal cavity (see FIG. 17-6 and FIGS. A3-7 and A3-9 in the Dissection Atlas). In addition to the two muscle layers already described, the stomach has a third, inner oblique (angled) layer that aids in grinding food and mixing it with digestive juices. The left-facing arch of the stomach is the **greater curvature**, whereas the smaller right surface forms the **lesser curvature.** The superior rounded portion under the left side of the diaphragm is the stomach's **fundus.** The region of the stomach leading into the small intestine is the **pylorus** (pi-LOR-us). The large region between the fundus and the pylorus is known as the stomach's body.

Sphincters A **sphincter** (SFINK-ter) is a muscular ring that regulates the size of an opening. There are two sphincters that separate the stomach from the organs above and below.

The **lower esophageal sphincter** (LES) controls the passage of food from the esophagus to the stomach. This muscle has also been called the *cardiac sphincter* because of its proximity to the heart. We are sometimes aware of the existence of this sphincter when it does not relax as it should, producing a feeling of being unable to swallow past that point. Between the distal, or far, end of the stomach and

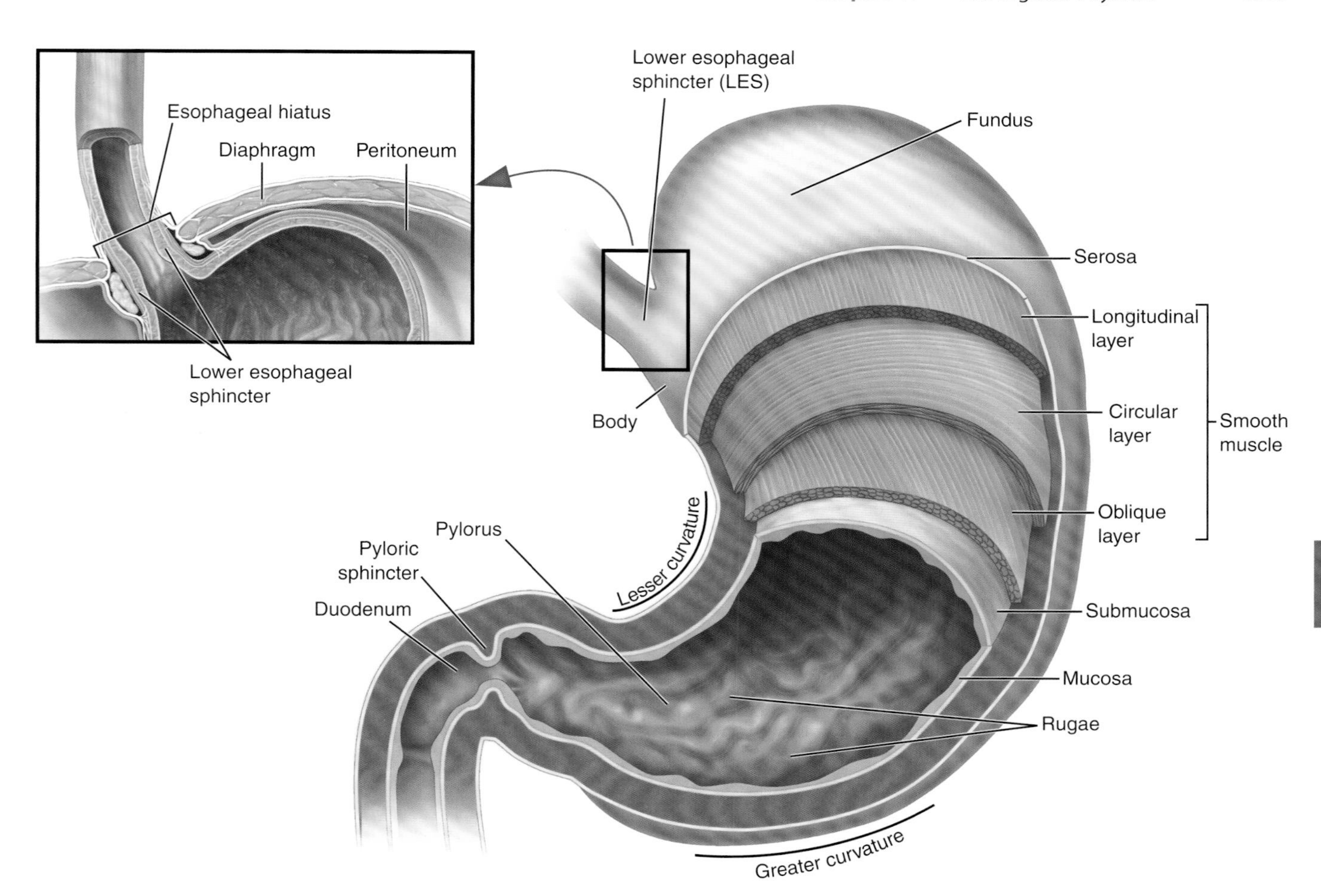

FIGURE 17-6 **Longitudinal section of the stomach.** KEY POINT The stomach's interior is shown along with a portion of the esophagus and the duodenum. Sphincters regulate the organ's entrance and exit openings. ZOOMING IN Which additional muscle layer is in the wall of the stomach that is not found in the rest of the digestive tract?

17

the small intestine is the **pyloric** (pi-LOR-ik) **sphincter**. This sphincter and the stomach's pylorus, which leads to it, are important in regulating how rapidly food moves into the small intestine.

Functions of the Stomach The stomach serves as a storage pouch, digestive organ, and churn. When the stomach is empty, the lining forms many folds called **rugae** (RU-je) (see FIG. 17-6). These folds disappear as the stomach expands. (The stomach can stretch to hold one-half of a gallon of food and liquid.) Special cells in the stomach's lining secrete substances that mix together to form gastric juice (*gastr/o* is the word root for "stomach.") Some of the cells secrete a great amount of mucus to protect the organ's lining from digestive secretions. Other cells produce the active components of the gastric juice, which are as follows:

- Hydrochloric acid (HCl), a strong acid that denatures (unwinds) proteins to prepare them for digestion and also destroys foreign organisms. HCl is produced in anticipation of eating and is produced in greater amounts when food enters the stomach.
- Pepsin, a protein-digesting enzyme. Pepsin is produced in an inactive form called *pepsinogen*, which is activated only when it contacts HCl or previously activated pepsin molecules.

Ingested food, gastric juice, and mucus (which is also secreted by cells of the gastric lining) are mixed to form **chyme** (*kime*), from a Greek word meaning "juice." This semiliquid, highly acidic substance is released gradually from the stomach through the pyloric sphincter into the small intestine for further digestion.

CASEPOINTS 

- [ ] **17-3** In the case study, Benjamin's stomach contents were escaping into his esophagus. Which sphincter normally keeps the stomach contents away from the esophagus?
- [ ] **17-4** In Benjamin's follow-up visit, he finds that his condition is complicated by a hiatal hernia. What organ passes through the hiatus?

THE SMALL INTESTINE

The small intestine is the longest part of the digestive tract (FIG. 17-7). It is known as the small intestine because, although it is longer than is the large intestine, it is smaller in diameter, with an average width of approximately 2.5 cm (1 in). After death, when relaxed to its full length, the small intestine is approximately 6 m (20 ft) long. In life, contraction of the longitudinal muscle shortens the small intestine to about 3 m (10 ft) in length. The first 25 cm (10 in) or so of the small intestine make up the **duodenum** (du-o-DE-num) (named for the Latin word for "twelve," based on its length of 12 finger widths). Beyond the duodenum are two more divisions: the **jejunum** (je-JU-num), which forms the next two-fifths of the small intestine, and the **ileum** (IL-e-um), which constitutes the remaining portion.

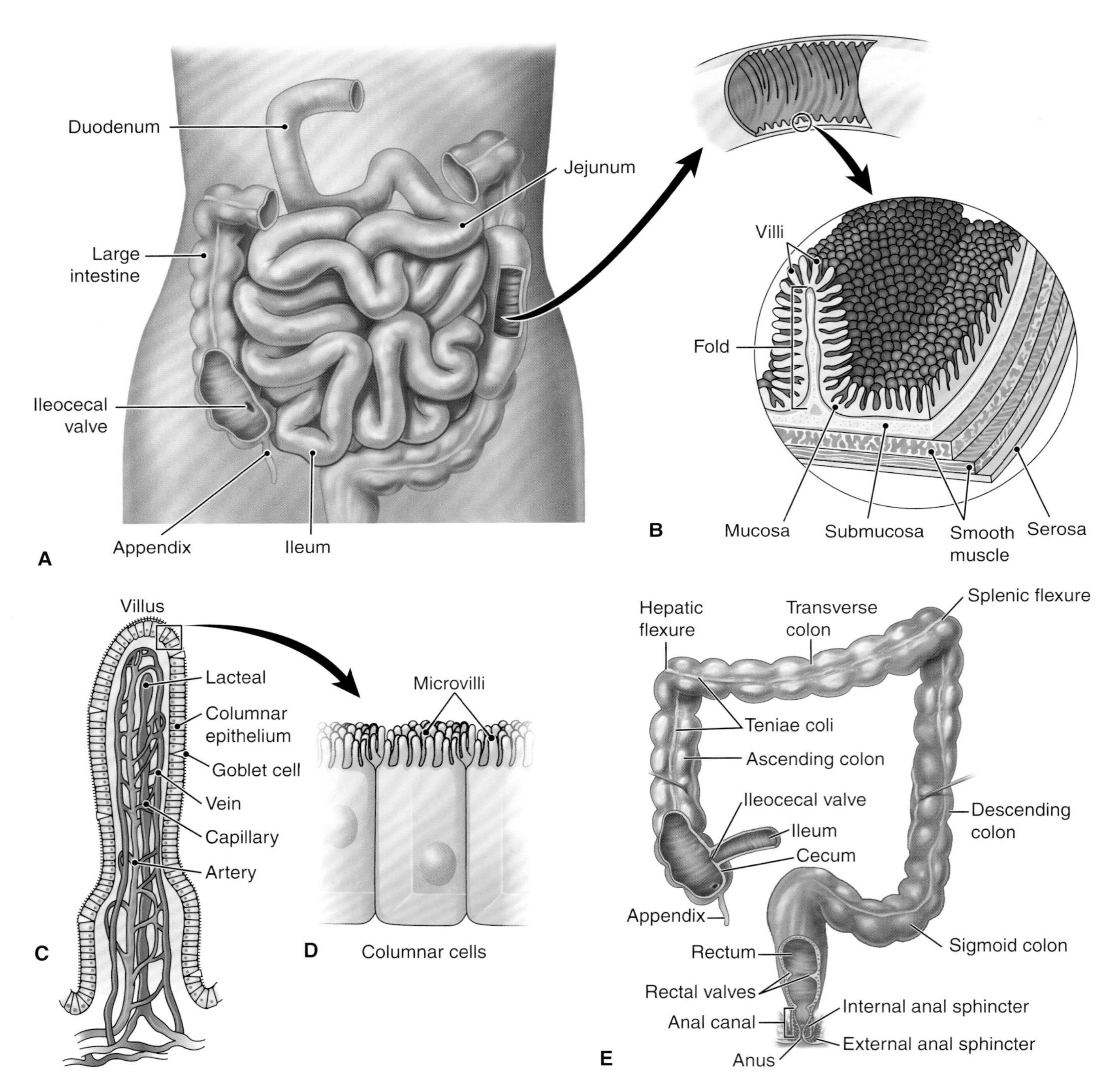

FIGURE 17-7 The intestines. KEY POINT The intestines have modifications that increase surface area. **A.** The small and large intestines. The colon is the main portion of the large intestine. **B.** The wall of the small intestine. **C.** Drawing of a villus. **D.** The plasma membrane of the columnar epithelial cells is folded into microvilli. **E.** Structure of the large intestine. ZOOMING IN Which portions of the small and large intestines join at the ileocecal valve?

Functions of the Small Intestine The small intestine participates in all aspects of digestive function: secretion, motility, digestion, and absorption.

Secretion Glands in the duodenal mucosa and submucosa secrete large amounts of mucus to protect the small intestine from the strongly acidic chyme entering from the stomach. Mucosal cells of the small intestine also produce enzymes that participate in the final stages of carbohydrate and protein digestion (discussed later under the digestive process). These enzymes are inserted into the cells' plasma membrane and act on nutrients that come in contact with the intestinal lining.

Motility Minimal peristalsis occurs in the small intestine. This form of motility is too rapid to allow for effective digestion and absorption in the small intestine. Instead, the process of segmentation (see FIG. 17-4) ensures that the food is thoroughly mixed with digestive juices and is placed in contact with the enzymes at the mucosal surface. Segmentation is regulated so that proximal segments contract before distal segments, slowly propelling the intestinal contents from the duodenum to the end of the ileum.

Digestion Most digestion takes place in the small intestine's lumen under the influence of secretions from the liver and pancreas as well as enzymes in the membranes of intestinal cells. These digestive processes are discussed in detail later.

Absorption As noted above, the means by which digested nutrients reach the blood is known as absorption. Most absorption takes place through the villi in the mucosa of the small intestine. FIGURE 17-7 shows three modifications of the small intestine lining that increase the surface area and thus maximize nutrient absorption.

1. The mucosa and submucosa are formed into large circular folds (see FIG. 17-7B).
2. The mucosa of each fold is formed into millions of tiny, finger-like projections, or **villi** (VIL-li), which give the inner surface a velvety appearance. Each villus contains a capillary and a specialized lymphatic capillary called a **lacteal** (LAK-tele) (see FIG. 17-7C).
3. The epithelial cells of the villi have small projecting folds of the plasma membrane known as **microvilli** (see FIG. 17-7D).

Simple sugars (monosaccharides), small peptides, amino acids, a few simple fatty acids, nucleic acids, water-soluble vitamins, and most of the water in the digestive tract are absorbed into the blood through these capillaries. From here, they pass by way of the portal vein to the liver, to be processed, stored, or released as needed. The role of the liver in digestion is discussed further below. Fats and fat-soluble vitamins, however, are absorbed into the lacteal, as discussed later.

Concept Mastery Alert

Make sure that you can distinguish between circular folds, villi, and microvilli.

BOX 17-1 has more information about the intestinal folds and how they promote nutrient absorption.

THE LARGE INTESTINE

The large intestine is approximately 6.5 cm (2.5 in) in diameter and approximately 1.5 m (5 ft) long (see FIG. 17-7E). It is named for its wide diameter, rather than its length. The outer longitudinal muscle fibers in its wall gather into three separate surface bands. These bands, known as **teniae** (TEN-e-e) **coli**, draw up the organ's wall to give it its distinctive puckered appearance. (The name is also spelled *taeniae*; the singular is *tenia* or *taenia*).

Subdivisions of the Large Intestine The large intestine begins in the lower right region of the abdomen. The first part is a small pouch called the **cecum** (SE-kum). The **ileocecal** (il-e-o-SE-kal) **valve** permits food passage from the ileum

A CLOSER LOOK

The Folded Intestine: More Absorption with Less Length

BOX 17-1

Whenever materials pass from one system to another, they must travel through a cellular membrane. A major factor in how much transport can occur per unit of time is the total surface area of the membrane; the greater the surface area, the higher the rate of transport. The problem of packing a large amount of surface into a small space is solved in the body by folding the membranes. We do the same thing in everyday life. Imagine trying to store a bed sheet in the closet without folding it!

In the small intestine, where digested food must absorb into the bloodstream, there is folding of membranes down to the level of single cells.

- The 6-m long organ is coiled to fit into the abdominal cavity.
- The inner wall of the organ is thrown into circular folds called plicae circulares, which not only increase surface area, but aid in mixing.
- The mucosal villi project into the lumen, providing more surface area than a flat membrane would.
- The individual cells that line the small intestine have microvilli, tiny finger-like folds of the plasma membrane that increase surface area tremendously.

Together, these structural features of the small intestine result in an absorptive surface area estimated to be about 250 m^2! Folding is present in other parts of the digestive system and in other areas of the body as well. Can you name other systems that show this folding pattern?

of the small intestine into the cecum, but not vice versa. Attached to the cecum is a small, blind tube containing lymphoid tissue; its full name is **vermiform** (VER-mih-form) **appendix** (*vermiform* means "wormlike") but usually just "appendix" is used.

The second portion, the **ascending colon**, extends superiorly along the right side of the abdomen toward the liver. It bends near the liver at the hepatic (right) flexure and extends across the abdomen as the **transverse colon**. It bends again sharply at the splenic (left) flexure and extends inferiorly on the left side of the abdomen into the pelvis, forming the **descending colon**. The distal colon bends backward into an S shape forming the **sigmoid colon** (named for the Greek letter *sigma*), which continues downward to empty into the **rectum**, a temporary storage area for indigestible or nonabsorbable food residue (see FIG. 17-7E). The narrow terminal portion of the large intestine is the **anal canal**, which leads to the outside of the body through an opening called the **anus** (A-nus).

Functions of the Large Intestine The colonic epithelium produces a great quantity of mucus, but no enzymes. Minimal digestion occurs in this organ, but some water is reabsorbed, and undigested food is stored, formed into solid waste material, called **feces** (FE-seze) or stool, and then eliminated.

Food waste moves through the large intestine by intermittent **mass movements**, waves of peristalsis that propel the contents up to several feet at a time toward the rectum. These movements often occur just after meals. Stretching of the rectum stimulates smooth muscle contraction in the rectal wall. Aided by voluntary contractions of the diaphragm and the abdominal muscles, the feces are eliminated from the body in a process called **defecation** (def-e-KA-shun). An anal sphincter provides voluntary control over defecation (see FIG. 17-7E).

Food waste remains quite stationary between mass movements. During these periods, bacteria that normally live in the colon act on it to produce vitamin K and some of the B complex vitamins. These vitamins are then absorbed by the mucosa of the large intestine. Systemic antibiotic therapy may destroy these symbiotic (helpful) bacteria living in the large intestine, causing undesirable side effects.

CHECKPOINTS 

- ☐ **17-4** Which form of motility occurs in the esophagus? In the small intestine?
- ☐ **17-5** What type of food is digested in the stomach?
- ☐ **17-6** What are the three divisions of the small intestine?
- ☐ **17-7** How does the small intestine function in the digestive process?
- ☐ **17-8** What are the functions of the large intestine?

CASEPOINT 

- ☐ **17-5** Benjamin's colonoscopy examined the health of his large intestine up to the cecum. Beginning with the anus, list the structures contacted by the instrument.

The Accessory Organs

The accessory organs release secretions through ducts into the digestive tract. Specifically, the salivary glands deliver their secretions into the mouth. The liver, gallbladder, and pancreas release secretions into the duodenum.

THE SALIVARY GLANDS

Saliva (sah-LI-vah) is a watery solution that moistens food and facilitates mastication (chewing) and deglutition (swallowing). Saliva helps keep the teeth and mouth clean. It also contains some antibodies and an enzyme (lysozyme) that help reduce bacterial growth.

This watery mixture contains mucus and an enzyme called **salivary amylase** (AM-ih-laze), which initiates carbohydrate digestion. Saliva is manufactured by three pairs of glands (FIG. 17-8):

- The **parotid** (pah-ROT-id) **glands**, the largest of the group, are located inferior and anterior to the ear.
- The **submandibular** (sub-man-DIB-u-lar) **glands**, also called *submaxillary* (sub-MAK-sih-ler-e) *glands*, are located near the body of the lower jaw.
- The **sublingual** (sub-LING-gwal) **glands** are under the tongue.

All these glands empty through ducts into the oral cavity.

THE LIVER

The **liver** (LIV-er), often referred to by the word root *hepat*, is the largest accessory organ (FIG. 17-9 and FIG. A3-10 **in the Dissection Atlas**). It is located in the superior right portion of the abdominal cavity under the dome of the diaphragm. The lower edge of a normal-sized liver is level with the ribs' lower margin. The human liver is the same reddish brown color as animal liver seen in the supermarket. It has a large right lobe and a smaller left lobe; the right lobe includes

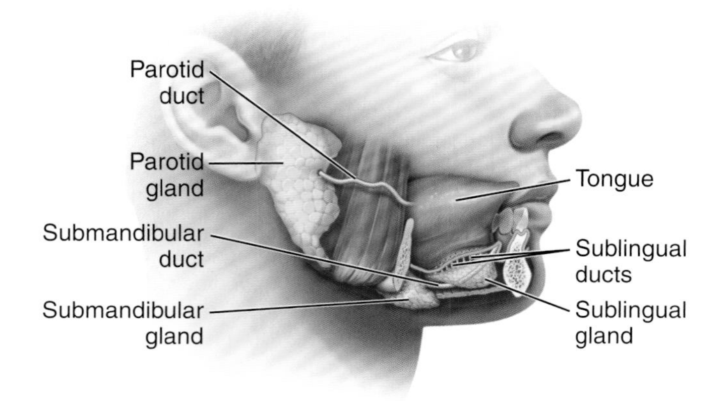

FIGURE 17-8 Salivary glands. KEY POINT Three pairs of glands secrete into the mouth through ducts. ZOOMING IN Which salivary glands are directly below the tongue?

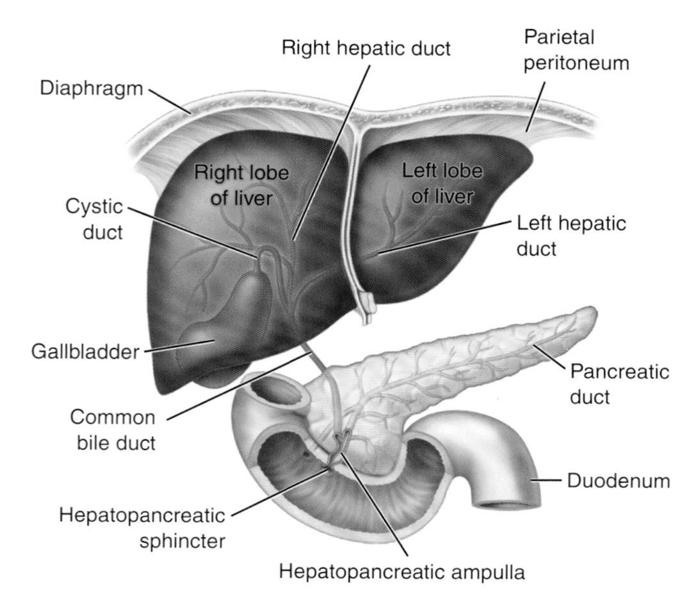

FIGURE 17-9 Accessory organs of digestion. **KEY POINT** The accessory organs secrete digestive substances into the small intestine. ZOOMING IN Into which part of the small intestine do these accessory organs secrete?

two inferior smaller lobes that are not illustrated here. The liver is supplied with blood through two vessels: the portal vein and the hepatic artery (the portal system and blood supply to the liver were described in Chapter 14). These vessels deliver about 1.5 qt (1.6 L) of blood to the liver every minute. The hepatic artery carries blood high in oxygen, whereas the venous portal system carries blood that is lower in oxygen and rich in digestive end products.

Liver Functions This most remarkable organ has many functions relating to digestion, metabolism, blood composition, and elimination of waste. Some of its major activities are:

- The manufacture of **bile**, a substance needed for the digestion of fats, discussed shortly.
- The control of blood glucose. As discussed in Chapter 11, the liver is an effector organ in feedback loops involving the hormones insulin and glucagon. The liver stores glucose in the form of glycogen, the animal equivalent of the starch found in plants. When the blood glucose level falls below normal, glucagon stimulates the conversion of glycogen to glucose and the synthesis of new glucose molecules from amino acids. The glucose enters blood to restore a normal concentration. When blood glucose levels are high, insulin stimulates the conversion of glucose into glycogen in order to reduce the blood glucose concentration.
- The storage of fat, iron, and some vitamins.
- The formation of blood plasma proteins, such as albumin, globulins, and clotting factors.
- The destruction of old red blood cells and the recycling or elimination of their breakdown products. One byproduct, a pigment called **bilirubin** (BIL-ih-ru-bin), is eliminated in bile and gives the stool its characteristic dark color.
- The synthesis of **urea** (u-RE-ah), a waste product of protein metabolism. Urea is released into the blood and transported to the kidneys for elimination.
- The **detoxification** (de-tok-sih-fih-KA-shun) (removal of toxicity) of harmful substances, such as alcohol and certain drugs.

Bile The liver's main digestive function is the production of bile, a substance needed for the processing of fats. Because of their hydrophobic nature, fats gather into droplets. The salts contained in bile act like a detergent to **emulsify** fat; that is, to break up fat into smaller droplets that can be acted on more effectively by digestive enzymes. Bile also prevents the products of fat digestion from forming droplets, thus aiding in their absorption from the small intestine.

Concept Mastery Alert

Bile does not participate in the actual digestion of fats, nor in the processing of other nutrient types.

Bile leaves the lobes of the liver by two ducts that merge to form the **common hepatic duct.** This duct joins with the **cystic** (SIS-tik) **duct** (from the gallbladder) to form the **common bile duct** (FIG. 17-9).

Concept Mastery Alert

Remember that the common bile duct carries bile from both the liver and the gallbladder, but the common hepatic duct carries bile only from the liver.

The common bile duct joins with the pancreatic duct to form a very short, wide channel called the **hepatopancreatic ampulla.** The **hepatopancreatic sphincter** controls the emptying of pancreatic fluids and bile from the ampulla into the duodenum, as discussed next.

See the animation "The Liver in Health and Disease" in the Student Resources on thePoint®.

THE GALLBLADDER

Although the liver may manufacture bile continuously, the body needs it only a few times a day, so the hepatopancreatic sphincter is usually closed. Consequently, bile from the liver flows up through the cystic duct into the **gallbladder** (GAWL-blad-er), a muscular sac on the inferior surface of the liver that stores bile (see FIG. 17-9 and FIG. A3-10 **in the Dissection Atlas**). When chyme enters the duodenum, the gallbladder contracts and the hepatopancreatic sphincter opens. Bile flows out of the gallbladder and liver into the duodenum.

THE PANCREAS

The **pancreas** (PAN-kre-as) is a long gland that extends from the duodenum to the spleen (see FIG. 17-9 and FIG. A3-9C **in**

17

the Dissection Atlas). The pancreas produces enzymes that digest fats, proteins, carbohydrates, and nucleic acids. The protein-digesting enzymes are produced in inactive forms, which must be converted to active forms in the small intestine by other enzymes.

The pancreas also releases large amounts of sodium bicarbonate ($NaHCO_3$), an alkaline (basic) fluid that neutralizes the acidic chyme in the small intestine, thus protecting the digestive tract's lining. These juices collect in the pancreatic duct and along with bile, enter the duodenum via the hepatopancreatic sphincter. Most people have an additional smaller pancreatic duct that opens directly into the duodenum and bypasses the hepatopancreatic ampulla.

As described in Chapter 11, the pancreas also functions as an endocrine gland, producing the hormones insulin and glucagon that regulate glucose metabolism. These islet cell secretions are released into the blood.

CHECKPOINTS

- ☐ **17-9** What are the names and locations of the salivary glands?
- ☐ **17-10** Which accessory organ secretes bile, and what is the function of bile in digestion?
- ☐ **17-11** What is the role of the gallbladder?
- ☐ **17-12** What accessory organ secretes sodium bicarbonate, and what is the function of this substance in digestion?

Table 17-2 Digestive Agents Produced by Digestive Tract Organs and Accessory Organs

Organ	Substance	Action
Salivary glands	Salivary amylase	Converts starch to maltose
Stomach	Hydrochloric acid (HCl)[a]	Denatures proteins; activates pepsin
	Pepsin	Digests proteins to peptides
Liver	Bile salts[a]	Emulsify fats
Pancreas	Sodium bicarbonate[a]	Neutralizes HCl
	Amylase	Digests starch to maltose
	Trypsin	Digests protein to peptides
	Lipase	Digests fats to fatty acids and monoglycerides
	Nucleases	Digest nucleic acids
Small intestine (brush border enzymes)	Peptidases	Digest peptides to amino acids
	Lactase, maltase, sucrase	Digest disaccharides to monosaccharides

[a]Not enzymes.

Nutrient Digestion and Absorption

7 The chemical digestion of fats, proteins, carbohydrates, and nucleic acids involves breaking the covalent bonds holding the subunits together (**see FIGS. 2-8 through 2-10**). Digestion and absorption occur throughout the digestive tract, as summarized in **TABLE 17-1**. Specific enzymes, as well as other substances, are used for the digestion of each nutrient (**TABLE 17-2**). Recall from Chapter 2 (**see FIG. 2-11**) that enzymes are catalysts, substances that speed the rate of specific chemical reactions, but are not themselves changed or used up in the reactions. The digestion and absorption of nucleic acids are relatively straightforward. **Nucleases** (NU-kle-ases) in the pancreatic juice digest DNA and RNA, and the products are absorbed into intestinal capillaries. Next, we follow the fate of proteins, carbohydrates, and fats as they pass through the digestive tract and enter the bloodstream.

CARBOHYDRATES

Carbohydrate digestion occurs in the mouth and the small intestine. Saliva contains the enzyme amylase, which digests some starch (consisting of many glucose molecules) into maltose (consisting of two glucose molecules). However, amylase can only access the starch molecules on the outer

Table 17-1 Activity of the Digestive Organs

Organ	Activity	Nutrients Digested	Active Substances
Mouth	Chews food and mixes it with saliva; forms it into bolus for swallowing	Starch	Salivary amylase
Esophagus	Moves food by peristalsis into stomach	—	—
Stomach	Stores food, churns food, and mixes it with digestive juices	Proteins	Hydrochloric acid, pepsin
Small intestine	Mixes food with secretions from pancreas and liver, digests food, absorbs nutrients and water into the blood or lymph	Fats, proteins, carbohydrates, nucleic acids	Intestinal enzymes (in cell membranes), pancreatic enzymes, bile from liver
Large intestine	Reabsorbs some water; forms, stores, and eliminates stool	—	—

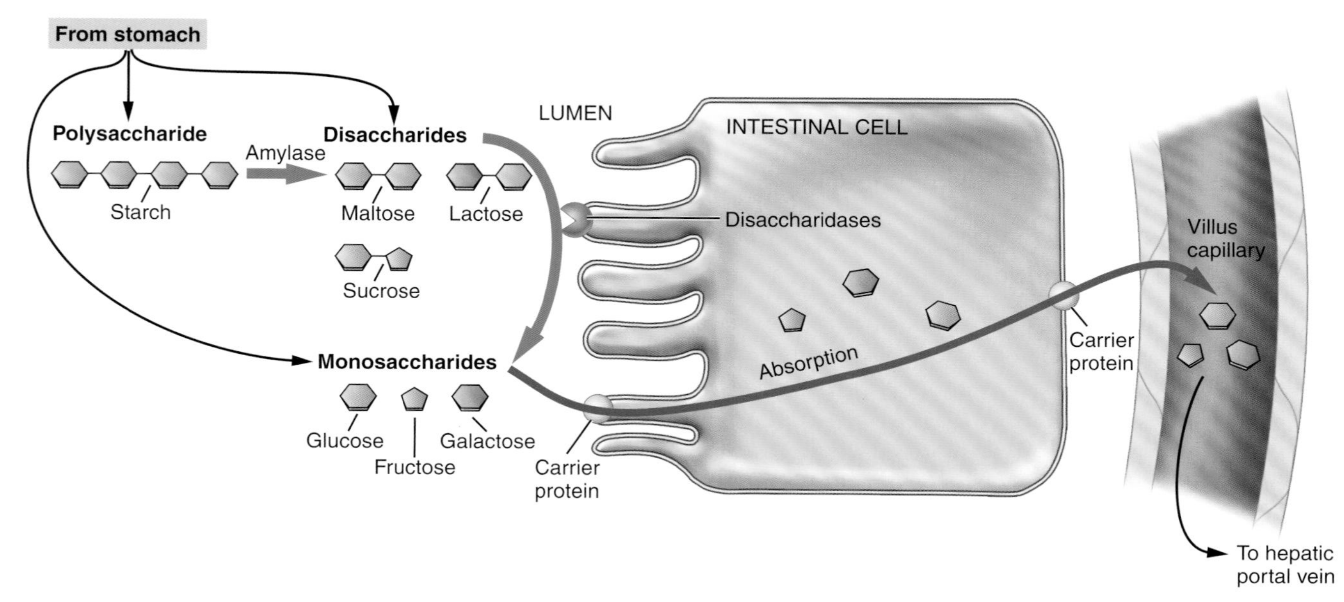

FIGURE 17-10 Carbohydrate digestion. KEY POINT Carbohydrates are digested into monosaccharides before absorption can occur. ZOOMING IN Name the two monosaccharide components of sucrose.

surface of the food bolus and only has limited time to work before it is inactivated by the stomach. So, most starch digestion occurs in the small intestine under the influence of pancreatic amylase. Enzymes in the intestinal cell membranes digest maltose and the other disaccharides into monosaccharides. These enzymes are **maltase**, **sucrase**, and **lactase**, which act on the disaccharides maltose, sucrose, and lactose, respectively (FIG. 17-10). The monosaccharides are then absorbed into the villus capillary. From here, they pass by way of the portal system to the liver, to be processed, stored, or released as needed.

PROTEINS

Protein digestion occurs in the stomach and small intestine. As discussed earlier, gastric juice contains HCl, which denatures (unwinds) proteins to prepare them for digestion. In addition, HCl activates the enzyme pepsin, which is secreted by gastric cells in an inactive form. Once activated by HCl, pepsin works to cleave long protein chains into shorter chains of amino acids.

Protein digestion is completed in the small intestine. Pancreatic enzymes, such as **trypsin** (TRIP-sin), split proteins into small peptides. Other pancreatic enzymes and enzymes in the intestinal cell plasma membranes digest these short peptides into individual amino acids. Amino acids and even some undigested small peptides are then absorbed into a villus capillary and travel to the liver.

FATS

While a small amount of fat digestion may occur in the mouth and stomach, we will limit our discussion to fat digestion in the small intestine. Most ingested fat is in the form of **triglycerides**, which are composed of glycerol and three fatty acids. First, bile emulsifies fats into tiny droplets. Next, the pancreatic enzyme **lipase** digests the triglycerides into free fatty acids (two from each triglyceride) and monoglycerides (glycerol combined with one fatty acid). If pancreatic lipase is absent, fats are expelled with the feces in an undigested form.

Fatty acids and monoglycerides diffuse into the intestinal cells, where most are reassembled into triglycerides in the smooth endoplasmic reticulum (ER). They are then packaged along with proteins into particles called **chylomicrons** (ki-lo-MI-kronz) that are soluble in blood. However, chylomicrons are too large to enter blood capillaries. Instead, they are absorbed by the villi's more permeable lymphatic capillaries, the lacteals. The absorbed fat droplets give the lymph a milky appearance. The mixture of lymph and fat globules that drains from the small intestine after fat has been digested is called **chyle** (kile). Chyle merges with the lymphatic circulation and eventually enters the blood when the lymph drains into veins near the heart. The absorbed fats then move to the liver for further processing.

Fat soluble vitamins, such as vitamins D (calciferol) and E (tocopherol), are absorbed alongside dietary fats. These vitamins are discussed in greater detail in Chapter 18.

THE ROLE OF WATER IN DIGESTION

6 Each of the digestive reactions discussed above requires water. Because water is added to nutrient molecules as they are split by enzymes, the process of digestion is referred to chemically as **hydrolysis** (hi-DROL-ih-sis), which means "splitting (lysis) by means of water (hydr/o)." In this chemical process, water's hydroxyl group (OH^-) is added to one fragment and the hydrogen ion (H^+) is added to the other, splitting the molecule. FIGURE 17-11 shows an example of hydrolysis.

Each hydrolysis reaction requires a specific enzyme and uses one molecule of water. About 7 L of water are secreted

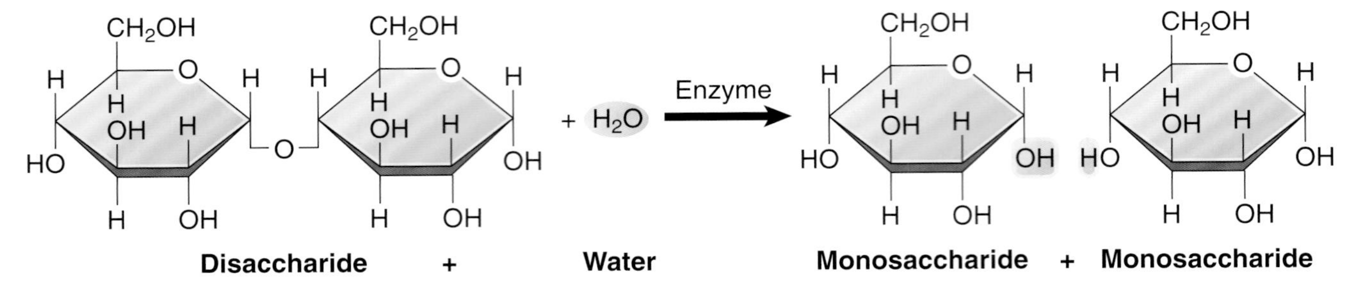

FIGURE 17-11 Hydrolysis. KEY POINT In digestion, water is added to compounds to split them into simpler building blocks. The figure shows the splitting of a disaccharide (double sugar) into two monosaccharides (simple sugars) by the addition of H^+ to one and OH^- to the other. The chemical bonds between the building blocks of fats and proteins are split in the same way. A specific enzyme is needed for each hydrolysis reaction.

into the digestive tract each day, in addition to the nearly 2 L taken in with food and drink. You can now understand why so large an amount of water is needed. Water is not only used to produce digestive juices and to dilute food so that it can move more easily through the digestive tract but is also used in the chemical process of digestion itself.

CHECKPOINTS

- [ ] **17-13** Name the four nutrients digested by enzymes in the intestinal cell plasma membranes.
- [ ] **17-14** What process means "splitting by means of water," as in digestion?
- [ ] **17-15** Which organ produces the enzymes found in the intestinal lumen?
- [ ] **17-16** Which nutrients enter the intestinal capillaries and which enter the lacteals?

CASEPOINTS

- [ ] **17-6** Benjamin's medication reduces acid production in his stomach. Which stomach enzyme could be affected by this change?
- [ ] **17-7** Which pancreatic enzymes could compensate for the decreased activity of this enzyme?

Control of Digestion

As food moves through the digestive tract, its rate of movement and the activity of each organ it passes through must be carefully regulated. If food moves too slowly or digestive secretions are inadequate, the body will not get enough nourishment. If food moves too rapidly or excess secretions are produced, digestion and absorption may be incomplete, or the digestive tract's lining may be damaged. 3 The body senses the amount and concentration of the digestive tract contents and modifies the activity of the digestive organs using hormonal and nervous signals.

THE ENTERIC NERVOUS SYSTEM

The nerves that control digestive activity are located in the submucosa and between the muscle layers of the organ walls (see FIG. 17-3). These nerves form a complete neural network known as the **enteric nervous system** (ENS), or less formally, the "gut brain." Like any nervous system, the ENS receives sensory stimuli, integrates information from many sources, and issues commands. More specifically, the ENS receives information from sensors in the digestive tract wall, and controls the glands and smooth muscle of the digestive system. The activity of the ENS can be modified by the autonomic (visceral) nervous system. In general, parasympathetic stimulation increases activity, and sympathetic stimulation decreases activity. Excess sympathetic stimulation, as can be caused by stress, can slow food's movement through the digestive tract and inhibit mucus secretion, which is crucial to protecting the digestive tract's lining.

PATHWAYS CONTROLLING DIGESTION

The body prepares for digestion even before food reaches the stomach. The sight, smell, thought, taste, or feel of food in the mouth stimulates, through the autonomic nervous system, the secretion of saliva and the release of gastric juice. This form of regulation is called *anticipatory* or *feed-forward* control, since it anticipates the need for these secretions. Once in the stomach, food stimulates the release of the hormone **gastrin** into the blood. Gastrin further promotes stomach secretions and motility.

Chyme entering the duodenum is sensed by receptors in the intestinal wall. These receptors activate nerve pathways that inhibit stomach motility, so that food will not move too rapidly into the small intestine. Chyme also stimulates the release of intestinal hormones that reduce gastric motility and acid production. These electrical and hormonal signals thus act together in a negative feedback loop controlling the volume and composition of the intestinal contents.

Each intestinal hormone is also involved in a specific feedback loop, as illustrated in FIGURE 17-12. For instance, fats and proteins in chyme are the regulated variables shown in the left panel. They stimulate the release of **cholecystokinin** (ko-le-sis-to-KI-nin) (CCK). CCK stimulates the secretion of pancreatic enzymes and the release of bile from the gallbladder. These secretions digest the fats and proteins, thereby lowering their concentration. The pH of the intestinal contents is the regulated variable in the middle panel of FIGURE 17-12. The signal in this loop, **secretin** (se-KRE-tin), is released in response to increased acidity. Secretin stimulates

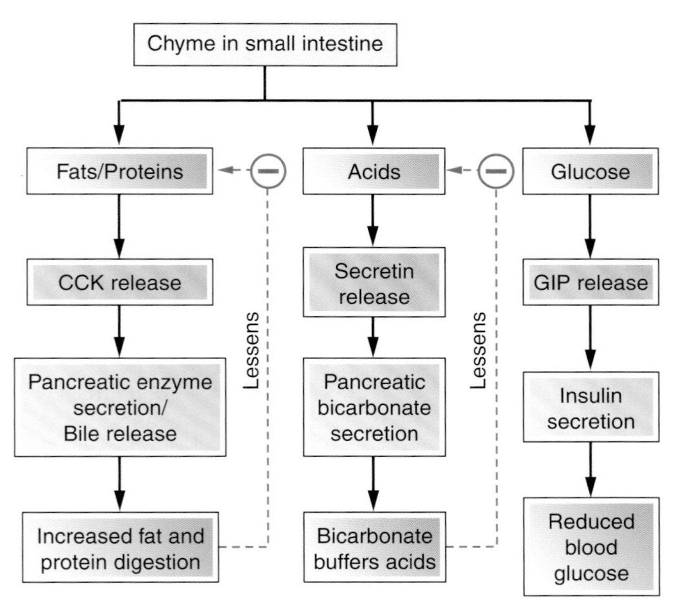

FIGURE 17-12 Hormonal control of digestion. KEY POINT Intestinal hormones act as signals in negative feedback loops controlling the intestinal contents. GIP also acts as an anticipatory signal in blood glucose control. In addition to the specific actions shown here, all three hormones inhibit gastric motility. Note that the different components of the feedback loop are indicated by different colors (regulated variable, green; signals, purple; effectors, blue). The sensors and control center for each loop are not illustrated. ZOOMING IN Name the hormone that increases the pH of the intestinal contents.

the pancreas to release bicarbonate, which acts as a buffer to raise the pH of the intestinal contents.

Elevated concentrations of glucose in the intestine stimulate the release of **gastric inhibitory peptide** (GIP) (right panel). GIP, in turn, stimulates insulin release, which reduces blood glucose in anticipation of glucose absorption from the chyme.

Control of Hunger and Appetite

Hunger is the desire for food, which can be satisfied by the ingestion of a filling meal. Appetite differs from hunger in that although it is basically a desire for food, it often has no relationship to the need for food. Even after an adequate meal that has relieved hunger, a person may still have an appetite for additional food. A variety of factors, such as emotional state, cultural influences, habit, and memories of past food intake, can affect appetite.

Hunger is regulated by hypothalamic centers that respond to nutrient levels, neural input from the digestive tract, and hormones.

SHORT-TERM REGULATION OF HUNGER

The hypothalamus induces hunger sensations when we haven't eaten in a while and fullness sensations when we should stop eating. Between meals, the empty stomach releases a hormone called **ghrelin** (GREL-in) that stimulates hunger (FIG. 17-13A). Low nutrient levels, particularly hypoglycemia, also activate the hypothalamic hunger center. On the other hand, food consumption distends the digestive tract, sending neural signals that decrease hunger. Also, remember that food arrival in the duodenum stimulates the release of the hormones CCK and GIP, and GIP stimulates insulin production. CCK and insulin further suppress hunger, as does the increased blood glucose level resulting from nutrient digestion (FIG. 17-13B). These neural and hormonal signals prevent the consumption of more food than the intestine can process.

LONG-TERM REGULATION OF BODY WEIGHT

Despite day-to-day variations in food intake and physical activity, a healthy individual maintains a constant body weight and energy reserves of fat over long periods. With the discovery of the hormone leptin (from the Greek word *leptos*, meaning "thin"), researchers have been able to piece together one long-term mechanism for regulating weight. Leptin is produced by adipocytes (fat cells). When fat stores increase because of excess food intake or inadequate activity, the cells release more leptin (FIG. 17-13C). Centers in the hypothalamus respond to the hormone by decreasing food intake and increasing energy expenditure, resulting in weight loss. If this feedback mechanism is disrupted, obesity will result. Early hopes of using leptin to treat human obesity have dimmed, however, because obese people do not have a leptin deficiency. This system's failure in humans appears to be caused by the hypothalamus's inability to respond to leptin rather than our inability to make the hormone.

17

CHECKPOINTS 

- ☐ **17-17** What are the two types of control over the digestive process?
- ☐ **17-18** What is the difference between hunger and appetite?

CASEPOINT

- ☐ **17-8** Benjamin would have fasted before his GED procedure. Would his ghrelin level have been higher or lower than normal?

Effects of Aging on the Digestive System

With age, receptors for taste and smell deteriorate, leading to a loss of appetite and decreased enjoyment of food. A decrease in saliva and poor gag reflex make swallowing more difficult. Tooth loss or poorly fitting dentures may make chewing food more difficult.

Activity of the digestive organs decreases. These changes can be seen in poor absorption of certain vitamins and poor protein digestion. Slowing of peristalsis in the large intestine

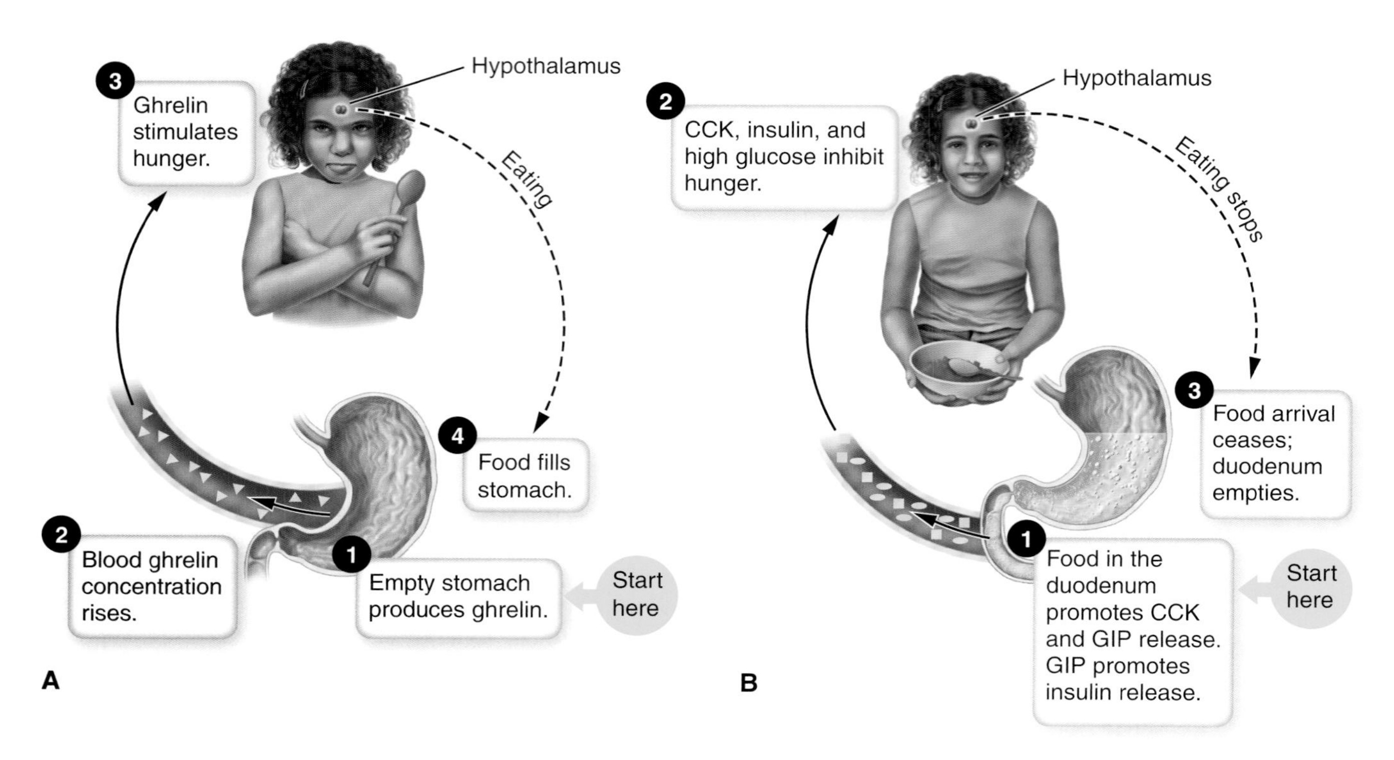

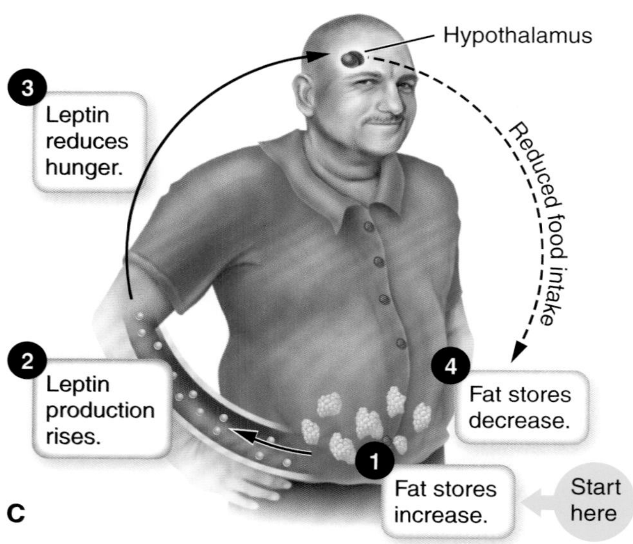

FIGURE 17-13 Regulation of food intake. A. Ghrelin initiates hunger. **B.** CCK (among other factors) inhibits hunger. **C.** Leptin governs food intake over the long term.

and increased consumption of easily chewed, refined foods contribute to the common occurrence of constipation.

The tissues of the digestive system require constant replacement. Slowing of this process contributes to a variety of digestive disorders, including gastritis, ulcers, and diverticulosis. As with many body systems, tumors and cancer occur more frequently with age.

As discussed in **BOX 17-2**, endoscopy is a nonsurgical medical technique that can detect digestive tract tumors. It can be an important part of healthcare in older adults. In the case study, Benjamin had a normal colonoscopy the year before consulting his physician for his GERD symptoms. His EGD examination involved endoscopy of the esophagus, stomach, and duodenum.

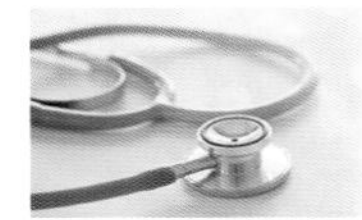

CLINICAL PERSPECTIVES
Endoscopy: A View from Within

BOX 17-2

Modern medicine has made great strides toward looking into the body without resorting to invasive surgery. An instrument that has made this possible in many cases is the **endoscope**, which is inserted into the body through an orifice or small incision and used to examine passageways, hollow organs, and body cavities. The first endoscopes were rigid lighted telescopes that could be inserted only a short distance into the body. Today, physicians are able to navigate the twists and turns of the digestive tract using long **fiberoptic endoscopes** composed of flexible bundles of glass or plastic that transmit light.

In the GI tract, endoscopy can detect structural abnormalities, bleeding ulcers, inflammation, and tumors. In addition, endoscopes can be used to remove fluid samples or tissue specimens. Some surgery can even be done with an endoscope, such as polyp removal from the colon or expansion of a sphincter. Endoscopy can also be used to examine and operate on joints (arthroscopy), the bladder (cystoscopy), respiratory passages (bronchoscopy), and the abdominal cavity (laparoscopy).

Capsular endoscopy, a recent technological advance, has made examination of the **gastrointestinal tract** even easier. It uses a pill-sized camera that can be swallowed. As the camera moves through the digestive tract, it transmits video images to a data recorder worn on the patient's belt. Each camera is used only once; after about 24 hours, it is eliminated with the stool and flushed away.

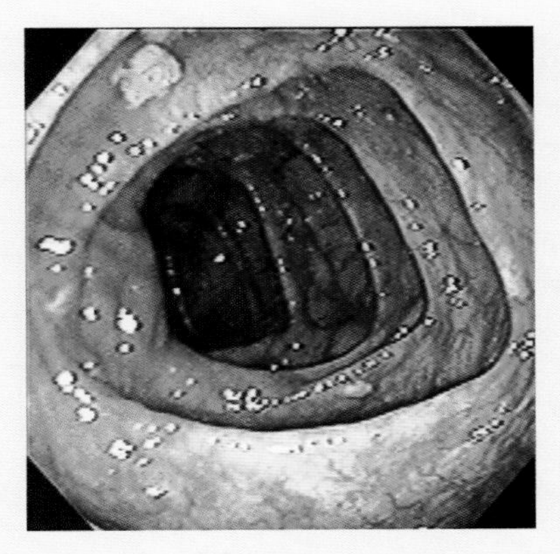

An endoscopic image of the cecum, the first portion of the large intestine.

17

A & P in Action Revisited: Benjamin's Gastroesophageal Reflux Disease (GERD)

Benjamin returned to Dr. Eliason's office 4 weeks later for his follow-up appointment. "So, Benjamin, has the medication helped?" he asked.

Looking sheepish, Benjamin replied, "I started feeling better after a few weeks, so I stopped taking the medication. Now the pain and the other symptoms have returned, and I'm waking up at night. I'm also having difficulty swallowing sometimes. I've never had that problem before."

Dr. Eliason explained, "The medication wasn't designed to be a fast cure. You need to take it consistently, because your endoscopy revealed moderate damage to your esophagus, as well as a small hiatal hernia. Without treatment, your reflux disease could damage your lungs and cause further damage to your esophagus. But, medication and some lifestyle changes should prevent further complications. You need to continue the medication for 3 continuous months. Then we'll see if you have healed and can manage with proper habits alone." He counseled Benjamin to decrease the fat in his meals, avoid lying down for a least two hours after meals, and limit alcohol intake.

CHAPTER 17 Chapter Wrap-Up

OVERVIEW

A detailed chapter outline with space for note-taking is on thePoint®. The figure below illustrates the main topics covered in this chapter.

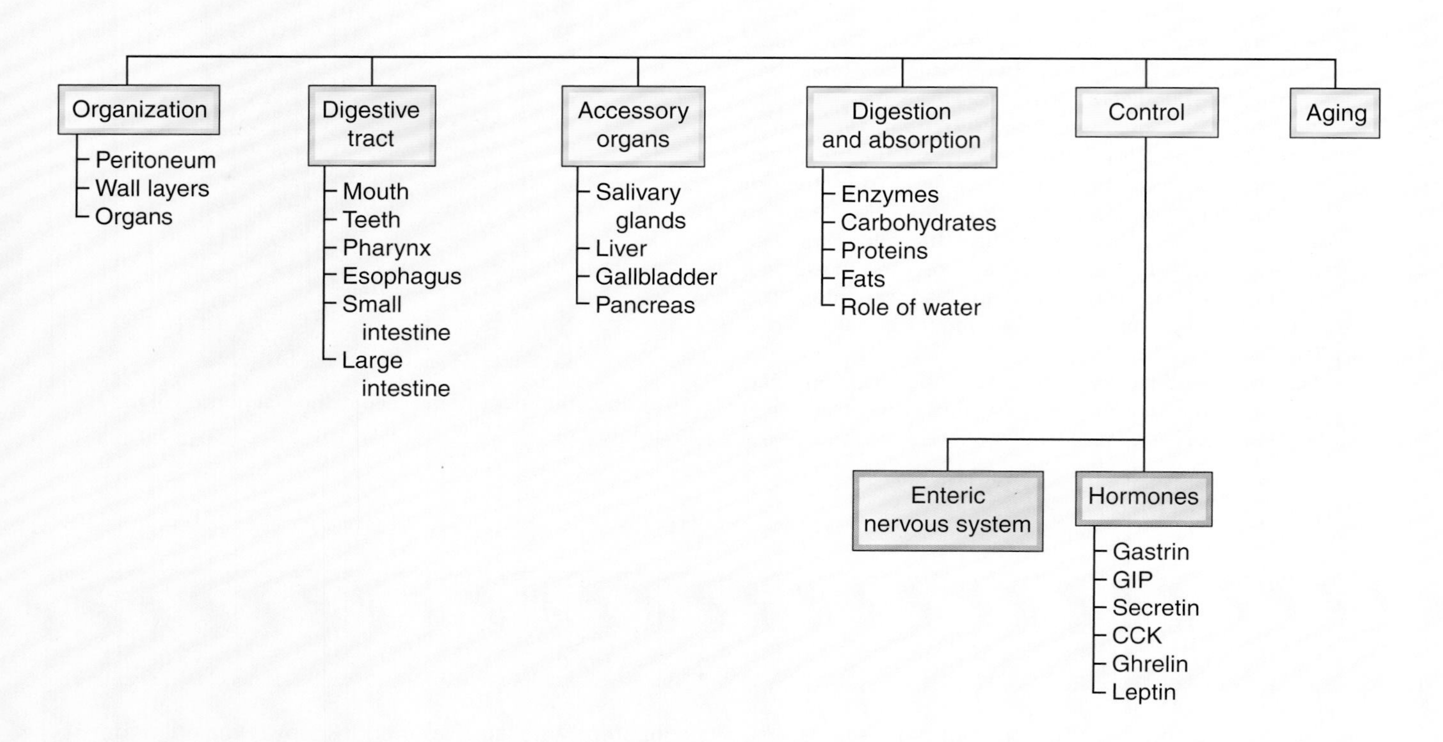

KEY TERMS

The terms listed below are emphasized in this chapter. Knowing them will help you organize and prioritize your learning. These and other bold face terms are defined in the Glossary with phonetic pronunciations.

absorption
bile
chyle
chyme
defecation
deglutition
digestion
duodenum
emulsify
esophagus
gallbladder
hydrolysis
ingestion
intestine
lacteal
liver
mastication
mesentery
pancreas
peristalsis
peritoneum
saliva
segmentation
sphincter
stomach
villi

WORD ANATOMY

Medical terms are built from standardized word parts (prefixes, roots, and suffixes). Learning the meanings of these parts can help you remember words and interpret unfamiliar terms.

WORD PART	MEANING	EXAMPLE
GENERAL STRUCTURE AND FUNCTION OF THE DIGESTIVE SYSTEM		
ab-	away from	In *absorption*, digested materials are taken from the digestive tract into the circulation.
enter/o	intestine	The *mesentery* is the portion of the peritoneum around the intestine.
mes/o-	middle	The *mesocolon*, like the mesentery, comes from the middle layer of cells in the embryo, the mesoderm.
DIGESTIVE TRACT ORGANS		
gastr/o	stomach	The *gastrointestinal* tract consists mainly of the stomach and intestine.
THE ACCESSORY ORGANS		
amyl/o	starch	The starch-digesting enzyme in saliva is salivary *amylase*.
bil/i	bile	*Bilirubin* is a pigment found in bile.
cyst/o	bladder, sac	The *cystic* duct carries bile into and out of the gallbladder.
hepat/o	liver	The *hepatic* portal system carries blood to the liver.
lingu/o	tongue	The *sublingual* salivary glands are under the tongue.
CONTROL OF DIGESTION		
chole	bile, gall	*Cholecystokinin* is a hormone that activates the gallbladder (cholecyst/o).

QUESTIONS FOR STUDY AND REVIEW

BUILDING UNDERSTANDING

Fill in the Blanks

1. The wavelike movement of the digestive tract wall is called ______.
2. The small intestine is connected to the posterior abdominal wall by ______.
3. The liver can store glucose in the form of ______.
4. The parotid glands secrete ______.
5. The membrane that lines the abdominal cavity and supports the organs it contains is the ______.

Matching Match each numbered item with the most closely related lettered item.

___ 6. Digests starch
___ 7. Begins protein digestion
___ 8. Digests fats
___ 9. Splits protein into amino acids
___ 10. Emulsifies fats

a. lipase
b. amylase
c. trypsin
d. pepsin
e. bile salt

Multiple Choice

___ 11. The teeth break up food into small parts by a process called
a. absorption
b. deglutition
c. ingestion
d. mastication

___ 12. Which organ secretes hydrochloric acid and pepsin?
a. salivary glands
b. stomach
c. pancreas
d. liver

___ 13. What is the main substance of a tooth?
a. gingiva
b. cuspid
c. dentin
d. cementum

___ 14. What is the soft, fleshy V-shaped mass of tissue that hangs from the soft palate called?
a. epiglottis
b. esophageal hiatus
c. uvula
d. gingiva

___ 15. What structure guards the entrance to the trachea during swallowing?
a. vocal fold
b. gingiva
c. bolus
d. epiglottis

UNDERSTANDING CONCEPTS

16. Refer to the Dissection Atlas in Appendix 3 to answer the following questions:
    a. Look at **FIGURE A3-9C**. Which organ is anterior to the aorta, and which organ is posterior?
    b. Look at **FIGURE A3-10**. List all of the labeled structures in the order in which they encounter bile in its trip from the liver to the gallbladder and then to the duodenum.
17. Referring to the digestive system in "The Body Visible" at the beginning of this book, give the name and number of the following:
    a. the first portion of the large intestine
    b. the muscle that controls movement of food from the stomach to the small intestine
    c. the largest salivary gland
    d. folds in the lining of the stomach
    e. the middle portion of the small intestine
18. Differentiate between the terms in each of the following pairs:
    a. deciduous and permanent teeth
    b. digestion and absorption
    c. parietal and visceral peritoneum
    d. gastrin and gastric inhibitory peptide
    e. secretin and cholecystokinin
19. Name the four layers of the digestive tract. Which tissue makes up each layer? What is the function of each layer?
20. Trace the path of a bolus of food through the digestive system.
21. Describe the structure and function of the liver.
22. Describe the system that delivers secretions from the liver, pancreas, and gallbladder to the duodenum.
23. What is hydrolysis? Give three examples of hydrolysis.
24. Where does absorption occur in the digestive tract, and what structures are needed for absorption? Which types of nutrients are absorbed into the blood? Into the lymph?
25. Give one example of negative feedback and one example of anticipatory control involved in the regulation of digestion.
26. Name two hormones that regulate eating, and give the function of each.

CONCEPTUAL THINKING

27. In the case study, Dr. Eliason is concerned that Benjamin's reflux might damage his lungs. Referring to Chapters 16 and 17, describe the path that the stomach acid would take from the stomach to the alveoli.
28. Overuse of antacids can inhibit protein digestion. How?

For more questions, see the Learning Activities on thePoint®.

CHAPTER 18

Metabolism, Nutrition, and Body Temperature

Learning Objectives

After careful study of this chapter, you should be able to:

1. Give examples of catabolic and anabolic reactions. *p. 374*
2. Differentiate between the anaerobic and aerobic phases of glucose catabolism and give the end products and the relative amount of energy released by each. *p. 374*
3. Explain how fats and proteins are metabolized for energy. *p. 376*
4. Define metabolic rate, and name factors that affect it. *p. 376*
5. Describe the different fates of ingested carbohydrates, fats, and proteins. *p. 376*
6. Describe the types and functions of dietary carbohydrates, proteins, and fats. *p. 377*
7. Explain the roles of minerals and vitamins in nutrition, and give examples of each. *p. 380*
8. Describe the recommendations found in the USDA dietary guidelines. *p. 382*
9. Discuss the implications of alcohol consumption on health. *p. 382*
10. Explain how heat is gained and lost in the body. *p. 384*
11. Describe the elements of the negative feedback loops regulating body temperature in response to hot or cold conditions. *p. 385*
12. Using the case study and the text, describe the adverse effects of undernutrition. *pp. 373, 387*
13. Show how word parts are used to build words related to metabolism, nutrition, and body temperature (see Word Anatomy at the end of the chapter). *p. 389*

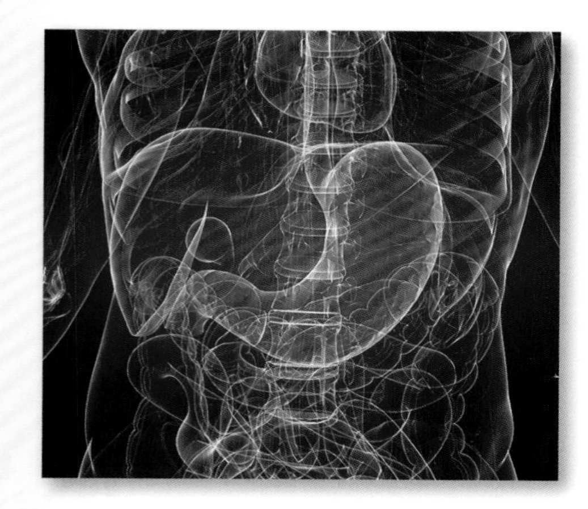

A & P in Action
Claudia's **Friends Are Concerned**

Dr. Wade, program advisor of Health Sciences at the university, was wrapping things up in her office for the day when Josie, a nursing student, stopped by.

"Hi Dr. Wade, do you have a minute to talk? I am worried about Claudia. She and I are roommates and close friends. After your lecture on eating disorders this week, well, I think she might have anorexia."

Dr. Wade told Josie to come in and sit down and then closed the door.

"I appreciate your concern for your classmate," she said. "What have you noticed?"

"We are in a couple of study groups together, and when we break for lunch, she always has an excuse for why she doesn't eat," Josie explained. "Either says she's not hungry, didn't bring a lunch, or doesn't have any money. We always offer to share our food but she won't eat. And yesterday when we all insisted she eat something, she had a few bites then went to the restroom. I think she may have made herself throw up. Also, after class, a bunch of us take the shuttle back to the dorm for dinner but Claudia has been jogging back and stops at the gym to work out. That might not sound like much but you should see her workout schedule; it's intense. She's always weighing herself, and I think she's overdoing the exercise. And recently when we were practicing measuring our BMIs for class, I noticed she didn't really participate in the exercise. She's pretty thin, and I know she thinks about her weight since she joined the cheerleading squad. She gets cold really easily, and I don't think she's had her period for months. I hate being a tattletale, but I'm really getting worried!"

"Josie, over half of teenage girls engage in unhealthy eating behaviors, and it can be hard to ask for help," Dr. Wade said. "You've made some good observations about your friend, and your timing is good in coming to me. I'll try to speak with Claudia tomorrow after class, but see if you can get her to seek help at the Student Health Center this evening. Also, let me give you some Web sites with tips on supporting friends with eating disorders."

Dr. Wade thought about Claudia after Josie left. She herself had notice Claudia's lack of concentration in class, and although a perfectionist, her academic performance was slipping. This student was showing clear signs of having an eating disorder.

Later, we will revisit Claudia and check on her condition. Nutrition and weight control will be discussed in this chapter, and we will see how malnutrition and being underweight can negatively affect one's health.

As you study this chapter, CasePoints will give you opportunities to apply your learning to this case.

Visit thePoint® to access the following resources. For guidance in using these resources most effectively, see pp. xv–xvii.

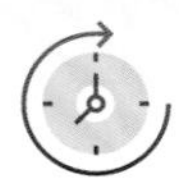

Preparing to Learn

- Tips for Effective Studying
- Pre-Quiz

While You Are Learning

- Web Figure: Effects of Alcoholism
- Chapter Notes Outline
- Audio Pronunciation Glossary

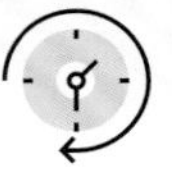

When You Are Reviewing

- Answers to Questions for Study and Review
- Health Professions: Dietitian and Nutritionist
- Interactive Learning Activities

A LOOK BACK

The concept of metabolism was introduced in Chapter 2 and applied to discussion of muscle function in Chapter 8. This chapter concentrates on the metabolic reactions that liberate energy 8 *from the nutrients that we eat, use, and store. The key ideas of mass balance* 13 *and enzymes* 7 *are relevant to this discussion. Finally, reviewing the key ideas of homeostasis* 3 *and flow down gradients* 5 *will help you understand the maintenance of normal body temperature.*

Metabolism

Chapter 17 discussed how nutrients absorbed from the digestive tract are chemically transformed into other substances for the body's use. These reactions, along with those involving stored body substances, are described as the body's **metabolism.** In Chapter 2, we said that there are two types of metabolic activities:

- **Catabolism,** the breakdown of complex compounds into simpler components. Catabolism includes the digestion of food into small molecules and the release of energy from these molecules within the cell.
- **Anabolism,** the building of simple substances into nutrient storage compounds, structural materials, and functional molecules such as enzymes and transporters.

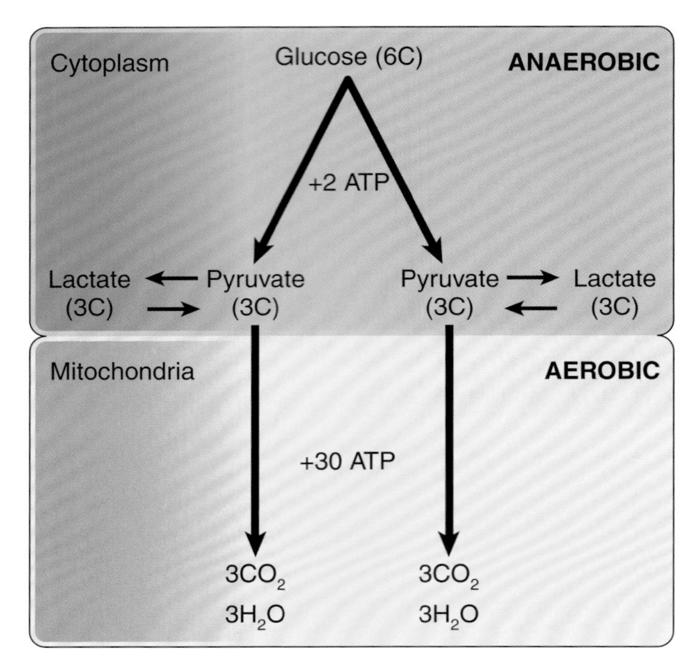

FIGURE 18-1 Cellular respiration. KEY POINT There are two stages of glucose catabolism, the first occurring without oxygen (anaerobic), followed by steps occurring with oxygen (aerobic). (The letter *c* represents a carbon atom, and the numbers show the number of carbon atoms in one molecule of the named substance.) ZOOMING IN How many pyruvate molecules can be generated from a single glucose molecule?

CELLULAR RESPIRATION

8 Recall from Chapter 2 that chemical bonds store potential energy. A series of catabolic reactions called **cellular respiration** (FIG. 18-1 and TABLE 18-1) transfers some of the potential energy in nutrients to the bonds of ATP, which can be used by all cells to do work. The rest of the energy becomes heat, a form of kinetic energy, which is used to maintain body temperature. These reactions can begin with glucose, fatty acids, or, more rarely, amino acids. We begin our explanation with glucose, a simple sugar that is a principal energy source for cells.

Glucose Catabolism: Anaerobic Phase The first steps in the breakdown of glucose do not require oxygen; that is, they are **anaerobic.** This phase of catabolism, known as **glycolysis** (gli-KOL-ih-sis), occurs in the cell's cytoplasm. Each glucose molecule yields enough energy by this process to produce two molecules of ATP.

Glycolysis ends with formation of an organic product called **pyruvate** (PI-ru-vate). This molecule is further metabolized in the next phase of cellular respiration, which occurs in the mitochondria and requires oxygen. In muscle cells that need to produce large amounts of ATP rapidly, for example, at the start of exercise or during very intense exercise, pyruvate cannot be metabolized fast enough and accumulates in the cells. In this case, pyruvate is converted to a related compound called lactate, which can be converted back into pyruvate once mitochondrial activity can keep pace with pyruvate production. As lactate accumulates in the cells, it can spill over into the blood. Other tissues, particularly the heart, can take up lactate and convert it to pyruvate for further catabolism. Lactate is thus a useful energy intermediate, and, contrary to previous thought, does not cause muscle damage or fatigue.

Table 18-1 Summary of Cellular Respiration of Glucose

Phase	Location in Cell	End Product(s)	Energy Yield/Glucose
Anaerobic (glycolysis)	Cytoplasm	Pyruvate	Two adenosine triphosphates (ATP)
Aerobic	Mitochondria	Carbon dioxide and water	30 ATP

Glucose Catabolism: Aerobic Phase To generate enough energy for survival, the body's cells must break pyruvate down more completely. As noted, these next **aerobic** (oxygen-requiring) reactions occur within the cell's mitochondria. On average, cells are able to form about 30 molecules of ATP aerobically per glucose molecule. Statements on ATP yields may differ slightly, because cells in different tissues vary in their metabolic pathways and in the amount of potential energy converted to heat. In any case, this additional yield is quite an increase over glycolysis alone, generally resulting in a total of 32 molecules of ATP per glucose as compared to two.

During the aerobic steps of cellular respiration, the mitochondria produce carbon dioxide and water. Because of the type of chemical reactions involved and because oxygen is used in the final steps, cellular respiration is described as nutrient **oxidation.** 7 Note that enzymes are required as catalysts in all these metabolic reactions. Many of the vitamins and minerals described later in this chapter are parts of these enzymes.

The net balanced equation for cellular respiration, starting with glucose, is as follows:

$$\underset{\text{glucose}}{C_6H_{12}O_6} + \underset{\text{oxygen}}{6O_2} \rightarrow \underset{\text{carbon dioxide}}{6CO_2} + \underset{\text{water}}{6H_2O}$$

See **BOX 18-1** for help in understanding this and other chemical equations.

Cellular Respiration of Fatty Acids and Amino Acids Fatty acids and, to a lesser extent, amino acids can also be oxidized to generate energy. Unlike glucose, these nutrients can generate ATP only by aerobic mechanisms, so they are of minimal use in high-intensity exercise and in cells with few mitochondria. As with pyruvate, fatty acids and amino acids are broken down completely to yield ATP, water, and carbon dioxide. The ATP yield and the precise sequence of chemical reactions vary among different fatty acids and amino acids.

Before amino acids are oxidized for energy, they must have their nitrogen (amine) groups removed. This removal,

ONE STEP AT A TIME

Chemical Equations

BOX 18-1

In the case study, Claudia failed to consume enough nutrients to meet her body's needs. Like everyone, Claudia needs to make ATP from food in order to survive.

The chemical equation below shows how glucose is used to generate ATP:

$$\underset{\text{glucose}}{C_6H_{12}O_6} + \underset{\text{oxygen}}{6O_2} \rightarrow \underset{\text{carbon dioxide}}{6CO_2} + \underset{\text{water}}{6H_2O} + \text{ATP} + \text{heat}$$

A chemical equation is a sort of shorthand, used to represent a chemical reaction. Familiarity with this shorthand will serve you well in your study of the human body.

Question

How many oxygen atoms are required to break down a single molecule of glucose? How many carbon dioxide molecules are produced?

Answer

Step 1. Identify the elements. As discussed in Chapter 2, each element has a one- or two-letter abbreviation. This equation includes three elements: carbon (C), hydrogen (H), and oxygen (O). The periodic table in Appendix 1 gives you the abbreviation for each element.

Step 2. Identify the molecules. The chemical formula of a molecule describes the atoms it contains. The subscript shows how many atoms of a particular element are found in the molecule. So, $C_6H_{12}O_6$ contains 6 carbon atoms, 12 hydrogen atoms, and 6 oxygen atoms. Each molecule in the equation above is identified under its chemical formula, but it can be useful to memorize common molecules such as water and carbon dioxide.

Step 3. Decipher the chemical equation. Like English sentences, most chemical equations are read left to right. The molecules to the left of the sideways arrow react together to form the products to the right of the arrow. So, glucose and oxygen combine to form carbon dioxide and water. Note that chemical equations only work in the direction of the arrow. That is, carbon dioxide and water do not combine to form glucose. Any equation that can work in both directions includes arrows going in both directions.

Step 4. Determine the quantities involved. Chemical equations also describe the relative amounts of each molecule. Normal chemical reactions cannot create or destroy atoms, so an equation must contain the same number of atoms for each element on either side of the arrow. For instance, 12 hydrogen atoms react (as part of glucose), and 12 hydrogen atoms are produced (as part of water).

So, the answer to the question is that a single glucose molecule reacts with 12 oxygen atoms (in the form of 6 oxygen molecules) to produce 6 carbon dioxide molecules.

Note that we see two elements on the right side of the equation that are not found on the left—ATP and heat. Without going into too much detail about this complex topic, the energy in the chemical bonds of the glucose molecule enables the mitochondria to make ATP from ADP, producing heat as a byproduct. Claudia's body has a fixed need for ATP. Since she cannot get enough ATP from her consumed nutrients, she needs to get it from her body stores.

See the Study Guide (available separately) for more practice analyzing chemical equations.

called **deamination** (de-am-ih-NA-shun), occurs in the liver, where the nitrogen groups are then formed into urea by combination with carbon dioxide. The blood transports urea to the kidneys to be eliminated.

13 The key idea of mass balance can be applied to the chemical reactions of metabolism. This idea requires that we balance the "ins" of a chemical reaction with the "outs." Remember that atoms are not created or destroyed in metabolic reactions. So, for example, if an individual loses 40 lb of adipose tissue (primarily stored triglycerides), where does the weight go? Figuring out the answer requires us to follow the carbon, oxygen, and hydrogen atoms that made up the triglycerides as they are catabolized. The atoms are not burned or lost, but converted into carbon dioxide and water. So, we actually breathe out (CO_2) and excrete (H_2O) the weight.

METABOLIC RATE

Metabolic rate refers to the rate at which cellular respiration metabolizes nutrients to produce ATP. Since the body makes ATP on demand, the metabolic rate relates to overall energy requirements. It is affected by a person's size, body fat, sex, age, activity, and hormones, especially thyroid hormone (thyroxine). Metabolic rate is high in children and adolescents and decreases with age. **Basal metabolism** is the amount of energy needed to maintain life functions while the body is at rest. Thus, your *basal metabolic rate* (*BMR*) is the energy you expend each day simply to stay alive. Any activity you perform, even tapping a toe, increases your energy expenditure above your BMR.

The unit used to measure energy is the **kilocalorie** (kcal), which is the amount of heat needed to raise 1 kg of water 1°C. Nutrition information for the general public typically replaces the word kilocalorie with *calorie* (C). To estimate how many calories you need each day taking into account your activity level, see **BOX 18-2**.

NUTRIENT METABOLISM

Instead of being processed for energy, nutrients may have other fates. They can be converted into storage forms, functional molecules, or even other nutrient types. 7 Enzymes catalyze all of these chemical reactions. The body controls these conversion reactions by controlling enzyme activity.

Carbohydrates We ingest carbohydrates in many different forms, as shown in **FIGURE 17-10**. As discussed in Chapter 17, dietary carbohydrates are digested into monosaccharides (simple sugars) before absorption into blood. The liver then converts the monosaccharides into glucose, which can be used for energy.

When nutrients are abundant and the body's energy needs are low, liver, and muscle cells convert glucose into

A CLOSER LOOK — BOX 18-2
Calorie Counting: Estimating Daily Energy Needs

Have you ever wondered how many calories you need to eat each day in order to avoid gaining or losing weight? To answer that question, you first need to calculate your basal metabolic rate (BMR) and then estimate the calories you burn each day in physical activity. Adding those two numbers together should give you your daily calorie needs.

You can estimate your BMR with a simple formula. An average woman requires 0.9 kcal/kg/h, and a man, 1.0 kcal/kg/h. If you need to convert your body weight from pounds to kilograms (kg), divide your weight in pounds by 2.2. Next, multiply your body weight in kilogram by 0.9 if you are female and by 1.0 if you are male. This gives you kcal burned per hour. Finally, multiply by 24 to find your BMR (the number of kcal you expend at rest per day).

For example, if you are female and weigh 132 lb, your equation would be as follows:

$$132\text{ lb} \div 2.2\text{ lb/kg} = 60\text{ kg}$$

$$0.9\text{ kcal/kg/h} \times 60\text{ kg} = 54\text{ kcal/h}$$

$$54\text{ kcal/h} \times 24\text{ h/day} = 1{,}296\text{ kcal/day}$$

Notice that, if you are male, you can skip step 2, since you'd simply be multiplying by 1.

To estimate your total energy needs for a day, you need to add to your BMR a percentage based on your current activity level ("couch potato" to serious athlete). These percentages are shown in the table that follows.

The equation to calculate total energy needs for a day is

$$\text{BMR} + (\text{BMR} \times \text{activity level})$$

Using the BMR from our previous example with different activity levels, the following equations apply:

At 25% activity:

$$1{,}296\text{ kcal/day} + (1{,}296\text{ kcal/day} \times 25\%) = 1{,}620\text{ kcal/day}$$

At 60% activity:

$$1{,}296\text{ kcal/day} + (1{,}296\text{ kcal/day} \times 60\%) = 2{,}073.6\text{ kcal/day}$$

As you can see, physical activity can help you maintain a healthful body weight. In this case, increased activity increased the individual's energy needs by more than 450 kcal/day!

Activity Level	Male	Female
Little activity ("couch potato")	25%–40%	25%–35%
Light activity (e.g., walking to and from class, but little or no intentional exercise)	50%–75%	40%–60%
Moderate activity (e.g., aerobics several times a week)	65%–80%	50%–70%
Heavy activity (serious athlete)	90%–120%	80%–100%

glycogen (GLI-ko-jen), the storage form of carbohydrates. Recall that the hormone insulin promotes this reaction. When glucose is needed for energy, glycogen is broken down under the influence of the hormone glucagon to yield glucose. Muscle glycogen is used only by the muscle cell that produced it. The glucose from liver glycogen can be released into the bloodstream to power other cells. However, the body's glycogen stores are limited. If we ingest more glucose than is needed for energy and glycogen storage, it is converted to fat and stored in adipose tissue and the liver.

Concept Mastery Alert

You can use your knowledge of word parts to distinguish between glycogen and glucose. Remember that glycogen generates (gen) glucose (glyco).

Fats Most tissues can use fatty acids for energy. Some organs, such as the liver, rely exclusively on fatty acids, and other tissues, such as muscle, use fatty acids during rest and low-intensity exercise. Brain tissue is one exception; it relies on glucose or, if glucose is not available, partially catabolized fatty acids called **ketone bodies**. Large amounts of ketone bodies are produced by the liver in cases of starvation or low carbohydrate intake, and the ketones can provide energy for the brain and other tissues. They are also produced in uncontrolled diabetes mellitus, in which insulin deficiency "tricks" the liver into thinking that glucose is unavailable. However, ketone bodies are acidic, and excesses disrupt the body's acid–base balance. We will say more about these changes in Chapter 19.

Fatty acids are a highly concentrated energy source and generate more ATP per molecule than does glucose. In fact, fat in the diet yields more than twice as much energy as do protein and carbohydrate; fat yields 9 kcal of energy per gram, whereas protein and carbohydrate each yields 4 kcal/g.

Because fats are such a concentrated energy source, they are the most efficient way to store excess calories. Excess caloric intake, be it carbohydrate, protein, or fat, can be converted into triglycerides and stored in adipose tissue. Then, in times of nutrient deficiency, the triglycerides can be converted to fatty acids for use throughout the body. Adipose tissue growth is virtually unlimited. According to the Guinness Book of World Records in 2017, the world's heaviest man, Jon Brower Minnoch, lost 924 lb from his peak weight of 1,400 lb, most of which would have been adipose tissue.

Proteins There are no specialized storage forms of proteins, as there are for carbohydrates (glycogen) and fats (triglycerides), because specific proteins are synthesized to meet specific body needs. For instance, trying to lift a heavy weight places a load on muscle tissue, stimulating the production of actin and myosin proteins. Protein consumed in excess of daily needs is not stored as protein, but is catabolized for energy or converted to triglycerides or glucose. Conversely, if protein intake is inadequate, the body generates amino acids by breaking down the proteins in muscle and other body tissues. Fats and carbohydrates are described as "protein sparing," because they are used for energy before proteins are and thus spare proteins for the synthesis of necessary body components.

CHECKPOINTS 

- ☐ **18-1** What are the two types of activities that make up metabolism?
- ☐ **18-2** What name is given to the series of cellular reactions that releases energy from nutrients?
- ☐ **18-3** What is the organic end product of glycolysis?
- ☐ **18-4** What element is required for aerobic cellular respiration but not for glycolysis?
- ☐ **18-5** Which of the three main nutrients is not stored as a reserve in the body?

CASEPOINTS 

- ☐ **18-1** If Claudia is not getting enough food and is metabolizing body fat for energy, what acidic compounds will be produced?
- ☐ **18-2** If Claudia is metabolizing body proteins for energy, what portion needs to be removed from their component amino acids before they are oxidized?
- ☐ **18-3** Claudia's ideal weight is 126 lb. What is her needed basal calorie intake per day?
- ☐ **18-4** According to her friends; description of her activities, what should her minimum daily calorie intake be?

Nutritional Guidelines

The relative amounts of carbohydrates, fats, and proteins that should be in the daily diet vary somewhat with the individual. Typical recommendations for the number of calories derived each day from the three types of food are as follows:

- Carbohydrate: 55% to 60%
- Fat: 30% or less
- Protein: 15% to 20%

It is important to realize that the type as well as the amount of each nutrient is a factor in good health. A weight loss diet should follow the same proportions as given above but with a reduction in portion sizes.

CARBOHYDRATES

A healthful diet provides abundant complex carbohydrates, whereas simple sugars are kept to a minimum. Simple sugars are monosaccharides, such as glucose and fructose (fruit sugar), and disaccharides, such as sucrose (table sugar) and lactose (milk sugar). Simple sugars are a source of fast energy because they are metabolized rapidly. However, they boost pancreatic insulin output, and as a result, they cause blood glucose levels to rise and fall rapidly. It is healthier to maintain steady glucose levels, which normally range from approximately 85 to 125 mg/dL throughout the day.

The **glycemic effect** is a measure of how rapidly a particular food raises the blood glucose level and stimulates the release of insulin. The effect is generally low for whole grains, fruit, vegetables, legumes, and dairy products, and high for refined sugars and refined ("white") grains. Note, however, that the glycemic effect of a food also depends on when it is eaten during the day and if or how it is combined with other foods.

Complex carbohydrates are polysaccharides. Examples are:

- Starches, found in grains, legumes, and potatoes and other starchy vegetables
- Fibers, such as cellulose, pectins, and gums, which are the structural materials of plants

Fiber cannot be used for energy, but it adds bulk to the stool and promotes elimination of toxins and waste. It also slows the digestion and absorption of other carbohydrates, thus regulating the release of glucose. It helps in weight control by providing a sense of fullness and limiting caloric intake. Adequate fiber in the diet lowers cholesterol and helps to prevent diabetes, colon cancer, hemorrhoids, appendicitis, and diverticulitis. Foods high in fiber, such as whole grains, fruits, and vegetables, are also rich in vitamins and minerals.

FATS

Moderate amounts of fats are necessary for health, providing energy and cell components, and adding taste to other critical nutrients. However, not all fats are the same. The healthiness of a triglyceride is determined by the characteristics of its three fatty acids.

Essential Fatty Acids While any fatty acid can provide energy, some cell components require specific fatty acids. Most fatty acid varieties can be synthesized by body cells. However, there are two essential fatty acids, **linoleic** (lin-o-LE-ik) **acid** and **alpha-linolenic** (lin-o-LEN-ik) **acid**, which must be taken in as food. Linoleic acid is easily obtained through a healthful, balanced diet that includes plenty of vegetables and vegetable oils. It is used to make prostaglandins, among other functions. In contrast, alpha-linolenic acid is found primarily in fatty fish and shellfish, dark green, leafy vegetables, and flaxseeds, soybeans, walnuts, and their oils. Thus, it is somewhat more difficult to obtain.

Saturated and Unsaturated Fats Fats are subdivided into saturated and unsaturated forms based on their chemical structures. The fatty acids in **saturated fats** have more hydrogen atoms in their molecules because they have no double bonds between carbons atoms. In other words, their carbon atoms are fully "saturated" with hydrogen (FIG. 18-2). Most saturated fats are from animal sources and are solid at room temperature, such as butter and lard. Also included in this group are the so-called tropical oils: coconut oil and palm oil.

Unsaturated fats are derived from plants. They are liquid at room temperature and are generally referred to as oils, such as corn, peanut, olive, and canola oils. Their fatty acids have one or more double carbon bonds, which exclude hydrogen (see FIG. 18-2). An unsaturated fatty acid is *monounsaturated* if it has a single double bond and polyunsaturated if it has more than one double bond. You may have heard or read about fatty acids described as "omega" acids with a number. This terminology refers to the position of the last double-bonded carbon in the chain with regard to the last carbon, named the omega carbon for the

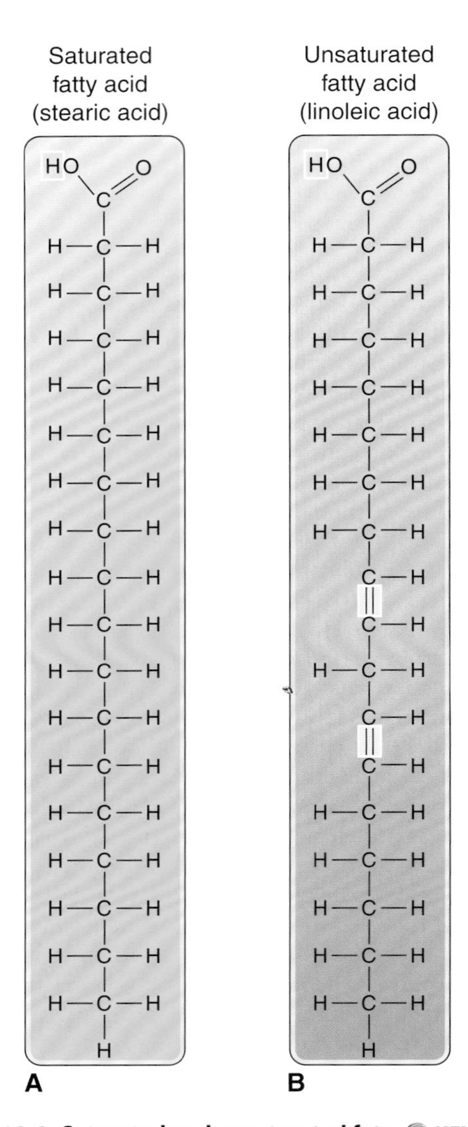

FIGURE 18-2 Saturated and unsaturated fats. KEY POINT Saturated and unsaturated fats differ in their bonding structures. **A.** Saturated fatty acids contain the maximum numbers of hydrogen atoms attached to carbons and no double bonds between carbon atoms. **B.** Unsaturated fatty acids have less than the maximum number of hydrogen atoms attached to carbons and one or more double bonds between carbon atoms (highlighted). The first carbon in the chain (*top*) is the alpha carbon; the last (*bottom*) is the omega carbon. The "omega" fatty acids are described according to how many carbons there are from the last double-bonded carbon in the chain to the omega carbon, such as omega-3 or omega-6. Linoleic acid is an omega-6 fatty acid.

last letter of the Greek alphabet. In the essential fatty acid linoleic acid, for example, the last double-bonded carbon is the sixth carbon from the omega carbon, so linoleic acid is an omega-6 fatty acid (see FIG. 18-2B). The essential fatty acid alpha-linolenic acid is an omega-3 fatty acid. As to fat intake, a predominance of unsaturated fats in general, in addition to the previously mentioned essential fatty acids, contributes to a healthful diet.

Saturated fats should make up less than one-third of the fat in the diet (less than 10% of total calories). Diets high in saturated fats are associated with a higher than normal incidence of cancer, heart disease, and cardiovascular problems, although the relation between these factors is not fully understood.

Because most unsaturated fats are liquid and easily become rancid (spoil), commercial food manufacturers have used fats that are artificially saturated to extend shelf life and improve consistency of foods. These fats are listed on food labels as partially hydrogenated (hi-DROJ-en-a-ted) vegetable oils and are found commonly in baked goods, processed peanut butter, vegetable shortening, and some solid margarines. Evidence shows that components of hydrogenated fats, known as *trans-fatty acids*, are more harmful than even natural saturated fats. They raise blood levels of LDL (the less healthful lipoproteins) and lower levels of HDLs (the healthful lipoproteins). Food manufacturers are responding to evidence that trans fats are harmful and are making efforts to remove them from processed foods. Many countries now require labeling of trans fat content on food labels. In the United States, trans fat contents of 0.5 g or more per serving must be listed. Bear in mind, though, that restaurants still may be using trans fats in cooking, especially deep-frying.

PROTEINS

The synthesis of each protein requires a specific set of amino acids. Of the 20 amino acids needed to build proteins, metabolic reactions can synthesize 11 internally. These 11 amino acids are described as *nonessential* because they need not be taken in as food (TABLE 18-2). The remaining nine amino acids cannot be made metabolically and therefore must be taken in as part of the diet; these are the **essential amino acids**. Note that some nonessential amino acids may become essential under certain conditions, as during extreme physical stress, or in certain hereditary metabolic diseases.

Because proteins, unlike carbohydrates and fats, are not stored in special reserves, protein foods should be consumed regularly with attention to obtaining the essential amino acids. Most animal proteins supply all of the essential amino acids and are described as complete proteins.

Concept Mastery Alert

Note that a complete protein does not need to contain every nonessential amino acid, because they can be synthesized by the body.

Most plant proteins are lacking in one or more of the essential amino acids. People on strict vegetarian diets must learn to combine foods, such as legumes (e.g., beans and peas) with grains (e.g., rice, corn, or wheat), to obtain all the essential amino acids each day. FIGURE 18-3 demonstrates the principles of combining two foods, legumes and grains, to supply essential amino acids that might be missing in one food or the other. Legumes are rich in isoleucine

Table 18-2 Amino Acids

Nonessential Amino Acids[a]		Essential Amino Acids[b]	
Name	**Pronunciation**	**Name[c]**	**Pronunciation**
Alanine	AL-ah-nene	Histidine	HIS-tih-dene
Arginine	AR-jih-nene	Isoleucine	i-so-LU-sene
Asparagine	ah-SPAR-ah-jene	Leucine	LU-sene
Aspartic acid	ah-SPAR-tik AH-sid	Lysine	LI-sene
Cysteine	SIS-teh-ene	Methionine	meh-THI-o-nene
Glutamic acid	glu-TAM-ik AH-sid	Phenylalanine	fen-il-AL-ah-nene
Glutamine	GLU-tah-mene	Threonine	THRE-o-nene
Glycine	GLY-sene	Tryptophan	TRIP-to-fan
Proline	PRO-lene	Valine	VA-lene
Serine	SERE-ene		
Tyrosine	TI-ro-sene		

[a]Nonessential amino acids can be synthesized by the body.
[b]Essential amino acids cannot be synthesized by the body; they must be taken in as part of the diet.
[c]If you are ever called upon to memorize the essential amino acids, the mnemonic (memory device) Pvt. T. M. Hill gives the first letter of each name.

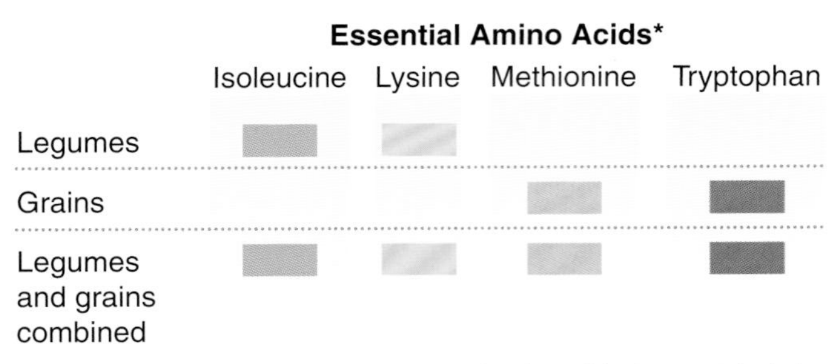

FIGURE 18-3 Combining foods to obtain the essential amino acids. KEY POINT There are nine essential amino acids. If someone does not eat animal proteins, most of which supply all of the essential amino acids, foods must be combined to supply all the needed amino acids within the day. In this illustration, legumes and grains are combined to provide four of the nine essential amino acids.

and lysine but poor in methionine and tryptophan, while grains are just the opposite. For illustration purposes, FIGURE 18-3 includes only the four missing essential amino acids (there are nine total). Traditional ethnic diets reflect these healthful combinations, for example, beans with corn or rice in Mexican dishes or chickpeas and lentils with wheat in Middle Eastern fare. Nuts and vegetables also provide plant proteins; thus, a peanut butter sandwich on whole grain bread or vegetables over rice are other complementary combinations.

MINERALS AND VITAMINS

In addition to needing carbohydrates, fats, and proteins, the body requires minerals and vitamins.

Minerals are chemical elements needed for body structure, fluid balance, and such activities as muscle contraction, nerve impulse conduction, and blood clotting. Some minerals are components of vitamins. A list of the main minerals needed in a proper diet is given in **TABLE 18-3**. Some additional minerals not listed are also required for good health. A mineral needed in extremely small amounts is referred to as a **trace element**.

Vitamins are complex organic substances needed in small quantities. Vitamins are parts of enzymes or other substances essential for metabolism, and vitamin deficiencies lead to a variety of nutritional diseases.

Table 18-3 Minerals

Mineral	Functions	Sources	Results of Deficiency
Calcium (Ca)	Formation of bones and teeth; blood clotting; nerve conduction; muscle contraction	Dairy products, eggs, green vegetables, legumes (peas and beans)	Rickets, tetany, osteoporosis
Phosphorus (P)	Formation of bones and teeth; found in ATP, nucleic acids	Meat, fish, poultry, egg yolk, dairy products	Osteoporosis, abnormal metabolism
Sodium (Na)	Fluid balance; nerve impulse conduction; muscle contraction	Most foods, especially processed foods, table salt	Weakness, cramps, diarrhea, dehydration
Potassium (K)	Fluid balance; nerve and muscle activity	Fruits, meats, seafood, milk, vegetables, grains	Muscular and neurologic disorders
Chloride (Cl)	Fluid balance; hydrochloric acid in stomach	Meat, milk, eggs, processed food, table salt	Rarely occurs
Iron (Fe)	Oxygen carrier (hemoglobin, myoglobin)	Meat, eggs, fortified cereals, legumes, dried fruit	Anemia, dry skin, indigestion
Iodine (I)	Thyroid hormones	Seafood, iodized salt	Hypothyroidism, goiter
Magnesium (Mg)	Catalyst for enzyme reactions; carbohydrate metabolism	Green vegetables, grains, nuts, legumes	Spasticity, arrhythmia, vasodilation
Manganese (Mn)	Catalyst in actions of calcium and phosphorus; facilitator of many cellular processes	Many foods	Possible reproductive disorders
Copper (Cu)	Necessary for absorption and use of iron in formation of hemoglobin; part of some enzymes	Meat, water	Anemia
Chromium (Cr)	Works with insulin to regulate blood glucose levels	Meat, unrefined food, fats and oils	Inability to use glucose
Cobalt (Co)	Part of vitamin B_{12}	Animal products	Pernicious anemia
Zinc (Zn)	Promotes carbon dioxide transport and energy metabolism; found in enzymes	Meat, fish, poultry, grains, vegetables	Alopecia (baldness); possibly related to diabetes
Fluoride (F)	Prevents tooth decay	Fluoridated water, tea, seafood	Dental caries

The water-soluble vitamins are the B vitamins and vitamin C. These are not stored and must be taken in regularly with food. The fat-soluble vitamins are A, D, E, and K. These vitamins are kept in reserve in fatty tissue. Excess intake of the fat-soluble vitamins can lead to toxicity. A list of vitamins is given in **TABLE 18-4**.

Certain substances are valuable in the diet as **antioxidants**. They defend against the harmful effects of reactive oxygen species (ROS), also described as *free radicals*, highly reactive and unstable molecules produced from oxygen in the normal course of metabolism (and also resulting from UV radiation, air pollution, and tobacco smoke). Free radicals contribute to aging and disease. Antioxidants react with ROS to stabilize them and minimize their harmful effects on cells. Vitamins C and E and beta-carotene, an orange pigment found in plants that is converted to vitamin A, are antioxidants. There are also many compounds found in plants (e.g., soybeans and tomatoes) that are antioxidants.

Mineral and Vitamin Supplements The benefits of adding mineral and vitamin supplements to the diet of healthy individuals is a subject of controversy. Some researchers maintain that adequate amounts of these substances can be obtained from a varied, healthful diet. Many commercial foods, including milk, cereal, and bread, are already fortified with minerals and vitamins. Others hold that pollution, depletion of the soils, and the storage, refining, and processing of foods make additional supplementation beneficial. At present, research does not support claims that multivitamins increase longevity, improve brain function, prevent cancer, or convey infection resistance in healthy adults, including in the elderly. While individuals with chronic illnesses often have vitamin deficiencies, the deficiencies may be the result rather than the cause of the disease.

Nevertheless, specific populations benefit from specific supplements. Pregnant women and their babies, for instance, benefit from supplemental iron and folic acid, and

18

Table 18-4 Vitamins

Vitamins	Functions	Sources	Results of Deficiency
A (retinol)	Required for healthy epithelial tissue and for eye pigments; involved in reproduction and immunity	Orange fruits and vegetables, liver, eggs, dairy products, dark green vegetables	Night blindness, dry, scaly skin, decreased immunity
B_1 (thiamin)	Required for enzymes involved in oxidation of nutrients; nerve function	Pork, cereal, grains, meats, legumes, nuts	Beriberi, a disease of nerves
B_2 (riboflavin)	In enzymes required for oxidation of nutrients	Milk, eggs, liver, green leafy vegetables, grains	Skin and tongue disorders
B_3 (niacin, nicotinic acid)	Involved in oxidation of nutrients	Yeast, meat, liver, grains, legumes, nuts	Pellagra with dermatitis, diarrhea, mental disorders
B_6 (pyridoxine)	Amino acid and fatty acid metabolism; formation of niacin; manufacture of red blood cells	Meat, fish, poultry, fruit grains, legumes, vegetables	Anemia, irritability, convulsions, muscle twitching, skin disorders
Pantothenic acid	Essential for normal growth; energy metabolism	Yeast, liver, eggs, and many other foods	Sleep disturbances, digestive upset
B_{12} (cyanocobalamin)	Production of cells; maintenance of nerve cells; fatty acid and amino acid metabolism	Animal products	Pernicious anemia
Biotin (a B vitamin)	Involved in fat and glycogen formation, amino acid metabolism	Peanuts, liver, tomatoes, eggs, oatmeal, soy, and many other foods	Lack of coordination, dermatitis, fatigue
Folate (folic acid, a B vitamin)	Required for amino acid metabolism; DNA synthesis; maturation of red blood cells	Vegetables, liver, legumes, seeds	Anemia, digestive disorders, neural tube defects in the embryo
C (ascorbic acid)	Maintains healthy skin and mucous membranes, involved in synthesis of collagen; antioxidant	Citrus fruits, green vegetables, potatoes, orange fruits	Scurvy, poor wound healing, anemia, weak bones
D (calciferol)	Aids in absorption of calcium and phosphorus from intestinal tract	Fatty fish, liver, eggs, fortified milk, cereal	Rickets, bone deformities, osteoporosis
E (tocopherol)	Protects cell membranes; antioxidant	Seeds, green vegetables, nuts, grains, oils	Anemia, muscle and liver degeneration, pain
K	Synthesis of blood clotting factors; bone formation	Liver, cabbage, and leafy green vegetables; bacteria in digestive tract	Hemorrhage

individuals with low bone mass improve when treated with vitamin D. When required, supplements should be selected by a physician or nutritionist to fit an individual's particular needs. Megavitamin dosages may cause unpleasant reactions and in some cases are hazardous. Fat-soluble vitamins have caused serious toxic effects when taken in excess of the established safe upper limit (UL).

USDA DIETARY GUIDELINES

The United States Department of Agriculture (USDA) has published dietary guidelines at regular intervals since 1916. They are currently revised every five years. The newest guidelines were published in 2015 and will be in effect until 2020. The graphic design that accompanies these recommendations promotes healthy eating choices from a variety of foods and beverages (**FIG. 18-4**). The various categories include fruits, vegetables, proteins, dairy products, grains, and oils. Foods high in proteins are meat, poultry, seafood, and eggs, but proteins are also found in plant products, such as seeds, nuts, legumes, and grains. The grains recommended are whole, unrefined grains. Saturated and trans fats, added sugars, and salt should be limited, the fats and sugars to less than 10% of daily total calorie intake and sodium to less than 2,300 mg/day, equivalent to about 1 teaspoon of salt. Note that much of the excess salt we eat comes from processed foods and restaurant fare. Solid fats and added sugars ("SoFAS") are described as "empty calories" that provide energy but little nutrition. These are "extras" that you can eat within your recommended daily energy limit after you meet your nutritional needs. Of course, you could also select additional foods from among the recommended nutrient-rich groups to satisfy your energy needs. The goal of the USDA guidelines is to help people maintain a healthy weight and reduce the risks of chronic diseases.

The USDA also helps to implement these guidelines through the website "choosemyplate.gov." Here, one can find resources to help make choices that align with the guidelines. There is information on planning diets at various daily calorie counts for different age groups and for those with special dietary needs or preferences. These include people with gluten allergies, those who are lactose intolerant, strict vegetarians (vegans) who eat no animal products, and vegetarians who eat dairy products and eggs (lacto-ovo vegetarians). On the website, you can get a personalized estimate of what and how much you should eat based on your height, weight, age, sex, and level of physical activity. You can also assess your diet and get advice on planning and preparing healthy meals. Another reference describes acceptable portion sizes, which are usually much smaller than we imagine.

A Healthy Eating Pattern Includes:

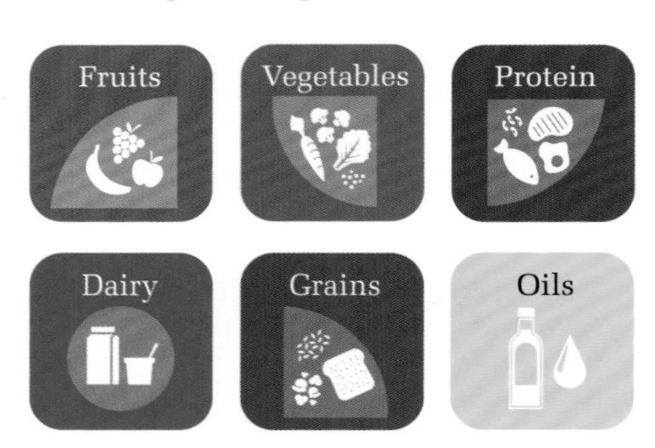

A Healthy Eating Pattern Limits:

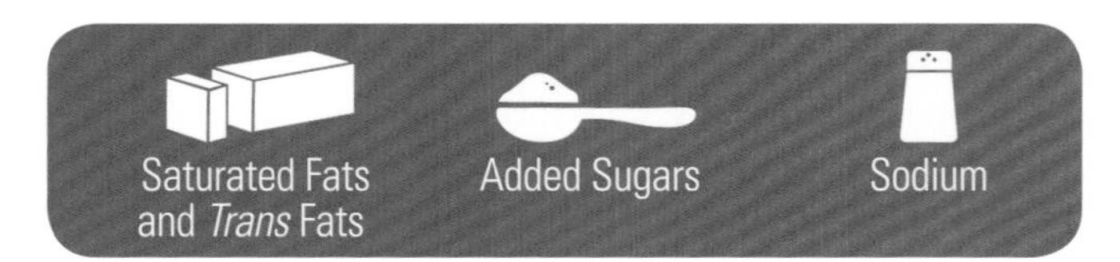

FIGURE 18-4 USDA dietary guidelines. The graphic shows the variety of nutrients recommended for a healthy diet.

See the Student Resources on thePoint® for information on dietitians and nutritionists, who study nutrition and metabolism and help people plan healthful diets.

ALCOHOL

Alcohol yields energy in the amount of 7 kcal/g, but it is not considered a nutrient because it does not yield useful end products. In fact, alcohol interferes with metabolism and contributes to a variety of disorders.

The body can metabolize about 1/2 oz of pure alcohol (ethanol) per hour. This amount translates into one glass of wine, one can of beer, or one shot of hard liquor. Consumed at a more rapid rate, alcohol accumulates in the bloodstream and affects many cells, notably in the brain.

Alcohol is rapidly absorbed through the stomach and small intestine and is detoxified by the liver. When delivered in excess to the liver, alcohol can lead to the accumulation of fat as well as inflammation and scarring of liver tissue. It can eventually cause cirrhosis (sih-RO-sis), which involves irreversible changes in liver structure. Alcohol metabolism ties up enzymes needed for oxidation of nutrients and also results in byproducts that acidify body fluids. Other effects of alcoholism include obesity, malnutrition, cancer, ulcers, and fetal alcohol syndrome. Health professionals advise pregnant women not to drink any alcohol. In addition, alcohol impairs judgment and leads to increased involvement in motor vehicle accidents, drownings, falls, and other accidental injuries, as well as suicides and homicides.

Moderate alcohol consumption is defined as no more than one drink per day for women and two drinks per day for men. Current research suggests that moderate alcohol consumption is compatible with good health and may even have a beneficial effect on the cardiovascular system.

See the Student Resources on thePoint® for figures showing the effects of alcoholism.

CHECKPOINTS 

- [ ] **18-6** What is the term for how rapidly a food raises the blood glucose level?
- [ ] **18-7** What is meant when an amino acid or a fatty acid is described as essential?
- [ ] **18-8** What is the difference between saturated and unsaturated fats?
- [ ] **18-9** What is the difference between vitamins and minerals?
- [ ] **18-10** How much pure alcohol can the average person metabolize per hour?

CASEPOINTS 

- [ ] **18-5** Because she was concerned about her weight, Claudia's diet was practically fat free and her fat stores were very low. Which important vitamins will she lack?
- [ ] **18-6** What are other important roles of adipose tissues that will be compromised in Claudia's case?

WEIGHT CONTROL

Body mass index (BMI) is a measurement used to evaluate body size. It is based on the ratio of weight to height (**FIG. 18-5**). BMI is calculated by dividing weight in kilograms by height in meters squared. (For those not accustomed to using the metric system, an alternate method is to divide weight in pounds by the square of height in inches and multiply by 703.) A healthy range for this measurement is 18.5 to 24.9.

BMI does not take into account the relative amount of muscle and fat in the body. For example, a bodybuilder might be healthy with a higher than typical BMI because muscle has a higher density than fat. Researchers have also noted that people of different ethnicities commonly have different levels of body fat at the same BMI. Finally, BMI does not give a fair indication of overweight or obesity in people over age 65 or in children. Alternative measures of adiposity include the body adiposity index (BAI), which compares hip circumference and height, and the waist-to-height ratio (WtH). However, none of these measures have received widespread acceptance.

Calculation of body mass index (BMI)

Formula:

$$BMI = \frac{\text{Weight (kg)}}{\text{Height (m)}^2}$$

Conversion:

Kilograms = pounds ÷ 2.2

Meters = inches ÷ 39.4

Example:

A woman who is 5′4″ tall and weighs 134 pounds has a BMI of 23.5.

Weight: 134 pounds ÷ 2.2 = 61 kg

Height: 64 inches ÷ 39.4 = 1.6 m; $(1.6)^2 = 2.6$

$$BMI = \frac{61 \text{ kg}}{2.6 \text{ m}} = 23.5$$

FIGURE 18-5 Calculation of body mass index (BMI). KEY POINT BMI is a measurement used to evaluate body size based on the ratio of weight to height. ZOOMING IN What is the BMI of a male 5 ft 10 in in height who weighs 170 lb? (Round off to one decimal place.)

Overweight and Obesity Overweight is defined as a BMI of 25 to 30 and obesity as a BMI greater than 30. Although the prevalence of overweight among Americans fluctuated between 31% and 34% after national surveys began in 1960, it is common knowledge that obesity has increased in recent decades. In 1962, the US obesity rate was below 14%. As of 2018, almost 40% of adult Americans were obese. Adding overweight to this figure brings the total to over 70%.

The causes of obesity are complex, involving social, economic, genetic, psychological, and metabolic factors. Obesity shortens the life span and is associated with cardiovascular disease, diabetes, some cancers, and other diseases. Unfortunately, obesity rates have increased over past decades among American children. Not surprisingly, the incidence of type 2 diabetes, once considered to have an adult onset, has also increased greatly among children. Some researchers hold that obesity has a closer correlation to chronic disease than poverty, smoking, or drinking alcohol.

Numerous prescription weight loss drugs have been developed, but all have risks and none are universally effective. For most people, a varied diet eaten in moderation and regular exercise are the surest ways to avoid obesity. Recent studies have questioned the belief that weight loss is best accomplished with large volume, low-intensity exercise. A number of investigators have found that high-intensity intermittent exercise (HIIE) results in much more fat loss (both subcutaneous and visceral) than low-intensity exercise. HIIE protocols are quite short, involving perhaps 20 minutes of exercise alternating between 8-second sprints and 12-second recovery periods. Researchers are still determining the mechanisms underlying this unexpected result, but they suspect decreased appetite, good adherence to the exercise program, and decreased insulin resistance are involved.

Underweight A BMI of less than 18.5 is defined as underweight. People who are underweight can have as much difficulty gaining weight as others have losing it. The problem of underweight may result from eating disorders, rapid growth, allergies, illness, or psychological factors. It is associated with low reserves of energy, reproductive disturbances, and nutritional deficiencies. In women, inadequate fat stores lead to a lack of estrogen production, which may cause menstrual periods to cease. To gain weight, people have to increase their intake of calories, but they should also engage in regular exercise to add muscle tissue and not just fat.

A chronic loss of appetite, called **anorexia** (an-o-REK-se-ah), may be caused by a great variety of physical and mental disorders. Because the hypothalamus and the higher brain centers are involved in the regulation of hunger, it is possible that emotional and social factors contribute to the development of anorexia.

Anorexia nervosa is a psychological disorder that predominantly afflicts young women, as illustrated in Claudia's case study. In a desire to be excessively thin, affected people literally starve themselves, sometimes to the point of death. As in all cases of starvation, the body resorts to breaking down body protein to generate amino acids, which can then be used for energy or to make glucose. The loss of bone protein weakens bone, the loss of cardiac muscle protein can lead to heart failure, and the loss of skeletal muscle protein results in muscle weakness.

CHECKPOINT 

☐ **18-11** What does the abbreviation BMI stand for?

CASEPOINT 

☐ **18-7** Claudia is 5′4″. If she weighs 106 lb, what would her BMI be?

Nutrition and Aging

With age, a person may find it difficult to maintain a balanced diet. Often, the elderly lose interest in buying and preparing food or are unable to do so. Because metabolism generally slows and less food is required to meet energy needs, nutritional deficiencies may develop. With age, the ability to synthesize vitamin D declines, and the kidneys are less able to convert it to its active form. Older adults may need supplements of vitamin D as well as calcium to prevent loss of bone density. The senses of smell and taste become less acute with age, diminishing appetite. Medications may interfere with appetite and with the absorption and use of specific nutrients.

It is important for older people to seek out foods that are "nutrient dense," that is, foods that have a high proportion of nutrients in comparison with the number of calories they provide. Exercise helps boost appetite and maintain muscle tissue, which is more metabolically active.

Body Temperature

Heat is an important byproduct of the many chemical activities constantly occurring in body tissues. At the same time, heat is always being lost through a variety of outlets. 3 While the amount of heat gained and lost by the body constantly varies, body temperature is a regulated variable that must be maintained within a narrow range. This section introduces the mechanisms of heat gain and loss before discussing the negative feedback loop involved in body temperature regulation.

HEAT LOSS

Although 15% to 20% of heat loss occurs through the respiratory system and with urine and feces, more than 80% of heat loss occurs through the skin. Networks of blood vessels in the skin's dermis (deeper part) can bring considerable quantities of blood near the surface, so that heat can be dissipated to the outside.

Mechanisms of Heat Loss Heat loss from the skin can occur in several ways (**FIG. 18-6**):

- In **radiation**, heat travels from its source as heat waves or rays. Radiation does not require contact between the heat source and the heat receiver.
- In **convection**, heat transfer is promoted by movement of a cooler contacting medium. For example, if the air around the skin is put in motion, as by an electric fan, the layer of heated air next to the body is constantly carried away and replaced with cooler air. Another

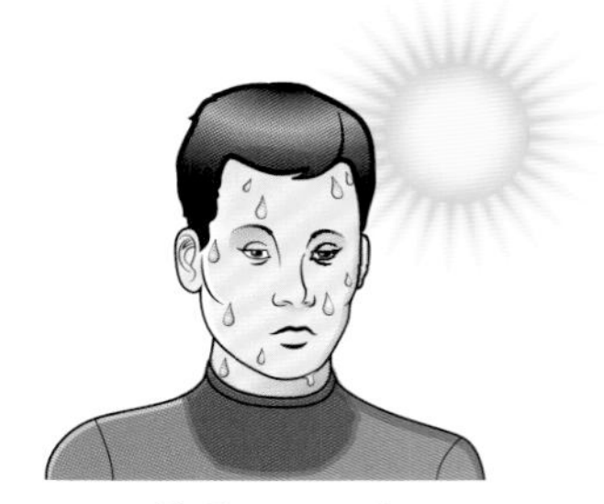

FIGURE 18-6 Mechanisms of heat loss. **KEY POINT** There are four processes that promote heat loss of body heat to the environment. **A.** Radiation—heat travels away as heat waves or rays. **B.** Convection—heat transfer is promoted by movement of a cooler medium. **C.** Evaporation—heat is used to change a liquid (such as sweat) to a vapor. **D.** Conduction—heat is transferred to a cooler object. **ZOOMING IN** What will happen in **(B)** if the fan speed is increased? What will happen in **(C)** if environmental humidity increases?

example of convection is the transfer of heat from muscle tissue to circulating blood.

- In **evaporation**, heat is lost in the process of changing a liquid to the vapor state. Heat from the skin provides the energy to evaporate sweat from the skin's surface, just as heat applied to a pot of water converts the water to steam.
- In **conduction**, a warm object transfers heat energy to a cooler object. For example, heat can be transferred from the skin directly to an ice pack. 5 Conduction, then, can be thought of as heat moving down a temperature gradient.

To illustrate evaporation, rub some alcohol on your skin; it evaporates rapidly, using so much heat that your skin feels cold. Perspiration does the same thing, although not as quickly.

Prevention of Heat Loss Factors that play a part in heat loss include the volume of tissue compared with the amount of skin surface. A child loses heat more rapidly than does an adult. Such parts as fingers and toes are affected most by exposure to cold because they have a great amount of skin compared with total tissue volume.

An effective natural insulation against heat loss is the layer of fat under the skin. The degree of insulation depends on the thickness of the subcutaneous layer. Even when skin temperature is low, this fatty tissue prevents the deeper tissues from losing much heat. On average, this layer is slightly thicker in females than in males.

HEAT GAIN

The body gains heat from the environment by two different mechanisms: radiation from the sun or conduction from contact with a warmer surface. Heat is also generated within the body, because a mitochondrion produces a certain amount of heat every time it makes a molecule of ATP. Therefore, heat production directly correlates with tissue activity. The activity of most tissues, including the brain and liver, remains constant. (An increase in nervous tissue activity is not associated with significantly increased ATP production, so intense studying doesn't burn more calories or generate additional heat.)

Muscle cell activity and heat production, though, increase enormously with exercise. Food intake also augments heat production, because of the high energy cost of processing and storing nutrients. This change is greatest with protein-rich meals and lowest with fat-rich meals. Circulating blood distributes heat from warmer body regions to cooler ones. For instance, strenuous exercise can make your head feel hot even though the heat was generated in the leg muscles.

In addition to exercise and digestion, heat production can be increased by hormones, such as thyroid hormones, epinephrine (adrenaline) from the adrenal medulla, and progesterone from the ovary. By increasing mitochondrial activity, these hormones increase both ATP production and heat production. 13 In a specialized form of adipose tissue called *brown fat*, hormones alter body temperature by changing the balance between these two metabolic outcomes. In cold conditions, they decrease the amount of ATP produced and increase the amount of heat produced for a given amount of glucose.

TEMPERATURE REGULATION

3 The control center in the negative feedback loop controlling body temperature is the **hypothalamus**, the area of the brain located just above the pituitary gland. The hypothalamus receives information from temperature receptors in the skin and within the hypothalamus itself and sends signals to numerous effectors, including muscles and blood vessels.

Normal Body Temperature The normal temperature range obtained by either a mercury or an electronic thermometer may vary from 36.2°C to 37.6°C (97°F to 100°F). Within this range, body temperature varies according to the time of day. Usually, it is lowest in the early morning because the muscles have been relaxed and no food has been taken in for several hours. Body temperature tends to be higher in the late afternoon and evening because of physical activity and food consumption.

Normal temperature also varies in different parts of the body. Skin temperature obtained in the axilla (armpit) is lower than mouth temperature, and mouth temperature is a degree or so lower than rectal temperature. It is believed that, if it were possible to place a thermometer inside the liver, it would register one or more degrees higher than rectal temperature. The temperature within a muscle might be even higher during activity.

Although the Fahrenheit scale is used in the United States, in most parts of the world, temperature is measured with the Celsius (SEL-se-us) thermometer. On this scale, the ice point is at 0° and the normal boiling point of water is at 100°, the interval between these two points being divided into 100 equal units. The Celsius scale is also called the centigrade scale (think of 100 cents in a dollar).

Responses to Cold Conditions If body temperature falls below the set point, the hypothalamus stimulates constriction of blood vessels in the skin to reduce heat loss (FIG. 18-7A). In addition, impulses sent to the skeletal muscles cause shivering, rhythmic contractions that result in increased metabolic heat production. Equally important are our conscious responses to cold conditions. The hypothalamus notifies the cerebral cortex that body temperature is unpleasantly low, inspiring behavioral responses such as finding a warmer environment or adding extra clothing. Clothes reduce heat loss by convection and radiation.

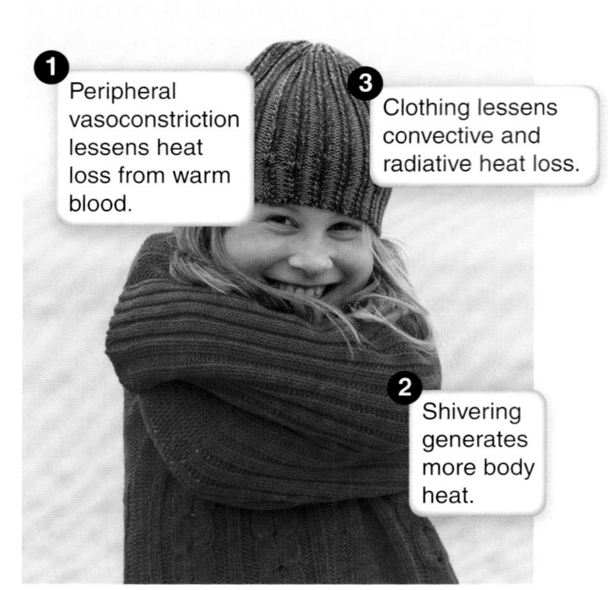

A Cold conditions

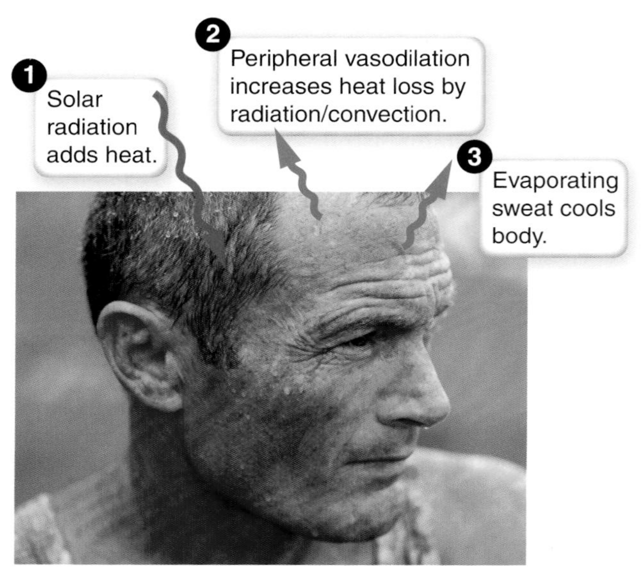

B Hot conditions

FIGURE 18-7 Thermoregulation. KEY POINT Negative feedback loops help maintain body temperature within normal limits. **A.** Under cold conditions, physiological and behavioral responses limit heat loss and generate additional metabolic heat. **B.** Under hot conditions, sweating and peripheral vasodilation help the body shed metabolic heat and heat absorbed from solar radiation. ZOOMING IN Is blood flow in the skin higher under cold conditions or hot conditions?

Responses to Hot Conditions Hot environmental conditions can provide a particular challenge to thermoregulation, because the body must disperse heat gained from solar radiation (FIG. 18-7B). Moreover, the smaller difference between body temperature and air temperature decreases the body's ability to lose heat by convection and conduction.

If body temperature increases above the set point, signals from the hypothalamus cause cutaneous blood vessels to dilate, so that increased blood flow will promote heat loss. The hypothalamus also stimulates the sweat glands to increase their activity, thus maximizing evaporation. However, the rate of heat loss through evaporation depends on the humidity of the surrounding air. When humidity exceeds 60% or so, perspiration does not evaporate as readily, making one feel uncomfortable unless some other means of heat loss is available, such as convection caused by a fan. It's important to remember that sweat only cools the body if it evaporates. Wiping away sweat eliminates its usefulness.

Muscles are especially important in temperature regulation because variations in the activity of these large tissue masses can readily increase or decrease heat generation. Because muscles form roughly one-third of the body, either an involuntary or an intentional increase in their activity can produce enough heat to offset a considerable decrease in the environmental temperature.

Age Factors Very young and very old people are limited in their ability to regulate body temperature when exposed to environmental extremes. A newborn infant's body temperature decreases if the infant is exposed to a cool environment for a long period. Elderly people are also not able to produce enough heat to maintain body temperature in a cool environment.

With regard to overheating in these age groups, heat loss mechanisms are not fully developed in the newborn. The elderly do not lose as much heat from their skin as do younger people. Both groups should be protected from extreme temperatures.

CHECKPOINTS 

- ☐ **18-12** What part of the brain is responsible for regulating body temperature?
- ☐ **18-13** What change occurs in cutaneous blood vessels under cold conditions? Under hot conditions?

CASEPOINT 

- ☐ **18-8** Why do you think Claudia was more sensitive to cold than her friends?

A & P in Action Revisited: Claudia's Anorexia Nervosa

Claudia approached Dr. Wade at the end of the next class, waiting until all of the other students had left.

"Do you have a minute to talk?" she asked. "I was hoping to speak with you today."

Dr. Wade replied, "Let's walk down toward the river where we can have some privacy. What's been happening with you?"

Claudia started to talk. "My friend Josie took me to the Student Health Center last night. They think I have anorexia nervosa! At first I didn't believe it, but they did some tests, and my results were similar to that case study we did in class about starving kids in war zones. My BMI is 16, and there were ketones in my urine. They did a heart test, and my ECG showed an arrhythmia. I don't know how this happened; I'm enjoying college life and I thought I was dealing with the stress, but I'm really in trouble. The psychologist wants me to keep a food diary, try to eat 2,000 calories a day, and limit my daily exercise to an hour. I have to meet with her every day, and if I can't turn things around in the next few weeks, they might refer me to an inpatient treatment center. I don't know if I can do it!"

Dr. Wade took a moment to think and said carefully, "Claudia, I'm so relieved that you are dealing with your illness. You have a support network of health professionals and friends, and I'm very hopeful for you. Please check in with me every week or so to let me know how you are doing, and let me know if you need to get some assignments or tests rescheduled."

CHAPTER 18 Chapter Wrap-Up

OVERVIEW

A detailed chapter outline with space for note-taking is on thePoint®. The figure below illustrates the main topics covered in this chapter.

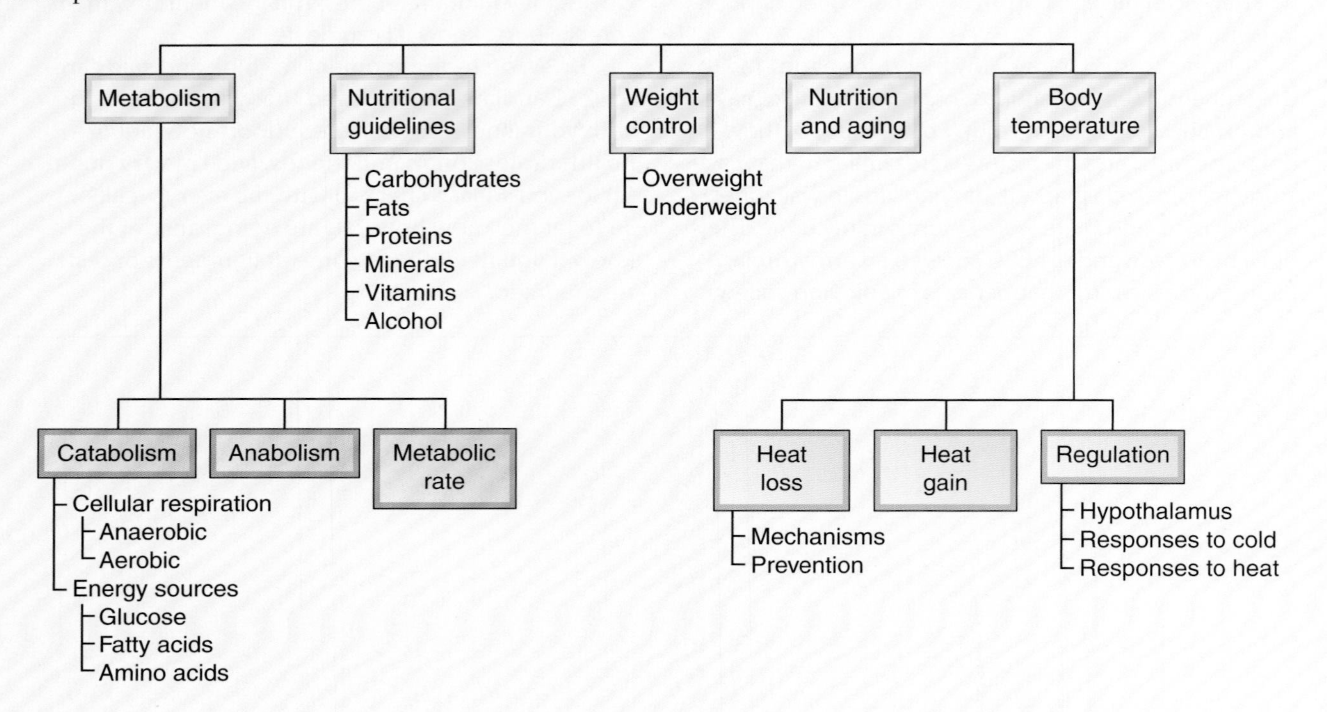

KEY TERMS

The terms listed below are emphasized in this chapter. Knowing them will help you organize and prioritize your learning. These and other boldface terms are defined in the Glossary with phonetic pronunciations.

anabolism
basal metabolism
catabolism
cellular respiration
glucose
glycogen
glycolysis
hypothalamus
kilocalorie
metabolic rate
mineral
oxidation
vitamin

WORD ANATOMY

Medical terms are built from standardized word parts (prefixes, roots, and suffixes). Learning the meanings of these parts can help you remember words and interpret unfamiliar terms.

WORD PART	MEANING	EXAMPLE
METABOLISM		
glyc/o	sugar, sweet	*Glycogen* yields glucose molecules when it breaks down.
-lysis	separating, dissolving	*Glycolysis* is the breakdown of glucose for energy.
BODY TEMPERATURE		
therm/o	heat	A *thermometer* is used to measure temperature.

QUESTIONS FOR STUDY AND REVIEW

BUILDING UNDERSTANDING

18

Fill in the Blanks

1. The amount of energy needed to maintain life functions while at rest is _______.
2. Reserves of glucose are stored in liver and muscle as _______.
3. The area of the brain most important for temperature regulation is the _______.
4. A mineral needed in extremely small amounts in the diet is termed a(n) _______.
5. The removal of nitrogen groups from amino acids in metabolism is called _______.

Matching Match each numbered item with the most closely related lettered item:

___ 6. Major energy source for the body
___ 7. Chemical element required for normal body function
___ 8. Complex organic substance required for normal body function
___ 9. A byproduct of fatty acid metabolism
___ 10. A simple fat

a. triglyceride
b. vitamin
c. mineral
d. ketone bodies
e. glucose

Multiple Choice

___ 11. Which of the following is NOT an example of catabolism?
a. glycolysis
b. deamination
c. cellular respiration
d. glycogen formation

___ 12. Which of the following would have the lowest glycemic effect?
a. glucose
b. sucrose
c. lactose
d. starch

___ 13. Which organ catabolizes alcohol?
a. small intestine
b. liver
c. pancreas
d. spleen

___ 14. A substance that protects against reactive oxygen species (free radicals) is a(n)
a. antioxidant
b. glycogen
c. fatty acid
d. pyruvate

UNDERSTANDING CONCEPTS

15. In what part of the cell does anaerobic respiration occur, and what are its end products? In what part of the cell does aerobic respiration occur? What are its end products?
16. About how many kilocalories are released from a tablespoon of butter (14 g)? A tablespoon of sugar (12 g)? A tablespoon of egg white (15 g)?
17. If you eat 2,000 kcal a day, how many kilocalories should come from carbohydrates? From fats? From protein?
18. How is heat produced in the body? What structures produce the most heat during increased activity?
19. During the course of a day, Emily's body temperature increased from 36.4°C to 37.1°C and then decreased to 36.4°C again. Describe the feedback mechanism regulating Emily's body temperature.
20. Differentiate between the terms in the following pairs:
    a. aerobic and anaerobic
    b. essential and nonessential amino acids
    c. saturated and unsaturated fats
    d. simple and complex carbohydrates
    e. radiation and conduction

CONCEPTUAL THINKING

21. The oxidation of glucose to form ATP is often compared to the burning of fuel. Why is this analogy inaccurate?
22. Richard M, a self-described couch potato, is 6 ft tall and weighs 240 lb. Calculate Richard's body mass index. Is Richard overweight or obese? List some diseases associated with obesity.
23. Referring to the case study in this chapter, explain how Claudia's body is compensating metabolically for her anorexia nervosa and what effects her disorder might have.
24. Jennifer's 85-year-old grandmother is a widow and lives alone. Should Jennifer be concerned about her nutritional well-being? What can she do to promote a healthful diet for her?

For more questions, see the Learning Activities on thePoint®.

CHAPTER 19

The Urinary System and Body Fluids

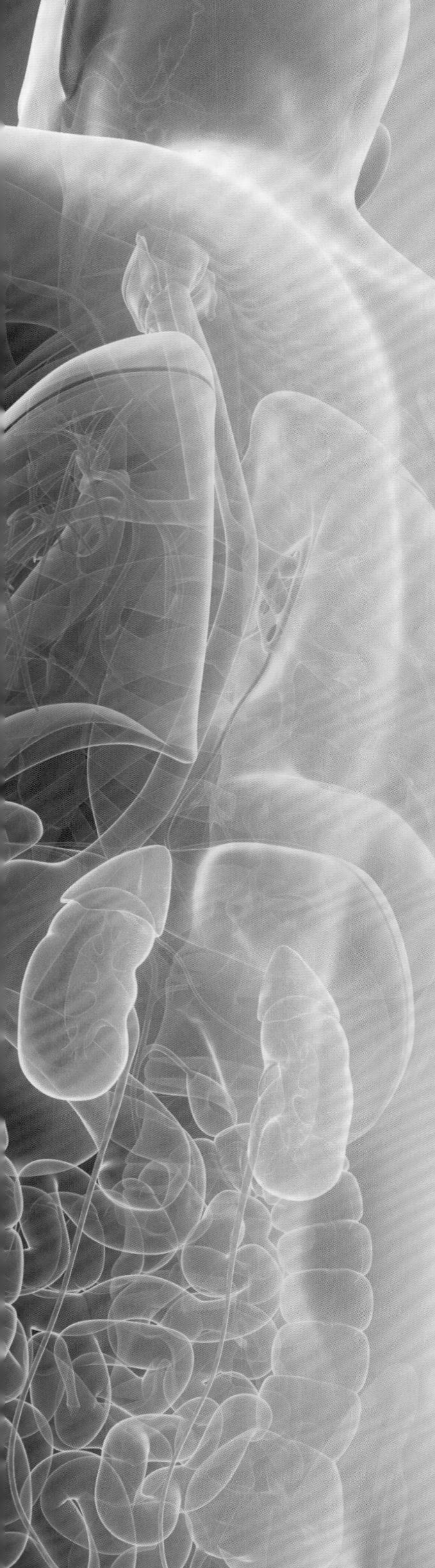

Learning Objectives

After careful study of this chapter, you should be able to:

1. Describe the organs of the urinary system, and give the functions of each. *p. 394*
2. List four systems that eliminate waste, and name the substances each eliminates. *p. 394*
3. List five important activities of the kidneys. *p. 395*
4. Describe the location and internal organization of the kidneys. *p. 395*
5. Describe a nephron. *p. 396*
6. Trace the path of a drop of blood as it flows through the kidney. *p. 396*
7. Name the four processes involved in urine formation, and describe what happens during each. *p. 396*
8. Explain the roles of juxtamedullary nephrons and antidiuretic hormone (ADH) in urine formation. *p. 399*
9. Describe the process of micturition. *p. 402*
10. List normal constituents of urine. *p. 402*
11. List the types of intracellular and extracellular fluids. *p. 403*
12. Define electrolytes, and describe some of their functions. *p. 404*
13. List the ways in which the body gains and loses water. *p. 405*
14. Explain how thirst is regulated. *p. 405*
15. List the steps involved in angiotensin II production, highlighting the role of the juxtaglomerular apparatus. *p. 407*
16. Diagram the feedback loops controlling body fluid osmolarity and blood pressure. *p. 409*
17. Describe three methods for regulating the pH of body fluids. *p. 409*
18. Referring to the opening case study, explain the dangers of overhydrating. *pp. 393, 410*
19. Show how word parts are used to build words related to the urinary system and body fluids (see Word Anatomy at the end of the chapter). *p. 412*

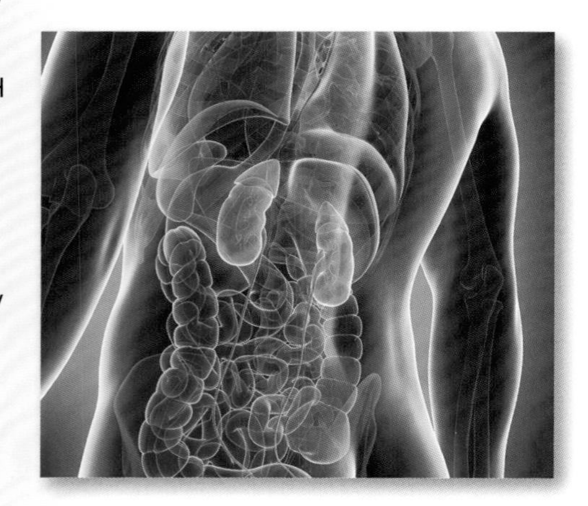

A & P in Action
Ethan's Triathlon Emergency

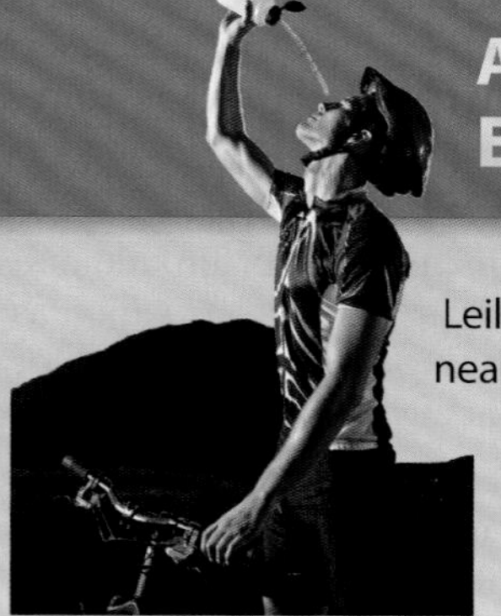

Leilani yawned as she waited near the finish line of the Ironman Triathlon in Kona, Hawaii. She had volunteered as a finish line attendant, which meant that she and her partner "caught" the athletes as they crossed the finish line and escorted them to get their medals, find their families, or, increasingly, get a bed in the medical tent. The race was over in three minutes, and there were still two athletes out on the course. They had already swum 2.4 mi in choppy surf, biked over 180 mi through lava fields, and had nearly finished the 26.2-mi run portion of the punishing event. Despite all of their efforts, they had to complete their run before midnight in order to qualify as Ironman finishers. The crowd noise increased as a tall man stumbled over to the finish line, to the announcer's cry, "Ethan Landry, you are an Ironman!" Leilani and her partner raced over to catch him.

"We better get this one straight to the medical tent," her partner gasped as they tried to keep the man from sinking to the pavement. They were met by Darren, a local physician assistant, who helped them lay the man down on a stretcher.

"What's your name? Do you know where you are?" Darren asked the man as he strapped on a blood pressure monitor.

"Um, Ethan? I'm in Hawaii? I don't feel well. I'm so tired and dizzy. I feel sick to my stomach, and I've had a terrible headache for the last few hours," the athlete replied. Darren pinched Ethan's arm and noted that the pinched skin returned immediately to its original state. His face seemed swollen, and his wedding ring was sunk deep into his engorged fingers.

"Should I get him some water?" Leilani asked.

"I don't think so," Darren answered, as he tilted Ethan's bed backward. "His blood pressure is low, which is why I'm lowering his head below his body, but he has some evidence of edema. I think that Ethan is actually overhydrated and suffering from exercise-associated hyponatremia, or EAH. This condition is used to be called water intoxication, and it happens sometimes in endurance events. Basically, consuming too much water can dilute your blood and other body fluids and lead to swelling. Look at his face and fingers; it's as if they are waterlogged. Let's get him to the hospital for some tests."

In Chapter 3, we discussed the effect of osmotic pressure on cells. We continue in this chapter with a more detailed study of body fluids and the need to keep the total volume and composition of these fluids within normal limits. Later, we'll find out if Darren's suspicions were correct.

As you study this chapter, CasePoints will give you opportunities to apply your learning to this case.

Visit thePoint® to access the following resources. For guidance in using these resources most effectively, see pp. xv–xvii.

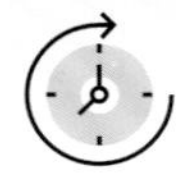

Preparing to Learn

- Tips for Effective Studying
- Pre-Quiz

While You Are Learning

- Web chart: Kidney Regulation of Blood Pressure
- Animation: Renal Function
- Animation: The Juxtaglomerular Apparatus
- Animation: Osmosis
- Chapter Notes Outline
- Audio Pronunciation Glossary

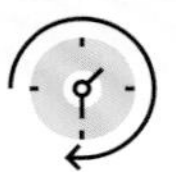

When You Are Reviewing

- Answers to Questions for Study and Review
- Health Professions: Emergency Medical Technician
- Interactive Learning Activities

A LOOK BACK

The importance of body fluid volume and composition has been stressed throughout previous chapters, beginning with the discussion of the chemistry of water 6, *electrolytes, and pH in Chapter 2. The urinary system plays a key role in body fluid homeostasis* 3, *largely by matching input with output* 13. *The key ideas of barriers* 4 *and gradients and flow* 5 *are also relevant to this chapter.*

Systems Involved in Excretion

The urinary system is also called the *excretory system* because one of its main functions is **excretion**, removal and elimination of unwanted substances from the blood. The urinary system also regulates the volume, acid–base balance (pH), and electrolyte composition of body fluids.

The main parts of the urinary system, shown in FIGURE 19-1, are as follows:

- Two **kidneys**. These organs extract wastes from the blood, balance body fluids, and form urine.
- Two **ureters** (U-re-ters). These tubes conduct urine from the kidneys to the urinary bladder.
- A single **urinary bladder**. This reservoir receives and stores the urine brought to it by the two ureters.
- A single **urethra** (u-RE-thrah). This tube conducts urine from the bladder to the outside of the body for elimination.

Although the focus of this chapter is the urinary system, some other systems are mentioned here as well, because they also function in excretion. These systems and some of the substances they eliminate are the following:

- The digestive system eliminates water, some salts, and bile, in addition to digestive residue, all of which are contained in the feces. The liver is important in eliminating the products of red blood cell destruction and in breaking down certain drugs and toxins.
- The respiratory system eliminates carbon dioxide and small amounts of water. The latter appears as vapor, as can be demonstrated by breathing on a windowpane or a mirror, where the water condenses.
- The skin, or integumentary system, excretes water, salts, and very small quantities of nitrogenous wastes. These all appear in perspiration, although water also evaporates continuously from the skin without our being conscious of it.

CHECKPOINTS 

- ☐ **19-1** What are the organs of the urinary system?
- ☐ **19-2** What are three systems other than the urinary system that eliminate waste?

The Kidneys

The kidneys interact with other systems, primarily the cardiovascular system, as prime regulators of homeostasis. After a brief overview of the kidney's many activities, we will describe its structure and specific roles in urine formation.

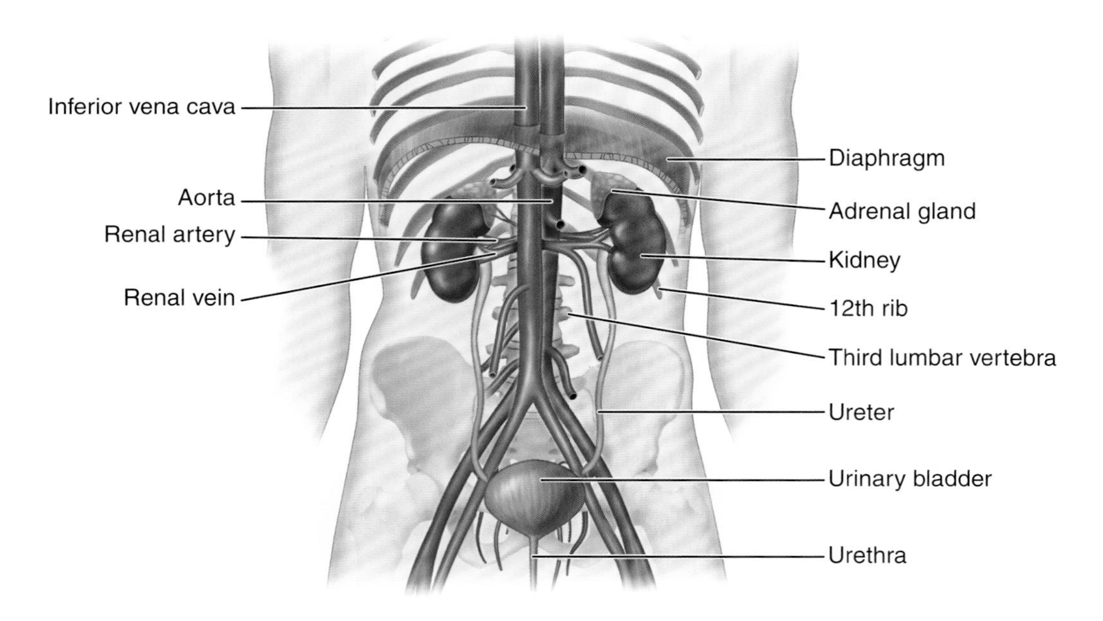

FIGURE 19-1 Urinary system. The urinary system consists of the kidneys, ureters, urinary bladder, and urethra. ZOOMING IN Identify the structures that carry urine to and from the bladder.

KIDNEY ACTIVITIES

The kidneys are involved in the following processes:

- Excretion of unwanted substances, such as cellular metabolic waste and toxins. One product of amino acid metabolism is nitrogen-containing waste material, chiefly **urea** (u-RE-ah). After synthesis in the liver, urea is transported in the blood to the kidneys for elimination. The kidneys have a specialized mechanism for the elimination of urea and other nitrogenous (ni-TROJ-en-us) wastes.
- Homeostasis of body fluids. As discussed later in this chapter, the kidneys control the sodium and water content of body fluids by altering the composition and volume of **urine**. The kidneys also contribute to blood pH control by excreting variable amounts of acids and bases and by synthesizing buffers.
- Blood pressure regulation. The kidneys help control blood pressure by altering fluid balance. This process is explained more completely later in this chapter.
- Hormone production. The kidneys produce erythropoietin (EPO), which stimulates red cell production in the bone marrow (see Chapter 12). They also activate vitamin D, which is necessary for bone health (see Chapter 11). Individuals with chronic kidney failure may not make enough of these two substances and thus require supplementation to prevent anemia (inadequate red blood cells) and osteomalacia (weak bones).
- Blood glucose control. The kidneys can synthesize glucose from noncarbohydrate sources, such as certain amino acids, glycerol, pyruvate, and lactate. Although the liver is the main site of this activity, known as gluconeogenesis, the kidneys can produce 10% to 20% of this yield as needed.

KIDNEY STRUCTURE

The kidneys lie against the back muscles in the upper abdomen at about the level of the last thoracic and first three lumbar vertebrae. The right kidney is slightly lower than the left to accommodate the liver (see FIG. 19-1). Each kidney is firmly enclosed in a membranous **renal capsule** made of fibrous connective tissue (FIG. 19-2). In addition, there is a protective layer of fat called the *adipose capsule* around the organ that is not illustrated. An outermost layer of fascia (connective tissue) anchors the kidney to the peritoneum and abdominal wall. The kidneys, as well as the ureters, lie posterior to the peritoneum (see Chapter 19). Thus, they are not in the peritoneal cavity but rather in an area known as the **retroperitoneal** (ret-ro-per-ih-to-NE-al) **space**.

19

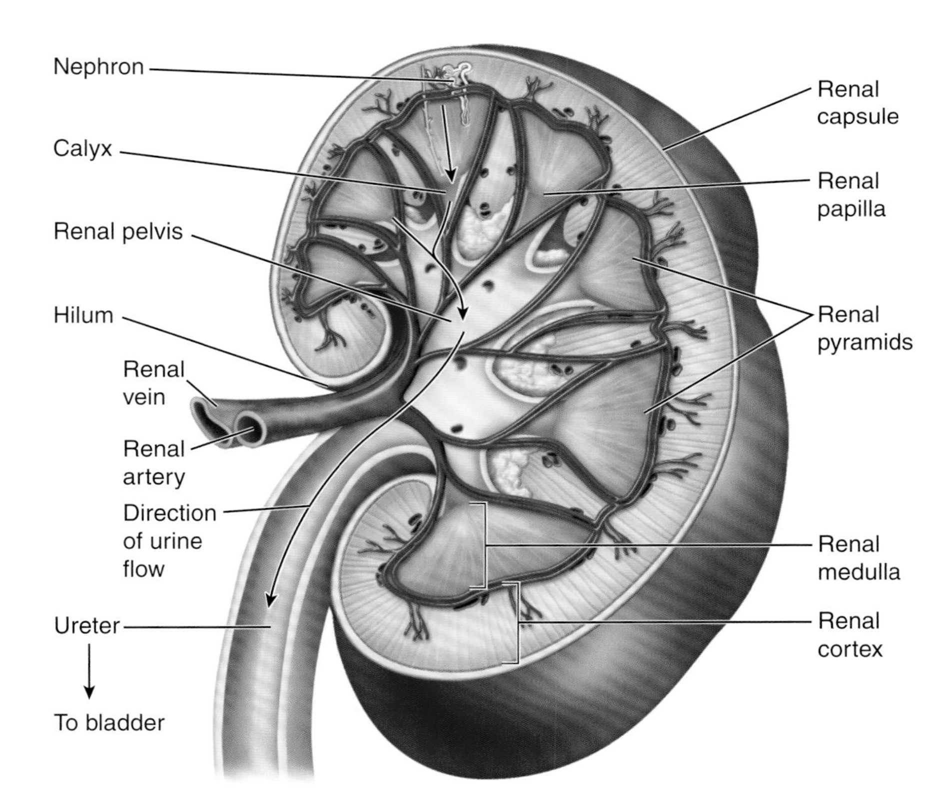

FIGURE 19-2 Kidney structure and the renal blood supply. A longitudinal section through the kidney shows its internal structure. Urine made in the nephrons drains into the renal pelvis and then the ureter. The renal artery branches to all regions of the kidney. Venous drainage combines to flow into the renal vein. ZOOMING IN What is the outer region of the kidney called? What is the inner region of the kidney called?

Blood is brought to the kidney by a short branch of the abdominal aorta called the **renal artery**. Blood leaves the kidney via the **renal vein**, which carries blood into the inferior vena cava for return to the heart. See **FIGURE A3-11A** in the Dissection Atlas for a photograph showing the kidneys and ureters, in place.

Organization The kidney is a somewhat flattened organ approximately 10-cm (4-in) long, 5-cm (2-in) wide, and 2.5-cm (1-in) thick. On the medial border, there is a notch called the **hilum**, where the renal artery, the renal vein, and the ureter connect with the kidney (**see FIG. 19-2**). The lateral border is convex (curved outward), giving the entire organ a bean-shaped appearance.

The kidney is divided into two regions: the renal cortex and the renal medulla (**see FIG. 19-2**). The **renal cortex** is the kidney's outer portion. The internal **renal medulla** contains tubes in which urine is formed and collected. These tubes form a number of cone-shaped structures called **renal pyramids**. The tips of the pyramids point toward the **renal pelvis**, a funnel-shaped basin that forms the upper end of the ureter. Cuplike extensions of the renal pelvis surround the tips of the pyramids and collect urine; these extensions are called **calyces** (KA-lih-seze; singular, **calyx**, KA-liks). The urine that collects in the pelvis then passes down the ureters to the bladder. See **FIGURE A3-11B** in the Dissection Atlas for a photograph of the internal kidney.

The Nephron and Its Blood Supply As is the case with most organs, the kidney's most fascinating aspect is too small to be seen with the naked eye. This basic unit, which actually creates urine, is the **nephron** (NEF-ron) (**FIG. 19-3A**). Each nephron begins with a hollow, cup-shaped bulb known as the **glomerular** (glo-MER-u-lar) **capsule** or *Bowman capsule*. This structure gets its name from the cluster of capillaries it surrounds, which is called the **glomerulus** (a word that comes from the Latin meaning "ball of yarn"). This combined unit of the glomerulus with its glomerular capsule is the nephron's filtering device. The remainder of the nephron is essentially a tiny coiled tube, the **renal tubule**, which consists of several parts. The portion leading from the glomerular capsule is called the **proximal tubule**. The renal tubule then untwists to form a hairpin-shaped segment called the **nephron loop**, or *loop of Henle*. The first part of the loop, which carries fluid toward the medulla, is the **descending limb**. The part that continues from the loop's turn and carries fluid away from the medulla is the **ascending limb**. Continuing from the ascending limb, the tubule folds into the **distal tubule**, so called because it is farther along the tubule from the glomerular capsule than the proximal tubule. Many renal tubules empty into each collecting duct, which then continues through the medulla toward the renal pelvis.

In most cases, the glomerulus, the glomerular capsule, and the proximal and distal tubules of the nephron are within the renal cortex. The nephron loop and collecting duct extend into the medulla (**see FIG. 19-3A**).

A small blood vessel, the **afferent arteriole**, supplies the glomerulus with blood; another small vessel, called the **efferent arteriole**, carries blood away from the glomerulus (**see FIG. 19-3B**). When blood leaves the glomerulus, it does not head immediately back toward the heart. Instead, it flows into a capillary network that surrounds the renal tubule. These **peritubular capillaries** are named for their location.

Each kidney contains about 1 million nephrons; if all these coiled tubes were separated, straightened out, and laid end to end, they would span some 120 km (75 miles)! **FIGURE 19-4** is a microscopic view of kidney tissue showing several glomeruli, each surrounded by a glomerular capsule. This figure also shows sections through renal tubules.

CHECKPOINTS 

- ☐ **19-3** Where is the retroperitoneal space?
- ☐ **19-4** What vessel supplies blood to the kidney, and what vessel drains blood from the kidney?
- ☐ **19-5** What is the name of the funnel-shaped collecting area that forms the upper end of the ureter?
- ☐ **19-6** What is the functional unit of the kidney called?
- ☐ **19-7** What name is given to the coil of capillaries in the glomerular capsule?

URINE FORMATION

The following explanation of urine formation describes a complex process, involving many back-and-forth exchanges between the bloodstream and the kidney tubules. As fluid filtered from the blood travels slowly through the nephron's twists and turns, there is ample time for exchanges to take place. These processes together allow the kidney to "fine-tune" body fluids as they adjust the urine's composition.

> **See the Student Resources on thePoint® for the animation "Renal Function," which shows the process of urine formation in action.**

Glomerular Filtration Chapter 14 explained how blood fluids cross the capillary wall and enter the interstitial fluid (IF) by the process of bulk flow, specifically filtration. The same process occurs in the nephron, but the fluid enters the glomerular capsule instead of the IF. In the nephron, fluid must pass through the walls of both the capillary and the glomerular capsule. Gaps between the cells making up these walls permit the passage of water and dissolved materials; however, the capillary walls retain blood cells and large proteins, and these components remain in the blood (**FIG. 19-5**).

Because the diameter of the afferent arteriole is slightly larger than that of the efferent arteriole (**see FIG. 19-5**), blood can enter the glomerulus more easily than it can leave. Thus, blood pressure in the glomerulus is about three to four times higher than that in other capillaries. To understand this effect, think of placing your thumb over the end of a garden

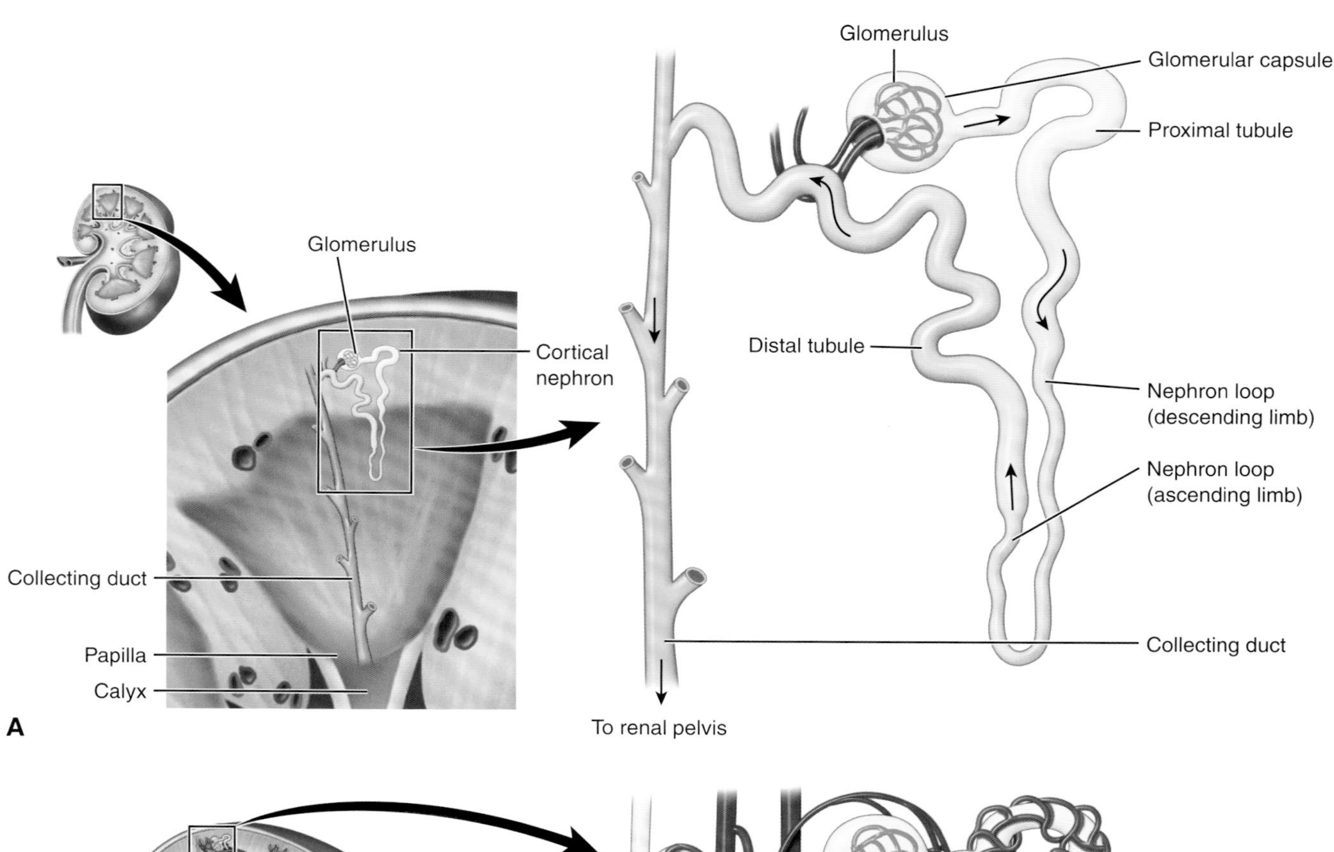

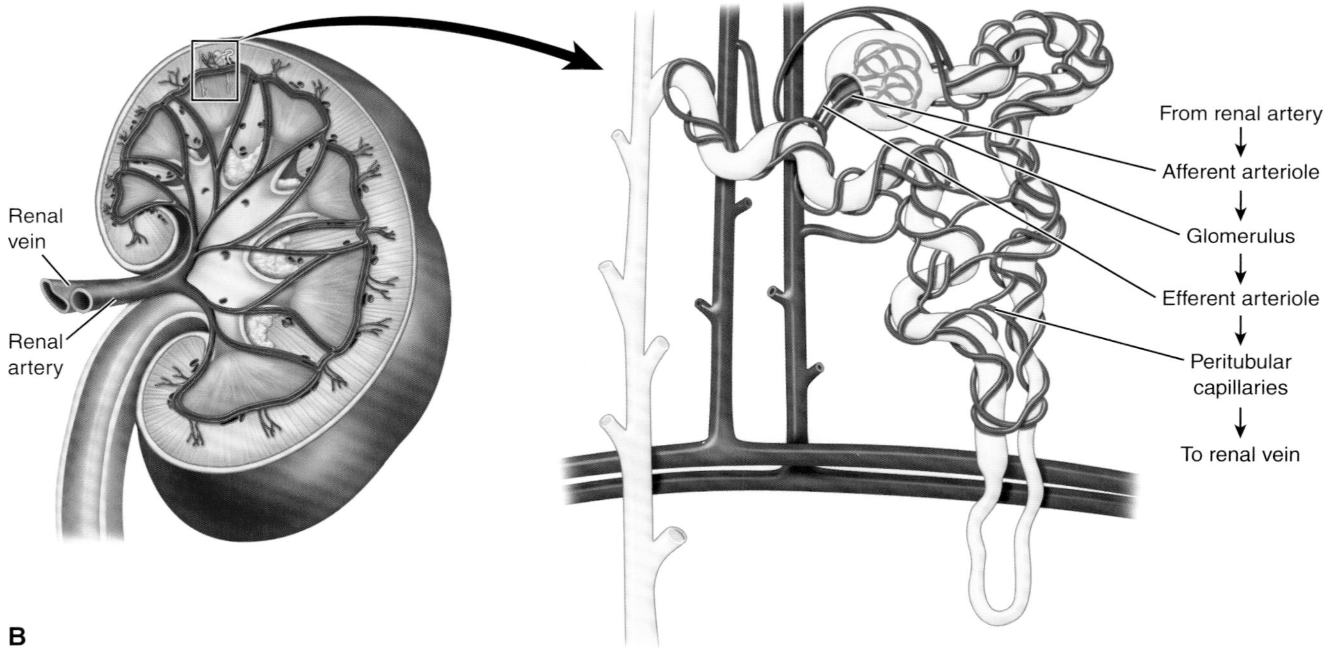

FIGURE 19-3 A nephron and its blood supply. KEY POINT The nephron regulates the proportions of urinary water, waste, and other materials according to the body's constantly changing needs. Materials that enter the nephron can be returned to the blood through the surrounding capillaries. **A.** A nephron with its blood vessels removed. **B.** A nephron and its associated blood vessels. ZOOMING IN The nephron is associated with two capillary beds. Which capillary bed receives blood first?

hose as water comes through. As you make the diameter of the opening smaller, water is forced out under higher pressure. As a result of increased fluid (hydrostatic) pressure in the glomerulus, materials are constantly being pushed out of the blood and into the nephron's glomerular capsule. This movement of materials under pressure from the blood into the capsule is therefore known as **glomerular filtration.** Note that only a small proportion of the blood's volume passes into the glomerular capsule. The rest continues out the efferent arteriole into the peritubular capillaries.

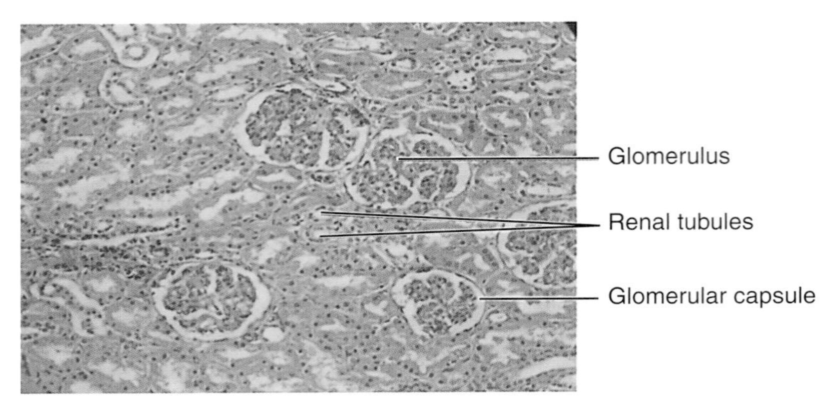

FIGURE 19-4 **Microscopic view of the kidney.** KEY POINT The glomeruli and glomerular capsules are visible along with cross-sections of the renal tubules.

The fluid that enters the glomerular capsule, called the **glomerular filtrate**, begins its journey along the renal tubule. In addition to water and the normal soluble substances in the blood, other substances, such as vitamins and drugs, also may be filtered and become part of the glomerular filtrate. As noted, blood cells and proteins are not normally found in the filtrate.

Tubular Reabsorption The kidneys form about 160 to 180 L of filtrate each day. However, only 1 to 1.5 L of urine is eliminated daily. Clearly, most of the water that enters the nephron is not excreted with the urine but rather is returned to the circulation. In addition to water, many other substances the body needs, such as nutrients and ions, pass into the nephron as part of the filtrate, and these also must be returned. Therefore, the process of filtration that occurs in the glomerular capsule is followed by a process of **tubular reabsorption.** As the filtrate travels through the renal tubule, water and other needed substances pass through tubular cells and enter the IF. They move by several processes previously described in Chapter 3, including:

- Diffusion. The movement of substances from an area of higher concentration to an area of lower concentration (following the concentration gradient).
- Osmosis. Diffusion of water through a semipermeable membrane, from an area of low solute concentration to an area of higher solute concentration.
- Active transport. Movement of materials through the plasma membrane against the concentration gradient using energy and transporters.

Reabsorbed substances move from the nephron to the IF and then enter the peritubular capillaries and return to the circulation. Substances that are filtered but not reabsorbed

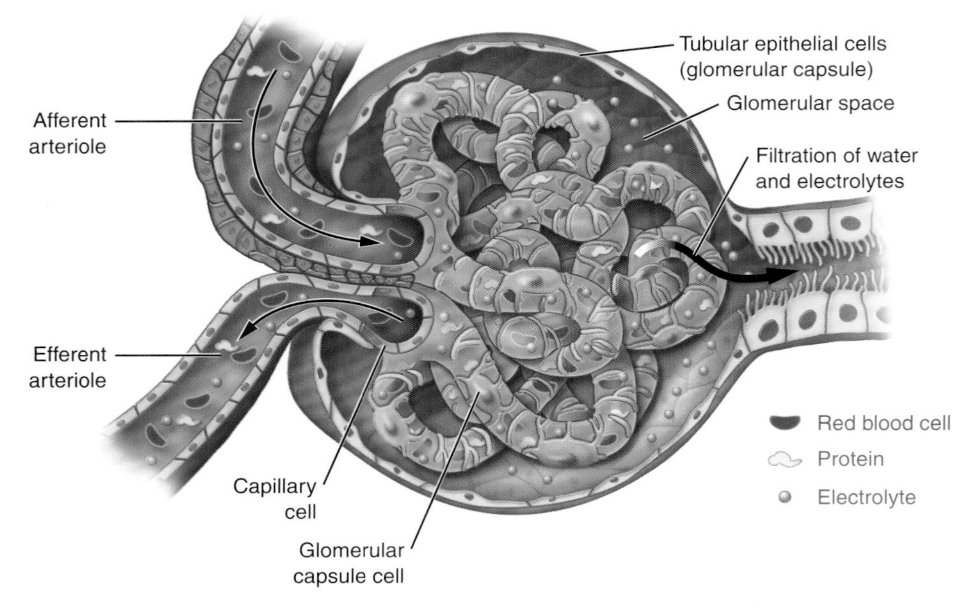

FIGURE 19-5 **Glomerular filtration: the first step in urine formation.** KEY POINT The cells forming the walls of the capillary and the glomerular capsule serve as a filter keeping cells and proteins in the blood and letting water and small solutes through. Blood pressure provides the gradient that enables filtration. Blood cells and large proteins remain behind in the blood.

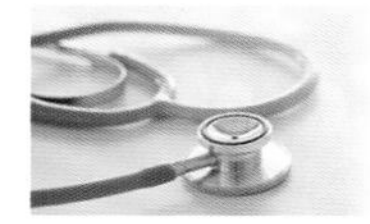

CLINICAL PERSPECTIVES
Transport Maximum

BOX 19-1

The kidneys work efficiently to return valuable substances to the blood after glomerular filtration. However, the carriers that are needed for active transport of these substances can become overloaded. Thus, there is a limit to the amount of each substance that can be reabsorbed in a given time period. The limit of this reabsorption rate is called the **transport maximum (Tm)**, or tubular maximum, and it is measured in milligrams (mg) per minute. For example, the Tm for glucose is approximately 375 mg/minute.

If a substance is present in excess in the blood, it may exceed its transport maximum, and then, because it cannot be totally reabsorbed, some will be excreted in the urine. Thus, the transport maximum determines the **renal threshold**—the plasma concentration at which a substance will begin to be excreted in the urine, which is measured in mg per deciliter (dL). For example, if the concentration of glucose in the blood exceeds its renal threshold (180 mg/dL), glucose will begin to appear in the urine, a condition called glycosuria. The most common cause of glycosuria is uncontrolled diabetes mellitus.

remain in the filtrate and are excreted in the urine. For example, about 50% of the urea and uric acid (products of amino acid and nucleic acid breakdown, respectively) and all of the creatinine (a product of muscle cell breakdown) are kept within the tubule to be eliminated. **BOX 19-1** presents additional information on tubular reabsorption in the nephron.

As discussed later in this chapter, the body controls the blood concentrations of certain ions by altering how much reabsorption occurs.

Concept Mastery Alert

Remember that reabsorption refers specifically to the movement of substances between the tubule and the blood. So, filtrate flow occurs from the proximal tubule to the nephron loop, but reabsorption occurs from the proximal tubule to the peritubular capillaries.

Tubular Secretion The process of **tubular secretion** actively transports specific substances from the peritubular capillary into the nephron tubule. These substances were not filtered from the blood or were reabsorbed in a more proximal portion of the tubule. Potassium and hydrogen ions are moved into the urine in this manner. The kidneys thus regulate the acid–base (pH) balance of body fluids by the active secretion of hydrogen ions. In addition, some drugs, such as penicillin, are actively secreted into the nephron tubule for elimination. As with reabsorption, regulation of tubular secretion aids in maintaining body fluid homeostasis, as discussed later in this chapter.

Concentration of the Urine The proximal tubule reabsorbs about 65% of filtered water by osmosis, as water follows reabsorbed solutes. The net result is a tubular fluid with the same **osmolarity** (solute concentration) as the IF and blood. However, the mammalian kidney has the remarkable ability to produce urine that is more concentrated than other body fluids. This ability relies on the actions of specialized **juxtamedullary nephrons** with exceptionally long nephron loops dipping deep into the renal medulla (*juxtamedullary* means "near the medulla") **(FIG. 19-6A)**. These long loops establish the **medullary osmotic gradient**, which is a difference in solute concentration (and thus osmotic pressure) in different regions of the medulla. The IF in deeper medullary regions has more solutes and thus greater osmotic pressure than does the IF in regions closer to the cortex.

All collecting ducts use the medullary gradient to produce concentrated urine. As the dilute filtrate passes through the collecting duct toward the renal pelvis, it encounters progressively more concentrated IF. Water moves by osmosis down the osmotic gradient, from the dilute filtrate into the more concentrated IF and then into the blood **(see FIG. 19-6B)**. In this manner, the urine becomes more concentrated as it leaves the nephron and its volume is reduced.

Water reabsorption from the collecting duct is controlled by **antidiuretic hormone (ADH)**, a hormone released from the posterior pituitary gland. 5 ADH does not alter the osmotic gradient; instead, it modifies the other parameter of flow: resistance. It makes the walls of the collecting duct more permeable to water by stimulating the insertion of aquaporins. As a result, more water will be reabsorbed and less water will be excreted with the urine **(see FIG. 19-6B)**. In the absence of ADH, collecting duct plasma membranes contain very few aquaporins, so very little water moves down the osmotic gradient from tubular fluid into medullary IF **(see FIG. 19-6C)**.

3 As discussed later in this chapter, ADH is a signal in feedback loops controlling blood osmolarity and volume **(see TABLE 19-1)**. As the blood becomes more concentrated or blood volume decreases, the hypothalamus triggers more ADH release from the posterior pituitary; as the blood becomes more dilute or increases in volume, less ADH is released.

Summary of Urine Formation The processes involved in urine formation are summarized below and illustrated in **FIGURE 19-7**.

1. Glomerular filtration moves water and solutes from the blood into the nephron tubule.
2. Tubular reabsorption moves water and other useful substances back into the blood.
3. Tubular secretion moves unfiltered or reabsorbed substances from the blood into the nephron for elimination.
4. In the presence of ADH, the collecting duct concentrates the urine and reduces the volume excreted.

19

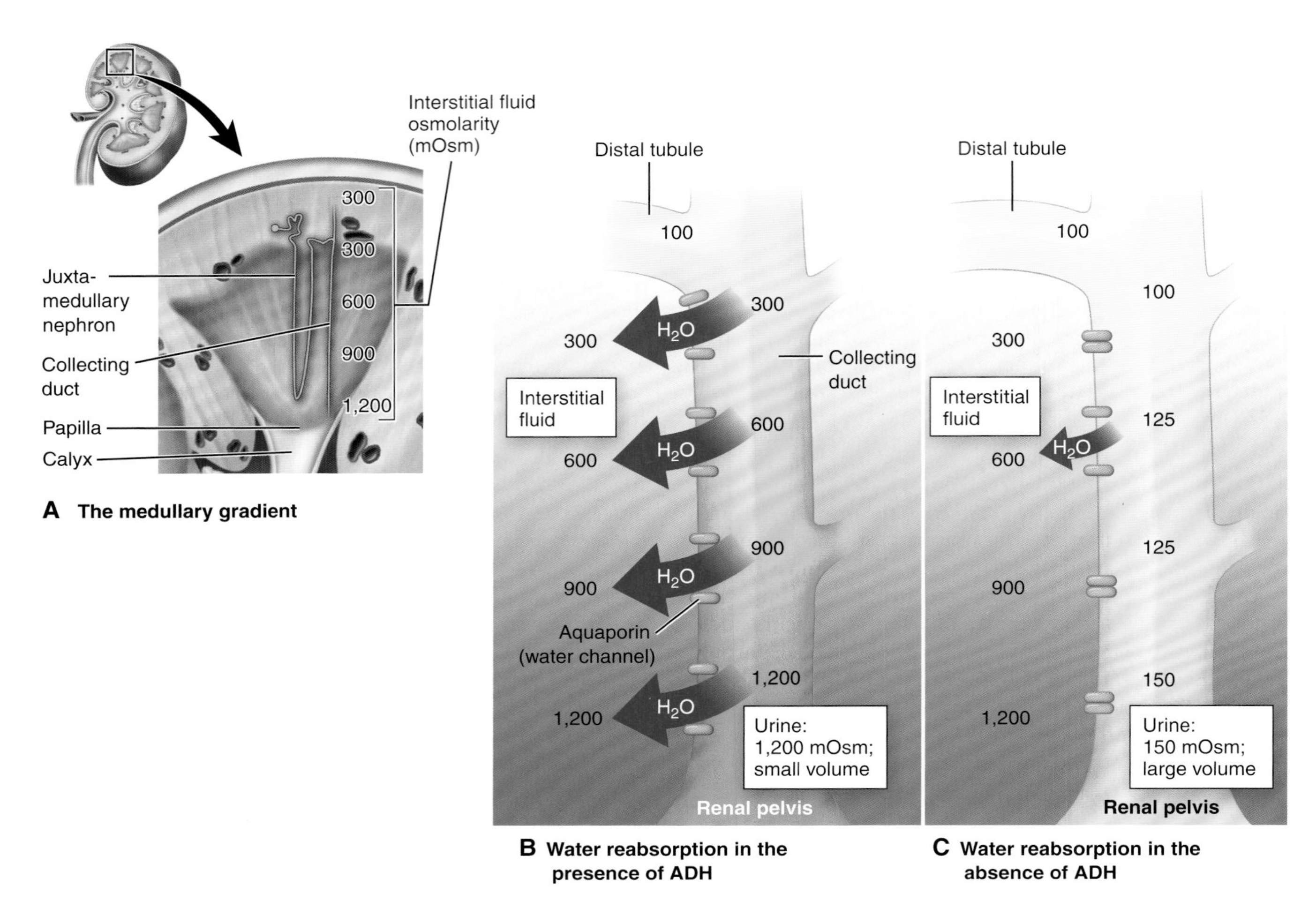

FIGURE 19-6 Urine concentration. **KEY POINT** The renal tubule establishes the medullary gradient, as measured in milliosmoles (mOsm). The collecting duct, under the influence of antidiuretic hormone (ADH), concentrates urine. **A.** The nephron loop establishes an osmotic gradient in the renal medulla. **B.** If ADH is present, water leaves the collecting duct by osmosis, resulting in concentrated urine. **C.** If ADH is absent, water cannot leave the collecting duct and the urine remains dilute. ZOOMING IN Where is the osmotic gradient stronger—near the distal tubule or near the renal pelvis?

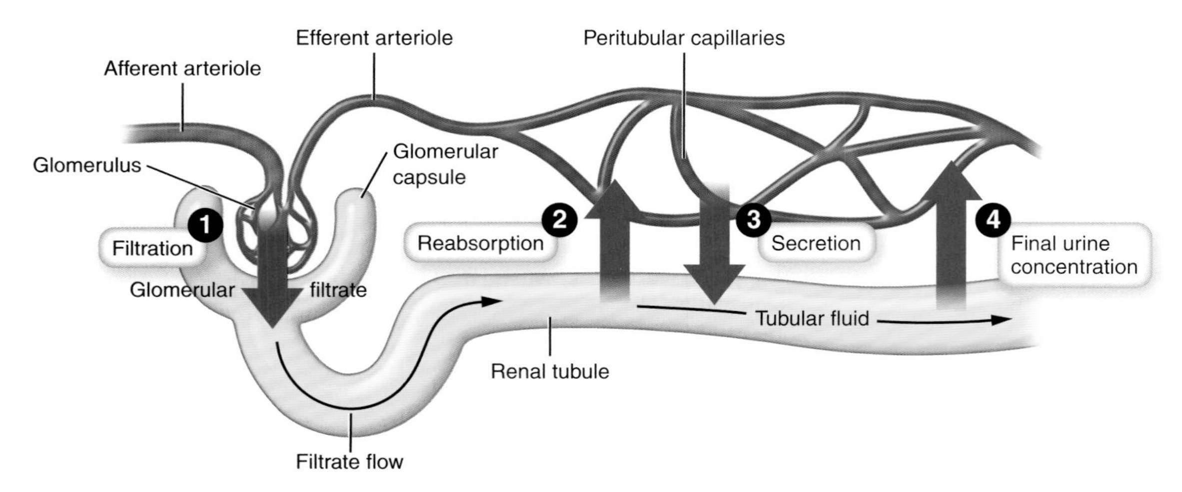

FIGURE 19-7 Summary of urine formation in a nephron. **KEY POINT** Four processes involved in urine formation are shown. ZOOMING IN Which process or processes move substances from the blood to the renal tubule?

The urine is not further modified once it enters the renal pelvis. Therefore, urine contains substances that were secreted as well as substances that were filtered but not reabsorbed.

CHECKPOINTS 

- ☐ **19-8** What process drives materials out of the glomerulus and into the glomerular capsule?
- ☐ **19-9** What is the name of the process that returns materials from the nephron back to the circulation?
- ☐ **19-10** What component of the filtrate is moved by tubular secretion to balance pH?
- ☐ **19-11** What hormone controls water reabsorption from the collecting duct of the nephron?
- ☐ **19-12** What parameter of flow is controlled in order to alter water reabsorption in the collecting duct: the gradient or the resistance?

Elimination of Urine

Urine is excreted from the kidneys into the two ureters, which transport urine to the bladder. It is then stored until eliminated from the body via the urethra. Let us take a closer look at each of these organs.

THE URETERS

Each of the two ureters is a long, slender, muscular tube that extends from the kidney down to and through the inferior portion of the urinary bladder (see FIG. 19-1). Like the other parts of the urinary system, the ureters are entirely extraperitoneal (outside the peritoneum). Their length naturally varies with the size of the individual; they may be anywhere from 25 to 32 cm (10- to 13-in) long. Nearly 2.5 cm (1 in) of the terminal distal ureter passes obliquely (at an angle) through the inferior bladder wall. The two ureteral openings are located just superior and lateral to the urethral opening on the inferior surface of the bladder and are protected by one-way valves (FIG. 19-8). A full bladder compresses the distal ureters and closes these valves, preventing the backflow of urine.

The ureteral wall includes a lining of epithelial cells, a relatively thick layer of involuntary muscle, and, finally, an outer coat of fibrous connective tissue. The epithelium is the transitional type, which flattens from a cuboidal shape as the tube stretches. This same type of epithelium lines the renal pelvis, the bladder, and the proximal portion of the urethra. The ureteral muscles are capable of the same rhythmic contraction (peristalsis) that occurs in the digestive system. Urine is moved along the ureter from the kidneys to the bladder by gravity and by peristalsis at frequent intervals.

THE URINARY BLADDER

When it is empty, the urinary bladder is located posterior to the pubic symphysis, as shown in FIGURE 20-1. The urinary bladder is a temporary reservoir for urine, just as the gallbladder is a storage sac for bile. A full (distended) bladder lies in an unprotected position in the lower abdomen, and a blow may rupture it, necessitating immediate surgical repair.

The bladder wall has many layers. It is lined with mucous membrane containing transitional epithelium. The bladder's lining, like that of the stomach, is thrown into folds called rugae (see FIG. 19-8) when the organ is empty. Beneath the mucosa is a layer of connective tissue, followed by a layer of involuntary smooth muscle that can stretch considerably. Finally, the parietal peritoneum covers the bladder's superior portion.

When the bladder is empty, the muscular wall thickens, and the entire organ feels firm. As the bladder fills, the muscular wall becomes thinner, and the organ may increase from a length of 5 cm (2 in) up to as much as 12.5 cm (5 in) or even more. A moderately full bladder holds about 470 mL (1 pint) of urine.

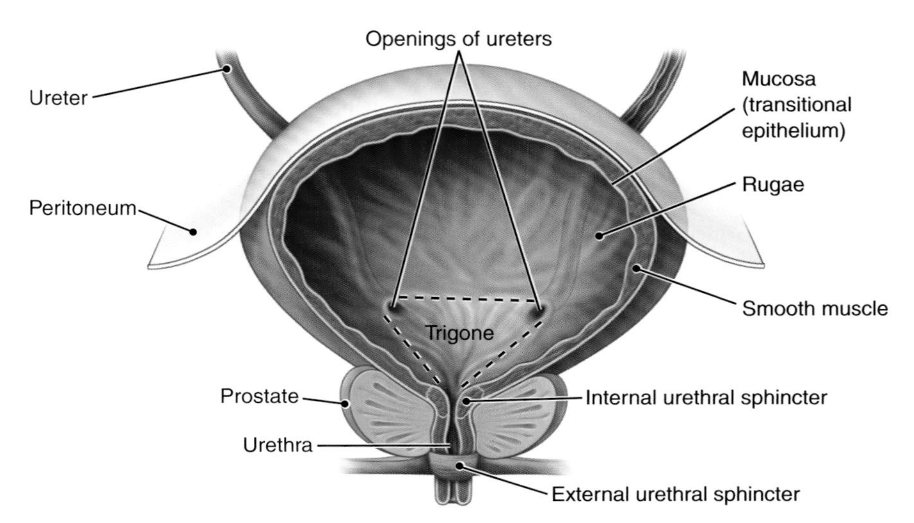

FIGURE 19-8 **The male urinary bladder.** The interior of the bladder is folded into rugae. Internal and external sphincters regulate urination. The trigone is a triangular region in the floor of the bladder marked by the openings of the ureters and the urethra. ZOOMING IN What gland does the urethra pass through in the male?

The **trigone** (TRI-gone) is a triangular-shaped region in the floor of the bladder. It is marked by the openings of the two ureters and the urethra (see FIG. 19-8). As the bladder fills with urine, it expands upward, leaving the trigone at the base stationary. This stability prevents stretching of the ureteral openings and the possible backflow of urine into the ureters.

THE URETHRA

The urethra carries urine from the bladder to the outside (see FIG. 19-1). The urethra differs in males and females; in the male, it is part of both the reproductive system and the urinary system, and it is much longer than is the female urethra. FIGURE A3-12 in the Dissection Atlas shows photographs of the bladder and urethra in both males and females.

The male urethra is approximately 20 cm (8 in) in length. Proximally, it passes through the prostate gland, where it is joined by two ducts carrying male reproductive cells (spermatozoa) from the testes and glandular secretions (see FIG. 20-1). From here, it leads to the outside through the **penis** (PE-nis), the male organ of copulation. The male urethra serves the dual purpose of draining the bladder and conveying semen with spermatozoa.

"Straddle" injuries to the urethra are common in males. This type of injury occurs when a man or child slips and lands with a narrow rod or beam between his legs, as when riding a bike or doing construction work. Such an accident may catch the urethra between the hard surface and the pubic arch and rupture the urethra. In accidents in which the bones of the pelvis are fractured, urethral rupture is fairly common.

The urethra in the female is a thin-walled tube about 4-cm (1.5-in) long. It is posterior to the pubic symphysis and is embedded in the muscle of the anterior vaginal wall (see FIG. 20-6A). The external opening, the urethral orifice, is located just anterior to the vaginal opening between the labia minora. The female urethra drains only the bladder and is entirely separate from the reproductive system.

URINATION

The process of expelling (voiding) urine from the bladder is called **urination** or **micturition** (mik-tu-RISH-un). This process is controlled both voluntarily and involuntarily with the aid of two muscular rings (sphincters) that surround the urethra (see FIG. 19-8). Near the bladder's outlet is an involuntary **internal urethral sphincter** formed by a continuation of the bladder's smooth muscle. Below this muscle is a voluntary **external urethral sphincter** formed by the muscles of the pelvic floor.

As the bladder fills with urine, stretch receptors in its wall send impulses to a center in the lower part of the spinal cord. Motor impulses from this center stimulate contraction of the bladder wall and relaxation of the urethral sphincters, forcing urine outward. In the infant, this emptying occurs automatically as a simple reflex. Impulses from higher brain centers can overcome this reflexive relaxation of the external urethral sphincter, blocking involuntary urination. Urination can also be voluntarily initiated by higher motor centers, even when the bladder is virtually empty. Early in life, a child learns to control urination, a process known as *toilet training*. The impulse to urinate will override conscious controls if the bladder becomes too full.

URINE

Urine is a yellowish liquid that is approximately 95% water and 5% dissolved solids and gases. The pH of freshly collected urine averages 6.0, with a range of 4.5 to 8.0. Diet may cause considerable variation in pH as well as color and odor. For example, some people's urine takes on an unusual odor when they eat asparagus, and beet consumption can give urine a worrying reddish hue.

The amount of dissolved substances in urine is indicated by its **specific gravity**. The specific gravity of pure water, used as a standard, is 1.000. Because of the dissolved materials it contains, urine has a specific gravity that normally varies from 1.002 (very dilute urine) to 1.040 (very concentrated urine). When the kidneys are diseased, they lose the ability to concentrate urine, and the specific gravity no longer varies as it does when the kidneys function normally.

Some of the dissolved substances normally found in the urine are the following:

- **Nitrogenous waste products**, including the following:
  - Urea, formed from amine groups released in protein catabolism
  - Uric acid from the breakdown of purines, which are found in some foods and nucleic acids
  - Creatinine (kre-AT-ih-nin), a breakdown product of muscle creatine
- **Electrolytes**, including sodium ions, chloride, and different kinds of sulfates and phosphates. Electrolytes are excreted in appropriate amounts to keep their blood concentration constant.
- **Pigments**. Urochrome, a yellow substance derived from the breakdown of hemoglobin, is the main pigment in urine. Small amount of bilirubin and other bile pigments are also found in normal urine. Beets and other dark foods can add color, as can B vitamins, vitamin C, and food dyes. Also, certain drugs can alter the color of the urine.

CHECKPOINTS 

- ☐ **19-13** What is the name of the tube that carries urine from the kidney to the bladder?
- ☐ **19-14** What openings form the bladder's trigone?
- ☐ **19-15** What is the name of the tube that carries urine from the bladder to the outside?

CASEPOINT 

- ☐ **19-1** If Ethan's body was functioning normally, what would happen to the specific gravity of his urine as the kidneys compensated for his excess water intake?

The Effects of Aging on the Urinary System

Even without renal disease, aging causes the kidneys to lose some of their ability to concentrate urine. With aging, progressively more water is needed to excrete the same amount of waste. Older people find it necessary to drink more water than do young people, and they eliminate larger amounts of urine (polyuria), even at night (nocturia).

Beginning at about 40 years of age, there is a decrease in the number and size of the nephrons. Often, more than half of them are lost before the age of 80 years. There may be an increase in blood urea nitrogen without serious symptoms. Elderly people are more susceptible than young people to UTIs. Childbearing may cause damage to the pelvic floor musculature, resulting in urinary tract problems in later years.

Enlargement of the prostate, common in older men, may cause obstruction and back pressure in the ureters and kidneys (see FIG. 20-1). If this condition is untreated, it will cause permanent damage to the kidneys. Changes with age, including decreased bladder capacity and decreased muscle tone in the bladder and urinary sphincters, may predispose people to incontinence. However, most elderly people (60% in nursing homes and up to 85% living independently) have no incontinence.

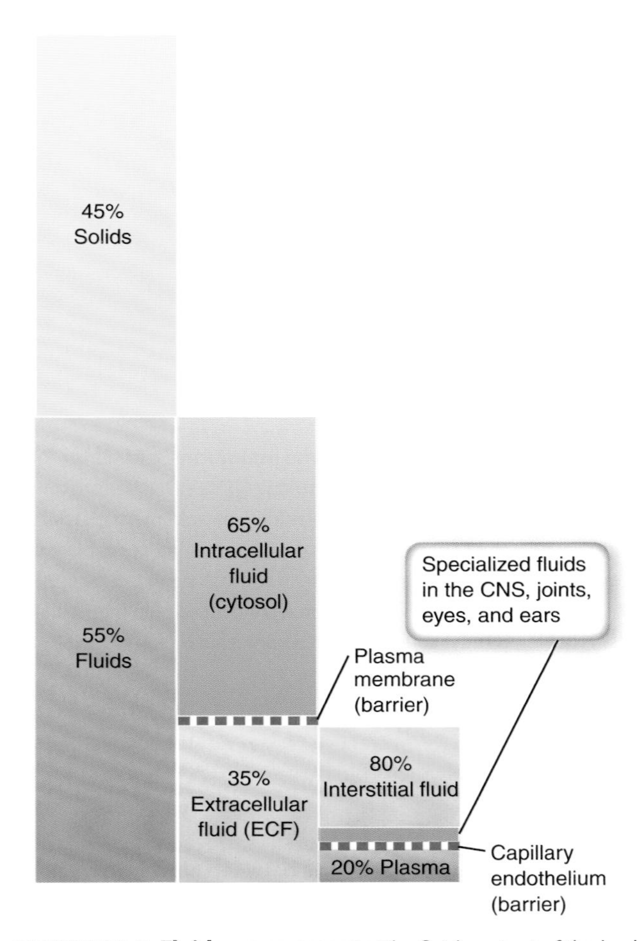

FIGURE 19-9 **Fluid compartments.** The fluid content of the body can be divided into numerous compartments. This figure shows the approximate values for a lean female. KEY POINT Semipermeable barriers separate some fluid compartments. ZOOMING IN Name the two most abundant extracellular fluids.

Body Fluids

Despite the body's apparent solidity, fluids make up over half of its mass. Composed primarily of water, body fluids also contain electrolytes (salts), nutrients, gases, waste, and special substances, such as enzymes and hormones. This chapter begins with a discussion of the location and composition of body fluids and then discusses some important aspects of body fluid regulation.

FLUID COMPARTMENTS

Although body fluids have much in common no matter where they are located, there are some important differences between fluid inside and outside cells. Accordingly, fluids are grouped into two main compartments (FIG. 19-9):

- **Intracellular fluid** (ICF) is contained within the cells. About two-thirds to three-fourths of all body fluids are in this category. Most of this fluid is found in the cytosol.
- **Extracellular fluid** (ECF) includes all body fluids outside of cells. In this group are the following:
    - **Interstitial** (in-ter-STISH-al) **fluid**, or more simply, tissue fluid. This fluid is located in the spaces between the cells in tissues throughout the body. It is estimated that tissue fluid constitutes about 15% of body weight.
    - **Blood plasma**, which constitutes about 4% of body weight.
    - **Lymph**, the fluid that drains from the tissues into the lymphatic system. This is about 1% of body weight.
    - **Fluid in special compartments**, such as cerebrospinal fluid, the aqueous and vitreous humors of the eye, serous fluid, and synovial fluid. Together, these make up about 1% to 3% of total body fluids, so less than 1% of body weight.

4 Selectively permeable barriers separate the fluids in the different compartments. For the most part, water can freely move through the capillary wall separating the blood from the interstitial fluid and across the plasma membrane separating interstitial fluid and from intracellular fluid (see FIG. 19-9). As a result, virtually all body fluids have the same osmolarity (osmotic concentration) (FIG. 19-10, right side).

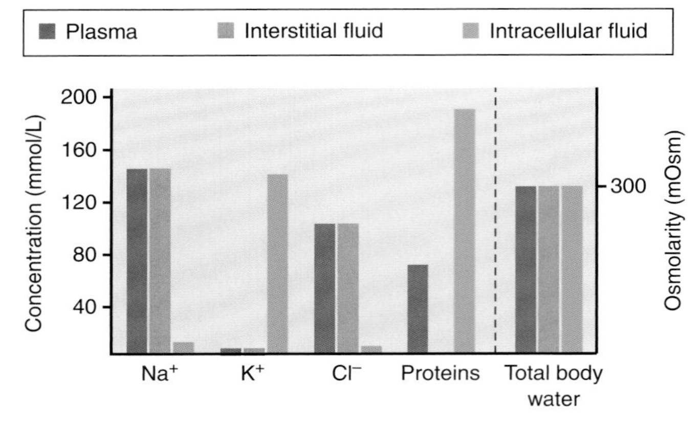

FIGURE 19-10 Distribution of some major ions and water in intracellular and extracellular fluids. **KEY POINT** Ion distribution varies between extracellular (plasma and interstitial fluid) and intracellular fluid components. Proteins are found only in plasma and cytosol. All fluid compartments are in osmotic equilibrium. **ZOOMING IN** What ions are the most concentrated in interstitial fluids? What ion is most concentrated in intracellular fluid?

However, these fluids differ in their concentrations of electrolytes and proteins (**FIG. 19-10, left side**), because not all substances can cross these barriers.

Concept Mastery Alert

Remember that interstitial fluid is between (inter-) cells and intracellular fluid is within (intra-) cells.

WATER AND ITS FUNCTIONS

6 Water is important to living cells as a solvent, a transport medium, and a participant in metabolic reactions. Importantly, the force of water in the vessels helps maintain blood pressure, which creates the gradient for blood flow and capillary exchange. Changes in water volume thus alter blood pressure. The normal proportion of body water varies from 50% to 70% of a person's weight (**see FIGURE 19-9**). It is highest in the young and in thin, muscular individuals. As the amount of fat increases, the percentage of water in the body decreases, because adipose tissue holds little water compared with muscle tissue. In infants, water makes up 75% of the total body mass. That's why infants are in greater danger from dehydration than are adults.

ELECTROLYTES AND THEIR FUNCTIONS

Electrolytes are important constituents of body fluids. These compounds separate into positively and negatively charged ions in solution. As discussed in Chapter 2, positively charged ions are called **cations**; negatively charged ions are called **anions**. Electrolytes are so named because they conduct an electric current in solution.

Major Ions A few of the most important cations and anions are reviewed next. **FIGURE 19-10** shows the distribution of common electrolytes in the different fluid compartments.

- Cations (positive ions):
  - **Sodium** is the most abundant cation in extracellular fluids. It plays a key role in maintaining osmotic balance and body fluid volume, as discussed later in this chapter. Other roles for the sodium ion (Na^+) include nerve impulse conduction and acid–base balance.
  - **Potassium** is the most abundant cation in intracellular fluids. It determines the excitability of neurons and muscle cells, including the heart muscle.
  - **Calcium** is required for bone formation, muscle contraction, neurotransmitter release, and blood clotting.
  - **Magnesium** is necessary for muscle contraction and for the action of some enzymes.
- Anions (negative ions):
  - **Bicarbonate** plays a key role in acid–base balance, as discussed later in this chapter.
  - **Chloride** is the most abundant anion in extracellular fluids. It is part of the hydrochloric acid formed by the stomach. It also helps regulate fluid balance and pH.
  - **Phosphate** is essential in carbohydrate metabolism, bone formation, and acid–base balance. Phosphates are found in plasma membranes, nucleic acids (DNA and RNA), and ATP.
  - **Proteins** are frequently negatively charged and are the most abundant anions in intracellular fluids. Proteins are also abundant in blood but not interstitial fluids, because they usually cannot cross the capillary wall.

CHECKPOINTS

- ☐ **19-16** Where are intracellular fluids located? Extracellular fluids?
- ☐ **19-17** What is interstitial fluid?
- ☐ **19-18** What are cations and anions?

CASEPOINTS

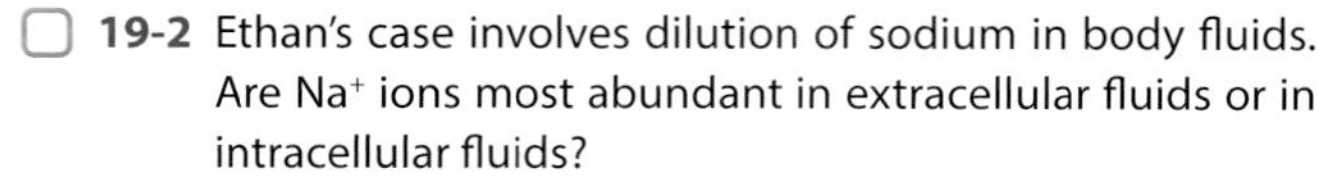

- ☐ **19-2** Ethan's case involves dilution of sodium in body fluids. Are Na^+ ions most abundant in extracellular fluids or in intracellular fluids?
- ☐ **19-3** What will happen to the osmotic pressure of Ethan's body fluids as the sodium concentration declines?

Regulation of Body Fluids

Multiple negative feedback mechanisms maintain the volume and composition of body fluids within narrow limits. Whenever these parameters vary even slightly outside normal ranges, disease results. This section explains how the body controls the overall fluid volume and the concentrations of sodium and potassium. Later in this chapter, we discuss the regulation of body fluid pH.

WATER GAIN AND LOSS

13 In a healthy person, a consistent body fluid volume is maintained by matching the quantity of water gained in a day (input) to the quantity lost (output) (FIG. 19-11). The quantity of water consumed in a day varies considerably among individuals and is typically increased in hot weather; however, the average adult in a comfortable environment takes in about 2,300 mL of water (about 2½ qt) daily. About two-thirds of this quantity comes from drinking water and other beverages; about one-third comes from foods, such as fruits, vegetables, and soups. Fluid intake is determined, in part, by the thirst mechanism, discussed shortly. About 200 mL of water is produced each day as a byproduct of cellular respiration (see Chapter 18). This water, described as *metabolic water*, brings the total average gain to 2,500 mL each day.

The same volume of water is constantly being lost from the body by the following routes:

- The **kidneys** excrete the largest quantity of water lost each day. About 1 to 1.5 L of water is eliminated daily in the urine, depending on fluid intake and fluid losses by other means.
- The **skin**. Although sebum and keratin help prevent dehydration, water is constantly evaporating from the skin's surface. Larger amounts of water are lost from the skin as sweat when it is necessary to cool the body, as was discussed in Chapter 18.

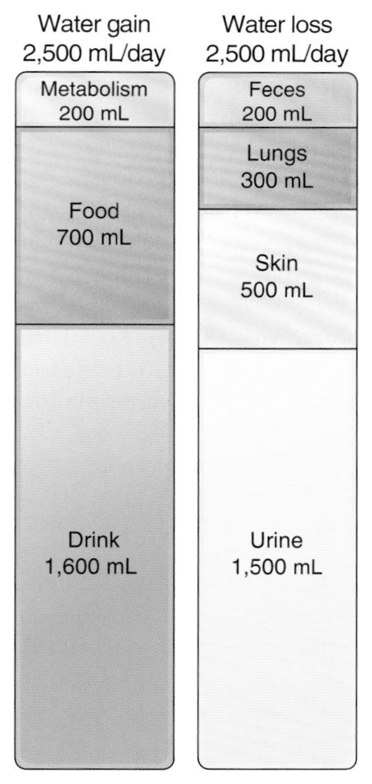

FIGURE 19-11 Daily gain and loss of water. **KEY POINT** In a healthy person, water gained in a day is approximately equal to the quantity lost. **ZOOMING IN** In what way is the most water lost in a day?

- The **lungs** expel water along with exhaled air.
- The **intestinal tract** eliminates water along with the feces.

In many disorders, it is important for the healthcare team to know whether a patient's intake and output are equal; in such a case, a 24-hour intake–output record is kept. The intake record includes *all* the liquid the patient has taken in. This means fluids administered intravenously as well as those consumed by mouth. The healthcare provider must account for water, other beverages, and liquid foods, such as soup and ice cream. The output record includes the quantity of urine excreted in the same 24-hour period as well as an estimation of fluid losses due to fever, vomiting, diarrhea, bleeding, wound discharge, or other causes.

THIRST AND DRINKING

As you can see by studying FIGURE 19-11, it is essential to take in enough fluid each day to replace physiologic losses. Our thirst mechanism prompts us to drink water. Thirst and drinking are our only defenses against decreased fluid volume, because the kidneys can only conserve existing body fluids. The control center for the sense of thirst is located in the brain's hypothalamus. This center plays a major role in the regulation of total fluid volume and composition. An increase in the osmolarity of extracellular fluids or a decrease in blood pressure stimulates the thirst center, causing a person to drink water or other fluids that will then dilute the blood and increase its volume. Dryness of the mouth also causes a sensation of thirst. Excessive thirst, such as that caused by excessive urine loss in cases of diabetes, is called **polydipsia** (pol-e-DIP-se-ah). See **BOX 19-2** for more information on receptors involved in sensing osmolarity.

Ideally, the thirst mechanism should stimulate just enough drinking to balance fluids, but this is not always the case. During vigorous exercise, especially in hot weather, people may not drink enough to replace fluids lost in sweat. While plain water is usually the best choice, prolonged dehydration should be treated with beverages containing sodium and other electrolytes in order to maximize water absorption from the intestine. Because we often consume beverages for reasons other than thirst (taste, social situations), excess fluid consumption also occurs. The kidneys usually compensate for excessive fluid intake by excreting more urine, as discussed later in this chapter. A failure of this homeostatic mechanism can cause serious problems, as illustrated in Ethan's case study.

CONTROL OF FLUID BALANCE

Fluid homeostasis relies on the coordinated efforts of the thirst center, which controls fluid intake, and the kidneys, which modify the volume and composition of urine. 3 Some of the regulated variables involved in fluid homeostasis include:

1. Blood pressure. Generally speaking, the body does not directly sense or regulate overall fluid volume. Instead, it senses and regulates blood pressure, which is determined

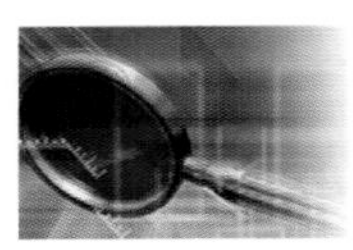

A CLOSER LOOK
Osmoreceptors: Thinking about Thirst

BOX 19-2

Osmoreceptors are specialized neurons that act as sensors in feedback loops controlling fluid balance. Located in the hypothalamus of the brain, they monitor the osmotic pressure (solute concentration) of the interstitial fluid and thus of circulating blood plasma.

Osmoreceptors respond primarily to small increases in sodium, the most common cation in ECF. As the blood becomes more concentrated, sodium draws water out of the cells, initiating nerve impulses. Traveling to different regions of the hypothalamus, these impulses may have two different but related effects:

- They stimulate the hypothalamus to produce antidiuretic hormone (ADH), which is then released from the posterior pituitary. ADH travels to the kidneys and causes these organs to conserve water.
- They stimulate the thirst center of the hypothalamus, causing increased consumption of water. Almost as soon as water consumption begins, however, the sensation of thirst disappears. Receptors in the throat and stomach send inhibitory signals to the thirst center, preventing overconsumption of water and allowing time for ADH to affect the kidneys.

Both of these mechanisms serve to dilute the blood and other body fluids. Both are needed to maintain optimal water balance. If either fails, a person soon becomes dehydrated.

in part by overall fluid volume. Thus, the thirst center and the kidneys are effectors in the control of blood pressure.

2. Plasma osmolarity. Osmoreceptors in the hypothalamus and elsewhere sense the overall solute concentration (osmolarity) of blood plasma, changes in which usually reflect changes in plasma Na^+ concentrations.
3. The plasma concentration of potassium.

Importantly, these variables are related to each other. As the body's sodium content rises (for instance, by the consumption of salty food), water intake and retention both increase. The resultant increase in blood volume can increase blood pressure if the cardiovascular system is unable to compensate for the change. This relationship explains why nutritionists recommend a low-salt diet in order to reduce blood pressure. Additionally, sodium and potassium excretion are tightly linked. As sodium excretion increases, potassium excretion decreases and vice versa.

A number of hormones act as signals in feedback loops regulating body fluids and blood pressure (**TABLE 19-1**). We begin by considering each hormone individually and then summarize briefly how they work together.

See the Student Resources on thePoint® for a chart summarizing the role of the kidney in regulating blood pressure.

Aldosterone As introduced in Chapter 11, **aldosterone** is an adrenal steroid hormone that alters Na^+ and K^+ handling by the kidney (**FIG. 19-12**). Its release is stimulated indirectly by low blood pressure; the hormone *angiotensin II* (ATII) mediates this effect, as discussed later. Aldosterone promotes the reabsorption of sodium (and thus water) in the distal tubule, so its actions tend to increase blood pressure.

Aldosterone is also involved in a second feedback loop involving K^+. Increased blood concentrations of K^+ act directly on the adrenal gland to promote aldosterone release. Aldosterone stimulates K^+ secretion from the blood into the distal tubule, so it increases K^+ excretion in the urine and decreases the amount of K^+ in the blood.

Antidiuretic Hormone (ADH) Recall that ADH is synthesized by the hypothalamus and released by the posterior pituitary. ADH secretion is stimulated by increased

Table 19-1 Hormonal Regulation of Fluid Homeostasis

Hormone	Stimulus for Secretion	Response
Aldosterone	Low blood pressure (via angiotensin II) or elevated blood potassium concentration	Increased sodium and water retention, resulting in increased blood pressure; increased potassium loss
Antidiuretic hormone (ADH)	Elevated blood sodium concentration or low blood pressure (via angiotensin II)	Increased water retention, resulting in more dilute body fluids and increased blood pressure
Angiotensin II	Low blood pressure	Increased blood pressure, via increased thirst, aldosterone secretion, ADH secretion, and vasoconstriction
Atrial natriuretic peptide (ANP)	High blood pressure in the atrium	Increased excretion of sodium and water, resulting in decreased blood pressure

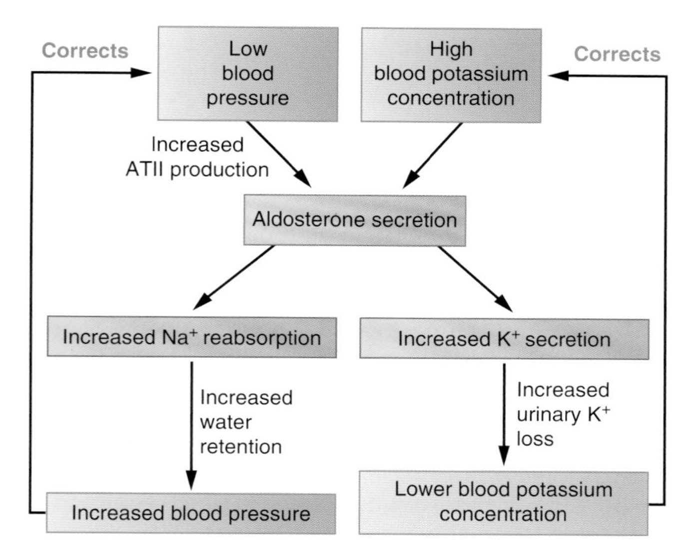

FIGURE 19-12 Aldosterone. **KEY POINT** Aldosterone increases water retention and potassium loss. Note that the different components of the feedback loop are indicated by different colors (regulated variable, *green*; signals, *purple*; effectors, *blue*). The sensors and control center are not illustrated. **ZOOMING IN** How would the overconsumption of potassium-rich foods like bananas alter aldosterone secretion?

blood osmolarity, which is detected by osmoreceptors in the hypothalamus. And, as with aldosterone, low blood pressure indirectly stimulates ADH production via ATII (discussed next). The kidney's collecting ducts respond to ADH by increasing water reabsorption into the blood, thereby concentrating the urine **(see FIG. 19-6)**. The reabsorbed water dilutes the excess sodium and increases blood pressure.

In the disease diabetes insipidus, there is inadequate secretion of ADH from the hypothalamus. The collecting duct is not very permeable to water in the absence of ADH, so large amounts of dilute urine are produced.

Angiotensin II (ATII) and the Juxtaglomerular Apparatus ATII is an important signal in the feedback loop regulating fluid balance and blood pressure, but both its production and action are relatively complex. In brief, decreased blood pressure stimulates ATII production, which then acts to increase blood pressure.

The sensors and control center for this feedback loop are found in a specialized region of the kidney known as the **juxtaglomerular (JG) apparatus** (the name means "near the glomerulus"). As seen in **FIGURE 19-13**, the first portion of the distal tubule curves backward toward the glomerulus to pass between the afferent and efferent arterioles. At the point where the distal tubule makes contact with the afferent arteriole, there are specialized cells in each that together make up the JG apparatus.

Receptors in the distal tubule cells of the JG apparatus are sensors for this feedback loop. These sensors measure filtrate volume as an indicator of blood pressure. They send signals to the afferent arteriole cells, which act as a control center. When blood pressure (and thus filtrate volume) decreases, the afferent arteriole cells secrete greater amounts of the enzyme **renin** (RE-nin) **(see TABLE 19-1)**. This enzyme participates in the production of ATII from inactive precursors. So, decreased blood pressure decreases filtrate volume, which increases renin production, which increases ATII production.

The actions of ATII are equally complex, but all lead to increased blood pressure. ATII has the immediate effect of stimulating vasoconstriction (narrowing) of the arterioles, which increases blood pressure. Over the longer term, it increases blood volume (and thus blood pressure) by stimulating thirst and, as mentioned earlier, by stimulating the release of both ADH and aldosterone. Because of its widespread effects, ATII inhibitors are useful treatments for hypertension (high blood pressure). **BOX 19-3** has more information on this topic.

See the Student Resources on thePoint® for the animation "The Juxtaglomerular Apparatus."

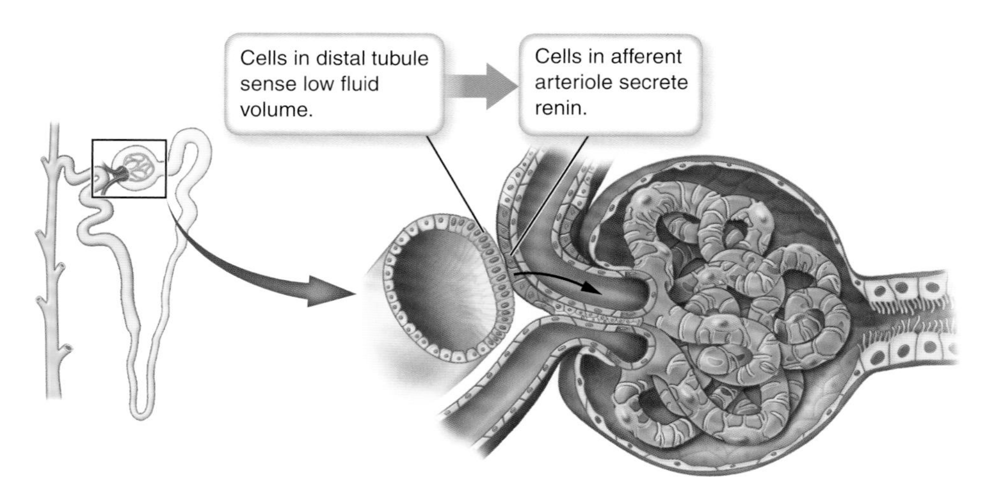

FIGURE 19-13 The juxtaglomerular (JG) apparatus. **KEY POINT** Note how the distal tubule contacts the afferent arteriole. When specialized distal tubule cells sense a decrease in filtrate volume, they send signals to specialized afferent arteriole cells to increase the production of renin. The resulting increase in ATII production increases blood pressure.

BOX 19-3

A CLOSER LOOK
The Renin–Angiotensin Pathway

In addition to forming urine, the kidneys play an integral role in regulating blood pressure. When blood pressure drops, cells of the JG apparatus secrete the enzyme renin into the blood. Renin acts on another blood protein, **angiotensinogen**, which is manufactured by the liver. Renin converts angiotensinogen into **angiotensin I** by cleaving off some amino acids from the end of the protein. Angiotensin I is then converted into **angiotensin II** (ATII) by yet another enzyme called angiotensin converting enzyme (ACE), which is manufactured by the capillary endothelium, especially in the lungs. ATII increases blood pressure in multiple ways:

- It acts on the cardiovascular system to increase cardiac output and stimulate vasoconstriction.
- It stimulates the release of aldosterone to increase sodium reabsorption and water reabsorption in the kidneys.
- It stimulates the release of ADH to increase water reabsorption in the kidneys.
- It stimulates thirst centers in the hypothalamus.

The combined effects of ATII produce a dramatic increase in blood pressure. In fact, ATII is estimated to be four to eight times more powerful than norepinephrine, another potent stimulator of hypertension, and thus, it is a good target for blood pressure–controlling drugs. One class of such drugs is the **ACE inhibitors**, which lower blood pressure by blocking the production of ATII via its converting enzyme. Another category is the ARBs, or angiotensin receptor blockers, which prevent ATII from binding to its receptors in blood vessels and other tissues.

Atrial Natriuretic Peptide (ANP) All of the hormones discussed thus far—ADH, aldosterone, and ATII—work together to increase blood pressure. This redundancy makes sense, as low blood pressure causes an immediate threat to life by reducing blood flow to critical organs, such as the brain. A single hormone has the opposite effect of reducing blood pressure—**atrial natriuretic peptide** (ANP). This hormone is secreted by specialized atrial myocardial cells when blood pressure rises too high. ANP causes the kidneys to excrete sodium and water, thus decreasing blood volume and lowering blood pressure. The name comes from *natrium*, the Latin name for sodium, and the adjective for *uresis*, which refers to urination.

Challenges to Fluid Homeostasis It can be helpful to consider how these hormones work together when faced with a homeostatic challenge. First, consider the response to excessive blood loss, which alters blood volume but not osmolarity (FIG. 19-14). The cardiovascular system reacts to the resulting decrease in blood pressure by increasing heart rate, stroke volume, and vasoconstriction. The JG apparatus reacts to the decreased blood pressure by releasing renin, which increases ATII production. ATII, in turn, increases vasoconstriction and water intake and reduces water loss. It does so directly and by stimulating the production of ADH and aldosterone. The net result is increased blood pressure.

In dehydration, the body's fluids decrease in volume but increase in osmolarity. In addition to the responses to decreased blood volume summarized above, the increased blood osmolarity stimulates ADH production and thirst to an even greater extent, thereby diluting the blood and returning blood pressure to normal.

CHECKPOINTS 

- ☐ **19-19** What are four routes for water loss from the body?
- ☐ **19-20** What hormone from the adrenal cortex promotes sodium reabsorption in the kidney?
- ☐ **19-21** What pituitary hormone increases water reabsorption in the kidney?

CASEPOINTS 

- ☐ **19-4** What body system would compensate most effectively for Ethan's excess water consumption?
- ☐ **19-5** Ethan's body fluids are excessively dilute. How would this normally affect ADH secretion?
- ☐ **19-6** What is the response of the hypothalamic thirst center to the dilution of body fluids? Did Ethan listen to his body signals?

Acid–Base Balance

The **pH scale** is a measure of how acidic or basic (alkaline) a solution is. As described in Chapter 2, the pH scale measures the hydrogen ion (H^+) concentration in a solution with higher numbers indicating a lower concentration. Body fluids are slightly alkaline, with a pH range of 7.35 to 7.45. These fluids must be kept within a narrow pH range, or damage, even death, will result. A shift in either direction by three-tenths of a point on the pH scale, to 7.0 or 7.7, is fatal.

REGULATION OF pH

Recall from Chapter 16 that carbon dioxide, generated by cellular respiration, dissolves in the blood and yields carbonic acid. Carbonic acid lowers blood pH. This change usually appears exclusively in venous blood, because healthy lungs exhale as much CO_2 as the body produces. However, any impairment of lung function can cause CO_2 to accumulate in arterial blood and thus lower blood pH.

The body constantly produces acids in the course of metabolism that cannot be eliminated in exhaled air. Catabolism of fats yields fatty acids and ketones, and intense exercise generates hydrogen ions. Moreover, significant amounts of base are lost in the feces in the form of bicarbonate ions. Normal metabolic activities thus tend to

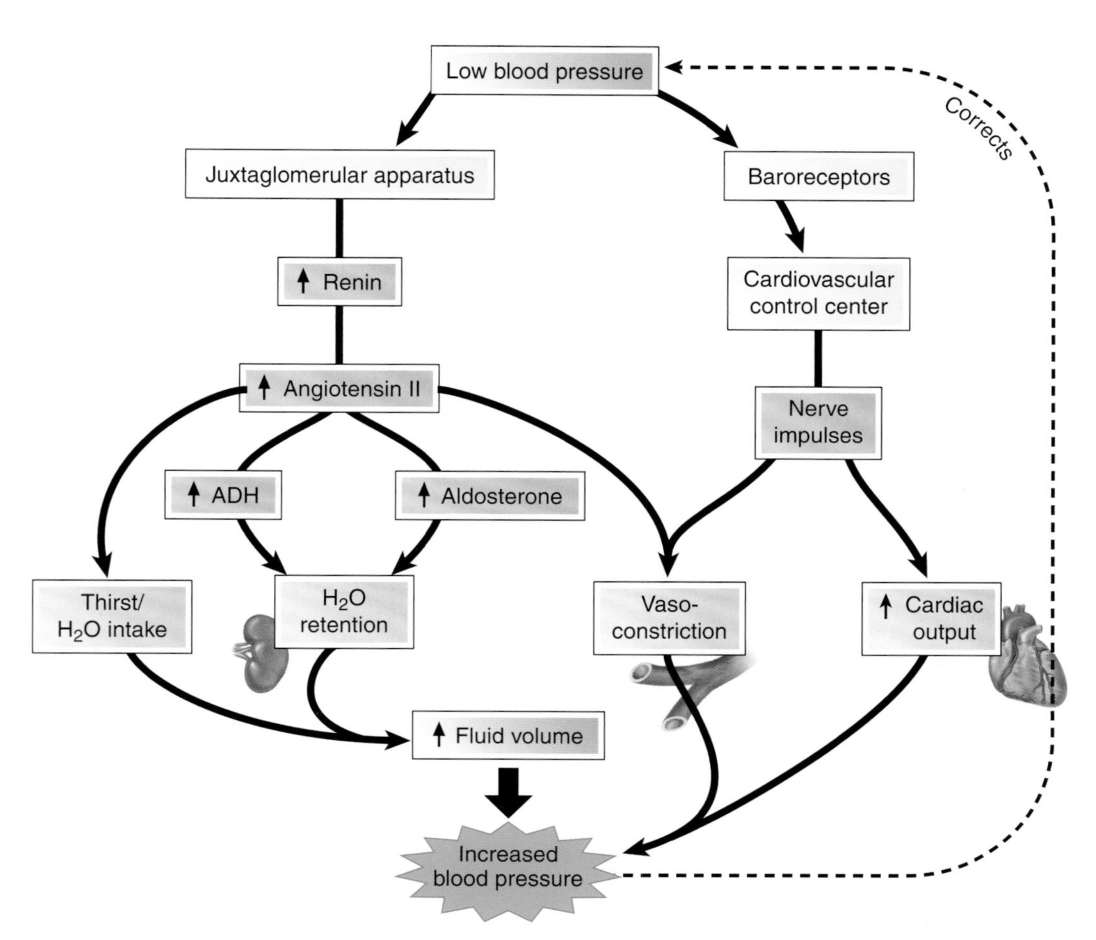

FIGURE 19-14 Integrated control of blood pressure. **KEY POINT** The kidneys, heart, and blood vessels are key effectors in blood pressure homeostasis. Note that the different components of the feedback loop are indicated by different colors (regulated variable and its determinants, *green*; sensors and control centers, *yellow*; signals, *purple*; effectors, blue). **ZOOMING IN** Which hormone stimulates thirst?

reduce blood pH. Conversely, a few abnormal conditions, such as excessive vomiting or excessive antacid consumption, may cause alkaline shifts in pH. Several systems act together to counteract these changes and maintain acid–base balance (**FIG. 19-15**):

- **Buffer systems.** Buffers are substances that prevent sharp changes in hydrogen ion (H^+) concentration and thus maintain a relatively constant pH. Buffers work by accepting or releasing these ions as needed to keep the pH steady. The main buffer systems in the body are bicarbonate buffers, phosphate buffers, and proteins, such as hemoglobin in red blood cells and plasma proteins.
- **Respiration.** The respiratory system can compensate for increased metabolic acid production, because faster, deeper breathing eliminates more carbon dioxide and makes the blood more alkaline. Conversely, metabolic alkalosis, as from excess vomiting of stomach acid, reduces the urge to breathe so that carbon dioxide is retained and blood becomes a bit more acidic. However, the respiratory system uses up buffers as it compensates for pH disturbances, so its actions are temporary.
- **Kidney function.** The kidneys regulate pH by reabsorbing or secreting hydrogen ions as needed and creating new buffer molecules, such as bicarbonate ions.

CHECKPOINT 

☐ **19-22** What are three mechanisms for maintaining the acid–base balance of body fluids?

See the Student Resources on thePoint® for the animation "Osmosis" and for information on emergency medical technicians, who often must administer fluids in providing healthcare.

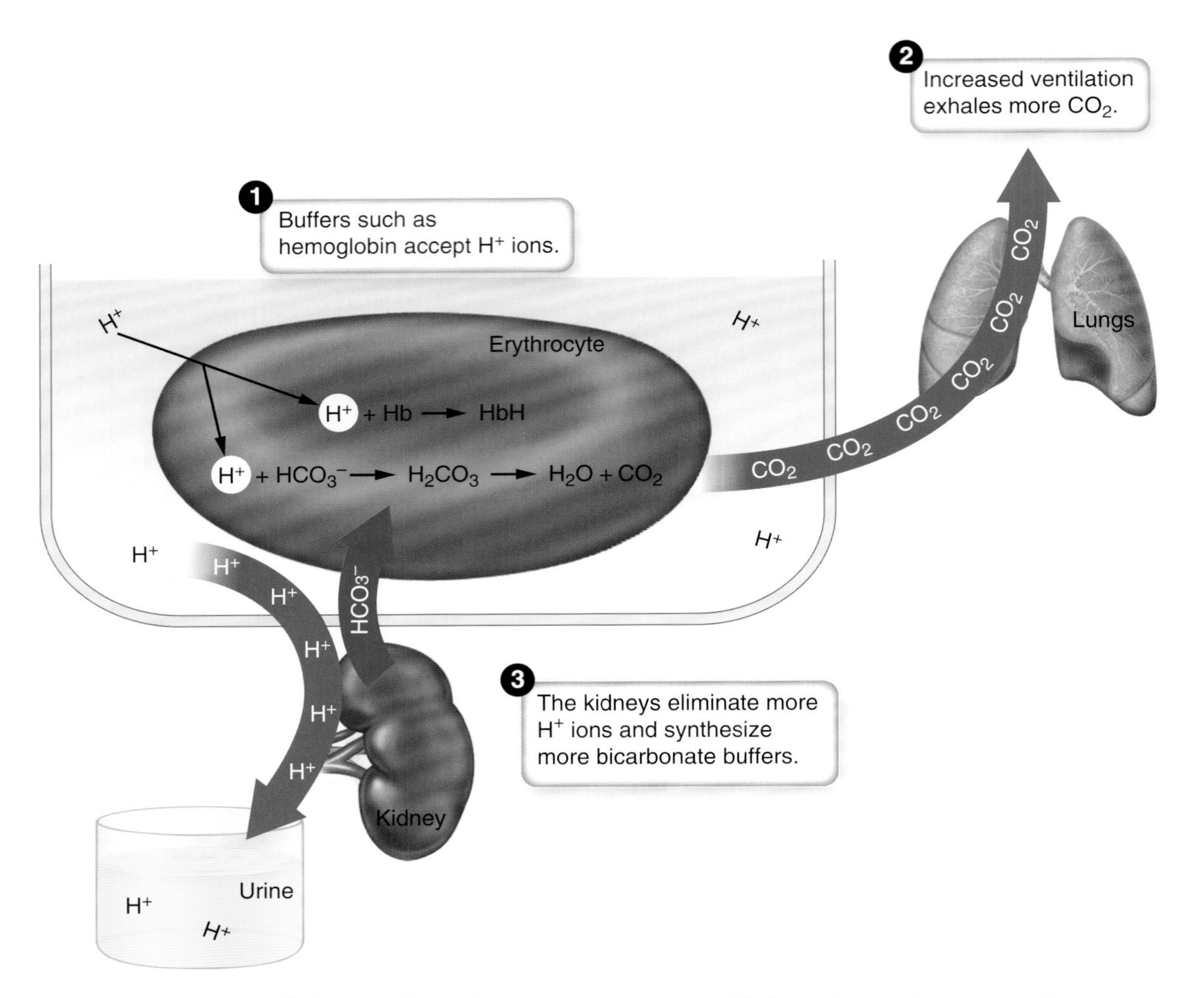

FIGURE 19-15 pH regulation. This figure illustrates the response to acids that are ingested or generated by metabolic processes. **KEY POINT** The body compensates for these acids using buffers and increased ventilation, but renal actions are required to eliminate the excess acids. **ZOOMING IN** Which organ generates bicarbonate ions?

A & P in Action Revisited: Ethan Overdid the Hydration

The next day, Leilani was heading down to the bay to go paddle boarding when she ran into Darren.

"Hey Darren, whatever happened to the late finisher, Ethan?" she asked.

"Just as I thought, he had exercise-associated hyponatremia," Darren replied. "Blood tests showed that his blood sodium level was only 129 mmol/L, and it should be between 135 and 145. We treated him with an oral hypertonic saline solution, and he was OK after about 8 hours."

"Do you know what caused this?" Leilani asked. "I know athletes have to guard against dehydration, but I never knew that overhydration could be a problem."

Darren smiled, "Ethan's coach back home told him to drink as much and as often as he could, and he followed the advice too enthusiastically. When he started to feel bad, he assumed that he was dehydrated, and so he drank more water. Most people respond to the increased water intake by urinating more, and they don't get sick. But for reasons we don't yet understand, some people who overdrink retain too much water and dangerously dilute their body fluids. We estimate that Ethan must have gained about 8 lb of water weight during the race, showing that he is susceptible to EAH and will have to really watch his hydration in future races. I'm not saying that dehydration isn't dangerous—it is—but people have to trust their sense of thirst more and let their body signals regulate their fluid intake. We can't know in advance who will develop EAH, so we advise all of our triathletes to drink no more than two cups of liquid per hour. We weigh them before they start the run portion, and if they haven't lost some weight, we ask them to stop drinking for a while."

CHAPTER 19 Chapter Wrap-Up

OVERVIEW

A detailed chapter outline with space for note-taking is on thePoint®. The figure below illustrates the main topics covered in this chapter.

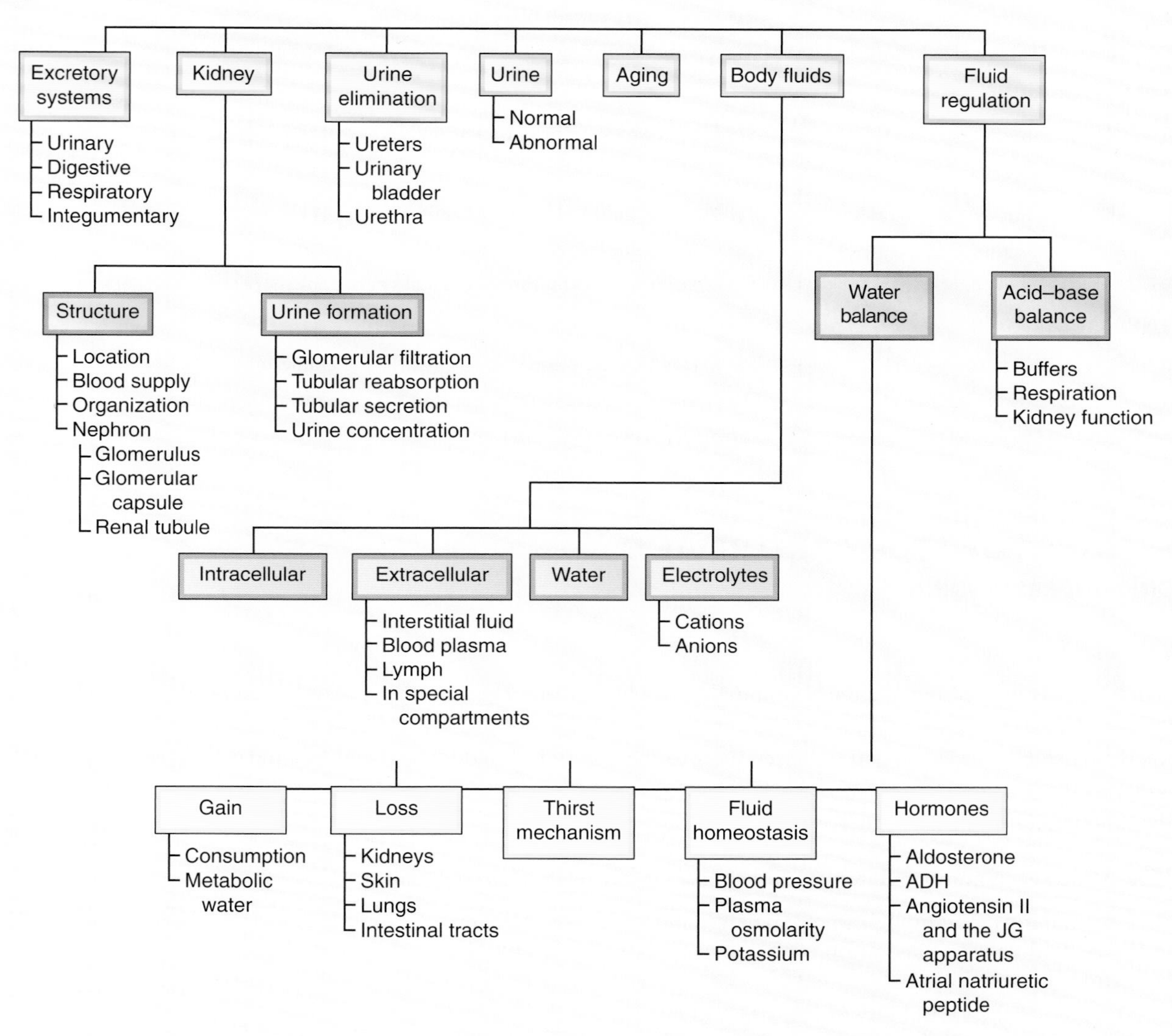

KEY TERMS

The terms listed below are emphasized in this chapter. Knowing them will help you organize and prioritize your learning. These and other boldface terms are defined in the Glossary with phonetic pronunciations.

aldosterone
angiotensin II
anion
antidiuretic hormone (ADH)
atrial natriuretic peptide
cation
excretion
extracellular
glomerular filtrate
glomerulus
interstitial
intracellular
kidney
micturition
nephron
osmolarity
pH scale
tubular reabsorption
urea
ureter
urethra
urinalysis
urinary bladder
urine

WORD ANATOMY

Medical terms are built from standardized word parts (prefixes, roots, and suffixes). Learning the meanings of these parts can help you remember words and interpret unfamiliar terms.

WORD PART	MEANING	EXAMPLE
THE KIDNEYS		
nephr/o	kidney	The *nephron* is the functional unit of the kidney.
ren/o	kidney	The *renal* artery carries blood to the kidney.
retro	backward, behind	The *retroperitoneal* space is posterior to the peritoneal cavity.
THE EFFECTS OF AGING		
noct/i	night	*Nocturia* is excessive urination at night.
FLUID COMPARTMENTS		
extra-	outside of, beyond	*Extracellular* fluid is outside the cells.
intra-	within	*Intracellular* fluid is within a cell.
semi-	partial, half	A *semipermeable* membrane is partially permeable.
WATER BALANCE		
juxta	next to	The *juxtaglomerular* apparatus is next to the glomerulus
osmo	osmosis	*Osmoreceptors* detect changes in osmotic concentration of fluids.
poly-	many	*Polydipsia* is excessive thirst.

QUESTIONS FOR STUDY AND REVIEW

BUILDING UNDERSTANDING

Fill in the Blanks

1. Each kidney is located outside the abdominal cavity in the _____ space.
2. The renal artery, renal vein, and ureter connect to the kidney at the _____.
3. The functional unit of the kidney is the _____.
4. The amount of dissolved substances in urine is indicated by its _____.
5. Fluid located between the cells in tissues is called _____.

Matching Match each numbered item with the most closely related lettered item.

___ 6. Essential for maintaining osmotic balance and body fluid volume, this cation is abundant in the extracellular fluid

___ 7. Determines the excitability of neurons and muscle cells, this cation is abundant in the intracellular fluid

___ 8. Required for bone formation, muscle contraction, and blood clotting

___ 9. Essential for bone formation and acid–base balance, this anion is found in plasma membranes, ATP, and nucleic acids

___ 10. A component of gastric acid, this anion is abundant in the extracellular fluid

a. calcium
b. chloride
c. phosphate
d. potassium
e. sodium

Multiple Choice

___ 11. What is the cluster of capillaries involved in renal filtration?
a. renal capsule
b. peritubular capillaries
c. glomerulus
d. glomerular capsule

___ 12. Which of the following is NOT a nitrogenous waste product?
a. urochrome
b. creatinine
c. urea
d. uric acid

___ 13. Reabsorbed or unfiltered materials can be moved into the nephron by what process?
a. filtration
b. tubular secretion
c. diffusion
d. osmosis

___ 14. Which structure is responsible for voluntary control of urination?
a. trigone
b. internal urethral sphincter
c. external urethral sphincter
d. urinary meatus

___ 15. Which of the following is an important buffer in body fluids?
a. chloride
b. bicarbonate
c. metabolic water
d. aldosterone

___ 16. Which system is responsible for long-term pH regulation?
a. buffer system
b. digestive system
c. respiratory system
d. urinary system

UNDERSTANDING CONCEPTS

17. List four organ systems active in excretion. What are the products eliminated by each?
18. Referring to The Body Visible in the front of the book, give the name and number of:
    a. the triangle at the base of the bladder
    b. the vessel that collects blood from the renal vein
    c. a vessel that travels along the base of a renal pyramid
    d. vessels that surround the ascending and descending limbs of the nephron
    e. the tube that empties into a minor calyx
19. Compare and contrast the following terms:
    a. glomerular filtration and tubular reabsorption
    b. afferent arteriole and efferent arteriole
    c. proximal tubule and distal tubule
    d. ureter and urethra
    e. aldosterone and antidiuretic hormone
    f. angiotensin II (ATII) and atrial natriuretic peptide (ANP)
20. Trace the pathway of a urea molecule from the afferent arteriole to the urinary meatus.
21. Describe the four processes involved in the formation of urine.
22. In a healthy person, what is the ratio of fluid intake to output?
23. Explain the role of the hypothalamus in water balance.
24. What are some metabolic sources of acids in body fluids?
25. How do the respiratory and urinary systems regulate pH?

CONCEPTUAL THINKING

26. In the opening case study, Ethan replaced body fluids lost in the Ironman competition with pure water. What physiologic changes resulted, and how did they relate to his symptoms?
27. A class of antihypertensive drugs called loop diuretics prevents sodium reabsorption in the nephron loop. How could a drug like this lower blood pressure?

For more questions, see the Learning Activities on thePoint®.

UNIT VI

Perpetuation of Life

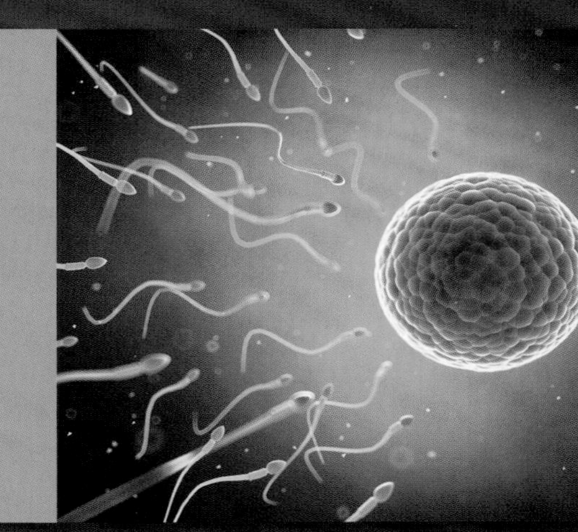

CHAPTER 20

The Male and Female Reproductive Systems

Learning Objectives

After careful study of this chapter, you should be able to:

1. Identify the male and female gametes, and state the purpose of meiosis. *p. 418*
2. Name the accessory organs and gonads of the male reproductive system, and cite the function of each. *p. 418*
3. Describe the composition and function of semen. *p. 419*
4. Draw and label a spermatozoon. *p. 421*
5. Identify the two hormones that regulate the production and development of the male gametes. *p. 421*
6. Explain how testosterone production is regulated and list its effects. *p. 421*
7. Name the accessory organs and gonads of the female reproductive system, and cite the function of each. *p. 423*
8. Describe changes during the course of the menstrual cycle with regard to hormone production, the ovary, and the uterus. *p. 426*
9. Describe the changes that occur during and after menopause. *p. 427*
10. Cite the main methods of birth control in use. *p. 429*
11. Using the text and information in the case study, discuss possible causes of infertility in men and women. *pp. 417, 430*
12. Show how word parts are used to build words related to the reproductive systems (see Word Anatomy at the end of the chapter). *p. 432*

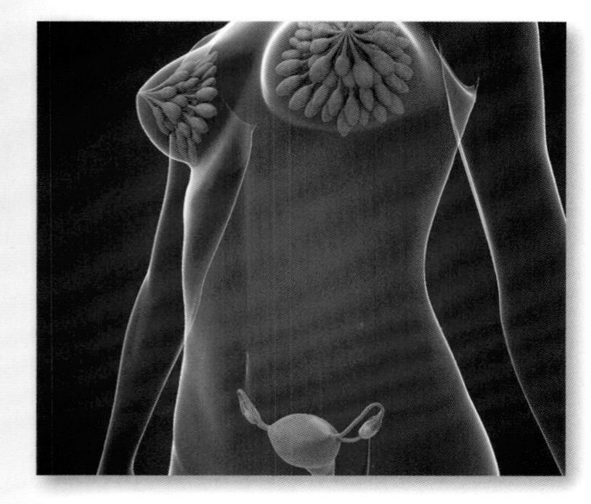

A & P in Action
Jessica and Brett's Infertility Problems

"Do you think my history of endometriosis is the reason I am unable to conceive? In my early 20s, I started taking the pill to help with my heavy and painful periods. My gynecologist said that I had evidence of endometriosis."

Jessica, a 34-year-old computer systems analyst, and her husband were conferring with Dr. Christensen, an infertility specialist recommended by her family physician.

"I stopped taking the pill about two years ago when Brett and I decided to start a family," Jessica said. "I know it can take a while for the effect of contraceptive pills to wear off, but two years seems really long."

"Your medical history is definitely relevant," the specialist replied and went on to explain. "As you are aware, endometriosis is a condition in which the lining of the uterus grows in locations outside of the uterus. These deposits break down and bleed every month, just like the endometrium within your uterus. This might cause inflammation and changes in the pelvic organs that would affect your ability to conceive. Your physician has already done the preliminary tests for both you and Brett. Your thyroid hormones and FSH levels are normal. Also I see that you monitored your LH levels over several months with urine tests, and the results showed that you are ovulating at about day 15 of your cycle. Brett, your sperm count was well within normal limits, and the cells were normal in shape and active. This means that you are both able to make healthy gametes. The next step is to look for structural problems in Jessica's reproductive tract that might interfere with fertilization or nourishing the fertilized egg. I'll schedule an endoscopic exam of your uterus and uterine tubes. This should reveal any abnormalities in those areas and any sign of blockage."

Brett chimed in, "What if we find out that there's something wrong? Do we have any other options?"

"Well, let's take this one step at a time," the doctor responded. "There's always the possibility of in vitro fertilization, or IVF, but first let's see what might be preventing you from becoming pregnant."

Later, we will see the results of Jessica's examination and the couple's options for starting a family.

As you study this chapter, CasePoints will give you opportunities to apply your learning to this case.

Visit thePoint® to access the following resources. For guidance in using these resources most effectively, see pp. xv–xvii.

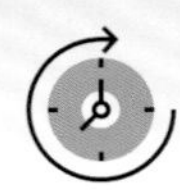

Preparing to Learn

- Tips for Effective Studying
- Pre-Quiz

While You Are Learning

- Web Figure: Descent of the Testes
- Web Figure: Laparoscopic Sterilization
- Web Chart: Reproductive Hormones
- Animation: Oogenesis and the Menstrual Cycle
- Chapter Notes Outline
- Audio Pronunciation Glossary

When You Are Reviewing

- Answers to Questions for Study and Review
- Health Professions: Physician Assistant
- Interactive Learning Activities

A LOOK BACK

In this chapter, we return once again to the concept of negative feedback 3, which regulates some reproductive activities in males and females. We also provide more details on the actions of the sex hormones and the hypothalamic and pituitary hormones that regulate their production, first introduced in Chapter 11. Other important key ideas include structure–function 1 and causation 12.

Introduction

The chapters in this unit deal with what is certainly one of the most interesting and mysterious attributes of life: the ability to reproduce. The simplest forms of life, one-celled organisms, usually need no partner to reproduce; they simply divide by themselves. This form of reproduction is known as **asexual** (nonsexual) reproduction.

In most animals, however, reproduction is **sexual**, meaning that there are two kinds of individuals, males and females, each of which has specialized cells designed specifically for the perpetuation of the species. These specialized sex cells are known as **gametes** (GAM-etes), or *germ cells*. In the male, they are called **spermatozoa** (sper-mah-to-ZO-ah; sing., spermatozoon) or simply sperm cells; in the female, they are called **ova** (O-vah; sing., ovum) or eggs.

Gametes are characterized by having half as many chromosomes as are found in any other body cell. During their formation, they go through a special process of cell division, called **meiosis** (mi-O-sis), which halves the number of chromosomes (see FIG. 21-2). In humans, meiosis reduces the chromosome number in a cell from 46 to 23. The role of meiosis in reproduction is explained in more detail in Chapter 21.

The male and female reproductive systems each include two groups of organs, primary and accessory:

- The primary organs are the **gonads** (GO-nads), or sex glands; they produce the gametes and manufacture hormones. The male gonad is the **testis** (pl., testes), and the female gonad is the **ovary**.
- The **accessory organs** include a series of ducts that transport the gametes as well as various exocrine glands.

The Male Reproductive System

The male reproductive system functions to manufacture spermatozoa (sperm cells) and to deliver them to the female. As illustrated in FIGURE 20-1, this system includes two testes, which are the sites of spermatozoa production, and the accessory organs. We begin our discussion with the ductal system that stores sperm cells and delivers them outside the body. See FIGURES A3-12A and A3-13 in the Dissection Atlas for photographs of the male reproductive system.

ACCESSORY ORGANS

The testes deliver spermatozoa into a greatly coiled tube called the **epididymis** (ep-ih-DID-ih-mis), which is 6 m (20 ft) long and is located on the surface of the testis (see FIGS. 20-1 and 20-3A). During their two-to-three day passage through the epididymis, the sperm cells mature and become motile, able to "swim" by themselves.

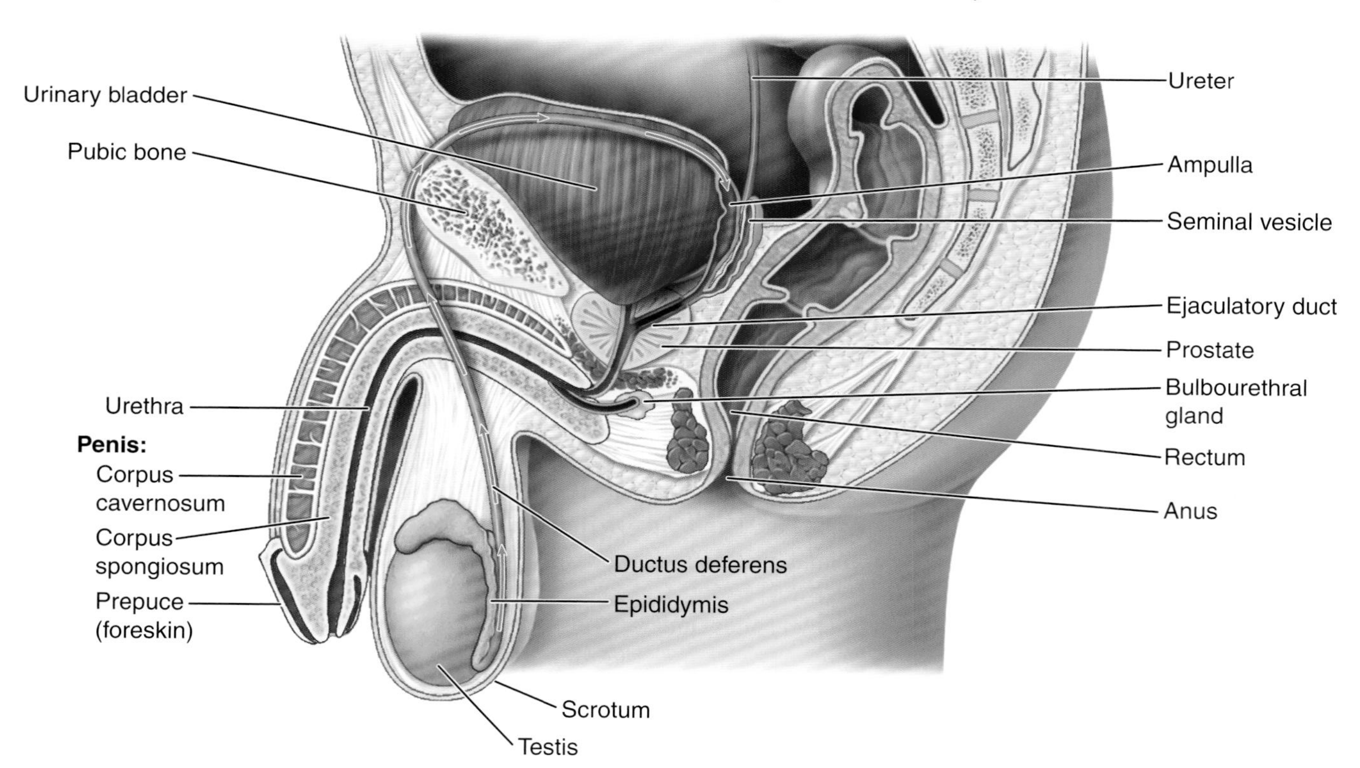

FIGURE 20-1 Male reproductive system. This sagittal section illustrates the organs of the male reproductive system. ZOOMING IN What four glands empty secretions into the urethra? What duct receives secretions from the epididymis?

The epididymis extends upward as the **ductus deferens** (DEF-er-enz), also called the *vas deferens*. This tube loops over the pubic bone and curves behind the urinary bladder. The ductus deferens widens to form an **ampulla** (dilation) just before it joins with the duct of the **seminal vesicle** (VES-ih-kl) on the same side to form the **ejaculatory** (e-JAK-u-lah-to-re) **duct.** The right and left ejaculatory ducts travel through the body of the prostate gland and then empty into the urethra.

SEMEN

Semen (SE-men) (meaning "seed") is the mixture of sperm cells and various secretions that is expelled from the body. It is a sticky fluid with a milky appearance. The pH is in the alkaline range of 7.2 to 7.8. The secretions in semen serve several functions:

- Nourish the spermatozoa
- Transport the spermatozoa
- Neutralize the acidity of the male urethra and the female vaginal tract
- Lubricate the reproductive tract during sexual intercourse
- Prevent infection by means of antibacterial enzymes and antibodies

The glands discussed next contribute secretions to the semen **(see FIG. 20-1)**.

The Seminal Vesicles The seminal vesicles are twisted muscular tubes with many small outpouchings. They are approximately 7.5 cm (3 in) long and are attached to the connective tissue at the posterior of the urinary bladder. The glandular lining produces a thick, yellow, alkaline secretion containing large quantities of simple sugars and other substances that provide nourishment for the spermatozoa. The seminal fluid makes up a large part of the semen's volume.

The Prostate Gland The **prostate gland** lies immediately inferior to the urinary bladder, where it surrounds the first part of the urethra. Ducts from the prostate carry its secretions into the urethra. The thin, alkaline prostatic secretion helps neutralize vaginal acidity and enhances the spermatozoa's motility. The prostate gland is also supplied with muscular tissue, which, upon signals from the nervous system, contracts to aid in the expulsion of the semen from the body.

Bulbourethral Glands The **bulbourethral** (bul-bo-u-RE-thral) **glands**, also called *Cowper glands*, are a pair of pea-sized organs located in the pelvic floor just inferior to the prostate gland. They secrete mucus to lubricate the urethra and tip of the penis during sexual stimulation. The ducts of these glands extend approximately 2.5 cm (1 in) from each side and empty into the urethra before it extends into the penis.

Other very small glands secrete mucus into the urethra as it passes through the penis.

THE URETHRA AND PENIS

The male urethra, as discussed in Chapter 19, serves the dual purpose of conveying urine from the bladder and semen from the ejaculatory duct to the outside. Semen ejection is made possible by **erection**, the stiffening and enlargement of the penis, through which the major portion of the urethra extends. The penis is made of spongy tissue containing many blood spaces that are relatively empty when the organ is flaccid but that fill with blood and distend when the penis is erect. This tissue is subdivided into three segments, each called a **corpus** (body) **(see FIGS. 20-1 and 20-2)**. A single, ventrally located **corpus spongiosum** contains the urethra. On either side is a larger **corpus cavernosum** (pl., corpora cavernosa). At the distal end of the penis, the corpus spongiosum enlarges to form the **glans penis**, which is covered with a loose fold of skin, the **prepuce** (PRE-puse), commonly called the *foreskin*. It is the end of the foreskin that is removed in a **circumcision** (sir-kum-SIZH-un), a surgery frequently performed on male infants for religious or cultural reasons. Experts vary in their opinions on the medical value of circumcision with regard to improved cleanliness and disease prevention.

Ejaculation (e-jak-u-LA-shun) is the forceful expulsion of semen through the urethra to the outside. The process is initiated by reflex centers in the spinal cord that stimulate smooth muscle contraction in the prostate. This is followed by contraction of skeletal muscle in the pelvic floor, which provides the force needed for expulsion. During ejaculation, the involuntary sphincter at the base of the bladder closes to prevent the release of urine.

A male typically ejaculates 2 to 5 mL of semen containing 50 to 150 million sperm cells per mL. Out of the millions of spermatozoa in an ejaculation, only one, if any, can fertilize an ovum. The remainder of the cells live from only a few hours up to a maximum of five to seven days.

THE TESTES

The testes (TES-teze) are located outside of the body proper, suspended between the thighs in a sac called the **scrotum**

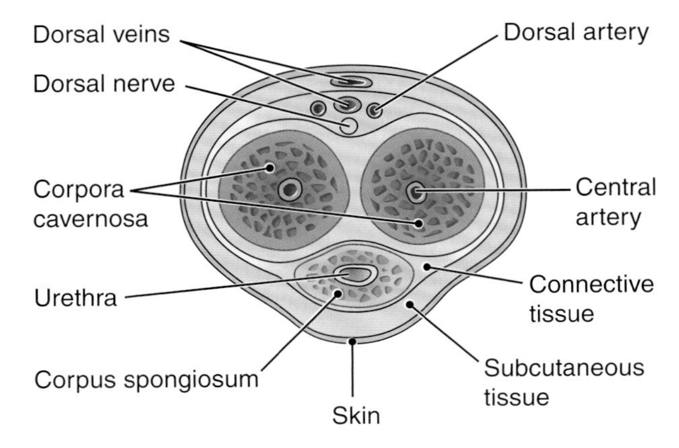

FIGURE 20-2 Cross-section of the penis. The subdivisions of the penis are shown along with associated vessels and a nerve. ZOOMING IN What subdivision of the penis contains the urethra?

20

(SKRO-tum) (**FIG. 20-3**). The penis and scrotum together make up the male **external genitalia** (jen-ih-TA-le-ah). The scrotum also contains the epididymis and the proximal portion of the ductus deferens. The testes are oval organs measuring approximately 4.0 cm (1.5 in) in length and approximately 2.5 cm (1 in) in each of the other two dimensions. During embryonic life, each testis develops from tissue near the kidney.

A month or two before birth, each testis normally descends (moves downward) through the **inguinal** (ING-gwih-nal) **canal** in the abdominal wall into the scrotum. There, the testis is suspended by a **spermatic cord** that extends through the inguinal canal (**see FIG. 20-3D**). This cord contains blood vessels, lymphatic vessels, nerves, and the ductus deferens. The gland must descend completely if it is to function normally; to produce spermatozoa, the testis must be kept at the temperature of the scrotum, which is several degrees lower than that of the abdominal cavity.

See the Student Resources on thePoint® for a figure showing descent of the testes.

Internal Structure Most of the specialized tissue of the testis consists of tiny, coiled **seminiferous** (seh-mih-NIF-er-us)

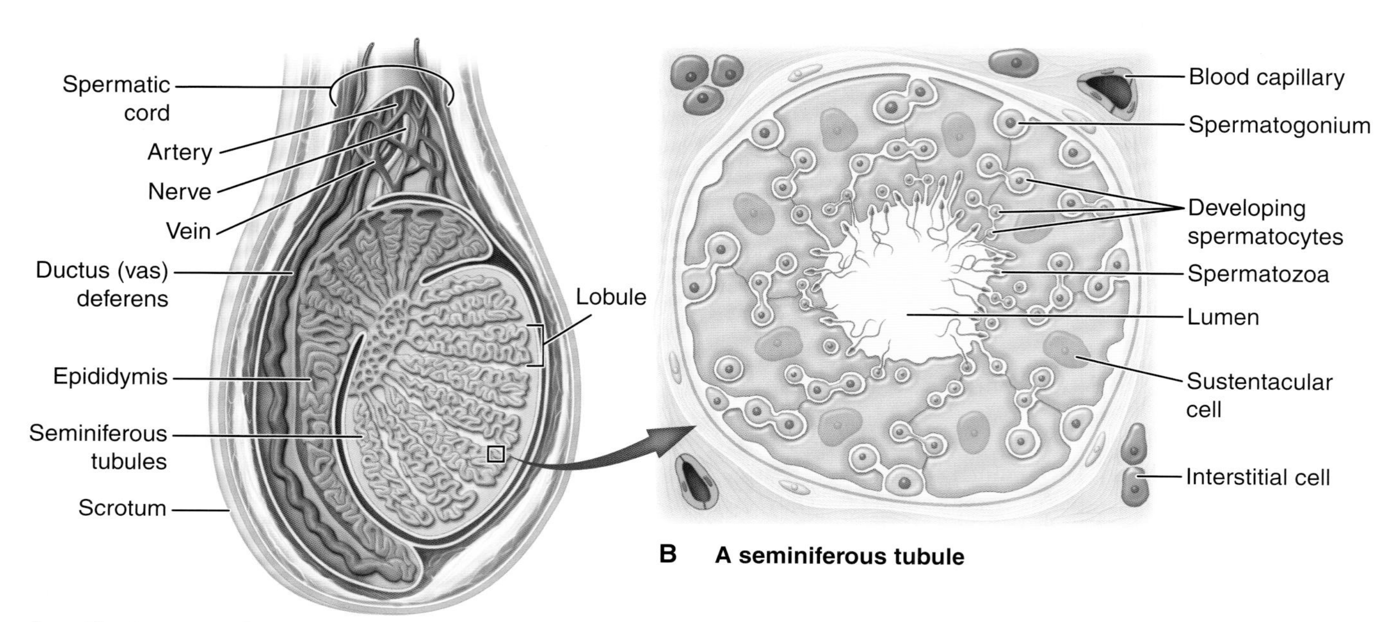

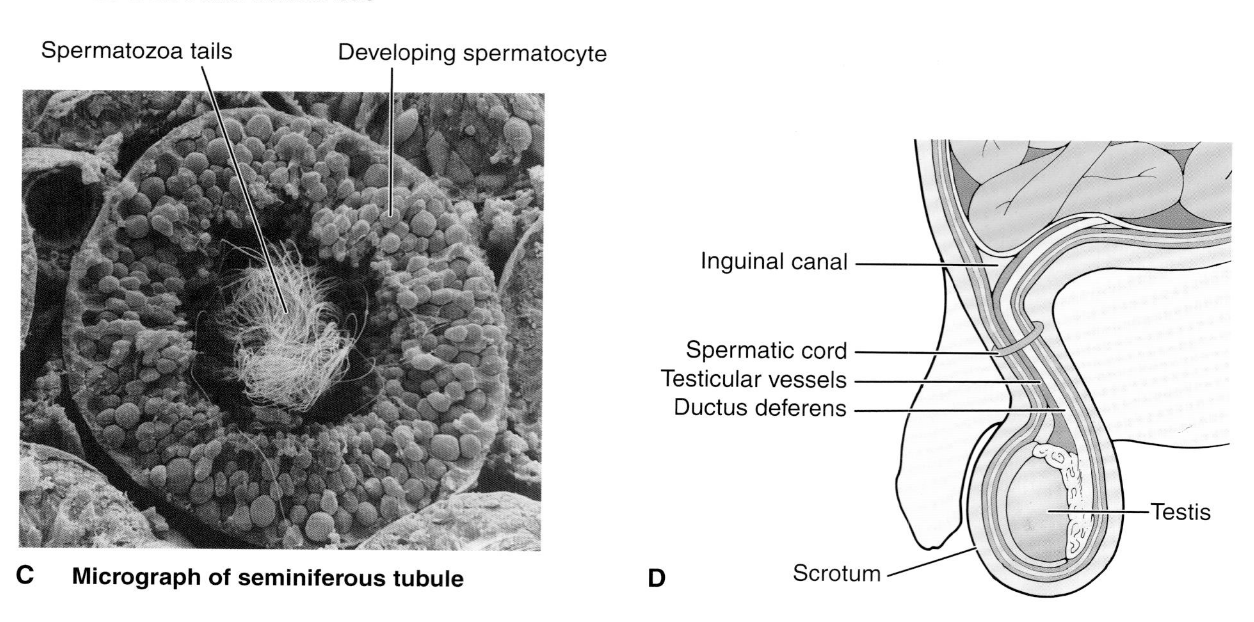

FIGURE 20-3 The testis. KEY POINT Spermatozoa develop within the testis's seminiferous tubules. **A.** The testes and scrotal sac. **B.** A seminiferous tubule. **C.** A micrograph illustrating a cross-section of a seminiferous tubule. **D.** The testis suspended by the spermatic cord that extends through the inguinal canal. ZOOMING IN Where are the interstitial cells located?

tubules (see FIG. 20-3A). Primitive cells in the walls of these tubules develop into spermatozoa, aided by neighboring cells called **sustentacular** (sus-ten-TAK-u-lar) (Sertoli) **cells** (see FIG. 20-3B). These so-called nurse cells nourish and protect the developing gametes. Once produced, spermatozoa pass into the epididymis for storage and final maturation.

Specialized **interstitial** (in-ter-STISH-al) **cells** that secrete the male sex steroid hormone **testosterone** (tes-TOS-teh-rone) are located between the seminiferous tubules. An older name for these cells is *Leydig* (LI-dig) cells. FIGURE 20-3C is a microscopic view of a seminiferous tubule in cross-section, showing the developing spermatozoa.

The Spermatozoa Spermatozoa are tiny individual cells illustrated in FIGURE 20-4. They are so small that at least 200 million can be contained in the average ejaculation. Beginning at puberty, sperm cells are manufactured continuously in the seminiferous tubules.

1 The spermatozoon is one of the most specialized cells in the body, reflecting its highly specific function. The spermatozoon has an oval head that is mostly a nucleus containing chromosomes. The **acrosome** (AK-ro-some), which covers the head like a cap, contains enzymes that help the sperm cell penetrate the ovum.

Whiplike movements of the tail (flagellum) propel the sperm through the female reproductive tract to the ovum. The cell's middle region (midpiece) contains many mitochondria that provide energy for movement.

Concept Mastery Alert

Look closely at the spermatozoon in FIGURE 20-4. Note that it contains very little cytoplasm.

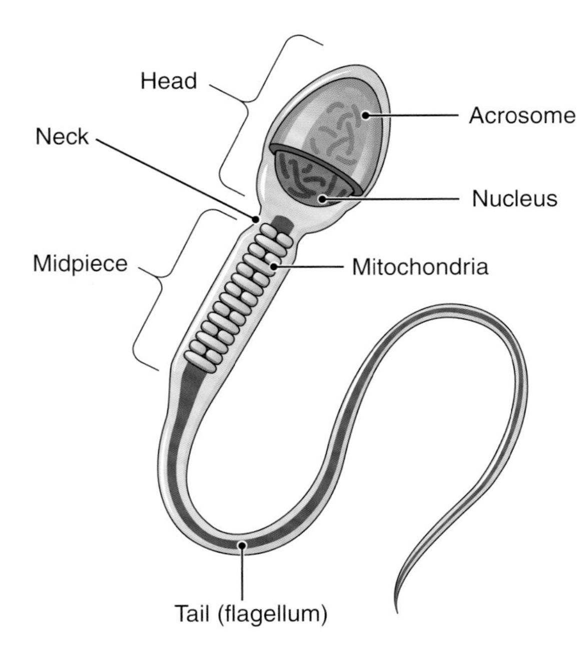

FIGURE 20-4 Human spermatozoon. The diagram shows major structural features of the male gamete. ZOOMING IN What organelles provide energy for sperm cell motility?

CHECKPOINTS 

- [ ] **20-1** What is the process of cell division that halves the chromosome number in a cell to produce a gamete?
- [ ] **20-2** What is the male gamete called?
- [ ] **20-3** What is the male gonad?
- [ ] **20-4** What is the structure on the surface of the testis that stores sperm?
- [ ] **20-5** What glands, aside from the testis, contribute secretions to semen?

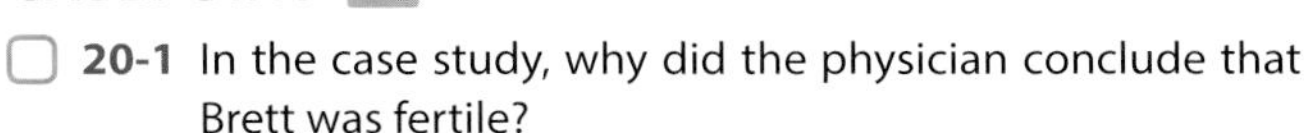

- [ ] **20-6** What are the main subdivisions of a spermatozoon?

CASEPOINT

- [ ] **20-1** In the case study, why did the physician conclude that Brett was fertile?

Hormonal Control of Male Reproduction

The activities of the testes are under the control of two hormones produced by the anterior pituitary as well as testicular hormones themselves (FIG. 20-5). These protein hormones are named for their activity in female reproduction (described later), although they are chemically the same in both males and females. Collectively known as the gonadotropins, they are:

- **Follicle-stimulating hormone** (**FSH**) stimulates the sustentacular cells to produce growth factors that promote the formation of spermatozoa.
- **Luteinizing hormone** (**LH**) stimulates the interstitial cells between the seminiferous tubules to produce testosterone, which is also needed for sperm cell development.

Starting at puberty, the hypothalamus begins to secrete a hormone (**gonadotropin-releasing hormone**, known as GnRH) that triggers the release of FSH and LH. These hormones are continuously secreted in the adult male.

3 The circulating concentration of testosterone is kept relatively constant by a negative feedback mechanism. As the blood level of testosterone increases, the hypothalamus secretes less GnRH; as the level of testosterone decreases, the hypothalamus secretes more GnRH.

TESTOSTERONE

From the interstitial cells, testosterone diffuses into surrounding fluids and is then absorbed into the bloodstream. This hormone has many functions, including (see FIG. 20-5):

- Development and maintenance of the male reproductive accessory organs
- Development of spermatozoa
- Development of **secondary sex characteristics**, traits that characterize males and females but are not directly concerned with reproduction. In males, these traits include a deeper voice, broader shoulders, narrower hips, a greater percentage of muscle tissue, and more body hair than found in females.

20

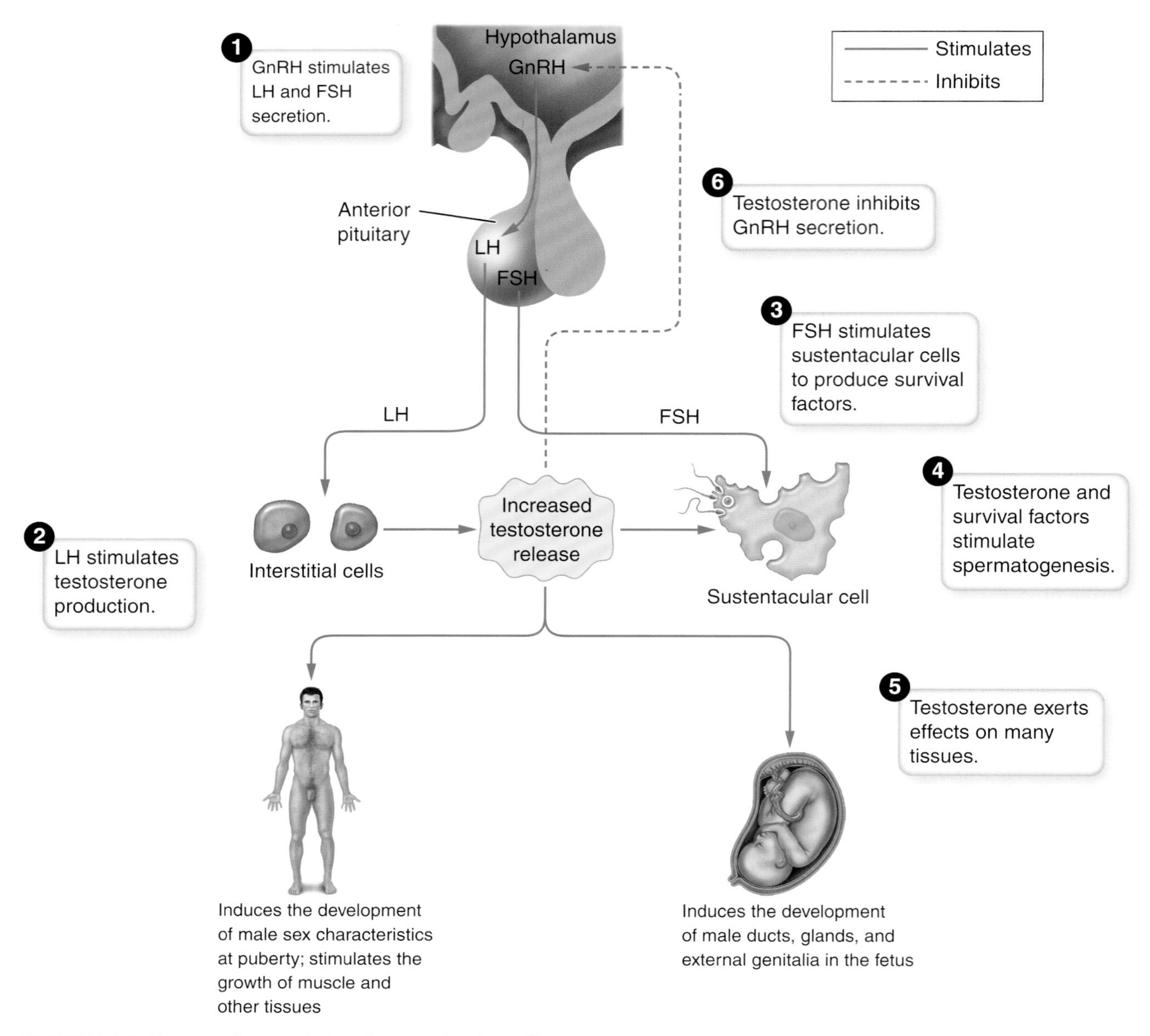

FIGURE 20-5 Hormonal control of male reproduction. **KEY POINT** The hypothalamus and anterior pituitary gland control testosterone synthesis by the testes. Testosterone regulates sperm production and other functions relating to male reproduction and also feeds back to inhibit its own secretion. **ZOOMING IN** Which hormone stimulates testosterone secretion—LH or FSH?

CHECKPOINTS

- ☐ **20-7** What two pituitary hormones regulate both male and female reproduction?
- ☐ **20-8** Which cell type in the testis produces the main male sex steroid hormone?

The Effects of Aging on Male Reproduction

A gradual decrease in the production of testosterone and spermatozoa begins at about the age of 40 and continues throughout life. Sperm motility and quality also decline in middle and later life, but some men retain fertility into their 90s. Secretions from the prostate and seminal vesicles decrease in amount and become less viscous.

Erectile dysfunction may affect men at any age, but its prevalence increases with age. It is defined as an inability to attain or maintain an erection adequate to engage in sexual intercourse. **BOX 20-1** has more information on the causes and treatments of this disorder.

Nonmalignant enlargement of the prostate, known as benign prostatic hyperplasia (BPH), exists in nearly all men by the age of 80 and causes symptoms in about half of them. As noted in Chapter 19, an enlarged prostate can put pressure on the urethra and interfere with urination.

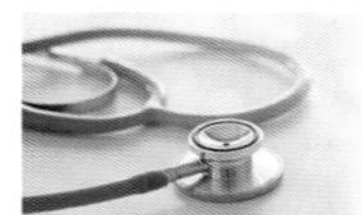

CLINICAL PERSPECTIVES
Treating Erectile Dysfunction

BOX 20-1

Approximately 25 million American men and their partners are affected by **erectile dysfunction** (ED), the inability to achieve or sustain an erection long enough to have satisfying sexual intercourse. Although ED is more common in men over the age of 65, it can occur at any age and can have many causes. Until recently, ED was believed to be caused by psychological factors, such as stress or depression. It is now known that many cases of ED are caused by physical factors, including cardiovascular disease, diabetes, spinal cord injury, and damage to penile nerves during prostate surgery. Antidepressant and antihypertensive medications also can produce ED.

Erection results from interaction between the autonomic nervous system and penile blood vessels. Sexual arousal stimulates parasympathetic nerves in the penis to release a compound called nitric oxide (NO), which activates the vascular smooth muscle enzyme guanylyl cyclase. This enzyme catalyzes production of cyclic GMP (cGMP), a potent vasodilator that increases blood flow into the penis to cause erection. Physical factors that cause ED prevent these physiologic occurrences.

Until recently, treatment options for ED, such as penile injections, vacuum pumps, and insertion of medications into the penile urethra, were inadequate, inconvenient, and painful. Today, drugs that target the physiologic mechanisms that underlie erection are giving men who suffer from ED new hope. The best known of these is sildenafil (Viagra), which works by inhibiting the enzyme that breaks down cGMP, thus prolonging the effects of NO. Because of its short duration of action, Viagra must be taken shortly before sexual intercourse. Other drugs, such as tadalafil (Cialis) can be taken once daily, removing the need to plan the timing of sexual activity.

Although effective in about 80% of all ED cases, Viagra can cause some relatively minor side effects, including headache, nasal congestion, stomach upset, and blue-tinged vision. Viagra should never be used by men who are taking nitrate drugs to treat angina. Because nitrate drugs elevate NO levels, taking them with Viagra, a drug that prolongs the effects of NO, can cause life-threatening hypotension.

The Female Reproductive System

The female gonads are the paired ovaries (O-vah-reze), where the female gametes, or ova, are formed. The remainder of the female reproductive tract consists of an organ (uterus) to hold and nourish a developing infant, various passageways, and the external genital organs (**FIG. 20-6**). As with our discussion of the male, we introduce the accessory organs before focusing on the structure and function of the ovary. See **FIGURES A3-12B and A3-14** in the Dissection Atlas for photographs of the female reproductive system.

ACCESSORY ORGANS

The accessory organs in the female are the uterus, the uterine tubes, the vagina, the greater vestibular glands (not illustrated), and the vulva (**see FIG. 20-6**). The breasts are also considered to be accessory reproductive organs in the female; because of their role in child nourishment, they are discussed in Chapter 21.

The Uterus The **uterus** (U-ter-us) is a pear-shaped, muscular organ approximately 7.5 cm (3 in) long, 5 cm (2 in) wide, and 2.5 cm (1 in) deep, in which the fetus develops to maturity (**see FIG. 20-6**). (The organ is typically larger in women who have borne children and smaller in postmenopausal women.) The uterus's superior portion rests on the upper surface of the urinary bladder; the inferior portion is embedded in the pelvic floor between the bladder and the rectum. Its wider upper region is called the **body** or corpus; the lower, narrower region is the **cervix** (SER-viks) or neck. The small, rounded region above the level of the tubal entrances is known as the **fundus** (FUN-dus).

Folds of the peritoneum called the *broad ligaments* support the uterus, extending from each side of the organ to the lateral body wall. Along with the uterus, these two membranes form a partition dividing the female pelvis into anterior and posterior areas. The ovaries are suspended from the broad ligaments, and the uterine tubes lie within the upper borders. Blood vessels that supply these organs are found between the layers of the broad ligaments.

The muscular wall of the uterus is called the **myometrium** (mi-o-ME-tre-um) (**see FIG. 20-6B**). The lining of the uterus is a specialized epithelium known as **endometrium** (en-do-ME-tre-um). This inner layer undergoes monthly changes during a woman's reproductive years, first building up to nourish a fertilized egg, then breaking down if no fertilization has occurred. This degenerated endometrial tissue is released as the menstrual flow. The cavity inside the uterus is shaped somewhat like a capital T, but it is capable of changing shape and enlarging as a fetus develops.

The Uterine Tubes Each **uterine** (U-ter-in) fallopian **tube** is a small, muscular structure, nearly 12.5 cm (5 in) long, extending from the uterus to a point near the ovary (**see FIG. 20-6**). There is no direct connection between the ovary and uterine tube. Instead, the ova are released from the ovary into the abdominal cavity. **Fimbriae** (FIM-bre-e), small, fringelike extensions of the tube's opening, create currents in the peritoneal fluid that sweep the ovum into the uterine tube.

Unlike the sperm cell, the ovum cannot move by itself. Its progress through the uterine tube toward the uterus depends on the sweeping action of cilia in the tube's lining and on peristalsis of the tube. It takes about five days for an ovum to reach the uterus from the ovary.

The Vagina The **vagina** is a muscular tube approximately 7.5 cm (3 in) long (see FIG. 20-6A). The cervix dips into the vagina's superior portion forming a circular recess known as the **fornix** (FOR-niks). The deepest area of the fornix, located behind the cervix is the **posterior fornix.** This recess in the posterior vagina lies adjacent to the most inferior portion of the peritoneal cavity, a narrow passage between the uterus and the rectum named the **rectouterine pouch.** A rather thin layer of tissue separates the posterior fornix from this region so that abscesses or tumors in the peritoneal cavity can sometimes be detected by vaginal examination.

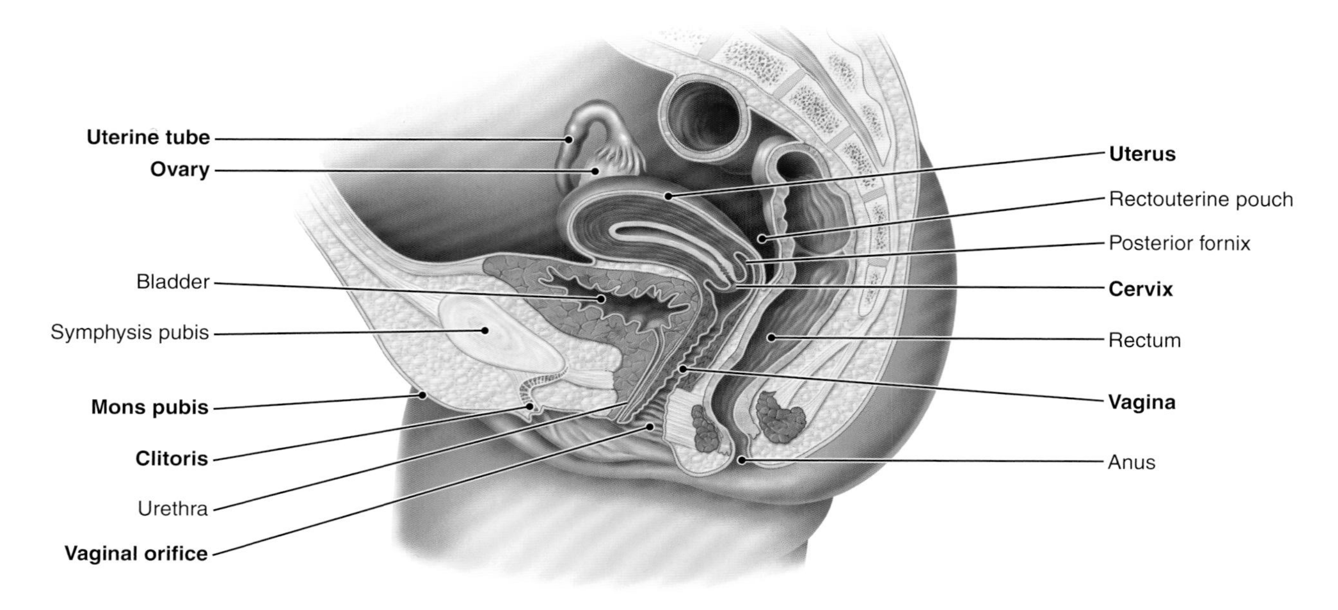

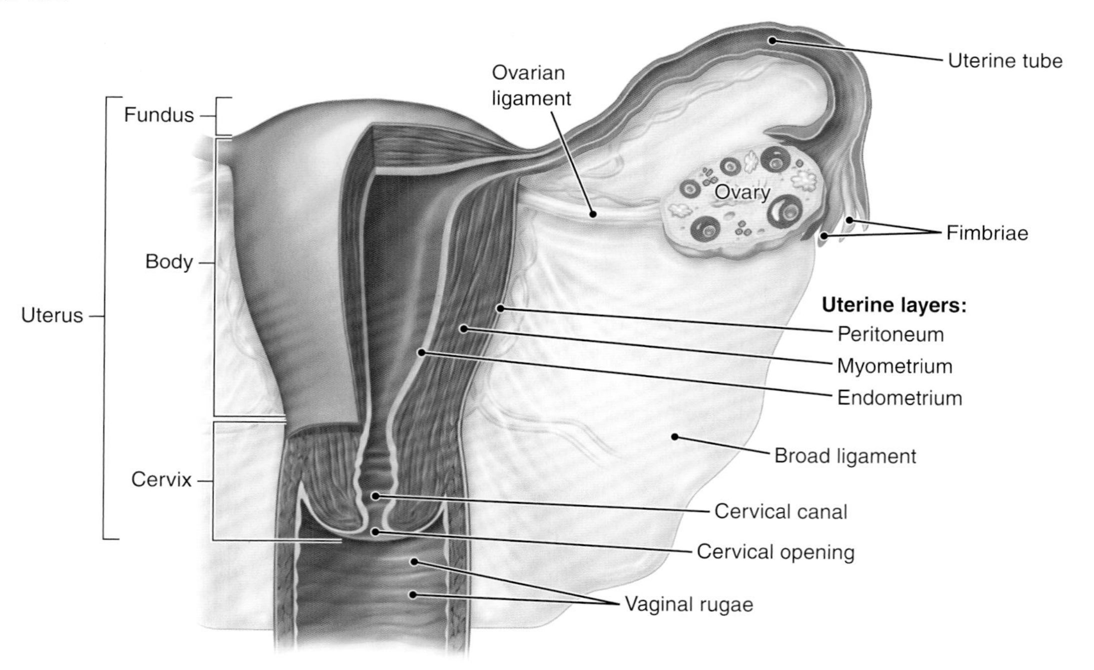

FIGURE 20-6 **Female reproductive system. A.** As shown in this sagittal view, the female internal genitalia are sandwiched between structures of the urinary and gastrointestinal systems, which are also shown. Structures of the reproductive system are labeled in bold type. **B.** Ligaments hold the uterus and ovaries in place. ZOOMING IN What is the deepest part of the uterus called? The most inferior portion?

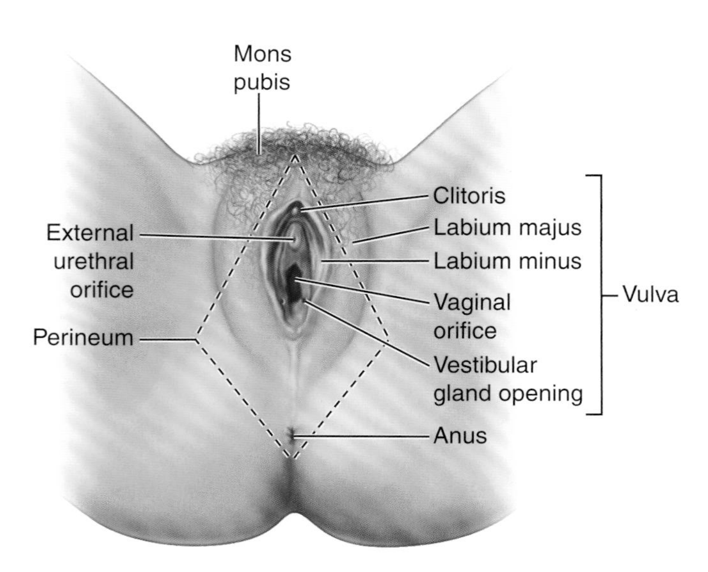

FIGURE 20-7 External parts of the female reproductive system. KEY POINT The external female genitalia, or vulva, includes the labia, clitoris, and mons pubis. Nearby structures are also shown.

The lining of the vagina is a mucous membrane with a surface layer of stratified squamous epithelium. Rugae (folds) in this lining permit enlargement to accommodate childbirth (see FIG. 20-6B). In addition to being a part of the birth canal, the vagina is the organ that receives the penis during sexual intercourse. A fold of membrane called the **hymen** (HI-men) may sometimes be found at or near the vaginal (VAJ-ih-nal) canal opening.

The Greater Vestibular Glands Just superior and lateral to the vaginal opening are the two mucus-producing **greater vestibular** (ves-TIB-u-lar) (Bartholin) **glands** (not illustrated). These glands secrete into an area near the vaginal opening known as the *vestibule*. Like the bulbourethral glands in males, these glands provide lubrication during intercourse. If a gland becomes infected, a surgical incision may be needed to reduce swelling and promote drainage.

The Vulva and the Perineum The external female genitalia make up the **vulva** (VUL-vah) (FIG. 20-7). This includes two pairs of **labia** (LA-be-ah), the larger *labia majora* (sing., labium majus) and smaller *labia minora* (sing., labium minus). It also includes the **clitoris** (KLIT-o-ris), a small organ of great sensitivity, as well as the openings of the urethra and vagina, and the *mons pubis*, a pad of fatty tissue over the pubic symphysis (joint) (see FIG. 20-6A). Although the entire pelvic floor in both the male and female is properly called the **perineum** (per-ih-NE-um), those who care for pregnant women usually refer to the limited area between the vaginal opening and the anus as the perineum or *obstetric perineum*.

THE OVARIES AND OVA

The ovary (O-vah-re) is a small, somewhat flattened oval body measuring approximately 4 cm (1.6 in) in length, 2 cm (0.8 in) in width, and 1 cm (0.4 in) in depth (FIG. 20-8). Like the testes, the ovaries descend, but only as far as the pelvic cavity. Here, they are held in place by ligaments, including the broad ligaments, the ovarian ligaments, and others, that attach them to the uterus and the body wall (see FIG. 20-6B).

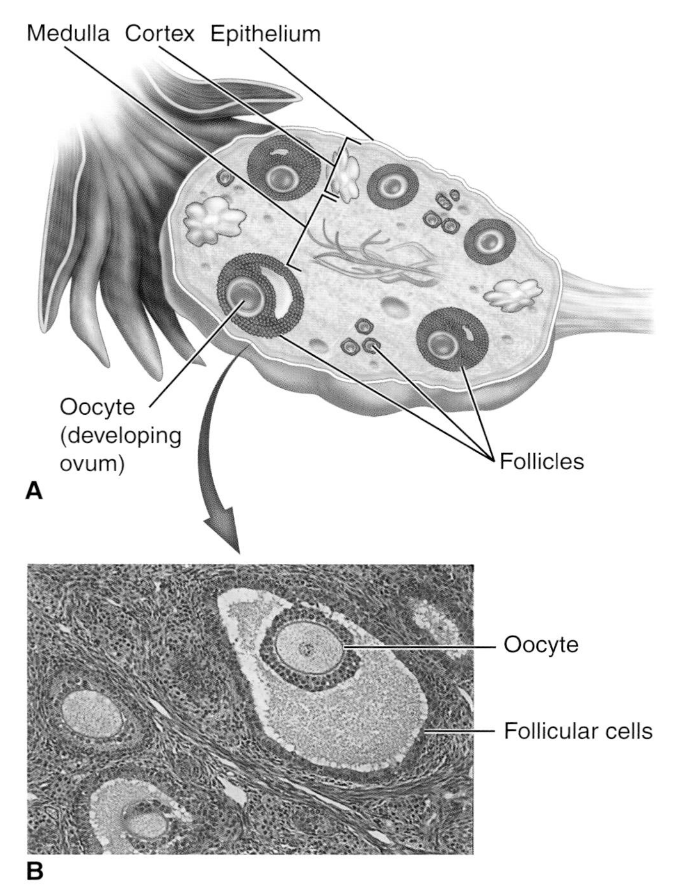

FIGURE 20-8 The ovary. KEY POINT The gametes and hormone-producing cells of the ovary are organized in follicles. **A.** The ovary, containing follicles in different stages of maturation. **B.** This micrograph shows an oocyte within a mature follicle.

The outer layer of each ovary is made of a single layer of epithelium. The ova are produced beneath this layer. Each developing ovum, or **oocyte**, is contained within a small cluster of cells called an **ovarian follicle** (o-VA-re-an FOL-ih-kl) (FIG. 20-8). The follicular cells protect the ovum and produce the ovarian hormones.

Unlike males, females are born with all of the gametes they will ever produce. The ovaries of a newborn female contain more than a million immature follicles, but only a few hundred thousand remain viable at puberty. As puberty approaches, a few follicles begin to develop. Their oocytes enlarge, and the follicular cells multiply. During each month of a woman's reproductive years, one developing follicle (usually) completes the maturation process and releases its ovum. FIGURE 20-8 shows follicles at different stages of development. The final stages of development and release are discussed shortly in the context of the female reproductive cycle.

CHECKPOINTS

- ☐ **20-9** What is the female gamete called?
- ☐ **20-10** What is the female gonad called?
- ☐ **20-11** In what organ does a fetus develop?
- ☐ **20-12** In what structure does an ovum mature?

The Female Reproductive Cycle

In the female, as in the male, reproductive function is controlled by gonadotropins from the anterior pituitary that are regulated by GnRH from the hypothalamus **(see FIG. 11-2)**. Female activity differs, however, in that it is cyclic; it shows regular patterns of increases and decreases in hormone levels. The most obvious sign of these changes is periodic vaginal bleeding, or **menstruation** (men-stru-A-shun), so this cycle is known as the **menstrual cycle**. The typical length of the menstrual cycle varies between 22 and 45 days, but 28 days is taken as an average, with the first day of menstrual flow being considered the first day of the cycle **(FIG. 20-9)**.

Before we delve into the details of the cycle, take a moment to scan **FIGURE 20-9**. **Ovulation** (ov-u-LA-shun), the release of the gamete from the ovary, separates the reproductive cycle into the preovulatory (follicular) and postovulatory (luteal) phases. The first row illustrates changes in the production of pituitary hormones. The second row illustrates how pituitary hormones alter follicular development in the ovary. As the follicle grows and transforms, it produces varying amounts of ovarian hormones, as shown in the third row. These hormones cause cyclic changes in the uterus, as shown at the bottom of the diagram. 12 Following the cause-and-effect relationships between the elements in the different rows is key to understanding the menstrual cycle.

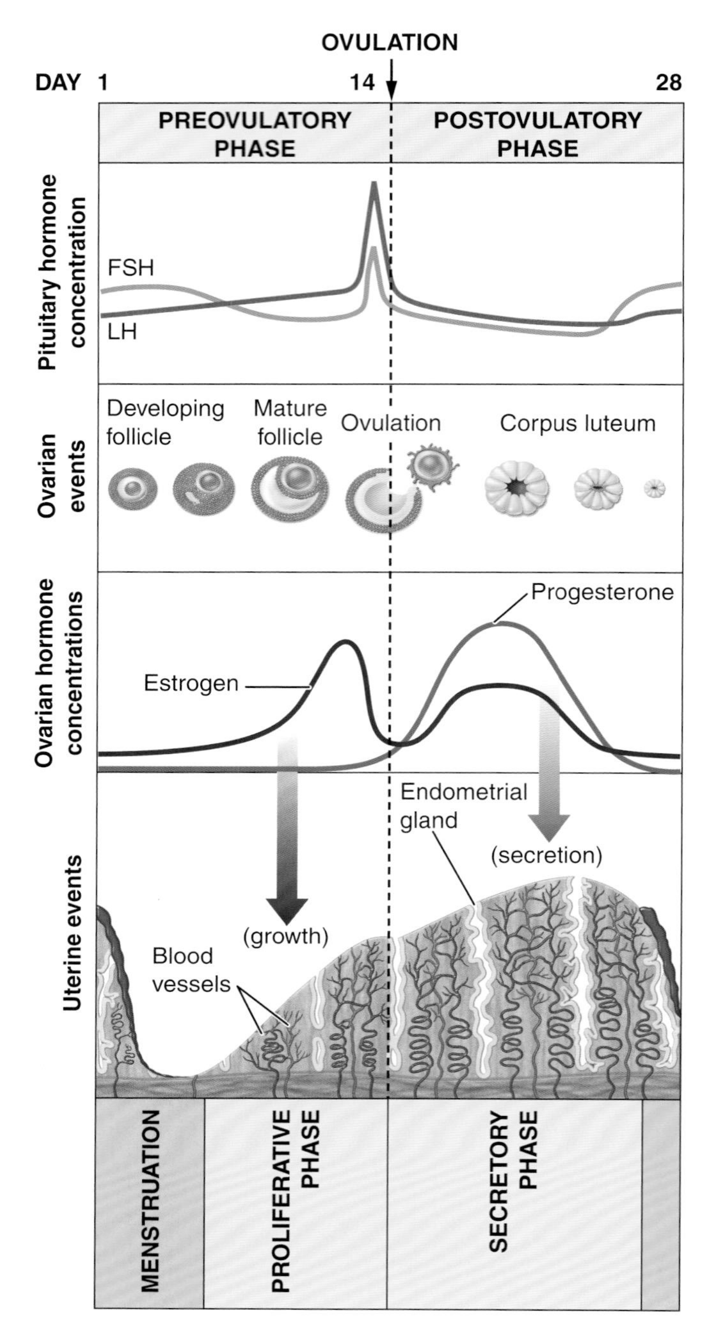

FIGURE 20-9 The female reproductive cycle. KEY POINT The anterior pituitary gland secretes FSH and LH, which control the follicle and corpus luteum. The follicle and corpus luteum secrete ovarian hormones, which control changes in the uterine endometrium. This figure illustrates an average 28-day menstrual cycle with ovulation on day 14. ZOOMING IN What ovarian hormone peaks closest to ovulation? What ovarian hormone peaks after ovulation?

PREOVULATORY PHASE

At the start of each cycle, under the influence of pituitary FSH, several follicles, each containing an ovum, enter the final stages of maturation. As each follicle matures, it enlarges, and fluid accumulates in its central cavity. While it grows, the follicle secretes increasing amounts of the ovarian hormone **estrogen** (ES-tro-jen), which stimulates further growth of the follicle **(see FIG. 20-9, second row)**. (*Estrogen* is the term used for a group of related hormones, the most active of which is estradiol.) Eventually, most of the developing follicles die off, and only a single follicle survives to release its ovum, as discussed shortly. Any maturing follicles that do not release their ova simply degenerate.

The estrogen produced by the follicles travels through the blood to the uterus (as well as to other tissues), where it starts preparing the endometrium for a possible pregnancy. This preparation includes thickening of the endometrium and elongation of the glands that produce uterine secretions. The time period between menstruation and ovulation is known as the **proliferative phase** of uterine development **(see FIG. 20-9, bottom)**.

3 During most of the menstrual cycle, estrogen acts as a negative feedback messenger to inhibit the release of GnRH, LH, and FSH. However, when estrogen reaches high

levels, it stimulates the production of both GnRH and LH. This stimulation results in the **LH surge**, a sharp rise of LH in the blood **(see FIG. 20-9, first row)**. (Note that there is also a small rise in FSH at this time caused by the increase in GnRH.) The LH surge triggers the next stage of the reproductive cycle—ovulation.

OVULATION AND THE POSTOVULATORY PHASE

Approximately one day after the LH surge, ovulation occurs. The wall of the mature follicle and the adjacent ovarian tissue both rupture, releasing the ovum into the peritoneum. Currents in the peritoneal fluid created by the fimbriae of the uterine tubes usually propel the ovum into the tube. In some women, ovulation causes a stabbing abdominal pain that can persist minutes or hours.

In an average 28-day cycle, ovulation occurs on day 14 and is followed two weeks later by the start of the menstrual flow (the beginning of a new cycle). However, an ovum can be released any time from days 7 to 21, thus accounting for the variation in the length of normal cycles.

In addition to causing ovulation, LH transforms the ruptured follicle into a solid mass called the **corpus luteum** (LU-te-um). This structure secretes estrogen and also **progesterone** (pro-JES-ter-one), a hormone that promotes the survival of a fertilized ovum and eventually an embryo. Under the influence of estrogen and progesterone, the endometrium continues to thicken, and the glands and blood vessels increase in size. The glands also secrete increasing amounts of fluid and nutrients, so the time period after ovulation and before menstruation is known as the **secretory phase** of the uterine cycle. The rising levels of estrogen and progesterone feed back to inhibit the release of FSH and LH from the pituitary.

During the postovulatory phase, the ovum makes its journey to the uterus by way of the uterine tube. If the ovum is fertilized, it begins to secrete hormones that maintain the corpus luteum for a few months. However, if the ovum is not fertilized, it dies within two to three days, and the corpus luteum degenerates by about 11 days postovulation. Sometimes, as a result of normal ovulation, the corpus luteum persists and forms a small ovarian cyst (fluid-filled sac). This condition usually resolves without treatment, but large cysts may rupture or leak, causing severe abdominal pain. These larger cysts may require surgery.

Concept Mastery Alert

Compare the ovarian events to the ovarian hormone concentrations. Note that estrogen production increases as the follicle enlarges, and both progesterone and estrogen production increase as the corpus luteum enlarges.

MENSTRUATION

If fertilization does not occur, the corpus luteum degenerates, and the levels of estrogen and progesterone decrease. Without the hormones to support growth, the endometrium degenerates. Small hemorrhages appear in this tissue, producing the bloody discharge known as the **menses** (MEN-seze) or *menstrual flow*. Bits of endometrium break away and accompany the blood flow during this period of menstruation. The average duration of menstruation is two to six days.

Even before the menstrual flow ceases, the endometrium begins to repair itself through the growth of new cells. The low levels of estrogen and progesterone allow the release of FSH from the anterior pituitary. FSH causes new follicles to begin to ripen within the ovaries, and the cycle begins anew.

The activity of ovarian hormones as negative feedback messengers is the basis of hormonal methods of contraception (birth control). The estrogen and progesterone in birth control pills inhibit the release of FSH and LH from the pituitary, preventing follicular development and ovulation. The menstrual period that follows withdrawal of this pharmaceutical estrogen and progesterone is **anovulatory** (an-OV-u-lah-tor-e); that is, it is not preceded by ovulation.

See the Student Resources on thePoint® for a summary chart on reproductive hormones and the animation "Oogenesis and the Menstrual Cycle."

20

CHECKPOINTS 

- ☐ **20-13** What are the two hormones produced in the ovaries?
- ☐ **20-14** What process releases an ovum from the ovary?
- ☐ **20-15** What does the follicle become after its ovum is released?

CASEPOINTS 

- ☐ **20-2** During which phase of the ovarian cycle and the uterine cycle would Jessica's endometrial tissue reach its maximum thickness?
- ☐ **20-3** Jessica took oral contraceptives (estrogen and progesterone pills) for two years. What effect did this have on her FSH levels during those years?
- ☐ **20-4** If Jessica ovulated on day 15 of her menstrual cycle, on what day can she expect her period?

Menopause

Menopause (MEN-o-pawz) is the time period during which menstruation ceases altogether. It ordinarily occurs gradually between the ages of 45 and 55 years and is caused by a normal decline in ovarian function. The ovary becomes chiefly scar tissue and no longer produces mature follicles or appreciable amounts of estrogen and progesterone. Eventually, the uterus, uterine tubes, vagina, and vulva all become somewhat atrophied, and the vaginal mucosa becomes thinner, dryer, and more sensitive.

Menopause is an entirely normal condition, but its onset sometimes brings about effects that are temporarily disturbing. The decrease in estrogen levels can cause nervous symptoms, such as anxiety and insomnia. Because estrogen also helps maintain the vascular dilation that promotes heat loss, low levels may result in "hot flashes."

Physicians may prescribe hormone replacement therapy (HRT) to relieve the discomforts associated with menopause. This medication is usually a combination of estrogen with a synthetic progesterone (progestin), which is included to prevent overgrowth of the endometrium and the risk of endometrial cancer. Studies with the most commonly prescribed form of HRT have shown it lowers the incidence of colorectal cancer. It also lowers the incidence of hip fractures, a sign of osteoporosis. However, in addition to an increased risk of breast cancer, HRT also carries a risk of

Table 20-1 Main Methods of Birth Control Currently in Use

Method	Description	Advantages	Disadvantages
Surgical			
Vasectomy/tubal ligation	Cutting and tying of tubes carrying gametes	Nearly 100% effective; involves no chemical or mechanical devices	Not usually reversible; rare surgical complications
Hormonal			
Birth control pills	Estrogen and progestin, or progestin alone, taken orally to prevent ovulation	Highly effective; requires no last-minute preparation	Alters physiology; return to fertility may be delayed; risk of cardiovascular disease in older women who smoke or have hypertension
Birth control shot	Injection of synthetic progestin every 3 months to prevent ovulation	Highly effective; lasts for 3–4 months	Alters physiology; same side effects as birth control pill; possible menstrual irregularity, amenorrhea
Birth control patch	Adhesive patch placed on body that administers estrogen and progestin through the skin; left on for 3 weeks and removed for a 4th week	Protects long term; less chance of incorrect use; no last-minute preparation	Alters physiology; same possible side effects as birth control pill
Birth control ring	Flexible ring inserted into vagina that releases progestin and estrogen; left in place for 3 weeks and removed for a 4th week	Long lasting; highly effective; no last-minute preparation	Possible infections, irritation; same possible side effects as birth control pill
Barrier			
Condom	Sheath that prevents semen from contacting the female reproductive tract	Readily available; does not affect physiology; does not require medical consultation; protects against STIs	
Male	Sheath that fits over erect penis and prevents release of semen	Inexpensive	Must be applied just before intercourse; may slip or tear
Female	Sheath that fits into vagina and covers cervix	Gives women control over last-minute contraception	Relatively expensive; may be difficult or inconvenient to insert
Diaphragm (with spermicide)	Rubber cap that fits over cervix and prevents entrance of sperm	Does not affect physiology; no side effects	Must be inserted before intercourse and left in place for 6 hours; requires fitting by physician
Contraceptive sponge (with spermicide)	Soft, disposable foam disk containing spermicide which is moistened with water and inserted into the vagina	Protects against pregnancy for 24 hours; nonhormonal; available without prescription; inexpensive	85%–90% effective depending on proper use; possible skin irritation
Intrauterine contraceptive	Metal or plastic device inserted into uterus through vagina; prevents fertilization and implantation by release of copper or birth control hormones	Highly effective for 5–10 years depending on type; reversible; economical; no last-minute preparation	Must be introduced by health professional; can result in heavy menstrual bleeding
Other			
Spermicide	Chemicals used to kill sperm; best when used in combination with a barrier method	Available without prescription; inexpensive; does not affect physiology	May cause local irritation; must be used just before intercourse
Fertility awareness	Abstinence during fertile part of cycle as determined by menstrual history, basal body temperature, or quality of cervical mucus	Does not affect physiology; accepted by certain religions	High failure rate; requires careful record keeping

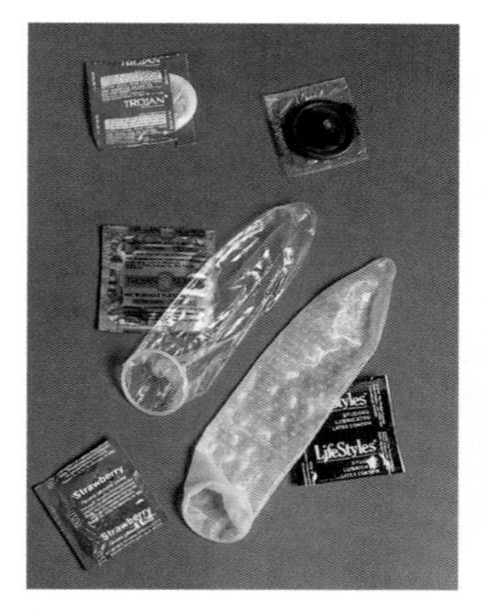

A **Male condoms**

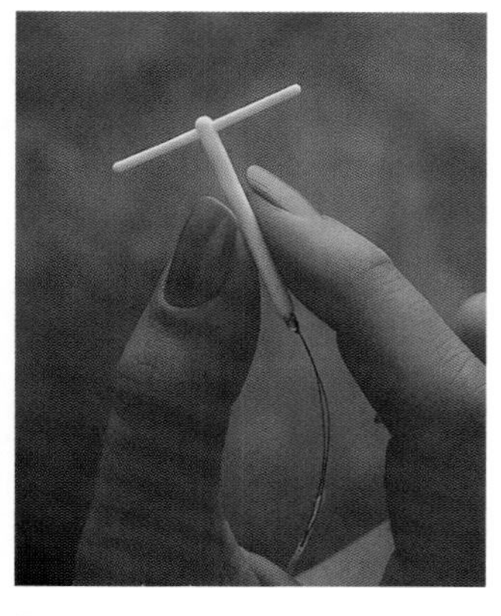

B **An intrauterine device (IUD)**

FIGURE 20-10 **Contraception. A**. The male condom consists of a flexible sheath. **B.** Intrauterine devices (IUDs) are implanted within the uterus. ZOOMING IN Of the two types of contraceptives shown here, which type prevents sexually transmitted diseases?

thrombosis and embolism, which is highest among women who smoke. All HRT risks increase with the duration of therapy. Therefore, treatment should be given for a short time and at the lowest effective dose. Women with a history or family history of breast cancer or circulatory problems should not take HRT.

Because of its beneficial effects, studies are continuing with estrogen alone, generally prescribed for women who have undergone a hysterectomy and do not have a uterus.

CHECKPOINT

- [ ] **20-16** What is the term describing the complete cessation of menstrual cycles?

Birth Control

Birth control is most commonly achieved by **contraception**, which is the use of artificial methods to prevent fertilization of the ovum. Birth control measures that prevent implantation of the fertilized ovum are also considered contraceptives, although technically they do not prevent conception and are more accurately called **abortifacients** (ah-bor-tih-FA-shents) (agents that cause abortion). Some birth control methods act by both mechanisms. **TABLE 20-1** presents a brief description of the main contraceptive methods currently in use along with some advantages and disadvantages of each. The list is given in a rough order of decreasing effectiveness.

The surest contraceptive method is surgical sterilization. Tubal ligation and vasectomy work by severing and tying off (or cauterizing) the ducts that carry the gametes: the uterine tubes in women and the ductus deferens in men. A man who has had a vasectomy retains the ability to produce hormones and semen as well as the ability to engage in sexual intercourse, but no fertilization can occur because the semen does not contain spermatozoa.

Condoms provide a physical barrier that prevents the union of a sperm and ovum (**FIG. 20-10**). The female condom is a sheath that fits into the vagina. A flexible inner ring inserted into the top of the vagina holds the condom in place; an outer ring remains outside the vaginal opening. While more expensive than the male condom, it gives a woman control over last-minute contraception and does not require any drugs or medical visits. Male and female condoms have an added benefit: they protect against STIs. See **BOX 20-2** for more information about these diseases and how to prevent them.

Hormonal methods of birth control in females exploit the negative feedback loop discussed earlier. They inhibit GnRH, FSH, and LH production (and thus ovulation) by providing a continuous supply of estrogen and progestin (a progesterone-like drug) or just progestin. The hormones can be delivered by pills, a skin patch, or a vaginal ring, in which case the treatment usually stops for one week every month so that menstruation occurs. Other methods include injections every three months or the implantation of a hormone-releasing rod under the skin. These longer-lasting methods both require the intervention of a health professional. The emergency contraceptive pill is a synthetic progesterone (progestin) taken within 72 hours after intercourse, usually

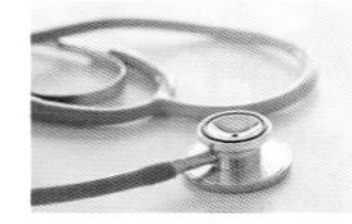

CLINICAL PERSPECTIVES

BOX 20-2

Sexually Transmitted Infections: Lowering Your Risks

Sexually transmitted infection (STIs), such as chlamydia, gonorrhea, genital herpes, HIV, and syphilis, are some of the most common infectious diseases in the United States, affecting more than 13 million men and women each year. These diseases are associated with complications, such as pelvic inflammatory disease, inflammation of the epididymis, infertility, liver failure, neurologic disorders, cancer, and AIDS. Women are more likely to contract STIs than are men. The same mechanisms that transport sperm cells through the female reproductive tract also move infectious organisms. The surest way to prevent STIs is to avoid sexual contact with others. If you are sexually active, the following techniques can lower your risks:

- Maintain a monogamous sexual relationship with an uninfected partner.
- Correctly and consistently use a male or female condom. Although not 100% effective, condoms greatly reduce the risk of contracting an STI.
- Avoid contact with body fluids, such as blood, semen, and vaginal fluids, all of which may harbor infectious organisms.
- Urinate and wash the genitals after sex. This may help remove infectious organisms before they cause disease.
- Have regular checkups for STIs. Most of the time, STIs cause no symptoms, particularly in women.

in two doses 12 hours apart. It reduces the risk of pregnancy following unprotected intercourse. This so-called morning-after pill is intended for emergency use and not as a regular birth control method.

Researchers have done trials with a male contraceptive pill, but none is on the market as yet. The male version of "the pill" also inhibits release of FSH and LH, which are important in spermatogenesis. Use of testosterone as a negative feedback messenger requires regular injections and has some undesirable side effects at the doses needed. Administration of the female hormone progesterone prevents spermatogenesis but also inhibits normal testosterone production. Studies are ongoing to find the best way to deliver the right male contraceptive hormones at safe and effective doses.

Mifepristone (RU 486) is a drug taken after conception to terminate an early pregnancy. It blocks the action of progesterone, causing the uterus to shed its lining and release the fertilized egg. It must be combined with administration of prostaglandins to expel the uterine tissue.

An **intrauterine contraceptive** (IUC), also known as an intrauterine device (IUD), is a small T-shaped structure inserted into the uterus by a healthcare provider **(see FIG. 20-10B)**. Some IUCs prevent ovulation by releasing progestin, as used in the emergency contraceptive pill. A different type contains copper. This element causes an inflammatory response that prevents conception by thickening cervical mucus and by immobilizing sperm.

See the Student Resources on thePoint® for an illustration of laparoscopic sterilization.

CHECKPOINTS 

- ☐ **20-17** What term describes the use of artificial means to prevent fertilization of an ovum?
- ☐ **20-18** Name the two methods of surgical sterilization.

Clinics and medical offices, like those of gynecologists and other doctors, may employ physician assistants. See the Student Resources on thePoint® for a description of this career.

A & P in Action Revisited: Jessica and Brett Discuss IVF

Dr. Christensen greeted the anxious couple as they waited in his office to get the results of Jessica's hysterosalpingoscopy.

"It appears that your uterine tubes have been slightly scarred and possibly blocked, perhaps because of your endometriosis," he told Jessica. "The good news is that your uterus looks healthy. I didn't see any fibroids or malformations that would interfere with your ability to carry a fetus to term. I suggest as a next step a laparoscopic examination to obtain a clearer diagnosis and, if possible, to repair any damage. If necessary, we can talk about further steps leading to pregnancy, including inducing ovulation and intrauterine insemination."

"Well, we're a little impatient at this point" said Brett. "Suppose we need to move on to in vitro fertilization. Can you tell us more about that?"

"Well first, Jessica, you would go on a schedule of hormone supplements to induce the ripening of multiple eggs," the doctor explained. "These eggs are then removed from your ovaries prior to ovulation. They are fertilized with Brett's sperm in a laboratory and then transferred into your uterus, bypassing the uterine tubes. This may have to be done more than once. I'll give you some information on costs and reimbursements for this procedure."

The couple thanked Dr. Christensen for his guidance and said they would think about their options.

Jessica's case illustrates a clinical condition associated with the female reproductive system and some procedures available to sidestep infertility. **BOX 21-1** in the next chapter has more information on assisted reproductive techniques currently in use.

CHAPTER 20

Chapter Wrap-Up

OVERVIEW

A detailed chapter outline with space for note-taking is on thePoint®. The figure below illustrates the main topics covered in this chapter.

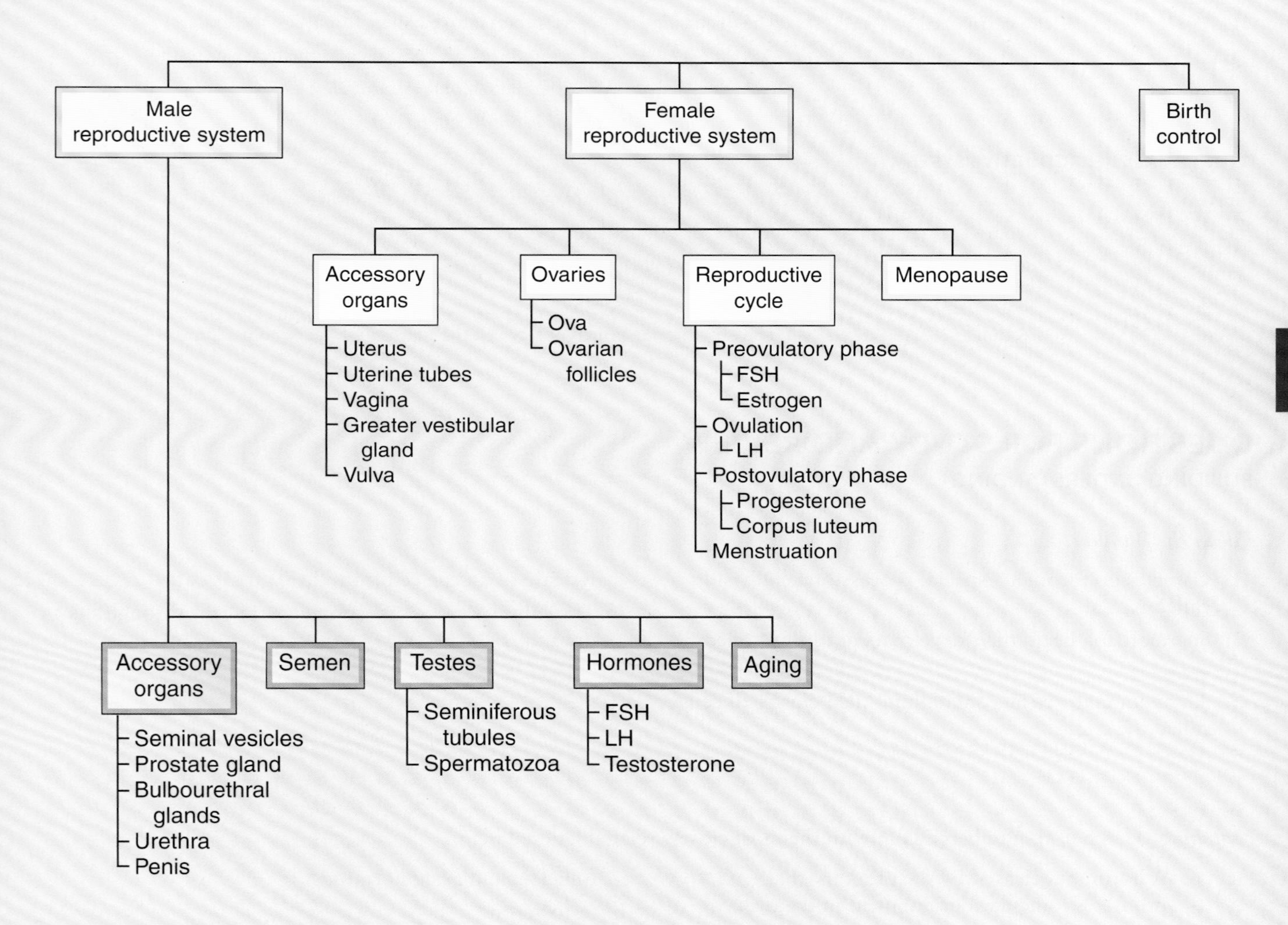

KEY TERMS

The terms listed below are emphasized in this chapter. Knowing them will help you organize and prioritize your learning. These and other boldface terms are defined in the Glossary with phonetic pronunciations.

corpus luteum
endometrium
estrogen
follicle-stimulating hormone (FSH)
gamete
gonadotropin-releasing hormone (GnRH)
infertility
luteinizing hormone (LH)
menopause
menses
menstruation
oocyte
ovarian follicle
ovary
ovulation
ovum (pl., ova)
progesterone
semen
spermatozoon (pl., spermatozoa)
testis (pl., testes)
testosterone
uterus

WORD ANATOMY

Medical terms are built from standardized word parts (prefixes, roots, and suffixes). Learning the meanings of these parts can help you remember words and interpret unfamiliar terms.

WORD PART	MEANING	EXAMPLE
THE MALE REPRODUCTIVE SYSTEM		
acr/o	extremity, end	The *acrosome* covers the head of a sperm cell.
circum-	around	A cut is made around the glans to remove part of the foreskin in a *circumcision*.
fer	to carry	The ductus *deferens* carries spermatozoa away from (de-) the testis.
gon	seed	A *gonad* is an organ that produces sex cells, such as the ovary or testis.
semin/o	semen, seed	Sperm cells are produced in the *seminiferous* tubules.
test/o	testis	The hormone *testosterone* is produced in the testis.
THE FEMALE REPRODUCTIVE SYSTEM		
metr/o	uterus	The *myometrium* is the muscular (my/o) layer of the uterus.
oo	ovum	An *oocyte* is a immature ovum
ovar, ovari/o	ovary	An *ovarian* follicle encloses an ovum.
ov/o, ov/i	egg	An *ovum* is an egg cell.
rect/o	rectum	The *rectouterine* pouch is between the uterus and the rectum.

QUESTIONS FOR STUDY AND REVIEW

BUILDING UNDERSTANDING

Fill in the Blanks

1. Gametes go through a special process of cell division called _____.
2. Spermatozoa begin their development in tiny coiled _____.
3. An ovum matures in a small fluid-filled cluster of cells called the _____.
4. Release of an egg cell from the ovary is called _____.
5. The substance that transports sperm cells through the urethra is _____.

Matching Match each numbered item with the most closely related lettered item.

___ 6. A hormone released by the pituitary that promotes follicular development in the ovary

___ 7. A hormone released by developing follicles that promotes thickening of the endometrium

___ 8. A hormone released by the pituitary that stimulates ovulation

___ 9. A hormone released by the corpus luteum that peaks after ovulation

___ 10. The main male sex hormone

a. follicle-stimulating hormone
b. testosterone
c. estrogen
d. luteinizing hormone
e. progesterone

Multiple Choice

___ 11. What structure does the fetal testis travel through to descend into the scrotum?
a. spermatic cord
b. inguinal canal
c. seminiferous tubule
d. vas deferens

___ 12. Where are the enzymes that help the sperm cell penetrate the ovum located?
a. acrosome
b. head
c. midpiece
d. flagellum

___ 13. Which structure contains the male urethra?
a. corpus cavernosum
b. corpus spongiosum
c. vas deferens
d. seminiferous tubules

___ 14. What structure(s) suspend(s) the uterus and ovaries in the pelvic cavity?
a. uterine tubes
b. broad ligaments
c. fimbriae
d. fornix

___ 15. What is the term for the area between the vaginal opening and the anus?
a. vestibule
b. vulva
c. hymen
d. perineum

___ 16. What phase of the uterine cycle is brought on by low estrogen and progesterone levels?
a. luteal phase
b. menstrual phase
c. proliferative phase
d. secretory phase

UNDERSTANDING CONCEPTS

17. Referring to the male and female reproductive systems in *The Body Visible* at the beginning of the book, give the name and number of.
    a. the duct that delivers semen to the urethra
    b. the section of the penis that surrounds the urethra
    c. a gland that secretes mucus into the urethra in males
    d. structure formed by the follicle after ovulation
    e. extensions of the uterine tube that sweep an ovum into the tube
    f. organ of sexual stimulation in females
    g. portion of the recess around the cervix
18. Compare and contrast the following terms:
    a. asexual reproduction and sexual reproduction
    b. spermatozoa and ova
    c. sustentacular cell and interstitial cell
    d. ovarian follicle and corpus luteum
    e. myometrium and endometrium
    f. contraceptive and abortifacient
19. Trace the pathway of sperm from the site of production to the urethra.
20. Referring to the Dissection Atlas in Appendix 3, name the four parts of the male urethra.
21. Describe the components of semen, their sites of production, and their functions.
22. List the hormones that control male reproduction, and state their functions.
23. Trace the pathway of an ovum from the site of production to the site of implantation.
24. Beginning with the first day of the menstrual flow, describe the events of one complete cycle, including the role of the hormones involved.
25. Describe methods of contraception that involve (1) barriers, (2) chemicals, (3) hormones, and (4) prevention of implantation.

CONCEPTUAL THINKING

26. Jodie, a middle-aged mother of three, is considering a tubal ligation, a contraceptive procedure that involves cutting the uterine tubes. Jodie is worried that this might cause her to enter early menopause. Should she be worried? Explain.
27. Some body builders consume large quantities of testosterone in order to increase their muscle mass. A common side effect of this regime is infertility. Use your knowledge of negative feedback and sperm development to explain why this is so.
28. In the opening case study, Jessica has fertility problems, possibly related to endometriosis. The word ending -sis means "condition of." What are the other word parts in this term and what do they mean?

For more questions, see the Learning Activities on thePoint®.

CHAPTER 21

Development and Heredity

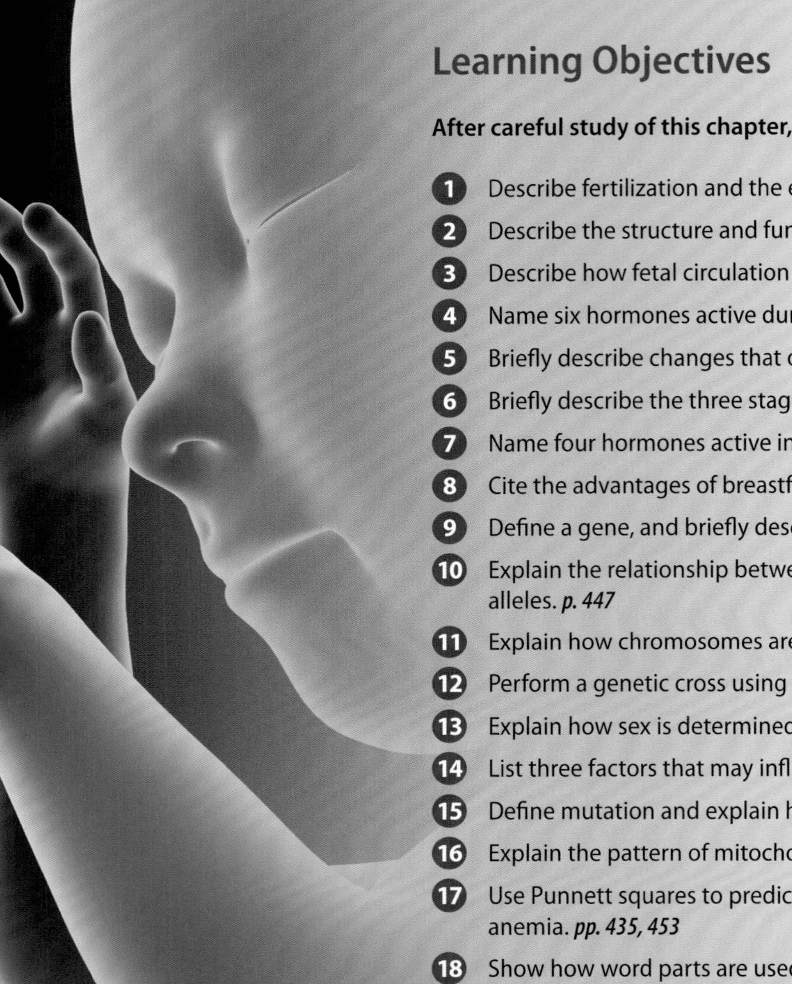

Learning Objectives

After careful study of this chapter, you should be able to:

1. Describe fertilization and the early development of the fertilized egg. ***p. 436***
2. Describe the structure and function of the placenta. ***p. 436***
3. Describe how fetal circulation differs from adult circulation. ***p. 436***
4. Name six hormones active during pregnancy, and describe the function of each. ***p. 438***
5. Briefly describe changes that occur in the embryo, fetus, and mother during pregnancy. ***p. 439***
6. Briefly describe the three stages of labor. ***p. 442***
7. Name four hormones active in lactation, and describe the action of each. ***p. 445***
8. Cite the advantages of breastfeeding. ***p. 446***
9. Define a gene, and briefly describe how genes function. ***p. 446***
10. Explain the relationship between genotype and phenotype for dominant and recessive alleles. ***p. 447***
11. Explain how chromosomes are distributed in meiosis. ***p. 448***
12. Perform a genetic cross using a Punnett square. ***p. 448***
13. Explain how sex is determined in humans and how sex-linked traits are inherited. ***p. 449***
14. List three factors that may influence the expression of a gene. ***p. 450***
15. Define mutation and explain how mutations can affect phenotype. ***p. 452***
16. Explain the pattern of mitochondrial inheritance. ***p. 452***
17. Use Punnett squares to predict the risk of inheriting genetic disorders, such as sickle cell anemia. ***pp. 435, 453***
18. Show how word parts are used to build words related to development and heredity (see "Word Anatomy" at the end of the chapter). ***p. 455***

A & P in Action
Baby Cole's Hereditary Blood Disorder

Both baby Cole and his mother Jada were napping in the maternity ward. Although Cole was only a few hours old, he had already practiced nursing with his mom and been bathed, fed, and given a thorough examination by his pediatrician, including a hearing test and Apgar score. As part of the hospital's routine neonatal procedures, a heel stick was done to obtain a sample of Cole's blood for testing in the hematology lab.

In the lab, Andrea, the technologist, ran a panel of tests to screen for congenital hypothyroidism, PKU, and a number of other hereditary metabolic diseases, as required by the state. When Cole's blood screen showed positive for abnormal hemoglobin S, Andrea confirmed her suspicions by examining the blood cells directly under the microscope.

"Looks like your patient Cole has sickle cell anemia," she reported to Dr. Hoffman, Cole's pediatrician. In response, the physician tracked down the hospital's hematologist, and together they went to the Armstrongs' room to tell them about the lab results. Jada and her husband Carl were surprised to see the two physicians enter.

Dr. Hoffman sat down with them for a moment and then explained. "We don't mean to frighten you, but Cole's lab tests have shown some abnormalities. It appears that he has a genetic disorder called sickle cell anemia. The hematologist is here with me to tell you more about it."

Dr. Evans introduced herself and continued, "Cole's disease is caused by a mutation in the gene that directs the manufacture of the blood's hemoglobin. You both carry two versions of the gene, one normal and one abnormal. Neither of you shows the disease because your normal gene masks the effect of the abnormal gene. But Cole inherited the abnormal copy of the gene from each of you and thus shows sickle cell anemia. That means, his red cells can take on a crescent shape when they are stressed, blocking blood flow and the delivery of oxygen to his tissues."

Seeing the parents' look of alarm, she tried to reassure them. "Starting immediately, a medical team will work with you to manage Cole's disease and minimize tissue damage and pain."

At first, Jada and Carl's attention was focused only on their child and the care he would need. Then, they began to wonder how they had passed on the disease and whether or not other children they might have would be at risk.

In this chapter, we'll learn pregnancy and childbirth, as well as genetic inheritance. We'll also go with the Armstrongs to visit a genetic counselor and find out the chances of their passing a disease gene to another child.

As you study this chapter, CasePoints will give you opportunities to apply your learning to this case.

Visit thePoint® to access the following resources. For guidance in using these resources most effectively, see pp. xv–xvii.

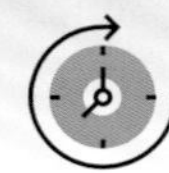

Preparing to Learn

- Tips for Effective Studying
- Pre-Quiz

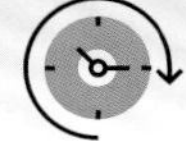

While You Are Learning

- Web Figure: Meiosis in Male and Female Gamete Formation
- Web Figure: Apgar Scoring System
- Web Chart: Placental Hormones
- Animation: Ovulation and Fertilization
- Animation: Fetal Circulation
- Animation: Single Gene Inheritance
- Chapter Notes Outline
- Audio Pronunciation Glossary

When You Are Reviewing

- Answers to Questions for Study and Review
- Health Professions: Midwives and Doulas
- Interactive Learning Activities

A LOOK BACK

This chapter concludes our discussion of the reproductive system by discussing the production of offspring. We reference the key idea of communication 11 in our discussion of the many hormones involved in pregnancy, childbirth, and lactation. Revisiting the key idea of gradients 5 will help you understand how substances move between the maternal and fetal circulation. Negative feedback mechanisms control many aspects of reproduction 3. However, childbirth and lactation represent a different type of feedback process—positive feedback—a system that accelerates rather than reverses an action. A thorough understanding of the female and male reproductive systems, as discussed in Chapter 20, will help you understand the topics in this chapter. We also rely on a key theme of Chapter 3, genes and proteins 9, that explains how genes determine hereditary traits. We examine this theme at multiple levels of organization 2, ranging from individual genes to entire individuals. The key ideas of causation 12 and mass balance 13 are also applied to the topics of genetics.

Pregnancy

Pregnancy begins with fertilization of an ovum and ends with childbirth. During this approximately 38-week period of development, known as **gestation** (jes-TA-shun), a single fertilized egg divides repeatedly, and the new cells differentiate into all the tissues of the developing offspring. Along the way, many changes occur in the mother as well as her baby. **Obstetrics** (ob-STET-riks) (OB) is the branch of medicine that is concerned with the care of women during pregnancy, childbirth, and the six weeks after childbirth. A physician who specializes in obstetrics is an *obstetrician*.

FERTILIZATION AND THE START OF PREGNANCY

When semen is deposited in the vagina, the millions of spermatozoa immediately wriggle about in all directions, some traveling into the uterus and uterine tubes. If an oocyte (immature egg cell) is present in the uterine tube, many spermatozoa cluster around it (FIG. 21-1). Enzymes released from acrosomes on the sperm cells dissolve the coating around the ovum so that the plasma membranes of the ovum and one sperm cell can fuse. The nuclei of the sperm and egg then combine in **fertilization** (see BOX 21-1 for information on assisted reproduction).

The result of this union is a single cell called a **zygote** (ZI-gote), which has the full human chromosome number of 46. The zygote divides rapidly into two cells and then four cells and soon forms a ball of identical cells called a **morula** (MOR-u-lah). During this time, the cell cluster is traveling toward the uterine cavity, pushed along by cilia lining the oviduct and by peristalsis (contractions) of the tube. Before it reaches the uterus, the morula develops into a partially hollow structure called a **blastocyst** (BLAS-to-sist). The blastocyst then burrows into the greatly thickened uterine lining and is soon implanted and completely covered. After **implantation** in the uterus, a group of cells within the blastocyst, called the **inner cell mass**, becomes an **embryo** (EM-bre-o), the term used for the growing offspring in the early stage of gestation. The rest of the blastocyst cells, known as **trophoblasts**, will differentiate into tissue that will support the developing **fetus** (FE-tus), the growing offspring from the beginning of the third month of gestation until birth.

THE PLACENTA

For a few days after implantation, the embryo gets nourishment from its surrounding fluids. As it grows, the embryo's increasing needs are met by the **placenta** (plah-SEN-tah), a flat, circular organ that consists of a spongy network of blood-filled sinuses and capillary-containing villi (FIG. 21-2A). (*Placenta* is from a Latin word meaning "pancake.") The placenta consists of both maternal and embryonic tissue. The maternal portion is simply a well-vascularized internal portion of the endometrium named the **decidua** (de-SID-u-ah). The embryonic portion (derived from trophoblasts) is called the **chorion** (KO-re-on), which forms projections called *chorionic villi*. The chorionic villi break down the endometrial tissue, creating a network of venous sinuses filled with maternal blood. The placenta is the organ of nutrition, respiration, and excretion for the developing offspring throughout gestation. 5 Although the blood of the mother and her offspring ideally do not mix—each has its own blood and cardiovascular system—diffusion takes place between maternal blood in the sinuses and embryonic blood in the capillaries of the chorionic villi. In this manner, gases (CO_2 and O_2) are exchanged, nutrients are provided to the developing offspring, and waste products diffuse into the maternal blood to be eliminated.

The Umbilical Cord The embryo is connected to the developing placenta by a stalk of tissue that eventually becomes the **umbilical** (um-BIL-ih-kal) **cord**. This structure carries blood to and from the embryo, later called the fetus. The cord encloses two arteries that carry blood low in oxygen from the fetus to the placenta and one vein that carries blood high in oxygen from the placenta to the fetus (see FIG. 21-2). (Note that, like the pulmonary vessels, these arteries carry blood low in oxygen, and this vein carries blood high in oxygen.)

Fetal Circulation The fetus has special circulatory adaptations to carry blood to and from the umbilical cord and to bypass its nonfunctional lungs (FIG. 21-2B). A small amount of the oxygen-rich blood traveling toward the fetus in the **umbilical vein** is delivered directly to the liver. However, most of the blood bypasses the liver and is added to the oxygen-poor blood in the inferior vena cava through a small vessel, the **ductus venosus**. Although mixed, this blood still contains enough oxygen to nourish

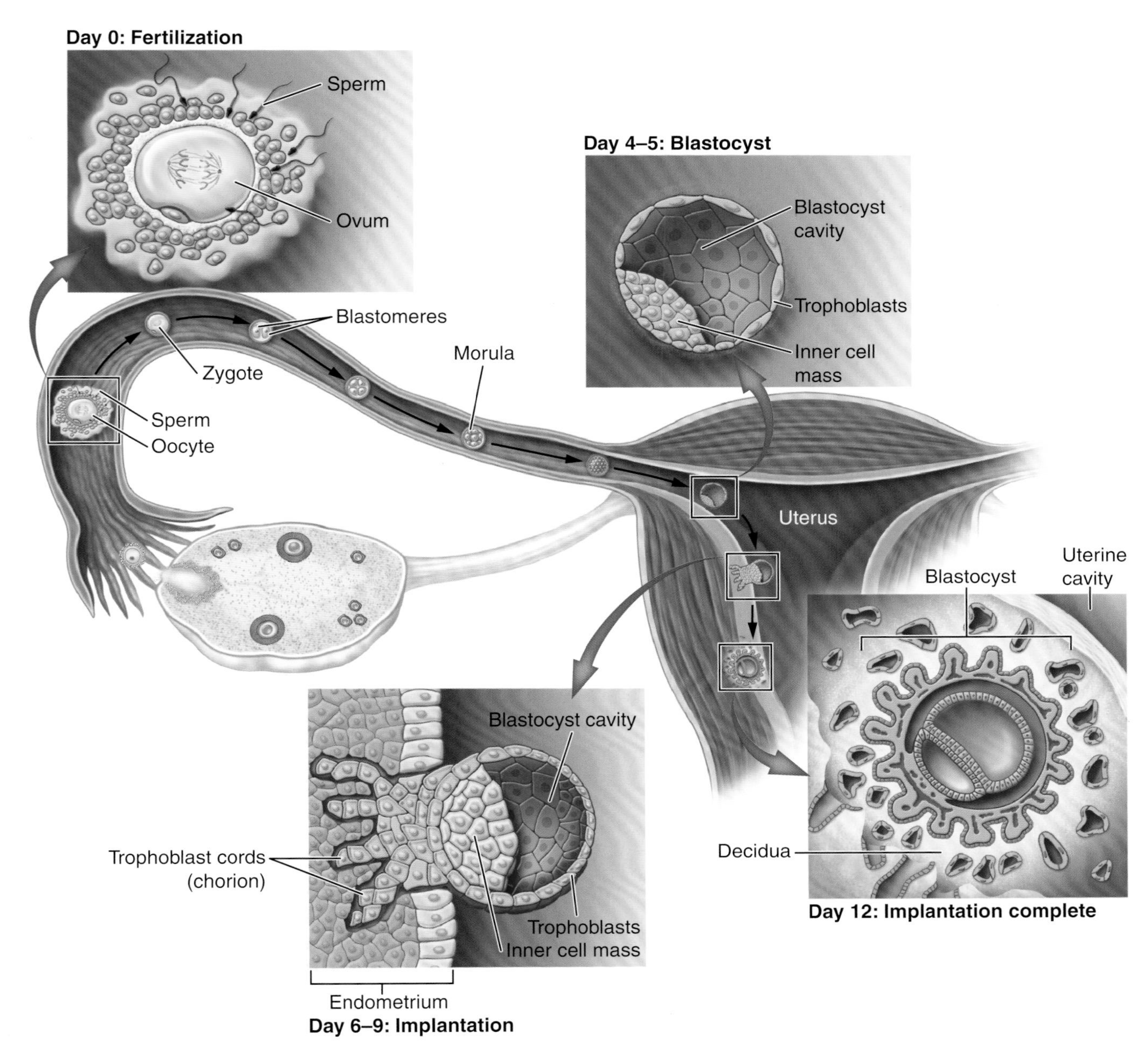

FIGURE 21-1 Fertilization and implantation. **KEY POINT** The union of oocyte and sperm produces a zygote, which develops into a blastocyst and implants into the endometrium. **ZOOMING IN** Where is the ovum fertilized?

fetal tissues. Once in the right atrium, most of the blood flows directly into the left atrium through a small hole in the atrial septum, the **foramen ovale** (o-VA-le). This blood has bypassed the right ventricle and the pulmonary circuit. Blood that does enter the right ventricle is pumped into the pulmonary artery. Although a small amount of this blood goes to the lungs, most of it shunts directly into the systemic circuit through a small vessel, the **ductus arteriosus**, which connects the pulmonary artery to the descending aorta. After traveling throughout fetal tissue, blood returns to the placenta to pick up oxygen through the two **umbilical arteries**.

After birth, when the baby's lungs are functioning, these adaptations begin to close. The foramen ovale gradually seals, and the various vessels constrict into fibrous cords, usually within minutes after birth (only the proximal parts of the umbilical arteries persist as arteries to the urinary bladder).

CASEPOINT

☐ **21-1** How do Cole's red cells pick up oxygen while he is developing in utero?

See the Student Resources on thePoint® to view the animations "Ovulation and Fertilization" and "Fetal Circulation."

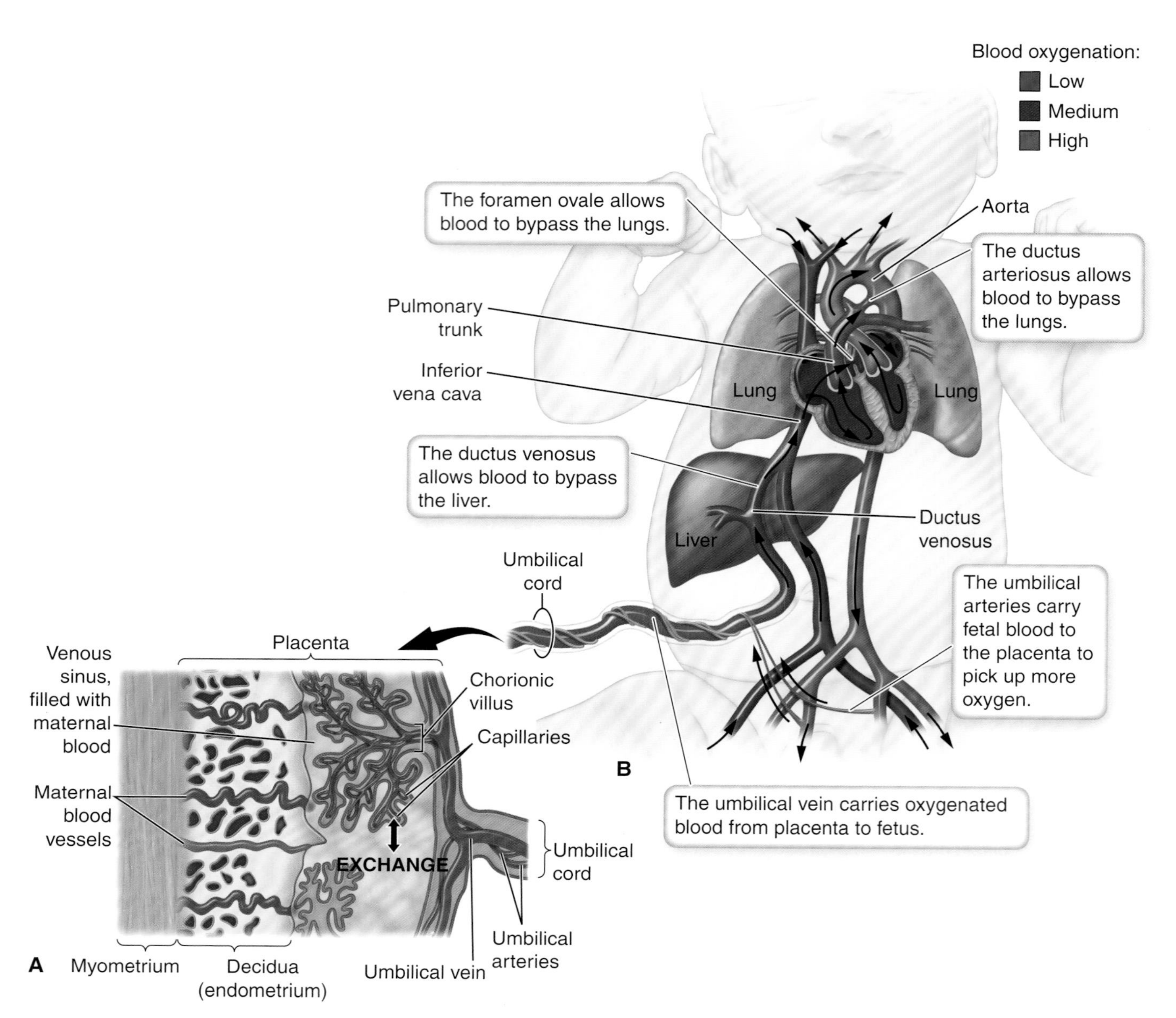

FIGURE 21-2 The placenta and the fetal circulation. KEY POINT The placenta acts as the organ of gas exchange and waste removal for the fetus. **A.** The placenta. The placenta is formed by the maternal endometrium and fetal chorion. Exchanges between fetal and maternal blood occur through the capillaries of the chorionic villi. **B.** Fetal circulation. During fetal life, specialized vessels and openings shunt blood away from nonfunctioning lungs and toward the placenta. Colors show relative oxygen content of blood. ZOOMING IN What is signified by the purple color in this illustration?

HORMONES AND PREGNANCY

Beginning soon after implantation, the blastocyst produces the hormone **human chorionic gonadotropin** (ko-re-ON-ikgon-ah-do-TRO-pin) (hCG). Because hCG production occurs only in fetal tissues, the presence of this hormone in blood or urine is used as an indicator of pregnancy. 11 hCG is very similar in structure to luteinizing hormone (LH) and can bind to LH receptors. By activating these receptors, hCG stimulates the growth and activity of the ovarian corpus luteum. The corpus luteum continues to grow and produce increasing amounts of progesterone and estrogen for about 11 or 12 weeks postfertilization, at which point it degenerates. By this time, the placenta has developed the capacity to secrete adequate amounts of progesterone and estrogen. Miscarriages (loss of an embryo or fetus) are most likely to occur during this critical time when hormone secretion is shifting from the corpus luteum to the placenta.

Progesterone is essential for the maintenance of pregnancy. It promotes endometrial secretions to nourish the embryo, maintains the endometrium, and decreases the uterine muscle's ability to contract, thus preventing the embryo from being expelled from the body. During pregnancy, progesterone also helps prepare the breasts for milk secretion. Estrogen promotes enlargement of the uterus and breasts.

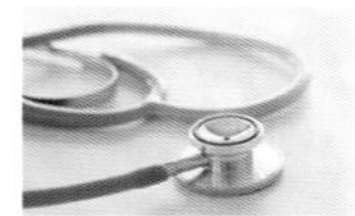

CLINICAL PERSPECTIVES
Assisted Reproductive Technology: The "ART" of Conception

BOX 21-1

At least 1 in 10 American couples is affected by infertility. Assisted reproductive technologies such as in vitro fertilization (IVF), gamete intrafallopian transfer (GIFT), and zygote intrafallopian transfer (ZIFT) can help these couples become pregnant.

In vitro fertilization refers to fertilization of an ovum outside the mother's body in a laboratory dish (in vitro means "in glass"). It is often used when a woman's uterine tubes are blocked or when a man has a low sperm count. The woman participating in IVF is given hormones to cause ovulation of several ova. These are then withdrawn with a needle and fertilized with the father's sperm. After a few divisions, some of the fertilized ova are placed in the uterus, thus bypassing the uterine tubes. Additional fertilized ova can be frozen to repeat the procedure in case of failure or for later pregnancies.

GIFT can be used when the woman has at least one normal uterine tube and the man has an adequate sperm count. As in IVF, the woman is given hormones to cause ovulation of several ova, which are collected. Then, the ova and the father's sperm are placed into the uterine tube using a catheter. Thus, in GIFT, fertilization occurs inside the woman, not in a laboratory dish.

ZIFT is a combination of both IVF and GIFT. Fertilization takes place in a laboratory dish, and then the zygote is placed into the uterine tube.

Because of a lack of guidelines or restrictions in the United States in the field of assisted reproductive technology, some problems have arisen. These issues concern the use of stored embryos and gametes, use of embryos without consent, and improper screening for disease among donors. In addition, the implantation of more than one fertilized ovum has resulted in a high incidence of multiple births, even up to seven or eight offspring in a single pregnancy, a situation that imperils the survival and health of the babies.

The placenta also produces a group of hormones that are very similar to pituitary growth hormone and bind to growth hormone receptors. One of these hormones is so similar that it is named placental growth hormone. Another is human chorionic somatomammotropin (hCS), also known as human placental lactogen (hPL). These hormones increase fatty acids and glucose availability for the fetus by modifying maternal metabolism.

Relaxin is a placental hormone that softens the cervix and relaxes the sacral joints and the pubic symphysis. These changes help widen the birth canal and aid in birth.

See the Student Resources on thePoint® for a summary chart on placental hormones.

CHECKPOINTS 

- ☐ **21-1** What structure is formed by the union of an ovum and a spermatozoon?
- ☐ **21-2** What structure nourishes the developing fetus?
- ☐ **21-3** What is the function of the umbilical cord?
- ☐ **21-4** Fetal circulation is adapted to bypass what organs?
- ☐ **21-5** What embryonic hormone maintains the corpus luteum early in pregnancy?

DEVELOPMENT OF THE EMBRYO

The developing offspring is referred to as an embryo for the first 8 weeks of life (**FIG. 21-3**), and the study of growth during this period is called **embryology** (em-bre-OL-o-je). The beginnings of all body systems are established during this time. The heart and the brain are among the first organs to develop. A primitive nervous system begins to form in the third week. The heart and blood vessels originate during the second week, and the first heartbeat appears during week 4 at the same time other muscles begin to develop.

By the end of the first month, the embryo is approximately 0.62 cm (0.25 in) long with four small swellings at the sides called **limb buds**, which will develop into the four extremities. At this time, the heart produces a prominent bulge at the anterior of the embryo.

By the end of the second month, the embryo takes on an appearance that is recognizably human. In male embryos, the primitive testes have formed and have begun to secrete testosterone, which will direct formation of the male reproductive organs as gestation continues. **FIGURE 21-4** shows photographs of embryonic and early fetal development. A developing embryo is especially sensitive to harmful substances and poor maternal nutrition during this early stage of pregnancy when so many important developmental events are occurring.

DEVELOPMENT OF THE FETUS

During the period of fetal development, from the beginning of the third month until birth, the organ systems continue to grow and mature. The ovaries form in the female early in this fetal period, and at this stage, they contain all the primitive cells (oocytes) that can later develop into mature ova (eggs).

The entire gestation period may be divided into three equal segments or **trimesters**. The fetus's most rapid growth occurs during the second trimester (months four to six). By the end of the fourth month, the fetus is almost 15 cm (6 in) long, and its external genitalia are sufficiently developed to reveal its sex. By the seventh month, the fetus is usually approximately 35 cm (14 in) long and weighs approximately 1.1 kg (2.4 lb). At the end of pregnancy, the normal length of the fetus is 45 to 56 cm (18 to 22.5 in), and the weight varies from 2.7 to 4.5 kg (6 to 10 lb).

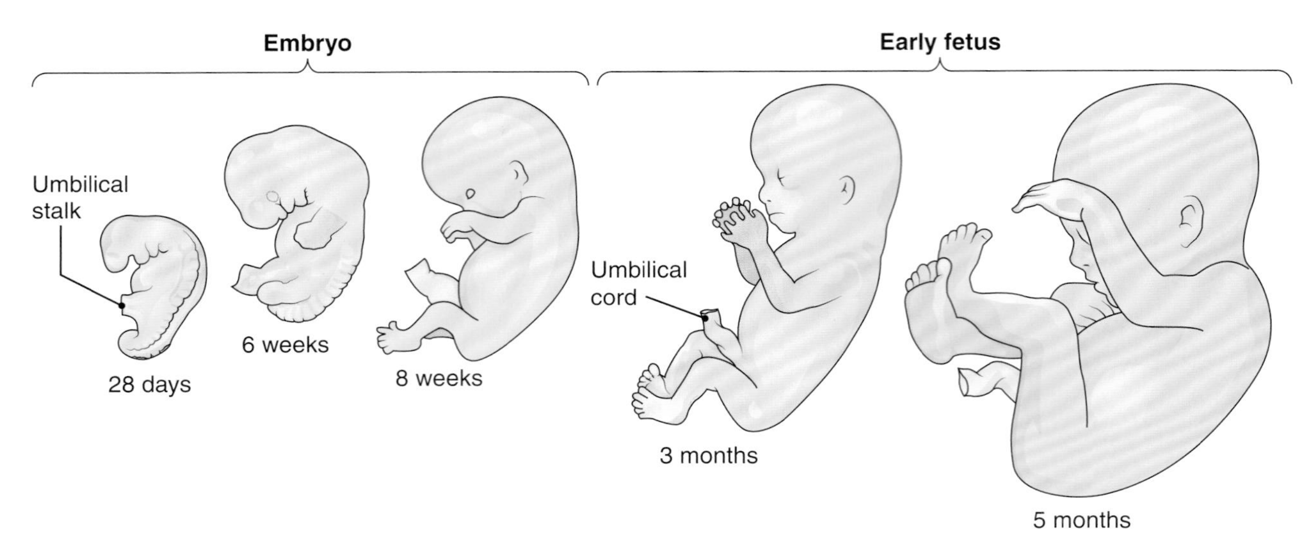

FIGURE 21-3 Development of an embryo and early fetus.

The **amniotic** (am-ne-OT-ik) **sac**, which is filled with a clear liquid known as **amniotic fluid**, surrounds the fetus and serves as a protective cushion for it (FIG. 21-5). The amniotic sac ruptures at birth, an event marked by the common expression that the mother's "water broke."

During development, the fetal skin is protected by a layer of cheeselike material called the **vernix caseosa** (VER-nikska-se-O-sah) (literally, "cheesy varnish").

THE MOTHER

The total period of pregnancy, from fertilization of the ovum to birth, is approximately 266 days, also given as 280 days or 40 weeks from the last menstrual period (LMP). During this time, the mother must supply all the food and oxygen for the fetus and eliminate its waste materials. To support the additional demands of the growing fetus, the

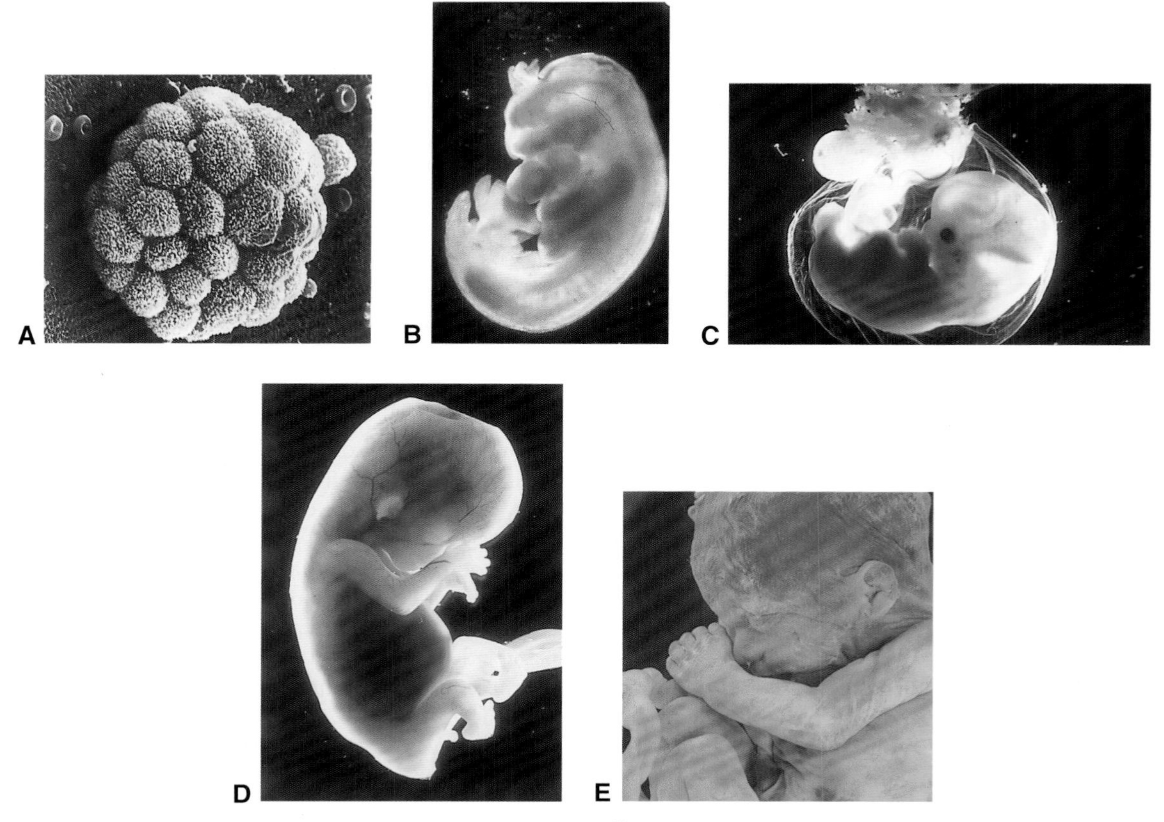

FIGURE 21-4 Human embryos at different stages and early fetus. KEY POINT The embryonic stage lasts for the first two months of pregnancy. Thereafter, the developing offspring is called a fetus. **A.** Implantation in uterus seven to eight days after conception. **B.** Embryo at 32 days. **C.** At 37 days. **D.** At 41 days. **E.** Fetus between 12 and 15 weeks.

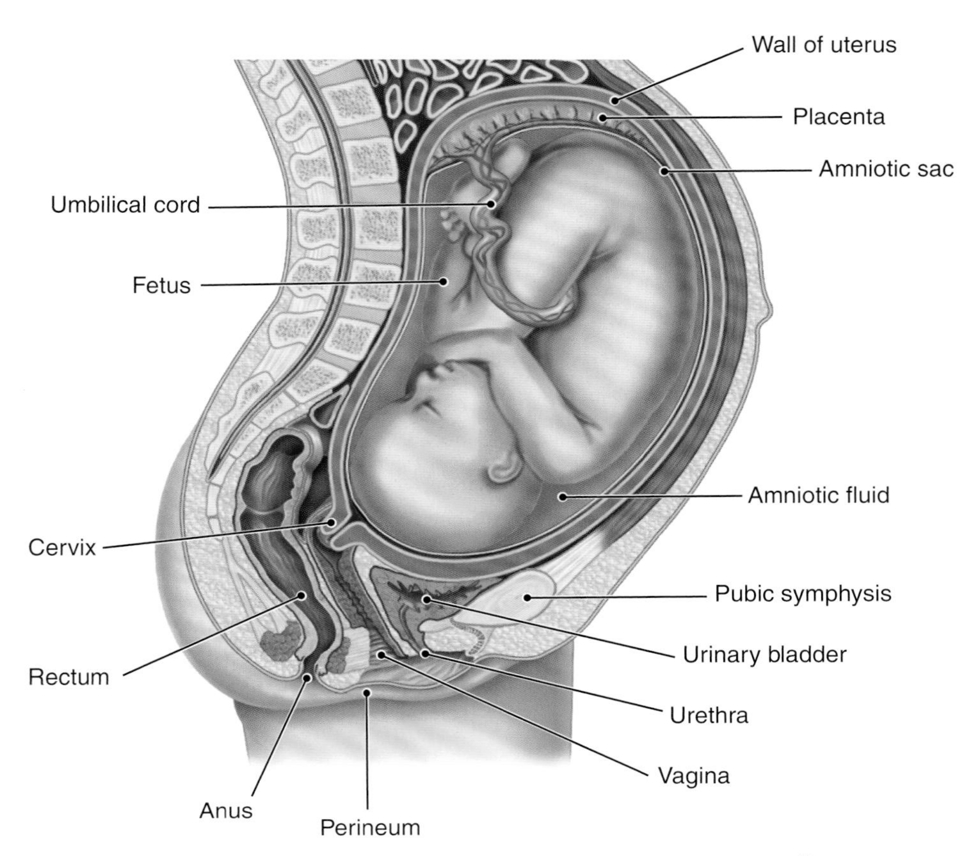

FIGURE 21-5 Midsagittal section of a pregnant uterus with intact fetus. KEY POINT The fetus is contained in an amniotic sac filled with amniotic fluid. ZOOMING IN What structure connects the fetus to the placenta?

mother's metabolism changes markedly, and several organ systems increase their output:

- The cardiac output and blood volume increase in order to deliver enough blood to the uterus and the fetus.
- The respiratory system provides more oxygen and eliminates more carbon dioxide by increasing the rate and depth of respiration.
- The kidneys excrete nitrogenous wastes from both the fetus and the mother.
- Food intake increases, so that the digestive system can supply additional nutrients for the growth of maternal organs (uterus and breasts) and fetal growth, as well as for subsequent labor and milk secretion.

Nausea and vomiting are common discomforts in early pregnancy. These most often occur upon arising or during periods of fatigue and are more common in women who smoke. The specific cause of these symptoms is not known, but they may be due to the great changes in hormone levels that occur at this time. The nausea and vomiting usually last for only a few weeks to several months but may persist for the entire gestational period.

Urinary frequency and constipation are often present during the early stages of pregnancy and then usually disappear. They may reappear late in pregnancy as the fetus's head drops from the abdominal region down into the pelvis, pressing on the urinary bladder and the rectum.

THE USE OF ULTRASOUND IN OBSTETRICS

Ultrasonography (ul-trah-son-OG-rah-fe) is a safe, painless, and noninvasive method for studying soft tissue. It has proved extremely valuable for monitoring pregnancies and childbirth.

An ultrasound image, called a *sonogram*, is made by sending high-frequency sound waves into the body (FIG. 21-6). Each time a wave meets an interface between two tissues of different densities, an echo is produced. An instrument called a *transducer* converts the reflected sound waves into electrical energy, and a computer is used to generate an image on a viewing screen.

Ultrasound scans can be used in obstetrics to diagnose pregnancy, judge fetal age, and determine the location of the placenta.

CHECKPOINTS 

- ☐ **21-6** At about what time in gestation does the heartbeat first appear?
- ☐ **21-7** What is the name of the fluid-filled sac that holds the fetus?
- ☐ **21-8** What is the approximate duration of pregnancy in days from the time of fertilization?

21

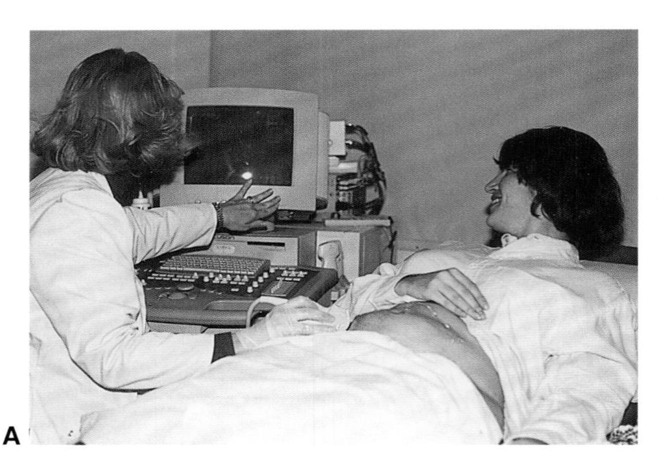

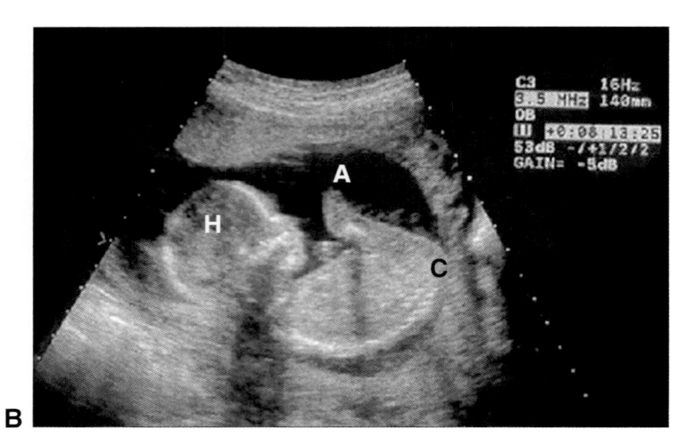

FIGURE 21-6 Sonography. KEY POINT Ultrasound is used to monitor pregnancy and childbirth. **A.** A sonogram is recorded as a mother views her baby on the monitor. **B.** Sonogram of a pregnant uterus at 10 to 11 weeks showing the amniotic cavity (*A*) filled with amniotic fluid. The fetus is seen in longitudinal section showing the head (*H*) and coccyx (*C*).

Childbirth

The exact mechanisms that trigger the beginning of uterine contractions for childbirth are still not completely known. Some fetal and maternal factors that probably work in combination to start labor are as follows:

- Stretching of the uterine muscle stimulates production of prostaglandins, which promote uterine contractions.
- Pressure on the cervix from the baby stimulates release of **oxytocin** (ok-se-TO-sin) from the posterior pituitary. Oxytocin stimulates uterine contractions, and the uterine muscle becomes increasingly sensitive to this hormone late in pregnancy.
- Changes in the placenta that occur with time may contribute to the start of labor.
- Cortisol from the fetal adrenal cortex inhibits the mother's progesterone production. Increase in the relative amount of estrogen as compared to progesterone stimulates uterine contractions.

POSITIVE FEEDBACK AND OXYTOCIN

The events that occur in childbirth involve **positive feedback**, a type of feedback mechanism we have not yet discussed. 3 Positive feedback is much less common than the negative feedback illustrated many times in previous chapters and is not involved in homeostasis. Whereas negative feedback reverses a condition to bring it back to a norm, positive feedback intensifies a response. (Think of a fist fight escalating among a group of people.) Activity continues until resources are exhausted, the stimulus is removed, or some outside force interrupts the activity. (In our fight analogy, people tire, the original fighters are separated, or the police arrive to break things up!) However, positive feedback still involves the same elements as negative feedback, including a regulated variable, sensors, a control center, signals, and effectors. The diagrams in FIGURE 21-7 illustrate positive feedback.

During childbirth, for example, cervical stretching activates stretch receptors in the uterine wall. These receptors send signals to the posterior pituitary (the control center) to release oxytocin. This hormone stimulates further contractions of the uterine muscle (the effector). As contractions increase in force, the cervix is stretched even more, causing further release of oxytocin. The escalating contractions and hormone release continue until the baby is born and the stimulus is removed. Oxytocin in the form of the trade name drug Pitocin can be used to hasten labor, as was about to happen in Amy's case study during childbirth.

THE THREE STAGES OF LABOR

The process by which the fetus is expelled from the uterus is known as **parturition** (par-tu-RISH-un) or *labor* (FIG. 21-8). It is divided into three stages:

1. The *first stage* begins with the onset of regular uterine contractions and dilation of the cervix. With each contraction, the cervix becomes thinner and the cervical canal larger. Rupture of the amniotic sac may occur at any time, with a release of fluid from the vagina.
2. The *second stage* begins when the cervix is completely dilated and ends with the baby's birth. This stage involves the passage of the fetus, usually head first, through the cervical canal and the vagina to the outside (see FIG. 21-8B and C). To prevent tissues of the pelvic floor from being torn during childbirth, as often happens, the obstetrician may cut the mother's perineum just before her infant is born; such an operation is called an **episiotomy** (eh-piz-e-OT-o-me). The area between the vagina and the anus that is cut in an episiotomy is referred to as the surgical or obstetrical perineum. (see FIG. 20-7 in Chapter 20.)
3. The *third stage* begins after the child is born and ends with the expulsion of the *afterbirth*, that is, the placenta, the membranes of the amniotic sac, and the umbilical cord, except for a small portion remaining attached to the baby's umbilicus (um-BIL-ih-kus) or

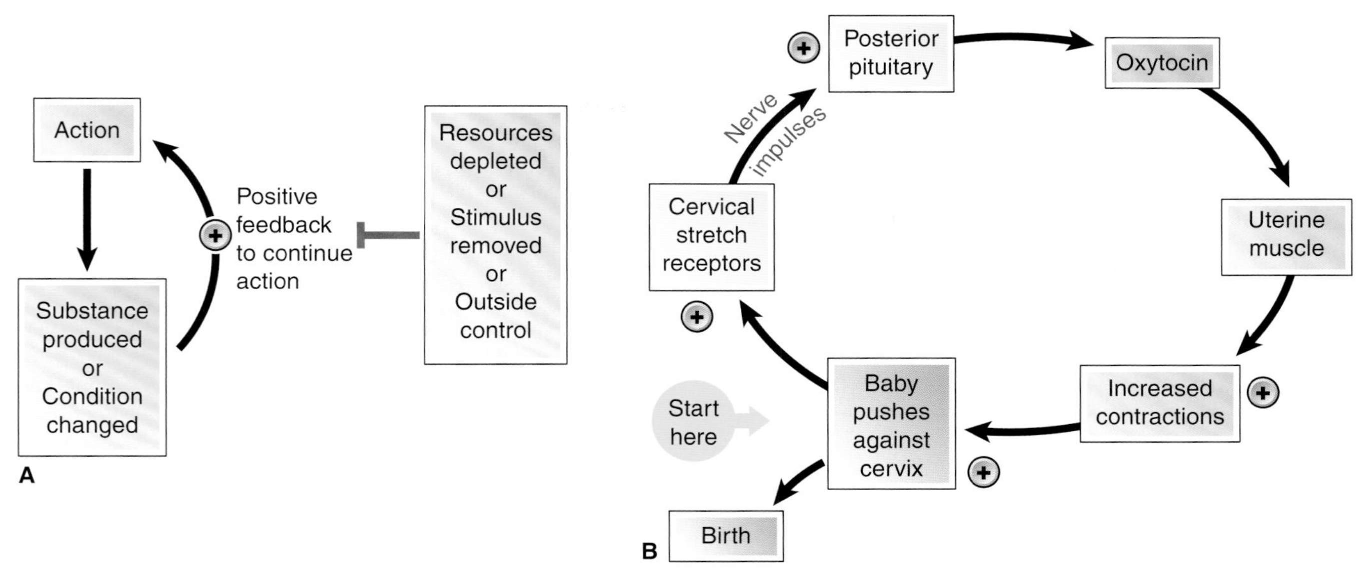

FIGURE 21-7 Positive feedback and childbirth. A. Positive feedback intensifies responses until the feedback loop is interrupted. **B.** Oxytocin is an important chemical signal in the positive feedback loop involved in childbirth. Note that the different components of the feedback loop are indicated by different colors (events relating to the regulated variable [cervical stretch], *green*; sensors and control centers, *yellow*; signals, *purple*; effectors, *blue*). ZOOMING IN What is the effector organ in the feedback loop illustrated in part B?

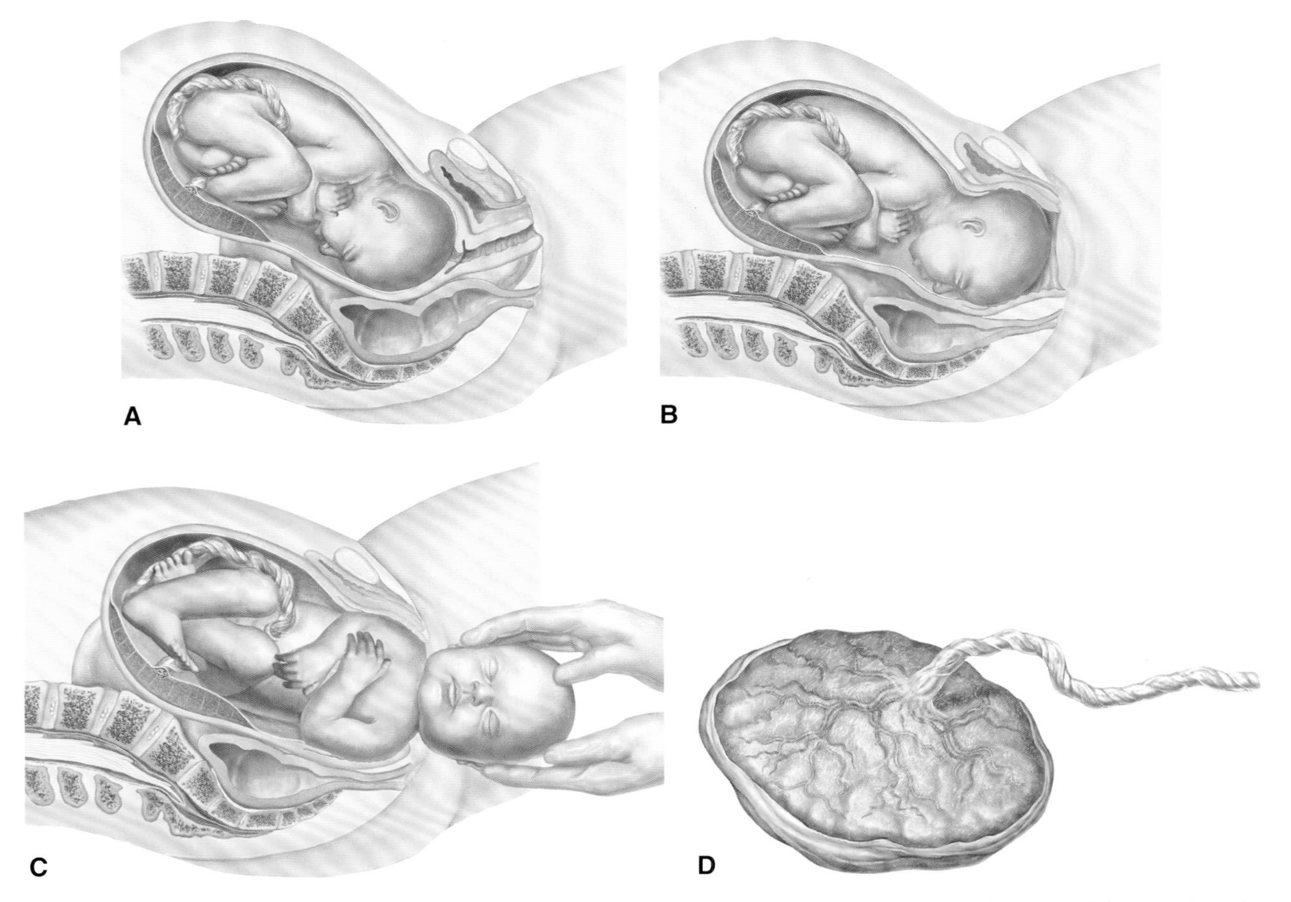

FIGURE 21-8 Stages of labor. The first stage **(A)** begins with the onset of uterine contractions; the second stage **(B,C)** begins when the cervix is completely dilated and ends with birth of the baby; the third stage **(D)** ends with expulsion of the afterbirth.

21

navel (see FIG. 21-8D). Contraction of the uterine muscle acts to close off the blood vessels leading to the placental site. Nursing a newborn after birth stimulates oxytocin release and helps to decrease postpartum bleeding.

Once labor is complete, the health professionals ensure that the placenta is intact and that the bleeding is controlled. If an episiotomy was performed, the obstetrician now repairs this clean cut.

See the Student Resources on thePoint® for information on birth assistants, midwives, and doulas.

CESAREAN SECTION

A **cesarean** (se-ZAR-re-an) **section** (C-section) is an incision made in the abdominal wall and uterine wall through which the fetus is manually removed from the mother's body. A cesarean section may be required for a variety of reasons, including placental abnormalities, abnormal fetal position, disproportion between the head of the fetus and the mother's pelvis that makes vaginal birth difficult or dangerous, and other problems that may arise during pregnancy and labor.

MULTIPLE BIRTHS

Until recently, statistics indicated that twins occurred in about one of every 80 to 90 births, varying somewhat in different countries. Triplets occurred much less frequently, usually once in several thousand births, whereas quadruplets occurred very rarely. The birth of quintuplets represented a historic event unless the mother had taken fertility drugs. Now these fertility drugs, usually gonadotropins, are given more commonly, and the number of multiple births has increased significantly. Multiple fetuses tend to be born prematurely and therefore have a high death rate. However, better care of infants and newer treatments have resulted in more living multiple births than ever.

Twins originate in two different ways and on this basis are divided into two types (FIG. 21-9):

- **Fraternal twins** result from the fertilization of two different ova by two spermatozoa. Two completely different individuals, as genetically distinct from each other as brothers and sisters of different ages, are produced. Each fetus has its own placenta and surrounding amniotic sac. Fraternal twins are described as *dizygotic*, as they develop from two (di-) separate zygotes.
- **Identical twins** develop from a single zygote formed from a single ovum fertilized by a single spermatozoon. Sometime, during the early stages of development, the embryonic cells separate into two units. Depending on the time of separation, the twins may or may not share a single placenta. However, there must always be a separate umbilical cord for each fetus. Identical twins are always the same sex and carry the same inherited traits. Identical twins are described as *monozygotic*, as they develop from a single (mono-) zygote.

Other multiple births may be fraternal, identical, or combinations of these. The tendency for fraternal, but not identical, multiple births seems to be hereditary.

PREGNANCY OUTCOMES

An infant born between 37 and 42 weeks of gestation is described as a **term infant**. However, a pregnancy may end before its full term has been completed. The term **live birth** is used if the baby breathes or shows any evidence of life such as heartbeat, pulsation of the umbilical cord, or movement of voluntary muscles. Infants born before the 37th week of gestation are considered **preterm**. Often these babies have low birth weights of less than 2,500 g (5.5 lb) and are immature (premature) in development. They are subject to a number of medical conditions, including anemia, jaundice, respiratory problems, and feeding difficulties. The underlying causes of preterm labor are not well understood, but risk factors include trauma, infections, poor maternal health, and chronic health conditions.

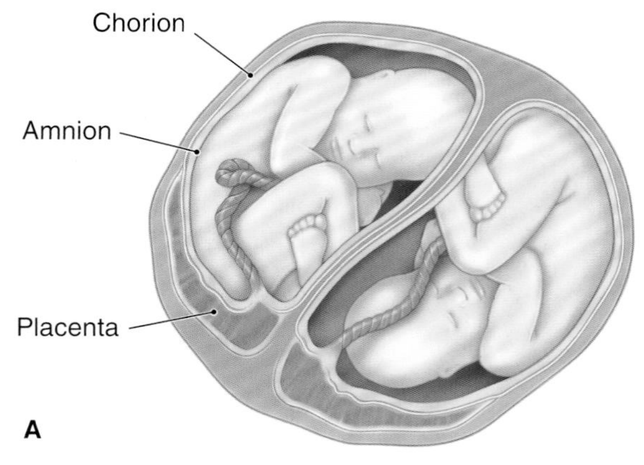

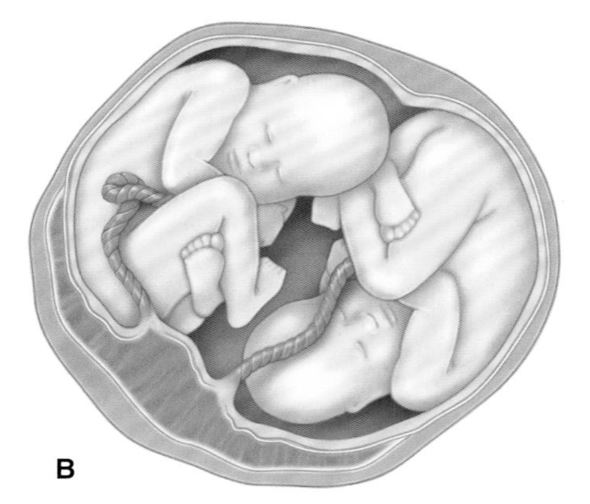

FIGURE 21-9 Twins. A. Fraternal twins. **B.** Identical twins. ZOOMING IN Which type of twins may share a single placenta?

Loss of the fetus is classified according to the duration of the pregnancy:

- The term **abortion** refers to loss of the embryo or fetus before the 20th week or weight of about 500 g (1.1 lb). This loss can be either spontaneous or induced.
  - **Spontaneous abortion,** commonly known as a **miscarriage,** occurs naturally with no interference. It is estimated that 22% of pregnancies end in spontaneous abortion. About 50% of these cases reflect a chromosomal abnormality in the embryo or fetus. Other causes include insufficient progesterone production by the corpus luteum or a structural abnormality of the mother's reproductive organs. Acute maternal infections can spread from the blood or from the urinary tract to infect the uterine cavity and cause miscarriage. Finally, chronic disorders, such as uncontrolled diabetes mellitus, are associated with increased risk of spontaneous abortion.
  - **Induced abortion** occurs as a result of deliberate interruption of pregnancy. A **therapeutic abortion** is an abortion performed by a physician as a treatment for a variety of reasons. More liberal access to this type of abortion has dramatically reduced the incidence of death related to illegal abortion.
- The term **fetal death** refers to loss of the fetus after the 20th week of pregnancy. **Stillbirth** refers to the birth of an infant who is lifeless. The loss of a fetus often indicates an infection or interference with the fetal blood supply by cord compression or placental failure.

Immaturity is a leading cause of death in the newborn. After the 20th week of pregnancy, the fetus is considered **viable**, that is, able to live outside the uterus. A fetus expelled before the 24th week or before reaching a weight of 1,000 g (2.2 lb) has little more than a 50% chance of survival; one born at a point closer to the full 38 weeks of gestation stands a much better chance of living. However, increasing numbers of immature infants are being saved because of advances in neonatal intensive care.

Hospitals use the Apgar score to assess a newborn's health and predict survival. Five features—heart rate, respiratory effort, muscle tone, response to stimulation, and skin coloration— are rated as 0, 1, or 2 at one minute and five minutes after birth. The maximum possible score on each test is 10. Infants with low scores require medical attention and have lower survival rates.

See the Student Resources on thePoint® for the Apgar scoring system.

CHECKPOINTS 

- 21-9 What pituitary hormone stimulates uterine contractions?
- 21-10 What is parturition?
- 21-11 During what stage of labor is the afterbirth expelled?
- 21-12 What surgical procedure is used to remove a fetus manually from the uterus?
- 21-13 What term describes a fetus able to live outside the uterus?

The Mammary Glands and Lactation

The **mammary glands**, contained in the female breasts, are associated organs of the reproductive system. They provide nourishment for the baby after its birth in a process termed **lactation**, or milk production. These exocrine glands are similar in construction to the sweat glands. Each gland is divided into a number of lobes composed of glandular tissue and fat, and each lobe is further subdivided. Secretions from the lobes are conveyed through **lactiferous** (lak-TIF-er-us) **ducts**, all of which converge at the papilla (nipple) (FIG. 21-10).

The mammary glands begin developing during puberty, but they do not become functional until the end of a pregnancy. As already noted, estrogen and progesterone during pregnancy help prepare the breasts for lactation, as does prolactin (PRL) from the anterior pituitary. The first mammary gland secretion is a thin liquid called **colostrum** (ko-LOS-trum). It is nutritious but has a somewhat different composition from milk. Milk secretion begins within a few days following birth and can continue for several years as long as milk is frequently removed by the suckling baby or by pumping. In another example of positive feedback, the nursing infant's suckling at

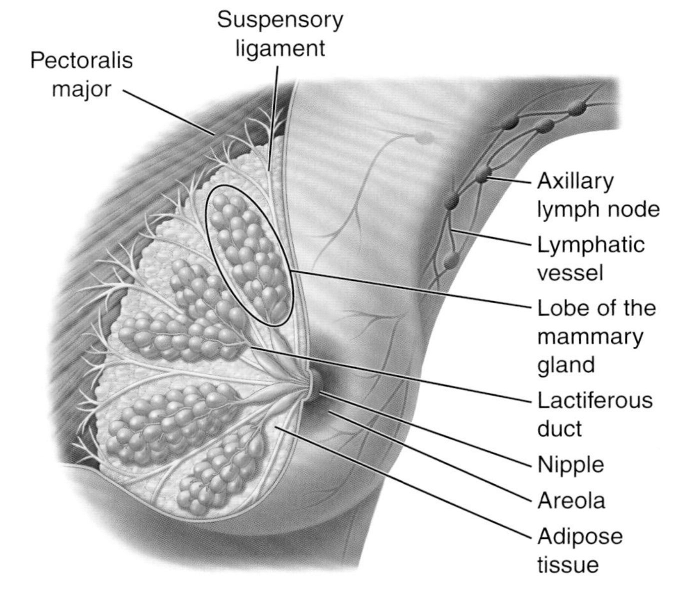

FIGURE 21-10 Section of the breast (mammary gland). KEY POINT The glands are divided into lobes containing milk ducts that converge at the nipple. ZOOMING IN What muscle underlies the breast?

21

the breast promotes release of PRL, which activates the mammary glands' secretory cells. Nursing also stimulates oxytocin release from the posterior pituitary. This hormone causes the milk ducts to contract, resulting in the ejection, or *letdown*, of milk.

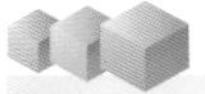
Concept Mastery Alert

Remember that prolactin stimulates milk synthesis, but oxytocin stimulates milk release.

The newborn baby's digestive tract is not ready for the usual adult mixed diet. Mother's milk is more desirable for the young infant than milk from other animals for several reasons, some of which are listed below:

- Infections that may be transmitted by foods exposed to the outside air are avoided by nursing.
- Both breast milk and colostrum contain maternal antibodies that help protect the baby against pathogens.
- The proportions of various nutrients and other substances in human milk are perfectly suited to the human infant, changing within a single feeding from a watery fluid that quenches the infant's thirst to a fluid rich in fat that satisfies the infant's hunger. The composition of breast milk also changes over time as the infant grows. Substitutes are not exact imitations of human milk. Nutrients are present in more desirable amounts if the mother's diet is well balanced.
- Breast milk contains complex carbohydrates that humans cannot digest but promote the growth of beneficial organisms in the intestinal microbiome.
- The psychological and emotional benefits of nursing are of infinite value to both the mother and the infant.

CHECKPOINTS 

- [ ] **21-14** What are the ducts that carry milk out of the breast?
- [ ] **21-15** What substance is secreted by the breast before milk production begins?

Heredity

We are often struck by the resemblance of a baby to one or both of its parents, yet rarely do we stop to consider *how* various traits are transmitted from parents to offspring. This subject—**heredity**—has fascinated humans for thousands of years. The Old Testament contains numerous references to heredity (although the word itself was unknown in ancient times). It was not until the 19th century, however, that methodical investigation into heredity was begun. At that time, an Austrian monk Gregor Mendel used garden peas to reveal a precise pattern in the appearance of characteristics among parents and their **progeny** (PROJ-eh-ne), their offspring, or descendants. Mendel's most important contribution to the understanding of heredity was the demonstration that there are independent units of heredity in the cells. Later, these independent units were given the name **genes**.

Genes and Chromosomes

Genes are actually segments of DNA contained in the threadlike **chromosomes** within the nucleus of each cell. (Only the mature red blood cell, which has lost its nucleus, lacks DNA.) Genes govern cells by controlling the manufacture of proteins, especially enzymes, which are necessary for all the chemical reactions that occur within the cell. They also contain the information to make the proteins that constitute structural materials, hormones, and growth factors. This is a good time to look back at Chapter 3, which has details on chromosomes, genes, and DNA (see FIG. 3-12).

When body cells divide by the process of mitosis, the DNA that makes up the chromosomes is replicated and distributed to the daughter cells, so that each daughter cell gets exactly the same kind and number of chromosomes as were in the original cell (see FIG. 3-15). 2 Each chromosome (aside from the Y chromosome, which determines male sex) may carry thousands of genes, and each gene carries the code for a specific **trait** (characteristic) (**FIG. 21-11**). These traits constitute the physical, biochemical, and physiologic makeup of every cell in the body (**BOX 21-2** on modern studies of the human genetic makeup).

In humans, every nucleated cell except the gametes (reproductive cells) contains 46 chromosomes. The chromosomes exist in pairs. One member of each pair was received at the time of fertilization from the offspring's father, and one was received from the mother. The paired chromosomes, except

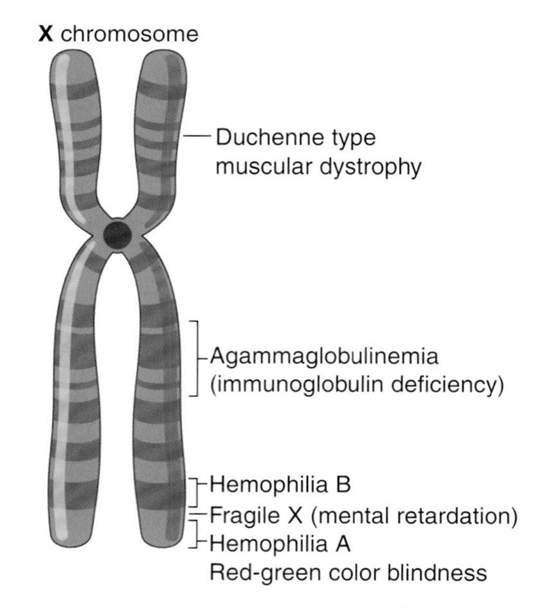

FIGURE 21-11 Genes and chromosomes. **KEY POINT** Genes are segments of DNA located at specific sites on a chromosome. This figure shows some genes on the X chromosome relevant to different diseases.

HOT TOPICS

BOX 21-2

The Human Genome Project: Reading the Book of Life

Packed tightly in nearly every one of your body cells (except the red blood cells) is a complete copy of your genome—the genetic instructions that direct all of your cellular activities. Written in the language of DNA, these instructions consist of genes parceled into 46 chromosomes that code for proteins. In 1990, a consortium of scientists from around the world set out to crack the genetic code and read the human genome, our "book of life." This monumental task, called the Human Genome Project, was completed in 2003 and succeeded in mapping the entire human genome—3 billion DNA base pairs arranged into about 20,000 to 25,000 genes. Now, scientists can pinpoint the exact location and chemical code of every gene in the body.

The human genome was decoded using a technique called sequencing. Samples of human DNA were fragmented into smaller pieces and then inserted into bacteria. As the bacteria multiplied, they produced more and more copies of the human DNA fragments, which the scientists extracted. The DNA copies were loaded into a sequencing machine capable of "reading" the string of DNA nucleotides that composed each fragment. Then, using computers, the scientists put all of the sequences from the fragments back together to get the entire human genome.

Now, scientists hope to use all these pages of the book of life to revolutionize the treatment of human disease. The information obtained from the Human Genome Project may lead to improved disease diagnosis, new drug treatments, and even gene therapy.

for the pair that determines sex, are alike in size and appearance. Thus, each body cell has one pair of sex chromosomes and 22 pairs (44 chromosomes) that are not involved in sex determination and are known as **autosomes** (AW-to-somes).

The paired autosomes carry the genes for a specific trait at the same physical locations, but they might have different versions of that gene (**FIG. 21-11**). The alternative forms of a given gene have slightly different DNA sequences and encode variants of a trait (e.g., curly vs. straight hair). Each version of a specific gene is called an **allele** (al-LEEL).

Concept Mastery Alert

Be careful not to confuse alleles with autosomes.

DOMINANT AND RECESSIVE ALLELES

Because our chromosomes are paired, we have two alleles for every gene. When both the alleles for a trait are the same, the alleles are said to be **homozygous** (ho-mo-ZI-gus); if they are different, they are described as **heterozygous** (het-er-o-ZI-gus). The concept of dominance explains how the two alleles interact to determine a particular trait. A **dominant** allele is one that expresses its effect in the cell regardless of the identity of the other allele. The allele needs to be received from only one parent for the trait to be expressed in the offspring.

The effect of a **recessive** allele is not evident unless its paired allele on the matching chromosome is also recessive. Thus, a recessive trait appears only if the recessive gene versions for that trait are received from both parents. A simple test, often done in biology labs to study genetic inheritance, is a test for the ability to taste phenylthiocarbamide (PTC), a harmless organic chemical not found in food. The ability to detect PTC's bitter taste is inherited as a dominant allele. Inability to taste it is carried by a recessive allele. Thus, nontasting appears in an offspring only if alleles for nontasting are received from both parents. A recessive trait only appears if the person's alleles are homozygous for that trait. In contrast, a dominant trait will appear whether the alleles are homozygous (both dominant) or heterozygous (one of each) because a dominant allele is always expressed if it is present.

Any characteristic that can be observed or can be tested for is part of a person's **phenotype** (FE-no-tipe). Eye color, for example, can be seen when looking at a person. Blood type is not visible but can be determined by testing and is also a part of a person's phenotype. When someone has the recessive phenotype, his or her genetic makeup, or **genotype** (JEN-o-tipe), is obviously homozygous recessive. When a dominant phenotype appears, the person's genotype can be either homozygous or heterozygous dominant, as noted above. Only genetic studies or family studies can reveal which it is.

A recessive allele is not expressed if it is present in a cell together with a dominant allele. However, the recessive allele can be passed onto offspring and may thus appear in future generations. An individual who shows no evidence of a trait but has a recessive allele for that trait is described as a **carrier** of the gene. Using genetic terminology, that person shows the dominant phenotype but has a heterozygous genotype for that trait. Using our PTC tasting example, a taster might be a carrier for the nontasting gene and could pass it on to his or her children.

Concept Mastery Alert

Carriers exist for recessive traits but not for dominant traits.

CASEPOINTS 

- ☐ **21-2** Sickle cell anemia is a recessive trait. How many alleles for the sickle cell trait does Cole have?
- ☐ **21-3** How would you describe his genotype for that trait?
- ☐ **21-4** Neither of Cole's parents has sickle cell anemia. How would you describe their genotype for sickle cell disease?

21

DISTRIBUTION OF CHROMOSOMES TO OFFSPRING

The reproductive cells (ova and spermatozoa) are produced by a special process of cell division called **meiosis** (mi-O-sis) (**FIG. 21-12**). This process divides the chromosome number in half, so that each reproductive cell has 23 chromosomes instead of the 46 in other body cells. The process of meiosis begins, as does mitosis, with replication of the chromosomes (see **FIG. 3-15** in Chapter 3). Then, these duplicated chromosomes line up across the center of the cell. However, instead of a random distribution of the 46 chromosomes, as occurs in mitosis, the chromosome pairs line up at the center of the cell (see **FIG. 21-12**). The first meiotic division (meiosis I) distributes the two members of each chromosome pair into separate cells. Then, a second meiotic division (meiosis II) separates the strands of the duplicated chromosomes and distributes each strand to an individual gamete. 13 Thus, in a single cell, 92 chromosomes enter meiosis, and each of the four gametes receives 23 different chromosomes, each representing one member of a chromosome pair.

Remember that for each chromosome pair, one chromosome was originally inherited from the father and one from the mother. It is very important to note that the separation of the chromosome pairs occurs at random, meaning that each gamete receives a mixture of maternal and paternal chromosomes. This reassortment of chromosomes in the gametes leads to increased variety within the population. (In a human cell with 46 chromosomes, the chances for varying combinations are much greater than in the example shown in **FIG. 21-12**.) Thus, children in a family resemble each other, but no two look exactly alike (unless they are identical twins), because each gamete contained a unique assortment of chromosomes from all four of their grandparents.

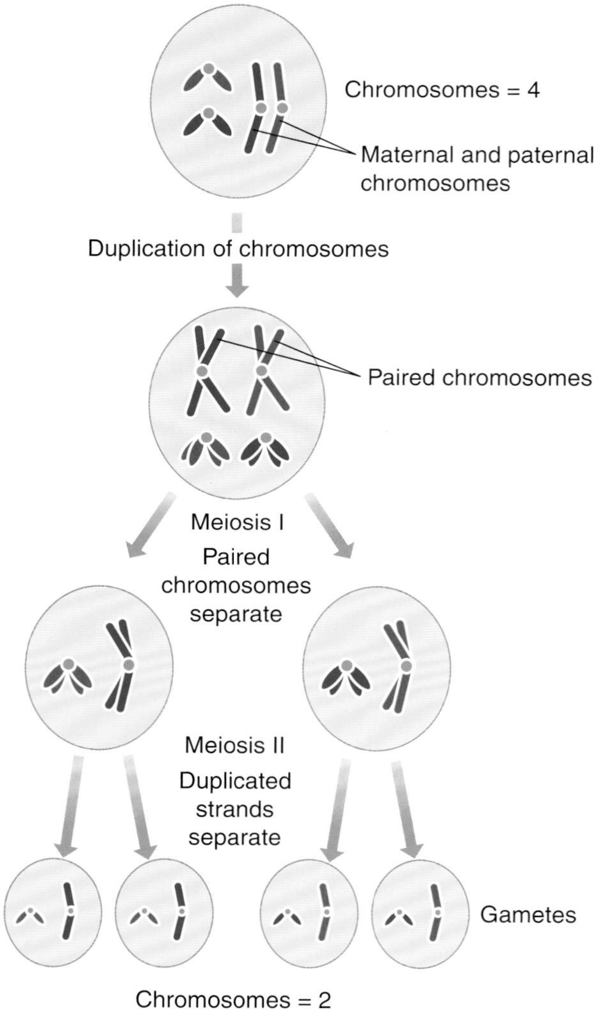

FIGURE 21-12 Meiosis. **KEY POINT** Meiosis halves the chromosome number in formation of gametes. Because the process begins with chromosome replication, two meiotic divisions are needed, first to reduce the chromosome number and then to separate the duplicated strands. The example shows a cell with a chromosome number of four, in contrast to a human cell, which has 46 chromosomes. **ZOOMING IN** How many cells are produced in one complete meiosis?

Concept Mastery Alert

Meiosis is a difficult concept for many students, so take time to study FIG. 21-12. At the end of this complex process, each of the four cells produced from a single cell contains 22 autosomes and one sex chromosome.

See the "Student Resources" on thePoint® for a figure on meiosis in male and female gamete formation.

PUNNETT SQUARES

Geneticists use a grid called a **Punnett square** to show all the combinations of alleles that can result from a given parental cross, that is, a mating that produces offspring (**FIG. 21-13**). In these calculations, a capital letter is used for the dominant allele, and the recessive allele is represented by the lower case of the same letter. For example, if T represents the allele for the dominant trait PTC taster, then t would be the recessive allele for nontaster. In the offspring, the genotype TT is homozygous dominant, and the genotype Tt is heterozygous, both of which will show the dominant phenotype taster. The homozygous recessive genotype tt will show the recessive phenotype nontaster.

A Punnett square shows all the possible gene combinations of a given cross and the theoretical ratios of all the genotypes produced. For example, in the cross shown in **FIGURE 21-13** between two heterozygous parents, the theoretical chance of producing a baby with the genotype TT is 25% (one in four). The chance of producing a baby with the genotype of tt and the recessive phenotype nontaster is also 25%. The chance of producing heterozygous Tt offspring for this trait is 50% (two in four). In all, 75% of the offspring will have the dominant phenotype taster, because they have at least one dominant gene for the trait. In Cole's case, because sickle cell anemia is determined by a recessive

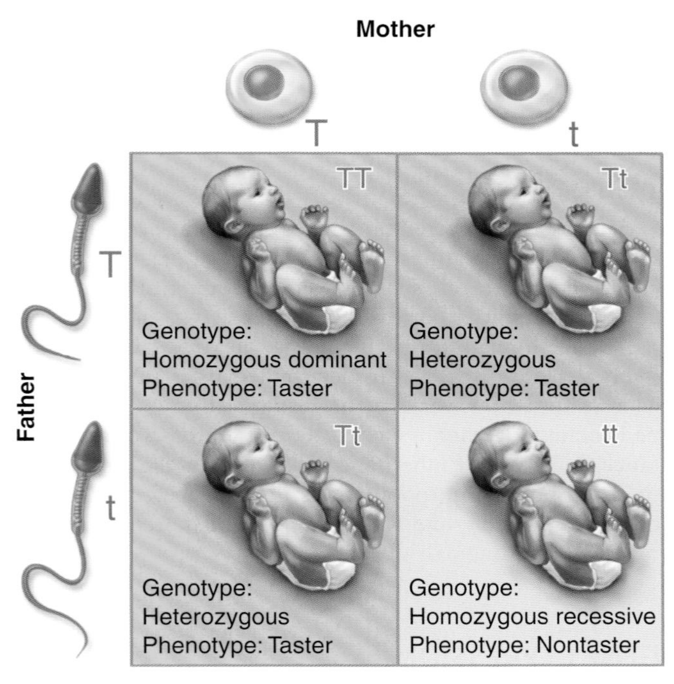

FIGURE 21-13 A Punnett square. **KEY POINT** Geneticists use this grid to show all the possible combinations of a given cross, in this case, Tt × Tt. **ZOOMING IN** What percentage of children from this cross will show the recessive phenotype? What percentage will be heterozygous?

allele, there was a 25% chance that he would have that disorder, because both his parents are carriers of the allele. If his parents have other children, the risk is the same with each birth.

In real life, genetic ratios may differ from the theoretical predictions, especially if the number of offspring is small. For example, the chance of having a male or female baby is 50/50 with each birth for reasons explained shortly, but a family might have several girls before having a boy, and vice versa. The chances of seeing the predicted ratios improve as the number of offspring increases.

CASEPOINTS

- ☐ **21-5** When Jada's gametes underwent meiosis, when did the sickle cell allele separate from the normal allele?
- ☐ **21-6** Based on predicted ratios, what percentage of children in Cole's family will be carriers of the sickle cell allele? What percentage will neither have sickle cell anemia nor be able to pass the recessive allele to offspring?

SEX DETERMINATION

The two chromosomes that determine the offspring's sex, unlike the autosomes (the other 22 pairs of chromosomes), are not matched in size and appearance. The female X chromosome is larger than most other chromosomes and carries genes for other characteristics in addition to that for sex. The male Y chromosome is smaller than other chromosomes and mainly determines sex. A female has two X chromosomes in each body cell; a male has one X from his mother and one Y from his father.

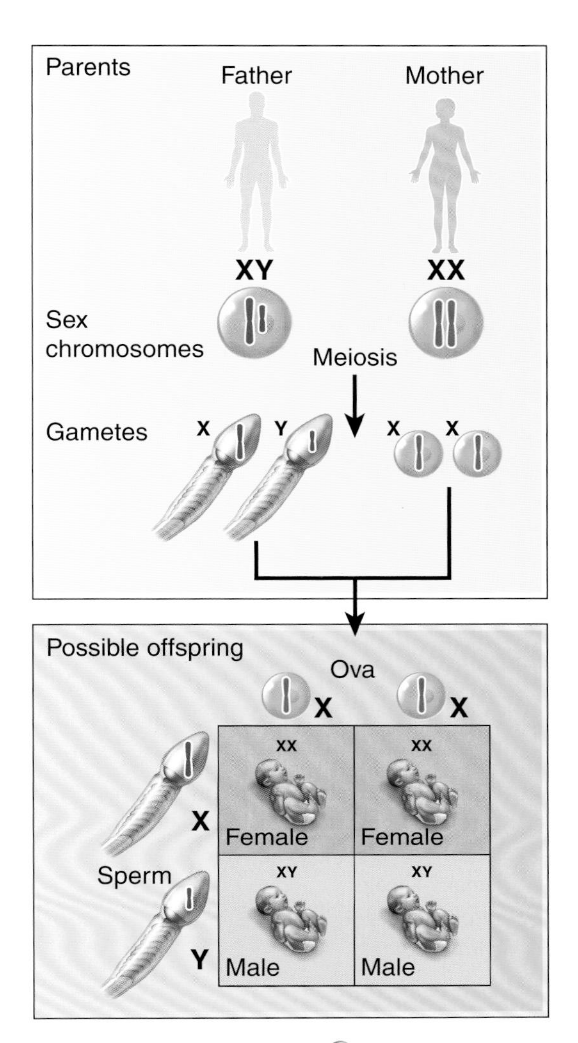

FIGURE 21-14 Sex determination. **KEY POINT** If an X chromosome from a male unites with an X chromosome from a female, the child is female (XX); if a Y chromosome from a male unites with an X chromosome from a female, the child is male (XY). **ZOOMING IN** What is the expected ratio of male to female offspring in a family?

By the process of meiosis, each sperm cell receives either an X or a Y chromosome, whereas every ovum receives only an X chromosome (**FIG. 21-14**). If a sperm cell with an X chromosome fertilizes an ovum, the resulting infant will be female; if a sperm with a Y chromosome fertilizes an ovum, the resulting infant will be male.

SEX-LINKED TRAITS

Any trait that is carried on a sex chromosome is said to be **sex-linked**. Because the Y chromosome carries few traits aside from sex determination, most sex-linked traits are carried on the X chromosome and are best described as *X-linked*. Examples are hemophilia, certain forms of baldness, and red–green color blindness.

Sex-linked traits appear almost exclusively in males. The reason for this is that most of these traits are recessive, and if a recessive allele is located on the X chromosome in a male, it cannot be masked by a matching dominant allele. (Remember that the Y chromosome with which the X chromosome pairs is very small and carries few genes.) Thus, a male who has only one recessive allele for a trait will exhibit that characteristic, whereas a female must have two recessive alleles to show the trait. The female must inherit a recessive allele for that trait from each parent and be homozygous recessive in order for the trait to appear. The daughter of a man with an X-linked disorder is always at least a carrier of the allele for that disorder, because a daughter always receives her father's single X chromosome. So, for example, if you are female and your father is color blind, then each of your sons will have a 50/50 chance of being color blind (see **FIGURE 21-15** for an example of sex-linked inheritance, illustrating a match between a normal male and a carrier female).

CHECKPOINTS 

- ☐ **21-16** What is a gene? What is a gene made of?
- ☐ **21-17** What term describes a gene that always expresses its effect?
- ☐ **21-18** What is the difference between a genotype and a phenotype?
- ☐ **21-19** What is the process of cell division that forms the gametes?
- ☐ **21-20** Human body cells have 46 chromosomes. How many chromosomes are in each gamete?
- ☐ **21-21** What sex chromosome combination determines a female? A male?
- ☐ **21-22** What term describes a trait carried on a sex chromosome?

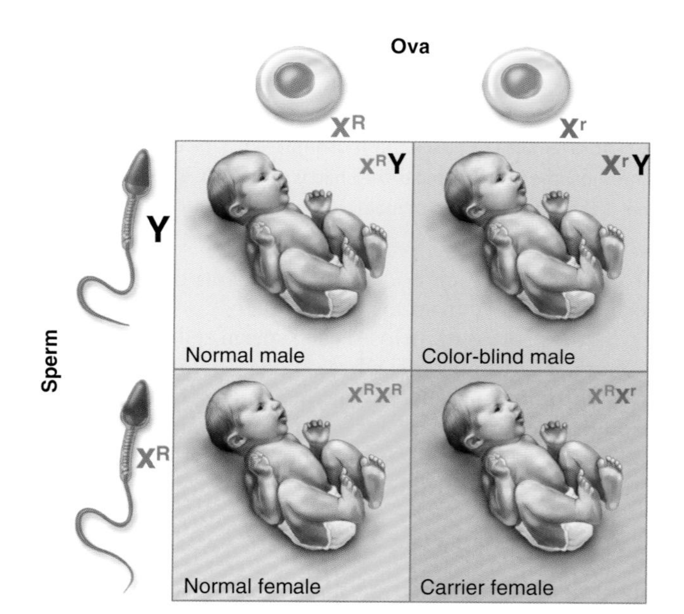

FIGURE 21-15 Inheritance of sex-linked traits. KEY POINT Sex-linked traits, such as color blindness, are encoded by genes on the X chromosome. This Punnett square illustrates the mating between a normal male and a female carrier. The "R" superscript is the dominant (normal) allele, and the "r" superscript represents the recessive allele for color blindness. ZOOMING IN What is the genotype of a carrier female?

See the "Student Resources" on thePoint® for the animation "Single Gene Inheritance."

Hereditary Traits

Some observable hereditary traits are skin, eye, and hair color and facial features. Also influenced by genetics are less clearly defined traits, such as weight, body build, life span, and susceptibility to disease.

Some human traits, including the traits involved in many genetic diseases, are determined by a single pair of genes; most, however, are the result of two or more gene pairs acting together in what is termed **multifactorial inheritance.** This type of inheritance accounts for the wide range of variations within populations in such characteristics as coloration, height, and weight, all of which are determined by more than one pair of genes.

GENE EXPRESSION

The phrase **gene expression** describes if, and to what extent, a gene has an effect on an individual's phenotype. Gene expression may be influenced by a variety of factors, including the individual's sex or the presence of other genes. For example, the genes for certain types of baldness may be inherited by either males or females, but the traits appear mostly in males under the effects of male sex hormone. Gene expression also varies during development and aging. For instance, there are actually two hemoglobin genes. One gene is expressed in the fetus, and a different gene is expressed after birth. Only the adult gene is affected in sickle-cell anemia, so Cole had normal blood cells in utero. See **BOX 21-3** to read about a possible treatment for Cole's disorder involving the fetal hemoglobin gene.

Environment also plays a part in gene expression. For example, a person might inherit the dominant gene for freckles, but the freckles appear only when the skin is exposed to sunlight. In the case of multifactorial inheritance, gene expression might be quite complex. You inherit a potential for a given size, for example, but your actual size is additionally influenced by such factors as nutrition, development, and general state of health. The same is true of life span and susceptibility to diseases.

We should also note here that dominance is not always clear cut. Sometimes, dominant alleles are expressed together as *codominance*. Take the inheritance of blood type, for example. Alleles for type A and type B are dominant over O. A person with the genotype AO is type A; a person with the genotype BO is type B. However, neither A nor B is dominant over the other. If a person inherits an allele for both A and B, his or her blood type is AB. Also, a person heterozygous for a certain trait may show some intermediate expression of the trait. The term for this type of gene interaction is *incomplete dominance*. Cole's disorder, sickle cell anemia, is determined by a recessive allele of the hemoglobin gene. However, a heterozygous carrier may have symptoms of sickle cell anemia under low oxygen conditions.

BOX 21-3

ONE STEP AT A TIME
The Proof is in the Statistics

Cole, the subject of the case study, faces a lifetime of treatments and a shortened life expectancy as a result of his sickle cell anemia. However, scientists are working on a cure—gene therapy—with a goal of producing fully functional hemoglobin and preventing red cell destruction.

How can we tell if the therapy works? The answer lies in statistics. Statistics tell us whether an apparent difference between the control group and the experimental group could have arisen by chance. This box presents results of a study in mice with sickle cell anemia. Scientists may choose to present their results in a table or in a graph, so this box provides strategies to interpret both.

Mice	Hct (%)	Average Hb content/ RBC (MCH, pg)
Normal (control)	44.2 ± 1.0	13.0 ± 0.4
Sickle-cell disorder	28.3 ± 1.9	12.3 ± 0.6
Sickle-cell disorder + gene therapy	46.2 ± 1.4 **	13.8 ± 0.2

A

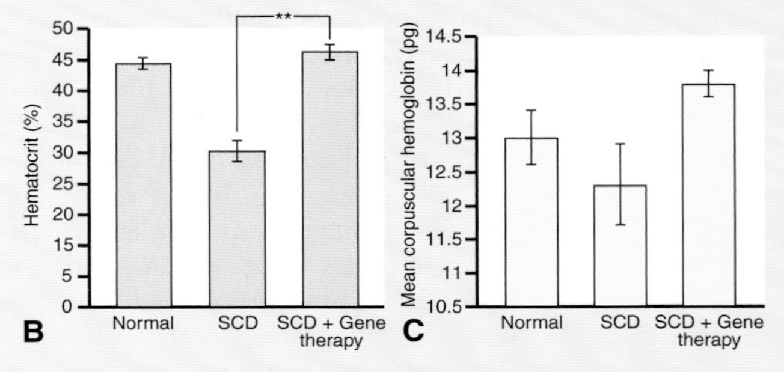

Blood parameters in normal mice, sickle-cell disease (SCD) mice, and SCD mice treated with gene therapy. All values represent means ± standard error of the mean (SEM). **A.** Changes in hematocrit (Hct) and average hemoglobin (Hb) content per red blood cell (mean corpuscular hemoglobin).**, $P < 0.00001$, compared with SCD mice. **B.** Hematocrit values.**, $P < 0.00001$. **C.** Mean corpuscular hemoglobin. (Data taken from: Xu, P. *et al.* Correction of sickle cell disease in adult mice by interference with fetal hemoglobin silencing. *Science*. 2011;334:993-996.)

Question

Is gene therapy effective in mice with sickle cell disorder? If so, which blood parameters are improved?

Answer

Step 1. Look at the mean values. The mean (average) value for each group can be determined by adding up all of the values and then dividing by the number of individuals. In the table, the mean of each value is the number before each plus–minus (±) sign. In the graphs, the mean is represented by the size of each column. Consider first the hematocrit, which indicates the proportion of blood volume taken up by red blood cells. We can see that the hematocrit is much lower in the SCD mice (28.3%) than in the normal mice (44.2%) or in the SCD mice receiving the therapy (46.2%). This finding indicates that fewer red blood cells are destroyed in the treated mice. A similar trend can be seen for the average hemoglobin content (MCH), but the differences are much smaller.

Step 2. Look at the error bars. The error bars in the graphs are the small black brackets at the top of each column. In this case, they represent a value called the *standard error of the mean (SEM)*, which measures the amount of variation in the data from each group. The SEM is the number after the ± sign in the table. For example, the SEM would be very large if individual values in a group were 5, 10, and 15 and very small if the individual values were 9, 10, and 11 (but the means would be identical). A large SEM can indicate that the study did not control enough variables, or it can simply indicate that the parameter varies considerably in the studied population. Compared with the size of the columns, the error bars are quite small in panel B but very large in panel C.

Step 3. Look for significance values. A significance value, such as $P < 0.05$, indicates the probability that the difference between the two groups could have arisen by chance (in this case, less than 5%). Generally, a significance value (also known as a P value) above 0.05 indicates that the difference between the two groups is not significant. In other words, it shows that the treatment did not significantly alter the studied parameter. The lower the P value, the more significant the difference. Tables and graphs often use one or more asterisks (*) to highlight significant differences. Graph B shows a significant difference in hematocrit between the SCD mice and the SCD+ gene therapy. The legend informs us that the probability of this difference arising by chance is less than 0.001%! We do not see a significant difference in MCH between the two SCD groups.

Step 4. Draw conclusions. Our analysis suggests that gene therapy can restore the hematocrit of sickle-cell mice to normal levels. The therapy did not alter the hemoglobin content of individual cells, so it must have increased the number of red blood cells.

See the Study Guide (available separately) for a similar question that you can answer yourself.

21

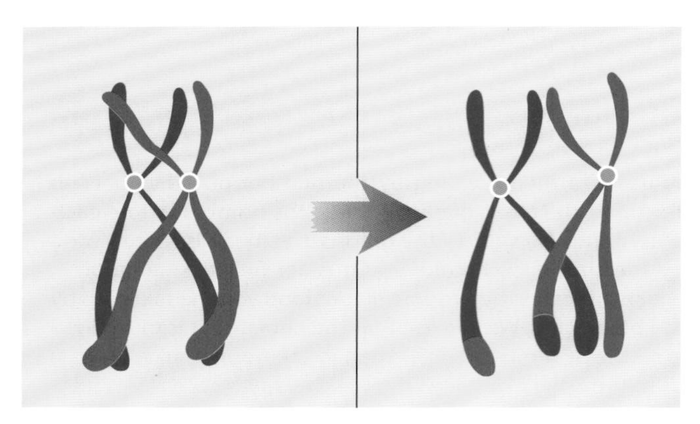

FIGURE 21-16 Genetic exchange. KEY POINT In meiosis, chromosomes tangle and exchange sections. These exchanges may be equal, maintaining a normal genetic makeup. They may also result in losses or duplications known as mutations.

GENETIC MUTATION

Any change in a gene or a chromosome is termed a genetic **mutation** (mu-TA-shun). Mutations may occur spontaneously or may be induced by some agent, such as ionizing radiation or a chemical, described as a **mutagen** (MU-tah-jen) or mutagenic agent. Often, small changes in gene sequence occur as copying errors during DNA synthesis. Larger mutations, in which portions of a chromosome or entire chromosomes are gained or lost, often occur during meiosis as chromosomes come together, reassort, and get distributed to two new cells. The long, threadlike chromosomes can intertwine and exchange segments (FIG. 21-16). Sometimes, these exchanges are equivalent, and the total genetic material remains normal, but sometimes genes are damaged, or there is a loss, duplication, or rearrangement of the genetic material. These changes may involve a single gene, a portion of a chromosome, or whole chromosomes.

If a mutation occurs in an ovum or a spermatozoon, the altered trait may be inherited by an offspring. The vast majority of harmful mutations are never expressed because the affected fetus dies and is spontaneously aborted. Most remaining mutations are so inconsequential that they have no detectable effect. Beneficial mutations, on the other hand, tend to survive and become increasingly common as a population evolves.

MUTATIONS AND PHENOTYPES

12 The different inheritance patterns discussed early in this chapter reflect the nature of the mutation causing the variation among alleles. Most recessive traits result from *loss of function* mutations; that is, the mutated gene encodes a protein that is not produced or that does not carry out its normal function. Heterozygous individuals have one normal copy of the gene and one mutated copy. The normal copy guides the synthesis of enough functional protein to result in the normal phenotype. For instance, a single normal copy of the hemoglobin gene is enough to ensure healthy blood cells, so homozygotes for the normal gene as well as heterozygotes exhibit the normal phenotype. However, individuals with two copies of the sickle cell allele only make abnormal hemoglobin, so they exhibit the sickle cell anemia phenotype.

In contrast, most dominant disorders result from *gain of function* mutations in which the resulting mutated protein exhibits excess activity. Cancer can result from this type of mutation in growth factor receptor genes. The mutated receptor protein is always active, triggering constant and abnormal cell divisions that can lead to cancer.

MITOCHONDRIAL INHERITANCE

Although we have emphasized the nucleus as the repository of genetic material in the cell, the mitochondria also contain some hereditary material. Mitochondria, you will recall, are the cells' powerhouse organelles, converting the energy in nutrients into ATP. Because mitochondria were separate organisms early in the evolution of life and later merged with other cells, they have their own DNA and multiply independently. Mitochondrial DNA can mutate, as does nuclear DNA, and disease can be the result. Such diseases can cause serious damage to metabolically active cells in the brain, liver, muscles, kidneys, and endocrine organs. These mitochondrial genes are passed only from a mother to her sons and daughters because almost all the cytoplasm in the zygote is contributed by the ovum. Because these cells must be smaller and lighter, the head of the spermatozoon carries very little cytoplasm and no mitochondria.

CHECKPOINTS

- 21-23 What is a mutagen?
- 21-24 From which parent does a child get mitochondrial genes?

CASEPOINT

- 21-7 Based on its inheritance pattern, does Cole's sickle cell anemia likely result from a loss-of-function or a gain-of-function mutation?

A & P in Action Revisited: The Armstrongs Visit a Genetic Counselor

Ms. Clarkson, a genetic counselor at the hospital where Cole was born, met with Jada and Carl later in the year to explain again the inheritance of their son's disorder.

"You each carry a recessive sickle cell gene for abnormal hemoglobin, and by pure chance, Cole inherited that gene from each of you, leading to his disorder," Ms. Clarkson explained.

"Where did this gene come from?" Jada asked. "And will all of our children have this disease? Should we stop here with our family?"

"Not necessarily," said Ms. Clarkson. "A mutation might have arisen in either of your reproductive cells, but more likely this trait has been carried somewhere in your families. We can try to develop a family pedigree, asking if there have been any relatives with unexplained illness in the past, but Dr. Hoffman has already reported that your blood tests confirm that you are both carriers."

"So how do the percentages work for future children?" Carl asked.

"Well, there's a one-in-four chance that any of your children will have sickle cell anemia," the counselor said. "But there is a 25% chance with each pregnancy that the child will be totally free of the mutation and a 50% chance with each pregnancy that the child will just be a carrier, like yourselves. Those percentages apply for each birth. Fortunately, there's a prenatal test for sickle cell disease, and you can choose to make family decisions based on those results."

CHAPTER 21

Chapter Wrap-Up

OVERVIEW

A detailed chapter outline with space for note-taking is on thePoint®. The figure below illustrates the main topics covered in this chapter.

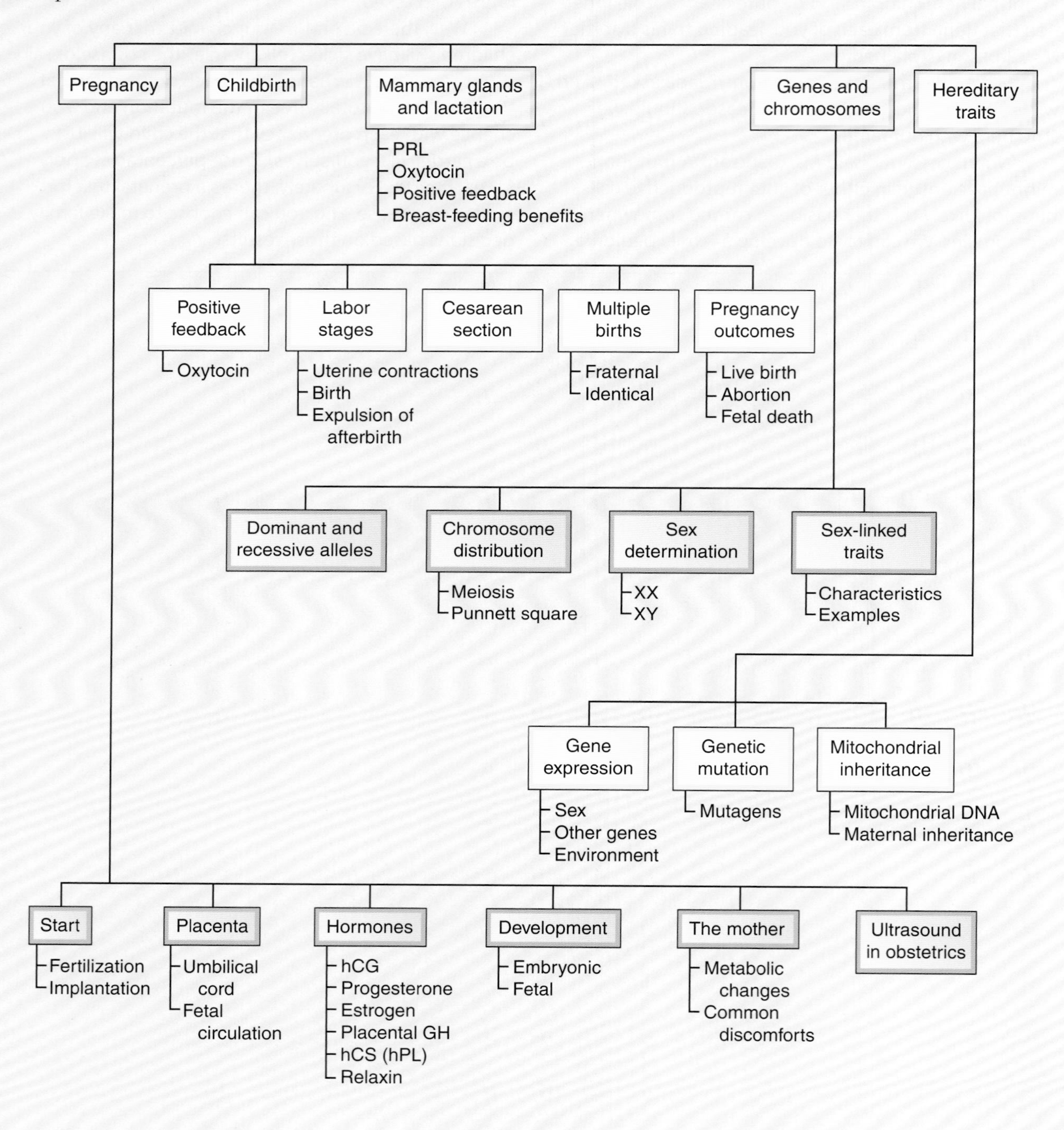

KEY TERMS

The terms listed below are emphasized in this chapter. Knowing them will help you organize and prioritize your learning. These and other boldface terms are defined in the Glossary with phonetic pronunciations.

abortion
allele
amniotic sac
autosome
carrier
chorion
chromosome
colostrum
decidua
dominant
embryo
familial
fertilization
fetus
gene
genetic
genotype
gestation
heredity
heterozygous
homozygous
human chorionic gonadotropin (hCG)
implantation
lactation
meiosis
mutagen
mutation
obstetrics
oxytocin
parturition
phenotype
placenta
progeny
recessive
relaxin
sex-linked
trait
umbilical cord
zygote

WORD ANATOMY

Medical terms are built from standardized word parts (prefixes, roots, and suffixes). Learning the meanings of these parts can help you remember words and interpret unfamiliar terms.

WORD PART	MEANING	EXAMPLE
PREGNANCY		
chori/o	membrane, chorion	Human *chorionic* gonadotropin is produced by the chorion (outermost cells) of the embryo.
zyg/o	joined	An ovum and spermatozoon join to form a *zygote*.
CHILDBIRTH		
ox/y	sharp, acute	*Oxytocin* is the hormone that stimulates labor.
toc/o	labor	See preceding example.
THE MAMMARY GLANDS AND LACTATION		
lact/o	milk	The *lactiferous* ducts carry milk from the mammary glands.
mamm/o	breast, mammary gland	A *mammogram* is radiographic study of the breast.
GENES AND CHROMOSOMES		
aut/o-	self	*Autosomes* are all the chromosomes aside from the two that determine sex.
chrom/o	color	*Chromosomes* color darkly with stains.
heter/o	other, different	*Heterozygous* paired genes (alleles) are different from each other.
homo-	same	*Homozygous* paired genes (alleles) are the same.
phen/o	to show	Traits that can be observed or tested for making up a person's *phenotype*.
HEREDITARY TRAITS		
multi-	many	*Multifactorial* traits are determined by multiple pairs of genes.

QUESTIONS FOR STUDY AND REVIEW

BUILDING UNDERSTANDING

Fill in the Blanks

1. Following ovulation, the mature follicle becomes the ______.
2. The first mammary gland secretion is called ______.
3. The basic unit of heredity is a(n) ______.
4. Chromosomes not involved in sex determination are known as ______.
5. The number of chromosomes in each human body cell is ______.

Matching Match each numbered item with the most closely related lettered item.

___ 6. An embryonic hormone that stimulates the ovaries to secrete progesterone

___ 7. A placental hormone that regulates maternal blood nutrient levels

___ 8. A placental hormone that softens the cervix to widen the birth canal

___ 9. A pituitary hormone that stimulates uterine contractions

___ 10. A pituitary hormone that stimulates maternal milk production

a. placental growth hormone
b. prolactin
c. oxytocin
d. relaxin
e. human chorionic gonadotropin

Multiple Choice

___ 11. What is a morula?
a. a ball of identical cells formed from a zygote
b. the fetal portion of the placenta
c. the vessel that carries blood from the embryo to the placenta
d. an abnormality of the placenta

___ 12. What is an acrosome?
a. the cheeselike material that protects fetal skin
b. the hollow structure that implants in the uterine lining
c. the enzyme-releasing cap over the head of a spermatozoon
d. a small vessel that bypasses the liver in fetal circulation

___ 13. How long is the average total period of gestation from fertilization to birth?
a. 30 weeks
b. 38 weeks
c. 40 weeks
d. 48 weeks

___ 14. With regard to identical twins, which statement is incorrect?
a. They develop from a single zygote.
b. They share one umbilical cord.
c. They are always the same sex.
d. They carry the same inherited traits.

___ 15. What do genes code for?
a. carbohydrates
b. lipids
c. proteins
d. electrolytes

___ 16. What is the term for all the variants of a particular gene?
a. chromosomes
b. ribosomes
c. nucleotides
d. alleles

UNDERSTANDING CONCEPTS

17. Describe the function and structure of the placenta, including the terms for its maternal and fetal portions.
18. Is blood in the umbilical arteries relatively high or low in oxygen? In the umbilical vein? Explain your answer.
19. Describe some of the events thought to initiate labor.
20. What is the major event of each of the three stages of parturition?
21. List several reasons why breast milk is beneficial for the baby.
22. Describe the process of meiosis, and explain how it results in genetic variation.
23. Dana has one dominant allele for freckles (F) and one recessive allele for no freckles (f). What is Dana's genotype? What is her phenotype?
24. Explain the great variation in the color of the skin and hair in humans.
25. Explain why the risk of miscarriage is highest at 12 weeks of pregnancy.

26. Compare and contrast the following terms:
    a. embryo and fetus
    b. foramen ovale and ductus arteriosus
    c. fertilization and implantation
    d. genotype and phenotype
    e. dominant and recessive genes
    f. homozygous and heterozygous genotypes

CONCEPTUAL THINKING

27. Although it is strongly suggested that a woman avoid harmful substances (like alcohol) during her entire pregnancy, why is this advice particularly important during the first trimester?
28. In the process of blood clotting, platelets adhere to damaged tissue and release chemicals that attract more platelets. This continues until a clot is formed and other blood factors stop the process. What type of control mechanism does this represent? Give two other examples of this type of action.
29. Ms. Clarkson, the genetic counselor, is explaining to Cole's parents the chances that any future children of theirs will have sickle cell anemia. Show how she explains these percentages using a Punnett square. Use S for normal hemoglobin and s for the abnormal hemoglobin gene.
30. Richard and Tamara are expecting their first child. Tamara has normal color vision. Richard is color blind and wonders if their baby might be. A family history reveals that Tamara's father is color blind too. What are Richard's and Tamara's genotypes and phenotypes? Using a Punnett square, calculate the possible genotypic and phenotypic combinations for their offspring.

For more questions, see the Learning Activities on thePoint®.

Glossary

A

abdominopelvic (ab-dom-ih-no-PEL-vik) Pertaining to the abdomen and pelvis.

abduction (ab-DUK-shun) Movement away from the midline.

abortifacient (ah-bor-tih-FA-shent) Agent that induces an abortion.

abortion (ah-BOR-shun) Loss of an embryo or fetus before the 20th week of pregnancy.

absorption (ab-SORP-shun) The process of taking up or assimilating a substance.

accommodation (ah-kom-o-DA-shun) Coordinated changes in the lens of the eye that enable one to focus on near and far objects.

acetylcholine (as-e-til-KO-lene) **(ACh)** Neurotransmitter; released at synapses within the nervous system and at the neuromuscular junction.

acid (AH-sid) Substance that can release a hydrogen ion when dissolved in water.

acidosis (as-ih-DO-sis) Condition that results from a decrease in the pH of body fluids.

acrosome (AK-ro-some) Caplike structure over the head of the sperm cell that helps the sperm to penetrate the ovum.

ACTH See adrenocorticotropic hormone.

actin (AK-tin) One of the two contractile proteins in muscle cells, the other being myosin.

action potential Rapid depolarization and repolarization of the plasma membrane capable of spreading down the neuron or muscle membrane; nerve impulse.

active transport Movement of molecules into or out of a cell from an area where they are in lower concentration to an area where they are in higher concentration. Such movement, which is opposite to the direction of normal flow by diffusion, requires energy and transporters.

adaptation (ad-ap-TA-shun) Change in an organ or tissue to meet new conditions.

adduction (ad-DUK-shun) Movement toward the midline.

adenoids (AD-eh-noyds) Popular name for pharyngeal tonsil located in the nasopharynx.

adenosine triphosphate (ah-DEN-o-sene tri-FOS-fate) **(ATP)** Energy-storing compound found in all cells.

ADH See antidiuretic hormone.

adhesion (ad-HE-zhun) Holding together of two surfaces or parts; band of connective tissue between parts that are normally separate.

adipose (AD-ih-pose) Referring to fats or a type of connective tissue that stores fat.

adrenal (ah-DRE-nal) **gland** Endocrine gland located above the kidney; suprarenal gland.

adrenaline (ah-DREN-ah-lin) See epinephrine.

adrenergic (ad-ren-ER-jik) An activity or structure that responds to epinephrine (adrenaline).

adrenocorticotropic (ah-dre-no-kor-tih-ko-TRO-pik) **hormone (ACTH)** Hormone produced by the anterior pituitary that stimulates the adrenal cortex.

aerobic (air-O-bik) Requiring oxygen.

afferent (AF-fer-ent) Carrying toward a given point, such as a sensory neuron that carries nerve impulses toward the central nervous system.

agglutination (ah-glu-tih-NA-shun) Clumping of cells due to an antigen–antibody reaction.

agglutinin (a-GLU-tih-nin) An antibody that causes clumping or agglutination of the agent that stimulated its formation.

agonist (AG-on-ist) Muscle that contracts to perform a given movement.

agranulocyte (a-GRAN-u-lo-site) Leukocyte without visible granules in the cytoplasm when stained; lymphocyte or monocyte.

albumin (al-BU-min) Protein in blood plasma and other body fluids; helps maintain the osmotic pressure of the blood.

aldosterone (al-DOS-ter-one) Hormone released by the adrenal cortex that promotes sodium, and indirectly, water reabsorption in the kidneys.

alkali (AL-kah-li) Substance that can accept a hydrogen ion (H^+); substance that releases a hydroxide ion (OH^-) when dissolved in water; a base.

alkalosis (al-kah-LO-sis) Condition that results from an increase in the pH of body fluids.

allele (al-LEEL) A version of a gene that controls a given trait.

allergen (AL-er-jen) Environmental substance that causes hypersensitivity; substance that induces allergy.

alveolus (al-VE-o-lus) Small sac or pouch; usually a tiny air sac in the lungs through which gases are exchanged between the outside air and the blood; tooth socket; pl., alveoli.

amino (ah-ME-no) **acid** Building block of protein.

amniotic (am-ne-OT-ik) Pertaining to the sac that surrounds and cushions the developing fetus or to the fluid that fills that sac.

amniotic (am-ne-OT-ik) **sac** Fluid-filled sac that surrounds and cushions the developing fetus.

amphiarthrosis (am-fe-ar-THRO-sis) Slightly movable joint.

ampulla (am-PUL-ah) A saccular dilation of a canal or duct.

amygdala (ah-MIG-dah-lah) Two clusters of nuclei deep in the temporal lobes that coordinate emotional responses to stimuli.

amyotrophic (ah-mi-o-TROF-ik) **lateral sclerosis (ALS)** Disorder of the nervous system in which motor neurons are destroyed.

anabolism (ah-NAB-o-lizm) Metabolic building of simple substances into materials needed by the body.

anaerobic (an-air-O-bik) Not requiring oxygen.

anaphase (AN-ah-faze) The third stage of mitosis in which chromosomes separate to opposite sides of the cell.

anastomosis (ah-nas-to-MO-sis) Communication between two structures, such as blood vessels.

anatomic position Standard position of the body for anatomic studies or designations: upright, face front, arms at side with palms forward, and feet parallel.

anatomy (ah-NAT-o-me) Study of body structure.

anemia (ah-NE-me-ah) Abnormally low level of hemoglobin or red cells in the blood, resulting in inadequate delivery of oxygen to the tissues.

androgen (AN-dro-jen) Any male sex hormone.

angiography (an-je-OG-rah-fe) Radiography of vessels after injection of contrast material.

angiotensin II (an-je-o-TEN-sin) (ATII) Substance produced from inactive precursors by the action of the renal enzyme renin and other enzymes; increases blood pressure by causing vascular constriction, stimulating the release of aldosterone from the adrenal cortex and ADH from the posterior pituitary, and increasing thirst.

anion (AN-i-on) Negatively charged particle (ion).

anorexia (an-o-REK-se-ah) Chronic loss of appetite. Anorexia nervosa is a psychological condition in which a person may become seriously and even fatally weakened from lack of food.

anoxia (ah-NOK-se-ah) See hypoxia.

ANP See atrial natriuretic peptide.

ANS See autonomic nervous system.

antagonist (an-TAG-o-nist) Muscle that has an action opposite that of a given movement or muscle; substance that opposes the action of another substance.

anterior (an-TE-re-or) Toward the front or belly surface; ventral.

antibody (AN-te-bod-e) **(Ab)** Substance produced in response to a specific antigen; immunoglobulin.

antidiuretic (an-ti-di-u-RET-ik) **hormone (ADH)** Hormone released from the posterior pituitary gland that increases water reabsorption in the kidneys, thus decreasing the urinary output.

antigen (AN-te-jen) **(Ag)** Substance that induces an immune response.

antioxidant (an-te-OX-ih-dant) Substance in the diet that protects against reactive oxygen species (ROS), also known as harmful free radicals.

antiserum (an-te-SE-rum) Serum containing antibodies that may be given to provide passive immunity; immune serum.

antitoxin (an-te-TOKS-in) Antibody that neutralizes a toxin.

antivenin (an-te-VEN-in) Antibody that neutralizes snake venom.

anus (A-nus) Inferior opening of the digestive tract.

aorta (a-OR-tah) The largest artery; carries blood out of the heart's left ventricle.

apex (A-peks) Pointed region of a cone-shaped structure.

apnea (AP-ne-ah) Temporary cessation of breathing.

apocrine (AP-o-krin) Referring to a gland that releases some cellular material along with its secretions.

aponeurosis (ap-o-nu-RO-sis) Broad sheet of fibrous connective tissue that attaches muscle to bone or to other muscle.

apoptosis (ah-pop-TO-sis) Programmed cell death, an orderly process necessary for growth and remodeling of tissue.

appendicular (ap-en-DIK-u-lar) **skeleton** Part of the skeleton that includes the bones of the upper extremities, lower extremities, shoulder girdle, and hips.

appendix (ah-PEN-diks) Finger-like tube of lymphoid tissue attached to the first portion of the large intestine; vermiform (wormlike) appendix.

aquaporin (ak-kwah-POR-in) A transmembrane channel protein that enables water movement across the plasma membrane.

aqueous (A-kwe-us) Pertaining to water; an aqueous solution is one in which water is the solvent.

aqueous (A-kwe-us) **humor** Watery fluid that fills much of the eyeball anterior to the lens.

arachnoid (ah-RAK-noyd) Middle layer of the meninges.

areolar (ah-RE-o-lar) Referring to loose connective tissue, to any small space, or to an areola, a circular area of marked color.

arrector pili (ah-REK-tor PI-li) Muscle attached to a hair follicle that raises the hair.

arrhythmia (ah-RITH-me-ah) Abnormal rhythm of the heartbeat; dysrhythmia.

arteriole (ar-TE-re-ole) Vessel between a small artery and a capillary.

artery (AR-ter-e) Vessel that carries blood away from the heart.

arthritis (arth-RI-tis) Inflammation of the joints.

articular (ar-TIK-u-lar) Pertaining to a joint.

articulation (ar-tik-u-LA-shun) A joint; an area of junction or union between two or more structures.

asexual (a-SEX-u-al) Nonsexual; referring to reproduction that does not require fusion of two different gametes.

asthma (AZ-mah) Inflammation and constriction of the air passageways.

astrocyte (AS-tro-site) A star-shaped type of neuroglial cell located in the CNS.

atom (AT-om) Smallest subunit of a chemical element.

atomic number The number of protons in the nucleus of an element's atoms; a number characteristic of each element.

ATP See adenosine triphosphate.

atrial natriuretic (na-tre-u-RET-ik) **peptide (ANP)** Hormone produced by the atria of the heart that lowers blood pressure by promoting excretion of sodium and water.

atrioventricular (a-tre-o-ven-TRIK-u-lar) **(AV) node** Part of the heart's conduction systemlocated in the interatrial septum at the bottom of the right atrium.

atrium (A-tre-um) One of the heart's two upper chambers; adj., atrial.

attenuated (ah-TEN-u-a-ted) Weakened.

auditory (AW-dih-tor-e) **tube** Tube that connects the middle ear cavity to the throat; eustachian tube.

autologous (aw-TOL-o-gus) Related to self, such as blood or tissue taken from one's own body.

autolysis (aw-TOL-ih-sis) Lysosomal self-digestion of a cell.

autonomic (aw-to-NOM-ik) **nervous system (ANS)** The part of the nervous system that controls smooth muscle, cardiac muscle, and glands; the visceral or involuntary nervous system.
autophagy (aw-TOF-ah-je) Lysosomal self-digestion of cellular structures.
autosome (AW-to-some) Any chromosome not involved in sex determination. There are 44 autosomes (22 pairs) in humans.
AV node See atrioventricular node.
axial (AK-se-al) **skeleton** The part of the skeleton that includes the skull, spinal column, ribs, and sternum.
axilla (ak-SIL-ah) Hollow beneath the arm where it joins the body; armpit.
axis (AK-sis) A central pole or line around which a structure may revolve. The second cervical vertebra.
axon (AK-son) Fiber of a neuron that conducts impulses away from the cell body.

B

band cell Immature neutrophil.
baroreceptor (bar-o-re-SEP-ter) Receptor that responds to pressure, such as those in vessel walls that respond to stretching and help regulate blood pressure; a type of mechanoreceptor.
basal ganglia (BA-sal GANG-le-ah) Gray masses in the lower part of the forebrain that aid in motor planning; basal nuclei.
basal (BA-sal) **metabolism** The amount of energy needed to maintain life functions while the body is at rest.
basal nuclei (BA-sal NU-kle-i) Interconnected masses of gray matter spread throughout the brain that modulate motor inputs and facilitate routine motor tasks; basal ganglia.
base A lower portion or foundation. Substance that can accept a hydrogen ion (H^+); substance that releases a hydroxide ion (OH^-) when dissolved in water; an alkali. The broad, superior portion of the heart.
basophil (BA-so-fil) Granular white blood cell that shows large, dark blue cytoplasmic granules when stained with basic stain.
B cell Agranular white blood cell that gives rise to antibody-producing plasma cells in response to an antigen; B lymphocyte.
benign (be-NINE) Denoting a mild condition, favorable for recovery, or a tumor that does not metastasize (spread).
bicarbonate ion (bi-KAR-bon-ate I-on) Molecule that combines with hydrogen ion to form carbonic acid, which separates into carbon dioxide and water; an important buffer in body fluids.
bile Substance produced in the liver that emulsifies fats.
bilirubin (BIL-ih-ru-bin) Pigment derived from the breakdown of hemoglobin and found in bile.
blastocyst (BLAS-to-sist) An early stage in embryonic development consisting of a hollow structure formed from a morula that implants in the uterine lining.
blood urea nitrogen (BUN) Amount of nitrogen from urea in the blood; test to evaluate kidney function.
body (BOD-e) The principal mass of any structure. The main portion of a vertebra. A corpus or subdivision of the penis.
bolus (BO-lus) A concentrated mass; the portion of food that is moved to the back of the mouth and swallowed.
bone Hard connective tissue that makes up most of the skeleton, or any structure composed of this type of tissue.
bone marrow Substance that fills the central cavity of a long bone (yellow marrow) and the spaces in spongy bone (red marrow).
bony labyrinth (LAB-ih-rinth) The outer, hollow bony shell of the inner ear that is filled with perilymph.
Bowman capsule See glomerular capsule.
bradycardia (brad-e-KAR-de-ah) Heart rate of less than 60 beats per minute.
brain (brane) The central controlling area of the central nervous system (CNS).
brain stem Portion of the brain that connects the cerebrum with the spinal cord; contains the midbrain, pons, and medulla oblongata.
broca (bro-KAH) **area** Area of the cerebral cortex concerned with motor control of speech; motor speech area.
bronchiole (BRONG-ke-ole) Microscopic branch of a bronchus.
bronchus (BRONG-kus) Large air passageway in the lung; pl., bronchi (BRONG-ki).
buffer (BUF-er) Substance that prevents sharp changes in a solution's pH.
bulbourethral (bul-bo-u-RE-thral) **gland** Gland that secretes mucus to lubricate the urethra and tip of the penis during sexual stimulation; Cowper gland.
bulk flow Movement of water and dissolved substances through the gaps in the capillary wall down a pressure gradient rather than a concentration gradient.
bulk transport Movement of large amounts of material through a cell's plasma membrane using vesicles; vesicular transport.
BUN See blood urea nitrogen.
bursa (BER-sah) Small, fluid-filled sac found in an area subject to stress around bones and joints; pl., bursae (BER-se).

C

calcitriol (kal-sih-TRI-ol) The active form of vitamin D; dihydroxycholecalciferol (di-hi-drok-se-ko-le-kal-SIF-eh-rol).
calcium (KAL-se-um) An element required for bone formation, muscle contraction, nerve impulse conduction, and blood clotting.
calyx (KA-liks) Cuplike extension of the renal pelvis that collects urine; pl., calyces (KA-lih-seze).
cancellous (KAN-sel-us) Referring to spongy bone tissue.
cancer (KAN-ser) Tumor that spreads to other tissues; a malignant neoplasm.
capillary (CAP-ih-lar-e) Microscopic vessel through which exchanges take place between the blood and the tissues.
carbohydrate (kar-bo-HI-drate) Simple sugar or compound made from simple sugars linked together, such as starch or glycogen.
carbon Element that is the basis of organic chemistry.
carbon dioxide (di-OX-ide) (CO_2) Gaseous waste product of cellular metabolism.

carbonic acid (kar-BON-ik) Acid formed when carbon dioxide dissolves in water; carbonic acid then separates into hydrogen ion and bicarbonate ion.

carbonic anhydrase (an-HI-drase) Enzyme that catalyzes the interconversion of carbon dioxide with bicarbonate ion and hydrogen ion.

cardiac (KAR-de-ak) Pertaining to the heart.

cardiac (KAR-de-ak) **output (CO)** Volume of blood pumped by each ventricle in 1 minute. CO is the product of stroke volume and heart rate.

cardiovascular system (kar-de-o-VAS-ku-lar) System consisting of the heart and blood vessels that transports blood throughout the body.

carrier Individual who has a recessive allele of a gene that is not expressed in the phenotype but that can be passed to offspring.

cartilage (KAR-tih-lij) Type of hard connective tissue found at the ends of bones, the tip of the nose, the larynx, the trachea, and the embryonic skeleton.

catabolism (kah-TAB-o-lizm) Metabolic breakdown of substances into simpler components; includes the digestion of food and the oxidation of nutrient molecules for energy.

catalyst (KAT-ah-list) Substance that speeds the rate of a chemical reaction.

cation (KAT-i-on) Positively charged particle (ion).

cauda equina (KAW-dah eh-KWI-nah) Bundle composed of the spinal nerves that arise from the terminal region of the spinal cord; the individual nerves gradually exit from their appropriate segments of the spinal column.

caudal (KAWD-al) Toward or nearer to the sacral region of the spinal column.

cavity (KAV-ih-te) Hollow space; hole.

CCK See cholecystokinin.

cecum (SE-kum) Small pouch at the beginning of the large intestine.

cell Basic unit of life.

cell membrane See plasma membrane.

cellular respiration Series of reactions by which nutrients are oxidized for energy within the mitochondria of body cells.

central canal Channel in the center of an osteon of compact bone; channel in the center of the spinal cord that contains CSF.

central nervous system (CNS) Part of the nervous system that includes the brain and spinal cord.

centrifuge (SEN-trih-fuje) An instrument that separates materials in a mixture based on density.

centriole (SEN-tre-ole) Rod-shaped body near the nucleus of a cell; functions in cell division.

cerebellum (ser-eh-BEL-um) Small section of the brain inferior to the cerebral hemispheres; functions in coordination, balance, and muscle tone.

cerebral (SER-e-bral) **cortex** Very thin outer layer of gray matter on the surface of the cerebral hemispheres.

cerebrospinal (ser-e-bro-SPI-nal) **fluid (CSF)** Fluid that circulates in and around the brain and spinal cord.

cerebrum (SER-e-brum) Largest part of the brain; composed of two cerebral hemispheres.

cerumen (seh-RU-men) Earwax; adj., ceruminous (seh-RU-min-us).

cervical (SER-vih-kal) Pertaining to the neck or the uterine cervix.

cervix (SER-vix) Constricted portion of an organ or part, such as the lower portion of the uterus; neck; adj., cervical.

cesarean (se-ZAR-re-en) **section** Incision in the abdominal and uterine walls for delivery of a fetus; C-section.

chemistry (KEM-is-tre) Study of the composition and properties of matter.

chemoreceptor (ke-mo-re-SEP-tor) Receptor that responds to chemicals in body fluids.

chloride (KLOR-ide) Ion of chlorine.

cholecystokinin (ko-le-sis-to-KI-nin) **(CCK)** Duodenal hormone that stimulates release of enzymes from the pancreas and bile from the gallbladder.

cholesterol (ko-LES-ter-ol) Lipid synthesized by the liver that is found in all plasma membranes, bile, myelin, steroid hormones, and elsewhere; circulates in the blood and is stored in liver and adipose tissue; implicated in the development of atherosclerosis.

cholinergic (ko-lin-ER-jik) Activity or structure that responds to acetylcholine.

chondrocyte (KON-dro-site) Cell that produces and maintains cartilage.

chordae tendineae (KOR-de ten-DIN-e-e) Fibrous threads that stabilize the heart's AV valve flaps.

chorion (KO-re-on) Outer embryonic layer that, together with a layer of the endometrium, forms the placenta.

choroid (KO-royd) Pigmented middle layer of the eye.

choroid plexus (KO-royd PLEKS-us) Vascular network in the brain's ventricles that forms cerebrospinal fluid.

chromosome (KRO-mo-some) Dark-staining, threadlike body in a cell's nucleus; contains genes that determine hereditary traits.

chyle (kile) Milky-appearing fluid absorbed into the lymphatic system from the small intestine. It consists of lymph and droplets of digested fat.

chyme (kime) Mixture of partially digested food, water, and digestive juices that forms in the stomach.

cilia (SIL-e-ah) Hairs or hairlike processes, such as eyelashes or microscopic extensions from a cell's surface; sing., cilium.

ciliary (SIL-e-ar-e) **muscle** Eye muscle that controls the shape of the lens.

cingulate gyrus (SIN-gu-late JI-rus) Portion of the cerebral cortex that loops over the corpus callosum; region that associates emotions with memories.

circulation (ser-ku-LA-shun) Movement or flow within a closed system, as of blood or CSF.

circumcision (sir-kum-SIJ-un) Surgery to remove the foreskin of the penis.

circumduction (ser-kum-DUK-shun) Circular movement at a joint.

cisterna chyli (sis-TER-nah KI-li) Initial portion of the thoracic lymph duct, which is enlarged to form a temporary storage area.

CK See creatine kinase.

clitoris (KLIT-o-ris) Small organ of great sensitivity in the external genitalia of the female.

CNS See central nervous system.

coagulation (ko-ag-u-LA-shun) Clotting, as of blood.

coccyx (KOK-siks) Terminal portion of the vertebral column; formed of four or five small fused bones.
cochlea (KOK-le-ah) Coiled portion of the inner ear that contains the organ of hearing.
collagen (KOL-ah-jen) Flexible white protein that gives strength and resilience to connective tissue, such as bone and cartilage.
colloid (kol-OYD) Mixture in which suspended particles do not dissolve but remain distributed in the solvent because of their small size (e.g., cytoplasm); colloidal suspension.
colon (KO-lon) Main portion of the large intestine.
colostrum (ko-LOS-trum) Secretion of the mammary glands released prior to secretion of milk.
complement (KOM-ple-ment) Group of blood proteins that helps antibodies and phagocytes to destroy foreign cells.
compliance (kom-PLI-ans) Ease with which a hollow structure, such as the thorax or alveoli of the lungs, can be expanded under pressure.
compound Substance composed of two or more chemical elements.
computed tomography (to-MOG-rah-fe) **(CT)** Imaging method in which multiple radiographic views taken from different angles are analyzed by computer to show an area cross section; used to detect tumors and other abnormalities.
concha (KON-ka) Shell-like bone in the nasal cavity; pl., conchae (KON-ke).
conduction (kon-DUK-shun) Transmission or conveyance of energy from one point to another as in transmission of a nerve impulse or transfer of heat from a warm object to a cooler one.
condyle (KON-dile) Rounded projection, as on a bone.
cone Receptor cell in the eye's retina; used for vision in bright light.
congenital (con-JEN-ih-tal) Present at birth.
conjunctiva (kon-junk-TI-vah) Membrane that lines the eyelid and covers the anterior part of the sclera (white of the eye).
connective tissue The supporting tissue of the body ranging in consistency from liquid (e.g., blood) to semisolid (e.g., loose and adipose tissue) to firm (e.g., ligaments, tendons, cartilage) to hard (e.g., bone).
contraception (con-trah-SEP-shun) Prevention of an ovum's fertilization or a fertilized ovum's implantation; birth control.
contractility (kon-trak-TIL-ih-te) Capacity to undergo shortening, as in muscle tissue.
convection (kon-VEK-shun) Heat transfer promoted by movement of a cooler contacting medium, as in cooling by use of a fan.
convergence (kon-VER-jens) Centering of both eyes on the same visual field.
cornea (KOR-ne-ah) Clear portion of the sclera that covers the anterior of the eye.
coronary (KOR-on-ar-e) Referring to the heart or to the arteries supplying blood to the heart.
corpus (KOR-pus) Body; main part of an organ or other structure.
corpus callosum (kal-O-sum) Thick bundle of myelinated nerve cell fibers deep within the brain that carries nerve impulses from one cerebral hemisphere to the other.
corpus luteum (LU-te-um) Yellow body formed from ovarian follicle after ovulation; produces estrogen and progesterone.
corpus spongiosum (KOR-pus spon-je-O-sum) Central subdivision of the penis that contains the urethra.
cortex (KOR-tex) Outer layer of an organ, such as the brain, kidney, or adrenal gland.
covalent (KO-va-lent) **bond** Chemical bond formed by the sharing of electrons between atoms.
CPR See cardiopulmonary resuscitation.
cranial (KRA-ne-al) Pertaining to the cranium, the part of the skull that encloses the brain; toward the head or nearer to the head.
craniosacral (kra-ne-o-SA-kral) Pertaining to the cranial and sacral region of the spinal cord, as describes the parasympathetic nervous system.
creatine (KRE-ah-tin) **phosphate** Compound in muscle tissue that stores energy in high-energy bonds.
creatine kinase (KRE-ah-tin KI-nase) **(CK)** Enzyme in muscle cells that is needed to synthesize creatine phosphate and is released in increased amounts when muscle tissue is damaged; the form specific to cardiac muscle cells is creatine kinase MB (CK-MB).
creatinine (kre-AT-ih-nin) Nitrogenous waste product eliminated in urine.
crenation (kre-NA-shun) Shrinking of a cell, as when placed in a hypertonic solution.
crista (KRIS-tah) Receptor for the sense of rotational equilibrium; pl., cristae.
cryoprecipitate (kri-o-pre-SIP-ih-tate) Precipitate formed when plasma is frozen and then thawed.
CSF See cerebrospinal fluid.
CT See computed tomography.
cutaneous (ku-TA-ne-us) Referring to the skin.
cuticle (KU-tih-kl) Extension of the stratum corneum that seals the space between the nail plate and the skin above the nail root.
cystic (SIS-tik) **duct** Duct that carries bile into and out of the gallbladder.
cytokine (SI-to-kine) Peptide produced by immune cells or other cells that acts as a cellular signal (e.g., interleukin).
cytology (si-TOL-o-je) Study of cells.
cytoplasm (SI-to-plazm) Substance that fills the cell, consisting of a liquid cytosol and organelles.
cytosol (SI-to-sol) Liquid portion of the cytoplasm, consisting of nutrients, minerals, enzymes, and other materials in water.

D

deamination (de-am-ih-NA-shun) Removal of amino groups from proteins in metabolism.
decidua (de-SID-u-ah) The vascularized internal portion of the endometrium in a pregnant uterus; the maternal portion of the placenta.
defecation (def-eh-KA-shun) Act of eliminating undigested waste from the digestive tract.
degeneration (de-jen-er-A-shun) Breakdown, as from age, injury, or disease.

deglutition (deg-lu-TISH-un) Act of swallowing.
dehydration (de-hi-DRA-shun) Excessive loss of body fluid.
denaturation (de-na-tu-RA-shun) Change in structure of a protein, such as an enzyme, so that it can no longer function.
dendrite (DEN-drite) Neuron fiber that conducts impulses toward the cell body.
dendritic (den-DRIT-ik) **cell** Phagocytic white blood cell active in the immune system as an antigen-presenting cell (APC).
dentin (DEN-tin) The main substance of a tooth.
deoxyribonucleic (de-OK-se-ri-bo-nu-kle-ik) **acid (DNA)** Genetic material of the cell; makes up the chromosomes in the cell's nucleus.
depolarization (de-po-lar-ih-ZA-shun) Reduction of the membrane potential (charge) resulting from the entry of cations or the exit of anions.
dermal papillae (pah-PIL-le) Extensions of the dermis that project up into the epidermis; they contain blood vessels that supply the epidermis.
dermatome (DER-mah-tome) Region of the skin supplied by a single spinal nerve.
dermis (DER-mis) True skin; deeper part of the skin.
detoxification (de-tok-sih-fih-KA-shun) Removal of the toxicity of a harmful substance, such as alcohol and certain drugs.
dextrose (DEK-strose) Glucose, as found in nature.
diabetes mellitus (di-ah-BE-teze mel-LI-tus) Disease of insufficient insulin or insufficient response to insulin in which excess glucose is found in the blood and the urine; characterized by abnormal metabolism of glucose, protein, and fat.
diaphragm (DI-ah-fram) Dome-shaped muscle under the lungs that flattens during inhalation; a separating membrane or structure.
diaphysis (di-AF-ih-sis) Shaft of a long bone.
diarthrosis (di-ar-THRO-sis) Freely movable joint; synovial joint.
diastole (di-AS-to-le) Relaxation; adj., diastolic (di-as-TOL-ik).
diencephalon (di-en-SEF-ah-lon) Region of the brain between the cerebral hemispheres and the midbrain; contains the thalamus, hypothalamus, and pituitary gland.
diffusion (dih-FU-zhun) Movement of solutes from a region where they are in higher concentration to a region where they are in lower concentration.
digestion (di-JEST-yun) Process of breaking down food into absorbable particles.
digestive system (di-JES-tiv) The system involved in taking in nutrients, converting them to a form the body can use, and absorbing them into the circulation.
dihydroxycholecalciferol (di-hi-drok-se-ko-le-kal-SIF-eh-rol) The active form of vitamin D.
dilation (di-LA-shun) Widening of a part, such as the pupil of the eye, a blood vessel, or the uterine cervix; dilatation.
disaccharide (di-SAK-ah-ride) Compound formed of two simple sugars linked together, such as sucrose and lactose.
dissect (dis-sekt) To cut apart or separate tissues for study.
distal (DIS-tal) Farther from a structure's origin or from a given reference point.
DNA See deoxyribonucleic acid.
dominant (DOM-ih-nant) Referring to an allele of a gene that is always expressed in the phenotype if present.
dopamine (DO-pah-mene) Neurotransmitter.
dorsal (DOR-sal) At or toward the back; posterior.
dorsiflexion (dor-sih-FLEK-shun) Bending the foot upward at the ankle.
duct Tube or vessel.
ductus arteriosus (DUK-tus ar-te-re-O-sus) Small vessel in the fetus that carries blood from the pulmonary artery to the descending aorta.
ductus deferens (DEF-er-enz) Tube that carries sperm cells from the testis to the urethra; vas deferens.
ductus venosus (ve-NO-sus) Small vessel in the fetus that carries blood from the umbilical vein to the inferior vena cava.
duodenum (du-o-DE-num) First portion of the small intestine.
dura mater (DU-rah MA-ter) Outermost layer of the meninges.

E

eccrine (EK-rin) Referring to sweat glands that regulate body temperature and vent sweat directly to the surface of the skin through a pore.
ECG See electrocardiograph.
echocardiograph (ek-o-KAR-de-o-graf) Instrument to study the heart by means of ultrasound; the record produced is an echocardiogram.
eclampsia (eh-KLAMP-se-ah) Serious and sometimes fatal condition of pregnancy associated with kidney failure, convulsions, and coma that can develop from preeclampsia.
ectopic (ek-TOP-ik) Apart from a normal location, as a pregnancy or heartbeat.
eczema (EK-ze-mah) Skin disorder characterized by intense itching and inflammation; acute eczematous dermatitis.
edema (eh-DE-mah) Accumulation of fluid in the tissue spaces.
EEG See electroencephalograph.
effector (ef-FEK-tor) Muscle or gland that responds to a signal; effector organ.
efferent (EF-fer-ent) Carrying away from a given point, such as a motor neuron that carries nerve impulses away from the central nervous system.
effusion (eh-FU-zhun) Escape of fluid into a cavity or space; the fluid itself.
ejaculation (e-jak-u-LA-shun) Expulsion of semen through the urethra.
EKG See electrocardiograph.
elasticity (e-las-TIS-ih-te) Capacity of a structure to return to its original shape after being stretched.
electrocardiograph (e-lek-tro-KAR-de-o-graf) **(ECG, EKG)** Instrument to study the heart's electric activity; record made is an electrocardiogram.
electroencephalograph (e-lek-tro-en-SEF-ah-lo-graf) **(EEG)** Instrument used to study the brain's electric activity; record made is an electroencephalogram.
electrolyte (e-LEK-tro-lite) Compound that separates into ions in solution; substance that conducts an electric current in solution.

electron (e-LEK-tron) Negatively charged particle located in an energy level outside an atom's nucleus.
element (EL-eh-ment) One of the substances from which all matter is made; substance that cannot be decomposed into a simpler substance.
embryo (EM-bre-o) Developing offspring during the first 8 weeks of gestation.
emulsify (e-MUL-sih-fi) To break up fats into small particles; n. emulsification.
enamel (e-NAM-el) Material that covers the crown of a tooth; the hardest substance in the body.
endocrine (EN-do-krin) Referring to a gland that secretes into the bloodstream.
endocrine system System composed of glands that secrete hormones.
endocytosis (en-do-si-TO-sis) Movement of large amounts of material into a cell using vesicles (e.g., phagocytosis and pinocytosis).
endolymph (EN-do-limf) Fluid that fills the membranous labyrinth of the inner ear.
endometrium (en-do-ME-tre-um) Inner layer of the uterus.
endomysium (en-do-MIS-e-um) Connective tissue around an individual muscle fiber.
endoplasmicreticulum(en-do-PLAS-mikre-TIK-u-lum)**(ER)** Network of membranes in the cellular cytoplasm; may be smooth or rough based on absence or presence of ribosomes.
endorphin (en-DOR-fin) Pain-relieving substance released naturally from the brain.
endoscope (EN-do-skope) Lighted flexible instrument used to examine the interior of body cavities or organs.
endosteum (en-DOS-te-um) Thin membrane that lines a bone marrow cavity.
endothelium (en-do-THE-le-um) Epithelium that lines the heart, blood vessels, and lymphatic vessels.
enzyme (EN-zime) A protein that accelerates a specific chemical reaction.
eosinophil (e-o-SIN-o-fil) Granular white blood cell that shows beadlike, bright pink cytoplasmic granules when stained with acid stain.
ependymal (eh-PEN-dih-mal) **cell** Neuroglial cell in the CNS used to form the choroid plexus, a network in a ventricle that forms CSF.
epicardium (ep-ih-KAR-de-um) Membrane that forms the heart wall's outermost layer and is continuous with the lining of the fibrous pericardium; visceral pericardium.
epicondyle (ep-ih-KON-dile) Small projection on a bone above a condyle.
epidermis (ep-ih-DER-mis) Outermost layer of the skin.
epididymis (ep-ih-DID-ih-mis) Coiled tube on the surface of the testis in which sperm cells are stored and in which they mature.
epigastric (ep-ih-GAS-trik) Pertaining to the abdominal region just inferior to the sternum (breastbone); most superior midline region of the abdomen.
epiglottis (ep-e-GLOT-is) Leaf-shaped cartilage that covers the larynx during swallowing.
epimysium (ep-ih-MIS-e-um) Sheath of fibrous connective tissue that encloses a muscle.
epinephrine (ep-ih-NEF-rin) Hormone released from the adrenal medulla; adrenaline.
epiphysis (eh-PIF-ih-sis) End of a long bone; adj., epiphyseal (ep-ih-FIZ-e-al).
episiotomy (eh-piz-e-OT-o-me) Cutting of the perineum between the vaginal opening and the anus to reduce tissue tearing in childbirth.
epithelium (ep-ih-THE-le-um) One of the four main types of tissues; forms glands, covers surfaces, and lines cavities; adj., epithelial.
EPO See erythropoietin.
equilibrium (e-kwih-LIB-re-um) Sense of balance.
ER See endoplasmic reticulum.
erection (e-REK-shun) Stiffening and enlargement of the penis.
erythrocyte (eh-RITH-ro-site) Red blood cell.
erythropoietin (eh-rith-ro-POY-eh-tin) **(EPO)** Hormone released from the kidney that stimulates red blood cell production in the red bone marrow.
esophagus (eh-SOF-ah-gus) Muscular tube that carries food from the throat to the stomach.
estrogen (ES-tro-jen) Group of female sex hormones that promotes development of the ovarian follicle and the uterine lining and maintains secondary sex characteristics; the main estrogen is estradiol.
eustachian (u-STA-shun) **tube** See auditory tube.
evaporation (e-vap-o-RA-shun) Process of changing a liquid to a vapor.
eversion (e-VER-zhun) Turning outward, with reference to movement of the foot.
excitability In cells, the ability to transmit an electric current along the plasma membrane.
excretion (eks-KRE-shun) Removal and elimination of metabolic waste products from the blood.
exfoliation (eks-fo-le-A-shun) Loss of cells from the surface of tissue, such as the skin.
exhalation (eks-hah-LA-shun) Expulsion of air from the lungs; expiration.
exocrine (EK-so-krin) Referring to a gland that secretes through a duct.
exocytosis (eks-o-si-TO-sis) Movement of large amounts of material out of the cell using vesicles.
extension (eks-TEN-shun) Motion that increases the angle at a joint, returning a body part to the anatomic position.
extracellular (EK-strah-sel-u-lar) Outside the cell.
extremity (ek-STREM-ih-te) Limb.

F

fallopian (fah-LO-pe-an) **tube** See uterine tube.
familial (fah-MIL-e-al) Hereditary; passed from parents to children in the genes.
fascia (FASH-e-ah) Band or sheet of fibrous connective tissue.
fascicle (FAS-ih-kl) Small bundle, as of muscle cells or nerve cell fibers.
fat Type of lipid composed of glycerol and fatty acids; triglyceride.

feces (FE-seze) Waste material discharged from the large intestine; excrement; stool.
feedback Return of information into a system, so that it can be used to regulate that system.
fertilization (fer-til-ih-ZA-shun) Union of an ovum and a spermatozoon.
fetus (FE-tus) Developing offspring from the start of the ninth week of gestation until birth.
fever (FE-ver) Abnormally high body temperature involving an increased temperature set point.
fibrillation (fih-brih-LA-shun) Very rapid, uncoordinated beating of the heart.
fibrin (FI-brin) Blood protein that forms a blood clot.
fibrinogen (fi-BRIN-o-jen) Plasma protein that is converted to fibrin in blood clotting.
fibroblast (FI-bro-blast) Cell that produces protein fibers and other components of the matrix in connective tissue.
fight-or-flight response Physiologic response to a threatening or stressful situation as activated by the sympathetic nervous system.
filtration (fil-TRA-shun) Movement of material through a semipermeable membrane down a pressure gradient.
fimbriae (FIM-bre-e) Fringelike extensions of the uterine tube that sweep a released ovum into the tube.
fissure (FISH-ure) Deep groove.
flagellum (flah-JEL-lum) Long whiplike extension from a cell used for locomotion; pl., flagella.
flexion (FLEK-shun) Bending motion that decreases the angle between bones at a joint, moving a body part away from the anatomic position.
follicle (FOL-lih-kl) Sac or cavity, such as the ovarian follicle or hair follicle.
follicle-stimulating hormone (FSH) Hormone produced by the anterior pituitary that stimulates the development of ova in the ovary and spermatozoa in the testes.
fontanel (fon-tah-NEL) Membranous area in the infant skull where bone has not yet formed; also spelled fontanelle; "soft spot."
foramen (fo-RA-men) Opening or passageway, as into or through a bone; pl., foramina (fo-RAM-in-ah).
foramen magnum Large opening in the skull's occipital bone through which the spinal cord passes to join the brain.
foramen ovale (o-VA-le) Small hole in the fetal atrial septum that allows blood to pass directly from the right atrium to the left atrium.
formed elements Cells and cell fragments in the blood.
fornix (FOR-niks) Recess or archlike structure.
fossa (FOS-sah) Hollow or depression, as in a bone; pl., fossae (FOS-se).
fovea (FO-ve-ah) Small pit or cup-shaped depression in a surface; fovea centralis.
fovea centralis (FO-ve-ah sen-TRA-lis) Tiny depressed area near the optic nerve that is the point of sharpest vision.
frontal (FRONT-al) Describing a plane that divides a structure into anterior and posterior parts. Pertaining to the anterior bone of the cranium.
FSH See follicle-stimulating hormone.
fulcrum (FUL-krum) Pivot point in a lever system; joint in the skeletal system.
fundus (FUN-dus) The deepest portion of an organ, such as the eye, the stomach, or the uterus.

G

gallbladder (GAWL-blad-er) Muscular sac on the inferior surface of the liver that stores bile.
GALT Gut-associated lymphoid tissue.
gamete (GAM-ete) Reproductive cell; ovum or spermatozoon.
gamma globulin (GLOB-u-lin) Protein fraction in the blood plasma that contains antibodies.
ganglion (GANG-le-on) Collection of nerve cell bodies located outside the central nervous system.
gastric inhibitory peptide (GIP) Duodenal hormone that inhibits release of gastric juice and stimulates insulin release.
gastrin (GAS-trin) Hormone released from the stomach that stimulates stomach activity.
gastroesophageal (gas-tro-eh-sof-ah-JE-al) **reflux disease (GERD)** Chronic reflux from the stomach into the distal esophagus caused by weakness in the lower esophageal sphincter (LES).
gastrointestinal (gas-tro-in-TES-tih-nal) **(GI)** Pertaining to the stomach and intestine or the digestive tract as a whole.
gene Hereditary factor; portion of the DNA on a chromosome encoding a specific protein.
gene expression Term that describes if, and to what extent, a gene has an effect on an individual's phenotype.
genetic (jeh-NET-ik) Pertaining to the genes or heredity.
genital (JEN-ih-tal) Pertaining to reproduction or the reproductive organs (genitalia).
genitalia (jen-ih-TA-le-ah) Reproductive organs, both external and internal.
genotype (JEN-o-tipe) Genetic makeup of an organism.
gestation (jes-TA-shun) Period of development from conception to birth.
GH See growth hormone.
GI See gastrointestinal.
gingiva (JIN-jih-vah) Tissue around the teeth; gum.
gland Organ or cell specialized to produce a substance that is sent to other parts of the body.
glans Enlarged distal portion of the penis.
glial cells (GLI-al) Cells that support and protect the nervous system; neuroglia.
glomerular (glo-MER-u-lar) **capsule** Enlarged portion of the nephron that surrounds the glomerulus; Bowman capsule.
glomerular (glo-MER-u-lar) **filtrate** Fluid and dissolved materials that leave the blood and enter the kidney tubule.
glomerulus (glo-MER-u-lus) Cluster of capillaries surrounded by the kidney tubule's glomerular capsule.
glottis (GLOT-is) Space between the vocal cords.
glucagon (GLU-kah-gon) Hormone from the pancreatic islets that raises blood glucose level.
glucocorticoid (glu-ko-KOR-tih-koyd) Steroid hormone from the adrenal cortex that increases the concentration of nutrients in the blood during times of stress (e.g., cortisol).
glucose (GLU-kose) Simple sugar; main energy source for the cells; dextrose.

glycemic (gli-SE-mik) **effect** Measure of how rapidly a food raises the blood glucose level and stimulates release of insulin.

glycogen (GLI-ko-jen) Compound built from glucose molecules that is stored for energy in the liver and muscles.

glycolysis (gli-KOL-ih-sis) First, anaerobic phase of glucose's metabolic breakdown for energy; converts glucose into pyruvate.

goblet cell Single-celled gland that secretes mucus.

Golgi (GOL-je) **apparatus** System of cellular membranes that modifies and sorts proteins; also called Golgi complex.

gonad (GO-nad) Organ producing gametes and sex steroids; ovary or testis; adj., gonadal.

gonadotropin (gon-ah-do-TRO-pin) Hormone that acts on a reproductive gland (ovary or testis) (e.g., FSH, LH).

gradient (GRA-de-ent) A difference in specific values between two areas, such as gradients of position, concentration, or pressure within a system.

gram (g) Basic unit of weight in the metric system.

granulocyte (GRAN-u-lo-site) Leukocyte with visible granules in the cytoplasm when stained.

gray commissure (KOM-ih-shure) Bridge of gray matter that connects the right and left horns of the spinal cord.

gray matter Nervous tissue composed of unmyelinated fibers and cell bodies.

greater vestibular (ves-TIB-u-lar) **gland** Gland that secretes mucus into the vagina; Bartholin gland.

growth hormone (GH) Hormone produced by anterior pituitary that promotes tissue growth; somatotropin.

gustation (gus-TA-shun) Sense of taste; adj., gustatory.

gyrus (JI-rus) Raised area of the cerebral cortex; pl., gyri (JI-ri).

H

hair cells Receptor cells for hearing and equilibrium.

hair follicle (FOL-lih-kl) Sheath that encloses a hair.

haversian (ha-VER-shan) **canal** See central canal.

haversian system See osteon.

hearing (HERE-ing) Ability to perceive sound; audition.

heart (hart) Organ that pumps blood through the cardiovascular system.

hemapheresis (hem-ah-fer-E-sis) Return of blood components to a donor following separation and removal of desired components.

hematocrit (he-MAT-o-krit) **(Hct)** Volume percentage of red blood cells in whole blood; packed cell volume.

hematopoiesis (hem-mah-to-poy-E-sis) Formation of blood cells.

hematopoietic (he-mah-to-poy-ET-ik) Pertaining to blood cell formation.

hemocytometer (he-mo-si-TOM-eh-ter) Device used to count blood cells under the microscope.

hemoglobin (he-mo-GLO-bin) **(Hb)** Iron-containing protein in red blood cells that binds oxygen.

hemolysis (he-MOL-ih-sis) Rupture of red blood cells; v., hemolyze (HE-mo-lize).

hemopoiesis (he-mo-poy-E-sis) Production of blood cells; hematopoiesis.

hemorrhage (HEM-eh-rij) Loss of blood.

hemostasis (he-mo-STA-sis) Stoppage of bleeding.

heparin (HEP-ah-rin) Substance that prevents blood clotting; anticoagulant.

hepatic (heh-PAT-ik) Pertaining to the liver.

heredity (he-RED-ih-te) Transmission of characteristics from parent to offspring by means of the genes.

hereditary (he-RED-ih-tar-e) Transmitted or transmissible through the genes; familial.

heterozygous (het-er-o-ZI-gus) Having unmatched alleles for a given trait; hybrid.

hilum (HI-lum) Indented region of an organ where vessels and nerves enter or leave.

hippocampus (hip-o-KAM-pus) Sea horse–shaped region of the limbic system that functions in learning and formation of long-term memory.

histamine (HIS-tah-mene) Substance released from tissues during an inflammatory reaction; promotes redness, pain, and swelling.

histology (his-TOL-o-je) Study of tissues.

homeostasis (ho-me-o-STA-sis) State of balance within the body; maintenance of body conditions within set limits.

homozygous (ho-mo-ZI-gus) Having identical alleles in a given gene pair.

hormone Chemical messenger secreted by a tissue that has specific regulatory effects on certain other cells.

human chorionic gonadotropin (ko-re-ON-ik gon-ah-do-TRO-pin) **(hCG)** Hormone produced by embryonic cells soon after implantation that maintains the corpus luteum and is diagnostic of pregnancy.

human placental lactogen (hPL) See human chorionic somatomammotropin.

human chorionic somatomammotropin (hCS) Placental hormone similar to growth hormone that increases nutrient availability for the fetus; human placental lactogen (hPL).

humoral (HU-mor-al) Pertaining to body fluids, such as immunity based on antibodies circulating in the blood.

hyaline (HI-ah-lin) Clear, glasslike; referring to a type of cartilage.

hydrolysis (hi-DROL-ih-sis) Splitting of large molecules by the addition of water, as in digestion.

hydrophilic (hi-dro-FIL-ik) Mixing with or dissolving in water, such as salts; literally "water loving."

hydrophobic (hi-dro-FO-bik) Repelling and not dissolving in water, such as fats; literally "water fearing."

hymen Fold of membrane near the opening of the vaginal canal.

hypercapnia (hi-per-KAP-ne-ah) Increased level of carbon dioxide in the blood.

hyperglycemia (hi-per-gli-SE-me-ah) Abnormal increase in the amount of glucose in the blood.

hyperkalemia (hi-per-kah-LE-me-ah) Excess potassium in body fluids.

hyperpnea (hi-PERP-ne-ah) Increase in the depth and rate of breathing, as during exercise.

hypertonic (hi-per-TON-ik) Describing a solution that is more concentrated than the fluids within a cell.

hypertrophy (hy-PER-tro-fe) Enlargement or overgrowth of an organ or part.

hyperventilation (hi-per-ven-tih-LA-shun) Increase in the rate and depth of breathing above optimum levels.

hypocapnia (hi-po-KAP-ne-ah) Decreased level of carbon dioxide in the blood.
hypochondriac (hi-po-KON-dre-ak) Pertaining to a region on either side of the abdomen just inferior to the ribs.
hypogastric (hi-po-GAS-trik) Pertaining to an area inferior to the stomach; most inferior midline region of the abdomen.
hypoglycemia (hi-po-gli-SE-me-ah) Abnormal decrease in the concentration of glucose in the blood.
hyponatremia (hi-po-nah-TRE-me-ah) Deficiency of sodium in body fluids.
hypophysis (hi-POF-ih-sis) Pituitary gland.
hypopnea (hi-POP-ne-ah) Decrease in the rate and depth of breathing.
hypotension (hi-po-TEN-shun) Low blood pressure.
hypothalamus (hi-po-THAL-ah-mus) Region of the brain that controls the pituitary; control center for numerous homeostatic negative feedback loops and for the autonomic nervous system.
hypotonic (hi-po-TON-ik) Describing a solution that is less concentrated than are the fluids within a cell.
hypoventilation (hi-po-ven-tih-LA-shun) Insufficient amount of air entering the alveoli.
hypoxemia (hi-pok-SE-me-ah) Lower than normal concentration of oxygen in arterial blood.
hypoxia (hi-POK-se-ah) Lower than normal level of oxygen in the tissues.

I

ileum (IL-e-um) Most distal portion of the small intestine.
Iliac (IL-e-ak) Pertaining to the ilium, the upper portion of the hipbone; pertaining to the most inferior, lateral regions of the abdomen.
immune (ih-MUNE) **system** Complex of cellular and molecular components that provides defense against foreign cells and substances as well as abnormal body cells.
immunity (ih-MU-nih-te) Power of an individual to resist or overcome the effects of a disease or other harmful agent.
immunization (ih-mu-nih-ZA-shun) Use of a vaccine to produce immunity; vaccination.
immunoglobulin (im-mu-no-GLOB-u-lin) **(Ig)** See antibody.
implantation (im-plan-TA-shun) Embedding of a fertilized ovum into the uterine lining.
inferior (in-FE-re-or) Below or lower.
inferior vena cava (VE-nah KA-vah) Large vein that drains the lower body and empties into the heart's right atrium.
infertility (in-fer-TIL-ih-te) Decreased ability to reproduce.
inflammation (in-flah-MA-shun) Response of tissues to injury or infection; characterized by heat, redness, swelling, and pain.
infundibulum (in-fun-DIB-u-lum) Stalk that connects the pituitary gland to the brain's hypothalamus.
ingestion (in-JES-chun) Intake of food.
inguinal (IN-gwih-nal) Pertaining to the groin region or the region of the inguinal canal.
inhalation (in-hah-LA-shun) Drawing of air into the lungs; inspiration.
insertion (in-SER-shun) Muscle attachment connected to a movable part.
insula (IN-su-lah) Lobe of the cerebral hemisphere located interiorly.
insulin (IN-su-lin) Hormone from the pancreatic islets that lowers blood glucose level and promotes tissue building.
integument (in-TEG-u-ment) Skin; adj., integumentary.
integumentary system The skin and all its associated structures.
intercalated (in-TER-cah-la-ted) **disk** A modified plasma membrane in cardiac tissue that allows rapid transfer of electric impulses between cells.
intercellular (in-ter-SEL-u-lar) Between cells.
intercostal (in-ter-KOS-tal) Between the ribs.
interferon (in-ter-FERE-on) **(IFN)** Group of substances released from virus-infected cells that prevent spread of infection to other cells; also nonspecifically boosts the immune system.
interleukin (in-ter-LU-kin) Substance released by a T cell, macrophage, or endothelial cell that regulates other immune system cells.
interneuron (in-ter-NU-ron) Nerve cell that transmits impulses within the central nervous system or enteric (gastrointestinal) nervous system.
interphase (IN-ter-faze) Stage in a cell's life cycle between one mitosis and the next; time period when a cell is not dividing.
interstitial (in-ter-STISH-al) Between; pertaining to an organ's spaces or structures between active tissues; also refers to the fluid between cells.
intestine (in-TES-tin) Organ of the digestive tract between the stomach and the anus, consisting of the small and large intestine.
intracellular (in-trah-SEL-u-lar) Within a cell.
inversion (in-VER-zhun) Turning inward, with reference to movement of the foot.
ion (I-on) Atom or molecule with an electrical charge; anion or cation.
ionic bond Chemical bond formed by the exchange of electrons between atoms.
iris (I-ris) Circular colored region of the eye around the pupil.
islets (I-lets) Groups of cells in the pancreas that produce hormones; islets of Langerhans (LAHNG-er-hanz).
isometric (i-so-MET-rik) **contraction** Muscle contraction in which there is no change in muscle length but an increase in muscle tension, as in pushing against an immovable force.
isotonic (i-so-TON-ik) Describing a solution that has the same concentration as the fluid within a cell.
isotonic contraction Muscle contraction in which the tone within the muscle remains the same but the muscle shortens to produce movement.
isotope (I-so-tope) Form of an element that has the same atomic number as another form of that element but a different atomic weight; isotopes differ in their numbers of neutrons.
isthmus (IS-mus) Narrow band, such as the band that connects the two lobes of the thyroid gland.

J

jejunum (je-JU-num) Second portion of the small intestine.

joint Area of junction between two or more bones; articulation.

juxtaglomerular (juks-tah-glo-MER-u-lar) **(JG) apparatus** Structure in the kidney composed of cells of the afferent arteriole and distal tubule that increases the secretion of the enzyme renin in response to decreased blood pressure.

juxtamedullary nephron (juks-tah-MED-u-lar-e NEF-ron) Nephron with exceptionally long loop that dips deep into the renal medulla.

K

keratin (KER-ah-tin) Protein that thickens and protects the skin; makes up hair and nails.

ketone (KE-tone) Acidic organic compound formed metabolically from the incomplete oxidation of fats.

kidney (KID-ne) Organ of excretion, hormone synthesis, and blood pressure regulation.

kilocalorie (kil-o-KAL-o-re) **(kcal)** Measure of the energy content of food; technically, the amount of heat needed to raise 1 kg of water 1°C; calorie (C).

kinesthesia (kin-es-THE-ze-ah) Sense of body movement.

kinetic (kih-NET-ik) Pertaining to or producing movement.

kinetic energy The energy of movement, such as the energy of molecular vibrations.

Kupffer (KOOP-fer) **cells** Macrophages in the liver that help to fight infection.

L

labium (LA-be-um) Lip; pl., labia (LA-be-ah).

labor (LA-bor) Parturition; childbirth.

labyrinth (LAB-ih-rinth) Inner ear, named for its complex shape; maze.

lacrimal (LAK-rih-mal) Referring to tears or the tear glands.

lacrimal (LAK-rih-mal) **apparatus** Lacrimal (tear) gland and its associated ducts.

lacrimal gland Gland above the eye that secretes tears.

lactase (LAK-tase) Enzyme that aids in the digestion of lactose.

lactate (LAK-tate) Organic compound produced from pyruvate during rapid anaerobic carbohydrate metabolism.

lactation (lak-TA-shun) Secretion of milk.

lacteal (LAK-te-al) Lymphatic capillary that drains digested fats from the villi of the small intestine.

laryngopharynx (lah-rin-go-FAR-inks) Lowest portion of the pharynx, opening into the larynx and esophagus.

larynx (LAR-inks) Structure between the pharynx and trachea that contains the vocal cords; voice box.

lateral (LAT-er-al) Farther from the midline; toward the side.

lens Biconvex structure of the eye that changes in thickness to accommodate near and far vision; crystalline lens.

leptin (LEP-tin) Hormone produced by adipocytes that aids in weight control by decreasing food intake and increasing energy expenditure.

leukocyte (LU-ko-site) White blood cell.

leukocytosis (lu-ko-si-TO-sis) Increase in the number of white cells in the blood, as occurs during infection.

LH See luteinizing hormone.

ligament (LIG-ah-ment) Band of connective tissue that connects a bone to another bone; thickened portion or fold of the peritoneum that supports an organ or attaches it to another organ.

ligand (LIG-and) Substance that binds to a receptor in the plasma membrane or within the cell.

limbic (LIM-bik) **system** Area between the brain's cerebrum and diencephalon that is involved in emotional states, memory, and behavior.

lingual (LING-gwal) Pertaining to the tongue.

lipase (LI-pase) Enzyme that aids in fat digestion.

lipid (LIP-id) Type of organic compound, one example of which is a fat.

liter (LE-ter) **(L)** Basic unit of volume in the metric system; 1,000 mL; 1.06 qt.

liver (LIV-er) Large organ inferior to the diaphragm in the superior right abdomen; has many functions, including bile secretion, detoxification, storage, and interconversion of nutrients.

lobe Subdivision of an organ, as of the cerebrum, liver, or lung.

loop of Henle (HEN-le) See nephron loop.

lumbar (LUM-bar) Pertaining to the region of the spine between the thoracic vertebrae and the sacrum.

lumen (LU-men) Central opening of an organ or vessel.

lung Organ of respiration.

lunula (LU-nu-la) Pale half-moon–shaped area at the proximal end of the nail.

luteinizing (LU-te-in-i-zing) **hormone (LH)** Hormone produced by the anterior pituitary that induces ovulation and formation of the corpus luteum in females; in males, it stimulates cells in the testes to produce testosterone.

lymph (limf) Fluid in the lymphatic system.

lymph node Mass of lymphoid tissue along the path of a lymphatic vessel that filters lymph and harbors white blood cells active in immunity.

lymphatic duct (lim-FAH-tic) One of two large vessels draining lymph from the lymphatic system into the venous system.

lymphatic system System consisting of the lymphatic vessels and lymphoid tissue; involved in immunity, digestion, and fluid balance.

lymphocyte (LIM-fo-site) Agranular white blood cell that functions in acquired immunity.

lysis (LI-sis) Loosening, dissolving, or separating; a gradual decline, as of a fever.

lysosome (LI-so-some) Cell organelle that contains digestive enzymes.

M

macrophage (MAK-ro-faj) Large phagocytic cell that develops from a monocyte; presents antigen to other leukocytes in immune response.

macula (MAK-u-lah) Spot; flat, discolored spot on the skin, such as a freckle or measles. Area of the retina that contains the point of sharpest vision; equilibrium receptor in the vestibule of the inner ear. Also macule.

macula lutea (MAK-u-lah LU-te-ah) Area of the retina that contains the fovea centralis, the point of sharpest vision.

magnetic resonance imaging (MRI) Method for studying tissue based on nuclear movement after exposure to radio waves in a powerful magnetic field.

major histocompatibility complex (MHC) Group of genes that codes for specific proteins (antigens) on cellular surfaces; these antigens are important in crossmatching for tissue transplantation; they are also important in immune reactions.

MALT Mucosal-associated lymphoid tissue; tissue in the mucous membranes that helps fight infection.

maltase (MAL-tase) Enzyme that aids in the digestion of maltose.

mammary (MAM-er-e) **gland** Milk-secreting portion of the breast.

mast cell White blood cell related to a basophil that is present in tissues; active in inflammatory and allergic reactions.

mastication (mas-tih-KA-shun) Act of chewing.

matrix (MA-triks) The acellular background material in a tissue; the intercellular material.

meatus (me-A-tus) Short channel or passageway, such as the external opening of a canal or a channel in bone.

medial (ME-de-al) Nearer the midline of the body.

mediastinum (me-de-as-TI-num) Region between the lungs and the organs and vessels it contains.

medulla (meh-DUL-lah) Inner region of an organ; marrow.

medullary (MED-u-lar-e) **cavity** Channel at the center of a long bone that contains yellow bone marrow.

medulla oblongata (ob-long-GAH-tah) Part of the brain stem that connects the brain to the spinal cord.

megakaryocyte (meg-ah-KAR-e-o-site) Very large bone marrow cell that gives rise to blood platelets.

meibomian (mi-BO-me-an) **gland** Gland that produces a secretion that lubricates the eyelashes.

meiosis (mi-O-sis) Process of cell division that halves the chromosome number in the formation of the gametes.

melanin (MEL-ah-nin) Dark pigment found in the skin, hair, parts of the eye, and certain parts of the brain.

melanocyte (MEL-ah-no-site) Cell that produces melanin.

melatonin (mel-ah-TO-nin) Hormone produced by the pineal gland.

membrane (MEM-brane) Thin sheet of tissue; lipid bilayer surrounding a cell or an organelle.

membrane attack complex (MAC) Channel in a pathogen's membrane caused by the action of complement and aiding in destruction of the cell.

membrane potential (po-TEN-shal) Difference in electric charge on either side of a plasma membrane; transmembrane potential.

membranous labyrinth (LAB-ih-rinth) The inner membranous portion of the inner ear that is filled with endolymph.

Mendelian (men-DE-le-en) **laws** Principles of heredity discovered by an Austrian monk named Gregor Mendel.

meninges (men-IN-jeze) Three layers of fibrous membranes that cover the brain and spinal cord.

menopause (MEN-o-pawz) Time during which menstruation ceases.

menses (MEN-seze) Monthly flow of blood from the female reproductive tract.

menstruation (men-stru-A-shun) Period of menstrual flow.

mesentery (MES-en-ter-e) Connective tissue membrane that attaches the small intestine to the dorsal abdominal wall.

mesocolon (mes-o-KO-lon) Connective tissue membrane that attaches the colon to the dorsal abdominal wall.

mesothelium (mes-o-THE-le-um) Epithelial tissue found in serous membranes.

metabolic rate Rate at which energy is released from nutrients in the cells.

metabolism (meh-TAB-o-lizm) All the physical and chemical processes by which an organism is maintained.

metaphase (MET-ah-faze) Second stage of mitosis, during which the chromosomes line up across the equator of the cell.

meter (ME-ter) **(m)** Basic unit of length in the metric system; 1.1 yards.

MHC See major histocompatibility complex.

microglia (mi-KROG-le-ah) Glial cells that act as phagocytes in the CNS.

micrometer (MI-kro-me-ter) **(mcm)** 1/1,000th of a millimeter; an instrument for measuring through a microscope (pronounced mi-KROM-eh-ter).

microscope (MI-kro-skope) Magnifying instrument used to examine cells and other structures not visible with the naked eye; examples are the compound light microscope, transmission electron microscope (TEM), and scanning electron microscope (SEM).

microvilli (mi-kro-VIL-li) Small projections of the plasma membrane that increase surface area; sing., microvillus.

micturition (mik-tu-RISH-un) Act of urination; voiding of the urinary bladder.

midbrain Upper portion of the brain stem.

milliosmole (mil-e-OZ-mole) One thousandth of an osmole, a measure of osmotic concentration.

mineral (MIN-er-al) Inorganic substance; in the diet, an element needed in small amounts for health.

mineralocorticoid (min-er-al-o-KOR-tih-koyd) Steroid hormone from the adrenal cortex that regulates electrolyte balance, for example, aldosterone.

miscarriage Loss of an embryo or fetus; spontaneous abortion.

mitochondria (mi-to-KON-dre-ah) Cellular organelles that manufacture ATP with the energy released from the oxidation of nutrients; sing., mitochondrion.

mitosis (mi-TO-sis) Type of cell division that produces two daughter cells exactly like the parent cell.

mitral (MI-tral) **valve** Valve between the heart's left atrium and left ventricle; left AV valve; bicuspid valve.

mixture Blend of two or more substances.

molecule (MOL-eh-kule) Particle formed by covalent bonding of two or more atoms; smallest subunit of a compound.

monocyte (MON-o-site) Phagocytic agranular white blood cell that differentiates into a macrophage.

monomer (MON-o-mer) Building block or single unit of a larger molecule.

monosaccharide (mon-o-SAK-ah-ride) Simple sugar; basic unit of carbohydrates.

morula (MOR-u-lah) An early stage in embryonic development; a ball of identical cells formed by cellular division from a zygote.
motor (MO-tor) Describing structures or activities involved in transmitting impulses away from the central nervous system; efferent; pertaining to or producing movement.
motor end plate Region of a muscle cell membrane that receives nervous stimulation.
motor unit Group consisting of a single neuron and all the muscle fibers it stimulates.
mouth Proximal opening of the digestive tract where food is ingested, chewed, mixed with saliva, and swallowed.
MRI See magnetic resonance imaging.
mucosa (mu-KO-sah) Epithelial membrane that produces mucus; mucous membrane.
mucus (MU-kus) Thick protective fluid secreted by mucous membranes and glands; adj., mucous.
murmur Abnormal heart sound.
muscle (MUS-l) Tissue that contracts to produce movement or tension; includes skeletal, smooth, and cardiac types; adj., muscular.
muscular (MUS-ku-lar) **system** The system of skeletal muscles that moves the skeleton, supports and protects the organs, and maintains posture.
mutagen (MU-tah-jen) Agent that causes mutation; adj., mutagenic (mu-tah-JEN-ik).
mutation (mu-TA-shun) Change in a gene or a chromosome.
myelin (MI-el-in) Fatty material that covers and insulates the axons of some neurons.
myocardium (mi-o-KAR-de-um) Middle layer of the heart wall; heart muscle.
myoglobin (MI-o-glo-bin) Compound that stores oxygen in muscle cells.
myometrium (mi-o-ME-tre-um) Muscular layer of the uterus.
myosin (MI-o-sin) One of the two contractile proteins in muscle cells, the other being actin.

N

nasopharynx (na-zo-FAR-inks) Upper portion of the pharynx located posterior to the nasal cavity.
natural killer (NK) cell Type of lymphocyte that can nonspecifically destroy abnormal cells.
negative feedback Self-regulating system in which the result of an action reverses that action; a method for keeping body conditions within a normal range and maintaining homeostasis.
nephron (NEF-ron) Microscopic functional unit of the kidney; consists of the glomerulus and the renal tubule.
nephron loop Hairpin-shaped segment of the renal tubule between the proximal and distal tubules; loop of Henle.
nerve Bundle of neuron fibers outside the central nervous system.
nerve impulse Electric charge that spreads along the membrane of a neuron; action potential.
nervous system (NER-vus) The system that transports information in the body by means of electric impulses and neurotransmitters.
neurilemma (nu-rih-LEM-mah) Thin sheath that covers certain peripheral axons; aids in axon regeneration.
neuroglia (nu-ROG-le-ah) Supporting and protective cells of the nervous system; glial cells.
neuromuscular junction Point at which a neuron's axon contacts a muscle cell.
neuron (NU-ron) Conducting cell of the nervous system.
neurotransmitter (nu-ro-TRANS-mit-er) Chemical released from the ending of an axon that enables a nerve impulse to cross a chemical synapse.
neutron (NU-tron) Noncharged particle in an atom's nucleus.
neutrophil (NU-tro-fil) Phagocytic granular white blood cell; polymorph; poly; PMN; seg.
nitrogen (NI-tro-jen) Chemical element found in all proteins.
node Small mass of tissue, such as a lymph node; space between cells in the myelin sheath.
norepinephrine (nor-epi-ih-NEF-rin) Neurotransmitter similar in composition and action to the hormone epinephrine; noradrenaline.
normal saline Isotonic or physiologic salt solution.
nucleic acid (nu-KLE-ik) Complex organic substance composed of nucleotides; DNA and RNA.
nucleolus (nu-KLE-o-lus) Small unit within the nucleus that assembles ribosomes.
nucleotide (NU-kle-o-tide) Building block of DNA and RNA; one is also a component of ATP.
nucleus (NU-kle-us) Largest cellular organelle, containing the DNA, which directs all cell activities; group of neurons in the central nervous system; in chemistry, the central part of an atom.

O

obstetrics (ob-STET-riks) Branch of medicine that is concerned with the care of women during pregnancy, childbirth, and the 6 weeks after childbirth.
occlusion (ok-LU-zhun) Closing, as of a vessel.
olfaction (ol-FAK-shun) Sense of smell; adj., olfactory.
oligodendrocyte (ol-ih-go-DEN-dro-site) Type of neuroglial cell that forms the myelin sheath in the CNS.
omentum (o-MEN-tum) Portion of the peritoneum; greater omentum extends over the anterior abdomen; lesser omentum extends between the stomach and liver.
oocyte (O-o-site) A developing egg cell, or ovum.
ophthalmic (of-THAL-mik) Pertaining to the eye.
ophthalmology (of-thal-MOL-o-je) Study of the eye and diseases of the eye.
ophthalmoscope (of-THAL-mo-skope) Instrument for examining the posterior (fundus) of the eye.
organ (OR-gan) Body part containing two or more tissues functioning together for specific purposes.
organ of Corti (KOR-te) See spiral organ.
organelle (or-gan-EL) Specialized structure within a cell.
organic (or-GAN-ik) Referring to the typically large and complex carbon compounds found in living things; contain hydrogen and usually oxygen as well as other elements.
organism (OR-gan-izm) Any organized living thing, such as a plant, animal, or microorganism.

origin (OR-ih-jin) Source; beginning; muscle attachment connected to a nonmoving part.

oropharynx (o-ro-FAR-inks) Middle portion of the pharynx, located behind the mouth.

osmolarity (os-mo-LAR-ih-te) Term that refers to the solute concentration of a solution; osmotic concentration.

osmosis (os-MO-sis) Passage of water through a semipermeable membrane from the region of lower solute concentration to the region of higher solute concentration.

osmotic (os-MOT-ik) **pressure** Tendency of a solution to draw water into it; directly related to a solution's concentration.

osseous (OS-e-us) Pertaining to bone tissue.

ossicle (OS-ih-kl) One of three small bones of the middle ear: malleus, incus, or stapes.

ossification (os-ih-fih-KA-shun) Process of bone formation.

osteoblast (OS-te-o-blast) Bone-forming cell.

osteoclast (OS-te-o-clast) Cell that breaks down bone.

osteocyte (OS-te-o-site) Mature bone cell; maintains bone but does not produce new bone tissue.

osteon (OS-te-on) Subunit of compact bone, consisting of concentric rings of bone tissue around a central channel; haversian system.

otolithic (o-to-LITH-ik) **membrane** Gelatinous material covering the tips of equilibrium receptor cells in the vestibule of the inner ear.

otoliths (O-to-liths) Crystals that add weight to the otolithic membrane of equilibrium receptors in the vestibule of the inner ear.

oval window Area of the inner ear where sound waves are transmitted from the footplate of the stapes to the fluids in the spiral organ.

ovarian follicle (o-VA-re-an FOL-ih-kl) Cluster of cells containing an ovum. A follicle can mature during a menstrual cycle and release its ovum.

ovary (O-vah-re) Female reproductive organ; produces ova and female sex steroids.

oviduct (O-vih-dukt) See uterine tube.

ovulation (ov-u-LA-shun) Release of an ovum from a mature ovarian follicle (graafian follicle).

ovum (O-vum) Female reproductive cell or gamete; pl., ova.

oxidation (ok-sih-DA-shun) Chemical breakdown of nutrients for energy usually using oxygen.

oxygen (OK-sih-jen) **(O_2)** Gas needed to break down nutrients completely for energy within the cell.

oxytocin (ok-se-TO-sin) Hormone from the posterior pituitary that causes uterine contraction and milk ejection ("letdown") from the breasts.

P

pacemaker Group of cells or artificial device that sets activity rate; in the heart, the sinoatrial (SA) node that normally initiates contractions.

palate (PAL-at) Roof of the oral cavity; anterior portion is the hard palate, posterior portion is the soft palate.

pancreas (PAN-kre-as) Large, elongated gland behind the stomach; produces digestive enzymes and hormones (e.g., insulin, glucagon).

papilla (pah-PIL-ah) Small nipple-like projection or elevation.

papillary (PAP-ih-lar-e) **muscles** Columnar muscles in the heart's ventricular walls that anchor and pull on the chordae tendineae to prevent the valve flaps from everting when the ventricles contract.

parasympathetic nervous system Craniosacral division of the autonomic nervous system; generally reverses the fight-or-flight (stress) response.

parathyroid (par-ah-THI-royd) **gland** Any of four to six small glands embedded in the capsule enclosing the thyroid gland; produces parathyroid hormone, which raises blood calcium level by causing calcium release from bones and calcium retention in the kidney.

parietal (pah-RI-eh-tal) Pertaining to the wall of a space or cavity.

parotid (pah-ROT-id) **gland** Salivary gland located inferior and anterior to the ear.

partial pressure Pressure of an individual gas within a mixture.

parturition (par-tu-RISH-un) Childbirth; labor.

pedigree (PED-ih-gre) **chart** Family history; record used in the study of heredity; family tree.

pelvis (PEL-vis) Basin-like structure, such as the lower portion of the abdomen or the upper flared portion of the ureter (renal pelvis).

penis (PE-nis) Male organ of urination and sexual intercourse.

perforating canal Channel across a long bone that contains blood vessels and nerves; Volkmann canal.

pericardium (per-ih-KAR-de-um) Fibrous sac lined with serous membrane that encloses the heart.

perichondrium (per-ih-KON-dre-um) Layer of connective tissue that covers cartilage.

perilymph (PER-e-limf) Fluid that fills the inner ear's bony labyrinth.

perimysium (per-ih-MIS-e-um) Connective tissue around a fascicle of muscle tissue.

perineum (per-ih-NE-um) Pelvic floor; external region between the anus and genital organs.

periosteum (per-e-OS-te-um) Connective tissue membrane covering a bone.

peripheral (peh-RIF-er-al) Located away from a center or central structure.

peripheral nervous system (PNS) All the nerves and nervous tissue outside the central nervous system.

peristalsis (per-ih-STAL-sis) Wavelike movements in the wall of an organ or duct that propel its contents forward.

peritoneum (per-ih-to-NE-um) Serous membrane that lines the abdominal cavity and forms the outer layer of the abdominal organs; forms supporting ligaments for some organs.

peroxisome (per-OK-sih-some) Cell organelle that enzymatically destroys harmful substances produced in metabolism.

Peyer (PI-er) **patches** Clusters of lymphoid nodules in the mucous membranes lining the distal small intestine.

pH Symbol indicating hydrogen ion (H^+) concentration; lower numbers indicate a higher H^+ concentration and higher acidity.

pH scale System for indicating the relative concentration of hydrogen and hydrozide ions in a solution. The scale ranges from 1 to 14, with 1 most acidic and 14 most alkaline (basic) and 7 as neutral.

phagocyte (FAG-o-site) Cell capable of engulfing large particles, such as foreign matter or cellular debris, through the plasma membrane.
phagocytosis (fag-o-si-TO-sis) Engulfing of large particles through the plasma membrane.
pharynx (FAR-inks) Throat; passageway between the mouth and esophagus.
phenotype (FE-no-tipe) All the characteristics of an organism that can be seen or tested for.
phospholipid (fos-fo-LIP-id) Complex lipid containing phosphorus; major component of the plasma membrane.
phrenic (FREN-ik) Pertaining to the diaphragm.
phrenic nerve Nerve that activates the diaphragm.
physiology (fiz-e-OL-o-je) Study of the function of living organisms.
pia mater (PI-ah MA-ter) Innermost layer of the meninges.
pineal (PIN-e-al) **gland** Gland in the brain that is regulated by light; involved in sleep–wake cycles.
pinna (PIN-nah) Outer projecting portion of the ear; auricle.
pinocytosis (pi-no-si-TO-sis) Intake of small particles and droplets by a cell's plasma membrane.
pituitary (pih-TU-ih-tar-e) **gland** Endocrine gland located under and controlled by the hypothalamus; releases hormones that control other glands; hypophysis.
placenta (plah-SEN-tah) Structure that nourishes and maintains the developing fetus during pregnancy.
plantar flexion (PLAN-tar FLEK-shun) Bending the foot so that the toes point downward.
plasma (PLAZ-mah) Liquid portion of the blood.
plasma cell Cell derived from a B cell that produces antibodies.
plasma membrane Outer covering of a cell; regulates what enters and leaves the cell; cell membrane.
plasmapheresis (plas-mah-fer-E-sis) Separation and removal of plasma from donated blood and return of the formed elements to the donor.
platelet (PLATE-let) Cell fragment that forms a plug to stop bleeding and acts in blood clotting; thrombocyte.
pleura (PLU-rah) Serous membrane that lines the chest cavity and covers the lungs.
plexus (PLEK-sus) Network of vessels or nerves.
PNS See peripheral nervous system.
polymorph (POL-e-morf) Term for a neutrophil; polymorphonuclear neutrophil.
polysaccharide (pol-e-SAK-ah-ride) Compound formed from many simple sugars linked together (e.g., starch, glycogen).
pons (ponz) Area of the brain between the midbrain and medulla; connects the cerebellum with the rest of the central nervous system.
portal system Venous system that carries blood to a second capillary bed through which it circulates before returning to the heart.
positive feedback Control system in which an action or the product of an action maintains or intensifies that action. The action stops when materials are depleted, the stimulus is removed, or an outside force interrupts the action.
posterior (pos-TE-re-or) Toward the back; dorsal.
postsynaptic (post-sin-AP-tik) Distal to the synaptic cleft.
potential (po-TEN-shal) Electric charge, as on the plasma membrane of a neuron or other cell; potential difference, membrane potential, or transmembrane potential.
potential energy Stored energy, such as gravitational energy, the energy of position in a gravitational field.
precipitation (pre-sip-ih-TA-shun) Settling out of a solid previously held in solution or suspension in a liquid; in immunity, clumping of small particles as a result of an antigen–antibody reaction; seen as a cloudiness.
prefix A word part that comes before a root and modifies its meaning.
pregnancy (PREG-nan-se) Period during which an embryo or fetus is developing in the body.
prepuce (PRE-puse) Loose fold of skin that covers the glans penis; foreskin.
presynaptic (pre-sin-AP-tik) Proximal to the synaptic cleft.
preterm (PRE-term) Referring to an infant born before the 37th week of gestation.
prime mover The main muscle that produces a given movement.
PRL See prolactin.
progeny (PROJ-eh-ne) Offspring, descendent.
progesterone (pro-JES-ter-one) Hormone produced by the corpus luteum and placenta; maintains the uterine lining for pregnancy.
prolactin (pro-LAK-tin) Hormone from the anterior pituitary that stimulates milk production in the mammary glands; PRL.
pronation (pro-NA-shun) Turning the palm down or backward.
prone Face down or palm down.
prophase (PRO-faze) First stage of mitosis, during which the chromosomes become visible and the organelles disappear.
proprioceptor (pro-pre-o-SEP-tor) Sensory receptor that aids in judging body position and changes in position; located in muscles, tendons, and joints.
prostaglandin (pros-tah-GLAN-din) Any of a group of hormones produced by many cells that usually act on neighboring cells; these hormones are involved in pain and inflammation, as well as many other functions.
prostate (PROS-tate) **gland** Gland that surrounds the urethra below the bladder in males and contributes secretions to semen.
protein (PRO-tene) Organic compound made of amino acids; found as structural materials and metabolically active compounds, such as enzymes, some hormones, pigments, antibodies, and others.
proteome (PRO-te-ome) All the proteins that can be expressed in a cell.
prothrombin (pro-THROM-bin) Clotting factor; converted to thrombin during blood clotting.
prothrombinase (pro-THROM-bih-nase) Blood clotting factor that converts prothrombin to thrombin.
proton (PRO-ton) Positively charged particle in an atom's nucleus.
proximal (PROK-sih-mal) Nearer to the point of origin or to a reference point.
puerperal (pu-ER-per-al) Related to childbirth.

pulmonary circuit Circulatory pathway that carries blood from the heart to the lungs to pick up oxygen and release carbon dioxide and then returns the blood to the heart.

pulse Wave of increased pressure in the vessels produced by heart contraction.

pulse pressure Difference between systolic and diastolic pressures.

pupil (PU-pil) Opening in the center of the eye through which light enters.

Purkinje (pur-KIN-je) **fibers** Part of the heart's conduction system that branches through the ventricular walls.

pyloric (pi-LOR-ik) Pertaining to the pylorus, the distal region of the stomach.

pylorus (pi-LOR-us) Distal region of the stomach that leads to the pyloric sphincter.

pyruvate (PI-ru-vate) Organic intermediate product in the breakdown of glucose for energy.

R

radiation (ra-de-A-shun) Emission of rays, as of light, radio, or heat waves, ultraviolet, or x-rays. Method of heat loss by heat waves traveling from their source.

radioactive (ra-de-o-AK-tive) Pertaining to isotopes that fall apart easily, giving off radiation.

radioactivity (ra-de-o-ak-TIV-ih-te) Emission of atomic particles from an element.

radiography (ra-de-OG-rah-fe) Production of an image by passage of x-rays through the body onto sensitized film; record produced is a radiograph.

reabsorption (re-ab-SORP-shun) Absorbing or taking up again.

receptor (re-SEP-tor) Specialized cell or ending of a sensory neuron that can be excited by a stimulus. Protein in the plasma membrane or other part of a cell that binds a chemical signal (e.g., hormone, neurotransmitter) resulting in a change in cellular activity.

recessive (re-SES-iv) Referring to an allele that is not expressed in the phenotype if a dominant allele for the same trait is present.

rectum (REK-tum) Distal region of the large intestine between the sigmoid colon and the anal canal.

reflex (RE-fleks) Simple, rapid, automatic response to a specific stimulus.

reflex arc (ark) Pathway through the nervous system from stimulus to response; commonly involves a sensory receptor, sensory neuron, central neuron(s), motor neuron, and effector.

refraction (re-FRAK-shun) Bending of light rays as they pass from one medium to another of a different density.

regulated variable Body condition that is kept homeostatically within a narrow range using negative feedback; examples are blood pressure and body temperature.

relaxin (re-LAKS-in) Placental hormone that softens the cervix and relaxes the pelvic joints.

renal tubule Coiled and looped portion of a nephron between the glomerular capsule and the collecting duct.

renin (RE-nin) Enzyme released from the kidney's juxtaglomerular apparatus that indirectly increases blood pressure by activating angiotensin.

repolarization (re-po-lar-ih-ZA-shun) A change in the membrane potential that brings it closer to the resting value.

resistance (re-ZIS-tans) Any or all factors that resist flow down a gradient.

resorption (re-SORP-shun) Loss of substance from a solid tissue, such as bone or a tooth, and return of the components to the blood.

respiration (res-pih-RA-shun) Process by which oxygen is obtained from the environment and delivered to the cells as carbon dioxide is removed from the tissues and released to the environment.

respiratory system System consisting of the lungs and breathing passages involved in exchange of oxygen and carbon dioxide between the outside air and the blood.

reticular (reh-TIK-u-lar) **formation** Network in the limbic system that governs wakefulness and sleep.

reticulocyte (reh-TIK-u-lo-site) Immature form of erythrocyte.

retina (RET-ih-nah) Innermost layer of the eye; contains light-sensitive cells (rods and cones).

retroperitoneal (ret-ro-per-ih-to-NE-al) Behind the peritoneum, as are the kidneys, pancreas, and abdominal aorta.

Rh factor Red cell antigen; D antigen.

rhodopsin (ro-DOP-sin) Light-sensitive pigment in the rods of the eye.

rib One of the slender curved bones that make up most of the thorax; costa; adj., costal.

ribonucleic (RI-bo-nu-kle-ik) **acid (RNA)** Substance needed for protein manufacture in the cell.

ribosome (RI-bo-some) Small body in the cell's cytoplasm that is a site of protein manufacture.

RNA See ribonucleic acid.

rod Receptor cell in the retina of the eye; used for vision in dim light.

roentgenogram (rent-GEN-o-gram) Image produced by means of x-rays; radiograph.

root (rute) Basic part of a word; attached or embedded part, such as a nail or hair root. The branched portion of a spinal nerve that joins with the spinal cord. See also word root.

rotation (ro-TA-shun) Twisting or turning of a bone on its own axis.

rugae (RU-je) Folds in the lining of an organ, such as the stomach or urinary bladder; sing., ruga (RU-gah).

S

sacrum (SA-krum) Portion of the vertebral column between the lumbar vertebrae and coccyx. It is formed of five fused bones and completes the posterior part of the bony pelvis.

sagittal (SAJ-ih-tal) Referring to a plane that divides the body into left and right portions.

saliva (sah-LI-vah) Secretion of the salivary glands; moistens food and contains an enzyme that digests starch.

salt Compound formed by reaction between an acid and a base (e.g., NaCl, table salt).

saltatory (SAL-tah-to-re) **conduction** Transmission of an electric impulse from node to node along a myelinated fiber; faster than continuous conduction along the entire membrane.

SA node See sinoatrial node.

sarcolemma (sar-ko-LEM-mah) The plasma membrane of a muscle cell.

sarcomere (SAR-ko-mere) Contracting subunit of skeletal muscle.

sarcoplasmic reticulum (sar-ko-PLAS-mik re-TIK-u-lum) **(SR)** Intracellular membranous organelle in muscle cells that is equivalent to the endoplasmic reticulum (ER) in other cells; stores calcium needed for muscle contraction.

satellite (SAT-eh-lite) **cell** Postnatal stem cell located at the surface of a muscle cell that can, within limits, produce a new muscle cell when activated.

saturated fat Fat that has more hydrogen atoms and fewer double bonds between carbons than do unsaturated fats.

Schwann (shvahn) **cell** Cell in the nervous system that produces the myelin sheath around peripheral axons.

sclera (SKLE-rah) Outermost layer of the eye; made of tough connective tissue; "white" of the eye.

scrotum (SKRO-tum) Sac in which the testes are suspended.

sebaceous (seh-BA-chus) Pertaining to sebum, an oily substance secreted by skin glands.

sebum (SE-bum) Oily secretion that lubricates the skin; adj., sebaceous (se-BA-shus).

secretin (se-KRE-tin) Hormone from the duodenum that stimulates pancreatic release of water and bicarbonate.

segmentation (seg-men-TA-shun) Alternating contraction and relaxation of the circular muscle in the small intestine's wall that mix its contents with digestive juices and move them through the organ.

selectively permeable Describing a membrane that regulates what can pass through (e.g., a cell's plasma membrane).

sella turcica (SEL-ah TUR-sih-ka) Saddle-like depression in the floor of the skull that holds the pituitary gland.

semen (SE-men) Mixture of sperm cells and secretions from several glands of the male reproductive tract.

semicircular canal One of three curved channels of the inner ear where receptors for rotational equilibrium are located.

semilunar (sem-e-LU-nar) Shaped like a half-moon, such as the flaps of the pulmonary and aortic valves.

seminal vesicle (VES-ih-kl) Gland that contributes secretions to the semen.

seminiferous (seh-mih-NIF-er-us) **tubules** Tubules in which sperm cells develop in the testis.

semipermeable (sem-e-PER-me-ah-bl) Permeable (passable) to some substances but not to others, as describes the cellular plasma membrane.

sensory (SEN-so-re) Describing cells or activities involved in transmitting impulses toward the central nervous system; afferent; pertaining to the senses.

sensory adaptation Gradual loss of sensation when sensory receptors are exposed to continuous stimulation.

sensory receptor Part of the nervous system that detects a stimulus.

septum (SEP-tum) Dividing wall, as between the chambers of the heart or the nasal cavities.

serosa (se-RO-sah) Serous membrane; epithelial membrane that secretes a thin, watery fluid.

serotonin (ser-o-TO-nin) Neurotransmitter involved in mood and other cognitive functions.

Sertoli cells See sustentacular cells.

serum (SE-rum) Liquid portion of blood without clotting factors; thin, watery fluid; adj., serous (SE-rus).

sex-linked Referring to a gene carried on a sex chromosome, usually the X chromosome.

sickle cell anemia Hereditary disease in which abnormal hemoglobin causes red blood cells to change shape (sickle) when they release oxygen; sickle cell disease.

sigmoidoscope (sig-MOY-do-skope) Endoscope used to examine the rectum and lower colon.

sign (sine) Manifestation of a disease as noted by an observer.

sinoatrial (si-no-A-tre-al) **(SA) node** Tissue in the right atrium's upper wall that sets the rate of heart contractions; the heart's pacemaker.

sinus (SI-nus) Cavity or channel, such as the paranasal sinuses in the skull bones.

sinus rhythm Normal heart rhythm originating at the SA node.

sinusoid (SI-nus-oyd) Enlarged capillary that serves as a blood channel.

skeletal (SKEL-eh-tal) **system** The body system that includes the bones and joints.

skeleton (SKEL-eh-ton) Bony framework of the body; adj., skeletal.

skull Bony framework of the head.

sodium (SO-de-um) Positive ion found in body fluids and important in fluid balance, nerve impulse conduction, and acid–base balance.

solute (SOL-ute) Substance that is dissolved in another substance (the solvent).

solution (so-LU-shun) Homogeneous mixture of one substance dissolved in another; the components in a mixture are evenly distributed and cannot be distinguished from each other.

solvent (SOL-vent) Substance in which another substance (the solute) is dissolved.

somatic (so-MAT-ik) **nervous system** Division of the nervous system that controls voluntary activities and stimulates skeletal muscle.

somatotropin (so-mah-to-TRO-pin) Growth hormone.

specific gravity Weight of a substance as compared to the weight of an equal volume of pure water.

spermatic (sper-MAT-ik) **cord** Cord that extends through the inguinal canal and suspends the testis; contains blood vessels, nerves, and ductus deferens.

spermatozoon (sper-mah-to-ZO-on) Male reproductive cell or gamete; pl., spermatozoa; sperm cell.

sphincter (SFINK-ter) Muscular ring that regulates the size of an opening.

sphygmomanometer (sfig-mo-mah-NOM-eh-ter) Device used to measure blood pressure; blood pressure apparatus or cuff.

spinal cord Nervous tissue contained in the spinal column; major relay area between the brain and the peripheral nervous system.

spine Vertebral column. Sharp projection from the surface of a bone.

spiral (SPI-ral) **organ** Receptor for hearing located in the cochlea of the internal ear; organ of Corti.

spirometer (spi-ROM-eh-ter) Instrument for recording lung volumes; tracing obtained is a spirogram.

spleen Lymphoid organ in the upper left region of the abdomen.

squamous (SKWA-mus) Flat and irregular, as in squamous epithelium.

SR See sarcoplasmic reticulum.

stain (stane) Dye that aids in viewing structures under the microscope.

stapes (STA-peze) Innermost ossicle of the middle ear; transmits sound waves to the oval window of the inner ear.

stem cell Cell that has the potential to divide and produce different types of cells.

sterilization (ster-ih-li-ZA-shun) Process of killing every living microorganism on or in an object; procedure that makes an individual incapable of reproduction.

steroid (STE-royd) Category of lipids that includes the hormones of the sex glands and the adrenal cortex.

stethoscope (STETH-o-skope) Instrument for conveying sounds from the patient's body to the examiner's ears.

stimulus (STIM-u-lus) Change in the external or internal environment that produces a response.

stomach (STUM-ak) Organ of the digestive tract that stores food, mixes it with digestive juices, and moves it into the small intestine.

stratified In multiple layers (strata).

stratum (STRA-tum) A layer; pl., strata.

stratum basale (bas-A-le) Deepest layer of the epidermis; layer that produces new epidermal cells; stratum germinativum.

stratum corneum (KOR-ne-um) The thick uppermost layer of the epidermis.

striations (stri-A-shuns) Stripes or bands, as seen in skeletal muscle and cardiac muscle.

stroke volume Amount of blood ejected from a ventricle with each beat.

subcutaneous (sub-ku-TA-ne-us) Under the skin.

submucosa (sub-mu-KO-sah) Layer of connective tissue beneath the mucosa.

substrate Substance on which an enzyme works.

sudoriferous (su-do-RIF-er-us) Producing sweat; referring to the sweat glands.

suffix A word part that follows a root and modifies its meaning.

sulcus (SUL-kus) Shallow groove as between convolutions of the cerebral cortex; pl., sulci (SUL-si).

superior (su-PE-re-or) Above; in a higher position.

superior vena cava (VE-nah KA-vah) Large vein that drains the upper part of the body and empties into the heart's right atrium.

supination (su-pin-A-shun) Turning the palm up or forward.

supine (SU-pine) Face up or palm up.

surfactant (sur-FAK-tant) Substance in the alveoli that prevents their collapse by reducing surface tension of the fluid lining them.

suspension (sus-PEN-shun) Heterogeneous mixture that will separate unless shaken.

suspensory ligaments Filaments attached to the ciliary muscle of the eye that hold the lens in place and can be used to change its shape.

sustentacular (sus-ten-TAK-u-lar) **cells** Cells in the seminiferous tubules that aid in the development of spermatozoa; Sertoli cells.

suture (SU-chur) Type of joint in which bone surfaces are closely united, as in the skull; stitch used in surgery to bring parts together; to stitch parts together in surgery.

sweat (swet) **gland** Skin gland that produces perspiration; sudoriferous gland.

sympathetic nervous system Thoracolumbar division of the autonomic nervous system; stimulates a fight-or-flight (stress) response.

symptom (SIMP-tom) Evidence of disease noted by the patient; such evidence noted by an examiner is called a sign or an objective symptom.

synapse (SIN-aps) Junction between two neurons or between a neuron and an effector.

synarthrosis (sin-ar-THRO-sis) Immovable joint.

synergist (SIN-er-jist) Substance or structure that enhances the work of another. A muscle that works with a prime mover to produce a given movement.

synovial (sin-O-ve-al) Pertaining to a thick, lubricating fluid found in joints, bursae, and tendon sheaths; pertaining to a freely movable (diarthrotic) joint.

system (SIS-tem) Group of organs functioning together for the same general purposes; any group of related components.

systemic (sis-TEM-ik) Referring to a generalized infection or condition.

systemic circuit Circulatory pathway that carries blood to all and from tissues of the body to deliver oxygen and nutrients and pick up carbon dioxide and metabolic wastes.

systole (SIS-to-le) Contraction; adj., systolic (sis-TOL-ik).

T

tachycardia (tak-e-KAR-de-ah) Heart rate more than 100 beats per minute in an adult.

tachypnea (tak-IP-ne-ah) Increased rate of respiration.

tactile (TAK-til) Pertaining to the sense of touch.

target tissue Tissue that is capable of responding to a specific hormone.

T cell Lymphocyte active in immunity that matures in the thymus gland; may destroy foreign cells directly or help in or regulate the immune response; T lymphocyte.

T-tubule Transverse tubule; a tubular extension of the sarcolemma into the sarcoplasm of a skeletal muscle cell. T tubules conduct action potentials to the sarcoplasmic reticulum.

tectorial (tek-TO-re-al) **membrane** Membrane overlying the sensory (hair) cells in the spiral organ of hearing.

telophase (TEL-o-faze) Final stage of mitosis, during which new nuclei form and the cell contents usually divide.
tendon (TEN-don) Cord of regular dense connective tissue that attaches a muscle to a bone.
teniae (TEN-e-e) **coli** Bands of smooth muscle in the wall of the large intestine.
testis (TES-tis) Male reproductive gland; pl., testes (TES-teze).
testosterone (tes-TOS-ter-one) Male sex hormone produced in the testes; promotes sperm cell development and maintains secondary sex characteristics.
thalamus (THAL-ah-mus) Region of the brain located in the diencephalon; chief relay center for sensory impulses traveling to the cerebral cortex.
thorax (THO-raks) Chest; adj., thoracic (tho-RAS-ik).
thrombin (THROM-bin) Clotting factor in the blood needed to convert fibrinogen to fibrin.
thrombocyte (THROM-bo-site) Blood platelet; cell fragment that participates in clotting.
thrombolytic (throm-bo-LIT-ik) Dissolving blood clots.
thrombosis (throm-BO-sis) Condition of having a thrombus (blood clot in a vessel).
thrombus (THROM-bus) Blood clot within a vessel.
thymus (THI-mus) Lymphoid organ in the upper portion of the chest; site of T cell development.
thyroid (THI-royd) Endocrine gland in the neck.
thyroid-stimulating hormone (TSH) Hormone produced by the anterior pituitary that stimulates the thyroid gland; thyrotropin.
thyroxine (thi-ROK-sin) Hormone produced by the thyroid gland; increases metabolic rate and needed for normal growth; T_4.
tissue Group of similar cells that performs a specialized function.
toll-like receptor (TRL) Receptor on a cell of the innate immune system that can recognize components of a pathogen as being foreign.
tongue (tung) Muscular organ in the mouth that contains the taste buds and functions in chewing, swallowing, and speech production.
tonsil (TON-sil) Mass of lymphoid tissue in the region of the pharynx.
tonus (TO-nus) Partially contracted state of muscle; also, tone.
toxin (TOK-sin) Poison.
toxoid (TOK-soyd) Altered toxin used to produce active immunity.
trachea (TRA-ke-ah) Tube that extends from the larynx to the bronchi; windpipe.
tract Bundle of neuron fibers within the central nervous system.
trait Characteristic.
transfusion (trans-FU-zhun) Introduction of blood or blood components directly into the bloodstream of a recipient.
transplantation (trans-plan-TA-shun) The grafting to a recipient of an organ or tissue from an animal or other human to replace an injured or incompetent body part.
transport maximum Maximum amount of a particular substance that can be reabsorbed from the nephron tubule in milligrams per minute; tubular maximum.
transverse Describing a plane that divides a structure into superior and inferior parts.
tricuspid (tri-KUS-pid) **valve** Valve between the heart's right atrium and right ventricle.
triglyceride (tri-GLIS-er-ide) Simple fat composed of glycerol and three fatty acids.
trigone (TRI-gone) Triangular-shaped region in the floor of the bladder that remains stable as the bladder fills.
triiodothyronine (tri-i-o-do-THI-ro-nin) Thyroid hormone that acts to raise cellular metabolism; T_3.
trophoblast (TRO-fo-blast) Region of the blastocyst in embryonic development that forms the fetal portion of the placenta.
tropomyosin (tro-po-MI-o-sin) Protein that works with troponin to regulate contraction in skeletal muscle.
troponin (tro-PO-nin) Protein that works with tropomyosin to regulate contraction in skeletal muscle.
trypsin (TRIP-sin) Pancreatic enzyme active in the digestion of proteins.
TSH See thyroid-stimulating hormone.
tympanic (tim-PAN-ik) **membrane** Membrane between the external and middle ear that transmits sound waves to the bones of the middle ear; eardrum.

U

ultrasound (UL-trah-sound) Very high-frequency sound waves; used in medical imaging to visualize soft structures.
umbilical (um-BIL-ih-kal) **cord** Structure that connects the fetus with the placenta; contains vessels that carry blood between the fetus and placenta.
umbilicus (um-BIL-ih-kus) Small scar on the abdomen that marks the former attachment of the umbilical cord to the fetus; navel.
universal solvent Term used for water because it dissolves more substances than any other solvent.
unsaturated fat Fat that has fewer hydrogen atoms and more double bonds between carbons than do saturated fats.
urea (u-RE-ah) Nitrogenous waste product excreted in the urine; end product of protein metabolism.
ureter (U-re-ter) Tube that carries urine from the kidney to the urinary bladder.
urethra (u-RE-thrah) Tube that carries urine from the urinary bladder to the outside of the body.
urinalysis (u-rin-AL-ih-sis) Laboratory examination of urine's physical and chemical properties.
urinary (U-rin-ar-e) **bladder** Hollow organ that stores urine until it is eliminated.
urinary system The system involved in elimination of soluble waste, water balance, and regulation of body fluids.
urination (u-rin-A-shun) Voiding of urine; micturition.
urine (U-rin) Liquid waste excreted by the kidneys.

uterine (U-ter-in) **tube** Tube that carries ova from the ovary to the uterus; fallopian tube.

uterus (U-ter-us) Muscular, pear-shaped organ in the female pelvis within which the fetus develops during pregnancy; adj., uterine.

uvula (U-vu-lah) Soft, fleshy, V-shaped mass that hangs from the soft palate.

V

vaccination (vak-sin-A-shun) Administration of a vaccine to protect against a specific disease; immunization.

vaccine (vak-SENE) Substance used to produce active immunity; usually, a suspension of attenuated or killed pathogens or some component of a pathogen given by inoculation to prevent a specific disease.

vagina (vah-JI-nah) Distal part of the birth canal that opens to the outside of the body; female organ of sexual intercourse.

vagus (VA-gus) **nerve** Tenth cranial nerve.

valence (VA-lens) Combining power of an atom; number of electrons lost, gained, or shared by atoms of an element in chemical reactions.

valve Structure that prevents fluid from flowing backward, as in the heart, veins, and lymphatic vessels.

variable (VAR-e-ah-bl) A value that is subject to change; capable of changing.

vas deferens (DEF-er-enz) Tube that carries sperm cells from the testis to the urethra; ductus deferens.

vascular (VAS-ku-lar) Pertaining to blood vessels.

vasectomy (vah-SEK-to-me) Surgical removal of part or all of the ductus (vas) deferens; usually done on both sides to produce sterility.

vasoconstriction (vas-o-kon-STRIK-shun) Decrease in a blood vessel's lumen diameter.

vasodilation (vas-o-di-LA-shun) Increase in a blood vessel's lumen diameter.

vasomotor (va-so-MO-tor) Pertaining to dilation or constriction of blood vessels.

vein (vane) Vessel that carries blood toward the heart.

vena cava (VE-nah KA-vah) Large vein that carries blood into the heart's right atrium; superior vena cava or inferior vena cava.

venous sinus (VE-nus SI-nus) Large channel that drains blood low in oxygen.

ventilation (ven-tih-LA-shun) Movement of air into and out of the lungs.

ventral (VEN-tral) At or toward the front or belly surface; anterior.

ventricle (VEN-trih-kl) Cavity or chamber. One of the heart's two lower chambers. One of the brain's four chambers in which cerebrospinal fluid is produced; adj., ventricular (ven-TRIK-u-lar).

venule (VEN-ule) Vessel between a capillary and a vein.

vernix caseosa (VER-niks ka-se-O-sah) Cheeselike sebaceous secretion that covers a newborn.

vertebra (VER-teh-brah) Bone of the spinal column; pl., vertebrae (VER-teh-bre).

vesicular transport Use of vesicles to move large amounts of material through a cell's plasma membrane; bulk transport.

vestibular apparatus (ves-TIB-u-lar) Part of the inner ear concerned with equilibrium; consists of the semicircular canals and vestibule.

vestibular (ves-TIB-u-lar) **folds** Folds of mucous membrane superior to the vocal cords in the larynx. They close off the glottis during swallowing and straining down; false vocal cords.

vestibule (VES-tih-bule) Any space at the entrance to a canal or organ; in the inner ear, area that contains some receptors for the sense of equilibrium.

vestibulocochlear (ves-tib-u-lo-KOK-le-ar) **nerve** Eighth cranial nerve concerned with hearing and equilibrium.

villi (VIL-li) Small finger-like projections from the surface of a membrane; projections in the lining of the small intestine through which digested food is absorbed; sing., villus.

viscera (VIS-er-ah) Organs in the ventral body cavities, especially the abdominal organs; adj., visceral.

viscosity (vis-KOS-ih-te) Thickness, as of the blood or other fluid.

vitamin (VI-tah-min) Organic compound needed in small amounts for health.

vitreous (VIT-re-us) **body** Soft, jelly-like substance that fills the eyeball and holds the shape of the eye; vitreous humor.

vocal folds Bands of mucous membrane in the larynx used in producing speech; vocal cords.

Volkmann canal See perforating canal.

vulva (VUL-va) External female genitalia.

W

Wernicke (VER-nih-ke) **area** Portion of the cerebral cortex concerned with speech recognition and the meaning of words.

white matter Nervous tissue composed of myelinated fibers.

word root The main part of a word to which prefixes and suffixes may be attached.

X

x-ray Ray or radiation of extremely short wavelength that can penetrate opaque substances and affect photographic plates and fluorescent screens.

Z

zygote (ZI-gote) Fertilized ovum; cell formed by the union of a sperm and an egg.

Glossary of Word Parts
Use of Word Parts in Medical Terminology

Medical terminology, the special language of the health occupations, is based on an understanding of a few relatively basic elements. These elements—roots, prefixes, and suffixes—form the foundation of almost all medical terms. A useful way to familiarize yourself with each term is to learn to pronounce it correctly and say it aloud several times. Soon it will become an integral part of your vocabulary.

The foundation of a word is the word root. Examples of word roots are *abdomin*, referring to the belly region, and *aden*, pertaining to a gland. A word root is often followed by a vowel to facilitate pronunciation when an ending is added, as in *abdomino* and *adeno*. We then refer to it as a "combining form," and it usually appears in texts with a slash before the vowel, as in *abdomin/o* and *aden/o*.

A prefix is a part of a word that precedes the word root and changes its meaning. For example, the prefix *pre-* in prenatal means "before," and the word means "before birth." A suffix, or word ending, is a part that follows the word root and adds to or changes its meaning. The suffix *-ism* means condition, so the term hypothyroidism means "an underactive (hypo-) thyroid condition."

Many medical words are compound words; that is, they are made up of more than one root or combining form. Examples of such compound words are *cardiovascular*, (pertaining to the heart and blood vessels), *lymphocyte* (a cell found in the lymphatic system), and *electroencephalograph* (an instrument for recording the electrical activity of the brain).

A general knowledge of language structure and spelling rules is also helpful in mastering medical terminology. For example, adjectives include words that end in *-al*, as in sternal (the noun is sternum), and words that end in *-ous*, as in mucous (the noun is mucus).

The following list includes some of the most commonly used word roots, prefixes, and suffixes, as well as examples of their use. Prefixes are followed by a hyphen; suffixes are preceded by a hyphen; and word roots have no hyphen. Commonly used combining vowels are added following a slash.

Word Parts

a-, an- not, without: *aphasia, atrophy, anemia, anuria*
ab- away from: *abduction, aboral*
abdomin/o belly or abdominal area: *abdominocentesis, abdominoscopy*
acous, acus hearing, sound: *acoustic, presbyacusis*
acr/o- end, extremity: *acromegaly, acromion*
actin/o, actin/i relation to raylike structures or, more commonly, to light or roentgen (x-) rays or some other type of radiation: *actiniform, actinodermatitis*
ad- (sometimes converted to *ac-, af-, ag-, ap-, as-, at-)* toward, added to, near: *adrenal, accretion, agglomerated, afferent*
aden/o gland: *adenectomy, adenitis, adenocarcinoma*
aer/o air, gas, oxygen: *aerobic, aerate*
-agogue inducing, leading, stimulating: *cholagogue, galactagogue*
-al pertaining to, resembling: *skeletal, surgical, ileal*
alb/i- white: *albinism, albinuria*
alge, alg/o, alges/i pain: *algetic, algophobia, analgesic*
-algia pain, painful condition: *myalgia, neuralgia*
amb/i- both, on two sides: *ambidexterity, ambivalent*
ambly- dimness, dullness: *amblyopia*
amphi- on both sides, around, double: *amphiarthrosis, amphibian*
amyl/o starch: *amylase, amyloid*
an- not, without: *anaerobic, anoxia, anemic*
ana- upward, back, again, excessive: *anatomy, anastomosis, anabolism*
andr/o male: *androgen, androgenous*
angi/o vessel: *angiogram, angiotensin*
ant/i- against; to prevent, suppress, or destroy: *antarthritic, antibiotic, anticoagulant*
ante- before, ahead of: *antenatal, antepartum*
anter/o- position ahead of or in front of (i.e., anterior to) another part: *anterolateral, anteroventral*
-apheresis take away, withdraw: *hemapheresis, plasmapheresis*
ap/o- separation, derivation from: *apocrine, apoptosis, apophysis*
aqu/e water: *aqueous, aquatic, aqueduct*
-ar pertaining to, resembling: *muscular, nuclear*
arthr/o joint or articulation: *arthrolysis, arthrostomy, arthritis*
-ary pertaining to, resembling: *salivary, dietary, urinary*
-ase enzyme: *lipase, protease*
-asis: See -sis.
atel/o- imperfect: *atelectasis*
ather/o gruel: *atherosclerosis, atheroma*
-ation process or condition: *discoloration, hydration*
audi/o sound, hearing: *audiogenic, audiometry, audiovisual*
aut/o- self: *autistic, autodigestion, autoimmune*

bar/o pressure, weight: *baroreceptor, barometer, bariatrics*
bas/o- alkaline: *basic, basophilic*
bi- two, twice: *bifurcate, bisexual*
bil/i bile: *biliary, bilirubin*
bio- life, living organism: *biopsy, antibiotic*
blast/o, -blast early stage of a cell, immature cell: *blastula, blastophore, erythroblast*
bleph, blephar/o eyelid, eyelash: *blepharism, blepharitis, blepharospasm*
brachi, brachi/o arm: *brachial, brachiocephalic, brachiotomy*

brachy- short: *brachydactylia, brachyesophagus*
brady- slow: *bradycardia*
bronch/o-, bronch/i bronchus: *bronchiectasis, bronchoscope*
bucc/o cheek: *buccal*

capn/o carbon dioxide: *hypocapnia, hypercapnia*
carcin/o cancer: *carcinogenic, carcinoma*
cardi/o, cardi/a heart: *carditis, cardiac, cardiologist*
cata- down: *catabolism, catalyst*
-cele swelling; enlarged space or cavity: *cystocele, meningocele, rectocele*
celi/o abdomen: *celiac, celiocentesis*
centi- relating to 100 (used in naming units of measurements): *centigrade, centimeter*
-centesis perforation, tapping: *amniocentesis, paracentesis*
cephal/o head: *cephalalgia, cephalopelvic*
cerebr/o brain: *cerebrospinal, cerebrum*
cervi neck: *cervical, cervix*
cheil/o lips; brim or edge: *cheilitis, cheilosis*
chem/o, chem/i chemistry, chemical: *chemotherapy, chemocautery, chemoreceptor*
chir/o, cheir/o hand: *cheiralgia, cheiromegaly, chiropractic*
chol/e, chol/o bile, gall: *cholagogue, cholecyst, cholelith*
cholecyst/o gallbladder: *cholecystitis, cholecystokinin*
chondr/o, chondri/o cartilage: *chondric, chondrocyte, chondroma*
chori/o membrane: *chorion, choroid, choriocarcinoma*
chrom/o, chromat/o color: *chromosome, chromatin, chromophilic*
-cid, -cide to cut, kill, destroy: *bactericidal, germicide, suicide*
circum- around, surrounding: *circumorbital, circumrenal, circumduction*
-clast break: *osteoclast*
clav/o, cleid/o clavicle: *cleidomastoid, subclavian*
co- with, together: *cofactor, cohesion, coinfection*
colp/o vagina: *colpectasia, colposcope, colpotomy*
con- with: *concentric, concentrate, conduct*
contra- opposed, against: *contraindication, contralateral*
corne/o horny: *corneum, cornified, cornea*
cortic/o cortex: *cortical, corticotropic, cortisone*
cost/a, cost/o- ribs: *intercostal, costosternal*
counter- against, opposite to: *counteract, counterirritation, countertraction*
crani/o skull: *cranium, craniotomy*
cry/o- cold: *cryalgesia, cryogenic, cryotherapy*
crypt/o- hidden, concealed: *cryptic, cryptogenic, cryptorchidism*
-cusis hearing: *acusis, presbyacusis*
cut skin: *subcutaneous*
cyan/o- blue: *cyanosis, cyanogen*
cyst/i, cyst/o sac, bladder: *cystitis, cystoscope*
cyt/o, -cyte cell: *cytology, cytoplasm, osteocyte*

dactyl/o digits (usually fingers, but sometimes toes): *dactylitis, polydactyly*
de- remove: *detoxify, dehydration*
dendr tree: *dendrite*
dent/o, dent/i tooth: *dentition, dentin, dentifrice*
derm/o, dermat/o skin: *dermatitis, dermatology, dermatosis*
di- twice, double: *dimorphism, dibasic, dihybrid*
dipl/o- double: *diplopia, diplococcus*
dia- through, between, across, apart: *diaphragm, diaphysis*
dis- apart, away from: *disarticulation, distal*
dors/i, dors/o- back (in the human, this combining form is the same as poster/o-): *dorsal, dorsiflexion, dorsonuchal*
dys- disordered, difficult, painful: *dysentery, dysphagia, dyspnea*

e- out: *enucleation, evisceration, ejection*
-ectasis expansion, dilation, stretching: *angiectasis, bronchiectasis*
ecto- outside, external: *ectoderm, ectogenous*
-ectomy surgical removal or destruction by other means: *appendectomy, thyroidectomy*
edem swelling: *edema*
-emia condition of blood: *glycemia, hyperemia*
encephal/o brain: *encephalitis, encephalogram*
end/o- in, within, innermost: *endarterial, endocardium, endothelium*
enter/o intestine: *enteritis, enterocolitis*
epi- on, upon: *epicardium, epidermis*
equi- equal: *equidistant, equivalent, equilibrium*
erg/o work: *ergonomic, energy, synergy*
eryth-, erythr/o- red: *erythema, erythrocyte*
-esthesia sensation: *anesthesia, paresthesia*
eu- well, normal, good: *euphoria, eupnea*
ex/o- outside, out of, away from: *excretion, exocrine, exophthalmic*
extra- beyond, outside of, in addition to: *extracellular, extrasystole, extravasation*

fasci fibrous connective tissue layers: *fascia, fascitis, fascicle*
fer, -ferent to bear, to carry: *afferent, efferent, transfer*
fibr/o threadlike structures, fibers: *fibrillation, fibroblast, fibrositis*

gastr/o stomach: *gastritis, gastroenterostomy*
-gen an agent that produces or originates: *allergen, pathogen, fibrinogen*
-genic produced from, producing: *neurogenic, pyogenic, psychogenic*
genit/o- organs of reproduction: *genitoplasty, genitourinary*
gen/o- related to reproduction or sex: *genealogy, generate, genetic, genotype*
-geny manner of origin, development, or production: *ontogeny, progeny*
gest/o gestation, pregnancy: *progesterone, gestagen*
glio, -glia gluey material; specifically, the support tissue of the central nervous system: *glioma, neuroglia*
gloss/o tongue: *glossitis, glossopharyngeal*
glyc/o- relating to sugar, glucose, sweet: *glycemia, glycosuria*
gnath/o related to the jaw: *prognathic, gnathoplasty*
gnos to perceive, recognize: *agnostic, prognosis, diagnosis*
gon seed, knee: *gonad, gonarthritis*
gonio angle: *goniometer, goniotomy*
-gram record, that which is recorded: *electrocardiogram, electroencephalogram*
graph/o, -graph instrument for recording, writing: *electrocardiograph, electroencephalograph, micrograph*

-graphy process of recording data: *photography, radiography*
gyn/o, gyne, gynec/o female, woman: *gynecology, gynecomastia, gynoplasty*
gyr/o circle: *gyroscope, gyrus, gyration*

hema, hemo, hemat/o blood: *hematoma, hematuria, hemorrhage*
hemi- one-half: *hemisphere, heminephrectomy, hemiplegia*
hepat/o- liver: *hepatitis, hepatogenous*
heter/o- other, different: *heterogeneous, heterosexual, heterochromia*
hist/o, histi/o tissue: *histology, histiocyte*
homeo-, homo- unchanging, the same: *homeostasis, homosexual*
hydr/o- water: *hydrolysis, hydrocephalus*
hyper- above, over, excessive: *hyperesthesia, hyperglycemia, hypertrophy*
hypo- deficient, below, beneath: *hypochondrium, hypodermic, hypogastrium*
hyster/o uterus: *hysterectomy*

-ia state of, condition of: *myopia, hypochondria, ischemia*
-iatrics, -trics medical specialty: *pediatrics, obstetrics*
iatr/o physician, medicine: *iatrogenic*
-ic pertaining to, resembling: *metric, psychiatric, geriatric*
idio- self, one's own, separate, distinct: *idiopathic, idiosyncrasy*
-ile pertaining to, resembling: *febrile, virile*
im-, in- in, into, lacking: *implantation, infiltration, inanimate*
infra- below, inferior: *infraspinous, infracortical*
insul/o pancreatic islet, island: *insulin, insulation, insulinoma*
inter- between: *intercostal, interstitial*
intra- within a part or structure: *intracranial, intracellular, intraocular*
isch suppression: *ischemia*
-ism state of: *alcoholism, hyperthyroidism*
iso- same, equal: *isotonic, isometric*
-ist one who specializes in a field of study: *cardiologist, gastroenterologist*
-itis inflammation: *dermatitis, keratitis, neuritis*

juxta- next to: *juxtaglomerular, juxtaposition*

kary/o nucleus: *karyotype, karyoplasm*
kerat/o cornea of the eye, certain cornified tissues: *keratin, keratitis, keratoplasty*
kine movement, motion: *kinesiology, kinesthesia*
kinet/o movement, motion: *kinetic, kinetogenic*

lacri- tear: *lacrimal*
lact/o milk: *lactation, lactogenic*
laryng/o larynx: *laryngeal, laryngectomy, laryngitis*
later/o- side: *lateral*
-lemma sheath: *neurilemma, sarcolemma*
leuk/o- (also written as *leuc-, leuco-*) white, colorless: *leukocyte, leukoplakia*
lip/o lipid, fat: *lipase, lipoma*
lig- bind: *ligament, ligature*
lingu/o tongue: *lingual, linguodental*
lith/o stone (calculus): *lithiasis, lithotripsy*
-logy study of: *physiology, gynecology*
lute/o yellow: *macula lutea, corpus luteum*
lymph/o lymph, lymphatic system, lymphocyte: *lymphoid, lymphedema*
lyso-, -lysis, -lytic loosening, dissolving, separating: *hemolysis, paralysis, lysosome*

macr/o- large, abnormal length: *macrophage, macroblast.* See also mega-, megal/o-.
mal- bad, diseased, disordered, abnormal: *malnutrition, malocclusion, malunion*
malac/o, -malacia softening: *malacoma, osteomalacia*
mamm/o- breast, mammary gland: *mammogram, mammoplasty, mammal*
man/o pressure: *manometer, sphygmomanometer*
mast/o breast: *mastectomy, mastitis*
meg/a-, megal/o, -megaly unusually or excessively large: *megacolon, megaloblast, splenomegaly, megakaryocyte*
melan/o dark, black: *melanin, melanocyte, melanoma*
men/o physiologic uterine bleeding, menses: *menstrual, menorrhagia, menopause*
mening/o membranes covering the brain and spinal cord: *meningitis, meningocele*
mes/a, mes/o- middle, midline: *mesencephalon, mesoderm*
meta- change, beyond, after, over, near: *metabolism, metacarpal, metaplasia*
-meter, metr/o measure: *hemocytometer, sphygmomanometer, spirometer, isometric*
metr/o uterus: *endometrium, metroptosis, metrorrhagia*
micro- very small: *microscope, microbiology, microsurgery, micrometer*
mon/o- single, one: *monocyte, mononucleosis*
morph/o- shape, form: *morphogenesis, morphology*
multi- many: *multiple, multifactorial, multipara*
my/o- muscle: *myenteron, myocardium, myometrium*
myc/o, mycet fungus: *mycid, mycete, mycology, mycosis, mycelium*
myel/o marrow (often used in reference to the spinal cord): *myeloid, myeloblast, osteomyelitis, poliomyelitis*
myring/o tympanic membrane: *myringotomy, myringitis*
myx/o mucus: *myxoma, myxovirus*

narc/o- stupor: *narcosis, narcolepsy, narcotic*
nas/o nose: *nasopharynx, paranasal*
natri sodium: *hyponatremia, natriuretic*
necr/o death, corpse: *necrosis*
neo- new: *neoplasm, neonatal*
neph, nephr/o kidney: *nephrectomy, nephron*
neur/o, neur/i nerve, nervous tissue: *neuron, neuralgia, neuroma*
neutr/o neutral: *neutrophil, neutropenia*
noct/i night: *noctambulation, nocturia, noctiphobia*

ocul/o- eye: *ocular, oculomotor*
odont/o- tooth, teeth: *odontalgia, orthodontics*
-odynia pain, tenderness: *myodynia, neurodynia*
-oid like, resembling: *lymphoid, myeloid*

olig/o- few, a deficiency: *oligospermia, oliguria*
-oma tumor, swelling: *hematoma, sarcoma*
-one ending for steroid hormone: *testosterone, progesterone*
onych/o nails: *paronychia, onychoma*
oo ovum, egg: *oocyte, oogenesis* (do not confuse with oophor-)
oophor/o ovary: *oophorectomy, oophoritis, oophorocystectomy.* See also ovar-.
ophthalm/o- eye: *ophthalmia, ophthalmologist, ophthalmoscope*
-opia condition of the eye or vision: *heterotropia, myopia, hyperopia*
opt/o eye: optic, optometrist
or/o mouth: *oropharynx, oral*
orchi/o, orchid/o testis: *orchitis, cryptorchidism*
orth/o- straight, normal: *orthopedics, orthopnea, orthosis*
-ory pertaining to, resembling: *respiratory, circulatory*
oscill/o to swing to and fro: *oscilloscope*
osmo- osmosis: *osmoreceptor, osmotic*
oss/i, osse/o, oste/o bone, bone tissue: *osseous, ossicle, osteocyte, osteomyelitis*
ot/o ear: *otalgia, otitis, otomycosis*
-ous pertaining to, resembling: *fibrous, venous, androgynous*
ov/o egg, ovum: *oviduct, ovulation*
ovar, ovari/o ovary: *ovariectomy.* See also oophor.
ox-, -oxia pertaining to oxygen: *hypoxemia, hypoxia, anoxia*
oxy sharp, acute: *oxygen, oxytocia*

pan- all: *pandemic, panacea*
papill/o nipple: *papilloma, papillary*
para- near, beyond, apart from, beside: *paramedical, parametrium, parathyroid, parasagittal*
pariet/o wall: *parietal*
path/o, -pathy disease, abnormal condition: *pathogen, pathology, neuropathy*
ped/o-, -pedia child, foot: *pedophobia, pediatrician, pedialgia*
-penia lack of: *leukopenia, thrombocytopenia*
per- through, excessively: *percutaneous, perfusion*
peri- around: *pericardium, perichondrium*
-pexy fixation: *nephropexy, proctopexy*
phag/o to eat, to ingest: *phagocyte, phagosome*
-phagia, -phagy eating, swallowing: *aphagia, dysphagia*
-phasia speech, ability to talk: a*phasia, dysphasia*
phen/o to show: *phenotype*
-phil, -philic to like, have an affinity for: *eosinophilia, hemophilia, hydrophilic*
phleb/o vein: *phlebitis, phlebotomy*
-phobia fear, dread, abnormal aversion: *phobic, acrophobia, hydrophobia*
phot/o- light: *photoreceptor, photophobia*
phren/o diaphragm: *phrenic, phrenicotomy*
physi/o- natural, physical: *physiology, physician*
pil/e-, pil/i-, pil/o- hair, resembling hair: *pileous, piliation, pilonidal*
pin/o to drink: *pinocytosis*
-plasty molding, surgical formation: *cystoplasty, gastroplasty, kineplasty*
-plegia stroke, paralysis: *paraplegia, hemiplegia*
pleur/o side, rib, pleura: *pleurisy, pleurotomy*
-pnea air, breathing: *dyspnea, eupnea*
pneum/o-, pneumat/o- air, gas, respiration: *pneumothorax, pneumograph, pneumatocele*
pneumon/o lung: *pneumonia, pneumonectomy*
pod/o foot: *podiatry, pododynia*
-poiesis making, forming: *erythropoiesis, hematopoiesis*
polio- gray, gray matter: *polioencephalitis, poliomyelitis*
poly- many: *polyarthritis, polycystic, polycythemia*
post- behind, after, following: *postnatal, postocular, postpartum*
pre- before, ahead of: *precancerous, preclinical, prenatal*
presby- old age: *presbycusis, presbyopia*
pro- before, in front of, in favor of: *prodromal, prosencephalon, prolapse, prothrombin*
proct/o rectum: *proctitis, proctocele, proctologist*
propri/o own: *proprioception*
pseud/o false: *pseudoarthrosis, pseudostratified, pseudopod*
psych/o mind: *psychosomatic, psychotherapy, psychosis*
-ptosis downward displacement, falling, prolapse: *blepharoptosis, enteroptosis, nephroptosis*
pulm/o, pulmon/o lung: *pulmonary, pulmonic, pulmonology*
py/o pus: *pyuria, pyogenic, pyorrhea*
pyel/o renal pelvis: *pyelitis, pyelogram, pyelonephrosis*
pyr/o fire, fever: *pyrogen, antipyretic, pyromania*

quadr/i- four: *quadriceps, quadriplegic*

rachi/o spine: *rachicentesis, rachischisis*
radio- emission of rays or radiation: *radioactive, radiography, radiology*
re- again, back: *reabsorption, reaction, regenerate*
rect/o rectum: *rectal, rectouterine*
ren/o kidney: *renal, renopathy*
reticul/o network: *reticulum, reticular*
retro- backward, located behind: *retrocecal, retroperitoneal*
rhin/o nose: *rhinitis, rhinoplasty*
-rhage, -rhagia* bursting forth, excessive flow: *hemorrhage, menorrhagia*
-rhaphy* suturing or sewing up a gap or defect in a part: *herniorrhaphy, gastrorrhaphy, cystorrhaphy*
-rhea* flow, discharge: *diarrhea, gonorrhea, seborrhea*

sacchar/o sugar: *monosaccharide, polysaccharide*
salping/o tube: *salpingitis, salpingoscopy*
sarc/o flesh: *sarcolemma, sarcoplasm, sarcomere*
scler/o- hard, hardness: *scleroderma, sclerosis*
scoli/o- twisted, crooked: *scoliosis, scoliometer*
-scope instrument used to look into or examine a part: *bronchoscope, endoscope, arthroscope*
semi- partial, half: *semipermeable, semicoma*
semin/o semen, seed: *seminiferous, seminal*
sep, septic poison, rot, decay: *sepsis, septicemia*
sin/o sinus: *sinusitis, sinusoid, sinoatrial*
-sis condition or process, usually abnormal: *dermatosis, osteoporosis*
soma-, somat/o, -some body: *somatic, somatotype, somatotropin*
son/o- sound: *sonogram, sonography*
sphygm/o- pulse: *sphygmomanometer*

spir/o breathing: *spirometer, inspiration, expiration*
splanchn/o- internal organs: *splanchnic, splanchnoptosis*
splen/o spleen: *splenectomy, splenic*
staphyl/o- grapelike cluster: *staphylococcus*
stat, -stasis stand, stoppage, remain at rest: *hemostasis, static, homeostasis*
sten/o- contracted, narrowed: *stenosis*
sthen/o, -sthenia, -sthenic strength: *asthenic, calisthenics, neurasthenia*
steth/o- chest: *stethoscope*
stoma, stomat/o mouth: *stomatitis*
-stomy surgical creation of an opening into a hollow organ or an opening between two organs: *colostomy, tracheostomy, gastroenterostomy*
strept/o chain: *streptococcus, streptobacillus*
sub- under, below, near, almost: *subclavian, subcutaneous, subluxation*
super- over, above, excessive: *superego, supernatant, superficial*
supra- above, over, superior: *supranasal, suprarenal*
sym-, syn- with, together *symphysis, synapse*
syring/o fistula, tube, cavity: *syringectomy, syringomyelia*

tach/o-, tachy- rapid: *tachycardia, tachypnea*
tars/o eyelid, foot: *tarsitis, tarsoplasty, tarsoptosis*
-taxia, -taxis order, arrangement: *ataxia, chemotaxis, thermotaxis*
tel/o- end, far, at a distance: *telophase, telomere, telegraph, telemetry*
tens- stretch, pull: *extension, tensor*
terat/o malformed fetus: *teratogen, teratogenic*
test/o testis: *testosterone, testicular*
tetr/a four: *tetralogy, tetraplegia*
therm/o-, -thermy heat: *thermalgesia, thermocautery, diathermy, thermometer*
thromb/o- blood clot: *thrombosis, thrombocyte*
toc/o labor: *eutocia, dystocia, oxytocin*
tom/o, -tomy incision of, cutting: *anatomy, phlebotomy, laparotomy*
ton/o tone, tension: *tonicity, tonic*
tox-, toxic/o- poison: *toxin, cytotoxic, toxemia, toxicology*
trache/o trachea, windpipe: *tracheal, tracheitis, tracheotomy*
trans- across, through, beyond: *transorbital, transpiration, transplant, transport*
tri- three: *triad, triceps*
trich/o hair: *trichiasis, trichosis, trichology*
troph/o, -trophic, -trophy nutrition, nurture: *atrophic, atrophy, hypertrophy*
trop/o, -tropin, -tropic turning toward, acting on, influencing, changing: *thyrotropin, adrenocorticotropic, gonadotropic*
tympan/o drum: *tympanic, tympanum*

ultra- beyond or excessive: *ultrasound, ultraviolent, ultrastructure*
uni- one: *unilateral, uniovular, unicellular*
-uria urine: *glycosuria, hematuria, pyuria*
ur/o urine, urinary tract: *urology, urogenital*

vas/o vessel, duct: *vascular, vasectomy, vasodilation*
viscer/o internal organs, viscera: *visceral, visceroptosis*
vitre/o- glasslike: *vitreous*
xer/o- dryness: *xeroderma, xerophthalmia, xerosis*

-y condition of: *tetany, atony, dysentery*

zyg/o joined: *zygote, heterozygous, monozygotic*

---

*When a suffix beginning with *rh* is added to a word root, the *r* is doubled.

Appendix 1 Periodic Table of the Elements

The periodic table lists the chemical elements according to their atomic numbers. The boxes in the table have information about the elements, as shown by the example at the top of the chart. The upper number in each box is the atomic number, which represents the number of protons in the nucleus of the atom. Under the name of the element is its chemical symbol, an abbreviation of its modern or Latin name. The Latin names of four common elements are shown below the chart. The bottom number in each box gives the atomic weight (mass) of that element's atoms compared with the weight of carbon atoms. Atomic weight is the sum of the weights of the protons and neutrons in the nucleus.

All the elements in a column share similar chemical properties based on the number of electrons in their outermost energy levels. Those in column VIII are nonreactive (inert) and are referred to as noble gases. The 26 elements found in the body are color coded according to quantity (see totals above the chart). Carbon, hydrogen, oxygen, and nitrogen make up 96% of body weight. The first three of these are present in all carbohydrates, lipids, proteins, and nucleic acids. Nitrogen is an additional component of all proteins. Nine other elements make up almost all the rest of the body weight. The remaining 13 elements are present in very small amounts and are referred to as trace elements. Although needed in very small quantities, they are essential for good health, as they are parts of enzymes and other compounds used in metabolism.

PERIODIC TABLE OF THE ELEMENTS

96% of body weight
3.9% of body weight
0.1% of body weight

Notation:

6 Carbon C 12.01	
6	Atomic number
Carbon	Name
C	Symbol
12.01	Atomic weight

I	II											III	IV	V	VI	VII	VIII
1 Hydrogen H 1.01																	2 Helium He 4.00
3 Lithium Li 6.94	4 Beryllium Be 9.01											5 Boron B 10.81	6 Carbon C 12.01	7 Nitrogen N 14.01	8 Oxygen O 16.00	9 Fluorine F 19.00	10 Neon Ne 20.18
11 Sodium Na 22.99	12 Magnesium Mg 24.31											13 Aluminum Al 26.98	14 Silicon Si 28.09	15 Phosphorus P 30.97	16 Sulfur S 32.07	17 Chlorine Cl 35.45	18 Argon Ar 39.95
19 Potassium K 39.10	20 Calcium Ca 40.08	21 Scandium Sc 44.96	22 Titanium Ti 47.88	23 Vanadium V 50.94	24 Chromium Cr 52.00	25 Manganese Mn 54.94	26 Iron Fe 55.85	27 Cobalt Co 58.93	28 Nickel Ni 58.69	29 Copper Cu 63.55	30 Zinc Zn 65.39	31 Gallium Ga 69.72	32 Germanium Ge 72.59	33 Arsenic As 74.92	34 Selenium Se 78.96	35 Bromine Br 79.90	36 Krypton Kr 83.80
37 Rubidium Rb 85.47	38 Strontium Sr 87.62	39 Yttrium Y 88.91	40 Zirconium Zr 91.22	41 Niobium Nb 92.91	42 Molybdenum Mo 95.94	43 Technetium Tc (98)	44 Ruthenium Ru 101.1	45 Rhodium Rh 102.9	46 Palladium Pd 106.4	47 Silver Ag 107.9	48 Cadmium Cd 112.4	49 Indium In 114.8	50 Tin Sn 118.7	51 Antimony Sb 121.8	52 Tellurium Te 127.6	53 Iodine I 126.9	54 Xenon Xe 131.3
55 Cesium Cs 132.91	56 Barium Ba 137.34		72 Hafnium Hf 178.5	73 Tantalum Ta 180.9	74 Tungsten W 183.9	75 Rhenium Re 186.2	76 Osmium Os 190.2	77 Iridium Ir 192.2	78 Platinum Pt 195.1	79 Gold Au 196.9	80 Mercury Hg 200.6	81 Thallium Tl 204.4	82 Lead Pb 207.2	83 Bismuth Bi 209.0	84 Polonium Po (210)	85 Astatine At (210)	86 Radon Rn (222)
87 Francium Fr (223)	88 Radium Ra (226)		104 Rutherfordium Rf (257)	105 Dubnium Db (260)	106 Seaborgium Sg (263)	107 Bohrium Bh (262)	108 Hassium Hs (265)	109 Meitnerium Mt (267)	110 Darmstadtium Ds (271)	111 Unnamed (272)	112 Unnamed (277)						

57–71 Lanthanides	57 Lanthanum La 138.9	58 Cerium Ce 140.1	59 Praseodymium Pr 140.9	60 Neodymium Nd 144.2	61 Promethium Pm (145)	62 Samarium Sm (150.4)	63 Europium Eu 152.0	64 Gadolinium Gd 157.3	65 Terbium Tb 158.9	66 Dysprosium Dy 162.5	67 Holmium Ho 164.9	68 Erbium Er 167.3	69 Thulium Tm 168.9	70 Ytterbium Yb 173.0	71 Lutetium Lu 175.0
89–103 Actinides	89 Actinium Ac (227)	90 Thorium Th 232.0	91 Protactinium Pa (231)	92 Uranium U (238)	93 Neptunium Np (237)	94 Plutonium Pu (244)	95 Americium Am (243)	96 Curium Cm (247)	97 Berkelium Bk (247)	98 Californium Cf (251)	99 Einsteinium Es (254)	100 Fermium Fm (257)	101 Mendelevium Md (256)	102 Nobelium No (259)	103 Lawrencium Lr (257)

Name	Latin name	Symbol
Copper	*cuprium*	Cu
Iron	*ferrum*	Fe
Potassium	*kalium*	K
Sodium	*natrium*	Na

Appendix  Answers to Chapter Checkpoint, Casepoint, and Zooming In Questions

Chapter 1

Answers to Checkpoint Questions

1-1 The study of body structure is anatomy; the study of body function is physiology.
1-2 Organs working together combine to form systems.
1-3 Homeostasis is a state of internal balance.
1-4 The components of a negative feedback loop are sensor, control center, and effector.
1-5 A barrier is any structure that separates differing environments.
1-6 A gradient represents a difference in a specific physical or chemical value between two regions.
1-7 Resistance is a factor that opposes flow along a gradient.
1-8 Distal describes a location farther from an origin. The wrist is distal to the elbow.
1-9 The three planes in which the body can be cut are sagittal, frontal (coronal), and transverse (horizontal).
1-10 The two main body cavities are the dorsal and ventral cavities.
1-11 The three central regions of the abdomen are the epigastric, umbilical, and hypogastric regions. The three left and right lateral regions are the hypochondriac, lumbar, and iliac (inguinal) regions.

Answers to Casepoint Questions

1-1 The spinal column is part of the skeletal system; the blood vessels are part of the cardiovascular system.
1-2 The major challenge to Mike's homeostasis was bleeding, leading to low blood pressure.
1-3 The heart is an effector in the feedback loop controlling blood pressure.
1-4 In *hypotensive*, the prefix is *hypo-* meaning deficiency, below, or beneath; the root is *tens*, which means to stretch or pull. In *tachycardic*, the prefix is *tachy-*, meaning *rapid*; the root is *cardi/o*, which pertains to the heart.

Answers to Zooming In Questions

1-5 Movement will increase if the angle of the ramp increases (which increases the gradient). Movement will slow if the surface of the ramp provides more resistance.
1-6 The figures are standing in the anatomic position.
1-7 The transverse (horizontal) plane divides the body into superior and inferior parts. The frontal (coronal) plane divides the body into anterior and posterior parts.
1-10 The ventral cavity contains the diaphragm.
1-13 The umbilical, hypogastric, left lumbar, and left iliac (inguinal) regions are represented in the left lower quadrant.

Chapter 2

Answers to Checkpoint Questions

2-1 Atoms are subunits of elements.
2-2 Three types of particles found in atoms are protons, neutrons, and electrons.
2-3 The atom with six electrons in its outermost energy level is more likely to participate in a chemical reaction, as the outermost energy level of the atom with eight electrons is complete.
2-4 Ionic bonds are formed by exchange of electrons between atoms. Covalent bonds are formed by the sharing of electrons between atoms.
2-5 When an electrolyte goes into solution, it separates into charged particles called ions (cations and anions).
2-6 Molecules are units composed of two or more covalently bonded atoms. Compounds are substances composed of two or more different elements.
2-7 $X + Y \rightleftarrows Z$.
2-8 In a solution, the solute dissolves and remains evenly distributed (the mixture is homogeneous); in a suspension, the material in suspension does not dissolve and settles out unless the mixture is shaken (the mixture is heterogeneous).
2-9 Water is the most abundant compound in the body.
2-10 A value of 7.0 is neutral on the pH scale. An acid measures lower than 7.0; a base measures higher than 7.0.
2-11 A buffer is a substance that maintains a steady pH of a solution.
2-12 Isotopes that give off radiation are termed radioactive and are called radioisotopes.
2-13 Carbon is the basis of organic chemistry.
2-14 The three main categories of organic compounds are carbohydrates, lipids, and proteins.
2-15 An enzyme is a catalyst that speeds the rate of chemical reactions in the body.
2-16 A nucleotide contains a sugar, a phosphate group, and a nitrogenous base. DNA, RNA, and ATP are examples of compounds composed of nucleotides.
2-17 Catabolism and anabolism are the two types of metabolic activities. In catabolism, complex substances are broken down into simpler components. In anabolism, simple substances are used to manufacture materials needed for growth, function, and tissue repair.
2-18 The energy in ATP is potential energy and chemical energy.

Answers to Casepoint Questions

2-1 The emergency team administered an aqueous solution with small amounts of glucose and salts.
2-2 Water is the solvent in this solution.
2-3 At a blood pH of 7.28, Margaret is suffering from acidosis.
2-4 The simple sugar glucose, a monosaccharide, was administered to Margaret in the case study.

Answers to Zooming In Questions

2-2 The number of protons is equal to the number of electrons. There are eight protons and eight electrons in the oxygen atom.
2-3 Oxygen needs two electrons to complete its outermost energy level.
2-4 A sodium atom has one electron in its outermost energy level. A sodium ion has eight electrons in its outermost energy level.
2-5 Two electrons are needed to complete the energy level of each hydrogen atom in a hydrogen molecule.

2-6 Two hydrogen atoms bond with an oxygen atom to form water.
2-7 The amount of hydroxide ion (OH^-) in a solution decreases when the amount of hydrogen ion (H^+) increases.
2-8 The building blocks (monomers) of disaccharides and polysaccharides are monosaccharides.
2-9 There are three carbon atoms in glycerol.
2-10 The amino group of an amino acid contains nitrogen.
2-11 The shape of an enzyme after the reaction is the same as it was before the reaction.
2-12 The prefix *tri-* means three.
2-14 Carbon, hydrogen, and oxygen combine to form glucose.

Chapter 3

Answers to Checkpoint Questions

3-1 The cell shows organization, metabolism, responsiveness, homeostasis, growth, and reproduction.
3-2 Three types of microscopes are the compound light microscope, transmission electron microscope (TEM), and scanning electron microscope (SEM).
3-3 The four substances found within the plasma membrane are phospholipids, cholesterol, carbohydrates (glycoproteins and glycolipids), and proteins.
3-4 The cell organelles are specialized structures that perform different tasks.
3-5 The nucleus is called the cell's control center because it contains the chromosomes, hereditary units that control all cellular activities.
3-6 The two types of organelles used for movement are cilia, which are small and hairlike, and the flagellum, which is long and whiplike.
3-7 Diffusion, osmosis, and filtration do not directly require chemical energy. Active transport and vesicular transport require chemical energy. Vesicular transport includes endocytosis (phagocytosis, pinocytosis, receptor-mediated endocytosis) and exocytosis.
3-8 An isotonic solution is the same concentration as the cytoplasm; a hypotonic solution is less concentrated; a hypertonic solution is more concentrated.
3-9 Nucleotides are the building blocks of nucleic acids.
3-10 DNA codes for proteins in the cell.
3-11 The three types of RNA active in protein synthesis are messenger RNA (rRNA), ribosomal RNA (rRNA), and transfer RNA (tRNA).
3-12 Before mitosis can occur, the DNA must replicate (double). Replication occurs during interphase.
3-13 The four stages of mitosis are prophase, metaphase, anaphase, and telophase.

Answers to Casepoint Questions

3-1 A channel is a protein that allows ions to pass through the plasma membrane.
3-2 Ribosomes, rough ER, the Golgi apparatus, and vesicles are organelles that synthesize, modify, and transport proteins to the plasma membrane.
3-3 Preventing chloride from entering cells would make the surrounding fluid hypertonic.
3-4 Deletion of three consecutive nucleotides in the CFTR gene would result in deletion of one amino acid.

Answers to Zooming In Questions

3-1 The transmission electron microscope (TEM) in B shows the most internal structure. The scanning electron microscope (SEM) in C shows the cilia in three dimensions.
3-2 Ribosomes attached to the ER make it look rough. Cytosol is the liquid part of the cytoplasm.
3-3 The plasma membrane is described as a bilayer because it is constructed of two layers of phospholipids.
3-4 Epithelial cells (B) would best cover a large surface area because they are flat.
3-6 A decrease in the number of transporters would decrease solute transport.
3-7 If the solute could pass through the membrane in this system, the solute and solvent molecules would equalize on the two sides of the membrane, and the fluid level would be the same on both sides.
3-8 If the concentration of solute was increased on side B of this system, the osmotic pressure would increase.
3-9 If lost blood were replaced with pure water, red blood cells would swell because the blood plasma would become hypotonic to the cells.
3-10 A lysosome would likely help destroy a particle taken in by phagocytosis.
3-12 The nucleotides pair up so that there is one larger nucleotide and one smaller nucleotide in each pair. Specifically, A pairs with T, and G pairs with C.
3-15 If the original cell has 46 chromosomes, each daughter cell will have 46 chromosomes after mitosis.

Chapter 4

Answers to Checkpoint Questions

4-1 The three basic shapes of epithelial cells are squamous (flat and irregular), cuboidal (square), and columnar (long and narrow).
4-2 Exocrine glands secrete into a nearby organ, cavity, or to the surface of the skin and generally secrete through ducts. Endocrine glands secrete directly into surrounding tissue fluids and into the blood.
4-3 The intercellular material in connective tissue is the matrix.
4-4 Collagen makes up the most abundant fibers in connective tissue.
4-5 Fibroblasts characterize dense connective tissue. Chondrocytes characterize cartilage. Osteoblasts and osteocytes characterize bone tissue.
4-6 The three types of muscle tissue are skeletal (voluntary), cardiac, and smooth (visceral) muscle.
4-7 The basic cell of the nervous system is the neuron, and it carries nerve impulses.
4-8 The nonconducting support cells of the nervous system are neuroglia (glial cells).
4-9 The three types of epithelial membranes are serous membranes, mucous membranes, and the cutaneous membrane (skin).
4-10 A parietal serous membrane is attached to the wall of a cavity or sac. A visceral serous membrane is attached to the surface of an organ.
4-11 Fascia is a fibrous band or sheet that supports and holds organs. Superficial fascia underlies the skin, and deep fascia covers muscles, nerves, and vessels.

Answers to Casepoint Questions

4-1 Squamous cells are flat and irregular.
4-2 The cutaneous membrane was involved in Paul's case study. It is an epithelial membrane composed of stratified squamous epithelium and dense irregular connective tissue.

Answers to Zooming In Questions

4-1 The epithelial cells are in a single layer.
4-2 Stratified epithelium protects underlying tissue from wear and tear.
4-4 Of the tissues shown, areolar connective tissue has the most fibers; adipose tissue is modified for storage.

Chapter 5

Answers to Checkpoint Questions

5-1 The integumentary system comprises the skin and all its associated structures.
5-2 The epidermis is the superficial layer of the skin; the dermis is the deeper layer.
5-3 The subcutaneous layer is composed of loose connective tissue and adipose (fat) tissue.
5-4 The sebaceous glands produce an oily secretion called sebum.
5-5 The sweat glands are the sudoriferous glands.
5-6 A hair follicle is the sheath in which a hair develops.
5-7 The active cells that produce a nail are located in the nail root at the proximal end of the nail.
5-8 Keratin and sebum help prevent dehydration.
5-9 Temperature is regulated through the skin by dilation (widening) and constriction (narrowing) of blood vessels and by evaporation of perspiration from the body surface.
5-10 Melanin, hemoglobin, and carotene are pigments that give color to the skin.
5-11 Four factors that affect healing are nutrition, blood supply, infection, and age.

Answers to Casepoint Questions

5-1 Hazels' hand burn involved the epidermis; the forearm burn involved both the epidermis and the dermis.
5-2 The stratum basale (germinativum) will produce new cells to replace Hazel's damaged epidermis.
5-3 The physician used antibacterial cream because without the skin as a barrier, the underlying tissue is subject to infection.
5-4 Pain indicated that the nerve endings in the dermis were not destroyed.
5-5 Hemoglobin gave Hazel's skin its color at the burn site.

Answers to Zooming In Questions

5-1 The epidermis is supplied with oxygen and nutrients from blood vessels in the dermis. The subcutaneous layer is located beneath the skin.
5-4 The sebaceous glands and apocrine sweat glands secrete to the outside through the hair follicles. The sweat glands are made of simple cuboidal epithelium.

Chapter 6

Answers to Checkpoint Questions

6-1 The diaphysis is the shaft of a long bone; the epiphysis is the end of a long bone.
6-2 Calcium compounds are deposited in the intercellular matrix of the embryonic skeleton to harden it.
6-3 The cells found in bone are osteoblasts, which build bone tissue; osteocytes, which maintain bone; and osteoclasts, which break down (resorb) bone.
6-4 Compact bone makes up the main shaft of a long bone and the outer layer of other bones; spongy (cancellous) bone makes up the ends of the long bones and the center of other bones. It also lines the medullary cavity of the long bones.
6-5 The epiphyseal plates are the secondary growth centers of a long bone.
6-6 Parathyroid hormone (PTH) stimulates osteoclast activities.
6-7 The markings on bones help form joints, are points for muscle attachments, and allow passage of nerves and blood vessels.
6-8 The skeleton of the trunk consists of the vertebral column and the bones of the thorax, which are the ribs and sternum.
6-9 The five regions of the vertebral column are the cervical vertebrae, thoracic vertebrae, lumbar vertebrae, sacrum, and coccyx.
6-10 The four regions of the appendicular skeleton are the shoulder girdle, upper extremity, pelvic bones, and lower extremity.
6-11 Phalanges constitute the bones of the fingers and toes.
6-12 The three types of joints based on the degree of movement allowed are synarthrosis, amphiarthrosis, and diarthrosis.
6-13 A synovial joint or diarthrosis is the most freely moveable type of joint.

Answers to Casepoint Questions

6-1 The end of a long bone is the epiphysis.
6-2 The fracture line would contain red marrow.
6-3 Osteoblasts would produce new bone tissue.
6-4 Reggie's fracture crosses the proximal end of the bone.
6-5 Reggie's fracture would not pass through the medullary cavity.
6-6 The hip joint that is proximal to Reggie's fracture is a diarthrotic joint, synovial joint, and ball-and-socket joint.
6-7 Abduction is the term that describes movement away from the midline of the body.

Answers to Zooming In Questions

6-2 The periosteum is the membrane on the outside of a long bone; the endosteum is the membrane on the inside of a long bone.
6-3 Osteocytes are located in the spaces (lacunae) of compact bone.
6-4 In fracture repair, a soft callus forms before the bony callus.
6-6 The maxilla and palatine bones make up each side of the hard palate. A foramen is a hole. The ethmoid makes up the superior and middle conchae.
6-7 The anterior fontanelle is the largest fontanelle.
6-8 The cervical and lumbar vertebrae form a convex curve. The lumbar vertebrae are the largest and heaviest because they bear the most weight.
6-9 The costal cartilages attach to the ribs.
6-10 The prefix *supra-* means above; the prefix *infra-* means below.
6-11 The ulna is the medial bone of the forearm.
6-13 The olecranon of the ulna forms the bony prominence of the elbow.
6-14 There are 14 phalanges on each hand.
6-15 The ischium is nicknamed the "sit bone."
6-17 The fibula is the lateral bone of the leg; the tibia is weight bearing.
6-18 The calcaneus is the heel bone; the talus forms a joint with the tibia.
6-19 The greater trochanter is a point for ligament attachment; articular (hyaline) cartilage covers the ends of a bone.

Chapter 7

Answers to Checkpoint Questions

7-1 The three types of muscle are smooth muscle, cardiac muscle, and skeletal muscle.
7-2 The three main functions of skeletal muscle are movement of the skeleton, maintenance of posture, and generation of heat.
7-3 Bundles of muscle fibers are called fascicles.
7-4 Myofibrils are the bundles of protein molecules within individual muscle fibers.
7-5 The membrane or transmembrane potential is the term for the difference in electrical charge on the two sides of a membrane.
7-6 The neuromuscular junction (NMJ) is the special synapse where a nerve cell makes contact with a muscle cell.
7-7 Acetylcholine (ACh) is the neurotransmitter involved in the stimulation of skeletal muscle cells.
7-8 Calcium is needed to allow the contractile filaments, actin and myosin, to interact.
7-9 ATP is the compound produced by the oxidation of nutrients that supplies the energy for muscle contraction.
7-10 Myoglobin stores reserves of oxygen in muscle cells.
7-11 The two main types of muscle contraction are isotonic and isometric contractions.
7-12 The origin is the muscle's attachment to a less movable part of the skeleton; the insertion is the attachment to a movable part of the skeleton. When a muscle contracts, the insertion is brought closer to the origin.
7-13 The muscle that produces a movement is the prime mover; the muscle that produces the opposite movement is the antagonist.
7-14 A third-class lever represents the action of most muscles. In this system, the fulcrum is behind the point of effort and the resistance (weight).
7-15 The diaphragm is the muscle most important in breathing.
7-16 The muscles of the abdominal wall are strengthened by having the muscle fibers run in different directions.

Answers to Casepoint Questions

7-1 Skeletal (striated) muscles are the effectors in voluntary movement.
7-2 The voluntary, or somatic, nervous system controls voluntary movements.
7-3 Actin and myosin are the main myofilaments in muscle cells' myofibrils.
7-4 The two largest muscles of the calf are the soleus and gastrocnemius.
7-5 The quadriceps femoris is the large muscle group of the anterior thigh.
7-6 The hamstring muscles form the muscle group of the posterior thigh.

Answers to Zooming In Questions

7-1 The endomysium is the innermost layer of connective tissue in a skeletal muscle. Perimysium surrounds a fascicle of muscle fibers.
7-4 The actin and myosin filaments do not change in length as muscle contracts; they simply overlap more.
7-6 Contraction of the brachialis produces flexion at the elbow.
7-7 In a third-class lever system, the fulcrum is behind the point of effort and the resistance (weight).
7-10 The frontalis, temporalis, nasalis, and zygomaticus are named for the bones they are near.
7-11 Carpi means wrist; digitorum means fingers.
7-13 Rectus means straight; oblique means at an angle.
7-15 Four muscles make up the quadriceps femoris.
7-16 The Achilles tendon inserts on the calcaneus (heel bone).

Chapter 8

Answers to Checkpoint Questions

8-1 The two structural divisions of the nervous system are the central nervous system and peripheral nervous system.
8-2 The somatic nervous system is voluntary and controls skeletal muscle; the autonomic (visceral) nervous system is involuntary and controls involuntary muscles (cardiac and smooth muscle) and glands.
8-3 A dendrite is a neuron fiber that carries impulses toward the cell body; an axon is a neuron fiber that carries impulses away from the cell body.
8-4 Myelinated fibers are white; unmyelinated tissue is gray.
8-5 Sensory (afferent) nerves convey impulses toward the CNS; motor (efferent) nerves convey impulses away from the CNS.
8-6 A nerve is a bundle of neuron fibers in the peripheral nervous system; a tract is a bundle of neuron fibers in the central nervous system.
8-7 Neuroglia (glial cells) are the nonconducting cells of the nervous system that protect, nourish, and support the neurons.
8-8 The two stages of an action potential are the rising phase or depolarization, when the charge on the membrane reverses, and the falling phase or repolarization, when the charge returns to the resting state.
8-9 Sodium ion (Na^+) and potassium ion (K^+) are the two ions involved in generating an action potential.
8-10 The myelin sheath speeds conduction along an axon.
8-11 A synapse is the junction between two neurons.
8-12 Neurotransmitters are the chemicals that carry information across the synaptic cleft.
8-13 The gray matter forms an H-shaped section in the center of the cord and extends in two pairs of columns called the dorsal and ventral horns. The white matter is located around the gray matter.
8-14 Tracts in the white matter of the spinal cord carry impulses to and from the brain. Ascending tracts conduct toward the brain; descending tracts conduct away from the brain.
8-15 There are 31 pairs of spinal nerves.
8-16 The dorsal root of a spinal nerve contains sensory fibers; the ventral root contains motor fibers.
8-17 A plexus is a network of spinal nerves.
8-18 A reflex arc is a pathway through the nervous system from a stimulus to an effector.
8-19 There are two motor neurons in each motor pathway of the autonomic nervous system (ANS).
8-20 The sympathetic system stimulates a stress response; the parasympathetic system reverses a stress response.

Answers to Casepoint Questions

8-1 The neuron fibers in the white matter are myelinated.
8-2 Descending tracts in the spinal cord carry motor impulses from the brain to muscles.
8-3 The fifth lumbar spinal nerve carried the sensory signal for touch from the top of her foot.
8-4 Ascending tracts in the spinal cord conduct the touch stimuli to the brain.
8-5 Sue's physician was studying simple spinal reflexes.

Answers to Zooming In Questions

8-1 The brain and spinal cord make up the central nervous system; the cranial nerves and spinal nerves make up the peripheral nervous system.
8-2 The neuron shown is a motor neuron because it is carrying information away from the CNS toward an effector organ. It is part of the somatic nervous system, because the effector is skeletal muscle.
8-5 The endoneurium is the deepest layer of connective tissue in a nerve; the epineurium is the outermost layer.
8-8 The charge on the membrane reverses at the point of an action potential.
8-11 The spinal cord is not as long as is the spinal column. There are seven cervical vertebrae and eight cervical spinal nerves.
8-12 The sacral spinal nerves (S1) carry impulses from the skin of the toes. The cervical spinal nerves (C6, C7, C8) carry impulses from the skin of the anterior hand and fingers.
8-13 The reflex arc shown is a somatic reflex arc, as it involves a skeletal muscle effector. An interneuron is located between the sensory and motor neurons in the CNS.
8-14 The parasympathetic system has ganglia closer to the effector organs.

Chapter 9

Answers to Checkpoint Questions

9-1 The main divisions of the brain are the cerebrum, diencephalon, brain stem, and cerebellum.
9-2 The three layers of the meninges are the dura mater, arachnoid, and pia mater.
9-3 CSF is produced in the ventricles of the brain. The two lateral ventricles are in the cerebral hemispheres, the third ventricle is in the diencephalon, and the fourth is between the brain stem and the cerebellum.
9-4 The frontal, parietal, temporal, and occipital lobes are the four surface lobes of each cerebral hemisphere.
9-5 The cerebral cortex is the outer layer of gray matter of the cerebral hemispheres where higher functions occur.
9-6 The thalamus and hypothalamus are the two main portions of the diencephalon. The thalamus directs sensory input to the cerebral cortex; the hypothalamus helps to maintain homeostasis.
9-7 The midbrain, pons, and medulla oblongata are the three subdivisions of the brain stem.
9-8 The cerebellum aids in coordination of voluntary muscles, maintenance of balance, and maintenance of muscle tone.
9-9 The cingulate gyrus, hippocampus, amygdala, and portions of the hypothalamus are four structures in the limbic system.
9-10 The basal nuclei modulate motor inputs and facilitate routine motor tasks.
9-11 The reticular activating system keeps one awake and attentive and screens out unnecessary sensory input.
9-12 There are 12 pairs of cranial nerves.
9-13 Cranial nerves may be sensory, motor, and mixed. A mixed nerve has both sensory and motor fibers.

Answers to Casepoint Questions

9-1 The occipital lobe of the brain processes visual stimuli.
9-2 Natalie's short-term memory has been affected.
9-3 The reticular system is involved with sleep–wake cycles. The limbic system is involved in emotional responses.
9-4 A CT scan evaluates lesions of bone, soft tissue, and brain cavities. Further studies might require an MRI.
9-5 Cranial nerves II, III, IV, or VI might be involved in visual symptoms.

Answers to Zooming In Questions

9-1 The cerebrum is the largest part of the brain. The brain stem, specifically the medulla oblongata, connects with the spinal cord.
9-2 The dural (venous) sinuses are formed where the dura mater divides into two layers. There are three layers of meninges.
9-3 The fourth ventricle is continuous with the central canal of the spinal cord.
9-4 The central sulcus separates the frontal from the parietal lobe; the lateral sulcus separates the temporal from the frontal and parietal lobes.
9-5 The primary somatosensory area is posterior to the central sulcus. The primary motor area is anterior to the central sulcus.
9-6 The pituitary gland is attached to the hypothalamus.
9-8 The cingulate gyrus is the part of the cerebral cortex that contributes to the limbic system.

Chapter 10

Answers to Checkpoint Questions

10-1 A sensory receptor is a part of the nervous system that detects a stimulus.
10-2 Based on types of stimulus, there are chemoreceptors that respond to chemicals in solution, photoreceptors that respond to light, thermoreceptors that respond to temperature, and mechanoreceptors that respond to movement.
10-3 The general senses are widely distributed; the special senses are in specialized sense organs.
10-4 When a sensory receptor adapts to a stimulus, it fails to respond.
10-5 Structures that protect the eye include the skull bones, eyelids, eyelashes and eyebrow, lacrimal gland, and conjunctiva.
10-6 The extrinsic eye muscles pull on the eyeball so that both eyes center on one visual field, a process known as convergence.
10-7 The optic nerve (cranial nerve II) carries visual impulses from the retina to the brain.
10-8 The fibrous tunic is the outermost tunic of the eye; the vascular tunic is the middle tunic; the nervous tunic is the innermost tunic.
10-9 The structures that refract light as it passes through the eye are the cornea, aqueous humor, lens, and vitreous body.
10-10 The ciliary muscle contracts and relaxes to regulate the thickness of the lens and allow accommodation for near and far vision.
10-11 The iris adjusts the size of the pupil to regulate the amount of light that enters the eye.
10-12 The rods and cones are the receptor cells of the retina.
10-13 The three divisions of the ear are the outer, middle, and inner ear.
10-14 The ear ossicles are the malleus, incus, and stapes. They transmit sound waves from the tympanic membrane to the inner ear.
10-15 The fluids in the inner ear are the endolymph and the perilymph. Endolymph is found in the membranous

labyrinth; perilymph is found around the membranous labyrinth in the bony labyrinth.

10-16 The hearing organ is the spiral organ (organ of Corti) located in the cochlear duct within the cochlea.

10-17 The equilibrium receptors are located in the vestibule and semicircular canals of the inner ear.

10-18 The senses of taste and smell are the special senses that respond to chemical stimuli.

10-19 Touch, pressure, temperature, position (proprioception), and pain are the general senses.

10-20 Proprioceptors are the receptors that respond to change in position. They are located in muscles, tendons, and joints.

Answers to Casepoint Questions

10-1 Mechanoreceptors concerned with hearing are involved in Evan's case. Hearing is a special sense.

10-2 The ciliated hair cells in the spiral organ were damaged in Evan's case.

10-3 A cochlear implant would not work if the eighth cranial nerve was damaged. Hearing stimuli would bypass the injured hair cells, but nerve impulses would not be conducted to the brain for interpretation.

10-4 Equilibrium is not affected by injury to the cochlea because equilibrium receptors are separate from the hearing receptors and are in a different location.

Answers to Zooming In Questions

10-3 The fibers in a rectus muscle are in a straight direction; the fibers in an oblique muscle run at an angle.

10-4 The cornea is the anterior structure continuous with the sclera.

10-5 The suspensory ligaments hold the lens in place.

10-6 The ciliary muscle is contracted when the lens is rounded.

10-7 The circular muscle fibers of the iris contract to make the pupil smaller. The radial muscles contract to make the pupil larger.

10-9 The tympanic membrane separates the outer ear from the middle ear.

10-10 Perilymph is in contact with the membrane lining the vestibular duct.

10-11 The tectorial membrane contains the cilia of the hair cells.

10-12 The cilia of the macular cells bend when the otolithic membrane moves.

10-15 The dendrite of an olfactory receptor cell interacts with an odorant.

Chapter 11

Answers to Checkpoint Questions

11-1 Hormones are chemicals that have specific regulatory effects on certain cells or organs in the body. Some of their effects are to regulate growth, metabolism, reproduction, and behavior.

11-2 The target tissue is the specific tissue that responds to a hormone.

11-3 Hormones chemically are amino acid compounds and steroids.

11-4 Negative feedback is the main method used to regulate hormone secretion.

11-5 The hypothalamus controls the pituitary.

11-6 Antidiuretic hormone (ADH) and oxytocin are released from the posterior pituitary.

11-7 Growth hormone (GH), thyroid-stimulating hormone (TSH), adrenocorticotropic hormone (ACTH), prolactin (PRL), follicle-stimulating hormone (FSH), and luteinizing hormone (LH) are the hormones released from the anterior pituitary.

11-8 Thyroid hormones increase the metabolic rate in cells.

11-9 Iodine is needed to produce thyroid hormones.

11-10 Calcium is regulated by parathyroid hormone and calcitriol.

11-11 Epinephrine (adrenaline) is the main hormone produced by the adrenal medulla.

11-12 Glucocorticoids, mineralocorticoids, and sex hormones are released by the adrenal cortex.

11-13 Cortisol raises blood glucose levels.

11-14 Insulin and glucagon produced by the pancreatic islets regulate glucose levels.

11-15 Secondary sex characteristics are features associated with gender other than reproductive activity.

11-16 The pineal gland secretes melatonin.

11-17 The kidneys produce erythropoietin. Bone tissue produces osteocalcin. The atria of the heart produce ANP.

11-18 Epinephrine, cortisol, ADH, and growth hormone are four hormones released in times of stress.

Answers to Casepoint Questions

11-1 The step between the control center and the effector is interrupted in Becky's case, as the pancreas does not produce insulin (which is the signal).

11-2 Insulin is a protein hormone composed of amino acids.

11-3 Becky is losing weight because her body cannot utilize glucose normally or convert nutrients into fat.

11-4 Glucagon secretion would be reduced because diabetes mellitus increases blood glucose, inhibiting glucagon secretion by negative feedback.

11-5 Becky has to inject insulin daily because her pancreas is not producing the hormone, so it is not possible to treat her by improving the hormone's effectiveness.

Answers to Zooming In Questions

11-2 The infundibulum connects the hypothalamus and the pituitary gland.

11-3 The anterior pituitary controls the thyroid gland.

11-4 The larynx is superior to the thyroid; the trachea is inferior to the thyroid.

11-5 The cortex is the outer region of the adrenal gland; the medulla is the inner region.

11-7 Skeletal muscles respond to insulin but not glucagon.

Chapter 12

Answers to Checkpoint Questions

12-1 Four types of substances transported in the blood are gases, nutrients, waste, and hormones.

12-2 7.35 to 7.45 is the pH range of the blood.

12-3 The two main components of blood are the liquid portion, or plasma, and the formed elements, which include the cells and cell fragments.

12-4 Protein is the most abundant type of substance in plasma aside from water.

12-5 All blood cells form in red bone marrow and are produced by hematopoietic stem cells.

12-6 The main function of hemoglobin is to carry oxygen in the blood.

12-7 Erythropoietin (EPO) is the signal in the feedback loop regulating blood cell production. Blood oxygen content is the regulated variable.

12-8 The granular leukocytes are neutrophils, eosinophils, and basophils. The agranular leukocytes are lymphocytes and monocytes.

12-9 The main function of leukocytes is to destroy pathogens. They also remove other foreign material and cellular debris.
12-10 Platelets are essential to blood clotting (coagulation).
12-11 Hemostasis is the process that stops blood loss.
12-12 Fibrin forms a clot.
12-13 Serum does not contain clotting factors; plasma does.
12-14 An antigen is any substance that activates an immune response.
12-15 The four ABO blood type groups are A, B, AB, and O.
12-16 The Rh factor is associated with incompatibility during pregnancy.
12-17 A centrifuge is used to separate blood into its component parts.
12-18 A hematocrit measures the relative volume of red cells in blood.
12-19 Hemoglobin is expressed as grams per deciliter (dL) of whole blood or as a percentage of a given standard, usually the average male normal of 15.6 grams hemoglobin per dL.

Answers to Casepoint Questions

12-1 Allen's pale-colored palate was caused by low hemoglobin levels, as hemoglobin imparts color to the skin and mucous membranes. He was tired because, as a result of low hemoglobin, his cells were not getting enough oxygen to generate the amount of ATP needed for everyday activities.
12-2 Allen was subject to repeated infections because he had fewer infection-fighting white blood cells.
12-3 Allen's nosebleeds and tiny skin hemorrhages indicated a low platelet count.
12-4 A person with type O blood has no ABO blood type antigens on red cells.
12-5 A person with type O blood has both A and B antibodies in blood plasma.
12-6 A person who is Rh positive has no Rh antibodies in plasma or the plasma would agglutinate his or her own red cells.
12-7 A platelet transfusion was needed to treat his bleeding problem.
12-8 A likely hematocrit value for Allen was 30%, as his red cell count was low.

Answers to Zooming In Questions

12-2 Erythrocytes (red blood cells) are the most numerous cells in the blood.
12-3 Erythrocytes are described as biconcave because they have an inward depression on both sides.
12-5 Simple squamous epithelium makes up the capillary wall.
12-7 The suffix *-ase* indicates that prothrombinase is an enzyme. The prefix *pro-* indicates that prothrombin is a precursor.
12-8 No. To test for Rh antigen, you have to use anti-Rh serum. The two types of antigens are independent.

Chapter 13

Answers to Checkpoint Questions

13-1 The endocardium is the innermost layer of the heart wall; the myocardium is the middle layer; the epicardium is the outermost layer.
13-2 The pericardium is the sac that encloses the heart.
13-3 The atrium is the heart's upper receiving chamber on each side. The ventricle is the lower pumping chamber on each side.
13-4 The valves direct the forward flow of blood through the heart.
13-5 The coronary circulation is the system that supplies blood to the myocardium.
13-6 Systole is the contraction phase of the cardiac cycle. Diastole is the relaxation phase.
13-7 Cardiac output is the amount of blood pumped by each ventricle in 1 minute. It is determined by the stroke volume, the amount of blood ejected from the ventricle with each beat, and by the heart rate, the number of times the heart beats per minute.
13-8 The scientific name for the pacemaker is the sinoatrial (SA) node.
13-9 The autonomic nervous system is the main influence on the rate and strength of heart contractions.
13-10 A heart murmur is an abnormal heart sound.
13-11 ECG (EKG) stands for electrocardiogram or electrocardiography.
13-12 Catheterization is the use of a thin tube threaded through a vessel for diagnosis or repair.
13-13 Coronary angiography and coronary computed tomography angiography (CTA) are techniques that use a dye and x-rays to visualize the coronary arteries.

Answers to Casepoint Questions

13-1 If the left coronary artery is blocked, the circumflex artery and the left anterior descending artery would not receive blood.
13-2 At a heart rate of 80 bpm and stroke volume of 50 mL, Jim's cardiac output is 4 L/min (4,000 mL).
13-3 The medulla oblongata of the brain stem has vital centers that control heart rate and respiration rate.
13-4 A heart rate of 160 bpm would be considered tachycardia.
13-5 Ventricular repolarization is occurring during the T wave.

Answers to Zooming In Questions

13-1 The myocardium is the thickest layer of the heart wall.
13-2 The left ventricle of the heart has the thickest wall.
13-3 The right AV valve has three cusps; the left AV valve has two cusps.
13-4 The coronary sinus is the largest cardiac vein. It leads into the right atrium.
13-6 The AV (tricuspid and mitral) valves close when the ventricles contract. The semilunar (pulmonary and aortic valves) open when the ventricles contract.
13-7 Atrial muscle fibers carry signals between the sinoatrial (SA) and atrioventricular (AV) nodes.
13-8 The vagus nerve (cranial nerve X) carries parasympathetic impulses to the heart.
13-9 The length of the cardiac cycle in this diagram is 0.8 seconds.

Chapter 14

Answers to Checkpoint Questions

14-1 The five types of blood vessels are arteries, arterioles, capillaries, venules, and veins.
14-2 The pulmonary circuit carries blood from the heart to the lungs and back to the heart; the systemic circuit carries blood to and from all remaining tissues in the body.
14-3 Smooth muscle makes up the middle layer of arteries and veins. Smooth muscle is involuntary muscle controlled by the autonomic nervous system.
14-4 A single layer of squamous epithelium makes up the wall of a capillary.

14-5 The aorta is divided into the ascending aorta, aortic arch, thoracic aorta, and abdominal aorta. The thoracic aorta and abdominal aorta make up the descending aorta.
14-6 The brachiocephalic, left common carotid, and left subclavian arteries are the three branches of the aortic arch.
14-7 The brachiocephalic artery supplies the arm and head on the right side.
14-8 An anastomosis is a communication between two vessels.
14-9 Superficial veins are near the surface; deep veins are closer to the interior and generally parallel the arteries.
14-10 The superior vena cava and inferior vena cava drain the systemic circuit and empty into the right atrium.
14-11 A venous sinus is a large channel that drains blood low in oxygen.
14-12 The hepatic portal system takes blood from the abdominal organs to the liver.
14-13 Blood pressure helps push materials out of a capillary; blood osmotic pressure helps draw materials into a capillary.
14-14 Vasodilation (widening) and vasoconstriction (narrowing) are the two types of vasomotor changes.
14-15 Vasomotor activities are regulated in the medulla of the brain stem.
14-16 Pulse is the wave of pressure that begins at the heart and travels along the arteries.
14-17 Blood pressure is the force exerted by blood against the walls of the vessels.
14-18 The heart and blood vessels are the effectors involved in the short-term regulation of blood pressure.
14-19 Baroreceptors are the sensors for blood pressure.
14-20 Systolic pressure, the maximum pressure in the arteries after heart contraction, and diastolic pressure, the lowest pressure in the arteries after heart relaxation, are measured when taking blood pressure.

Answers to Casepoint Questions

14-1 A blood clot in a leg vein would block blood flow toward the heart.
14-2 The inner layer of a blood vessel is the endothelium, made of simple squamous epithelium.
14-3 From the popliteal vein, the clot would enter the femoral vein.
14-4 The inferior vena cava would carry the clot into the heart.
14-5 The clot would enter the right atrium of the heart.
14-6 The clot would not enter the hepatic portal system, because that system receives blood from the abdominal organs.
14-7 Increased blood pressure in the capillaries would force fluid out of the capillaries and into the tissues, causing swelling of the tissues.

Answers to Zooming In Questions

14-1 Pulmonary arteries contain oxygen-poor blood. Pulmonary veins contain oxygen-rich blood.
14-2 Veins have valves that control the direction of blood flow.
14-3 An artery has a thicker wall than does a vein of comparable size.
14-4 There is one brachiocephalic artery.
14-5 The left and right common iliac arteries branch from the terminal aorta.
14-7 There are two brachiocephalic veins.
14-8 The internal jugular vein receives blood from the transverse sinus.
14-9 The hepatic veins drain into the inferior vena cava.
14-11 The proximal valve is closer to the heart.
14-12 Pulse pressure drops to 0 in the arterioles.
14-13 Blood pressure will fall if arteriolar diameter increases.
14-14 Heart rate increases when blood pressure falls.
14-15 The systolic pressure in part B is 120 mm Hg.

Chapter 15

Answers to Checkpoint Questions

15-1 The lymphatic system helps maintain fluid balance by bringing excess fluid and proteins from the tissues to the blood; protects against foreign material and foreign or abnormal cells; and absorbs fats from the small intestine.
15-2 The lymphatic capillaries are more permeable than are blood capillaries and begin blindly. They are closed at one end and do not bridge two vessels.
15-3 The right lymphatic duct and the thoracic duct are the two main lymphatic vessels.
15-4 Lymph nodes filter lymph. They also have lymphocytes to fight infection.
15-5 The spleen filters blood.
15-6 T cells of the immune system develop in the thymus.
15-7 MALT stands for mucosa-associated lymphoid tissue; GALT stands for gut-associated lymphoid tissue.
15-8 Tonsils are located in the vicinity of the pharynx (throat).
15-9 Innate barriers such as the skin, mucous membranes, body secretions, and reflexes are the first line of defense against the invasion of pathogens.
15-10 Innate cells and chemicals are two components in the second line of defense against infection.
15-11 Heat, redness, swelling, and pain are the four signs of inflammation.
15-12 Fever boosts the immune system by stimulating phagocytes, increasing metabolism, and decreasing the ability of certain microorganisms to multiply.
15-13 Adaptive immunity is an individual's power to resist or overcome any particular disease or its products. Adaptive immunity is acquired during a person's lifetime, usually from contact with a disease organism or a vaccine.
15-14 An antigen is any foreign substance, usually a protein, that induces an immune response.
15-15 The four types of T cells are cytotoxic, helper, regulatory, and memory T cells.
15-16 Antigen-presenting cells (APC) take in and digest a foreign antigen. They then display fragments of the antigen in their plasma membrane along with self (MHC) antigens that a T cell can recognize.
15-17 An antibody is a substance produced in response to an antigen.
15-18 Plasma cells, derived from B cells, produce antibodies.
15-19 Active adaptive immunity involves a person's own immune system. Passive adaptive immunity depends on antibodies from an outside source.
15-20 Natural adaptive immunity results from contact with a disease organism or obtaining antibodies from an outside source without medical intervention. Artificial adaptive immunity results from administration of a vaccine or antiserum.
15-21 A vaccine is a prepared substance that induces an immune response.
15-22 A booster is a repeated inoculation given to increase antibodies to a disease.
15-23 An antiserum is an antibody prepared in an outside source, usually an animal. Antisera are used in emergencies to provide quick passive immunization against a disease organism, toxin, or venom.

Answers to Casepoint Questions

15-1 The cervical lymph nodes are in the neck, axillary in the armpit, and inguinal in the groin.

15-2 Lymph from the left side travels to the thoracic duct and then drains into the left subclavian vein near the left internal jugular vein.
15-3 Enlargement of the germinal centers of the lymph nodes indicates multiplication and activity of lymphocytes.
15-4 The doctor would have seen the palatine tonsils in the throat.
15-5 The lymph nodes filter lymph; the spleen filters blood.
15-6 Signs of inflammation in Lucas's study are pain, heat, redness, and swelling.
15-7 Other signs that Lucas was fighting an infection were fatigue, weakness, malaise, enlarged lymph nodes and spleen.
15-8 Cytotoxic T cells (Tc) destroy virus-infected cells.
15-9 Memory T and B cells confer immunity to subsequent infections with the same organism.
15-10 Lucas will develop natural active immunity to mono because he was infected with and made antibodies against the agent that causes mono, EBV.

Answers to Zooming In Questions

15-1 A vein receives lymph collected from the body.
15-3 Axillary nodes receive lymph drainage from the mammary vessels. Inguinal nodes receive lymph drainage from the leg vessels.
15-4 An afferent vessel carries lymph into a node; an efferent vessel carries lymph out of a node.
15-9 Increased blood flow causes heat and redness. Excess fluid in the tissues causes swelling and pain.
15-10 Digestive enzymes are contained in the lysosome that joins the phagocytic vesicle.
15-11 Activated B cells produce plasma cells that produce antibodies and memory cells that respond to reinfection

Chapter 16

Answers to Checkpoint Questions

16-1 The four phases of respiration are pulmonary ventilation, external gas exchange, gas transport in the blood, and internal gas exchange.
16-2 As air passes over the nasal mucosa, it is filtered, warmed, and moistened.
16-3 The three regions of the pharynx are the nasopharynx, oropharynx, and laryngopharynx.
16-4 The scientific name for the throat is the pharynx; the voice box is the larynx; the windpipe is the trachea.
16-5 The two mainstem (primary) bronchi are formed by the inferior branching of the trachea.
16-6 Cilia on the cells that line the respiratory passageways move fluid with entrapped impurities upward to be eliminated from the respiratory system.
16-7 Gas exchange in the lungs occurs in the alveoli.
16-8 The pleura is the membrane that encloses the lung.
16-9 Inhalation and exhalation are the two phases of quiet breathing. Inhalation is active, and exhalation is passive.
16-10 Surfactant produced by lung cells aids in compliance.
16-11 A pressure gradient drives diffusion across the respiratory membrane.
16-12 Hemoglobin in red blood cells holds almost all of the oxygen carried in the blood.
16-13 The main form in which carbon dioxide is carried in the blood is as bicarbonate ion.
16-14 The medulla of the brain stem has the centers that set the basic pattern of respiration.
16-15 The phrenic nerve is the motor nerve that controls the diaphragm.
16-16 Carbon dioxide is the main chemical controller of respiration.
16-17 The word ending *-pnea* means breathing.

Answers to Casepoint Questions

16-1 Pulmonary ventilation is affected by Emily's asthma.
16-2 Swelling is the characteristic of inflammation that affects Emily's breathing.
16-3 Activation of the adrenergic receptors causes relaxation of the smooth muscle cells, resulting in dilation of the airways.
16-4 The diaphragm relaxes during exhalation.
16-5 Emily can use some intercostal muscles and the muscles of the abdominal wall to aid exhalation.
16-6 If Emily has trouble exhaling carbon dioxide, her breathing rate will increase.

Answers to Zooming In Questions

16-1 Blood travels to the lungs from the right side of the heart; blood returns to the left side of the heart.
16-2 The heart is located in the medial depression of the left lung.
16-4 The epiglottis is named for its position above the glottis.
16-6 The intercostal muscles are located between the ribs.
16-7 Intrapulmonary pressure is greater than atmospheric pressure during exhalation.
16-8 Residual volume cannot be measured with a spirometer; total lung capacity and functional residual capacity cannot be measured with a spirometer.

Chapter 17

Answers to Checkpoint Questions

17-1 Food must be broken down by digestion into particles small enough to pass through the plasma membrane and enter cells.
17-2 The peritoneum is the large serous membrane that lines the abdominopelvic cavity and covers the organs it contains.
17-3 The digestive tract typically has a wall composed of a mucous membrane (mucosa), a submucosa, smooth muscle (muscularis externa), and a serous membrane (serosa).
17-4 Peristalsis occurs in the esophagus. Segmentation occurs in the small intestine.
17-5 Proteins are digested in the stomach.
17-6 The duodenum, jejunum, and ileum are the three divisions of the small intestine.
17-7 Most digestion takes place in the small intestine under the effects of digestive juices from the small intestine and the accessory organs. Most absorption of digested food and water also occurs in the small intestine.
17-8 The large intestine reabsorbs some water and stores, forms, and eliminates the stool. It also houses bacteria that provide some vitamins.
17-9 The salivary glands are the parotid, anterior and inferior to the ear; submandibular (submaxillary), near the body of the lower jaw; and sublingual, under the tongue.
17-10 The liver secretes bile, which emulsifies fats, that is, breaks it down into small particles. Bile also prevents formation of fat droplets, thus aiding in absorption.
17-11 The gallbladder stores bile and contracts to release it into the duodenum.
17-12 The pancreas secretes sodium bicarbonate, which neutralizes the acidic chyme from the stomach.
17-13 The disaccharides maltose, sucrose, and lactose and also small peptides are the nutrients digested by enzymes in the intestinal cell membranes.

17-14 Hydrolysis means "splitting by means of water" as in digestion.
17-15 The pancreas produces the enzymes found in the intestinal lumen.
17-16 The products of nucleic acid digestion, monosaccharides, amino acids, and small peptides are absorbed into the intestinal capillaries. Fatty acids, monoglycerides, and fat-soluble vitamins are absorbed into the lacteals.
17-17 The two types of control over the digestive process are nervous control and hormonal control.
17-18 Hunger is the desire for food that can be satisfied by the ingestion of a filling meal. Appetite is a desire for food that is unrelated to a need for food.

Answers to Casepoint Questions

17-1 In an EGD study, the endoscope will encounter the lips, teeth, tongue, oral cavity, pharynx, esophagus, stomach, and duodenum of the small intestine.
17-2 The endoscope will contact the mucosa of the digestive tract wall in an EGD exam.
17-3 The lower esophageal sphincter (LES) normally keeps stomach contents from passing upward into the esophagus.
17-4 The esophagus passes through the hiatus (space) in the diaphragm.
17-5 The colonoscope contacts the anus, rectum, sigmoid colon, descending colon, transverse colon, and ascending colon.
17-6 Acid reduction could affect production of pepsin, as the conversion of pepsinogen into pepsin is catalyzed by the hydrochloric acid in the stomach.
17-7 Trypsin and other protein-digesting enzymes, such as peptidases, could compensate for decreased activity of pepsin.
17-8 Ghrelin is produced by the empty stomach to stimulate hunger, so a fasting person would have a higher level than normal.

Answers to Zooming In Questions

17-1 The salivary glands are the accessory organs that secrete into the mouth.
17-2 The mesentery is the part of the peritoneum around the small intestine.
17-3 Smooth muscle (circular and longitudinal) is between the submucosa and the serous membrane in the digestive tract wall.
17-4 Peristalsis to move food rapidly is most useful in the esophagus.
17-5 The common name for the gingiva is the gum.
17-6 The oblique muscle layer is an additional muscle layer in the stomach that is not found in the rest of the digestive tract.
17-7 The ileum of the small intestine joins the cecum of the large intestine at the ileocecal valve.
17-8 The sublingual salivary glands are directly below the tongue.
17-9 The accessory organs shown secrete into the duodenum.
17-10 Glucose and fructose are the two monosaccharide components of sucrose.
17-12 Secretin is the hormone that raises the pH of the intestinal contents by stimulating the release of bicarbonate, a buffer.

Chapter 18

Answers to Checkpoint Questions

18-1 The two types of activities that make up metabolism are catabolism, the breakdown of compounds into simpler components, and anabolism, the building of simple substances into materials needed by the body, such as nutrient storage compounds, structural materials, and functional molecules, including enzymes and transporters.
18-2 Cellular respiration is the series of cellular reactions that releases energy from nutrients.
18-3 Pyruvate is the end product of glycolysis.
18-4 Oxygen is the element required for aerobic cellular respiration but not for glycolysis.
18-5 Proteins are not stored as reserves in the body.
18-6 The glycemic index is the term for how rapidly a food raises the blood glucose level.
18-7 An essential amino acid or fatty acid cannot be made metabolically and must be taken in as part of the diet.
18-8 Saturated fats have no double-bonded carbons and have the maximum number of hydrogen atoms possible. Unsaturated fats have one or more double-bonded carbons and less than the maximum number of hydrogen atoms possible.
18-9 Minerals are chemical elements, and vitamins are complex organic substances.
18-10 The average person can metabolize 1/2 oz of pure alcohol per hour.
18-11 BMI stands for body mass index.
18-12 The hypothalamus is responsible for regulating body temperature.
18-13 Cutaneous blood vessels constrict under cold conditions. Under hot conditions, they dilate.

Answers to Casepoint Questions

18-1 If Claudia is metabolizing body fat for energy, acidic ketones will be produced.
18-2 To metabolize amino acids for energy, the amine (nitrogen-containing) portions must be removed.
18-3 Claudia's basal needs are 1,237 kcal/day ($[126 \div 2.2] \times 0.9 \times 24$).
18-4 According to her friends, Claudia engages in heavy activity. At a minimum addition of 80% of her BMR, her daily need is 2,227 kcal/day.
18-5 Claudia will lack the fat-soluble vitamins, such as A, D, and E, found in fatty foods.
18-6 Claudia will lose insulation, padding, and energy reserves if her fat tissues are diminished.
18-7 Claudia's BMI is 18.5 ($106 \div 2.2 = 48 \div 2.6$).
18-8 Claudia lacks insulating fat tissue, so she is more sensitive to cold than her friends.

Answers to Zooming In Questions

18-1 Two pyruvate molecules can be generated from a single molecule of glucose.
18-5 The BMI is 24 ($77 \div 3.2 = 24$).
18-6 If the fan speed is increased in (B), convection, heat loss will increase. If environmental humidity increases in (C), evaporation, heat loss will decrease.
18-7 Blood flow in the skin is higher under hot conditions.

Chapter 19

Answers to Checkpoint Questions

19-1 The urinary system consists of two kidneys, two ureters, the bladder, and the urethra.
19-2 Systems other than the urinary system that eliminate waste include the digestive, respiratory, and integumentary systems.
19-3 The retroperitoneal space is posterior to the peritoneum.
19-4 The renal artery supplies blood to the kidney, and the renal vein drains blood from the kidney.

19-5 The renal pelvis is the funnel-shaped collecting area that forms the upper end of the ureter.
19-6 The nephron is the functional unit of the kidney.
19-7 The glomerulus is the coil of capillaries in the glomerular capsule.
19-8 Filtration is the process that drives materials out of the glomerulus and into the glomerular capsule.
19-9 Tubular reabsorption is the process that returns materials from the nephron back to the circulation.
19-10 Hydrogen ions (H^+) are moved by tubular secretion to balance pH.
19-11 Antidiuretic hormone (ADH) controls water reabsorption from the collecting duct of the nephron.
19-12 Resistance is controlled to alter water reabsorption.
19-13 The ureter carries urine from the kidney to the bladder.
19-14 The openings of the ureters and the urethra form the bladder's trigone.
19-15 The urethra carries urine from the bladder to the outside.
19-16 Intracellular fluids are located within cells. Extracellular fluids are located outside of cells.
19-17 Interstitial fluid is fluid located between cells in tissues.
19-18 Cations are positively charged ions. Anions are negatively charged ions.
19-19 Water is lost from the body through the kidneys, the skin, the lungs, and the intestinal tract.
19-20 Aldosterone is the hormone from the adrenal cortex that promotes sodium reabsorption in the kidneys.
19-21 Antidiuretic hormone (ADH) is the pituitary hormone that directly increases water reabsorption in the kidney.
19-22 The acid–base balance of body fluids is maintained by buffer systems, respiration, and kidney function.

Answers to Casepoint Questions

19-1 As the kidneys excreted more water, urine would be diluted and its specific gravity would decrease.
19-2 Sodium (Na^+) ions are most abundant in extracellular fluids.
19-3 The osmotic pressure of Ethan's body fluids will decrease as sodium concentration declines.
19-4 The urinary system would compensate most effectively for Ethan's excess water consumption.
19-5 Dilution of body fluids would normally inhibit secretion of ADH so excess fluid could be eliminated.
19-6 Dilution of body fluids will cause the hypothalamic thirst center to reduce thirst and limit water intake. Ethan did not listen to his body signals, as he continued to drink when he was not thirsty.

Answers to Zooming In Questions

19-1 The ureters carry urine to the bladder, and the urethra carries urine from the bladder.
19-2 The outer region of the kidney is the renal cortex. The inner region of the kidney is the renal medulla.
19-3 The glomerulus is the first capillary bed to receive blood.
19-6 The osmotic gradient is stronger by the renal pelvis than by the distal tubule.
19-7 Filtration and tubular secretion move substances from the blood to the renal tubule.
19-8 The urethra passes through the prostate gland in the male.
19-9 The two most abundant extracellular fluids are interstitial fluid and blood plasma
19-10 Sodium and chloride are highest in interstitial fluids; potassium is most concentrated in intracellular fluid.
19-11 Most water is lost in a day in urine.
19-12 High blood potassium increases aldosterone secretion and promotes loss of potassium.
19-14 Angiotensin II stimulates thirst.
19-15 The kidneys generate bicarbonate ions (buffers).

Chapter 20

Answers to Checkpoint Questions

20-1 Meiosis is the process of cell division that halves the chromosome number in a cell to produce a gamete.
20-2 The spermatozoon, or sperm cell, is the male gamete (sex cell).
20-3 The testis is the male gonad.
20-4 The epididymis is the structure on the surface of the testis that stores sperm.
20-5 Glands that contribute secretions to the semen, aside from the testes, are the seminal vesicles, prostate, and bulbourethral glands.
20-6 The main subdivisions of a spermatozoon are the head, midpiece, and tail (flagellum).
20-7 Follicle-stimulating hormone (FSH) and luteinizing hormone (LH) are the pituitary hormones that regulate male and female reproduction.
20-8 Specialized interstitial cells (Leydig cells) in the testis secrete the main male sex steroid hormone (testosterone).
20-9 Ovum is the female gamete.
20-10 The ovary is the female gonad.
20-11 A fetus develops in the uterus.
20-12 An ovum matures in an ovarian follicle.
20-13 Estrogen and progesterone are the two hormones produced in the ovaries.
20-14 Ovulation releases an ovum from the ovary.
20-15 The follicle becomes a corpus luteum after its ovum is released.
20-16 Menopause is the term for complete cessation of menstrual cycles.
20-17 Contraception is the use of artificial methods to prevent fertilization of an ovum.
20-18 Vasectomy and tubal ligation are two methods of surgical sterilization.

Answers to Casepoint Questions

20-1 The physician concluded that Brett was fertile because his sperm count was within normal limits, and the cells were normal in shape and were active.
20-2 Endometrial tissue would reach maximum thickness during the postovulatory phase of the ovarian cycle and the secretory phase of the uterine cycle.
20-3 Taking oral contraceptives would decrease FSH levels in a negative feedback response.
20-4 Ovulation on day 15 should result in a period on day 29 of the menstrual cycle.

Answers to Zooming In Questions

20-1 The four glands that empty secretions into the urethra are the testes, seminal vesicles, prostate, and bulbourethral glands. The ductus (vas) deferens receives secretions from the epididymis.
20-2 The corpus spongiosum of the penis contains the urethra.
20-3 The interstitial cells are located between the seminiferous tubules.
20-4 Mitochondria are the organelles that provide energy for sperm cell motility.
20-5 LH stimulates testosterone secretion.
20-6 The fundus of the uterus is the deepest part. The cervix is the most inferior portion.

20-9 Estrogen peaks closest to ovulation. Progesterone peaks after ovulation.
20-10 The male condom prevents sexually transmitted diseases.

Chapter 21

Answers to Checkpoint Questions

21-1 A zygote is formed by the union of an ovum and a spermatozoon.
21-2 The placenta nourishes the developing fetus.
21-3 The umbilical cord carries blood between the fetus and the placenta.
21-4 Fetal circulation is adapted to bypass the lungs.
21-5 The embryonic hormone hCG (human chorionic gonadotropin) maintains the corpus luteum early in pregnancy.
21-6 The heartbeat first appears during the fourth week of gestation.
21-7 The amniotic sac is the fluid-filled sac that holds the fetus.
21-8 The approximate length of pregnancy in days from the time of fertilization is 266.
21-9 Oxytocin is the pituitary hormone that stimulates uterine contractions.
21-10 Parturition is the process of labor and childbirth.
21-11 The afterbirth is expelled during the third stage of labor.
21-12 A cesarean section is used to remove a fetus manually from the uterus.
21-13 The term *viable* describes a fetus able to live outside the uterus.
21-14 The lactiferous ducts carry milk out of the breast.
21-15 The breast secretes colostrum before milk production begins.
21-16 A gene is an independent unit of heredity. Each gene is a segment of DNA contained in a chromosome.
21-17 A dominant gene is always expressed, regardless of the gene on the matching chromosome.
21-18 A genotype is the genetic makeup of an individual; a phenotype is all the traits that can be observed or tested for.
21-19 Meiosis is the process of cell division that forms the gametes.
21-20 There are 23 chromosomes in each human gamete.
21-21 The sex chromosome combination that determines a female is XX; a male is XY.
21-22 A trait carried on a sex chromosome is described as sex-linked.
21-23 A mutagen is any agent that causes a mutation, a change in a gene or a chromosome.
21-24 A child gets mitochondrial genes from the mother, as mitochondria are inherited in the ovum.

Answers to Casepoint Questions

21-1 Cole's red cells pick up oxygen from his mother's circulation by diffusion from the venous sinuses of the placenta into the capillaries of the chorionic villi.
21-2 Cole must have two sickle cell alleles in order for the recessive trait to appear.
21-3 His genotype for sickle cell is homozygous recessive.
21-4 The parents are heterozygous for the sickle cell trait. They have one normal and one abnormal allele and are carriers of sickle cell.
21-5 The sickle cell allele separated from the normal allele in the first meiotic division.
21-6 Based on predicted ratios, it is expected that 50% of the children in Cole's family will be carriers of the sickle cell allele. Theoretically, 25% of the children will be homozygous dominant and will neither have sickle cell anemia nor be able to pass the allele to offspring.
21-7 Sickle cell anemia likely results from a loss-of-function mutation as it is a recessive disorder.

Answers to Zooming In Questions

21-1 The ovum is fertilized in the uterine tube.
21-2 The purple color signifies a mixture of blood that is low in oxygen with blood that is high in oxygen.
21-5 The umbilical cord connects the fetus to the placenta.
21-7 The uterine muscle is the effector in the feedback loop shown in part B.
21-9 Identical twins may share a single placenta.
21-10 The pectoralis major muscle underlies the breast.
21-12 Four cells are produced in one complete meiosis.
21-13 Twenty-five percent of children will show the recessive phenotype nontaster. Fifty percent of children will be heterozygous.
21-14 The expected ratio of male to female offspring in a family is 50/50.
21-15 The genotype of a female carrier is X^RX^r.

Appendix 3 Dissection Atlas

A. = artery   V. = vein   M. = muscle   N. = nerve

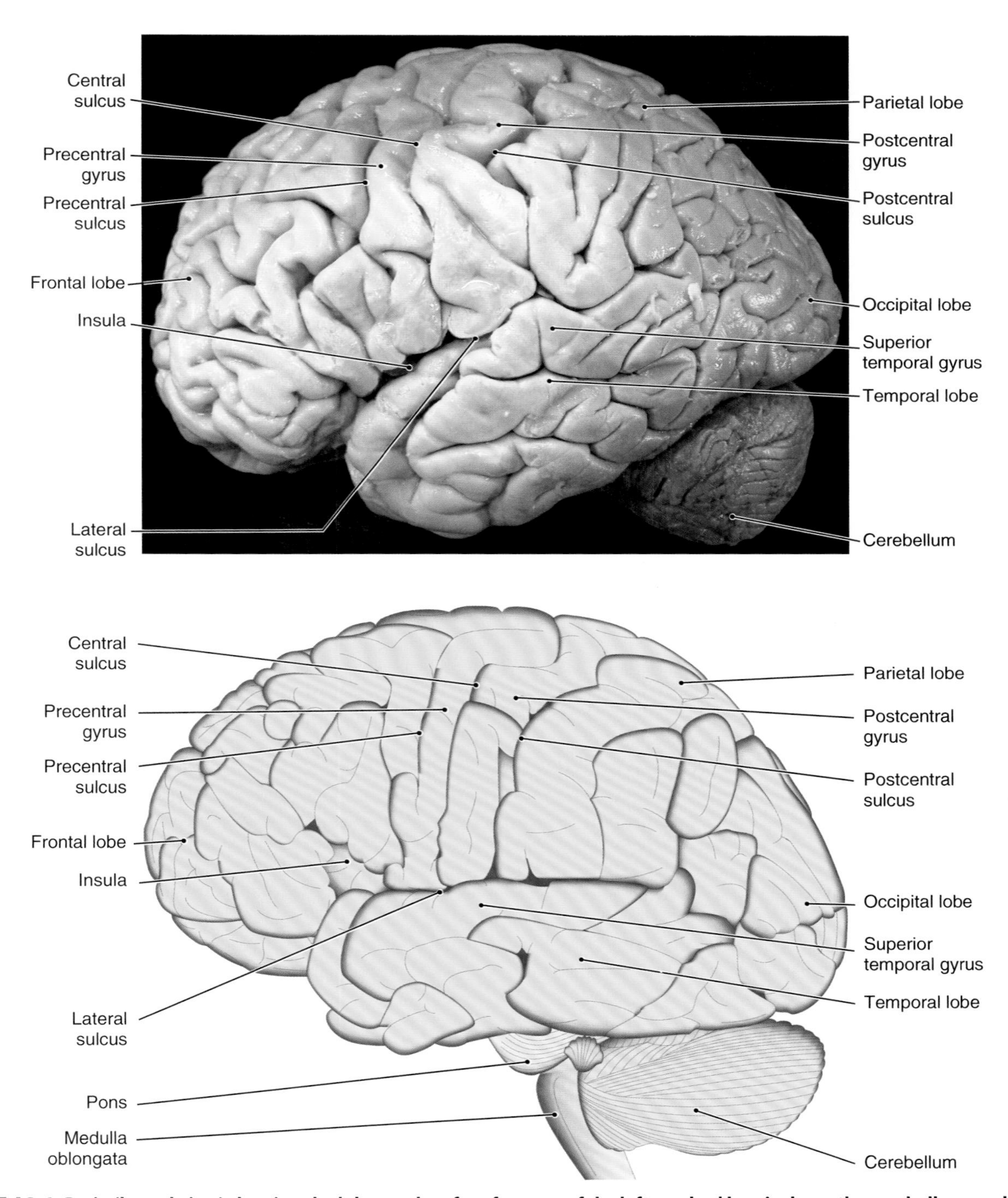

FIGURE A3-1 **Brain (lateral view) showing the lobes and surface features of the left cerebral hemisphere, the cerebellum, and parts of the brain stem.** (Modified from with permission Olinger AB. *Human Gross Anatomy*. Baltimore, MD: Wolters Kluwer, 2016.)

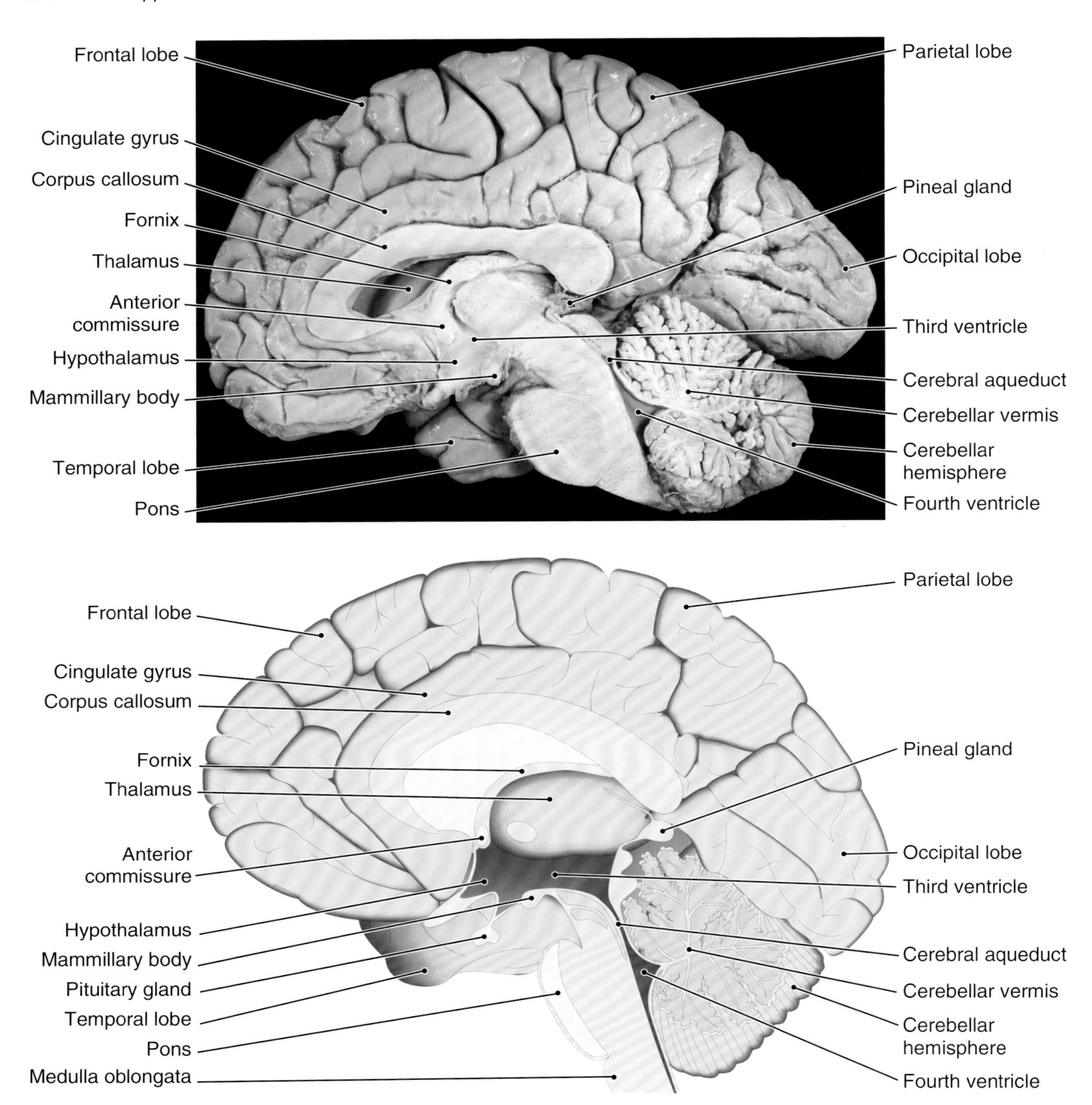

FIGURE A3-2 **Interior of the brain (sagittal section).** (Modified with permission from Olinger AB. *Human Gross Anatomy*. Baltimore, MD: Wolters Kluwer, 2016.)

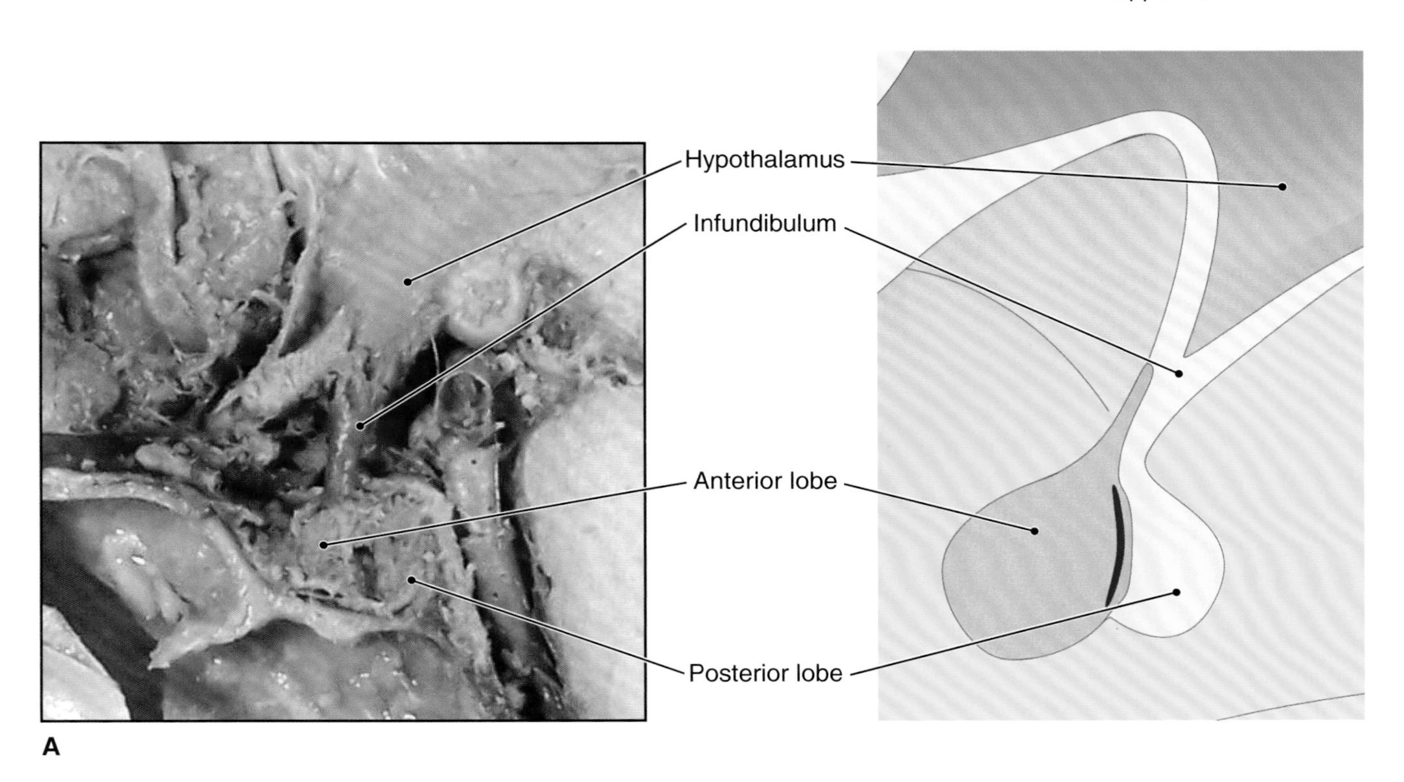

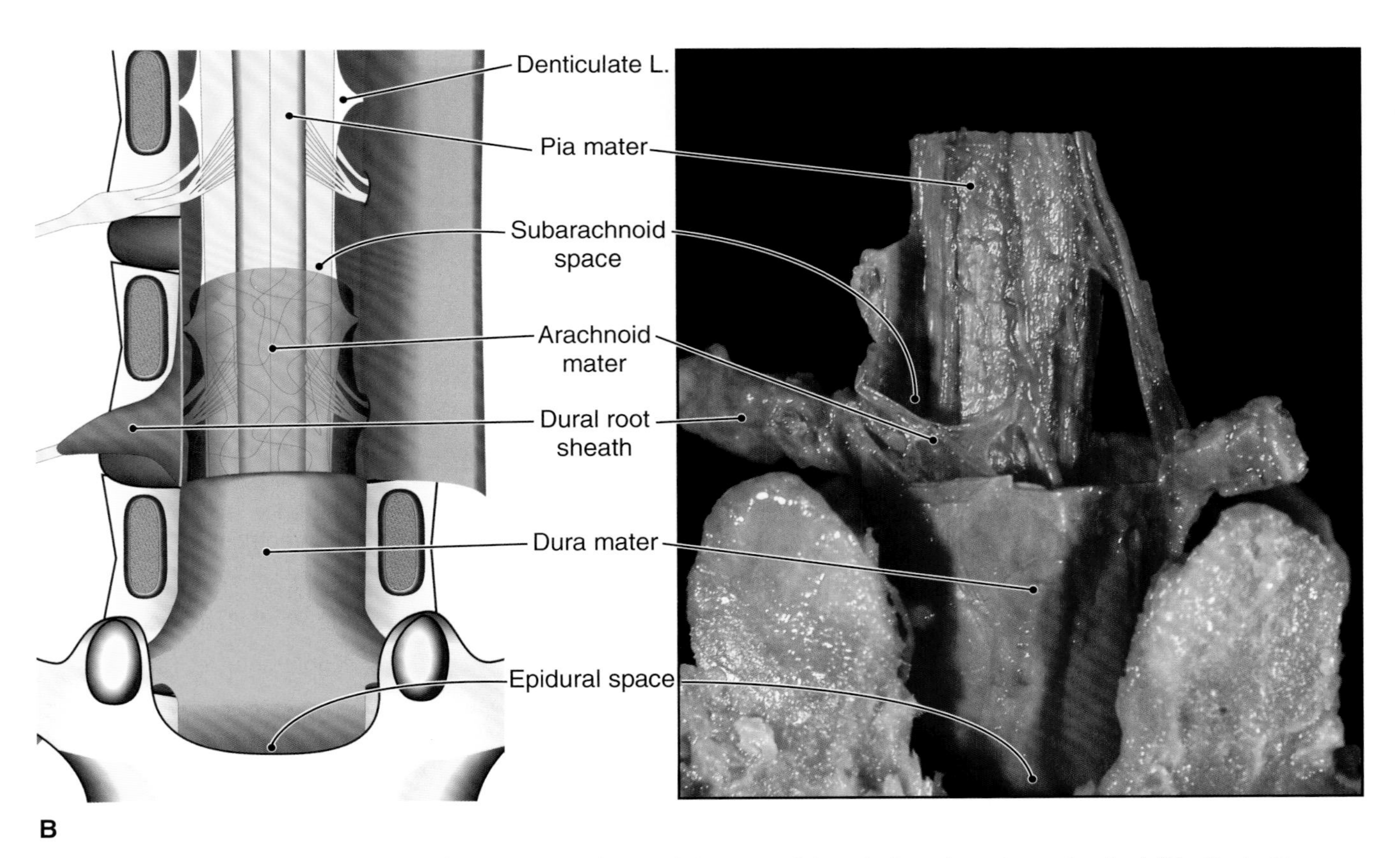

FIGURE A3-3 **A.** Pituitary gland and hypothalamus (sagittal section). **B.** Layers of the spinal meninges (posterior view). (Modified with permission from Olinger AB. *Human Gross Anatomy*. Baltimore, MD: Wolters Kluwer, 2016.)

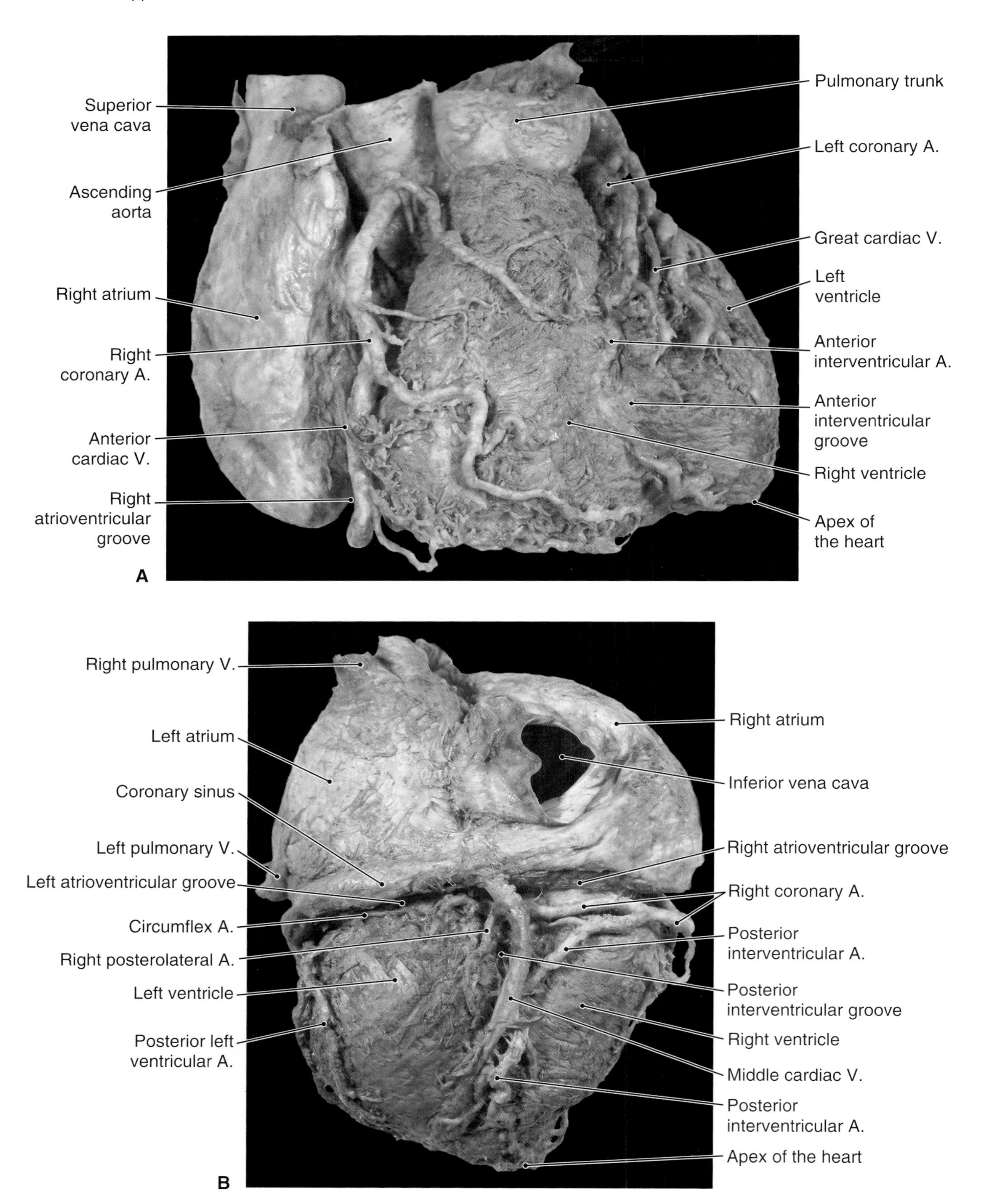

FIGURE A3-4 **A.** External features of the heart (anterior view). **B.** External features of the heart (posterior view). (Modified with permission from Olinger AB. *Human Gross Anatomy*. Baltimore, MD: Wolters Kluwer, 2016.)

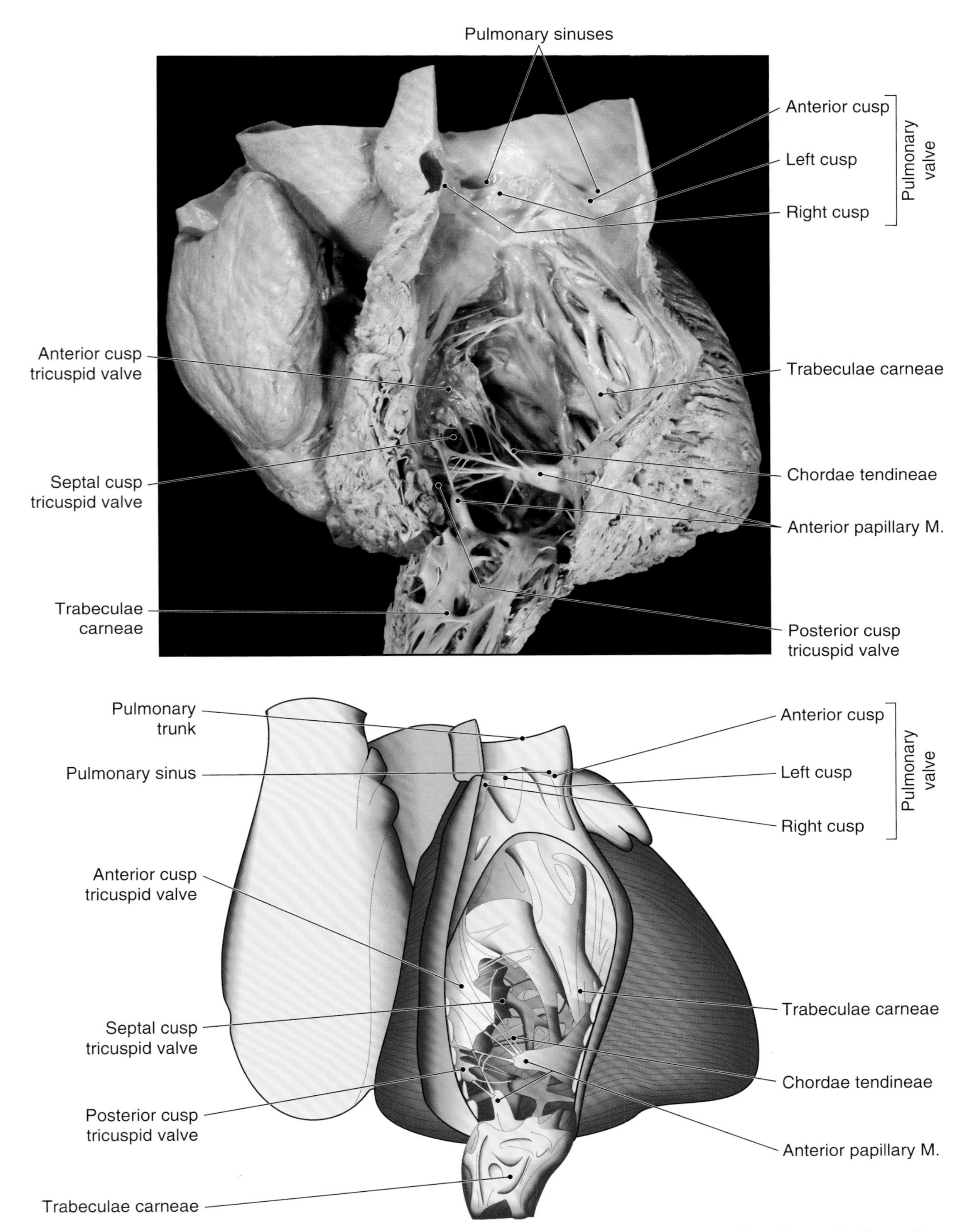

FIGURE A3-5 Internal features of the heart, right ventricle (anterior view). (Modified with permission from Olinger AB. *Human Gross Anatomy*. Baltimore, MD: Wolters Kluwer, 2016.)

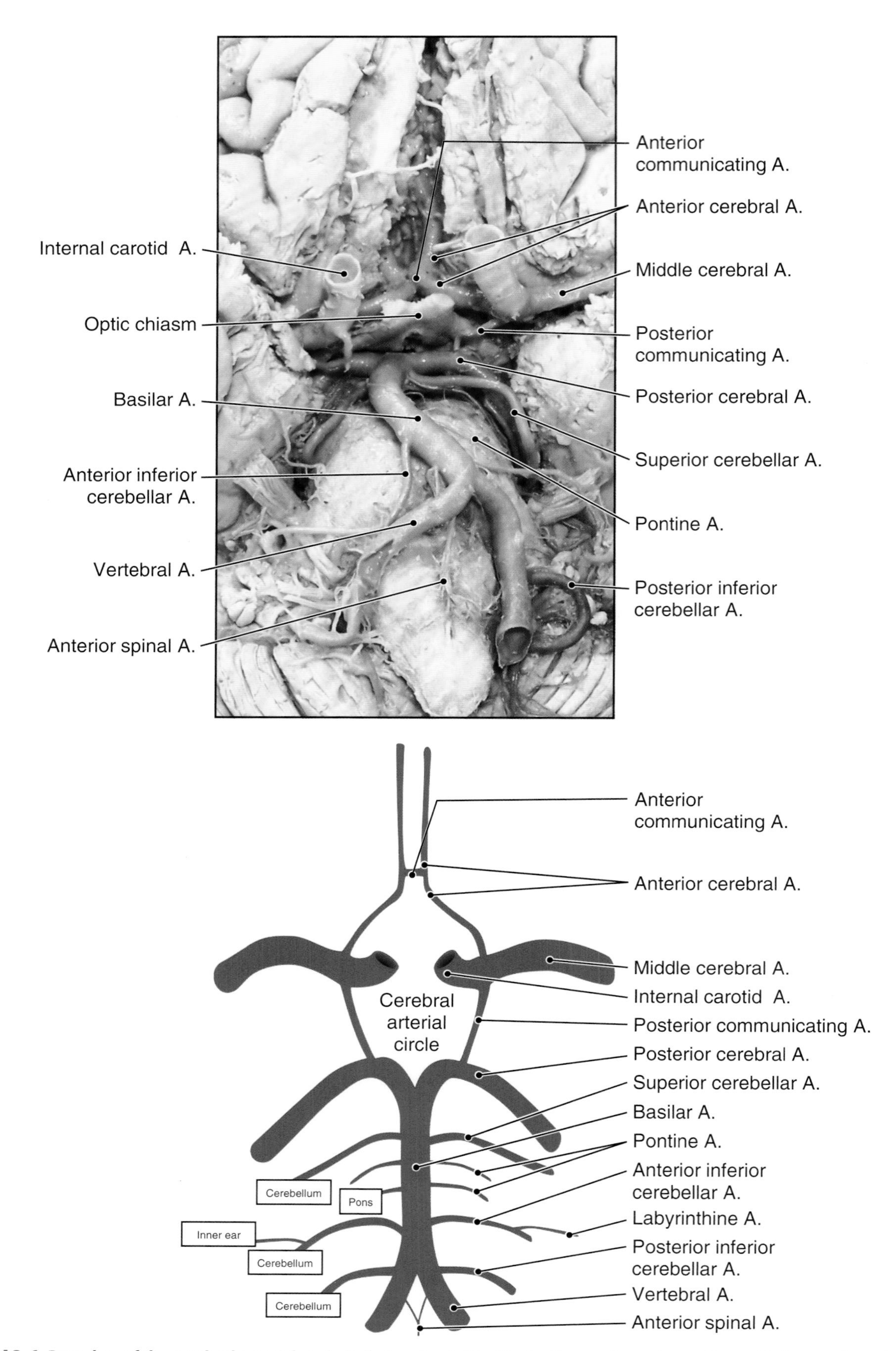

FIGURE A3-6 **Branches of the cerebral arterial circle (inferior view).** (Modified with permission from Olinger AB. *Human Gross Anatomy*. Baltimore, MD: Wolters Kluwer, 2016.)

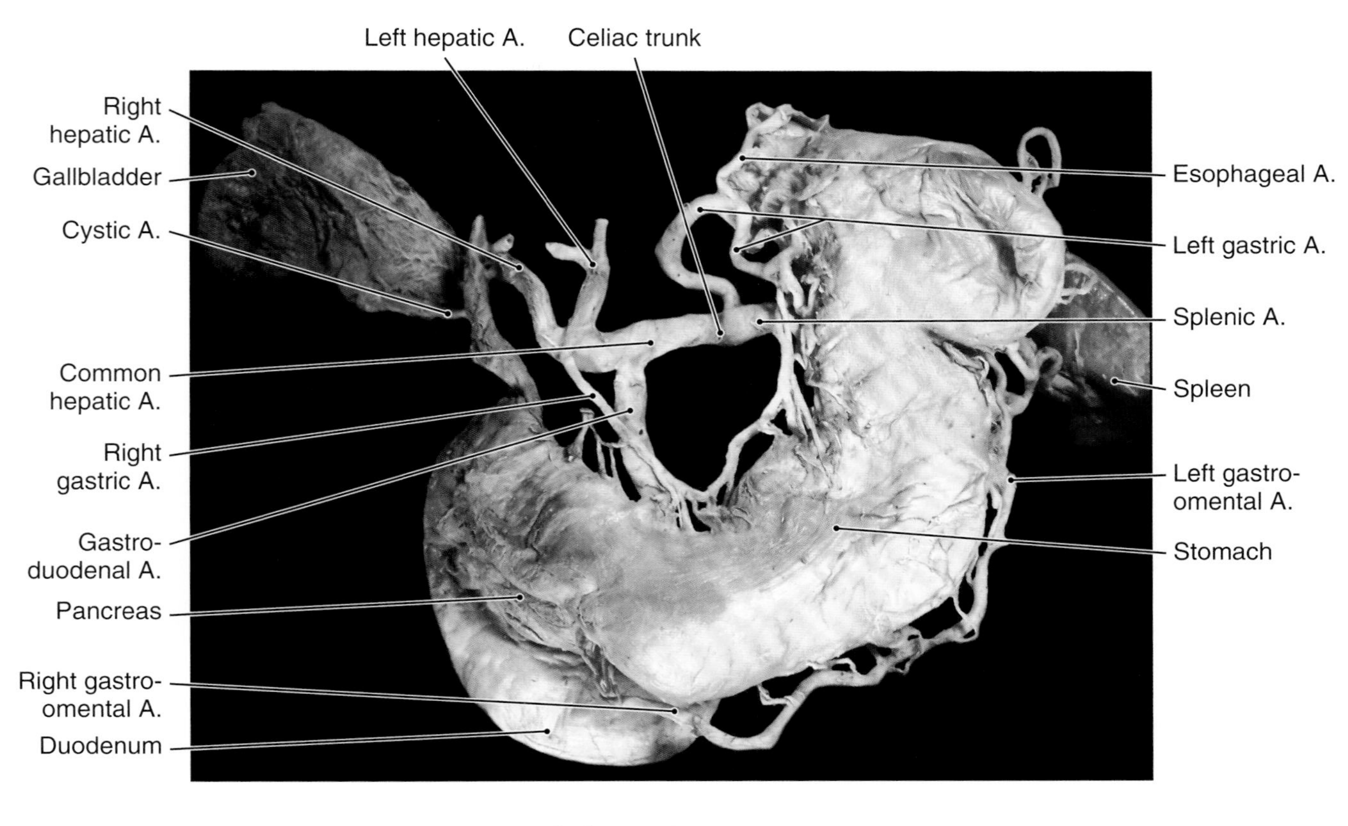

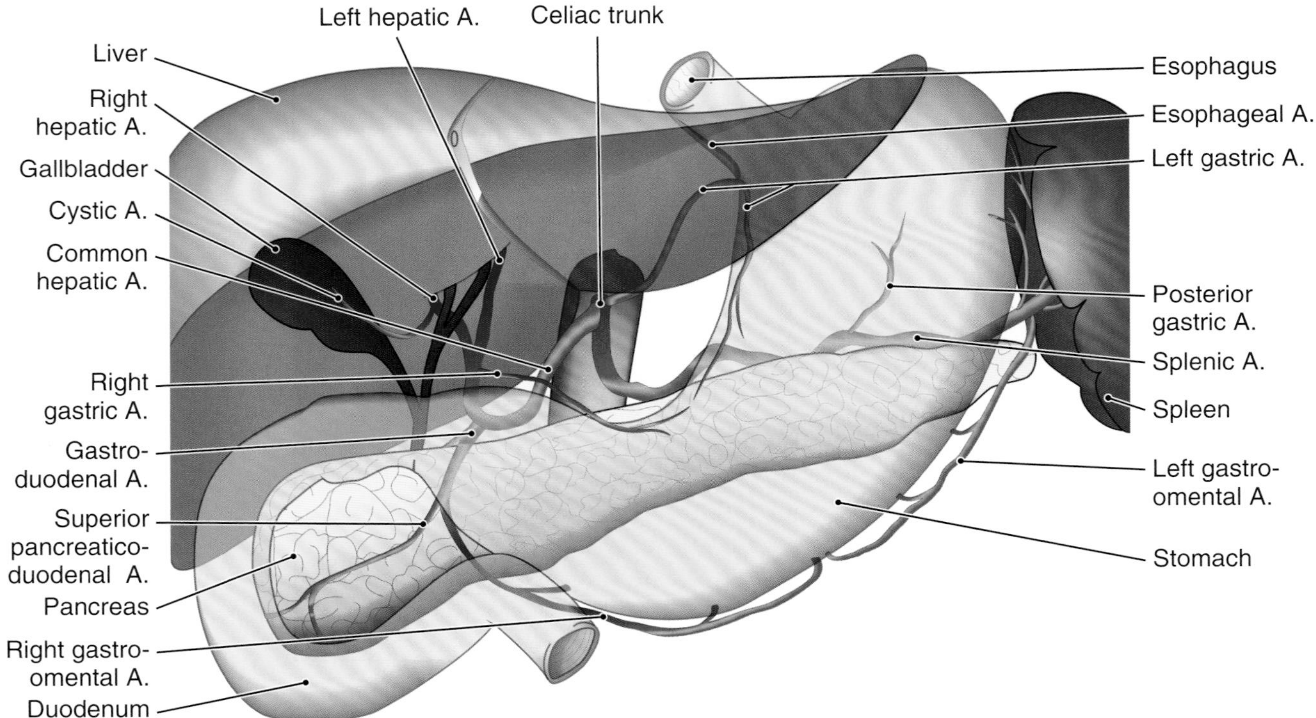

FIGURE A3-7 The celiac trunk and its branches to the abdominal organs. (Modified with permission from Olinger AB. *Human Gross Anatomy*. Baltimore, MD: Wolters Kluwer, 2016.)

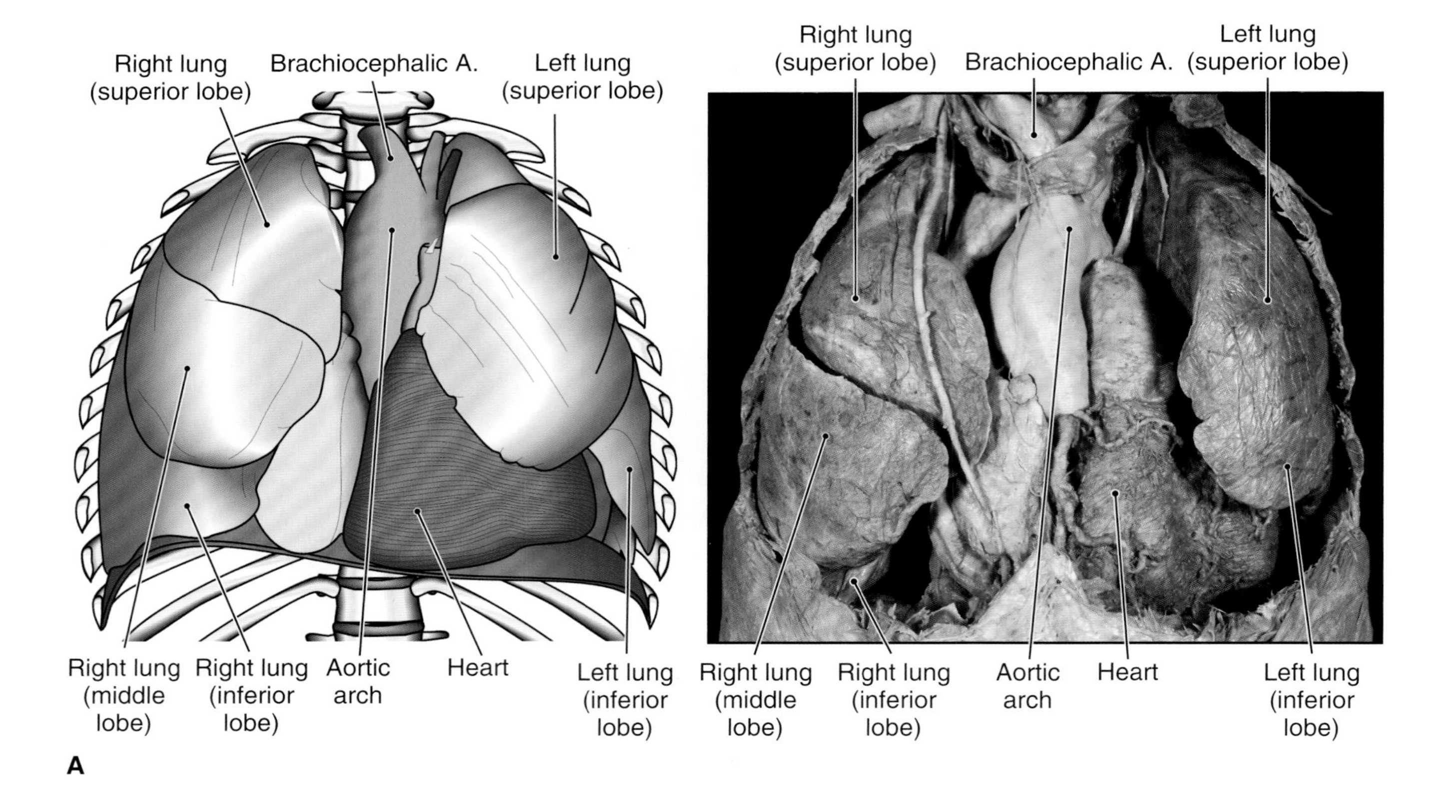

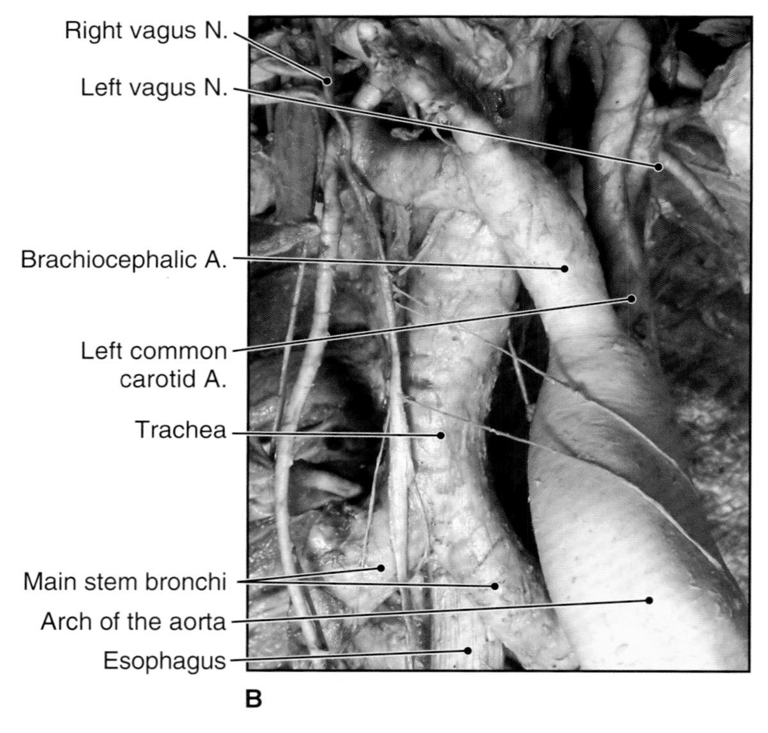

FIGURE A3-8 **A.** Thoracic cavity showing the pulmonary cavities, containing the lungs, and the mediastinum, containing the heart and other organs of the thorax. **B.** Respiratory passageways of the neck and upper thorax with nearby structures (anterior view). (Modified with permission from Olinger AB. *Human Gross Anatomy*. Baltimore, MD: Wolters Kluwer, 2016.)

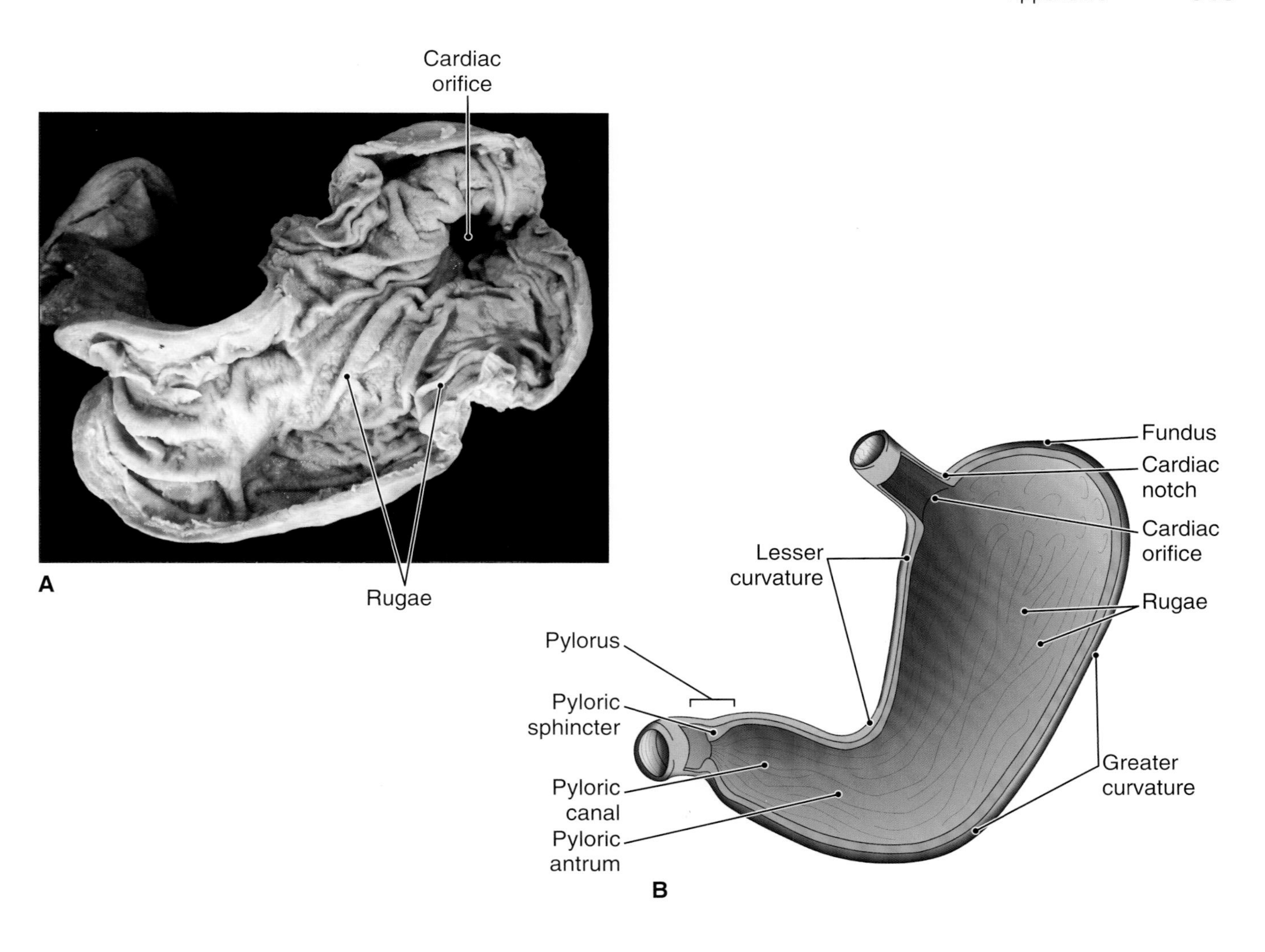

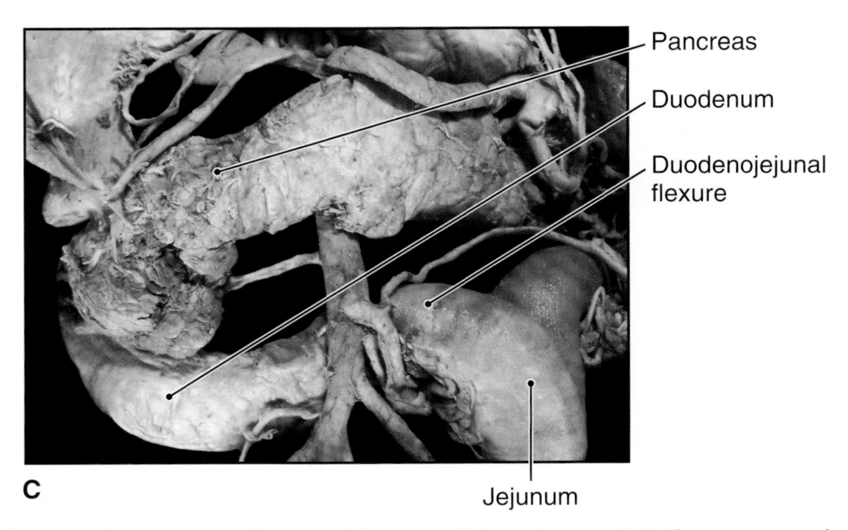

FIGURE A3-9 **A.** Internal parts of the stomach. **B.** Anatomic features of the stomach (anterior aspect). **C.** The pancreas and portions of the small intestine. (Modified with permission from Olinger AB. *Human Gross Anatomy*. Baltimore, MD: Wolters Kluwer, 2016.)

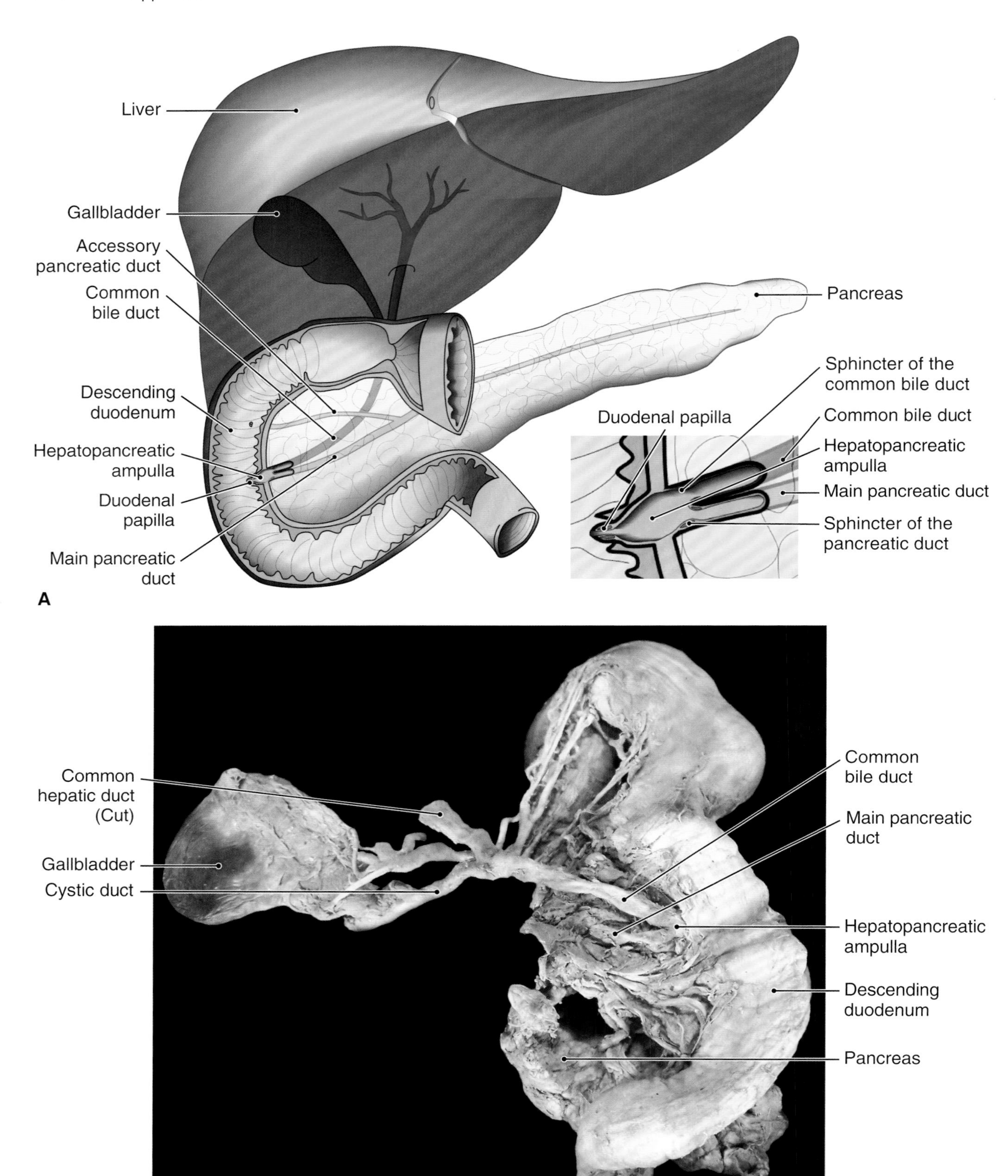

FIGURE A3-10 **A.** The pancreas and biliary system showing the passageways leading to the duodenum. **B.** Dissection showing the biliary system and pancreas. (Modified with permission from Olinger AB. *Human Gross Anatomy*. Baltimore, MD: Wolters Kluwer, 2016.)

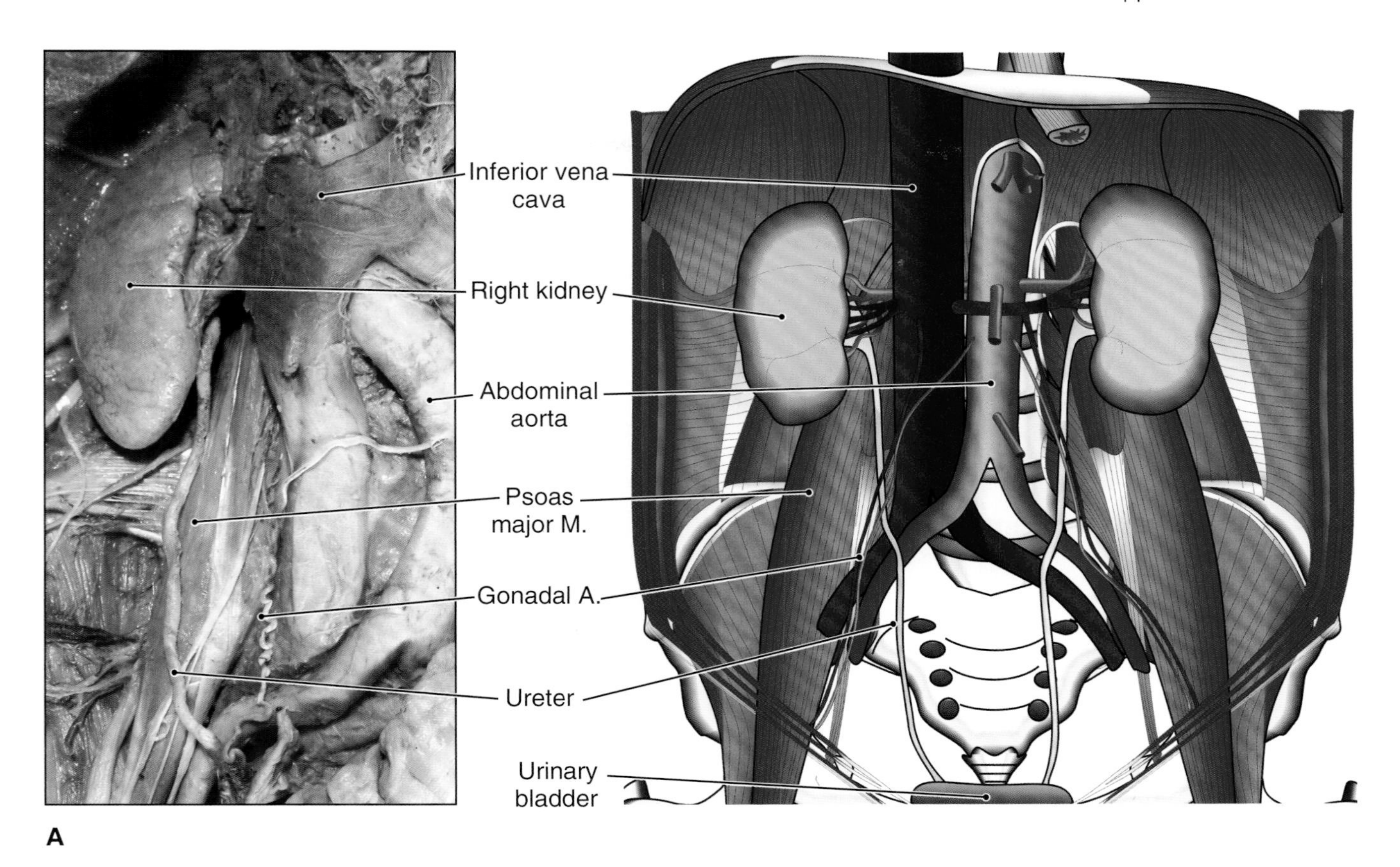

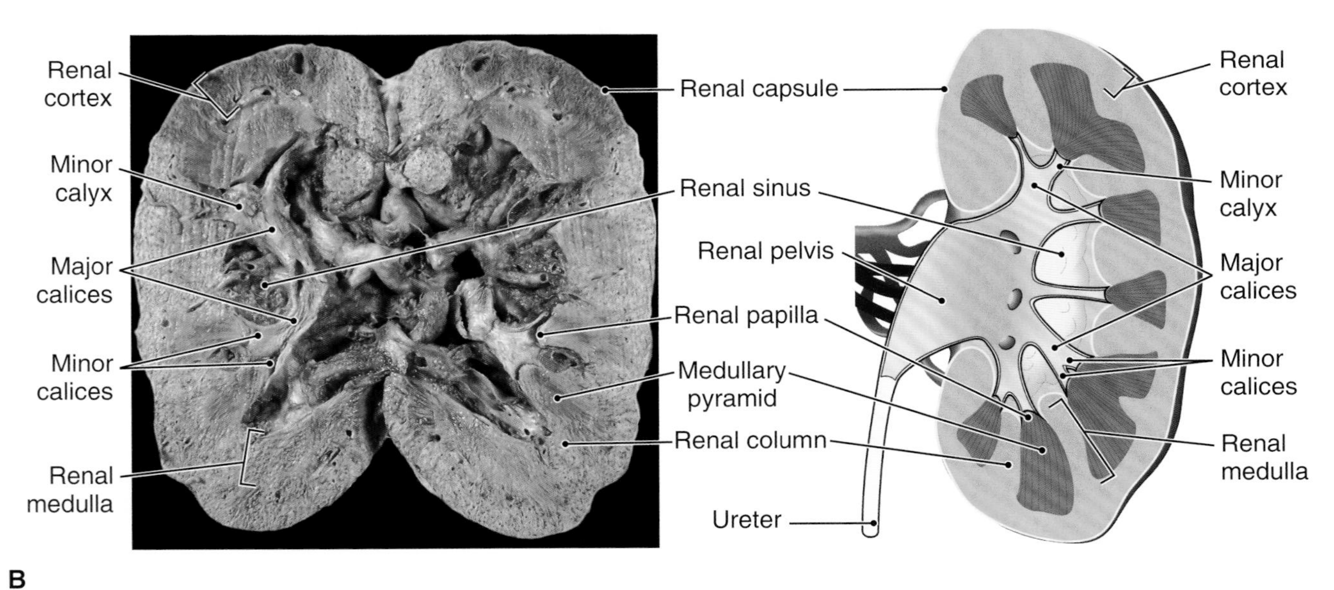

FIGURE A3-11 **A.** Components of the urinary system showing nearby vessels (anterior view). **B.** Internal structure of the kidney (coronal section). (Modified with permission from Olinger AB. *Human Gross Anatomy*. Baltimore, MD: Wolters Kluwer, 2016.)

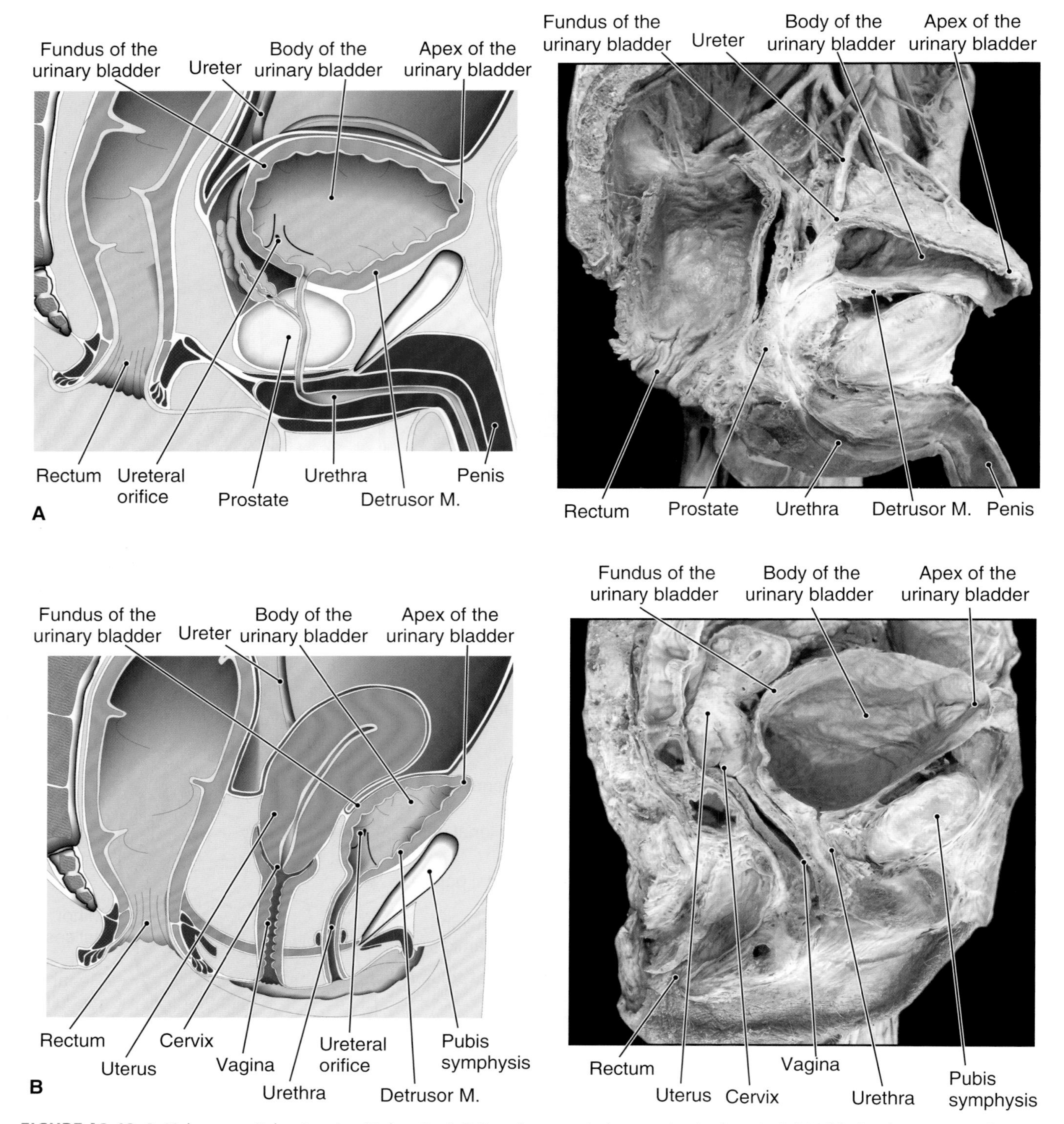

FIGURE A3-12 A. Male urogenital system (sagittal section). **B.** Female urogenital system (sagittal section). (Modified with permission from Olinger AB. *Human Gross Anatomy*. Baltimore, MD: Wolters Kluwer, 2016.)

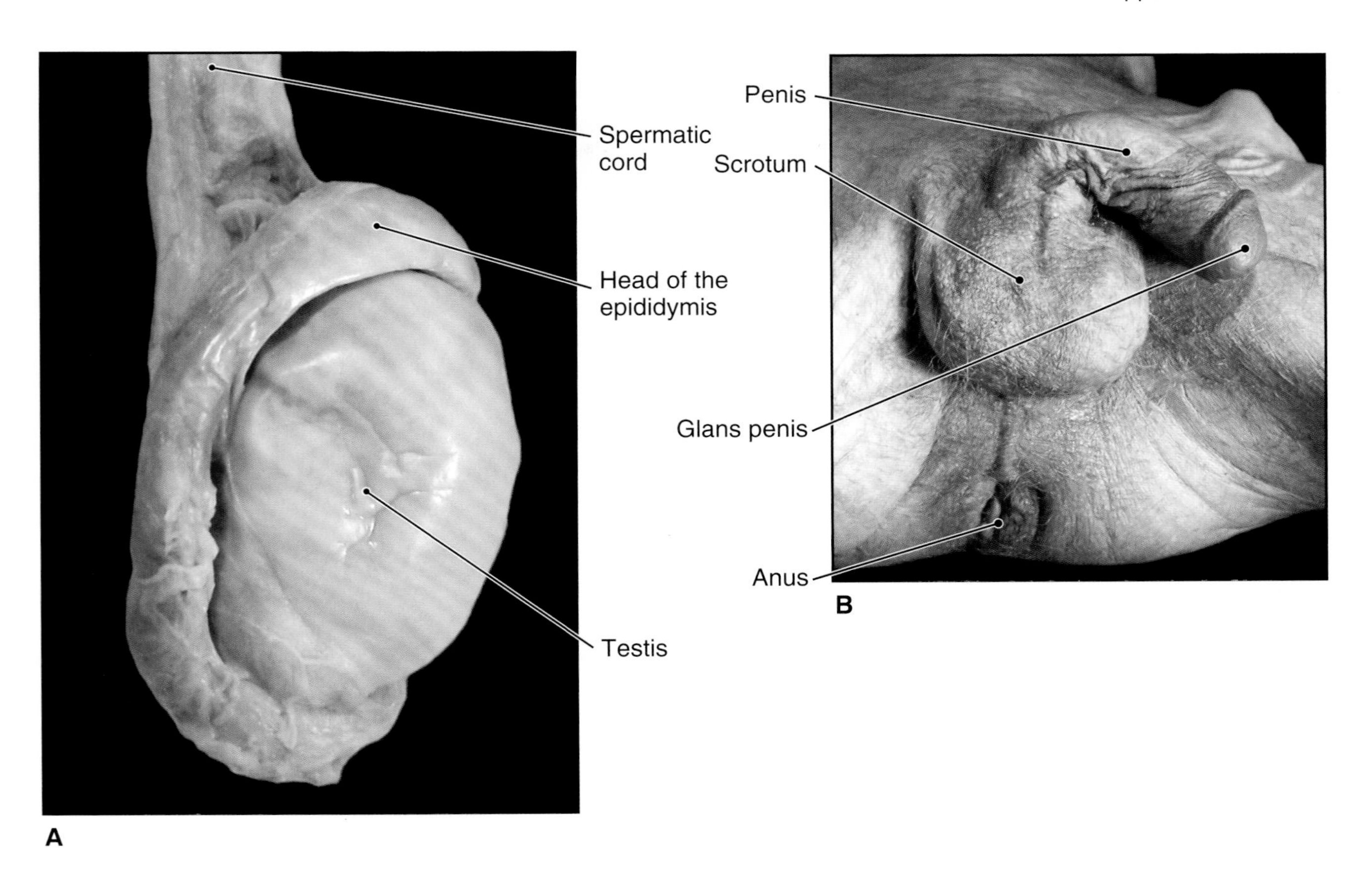

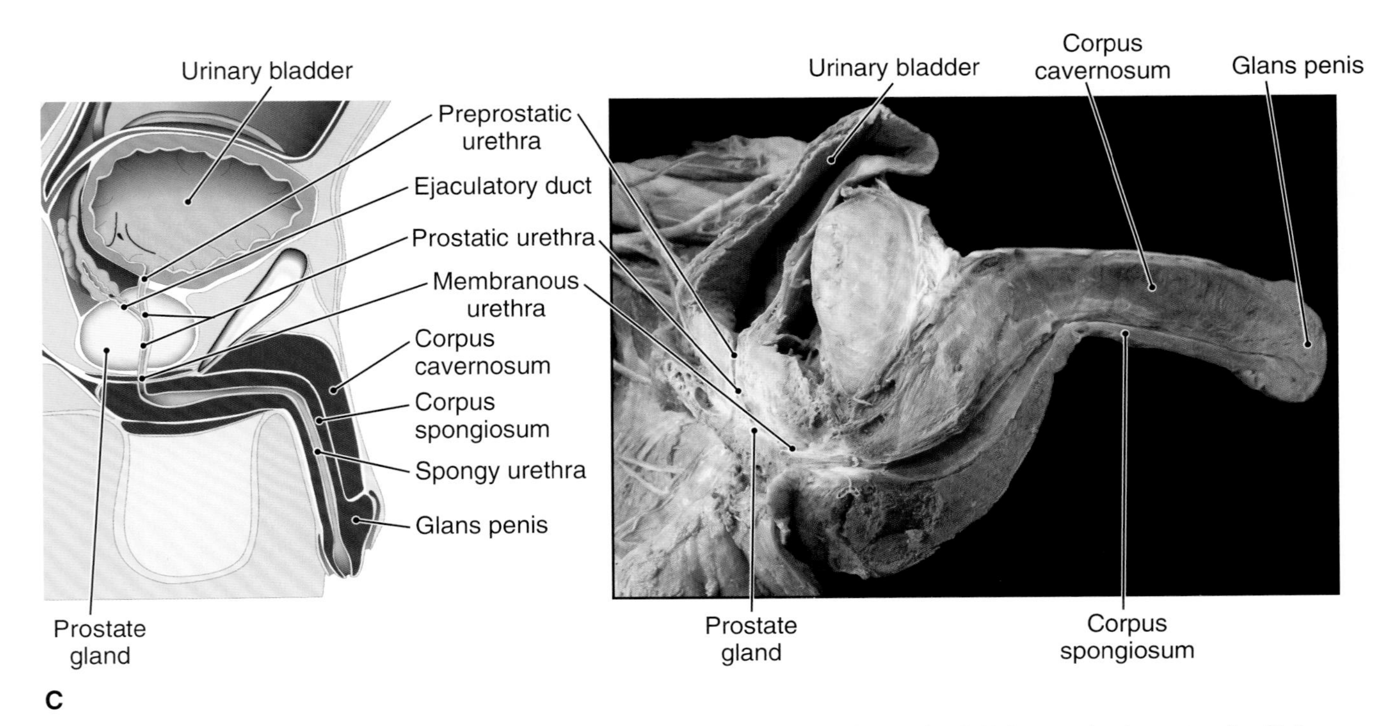

FIGURE A3-13 **A.** Testis showing the epididymis and spermatic cord. **B.** Male external genitalia. **C.** Male reproductive system (sagittal section). (Modified with permission from Olinger AB. *Human Gross Anatomy*. Baltimore, MD: Wolters Kluwer, 2016.)

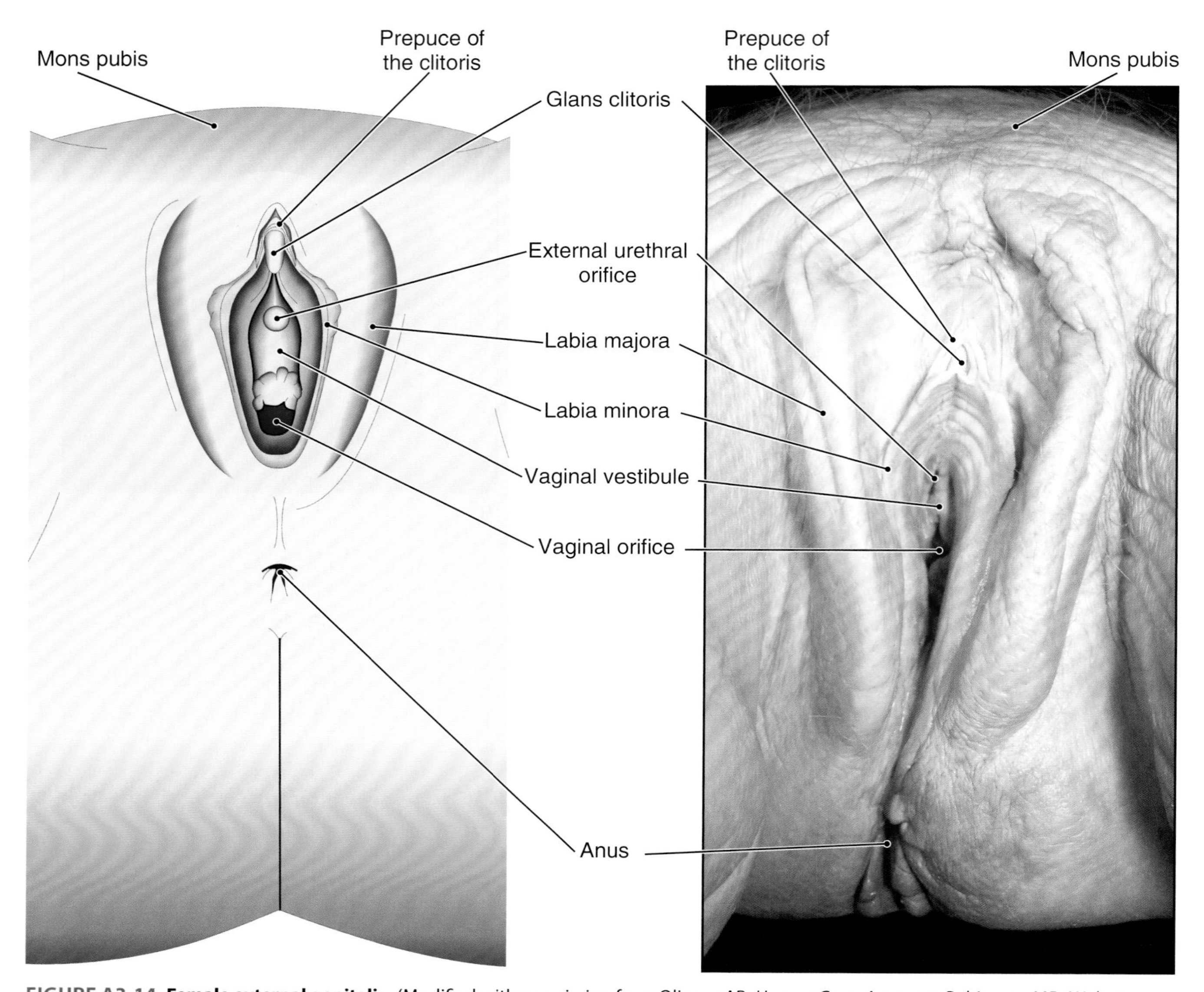

FIGURE A3-14 Female external genitalia. (Modified with permission from Olinger AB. *Human Gross Anatomy*. Baltimore, MD: Wolters Kluwer, 2016.)

Figure Credits

Chapter 1

1-5. Reprinted with permission from McConnell T, Hull K. *Human Form, Human Function*. Philadelphia, PA: Lippincott Williams & Wilkins, 2011.

1-9. Reprinted with permission from Erkonen WE, Smith WL. *Radiology 101: Basics and Fundamentals of Imaging*, 3rd ed. Philadelphia, PA: Lippincott Williams & Wilkins, 2010.

1-13. Reprinted with permission from McConnell T, Hull K. *Human Form, Human Function*. Philadelphia, PA: Lippincott Williams & Wilkins, 2011.

1-14. Reprinted with permission from McConnell T, Hull K. *Human Form, Human Function*. Philadelphia, PA: Lippincott Williams & Wilkins, 2011.

Chapter 2

2-14. Adapted with permission from McConnell T, Hull K. *Human Form, Human Function*. Philadelphia, PA: Lippincott Williams & Wilkins, 2011.

Chapter 3

3-1. **A.** Reprinted with permission from Cormack DH. *Essential Histology*, 2nd ed. Philadelphia, PA: Lippincott Williams & Wilkins, 2001; **B.** Reprinted with permission from Quinton P, Martinez R, eds. *Fluid and Electrolyte Transport in Exocrine Glands in Cystic Fibrosis*. San Francisco, CA: San Francisco Press, 1982; **C.** Reprinted with permission from Hafez ESE. *Scanning Electron Microscopic Atlas of Mammalian Reproduction*. Tokyo, Japan: Igaku-Shoin, 1975.

3-9. Reprinted with permission from McConnell T, Hull K. *Human Form, Human Function*. Philadelphia, PA: Lippincott Williams & Wilkins, 2011.

3-15. Photomicrographs reprinted with permission from Cormack DH. *Essential Histology*, 2nd ed. Philadelphia, PA: Lippincott Williams & Wilkins, 2001.

Chapter 4

4-1. Photomicrographs reprinted with permission from Cormack DH. *Essential Histology*, 2nd ed. Philadelphia, PA: Lippincott Williams & Wilkins, 2001.

4-4. **A.** Reprinted with permission from McClatchey KD. *Clinical Laboratory Medicine*, 2nd ed. Baltimore, MD: Lippincott Williams & Wilkins, 2001; **B.** Reprinted with permission from Cormack DH. *Essential Histology*, 2nd ed. Philadelphia, PA: Lippincott Williams & Wilkins, 2001; **C.** Reprinted with permission from Mills SE. Histology for Pathologists, 3rd ed. Philadelphia, PA: Lippincott Williams & Wilkins, 2006.

4-5. Reprinted with permission from Cormack DH. *Essential Histology*, 2nd ed. Philadelphia, PA: Lippincott Williams & Wilkins, 2001; **A–C.** Reprinted with permission from Mills SE. *Histology for Pathologists*, 3rd ed. Philadelphia, PA: Lippincott Williams & Wilkins, 2006; **D.** Reprinted with permission from Gartner LP, Hiatt JL. *Color Atlas of Histology*, 4th ed. Baltimore, MD: Lippincott Williams & Wilkins, 2005.

4-6. **A and C.** Reprinted with permission from Cormack DH. *Essential Histology*, 2nd ed. Philadelphia, PA: Lippincott Williams & Wilkins, 2001; **B.** Reprinted with permission from Gartner LP, Hiatt JL. *Color Atlas of Histology*, 4th ed. Baltimore, MD: Lippincott Williams & Wilkins, 2005.

4-7. Reprinted with permission from Cormack DH. *Essential Histology*, 2nd ed. Philadelphia, PA: Lippincott Williams & Wilkins, 2001.

4-9. Reprinted with permission from Okazaki H, Scheithauer BW. *Atlas of Neuropathology*. New York, NY: Gower Medical Publishing, 1988.

Chapter 5

5-2. Reprinted with permission from Cormack DH. *Essential Histology*, 2nd ed. Philadelphia, PA: Lippincott Williams & Wilkins, 2001.

5-3. Reprinted with permission from Cormack DH. *Essential Histology*, 2nd ed. Philadelphia, PA: Lippincott Williams & Wilkins, 2001.

5-4. **A and B.** Reprinted with permission from Cormack DH. *Essential Histology*, 2nd ed. Philadelphia, PA: Lippincott Williams & Wilkins, 2001.

5-5. **A.** Reprinted with permission from Bickley LS. *Bates' Guide to Physical Examination and History Taking*, 8th ed. Philadelphia, PA: Lippincott Williams & Wilkins, 2003.

5-6. **A**, Shutterstock. **B**, Reprinted with permission from Fleisher GR, Ludwig S, Baskin MN. Atlas of Pediatric Emergency Medicine. Philadelphia, PA: Lippincott Williams & Wilkins, 2004.

Chapter 6

6-1. Reprinted with permission from Erkonen WE, Smith WL. *Radiology 101*, 3rd ed. Philadelphia, PA: Lippincott Williams & Wilkins, 2010.

6-2. **B.** Reprinted with permission from Rubin R, Strayer DS. *Rubin's Pathology: Clinicopathologic Foundations of Medicine*, 5th ed. Baltimore, MD: Lippincott Williams & Wilkins, 2008.

6-3. **A.** Reprinted with permission from McConnell T, Hull K. *Human Form, Human Function*. Philadelphia, PA: Lippincott Williams & Wilkins, 2011; **B.** Reprinted with permission from Gartner LP, Hiatt JL. *Color Atlas of Histology*, 4th ed. Baltimore, MD: Lippincott Williams & Wilkins, 2005.

6-4. Modified with permission from McConnell T, Hull K. *Human Form, Human Function*. Philadelphia, PA: Lippincott Williams & Wilkins, 2011.

6-5. Reprinted with permission from McConnell T, Hull K. *Human Form, Human Function*. Philadelphia, PA: Lippincott Williams & Wilkins, 2011.

6-8. **D.** Reprinted with permission from McConnell T, Hull K. *Human Form, Human Function*. Philadelphia, PA: Lippincott Williams & Wilkins, 2011.

6-9. Reprinted with permission from McConnell T, Hull K. *Human Form, Human Function*. Philadelphia, PA: Lippincott Williams & Wilkins, 2011.

6-10. Reprinted with permission from McConnell T, Hull K. *Human Form, Human Function*. Philadelphia, PA: Lippincott Williams & Wilkins, 2011.

6-19. Reprinted with permission from Rubin E, Farber JL. *Pathology*, 5th ed. Philadelphia, PA: Lippincott Williams & Wilkins, 2008.

6-20. **A and C.** Reprinted with permission from Rubin E, Farber JL. *Pathology*, 3rd ed. Philadelphia, PA: Lippincott Williams & Wilkins, 2008; **B.** Becker KL, Bilezikian JP, Brenner WJ, et al. *Principles and Practice of Endocrinology and Metabolism*, 3rd ed. Philadelphia, PA: Lippincott Williams & Wilkins, 2001.

6-23. Reprinted with permission from McConnell T, Hull K. *Human Form, Human Function*. Philadelphia, PA: Lippincott Williams & Wilkins, 2011.

6-24. Reprinted with permission from McConnell T, Hull K. *Human Form, Human Function*. Philadelphia, PA: Lippincott Williams & Wilkins, 2011.

6-25. Modified with permission from McConnell T, Hull K. *Human Form, Human Function*. Philadelphia, PA: Lippincott Williams & Wilkins, 2011.

Chapter 7

7-1. **A and B.** Reprinted with permission from McConnell T, Hull K. *Human Form, Human Function*. Philadelphia, PA: Lippincott Williams & Wilkins, 2011; **C.** Reprinted with permission from Gartner LP, Hiatt JL. *Color Atlas of Histology*, 3rd ed. Philadelphia, PA: Lippincott Williams & Wilkins, 2000.

7-2. **A.** Reprinted with permission from Cormack DH. *Essential Histology*, 2nd ed. Philadelphia, PA: Lippincott Williams & Wilkins, 2001; **E.** Courtesy of A. Sima.

7-3. **A.** Reprinted with permission from Mills SE. *Histology for Pathologists*, 3rd ed. Philadelphia, PA: Lippincott Williams & Wilkins, 2006.

Chapter 8

8-5. **B.** Reprinted with permission from Gartner LP, Hiatt JL. *Color Atlas of Histology*, 3rd ed. Philadelphia, PA: Lippincott Williams & Wilkins, 2000.

8-6. Reprinted with permission from McConnell T, Hull K. *Human Form, Human Function*. Philadelphia, PA: Lippincott Williams & Wilkins, 2011.

8-9. Reprinted with permission from Bear MF, et al. *Neuroscience, Exploring the Brain*, 3rd ed. Philadelphia, PA: Lippincott Williams & Wilkins, 2007.

8-11. **B and C.** Reprinted with permission from McConnell T, Hull K. *Human Form, Human Function*. Philadelphia, PA: Lippincott Williams & Wilkins, 2011.

8-12. **B.** Reprinted with permission from Mills SE. *Histology for Pathologists*, 3rd ed. Philadelphia, PA: Lippincott Williams & Wilkins, 2006.

Chapter 9

9-2. Modified with permission from McConnell T, Hull K. *Human Form, Human Function*. Philadelphia, PA: Lippincott Williams & Wilkins, 2011.

9-3. Modified with permission from McConnell T, Hull K. *Human Form, Human Function*. Philadelphia, PA: Lippincott Williams & Wilkins, 2011.

9-9. **A and B.** Reprinted with permission from Erkonen WE. *Radiology 101*. Philadelphia, PA: Lippincott Williams & Wilkins, 1998; **C.** Courtesy of Newport Diagnostic Center, Newport Beach, CA.

9-11. Reprinted with permission from McConnell T, Hull K. *Human Form, Human Function*. Philadelphia, PA: Lippincott Williams & Wilkins, 2011.

Chapter 10

10-1. Reprinted with permission from McConnell T, Hull K. *Human Form, Human Function*. Philadelphia, PA: Lippincott Williams & Wilkins, 2011.

10-2. Reprinted with permission from McConnell T, Hull K. *Human Form, Human Function*. Philadelphia, PA: Lippincott Williams & Wilkins, 2011.

10-3. Reprinted with permission from McConnell T, Hull K. *Human Form, Human Function*. Philadelphia, PA: Lippincott Williams & Wilkins, 2011.

10-4. Reprinted with permission from McConnell T, Hull K. *Human Form, Human Function*. Philadelphia, PA: Lippincott Williams & Wilkins, 2011.

10-6. Reprinted with permission from McConnell T, Hull K. *Human Form, Human Function*. Philadelphia, PA: Lippincott Williams & Wilkins, 2011.

10-7. Reprinted with permission from McConnell T, Hull K. *Human Form, Human Function*. Philadelphia, PA: Lippincott Williams & Wilkins, 2011.

10-8. **A.** Reprinted with permission from McConnell T, Hull K. *Human Form, Human Function*. Philadelphia, PA: Lippincott Williams & Wilkins, 2011.

10-9. Reprinted with permission from McConnell T, Hull K. *Human Form, Human Function*. Philadelphia, PA: Lippincott Williams & Wilkins, 2011.

10-10. Reprinted with permission from McConnell T, Hull K. *Human Form, Human Function*. Philadelphia, PA: Lippincott Williams & Wilkins, 2011.

10-11. Reprinted with permission from McConnell T, Hull K. *Human Form, Human Function*. Philadelphia, PA: Lippincott Williams & Wilkins, 2011.

10-14. Reprinted with permission from McConnell T, Hull K. *Human Form, Human Function*. Philadelphia, PA: Lippincott Williams & Wilkins, 2011.

10-15 Reprinted with permission from McConnell T, Hull K. *Human Form, Human Function*. Philadelphia, PA: Lippincott Williams & Wilkins, 2011.

Unnumbered Fig. Box 10-1. Modified with permission from McConnell T, Hull K. Human Form, Human Function. Philadelphia, PA: Lippincott Williams & Wilkins, 2011.

Unnumbered Fig. p. 218. Reprinted with permission from McConnell T, Hull K. *Human Form, Human Function*. Philadelphia, PA: Lippincott Williams & Wilkins, 2011.

Chapter 11

11-4. Modified with permission from McConnell T, Hull K. *Human Form, Human Function*. Philadelphia, PA: Lippincott Williams & Wilkins, 2011.

11-5. Reprinted with permission from Cohen BJ. *Medical Terminology*, 8th ed. Philadelphia, PA: Lippincott Williams & Wilkins, 2017.

11-6. **A and B.** Reprinted with permission from McConnell TH. *The Nature of Disease*. Philadelphia, PA: Lippincott Williams & Wilkins, 2007; **C.** Courtesy of Dana Morse Bittus and B. J. Cohen.

Chapter 12

12-4. Reprinted with permission from Gartner LP, Hiatt JL. *Color Atlas of Histology*, 5th ed. Philadelphia, PA: Lippincott Williams & Wilkins, 2009.

12-5. **B.** Reprinted with permission from McConnell T, Hull K. *Human Form, Human Function*. Philadelphia, PA: Lippincott Williams & Wilkins, 2010; **C.** SEM courtesy of Erlandsen S, Engelkirk P, Kengelkirk PG, et al. *Burton's Microbiology for the Health Sciences*, 8th ed. Philadelphia, PA: Lippincott Williams & Wilkins, 2007.

12-6. **B.** Reprinted with permission from Gartner LP, Hiatt JL. *Color Atlas of Histology*, 5th ed. Philadelphia, PA: Lippincott Williams & Wilkins, 2009.

12-7. **A and C.** Reprinted with permission from McConnell TH, Hull KH. *Human Form, Human Function*. Philadelphia, PA: Lippincott Williams & Wilkins, 2011.

12-10. Reprinted with permission from McConnell T, Hull K. *Human Form, Human Function*. Philadelphia, PA: Lippincott Williams & Wilkins, 2011.

Unnumbered Fig. Box 12-3. Reprinted with permission from Cormack DH. *Essential Histology*, 2nd ed. Philadelphia, PA: Lippincott Williams & Wilkins, 2001.

Chapter 13

13-1. **A and B.** Modified with permission from McConnell T, Hull K. *Human Form, Human Function*. Philadelphia, PA: Lippincott Williams & Wilkins, 2011; **C.** Reprinted with permission from Gartner LP, Hiatt JL. *Color Atlas of Histology*, 5th ed. Philadelphia, PA: Lippincott Williams & Wilkins, 2009.

13-2. Modified with permission from McConnell T, Hull K. *Human Form, Human Function*. Philadelphia, PA: Lippincott Williams & Wilkins, 2011.

13-4. Modified with permission from McConnell T, Hull K. *Human Form, Human Function*. Philadelphia, PA: Lippincott Williams & Wilkins, 2011.

13-7. Modified with permission from McConnell T, Hull K. *Human Form, Human Function*. Philadelphia, PA: Lippincott Williams & Wilkins, 2011.

13-8. Modified with permission from McConnell T, Hull K. *Human Form, Human Function*. Philadelphia, PA: Lippincott Williams & Wilkins, 2011.

13-9. Modified with permission from McConnell T, Hull K. *Human Form, Human Function*. Philadelphia, PA: Lippincott Williams & Wilkins, 2011.

13-10. Reprinted with permission from Baim DS. *Grossman's Cardiac Catheterization, Angiography, and Intervention*, 7th ed. Philadelphia, PA: Lippincott Williams & Wilkins, 2006.

Chapter 14

14-1. Reprinted with permission from McConnell T, Hull K. *Human Form, Human Function*. Philadelphia, PA: Lippincott Williams & Wilkins, 2011.

14-2. Modified with permission from McConnell T, Hull K. *Human Form, Human Function*. Philadelphia, PA: Lippincott Williams & Wilkins, 2011.

14-3. Reprinted with permission from Cormack DH. *Essential Histology*, 2nd ed. Philadelphia, PA: Lippincott Williams & Wilkins, 2001.

14-5. Reprinted with permission from McConnell T, Hull K. *Human Form, Human Function*, Philadelphia, PA: Lippincott Williams & Wilkins, 2011.

14-6. Reprinted with permission from McConnell T, Hull K. *Human Form, Human Function*. Philadelphia, PA: Lippincott Williams & Wilkins, 2011.

14-7. Reprinted with permission from McConnell T, Hull K. *Human Form, Human Function*. Philadelphia, PA: Lippincott Williams & Wilkins, 2011.

14-8. Reprinted with permission from McConnell T, Hull K. *Human Form, Human Function*. Philadelphia, PA: Lippincott Williams & Wilkins, 2011.

14-9. Modified with permission from McConnell T, Hull K. *Human Form, Human Function*. Philadelphia, PA: Lippincott Williams & Wilkins, 2011.

14-10. Modified with permission from McConnell T, Hull K. *Human Form, Human Function*. Philadelphia, PA: Lippincott Williams & Wilkins, 2011.

14-11. Reprinted with permission from McConnell T, Hull K. *Human Form, Human Function*. Philadelphia, PA: Lippincott Williams & Wilkins, 2011.

14-12. Reprinted with permission from McConnell T, Hull K. *Human Form, Human Function*. Philadelphia, PA: Lippincott Williams & Wilkins, 2011.

14-13. Modified with permission from McConnell T, Hull K. *Human Form, Human Function*. Philadelphia, PA: Lippincott Williams & Wilkins, 2011.

14-14. Reprinted with permission from McConnell T, Hull K. *Human Form, Human Function*. Philadelphia, PA: Lippincott Williams & Wilkins, 2011.

14-15. Reprinted with permission from McConnell T, Hull K. *Human Form, Human Function*. Philadelphia, PA: Lippincott Williams & Wilkins, 2011.

Unnumbered Fig. Box 14-1. Reprinted with permission from *The Massage Connection Anatomy and Physiology*. Philadelphia, PA: Lippincott Williams & Wilkins, 2004.

Chapter 15

15-1. Reprinted with permission from McConnell T, Hull K. *Human Form, Human Function*. Philadelphia, PA: Lippincott Williams & Wilkins, 2011.

15-3. Reprinted with permission from McConnell T, Hull K. *Human Form, Human Function*. Philadelphia, PA: Lippincott Williams & Wilkins, 2011.

15-4. **A.** Reprinted with permission from McConnell T, Hull K. *Human Form, Human Function*. Philadelphia, PA: Lippincott Williams & Wilkins, 2011. **B.** Reprinted with permission from Cormack DH. *Essential Histology*, 2nd ed. Philadelphia, PA: Lippincott Williams & Wilkins, 2001.

15-5. Reprinted with permission from McConnell T, Hull K. *Human Form, Human Function*. Philadelphia, PA: Lippincott Williams & Wilkins, 2011.

15-6. **A.** Reprinted with permission from Archer P, Nelson LA. *Applied Anatomy and Physiology for Manual Therapists*. Philadelphia, PA: Lippincott Williams & Wilkins; **B.** Reprinted with permission from

Bickley LS. *Bates' Guide to Physical Examination and History Taking*, 8th ed. Philadelphia, PA: Lippincott Williams & Wilkins, 2003.

15-7. Reprinted with permission from McConnell T, Hull K. *Human Form, Human Function*. Philadelphia, PA: Lippincott Williams & Wilkins, 2011.

15-8. Reprinted with permission from McConnell T, Hull K. *Human Form, Human Function*. Philadelphia, PA: Lippincott Williams & Wilkins, 2011.

15-9. Reprinted with permission from Braun C, Anderson CM. *Pathophysiology*, 2nd ed. Philadelphia, PA: Lippincott Williams & Wilkins, 2012.

15-11. Reprinted with permission from McConnell T, Hull K. *Human Form, Human Function*. Philadelphia, PA: Lippincott Williams & Wilkins, 2011.

15-13. Modified with permission from McConnell T, Hull K. *Human Form, Human Function*. Philadelphia, PA: Lippincott Williams & Wilkins, 2011.

15-14. Reprinted with permission from Doan T, Melvold R, Viselli S, et al. *Immunology*, 2nd ed. Philadelphia, PA: Lippincott Williams & Wilkins, 2012.

Chapter 16

16-2. Reprinted with permission from McConnell T, Hull K. *Human Form, Human Function*. Philadelphia, PA: Lippincott Williams & Wilkins, 2011.

16-5. **A and B.** Reprinted with permission from McConnell TH, Hull KL. *Human Form, Human Function*. Philadelphia, PA: Lippincott Williams & Wilkins, 2011; **C.** Reprinted with permission from Cui D, et al. *Atlas of Histology with Functional and Clinical Correlations*. Philadelphia, PA: Lippincott Williams & Wilkins, 2010; **D.** Reprinted with permission from Snell RS. *Clinical Anatomy*, 7th ed. Philadelphia, PA: Lippincott Williams & Wilkins, 2003.

16-7. Modified with permission from McConnell T, Hull K. *Human Form, Human Function*. Philadelphia, PA: Lippincott Williams & Wilkins, 2011.

16-8. Modified with permission from McConnell T, Hull K. *Human Form, Human Function*. Philadelphia, PA: Lippincott Williams & Wilkins, 2011.

16-10. Reprinted with permission from McConnell T, Hull K. *Human Form, Human Function*. Philadelphia, PA: Lippincott Williams & Wilkins, 2011.

16-11. Reprinted with permission from McConnell T, Hull K. *Human Form, Human Function*. Philadelphia, PA: Lippincott Williams & Wilkins, 2011.

Chapter 17

17-1. Reprinted with permission from McConnell T, Hull K. *Human Form, Human Function*. Philadelphia, PA: Lippincott Williams & Wilkins, 2011.

17-2. Reprinted with permission from McConnell T, Hull K. *Human Form, Human Function*. Philadelphia, PA: Lippincott Williams & Wilkins, 2011.

17-3. Reprinted with permission from McConnell T, Hull K. *Human Form, Human Function*. Philadelphia, PA: Lippincott Williams & Wilkins, 2011.

17-4. Reprinted with permission from McConnell T, Hull K. *Human Form, Human Function*. Philadelphia, PA: Lippincott Williams & Wilkins, 2011.

17-5. Reprinted with permission from McConnell T, Hull K. *Human Form, Human Function*. Philadelphia, PA: Lippincott Williams & Wilkins, 2011.

17-6. Modified with permission from McConnell T, Hull K. *Human Form, Human Function*. Philadelphia, PA: Lippincott Williams & Wilkins, 2011. Insert modified with permission from Mulholland M, Albo D, Dalman R, et al. *Operative Techniques in Surgery*. Philadelphia, PA: Wolters Kluwer Health, 2014.

17-7. Modified with permission from McConnell T, Hull K. *Human Form, Human Function*. Philadelphia, PA: Lippincott Williams & Wilkins, 2011.

17-8. Modified with permission from McConnell T, Hull K. *Human Form, Human Function*. Philadelphia, PA: Lippincott Williams & Wilkins, 2011.

17-9. Modified with permission from McConnell T, Hull K. *Human Form, Human Function*. Philadelphia, PA: Lippincott Williams & Wilkins, 2011.

17-11. Modified with permission from McConnell T, Hull K. *Human Form, Human Function*. Philadelphia, PA: Lippincott Williams & Wilkins, 2011.

17-12. Modified with permission from McConnell T, Hull K. *Human Form, Human Function*. Philadelphia, PA: Lippincott Williams & Wilkins, 2011.

17-13. Modified with permission from McConnell T, Hull K. *Human Form, Human Function*. Philadelphia, PA: Lippincott Williams & Wilkins, 2011.

Unnumbered Fig. Box 17-2. Reprinted with permission from Cohen BJ. *Medical Terminology*, 8th ed. Philadelphia, PA: Lippincott Williams & Wilkins, 2017.

Chapter 18

18-4. Dietary Guidelines 2015. Department of Agriculture/ Center for Nutrition Policy and Promotion, https://health.gov/dietaryguidelines/2015/guidelines/executive-summary/#figure-es-1

18-6. Reprinted with permission from Taylor C, et al. *Fundamentals of Nursing*, 5th ed. Philadelphia, PA: Lippincott Williams & Wilkins, 2005.

18-7. Reprinted with permission from McConnell T, Hull K. *Human Form, Human Function*. Philadelphia, PA: Lippincott Williams & Wilkins, 2011.

Chapter 19

19-1. Reprinted with permission from McConnell T, Hull K. *Human Form, Human Function*. Philadelphia, PA: Lippincott Williams & Wilkins, 2011.

19-2. Modified with permission from McConnell T, Hull K. *Human Form, Human Function*. Philadelphia, PA: Lippincott Williams & Wilkins, 2011.

19-3. Reprinted with permission from McConnell T, Hull K. *Human Form, Human Function*. Philadelphia, PA: Lippincott Williams & Wilkins, 2011.

19-4. Courtesy of Dana Morse Bittus and BJ Cohen.

19-5. Modified with permission from McConnell TH. *The Nature of Disease*. Philadelphia, PA: Lippincott Williams & Wilkins, 2011

19-6. Modified with permission from McConnell T, Hull K. *Human Form, Human Function*. Philadelphia, PA: Lippincott Williams & Wilkins, 2011.

19-7. Reprinted with permission from McConnell TH. *The Nature of Disease*. Philadelphia, PA: Lippincott Williams & Wilkins, 2011

19-8. Modified with permission from McConnell T, Hull K. *Human Form, Human Function*. Philadelphia, PA: Lippincott Williams & Wilkins, 2011.

19-9. Reprinted with permission from McConnell TH. *The Nature of Disease*. Philadelphia, PA: Lippincott Williams & Wilkins, 2011.

19-10. Reprinted with permission from McConnell TH. *The Nature of Disease*. Philadelphia, PA: Lippincott Williams & Wilkins, 2011.

19-12. Modified with permission from McConnell T, Hull K. *Human Form, Human Function*. Philadelphia, PA: Lippincott Williams & Wilkins, 2011.

19-13. Reprinted with permission from McConnell TH. *The Nature of Disease*. Philadelphia, PA: Lippincott Williams & Wilkins, 2011.

Chapter 20

20-1. Modified with permission from McConnell T, Hull K. *Human Form, Human Function*. Philadelphia, PA: Lippincott Williams & Wilkins, 2011.

20-3. **A and B.** Reprinted with permission from McConnell T, Hull K. *Human Form, Human Function*. Philadelphia, PA: Lippincott Williams & Wilkins, 2011; **C.** Reprinted with permission from Eroschenko VP. *Di Fiore's Atlas of Histology with Functional Correlations*, 8th ed. Philadelphia, PA: Lippincott Williams & Wilkins, 1995; **D.** Adapted from Moore KL, Agur AM. *Essentials of Clinical Anatomy*, 2nd ed. Philadelphia, PA: Lippincott Williams & Wilkins, 2002.

20-5. Modified with permission from McConnell T, Hull K. *Human Form, Human Function*. Philadelphia, PA: Lippincott Williams & Wilkins, 2011.

20-6. Modified with permission from McConnell T, Hull K. *Human Form, Human Function*. Philadelphia, PA: Lippincott Williams & Wilkins, 2011.

20-7. Reprinted with permission from McConnell T, Hull K. *Human Form, Human Function*. Philadelphia, PA: Lippincott Williams & Wilkins, 2011.

20-8. **A.** Reprinted with permission from McConnell T, Hull K. *Human Form, Human Function*. Philadelphia, PA: Lippincott Williams & Wilkins, 2011; **C.** Courtesy of Dana Morse Bittus and Barbara Cohen.

20-9. Modified with permission from McConnell T, Hull K. *Human Form, Human Function*. Philadelphia, PA: Lippincott Williams & Wilkins, 2011.

20-10. Courtesy of Ansell Healthcare, Inc., Personal Products Group, and Carter-Wallace Inc. Reprinted from Westheimer R, Lopater S. *Human Sexuality: A Psychosocial Perspective*. Philadelphia, PA: Lippincott Williams & Wilkins, 2002; **B.** Courtesy of ALZA Pharmaceuticals, Palo Alto, CA.

Chapter 21

21-1. Reprinted with permission from McConnell T, Hull K. *Human Form, Human Function*. Philadelphia, PA: Lippincott Williams & Wilkins, 2011.

21-2. Modified with permission from McConnell T, Hull K. *Human Form, Human Function*. Philadelphia, PA: Lippincott Williams & Wilkins, 2011.

21-4. Reprinted with permission from Pillitteri A. *Maternal and Child Health Nursing*, 6th ed. Philadelphia, PA: Lippincott Williams & Wilkins, 2009.

21-5. Modified with permission from McConnell TH. *The Nature of Disease*. Philadelphia, PA: Lippincott Williams & Wilkins, 2011.

21-6. **A.** Reprinted with permission from Pillitteri A. *Maternal & Child Health Nursing*, 6th ed. Philadelphia, PA: Lippincott Williams & Wilkins, 2009; **B.** Reprinted with permission from Bushberg JT, Seibert JA, Leidholdt EM, et al. *Essential Physics of Medical Imaging*, 3rd ed. Philadelphia, PA: Lippincott Williams & Wilkins, 2011.

21-8. Reprinted with permission from Anatomical Chart Company. *OB/GYN Disorders*. Philadelphia, PA: Lippincott Williams & Wilkins, 2008.

21-9. Reprinted with permission from Snyder R, Dent N, Fowler W, Ling F. *Step-up to Obstetrics and Gynecology*. Philadelphia, PA: Wolters Kluwer Health, 2014.

21-10. Reprinted with permission from McConnell TH. *The Nature of Disease*. Philadelphia, PA: Lippincott Williams & Wilkins, 2011.

21-11. Modified with permission from McConnell TH. *The Nature of Disease*, 2nd ed. Philadelphia, PA: Lippincott Williams & Wilkins, 2013.

21-13. Modified with permission from McConnell T, Hull K. *Human Form, Human Function*. Philadelphia, PA: Lippincott Williams & Wilkins, 2011.

21-15. Modified with permission from McConnell T, Hull K. *Human Form, Human Function*. Philadelphia, PA: Lippincott Williams & Wilkins, 2011.

Index of Boxes

Clinical Perspectives

A Closer Look

Hot Topics

One Step at a Time

Index

Note: Page numbers followed by the letter "*f*" refer to figures; those followed by the letter "*t*" refer to tables; and those followed by the letter "*b*" refer to boxes.

S

T